Lecture Notes in Computer Science 12354

Founding Editors

Gerhard Goos
Karlsruhe Institute of Technology, Karlsruhe, Germany

Juris Hartmanis
Cornell University, Ithaca, NY, USA

Editorial Board Members

Elisa Bertino
Purdue University, West Lafayette, IN, USA

Wen Gao
Peking University, Beijing, China

Bernhard Steffen
TU Dortmund University, Dortmund, Germany

Gerhard Woeginger
RWTH Aachen, Aachen, Germany

Moti Yung
Columbia University, New York, NY, USA

More information about this series at http://www.springer.com/series/7412

Andrea Vedaldi · Horst Bischof ·
Thomas Brox · Jan-Michael Frahm (Eds.)

Computer Vision – ECCV 2020

16th European Conference
Glasgow, UK, August 23–28, 2020
Proceedings, Part IX

Editors
Andrea Vedaldi
University of Oxford
Oxford, UK

Horst Bischof
Graz University of Technology
Graz, Austria

Thomas Brox
University of Freiburg
Freiburg im Breisgau, Germany

Jan-Michael Frahm
University of North Carolina at Chapel Hill
Chapel Hill, NC, USA

ISSN 0302-9743 ISSN 1611-3349 (electronic)
Lecture Notes in Computer Science
ISBN 978-3-030-58544-0 ISBN 978-3-030-58545-7 (eBook)
https://doi.org/10.1007/978-3-030-58545-7

LNCS Sublibrary: SL6 – Image Processing, Computer Vision, Pattern Recognition, and Graphics

© Springer Nature Switzerland AG 2020, corrected publication 2020
This work is subject to copyright. All rights are reserved by the Publisher, whether the whole or part of the material is concerned, specifically the rights of translation, reprinting, reuse of illustrations, recitation, broadcasting, reproduction on microfilms or in any other physical way, and transmission or information storage and retrieval, electronic adaptation, computer software, or by similar or dissimilar methodology now known or hereafter developed.
The use of general descriptive names, registered names, trademarks, service marks, etc. in this publication does not imply, even in the absence of a specific statement, that such names are exempt from the relevant protective laws and regulations and therefore free for general use.
The publisher, the authors and the editors are safe to assume that the advice and information in this book are believed to be true and accurate at the date of publication. Neither the publisher nor the authors or the editors give a warranty, expressed or implied, with respect to the material contained herein or for any errors or omissions that may have been made. The publisher remains neutral with regard to jurisdictional claims in published maps and institutional affiliations.

This Springer imprint is published by the registered company Springer Nature Switzerland AG
The registered company address is: Gewerbestrasse 11, 6330 Cham, Switzerland

Foreword

Hosting the European Conference on Computer Vision (ECCV 2020) was certainly an exciting journey. From the 2016 plan to hold it at the Edinburgh International Conference Centre (hosting 1,800 delegates) to the 2018 plan to hold it at Glasgow's Scottish Exhibition Centre (up to 6,000 delegates), we finally ended with moving online because of the COVID-19 outbreak. While possibly having fewer delegates than expected because of the online format, ECCV 2020 still had over 3,100 registered participants.

Although online, the conference delivered most of the activities expected at a face-to-face conference: peer-reviewed papers, industrial exhibitors, demonstrations, and messaging between delegates. In addition to the main technical sessions, the conference included a strong program of satellite events with 16 tutorials and 44 workshops.

Furthermore, the online conference format enabled new conference features. Every paper had an associated teaser video and a longer full presentation video. Along with the papers and slides from the videos, all these materials were available the week before the conference. This allowed delegates to become familiar with the paper content and be ready for the live interaction with the authors during the conference week. The live event consisted of brief presentations by the oral and spotlight authors and industrial sponsors. Question and answer sessions for all papers were timed to occur twice so delegates from around the world had convenient access to the authors.

As with ECCV 2018, authors' draft versions of the papers appeared online with open access, now on both the Computer Vision Foundation (CVF) and the European Computer Vision Association (ECVA) websites. An archival publication arrangement was put in place with the cooperation of Springer. SpringerLink hosts the final version of the papers with further improvements, such as activating reference links and supplementary materials. These two approaches benefit all potential readers: a version available freely for all researchers, and an authoritative and citable version with additional benefits for SpringerLink subscribers. We thank Alfred Hofmann and Aliaksandr Birukou from Springer for helping to negotiate this agreement, which we expect will continue for future versions of ECCV.

August 2020

Vittorio Ferrari
Bob Fisher
Cordelia Schmid
Emanuele Trucco

Preface

Welcome to the proceedings of the European Conference on Computer Vision (ECCV 2020). This is a unique edition of ECCV in many ways. Due to the COVID-19 pandemic, this is the first time the conference was held online, in a virtual format. This was also the first time the conference relied exclusively on the Open Review platform to manage the review process. Despite these challenges ECCV is thriving. The conference received 5,150 valid paper submissions, of which 1,360 were accepted for publication (27%) and, of those, 160 were presented as spotlights (3%) and 104 as orals (2%). This amounts to more than twice the number of submissions to ECCV 2018 (2,439). Furthermore, CVPR, the largest conference on computer vision, received 5,850 submissions this year, meaning that ECCV is now 87% the size of CVPR in terms of submissions. By comparison, in 2018 the size of ECCV was only 73% of CVPR.

The review model was similar to previous editions of ECCV; in particular, it was double blind in the sense that the authors did not know the name of the reviewers and vice versa. Furthermore, each conference submission was held confidentially, and was only publicly revealed if and once accepted for publication. Each paper received at least three reviews, totalling more than 15,000 reviews. Handling the review process at this scale was a significant challenge. In order to ensure that each submission received as fair and high-quality reviews as possible, we recruited 2,830 reviewers (a 130% increase with reference to 2018) and 207 area chairs (a 60% increase). The area chairs were selected based on their technical expertise and reputation, largely among people that served as area chair in previous top computer vision and machine learning conferences (ECCV, ICCV, CVPR, NeurIPS, etc.). Reviewers were similarly invited from previous conferences. We also encouraged experienced area chairs to suggest additional chairs and reviewers in the initial phase of recruiting.

Despite doubling the number of submissions, the reviewer load was slightly reduced from 2018, from a maximum of 8 papers down to 7 (with some reviewers offering to handle 6 papers plus an emergency review). The area chair load increased slightly, from 18 papers on average to 22 papers on average.

Conflicts of interest between authors, area chairs, and reviewers were handled largely automatically by the Open Review platform via their curated list of user profiles. Many authors submitting to ECCV already had a profile in Open Review. We set a paper registration deadline one week before the paper submission deadline in order to encourage all missing authors to register and create their Open Review profiles well on time (in practice, we allowed authors to create/change papers arbitrarily until the submission deadline). Except for minor issues with users creating duplicate profiles, this allowed us to easily and quickly identify institutional conflicts, and avoid them, while matching papers to area chairs and reviewers.

Papers were matched to area chairs based on: an affinity score computed by the Open Review platform, which is based on paper titles and abstracts, and an affinity score computed by the Toronto Paper Matching System (TPMS), which is based on the paper's full text, the area chair bids for individual papers, load balancing, and conflict avoidance. Open Review provides the program chairs a convenient web interface to experiment with different configurations of the matching algorithm. The chosen configuration resulted in about 50% of the assigned papers to be highly ranked by the area chair bids, and 50% to be ranked in the middle, with very few low bids assigned.

Assignments to reviewers were similar, with two differences. First, there was a maximum of 7 papers assigned to each reviewer. Second, area chairs recommended up to seven reviewers per paper, providing another highly-weighed term to the affinity scores used for matching.

The assignment of papers to area chairs was smooth. However, it was more difficult to find suitable reviewers for all papers. Having a ratio of 5.6 papers per reviewer with a maximum load of 7 (due to emergency reviewer commitment), which did not allow for much wiggle room in order to also satisfy conflict and expertise constraints. We received some complaints from reviewers who did not feel qualified to review specific papers and we reassigned them wherever possible. However, the large scale of the conference, the many constraints, and the fact that a large fraction of such complaints arrived very late in the review process made this process very difficult and not all complaints could be addressed.

Reviewers had six weeks to complete their assignments. Possibly due to COVID-19 or the fact that the NeurIPS deadline was moved closer to the review deadline, a record 30% of the reviews were still missing after the deadline. By comparison, ECCV 2018 experienced only 10% missing reviews at this stage of the process. In the subsequent week, area chairs chased the missing reviews intensely, found replacement reviewers in their own team, and managed to reach 10% missing reviews. Eventually, we could provide almost all reviews (more than 99.9%) with a delay of only a couple of days on the initial schedule by a significant use of emergency reviews. If this trend is confirmed, it might be a major challenge to run a smooth review process in future editions of ECCV. The community must reconsider prioritization of the time spent on paper writing (the number of submissions increased a lot despite COVID-19) and time spent on paper reviewing (the number of reviews delivered in time decreased a lot presumably due to COVID-19 or NeurIPS deadline). With this imbalance the peer-review system that ensures the quality of our top conferences may break soon.

Reviewers submitted their reviews independently. In the reviews, they had the opportunity to ask questions to the authors to be addressed in the rebuttal. However, reviewers were told not to request any significant new experiment. Using the Open Review interface, authors could provide an answer to each individual review, but were also allowed to cross-reference reviews and responses in their answers. Rather than PDF files, we allowed the use of formatted text for the rebuttal. The rebuttal and initial reviews were then made visible to all reviewers and the primary area chair for a given paper. The area chair encouraged and moderated the reviewer discussion. During the discussions, reviewers were invited to reach a consensus and possibly adjust their ratings as a result of the discussion and of the evidence in the rebuttal.

After the discussion period ended, most reviewers entered a final rating and recommendation, although in many cases this did not differ from their initial recommendation. Based on the updated reviews and discussion, the primary area chair then made a preliminary decision to accept or reject the paper and wrote a justification for it (meta-review). Except for cases where the outcome of this process was absolutely clear (as indicated by the three reviewers and primary area chairs all recommending clear rejection), the decision was then examined and potentially challenged by a secondary area chair. This led to further discussion and overturning a small number of preliminary decisions. Needless to say, there was no in-person area chair meeting, which would have been impossible due to COVID-19.

Area chairs were invited to observe the consensus of the reviewers whenever possible and use extreme caution in overturning a clear consensus to accept or reject a paper. If an area chair still decided to do so, she/he was asked to clearly justify it in the meta-review and to explicitly obtain the agreement of the secondary area chair. In practice, very few papers were rejected after being confidently accepted by the reviewers.

This was the first time Open Review was used as the main platform to run ECCV. In 2018, the program chairs used CMT3 for the user-facing interface and Open Review internally, for matching and conflict resolution. Since it is clearly preferable to only use a single platform, this year we switched to using Open Review in full. The experience was largely positive. The platform is highly-configurable, scalable, and open source. Being written in Python, it is easy to write scripts to extract data programmatically. The paper matching and conflict resolution algorithms and interfaces are top-notch, also due to the excellent author profiles in the platform. Naturally, there were a few kinks along the way due to the fact that the ECCV Open Review configuration was created from scratch for this event and it differs in substantial ways from many other Open Review conferences. However, the Open Review development and support team did a fantastic job in helping us to get the configuration right and to address issues in a timely manner as they unavoidably occurred. We cannot thank them enough for the tremendous effort they put into this project.

Finally, we would like to thank everyone involved in making ECCV 2020 possible in these very strange and difficult times. This starts with our authors, followed by the area chairs and reviewers, who ran the review process at an unprecedented scale. The whole Open Review team (and in particular Melisa Bok, Mohit Unyal, Carlos Mondragon Chapa, and Celeste Martinez Gomez) worked incredibly hard for the entire duration of the process. We would also like to thank René Vidal for contributing to the adoption of Open Review. Our thanks also go to Laurent Charling for TPMS and to the program chairs of ICML, ICLR, and NeurIPS for cross checking double submissions. We thank the website chair, Giovanni Farinella, and the CPI team (in particular Ashley Cook, Miriam Verdon, Nicola McGrane, and Sharon Kerr) for promptly adding material to the website as needed in the various phases of the process. Finally, we thank the publication chairs, Albert Ali Salah, Hamdi Dibeklioglu, Metehan Doyran, Henry Howard-Jenkins, Victor Prisacariu, Siyu Tang, and Gul Varol, who managed to compile these substantial proceedings in an exceedingly compressed schedule. We express our thanks to the ECVA team, in particular Kristina Scherbaum for allowing open access of the proceedings. We thank Alfred Hofmann from Springer who again serve as the publisher. Finally, we thank the other chairs of ECCV 2020, including in particular the general chairs for very useful feedback with the handling of the program.

August 2020

Andrea Vedaldi
Horst Bischof
Thomas Brox
Jan-Michael Frahm

Organization

General Chairs

Vittorio Ferrari	Google Research, Switzerland
Bob Fisher	University of Edinburgh, UK
Cordelia Schmid	Google and Inria, France
Emanuele Trucco	University of Dundee, UK

Program Chairs

Andrea Vedaldi	University of Oxford, UK
Horst Bischof	Graz University of Technology, Austria
Thomas Brox	University of Freiburg, Germany
Jan-Michael Frahm	University of North Carolina, USA

Industrial Liaison Chairs

Jim Ashe	University of Edinburgh, UK
Helmut Grabner	Zurich University of Applied Sciences, Switzerland
Diane Larlus	NAVER LABS Europe, France
Cristian Novotny	University of Edinburgh, UK

Local Arrangement Chairs

Yvan Petillot	Heriot-Watt University, UK
Paul Siebert	University of Glasgow, UK

Academic Demonstration Chair

Thomas Mensink	Google Research and University of Amsterdam, The Netherlands

Poster Chair

Stephen Mckenna	University of Dundee, UK

Technology Chair

Gerardo Aragon Camarasa	University of Glasgow, UK

Tutorial Chairs

Carlo Colombo	University of Florence, Italy
Sotirios Tsaftaris	University of Edinburgh, UK

Publication Chairs

Albert Ali Salah	Utrecht University, The Netherlands
Hamdi Dibeklioglu	Bilkent University, Turkey
Metehan Doyran	Utrecht University, The Netherlands
Henry Howard-Jenkins	University of Oxford, UK
Victor Adrian Prisacariu	University of Oxford, UK
Siyu Tang	ETH Zurich, Switzerland
Gul Varol	University of Oxford, UK

Website Chair

Giovanni Maria Farinella	University of Catania, Italy

Workshops Chairs

Adrien Bartoli	University of Clermont Auvergne, France
Andrea Fusiello	University of Udine, Italy

Area Chairs

Lourdes Agapito	University College London, UK
Zeynep Akata	University of Tübingen, Germany
Karteek Alahari	Inria, France
Antonis Argyros	University of Crete, Greece
Hossein Azizpour	KTH Royal Institute of Technology, Sweden
Joao P. Barreto	Universidade de Coimbra, Portugal
Alexander C. Berg	University of North Carolina at Chapel Hill, USA
Matthew B. Blaschko	KU Leuven, Belgium
Lubomir D. Bourdev	WaveOne, Inc., USA
Edmond Boyer	Inria, France
Yuri Boykov	University of Waterloo, Canada
Gabriel Brostow	University College London, UK
Michael S. Brown	National University of Singapore, Singapore
Jianfei Cai	Monash University, Australia
Barbara Caputo	Politecnico di Torino, Italy
Ayan Chakrabarti	Washington University, St. Louis, USA
Tat-Jen Cham	Nanyang Technological University, Singapore
Manmohan Chandraker	University of California, San Diego, USA
Rama Chellappa	Johns Hopkins University, USA
Liang-Chieh Chen	Google, USA

Yung-Yu Chuang	National Taiwan University, Taiwan
Ondrej Chum	Czech Technical University in Prague, Czech Republic
Brian Clipp	Kitware, USA
John Collomosse	University of Surrey and Adobe Research, UK
Jason J. Corso	University of Michigan, USA
David J. Crandall	Indiana University, USA
Daniel Cremers	University of California, Los Angeles, USA
Fabio Cuzzolin	Oxford Brookes University, UK
Jifeng Dai	SenseTime, SAR China
Kostas Daniilidis	University of Pennsylvania, USA
Andrew Davison	Imperial College London, UK
Alessio Del Bue	Fondazione Istituto Italiano di Tecnologia, Italy
Jia Deng	Princeton University, USA
Alexey Dosovitskiy	Google, Germany
Matthijs Douze	Facebook, France
Enrique Dunn	Stevens Institute of Technology, USA
Irfan Essa	Georgia Institute of Technology and Google, USA
Giovanni Maria Farinella	University of Catania, Italy
Ryan Farrell	Brigham Young University, USA
Paolo Favaro	University of Bern, Switzerland
Rogerio Feris	International Business Machines, USA
Cornelia Fermuller	University of Maryland, College Park, USA
David J. Fleet	Vector Institute, Canada
Friedrich Fraundorfer	DLR, Austria
Mario Fritz	CISPA Helmholtz Center for Information Security, Germany
Pascal Fua	EPFL (Swiss Federal Institute of Technology Lausanne), Switzerland
Yasutaka Furukawa	Simon Fraser University, Canada
Li Fuxin	Oregon State University, USA
Efstratios Gavves	University of Amsterdam, The Netherlands
Peter Vincent Gehler	Amazon, USA
Theo Gevers	University of Amsterdam, The Netherlands
Ross Girshick	Facebook AI Research, USA
Boqing Gong	Google, USA
Stephen Gould	Australian National University, Australia
Jinwei Gu	SenseTime Research, USA
Abhinav Gupta	Facebook, USA
Bohyung Han	Seoul National University, South Korea
Bharath Hariharan	Cornell University, USA
Tal Hassner	Facebook AI Research, USA
Xuming He	Australian National University, Australia
Joao F. Henriques	University of Oxford, UK
Adrian Hilton	University of Surrey, UK
Minh Hoai	Stony Brooks, State University of New York, USA
Derek Hoiem	University of Illinois Urbana-Champaign, USA

Timothy Hospedales	University of Edinburgh and Samsung, UK
Gang Hua	Wormpex AI Research, USA
Slobodan Ilic	Siemens AG, Germany
Hiroshi Ishikawa	Waseda University, Japan
Jiaya Jia	The Chinese University of Hong Kong, SAR China
Hailin Jin	Adobe Research, USA
Justin Johnson	University of Michigan, USA
Frederic Jurie	University of Caen Normandie, France
Fredrik Kahl	Chalmers University, Sweden
Sing Bing Kang	Zillow, USA
Gunhee Kim	Seoul National University, South Korea
Junmo Kim	Korea Advanced Institute of Science and Technology, South Korea
Tae-Kyun Kim	Imperial College London, UK
Ron Kimmel	Technion-Israel Institute of Technology, Israel
Alexander Kirillov	Facebook AI Research, USA
Kris Kitani	Carnegie Mellon University, USA
Iasonas Kokkinos	Ariel AI, UK
Vladlen Koltun	Intel Labs, USA
Nikos Komodakis	Ecole des Ponts ParisTech, France
Piotr Koniusz	Australian National University, Australia
M. Pawan Kumar	University of Oxford, UK
Kyros Kutulakos	University of Toronto, Canada
Christoph Lampert	IST Austria, Austria
Ivan Laptev	Inria, France
Diane Larlus	NAVER LABS Europe, France
Laura Leal-Taixe	Technical University Munich, Germany
Honglak Lee	Google and University of Michigan, USA
Joon-Young Lee	Adobe Research, USA
Kyoung Mu Lee	Seoul National University, South Korea
Seungyong Lee	POSTECH, South Korea
Yong Jae Lee	University of California, Davis, USA
Bastian Leibe	RWTH Aachen University, Germany
Victor Lempitsky	Samsung, Russia
Ales Leonardis	University of Birmingham, UK
Marius Leordeanu	Institute of Mathematics of the Romanian Academy, Romania
Vincent Lepetit	ENPC ParisTech, France
Hongdong Li	The Australian National University, Australia
Xi Li	Zhejiang University, China
Yin Li	University of Wisconsin-Madison, USA
Zicheng Liao	Zhejiang University, China
Jongwoo Lim	Hanyang University, South Korea
Stephen Lin	Microsoft Research Asia, China
Yen-Yu Lin	National Chiao Tung University, Taiwan, China
Zhe Lin	Adobe Research, USA

Haibin Ling	Stony Brooks, State University of New York, USA
Jiaying Liu	Peking University, China
Ming-Yu Liu	NVIDIA, USA
Si Liu	Beihang University, China
Xiaoming Liu	Michigan State University, USA
Huchuan Lu	Dalian University of Technology, China
Simon Lucey	Carnegie Mellon University, USA
Jiebo Luo	University of Rochester, USA
Julien Mairal	Inria, France
Michael Maire	University of Chicago, USA
Subhransu Maji	University of Massachusetts, Amherst, USA
Yasushi Makihara	Osaka University, Japan
Jiri Matas	Czech Technical University in Prague, Czech Republic
Yasuyuki Matsushita	Osaka University, Japan
Philippos Mordohai	Stevens Institute of Technology, USA
Vittorio Murino	University of Verona, Italy
Naila Murray	NAVER LABS Europe, France
Hajime Nagahara	Osaka University, Japan
P. J. Narayanan	International Institute of Information Technology (IIIT), Hyderabad, India
Nassir Navab	Technical University of Munich, Germany
Natalia Neverova	Facebook AI Research, France
Matthias Niessner	Technical University of Munich, Germany
Jean-Marc Odobez	Idiap Research Institute and Swiss Federal Institute of Technology Lausanne, Switzerland
Francesca Odone	Università di Genova, Italy
Takeshi Oishi	The University of Tokyo, Tokyo Institute of Technology, Japan
Vicente Ordonez	University of Virginia, USA
Manohar Paluri	Facebook AI Research, USA
Maja Pantic	Imperial College London, UK
In Kyu Park	Inha University, South Korea
Ioannis Patras	Queen Mary University of London, UK
Patrick Perez	Valeo, France
Bryan A. Plummer	Boston University, USA
Thomas Pock	Graz University of Technology, Austria
Marc Pollefeys	ETH Zurich and Microsoft MR & AI Zurich Lab, Switzerland
Jean Ponce	Inria, France
Gerard Pons-Moll	MPII, Saarland Informatics Campus, Germany
Jordi Pont-Tuset	Google, Switzerland
James Matthew Rehg	Georgia Institute of Technology, USA
Ian Reid	University of Adelaide, Australia
Olaf Ronneberger	DeepMind London, UK
Stefan Roth	TU Darmstadt, Germany
Bryan Russell	Adobe Research, USA

Mathieu Salzmann	EPFL, Switzerland
Dimitris Samaras	Stony Brook University, USA
Imari Sato	National Institute of Informatics (NII), Japan
Yoichi Sato	The University of Tokyo, Japan
Torsten Sattler	Czech Technical University in Prague, Czech Republic
Daniel Scharstein	Middlebury College, USA
Bernt Schiele	MPII, Saarland Informatics Campus, Germany
Julia A. Schnabel	King's College London, UK
Nicu Sebe	University of Trento, Italy
Greg Shakhnarovich	Toyota Technological Institute at Chicago, USA
Humphrey Shi	University of Oregon, USA
Jianbo Shi	University of Pennsylvania, USA
Jianping Shi	SenseTime, China
Leonid Sigal	University of British Columbia, Canada
Cees Snoek	University of Amsterdam, The Netherlands
Richard Souvenir	Temple University, USA
Hao Su	University of California, San Diego, USA
Akihiro Sugimoto	National Institute of Informatics (NII), Japan
Jian Sun	Megvii Technology, China
Jian Sun	Xi'an Jiaotong University, China
Chris Sweeney	Facebook Reality Labs, USA
Yu-wing Tai	Kuaishou Technology, China
Chi-Keung Tang	The Hong Kong University of Science and Technology, SAR China
Radu Timofte	ETH Zurich, Switzerland
Sinisa Todorovic	Oregon State University, USA
Giorgos Tolias	Czech Technical University in Prague, Czech Republic
Carlo Tomasi	Duke University, USA
Tatiana Tommasi	Politecnico di Torino, Italy
Lorenzo Torresani	Facebook AI Research and Dartmouth College, USA
Alexander Toshev	Google, USA
Zhuowen Tu	University of California, San Diego, USA
Tinne Tuytelaars	KU Leuven, Belgium
Jasper Uijlings	Google, Switzerland
Nuno Vasconcelos	University of California, San Diego, USA
Olga Veksler	University of Waterloo, Canada
Rene Vidal	Johns Hopkins University, USA
Gang Wang	Alibaba Group, China
Jingdong Wang	Microsoft Research Asia, China
Yizhou Wang	Peking University, China
Lior Wolf	Facebook AI Research and Tel Aviv University, Israel
Jianxin Wu	Nanjing University, China
Tao Xiang	University of Surrey, UK
Saining Xie	Facebook AI Research, USA
Ming-Hsuan Yang	University of California at Merced and Google, USA
Ruigang Yang	University of Kentucky, USA

Kwang Moo Yi — University of Victoria, Canada
Zhaozheng Yin — Stony Brook, State University of New York, USA
Chang D. Yoo — Korea Advanced Institute of Science and Technology, South Korea
Shaodi You — University of Amsterdam, The Netherlands
Jingyi Yu — ShanghaiTech University, China
Stella Yu — University of California, Berkeley, and ICSI, USA
Stefanos Zafeiriou — Imperial College London, UK
Hongbin Zha — Peking University, China
Tianzhu Zhang — University of Science and Technology of China, China
Liang Zheng — Australian National University, Australia
Todd E. Zickler — Harvard University, USA
Andrew Zisserman — University of Oxford, UK

Technical Program Committee

Sathyanarayanan N. Aakur
Wael Abd Almgaeed
Abdelrahman Abdelhamed
Abdullah Abuolaim
Supreeth Achar
Hanno Ackermann
Ehsan Adeli
Triantafyllos Afouras
Sameer Agarwal
Aishwarya Agrawal
Harsh Agrawal
Pulkit Agrawal
Antonio Agudo
Eirikur Agustsson
Karim Ahmed
Byeongjoo Ahn
Unaiza Ahsan
Thalaiyasingam Ajanthan
Kenan E. Ak
Emre Akbas
Naveed Akhtar
Derya Akkaynak
Yagiz Aksoy
Ziad Al-Halah
Xavier Alameda-Pineda
Jean-Baptiste Alayrac
Samuel Albanie
Shadi Albarqouni
Cenek Albl
Hassan Abu Alhaija
Daniel Aliaga
Mohammad S. Aliakbarian
Rahaf Aljundi
Thiemo Alldieck
Jon Almazan
Jose M. Alvarez
Senjian An
Saket Anand
Codruta Ancuti
Cosmin Ancuti
Peter Anderson
Juan Andrade-Cetto
Alexander Andreopoulos
Misha Andriluka
Dragomir Anguelov
Rushil Anirudh
Michel Antunes
Oisin Mac Aodha
Srikar Appalaraju
Relja Arandjelovic
Nikita Araslanov
Andre Araujo
Helder Araujo
Pablo Arbelaez
Shervin Ardeshir
Sercan O. Arik
Anil Armagan
Anurag Arnab
Chetan Arora
Federica Arrigoni
Mathieu Aubry
Shai Avidan
Angelica I. Aviles-Rivero
Yannis Avrithis
Ismail Ben Ayed
Shekoofeh Azizi
Ioan Andrei Bârsan
Artem Babenko
Deepak Babu Sam
Seung-Hwan Baek
Seungryul Baek
Andrew D. Bagdanov
Shai Bagon
Yuval Bahat
Junjie Bai
Song Bai
Xiang Bai
Yalong Bai
Yancheng Bai
Peter Bajcsy
Slawomir Bak

Mahsa Baktashmotlagh
Kavita Bala
Yogesh Balaji
Guha Balakrishnan
V. N. Balasubramanian
Federico Baldassarre
Vassileios Balntas
Shurjo Banerjee
Aayush Bansal
Ankan Bansal
Jianmin Bao
Linchao Bao
Wenbo Bao
Yingze Bao
Akash Bapat
Md Jawadul Hasan Bappy
Fabien Baradel
Lorenzo Baraldi
Daniel Barath
Adrian Barbu
Kobus Barnard
Nick Barnes
Francisco Barranco
Jonathan T. Barron
Arslan Basharat
Chaim Baskin
Anil S. Baslamisli
Jorge Batista
Kayhan Batmanghelich
Konstantinos Batsos
David Bau
Luis Baumela
Christoph Baur
Eduardo Bayro-Corrochano
Paul Beardsley
Jan Bednavr'ik
Oscar Beijbom
Philippe Bekaert
Esube Bekele
Vasileios Belagiannis
Ohad Ben-Shahar
Abhijit Bendale
Róger Bermúdez-Chacón
Maxim Berman
Jesus Bermudez-cameo
Florian Bernard
Stefano Berretti
Marcelo Bertalmio
Gedas Bertasius
Cigdem Beyan
Lucas Beyer
Vijayakumar Bhagavatula
Arjun Nitin Bhagoji
Apratim Bhattacharyya
Binod Bhattarai
Sai Bi
Jia-Wang Bian
Simone Bianco
Adel Bibi
Tolga Birdal
Tom Bishop
Soma Biswas
Mårten Björkman
Volker Blanz
Vishnu Boddeti
Navaneeth Bodla
Simion-Vlad Bogolin
Xavier Boix
Piotr Bojanowski
Timo Bolkart
Guido Borghi
Larbi Boubchir
Guillaume Bourmaud
Adrien Bousseau
Thierry Bouwmans
Richard Bowden
Hakan Boyraz
Mathieu Brédif
Samarth Brahmbhatt
Steve Branson
Nikolas Brasch
Biagio Brattoli
Ernesto Brau
Toby P. Breckon
Francois Bremond
Jesus Briales
Sofia Broomé
Marcus A. Brubaker
Luc Brun
Silvia Bucci
Shyamal Buch
Pradeep Buddharaju
Uta Buechler
Mai Bui
Tu Bui
Adrian Bulat
Giedrius T. Burachas
Elena Burceanu
Xavier P. Burgos-Artizzu
Kaylee Burns
Andrei Bursuc
Benjamin Busam
Wonmin Byeon
Zoya Bylinskii
Sergi Caelles
Jianrui Cai
Minjie Cai
Yujun Cai
Zhaowei Cai
Zhipeng Cai
Juan C. Caicedo
Simone Calderara
Necati Cihan Camgoz
Dylan Campbell
Octavia Camps
Jiale Cao
Kaidi Cao
Liangliang Cao
Xiangyong Cao
Xiaochun Cao
Yang Cao
Yu Cao
Yue Cao
Zhangjie Cao
Luca Carlone
Mathilde Caron
Dan Casas
Thomas J. Cashman
Umberto Castellani
Lluis Castrejon
Jacopo Cavazza
Fabio Cermelli
Hakan Cevikalp
Menglei Chai
Ishani Chakraborty
Rudrasis Chakraborty
Antoni B. Chan

Kwok-Ping Chan
Siddhartha Chandra
Sharat Chandran
Arjun Chandrasekaran
Angel X. Chang
Che-Han Chang
Hong Chang
Hyun Sung Chang
Hyung Jin Chang
Jianlong Chang
Ju Yong Chang
Ming-Ching Chang
Simyung Chang
Xiaojun Chang
Yu-Wei Chao
Devendra S. Chaplot
Arslan Chaudhry
Rizwan A. Chaudhry
Can Chen
Chang Chen
Chao Chen
Chen Chen
Chu-Song Chen
Dapeng Chen
Dong Chen
Dongdong Chen
Guanying Chen
Hongge Chen
Hsin-yi Chen
Huaijin Chen
Hwann-Tzong Chen
Jianbo Chen
Jianhui Chen
Jiansheng Chen
Jiaxin Chen
Jie Chen
Jun-Cheng Chen
Kan Chen
Kevin Chen
Lin Chen
Long Chen
Min-Hung Chen
Qifeng Chen
Shi Chen
Shixing Chen
Tianshui Chen
Weifeng Chen
Weikai Chen
Xi Chen
Xiaohan Chen
Xiaozhi Chen
Xilin Chen
Xingyu Chen
Xinlei Chen
Xinyun Chen
Yi-Ting Chen
Yilun Chen
Ying-Cong Chen
Yinpeng Chen
Yiran Chen
Yu Chen
Yu-Sheng Chen
Yuhua Chen
Yun-Chun Chen
Yunpeng Chen
Yuntao Chen
Zhuoyuan Chen
Zitian Chen
Anchieh Cheng
Bowen Cheng
Erkang Cheng
Gong Cheng
Guangliang Cheng
Jingchun Cheng
Jun Cheng
Li Cheng
Ming-Ming Cheng
Yu Cheng
Ziang Cheng
Anoop Cherian
Dmitry Chetverikov
Ngai-man Cheung
William Cheung
Ajad Chhatkuli
Naoki Chiba
Benjamin Chidester
Han-pang Chiu
Mang Tik Chiu
Wei-Chen Chiu
Donghyeon Cho
Hojin Cho
Minsu Cho
Nam Ik Cho
Tim Cho
Tae Eun Choe
Chiho Choi
Edward Choi
Inchang Choi
Jinsoo Choi
Jonghyun Choi
Jongwon Choi
Yukyung Choi
Hisham Cholakkal
Eunji Chong
Jaegul Choo
Christopher Choy
Hang Chu
Peng Chu
Wen-Sheng Chu
Albert Chung
Joon Son Chung
Hai Ci
Safa Cicek
Ramazan G. Cinbis
Arridhana Ciptadi
Javier Civera
James J. Clark
Ronald Clark
Felipe Codevilla
Michael Cogswell
Andrea Cohen
Maxwell D. Collins
Carlo Colombo
Yang Cong
Adria R. Continente
Marcella Cornia
John Richard Corring
Darren Cosker
Dragos Costea
Garrison W. Cottrell
Florent Couzinie-Devy
Marco Cristani
Ioana Croitoru
James L. Crowley
Jiequan Cui
Zhaopeng Cui
Ross Cutler
Antonio D'Innocente

Rozenn Dahyot
Bo Dai
Dengxin Dai
Hang Dai
Longquan Dai
Shuyang Dai
Xiyang Dai
Yuchao Dai
Adrian V. Dalca
Dima Damen
Bharath B. Damodaran
Kristin Dana
Martin Danelljan
Zheng Dang
Zachary Alan Daniels
Donald G. Dansereau
Abhishek Das
Samyak Datta
Achal Dave
Titas De
Rodrigo de Bem
Teo de Campos
Raoul de Charette
Shalini De Mello
Joseph DeGol
Herve Delingette
Haowen Deng
Jiankang Deng
Weijian Deng
Zhiwei Deng
Joachim Denzler
Konstantinos G. Derpanis
Aditya Deshpande
Frederic Devernay
Somdip Dey
Arturo Deza
Abhinav Dhall
Helisa Dhamo
Vikas Dhiman
Fillipe Dias Moreira de Souza
Ali Diba
Ferran Diego
Guiguang Ding
Henghui Ding
Jian Ding
Mingyu Ding
Xinghao Ding
Zhengming Ding
Robert DiPietro
Cosimo Distante
Ajay Divakaran
Mandar Dixit
Abdelaziz Djelouah
Thanh-Toan Do
Jose Dolz
Bo Dong
Chao Dong
Jiangxin Dong
Weiming Dong
Weisheng Dong
Xingping Dong
Xuanyi Dong
Yinpeng Dong
Gianfranco Doretto
Hazel Doughty
Hassen Drira
Bertram Drost
Dawei Du
Ye Duan
Yueqi Duan
Abhimanyu Dubey
Anastasia Dubrovina
Stefan Duffner
Chi Nhan Duong
Thibaut Durand
Zoran Duric
Iulia Duta
Debidatta Dwibedi
Benjamin Eckart
Marc Eder
Marzieh Edraki
Alexei A. Efros
Kiana Ehsani
Hazm Kemal Ekenel
James H. Elder
Mohamed Elgharib
Shireen Elhabian
Ehsan Elhamifar
Mohamed Elhoseiny
Ian Endres
N. Benjamin Erichson
Jan Ernst
Sergio Escalera
Francisco Escolano
Victor Escorcia
Carlos Esteves
Francisco J. Estrada
Bin Fan
Chenyou Fan
Deng-Ping Fan
Haoqi Fan
Hehe Fan
Heng Fan
Kai Fan
Lijie Fan
Linxi Fan
Quanfu Fan
Shaojing Fan
Xiaochuan Fan
Xin Fan
Yuchen Fan
Sean Fanello
Hao-Shu Fang
Haoyang Fang
Kuan Fang
Yi Fang
Yuming Fang
Azade Farshad
Alireza Fathi
Raanan Fattal
Joao Fayad
Xiaohan Fei
Christoph Feichtenhofer
Michael Felsberg
Chen Feng
Jiashi Feng
Junyi Feng
Mengyang Feng
Qianli Feng
Zhenhua Feng
Michele Fenzi
Andras Ferencz
Martin Fergie
Basura Fernando
Ethan Fetaya
Michael Firman
John W. Fisher

Matthew Fisher
Boris Flach
Corneliu Florea
Wolfgang Foerstner
David Fofi
Gian Luca Foresti
Per-Erik Forssen
David Fouhey
Katerina Fragkiadaki
Victor Fragoso
Jean-Sébastien Franco
Ohad Fried
Iuri Frosio
Cheng-Yang Fu
Huazhu Fu
Jianlong Fu
Jingjing Fu
Xueyang Fu
Yanwei Fu
Ying Fu
Yun Fu
Olac Fuentes
Kent Fujiwara
Takuya Funatomi
Christopher Funk
Thomas Funkhouser
Antonino Furnari
Ryo Furukawa
Erik Gärtner
Raghudeep Gadde
Matheus Gadelha
Vandit Gajjar
Trevor Gale
Juergen Gall
Mathias Gallardo
Guillermo Gallego
Orazio Gallo
Chuang Gan
Zhe Gan
Madan Ravi Ganesh
Aditya Ganeshan
Siddha Ganju
Bin-Bin Gao
Changxin Gao
Feng Gao
Hongchang Gao
Jin Gao
Jiyang Gao
Junbin Gao
Katelyn Gao
Lin Gao
Mingfei Gao
Ruiqi Gao
Ruohan Gao
Shenghua Gao
Yuan Gao
Yue Gao
Noa Garcia
Alberto Garcia-Garcia
Guillermo Garcia-Hernando
Jacob R. Gardner
Animesh Garg
Kshitiz Garg
Rahul Garg
Ravi Garg
Philip N. Garner
Kirill Gavrilyuk
Paul Gay
Shiming Ge
Weifeng Ge
Baris Gecer
Xin Geng
Kyle Genova
Stamatios Georgoulis
Bernard Ghanem
Michael Gharbi
Kamran Ghasedi
Golnaz Ghiasi
Arnab Ghosh
Partha Ghosh
Silvio Giancola
Andrew Gilbert
Rohit Girdhar
Xavier Giro-i-Nieto
Thomas Gittings
Ioannis Gkioulekas
Clement Godard
Vaibhava Goel
Bastian Goldluecke
Lluis Gomez
Nuno Gonçalves
Dong Gong
Ke Gong
Mingming Gong
Abel Gonzalez-Garcia
Ariel Gordon
Daniel Gordon
Paulo Gotardo
Venu Madhav Govindu
Ankit Goyal
Priya Goyal
Raghav Goyal
Benjamin Graham
Douglas Gray
Brent A. Griffin
Etienne Grossmann
David Gu
Jiayuan Gu
Jiuxiang Gu
Lin Gu
Qiao Gu
Shuhang Gu
Jose J. Guerrero
Paul Guerrero
Jie Gui
Jean-Yves Guillemaut
Riza Alp Guler
Erhan Gundogdu
Fatma Guney
Guodong Guo
Kaiwen Guo
Qi Guo
Sheng Guo
Shi Guo
Tiantong Guo
Xiaojie Guo
Yijie Guo
Yiluan Guo
Yuanfang Guo
Yulan Guo
Agrim Gupta
Ankush Gupta
Mohit Gupta
Saurabh Gupta
Tanmay Gupta
Danna Gurari
Abner Guzman-Rivera

JunYoung Gwak
Michael Gygli
Jung-Woo Ha
Simon Hadfield
Isma Hadji
Bjoern Haefner
Taeyoung Hahn
Levente Hajder
Peter Hall
Emanuela Haller
Stefan Haller
Bumsub Ham
Abdullah Hamdi
Dongyoon Han
Hu Han
Jungong Han
Junwei Han
Kai Han
Tian Han
Xiaoguang Han
Xintong Han
Yahong Han
Ankur Handa
Zekun Hao
Albert Haque
Tatsuya Harada
Mehrtash Harandi
Adam W. Harley
Mahmudul Hasan
Atsushi Hashimoto
Ali Hatamizadeh
Munawar Hayat
Dongliang He
Jingrui He
Junfeng He
Kaiming He
Kun He
Lei He
Pan He
Ran He
Shengfeng He
Tong He
Weipeng He
Xuming He
Yang He
Yihui He
Zhihai He
Chinmay Hegde
Janne Heikkila
Mattias P. Heinrich
Stéphane Herbin
Alexander Hermans
Luis Herranz
John R. Hershey
Aaron Hertzmann
Roei Herzig
Anders Heyden
Steven Hickson
Otmar Hilliges
Tomas Hodan
Judy Hoffman
Michael Hofmann
Yannick Hold-Geoffroy
Namdar Homayounfar
Sina Honari
Richang Hong
Seunghoon Hong
Xiaopeng Hong
Yi Hong
Hidekata Hontani
Anthony Hoogs
Yedid Hoshen
Mir Rayat Imtiaz Hossain
Junhui Hou
Le Hou
Lu Hou
Tingbo Hou
Wei-Lin Hsiao
Cheng-Chun Hsu
Gee-Sern Jison Hsu
Kuang-jui Hsu
Changbo Hu
Di Hu
Guosheng Hu
Han Hu
Hao Hu
Hexiang Hu
Hou-Ning Hu
Jie Hu
Junlin Hu
Nan Hu
Ping Hu
Ronghang Hu
Xiaowei Hu
Yinlin Hu
Yuan-Ting Hu
Zhe Hu
Binh-Son Hua
Yang Hua
Bingyao Huang
Di Huang
Dong Huang
Fay Huang
Haibin Huang
Haozhi Huang
Heng Huang
Huaibo Huang
Jia-Bin Huang
Jing Huang
Jingwei Huang
Kaizhu Huang
Lei Huang
Qiangui Huang
Qiaoying Huang
Qingqiu Huang
Qixing Huang
Shaoli Huang
Sheng Huang
Siyuan Huang
Weilin Huang
Wenbing Huang
Xiangru Huang
Xun Huang
Yan Huang
Yifei Huang
Yue Huang
Zhiwu Huang
Zilong Huang
Minyoung Huh
Zhuo Hui
Matthias B. Hullin
Martin Humenberger
Wei-Chih Hung
Zhouyuan Huo
Junhwa Hur
Noureldien Hussein
Jyh-Jing Hwang
Seong Jae Hwang

Sung Ju Hwang
Ichiro Ide
Ivo Ihrke
Daiki Ikami
Satoshi Ikehata
Nazli Ikizler-Cinbis
Sunghoon Im
Yani Ioannou
Radu Tudor Ionescu
Umar Iqbal
Go Irie
Ahmet Iscen
Md Amirul Islam
Vamsi Ithapu
Nathan Jacobs
Arpit Jain
Himalaya Jain
Suyog Jain
Stuart James
Won-Dong Jang
Yunseok Jang
Ronnachai Jaroensri
Dinesh Jayaraman
Sadeep Jayasumana
Suren Jayasuriya
Herve Jegou
Simon Jenni
Hae-Gon Jeon
Yunho Jeon
Koteswar R. Jerripothula
Hueihan Jhuang
I-hong Jhuo
Dinghuang Ji
Hui Ji
Jingwei Ji
Pan Ji
Yanli Ji
Baoxiong Jia
Kui Jia
Xu Jia
Chiyu Max Jiang
Haiyong Jiang
Hao Jiang
Huaizu Jiang
Huajie Jiang
Ke Jiang
Lai Jiang
Li Jiang
Lu Jiang
Ming Jiang
Peng Jiang
Shuqiang Jiang
Wei Jiang
Xudong Jiang
Zhuolin Jiang
Jianbo Jiao
Zequn Jie
Dakai Jin
Kyong Hwan Jin
Lianwen Jin
SouYoung Jin
Xiaojie Jin
Xin Jin
Nebojsa Jojic
Alexis Joly
Michael Jeffrey Jones
Hanbyul Joo
Jungseock Joo
Kyungdon Joo
Ajjen Joshi
Shantanu H. Joshi
Da-Cheng Juan
Marco Körner
Kevin Köser
Asim Kadav
Christine Kaeser-Chen
Kushal Kafle
Dagmar Kainmueller
Ioannis A. Kakadiaris
Zdenek Kalal
Nima Kalantari
Yannis Kalantidis
Mahdi M. Kalayeh
Anmol Kalia
Sinan Kalkan
Vicky Kalogeiton
Ashwin Kalyan
Joni-kristian Kamarainen
Gerda Kamberova
Chandra Kambhamettu
Martin Kampel
Meina Kan
Christopher Kanan
Kenichi Kanatani
Angjoo Kanazawa
Atsushi Kanehira
Takuhiro Kaneko
Asako Kanezaki
Bingyi Kang
Di Kang
Sunghun Kang
Zhao Kang
Vadim Kantorov
Abhishek Kar
Amlan Kar
Theofanis Karaletsos
Leonid Karlinsky
Kevin Karsch
Angelos Katharopoulos
Isinsu Katircioglu
Hiroharu Kato
Zoltan Kato
Dotan Kaufman
Jan Kautz
Rei Kawakami
Qiuhong Ke
Wadim Kehl
Petr Kellnhofer
Aniruddha Kembhavi
Cem Keskin
Margret Keuper
Daniel Keysers
Ashkan Khakzar
Fahad Khan
Naeemullah Khan
Salman Khan
Siddhesh Khandelwal
Rawal Khirodkar
Anna Khoreva
Tejas Khot
Parmeshwar Khurd
Hadi Kiapour
Joe Kileel
Chanho Kim
Dahun Kim
Edward Kim
Eunwoo Kim
Han-ul Kim

Hansung Kim
Heewon Kim
Hyo Jin Kim
Hyunwoo J. Kim
Jinkyu Kim
Jiwon Kim
Jongmin Kim
Junsik Kim
Junyeong Kim
Min H. Kim
Namil Kim
Pyojin Kim
Seon Joo Kim
Seong Tae Kim
Seungryong Kim
Sungwoong Kim
Tae Hyun Kim
Vladimir Kim
Won Hwa Kim
Yonghyun Kim
Benjamin Kimia
Akisato Kimura
Pieter-Jan Kindermans
Zsolt Kira
Itaru Kitahara
Hedvig Kjellstrom
Jan Knopp
Takumi Kobayashi
Erich Kobler
Parker Koch
Reinhard Koch
Elyor Kodirov
Amir Kolaman
Nicholas Kolkin
Dimitrios Kollias
Stefanos Kollias
Soheil Kolouri
Adams Wai-Kin Kong
Naejin Kong
Shu Kong
Tao Kong
Yu Kong
Yoshinori Konishi
Daniil Kononenko
Theodora Kontogianni
Simon Korman
Adam Kortylewski
Jana Kosecka
Jean Kossaifi
Satwik Kottur
Rigas Kouskouridas
Adriana Kovashka
Rama Kovvuri
Adarsh Kowdle
Jedrzej Kozerawski
Mateusz Kozinski
Philipp Kraehenbuehl
Gregory Kramida
Josip Krapac
Dmitry Kravchenko
Ranjay Krishna
Pavel Krsek
Alexander Krull
Jakob Kruse
Hiroyuki Kubo
Hilde Kuehne
Jason Kuen
Andreas Kuhn
Arjan Kuijper
Zuzana Kukelova
Ajay Kumar
Amit Kumar
Avinash Kumar
Suryansh Kumar
Vijay Kumar
Kaustav Kundu
Weicheng Kuo
Nojun Kwak
Suha Kwak
Junseok Kwon
Nikolaos Kyriazis
Zorah Lähner
Ankit Laddha
Florent Lafarge
Jean Lahoud
Kevin Lai
Shang-Hong Lai
Wei-Sheng Lai
Yu-Kun Lai
Iro Laina
Antony Lam
John Wheatley Lambert
Xiangyuan lan
Xu Lan
Charis Lanaras
Georg Langs
Oswald Lanz
Dong Lao
Yizhen Lao
Agata Lapedriza
Gustav Larsson
Viktor Larsson
Katrin Lasinger
Christoph Lassner
Longin Jan Latecki
Stéphane Lathuilière
Rynson Lau
Hei Law
Justin Lazarow
Svetlana Lazebnik
Hieu Le
Huu Le
Ngan Hoang Le
Trung-Nghia Le
Vuong Le
Colin Lea
Erik Learned-Miller
Chen-Yu Lee
Gim Hee Lee
Hsin-Ying Lee
Hyungtae Lee
Jae-Han Lee
Jimmy Addison Lee
Joonseok Lee
Kibok Lee
Kuang-Huei Lee
Kwonjoon Lee
Minsik Lee
Sang-chul Lee
Seungkyu Lee
Soochan Lee
Stefan Lee
Taehee Lee
Andreas Lehrmann
Jie Lei
Peng Lei
Matthew Joseph Leotta
Wee Kheng Leow

Gil Levi
Evgeny Levinkov
Aviad Levis
Jose Lezama
Ang Li
Bin Li
Bing Li
Boyi Li
Changsheng Li
Chao Li
Chen Li
Cheng Li
Chenglong Li
Chi Li
Chun-Guang Li
Chun-Liang Li
Chunyuan Li
Dong Li
Guanbin Li
Hao Li
Haoxiang Li
Hongsheng Li
Hongyang Li
Houqiang Li
Huibin Li
Jia Li
Jianan Li
Jianguo Li
Junnan Li
Junxuan Li
Kai Li
Ke Li
Kejie Li
Kunpeng Li
Lerenhan Li
Li Erran Li
Mengtian Li
Mu Li
Peihua Li
Peiyi Li
Ping Li
Qi Li
Qing Li
Ruiyu Li
Ruoteng Li
Shaozi Li
Sheng Li
Shiwei Li
Shuang Li
Siyang Li
Stan Z. Li
Tianye Li
Wei Li
Weixin Li
Wen Li
Wenbo Li
Xiaomeng Li
Xin Li
Xiu Li
Xuelong Li
Xueting Li
Yan Li
Yandong Li
Yanghao Li
Yehao Li
Yi Li
Yijun Li
Yikang LI
Yining Li
Yongjie Li
Yu Li
Yu-Jhe Li
Yunpeng Li
Yunsheng Li
Yunzhu Li
Zhe Li
Zhen Li
Zhengqi Li
Zhenyang Li
Zhuwen Li
Dongze Lian
Xiaochen Lian
Zhouhui Lian
Chen Liang
Jie Liang
Ming Liang
Paul Pu Liang
Pengpeng Liang
Shu Liang
Wei Liang
Jing Liao
Minghui Liao
Renjie Liao
Shengcai Liao
Shuai Liao
Yiyi Liao
Ser-Nam Lim
Chen-Hsuan Lin
Chung-Ching Lin
Dahua Lin
Ji Lin
Kevin Lin
Tianwei Lin
Tsung-Yi Lin
Tsung-Yu Lin
Wei-An Lin
Weiyao Lin
Yen-Chen Lin
Yuewei Lin
David B. Lindell
Drew Linsley
Krzysztof Lis
Roee Litman
Jim Little
An-An Liu
Bo Liu
Buyu Liu
Chao Liu
Chen Liu
Cheng-lin Liu
Chenxi Liu
Dong Liu
Feng Liu
Guilin Liu
Haomiao Liu
Heshan Liu
Hong Liu
Ji Liu
Jingen Liu
Jun Liu
Lanlan Liu
Li Liu
Liu Liu
Mengyuan Liu
Miaomiao Liu
Nian Liu
Ping Liu
Risheng Liu

Sheng Liu
Shu Liu
Shuaicheng Liu
Sifei Liu
Siqi Liu
Siying Liu
Songtao Liu
Ting Liu
Tongliang Liu
Tyng-Luh Liu
Wanquan Liu
Wei Liu
Weiyang Liu
Weizhe Liu
Wenyu Liu
Wu Liu
Xialei Liu
Xianglong Liu
Xiaodong Liu
Xiaofeng Liu
Xihui Liu
Xingyu Liu
Xinwang Liu
Xuanqing Liu
Xuebo Liu
Yang Liu
Yaojie Liu
Yebin Liu
Yen-Cheng Liu
Yiming Liu
Yu Liu
Yu-Shen Liu
Yufan Liu
Yun Liu
Zheng Liu
Zhijian Liu
Zhuang Liu
Zichuan Liu
Ziwei Liu
Zongyi Liu
Stephan Liwicki
Liliana Lo Presti
Chengjiang Long
Fuchen Long
Mingsheng Long
Xiang Long
Yang Long
Charles T. Loop
Antonio Lopez
Roberto J. Lopez-Sastre
Javier Lorenzo-Navarro
Manolis Lourakis
Boyu Lu
Canyi Lu
Feng Lu
Guoyu Lu
Hongtao Lu
Jiajun Lu
Jiasen Lu
Jiwen Lu
Kaiyue Lu
Le Lu
Shao-Ping Lu
Shijian Lu
Xiankai Lu
Xin Lu
Yao Lu
Yiping Lu
Yongxi Lu
Yongyi Lu
Zhiwu Lu
Fujun Luan
Benjamin E. Lundell
Hao Luo
Jian-Hao Luo
Ruotian Luo
Weixin Luo
Wenhan Luo
Wenjie Luo
Yan Luo
Zelun Luo
Zixin Luo
Khoa Luu
Zhaoyang Lv
Pengyuan Lyu
Thomas Möllenhoff
Matthias Müller
Bingpeng Ma
Chih-Yao Ma
Chongyang Ma
Huimin Ma
Jiayi Ma
K. T. Ma
Ke Ma
Lin Ma
Liqian Ma
Shugao Ma
Wei-Chiu Ma
Xiaojian Ma
Xingjun Ma
Zhanyu Ma
Zheng Ma
Radek Jakob Mackowiak
Ludovic Magerand
Shweta Mahajan
Siddharth Mahendran
Long Mai
Ameesh Makadia
Oscar Mendez Maldonado
Mateusz Malinowski
Yury Malkov
Arun Mallya
Dipu Manandhar
Massimiliano Mancini
Fabian Manhardt
Kevis-kokitsi Maninis
Varun Manjunatha
Junhua Mao
Xudong Mao
Alina Marcu
Edgar Margffoy-Tuay
Dmitrii Marin
Manuel J. Marin-Jimenez
Kenneth Marino
Niki Martinel
Julieta Martinez
Jonathan Masci
Tomohiro Mashita
Iacopo Masi
David Masip
Daniela Massiceti
Stefan Mathe
Yusuke Matsui
Tetsu Matsukawa
Iain A. Matthews
Kevin James Matzen
Bruce Allen Maxwell
Stephen Maybank

Helmut Mayer
Amir Mazaheri
David McAllester
Steven McDonagh
Stephen J. Mckenna
Roey Mechrez
Prakhar Mehrotra
Christopher Mei
Xue Mei
Paulo R. S. Mendonca
Lili Meng
Zibo Meng
Thomas Mensink
Bjoern Menze
Michele Merler
Kourosh Meshgi
Pascal Mettes
Christopher Metzler
Liang Mi
Qiguang Miao
Xin Miao
Tomer Michaeli
Frank Michel
Antoine Miech
Krystian Mikolajczyk
Peyman Milanfar
Ben Mildenhall
Gregor Miller
Fausto Milletari
Dongbo Min
Kyle Min
Pedro Miraldo
Dmytro Mishkin
Anand Mishra
Ashish Mishra
Ishan Misra
Niluthpol C. Mithun
Kaushik Mitra
Niloy Mitra
Anton Mitrokhin
Ikuhisa Mitsugami
Anurag Mittal
Kaichun Mo
Zhipeng Mo
Davide Modolo
Michael Moeller
Pritish Mohapatra
Pavlo Molchanov
Davide Moltisanti
Pascal Monasse
Mathew Monfort
Aron Monszpart
Sean Moran
Vlad I. Morariu
Francesc Moreno-Noguer
Pietro Morerio
Stylianos Moschoglou
Yael Moses
Roozbeh Mottaghi
Pierre Moulon
Arsalan Mousavian
Yadong Mu
Yasuhiro Mukaigawa
Lopamudra Mukherjee
Yusuke Mukuta
Ravi Teja Mullapudi
Mario Enrique Munich
Zachary Murez
Ana C. Murillo
J. Krishna Murthy
Damien Muselet
Armin Mustafa
Siva Karthik Mustikovela
Carlo Dal Mutto
Moin Nabi
Varun K. Nagaraja
Tushar Nagarajan
Arsha Nagrani
Seungjun Nah
Nikhil Naik
Yoshikatsu Nakajima
Yuta Nakashima
Atsushi Nakazawa
Seonghyeon Nam
Vinay P. Namboodiri
Medhini Narasimhan
Srinivasa Narasimhan
Sanath Narayan
Erickson Rangel Nascimento
Jacinto Nascimento
Tayyab Naseer
Lakshmanan Nataraj
Neda Nategh
Nelson Isao Nauata
Fernando Navarro
Shah Nawaz
Lukas Neumann
Ram Nevatia
Alejandro Newell
Shawn Newsam
Joe Yue-Hei Ng
Trung Thanh Ngo
Duc Thanh Nguyen
Lam M. Nguyen
Phuc Xuan Nguyen
Thuong Nguyen Canh
Mihalis Nicolaou
Andrei Liviu Nicolicioiu
Xuecheng Nie
Michael Niemeyer
Simon Niklaus
Christophoros Nikou
David Nilsson
Jifeng Ning
Yuval Nirkin
Li Niu
Yuzhen Niu
Zhenxing Niu
Shohei Nobuhara
Nicoletta Noceti
Hyeonwoo Noh
Junhyug Noh
Mehdi Noroozi
Sotiris Nousias
Valsamis Ntouskos
Matthew O'Toole
Peter Ochs
Ferda Ofli
Seong Joon Oh
Seoung Wug Oh
Iason Oikonomidis
Utkarsh Ojha
Takahiro Okabe
Takayuki Okatani
Fumio Okura
Aude Oliva
Kyle Olszewski

Björn Ommer
Mohamed Omran
Elisabeta Oneata
Michael Opitz
Jose Oramas
Tribhuvanesh Orekondy
Shaul Oron
Sergio Orts-Escolano
Ivan Oseledets
Aljosa Osep
Magnus Oskarsson
Anton Osokin
Martin R. Oswald
Wanli Ouyang
Andrew Owens
Mete Ozay
Mustafa Ozuysal
Eduardo Pérez-Pellitero
Gautam Pai
Dipan Kumar Pal
P. H. Pamplona Savarese
Jinshan Pan
Junting Pan
Xingang Pan
Yingwei Pan
Yannis Panagakis
Rameswar Panda
Guan Pang
Jiahao Pang
Jiangmiao Pang
Tianyu Pang
Sharath Pankanti
Nicolas Papadakis
Dim Papadopoulos
George Papandreou
Toufiq Parag
Shaifali Parashar
Sarah Parisot
Eunhyeok Park
Hyun Soo Park
Jaesik Park
Min-Gyu Park
Taesung Park
Alvaro Parra
C. Alejandro Parraga
Despoina Paschalidou
Nikolaos Passalis
Vishal Patel
Viorica Patraucean
Badri Narayana Patro
Danda Pani Paudel
Sujoy Paul
Georgios Pavlakos
Ioannis Pavlidis
Vladimir Pavlovic
Nick Pears
Kim Steenstrup Pedersen
Selen Pehlivan
Shmuel Peleg
Chao Peng
Houwen Peng
Wen-Hsiao Peng
Xi Peng
Xiaojiang Peng
Xingchao Peng
Yuxin Peng
Federico Perazzi
Juan Camilo Perez
Vishwanath Peri
Federico Pernici
Luca Del Pero
Florent Perronnin
Stavros Petridis
Henning Petzka
Patrick Peursum
Michael Pfeiffer
Hanspeter Pfister
Roman Pflugfelder
Minh Tri Pham
Yongri Piao
David Picard
Tomasz Pieciak
A. J. Piergiovanni
Andrea Pilzer
Pedro O. Pinheiro
Silvia Laura Pintea
Lerrel Pinto
Axel Pinz
Robinson Piramuthu
Fiora Pirri
Leonid Pishchulin
Francesco Pittaluga
Daniel Pizarro
Tobias Plötz
Mirco Planamente
Matteo Poggi
Moacir A. Ponti
Parita Pooj
Fatih Porikli
Horst Possegger
Omid Poursaeed
Ameya Prabhu
Viraj Uday Prabhu
Dilip Prasad
Brian L. Price
True Price
Maria Priisalu
Veronique Prinet
Victor Adrian Prisacariu
Jan Prokaj
Sergey Prokudin
Nicolas Pugeault
Xavier Puig
Albert Pumarola
Pulak Purkait
Senthil Purushwalkam
Charles R. Qi
Hang Qi
Haozhi Qi
Lu Qi
Mengshi Qi
Siyuan Qi
Xiaojuan Qi
Yuankai Qi
Shengju Qian
Xuelin Qian
Siyuan Qiao
Yu Qiao
Jie Qin
Qiang Qiu
Weichao Qiu
Zhaofan Qiu
Kha Gia Quach
Yuhui Quan
Yvain Queau
Julian Quiroga
Faisal Qureshi
Mahdi Rad

Filip Radenovic
Petia Radeva
Venkatesh
B. Radhakrishnan
Ilija Radosavovic
Noha Radwan
Rahul Raguram
Tanzila Rahman
Amit Raj
Ajit Rajwade
Kandan Ramakrishnan
Santhosh
K. Ramakrishnan
Srikumar Ramalingam
Ravi Ramamoorthi
Vasili Ramanishka
Ramprasaath R. Selvaraju
Francois Rameau
Visvanathan Ramesh
Santu Rana
Rene Ranftl
Anand Rangarajan
Anurag Ranjan
Viresh Ranjan
Yongming Rao
Carolina Raposo
Vivek Rathod
Sathya N. Ravi
Avinash Ravichandran
Tammy Riklin Raviv
Daniel Rebain
Sylvestre-Alvise Rebuffi
N. Dinesh Reddy
Timo Rehfeld
Paolo Remagnino
Konstantinos Rematas
Edoardo Remelli
Dongwei Ren
Haibing Ren
Jian Ren
Jimmy Ren
Mengye Ren
Weihong Ren
Wenqi Ren
Zhile Ren
Zhongzheng Ren
Zhou Ren
Vijay Rengarajan
Md A. Reza
Farzaneh Rezaeianaran
Hamed R. Tavakoli
Nicholas Rhinehart
Helge Rhodin
Elisa Ricci
Alexander Richard
Eitan Richardson
Elad Richardson
Christian Richardt
Stephan Richter
Gernot Riegler
Daniel Ritchie
Tobias Ritschel
Samuel Rivera
Yong Man Ro
Richard Roberts
Joseph Robinson
Ignacio Rocco
Mrigank Rochan
Emanuele Rodolà
Mikel D. Rodriguez
Giorgio Roffo
Grégory Rogez
Gemma Roig
Javier Romero
Xuejian Rong
Yu Rong
Amir Rosenfeld
Bodo Rosenhahn
Guy Rosman
Arun Ross
Paolo Rota
Peter M. Roth
Anastasios Roussos
Anirban Roy
Sebastien Roy
Aruni RoyChowdhury
Artem Rozantsev
Ognjen Rudovic
Daniel Rueckert
Adria Ruiz
Javier Ruiz-del-solar
Christian Rupprecht
Chris Russell
Dan Ruta
Jongbin Ryu
Ömer Sümer
Alexandre Sablayrolles
Faraz Saeedan
Ryusuke Sagawa
Christos Sagonas
Tonmoy Saikia
Hideo Saito
Kuniaki Saito
Shunsuke Saito
Shunta Saito
Ken Sakurada
Joaquin Salas
Fatemeh Sadat Saleh
Mahdi Saleh
Pouya Samangouei
Leo Sampaio
Ferraz Ribeiro
Artsiom Olegovich
Sanakoyeu
Enrique Sanchez
Patsorn Sangkloy
Anush Sankaran
Aswin Sankaranarayanan
Swami Sankaranarayanan
Rodrigo Santa Cruz
Amartya Sanyal
Archana Sapkota
Nikolaos Sarafianos
Jun Sato
Shin'ichi Satoh
Hosnieh Sattar
Arman Savran
Manolis Savva
Alexander Sax
Hanno Scharr
Simone Schaub-Meyer
Konrad Schindler
Dmitrij Schlesinger
Uwe Schmidt
Dirk Schnieders
Björn Schuller
Samuel Schulter
Idan Schwartz

William Robson Schwartz
Alex Schwing
Sinisa Segvic
Lorenzo Seidenari
Pradeep Sen
Ozan Sener
Soumyadip Sengupta
Arda Senocak
Mojtaba Seyedhosseini
Shishir Shah
Shital Shah
Sohil Atul Shah
Tamar Rott Shaham
Huasong Shan
Qi Shan
Shiguang Shan
Jing Shao
Roman Shapovalov
Gaurav Sharma
Vivek Sharma
Viktoriia Sharmanska
Dongyu She
Sumit Shekhar
Evan Shelhamer
Chengyao Shen
Chunhua Shen
Falong Shen
Jie Shen
Li Shen
Liyue Shen
Shuhan Shen
Tianwei Shen
Wei Shen
William B. Shen
Yantao Shen
Ying Shen
Yiru Shen
Yujun Shen
Yuming Shen
Zhiqiang Shen
Ziyi Shen
Lu Sheng
Yu Sheng
Rakshith Shetty
Baoguang Shi
Guangming Shi
Hailin Shi
Miaojing Shi
Yemin Shi
Zhenmei Shi
Zhiyuan Shi
Kevin Jonathan Shih
Shiliang Shiliang
Hyunjung Shim
Atsushi Shimada
Nobutaka Shimada
Daeyun Shin
Young Min Shin
Koichi Shinoda
Konstantin Shmelkov
Michael Zheng Shou
Abhinav Shrivastava
Tianmin Shu
Zhixin Shu
Hong-Han Shuai
Pushkar Shukla
Christian Siagian
Mennatullah M. Siam
Kaleem Siddiqi
Karan Sikka
Jae-Young Sim
Christian Simon
Martin Simonovsky
Dheeraj Singaraju
Bharat Singh
Gurkirt Singh
Krishna Kumar Singh
Maneesh Kumar Singh
Richa Singh
Saurabh Singh
Suriya Singh
Vikas Singh
Sudipta N. Sinha
Vincent Sitzmann
Josef Sivic
Gregory Slabaugh
Miroslava Slavcheva
Ron Slossberg
Brandon Smith
Kevin Smith
Vladimir Smutny
Noah Snavely
Roger D. Soberanis-Mukul
Kihyuk Sohn
Francesco Solera
Eric Sommerlade
Sanghyun Son
Byung Cheol Song
Chunfeng Song
Dongjin Song
Jiaming Song
Jie Song
Jifei Song
Jingkuan Song
Mingli Song
Shiyu Song
Shuran Song
Xiao Song
Yafei Song
Yale Song
Yang Song
Yi-Zhe Song
Yibing Song
Humberto Sossa
Cesar de Souza
Adrian Spurr
Srinath Sridhar
Suraj Srinivas
Pratul P. Srinivasan
Anuj Srivastava
Tania Stathaki
Christopher Stauffer
Simon Stent
Rainer Stiefelhagen
Pierre Stock
Julian Straub
Jonathan C. Stroud
Joerg Stueckler
Jan Stuehmer
David Stutz
Chi Su
Hang Su
Jong-Chyi Su
Shuochen Su
Yu-Chuan Su
Ramanathan Subramanian
Yusuke Sugano

Masanori Suganuma
Yumin Suh
Mohammed Suhail
Yao Sui
Heung-Il Suk
Josephine Sullivan
Baochen Sun
Chen Sun
Chong Sun
Deqing Sun
Jin Sun
Liang Sun
Lin Sun
Qianru Sun
Shao-Hua Sun
Shuyang Sun
Weiwei Sun
Wenxiu Sun
Xiaoshuai Sun
Xiaoxiao Sun
Xingyuan Sun
Yifan Sun
Zhun Sun
Sabine Susstrunk
David Suter
Supasorn Suwajanakorn
Tomas Svoboda
Eran Swears
Paul Swoboda
Attila Szabo
Richard Szeliski
Duy-Nguyen Ta
Andrea Tagliasacchi
Yuichi Taguchi
Ying Tai
Keita Takahashi
Kouske Takahashi
Jun Takamatsu
Hugues Talbot
Toru Tamaki
Chaowei Tan
Fuwen Tan
Mingkui Tan
Mingxing Tan
Qingyang Tan
Robby T. Tan
Xiaoyang Tan
Kenichiro Tanaka
Masayuki Tanaka
Chang Tang
Chengzhou Tang
Danhang Tang
Ming Tang
Peng Tang
Qingming Tang
Wei Tang
Xu Tang
Yansong Tang
Youbao Tang
Yuxing Tang
Zhiqiang Tang
Tatsunori Taniai
Junli Tao
Xin Tao
Makarand Tapaswi
Jean-Philippe Tarel
Lyne Tchapmi
Zachary Teed
Bugra Tekin
Damien Teney
Ayush Tewari
Christian Theobalt
Christopher Thomas
Diego Thomas
Jim Thomas
Rajat Mani Thomas
Xinmei Tian
Yapeng Tian
Yingli Tian
Yonglong Tian
Zhi Tian
Zhuotao Tian
Kinh Tieu
Joseph Tighe
Massimo Tistarelli
Matthew Toews
Carl Toft
Pavel Tokmakov
Federico Tombari
Chetan Tonde
Yan Tong
Alessio Tonioni
Andrea Torsello
Fabio Tosi
Du Tran
Luan Tran
Ngoc-Trung Tran
Quan Hung Tran
Truyen Tran
Rudolph Triebel
Martin Trimmel
Shashank Tripathi
Subarna Tripathi
Leonardo Trujillo
Eduard Trulls
Tomasz Trzcinski
Sam Tsai
Yi-Hsuan Tsai
Hung-Yu Tseng
Stavros Tsogkas
Aggeliki Tsoli
Devis Tuia
Shubham Tulsiani
Sergey Tulyakov
Frederick Tung
Tony Tung
Daniyar Turmukhambetov
Ambrish Tyagi
Radim Tylecek
Christos Tzelepis
Georgios Tzimiropoulos
Dimitrios Tzionas
Seiichi Uchida
Norimichi Ukita
Dmitry Ulyanov
Martin Urschler
Yoshitaka Ushiku
Ben Usman
Alexander Vakhitov
Julien P. C. Valentin
Jack Valmadre
Ernest Valveny
Joost van de Weijer
Jan van Gemert
Koen Van Leemput
Gul Varol
Sebastiano Vascon
M. Alex O. Vasilescu

Subeesh Vasu
Mayank Vatsa
David Vazquez
Javier Vazquez-Corral
Ashok Veeraraghavan
Erik Velasco-Salido
Raviteja Vemulapalli
Jonathan Ventura
Manisha Verma
Roberto Vezzani
Ruben Villegas
Minh Vo
MinhDuc Vo
Nam Vo
Michele Volpi
Riccardo Volpi
Carl Vondrick
Konstantinos Vougioukas
Tuan-Hung Vu
Sven Wachsmuth
Neal Wadhwa
Catherine Wah
Jacob C. Walker
Thomas S. A. Wallis
Chengde Wan
Jun Wan
Liang Wan
Renjie Wan
Baoyuan Wang
Boyu Wang
Cheng Wang
Chu Wang
Chuan Wang
Chunyu Wang
Dequan Wang
Di Wang
Dilin Wang
Dong Wang
Fang Wang
Guanzhi Wang
Guoyin Wang
Hanzi Wang
Hao Wang
He Wang
Heng Wang
Hongcheng Wang
Hongxing Wang
Hua Wang
Jian Wang
Jingbo Wang
Jinglu Wang
Jingya Wang
Jinjun Wang
Jinqiao Wang
Jue Wang
Ke Wang
Keze Wang
Le Wang
Lei Wang
Lezi Wang
Li Wang
Liang Wang
Lijun Wang
Limin Wang
Linwei Wang
Lizhi Wang
Mengjiao Wang
Mingzhe Wang
Minsi Wang
Naiyan Wang
Nannan Wang
Ning Wang
Oliver Wang
Pei Wang
Peng Wang
Pichao Wang
Qi Wang
Qian Wang
Qiaosong Wang
Qifei Wang
Qilong Wang
Qing Wang
Qingzhong Wang
Quan Wang
Rui Wang
Ruiping Wang
Ruixing Wang
Shangfei Wang
Shenlong Wang
Shiyao Wang
Shuhui Wang
Song Wang
Tao Wang
Tianlu Wang
Tiantian Wang
Ting-chun Wang
Tingwu Wang
Wei Wang
Weiyue Wang
Wenguan Wang
Wenlin Wang
Wenqi Wang
Xiang Wang
Xiaobo Wang
Xiaofang Wang
Xiaoling Wang
Xiaolong Wang
Xiaosong Wang
Xiaoyu Wang
Xin Eric Wang
Xinchao Wang
Xinggang Wang
Xintao Wang
Yali Wang
Yan Wang
Yang Wang
Yangang Wang
Yaxing Wang
Yi Wang
Yida Wang
Yilin Wang
Yiming Wang
Yisen Wang
Yongtao Wang
Yu-Xiong Wang
Yue Wang
Yujiang Wang
Yunbo Wang
Yunhe Wang
Zengmao Wang
Zhangyang Wang
Zhaowen Wang
Zhe Wang
Zhecan Wang
Zheng Wang
Zhixiang Wang
Zilei Wang
Jianqiao Wangni

Anne S. Wannenwetsch
Jan Dirk Wegner
Scott Wehrwein
Donglai Wei
Kaixuan Wei
Longhui Wei
Pengxu Wei
Ping Wei
Qi Wei
Shih-En Wei
Xing Wei
Yunchao Wei
Zijun Wei
Jerod Weinman
Michael Weinmann
Philippe Weinzaepfel
Yair Weiss
Bihan Wen
Longyin Wen
Wei Wen
Junwu Weng
Tsui-Wei Weng
Xinshuo Weng
Eric Wengrowski
Tomas Werner
Gordon Wetzstein
Tobias Weyand
Patrick Wieschollek
Maggie Wigness
Erik Wijmans
Richard Wildes
Olivia Wiles
Chris Williams
Williem Williem
Kyle Wilson
Calden Wloka
Nicolai Wojke
Christian Wolf
Yongkang Wong
Sanghyun Woo
Scott Workman
Baoyuan Wu
Bichen Wu
Chao-Yuan Wu
Huikai Wu
Jiajun Wu
Jialin Wu
Jiaxiang Wu
Jiqing Wu
Jonathan Wu
Lifang Wu
Qi Wu
Qiang Wu
Ruizheng Wu
Shangzhe Wu
Shun-Cheng Wu
Tianfu Wu
Wayne Wu
Wenxuan Wu
Xiao Wu
Xiaohe Wu
Xinxiao Wu
Yang Wu
Yi Wu
Yiming Wu
Ying Nian Wu
Yue Wu
Zheng Wu
Zhenyu Wu
Zhirong Wu
Zuxuan Wu
Stefanie Wuhrer
Jonas Wulff
Changqun Xia
Fangting Xia
Fei Xia
Gui-Song Xia
Lu Xia
Xide Xia
Yin Xia
Yingce Xia
Yongqin Xian
Lei Xiang
Shiming Xiang
Bin Xiao
Fanyi Xiao
Guobao Xiao
Huaxin Xiao
Taihong Xiao
Tete Xiao
Tong Xiao
Wang Xiao
Yang Xiao
Cihang Xie
Guosen Xie
Jianwen Xie
Lingxi Xie
Sirui Xie
Weidi Xie
Wenxuan Xie
Xiaohua Xie
Fuyong Xing
Jun Xing
Junliang Xing
Bo Xiong
Peixi Xiong
Yu Xiong
Yuanjun Xiong
Zhiwei Xiong
Chang Xu
Chenliang Xu
Dan Xu
Danfei Xu
Hang Xu
Hongteng Xu
Huijuan Xu
Jingwei Xu
Jun Xu
Kai Xu
Mengmeng Xu
Mingze Xu
Qianqian Xu
Ran Xu
Weijian Xu
Xiangyu Xu
Xiaogang Xu
Xing Xu
Xun Xu
Yanyu Xu
Yichao Xu
Yong Xu
Yongchao Xu
Yuanlu Xu
Zenglin Xu
Zheng Xu
Chuhui Xue
Jia Xue
Nan Xue

Tianfan Xue
Xiangyang Xue
Abhay Yadav
Yasushi Yagi
I. Zeki Yalniz
Kota Yamaguchi
Toshihiko Yamasaki
Takayoshi Yamashita
Junchi Yan
Ke Yan
Qingan Yan
Sijie Yan
Xinchen Yan
Yan Yan
Yichao Yan
Zhicheng Yan
Keiji Yanai
Bin Yang
Ceyuan Yang
Dawei Yang
Dong Yang
Fan Yang
Guandao Yang
Guorun Yang
Haichuan Yang
Hao Yang
Jianwei Yang
Jiaolong Yang
Jie Yang
Jing Yang
Kaiyu Yang
Linjie Yang
Meng Yang
Michael Ying Yang
Nan Yang
Shuai Yang
Shuo Yang
Tianyu Yang
Tien-Ju Yang
Tsun-Yi Yang
Wei Yang
Wenhan Yang
Xiao Yang
Xiaodong Yang
Xin Yang
Yan Yang
Yanchao Yang
Yee Hong Yang
Yezhou Yang
Zhenheng Yang
Anbang Yao
Angela Yao
Cong Yao
Jian Yao
Li Yao
Ting Yao
Yao Yao
Zhewei Yao
Chengxi Ye
Jianbo Ye
Keren Ye
Linwei Ye
Mang Ye
Mao Ye
Qi Ye
Qixiang Ye
Mei-Chen Yeh
Raymond Yeh
Yu-Ying Yeh
Sai-Kit Yeung
Serena Yeung
Kwang Moo Yi
Li Yi
Renjiao Yi
Alper Yilmaz
Junho Yim
Lijun Yin
Weidong Yin
Xi Yin
Zhichao Yin
Tatsuya Yokota
Ryo Yonetani
Donggeun Yoo
Jae Shin Yoon
Ju Hong Yoon
Sung-eui Yoon
Laurent Younes
Changqian Yu
Fisher Yu
Gang Yu
Jiahui Yu
Kaicheng Yu
Ke Yu
Lequan Yu
Ning Yu
Qian Yu
Ronald Yu
Ruichi Yu
Shoou-I Yu
Tao Yu
Tianshu Yu
Xiang Yu
Xin Yu
Xiyu Yu
Youngjae Yu
Yu Yu
Zhiding Yu
Chunfeng Yuan
Ganzhao Yuan
Jinwei Yuan
Lu Yuan
Quan Yuan
Shanxin Yuan
Tongtong Yuan
Wenjia Yuan
Ye Yuan
Yuan Yuan
Yuhui Yuan
Huanjing Yue
Xiangyu Yue
Ersin Yumer
Sergey Zagoruyko
Egor Zakharov
Amir Zamir
Andrei Zanfir
Mihai Zanfir
Pablo Zegers
Bernhard Zeisl
John S. Zelek
Niclas Zeller
Huayi Zeng
Jiabei Zeng
Wenjun Zeng
Yu Zeng
Xiaohua Zhai
Fangneng Zhan
Huangying Zhan
Kun Zhan

Xiaohang Zhan
Baochang Zhang
Bowen Zhang
Cecilia Zhang
Changqing Zhang
Chao Zhang
Chengquan Zhang
Chi Zhang
Chongyang Zhang
Dingwen Zhang
Dong Zhang
Feihu Zhang
Hang Zhang
Hanwang Zhang
Hao Zhang
He Zhang
Hongguang Zhang
Hua Zhang
Ji Zhang
Jianguo Zhang
Jianming Zhang
Jiawei Zhang
Jie Zhang
Jing Zhang
Juyong Zhang
Kai Zhang
Kaipeng Zhang
Ke Zhang
Le Zhang
Lei Zhang
Li Zhang
Lihe Zhang
Linguang Zhang
Lu Zhang
Mi Zhang
Mingda Zhang
Peng Zhang
Pingping Zhang
Qian Zhang
Qilin Zhang
Quanshi Zhang
Richard Zhang
Rui Zhang
Runze Zhang
Shengping Zhang
Shifeng Zhang
Shuai Zhang
Songyang Zhang
Tao Zhang
Ting Zhang
Tong Zhang
Wayne Zhang
Wei Zhang
Weizhong Zhang
Wenwei Zhang
Xiangyu Zhang
Xiaolin Zhang
Xiaopeng Zhang
Xiaoqin Zhang
Xiuming Zhang
Ya Zhang
Yang Zhang
Yimin Zhang
Yinda Zhang
Ying Zhang
Yongfei Zhang
Yu Zhang
Yulun Zhang
Yunhua Zhang
Yuting Zhang
Zhanpeng Zhang
Zhao Zhang
Zhaoxiang Zhang
Zhen Zhang
Zheng Zhang
Zhifei Zhang
Zhijin Zhang
Zhishuai Zhang
Ziming Zhang
Bo Zhao
Chen Zhao
Fang Zhao
Haiyu Zhao
Han Zhao
Hang Zhao
Hengshuang Zhao
Jian Zhao
Kai Zhao
Liang Zhao
Long Zhao
Qian Zhao
Qibin Zhao
Qijun Zhao
Rui Zhao
Shenglin Zhao
Sicheng Zhao
Tianyi Zhao
Wenda Zhao
Xiangyun Zhao
Xin Zhao
Yang Zhao
Yue Zhao
Zhichen Zhao
Zijing Zhao
Xiantong Zhen
Chuanxia Zheng
Feng Zheng
Haiyong Zheng
Jia Zheng
Kang Zheng
Shuai Kyle Zheng
Wei-Shi Zheng
Yinqiang Zheng
Zerong Zheng
Zhedong Zheng
Zilong Zheng
Bineng Zhong
Fangwei Zhong
Guangyu Zhong
Yiran Zhong
Yujie Zhong
Zhun Zhong
Chunluan Zhou
Huiyu Zhou
Jiahuan Zhou
Jun Zhou
Lei Zhou
Luowei Zhou
Luping Zhou
Mo Zhou
Ning Zhou
Pan Zhou
Peng Zhou
Qianyi Zhou
S. Kevin Zhou
Sanping Zhou
Wengang Zhou
Xingyi Zhou

Yanzhao Zhou
Yi Zhou
Yin Zhou
Yipin Zhou
Yuyin Zhou
Zihan Zhou
Alex Zihao Zhu
Chenchen Zhu
Feng Zhu
Guangming Zhu
Ji Zhu
Jun-Yan Zhu
Lei Zhu
Linchao Zhu
Rui Zhu
Shizhan Zhu
Tyler Lixuan Zhu
Wei Zhu
Xiangyu Zhu
Xinge Zhu
Xizhou Zhu
Yanjun Zhu
Yi Zhu
Yixin Zhu
Yizhe Zhu
Yousong Zhu
Zhe Zhu
Zhen Zhu
Zheng Zhu
Zhenyao Zhu
Zhihui Zhu
Zhuotun Zhu
Bingbing Zhuang
Wei Zhuo
Christian Zimmermann
Karel Zimmermann
Larry Zitnick
Mohammadreza Zolfaghari
Maria Zontak
Daniel Zoran
Changqing Zou
Chuhang Zou
Danping Zou
Qi Zou
Yang Zou
Yuliang Zou
Georgios Zoumpourlis
Wangmeng Zuo
Xinxin Zuo

Additional Reviewers

Victoria Fernandez Abrevaya
Maya Aghaei
Allam Allam
Christine Allen-Blanchette
Nicolas Aziere
Assia Benbihi
Neha Bhargava
Bharat Lal Bhatnagar
Joanna Bitton
Judy Borowski
Amine Bourki
Romain Brégier
Tali Brayer
Sebastian Bujwid
Andrea Burns
Yun-Hao Cao
Yuning Chai
Xiaojun Chang
Bo Chen
Shuo Chen
Zhixiang Chen
Junsuk Choe
Hung-Kuo Chu
Jonathan P. Crall
Kenan Dai
Lucas Deecke
Karan Desai
Prithviraj Dhar
Jing Dong
Wei Dong
Turan Kaan Elgin
Francis Engelmann
Erik Englesson
Fartash Faghri
Zicong Fan
Yang Fu
Risheek Garrepalli
Yifan Ge
Marco Godi
Helmut Grabner
Shuxuan Guo
Jianfeng He
Zhezhi He
Samitha Herath
Chih-Hui Ho
Yicong Hong
Vincent Tao Hu
Julio Hurtado
Jaedong Hwang
Andrey Ignatov
Muhammad Abdullah Jamal
Saumya Jetley
Meiguang Jin
Jeff Johnson
Minsoo Kang
Saeed Khorram
Mohammad Rami Koujan
Nilesh Kulkarni
Sudhakar Kumawat
Abdelhak Lemkhenter
Alexander Levine
Jiachen Li
Jing Li
Jun Li
Yi Li
Liang Liao
Ruochen Liao
Tzu-Heng Lin
Phillip Lippe
Bao-di Liu
Bo Liu
Fangchen Liu

Hanxiao Liu
Hongyu Liu
Huidong Liu
Miao Liu
Xinxin Liu
Yongfei Liu
Yu-Lun Liu
Amir Livne
Tiange Luo
Wei Ma
Xiaoxuan Ma
Ioannis Marras
Georg Martius
Effrosyni Mavroudi
Tim Meinhardt
Givi Meishvili
Meng Meng
Zihang Meng
Zhongqi Miao
Gyeongsik Moon
Khoi Nguyen
Yung-Kyun Noh
Antonio Norelli
Jaeyoo Park
Alexander Pashevich
Mandela Patrick
Mary Phuong
Bingqiao Qian
Yu Qiao
Zhen Qiao
Sai Saketh Rambhatla
Aniket Roy
Amelie Royer
Parikshit Vishwas Sakurikar
Mark Sandler
Mert Bülent Sarıyıldız
Tanner Schmidt
Anshul B. Shah
Ketul Shah
Rajvi Shah
Hengcan Shi
Xiangxi Shi
Yujiao Shi
William A. P. Smith
Guoxian Song
Robin Strudel
Abby Stylianou
Xinwei Sun
Reuben Tan
Qingyi Tao
Kedar S. Tatwawadi
Anh Tuan Tran
Son Dinh Tran
Eleni Triantafillou
Aristeidis Tsitiridis
Md Zasim Uddin
Andrea Vedaldi
Evangelos Ververas
Vidit Vidit
Paul Voigtlaender
Bo Wan
Huanyu Wang
Huiyu Wang
Junqiu Wang
Pengxiao Wang
Tai Wang
Xinyao Wang
Tomoki Watanabe
Mark Weber
Xi Wei
Botong Wu
James Wu
Jiamin Wu
Rujie Wu
Yu Wu
Rongchang Xie
Wei Xiong
Yunyang Xiong
An Xu
Chi Xu
Yinghao Xu
Fei Xue
Tingyun Yan
Zike Yan
Chao Yang
Heran Yang
Ren Yang
Wenfei Yang
Xu Yang
Rajeev Yasarla
Shaokai Ye
Yufei Ye
Kun Yi
Haichao Yu
Hanchao Yu
Ruixuan Yu
Liangzhe Yuan
Chen-Lin Zhang
Fandong Zhang
Tianyi Zhang
Yang Zhang
Yiyi Zhang
Yongshun Zhang
Yu Zhang
Zhiwei Zhang
Jiaojiao Zhao
Yipu Zhao
Xingjian Zhen
Haizhong Zheng
Tiancheng Zhi
Chengju Zhou
Hao Zhou
Hao Zhu
Alexander Zimin

Contents – Part IX

Search What You Want: Barrier Panelty NAS for Mixed Precision Quantization

Haibao Yu[1], Qi Han[1], Jianbo Li[1,2], Jianping Shi[1], Guangliang Cheng[1](✉), and Bin Fan[3](✉)

[1] SenseTime Research, Beijing, China
{yuhaibao,hanqi,shijianping,chengguangliang}@sensetime.com
[2] Peking University, Beijing, China
jianbo.li@pku.edu.cn
[3] University of Science and Technology Beijing, Beijing, China
bin.fan@ieee.org

Abstract. Emergent hardwares can support mixed precision CNN models inference that assign different bitwidths for different layers. Learning to find an optimal mixed precision model that can preserve accuracy and satisfy the specific constraints on model size and computation is extremely challenge due to the difficult in training a mixed precision model and the huge space of all possible bit quantizations.

In this paper, we propose a novel soft Barrier Penalty based NAS (BP-NAS) for mixed precision quantization, which ensures all the searched models are inside the valid domain defined by the complexity constraint, thus could return an optimal model under the given constraint by conducting search only one time. The proposed soft Barrier Penalty is differentiable and can impose very large losses to those models outside the valid domain while almost no punishment for models inside the valid domain, thus constraining the search only in the feasible domain. In addition, a differentiable Prob-1 regularizer is proposed to ensure learning with NAS is reasonable. A distribution reshaping training strategy is also used to make training more stable. BP-NAS sets new state of the arts on both classification (Cifar-10, ImageNet) and detection (COCO), surpassing all the efficient mixed precision methods designed manually and automatically. Particularly, BP-NAS achieves higher mAP (up to 2.7% mAP improvement) together with lower bit computation cost compared with the existing best mixed precision model on COCO detection.

Keywords: Mixed precision quantization · NAS · Optimization problem with constraint · Soft barrier penalty

H. Yu and Q. Han—Indicates equal contributions.

Electronic supplementary material The online version of this chapter (https://doi.org/10.1007/978-3-030-58545-7_1) contains supplementary material, which is available to authorized users.

© Springer Nature Switzerland AG 2020
A. Vedaldi et al. (Eds.): ECCV 2020, LNCS 12354, pp. 1–16, 2020.
https://doi.org/10.1007/978-3-030-58545-7_1

1 Introduction

Deep convolutional neural networks (CNNs) have achieved remarkable performance in a wide range of computer vision tasks, such as image classification [6,10], semantic segmentation [14,26] and object detection [17,18]. To deploy CNN models into resource-limited edge devices for real-time inference, one classical way is to quantize the floating-point weights and activations with fewer bits, where all the CNN layers share the same quantization bitwidth. However, different layers and structures may hold different sensibility to bitwidth for quantization, thus this kind of methods often causes serious performance drop, especially in low-bit quantization, for instance, 4-bit.

Currently, more and more hardware platforms, such as Turning GPUs [16] and FPGAs, have supported the mixed precision computation that assigns different quantization bitwidths to different layers. This motivates the research of quantizing different CNN layers with various bitwidths to pursue higher accuracy of quantized models while consuming much fewer bit operations (BOPs). Furthermore, many real applications usually set a hard constraint of complexity on mixed precision models that their BOPs can only be less than a preset budget. However, it is computational expensive to determine an appropriate mixed precision configuration from all the possible candidates so as to satisfy the BOPs constraint and achieve high accuracy. Assuming that the quantized model has N layers and each layer has M candidate options to quantize the weights and activations, the total search space is as large as $(M)^N$.

To alleviate the search cost, an attractive solution is to adopt weight-sharing approach like DARTS [13] which incorporates the BOPs cost as an additional loss term to search the mixed precision models. However, there exist some problems for this kind of methods: i) Invalid search. The framework actually seeks a balance between the accuracy and the BOPs cost, may causing the returned mixed precision models outside the BOPs constraint. To satisfy the application requirement, the search process has to be repeated many times by trial-and-error through tuning the BOPs cost loss weight. This is time consuming and often leads to suboptimal quantization. ii) Very similar importance factors. The importance factors corresponding to different bitwidths in each CNN layer often have similar values, which makes it hard to select a proper bitwidth for each layer. iii) Unstable training. Empirically, high-bit quantization and low-bit quantization prefer different training strategies. To balance the favors between high-bit and low-bit, the training of mixed precision models maybe fluctuate a lot and cannot achieve high accuracy.

To ensure the searched models to satisfy the BOPs constraint, we model the mixed precision quantization task as an constrained optimization problem, and propose a novel loss term called soft barrier penalty to model the complexity constraint. With the differentiable soft barrier penalty, it bounds the search in the valid domain and imposes large punishments to those mixed precision models near the constraint barrier, thus significantly reducing the risk that the returned models do not fulfill the final deploy requirement. Compared to existing methods [2,21,22] which incorporate the complexity cost into loss function and tune the corresponding balance weight, our soft barrier penalty method does not need to conduct numerous trial-and-error experiments. Moreover, our search process focuses on the candidates in the valid search space, then the search will be more efficient to approach much more optimal mixed precision model while satisfying the BOPs cost constraint.

To make the importance factors more distinguishable, we propose a differentiable Prob-1 regularizer to make the importance factors converge to 0–1 type. With the convergence of the importance factors in each CNN layer, the mixed precision configuration will be gradually determined. To deal with the unstable training issues and improve the accuracy of mixed precision model, we use the distribution reshaping method to train a specific float-point model with uniformly-distributed weights and no long-tailed activations, and use the float-point model as the pretrained mixed precision model. Experiments show that the pretrained model and uniformization method can make the mixed precision training much more robust and achieve higher accuracy.

In particular, our contributions are summarized as the followings:

- We propose a novel and deterministic method in BP-NAS to incorporate the BOPs constraint into loss function and bound the search process within the constraint, which saves numerous trial and error to search the mixed precision with high accuracy while satisfying the constraint.
- A novel distribution reshaping method is introduced to mixed precision quantization training, which can address the unstable mixed precision training problems while achieve high accuracy.
- BP-NAS surpasses the state-of-the-art efficient mixed precision methods designed manually and automatically on three public benchmarks. Specifically, BP-NAS achieves higher accuracy (up to 2.7% mAP) with lower bit computation compared with state-of-arts methods on COCO detection dataset.

2 Related Work

Mixed precision quantization techniques aims to allocate different weight and activation bitwidth for different CNN layers. Compared to the traditional fixed-point quantization methods, the mixed precision manner can efficiently recover the quantization performance while consuming less bit operations (BOPs). It mainly involves mixed precision quantization training and mixed precision search from the huge search space.

Quantization involves high-bit quantization training and low-bit quantization training. For the former, ELQ [27] adopts incremental strategy to fixes part of weights and update the rest in the training, HWGQ [3] introduces clipped ReLU to remove outliers, PACT [5] further proposes adaptatively learns and determines the clipping threshold, Tensorflow Lite [11] proposes asymmetric linear scheme to quantize the weights and activations with Integer-arithmetic-only. For the later, Dorefa [28] and TTN [29] focus on extremely low bit such as one or two bit. Wang et al. [20] proposes two-step to train the low-bit quantization. Gu et al. [8] proposes Bayesian-based method to optimize the 1-bit CNNs. These quantization methods simply assign the same bit for all layers, while we consider to apply the uniformization to robustly train the mixed precision quantization model with different bitwidth. Shkolnik et al. [15] also consider the uniformization for robustness but for post-training quantization.

Mixed precision search focuses on efficiently searching the mixed precision model with high accuracy while consuming less bit operations(BOPs). By treating the bitwidth as operations, several NAS methods could be used for mixed precision search: RL-based method [19] consider to collect the action-reward involving the hard-ware architecture into the loop, and then use these data to train the agent to search better mixed precision model. But it does not consider to use the loss-aware training model data to search the mixed precision. Single-Path [9] proposes to use evolutionary algorithm to search the mixed precision to achieve high precision and satisfy the constraint. But the search will cost huge resources to train numerous of mixed precision models in the early stage.

In this paper, we apply the weight-sharing NAS method into mixed precision search. To the best of our knowledge, our work is mostly related to DNAS [22], which also first introduces the weight-sharing NAS into mixed precision search. However there are significant differences: 1) We propose a novel soft barrier penalty strategy to ensure all the search model will be inside the valid domain. 2) We propose an effective differentiable Prob-1 regularizer to make the importance factors more distinguishable. 3) We incorporate distribution reshaping strategy to make the training much more robust.

3 Method

In this paper, we model the mixed precision quantization task as an constrained optimization problem. Our basic solution is to apply DARTS [13] for addressing such optimization problem. A similar solution [22] for this problem is also to apply DARTS with gumbel softmax trick by incorporating the complexity cost into loss term. However, it actually seeks a balance between the accuracy of the mixed-precision model and the BOPs constraint, making the search process repeate many times by tuning the complexity cost weight until the searched model can satisfy the application requirement. This is time consuming and often leads to suboptimal quantization. To address this issue, we propose a novel loss term to model the BOPs constraint so that it encourages the search conducting in the valid domain and imposes large punishment to those quantizations outside the valid domain, thus significantly reducing the risk that the returned models do not fulfill the final deploy requirement. Our method is termed as BP-NAS since the new loss term is based on the barrier penalty which will be elaborated in Sect. 3.2. In addition, a Prob-1 regularizer is proposed along with the barrier penalty to facilitate node selection in the supernet constructed by DARTS. Finally, we propose how to train a uniform-like pretrained model to stabilize the training of mixed precision quantization. In the following, we first describe the problem formulation (Sect. 3.1) and then detail our solution (Sect. 3.2).

3.1 Mixed Precision Quantization Search

Quantization aims to represent the float weights and activations with linearly discrete values. We denote ith convolutional layer weights as $\mathbf{W}_i$. For each weight atom $w \in \mathbf{W}_i$, we linearly quantize it into b_i-bit as

$$Q(w; b_i) = [\frac{\text{clamp}(w, \alpha)}{\alpha/(2^{b_i-1}-1)}] \cdot \alpha/(2^{b_i-1}-1), \tag{1}$$

where α is a learnable parameter, and clamp$(\cdot, \alpha)$ is to truncate the values into $[-\alpha, \alpha]$, $[\cdot]$ is the rounding operation. For activations, we truncates the values into the range $[0, \alpha]$ in a similar way since the activation values are non-negative after the ReLU layer. For simplicity, we use (m, n)-bit to denote quantizing a layer with m-bit for its weights and n-bit for activations.

Since different layers in a model may prefer different bitwidths for quantization, we consider the problem of mixed precision quantization. For a given network $\mathcal{N} = \{L_1, L_2, \cdots, L_N\}$, the mixed precision quantization aims at determining the optimal weight bitwidth $\{b_1, b_2, \cdots, b_N\}$ and activation bitwidth $\{a_1, a_2, \cdots, a_N\}$ for all layers from a set of predefined configurations. This is cast into an optimization problem as following,

$$\text{MP}^* = \underset{\text{MP}}{\text{argmin}}\, \text{Loss}_{val}(\mathcal{N}; \text{MP}) \tag{2}$$

where MP denotes the mixed precision configuration for $\mathcal{N}$, and Loss_{val} denotes the loss on the validation set.

Usually, there are additional requirements on a mixed precision model about its complexity and memory consumption. Bit operations(BOPs[1]) is a popular indicator to formulate these issues. Mathematically, BOPs is formulated as

$$\mathcal{B}(\mathcal{N}; \text{MP}) = \sum_i \text{FLOP}(L_i) * b_i * a_i, \tag{3}$$

where $\text{FLOP}(L_i)$ denotes the number of float point operations in layer L_i, (b_i, a_i)-bit denotes weight and activation bitwidths for L_i respectively. For convenience, we also refer BOPs to average bit operations which is actually used in this paper,

$$\mathcal{B}(\mathcal{N}; \text{MP}) = ((\sum_i \text{FLOP}(L_i) * b_i * a_i)/\sum_i \text{FLOP}(L_i))^{1/2}. \tag{4}$$

A typical requirement in mixed precision quantization search is called BOPs constraint to restrict the searched model within a preset budget,

$$\mathcal{B}(\mathcal{N}; \text{MP}) \leq \mathcal{B}_{max}. \tag{5}$$

As a result, the problem of mixed precision quantization search is to find a set of quantization parameters for different layers so as to keep the original model's accuracy as much as possible while fulfilling some complexity constraints, such as the one listed above.

[1] We also provide an example to demonstrate this BOPs calculation process in supplementary materials. More formulation of BOPs could be seen in [4,23], which consider the memory bandwidth and could be more suitable for real application.

3.2 BP-NAS for Mixed Precision Quantization Search

NAS for Mixed Precision Quantization. Due to the large cost of training a mixed precision model and the huge mixed precision configurations, it is impractical to train various mixed precision models and select one to satisfy Eq. (2) and Eq. (5). Similar to [22], we adopt the weight-sharing NAS, i.e., DARTS [13], to search mixed precision quantization models.

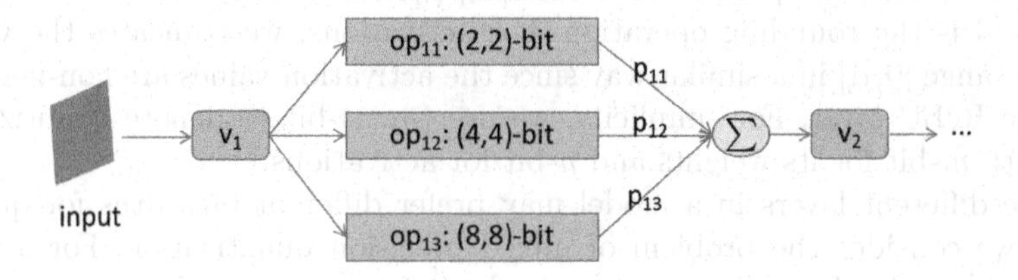

Fig. 1. Supernet architecture illustration with (2, 2)-bit, (4, 4)-bit, (8, 8)-bit as candidate bitwidth. v_i represents the input data tensor and each operation share the same input data but different bidwidth

Specifically, we first construct and train a supernet $\mathcal{SN}$, from which the mixed precision model is then sampled. The supernet is illustrated in Fig. 1. In the supernet $\mathcal{SN}$, the edge between node v_i (corresponds to the i-th layer in the mixed precision model) and v_{i+1} is composed of several candidate bit operations $\{op_{i,j}|_{j=1,\cdots,m}\}$ such as with learnable parameters $\{\theta_{i,j}|_{j=1,\cdots,m}\}$. The output of node v_{i+1} is calculated by assembling all edges as

$$v_{i+1} = \sum_j p_{i,j} * op_{i,j}(v_i), \tag{6}$$

where p_{ij} denotes the importance factor of the edge $op_{i,j}$ and is calculated with the learnable parameter $\theta_{i,j}$ as

$$p_{i,j} = \frac{exp(\theta_{i,j})}{\sum_l exp(\theta_{i,l})}. \tag{7}$$

Once the supernet has been trained, the mixed precision model can be sampled

$$\text{MP}^* = \{(b_i^*, a_i^*)\text{-bit}, i = 1, 2, \cdots, N\} = \{\text{argmax}_j\ p_{i,j}\} \tag{8}$$

Finally, the sampled mixed precision model is retrained to achieve much better performance.

To take the BOPs cost of the sampled model into consideration, the supernet is trained with the following loss

$$Loss = Loss_{val} + \lambda * E(\mathcal{SN}). \tag{9}$$

where BOPs cost of the sampled model is estimated with the supernet as

$$E(\mathcal{SN}) = \sum_i \sum_j p_{i,j} * \mathcal{B}(op_{i,j}). \tag{10}$$

Such an estimation is reasonable if the importance factors $p_{i,j}$ for each layer is highly selective, i.e., only one element approaches 1 and others being very small approaching 0. We will introduce later the proposed Prob-1 regularizer to ensure such property of $p_{i,j}$.

However, optimizing over the above loss only encourages the BOPs being as small as possible while maximizing the accuracy. With a large λ, it will make the learning focus on the complexity but is very likely to return a less accurate model with much less BOPs than what we want. On the contrary, a small λ makes the learning focus on the accuracy and will possibly return a model with high accuracy but with much larger BOPs than what we want. Therefore, it requires numerous trial and error to search the proper mixed precision. This will definitely consume much more time and GPU resources to find a satisfactory mixed precision model. Even though, the obtained model could be suboptimal.

Alternatively, to make the returned model satisfying the hard constraint in (5), the mixed precision search is expected to be focused on the feasible space defined by the imposed constraint. In this case, the search is able to return a model fulfilling the constraint and with the better accuracy. Since the search is only conducted in the feasible space, the searching space is also reduced, and so the search is more efficient. We achieve this goal by proposing a novel approach called BP-NAS, which incorporates a barrier penalty in NAS to bound the mixed precision search within the feasible solution space with a differentiable form.

Barrier Penalty Regularizer. The key idea of making the search within the feature space is to develop a regularizer so that it punishes the training with an extremely large loss once the search reaches the bound or is outside the feasible space. For the BOPs constraint in Eq. (5), it can be rewritten as a regularization term

$$\mathcal{L}_c(\theta) = \begin{cases} 0 & if \quad E(\mathcal{SN}) \leq \mathcal{B}_{max} \\ \infty & otherwise \end{cases} \tag{11}$$

It imposes an ∞ loss once the BOPs of the trained supernet falls outside the constraint, while no punishment in the other case. This could be an ideal regularizer as we analyzed before, however, it is non-differentiable and thus can not be used.

Inspired by the innerior method [1] for solving the constrained optimization problem, we approximate $\mathcal{L}_c(\theta)$ by

$$\mathcal{L}_c^*(\theta) = -\mu log(log(\mathcal{B}_{max} + 1 - E(\mathcal{SN}))), \tag{12}$$

where μ should be smaller in the early search stage because there is a large gap between the estimation BOPs of the initial supernet and the BOPs of the final sampled mixed precision model.

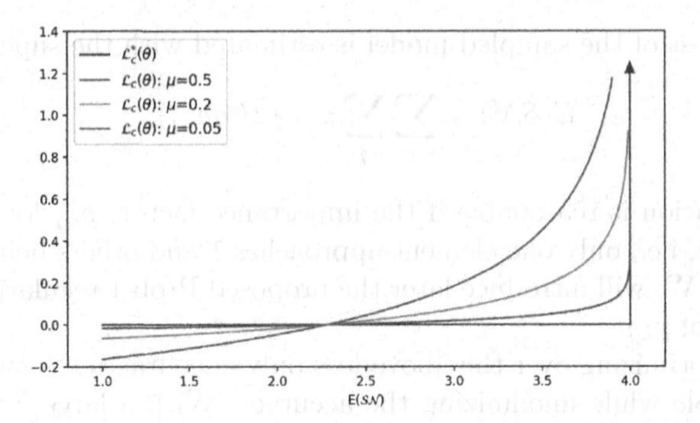

Fig. 2. Illustration of the barrier penalty regularizer under $\mathcal{B}_{max} = 4$. Blue curves denotes the function $\mathcal{L}_c(\theta)$. Other curves denote $\mathcal{L}_c^*(\theta)$ with different $\mu = 0.5, 0.2, 0.05$. With μ decreasing, the curve of $\mathcal{L}_c^*(\theta)$ will be more flatten and is near to 0 in valid interval, and it will be infinity when $E(\mathcal{SN})$ tends to barrier at $E(\mathcal{SN}) = \mathcal{B}_{max}$.

Equation (12) is differentiable with respect to $E(\mathcal{SN})$ in valid interval $(-\infty, \mathcal{B}_{max})$. As shown in Fig. 2, it approaches the hard constraint loss in Eq. (11) by decreasing μ and looks like setting a barrier at $E(\mathcal{SN}) = \mathcal{B}_{max}$. That why we call it the barrier penalty regularizer. For other values of $E(\mathcal{SN})$ in the valid interval, $\mathcal{L}_c^*(\theta)$ is near zero, so has very little impact when searching in the feasible solution space.

Prob-1 Regularizer. Since we use the expected BOPs of the supernet to estimate that of the sampled mixed precision model as in Eq. (10), this approximation is reasonable only when the supernet gradually converge to the sampled mixed precision model. In other words, the importance factors should gradually be 0–1 type. For this purpose, we introduce the differentiable Prob-1 regularizer.

As the preliminary, we first introduce the Prob-1 function that can be formulated as

$$f(\mathbf{x}) = \prod_j (1 - x_j), \qquad s.t. \quad \sum x_j = 1, 0 \leq x_j \leq 1, \tag{13}$$

where $\mathbf{x} = (x_1, \cdots, x_m)$ is m-dimension vector. For the Prob-1 function, it can be proved that it has the following property,

Property 1. Prob-1 function $f(x)$ achieves the minimal value if and only if there exists unique x_j to reach 1.

The detailed proof is in the supplementary. Since the importance factors of each layer in the supernet satisfy the constraint of Prob-1 function, we design a Prob-1 regularizer for them with the formula

$$\mathcal{L}_{prob-1} = \sum_i \prod_j (1 - p_{i,j}). \tag{14}$$

Due to the *Property 1*, this regularizer encourages to learn the 0–1 type importance factors.

BP-NAS Algorithm. The proposed BP-NAS algorithm is summarized in Algorithm 1. Basically, it contains three stages: training a supernet, sampling mixed precision from the supernet and retrain the mixed precision model. In the first stage, the network weights W and architecture parameter θ are trained alternately. The network weights in the supernet are updated by the gradient of loss function $\mathcal{L}_1$ on training set,

$$\mathcal{L}_1 = \mathcal{L}_{train}(W;\theta) \tag{15}$$

While θ is updated with the gradient of loss function $\mathcal{L}_2$ on validation set

$$\mathcal{L}_2 = \mathcal{L}_{val}(W;\theta) + \mathcal{L}_c^* + \mathcal{L}_{Prob-1} \tag{16}$$

The differentiation form of loss functions of Eq. (15)–(16) will be described in the supplementary. After training some epochs, we sample the mixed precision quantization model from the trained supernet according to the importance factors as Eq. (8). We retrain the mixed precision models for some epochs to achieve better accuracy.

Another issue about training mixed precision model is that the training often crashes since training a quantization (especially low-bit quantization) model is very sensitive and different bit quantization usually favors different training strategies [5,11]. To alleviate this problem, we use distribution reshaping to train a float-point model with uniformly-distributed weights and no long-tailed activations as the pretrained model, which is helpful for robust training under different bit quantizations. For each weight atom w in the ith layer's weight $\mathbf{W}_i$, the distribution reshaping simply uses the clipping method with the formula

$$\text{clip}(w) = \begin{cases} T, & w \geq T \\ w, & w \in (-T, T) \\ -T, & w \leq -T \end{cases} \tag{17}$$

where

$$T = k \cdot \text{mean}(|\mathbf{W}_i|). \tag{18}$$

Here we set $k = 2$ to reshape the weight distribution uniform-like. We use the Clip-ReLU as in PACT [5] to quantize the activations. Similar robust quantization work with distribution reshaping could be seen in [15,24].

4 Experiments

We implement the proposed BP-NAS for mixed precision search on image classification (Cifar-10, ImageNet) and object detection (COCO). The search adopts a block-wise manner, that all layers in the block share the same quantization bitwidth. We present all the block-wise configurations for mixed-precision quantization model in the supplementary material.

Algorithm 1. Barrier Penalty for Neural Network Search

Input: *supernet with weight* $\mathbf{W}$ *and architecture parameter* θ*, BOPs constraint* $\mathcal{B}_{max}$*, maximal epoch* $epoch_{max}$

Output: *the architecture with high accuracy under architecture constraints*

1: θ : initialize the architecture parameter θ

2: **for** epoch = 1:$epoch_{max}$ **do**

3: update the weight $\mathbf{W}$ of the supernet with $\mathcal{L}$ in Eq. (15)

4: update the architecture parameter θ in Eq. (16) by computing $\partial(\mathcal{L}_{val}(\mathbf{W};\theta) + \mathcal{L}_c^* + \mathcal{L}_{Prob-1})/\partial\theta$

5: **end for**

6: **return** mixed precision model MP* sampled from the supernet

4.1 Cifar-10

We implement the mixed precision search with ResNet20 on Cifar-10, and conduct three sets of experiments under different BOPs constraints $\mathcal{B}_{max}$ = 3-bit, 3.5-bit and 4-bit. Compared to 32-bit float point model, the three constraints respectively have 114×, 85× and 64× bit operations compression ratio($\mathcal{B}$-Comp).

We construct the supernet whose macro architecture is the same as ResNet20. Each block in the supernet contains {(2, 3), (2, 4), (3, 3), (3, 4), (4, 4), (4, 6), (6,4), (8, 4)}-bit quantization operations. To train the supernet, we randomly split 60% of the Cifar-10 training samples as training set and others as validation set, with batch size of 512. To train the weight W, we use SGD optimizer with an initial learning rate 0.2 (decayed by cosine schedule), momentum 0.9 and weight decay 5e-4. To train the architecture parameter θ, we use Adam optimizer with an initial learning rate 5e-3 and weight decay 1e−3. We sample the mixed precision from supernet as Eq. (8), and then retrain the sampled mixed precision for 160 epochs and use cutout in data augmentation. Other training settings are the same as the weight training stage. To reduce the deviation caused by different quantization settings and training strategies, and to make a fair comparison with other mixed precision methods like HAWQ [7], we also train {(32, 32), (2.43MP, 4)[2], (3, 3)}-bit ResNet20 with our quantization training strategies. We remark these results as "Baseline".

The results are shown in Table 1. Compared with PACT [5] and HAWQ [7], though our Baselines achieves lower accuracy with (3, 3)-bit and (2 MP,4)-bit on ResNet20, the BP-NAS still surpasses other state-of-the-art methods. Specifically, HAWQ [7] achieves 92.22% accuracy with 64.49× compression ratio, whereas BP-NAS achieves 92.30% accuracy with much higher compression ratio up to 81.53×. In addition, our faster version (i.e. BP-NAS with $\mathcal{B}_{max}$=3-bit) outperforms PACT [5] and LQ-Nets [25] more than 0.4%. Note that with different bit computation cost constraints, BP-NAS can search the mixed precision that could not only satisfy the constraint but also achieve even better performance.

[2] 2.43MP uses the mixed precision quantizations searched by HAWQ [7].

Table 1. Quantization results of ResNet20 on Cifar-10. "w-bits" and "a-bits" represent the bitwidth for weights and activations. "W-Comp", "B-Comp" and "Ave-bit' denote the model size compression ratio, the bit operations compression ratio, and the average bit operations of mixed precision model, respectively. "MP" denotes mixed precision.

Quantization	w-bits	a-bits	Top-1 Acc	W-Comp	B-Comp	Ave-bit
Baseline	32	32	92.61	1.00×	1.00×	32-bit
Baseline	2.43 MP	4	92.12	13.11×	64.49×	4.0-bit
DNAS [22]	MP	32	92.00	16.60×	16.60×	7.9-bit
DNAS [22]	MP	32	92.72	11.60×	11.60×	9.4-bit
HAWQ [7]	2.43 MP	4	92.22	13.11×	64.49×	4.0-bit
BP-NAS $\mathcal{B}_{max}$=4-bit	3.14 MP	MP	**92.30**	10.19×	**81.53**×	**3.5**-bit
BP-NAS $\mathcal{B}_{max}$=3.5-bit	2.86 MP	MP	**92.12**	10.74×	**95.61**×	**3.3**-bit
Baseline	3	3	91.80	10.67×	113.78×	3.0-bit
Dorefa [28]	3	3	89.90	10.67×	113.78×	3.0-bit
PACT [5]	3	3	91.10	10.67×	113.78×	3.0-bit
LQ-Nets [25]	3	3	91.60	10.67×	113.78×	3.0-bit
BP-NAS $\mathcal{B}_{max}$=3-bit	2.65 MP	MP	**92.04**	12.08×	**116.89**×	**2.9**-bit

4.2 ImageNet

In the ImageNet experiment, we search the mixed precision for the prevalent ResNet50. To speedup the supernet training, we randomly sample 10 categories with 5000 images as training set and with 5000 images as validation set, and set candidate operations as {(2, 4), (3, 3), (3, 4), (4, 3), (4, 4), (4, 6), (6, 4)}-bit. We sample the mixed precision from the supernet and then transfer it to ResNet50 with 1000 categories. We retrain the mixed precision for 150 epochs with batch size of 1024 and label smooth on 16 GPUs. We also train the (32, 32)-bit (3, 3)-bit and (4, 4)-bit ResNet50 with the same quantization training strategies, and remark the results as "Baseline". We report the quantization results in Table 2.

Under the constraint of $\mathcal{B}_{max}$ = 3-bit, the mixed precision ResNet50 searched by our BP-NAS achieves 75.71% Top-1 accuracy with compression ratio of 118.98×. Though our baseline Top-1 accuracy is inferior to PACT [5], our mixed precision result surpasses PACT [5] in (3, 3)-bit with similar compression ratio. Furthermore, we also present the mixed precision ResNet50 under the constraint $\mathcal{B}_{max}$ = 4-bit. Compared to the fixed (4, 4)-bit quantization, our mixed precision ResNet50 achieves 76.67% Top-1 accuracy with similar average bit.

We also compare the quantization results with HAQ [19] and HAWQ [7], which also use mixed precision for quantization. With similar compression ratio, our mixed precision on ResNet50 under $\mathcal{B}_{max}$ = 3-bit outperforms HAQ [19] and HAWQ [7] by 1% Top-1 accuracy.

Table 2. Quantization results of ResNet50 on ImageNet dataset.

Quantization	w-bits	a-bits	Acc.(Top-1)	Acc.(Top-5)	$\mathcal{B}$-Comp	Ave-bit
Baseline	32	32	77.56%	94.15%	1.00×	32-bit
Baseline	4	4	76.02%	93.01%	64.00×	4-bit
PACT [5]	4	4	76.50%	93.20%	64×	4-bit
HAWQ [7]	MP	MP	75.30%	92.37%	64.49×	4-bit
HAQ [19]	MP	MP	75.48%	92.42%	78.60×	3.6-bit
BP-NAS $_{\mathcal{B}=4\text{-bit}}$	MP	MP	**76.67%**	93.55%	**71.65**×	**3.8**-bit
Baseline	3	3	75.17%	92.31%	113.78×	3-bit
Dorefa [28]	3	3	69.90%	89.20%	113.78×	3-bit
PACT [5]	3	3	75.30%	92.60%	113.78×	3-bit
LQ-Nets [25]	3	3	74.20%	91.60%	113.78×	3-bit
BP-NAS $_{\mathcal{B}=3\text{-bit}}$	MP	MP	**75.71%**	**92.83%**	**118.98**×	**2.9**-bit

4.3 COCO Detection

We implement the mixed precision search with ResNet50 Faster R-CNN on COCO detection dataset, which contains 80 object categories and 330K images with 1.5 million object instances. Similar to state-of-the-art quantization algorithm FQN [12], we also quantize all the convolutional weights and activations including FC layer with fully quantized mode. We use the ImageNet to pretrain the backbone and finetune the quantized Faster R-CNN for 27 epochs with batch size of 16 on 8 GPUs. We use SGD optimizer with an initial learning rate 0.1 (decayed by MultiStepLR schedule similar to milestones [16,22,25]), momentum 0.9 and weight decay 1e-4. We report our 4-bit quantization results as well as FQN [12] in Table 3. FQN [12] achieves 0.331 mAP with 4-bit Faster R-CNN and results in 4.6% mAP loss compared to its float model. According to the "Baseline", our 4-bit Faster R-CNN achieves 0.343 mAP and outperforms FQN [12] by 1.2%. Note that our quantization method achieves new state-of-the-art performance on COCO dataset with 4-bit quantization.

Table 3. Quantization results of Faster R-CNN on COCO.

Quantization	Input	w-bits	a-bits	$\mathcal{B}$-Comp	mAP					
					AP	$AP^{0.5}$	$AP^{0.75}$	AP^{S}	AP^{M}	AP^{L}
Baseline	800	32	32	1.00×	0.373	0.593	0.400	0.222	0.408	0.475
Baseline	800	4	4	64.00×	**0.343**	0.558	0.364	0.210	0.385	0.427
FQN [12]	800	4	4	64.00×	0.331	0.540	0.355	0.182	0.362	0.436
BP-NAS$_{\mathcal{B}_{max}=4}$	800	MP	MP	64.76×	**0.358**	0.579	0.383	0.217	0.398	0.474

As we can see, 4-bit Faster R-CNN still results in significant performance degradation (i.e. 3% drop). To alleviate this, we implement BP-NAS to search the mixed precision for Faster R-CNN. We utilize the candidate operations from $\{(2, 4), (3, 3), (3, 4), (3, 5), (4, 4), (4, 6), (6, 4), (4, 8)\}$. Following [2], we sample paths by probability to participate the training. To train the supernet, we set the bit operations constraint $\mathcal{B}_{max}$ as 4-bit, which has the same compression ratio as 4-bit quantization. We randomly samples 5 categories with 6K images as training set and 4K images as validation set. We train the supernet for 40 epochs with batch size of 1. Other training settings are the same as the training with ResNet20 on Cifar-10. We train the sampled mixed precision Faster R-CNN with the same training settings as 4-bit quantizaiton. As illustrated in Table 3, the mixed 4-bit Faster R-CNN can achieve 0.358 mAP with bit computation cost compression ratio up to 64.76×, which outperforms the original 4-bit Faster R-CNN by 1.5% mAP. In addition, the mixed 4-bit model only brings in 1.5% mAP degradation compared to its float-point version, which is a new state-of-the-art on mixed 4-bit community.

5 Ablation Study

We also conduct experiments to illustrate the statistical significance for the mixed precision search, and the effectiveness of Prob-1 regularizer. All experiments are implemented with ResNet20 on Cifar-10 dataset.

5.1 Efficient Search

BP-NAS can make the mixed precision search focus on the feasible solution space and return the valid search results with high accuracy. We conduct two sets of search experiments for ResNet20 under different BOPs constraint $\mathcal{B}_{max}$ = 3-bit and $\mathcal{B}_{max}$ = 4-bit. We implement each set of experiments for 20 times respectively, and then we retrain the searched mixed precision models for 160 epoches. The accuracy and average bit of the search results are shown in Fig. 3.

In Fig. 3, the average bit of the search results are all smaller than the preset BOPs constraint. And after retaining for some epoches, many of these mixed precision model can achieve comparable, even higher accuracy than the fixed bit quantization with $\mathcal{B}_{max}$. Take the red circle as an example. They are mixed precision searched with BP-NAS under preset constraint $\mathcal{B}_{max}$=3-bit, and their average bit are all smaller than 3-bit. However, some of them can achieve 91.8% Top-1 accuracy, which is 3-bit quantization model's accuracy. Even one mixed precision achieves 92.04% Top-1 accuracy, which surpasses the 3-bit model up to 0.24% Top-1 accuracy. The results show that our BP-NAS is effective for mixed precision quantization with constraint.

Note that if without Barrier Penalty Regularizer, most of the searched bit-widths will tend to be higher bit-width like 8-bit, and the searched mixed-precision models achieve higher Top-1 accuracy up to 92.6%. Actually, it is natural to allocate higher bit width for each layer without BOPs constraint as higher bit-width leads to higher accuracy.

5.2 Differentiated Importance Factors

We conduct two mixed precision search experiments with and without Prob-1 regularizer, to verify whether the Prob-1 regularizer will differentiate the importance factors. Both mixed precision search is under the constraint $\mathcal{B}_{max}$ = 3-bit. In Fig. 4, we present the importance factors evolution of the first layer in the two supernets. The red curve in Fig. 4(a) tends to be 1 and other curves tend to 0 after several epoches. While in Fig. 4(b), all the curves have similar value, indicating that these importance factors are similar. Note that our Prob-1 regularizer does not disturb the mixed precision search in the early search stage.

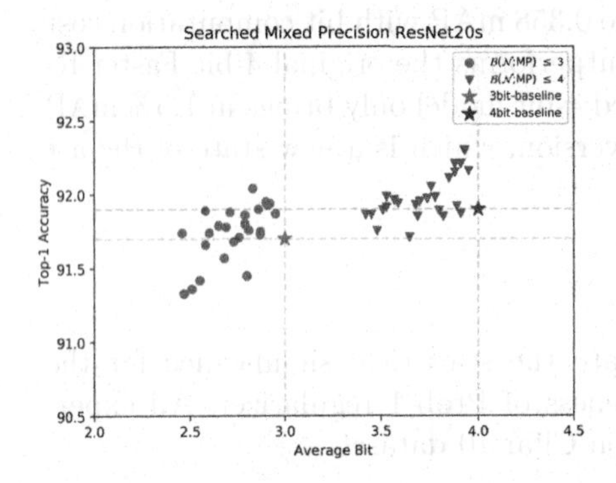

Fig. 3. Accuracy and average bit of the search results under different BOPs constraint. Red cicles and green triplet respectively denote the search results under constraints $\mathcal{B}_{max}$ = 3-bit and $\mathcal{B}_{max}$ = 4-bit

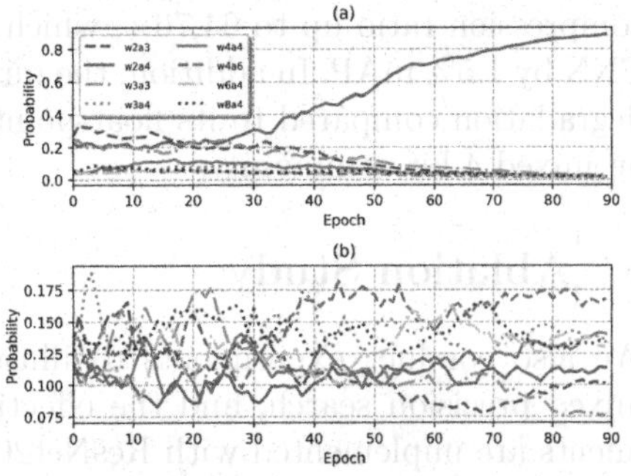

Fig. 4. Importance factors evolution. (a) The factors evolution of the first layer with Prob-1 regularizer; (b) The factors evolution of the second layer without the regularizer

6 Conclusion

In this paper, a novel soft Barrier Penalty Neural Architecture Search (BP-NAS) framework is proposed for mixed precision quantization, which ensures all the search models will satisfy the application meet or predefined constraint. In addition, an effective differentiable Prob-1 regularizer is proposed to differeentiate the importance factors in each layer, and the distribution reshaping strategy will make the training much more robust and achieve higher performance. Compared with existing gradient-based NAS, BP-NAS can guarantee that the searched mixed precision model within the constraint, while speedup the search to access to the better accuracy. Extensive experiments demonstrate that BP-NAS outperforms existing state-of-the-art mixed-precision algorithms by a large margin on public datasets.

Acknowledgments. We thank Ligeng Zhu for the supportive feedback, and Kuntao Xiao for the meaningful discussion on solving constrained-optimization problem. This work is supported by the National Natural Science Foundation of China (61876180), the Beijing Natural Science Foundation (4202073), the Young Elite Scientists Sponsorship Program by CAST (2018QNRC001).

References

1. Alizadeh, F.: Interior point methods in semidefinite programming with applications to combinatorial optimization. SIAM J. Optim. **5**(1), 13–51 (1995)
2. Cai, H., Zhu, L., Han, S.: ProxyLessnas: direct neural architecture search on target task and hardware. In: ICLR (2019)
3. Cai, Z., He, X., Sun, J., Vasconcelos, N.: Deep learning with low precision by half-wave Gaussian quantization. In: The IEEE Conference on Computer Vision and Pattern Recognition (CVPR), July 2017
4. Chaim, B., Eli, S., Evgenii, Z., Natan, L., Raja, G., Alex, M.B., Avi, M.: UNIQ: uniform noise injection for non-uniform quantization of neural networks. arXiv preprint arXiv:1804.10969 (2018)
5. Choi, J., Wang, Z., Venkataramani, S., Chuang, P.I.J., Srinivasan, V., Gopalakrishnan, K.: PACT: parameterized clipping activation for quantized neural networks. arXiv preprint arXiv:1805.06085 (2018)
6. Deng, J., Dong, W., Socher, R., Li, L.J., Li, K., Fei-Fei, L.: ImageNet: a large-scale hierarchical image database. In: IEEE Conference on Computer Vision and Pattern Recognition, pp. 248–255. IEEE (2009)
7. Dong, Z., Yao, Z., Gholami, A., Mahoney, M., Keutzer, K.: HAWQ: hessian aware quantization of neural networks with mixed-precision. In: International Conference on Computer Vision (ICCV) (2019)
8. Gu, J., Zhao, J., Jiang, X., Zhang, B., Liu, J., Guo, G., Ji, R.: Bayesian optimized 1-bit CNNs. In: ICCV (2019)
9. Guo, Z., et al.: Single path one-shot neural architecture search with uniform sampling. In: ECCV (2020)
10. He, K., Zhang, X., Ren, S., Sun, J.: Deep residual learning for image recognition (2016)
11. Jacob, B., et al.: Quantization and training of neural networks for efficient integer-arithmetic-only inference. arXiv preprint arXiv:1712.05877 (2017)
12. Li, R., Wang, Y., Liang, F., Qin, H., Yan, J., Fan, R.: Fully quantized network for object detection (2019)
13. Liu, H., Simonyan, K., Yang, Y.: DARTS: differentiable architecture search. In: ICLR (2019)
14. Long, J., Shelhamer, E., Darrell, T.: Fully convolutional networks for semantic segmentation. In: Proceedings of the IEEE Conference on Computer Vision and Pattern Recognition, pp. 3431–3440 (2015)
15. Moran, S., et al.: Robust quantization: one model to rule them all. ArXiv, abs/2002.07686 (2020)
16. Nvidia: Nvidia tensor cores (2018)
17. Redmon, J., Divvala, S., Girshick, R., Farhadi, A.: You only look once: unified, real-time object detection. In: Proceedings of the IEEE Conference on Computer Vision and Pattern Recognition, pp. 779–788 (2016)

18. Ren, S., He, K., Girshick, R., Sun, J.: Faster R-CNN: towards real-time object detection with region proposal networks. In: Advances in Neural Information Processing Systems, pp. 91–99 (2015)
19. Wang, K., Liu, Z., Lin, Y., Lin, J., Han, S.: HAQ: hardware-aware automated quantization. In: The IEEE Conference on Computer Vision and Pattern Recognition (CVPR) (2019)
20. Wang, P., et al.: Two-step quantization for low-bit neural networks. In: Proceedings of the IEEE Conference on Computer Vision and Pattern Recognition, pp. 4376–4384 (2018)
21. Wu, B., et al.: FBNet: hardware-aware efficient convnet design via differentiable neural architecture search. In: Proceedings of the IEEE Conference on Computer Vision and Pattern Recognition, pp. 10734–10742 (2019)
22. Wu, B., Wang, Y., Zhang, P., Tian, Y., Vajda, P., Keutzer, K.: Mixed precision quantization of ConvNets via differentiable neural architecture search. In: ICLR (2019)
23. Yochai, Z., et al.: Towards learning of filter-level heterogeneous compression of convolutional neural networks. In: ICML Workshop on AutoML (2019)
24. Yu, H., Wen, T., Cheng, G., Sun, J., Han, Q., Shi, J.: Low-bit quantization needs good distribution. In: CVPR Workshop on Efficient Deep Learning in Computer Vision (2020)
25. Zhang, D., Yang, J., Ye, D., Hua, G.: LQ-nets: learned quantization for highly accurate and compact deep neural networks. In: Proceedings of the European Conference on Computer Vision (ECCV), pp. 365–382 (2018)
26. Zhao, H., Shi, J., Qi, X., Wang, X., Jia, J.: Pyramid scene parsing network. In: IEEE Confernce on Computer Vision and Pattern Recognition (CVPR), pp. 2881–2890 (2017)
27. Zhou, A., Yao, A., Guo, Y., Xu, L., Chen, Y.: Incremental network quantization: towards lossless CNNs with low-precision weights. arXiv preprint arXiv:1702.03044 (2017)
28. Zhou, S., Wu, Y., Ni, Z., Zhou, X., Wen, H., Zou, Y.: DoReFa-net: training low bitwidth convolutional neural networks with low bitwidth gradients. arXiv preprint arXiv:1606.06160 (2016)
29. Zhu, C., Han, S., Mao, H., Dally, W.J.: Trained ternary quantization. arXiv preprint arXiv:1612.01064 (2016)

Monocular 3D Object Detection via Feature Domain Adaptation

Xiaoqing Ye[1(✉)], Liang Du[2,3(✉)], Yifeng Shi[1], Yingying Li[1], Xiao Tan[1], Jianfeng Feng[2,3], Errui Ding[1], and Shilei Wen[1]

[1] Baidu Inc., Beijing, China
yexiaoqing@baidu.com
[2] Institute of Science and Technology for Brain-Inspired Intelligence, Fudan University, Shanghai, China
duliang@mail.ustc.edu.cn
[3] Key Laboratory of Computational Neuroscience and Brain-Inspired Intelligence, Fudan University, Ministry of Education, Shanghai, China

Abstract. Monocular 3D object detection is a challenging task due to unreliable depth, resulting in a distinct performance gap between monocular and LiDAR-based approaches. In this paper, we propose a novel domain adaptation based monocular 3D object detection framework named DA-3Ddet, which adapts the feature from unsound image-based pseudo-LiDAR domain to the accurate real LiDAR domain for performance boosting. In order to solve the overlooked problem of inconsistency between the foreground mask of pseudo and real LiDAR caused by inaccurately estimated depth, we also introduce a context-aware foreground segmentation module which helps to involve relevant points for foreground masking. Extensive experiments on KITTI dataset demonstrate that our simple yet effective framework outperforms other state-of-the-arts by a large margin.

Keywords: Monocular · 3D Object detection · Domain adaptation · Pseudo-Lidar

1 Introduction

3D object detection is in a period of rapid development and plays a critical role in autonomous driving [16] and robot vision [4]. Currently, methods [34,39,47] based on LiDAR devices have shown favorable performance. However, the disadvantages of these approaches are also obvious due to the high cost of 3D

X. Ye and L. Du—The first two authors contributed equally to this work.

Electronic supplementary material The online version of this chapter (https://doi.org/10.1007/978-3-030-58545-7_2) contains supplementary material, which is available to authorized users.

© Springer Nature Switzerland AG 2020
A. Vedaldi et al. (Eds.): ECCV 2020, LNCS 12354, pp. 17–34, 2020.
https://doi.org/10.1007/978-3-030-58545-7_2

sensors. Alternatively, the much cheaper monocular cameras are drawing increasing attention of researchers to dig into the problem of monocular 3D detection [3,9,10,32,37,44]. Monocular-based methods can be roughly divided into two categories, one is RGB image-based approaches incorporating with geometry constraints [32] or semantic knowledge [9]. Unsatisfactory precision is observed due to the variance of the scale for an object caused by perspective projection and the lack of depth information. The other category leverages depth estimation to convert pixels into artificial point clouds, namely, pseudo-LiDAR [42,43], so as to boost the performance by borrowing benefits of approaches working for real-LiDAR points. In this case, 3D point cloud detection approaches [34] can be adopted for pseudo-LiDAR. Although recent works [42,43,46] have proven the superiority of pseudo-LiDAR in 3D object detection, the domain gap between pseudo-LiDAR and real LiDAR remains substantial due to their physical differences, which limits the higher performance of pseudo-LiDAR.

To solve the problem, most approaches [41,46] investigate more accurate depth estimation methods such as DenseDepth [2], DispNet [31], PSM-Net [8], or take advantage of semantic segmentation [9] as well as instance segmentation [37] to obtain an unalloyed object point cloud with background filtered out. However, the extra information makes the framework too heavy for real scene application. As shown in Fig. 1, compared against real LiDAR, pseudo-LiDAR point cloud has a weaker expression of the object structure.

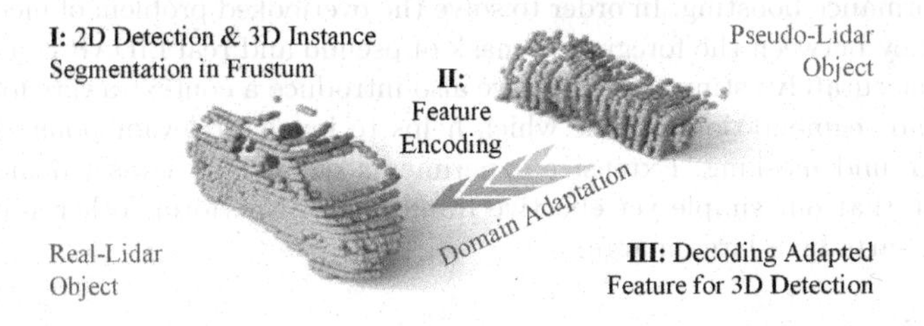

Fig. 1. A comparison of real and pseudo-LiDAR object point clouds. The real-LiDAR point cloud of the object has more accurate and crisper representation than pseudo-LiDAR, leading to a performance discrepancy. Domain adaptation approach is utilized to bridge the domain gap between these two modalities for further boosting the performance of monocular 3D object detection.

In contrast, we propose a simple yet effective way to boost the performance of pseudo-LiDAR by bridging the domain inconsistency between real LiDAR and pseudo-LiDAR with adaption. We build up a siamese network based on the off-the-shelf LiDAR based 3D object detection framework [34]. Two branches of the siamese network take real and pseudo-LiDAR as input, respectively. The difference between the two feature domains is minimized by the proposed network so as to encourage the pseudo-LiDAR feature being similar to the real-LiDAR feature. By narrowing the gap between high-dimensional feature representations of LiDAR and pseudo-LiDAR, our work surpasses previous state-of-the-arts on the KITTI 3D detection benchmark. The main contributions of our work are summarized as follows:

- We are the first to leverage domain adaptation approach to customize the features from pseudo-LiDAR domain to real-LiDAR domain, so as to bridge the performance gap between monocular-based and LiDAR point-based 3D detection approaches.
- To fully exploit the context information for feature generation, we investigate a context-aware foreground segmentation module (CAFS), which allows network using the both foreground and the context point clouds to map the pseudo features to discriminative LiDAR features.
- We achieve new state-of-the-art monocular 3D object detection performance on KITTI benchmark.

2 Related Works

2.1 LiDAR-Based 3D Object Detection

Current LiDAR Based 3D object detection methods can be divided into three categories: (1) Multi-view based methods [11,23] project the LiDAR point clouds into bird's eye view (BEV) or front view to extract features, and a fusion process is applied to merge features. (2) Voxel-based. LiDAR point clouds are first divided into voxels and then learned by 3D convolutions [13,47]. However, due to the sparsity and non-uniformity of point clouds, voxel-based methods suffer from high computation cost. To tackle this problem, sparse convolutions are applied on this form of data [12,45]. Besides, (3) direct operation on raw points are also investigated recently [34–36,39], since some researchers believe that data representation transformation may cause data variance and geometric information loss. For instance, F-PointNet [34] leverages both mature 2D object detectors and advanced 3D pointnet-based approach for robust object localization.

2.2 Monocular 3D Object Detection

Monocular-based 3D detection [5,25,28,30,33,38,40] is a more challenging task due to a lack of accurate 3D location information. Most prior works [3,9,10,24,37,44] on monocular 3D detection were RGB image-based, with auxiliary information like the semantic knowledge or geometry constraints and so on. AM3D [29] designed two modules for background points segmentation and RGB information aggregation respectively in order to improve 3D box estimation. Some other works involved 2D-3D geometric constraints to alleviate the difficulty caused by scale variety. Mousavian et al. [32] argued that 3D bounding box should fit tightly into 2D detection bounding box according to geometric constraints. Deep MANTA [7] encoded 3D vehicle information using key points of vehicles.

Another recently introduced approach for monocular 3D detection is based on pseudo-LiDAR [42,43,46], which utilizes depth information to convert image pixels into artificial point clouds, i.e., pseudo-LiDAR, and employs LiDAR-based frameworks for further detection. PL-MONO [42] was the pioneer work that pointed out the main reason for the performance gap is not attributed to the inaccurate depth information, but data representation. Their work achieved impressive improvements by converting image-based depth maps to pseudo-LiDAR representations. Mono3D-PLiDAR [43] trained a LiDAR-based 3D detection network with pseudo-LiDAR; therefore the LiDAR-based methods can work with a single image input. They also point out the noise in pseudo-LiDAR data is a bottleneck to improve performance. You et al. [46] believed pseudo-LiDAR relies heavily on the quality of depth estimation, so a stereo network architecture to achieve more accurate depth estimation is proposed. While pseudo-LiDAR has largely improved the performance of monocular 3D detection, there is still a notable performance gap between pseudo-LiDAR and real LiDAR.

2.3 Domain Adaptation

As shown in Fig. 1, given the same object, distinct point distribution discrepancy can be noticed between the depth-transformed pseudo-LiDAR and real LiDAR, which leads to a large domain gap between two modalities. Domain adaptation [1] is a machine learning paradigm aiming at bridging different domains. The critical point lies in how to reduce the distribution discrepancy across different domains while avoiding over-fitting. Thanks to deep neural networks that are able to extract high-level representations behind the data, domain adaptation [22] has made tremendous progress in object detection [19] and semantic segmentation [18]. In order to keep from overly dependent on the accuracy of depth estimation, we take advantage of LiDAR guidance in training stage to adapt the pseudo-LiDAR features to act as real-LiDAR features do. Following [27], our DA-3Ddet employs the L2-norm regularization to calculate the feature similarity for domain adaptation.

3 Methodology

3.1 Overview

The proposed framework DA-3Ddet is depicted in Fig. 2. It consists of two main components, the siamese branch for feature domain adaption and the context-aware foreground segmentation module. First the overall pipeline is introduced, then the two critical modules are elaborated in detail, and finally the training loss is given.

Recent research works [42] have verified the superiority of adopting pseudo-LiDAR representations from estimated depth to mimicking LiDAR point clouds for 3D object detection. However, there is still a large performance gap between pseudo-based and real LiDAR detection methods due to two reasons. For one thing, the generated pseudo-LiDAR representations heavily rely on the accuracy of the estimated depth. For another, the distribution and density are physically different between the two representations, as shown in Fig. 1. In light of such domain inconsistency issue, we propose a simple yet effective method to boost the performance of 3D object detection from only a monocular image. Rather than hunting for expensive multi-modal fusion strategies to improve the pseudo-LiDAR approaches, we build up a siamese network leveraging off-the-shelf LiDAR-based 3D detection frameworks as the backbone. Further, domain adaptation approach between different modal features is adopted to guide the pseudo-LiDAR representations to be closer to real-LiDAR representations.

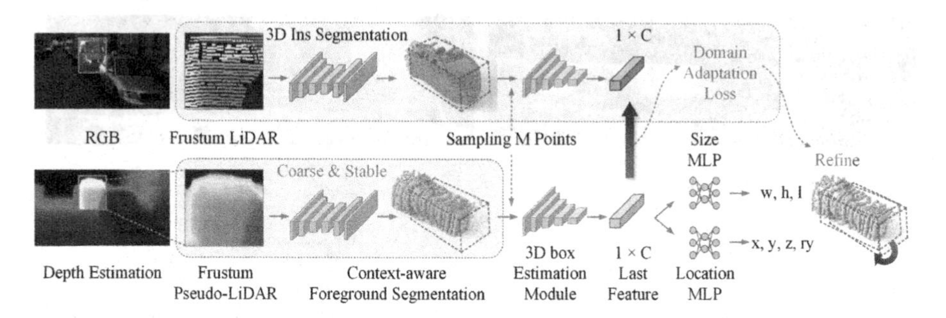

Fig. 2. Illustration of the proposed DA-3Ddet. A 2D object detector is first utilized to recognize and localize the objects before lifting the concerned 2D regions to 3D frustum proposals. Given a point cloud in a frustum ($N \times 3$ with N points and three channels of XYZ for each point), the object instance is segmented by binary classification of each point. The segmented foreground point cloud ($M \times 3$) is then encoded by 3D box estimation network. Domain adaptation is performed between the last layer ($1 \times C$) of real and pseudo encoded features. Finally, the box estimation net is employed to predict 3D bounding box parameters.

The input of our framework is a monocular image, and during the training process, real-LiDAR data is also utilized for feature domain adaption. Only a single image is required during the inference stage. First, the depth map is estimated given a monocular image and then transformed into point clouds in the LiDAR coordinate system, namely, pseudo-LiDAR. After that, real and pseudo LiDAR data of the same scene are simultaneously fed into the siamese branches, respectively, to obtain their high-dimensional feature representations. The features of pseudo-LiDAR domain are adapted to real-LiDAR feature domain during the training process. Finally, the aligned pseudo feature is decoded to regress the 3D parameters of detected objects. More technical details of our approach will be explained in the following sections.

3.2 Siamese Framework for Adapting Pseudo-LiDAR to LiDAR

To narrow the gap between pseudo-LiDAR generated by depth maps and real LiDAR based methods, we propose a naive yet effective adaption method. Any off-the-shelf LiDAR-based 3D object detection networks can be utilized as the backbone for encoding 3D points data. To fairly compare with most existing

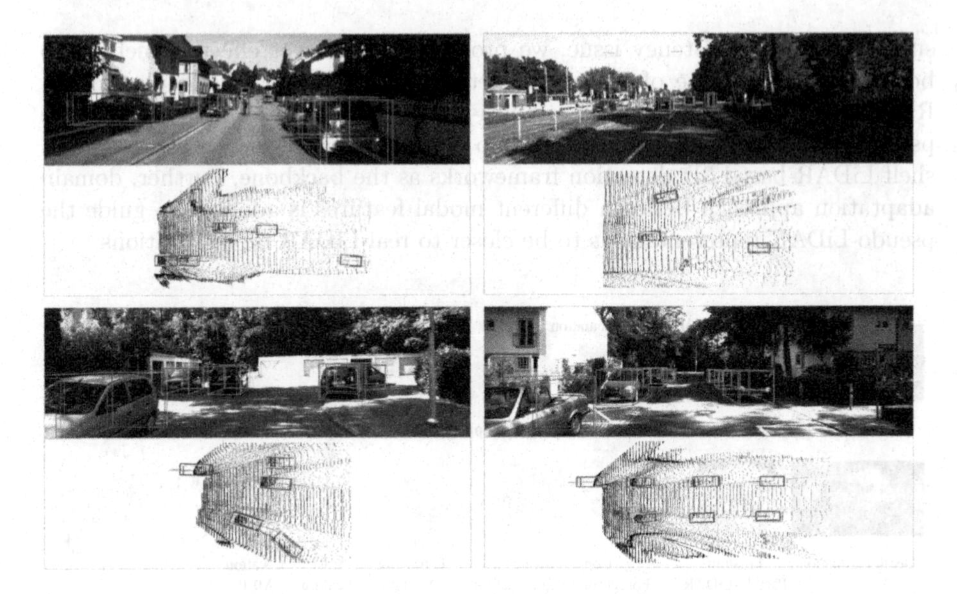

Fig. 3. The qualitative comparison of the ground truth (green), the baseline (yellow) and our method (red) on KITTI val set. The first row shows RGB images, and the second row shows the bird's-eye view, respectively. Our method effectively predicts reliable 3D bounding box of objects even with inaccurate depth estimation. (Color figure online)

pseudo approaches like [42], we adopt the same framework frustum PointNet (F-PointNet) [34] as our baseline.

First, we briefly review the pipeline of frustum PointNet [34]. A 2D object detection network is applied to the monocular image to detect the objects. Next, each region within the 2D bounding box is lifted to 3D frustum proposals. Each frustum point cloud is then fed into a PointNet encoder for 3D instance segmentation to perform foreground and background classification. Based on the masked object point cloud after binary segmentation, a simplified regression PointNet (T-Net) is further applied to translate the mask center to amodal box center. Finally, another PointNet module is followed to regress 3D box parameters. More details can be found in [34]. The advantage of the chosen baseline lies in its employment of 2D detector to restrain interested regions as well as operation on raw point clouds, which makes it robust to strong occlusion and sparsity at low cost (Fig. 3).

In our work, we adopt a siamese network consisting of two branches of frustum PointNet, as depicted in Fig. 2. For the upper branch corresponding to the real LiDAR, we utilize the pretrained model as prior, and it is only forwarded during the training process with parameters fixed. For the lower depth-transformed pseudo-LiDAR branch, the extracted frustum is fed into the frustum PointNet-based module, which is exactly the same architecture with the upper branch. Similar to F-PointNet [34], we also adopt a PointNet-based 3D instance segmentation network to filter out background or irrelevant instance point clouds in the frustum. Differently, due to inaccurate estimated depth, pseudo-LiDAR points within the ground truth 3D bounding box (provided by 3D LiDAR) can be inconsistent with the foreground points derived from 2D images. In consequence, a context-aware foreground segmentation module is proposed to alleviate the adverse effects caused by inaccurate depth estimation, which will be further explained in the following subsection.

The generated features of the two branches before the final head of 3D bounding box regression are encoded into $(1 \times C)$, respectively, where C is the channel number of the encoded feature. To make the domain adaptation more focused, it is restricted to segmented foreground points only. We calculate the $\mathcal{L}2$ distance between the pseudo and real LiDAR high-dimensional features to perform domain adaptation so that the engendered representations of pseudo-LiDAR resemble real-LiDAR features. After that, the amodal 3D box estimation network is adopted to decode the features after adaptation, so as to regress the 3D box parameters of the object. After the pseudo-LiDAR branch network is tuned to achieve an aligned feature domain, we can simply discard the upper real-LiDAR branch at inference time.

3.3 Context-Aware Foreground Segmentation

In real LiDAR-based 3D detection approaches, LiDAR points within the ground truth 3D bounding box are utilized as the supervising signal for 3D instance segmentation to filter out background points. However, for pseudo-LiDAR point clouds, points computed from 2D foreground instance pixels may be inconsistent with 3D foreground mask ground truth due to inaccurate depth estimation, as shown in Fig. 4. To be more specific, if the estimated depth differs from the ground truth depth to some degree, there can be fewer points within the 3D ground truth box, increasing the difficulty for regress the 3D object parameters. In contrast, relevant points (regions colored in light green in Fig. 4) that contain useful structural information are excluded by the ground truth 3D foreground mask. An extreme case can be no valid pseudo-LiDAR object points found in the ground truth 3D box in far-away regions due to large depth estimation offset to actual distance.

Although neglected by previous works, we argue that it must not be overlooked. To tackle this problem, we investigate a context-aware foreground segmentation module (CAFS). We first train a baseline model with 3D instance segmentation loss like F-PointNet [34] does. The pretrained model serves as a prior to help CAFS module recognize and segment the foreground point clouds in a coarse manner. During the end-to-end whole siamese network training, the foreground segmentation loss is then discarded to let the CAFS select both foreground and background points to generate stable and abundant features for domain adaptation.

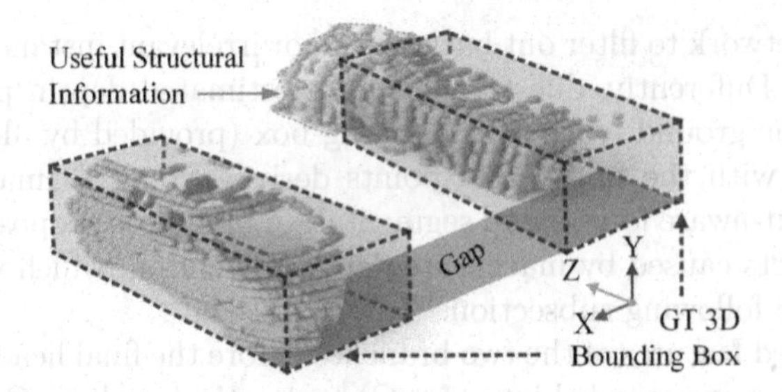

Fig. 4. An example of pseudo-LiDAR and real-LiDAR instance point cloud. Adopting GT 3D bounding box as supervising signal of pseudo-LiDAR can lead to the lost of structural information due to inaccurate depth estimation. (Color figure online)

3.4 Training Loss

In order to align the 3D parameters with projected 2D bounding boxes according to projective transformation, we define the ground truth 2D bounding box for our 2D detector: $[x_1, y_1, x_2, y_2]$ according to projected 3D ground truth boxes. In specific, we project all ground truth 3D bounding boxes onto image space given the camera intrinsic, as Eq. 1 shows.

$$Z \begin{bmatrix} u \\ v \\ 1 \end{bmatrix} = K \begin{bmatrix} X \\ Y \\ Z \\ 1 \end{bmatrix}_{P_c} \tag{1}$$

Where the left point $[u, v, 1]$ denotes the projected 2D image coordinate and Z is the depth, whereas the right part denotes 3D point in camera coordinate and K is the camera intrinsic computed in advance and used during the training and inference stage. Rather than directly adopting 2D annotations provided by KITTI, the computed minimum bounding rectangle of the projected eight 3D vertices are served as the ground truth of 2D detector. By means of 2D-3D alignment, the 3D prediction is jointly optimized.

For each input frustum point cloud, the outputs are parameterized as follows: $[c_x, c_y, c_z]_{3D}$, $[h, w, l]_{3D}$, $[R_y^{(m)}, R_{offset}^{(m)}]$, $C^{(s)}$, Corner$_i$,$i \in 1, ..., 8$, where $[c_x, c_y, c_z]_{3D}$ and $[h, w, l]_{3D}$ are the regressed center and size of 3D box, respectively. As is proved in previous works [34], a hybrid of classification and regression formulations makes heading angle estimation more robust. Thus the network predicts the scores of each equally split angle bins as well as their offset to the center of each bin, i.e., $[\mathrm{R}_y^{(m)}, R_{offset}^{(m)}]$, where m is the predefined number of bins. $C^{(s)}$ denotes the class of the given object, with s refers to the categories.

Following [34], for the sub-network of training pseudo-LiDAR data we adopt the same loss function, including cross-entropy foreground segmentation loss $\mathcal{L}_{seg}$, smooth-L1 3D box regression loss $\mathcal{L}_{3D_{reg}}$, cross-entropy classification loss $\mathcal{L}_{cls}$, as well as corner loss $\mathcal{L}_{corner}$. For corner loss, we compute the minimum one of the two mean distances between the eight corners derived from the predicted angle and its flipped angle with respect to ground truth:

$$\mathcal{L}_{corner} = \frac{1}{8}min(\sum_{i=1}^{8} |C_i - C_i^*|, \sum_{i=1}^{8} |C_i - C_i^{*'}|) \tag{2}$$

where C_i^* and $C_i^{*'}$ are predicted corners and the corners with flipped angle, C_i is the ground truth corner.

$$\mathcal{L}_{3D} = \mathcal{L}_{seg} + \mathcal{L}_{3D_{reg}} + \mathcal{L}_{cls} + \mathcal{L}_{corner} \tag{3}$$

For the domain adaption loss, we compute the $\mathcal{L}_2$ loss for feature alignment:

$$\mathcal{L}_{DA} = \mathcal{L}_2(\mathcal{F}_{real}, \mathcal{F}_{pseudo}) \tag{4}$$

The overall loss is then computed by Eq. 5, where α is a hyperparameter to balance these two terms.

$$\mathcal{L}_{all} = \mathcal{L}_{3D} + \alpha\mathcal{L}_{DA} \tag{5}$$

4 Experiment

4.1 Implementation

Dataset. The proposed approach is evaluated on the KITTI 3D object detection benchmark [15,16], which contains 7,481 images for training and 7,518 images for testing. We follow the same training and validation splits as suggested by [11], i.e., 3,712 and 3,769 images for train and val, respectively. For each training image, KITTI provides the corresponding LiDAR point cloud, right image from stereo cameras, as well as camera intrinsics and extrinsics.

Metric. We focus on 3D and bird's-eye-view (BEV) object detection and report the average precision (AP) results on validation and test set. Specifically, for "Car" category, we adopt IoU = 0.7 as threshold following [11]. Besides, to validate the effectiveness on the other two categories - "Pedestrian" and "Cyclist", we also include corresponding experiments on the two categories with IoU = 0.5 for fair comparison. AP for 3D and BEV tasks are denoted by AP_{3D} and AP_{BEV}, respectively. Note that there are three levels of difficulty defined in the benchmark according to the 2D bounding box height, occlusion and truncation degree, namely, easy, moderate and hard. The KITTI benchmark ranks algorithms mainly based on the moderate AP.

Monocular Depth Estimation. Different depth estimation approaches can have influence on the transformed pseudo-LiDAR. Thanks to the proposed feature domain adaptation and context-aware foreground segmentation module, our framework works on various depth predictors regardless of the degree of accuracy. For fair comparison with other works, we adopt the open-sourced monocular depth estimator DORN [14] to obtain depth maps. Note that our proposed

Table 1. Comparison on KITTI val and test set. The average precision (in %) of "Car" on 3D object detection (AP_{3D}) at IoU = 0.7 is reported. Our proposed DA-3Ddet achieves new state-of-the-art performance.

Method	Val			Test		
	Mod	Easy	Hard	Mod	Easy	Hard
SS3D [20]	13.2	14.5	11.9	7.7	10.8	6.5
RT-M3D [26]	16.9	20.8	16.6	10.1	13.6	8.2
Pseudo-LiDAR [42]	17.2	19.5	16.2	/	/	/
M3D-RPN [5]	17.1	20.3	15.2	9.7	14.8	7.4
Decoupled-3D [6]	18.7	27.0	15.8	7.3	11.7	5.7
Mono3D-PLiDAR [43]	21.0	31.5	17.5	7.5	10.8	6.1
AM3D [29]	21.1	32.2	17.3	10.7	16.5	**9.5**
Ours	**24.0**	**33.4**	**19.9**	**11.5**	**16.8**	8.9

framework observes improvements on various depth estimation strategies and experiment result in Table 7 validates the effectiveness.

Results on Other Categories. Due to the small sizes and non-rigid structures of the other two classes - "Pedestrian" and "Cyclist", it is much more challenging to perform 3D object detection from monocular image than cars. We guess it could be the reason that most of the previous monocular methods simply report their results on "Car" only. Nevertheless, we still conduct self-compared experiments with respect to the baseline on the given two classes. Table 2 reports the AP_{3D} results on KITTI val set at IoU = 0.5. Although the results seem worse than "Car", compared to the baseline pseudo-LiDAR, we observe an improvement on both categories at all difficulties due to our domain adaptation and context-aware 3D foreground segmentation module.

Table 2. Bird's eye view detection (AP_{BEV}) / 3D object detection (AP_{3D}) performance for "Pedestrian" and "Cyclist" on KITTI val split set at IoU = 0.5.

Method	Cyclist			Pedestrian		
	Mod	Easy	Hard	Mod	Easy	Hard
Baseline	10.8/10.6	11.7/11.4	10.8/10.6	9.3/6.0	11.7/7.2	7.8/5.4
Ours	**12.2/11.5**	**15.5/14.5**	**11.8/11.5**	**10.6/7.1**	**13.1/8.7**	**9.2/6.7**

Pseudo-LiDAR Frustum Generation. First the estimated depth map is back-projected into 3D points in LiDAR's coordinate system by the provided calibration matrices. Second, utilizing the 2D detector which is trained with the minimum bounding rectangle of the projected vertices of 3D boxes as ground truth, the frustum is lifted to serve as the input of our siamese network.

Training Details. The network is optimized by Adam optimizer [21] with initial learning rate 0.001 and a mini-batch size of 32 on TITAN RTX GPU. The number of points of the network input is fixed to 1024. Frustum with less than 1024 points will be sampled repeatedly and otherwise be randomly down-sampled. For training the coarse foreground segmentation module with loss, we trained for 150 epochs, and after that we further trained for 150 epochs with segmentation loss discarded. α in Eq. 5 is set to 1.0. And for final output, in order to decouple the size and location properties, we add a three-layer MLP to regress the size and location parameters of 3D box, respectively.

4.2 Comparison with State-of-the-art Methods

Results on KITTI Test and Val. The 3D object detection results on KITTI val and test set are summarized Table 1 at IoU threshold = 0.7. Compared with the top-ranked monocular-based 3D detection approaches in KITTI leader board, our method consistently outperforms the other methods and ranks 1st. In specific, (1) Both on validation and test set, our method achieves the highest performance on the moderate set, which is the main setting for ranking on the KITTI benchmark. Large margins are observed over the second top-performed method (AM3D [29]), that is, 13.6% and 4.7% on val and test set, respectively. (2) Some top-ranked monocular methods utilize extra information for 3D detection. For example, Mono3D-PliDAR [43] uses the instance mask instead of the bounding box as the representation of 2D proposals and AM3D [29] designs two extra modules for background points segmentation and RGB information aggregation. In contrast to the above-mentioned methods that utilize extra information, our simple yet effective method achieves appealing results.

4.3 Ablation Study

Main Ablative Analysis. We conduct ablation study by making comparison among ten variants of the proposed method as shown in Table 3. "DA" means applying our domain adaptation approach for pseudo-LiDAR. "SP" and "UnSP" represent the network with or without 3D instance segmentation loss when selecting foreground points during the whole training process, respectively. CAFS denotes the proposed module that segment the foreground points with context information for further domain adaptation and 3D bounding box regression. Furthermore, to compare the effect of decoupling size and location parameters of 3D object, three different head decoders are experimented. "D1", "D2" and "D3" indicate adopting single, double ("$xyzr$" and "whl") and triple ("xyz", "whl", "r") decoding branches and each branch is composed of a three-layer MLP, respectively. The baseline (a) is constructed following F-PointNet [34], with supervised instance segmentation loss during the whole training process and only one decoder branch, no domain adaptation is utilized. Note that for the baseline, we use the modified 2D detector with a minimum bounding rectangle of the projected eight 3D corners as ground truth.

Table 3. Ablative analysis on the KITTI val split set for AP_{3D} at IoU = 0.7. Experiment group (a) is our baseline method. Different experiment settings are applied: using domain adaptation (DA), using GT bounding box to supervise the 3D instance segmentation (SP), unsupervised segmentation (UnSp), using our context-aware foreground segmentation method (CAFS), single decoder (D1), two-branch decoder (D2) and three-branch decoder (D3).

Group	DA	SP	UnSP	CAFS	D1	D2	D3	Mod	Easy	Hard
(a)		✓			✓			21.9	28.6	18.4
(b)			✓		✓			15.7	18.5	14.8
(c)				✓	✓			22.4	29.4	18.8
(d)		✓				✓		22.1	28.8	18.5
(e)		✓					✓	21.4	27.6	18.2
(f)	✓	✓			✓			23.1	31.8	19.2
(g)	✓		✓		✓			16.1	19.0	14.9
(h)	✓			✓	✓			23.6	32.9	19.7
(i)	✓			✓		✓		**24.0**	**33.4**	**19.9**
(j)	✓			✓			✓	23.2	31.7	19.4

As depicted in Table 3, we can observe that: (1) Compared (a) with (b), (c), we found that without domain adaptation, the performance can be deteriorated by simply removing the foreground segmentation loss, since the regression relies on the useful foreground points rather than background noises. If we divide the whole training process into two stages and apply our context-aware foreground segmentation (CAFS) module, the performance of the baseline can be improved from 21.9 to 22.4 in moderate setting. (2) Compared (a) with (f), a noticeable gain is achieved due to our feature domain adaptation (21.9 vs. 23.1). (3) In addition, changing from supervised instance segmentation loss to our adaptive context-aware foreground segmentation further improves the performance, reflected from (g) to (h). (4) Finally, we also compare different decoder branch settings and find that employing separate heads for "*xyzr*" and "*whl*" achieves the best performance. We believe that "*whl*" are of size parameters, whereas "*xyzr*" is related to location and is estimated in residual form. Decoupling the two groups can slightly benefit to the regression task.

Statistic Analysis on 3D Metric. For monocular 3D detection, the depth (i.e., "z" in distance) and rotation "ry" around Y-axis in camera coordinates are the most challenging parameters, which have significant influence on the 3D detection precision. As a result, for further detailed explanation of the improvement over the baseline, we compare the errors on the above-mentioned two metrics of the baseline and our proposed DA-3Ddet. As shown in Fig. 5, we can clearly see that our proposed method improves the baseline method in "z" and "ry", which results in more accurate monocular 3D object detection.

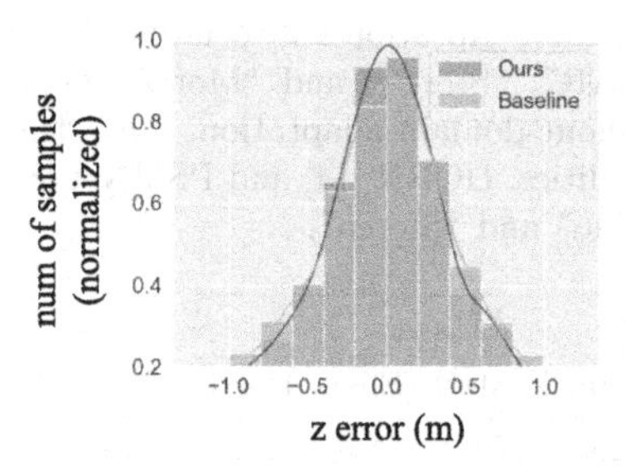

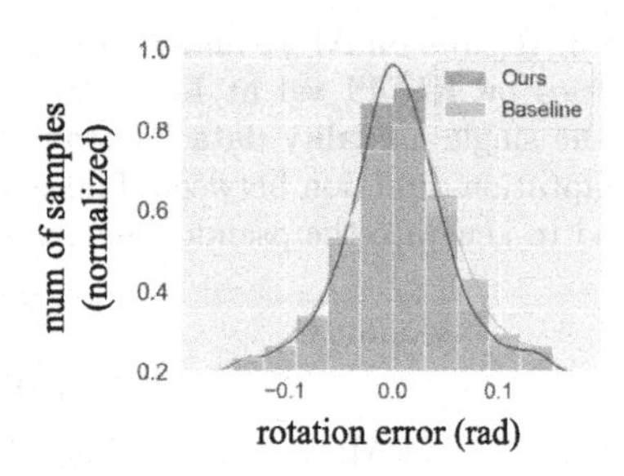

Fig. 5. The statistic analysis and comparison of the baseline method (blue) and our DA-3Ddet (red). The y axis of the chart represents the number of samples after normalization. Our method effectively improves the critical metrics "z" and rotation "ry" in 3D object detection. (Color figure online)

Table 4. Comparison of different sampling rates on val set at IoU = 0.7. The number of sampling points during the training and testing process is the same.

Sampling num	AP_{3D}			AP_{BEV}		
	Mod.	Easy	Hard	Mod.	Easy	Hard
768	23.5	32.4	19.6	32.1	45.0	26.6
1024	**24.0**	**33.4**	**19.9**	32.7	45.5	27.1
1536	23.9	32.3	19.8	33.1	45.6	27.2
2048	23.8	32.4	19.7	**33.1**	**46.0**	**27.2**

Impact of Different Point Cloud Densities. Real LiDAR point clouds are sparse and non-uniform whereas the depth-transformed pseudo-LiDAR data can be denser. As a result, to compare the influence of points number within the lifted frustum, we conduct the experiments with different point sampling rates from the frustum. As shown in Table 4, for 3D detection task, 1024 points perform best. For detection in bird's eye view, more points (2048) achieves better results. We claim that our framework is robust to point numbers to some degree since the performance gap of different densities is small.

Impact of Different Loss Functions for Domain Adaptation. For feature domain adaptation, we experimented with two kinds of losses, namely, the $\mathcal{L}1$ and $\mathcal{L}2$. Table 5 shows that $\mathcal{L}2$ performs better than $\mathcal{L}1$.

Table 5. Comparison of different loss functions for adaptation on val set at IoU = 0.7.

Adaptation loss	AP_{3D}			AP_{BEV}		
	Mod	Easy	Hard	Mod	Easy	Hard
$\mathcal{L}1$	23.3	32.1	19.4	32.5	44.8	26.9
$\mathcal{L}2$	**24.0**	**33.4**	**19.9**	**32.7**	**45.5**	**27.1**

Table 6. Results on AP_{3D} and AP_{BEV} via domain adaptation between different data modalities on KITTI val at IoU = 0.7. "LiDAR", "Stereo" and "Mono" represent using the single modality data for training without domain adaptation. "⇒" denotes the adaptation direction between different modalities. DORN [14] and PSMNet [8] are adopted to generate the pseudo-LiDAR of "Mono" and "Stereo".

Modality	AP_{3D}			AP_{BEV}		
	Mod	Easy	Hard	Mod	Easy	Hard
LiDAR	67.9	84.8	58.8	79.0	88.5	69.5
Stereo	44.0	59.2	36.4	55.2	73.0	46.3
Stereo ⇒ LiDAR	**46.1**	**66.7**	**38.2**	**56.1**	**73.8**	**47.3**
Mono	21.9	28.4	18.4	30.7	42.7	25.5
Mono ⇒ Stereo	23.4	32.0	19.5	32.0	44.8	26.7
Mono ⇒ LiDAR	**24.0**	**33.4**	**19.9**	**32.7**	**45.5**	**27.1**

Table 7. Comparison of different depth estimators on val split set at IoU = 0.7.

Depth	Method	AP_{3D}			AP_{BEV}		
		Mod	Easy	Hard	Mod	Easy	Hard
MonoDepth [17]	Baseline	16.4	20.4	15.2	22.6	31.2	18.6
	Ours	**18.1**	**23.9**	**16.6**	**23.8**	**33.5**	**19.7**
DORN [14]	Baseline	21.9	28.4	18.4	30.7	42.7	25.5
	Ours	**24.0**	**33.4**	**19.9**	**32.7**	**45.5**	**27.1**

4.4 Generalization Ability

For generalization ability validation, we include two kinds of experiments. The first aims to verify that feature adaptation can generalize to other data modalities, such as monocular to stereo, stereo to LiDAR. The second aims to prove that our approach gains improvement on different depth estimation methods.

Domain Adaptation Between Different Modalities. To validate the effectiveness and generalization ability of our domain adaptation based method, we perform feature adaptation between different data modalities. As illustrated in Table 6, the "LiDAR" and "Stereo" indicate adopting the same baseline for real-LiDAR and stereo-based pseudo-LiDAR method. Stereo ⇒ LiDAR, "Mono" ⇒ "Stereo" and "Mono" ⇒ "LiDAR" denote the feature adaptation between different modalities, respectively. The experiment results demonstrate that the feature domain adaptation from a less accurate feature representation to a more reliable feature representation could largely improve the 3D detection performance for both stereo and monocular approaches.

Impact of Different Depth Estimators. In this experiment, we choose the unsupervised depth estimator MonoDepth [17] as well as the supervised monocular DORN [14] for comparison. As is known, supervised approaches have higher accuracy in depth prediction than unsupervised methods. As shown in Table 7, the 3D detection precision is positively correlated with the accuracy of estimated depth. Besides, improvements can be noticed both in unsupervised and supervised depth predictors.

5 Conclusions

In this paper, we present a monocular 3D object detection framework based on domain adaptation to adapt features from the noisy pseudo-LiDAR domain to accurate real LiDAR domain. Motivated by the overlooked problem of foreground inconsistency between pseudo and real LiDAR caused by inaccurate estimated depth, we also introduce a context-aware foreground segmentation module which uses both foreground and the certain context points for foreground feature extraction. In future work, Generative Adversarial Networks is considered for feature domain adaptation instead of simple $\mathcal{L}2$ and RGB information may be incorporated with pseudo-LiDAR.

Acknowledgments. This work was supported by National Key R&D Program of China (2019YFA0709502), the 111 Project (NO. B18015), the key project of Shanghai Science & Technology (No. 16JC1420402), Shanghai Municipal Science and Technology Major Project (No. 2018SHZDZX01) and ZJLab, National Key R&D Program of China (No. 2018YFC1312900), National Natural Science Foundation of China (NSFC 91630314).

References

1. Achlioptas, P., Diamanti, O., Mitliagkas, I., Guibas, L.: Learning representations and generative models for 3D point clouds. arXiv preprint arXiv:1707.02392 (2017)
2. Alhashim, I., Wonka, P.: High quality monocular depth estimation via transfer learning. arXiv preprint arXiv:1812.11941 (2018)
3. Atoum, Y., Roth, J., Bliss, M., Zhang, W., Liu, X.: Monocular video-based trailer coupler detection using multiplexer convolutional neural network. In: Proceedings of the IEEE International Conference on Computer Vision, pp. 5477–5485 (2017)
4. Biegelbauer, G., Vincze, M.: Efficient 3D object detection by fitting superquadrics to range image data for robot's object manipulation. In: Proceedings 2007 IEEE International Conference on Robotics and Automation, pp. 1086–1091. IEEE (2007)
5. Brazil, G., Liu, X.: M3D-RPN: monocular 3D region proposal network for object detection. In: Proceedings of the IEEE International Conference on Computer Vision, pp. 9287–9296 (2019)
6. Cai, Y., Li, B., Jiao, Z., Li, H., Zeng, X., Wang, X.: Monocular 3D object detection with decoupled structured polygon estimation and height-guided depth estimation. arXiv preprint arXiv:2002.01619 (2020)
7. Chabot, F., Chaouch, M., Rabarisoa, J., Teuliere, C., Chateau, T.: Deep MANTA: a coarse-to-fine many-task network for joint 2D and 3D vehicle analysis from monocular image. In: Proceedings of the IEEE Conference on computer vision and Pattern Recognition, pp. 2040–2049 (2017)

8. Chang, J.R., Chen, Y.S.: Pyramid stereo matching network. In: Proceedings of the IEEE Conference on Computer Vision and Pattern Recognition, pp. 5410–5418 (2018)
9. Chen, X., Kundu, K., Zhang, Z., Ma, H., Fidler, S., Urtasun, R.: Monocular 3D object detection for autonomous driving. In: Proceedings of the IEEE Conference on Computer Vision and Pattern Recognition, pp. 2147–2156 (2016)
10. Chen, X., et al.: 3D object proposals for accurate object class detection. In: Advances in Neural Information Processing Systems, pp. 424–432 (2015)
11. Chen, X., Ma, H., Wan, J., Li, B., Xia, T.: Multi-view 3D object detection network for autonomous driving. In: Proceedings of the IEEE Conference on Computer Vision and Pattern Recognition, pp. 1907–1915 (2017)
12. Du, L., et al.: Associate-3Ddet: perceptual-to-conceptual association for 3D point cloud object detection. In: The IEEE/CVF Conference on Computer Vision and Pattern Recognition (CVPR), June 2020
13. Engelcke, M., Rao, D., Wang, D.Z., Tong, C.H., Posner, I.: Vote3Deep: fast object detection in 3D point clouds using efficient convolutional neural networks. In: 2017 IEEE International Conference on Robotics and Automation (ICRA), pp. 1355–1361. IEEE (2017)
14. Fu, H., Gong, M., Wang, C., Batmanghelich, K., Tao, D.: Deep ordinal regression network for monocular depth estimation. In: Proceedings of the IEEE Conference on Computer Vision and Pattern Recognition, pp. 2002–2011 (2018)
15. Geiger, A., Lenz, P., Stiller, C., Urtasun, R.: Vision meets robotics: the kitti dataset. Int. J. Robot. Res. **32**(11), 1231–1237 (2013)
16. Geiger, A., Lenz, P., Urtasun, R.: Are we ready for autonomous driving? The kitti vision benchmark suite. In: 2012 IEEE Conference on Computer Vision and Pattern Recognition, pp. 3354–3361. IEEE (2012)
17. Godard, C., Mac Aodha, O., Brostow, G.J.: Unsupervised monocular depth estimation with left-right consistency. In: Proceedings of the IEEE Conference on Computer Vision and Pattern Recognition, pp. 270–279 (2017)
18. Hoffman, J., Wang, D., Yu, F., Darrell, T.: FCNs in the wild: pixel-level adversarial and constraint-based adaptation. arXiv preprint arXiv:1612.02649 (2016)
19. Inoue, N., Furuta, R., Yamasaki, T., Aizawa, K.: Cross-domain weakly-supervised object detection through progressive domain adaptation. In: Proceedings of the IEEE conference on computer vision and pattern recognition, pp. 5001–5009 (2018)
20. Jörgensen, E., Zach, C., Kahl, F.: Monocular 3D object detection and box fitting trained end-to-end using intersection-over-union loss. arXiv preprint arXiv:1906.08070 (2019)
21. Kingma, D.P., Ba, J.: Adam: a method for stochastic optimization. arXiv preprint arXiv:1412.6980 (2014)
22. Kouw, W.M., Loog, M.: An introduction to domain adaptation and transfer learning. arXiv preprint arXiv:1812.11806 (2018)
23. Ku, J., Mozifian, M., Lee, J., Harakeh, A., Waslander, S.L.: Joint 3D proposal generation and object detection from view aggregation. In: 2018 IEEE/RSJ International Conference on Intelligent Robots and Systems (IROS), pp. 1–8. IEEE (2018)
24. Ku, J., Pon, A.D., Waslander, S.L.: Monocular 3D object detection leveraging accurate proposals and shape reconstruction. In: Proceedings of the IEEE Conference on Computer Vision and Pattern Recognition, pp. 11867–11876 (2019)
25. Li, B., Ouyang, W., Sheng, L., Zeng, X., Wang, X.: GS3D: an efficient 3D object detection framework for autonomous driving. In: Proceedings of the IEEE Conference on Computer Vision and Pattern Recognition, pp. 1019–1028 (2019)

26. Li, P., Zhao, H., Liu, P., Cao, F.: RTM3D: real-time monocular 3D detection from object keypoints for autonomous driving. arXiv preprint arXiv:2001.03343 (2020)
27. Li, X., Grandvalet, Y., Davoine, F.: Explicit inductive bias for transfer learning with convolutional networks. arXiv preprint arXiv:1802.01483 (2018)
28. Liu, L., Lu, J., Xu, C., Tian, Q., Zhou, J.: Deep fitting degree scoring network for monocular 3D object detection. In: Proceedings of the IEEE Conference on Computer Vision and Pattern Recognition, pp. 1057–1066 (2019)
29. Ma, X., Wang, Z., Li, H., Zhang, P., Ouyang, W., Fan, X.: Accurate monocular 3D object detection via color-embedded 3D reconstruction for autonomous driving. In: Proceedings of the IEEE International Conference on Computer Vision, pp. 6851–6860 (2019)
30. Manhardt, F., Kehl, W., Gaidon, A.: ROI-10D: monocular lifting of 2D detection to 6d pose and metric shape. In: Proceedings of the IEEE Conference on Computer Vision and Pattern Recognition, pp. 2069–2078 (2019)
31. Mayer, N., et al.: A large dataset to train convolutional networks for disparity, optical flow, and scene flow estimation. In: Proceedings of the IEEE Conference on Computer Vision and Pattern Recognition, pp. 4040–4048 (2016)
32. Mousavian, A., Anguelov, D., Flynn, J., Kosecka, J.: 3D bounding box estimation using deep learning and geometry. In: Proceedings of the IEEE Conference on Computer Vision and Pattern Recognition, pp. 7074–7082 (2017)
33. Naiden, A., Paunescu, V., Kim, G., Jeon, B., Leordeanu, M.: Shift R-CNN: deep monocular 3D object detection with closed-form geometric constraints. In: 2019 IEEE International Conference on Image Processing (ICIP), pp. 61–65. IEEE (2019)
34. Qi, C.R., Liu, W., Wu, C., Su, H., Guibas, L.J.: Frustum pointnets for 3D object detection from RGB-D data. In: Proceedings of the IEEE Conference on Computer Vision and Pattern Recognition, pp. 918–927 (2018)
35. Qi, C.R., Su, H., Mo, K., Guibas, L.J.: PointNet: deep learning on point sets for 3D classification and segmentation. In: Proceedings of the IEEE Conference on Computer Vision and Pattern Recognition, pp. 652–660 (2017)
36. Qi, C.R., Yi, L., Su, H., Guibas, L.J.: Pointnet++: deep hierarchical feature learning on point sets in a metric space. In: Advances in Neural Information Processing Systems, pp. 5099–5108 (2017)
37. Qin, Z., Wang, J., Lu, Y.: MonoGRNet: a geometric reasoning network for monocular 3D object localization. In: Proceedings of the AAAI Conference on Artificial Intelligence, vol. 33, pp. 8851–8858 (2019)
38. Roddick, T., Kendall, A., Cipolla, R.: Orthographic feature transform for monocular 3D object detection. arXiv preprint arXiv:1811.08188 (2018)
39. Shi, S., Wang, X., Li, H.: PointRCNN: 3D object proposal generation and detection from point cloud. In: Proceedings of the IEEE Conference on Computer Vision and Pattern Recognition, pp. 770–779 (2019)
40. Simonelli, A., Bulo, S.R., Porzi, L., López-Antequera, M., Kontschieder, P.: Disentangling monocular 3D object detection. In: Proceedings of the IEEE International Conference on Computer Vision, pp. 1991–1999 (2019)
41. Vianney, J.M.U., Aich, S., Liu, B.: RefinedMPL: refined monocular PseudoLiDAR for 3D object detection in autonomous driving. arXiv preprint arXiv:1911.09712 (2019)
42. Wang, Y., Chao, W.L., Garg, D., Hariharan, B., Campbell, M., Weinberger, K.Q.: Pseudo-LiDAR from visual depth estimation: Bridging the gap in 3D object detection for autonomous driving. In: Proceedings of the IEEE Conference on Computer Vision and Pattern Recognition, pp. 8445–8453 (2019)

43. Weng, X., Kitani, K.: Monocular 3D object detection with pseudo-LiDAR point cloud. In: Proceedings of the IEEE International Conference on Computer Vision Workshops (2019)
44. Xu, B., Chen, Z.: Multi-level fusion based 3D object detection from monocular images. In: Proceedings of the IEEE Conference on Computer Vision and Pattern Recognition, pp. 2345–2353 (2018)
45. Yan, Y., Mao, Y., Li, B.: Second: sparsely embedded convolutional detection. Sensors **18**(10), 3337 (2018)
46. You, Y., et al.: Pseudo-LiDAR++: accurate depth for 3D object detection in autonomous driving. arXiv preprint arXiv:1906.06310 (2019)
47. Zhou, Y., Tuzel, O.: VoxelNet: end-to-end learning for point cloud based 3D object detection. In: Proceedings of the IEEE Conference on Computer Vision and Pattern Recognition, pp. 4490–4499 (2018)

Talking-Head Generation with Rhythmic Head Motion

Lele Chen[1(✉)], Guofeng Cui[1], Celong Liu[2], Zhong Li[2], Ziyi Kou[1], Yi Xu[2], and Chenliang Xu[1]

[1] University of Rochester, Rochester, USA
lchen63@cs.rochester.edu
[2] OPPO US Research Center, Palo Alto, USA

Abstract. When people deliver a speech, they naturally move heads, and this rhythmic head motion conveys prosodic information. However, generating a lip-synced video while moving head naturally is challenging. While remarkably successful, existing works either generate still talking-face videos or rely on landmark/video frames as sparse/dense mapping guidance to generate head movements, which leads to unrealistic or uncontrollable video synthesis. To overcome the limitations, we propose a 3D-aware generative network along with a hybrid embedding module and a non-linear composition module. Through modeling the head motion and facial expressions (In our setting, facial expression means facial movement (e.g., blinks, and lip & chin movements).) explicitly, manipulating 3D animation carefully, and embedding reference images dynamically, our approach achieves controllable, photo-realistic, and temporally coherent talking-head videos with natural head movements. Thoughtful experiments on several standard benchmarks demonstrate that our method achieves significantly better results than the state-of-the-art methods in both quantitative and qualitative comparisons. The code is available on https://github.com/lelechen63/Talking-head-Generation-with-Rhythmic-Head-Motion.

1 Introduction

People naturally emit head movements when they speak, which contain non-verbal information that helps the audience comprehend the speech content [4,15]. Modeling the head motion and then generating a controllable talking-head video are valuable problems in computer vision. For example, it can benefit the research of adversarial attacks in security or provide more training samples for supervised learning approaches. Meanwhile, it is also crucial to real-world applications, such as enhancing speech comprehension for hearing-impaired people, and generating virtual characters with synchronized facial movements to speech audio in movies/games.

Electronic supplementary material The online version of this chapter (https://doi.org/10.1007/978-3-030-58545-7_3) contains supplementary material, which is available to authorized users.

© Springer Nature Switzerland AG 2020
A. Vedaldi et al. (Eds.): ECCV 2020, LNCS 12354, pp. 35–51, 2020.
https://doi.org/10.1007/978-3-030-58545-7_3

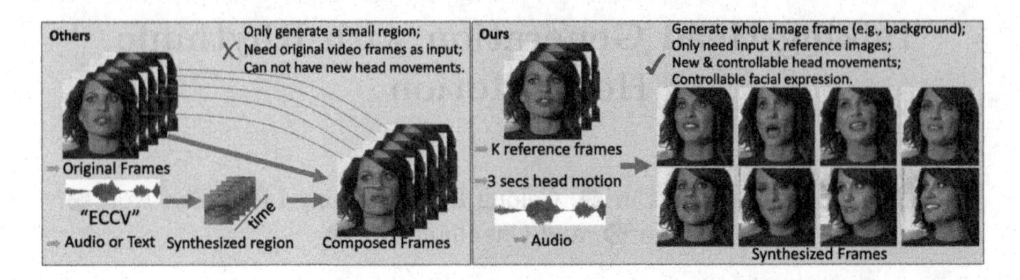

Fig. 1. The comparisons of other methods [13,27] and our method. The green arrows denote inputs. [13,27] require whole original video frames as input and can only edit a small region (e.g., lip region) on the original video frames. In contrast, through learning the appearance embedding from K reference frames and predicting future head movements using the head motion vector extracted from the 3-s reference video, our method generates whole future video frames with controllable head movements and facial expressions.

We humans can infer the visual prosody from a short conversation with speakers [22]. Inspired by that, in this paper, we consider such a task: given a short video[1] of the target subject and an arbitrary reference audio, generating a photo-realistic and lip-synced talking-head video with natural head movements. Similar problems [7,8,26,30,31,35,37,39] have been explored recently. However, several challenges on how to explicitly use the given video and then model the apparent head movements remain unsolved. In the rest of this section, we discuss our technical contributions concerning each challenge.

The deformation of a talking-head consists of his/her intrinsic subject traits, head movements, and facial expressions, which are highly convoluted. This complexity stems not only from modeling face regions but also from modeling head motion and background. While previous audio-driven talking-face generation methods [6,7,9,26,35,39] could generate lip-synced videos, they omit the head motion modeling, thus can only generate still talking-face with expressions under a fixed facial alignment. Other landmark/image-driven generation methods [31–33,35,37] can synthesize moving-head videos relying on the input facial landmarks or video frames as guidance to infer the convoluted head motion and facial expressions. However, those methods fail to output a controllable talking-head video (e.g., control facial expressions with speech), which greatly limits their use. To address the convoluted deformation problem, we propose a simple but effective method to decompose the head motion and facial expressions.

Another challenge is exploiting the information contained in the reference image/video. While the few-shot generation methods [20,31,36,37] can synthesize videos of unseen subjects by leveraging K reference images, they only utilize global appearance information. There remains valuable information underexplored. For example, we can infer the individual's head movement characteristics by leveraging the given short video. Therefore, we propose a novel method

[1] E.g., a 3-s video. We use it to learn head motion and only sample K images as reference frames.

to extrapolate rhythmic head motion based on the short input video and the conditioned audio, which enables generating a talking-head video with natural head poses. Besides, we propose a novel hybrid embedding network dynamically aggregating information from the reference images by approximating the relation between the target image with the reference images.

People are sensitive to any subtle artifacts and perceptual identity changes in a synthesized video, which are hard to avoid in GAN-based methods. 3D graphics modeling has been introduced by [13,17,38] in GAN-based methods due to its stability. In this paper, we employ the 3D modeling along with a novel non-linear composition module to alleviate the visual discontinuities caused by apparent head motion or facial expressions.

Combining the above features, which are designed to overcome limitations of existing methods, our framework can generate a talking-head video with natural head poses that conveys the given audio signal. We conduct extensive experimental validation with comparisons to various state-of-the-art methods on several benchmark datasets (e.g., VoxCeleb2 [9] and LRS3-TED [1] datasets) under several different settings (e.g., audio-driven and landmark-driven). Experimental results show that the proposed framework effectively addresses the limitations of those existing methods.

2 Related Work

2.1 Talking-Head Image Generation

The success of graphics-based approaches has been mainly limited to synthesizing talking-head videos for a specific person [2,5,13,14,19,27]. For instance, Suwajanakorn et al. [27] generate a small region (e.g., lip region, see Fig. 1)[2] and compose it with a retrieved frame from a large video corpus of the target person to produce photo-realistic videos. Although it can synthesize fairly accurate lip-synced videos, it requires a large amount of video footage of the target person to compose the final video. Moreover, this method can not be generalized to an unseen person due to the rigid matching scheme. More recently, video-to-video translation has been shown to be able to generate arbitrary faces from arbitrary input data. While the synthesized video conveys the input speech signal, the talking-face generation methods [7,8,24,26,30,39] can only generate facial expressions without any head movements under a fixed alignment since the head motion modeling has been omitted. Talking-head methods [31,35,37] can generate high-quality videos guided by landmarks/images. However, these methods can not generate controllable video since the facial expressions and head motion are convoluted in the guidance, e.g., using audio to drive the generation. In contrast, we explicitly model the head motion and facial expressions in a disentangled manner, and propose a method to extrapolate rhythmic head motion to enable generating future frames with natural head movements.

[2] We intent to highlight the different between [13,27] and our work. While they generate high-quality videos, they can not disentangle the head motion and facial movement due to their intrinsic limitations.

2.2 Related Techniques

Embedding Network refers to the external memory module to learn common feature over few reference images of the target subjects. And this network is critical to identity-preserving performance in generation task. Previous works [7,8,26,30] directly use CNN encoder to extract the feature from reference images, then concatenate with other driven vectors and decode it to new images. However, this encoder-decoder structure suffers identity-preserving problem caused by the deep convolutional layers. Wiles et al. [35] propose an embedding network to learn a bilinear sampler to map the reference frames to a face representation. More recently, [31,36,37] compute part of the network weights dynamically based on the reference images, which can adapt to novel cases quickly. Inspired by [31], we propose a hybrid embedding network to aggregate appearance information from reference images with apparent head movements.

Image matting function has been well explored in image/video generation task [7,24,29,31,32]. For instance, Pumarola et al. [24] compute the final output image by $\hat{\mathbf{I}} = (1-\mathbf{A})*\mathbf{C}+\mathbf{A}*\mathbf{I}_r$, where $\mathbf{I}_r$, $\mathbf{A}$ and $\mathbf{C}$ are input reference image, attention map and color mask, respectively. The attention map $\mathbf{A}$ indicates to what extend each pixel of $\mathbf{I}_r$ contributes to the output image $\hat{\mathbf{I}}$. However, this attention mechanism may not perform well if there exits a large deformation between reference frame $\mathbf{I}_r$ and target frame $\hat{\mathbf{I}}$. Wang et al. [33] use estimated optic flow to warp the $\mathbf{I}_r$ to align with $\hat{\mathbf{I}}$, which is computationally expensive and can not estimate the rotations in talking-head videos. In this paper, we propose a 3D-aware solution along with a non-linear composition module to better tackle the misalignment problem caused by apparent head movements.

3 Method

3.1 Problem Formulation

We introduce a neural approach for talking-head video generation, which takes as input sampled video frames, $\mathbf{y}_{1:\tau} \equiv \mathbf{y}_1, ..., \mathbf{y}_\tau$, of the target subject and a sequence of driving audio, $\mathbf{x}_{\tau+1:T} \equiv \mathbf{x}_{\tau+1}, ..., \mathbf{x}_T$, and synthesizes target video frames, $\hat{\mathbf{y}}_{\tau+1:T} \equiv \hat{\mathbf{y}}_{\tau+1}, ..., \hat{\mathbf{y}}_T$, that convey the given audio signal with realistic head movements. To explicitly model facial expressions and head movements, we decouple the full model into three sub-models: a facial expression learner (Ψ), a head motion learner (Φ), and a 3D-aware generative network (Θ). Specifically, given an audio sequence $\mathbf{x}_{\tau+1:T}$ and an example image frame $\mathbf{y}_{t_\mathcal{M}}$, Ψ generates the facial expressions $\hat{\mathbf{p}}_{\tau+1:T}$. Meanwhile, given a short reference video $\mathbf{y}_{1:\tau}$ and the driving audio $\mathbf{x}_{\tau+1:T}$, Φ extrapolates natural head motion $\hat{\mathbf{h}}_{\tau+1:T}$ to manipulate the head movements. Then, the Θ generates video frames $\hat{\mathbf{y}}_{\tau+1:T}$ using $\hat{\mathbf{p}}_{\tau+1:T}$, $\hat{\mathbf{h}}_{\tau+1:T}$, and $\mathbf{y}_{1:\tau}$. Thus, the full model is given by:

$$\hat{\mathbf{y}}_{\tau+1:T} = \Theta(\mathbf{y}_{1:\tau}, \hat{\mathbf{p}}_{\tau+1:T}, \hat{\mathbf{h}}_{\tau+1:T}) = \Theta(\mathbf{y}_{1:\tau}, \Psi(\mathbf{y}_{t_\mathcal{M}}, \mathbf{x}_{\tau+1:T}), \Phi(\mathbf{y}_{1:\tau}, \mathbf{x}_{\tau+1:T})).$$

The proposed framework (see Fig. 2) aims to exploit the facial texture and head motion information in $\mathbf{y}_{1:\tau}$ and maps the driving audio signal to a sequence of generated video frames $\hat{\mathbf{y}}_{\tau+1:T}$.

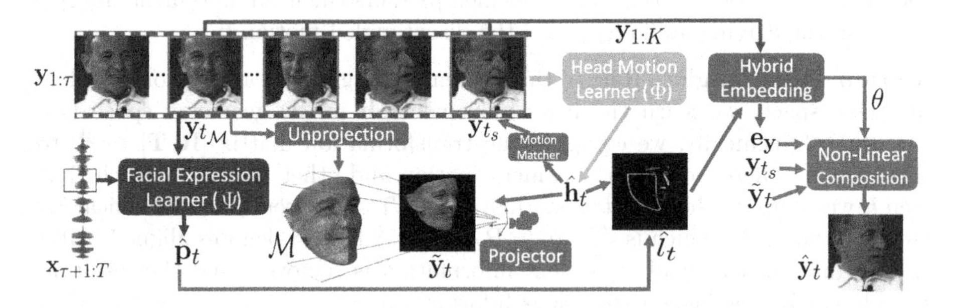

Fig. 2. The overview of the framework. Ψ and Φ are introduced in Sect. 3.2 and Sect. 3.3, respectively. The 3D-aware generative network consists of a 3D-Aware module (green part, see Sect. 4.1), a hybrid embedding module (blue part, see Sect. 4.2), and a non-linear composition module (orange part, see Sect. 4.3). (Color figure online)

3.2 The Facial Expression Learner

The facial expression learner Ψ (see Fig. 2 red block) receives as input the raw audio $\mathbf{x}_{\tau+1:T}$, and a subject-specific image $\mathbf{y}_{t_\mathcal{M}}$ (see Sect. 4.1 about how to select $\mathbf{y}_{t_\mathcal{M}}$), from which we extract the landmark identity feature $\mathbf{p}_{t_\mathcal{M}}$. The desired outputs are PCA components $\hat{\mathbf{p}}_{\tau+1:T}$ that convey the facial expression. During training, we mix $\mathbf{x}_{\tau+1:T}$ with a randomly selected noise file in 6 to 30 dB SNRs with 3 dB increments to increase the network robustness. At time step t, we encode audio clip $\mathbf{x}_{t-3:t+4}$ (0.04×7 s) into an audio feature vector, concatenate it with the encoded reference landmark feature, and then decode the fused feature into PCA components of target expression $\hat{\mathbf{p}}_t$. During inference, we add the identity feature $\mathbf{p}_{t_\mathcal{M}}$ back to the resultant facial expressions $\hat{\mathbf{p}}_{\tau+1:T}$ to keep the identity information.

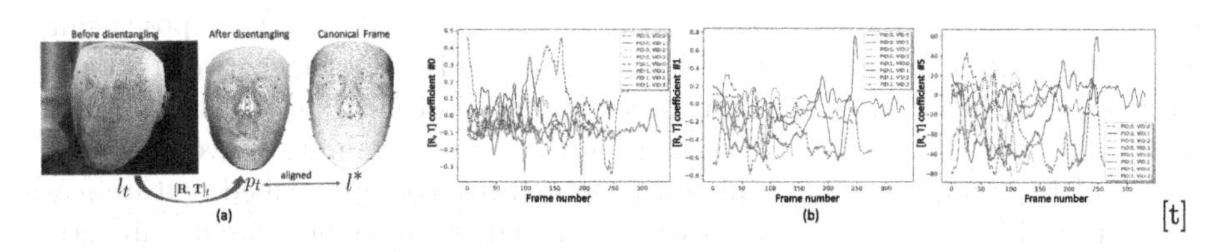

Fig. 3. (a) shows the motion disentangling process. (b) shows the head movement coefficients of two randomly-selected identities. PID, VID indicate identity id and video id, respectively.

3.3 The Head Motion Learner

To generate talking-head videos with natural head movements, we introduce a head motion learner Φ, which disentangles the reference head motion $\mathbf{h}_{1:\tau}$ from the input reference video clip $\mathbf{y}_{1:\tau}$ and then predicts the head movements $\hat{\mathbf{h}}_{\tau+1:T}$ based on the driving audio $\mathbf{x}_{1:T}$ and the disentangled motion $\mathbf{h}_{1:\tau}$.

Motion Disentangler. Rather than disentangling the head motion problem in image space, we learn the head movements $\mathbf{h}$ in 3D geometry space (see Fig. 3(a)). Specifically, we compute the transformation matrix $[\mathbf{R}, \mathbf{T}] \in \mathbb{R}^6$ to represent $\mathbf{h}$, where we omit the camera motion and other factors. At each time step t, we compute the rigid transformation $[\mathbf{R}, \mathbf{T}]_t$ between l_t and canonical 3D facial landmark l^*, which is $\mathit{l}^* \approx \mathbf{p}_t = \mathbf{R}_t \mathit{l}_t + \mathbf{T}_t$, where $\approx$ denotes aligned. After transformation, the head movement information is removed, and the resultant $\mathbf{p}_t$ only carries the facial expressions information.

The Natural Head Motion Extrapolation. We randomly select two subjects (each with four videos) from VoxCeleb2 [9] and plot the motion coefficients in Fig. 3(b). We can find that one subject (same PID, different VID) has similar head motion patterns in different videos, and different subjects (different PID) have much different head motion patterns. To explicitly learn the individual's motion from $\mathbf{h}_{1:\tau}$ and the audio-correlated head motion from $\mathbf{x}_{1:T}$, we propose a few-shot temporal convolution network Φ, which can be formulated as: $\hat{\mathbf{h}}_{\tau+1:T} = \Phi(f(\mathbf{h}_{1:\tau}, \mathbf{x}_{1:\tau}), \mathbf{x}_{\tau+1:T})$. In details, the encoder f first encodes $\mathbf{h}_{1:\tau}$ and $\mathbf{x}_{1:\tau}$ to a high-dimensional reference feature vector, and then transforms the reference feature to network weights $\mathbf{w}$ using a multi-layer perception. Meanwhile, $\mathbf{x}_{\tau+1:T}$ is encoded by several temporal convolutional blocks. Then, the encoded audio feature is processed by a multi-layer perception, where the weights are dynamically learnable $\mathbf{w}$, to generate $\hat{\mathbf{h}}_{\tau+1:T}$.

3.4 The 3D-Aware Generative Network

To generate controllable videos, we propose a 3D-aware generative network Θ, which fuses the head motion $\mathbf{h}_{\tau+1:T}$, and facial expressions $\mathbf{p}_{\tau+1:T}$ with the appearance information in $\mathbf{y}_{1:\tau}$ to synthesise target video frames, $\hat{\mathbf{y}}_{\tau+1:T}$. The Θ consists of three modules: a *3D-Aware* module, which manipulates the head motion to reconstruct an intermediate image that carries desired head pose using a differentiable unprojection-projection step to bridge the gap between images with different head poses. Comparing to landmark-driven methods [31,33,37], it can alleviate the overfitting problem and make the training converge faster; a *Hybrid Embedding* module, which is proposed to explicitly embed texture features from different appearance patterns carried by different reference images; a *Non-Linear Composition* module, which is proposed to stabilize the background and better preserve the individual's appearance. Comparing to image matting function [7,24,33,37,39], our non-linear composition module can synthesize videos with natural facial expression, smoother transition, and a more stable background.

4 3D-Aware Generation

4.1 3D-Aware Module

3D Unprojection. We assume that the image with frontal face contains more appearance information. Given a short reference video clip $\mathbf{y}_{1:\tau}$, we compute the head rotation $\mathbf{R}_{1:\tau}$ and choose the most visible frame $\mathbf{y}_{t_\mathcal{M}}$ with minimum rotation, such that $\mathbf{R}_{t_\mathcal{M}} \rightarrow 0$. Then we feed it to an unprojection network to obtain a 3D mesh $\mathcal{M}$. The Unprojection network is a U-net structure network [12] and we pretrain it on 300W-LP [40] dataset. After pretrianing, we fix the weights and use it to transfer the input image $\mathbf{y}_{t_\mathcal{M}}$ into the textured 3D mesh $\mathcal{M}$. The topology of $\mathcal{M}$ will be fixed for all frames in one video, we denote it as $\mathcal{F}^\mathcal{M}$.

3D Projector. In order to get the projected image $\tilde{\mathbf{y}}_t$ from the 3D face mesh $\mathcal{M}$, we need to compute the correct pose for $\mathcal{M}$ at time t. Suppose $\mathcal{M}$ is reconstructed from frame $\mathbf{y}_{t_\mathcal{M}}, (1 \leq t_\mathcal{M} \leq \tau)$, the position of vertices of $\mathcal{M}$ at time t can be computed by:

$$\mathcal{V}_t^\mathcal{M} = \mathbf{R}_t^{-1}(\mathbf{R}_{t_\mathcal{M}} \mathcal{V}_{t_\mathcal{M}}^\mathcal{M} + \mathbf{T}_{t_\mathcal{M}} - \mathbf{T}_t), \tag{1}$$

where $\mathcal{V}_{t_\mathcal{M}}^\mathcal{M}$ denotes the position of vertices of $\mathcal{M}$ at time $t_\mathcal{M}$. Hence, $\mathcal{M}$ at time t can be represented as $[\mathcal{V}_t^\mathcal{M}, \mathcal{F}^\mathcal{M}]$. Then, a differentiable rasterizer [21] is applied here to render $\tilde{\mathbf{y}}_t$ from $[\mathcal{V}_t^\mathcal{M}, \mathcal{F}^\mathcal{M}]$. After rendering, $\tilde{\mathbf{y}}_t$ carries the same head pose as the target frame $\mathbf{y}_t$ and the same expression as $\mathbf{y}_{t_\mathcal{M}}$.

Motion Matcher. We randomly select K frames $\mathbf{y}_{1:K}$ from $\mathbf{y}_{1:\tau}$ as reference images. To infer the prior knowledge about the background of the target image from $\mathbf{y}_{1:K}$, we propose a motion matcher, $M(\mathbf{h}_t, \mathbf{h}_{1:K}, \mathbf{y}_{1:K})$, which matches a frame $\mathbf{y}_{t_s}$ from $\mathbf{y}_{1:K}$ with the nearest background as $\mathbf{y}_t$ by comparing the similarities between $\mathbf{h}_t$ with $\mathbf{h}_{1:K}$. To solve this background retrieving problem, rather than directly computing the similarities between facial landmarks, we compute the geometry similarity cost c_t using the head movement coefficients and choose the $\mathbf{y}_{t_s}$ with the least cost c_t. That is:

$$c_t = \min_{k=1,2,\ldots K} \|\mathbf{h}_t - \mathbf{h}_k\|_2^2. \tag{2}$$

The matched $\mathbf{y}_{t_s}$ is passed to the non-linear composition module (see Sect. 4.3) since it carries the similar background pattern as $\mathbf{y}_t$. During training, we manipulate the selection of $\mathbf{y}_{t_s}$ by increasing the cost c_t, which makes the non-linear composition module more robust to different $\mathbf{y}_{t_s}$.

4.2 Hybrid Embedding Module

Instead of averaging the K reference image features [8,35,37], we design a hybrid embedding mechanism to dynamically aggregate the K visual features.

Recall that we obtain the facial expression $\mathbf{p}_t$ (Sect. 3.2) and the head motion $\mathbf{h}_t$ (Sect. 3.3) for time t. Here, we combine them to landmark l_t and transfer it to

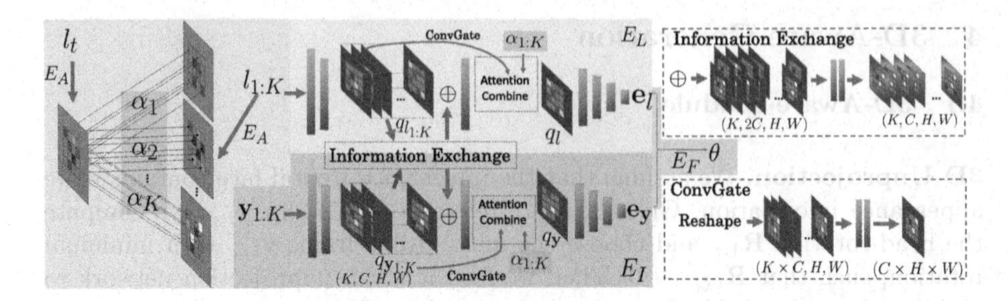

Fig. 4. The network structure of hybrid embedding module. Green, yellow, blue, pink parts are activation encoder E_A, landmark encoder E_L, image encoder E_I, and fusion network E_F, respectively. (Color figure online)

a gray-scale image. Our hybrid embedding network (Fig. 4) consists of four sub-networks: an Activation Encoder (E_A, green part) to generate the activation map between the query landmark $\boldsymbol{l}_t$ and reference landmarks $\boldsymbol{l}_{1:K}$; an Image Encoder (E_I, blue part) to extract the image feature $\mathbf{e}_{\mathbf{y}}$; a Landmark Encoder (E_L, yellow part) to extract landmark feature $\mathbf{e}_l$; a fusion network (E_F, pink part) to aggregate image feature and landmark feature into parameter set θ. The overview of the embedding network is performed by:

$$\boldsymbol{\alpha}_k = \text{softmax}(E_A(\boldsymbol{l}_k) \odot E_A(\boldsymbol{l}_t)), \ \ k \in [1, K], \tag{3}$$

$$\mathbf{e}_{\mathbf{y}} = E_I(\mathbf{y}_{1:K}, \boldsymbol{\alpha}_{1:K}), \quad \mathbf{e}_l = E_L(\boldsymbol{l}_{1:K}, \boldsymbol{\alpha}_{1:K}), \tag{4}$$

$$\theta = E_F(\mathbf{e}_l, \mathbf{e}_{\mathbf{y}}), \tag{5}$$

where $\odot$ denotes element-wise multiplication. We regard the activation map $\boldsymbol{\alpha}_k$ as the activation energy between $\boldsymbol{l}_k$ and $\boldsymbol{l}_t$, which approximates the similarity between $\mathbf{y}_k$ and $\mathbf{y}_t$.

We observe that different referent images may carry different appearance patterns, and those appearance patterns share some common features with the reference landmarks (e.g., head pose, and edges). Assuming that knowing the information of one modality can better encode the feature of another modality [28], we hybrid the information between E_I and E_L. Specifically, we use two convolutional layers to extract the feature map $\mathbf{q}_{\mathbf{y}_{1:K}}$ and $\mathbf{q}_{\boldsymbol{l}_{1:K}}$, and then forward them to an information exchange block to refine the features. After exchanging the information, the hybrid feature is concatenated with the original feature $\mathbf{q}_{\mathbf{y}_{1:K}}$. Then, we pass the concatenated feature to a bottleneck convolution layer to produce the refined feature $\mathbf{q}'_{\mathbf{y}_{1:K}}$. Meanwhile, we apply a ConvGate block on $\mathbf{q}_{\mathbf{y}_{1:K}}$ to self-aggregate the features from K references, and then combine the gated feature with the refined feature using learnable activation map $\boldsymbol{\alpha}_k$. This attention combine step can be formulated as:

$$\mathbf{q}_{\mathbf{y}} = \sum_{i=1}^{K} (\boldsymbol{\alpha}_{1:K} \odot \mathbf{q}'_{\mathbf{y}_{1:K}}) + \text{ConvGate}(\mathbf{q}_{\mathbf{y}_{1:K}}). \tag{6}$$

Then, by applying several convolutional layers to the aggregated feature $\mathbf{q_y}$, we obtain the image feature vector $\mathbf{e_y}$. Similarly, the landmark feature vector $\mathbf{e}_l$ can also be produced.

The Fusion network E_F consists of three blocks: a N-layer image feature encoding block E_{F_I}, a landmark feature encoding block E_{F_L}, and a multi-layer perception block E_{F_P}. Thus, we can rewrite Eq. 5 as:

$$\{\theta_\gamma^i, \theta_\beta^i, \theta_S^i\}_{i\in[1,N]} = E_{F_P}(\mathrm{softmax}(E_{F_L}(\mathbf{e}_l)) \odot E_{F_I}(\mathbf{e_y})), \quad (7)$$

where $\{\theta_\gamma^i, \theta_\beta^i, \theta_S^i\}_{i\in[1,N]}$ are the learnable network weights in the non-linear composition module in Sect. 4.3.

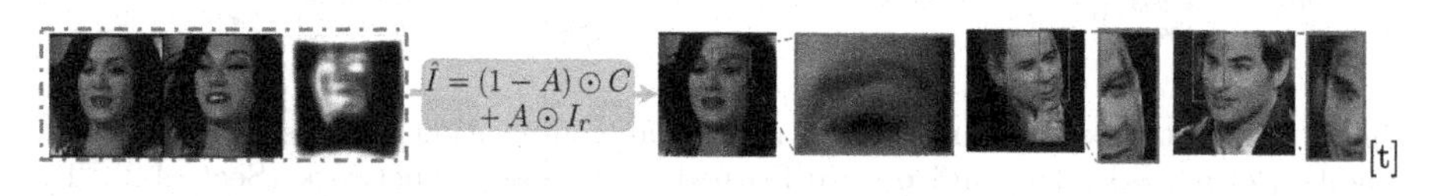

Fig. 5. The artifact examples produced by the image matting function. The blue dashed box shows the color map $\mathbf{C}$, reference image $\mathbf{I_r}$, and attention map $\mathbf{A}$, respectively. The blue box shows the image matting function (see Sect. 2.2) followed by the final result with its zoom-in artifact area. Two other artifact examples are on the right side.

4.3 Non-linear Composition Module

Since the projected image $\tilde{\mathbf{y}}_t$ carries valuable facial texture with better alignment to the ground-truth, and the matched image $\mathbf{y}_{t_s}$ contains similar background pattern, we input them to our composition module to ease the burden of the generation. We use FlowNet [32] to estimate the flow between $\tilde{l}_t$ and $\hat{l}_t$, and warp the projected image to obtain $\tilde{\mathbf{y}}'_t$. Meanwhile, the FlowNet also outputs an attention map $\boldsymbol{\alpha}_{\tilde{y}_t}$. Similarly, we obtain the warped matched image and its attention map $[\mathbf{y'}_{t_s}, \boldsymbol{\alpha}_{\mathbf{y}_{t_s}}]$. Although image matting function is a popular way to combine the warped image and the raw output of the generator [7,24,31,32] using these attention maps, it may generate apparent artifacts (Fig. 5) caused by misalignment, especially in the videos with apparent head motion. To solve this problem, we propose a nonlinear combination module (Fig. 6) based on the network structure proposed in [31].

The decoder G consists of N layers and each layer contains three parallel SPADE blocks [23]. Inspired by [31], in the ith layer of G, we first convolve $[\theta_\gamma^i, \theta_\beta^i]$ with $\phi_{l_t}^i$ to generate the scale and bias map $\{\gamma^i, \beta^i\}_{\phi_{l_t}^i}$ of SPADE (green box) to denormalize the appearance vector $\mathbf{e}_\mathbf{y}^i$. This step aims to sample the face representation $\mathbf{e}_\mathbf{y}^i$ towards the target pose and expression. In other two SPADE blocks (orange box), the scale and bias map $\{\gamma^i, \beta^i\}_{\phi_{\mathbf{y}_{t_s}}, \phi_{\tilde{\mathbf{y}}_t}}$ are generated by fixed weights convoluted on $\phi_{\mathbf{y}_{t_s}}$ and $\phi_{\tilde{\mathbf{y}}_t}$, respectively. Then, we sum up the three denomalized features and upsample them with a transposed convolution. We repeat this non-linear combination module in all layers of G and apply a hyperbolic tangent activation layer at the end of G to output fake image $\hat{\mathbf{y}}_t$. Our experiment results (ablation study in Sect. 6) show that this parallel non-linear combination is more robust to the spatial-misalignment among the input images, especially when there is an apparent head motion.

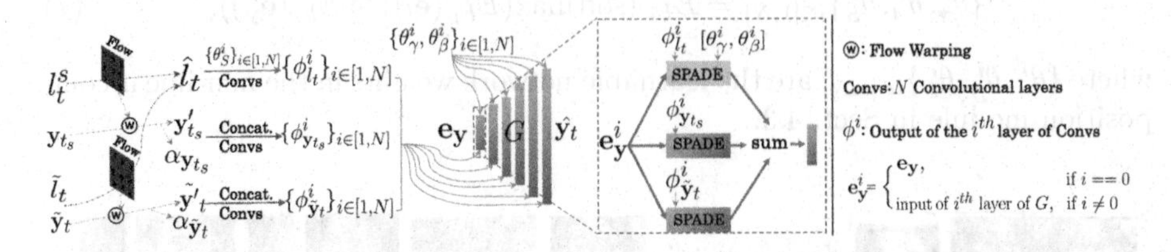

Fig. 6. The network structure of the non-linear composition module. The network weights $\{\theta^i_S, \theta^i_\gamma, \theta^i_\beta\}_{i\in[1,N]}$ and $\mathbf{e}_\mathbf{y}$ are learned in embedding network (Sect. 4.2). To better encode $\hat{l}_t$ with extra knowledge learned from the embedding network, we leverage $\{\theta^i_S\}_{i\in[1,N]}$ as the weights of the convolutional layers (green arrow) and save all the intermediate features $\{\phi^i_{\hat{l}_t}\}_{i\in[1,N]}$. Similarly, we obtain $\{\phi^i_{\tilde{\mathbf{y}}_t}, \phi^i_{\mathbf{y}_{t_s}}\}_{i\in[1,N]}$ with fixed convolutional weights (red arrow). Then, we input $\{\phi^i_{\tilde{\mathbf{y}}_t}, \phi^i_{\mathbf{y}_{t_s}}, \phi^i_{\hat{l}_t}\}_{i\in[1,N]}$ into the corresponding block of the decoder G to guide the image generation. (Color figure online)

4.4 Objective Function

The Multi-Scale Discriminators [33] are employed to differentiate the real and synthesized video frames. The discriminators D_1 and D_2 are trained on two different input scales, respectively. Thus, the best generator G^* is found by solving the following optimization problem:

$$G^* = \min_G \Big(\max_{D_1,D_2} \sum_{k=1,2} \mathcal{L}_{\text{GAN}}(G, D_k) \\ + \lambda_{\text{FM}} \sum_{k=1,2} \mathcal{L}_{\text{FM}}(G, D_k)\Big) + \lambda_{\text{PCT}}\mathcal{L}_{\text{PCT}}(G) + \lambda_W \mathcal{L}_W(G), \tag{8}$$

where $\mathcal{L}_{\text{FM}}$ and $\mathcal{L}_W$ are the feature matching loss and flow loss proposed in [32]. The $\mathcal{L}_{\text{PCT}}$ is the VGG19 [25] perceptual loss term, which measures the perceptual similarity. The λ_{FM}, λ_{PCT}, and λ_W control the importance of the four loss terms.

5 Experiments Setup

Dataset. We quantitatively and qualitatively evaluate our approach on three datasets: LRW [10], VoxCeleb2 [9], and LRS3-TED [1]. The LRW dataset consists of 500 different words spoken by hundreds of different speakers in the wild and the VoxCeleb2 contains over 1 million utterances for 6,112 celebrities. The videos in LRS3-TED [1] contain more diverse and challenging head movements than the others. We follow same data split as [1,9,10].

Implementation Details[3]**.** We follow the same training procedure as [32]. We adopt ADAM [18] optimizer with lr $= 2 \times 10^{-4}$ and $(\beta_1, \beta_2) = (0.5, 0.999)$. During training, we select $K = 1, 8, 32$ in our embedding network. The τ in all experiments is 64. The $\lambda_{\text{FM}} = \lambda_{\text{PCT}} = \lambda_{\text{PCT}} = 10$ in Eq. 8.

Table 1. Quantitative results of different audio to video methods on LRW dataset and VoxCeleb2 dataset. We bold each leading score.

Method	Audio-driven							
Dataset	LRW				VoxCeleb2			
	LMD↓	CSIM↑	SSIM↑	FID↓	LMD↓	CSIM↑	SSIM↑	FID↓
Chung et al. [8]	3.15	0.44	**0.91**	132	5.4	0.29	0.79	159
Chen et al. [7]	3.27	0.38	0.84	151	4.9	0.31	0.82	142
Vougioukas et al. [30]	3.62	0.35	0.88	116	6.3	0.33	**0.85**	127
Ours (K=1)	**3.13**	**0.49**	0.76	**62**	**3.37**	**0.42**	0.74	**47**

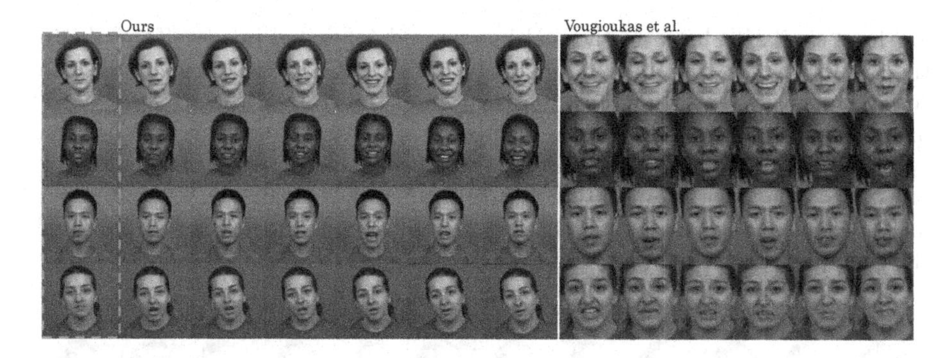

Fig. 7. Comparison of our model with Vougioukas et al. [30] on CREMA-D testing set.

6 Results and Analysis

Evaluation Metrics. We use several criteria for quantitative evaluation. We use Fréchet Inception Distance (FID) [16], mostly quantifying the fidelity of synthesized images, structured similarity (SSIM) [34], commonly used to measure the

[3] All experiments are conducted on an NVIDIA DGX-1 machine with 8 32 GB V100 GPUs. It takes 5 days to converge the training on VoxCeleb2/LRS3-TED since they contain millions of video clips. It takes less than 1 day to converge the training on the GRID/CREMA. For more details, please refer to supplemental materials.

Table 2. Quantitative results of different landmark to video methods on LRS3-TED and VoxCeleb2 datasets. We bold each leading score. K is the number of reference frames.

Method	Landmark-driven					
Dataset	LRS3-TED			VoxCeleb2		
	SSIM↑	CSIM↑	FID↓	SSIM↑	CSIM↑	FID↓
Wiles et al. [35]	0.57	0.28	172	0.65	0.31	117.3
Chen et al. [7]	0.66	0.31	294	0.61	0.39	107.2
Zakharov et al. [37] (K = 1)	0.56	0.27	361	0.64	0.42	88.0
Wang et al. [31] (K = 1)	0.59	0.33	**161**	0.69	**0.48**	59.4
Ours (K = 1)	**0.71**	**0.37**	354	**0.71**	0.44	**40.8**
Zakharov et al. [37] (K = 8)	0.65	0.32	284	0.71	**0.54**	62.6
Wang et al. [31] (K = 8)	0.68	0.36	**144**	**0.72**	0.53	42.6
Ours(K=8)	**0.76**	**0.41**	324	0.71	0.50	**37.1**
Zakharov et al. [37] (K = 32)	0.72	0.41	154	0.75	0.54	51.8
Wang et al. [31] (K = 32)	0.69	0.40	132	0.73	**0.60**	41.4
Ours (K=32)	**0.79**	**0.44**	**122**	**0.78**	0.58	**35.7**

low-level similarity between the real images and generated images. To evaluate the identity preserving ability, same as [37], we use CSIM, which computes the cosine similarity between embedding vectors of the state-of-the-art face recognition network [11] for measuring identity mismatch. To evaluate whether the synthesized video contains accurate lip movements that correspond to the input condition, we adopt the evaluation matrix Landmarks Distance (LMD) proposed in [6]. Additionally, we conduct user study to investigate the visual quality of the generated videos including lip-sync performance.

Fig. 8. We generate video frames on two real world images.

Comparison with Audio-driven Methods. We first consider such a scenario that takes audio and one frame as inputs and synthesizes a talking head saying the speech, which has been explored in Chung et al. [8], Wiles et al. [35], Chen et al. [7], Song et al. [26], Vougioukas et al. [30], and Zhou et al. [39]. Comparing with their methods, our method is able to generate vivid videos including controllable head movements and natural facial expressions. We show the videos driven by audio in the supplemental materials. Table 1 shows the quantitative results. For a fair comparison, we only input one reference frame. Note that other methods require pre-processing including affine transformation and cropping under a fixed alignment. And we do not add such constrains, which leads to a lower SSIM matrix value (e.g., complex backgrounds). We also compare the facial expression generation (see Fig. 7)[4] from audio with Vougioukas et al. [30] on CREMA-D dataset [3]. Furthermore, we show two generation examples in Fig. 8 driven by audio input, which demonstrates the generalizability of our network.

Fig. 9. Comparison on LRS3-TED and VoxCeleb2 testing set. We can find that our method can generate lip-synced video frames while preserving the identity information.

Comparison with Visual-driven Methods. We also compare our approach with state-of-the-art landmark/image-driven generation methods: Zakharov et al. [37], Wiles et al. [35], and Wang et al. [31] on VoxCeleb2 and LRS3-TED datasets. Figure 9 shows the visual comparison, from where we can find that our methods can synthesize accurate lip movement while moving the head accurately. Comparing with other methods, ours performs much better especially when there is a apparent deformation between the target image and reference image. We attribute it to the 3D-aware module, which can guide the network with rough geometry information. We also show the quantitative results in Table 2, which shows that our approach achieves best performance in most of the evaluation matrices. It is worth to mention that our method outperforms other methods in terms of the CSIM score, which demonstrates the generalizability of our method. We choose different K here to better understand our matching scheme. We can see that with larger K value, the inception performance (FID and SSIM) can be improved. Note that we only select K images from the first 64 frames, and Zakharov et al. [37] and Wang et al. [31] select K images from the whole video sequence, which arguably gives other methods an advantage.

[4] We show the raw output from the network in Fig. 7. Due to intrinsic limitations, [30] only generates facial regions with a fixed scale.

(a) Visualization of ablation studies. We show the artifacts using red circles/boxes.

(b) The statistics of user studies.

Fig. 10. (a)Ablation studies. (b) User studies.

Ablation Studies. We conduct thoughtful ablation experiments to study the contributions of each module we introduced in Sect. 3 and Sect. 4. The ablation studies are conducted on VoxCeleb2 dataset. As shown in Fig. 10(a), each component contributes to the full model. For instance, from the third column, we can find that the identity preserving ability drops dramatically if we remove the 3d-aware module. We attribute this to the valuable highly-aligned facial texture information provided 3D-aware module, which could stabilize the GAN training and lead to a faster convergence. Another interesting case is that if we remove our non-linear composition block or warping operation, the synthesized images may contain artifacts in the area near face edges or eyes. We attribute to the alignment issue caused by head movements. Please refer to supplemental materials for more quantitative results on ablation studies.

User Studies. To compare the photo-realism and faithfulness of the translation outputs, we perform human evaluation on videos generated by both audio and landmarks on videos randomly sampled from different testing sets, including LRW [10], VoxCeleb2 [9], and LRS3-TED [1]. Our results including: synthesized videos conditioned on audio input, videos conditioned on landmark guidance, and the generated videos that facial expressions are controlled by audio signal and the head movements are controlled by our motion learner. Human subjects evaluation is conducted to investigate the visual qualities of our generated results compared with Wiles et al. [35], Wang et al. [31], Zakharov et al. [37] in terms of two criteria: whether participants could regard the generated talking faces as realistic and whether the generated talking head temporally sync with the corresponding audio. The details of User Studies are described in the supplemental materials. From Fig. 10(b), we can find that all our methods outperform other two methods in terms of the extent of synchronization and authenticity. It is worth to mention that the videos synthesized with head movements (ours, K = 1) learned by the motion learner achieve much better performance that the one without head motion, which indicate that the participants prefer videos with natural head movements more than videos with still face.

7 Conclusion and Discussion

In this paper, we propose a novel approach to model the head motion and facial expressions explicitly, which can synthesize talking-head videos with natural head movements. By leveraging a 3d aware module, a hybrid embedding module, and a non-linear composition module, the proposed method can synthesize photo-realistic talking-head videos.

Although our approach outperforms previous methods, our model still fails in some situations. For example, our method struggles synthesizing extreme poses, especially there is no visual clues in the given reference frames. Furthermore, we omit modeling the camera motions, light conditions, and audio noise, which may affect our synthesizing performance.

Acknowledgement. This work was supported in part by NSF 1741472, 1813709, and 1909912. The article solely reflects the opinions and conclusions of its authors but not the funding agents.

References

1. Afouras, T., Chung, J.S., Zisserman, A.: LRS3-TED: a large-scale dataset for visual speech recognition. In: arXiv preprint arXiv:1809.00496 (2018)
2. Bregler, C., Covell, M., Slaney, M.: Video rewrite: driving visual speech with audio. In: Proceedings of the 24th Annual Conference on Computer Graphics and Interactive Techniques, pp. 353–360 (1997)
3. Cao, H., Cooper, D.G., Keutmann, M.K., Gur, R.C., Nenkova, A., Verma, R.: CREMA-D: crowd-sourced emotional multimodal actors dataset. IEEE Trans. Affect. Comput. **5**(4), 377–390 (2014)
4. Cassell, J., McNeill, D., McCullough, K.E.: Speech-gesture mismatches: evidence for one underlying representation of linguistic and nonlinguistic information. Pragmat. Cogn. **7**(1), 1–34 (1999)
5. Chang, Y.J., Ezzat, T.: Transferable videorealistic speech animation. In: Proceedings of the 2005 ACM SIGGRAPH/Eurographics Symposium on Computer Animation, pp. 143–151. ACM (2005)
6. Chen, L., Li, Z., K Maddox, R., Duan, Z., Xu, C.: Lip movements generation at a glance. In: Proceedings of the European Conference on Computer Vision (ECCV), pp. 520–535 (2018)
7. Chen, L., Maddox, R.K., Duan, Z., Xu, C.: Hierarchical cross-modal talking face generation with dynamic pixel-wise loss. In: Proceedings of the IEEE Conference on Computer Vision and Pattern Recognition, pp. 7832–7841 (2019)
8. Chung, J.S., Jamaludin, A., Zisserman, A.: You said that? In: British Machine Vision Conference (2017)
9. Chung, J.S., Nagrani, A., Zisserman, A.: VoxCeleb2: deep speaker recognition. In: INTERSPEECH (2018)
10. Chung, J.S., Zisserman, A.: Lip reading in the wild. In: Lai, S.-H., Lepetit, V., Nishino, K., Sato, Y. (eds.) ACCV 2016. LNCS, vol. 10112, pp. 87–103. Springer, Cham (2017). https://doi.org/10.1007/978-3-319-54184-6_6
11. Deng, J., Guo, J., Xue, N., Zafeiriou, S.: ArcFace: additive angular margin loss for deep face recognition. In: Proceedings of the IEEE Conference on Computer Vision and Pattern Recognition, pp. 4690–4699 (2019)

12. Feng, Y., Wu, F., Shao, X., Wang, Y., Zhou, X.: Joint 3D face reconstruction and dense alignment with position map regression network. In: Proceedings of the European Conference on Computer Vision (ECCV), pp. 534–551 (2018)
13. Fried, O., et al.: Text-based editing of talking-head video. ACM Trans. Graph. (TOG) **38**(4), 1–14 (2019)
14. Garrido, P., et al.: VDub: modifying face video of actors for plausible visual alignment to a dubbed audio track. In: Computer Graphics Forum, vol. 34, pp. 193–204. Wiley Online Library (2015)
15. Ginosar, S., Bar, A., Kohavi, G., Chan, C., Owens, A., Malik, J.: Learning individual styles of conversational gesture. In: Computer Vision and Pattern Recognition (CVPR). IEEE (2019)
16. Heusel, M., Ramsauer, H., Unterthiner, T., Nessler, B., Hochreiter, S.: GANs trained by a two time-scale update rule converge to a local nash equilibrium. In: Advances in Neural Information Processing Systems, pp. 6626–6637 (2017)
17. Kim, H., et al.: Deep video portraits. ACM Trans. Graph. (TOG) **37**(4), 1–14 (2018)
18. Kingma, D.P., Ba, J.: Adam: a method for stochastic optimization. arXiv preprint arXiv:1412.6980 (2014)
19. Liu, K., Ostermann, J.: Realistic facial expression synthesis for an image-based talking head. In: 2011 IEEE International Conference on Multimedia and Expo, pp. 1–6. IEEE (2011)
20. Liu, M.Y., et al.: Few-shot unsupervised image-to-image translation. In: IEEE International Conference on Computer Vision (ICCV) (2019)
21. Liu, S., Li, T., Chen, W., Li, H.: Soft rasterizer: a differentiable renderer for image-based 3d reasoning. The IEEE International Conference on Computer Vision (ICCV), October 2019
22. Munhall, K.G., Jones, J.A., Callan, D.E., Kuratate, T., Vatikiotis-Bateson, E.: Visual prosody and speech intelligibility: head movement improves auditory speech perception. Psychol. Sci. **15**(2), 133–137 (2004)
23. Park, T., Liu, M.Y., Wang, T.C., Zhu, J.Y.: Semantic image synthesis with spatially-adaptive normalization. In: Proceedings of the IEEE Conference on Computer Vision and Pattern Recognition, pp. 2337–2346 (2019)
24. Pumarola, A., Agudo, A., Martinez, A.M., Sanfeliu, A., Moreno-Noguer, F.: GANimation: one-shot anatomically consistent facial animation. Int. J. Comput. Vis. 1–16 (2019)
25. Simonyan, K., Zisserman, A.: Very deep convolutional networks for large-scale image recognition. In: International Conference on Learning Representations (2015)
26. Song, Y., Zhu, J., Li, D., Wang, A., Qi, H.: Talking face generation by conditional recurrent adversarial network. In: Proceedings of the Twenty-Eighth International Joint Conference on Artificial Intelligence, IJCAI-19, pp. 919–925. International Joint Conferences on Artificial Intelligence Organization, July 2019. https://doi.org/10.24963/ijcai.2019/129
27. Suwajanakorn, S., Seitz, S.M., Kemelmacher-Shlizerman, I.: Synthesizing obama: learning lip sync from audio. ACM Trans. Graph. (TOG) **36**(4), 95 (2017)
28. Tian, Y., Shi, J., Li, B., Duan, Z., Xu, C.: Audio-visual event localization in unconstrained videos. In: Proceedings of the European Conference on Computer Vision (ECCV), pp. 247–263 (2018)
29. Vondrick, C., Pirsiavash, H., Torralba, A.: Generating videos with scene dynamics. In: Advances in Neural Information Processing Systems, pp. 613–621 (2016)

30. Vougioukas, K., Petridis, S., Pantic, M.: Realistic speech-driven facial animation with GANs. Int. J. Comput. Vis. 1–16 (2019)
31. Wang, T.C., Liu, M.Y., Tao, A., Liu, G., Kautz, J., Catanzaro, B.: Few-shot video-to-video synthesis. In: Advances in Neural Information Processing Systems (NeurIPS) (2019)
32. Wang, T.C., et al.: Video-to-video synthesis. In: Advances in Neural Information Processing Systems (NeurIPS) (2018)
33. Wang, T.C., Liu, M.Y., Zhu, J.Y., Tao, A., Kautz, J., Catanzaro, B.: High-resolution image synthesis and semantic manipulation with conditional GANs. In: Proceedings of the IEEE Conference on Computer Vision and Pattern Recognition, pp. 8798–8807 (2018)
34. Wang, Z., Bovik, A.C., Sheikh, H.R., Simoncelli, E.P., et al.: Image quality assessment: from error visibility to structural similarity. IEEE Trans. Image Process. **13**(4), 600–612 (2004)
35. Wiles, O., Sophia Koepke, A., Zisserman, A.: X2Face: a network for controlling face generation using images, audio, and pose codes. In: Proceedings of the European Conference on Computer Vision (ECCV), pp. 670–686 (2018)
36. Yoo, S., Bahng, H., Chung, S., Lee, J., Chang, J., Choo, J.: Coloring with limited data: few-shot colorization via memory augmented networks. In: Proceedings of the IEEE Conference on Computer Vision and Pattern Recognition, pp. 11283–11292 (2019)
37. Zakharov, E., Shysheya, A., Burkov, E., Lempitsky, V.: Few-shot adversarial learning of realistic neural talking head models. In: The IEEE International Conference on Computer Vision (ICCV), October 2019
38. Zhou, H., Liu, J., Liu, Z., Liu, Y., Wang, X.: Rotate-and-render: unsupervised photorealistic face rotation from single-view images. In: IEEE Conference on Computer Vision and Pattern Recognition (CVPR) (2020)
39. Zhou, H., Liu, Y., Liu, Z., Luo, P., Wang, X.: Talking face generation by adversarially disentangled audio-visual representation. In: Proceedings of the AAAI Conference on Artificial Intelligence, vol. 33, pp. 9299–9306 (2019)
40. Zhu, X., Lei, Z., Liu, X., Shi, H., Li, S.Z.: Face alignment across large poses: a 3D solution. In: Proceedings of the IEEE Conference on Computer Vision and Pattern Recognition, pp. 146–155 (2016)

AUTO3D: Novel View Synthesis Through Unsupervisely Learned Variational Viewpoint and Global 3D Representation

Xiaofeng Liu[1,4(✉)], Tong Che[2], Yiqun Lu[3], Chao Yang[5], Site Li[6], and Jane You[7]

[1] HMS, Harvard University, Boston, MA 03315, USA
liuxiaofengcmu@gmail.com
[2] MILA, Université de Montréal, Montréal, Canada
[3] Nanjing University of Information Science and Technology, Nanjing 210044, China
[4] Fanhan Tech. Inc., Suzhou 215128, China
[5] Facebook AI, Boston, MA 03315, USA
[6] Carnegie Mellon University, Pittsburgh, PA 15213, USA
[7] Department of Computing, The Hong Kong Polytechnic University, Hung Hom, Hong Kong

Abstract. This paper targets on learning-based novel view synthesis from a single or limited 2D images without the pose supervision. In the viewer-centered coordinates, we construct an end-to-end trainable conditional variational framework to disentangle the unsupervisely learned relative-pose/rotation and implicit global 3D representation (shape, texture and the origin of viewer-centered coordinates, etc.). The global appearance of the 3D object is given by several appearance-describing images taken from any number of viewpoints. Our spatial correlation module extracts a global 3D representation from the appearance-describing images in a permutation invariant manner. Our system can achieve implicitly 3D understanding without explicitly 3D reconstruction. With an unsupervisely learned viewer-centered relative-pose/rotation code, the decoder can hallucinate the novel view continuously by sampling the relative-pose in a prior distribution. In various applications, we demonstrate that our model can achieve comparable or even better results than pose/3D model-supervised learning-based novel view synthesis (NVS) methods with any number of input views.

Keywords: Unsupervised novel view synthesis · Viewer-centered coordinates · Variational viewpoints · Global 3D representation

X. Liu, T. Che and Y. Lu—Contribute Equally.

Electronic supplementary material The online version of this chapter (https://doi.org/10.1007/978-3-030-58545-7_4) contains supplementary material, which is available to authorized users.

© Springer Nature Switzerland AG 2020
A. Vedaldi et al. (Eds.): ECCV 2020, LNCS 12354, pp. 52–71, 2020.
https://doi.org/10.1007/978-3-030-58545-7_4

1 Introduction

Novel view synthesis (NVS) [79] aims at generating novel images with arbitrary viewpoints given one or a few description images of an object. NVS has great potential in computer vision, computer graphics and virtual reality.

Current NVS methods can be grouped into two categories, i.e., geometry- and learning-based methods. The geometry-based methods [14,30] are usually challenging to estimate the geometric structure in 3D space with single or very limited 2D input [15,79], and need render for appearance mapping.

On the other hand, with the popularity of deep generative model [20], learning-based solutions directly generate the image in target view, without the explicitly 3D structure and the 2D rendering. As the 3D model estimation and render module are not necessary, it is promising in a wide range of scenarios [79].

The generative adversarial network (GAN) [20] can be used for NVS by discretizing the camera views and learn the view-to-view mapping functions between any two pre-defined views [4,72,73]. Without 3D understanding, these models cannot generalize unseen views effectively, e.g., trained with $10°, 20°$ and the model is asked to take a $15°$ input or generate the viewpoint of $25°$ [79].

To address this issue, [14,30] resort to the extra 3D information e.g., CAD labels, which are usually expensive or inaccessible. [79] introduces the Cycle GAN [86] to extract pose-invariant feature as implicitly 3D representation. However, all of the aforementioned learning-based methods rely on human-labeled camera pose/viewpoint in their training. Getting these viewpoint labels is costly because the position of camera and object both need to be measured. Besides, the results are usually noisy [53]. A more challenging issue of this approach is that it is sometimes difficult to define the origin of pose for unseen, complex new objects.

Actually, previous NVS works adopt the *object-centered coordinates* [66], where the shape of objects is represented with a canonical view. For example, shown either a front view or side view of a car, these approaches set the pre-defined frontal view as the origin and synthesize a view in this pose coordinates. Defining canonical poses can simplify some specific scenarios (e.g., face [13]), while it is problematic on many real-world tasks. It requires all the 3D objects to be aligned to a canonical pose, which is hard for a novel object that has not been encountered in the training set [53].

In contrast, *viewer-centered coordinates* [66,83] propose to represent the shape in a coordinate system that aligns with the viewing perspective of input image. We propose that the origin of NVS can be defined as the input view. In this setting, novel objects and poses can be generalized since it is not required to align canonical poses to 3D models. The manipulation code of relative-pose would be the difference between appearance-describing input and target view, rather than an absolute value in object-centered coordinates.

Besides, for complex objects, a single image is intrinsically ill-posed to describe the entire appearance information of their objects. Recent learning-based NVS works either hallucinate the blurry results [10] or use CAD model in training [58]. A straightforward solution to improve NVS quality is to collect several images of the same object taken from different viewpoints. Most learning-based works [59,73] directly average the representation of inputs with the help of pose label. While the multiple inputs can be aligned without pose supervision according to the texture in geometry-based methods.

Motivated by the aforementioned insights, we propose an unsupervised conditional variational autoencoder framework to achieve NVS in learned viewer-centered coordinates (abbreviated as AUTO3D). In this paper, we propose a method to benefit from both learning- and geometry-based methods while ameliorating their drawback. Our method is essentially a learning-based strategy without the need of the explicitly 3d reconstruction and render, and yet still infers 3D knowledge implicitly. It can automatically disentangle the relative-pose/rotation and a global 3D representation to summarize the other factors (e.g., shape, texture, illumination and the origin of viewer-centered coordinates) without any extra supervision of pose, 3D model or geometry priors of symmetry [2,31], and synthesize images of continuous viewpoints.

Our basic idea coincides with human's way of novel view imagination that we can perform virtual rotation of an implicitly 3D world understanding start from the given view in our mentality [66]. We do not need to define frontal view, have input pose label, and extract view-point independent representation as [73,79].

Besides, the disentanglement based on GANs can be unreliable for its unstable training dynamics what is known as mode collapse [6,18,51,54]. Unsupervised conditional β-variational autoencoder (VAE) adopted here for viewer-centered pose encoding offers a much easier and stable training than GANs [18]. Although GAN loss can always be added to enrich the generation details [36]. With end-to-end training, our model simultaneously learns to extract 3D information from appearance-describing images, to disentangle latent pose code, and to synthesize target image with a relative-pose code sampled on a prior distribution (e.g., Gaussian). All of these are achieved in a pose-unsupervised manner.

Our spatial correlation module (SCM) can take multiple images in a permutation invariant manner to generate a global 3D encoding. Based on the non-local mechanism [75,84], we further explore the spatial clues with Gaussian similarity metric and local diffusion-based complementary-aware formulation.

Since these images provide a complete description of the appearance of the object, we name them as "appearance-describing" images. Our model extracts the implicitly global 3D representation which provides a global overview of the objects from these appearance describing images. The representation is combined with the latent relative-pose code to synthesize the target image with the viewpoint. In our model, no explicit notion of "canonical pose" is given by the human labeler. Instead, it infers an implicit origin of viewer-centered coordinates from the appearance describing images, which is usually the average pose of these input images in our experiment observations. Besides, the input pose detection in testing is not required When synthesizing the view with a user-defined degree of rotation. Our contributions can be summarized as:

- We propose a novel learning-based NVS system to synthesize new images in arbitrary views without the supervision of pose. AUTO3D is the first attempt at adapting unsupervisely learned viewer-centered coordinates for NVS.
- A unified conditional variational framework is designed to achieve unsupervisely learned viewer-centered relative-pose encoding and global 3D representation (shape, texture, illumination and the origin of viewer-centered coordinates, etc.).
- Our model is general to take any number of images (from one to many) in a permutation-invariant manner. The complementary information is organized with a pose-unsupervised non-local mechanism beyond simply average.

We extensively evaluate our method on both objects and face NVS benchmarks and obtained comparable or even better performance than the pose/3D model-supervised methods. It can be applied to either a single or multiple inputs.

2 Related Work

Geometry-based NVS tries to explicitly model the 3D structure of objects and project it to 2D space [14,16,30,62,67]. However, the estimated point clouds are often not dense enough, especially when handling complicated texture [37,62]. [16,78] estimated the depth instead, but they are designed for binocular situations only. [32,64] proposed exemplar-based models that use large-scale collections of 3D models, and the accuracy largely depends on the variation and complexity of 3D models. [24] proposes to reconstruct the 3D model from a single 2D image without pose annotation, but its voxel setting does not consider the appearance. In contrast, our proposed framework is essentially learning-based without the need for explicit 3D reconstruction [28,69,77].

Learning-based NVS emerges with the development of convolutional neural networks (CNN) [21,42,43,46–48,52,55,80]. Early attempts directly map an input image to a paired target image with an encoder-decoder structure [11]. [85] predicts appearance flow instead of synthesizing pixels from scratch. But it is not able to hallucinate the pixels not contained in the appearance-describing view. [60] concatenates an additional image completion network, but its 3D annotation for training is not necessary for our setting.

Recently, GAN [18,22,23,40] has been utilized to improve the realism of synthesized images [49,50,81]. The generator learns to hallucinate the missing pixels to make the output realistic. Most methods essentially learn an view-to-view translator [29,38,86] between any two pre-defined discrete poses. Without taking the 3D knowledge into account, these methods can only synthesize decent results in several views presented in a training set with pose labels. In contrast, our AUTO3D can synthesize novel viewpoints even if they never appear in the training set and no pose label is given. [79] proposes to extract view-independent features to implicitly infer the 3D structure with pose supervision in the CycleGAN [86]. Indeed, all previous mentioned learning-based NVS require either 3D model or pose label in their training [7,59,68,71,79,85]. Besides, some methods introduce explicit 3D induction bias, e.g., surfel representation [63] and rigid-body transformation [57], but do not work on unseen objects in testing. However, based on a unified conditional variational framework, our AUTO3D learns an implicit global 3D representation on the unsupervisely learned viewer-centered coordinates without any 3D shape and pose supervisions, performing well with unseen objects and views.

Multiple-description NVS has also been investigated to provide more information about the object. Most works [41,44,73] directly average the representation of each appearance-describing input. [68] proposes a sophisticate 3D statistic model to integrate different views. Our spatial-aware self-attention can be a simple and efficient learning-based unified solution to tackle this problem.

Self-attention and Non-local Filtering. As attention models gain in popularity, [74] develops a self-attention mechanism for machine translation. A similar idea is inherited in the non-local algorithm [3], which is a classical image denoising technique. The interaction networks are also developed for modeling pair-wise interactions [45]. Moreover, [75] proposes to bridge self-attention to the more general non-local filtering operations and use it for action recognition in videos. [84] proposes to learn temporal dependencies between video frames at multiple time scales. However, we argue that it is essentially tailored for unordered image sets. We further incorporating spatial clues with Gaussian similarity matrix, and local diffusion-based complementary-aware formulation.

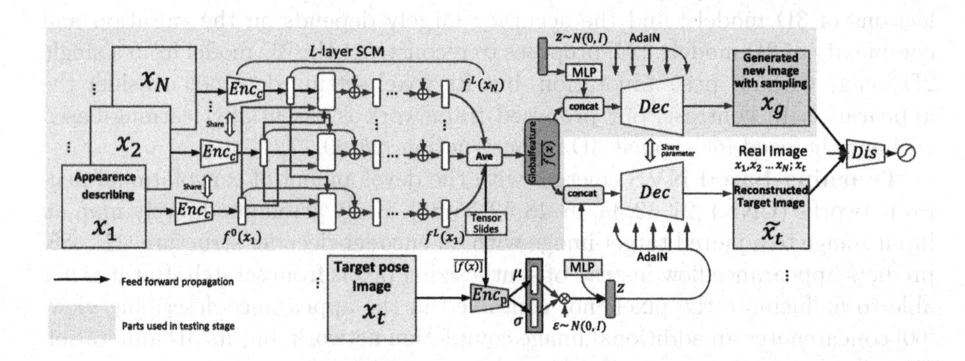

Fig. 1. Illustration of our proposed AUTO3D framework. It is based on VAE-GAN and consists with an unsupervised viewer-centered relative-pose encoding framework, and a spatial-aware self attention module for global 3D encoding to summarize the other factors. e.g., shape, texture, illumination and the origin of viewer-centered coordinates.

3 Methodology

Our goal is to generate a novel view image x_g with the controllable viewer-centered relative-pose code z given a global description of the object or scene.

The global 3D representation is a vector representation computed from a single or multiple appearance-describing images $\{x_1, x_2 \cdots x_N\}, N = 1, 2 \cdots$ which provides a partial or complete view of the 3D object. Our implicit global 3D representation does not pose-invariant as [79], since it is used to define the origin of unsupervisely learned viewer-centered coordinates.

The overall framework of our AUTO3D is shown in Fig. 1, which is based on the conditional β-variational autoencoder. Note that the GAN module is only applied to enrich the details rather than disentanglement. The system is composed of four modules for 1) global 3D feature encoding, 2) unsupervised viewer-centered relative-pose encoding, 3) conditional decoding and 4) discriminating the reconstructed target image with the generated image with z sampling respectively. The disentanglement of relative-viewpoints/rotation and 3D representations can be achieved via the variational framework without the supervision of the 3D model or view-point label, and not relies on adversarial training. Compared with the sophisticate triplet-based adversarial unsupervised disentanglement [56], our solution is simple but sufficient here.

3.1 Global 3D Encoding with Arbitrary Number of Appearance Describing Images

Previous works usually focused on generating 3D model from only a single image [79], but it is intrinsically hard to infer the hidden parts from one image for many complex 3D objects. Rather than simply using the average operation to aggregate multiple views [59,68,71,85] without alignment of different views, we propose to use the global 3D encoder to collect the global information of the object.

The inputs to our global 3D encoder network can be arbitrary number (one to many) of images of the same 3D object taken from different viewpoints, to provide the global information of the 3D object, namely shape, color, texture and the origin of viewer-centered coordinates, etc.

To organize multi-view inputs without the pose label, we first apply the fully convolutional content encoder $Enc_c : x_i \rightarrow \mathbb{R}^{H \times W \times D}$ on each 2D appearance-describing image x_i to extract a compressed representation, where H, W and D are the height, width and channel dimension of output feature respectively. In general, the extracted feature is expected to maintain the spatial relationship of each pixel in a 2D image. However, CNN is famous for its spatial invariant property. Following the CoordConv operation [39], we concatenate the location of the pixel as two additional channels to the feature map.

Since Enc_c is view-agnostic, simply averaging $Enc_c(x_i)$ does not give a chance for each input to be aware of the others, in order to build links and correspondences between different images, etc. We propose to harvest the spatially-aware inner-set correlations by exploiting the affinity of point-wise feature vectors. We use $i = 1, \cdots, H \times W$ to index the position in HW plane and the j is the index for all D-dimensional feature vectors other than the ith vector ($j = 1, \cdots, H \times W \times (N-1)$). Specifically, our non-local block can be formulated as

$$x_{n\_i}^{l} = x_{n\_i}^{l-1} + \frac{\Omega^l}{C_{n\_i}} \sum_{\forall n\_j} \omega(x_{n\_i}^{0}, x_{n\_j}^{0})(x_{n\_j}^{l-1} - x_{n\_i}^{l-1})\Delta_{i,j}$$

$$C_{n\_i} = \sum_{\forall n\_j} \omega(x_{n\_i}^{l-1}, x_{n\_j}^{l-1})\Delta_{i,j}; l = 0, 1, \cdots, L \quad (1)$$

where $\Omega^l \in \mathbb{R}^{1\times1\times D}$ is the weight vector to be learned, L being the number of stacked sub-self attention blocks and $x_n^0 = x_n$. The pairwise affinity $\omega(\cdot,\cdot)$ is an scalar. The response is normalized by C_{n_i}. The operation of ω in Eq. (1) is not sensitive to many function choices [75,84]. We simply choose the embedded Gaussian given by $\omega(x_{n\_i}^{l-1}, x_{n\_j}^{l-1}) = e^{\psi(x_{n\_i}^{l-1})^T \phi(x_{n\_j}^{l-1})}$, where $\psi(x_{n\_i}^{l-1}) = \Psi x_{n\_i}^{l-1}$ and $\phi(x_{n\_j}^{l-1}) = \Phi x_{n\_j}^{l-1}$ are two embeddings, and Ψ, Φ are matrices to be learned.

To explore the spatial clues, we further propose to use Gaussian kernel as a similarity measure $\Delta_{i,j} = \exp(\frac{\|hw_{n\_i} - hw_{n\_j}\|_2^2}{\sigma})$, where hw_{n_i}, $hw_{n\_j} \in \mathbb{R}^2$ represent the position of ith and j^{th} vectors in the HW-plane of x_n, respectively. The residual term is the difference between the neighboring feature (*i.e.*, $x_{n\_j}^{l-1}$) and the computed feature $x_{n\_i}^{l-1}$. If $x_{n\_j}^{l-1}$ incorporates complementary information and has better imaging/content quality compared to $x_{n\_i}^{l-1}$, then RSA will erase some information of the inferior $x_{n\_i}^{l-1}$ and replaces it by the more discriminative feature representation $x_{n\_j}^{l-1}$. Compared to the method of using only $x_{n\_j}^{l-1}$ [75], our setting shares more common features with diffusion maps [70], graph Laplacian [9] and non-local image processing [17]. All of them are non-local analogues [12] of local diffusions, which are expected to be more stable than its original non-local counterpart [75] due to the nature of its inherit Hilbert-Schmidt operator [12].

3.2 Unsupervised Viewer-Centered Relative-Pose Encoding

In the viewer-centered coordinates, the "average" viewpoint of all the appearance-describing images is defined as origin, while the relative-pose code z indicates the "rotation" from the origin to the pose of to be synthesized image.

Instead of inferring the viewpoint code only from a target image x_t, the viewer-centered relative-pose encoder Enc_p takes both x_t and $\overline{[f(x)]}$ as inputs. $\overline{[f(x)]}$ is a slice of $\overline{f(x)}$. In testing, our latent code z controls how the generated viewpoint is different from the origin w.r.t. a small set of input appearance-describing images.

The *Dec* maps global 3d feature $\overline{f(x)}$ to image domain with a reversed structure of *Enc* and conditional to the relative-pose code z. Instead of only resize z to match $\overline{f(x)}$ with a multi-layer perceptron (MLP) and concatenate them as the input of *Dec*, we also adopt the adaptive instance normalization (AdaIN) [27] after each convolution layer as previous conditional generation works [26,57,79,82]. Specifically, the mean (μ) and variance σ of AdaIN layers are normalized to match the relative-pose code z instead of the feature map itself. Here, it a injects stronger inductive bias of z to *Dec*.

The optimization objective of β-VAE [25] is to maximize the regularized evidence lower bound (ELBO) of $p(x_t|x_1, \cdots x_N)$. Specifically, $\log p(x_t|x_1, \cdots x_N) \geq E_{q(z|x_t, \overline{[f(x)]})} \log p(\tilde{x}_t|z, \overline{f(x)}) - \beta D_{KL}(q(z|x_t, \overline{[f(x)]})||p(z))$, where $q(z|x_t, \overline{f(x)})$ and $p(\tilde{x}_t|z, \overline{f(x)})$ are the parameterized *Enc* and *Dec* respectively, $p(z)$ is a prior distribution (e.g., Gaussian), D_{KL} is the Kullback-Leibler (KL) divergence. The regularization coefficient $\beta \geq 1$ constraints the capacity of the latent information bottleneck z [1,65]. Therefore, the higher β can put a stronger information bottleneck pressure on the latent posterior $q(z|x_t, \overline{[f(x)]})$. In this way, z is forced to contain as little information of x_t as possible, thus it drops all the appearance information and carries only the relative-pose information. Both latent z and $\overline{f(x)}$ are the inputs to the *Dec*. With the information bottleneck on z, the decoder is encouraged to get all its appearance information from $\overline{f(x)}$, thus the relative-pose and appearance information are automatically disentangled, without any pose supervision or adversarial training.

We follow the original VAEs [19] that the inference model has two output variables, *i.e.*, μ and σ. Then utilize the reparametric trick $z = \mu + \sigma \odot \epsilon$, where $\epsilon \in N(0, I)$. The posterior distribution is $q(z|x_t, \overline{[f(x)]}) \sim N(z; \mu, \sigma^2)$. In practice, the KL-divergence can be computed as

$$L_{KL}(z; \mu, \sigma) = \frac{1}{2} \sum_{j=1}^{M_z} (1 + \log(\sigma_j^2) - \mu_j^2 - \sigma_j^2) \tag{2}$$

where M_z the dimension of the latent code z. For the reconstruction error, we simply adopt the pixel-wise mean square error (MSE), i.e., L_2 loss. Let $\tilde{x}_t$ be the reconstructed x_t, their $L2$ loss can be formulated as

$$L_{REC}(x_t, \tilde{x}_t) = \frac{1}{2} \sum_{j=1}^{M_{rz}} ||x_{t,j} - \tilde{x}_{t,j}||_F^2 \tag{3}$$

where M_{rz} indicates the channel dimension of x_t or $\tilde{x}_t$.

3.3 Overall Framework and Optimization Objective

A limitation of VAEs is that the generated samples tend to be blurry. This is often result of the limited expressiveness of the inference models, the injected noise and imperfect element-wise criteria such as the squared error [36]. Although recent studies [34] have greatly improved the predicted log-likelihood, the VAE image generation quality still lags behind GAN.

In order to improve generation quality, we adopt the following adversarial training procedure. Similar to VAE-GAN [36], we train AUTO3D to discriminate real samples from both the reconstructions and the generated examples with sampling z. As shown in Fig. 1, these two types of samples are the reconstruction samples x_r and the new samples $\tilde{x}_t$. The adversarial game of GAN can be

$$L_{Adv} = \log(Dis(x_r)) + \log(1 - Dis(Dec(x_g))) + \log(1 - Dis(Dec(\tilde{x}_t))) \tag{4}$$

where $x_r \in \{x_1, x_2 \cdots x_N, x_t\}$ is the real image from either appereance describing set or target pose image. Actually, given a real x_r, the reconstructed sample $Dec(\tilde{x_t})$ can always be more realistic than the sampling image x_g. We usually use similar number of reconstructed and sampled image in training [36].

When the KL-divergence object of VAEs is adequately optimized, the posterior $q(z|x_t, \overline{f(x)})$ matches the prior $p(z) = N(z; 0, I)$ approximately and the samples are similar to each other. The combined use of samples from $p(z)$ and $q(z|x_t, [\overline{f(x)}])$ is also expected to mitigate the observation gap of z in training and testing stage, and empirically synthesize more realistic samples in the testing. The to be minimized objective of each module are respectively defined as

$$\begin{aligned} \mathcal{L}_{Enc_p} &= (L_{Rec} + L_{KL} + L_{Adv}); \ \mathcal{L}_{Dec} = (L_{Rec} + L_{Adv}) \\ \mathcal{L}_{Enc_c/SCM} &= (L_{Rec} + L_{Adv}); \quad \mathcal{L}_{Dis} = -L_{Adv} \end{aligned} \tag{5}$$

After the aforementioned modules are trained, we use Enc_c, spatial-aware self attention module (SCM) and Dec for the testing. Give a set of appearance-describing image, we can sampling on a prior $p(z)$ to control the projection view with user defined rotation. Note that the network mapping of z and the relative-pose difference is deterministic after the training.

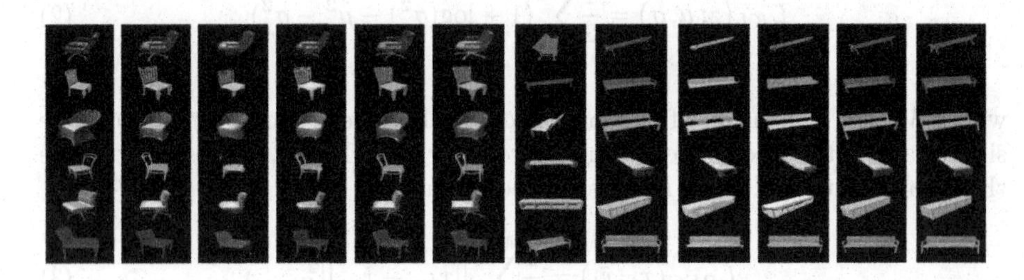

Fig. 2. Comparison of "chair, bench" category on ShapeNet with a single 2D input. From left to right: 2D-input, ground-of-truth, MV3D [71], AF [85], pose-supervised VIGAN [79], Our unsupervised AUTO3D. AUTO3D is comparable to the pose-supervised VIGAN and significantly better than MV3D/AF.

4 Experiments

We conduct a series of experiments on both large scale objects (ShapeNet [5]) and face (300W-LP) [87] datasets to evaluate the qualitative and quantitative performance of AUTO3D, along with the detailed ablation study. Note that the compared methods use the absolute pose value while our z defines the relative-pose/rotation. For the fair comparison, we calculate the difference of input and target pose label in the testing as our relative-pose. Note that AUTO3D can generate any pose continuously without the pose label in both training and NVS implementation.

For fair comparisons in the objects and continuous face rotation tasks, we choose the same Enc_c, Dec, Dis, MLP backbones and AdaIN setting as VI-GAN [79]. We set $|z|$=128 for all datasets except for Cars, where we use $|z|$=200. We train AUTO3D from scratch with Adam [33] solver and implemented on Pytorch [61]. Let $\mathcal{C}_{s,k,c}$ denote a convolutional layer with a stride s, kernel size k, and an output channel c. Then, the discriminator architecture can be expressed as $\mathcal{C}_{2,4,32} \rightarrow \mathcal{C}_{2,4,64} \rightarrow \mathcal{C}_{2,4,128} \rightarrow \mathcal{C}_{2,4,256} \rightarrow \mathcal{C}_{1,1,3}$. Note that we use a local discriminator similar to that of [29]. We use a Leaky ReLU activation function with slope of 0.2 on every layer, except for the last layer. Normalization layer is not applied. This architecture is shared across all experiments.

We implemented our model on Pytorch [61]. Our model is trained end-to-end using using ADAM [33] optimization with hyper-parameters β_1=0.9 and β_2=0.999. We used a batch size of 8 for ShapeNet objects. The encoder network is trained using a learning rate of 5×10^{-5} and the generator is trained using a learning rate 10^{-4}.

4.1 Datasets

ShapeNet [5] is a large collection of textured 3D CAD models of a variety of object categories. There are both single input setting and multiple inputs setting. For single image only, we use the image rendered by [8] following [79]. The chair, bench, and sofa are selected, and 80% models are used for training while 20% for testing [79]. Noticing the testing models are not seen by the network in the training stage. For the multiple viewpoint inputs, we follow the standard training and test data splits [59,60,68,71,85], and train a separate network for each object category (also standard), using 1 to 4 input images to synthesize the target view. The network architecture and training methods were fixed across categories.

300W-LP [87] is a synthesized large-pose face images from 300W. It generates 61,225 samples across large poses with the 3D Image meshing and rotation of in-the-wild face images, which is further expanded to 122,450 samples with flipping. Following [79], we use 80% identities for training and 20% for testing.

4.2 Qualitative Results

Object rotation targets on synthesizing novel views of certain categories for unseen objects. It is challenging, since different objects may have diverse structure and appearance. To demonstrate the capacity of our model, we evaluate our model on the ShapeNet [5] dataset using samples from "chair", "bench" and "sofa" categories. The results are given in Fig. 2.

MV3D [71] and Appearance-Flow (AF) [85] are two popular methods that perform well on this task, while VI-GAN [79] is the recent pose-supervised state-of-the-art. MV3D and AF deal with continuous camera pose by taking the difference between the 3×4 transformation matrices of the input and target views as the pose vector. We compare AUTO3D with them both qualitatively and quantitatively. As shown in Fig. 2, MV3D [71] and AF [85] usually miss small parts, while our results are closer to the ground truth and recent pose-supervised NVS method.

Table 1. Using a single input, the mean pixel-wise L_1 error (lower is better) and SSIM (higher is better) between ground truth and predictions generated by previous pose-supervised methods and different AUTO3D settings. When computing the L_1 error, pixel values are in range of [0, 255]. The best are bolded, while the second best are underlined.

Method	Chair		Bench		Sofa	
	$L_1 \downarrow$	SSIM$\uparrow$	$L_1 \downarrow$	SSIM$\uparrow$	$L_1 \downarrow$	SSIM$\uparrow$
MV3D [71] [need pose label]	24.25	0.76	20.24	0.75	17.52	0.73
AF [85] [need pose label]	18.44	0.82	14.42	0.85	13.26	0.77
VIGAN [79] [need pose label]	**12.56**	**0.87**	**11.52**	**0.88**	**10.13**	**0.83**
AUTO3D w/o AdaIN	12.65	0.83	11.88	0.85	10.39	0.79
AUTO3D w/o GAN	12.64	0.83	11.86	0.85	10.40	0.78
AUTO3D w/o TS	12.65	0.85	11.83	0.86	10.35	0.80
AUTO3D w/o SCM	12.62	0.86	11.80	0.87	10.31	0.82
AUTO3D	12.62	**0.87**	11.80	0.87	10.30	0.82

In the face rotation task, PRNet [13] uses the UV position map in 3DMM to record 3D coordinates and trains CNN to regress them from single views. Figure 3 qualitatively compares our method with PRNet [13] and pose-supervised VI-GAN [79]. Following [13,79], we choose the standard training protocol of 300W-LP, but not use the pose label. As shown in Fig. 3, PRNet [13] may introduce artifacts when information of certain regions is missing. This issue is severe when turning a profile into a frontal face. In contrast, our model produces more realistic images than PRNet [13] and comparable to pose-supervised VI-GAN [79].

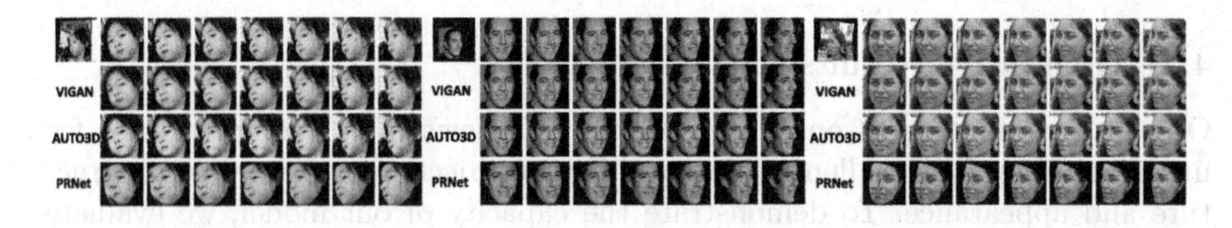

Fig. 3. Comparison with VIGAN [79], PRNet [13] on 300W-LP face dataset.

4.3 Quantitative Results

For quantitative evaluation, the mean pixel-wise L_1 error and the structural similarity index measure (SSIM) [76,78] between synthesized results and the ground truth are calculated following previous methods. We measure the capability of our approach to synthesize new views of objects under large transformations following the standard evaluation protocol. Table 1 shows that our model has on-par performance with pose-supervised VI-GAN in single-input setting following their experiment setting. AUTO3D achieves much lower L_1 error and higher SSIM than MV3D [71] and AF [85].

Table 2. The mean pixel-wise L_1 error (lower is better) and SSIM (higher is better) of AUTO3D and pose-supervised methods with 1 to 4 views, on Chair and Car categories of ShapeNet. Noticing that the setting is different from Table 1 as detailed in Sec 4.2.

Views	Method	Chair		Car		Views	Method	Chair		Car	
		$L_1\downarrow$	SSIM↑	$L_1\downarrow$	SSIM↑			$L_1\downarrow$	SSIM↑	$L_1\downarrow$	SSIM↑
1	MV3D [71] [pose]	0.223	0.882	0.139	0.875	3	MV3D [71] [pose]	0.197	0.898	0.116	0.887
	AF [85] [pose]	0.229	0.871	0.148	0.877		AF [85] [pose]	0.188	0.887	0.089	0.915
	MNV [68] [pose]	0.181	**0.895**	0.098	0.923		MNV [68] [pose]	0.122	0.919	0.068	0.941
	TBN [59] [pose]	**0.046**	**0.895**	**0.025**	**0.927**		TBN [59] [pose]	**0.023**	**0.936**	**0.017**	**0.943**
	AUTO3D w/o SCM	0.052	0.893	0.031	0.916		AUTO3D w/o SCM	0.029	0.930	0.024	0.935
	AUTO3D [SCM-SG]	0.053	0.892	0.031	0.916		AUTO3D [SCM-SG]	0.026	0.932	0.020	0.939
	AUTO3D [SCM-LDC]	0.052	0.893	0.031	0.917		AUTO3D [SCM-LDC]	0.027	0.934	0.019	0.939
	AUTO3D	0.053	0.893	0.030	0.916		AUTO3D	0.025	**0.936**	**0.017**	0.942
2	MV3D [71] [pose]	0.209	0.890	0.124	0.883	4	MV3D [71] [pose]	0.192	0.900	0.112	0.890
	AF [85] [pose]	0.207	0.881	0.107	0.901		AF [85] [pose]	0.165	0.891	0.081	0.924
	MNV [68] [pose]	0.141	0.911	0.078	0.935		MNV [68] [pose]	0.111	0.925	0.062	**0.946**
	TBN [59] [pose]	**0.027**	**0.928**	**0.019**	**0.939**		TBN [59] [pose]	0.022	**0.939**	**0.015**	**0.946**
	AUTO3D w/o SCM	0.036	0.918	0.028	0.929		AUTO3D w/o SCM	0.030	0.929	0.022	0.938
	AUTO3D [SCM-SG]	0.034	0.921	0.025	0.933		AUTO3D [SCM-SG]	0.024	0.935	0.019	0.942
	AUTO3D [SCM-LDC]	0.033	0.922	0.023	0.934		AUTO3D [SCM-LDC]	0.022	0.936	0.018	0.944
	AUTO3D	0.031	0.924	0.020	0.937		AUTO3D	**0.020**	0.938	0.016	**0.946**

Table 3. Turning into frontal face task on 300W-LP dataset.

Pre-training encoder	PRNet [ECCV2018] [13]	VIGAN [ICCV2019] [79]	Our AUTO3D (unsupervised)
$L_1\downarrow$	22.65	**15.32**	16.25± 0.005
SSIM↑	0.65	**0.73**	0.71± 0.003

Then, we demonstrate AUTO3D can infer high-quality views flexibly using limited (1–4) input views at testing. We following the experimental protocol of [59,68] to use up to 4 input images to infer a target image, which is usually challenging for geometry-based NVS. We report the quantitative results on Table 2, and compare our AUTO3D with other works that can take multiple inputs [59,68,71,85], as well as those only accepting single inputs [60]. AUTO3D is comparable or even better than previous pose-supervised methods, especially when more views available. Besides, the gap between AUTO3D and its SCM-free version is usually larger when views increase.

We also give a quantitative evaluation scheme when turning into frontal faces following [79]. Given a synthesized frontal image, it is aligned to its ground truth followed by cropping into the facial area. Its ground truth is also cropped with the same operation. L_1 error and SSIM are calculated between two facial areas and reported in Table 3. AUTO3D yields higher precision than PRNet [13] and is comparable to pose-supervised VIGAN [79] on the 300W-LP dataset.

4.4 Ablation Study of Each Module

Based on conditional β-VAE, our AdaIN, tensor slides (TS), spatial correlation module (SCM) and adversarial loss (GAN) also contribute to the final results.

From Table 1, 2, we can see that the SCM does not affect the performance of AUTO3D when only a single input is available. While it is critical to achieve better performance in multiple inputs cases as shown in Table 2. Adding SCM can consistently improve the appearance reconstruction. Besides, SCM without spatial-aware Gaussian (SCM-SG) or local diffusion-based complementary-aware formulation (SCM-LDC) is consistently inferior to the normal SCM, indicating the effectiveness of our modification on vanilla non-local.

The adversarial loss is utilized to enrich the details and sharpen the appearance. We do not manage to use it for disentanglement as previous unsupervised adversarial training works [56].

AdaIN also contributes to disentanglement, and improve the generation quality w.r.t. appearance. Noticing that the NVS is usually not sensitive to the tensor slides, while can speed up the training speed by 1.5 times.

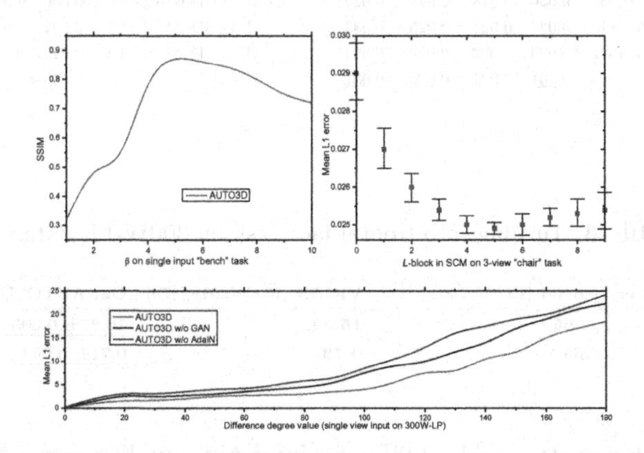

Fig. 4. Sensibility analysis of [Top left] different β (single view bench on ShapeNet), [Top right] the number of SCM blocks (3-view inputs of chair on ShapeNet) and [Bottom] rotation values (single view 300W-LP).

4.5 Sensitive Analysis

The value of β is critical to the performance. We use automatic selection with the disentanglement metric following [25], and fine-tune it according to visual quality. The sensitive analysis is shown in top left of Fig. 4.

The number of layers in our spatial correlation module (SCM) is also critical to the synthesis quality in multiple inputs cases. Here, we give a sensibility analysis in the top right of Fig. 4 (b). We can see that the performance is stable within the range of [4,7]. For simple operation, we choose 4-layer for all of our experiments with multiple inputs.

We also analyse the interaction between the conditional viewer-centered pose code z and generation quality. The bottom of Fig. 4 shows the comparisons of L_1 error as a function of view rotation on the face dataset. Noticing that z indicates the difference of appearance-describing and target view, and $0°$ means no viewpoint change. This illustrates that our AUTO3D can well tackle the extreme pose rotations even without the 3D model or pose label in the training.

4.6 Investigating the Global 3D Feature

We expect that the implicitly 3D structure information of objects can be captured. To evidence this, we implement the experiment of using the latent global 3D representation encoding for learning of 3D tasks.

Following [79], we adopt the 3D face landmark estimation task. The network has two parts where the encoder is the same as the encoder in AUTO3D and Multilayer Perceptron (MLP) is with 2-layers for estimating the coordinate of landmarks based on features extracted by the encoder. Noticing that the backbone of AUTO3D is identical to VIGAN [79]. We also choose 300W-LP [87] for training, in which 3D landmarks are obtained by using their 3DMM parameters.

We configure three training settings to extract the feature for 3D face landmark estimation. The *first* is to train the overall network from scratch to learn 3D features directly. The *second* is pre-train the encoder using the view-independent constraint of VI-GAN, then the 3D supervised data is then used to train the overall network. The *third* setting is to pre-train the Enc_c with our AUTO3D.

Following [79], testing involves 2,000 images from AFLW2000-3D [35] with 68 landmarks. Besides, the mean Normalized Mean Error (NME) [87] is employed for evaluation. We report the results of three settings in Table. 4, until the training loss of both settings no longer changes. The pose-supervised implicitly 3D feature extraction method [79] and our unsupervised AUTO3D get the mean NMEs of 6.8% and 6.9% respectively, which is significantly lower than the training from scratch. This demonstrates that the feature learned by the encoder of AUTO3D is 3D-related. It gives a good initialization for 3D tasks.

Table 4. The NME for 3D face landmark estimation.

Pre-train Enc_c	Scratch	VIGAN [ICCV2019] [79]	Our AUTO3D (unsupervised)
Mean NMEs↓	12.7%	**6.8%**	6.9%±0.12%

4.7 The Effect of Source Image Ordering

The sum operation used in AUTO3D is essentially permutation invariant. We conduct a simple experiment where we test the model on all possible order. We randomly sampled 1000 tuple of source (image, camera pose) pairs from ShapeNet cars and chairs, and evaluated on all 24 ordering. We have found that feeding the different order does not affect the performance of proposed AUTO3D. Our model shows robustness to ordering.

5 Conclusions

This paper presents a novel learning-based framework (AUTO3D) to achieve NVS without the supervision of pose labels and 3D models. It is essentially based on a conditional β-VAE which can be easily and stably trained to disentangle the relative viewpoint information from the other factors in global 3D representation (shape, appearance, lighting and the origin of viewer-centered coordinates, etc.). Instead of the conventional object-centered coordinates, we define the relative-pose/rotation in viewer-centered coordinates, for the first time, on NVS task. Therefore, we do not need to align both training exemplars and unseen objects in testing to a pre-defined canonical pose. Both single or multiple inputs can be naturally integrated with a spatial-aware self-attention (SCM) module. Our results evidenced that AUTO3D is a powerful and versatile unsupervised method for NVS. In the future, we plan to explore more 3D tasks with AUTO3D.

Acknowledgments. This work was supported by the Jangsu Youth Programme [grant number SBK2020041180], National Natural Science Foundation of China, Younth Programme [grant number 61705221], the Fundamental Research Funds for the Central Universities [grant number GK2240260006], NIH [NS061841, NS095986], Fanhan Technology, and Hong Kong Government General Research Fund GRF (Ref. No.152202/14E) are greatly appreciated.

References

1. Alemi, A.A., Fischer, I., Dillon, J.V., Murphy, K.: Deep variational information bottleneck. arXiv preprint arXiv:1612.00410 (2016)
2. Barron, J.T., Malik, J.: Shape, illumination, and reflectance from shading. IEEE Trans. Pattern Anal. Mach. Intell. **37**(8), 1670–1687 (2014)
3. Buades, A., Coll, B., Morel, J.M.: A non-local algorithm for image denoising. In: IEEE Computer Society Conference on Computer Vision and Pattern Recognition, CVPR 2005, vol. 2, pp. 60–65. IEEE (2005)
4. Cao, J., Hu, Y., Yu, B., He, R., Sun, Z.: Load balanced GANs for multi-view face image synthesis. arXiv preprint arXiv:1802.07447 (2018)
5. Chang, A.X., et al.: ShapeNet: an information-rich 3D model repository. arXiv preprint arXiv:1512.03012 (2015)
6. Che, T., et al.: Deep verifier networks: Verification of deep discriminative models with deep generative models. arXiv preprint arXiv:1911.07421 (2019)

7. Chen, X., Song, J., Hilliges, O.: Monocular neural image based rendering with continuous view control. In: Proceedings of the IEEE International Conference on Computer Vision. pp. 4090–4100 (2019)
8. Choy, C.B., Xu, D., Gwak, J.Y., Chen, K., Savarese, S.: 3D-R2N2: a unified approach for single and multi-view 3D object reconstruction. In: Leibe, B., Matas, J., Sebe, N., Welling, M. (eds.) ECCV 2016. LNCS, vol. 9912, pp. 628–644. Springer, Cham (2016). https://doi.org/10.1007/978-3-319-46484-8_38
9. Chung, F.R., Graham, F.C.: Spectral Graph Theory. No. 92. American Mathematical Soc. (1997)
10. Dosovitskiy, A., Springenberg, J.T., Tatarchenko, M., Brox, T.: Learning to generate chairs, tables and cars with convolutional networks. IEEE Trans. Pattern Anal. Mach. Intell. **39**(4), 692–705 (2016)
11. Dosovitskiy, A., Tobias Springenberg, J., Brox, T.: Learning to generate chairs with convolutional neural networks. In: Proceedings of the IEEE Conference on Computer Vision and Pattern Recognition, pp. 1538–1546 (2015)
12. Du, Q., Gunzburger, M., Lehoucq, R.B., Zhou, K.: Analysis and approximation of nonlocal diffusion problems with volume constraints. SIAM Rev. **54**(4), 667–696 (2012)
13. Feng, Y., Wu, F., Shao, X., Wang, Y., Zhou, X.: Joint 3D face reconstruction and dense alignment with position map regression network. In: Proceedings of the European Conference on Computer Vision (ECCV), pp. 534–551 (2018)
14. Flynn, J., Neulander, I., Philbin, J., Snavely, N.: DeepStereo: learning to predict new views from the world's imagery. In: Proceedings of the IEEE Conference on Computer Vision and Pattern Recognition, pp. 5515–5524 (2016)
15. Forsyth, D.A., Ponce, J.: Computer Vision: A Modern Approach. Prentice Hall Professional Technical Reference (2002)
16. Garg, R., B.G., V.K., Carneiro, G., Reid, I.: Unsupervised CNN for single view depth estimation: geometry to the rescue. In: Leibe, B., Matas, J., Sebe, N., Welling, M. (eds.) ECCV 2016. LNCS, vol. 9912, pp. 740–756. Springer, Cham (2016). https://doi.org/10.1007/978-3-319-46484-8_45
17. Gilboa, G., Osher, S.: Nonlocal linear image regularization and supervised segmentation. Multisc. Model. Simul. **6**(2), 595–630 (2007)
18. Goodfellow, I.: Nips 2016 tutorial: generative adversarial networks. arXiv preprint arXiv:1701.00160 (2016)
19. Goodfellow, I., Bengio, Y., Courville, A.: Deep Learning. MIT Press, Cambridge (2016)
20. Goodfellow, I., et al.: Generative adversarial nets. In: Advances in neural information processing systems, pp. 2672–2680 (2014)
21. Han, Y., et al.: Wasserstein loss-based deep object detection. In: Proceedings of the IEEE/CVF Conference on Computer Vision and Pattern Recognition Workshops, pp. 998–999 (2020)
22. He, G., Liu, X., Fan, F., You, J.: Classification-aware semi-supervised domain adaptation. In: Proceedings of the IEEE/CVF Conference on Computer Vision and Pattern Recognition Workshops, pp. 964–965 (2020)
23. He, G., Liu, X., Fan, F., You, J.: Image2Audio: facilitating semi-supervised audio emotion recognition with facial expression image. In: Proceedings of the IEEE/CVF Conference on Computer Vision and Pattern Recognition Workshops, pp. 912–913 (2020)
24. Henderson, P., Ferrari, V.: Learning single-image 3D reconstruction by generative modelling of shape, pose and shading. Int. J. Comput. Vis. 1–20 (2019)

25. Higgins, I., et al.: beta-vae: Learning basic visual concepts with a constrained variational framework. In: ICLR, vol. 2, no. 5, p. 6 (2017)
26. Huang, H., He, R., Sun, Z., Tan, T., et al.: IntroVAE: introspective variational autoencoders for photographic image synthesis. In: Advances in Neural Information Processing Systems, pp. 52–63 (2018)
27. Huang, X., Belongie, S.: Arbitrary style transfer in real-time with adaptive instance normalization. In: Proceedings of the IEEE International Conference on Computer Vision, pp. 1501–1510 (2017)
28. Insafutdinov, E., Dosovitskiy, A.: Unsupervised learning of shape and pose with differentiable point clouds. In: Advances in Neural Information Processing Systems, pp. 2802–2812 (2018)
29. Isola, P., Zhu, J.Y., Zhou, T., Efros, A.A.: Image-to-image translation with conditional adversarial networks. In: Proceedings of the IEEE Conference on Computer Vision and Pattern Recognition, pp. 1125–1134 (2017)
30. Ji, D., Kwon, J., McFarland, M., Savarese, S.: Deep view morphing. In: Proceedings of the IEEE Conference on Computer Vision and Pattern Recognition, pp. 2155–2163 (2017)
31. Kanazawa, A., Tulsiani, S., Efros, A.A., Malik, J.: Learning category-specific mesh reconstruction from image collections. In: Proceedings of the European Conference on Computer Vision (ECCV), pp. 371–386 (2018)
32. Kholgade, N., Simon, T., Efros, A., Sheikh, Y.: 3D object manipulation in a single photograph using stock 3D models. ACM Trans. Graph. (TOG) **33**(4), 1–12 (2014)
33. Kingma, D.P., Ba, J.: Adam: a method for stochastic optimization. arXiv preprint arXiv:1412.6980 (2014)
34. Kingma, D.P., Salimans, T., Jozefowicz, R., Chen, X., Sutskever, I., Welling, M.: Improved variational inference with inverse autoregressive flow. In: Advances in Neural Information Processing Systems, pp. 4743–4751 (2016)
35. Koestinger, M., Wohlhart, P., Roth, P.M., Bischof, H.: Annotated facial landmarks in the wild: a large-scale, real-world database for facial landmark localization. In: 2011 IEEE International Conference on Computer Vision Workshops (ICCV Workshops), pp. 2144–2151. IEEE (2011)
36. Larsen, A.B.L., Sønderby, S.K., Larochelle, H., Winther, O.: Autoencoding beyond pixels using a learned similarity metric. In: ICML (2016)
37. Lin, C.H., Kong, C., Lucey, S.: Learning efficient point cloud generation for dense 3D object reconstruction. In: Thirty-Second AAAI Conference on Artificial Intelligence (2018)
38. Liu, M.Y., Breuel, T., Kautz, J.: Unsupervised image-to-image translation networks. In: Advances in Neural Information Processing Systems, pp. 700–708 (2017)
39. Liu, R., et al.: An intriguing failing of convolutional neural networks and the CoordConv solution. In: Advances in Neural Information Processing Systems, pp. 9605–9616 (2018)
40. Liu, X.: Disentanglement for discriminative visual recognition. arXiv preprint arXiv:2006.07810 (2020)
41. Liu, X., B.V.K., K., Yang, C., Tang, Q., You, J.: Dependency-aware attention control for unconstrained face recognition with image sets. In: European Conference on Computer Vision (2018)
42. Liu, X., Fan, F., Kong, L., Diao, Z., Xie, W., Lu, J., You, J.: Unimodal regularized neuron stick-breaking for ordinal classification. Neurocomputing (2020)
43. Liu, X., Ge, Y., Yang, C., Jia, P.: Adaptive metric learning with deep neural networks for video-based facial expression recognition. J. Electron. Imaging **27**(1), 013022 (2018)

44. Liu, X., Guo, Z., Jia, J., Kumar, B.: Dependency-aware attention control for imageset-based face recognition. In: IEEE Transactions on Information Forensics and Security (2019)
45. Liu, X., Guo, Z., Li, S., Kong, L., Jia, P., You, J., Kumar, B.V.: Permutation-invariant feature restructuring for correlation-aware image set-based recognition. In: The IEEE International Conference on Computer Vision (ICCV), October 2019
46. Liu, X., et al.: Importance-aware semantic segmentation in self-driving with discrete wasserstein training. In: AAAI, pp. 11629–11636 (2020)
47. Liu, X., Ji, W., You, J., Fakhri, G.E., Woo, J.: Severity-aware semantic segmentation with reinforced wasserstein training. In: Proceedings of the IEEE/CVF Conference on Computer Vision and Pattern Recognition, pp. 12566–12575 (2020)
48. Liu, X., Kong, L., Diao, Z., Jia, P.: Line-scan system for continuous hand authentication. Opt. Eng. **56**(3), 033106 (2017)
49. Liu, X., Kumar, B.V., Ge, Y., Yang, C., You, J., Jia, P.: Normalized face image generation with perceptron generative adversarial networks. In: 2018 IEEE 4th International Conference on Identity, Security, and Behavior Analysis (ISBA), pp. 1–8 (2018)
50. Liu, X., Kumar, B.V., Jia, P., You, J.: Hard negative generation for identity-disentangled facial expression recognition. Pattern Recogn. **88**, 1–12 (2019)
51. Liu, X., Li, S., Kong, L., Xie, W., Jia, P., You, J., Kumar, B.: Feature-level Frankenstein: eliminating variations for discriminative recognition. In: Proceedings of the IEEE Conference on Computer Vision and Pattern Recognition, pp. 637–646 (2019)
52. Liu, X., Vijaya Kumar, B., You, J., Jia, P.: Adaptive deep metric learning for identity-aware facial expression recognition. In: CVPR Workshops, pp. 20–29 (2017)
53. Liu, X., et al.: Conservative wasserstein training for pose estimation. In: The IEEE International Conference on Computer Vision (ICCV), October 2019
54. Liu, X., et al.: Data augmentation via latent space interpolation for image classification. In: 24th International Conference on Pattern Recognition (ICPR), pp. 728–733 (2018)
55. Liu, X., Zou, Y., Song, Y., Yang, C., You, J., K Vijaya Kumar, B.: Ordinal regression with neuron stick-breaking for medical diagnosis. In: Proceedings of the European Conference on Computer Vision (ECCV) (2018)
56. Mathieu, M.F., Zhao, J.J., Zhao, J., Ramesh, A., Sprechmann, P., LeCun, Y.: Disentangling factors of variation in deep representation using adversarial training. In: Advances in Neural Information Processing Systems, pp. 5040–5048 (2016)
57. Nguyen-Phuoc, T., Li, C., Theis, L., Richardt, C., Yang, Y.L.: HoloGAN: unsupervised learning of 3D representations from natural images. arXiv preprint arXiv:1904.01326 (2019)
58. Nguyen-Phuoc, T.H., Li, C., Balaban, S., Yang, Y.: RenderNet: a deep convolutional network for differentiable rendering from 3D shapes. In: Advances in Neural Information Processing Systems, pp. 7891–7901 (2018)
59. Olszewski, K., Tulyakov, S., Woodford, O., Li, H., Luo, L.: Transformable bottleneck networks. arXiv preprint arXiv:1904.06458 (2019)
60. Park, E., Yang, J., Yumer, E., Ceylan, D., Berg, A.C.: Transformation-grounded image generation network for novel 3D view synthesis. In: Proceedings of the IEEE Conference on Computer Vision and Pattern Recognition, pp. 3500–3509 (2017)
61. Paszke, A., et al.: Automatic differentiation in PyTorch (2017)

62. Pontes, J.K., Kong, C., Sridharan, S., Lucey, S., Eriksson, A., Fookes, C.: Image2Mesh: a learning framework for single image 3D reconstruction. In: Jawahar, C.V., Li, H., Mori, G., Schindler, K. (eds.) ACCV 2018. LNCS, vol. 11361, pp. 365–381. Springer, Cham (2019). https://doi.org/10.1007/978-3-030-20887-5_23
63. Rajeswar, S., Mannan, F., Golemo, F., Vazquez, D., Nowrouzezahrai, D., Courville, A.: Pix2Scene: learning implicit 3D representations from images (2018)
64. Rematas, K., Nguyen, C.H., Ritschel, T., Fritz, M., Tuytelaars, T.: Novel views of objects from a single image. IEEE Trans. Pattern Anal. Mach. Intell. **39**(8), 1576–1590 (2016)
65. Saxe, A.M., et al.: On the information bottleneck theory of deep learning (2018)
66. Shin, D., Fowlkes, C.C., Hoiem, D.: Pixels, voxels, and views: A study of shape representations for single view 3D object shape prediction. In: Proceedings of the IEEE Conference on Computer Vision and Pattern Recognition, pp. 3061–3069 (2018)
67. Sturm, P., Triggs, B.: A factorization based algorithm for multi-image projective structure and motion. In: Buxton, B., Cipolla, R. (eds.) ECCV 1996. LNCS, vol. 1065, pp. 709–720. Springer, Heidelberg (1996). https://doi.org/10.1007/3-540-61123-1_183
68. Sun, S.H., Huh, M., Liao, Y.H., Zhang, N., Lim, J.J.: Multi-view to novel view: synthesizing novel views with self-learned confidence. In: Proceedings of the European Conference on Computer Vision (ECCV), pp. 155–171 (2018)
69. Szabó, A., Favaro, P.: Unsupervised 3D shape learning from image collections in the wild. arXiv preprint arXiv:1811.10519 (2018)
70. Tao, Y., Sun, Q., Du, Q., Liu, W.: Nonlocal neural networks, nonlocal diffusion and nonlocal modeling. arXiv preprint arXiv:1806.00681 (2018)
71. Tatarchenko, M., Dosovitskiy, A., Brox, T.: Multi-view 3D models from single images with a convolutional network. In: Leibe, B., Matas, J., Sebe, N., Welling, M. (eds.) ECCV 2016. LNCS, vol. 9911, pp. 322–337. Springer, Cham (2016). https://doi.org/10.1007/978-3-319-46478-7_20
72. Tian, Y., Peng, X., Zhao, L., Zhang, S., Metaxas, D.N.: CR-GAN: learning complete representations for multi-view generation. arXiv preprint arXiv:1806.11191 (2018)
73. Tran, L., Yin, X., Liu, X.: Disentangled representation learning GAN for pose-invariant face recognition. In: CVPR, vol. 3, p. 7 (2017)
74. Vaswani, A., et al.: Attention is all you need. In: Advances in Neural Information Processing Systems, pp. 5998–6008 (2017)
75. Wang, X., Girshick, R., Gupta, A., He, K.: Non-local neural networks. In: The IEEE Conference on Computer Vision and Pattern Recognition (CVPR) (2018)
76. Wang, Z., Bovik, A.C., Sheikh, H.R., Simoncelli, E.P., et al.: Image quality assessment: from error visibility to structural similarity. IEEE Trans. Image Process. **13**(4), 600–612 (2004)
77. Wu, J., Zhang, C., Zhang, X., Zhang, Z., Freeman, W.T., Tenenbaum, J.B.: Learning shape priors for single-view 3D completion and reconstruction. In: Proceedings of the European Conference on Computer Vision (ECCV), pp. 646–662 (2018)
78. Xie, J., Girshick, R., Farhadi, A.: Deep3D: fully automatic 2D-to-3D video conversion with deep convolutional neural networks. In: Leibe, B., Matas, J., Sebe, N., Welling, M. (eds.) ECCV 2016. LNCS, vol. 9908, pp. 842–857. Springer, Cham (2016). https://doi.org/10.1007/978-3-319-46493-0_51
79. Xu, X., Chen, Y.C., Jia, J.: View independent generative adversarial network for novel view synthesis. In: Proceedings of the IEEE International Conference on Computer Vision, pp. 7791–7800 (2019)

80. Yang, C., Liu, X., Tang, Q., Kuo, C.C.J.: Towards disentangled representations for human retargeting by multi-view learning. arXiv preprint arXiv:1912.06265 (2019)
81. Yang, C., Song, Y., Liu, X., Tang, Q., Kuo, C.C.J.: Image inpainting using block-wise procedural training with annealed adversarial counterpart. arXiv preprint arXiv:1803.08943 (2018)
82. Zakharov, E., Shysheya, A., Burkov, E., Lempitsky, V.: Few-shot adversarial learning of realistic neural talking head models. arXiv preprint arXiv:1905.08233 (2019)
83. Zhang, X., Zhang, Z., Zhang, C., Tenenbaum, J., Freeman, B., Wu, J.: Learning to reconstruct shapes from unseen classes. In: Advances in Neural Information Processing Systems, pp. 2257–2268 (2018)
84. Zhou, B., Andonian, A., Torralba, A.: Temporal relational reasoning in videos. In ECCV (2018)
85. Zhou, T., Tulsiani, S., Sun, W., Malik, J., Efros, A.A.: View synthesis by appearance flow. In: Leibe, B., Matas, J., Sebe, N., Welling, M. (eds.) ECCV 2016. LNCS, vol. 9908, pp. 286–301. Springer, Cham (2016). https://doi.org/10.1007/978-3-319-46493-0_18
86. Zhu, J.Y., Park, T., Isola, P., Efros, A.A.: Unpaired image-to-image translation using cycle-consistent adversarial networks. In: Proceedings of the IEEE International Conference on Computer Vision, pp. 2223–2232 (2017)
87. Zhu, X., Lei, Z., Liu, X., Shi, H., Li, S.Z.: Face alignment across large poses: a 3D solution. In: Proceedings of the IEEE Conference on Computer Vision and Pattern Recognition, pp. 146–155 (2016)

VPN: Learning Video-Pose Embedding for Activities of Daily Living

Srijan Das[(✉)], Saurav Sharma, Rui Dai, François Brémond, and Monique Thonnat

INRIA Université Nice Côte d'Azur, Nice, France
{srijan.das,saurav.sharma,rui.dai,francois.bremond, monique.thonnat}@inria.fr

Abstract. In this paper, we focus on the spatio-temporal aspect of recognizing Activities of Daily Living (ADL). ADL have two specific properties (i) subtle spatio-temporal patterns and (ii) similar visual patterns varying with time. Therefore, ADL may look very similar and often necessitate to look at their fine-grained details to distinguish them. Because the recent spatio-temporal 3D ConvNets are too rigid to capture the subtle visual patterns across an action, we propose a novel Video-Pose Network: **VPN**. The 2 key components of this VPN are a spatial embedding and an attention network. The spatial embedding projects the 3D poses and RGB cues in a common semantic space. This enables the action recognition framework to learn better spatio-temporal features exploiting both modalities. In order to discriminate similar actions, the attention network provides two functionalities - (i) an end-to-end learnable pose backbone exploiting the topology of human body, and (ii) a coupler to provide joint spatio-temporal attention weights across a video. Experiments (Code/models: https://github.com/srijandas07/VPN) show that VPN outperforms the state-of-the-art results for action classification on a large scale human activity dataset: **NTU-RGB+D 120**, its subset **NTU-RGB+D 60**, a real-world challenging human activity dataset: **Toyota Smarthome** and a small scale human-object interaction dataset **Northwestern UCLA**.

Keywords: Action recognition · Video · Pose · Embedding · Attention

1 Introduction

Monitoring human behavior requires fine-grained understanding of actions. Activities of Daily Living (ADL) may look simple but their recognition is often more challenging than activities present in sport, movie or Youtube videos. ADL often have very low inter-class variance making the task of discriminating them

Electronic supplementary material The online version of this chapter (https://doi.org/10.1007/978-3-030-58545-7_5) contains supplementary material, which is available to authorized users.

© Springer Nature Switzerland AG 2020
A. Vedaldi et al. (Eds.): ECCV 2020, LNCS 12354, pp. 72–90, 2020.
https://doi.org/10.1007/978-3-030-58545-7_5

from one another very challenging. The challenges characterizing ADL are illustrated in Fig. 1: (i) short and subtle actions like *pouring water* and *pouring grain* while *making coffee*; (ii) actions exhibiting similar visual patterns while differing in motion patterns like *rubbing hands* and *clapping*; and finally, (iii) actions observed from different camera views. In the recent literature, the main focus is the recognition of actions from internet videos [5,12,13,53,54] and very few studies have attempted to recognize ADL in indoor scenarios [4,8,16].

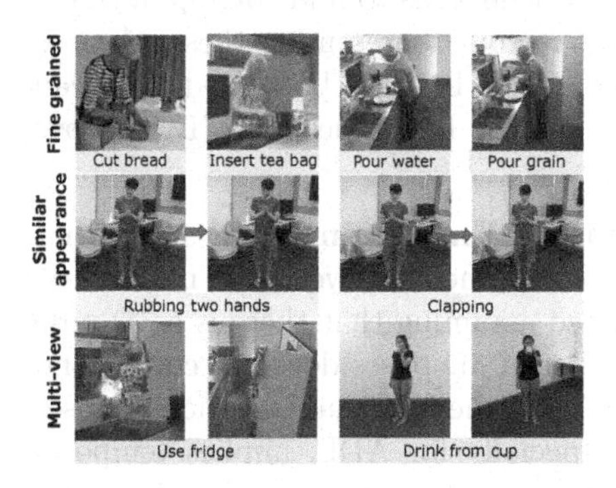

Fig. 1. Illustration of the challenges in Activities of Daily Living: fine-grained actions (top), actions with similar visual pattern (middle) and actions viewed from different cameras (below).

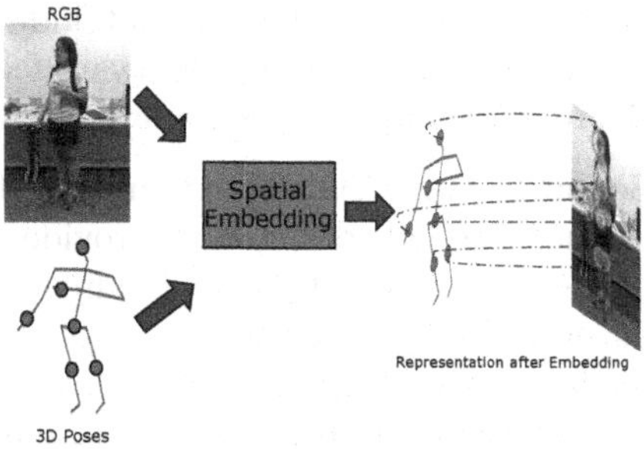

Fig. 2. Illustration of spatial embedding. Input is a RGB image and its corresponding 3D poses. For convenience, we only show 6 relevant human joints. The embedding enforces the human joints to represent the relevant regions in the image.

For instance, state-of-the-art 3D convolutional networks like I3D [5] pre-trained on huge video datasets [17,21,45] have successfully boosted the recognition of actions from internet videos. But, these networks with similar spatio-temporal kernels applied across the whole space-time volume cannot address the complex challenges exhibited by ADL. Attention mechanisms have thus been proposed on top of these 3D convolutional networks to guide them along the regions of interest of the targeted actions [13,16,54].

Following a different direction, action recognition for ADL has been dominated by the use of human 3D poses [56,57]. They provide a strong clue for understanding the visual patterns of an action over time. 3D poses are robust to illumination changes, view adaptive and provide critical geometric information about human actions. However, they lack incorporating the appearance information which is an essential property in ADL (especially for human-object interaction).

Consequently, attempts have been made to utilize 3D poses to weight the discriminative parts of a RGB feature map [2–4,7,8]. These methods have improved the action recognition performance but they do not take into account the alignment of the RGB cues and the corresponding 3D poses. Therefore, we propose a spatial embedding to project the visual features and the 3D poses in the same referential. Before describing our contribution, we answer two intuitive questions below.

First, Why is Spatial Embedding Important? Previous pose driven attention networks can be perceived as guiding networks to help the RGB cues focus on the salient information for action classification. For these guiding networks, it is important to have an accurate correspondences between the poses and RGB data. So, the objective of the spatial embedding is to find correspondences between the 3D human joints and the image regions representing these joints as illustrated in Fig. 2. This task of finding correlation between both modalities can (i) provide informative pose aware feedback to the RGB cues, and (ii) improve the functionalities of the guiding network.

Second, Why Not Performing Temporal Embedding? We argue that the need of embedding is to provide proper alignment between the modalities. Across time, the 3D poses are already aligned assuming that there is a 3D pose for every images. However, even if the number of 3D poses does not correspond to the number of image frames (as in [2–4,7,8]), the fact that variance in poses for few consecutive frames is negligible, especially for ADL, implies temporal embedding is not needed.

We propose a recognition model based on a Video-Pose Network, **VPN** to recognize a large variety of human actions. VPN consists of a spatial embedding and an attention network. VPN exhibits the following novelties: (i) a spatial embedding learns an accurate video-pose embedding to enforce the relationships between the visual content and 3D poses, (ii) an attention network learns the attention weights with a tight spatio-temporal coupling for better modulating the RGB feature map, (iii) the attention network takes the spatial layout of the human body into account by processing the 3D poses through Graph Convolutional Networks (GCNs).

The proposed recognition model is end-to-end trainable and our proposed VPN can be used as a layer on top of any 3D ConvNets.

2 Related Work

Below, we discuss the relevant action recognition algorithms w.r.t. their input modalities.

RGB. Traditionally, image level features [49,50] have been aggregated over time using encoding techniques like Fisher Vector [34] and NetVLAD [1]. But these video descriptors do not encode long-range temporal information. Then, temporal patterns of actions have been modelled in videos using sequential networks. These sequential networks like LSTMs are fed with convolutional features from images [10] and thus, they model the temporal information based on the evolution of appearance of the human actions. However, these methods first process the image level features and then capture their temporal evolution preventing the computation of joint spatio-temporal patterns over time.

Due to this reason, Du et al. [48] have proposed 3D convolution to model the spatio-temporal patterns within an action. The 3D kernels provide tight coupling of space and time towards better action classification. Later on, holistic methods like I3D [5], slow-fast network [12], MARS [6] and two-in-one stream network [59] have been fabricated for generic datasets like Kinetics [17] and UCF-101 [45]. But these networks are trained globally over the whole 3D volume of a video and thus, are too rigid to capture salient features for subtle spatio-temporal patterns for ADL.

Recently several attention mechanisms have been proposed on top of the aforementioned 3D ConvNets to extract salient spatio-temporal patterns. For instance, Wang et al. [54] have proposed a non-local module on top of I3D which computes the attention of each pixel as a weighted sum of the features of all pixels in the space-time volume. But this module relies too much on the appearance of the actions, i.e., pixel position within the space-time volume. As a consequence, this module though effective for the classification of actions in internet videos, fails to disambiguate ADL with similar motion and fails to address view invariant challenges.

3D Poses. To focus on the view-invariant challenge, temporal evolution of 3D poses have been leveraged through sequential networks like LSTM and GRU for skeleton based action recognition [26,56,57]. Taking a step ahead, LSTMs have also been used for spatial and temporal attention mechanisms to focus on the salient human joints and key temporal frames [44]. Another framework represents 3D poses as pseudo image to leverage the successful image classification CNNs for action classification [11,27]. Recently, graph-based methods model the data as a graph with joints as vertexes and bones as edges [40,47,55]. Compared to sequential networks and pseudo image based methods, graph-based methods make use of the spatial configuration of the human body joints and thus, are more effective. However, the skeleton based action recognition lacks in encoding the appearance information which is critical for ADL recognition.

RGB + 3D Poses. In order to make use of the pros of both modalities, i.e. RGB and 3D Poses, it is desirable to fuse these multi-modal information into an integrated set of discriminative features. As these modalities are heterogeneous, they must be processed by different kinds of network to show their effectiveness. This limits their performance in simple multi-modal fusion strategy [23,30,38]. As a consequence, many pose driven attention mechanisms have been proposed to guide the RGB cues for action recognition. In [2–4], the pose driven attention networks implemented through LSTMs, focus on the salient image features and the key frames. Then, with the success of 3D CNNs, 3D poses have been exploited to compute the attention weights of a spatio-temporal feature map. Das et al. [7] have proposed a spatial attention mechanism on top of 3D ConvNets to weight the pertinent human body parts relevant for an action. Then, authors in [8] have proposed a more general spatial and temporal attention mechanism in a dissociated manner. But these methods have the following drawbacks: (i) there is no accurate correspondence between the 3D poses and the RGB cues in the process of computing the attention weights [2–4,7,8]; (ii) the attention sub-networks [2–4,7,8] neglect the topology of the human body while computing the attention weights; (iii) the attention weights in [7,8] provide identical spatial attention along the video. As a result, action pairs with similar appearance like *jumping* and *hopping* are mis-classified.

In contrast, we propose a new spatial embedding to enforce the correspondences between RGB and 3D pose which has been missing in the state-of-the-art methods. The embedding is built upon an end-to-end learnable attention network. The attention network considers the human topology to better activate the relevant body joints for computing the attention weights. To the best of our knowledge, none of the previous action recognition methods have combined human topology with RGB cues. In addition, the proposed attention network couples the spatial and temporal attention weights in order to provide spatial attention weights varying along time.

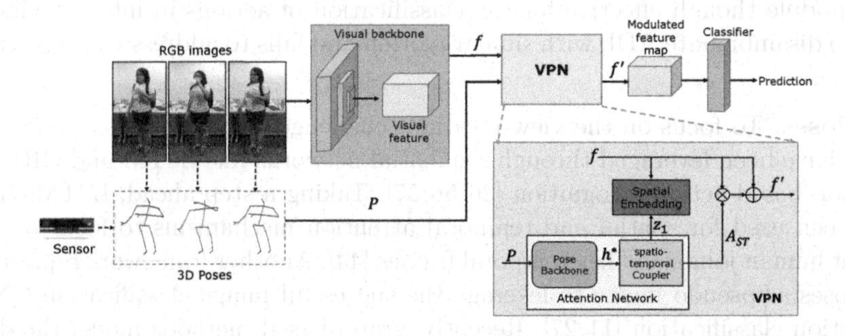

Fig. 3. Proposed Action Recognition Model: Our model takes as input RGB images with their corresponding 3D poses. The RGB images are processed by a visual backbone which generates a spatio-temporal feature map (f). The proposed **VPN** takes as input the feature map (f) and the 3D poses (P). VPN consists of two components: an attention network and a spatial embedding. The attention network further consists of a Pose Backbone and a spatio-temporal Coupler. VPN computes a modulated feature map f'. This modulated feature map f' is then used for classification.

3 Proposed Action Recognition Model

Our objective is to design an accurate spatial embedding of poses and visual content to better extract the discriminative spatio-temporal patterns. As shown in Fig. 3, the input of our proposed recognition model are the RGB images and their 3D poses. The 3D poses are either extracted from depth sensor or from RGB using LCRNet [37]. The proposed Video-Pose Network **VPN** takes as input the visual feature map and the 3D poses. Below, we discuss the action recognition model in details.

3.1 Video Representation

Taking as input a stack of human cropped images from a video clip, the spatio-temporal representation f is computed by a 3D convolutional network (the visual backbone in Fig. 3). f is a feature map of dimension $t_c \times m \times n \times c$, where t_c denotes the temporal dimension, $m \times n$ the spatial scale and c the channels. Then, the feature map f and the corresponding poses P are processed by the proposed network.

3.2 VPN

VPN can be thought as a layer which can be placed on top of any 3D convolutional backbone. VPN takes as input a 3D feature map (f) and its corresponding 3D poses (P) to perform two functionalities. First, to provide an accurate alignment of the human joints with the feature map f. Second, to compute a modulated feature map (f') which is further classified for action recognition. The modulated feature map (f') is weighted along space and time as per its relevance. VPN exploits the highly informative 3D pose information to transform the visual feature map f and finally, compute the attention weights. This network has two major components as shown in Fig. 4: (I) an attention network and (II) a spatial embedding. Though the intrinsic parameters of the attention network and the spatial embedding learns in parallel, we present these two components in the following order for better understanding.

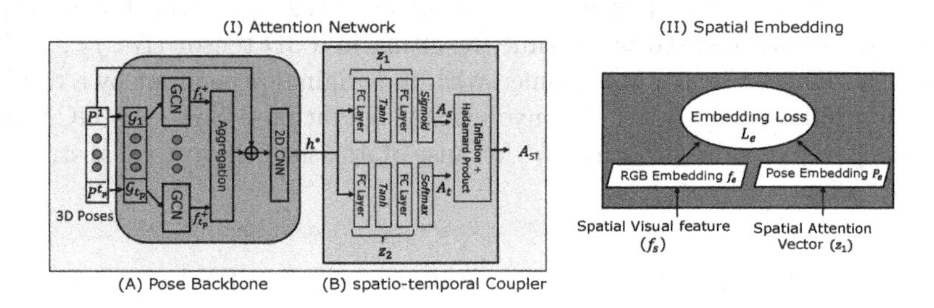

Fig. 4. The components in VPN: (I) Attention Network (left) and (II) Spatial Embedding (right). We present a zoom of the attention Network with: (A) a GCN Pose Backbone, and (B) a spatio-temporal Coupler to generate spatio-temporal attention weights A_{ST}.

(I) Attention Network. The attention network consists of a Pose Backbone and a spatio-temporal Coupler. Such a framework for pose driven attention network is unique compared to the other state-of-the-art methods using poses and RGB. The proposed attention network unlike [3,4,7,8] takes into account the human spatial configuration and it also learns coupled spatio-temporal attention weights for the visual feature map f.

Pose Backbone. The input poses along the video are processed in a Pose Backbone. The pose based input of VPN are the 3D human joint coordinates $P \in \mathbb{R}^{3 \times J \times t_p}$ stacked along t_p temporal dimension, where J is the number of skeleton joints. The Pose Backbone processes these 3D poses to compute pose features h^* which are used further in the attention network for computing the spatio-temporal attention weights. They carry meaningful information in a compact way, so the proposed attention network can efficiently focus on salient action parts.

For the Pose Backbone, we use **GCNs** to learn the spatial relationships between the 3D human joints to provide attention weights to the visual feature map (f). We aim at exploiting the graphical structure of the 3D poses. In Fig. 4(I), we illustrate our GCN pose backbone (marked (A)). For each pose input $P_t \in \mathbb{R}^{3 \times J}$ with J joints, we first construct a graph $\mathcal{G}_t(P_t, E)$ where E is the $J \times J$ weighted adjacency matrix:

$$e_{ij} = \begin{cases} 0, & \text{if } i = j \\ \alpha, & \text{if joint i and joint j are connected} \\ \beta, & \text{if joint i and joint j are disconnected} \end{cases}$$

Each graph $\mathcal{G}_t$ at time t is processed by a GCN to compute feature f_t^+:

$$f_t^+ = D^{-\frac{1}{2}}(E + I)D^{-\frac{1}{2}}\mathcal{G}_t W_t, \tag{1}$$

where W_t is the weight matrix and D is the diagonal degree matrix with $D_{ii} = \Sigma_j(E_{ij} + I_{ij})$ its diagonal elements. For all $t = 1, 2, ..., t_p$, the GCN output features f_t^+ are aggregated along time, resulting in a 3D tensor $[f_1^+, f_2^+, ..., f_{t_p}^+]$. Finally, the 3D pose tensor is combined with the original pose input by a residual connection followed by a set of convolutional operations. Now, the GCN pose backbone provides salient features h^* because of its use of the graphical structure of the 3D joints.

Spatio-Temporal Coupler. The attention network in VPN learns the spatio-temporal attention weights from the output of Pose Backbone in two steps as shown in Fig. 4(I)(B). In the first step, the spatial and temporal attention weights (A_S and A_T) are classically trained as in [44] to get the most important body part and key frames for an action. The output feature h^* of Pose Backbone follows two separate non-linear mapping functions to compute the spatial and temporal attention weights. These spatial A_S and temporal A_T weights are defined as

$$A_S = \sigma(z_1); \qquad A_T = softmax(z_2) \tag{2}$$

where $z_r = W_{z_r} tanh(W_{h_r} h^* + b_{h_r}) + b_{z_r}$ (for $r = 1, 2$) with subscripted W and b, the corresponding weights and biases are the latent spatial and temporal attention vectors. The dissociated attention weights A_S and A_T having dimension $m \times n$ and t_c respectively, can undergo a linear mapping to obtain spatially and temporally modulated feature maps. The resultant model is equivalent to the separable STA model [8]. In contrast, we propose to further perform a coupling of the spatial and temporal attention weights. Thus in the second step, joint spatio-temporal attention weights are computed by performing a Hadamard product on the spatial and temporal attention weights. In order to perform this matrix multiplication, the spatial and temporal attention weights are inflated by duplicating the same attention weights in temporal and spatial dimension respectively. Hence, the $m \times n \times t_c$ dimensional spatio-temporal attention weights A_{ST} are obtained by $A_{ST} = inflate(A_S) \circ inflate(A_T)$. This two-step attention learning process enables the attention network to compute spatio-temporal attention weights in which the spatial saliency varies with time. The obtained attention weights are crucial to disambiguate actions with similar appearance as they may have dissimilar motion over time.

Finally, the spatio-temporal attention weights A_{ST} are linearly multiplied with the input video feature map f, followed by a residual connection with the original feature map f to output the modulated feature map f'. The residual connection enables the network to retain the properties of the original visual features.

(II) Spatial Embedding of RGB and Pose. The objective of the embedding model is to provide tight correspondences between both pose and RGB modalities used in VPN. The state-of-the-art methods [7,8] attempt to provide the attention weights on the RGB feature map using 3D pose information without projecting them into the same 3D referential. The mapping with the pose is only done by cropping the person within the input RGB images. The spatial attention computed through the 3D joint coordinates does not correspond to the part of the image (no pixel to pixel correspondence), although it is crucial for recognizing fine-grained actions. To correlate both modalities, an embedding technique inspired from image captioning task [32,33] is used to build an accurate RGB-Pose embedding in order to enable the poses to represent the visual content of the actions (see Fig. 4(II)).

We assume that a low dimensional embedding exists for the global spatial representation of video feature map $f_s = \Sigma_{i=1}^{t_c} f(i,:,:,:)$ (a D_v dimensional vector) and its corresponding pose based latent spatial attention vector z_1 (a D_p dimensional vector). The mapping function can be derived from this embedding by

$$f_e = T_v f_s \quad and \quad P_e = T_p z_1, \tag{3}$$

where $T_v \in R^{D_e \times D_v}$ and $T_p \in R^{D_e \times D_p}$ are the transformation matrices that project the video content and the 3D poses into the common D_e dimensional embedding space. This mapping function is applied on the global spatial representation of the visual feature map and the pose based features in order to attain the aforementioned objective of the spatial embedding.

To measure the correspondence between the video content and the 3D poses, we compute the distance between their mappings in the embedding space. Thus, we define an embedding loss as a hypersphere feature metric space

$$L_e = ||\widehat{T_v f_s} - \widehat{T_p z_1}||_2^2 \qquad s.t. \quad ||T_v||_2 = ||T_p||_2 = 1 \tag{4}$$

$\widehat{T_v f_s} = \frac{T_v f_s}{||T_v f_s||_2}$ and $\widehat{T_p z_1} = \frac{T_p z_1}{||T_p z_1||_2}$ are the feature representations projected to the unit hypersphere. The norm constraint $||T_v||_2 = 1$ & $||T_p||_2 = 1$ simply prevents the trivial solution $\hat{T_v} = \hat{T_p} = 0$. This embedding loss along with the global classification loss provides a linear transformation on the RGB feature map that preserves the low-rank structure for the action representation and introduces a maximally separated features for different actions. Now, the kernels at the visual backbone are updated with a gradient proportional to $(f_e - P_e)$, which in turn transforms the visual feature map to learn pose aware characteristics. Consequently, we strengthen the correspondences between video and poses by minimizing the embedding loss. This embedding ensures that the pose information to be used for computing the spatial attention weights aligns with the content of the video.

Note that the embedding loss also provides feedback to the pose based latent spatial attention vectors (z_1), which in turn transfers knowledge from the 2D image space to pose 3D referential. This allows the attention network to provide better and meaningful spatial attention weights (A_s) compared to the attention network without the embedding. We will quantify this observation in the experiments.

3.3 Training Jointly the 3D ConvNet and VPN

VPN can be trained as a layer on top of any 3D ConvNet. The 3D ConvNet can be pre-trained for the action classification task for faster convergence. Finally, VPN is plugged into the 3D ConvNet for an end-to-end training with a regularized loss L formulated as

$$L = \lambda_1 L_C + (1 - \lambda_1) L_e + \lambda_2 L_a \tag{5}$$

Here, L_C is the cross-entropy loss, L_e is the embedding loss; the trade-off between these two losses is captured by linear fusion with a positive parameter λ_1; L_a is the attention regularizer with λ_2 weighting factor. The attention regularizer consists of the spatial and temporal attention weight regularizer and is formulated as

$$L_a = \sum_{j=1}^{m \times n} ||A_s(j)||_2 + \sum_{j=1}^{t_c} (1 - A_{t_c}(j))^2 \tag{6}$$

This additional regularization term L_a ensures that the attention weights are not biased to provide extremely high values to the parts of the spatio-temporal feature map with more relevance and completely neglecting the other parts.

4 Experiments

We evaluate the effectiveness of our model for action classification. We consider four public datasets which are the popular datasets for ADL: NTU-60 [39], NTU-120 [25], Toyota-Smarthome [8] and Northwestern-UCLA [51].

NTU RGB+D. (NTU-60 & NTU-120) NTU-60 is acquired with a Kinect v2 camera and consists of 56880 video samples with 60 activity classes. The activities were performed by 40 subjects and recorded from 80 viewpoints. For each frame, the dataset provides RGB, depth and a 25-joint skeleton of each subject in the frame. For evaluation, we follow the two protocols proposed in [39]: cross-subject (CS) and cross-view (CV). NTU-120 is a super-set of NTU-60 adding a lot of new similar actions. NTU-120 dataset contains 114k video clips of 106 distinct subjects performing 120 actions in a laboratory environment with 155 camera views. For evaluation, we follow a cross-subject (CS_1) protocol and a cross-setting (CS_2) protocol proposed in [25].

Toyota-Smarthome. (Smarthome) is a recent ADL dataset recorded in an apartment where 18 older subjects carry out tasks of daily living during a day. The dataset contains 16.1k video clips, 7 different camera views and 31 complex activities performed in a natural way without strong prior instructions. This dataset provides RGB data and 3D skeletons which are extracted from LCRNet [37]. For evaluation on this dataset, we follow cross-subject (CS) and cross-view (CV_1 and CV_2) protocols proposed in [8].

Northwestern-UCLA Multiview Activity 3D Dataset. (N-UCLA) is acquired simultaneously by three Kinect v1 cameras. The dataset consists of 1194 video samples with 10 activity classes. The activities were performed by 10 subjects, and recorded from three viewpoints. We performed experiments on N-UCLA using the cross-view (CV) protocol proposed in [51]: we trained our model on samples from two camera views and tested on the samples from the remaining view. For instance, the notation $V_{1,2}^{3}$ indicates that we trained on samples from view 1 and 2, and tested on samples from view 3.

The presence of ADL challenges like fine-grained and similar appearance activities is in higher magnitude in NTU-120 and Smarthome datasets. So, we perform all our ablation studies on these two datasets. We abbreviate Smarthome as SH in Tables 1, 2, 3 and 4.

4.1 Implementation Details

Training. In our experiments, the selected **visual backbone** is I3D [5] network pre-trained on ImageNet [9] and Kinetics-400 [17]. The visual backbone takes 64 video frames as input. The input of the **VPN** consists of the feature map extracted from `Mixed_5c` layer of I3D and the corresponding 3D poses.

The **pose backbone** takes as input a sequence of t_p 3D poses uniformly sampled from each clip. Hyper-parameter $t_p = 20, 20, 30$ and 5 for NTU-60, NTU-120, Smarthome and N-UCLA respectively. For the pose backbone, we use t_p number of GCNs, each processing a pose from the sequence. The weighting parameters α and β for computing the adjacency matrix of the pose based graph are set to 5 and 2 respectively. GCN projects the input joint coordinates to a 64$-$*dimensional* space. The output of the GCN is passed to a set of convolutional operations (see Fig. 4(I)(A)) which consists of three 2D convolutional layers each are followed by a Batch Normalization layer and a ReLU layer. The output channels of the convolutional layers are 64, 64 and 128.

For classification, a global-average pooling layer followed by a dropout [46] of 0.3 and a *softmax* layer are added at the end of the recognition model for class prediction. Our recognition model is trained with a 4-GPU machine where each GPU has 4 video clips in a mini-batch. Our model is trained for 30 epochs in total, with SGD optimizer having initial learning rate of 0.01 and decay rate of 0.1 after every 10 epochs. The trade off (λ_1) and regularizer (λ_2) parameters are set to 0.8 and 0.00001 respectively for all the experiments.

Inference. For the recognition model, we perform fully convolutional inference in space as in [54]. The final classification is obtained by max-pooling the softmax scores.

Table 1. Ablation study to show the effectiveness of each VPN component.

VPN components	NTU-120	NTU-120	SH	SH
	CS_1	CS_2	CS	CV_2
l_1: visual backbone	77.0	80.1	53.4	45.1
l_2: l_1 + attention network	85.4	86.9	56.4	50.5
l_3: l_2 + spatial embedding	**86.3**	**87.8**	**60.8**	**53.5**

Table 2. Performance of VPN with different choices of attention network.

Model	Pose backbone	Coupler	NTU-120	NTU-120	SH	SH
			CS_1	CS_2	CS	CV_2
l_4: VPN	LSTM	×	84.7	83.6	57.1	50.6
l_5: VPN	GCN	×	85.6	86.8	60.1	53.1
l_6: VPN	LSTM	✓	85.3	84.1	57.6	51.5
l_7: VPN	GCN	✓	**86.3**	**87.8**	**60.8**	**53.5**

Table 3. Performance of VPN with different embedding losses l_e.

Loss	NTU-120	NTU-120	SH	SH
	CS_1	CS_2	CS	CV_2
KL-divergence $D_{KL}(f_e\|\|P_e)$	85.5	87.1	57.2	50.9
KL-divergence $D_{KL}(P_e\|\|f_e)$	85.6	86.9	57.0	51.1
Bi-directional KL-divergence	86.1	87.2	57.2	51.7
Normalized Euclidean loss	**86.3**	**87.8**	**60.8**	**53.5**

Table 4. Impact of spatial embedding on spatial attention.

Model	Pose backbone	Spatial embedding	NTU-120	NTU-120	SH	SH
			CS_1	CS_2	CS	CV_2
VPN	LSTM	×	81.7	81.2	45.5	50.0
VPN	LSTM	✓	82.7	82.0	56.5	52.6
VPN	GCN	×	82.6	84.3	49.1	51.7
VPN	GCN	✓	**83.1**	**85.3**	**58.4**	**53.1**

4.2 Ablation Study

Our model includes two novel components, the spatial embedding and the attention network. Both of them are critical for good performance on ADL recognition. We show the importance of the attention network and the spatial embedding of VPN in Table 1. We also show the effectiveness of the spatial embedding with different instantiation of the attention network in Table 2.

How Effective is VPN? In order to answer this point, we show the action classification accuracy with baseline I3D (l_1) which is the visual backbone and then incorporate the VPN components: the attention network (l_2) and the spatial embedding (l_3) one-by-one in Table 1. The attention network (l_2) improves significantly the classification of the actions (upto 8.4% on NTU-120 and 5.4% on Smarthome) by providing spatio-temporal saliency to the I3D feature maps.

With the spatial embedding (l_3), the action classification further improves (upto 0.9% on NTU-120 and 4.4% on Smarthome).

Diagnosis of the Attention Network. In Table 2, we further illustrate the importance of each component in the attention network, i.e. the Pose Backbone and the spatio-temporal coupler. We have designed a baseline attention network with LSTM as pose backbone following [8]. We compare the LSTM pose backbone in l_4 and l_6 with our proposed GCN instantiation in l_5 and l_7. The attention network without a spatio-temporal coupler provides dissociated spatial and temporal attention weights in l_4 and l_5 in contrast to our proposed coupler in l_6 and l_7. Firstly, we observe that the GCN pose backbone makes use of the human joint topology, thus improves the classification accuracy in all scenarios with or without the coupler. Consequently, actions like *Snapping Finger* (+24.5%) and *Apply cream on face* (+23.9%) improves significantly with GCN instantiation (l_6) compared to LSTM (l_7). Secondly, we observe that the spatio-temporal coupler provides fine spatial attention weights for the most important frames in a video, which enables the model to disambiguate actions with similar appearance but dissimilar motion. Consequently, the coupler (l_7) improves the classification accuracy up to 1% on NTU-120 and 0.7% on Smarthome w.r.t. dissociating the attention weights (l_5). For instance, with dissociation of the attention weights, *rubbing two hands* was confused with *clapping* and *flicking hair* was confused with *putting on headphone.* With VPN, the coupler improves the classification accuracy of actions *rubbing two hands* and *flicking hair* by 25% and 19.6% respectively.

Which Loss is Better for Learning the Spatial Embedding? In this ablation study (Table 3), we compare different losses for projecting the 3D poses and RGB cues in a common semantic space. First, we compare the KL-divergence losses [15,19] ($D_{KL}(f_e||P_e)$ and $D_{KL}(P_e||f_e)$) from P_e to f_e and vice-versa. Then, we compare a bi-directional KL-divergence loss [29,52,58] ($D_{KL}(f_e||P_e) + D_{KL}(P_e||f_e)$) to our normalized euclidean loss. We observe that (i) the loss using $D_{KL}(f_e||P_e)$ and $D_{KL}(P_e||f_e)$ deteriorates the action classification accuracy as the feedback is in one direction either towards RGB or poses, implying two-way feedback for the visual features and the attention network is necessary, (ii) our normalized euclidean loss outperforms the bi-directional KL divergence loss, exhibit its superiority.

Impact of Embedding on Spatial attention. In Table 4, we show the impact of spatial embedding on the attention network providing spatial attention only. We perform the experiments with different choice of Pose Backbone, i.e. LSTM as discussed above and our proposed GCN. The spatial embedding provides a tight correspondence between the RGB data and poses. As a result, it boosts the classification accuracy in all the experiments. It is worth noting that the improvement is significant for Smarthome as it contains many fine-grained actions with videos captured by fixed cameras in an unconstrained Field of View. Thus, enforcing the embedding loss enhances the spatial precision during inference. As a result, the classification accuracy of fine-grained actions like *pouring water* (+77.7%), *pouring grains* (+76.1%) for making coffee, *cutting bread* (+50%), *pouring from kettle* (+42.8%) and *inserting teabag* (+35%) improves VPN with GCN pose backbone compared to its counterpart without embedding.

4.3 Qualitative Analysis

Figure 5(a) visualizes the activation of the human joints at the output of pose backbone (with GCNs) in VPN. The figure depicts the activations of the 3D joints. They are presented in a sequence of the human body topological order (follow first row of Fig. 5(a)) for convenient visualization. VPN is able to disambiguate actions with similar appearance like *hopping* and *jumping* due to high order activation at relevant joints of the human legs. The discriminative leg joints with high activation have been marked with a red bounding box in Fig. 5(a) (third row). Similarly, for actions like *put on headphone* with two hands and *flicking hair* with one hand, the blue bounding boxes demonstrate high activation of both the hand joints for the former action as compared to high activation of a single hand joints for the latter. For a very fine-grained action like *thumbs up*, the thumb joint is highly activated as compared to the other joints. This shows that the GCN pose backbone in VPN is a crucial ingredient for better action recognition.

In Fig. 5(b), for similar actions like *put on headphone* and *flicking hair*, along with salience precision of the VPN feature maps, the activations of their corresponding kernels show the discriminative power of VPN.

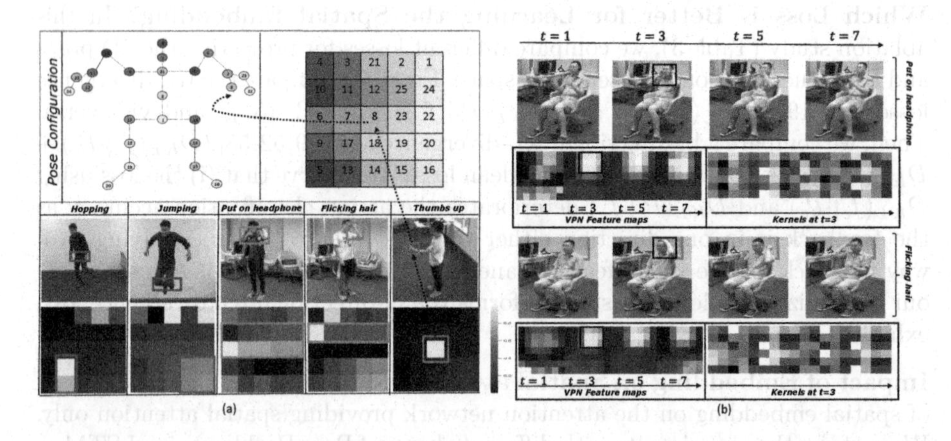

Fig. 5. (a) The heatmaps of the activations of the 3D joint coordinates (output of GCN) in the attention network of VPN. The area in the colored bounding boxes shows that different joints are activated for similar actions. (b) Visual feature maps of VPN across different time stamps for actions *put on headphone* and *flicking hair* on the left. Activated kernels corresponding to the **black** bounding box at $t = 3$ on the right.

In Fig. 6(a), we illustrate the performance of VPN w.r.t. I3D baseline for the dynamicity of an action along the videos. This dynamicity is computed by averaging the Procrustes distance [20] between subsequent 3D poses along the videos. If the average distance is large, it means the poses change a lot in an action. VPN significantly improves for actions with subtle motion like *hush* (+52.7%), *staple book* (+40.7%) and *reading* (+36.2%) which indicates the efficacy of VPN for fine-grained actions. The degradation of the VPN performance for high action dynamicity is negligible (−0.8%). In Fig. 6(b), we show the t-SNE plots of the feature spaces produced by I3D and VPN for some selected actions with similar appearance. It clearly shows the discriminative power of VPN for actions with similar appearance which is a frequent challenge in ADL.

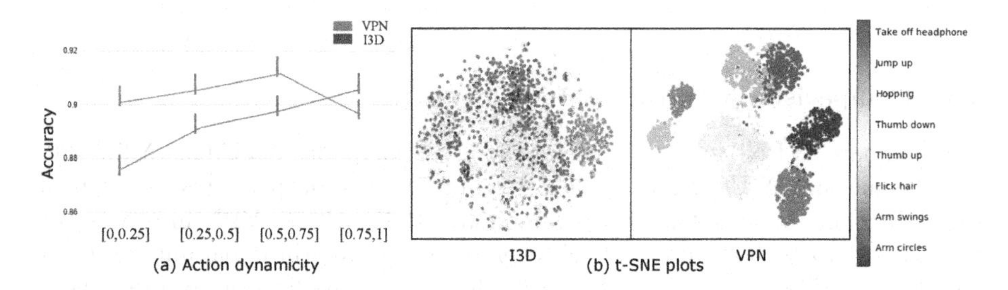

Fig. 6. (a) We compare our model against baseline I3D across action dynamicity. Our model significantly improves for most actions. (b) t-SNE plots of feature spaces produced by I3D and VPN for similar appearance actions.

Table 5. Results (accuracies in %) on NTU-60 with cross-subject (CS) and cross-view (CV) settings (at left) and NTU-120 with cross-subject (CS_1) and cross-setup (CS_2) settings (at right); Att indicates attention mechanism, ∘ indicates that the modality has only been used for training, the methods indicated with * are reproduced on this dataset. 3D ResNeXt-101 is abbreviated as RNX3D101.

Methods	Pose	RGB	Att	CS	CV
AGC-LSTM [42]	✓	×	✓	89.2	95.0
DGNN [40]	✓	×	×	89.9	96.1
STA-Hands [2]	✓	✓	✓	82.5	88.6
altered STA-Hands [3]	✓	✓	✓	84.8	90.6
Glimpse Cloud [4]	∘	✓	✓	86.6	93.2
PEM [28]	✓	✓	×	91.7	95.2
Separable STA [8]	✓	✓	✓	92.2	94.6
P-I3D [7]	✓	✓	✓	93	95.4
VPN	✓	✓	✓	**93.5**	**96.2**
VPN (RNX3D101)	✓	✓	✓	**95.5**	**98.0**

Methods	Pose	RGB	Att	CS_1	CS_2
ST-LSTM [26]	✓	×	✓	55.7	57.9
Two stream Att LSTM [24]	✓	×	✓	61.2	63.3
Multi-Task CNN [18]	✓	×	×	62.2	61.8
PEM [28]	✓	×	✓	64.6	66.9
2s-AGCN [41]	✓	×	✓	82.9	84.9
Two-streams [43]	×	✓	×	58.5	54.8
I3D* [5]	×	✓	×	77.0	80.1
Two-streams + ST-LSTM [25]	✓	✓	×	61.2	63.1
Separable STA* [8]	✓	✓	✓	83.8	82.5
VPN	✓	✓	✓	**86.3**	**87.8**

Table 6. Results on smarthome dataset with cross-subject (CS) and cross-view (CV_1 and CV_2) settings (accuracies in %). Att indicates attention mechanism.

Methods	Pose	RGB	Att	CS	CV_1	CV_2
DT [49]	×	✓	×	41.9	20.9	23.7
LSTM [31]	✓	×	×	42.5	13.4	17.2
I3D [5]	×	✓	×	53.4	34.9	45.1
I3D+NL [54]	×	✓	✓	53.6	34.3	43.9
P-I3D [7]	✓	✓	✓	54.2	35.1	50.3
Separable STA [8]	✓	✓	✓	54.2	35.2	50.3
VPN	✓	✓	✓	**60.8**	**43.8**	**53.5**

Table 7. Results on N-UCLA dataset with cross-view $V_{1,2}^3$ settings (accuracies in %); $\overline{Pose}$ indicate its usage only in the training phase.

Methods	Data	Att	$V_{1,2}^3$
HPM+TM [36]	Depth	×	91.9
Ensemble TS-LSTM [22]	Pose	×	89.2
NKTM [35]	RGB	×	85.6
Glimpse Cloud [4]	RGB+ $\overline{Pose}$	✓	90.1
Separable STA [8]	RGB+Pose	✓	92.4
P-I3D [7]	RGB+Pose	✓	93.1
VPN	RGB+Pose	✓	**93.5**

4.4 Comparison with the State-of-the-art

We compare VPN to the state-of-the-art (SoA) on NTU-60, NTU-120, Smarthome and N-UCLA in table 5, 6 and 7. VPN outperforms on each of them. In Table 5 (at left), for input modality RGB+Poses, VPN improves the SoA [7] by up to 0.8% on NTU-60 even by using one-third parameters compared to [7]. The SoA using Poses only [40] yields classification accuracy near to VPN for cross-view protocol (with 0.1% difference) due to their robustness to view changes. However, the lack of appearance information restricts these methods [40,42] to disambiguate actions with similar visual appearance, thus resulting in lower accuracy for cross-subject protocol. We have also tested VPN with 3D ResNeXt-101 [14] on NTU-60 dataset. The results in Table 5 show that VPN can be adapted with other existing video backbones.

Compared to the SoA results, the improvement by 3.9% and 4.9% (averaging over the protocols) on NTU-120 and Smarthome respectively are significant. It is worth noting that VPN improves further the classification of actions with similar appearance as compared to Separable STA [8]. For example, actions like *clapping* (+44.3%) and *flicking hair* (+19.1%) are now discriminated with better accuracy. In addition, the superior performance of VPN in cross-view protocol for both NTU-120 and Smarthome implies that it provides better view-adaptive characterization compared to all the prior methods.

For N-UCLA which is a small-scale dataset, we pre-train the visual backbone with NTU-60 for a fair comparison with [4,7,8]. We also outperform the SoA [7] by 0.4% on this dataset.

5 Conclusion

This paper addresses the challenges of ADL classification. We have proposed a novel Video-Pose Network **VPN** which provides an accurate video-pose embedding. We show that the embedding along with attention network yields a more discriminative feature map for action classification. The attention network leverages the topology of the human joints and with the coupler provides precise spatio-temporal attention weights along the video.

Our recognition model outperforms the state of-the-art results for action classification on 4 public datasets. This is a first step towards combining RGB and Pose through an explicit embedding. A future perspective of this work is to exploit this embedding even in case of noisy 3D poses in order to also boost action recognition for internet videos. This embedding could even help to refine these noisy 3D poses in a weakly supervised manner.

References

1. Arandjelović, R., Gronat, P., Torii, A., Pajdla, T., Sivic, J.: NetVLAD: CNN architecture for weakly supervised place recognition. In: IEEE Conference on Computer Vision and Pattern Recognition (2016)
2. Baradel, F., Wolf, C., Mille, J.: Human action recognition: pose-based attention draws focus to hands. In: 2017 IEEE International Conference on Computer Vision Workshops (ICCVW), pp. 604–613, October 2017. https://doi.org/10.1109/ICCVW.2017.77
3. Baradel, F., Wolf, C., Mille, J.: Human activity recognition with pose-driven attention to RGB. In: The British Machine Vision Conference (BMVC), September 2018
4. Baradel, F., Wolf, C., Mille, J., Taylor, G.W.: Glimpse clouds: human activity recognition from unstructured feature points. In: The IEEE Conference on Computer Vision and Pattern Recognition (CVPR), June 2018
5. Carreira, J., Zisserman, A.: Quo vadis, action recognition? A new model and the kinetics dataset. In: 2017 IEEE Conference on Computer Vision and Pattern Recognition (CVPR), pp. 4724–4733. IEEE (2017)
6. Crasto, N., Weinzaepfel, P., Alahari, K., Schmid, C.: MARS: motion-augmented RGB stream for action recognition. In: CVPR (2019)
7. Das, S., Chaudhary, A., Bremond, F., Thonnat, M.: Where to focus on for human action recognition? In: 2019 IEEE Winter Conference on Applications of Computer Vision (WACV), pp. 71–80, January 2019. https://doi.org/10.1109/WACV.2019.00015
8. Das, S., et al.: Toyota smarthome: real-world activities of daily living. In: ICCV (2019)
9. Deng, J., Dong, W., Socher, R., Li, L.J., Li, K., Fei-Fei, L.: ImageNet: a large-scale hierarchical image database. In: CVPR (2009)
10. Donahue, J., et al.: Long-term recurrent convolutional networks for visual recognition and description. In: The IEEE Conference on Computer Vision and Pattern Recognition (CVPR), June 2015
11. Du, Y., Fu, Y., Wang, L.: Skeleton based action recognition with convolutional neural network. In: 2015 3rd IAPR Asian Conference on Pattern Recognition (ACPR), pp. 579–583, November 2015. https://doi.org/10.1109/ACPR.2015.7486569
12. Feichtenhofer, C., Fan, H., Malik, J., He, K.: Slowfast networks for video recognition. In: The IEEE International Conference on Computer Vision (ICCV), October 2019
13. Girdhar, R., Carreira, J., Doersch, C., Zisserman, A.: Video action transformer network. CoRR abs/1812.02707 (2018). arxiv.org/abs/1812.02707

14. Hara, K., Kataoka, H., Satoh, Y.: Can spatiotemporal 3D CNNs retrace the history of 2D CNNs and ImageNet? In: The IEEE Conference on Computer Vision and Pattern Recognition (CVPR), June 2018
15. Hinton, G., Vinyals, O., Dean, J.: Distilling the knowledge in a neural network (2015)
16. Hussein, N., Gavves, E., Smeulders, A.W.M.: Timeception for complex action recognition. CoRR abs/1812.01289 (2018). http://arxiv.org/abs/1812.01289
17. Kay, W., et al.: The kinetics human action video dataset. arXiv preprint arXiv:1705.06950 (2017)
18. Ke, Q., Bennamoun, M., An, S., Sohel, F., Boussaid, F.: Learning clip representations for skeleton-based 3D action recognition. IEEE Trans. Image Process. **27**(6), 2842–2855 (2018). https://doi.org/10.1109/TIP.2018.2812099
19. Kim, S., Seltzer, M., Li, J., Zhao, R.: Improved training for online end-to-end speech recognition systems. In: Proceedings of Interspeech 2018, pp. 2913–2917 (2018). https://doi.org/10.21437/Interspeech.2018-2517. http://dx.doi.org/10.21437/Interspeech.2018-2517
20. Krzanowski, W.J.: Principles of Multivariate Analysis: A User's Perspective. Oxford University Press Inc., USA (1988)
21. Kuehne, H., Jhuang, H., Garrote, E., Poggio, T., Serre, T.: HMDB: a large video database for human motion recognition. In: 2011 International Conference on Computer Vision, pp. 2556–2563. IEEE (2011)
22. Lee, I., Kim, D., Kang, S., Lee, S.: Ensemble deep learning for skeleton-based action recognition using temporal sliding LSTM networks. In: Proceedings of the IEEE International Conference on Computer Vision (2017)
23. Liu, G., Qian, J., Wen, F., Zhu, X., Ying, R., Liu, P.: Action recognition based on 3D skeleton and RGB frame fusion. In: 2019 IEEE/RSJ International Conference on Intelligent Robots and Systems (IROS), pp. 258–264 (2019). https://doi.org/10.1109/IROS40897.2019.8967570
24. Liu, J., Wang, G., Hu, P., Duan, L., Kot, A.C.: Global context-aware attention LSTM networks for 3D action recognition. In: 2017 IEEE Conference on Computer Vision and Pattern Recognition (CVPR), pp. 3671–3680, July 2017. https://doi.org/10.1109/CVPR.2017.391
25. Liu, J., Shahroudy, A., Perez, M., Wang, G., Duan, L.Y., Kot, A.C.: NTU RGB+D 120: a large-scale benchmark for 3D human activity understanding. IEEE Trans. Pattern Anal. Mach. Intell. (2019). https://doi.org/10.1109/TPAMI.2019.2916873
26. Liu, J., Shahroudy, A., Xu, D., Wang, G.: Spatio-temporal LSTM with trust gates for 3D human action recognition. In: Leibe, B., Matas, J., Sebe, N., Welling, M. (eds.) ECCV 2016. LNCS, vol. 9907, pp. 816–833. Springer, Cham (2016). https://doi.org/10.1007/978-3-319-46487-9_50
27. Liu, M., Liu, H., Chen, C.: Enhanced skeleton visualization for view invariant human action recognition. Pattern Recogn. **68**, 346–362 (2017). https://doi.org/10.1016/j.patcog.2017.02.030. http://www.sciencedirect.com/science/article/pii/S0031320317300936
28. Liu, M., Yuan, J.: Recognizing human actions as the evolution of pose estimation maps. In: The IEEE Conference on Computer Vision and Pattern Recognition (CVPR), June 2018
29. Liu, Y., Guo, Y., Bakker, E.M., Lew, M.S.: Learning a recurrent residual fusion network for multimodal matching. In: 2017 IEEE International Conference on Computer Vision (ICCV), pp. 4127–4136, October 2017. https://doi.org/10.1109/ICCV.2017.442

30. Luo, Z., Hsieh, J.T., Jiang, L., Carlos Niebles, J., Fei-Fei, L.: Graph distillation for action detection with privileged modalities. In: The European Conference on Computer Vision (ECCV), September 2018
31. Mahasseni, B., Todorovic, S.: Regularizing long short term memory with 3D human-skeleton sequences for action recognition. In: Proceedings of the IEEE Conference on Computer Vision and Pattern Recognition, pp. 3054–3062 (2016)
32. Miech, A., Laptev, I., Sivic, J.: Learning a text-video embedding from incomplete and heterogeneous data. CoRR abs/1804.02516 (2018). http://arxiv.org/abs/1804.02516
33. Pan, Y., Mei, T., Yao, T., Li, H., Rui, Y.: Jointly modeling embedding and translation to bridge video and language. In: The IEEE Conference on Computer Vision and Pattern Recognition (CVPR), June 2016
34. Perronnin, F., Sánchez, J., Mensink, T.: Improving the fisher kernel for large-scale image classification. In: Daniilidis, K., Maragos, P., Paragios, N. (eds.) ECCV 2010. LNCS, vol. 6314, pp. 143–156. Springer, Heidelberg (2010). https://doi.org/10.1007/978-3-642-15561-1_11
35. Rahmani, H., Mian, A.: Learning a non-linear knowledge transfer model for cross-view action recognition. In: 2015 IEEE Conference on Computer Vision and Pattern Recognition (CVPR), pp. 2458–2466, June 2015. https://doi.org/10.1109/CVPR.2015.7298860
36. Rahmani, H., Mian, A.: 3D action recognition from novel viewpoints. In: 2016 IEEE Conference on Computer Vision and Pattern Recognition (CVPR), pp. 1506–1515, June 2016. https://doi.org/10.1109/CVPR.2016.167
37. Rogez, G., Weinzaepfel, P., Schmid, C.: LCR-Net++: multi-person 2D and 3D pose detection in natural images. IEEE Trans. Pattern Anal. Mach. Intell. (2019). https://doi.org/10.1109/TPAMI.2019.2916873
38. Shahroudy, A., Wang, G., Ng, T.: Multi-modal feature fusion for action recognition in RGB-D sequences. In: 2014 6th International Symposium on Communications, Control and Signal Processing (ISCCSP), pp. 1–4, May 2014. https://doi.org/10.1109/ISCCSP.2014.6877819
39. Shahroudy, A., Liu, J., Ng, T.T., Wang, G.: NTU RGB+D: a large scale dataset for 3D human activity analysis. In: The IEEE Conference on Computer Vision and Pattern Recognition (CVPR), June 2016
40. Shi, L., Zhang, Y., Cheng, J., Lu, H.: Skeleton-based action recognition with directed graph neural networks. In: The IEEE Conference on Computer Vision and Pattern Recognition (CVPR), June 2019
41. Shi, L., Zhang, Y., Cheng, J., Lu, H.: Two-stream adaptive graph convolutional networks for skeleton-based action recognition. In: CVPR (2019)
42. Si, C., Chen, W., Wang, W., Wang, L., Tan, T.: An attention enhanced graph convolutional LSTM network for skeleton-based action recognition. In: 2019 IEEE/CVF Conference on Computer Vision and Pattern Recognition (CVPR), pp. 1227–1236, June 2019. https://doi.org/10.1109/CVPR.2019.00132
43. Simonyan, K., Zisserman, A.: Two-stream convolutional networks for action recognition in videos. In: Advances in Neural Information Processing Systems, pp. 568–576 (2014)
44. Song, S., Lan, C., Xing, J., Zeng, W., Liu, J.: An end-to-end spatio-temporal attention model for human action recognition from skeleton data. In: AAAI Conference on Artificial Intelligence, pp. 4263–4270 (2017)
45. Soomro, K., Roshan Zamir, A., Shah, M.: UCF101: a dataset of 101 human actions classes from videos in the wild (2012)

46. Srivastava, N., Hinton, G., Krizhevsky, A., Sutskever, I., Salakhutdinov, R.: Dropout: A simple way to prevent neural networks from overfitting. J. Mach. Learn. Res. **15**(1), 1929–1958 (2014). http://dl.acm.org/citation.cfm?id=2627435.2670313
47. Tang, Y., Tian, Y., Lu, J., Li, P., Zhou, J.: Deep progressive reinforcement learning for skeleton-based action recognition. In: 2018 IEEE/CVF Conference on Computer Vision and Pattern Recognition, pp. 5323–5332, June 2018. https://doi.org/10.1109/CVPR.2018.00558
48. Tran, D., Bourdev, L., Fergus, R., Torresani, L., Paluri, M.: Learning spatiotemporal features with 3D convolutional networks. In: Proceedings of the 2015 IEEE International Conference on Computer Vision (ICCV), ICCV 2015, pp. 4489–4497. IEEE Computer Society, Washington, DC (2015). https://doi.org/10.1109/ICCV.2015.510
49. Wang, H., Kläser, A., Schmid, C., Liu, C.L.: Action recognition by dense trajectories. In: IEEE Conference on Computer Vision & Pattern Recognition, pp. 3169–3176. Colorado Springs, USA, June 2011. http://hal.inria.fr/inria-00583818/en
50. Wang, H., Schmid, C.: Action recognition with improved trajectories. In: IEEE International Conference on Computer Vision, Sydney, Australia (2013). http://hal.inria.fr/hal-00873267
51. Wang, J., Nie, X., Xia, Y., Wu, Y., Zhu, S.C.: Cross-view action modeling, learning, and recognition. In: 2014 IEEE Conference on Computer Vision and Pattern Recognition, pp. 2649–2656, June 2014. https://doi.org/10.1109/CVPR.2014.339
52. Wang, L., Li, Y., Lazebnik, S.: Learning deep structure-preserving image-text embeddings. In: 2016 IEEE Conference on Computer Vision and Pattern Recognition (CVPR), pp. 5005–5013, June 2016. https://doi.org/10.1109/CVPR.2016.541
53. Wang, L., et al.: Temporal segment networks: towards good practices for deep action recognition. In: Leibe, B., Matas, J., Sebe, N., Welling, M. (eds.) ECCV 2016. LNCS, vol. 9912, pp. 20–36. Springer, Cham (2016). https://doi.org/10.1007/978-3-319-46484-8_2
54. Wang, X., Girshick, R.B., Gupta, A., He, K.: Non-local neural networks. In: 2018 IEEE/CVF Conference on Computer Vision and Pattern Recognition, pp. 7794–7803 (2018)
55. Yan, S., Xiong, Y., Lin, D.: Spatial temporal graph convolutional networks for skeleton-based action recognition. In: AAAI (2018)
56. Zhang, P., Lan, C., Xing, J., Zeng, W., Xue, J., Zheng, N.: View adaptive recurrent neural networks for high performance human action recognition from skeleton data. In: The IEEE International Conference on Computer Vision (ICCV), October 2017
57. Zhang, S., Liu, X., Xiao, J.: On geometric features for skeleton-based action recognition using multilayer LSTM networks. In: 2017 IEEE Winter Conference on Applications of Computer Vision (WACV), pp. 148–157, March 2017. https://doi.org/10.1109/WACV.2017.24
58. Zhang, Y., Lu, H.: Deep cross-modal projection learning for image-text matching. In: ECCV (2018)
59. Zhao, J., Snoek, C.G.M.: Dance with flow: two-in-one stream action detection. In: The IEEE Conference on Computer Vision and Pattern Recognition (CVPR), June 2019

Soft Anchor-Point Object Detection

Chenchen Zhu(✉), Fangyi Chen, Zhiqiang Shen, and Marios Savvides

Carnegie Mellon University, Pittsburgh, PA 15213, USA
{chenchez,fangyic,zhiqians,marioss}@andrew.cmu.edu

Abstract. Recently, anchor-free detection methods have been through great progress. The major two families, anchor-point detection and key-point detection, are at opposite edges of the speed-accuracy trade-off, with anchor-point detectors having the speed advantage. In this work, we boost the performance of the anchor-point detector over the key-point counterparts while maintaining the speed advantage. To achieve this, we formulate the detection problem from the anchor point's perspective and identify ineffective training as the main problem. Our key insight is that anchor points should be optimized jointly as a group both within and across feature pyramid levels. We propose a simple yet effective training strategy with soft-weighted anchor points and soft-selected pyramid levels to address the false attention issue within each pyramid level and the feature selection issue across all the pyramid levels, respectively. To evaluate the effectiveness, we train a single-stage anchor-free detector called Soft Anchor-Point Detector (SAPD). Experiments show that our concise SAPD pushes the envelope of speed/accuracy trade-off to a new level, outperforming recent state-of-the-art anchor-free and anchor-based detectors. Without bells and whistles, our best model can achieve a single-model single-scale AP of 47.4% on COCO.

Keywords: Object detection · Anchor-point detector · Soft-weighted anchor points · Soft-selected pyramid levels

1 Introduction

Recently, anchor-free object detectors have drawn a lot of attention [6,11,12, 27–29,35–37]. They don't rely on anchor boxes. Predictions are generated in a point(s)-to-box style. Compared to conventional anchor-based approaches [1,3, 4,9,14,15,18,22–24,31], anchor-free detectors have a few advantages in general: 1) no manual tuning of hyperparameters for the anchor configuration; 2) usually simpler architecture of detection head; 3) less training memory cost.

The anchor-free detectors can be roughly divided into two categories, i.e. anchor-point detection and key-point detection. Anchor-point detectors, such as [10,11,27–30,37], encode and decode object bounding boxes as anchor points

Electronic supplementary material The online version of this chapter (https://doi.org/10.1007/978-3-030-58545-7_6) contains supplementary material, which is available to authorized users.

© Springer Nature Switzerland AG 2020
A. Vedaldi et al. (Eds.): ECCV 2020, LNCS 12354, pp. 91–107, 2020.
https://doi.org/10.1007/978-3-030-58545-7_6

with corresponding point-to-boundary distances, where the anchor points are the pixels on the pyramidal feature maps and they are associated with the features at their locations just like the anchor boxes. Key-point detectors, such as [6,12,36], predict the locations of key points of the bounding box, e.g. corners, center, or extreme points, using a high-resolution feature map and repeated bottom-up top-down inference [20], and group those key points to form a box. Compared to key-point detectors, anchor-point detectors have several advantages: 1) simpler network architecture; 2) faster training and inference speed; 3) potential to benefit from augmentations on feature pyramids [21,26,33]; 4) flexible feature level selection. However, they cannot be as accurate as key-point-based methods under the same image scale of testing.

A natural question to ask is: what hinders a simple anchor-point detector from achieving similar accuracy as key-point detectors? In this work, we push the envelope further: we present Soft Anchor-Point Detector (SAPD), a concise single-stage anchor-point detector with both faster speed and higher accuracy. To achieve this, we formulate the detection problem from the anchor point's perspective and identify ineffective training as the major obstacle impeding anchor-point detector from exploring more potentials of network power both within and across the feature pyramid levels. Specifically, the conventional training strategy has two overlooked issues, i.e. false attention within each pyramid level and feature selection across all pyramid levels. For anchor points on the same pyramid level, those receiving false attention in training will generate detections with unnecessarily high confidence scores but poor localization during inference, suppressing some anchor points with accurate localization but lower score. This can confuse the post-processing step since high-score detections usually have priority to be kept over the low-score ones in non-maximum suppression, resulting in low AP scores at strict IoU thresholds. For anchor points at the same spatial location across different pyramid levels, their associated features are similar but how much they contribute to the network loss is decided without careful consideration. Current methods make the selection based on ad-hoc heuristics like instance scale and usually limited to a single level per instance. This causes a waste of unselected features.

These issues motivate us to propose a novel training strategy with two softened optimization techniques, i.e. soft-weighted anchor points and soft-selected pyramid levels. For anchor points on the same pyramid level, we reduce the false attention by reweighting their contributions to the network loss according to their geometrical relation with the instance box. We argue that the more close to the instance boundaries, the harder for anchor points to localize objects precisely due to feature misalignment, the less they should contribute to the network loss. Additionally, we further reweight an anchor point by the instance-dependent "participation" degree of its pyramid level. We implement a light-weight feature selection network to learn the per-level "participation" degrees given the object instances. The feature selection network is jointly optimized with the detector and not involved in detector inference.

Comprehensive experiments show that the proposed training strategy consistently improves the baseline FSAF [37] module by a large margin *without inference slowdown*, e.g. 2.1% AP increase on COCO [16] detection benchmark with

ResNet-50 [8]. The improvements are robust and insensitive to specific hyper-parameters and implementations, including advanced feature pyramid designs. With Balanced Feature Pyramid [21], our complete detector achieves the best speed-accuracy balance among recent state-of-the-art anchor-free detectors, see Fig. 1. We report single-model single-scale speed/accuracy of SAPD with different backbones, and with or without DCN [4]. The fast variant without DCN outperforms the best key-point detector, CenterNet [6] (45.4% vs. 44.9%) while running about 2× times faster. The accurate variant with DCN forms an upper bound of speed/accuracy trade-offs for recent single-stage and multi-stage detectors, surpassing the accurate TridentNet [13] (47.4% vs. 46.8%) and being more than 3× faster.

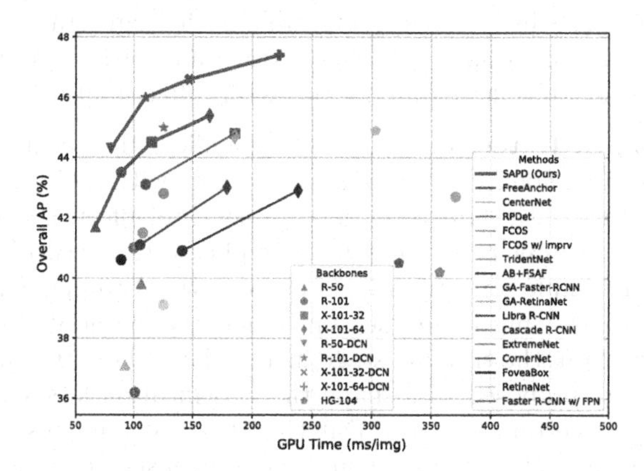

Fig. 1. Single-model single-scale speed (ms) vs. accuracy (AP) on COCO `test-dev`. We show variants of our SAPD with and without DCN [4]. Without DCN, our fastest version can run up to 5× faster than other methods with comparable accuracy. With DCN, our SAPD forms an upper envelop of all recent detectors.

2 Related Work

Anchor-Free Detectors. Despite the dominance of anchor-based methods, anchor-free detectors are continuously under development. Earlier works like DenseBox [10] and UnitBox [30] explore an alternate direction of region proposal. And it has been used in tasks such as scene text detection [34] and pedestrian detection [19]. Recent efforts have pushed the anchor-free detection outperforming the anchor-based counterparts. Most of them are single-stage detectors. For instance, CornerNet [12], ExtremeNet [36] and CenterNet [6] reformulate the detection problem as locating several key points of the bounding boxes. FSAF [37], Guided Anchoring [28], FCOS [27] and FoveaBox [11] encode and decode the bounding boxes as anchor points and point-to-boundary distances. Anchor-free methods can also be in the form of two-stage detectors, such as Guided Anchoring [28] and RPDet [29].

Feature Selection in Detection. Modern object detectors often construct the feature pyramid to alleviate the scale variation problem. With multiple levels in the feature pyramid, selecting the suitable feature level for each instance is a crucial problem. Anchor-based methods make the implicit selection by the anchor matching mechanism, which is based on ad-hoc heuristics like scales and aspect ratios. Similarly, most anchor-free approaches [11,27–29] assign the instances according to scale. The FSAF module [37], on the other hand, makes the assignment by choosing the pyramid level with the minimal instance-dependent loss in a dynamic style during training but limited to one level per instance. In two-stage detectors, some methods consider feature selection in the second stage by feature fusion. PANet [17] proposed the adaptive feature pooling with the element-wise maximize operation. But this requires the input of region proposals for both training and testing, which is not compatible with single-stage methods. Our soft feature selection is designed for single-stage anchor-free methods and can dynamically choose multiple pyramid levels with differentiation.

Soft Weighting in Detection. FCOS [27] predicts the "center-ness" masks and multiplies the confidence scores of anchor points with the masks. But this is in the inference stage so the false attention problem still affects the network training and the extra "center-ness" branches complicate the network architecture. We show that our simple soft-weighting scheme during training is more effective than the "center-ness" masks in the supplementary material. Previous works doing soft weighting in the training stage include Focal Loss [15] and Consistent Loss [25]. They are proposed to reshape the classification loss so that the network down-weights certain samples. But they treat all samples independently. Our training strategy is more direct and comprehensive since we reshape the combination of classification and localization losses and consider jointly weighting a group of anchor points spreading both within and across feature pyramid levels.

3 Soft Anchor-Point Detector

In this section, we present our Soft Anchor-Point Detector (SAPD). First, we formulate the detection problem from the anchor point's perspective in the setting of a vanilla anchor-point detector with a simple head architecture (Sect. 3.1). Then we introduce our novel training strategy (Fig. 3) including soft-weighted anchor points (Sect. 3.2) and soft-selected pyramid levels (Sect. 3.3) to address the false attention within pyramid level and feature selection across pyramid levels respectively.

3.1 Detection Formulation with Anchor Points

The first anchor-point detector can be traced back to DenseBox [10]. The recent modern anchor-point detectors are more or less attaching the detection head of DenseBox with additional convolution layers to multiple levels in the feature pyramids. Here we introduce the general concept of a representative in terms of network architecture, supervision targets, and loss functions.

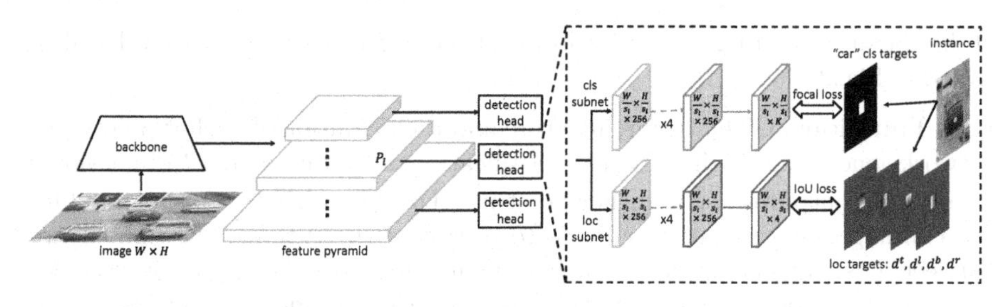

Fig. 2. The network architecture of a vanilla anchor-point detector with a simple detection head.

Network Architecture. As shown in Fig. 2, the network consists of a backbone, a feature pyramid, and one detection head per pyramid level, in a fully convolutional style. A pyramid level is denoted as P_l where l indicates the level number and it has $1/s_l$ resolution of the input image size $W \times H$. s_l is the feature stride and $s_l = 2^l$. A typical range of l is 3 to 7. A detection head has two task-specific subnets, i.e. classification subnet and localization subnet. They both have five 3×3 conv layers. The classification subnet predicts the probability of objects at each anchor point location for each of the K object classes. The localization subnet predicts the 4-dimensional class-agnostic distance from each anchor point to the boundaries of a nearby instance if the anchor point is positive (defined next).

Supervision Targets. We first define the concept of anchor points. An anchor point p_{lij} is a pixel on the pyramid level P_l located at (i, j) with $i = 0, 1, \ldots, W/s_l - 1$ and $j = 0, 1, \ldots, H/s_l - 1$. Each p_{lij} has a corresponding image space location (X_{lij}, Y_{lij}) where $X_{lij} = s_l(i + 0.5)$ and $Y_{lij} = s_l(j + 0.5)$. Next we define the valid box B_v of a ground-truth instance box $B = (c, x, y, w, h)$ where c is the class id, (x, y) is the box center, and w, h are box width and height respectively. B_v is a central shrunk box of B, i.e. $B_v = (c, x, y, \epsilon w, \epsilon h)$, where ϵ is the shrunk factor. An anchor point p_{lij} is positive if and only if some instance B is assigned to P_l and the image space location (X_{lij}, Y_{lij}) of p_{lij} is inside B_v, otherwise it is a negative anchor point. For a positive anchor point, its classification target is c and localization targets are calculated as the normalized distances $\mathbf{d} = (d^l, d^t, d^r, d^b)$ from the anchor point to the left, top, right, bottom boundaries of B respectively (1),

$$
\begin{aligned}
d^l &= \frac{1}{zs_l}[X_{lij} - (x - w/2)] \quad d^t = \frac{1}{zs_l}[Y_{lij} - (y - h/2)] \\
d^r &= \frac{1}{zs_l}[(x + w/2) - X_{lij}] \quad d^b = \frac{1}{zs_l}[(y + h/2) - Y_{lij}]
\end{aligned}
\tag{1}
$$

where z is the normalization scalar. For negative anchor points, their classification targets are background ($c = 0$), and localization targets are set to null because they don't need to be learned. To this end, we have a classification target c_{lij} and a localization target $\mathbf{d}_{lij}$ for all of each anchor point p_{lij}. A visualization of the classification targets and the localization targets of one feature level is illustrated in Fig. 2.

Loss Functions. Given the architecture and the definition of anchor points, the network generates a K-dimensional classification output $\hat{\mathbf{c}}_{lij}$ and a 4-dimensional localization output $\hat{\mathbf{d}}_{lij}$ per anchor point p_{lij}. Focal loss [15] (l_{FL}) is adopted for the training of classification subnets to overcome the extreme class imbalance between positive and negative anchor points. IoU loss [30] (l_{IoU}) is used for the training of localization subnets. Therefore, the per anchor point loss L_{lij} is calculated as Eq. (2).

$$L_{lij} = \begin{cases} l_{FL}(\hat{\mathbf{c}}_{lij}, c_{lij}) + l_{IoU}(\hat{\mathbf{d}}_{lij}, \mathbf{d}_{lij}), & p_{lij} \in p^+ \\ l_{FL}(\hat{\mathbf{c}}_{lij}, c_{lij}), & p_{lij} \in p^- \end{cases} \tag{2}$$

where p^+ and p^- are the sets of positive and negative anchor points respectively. The loss for the whole network is the summation of all anchor point losses divided by the number of positive anchor points (3).

$$L = \frac{1}{N_{p^+}} \sum_{l} \sum_{ij} L_{lij} \tag{3}$$

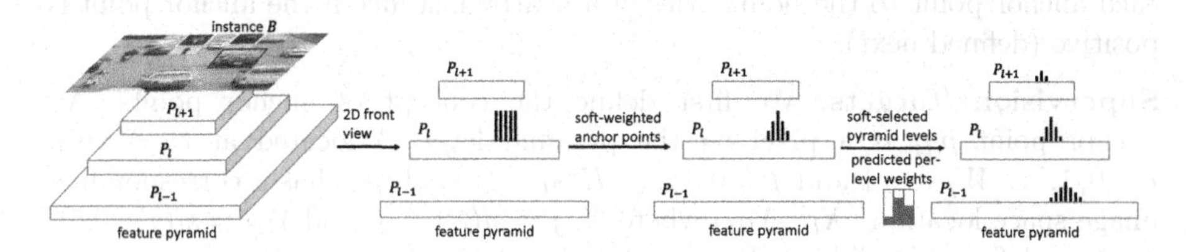

Fig. 3. Illustrative overview of our training strategy with soft-weighted anchor points and soft-selected pyramid levels. The black bars indicate the assigned weights of positive anchor points' contribution to the network loss. The key insight is the joint optimization of anchor points as a group both within and across feature pyramid levels.

3.2 Soft-Weighted Anchor Points

False Attention. Under the conventional training strategy, we observe that during inference some anchor points generate detection boxes with poor localization but high confidence score, which suppress the boxes with more precise localization but lower score. As a result, the non-maximum suppression (NMS) tends to keep the poorly localized detections, leading to low AP at a strict IoU threshold. We visualize an example of this observation in Fig. 4(a). We plot the detection boxes before NMS with confidence scores indicated by the color. The box with more precise localization of the person is suppressed by other boxes not so accurate but having high scores. Then the final detection (bold box) after NMS doesn't have high IoU with the ground-truth.

So why this is the case? The conventional training strategy treats anchor points independently in Eq. (3), i.e. they receive equal attention. For a group of anchor points inside B_v, their spatial locations and associated features are different. So their abilities to localize B are also different. We argue that anchor points located close to instance boundaries don't have features well aligned with the instance. Their features tend to be hurt by content outside the instance because their receptive fields include too much information from the background, resulting in less representation power for precise localization. Thus, forcing these anchor points to perform as well as those with powerful feature representation is misleading the network. Less attention should be paid to anchor points close to instance boundaries than those surrounding the center in training. In other words, the network should focus more on optimizing the anchor points with powerful feature representation and reduce the false attention to others.

Our Solution. To address the false attention issue, we propose a simple and effective soft-weighting scheme. The basic idea is to assign an attention weight w_{lij} for each anchor point's loss L_{lij}. For each positive anchor point, the weight depends on the distance between its image space location and the corresponding instance boundaries. The closer to a boundary, the more down-weighted the anchor point gets. Thus, anchor points close to boundaries are receiving less attention and the network focuses more on those surrounding the center. For negative anchor points, they are kept unchanged since they are not involved in localization, i.e. their weights are all set to 1. Mathematically, w_{lij} is defined in Eq. (4):

$$w_{lij} = \begin{cases} f(p_{lij}, B), & \exists B, p_{lij} \in B_v \\ 1, & \text{otherwise} \end{cases} \tag{4}$$

where f is a function reflecting how close p_{lij} is to the boundaries of B. Closer distance yields less attention weight. We instantiate f using a generalized version of centerness function [27], i.e. $f(p_{lij}, B) = [\frac{\min(d^l_{lij}, d^r_{lij})}{\max(d^l_{lij}, d^r_{lij})} \frac{\min(d^t_{lij}, d^b_{lij})}{\max(d^t_{lij}, d^b_{lij})}]^\eta$, where η controls the decreasing steepness. An illustration of the soft-weighted anchor points is shown in Fig. 3.

Fig. 4. (a) Poorly localized detection boxes with high scores are generated by anchor points receiving false attention. (b) Our soft-weighting scheme effectively improves localization. Box score is indicated by the color bar.

Fig. 5. Feature responses from P_3 to P_7. They look similar but the details gradually vanish as the resolution becomes smaller. Selecting a single level per instance causes the waste of network power.

3.3 Soft-Selected Pyramid Levels

Feature Selection. Unlike anchor-based detectors, anchor-free methods don't have constraints from anchor matching to select feature levels for instances from the feature pyramid. In other words, we can assign each instance to arbitrary feature level(s) in anchor-free methods during training. And selecting the right feature levels can make a big difference [37].

We approach the issue of feature selection by looking into the properties of the feature pyramid. Indeed, feature maps from different pyramid levels are somewhat similar to each other, especially the adjacent levels. We visualize the response of all pyramid levels in Fig. 5. It turns out that if one level of feature is activated in a certain region, the same regions of adjacent levels may also be activated in a similar style. But the similarity fades as the levels are farther apart. This means that features from more than one pyramid level can participate together in the detection of a particular instance, but the degrees of participation from different levels should be somewhat different.

Inspired by the above analysis, we argue there should be two principles for proper pyramid level selection. Firstly, the selection should be related to the pattern of feature response, rather than some ad-hoc heuristics. And the instance-dependent loss can be a good reflection of whether a pyramid level is suitable for detecting some instances. This principle is also supported by [37]. Secondly, we should allow features from multiple levels involved in the training and testing for each instance, and each level should make distinct contributions. FoveaBox [11] has shown that assigning instances to multiple feature levels can improve the performance to some extent. But assigning to too many levels may instead hurt the performance severely. We believe this limitation is caused by the hard selection of pyramid levels. For each instance, the pyramid levels in FoveaBox are either selected or discarded. The selected levels are treated equally no matter how different their feature responses are.

Therefore, the solution lies in reweighting the pyramid levels for each instance. In other words, a weight is assigned to each pyramid level according to the feature response, making the selection soft. This can also be viewed as assigning a proportion of the instance to a level.

Table 1. Architecture of the feature selection network. The conv layers have no padding.

Layer type	Output size	Layer setting	Activation
Input	$1280 \times 7 \times 7$	n/a	n/a
Conv	$256 \times 5 \times 5$	$3 \times 3, 256$	Relu
Conv	$256 \times 3 \times 3$	$3 \times 3, 256$	Relu
Conv	$256 \times 1 \times 1$	$3 \times 3, 256$	Relu
Fc	5	n/a	softmax

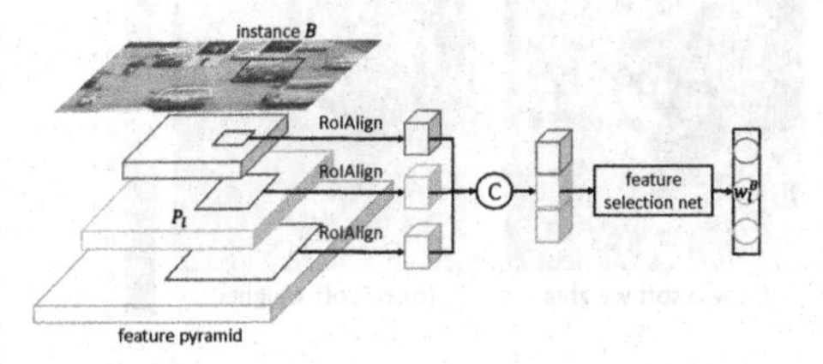

Fig. 6. The weights prediction for soft-selected pyramid levels. "C" indicates the concatenation operation.

Our Solution. So how to decide the weight of each pyramid level per instance? We propose to train a feature selection network to predict the weights for soft feature selection. The input to the network is instance-dependent feature responses extracted from all the pyramid levels. This is realized by applying the RoIAlign layer [7] to each pyramid feature followed by concatenation, where the RoI is the instance ground-truth box. Then the extracted feature goes through a feature selection network to output a vector of the probability distribution, as shown in Fig. 6. We use the probabilities as the weights of soft feature selection.

There are multiple architecture designs for the feature selection network. For simplicity, we present a light-weight instantiation. It consists of three 3×3 conv layers with no padding, each followed by the ReLU function, and a fully-connected layer with softmax. Table 1 details the architecture. The feature selection network is jointly trained with the detector. Cross entropy loss is used for optimization and the ground-truth is a one-hot vector indicating which pyramid level has minimal loss as defined in the FSAF module [37].

So far, each instance B is associated with a per level weight w_l^B via the feature selection network. Together with the soft-weighting scheme in Sect. 3.2, the anchor point loss L_{lij} is down-weighed further if B is assigned to P_l and p_{lij} is inside B_v. We assign each instance B to top$k$ feature levels with k minimal instance-dependent losses during training. Thus, Eq. (4) is augmented into Eq. (5).

$$w_{lij} = \begin{cases} w_l^B f(p_{lij}, B), & \exists B, p_{lij} \in B_v \\ 1, & \text{otherwise} \end{cases} \tag{5}$$

The total loss of the whole model is the weighted sum of anchor point losses plus the classification loss ($L_{\text{select-net}}$) from the feature selection network, as in Eq. (6). Figure 3 shows the effect of applying soft-selection weights.

$$L = \frac{1}{\sum_{p_{lij} \in p^+} w_{lij}} \sum_{lij} w_{lij} L_{lij} + \lambda L_{\text{select-net}} \tag{6}$$

where λ is the hyperparameter that controls the proportion of classification loss $L_{\text{select-net}}$ for feature selection.

3.4 Implementation Details

Initialization. We follow [37] for the initialization of the detection network. Specifically, the backbone networks are pre-trained on ImageNet1k [5]. The classification layers in the detection head are initialized with bias $-\log((1-\pi)/\pi)$ where $\pi = 0.01$, and a Gaussian weight. The localization layers in the detection head are initialized with bias 0.1, and also a Gaussian weight. For the newly added feature selection network, we initialize all layers in it using a Gaussian weight. All the Gaussian weights are filled with $\sigma = 0.01$.

Optimization. The entire detection network and the feature selection network are jointly trained with stochastic gradient descent on 8 GPUs with 2 images per

GPU using the COCO `train2017` set [16]. Unless otherwise noted, all models are trained for 12 epochs (~90k iterations) with an initial learning rate of 0.01, which is divided by 10 at the 9th and the 11th epoch. Horizontal image flipping is the only data augmentation unless otherwise specified. For the first 6 epochs, we don't use the output from the feature selection network. The detection network is trained with the same online feature selection strategy as in the FSAF module [37], i.e. each instance is assigned to only one feature level yielding the minimal loss. We plug in the soft selection weights and choose the top$k$ levels for the second 6 epochs. This is to stabilize the feature selection network first and to make the learning smoother in practice. We use the same training hyper-parameters for the shrunk factor $\epsilon = 0.2$ and the normalization scalar $z = 4.0$ as [37]. We set $\lambda = 0.1$ although results are robust to the exact value.

Inference. At the time of inference, the network architecture is as simple as in Fig. 2. The feature selection network is *not* involved in the inference so the runtime speed is not affected. An image is forwarded through the network in a fully convolutional style. Then classification prediction $\hat{\mathbf{c}}_{lij}$ and localization prediction $\hat{\mathbf{d}}_{lij}$ are generated for all each anchor point p_{lij}. Bounding boxes can be decoded using the reverse of Eq. (1). We only decode box predictions from at most 1k top-scoring anchor points in each pyramid level, after thresholding the confidence scores by 0.05. These top predictions from all feature levels are merged, followed by non-maximum suppression with a threshold of 0.5, yielding the final detections.

4 Experiments

We conduct experiments on the COCO [16] detection track using the MMDetection [2] codebase. All models are trained on the `train2017` split including around 115k images. We analyze our method by ablation studies on the `val2017` split containing 5k images. When comparing to the state-of-the-art detectors, we report the Average Precision (AP) scores on the `test-dev` split.

Table 2. Ablative experiments for the SAPD on the COCO `val2017`. ResNet-50 is the backbone network for all experiments in this table. We study the effect of **SW**: soft-weighted anchor points, **SS**: soft-selected pyramid levels, and **BFP** [21]: augmented feature pyramids.

	SW	SS	BFP	AP	AP_{50}	AP_{75}	AP_S	AP_M	AP_L
FSAF [37]				35.9	55.0	37.9	19.8	39.6	48.2
	✓			37.0	55.8	39.5	20.5	40.1	48.5
	✓	✓		38.0	56.9	40.5	21.2	41.2	50.2
			✓	36.8	57.4	38.8	21.6	40.9	47.6
SAPD	✓	✓	✓	**38.8**	**58.7**	**41.3**	**22.5**	**42.6**	**50.8**

4.1 Ablation Studies

All results in ablation studies are based on models trained and tested with an image scale of 800 pixels. We study the contribution of each proposed component by gradually applying these components to the baseline FSAF [37] module. For the soft-weighted anchor points and soft-selected pyramid levels, we first study the effect of varying hyperparameters on them and then apply each component with the best hyperparameter to the baseline. We also report more ablative experiments in the supplementary material.

Table 3. Varying η for the generalized centerness function in soft-weighted anchor points.

η	AP	AP_{50}	AP_{75}	AP_S	AP_M	AP_L
0.10	36.8	56.2	39.0	20.6	40.3	48.3
0.50	36.9	55.8	39.4	20.2	40.1	48.7
1.0	**37.0**	55.8	39.5	20.5	40.1	48.5
2.0	36.6	55.1	39.1	19.8	40.3	47.8

Table 4. Varying k for number of selected levels in soft-selected pyramid level with $\eta = 1.0$.

top$k$	AP	AP_{50}	AP_{75}	AP_S	AP_M	AP_L
2	37.9	56.9	40.5	21.1	41.0	50.1
3	**38.0**	56.9	40.5	21.2	41.2	50.2
4	37.9	56.9	40.3	21.2	41.1	50.2
5	37.9	56.8	40.5	21.0	41.1	50.2

Soft-Weighted Anchor Points Improve the Localization. We first apply the soft-weighting scheme (Eq. (4)) to the training of the baseline FSAF module. Results are reported in Table 2 and 3. Soft-weighted anchor points offer a significant improvement (up to 1.1% AP) over the baseline, being insensitive to various hyperparameters. More importantly, AP_{75} is increased by 1.6%, indicating better localization accuracy at a strict IoU threshold. We also visualize the effect of our soft-weighting scheme in Fig. 4(b). The precise box (marked as bold) is kept while the other poorly localized boxes are suppressed, reducing the false attention issue effectively.

Soft-Selected Pyramid Levels Utilize the Feature Power Better. Next, we further apply the soft feature selection on top of the soft-weighting scheme, so that each anchor point is down-weighted as in Eq. (5). Table 2 and 4 reports the ablative results. We find that as long as each instance is assigned to more than one pyramid level, we can observe robust $\sim$1.0% absolute AP improvements over the FSAF module plus soft-weighted anchor points. This indicates that allowing instances to optimize multiple pyramid levels is essential to utilize the feature power as much as possible. Empirically, we assign each instance to the top 3 feature levels with the minimal instance-dependent losses according to Table 4. To understand how does the feature selection network assign instances, we visualize the predicted soft selection weights in Fig. 7. It turns out that larger instances tend to be assigned high weights for higher pyramid levels. The majority of instances can be learned with no more than two levels. Very rare instances need to be modeled by more than two levels, e.g. the sofa in the top right sub-figure of Fig. 7. This is consistent with the results in Table 4.

Fig. 7. Visualization of the soft feature selection weights from the feature selection network. Weights (the top-left red bars) ranging from 0 to 1 of five pyramid levels (P_3 to P_7) are predicted for each instance (blue box). The more filled a red bar is, the higher the weight. *Best viewed in color when zoomed in.* (Color figure online)

Joint Training of the Feature Selection Network has a Negligible Effect on Performance. As shown in Fig. 6, the feature selection network takes in feature extracted from the shared feature pyramid and is jointly trained with the detector. One may argue that the performance improvement by the soft-selected pyramid levels is due to the multi-task learning of the feature selection network and the detector. We prove this is not the case. We conduct an experiment in which the feature selection network is jointly trained with the detector but its predicted soft selection weights are not used. In other words, the weights for anchor points are still following Eq. (4) and the feature selection strategy is the same as the baseline FSAF module. It turns out the final AP is 37.1%, only 0.1% higher than the 2nd entry and 0.9% lower than the 3rd entry in Table 2. This means that the major contribution of the soft-selected pyramid levels is actually from *softly selecting multiple levels* rather than the multi-task learning effect from the feature selection network.

Our Training Strategy Works Well with Augmented Feature Pyramids. Different from key-point detectors that use a single high-resolution feature map, the SAPD can enjoy the merits brought by the advanced designs of feature pyramids. Here we adopt the Balanced Feature Pyramid (BFP) [21] and achieve further improvement. The BFP pushes our model with ResNet-50 to a 38.8% AP, which is 2.9% higher than the baseline FSAF module. More importantly, our proposed training strategy can robustly work with advanced feature pyramids, offering a steady 2% AP gain (see 4th and 5th entries in Table 2).

SAPD is Robust and Efficient. Our SAPD can consistently provide robust performance using deeper and better backbone networks, while at the same time keeping the detection head as simple as possible. We report the head-to-head

Table 5. Head-to-head comparisons of anchor-based RetinaNet, anchor-based plus FSAF module, and our purely anchor-free SAPD with different backbone networks on the COCO `val2017` set. **AB:** Anchor-based branches. **R:** ResNet. **X:** ResNeXt.

Backbone	Method	AP	AP_{50}	FPS
R-50	RetinaNet	35.7	54.7	11.6
	AB+FSAF	37.2	57.2	9.0
	SAPD (Ours)	**38.8**	**58.7**	**14.9**
R-101	RetinaNet	37.7	57.2	8.0
	AB+FSAF	39.3	59.2	7.1
	SAPD (Ours)	**41.0**	**60.7**	**11.2**
X-101-64x4d	RetinaNet	39.8	59.5	4.5
	AB+FSAF	41.6	62.4	4.2
	SAPD (Ours)	**43.1**	**63.7**	**6.1**

comparisons with anchor-based RetinaNet and the more complex anchor-based plus FSAF detector in terms of detection accuracy and speed in Table 5. Except for the head architectures, all other settings are the same. All detectors run on a single GTX 1080Ti GPU with CUDA 10 using a batch size of 1. It turns out that our SAPD gets both sides of two worlds. Our SAPD with purely anchor-free heads can not only run faster than the anchor-based counterparts due to simpler head architecture, but also outperform the *combination* of anchor-based and anchor-free heads by significant margins, i.e. 1.6%, 1.7%, and 1.5% absolute AP increases on ResNet-50, ResNet-101, and ResNeXt-101-64x4d backbones respectively.

4.2 Comparison to State of the Art

We evaluate our complete SAPD on the COCO `test-dev` set to compare with recent state-of-the-art anchor-free and anchor-based detectors. All of our models are trained using scale jitter by randomly scaling the shorter side of images in the range from 640 to 800 and for 2× number of epochs as the models in Sect. 4.1 with the learning rate change points scaled proportionally. Other settings are the same as Sect. 4.1.

For a fair comparison, we report the results of single-model single-scale testing for all methods, as well as their corresponding inference speeds in Table 6. A visualization of the accuracy-speed trade-off is shown in Fig. 1. The inference speeds are measured by Frames-per-Second (FPS) on the same machine with a GTX 1080Ti GPU using a batch size of 1 whenever possible. A "n/a" indicates the case that the method doesn't provide trained models nor self-timing results from the original paper.

Our proposed SAPD pushes the envelope of accuracy-speed boundary to a new level. We report the results of two series of the backbone models, one without

Table 6. *Single-model and single-scale* accuracy and inference speed of SAPD vs. recent state-of-the-art detectors on the COCO `test-dev` set. FPS is measured on the same machine with a single *GTX 1080Ti* GPU using the official source code whenever possible. "n/a" means that both trained models and timing results from original papers are not available. **R:** ResNet. **X:** ResNeXt. **HG:** Hourglass. For a visualized comparison, please refer to Fig. 1.

Method	Backbone	Anchor free?	FPS	AP	AP_{50}	AP_{75}	AP_S	AP_M	AP_L
Multi-stage detectors									
Faster R-CNN w/FPN [14]	R-101		9.9	36.2	59.1	39.0	18.2	39.0	48.2
Cascade R-CNN [1]	R-101		8.0	42.8	62.1	46.3	23.7	45.5	55.2
GA-Faster-RCNN [28]	R-50	✓	9.4	39.8	59.2	43.5	21.8	42.6	50.7
Libra R-CNN [21]	R-101		9.5	41.1	62.1	44.7	23.4	43.7	52.5
Libra R-CNN [21]	X-101-64x4d		5.6	43.0	64.0	47.0	25.3	45.6	54.6
RPDet [29]	R-101	✓	10.0	41.0	62.9	44.3	23.6	44.1	51.7
RPDet [29]	R-101-DCN	✓	8.0	45.0	66.1	49.0	26.6	48.6	57.5
TridentNet [13]	R-101		2.7	42.7	63.6	46.5	23.9	46.6	56.6
TridentNet [13]	R-101-DCN		1.3	46.8	67.6	51.5	28.0	51.2	60.5
Single-stage detectors									
RetinaNet [15]	R-101		8.0	39.1	59.1	42.3	21.8	42.7	50.2
CornerNet [12]	HG-104	✓	3.1	40.5	56.5	43.1	19.4	42.7	53.9
AB+FSAF [37]	R-101		7.1	40.9	61.5	44.0	24.0	44.2	51.3
AB+FSAF [37]	X-101-64x4d		4.2	42.9	63.8	46.3	26.6	46.2	52.7
GA-RetinaNet [28]	R-50	✓	10.8	37.1	56.9	40.0	20.1	40.1	48.0
ExtremeNet [36]	HG-104	✓	2.8	40.2	55.5	43.2	20.4	43.2	53.1
FoveaBox [11]	X-101	✓	n/a	42.1	61.9	45.2	24.9	46.8	55.6
FCOS [27]	R-101	✓	9.3	41.5	60.7	45.0	24.4	44.8	51.6
FCOS [27] w/imprv	X-101-64x4d	✓	5.4	44.7	64.1	48.4	27.6	47.5	55.6
CenterNet [6]	HG-104	✓	3.3	44.9	62.4	48.1	25.6	47.4	57.4
FreeAnchor [32]	R-101		9.1	43.1	62.2	46.4	24.5	46.1	54.8
FreeAnchor [32]	X-101-32x8d		5.4	44.8	64.3	48.4	27.0	47.9	56.0
SAPD (Ours)	R-50	✓	14.9	41.7	61.9	44.6	24.1	44.6	51.6
SAPD (Ours)	R-101	✓	11.2	43.5	63.6	46.5	24.9	46.8	54.6
SAPD (Ours)	X-101-64x4d	✓	6.1	45.4	65.6	48.9	27.3	48.7	56.8
SAPD (Ours)	R-50-DCN	✓	12.4	44.3	64.4	47.7	25.5	47.3	57.0
SAPD (Ours)	R-101-DCN	✓	9.1	46.0	65.9	49.6	26.3	49.2	59.6
SAPD (Ours)	X-101-64x4d-DCN	✓	4.5	47.4	67.4	51.1	28.1	50.3	61.5

DCN and the other with DCN. Without DCN, our fastest SAPD version based on ResNet-50 can reach a 14.9 FPS while maintaining a 41.7% AP, outperforming some of the methods [11,14,15,21,27] using ResNet-101. With DCN, our SAPD forms an upper envelope of state-of-the-art anchor-free detectors and some recent anchor-based detectors. The closest competitor, RPDet [29], is 1.0% AP worse and 15ms slower than ours. Compared to key-point anchor-free detectors [6,12, 36] using Hourglass, our SAPD enjoys significantly faster inference speed (up to 5× times) and a 2.5% AP improvement (47.4% vs. 44.9%) over the best key-point detector, CenterNet [6].

5 Conclusion

This work studied the anchor-point object detection and discovered the key insight lies in the joint optimization of a group of anchor points both within and across the feature pyramid levels. We proposed a novel training strategy addressing two underexplored issues of anchor-point detection approaches, i.e. the false attention issue within each pyramid level and the feature selection issue across all pyramid levels. Applying our training strategy to a simple anchor-point detector leads to a new upper envelope of the speed-accuracy trade-off.

References

1. Cai, Z., Vasconcelos, N.: Cascade R-CNN: delving into high quality object detection. In: Proceedings of the IEEE Conference on Computer Vision and Pattern Recognition, pp. 6154–6162 (2018)
2. Chen, K., et al.: MMDetection: open MMLab detection toolbox and benchmark. arXiv preprint arXiv:1906.07155 (2019)
3. Dai, J., Li, Y., He, K., Sun, J.: R-FCN: object detection via region-based fully convolutional networks. In: Advances in Neural Information Processing Systems, pp. 379–387 (2016)
4. Dai, J., Qi, H., Xiong, Y., Li, Y., Zhang, G., Hu, H., Wei, Y.: Deformable convolutional networks. In: Proceedings of the IEEE International Conference on Computer Vision, pp. 764–773 (2017)
5. Deng, J., Dong, W., Socher, R., Li, L.J., Li, K., Fei-Fei, L.: ImageNet: a large-scale hierarchical image database. In: 2009 IEEE Conference on Computer Vision and Pattern Recognition, pp. 248–255. IEEE (2009)
6. Duan, K., Bai, S., Xie, L., Qi, H., Huang, Q., Tian, Q.: CenterNet: object detection with keypoint triplets. In: Proceedings of the IEEE International Conference on Computer Vision (2019)
7. He, K., Gkioxari, G., Dollár, P., Girshick, R.: Mask R-CNN. In: Proceedings of the IEEE International Conference on Computer Vision, pp. 2961–2969 (2017)
8. He, K., Zhang, X., Ren, S., Sun, J.: Deep residual learning for image recognition. In: Proceedings of the IEEE Conference on Computer Vision and Pattern Recognition, pp. 770–778 (2016)
9. He, Y., Zhu, C., Wang, J., Savvides, M., Zhang, X.: Bounding box regression with uncertainty for accurate object detection. In: Proceedings of the IEEE/CVF Conference on Computer Vision and Pattern Recognition (CVPR), June 2019
10. Huang, L., Yang, Y., Deng, Y., Yu, Y.: DenseBox: unifying landmark localization with end to end object detection. arXiv preprint arXiv:1509.04874 (2015)
11. Kong, T., Sun, F., Liu, H., Jiang, Y., Shi, J.: FoveaBox: beyond anchor-based object detector. arXiv preprint arXiv:1904.03797 (2019)
12. Law, H., Deng, J.: CornerNet: detecting objects as paired keypoints. In: Proceedings of the European Conference on Computer Vision (ECCV), pp. 734–750 (2018)
13. Li, Y., Chen, Y., Wang, N., Zhang, Z.: Scale-aware trident networks for object detection. In: Proceedings of the IEEE International Conference on Computer Vision (2019)
14. Lin, T.Y., Dollár, P., Girshick, R., He, K., Hariharan, B., Belongie, S.: Feature pyramid networks for object detection. In: Proceedings of the IEEE Conference on Computer Vision and Pattern Recognition, pp. 2117–2125 (2017)

15. Lin, T.Y., Goyal, P., Girshick, R., He, K., Dollár, P.: Focal loss for dense object detection. In: Proceedings of the IEEE International Conference on Computer Vision, pp. 2980–2988 (2017)
16. Lin, T.-Y., et al.: Microsoft COCO: common objects in context. In: Fleet, D., Pajdla, T., Schiele, B., Tuytelaars, T. (eds.) ECCV 2014. LNCS, vol. 8693, pp. 740–755. Springer, Cham (2014). https://doi.org/10.1007/978-3-319-10602-1_48
17. Liu, S., Qi, L., Qin, H., Shi, J., Jia, J.: Path aggregation network for instance segmentation. In: Proceedings of the IEEE Conference on Computer Vision and Pattern Recognition, pp. 8759–8768 (2018)
18. Liu, W., et al.: SSD: single shot multibox detector. In: Leibe, B., Matas, J., Sebe, N., Welling, M. (eds.) ECCV 2016. LNCS, vol. 9905, pp. 21–37. Springer, Cham (2016). https://doi.org/10.1007/978-3-319-46448-0_2
19. Liu, W., Liao, S., Ren, W., Hu, W., Yu, Y.: High-level semantic feature detection: a new perspective for pedestrian detection. In: The IEEE Conference on Computer Vision and Pattern Recognition (CVPR), June 2019
20. Newell, A., Yang, K., Deng, J.: Stacked hourglass networks for human pose estimation. In: Leibe, B., Matas, J., Sebe, N., Welling, M. (eds.) ECCV 2016. LNCS, vol. 9912, pp. 483–499. Springer, Cham (2016). https://doi.org/10.1007/978-3-319-46484-8_29
21. Pang, J., Chen, K., Shi, J., Feng, H., Ouyang, W., Lin, D.: Libra R-CNN: towards balanced learning for object detection. In: Proceedings of the IEEE Conference on Computer Vision and Pattern Recognition, pp. 821–830 (2019)
22. Redmon, J., Farhadi, A.: Yolo9000: better, faster, stronger. In: Proceedings of the IEEE Conference on Computer Vision and Pattern Recognition, pp. 7263–7271 (2017)
23. Ren, S., He, K., Girshick, R., Sun, J.: Faster R-CNN: towards real-time object detection with region proposal networks. In: Advances in Neural Information Processing Systems, pp. 91–99 (2015)
24. Shen, Z., Liu, Z., Li, J., Jiang, Y.G., Chen, Y., Xue, X.: DSOD: learning deeply supervised object detectors from scratch. In: Proceedings of the IEEE International Conference on Computer Vision, pp. 1919–1927 (2017)
25. Sun, F., Kong, T., Huang, W., Tan, C., Fang, B., Liu, H.: Feature pyramid reconfiguration with consistent loss for object detection. IEEE Trans. Image Process. **28**(10), 5041–5051 (2019)
26. Sun, K., Xiao, B., Liu, D., Wang, J.: Deep high-resolution representation learning for human pose estimation. In: Proceedings of the IEEE Conference on Computer Vision and Pattern Recognition, pp. 5693–5703 (2019)
27. Tian, Z., Shen, C., Chen, H., He, T.: FCOS: fully convolutional one-stage object detection. In: Proceedings of the IEEE International Conference on Computer Vision (2019)
28. Wang, J., Chen, K., Yang, S., Loy, C.C., Lin, D.: Region proposal by guided anchoring. In: Proceedings of the IEEE Conference on Computer Vision and Pattern Recognition, pp. 2965–2974 (2019)
29. Yang, Z., Liu, S., Hu, H., Wang, L., Lin, S.: RepPoints: point set representation for object detection. In: Proceedings of the IEEE International Conference on Computer Vision (2019)
30. Yu, J., Jiang, Y., Wang, Z., Cao, Z., Huang, T.: UnitBox: an advanced object detection network. In: Proceedings of the 24th ACM International Conference on Multimedia, pp. 516–520. ACM (2016)

31. Zhang, S., Wen, L., Bian, X., Lei, Z., Li, S.Z.: Single-shot refinement neural network for object detection. In: Proceedings of the IEEE Conference on Computer Vision and Pattern Recognition, pp. 4203–4212 (2018)
32. Zhang, X., Wan, F., Liu, C., Ji, R., Ye, Q.: FreeAnchor: learning to match anchors for visual object detection. In: Advances in Neural Information Processing Systems (2019)
33. Zhao, Q., et al.: M2Det: a single-shot object detector based on multi-level feature pyramid network. In: Proceedings of the AAAI Conference on Artificial Intelligence, vol. 33, pp. 9259–9266 (2019)
34. Zhong, Z., Sun, L., Huo, Q.: An anchor-free region proposal network for faster R-CNN based text detection approaches. arXiv preprint arXiv:1804.09003 (2018)
35. Zhou, X., Wang, D., Krähenbühl, P.: Objects as points. arXiv preprint arXiv:1904.07850 (2019)
36. Zhou, X., Zhuo, J., Krahenbuhl, P.: Bottom-up object detection by grouping extreme and center points. In: Proceedings of the IEEE Conference on Computer Vision and Pattern Recognition, pp. 850–859 (2019)
37. Zhu, C., He, Y., Savvides, M.: Feature selective anchor-free module for single-shot object detection. In: Proceedings of the IEEE/CVF Conference on Computer Vision and Pattern Recognition (CVPR), June 2019

Beyond Fixed Grid: Learning Geometric Image Representation with a Deformable Grid

Jun Gao[1,2,3](✉), Zian Wang[1,2], Jinchen Xuan[4], and Sanja Fidler[1,2,3]

[1] University of Toronto, Toronto, Canada
{jungao,zianwang,fidler}@cs.toronto.edu
[2] Vector Institute, Toronto, Canada
[3] NVIDIA, Toronto, Canada
[4] Peking University, Beijing, China
1600012865@pku.edu.cn

Abstract. In modern computer vision, images are typically represented as a fixed uniform grid with some stride and processed via a deep convolutional neural network. We argue that deforming the grid to better align with the high-frequency image content is a more effective strategy. We introduce *Deformable Grid* (DefGrid), a learnable neural network module that predicts location offsets of vertices of a 2-dimensional triangular grid, such that the edges of the deformed grid align with image boundaries. We showcase our DefGrid in a variety of use cases, i.e., by inserting it as a module at various levels of processing. We utilize DefGrid as an end-to-end *learnable geometric downsampling* layer that replaces standard pooling methods for reducing feature resolution when feeding images into a deep CNN. We show significantly improved results at the same grid resolution compared to using CNNs on uniform grids for the task of semantic segmentation. We also utilize DefGrid at the output layers for the task of object mask annotation, and show that reasoning about object boundaries on our predicted polygonal grid leads to more accurate results over existing pixel-wise and curve-based approaches. We finally showcase DefGrid as a standalone module for unsupervised image partitioning, showing superior performance over existing approaches. Project website: http://www.cs.toronto.edu/ jungao/def-grid.

1 Introduction

In modern computer vision approaches, an image is treated as a fixed uniform grid with a stride and processed through a deep convolutional neural network. Very high resolution images are typically processed at a lower resolution for increased efficiency, whereby the image is essentially blurred and subsampled. When fed to a neural network, each pixel thus contains a blurry version of the

Electronic supplementary material The online version of this chapter (https://doi.org/10.1007/978-3-030-58545-7_7) contains supplementary material, which is available to authorized users.

© Springer Nature Switzerland AG 2020
A. Vedaldi et al. (Eds.): ECCV 2020, LNCS 12354, pp. 108–125, 2020.
https://doi.org/10.1007/978-3-030-58545-7_7

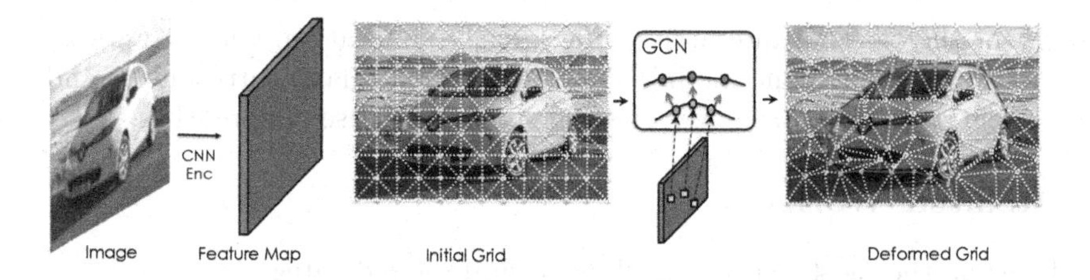

Fig. 1. DefGrid is a neural module that represents an image with a triangular grid. Initialized with an uniform grid, DefGrid deforms grid's vertices, such that grid's edges align with image boundaries, while keeping topology fixed.

original signal mixing information from both the foreground and background, possibly causing higher sensitivity and reliance of the network to objects and their context. In contrast, in many of the traditional computer vision pipelines the high resolution image was instead partitioned into a smaller set of superpixels that conform to image boundaries, leading to more effective reasoning in downstream tasks. We follow this line of thought and argue that deforming the grid to better align with the high-frequency information content in the input is a more effective representation strategy. This is conceptually akin to superpixels but conforming to a regular topology with geometric constraints thus still easily amenable for use with deep convolutional networks for downstream tasks.

Furthermore, tasks such as object mask annotation naturally require the output to be in the form of polygons with a manageable number of control points that a human annotator can edit. Previous work either parametrized the output as a closed curve with a fixed number of control points [27] or performed pixel-wise labeling followed by a (non-differentiable) polygonization step [26,29,39]. In the former, the predicted curves typically better utilize shape priors leading to "well behaved" predictions, however, the output is inherently limited in the genus and complexity of the shape it is able to represent. In contrast, pixel-wise methods can represent shapes of arbitrary genus, however, typically large input/output resolutions are required to produce accurate labeling around object boundaries. We argue that reasoning on a low-resolution polygonal grid that well aligns with image boundaries combines the advantages of the two approaches.

We introduce *Deformable Grid* (DefGrid), a neural network module that represents an image with a 2-dimensional triangular grid. The basic element of the grid is a triangular cell with vertices that place the triangle in the image plane. DefGrid is initialized with a uniform grid and utilizes a neural network that predicts location offsets of the triangle vertices such that the edges and vertices of the deformed grid align with image boundaries (Fig. 1). We propose several carefully designed loss functions that encourage this behaviour. Due to the differentiability of the deformation operations, DefGrid can be trained end-to-end with downstream neural networks as a plug-and-play module at various levels of deep processing. We showcase DefGrid in various use cases: as a learnable geometric image downsampling layer that affords high accuracy semantic segmentation at significantly reduced grid resolutions. Furthermore, when used to parametrize the output, we show that it leads to more effective and accurate results for the tasks of interactive object mask annotation. Our DefGrid can also be used a standalone module for unsupervised image partitioning, and we show superior performance over existing superpixel-based approaches.

2 Related Works

We focus on the most relevant work in several related categories.

Deformable Structure: Deformable convolutions [10] predict position offsets of each cell in the convolutional kernel's grid with the aim to better capture object deformation. This is in contrast to our approach which deforms a triangular grid which is then exploited in downstream processing. Note that our approach does not imply any particular downstream neural architecture and would further benefit by employing deformable convolutions. Related to our work, [23,31] learn to deform an image such that the corresponding warped image, when fed to a neural network, leads to improved downstream tasks. Our DefGrid, which is trained with both unsupervised and supervised loss functions to explicitly align with image boundaries, allows downstream tasks such as object segmentation to perform reasoning directly on the low-dimensional grid.

Image Partitioning: Polygonal image partitioning plays an important role in certain applications such as multi-view 3D object reconstruction [6] and graphics-based image manipulation [22]. Existing works tried to polygonize an image with triangles [6] or convex polygons [11]. [6] used Constrained Delaunay triangulation to get the triangular mesh. However, their method heavily relies either on having good key point locations or good edges in order to get boundary aligned triangles. [11] also relies on the initial line segment detection. In [2], a non-iterative method was proposed to obtain superpixels, followed by a polygonization method using a contour tracing algorithm. Our DefGrid produces a triangular grid that conforms to image boundaries and is end-to-end trainable.

Superpixels: Superpixel methods aim to partition the image into regions of homogenous color while regularizing their shape and preserving image boundaries [1,2,5,12,17,25,28,32]. Many algorithms have been proposed mainly differing in the energy function they optimize and the optimization technique they utilize [1,2,12,17,28,32,37,40,41]. Most of these approaches produce superpixels with irregular topology, and the final segmentation map is often disconnected and needs postprocessing. Note also that energy is often hand designed, and inference is optimization based. Recently, SSN [24] made clustering-based approaches differentiable by softly assigning pixels to superpixels with the exponential function. SEAL [36] learns superpixels by exploiting segmentation affinities. Both of these methods produce superpixels with highly irregular boundaries and region topology, and thus they may not be trivially embedded in existing convolutional neural architectures [13]. To produce regular grid-like topology, superpixel lattices [30] partition the image recursively, finding horizontal and vertical paths with minimal boundary cost at each iteration. Unlike their approach, DefGrid utilizes differentiable operations to predict boundary aligned triangular grids and is end-to-end trainable with both unsupervised and supervised loss functions.

3 Deformable Grid

Our DefGrid is a 2-dimensional triangular grid defined on an image plane. The basic cell in the grid is a triangle with three vertices, each with a location that position the triangle in the image. Edges of the triangle thus represent line segments and are expected not to self-intersect across triangles. The topology of the grid is fixed and does not depend on the input image. The geometric grid thus naturally partitions an image into regular segments, as shown in Fig. 1.

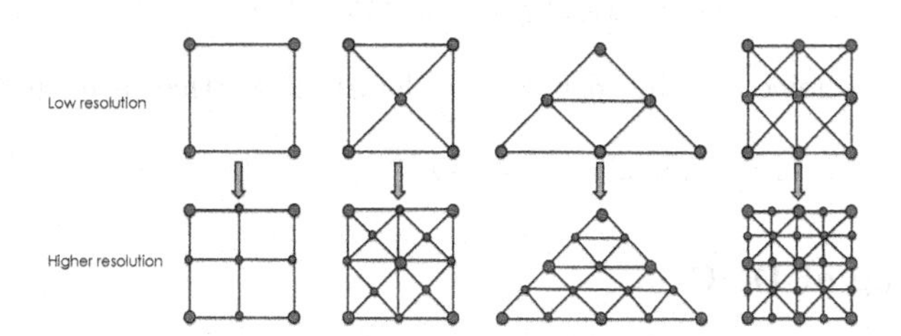

Fig. 2. Different grid topologies. We choose the last column for its flexibility in representing a variety of different edge orientations.

We formulate our approach as deforming a triangular grid with uniformly initialized vertex positions to better align with image boundaries. The grid is deformed via a neural network that a predicts position offset for each vertex while ensuring the topology does not change (no self-intersections occur).

Our main intuition is that when the edges of the grid align with image boundaries, the pixels inside each grid cell have minimal variance of RGB values (or one-hot masks when we have supervision), and vice versa. We aim to minimize this variance in a differentiable way with respect to the positions of the vertices, to make it amenable to deep learning. We describe our DefGrid formulation along with its training method in detail next. In Sect. 4, we show applications to different downstream tasks.

3.1 Grid Parameterization

Grid Topology: Choosing the right topology of the grid is an important aspect of our work. Since objects (and their parts) can appear at different scales in images, we ideally want a topology that can easily be subdivided to accommodate for this diversity. Furthermore, boundaries can be found in any orientation and thus the grid edges should be flexible enough to well align to any real edge. We experimentally tried four different topologies which are visualized in Fig. 2. We found the topology in the last column to outperform alternatives for its flexibility in representing different edge orientations. Note that our method is agnostic to the choice of topology and we provide a detailed comparison in the appendix.

Grid Representation: Let I be an input image. We denote each vertex of the grid in the image plane as $v_i = [x_i, y_i]^T$, where $i \in \{1, \cdots, n\}$ and n is the total number of vertices in the grid. Since the grid topology is fixed, the grid in the image is entirely specified by the positions of its vertices $\mathbf{v}$. We denote each triangular cell in the grid with its three vertices as $C_k = [v_{a_k}, v_{b_k}, v_{c_k}]$, with $k \in \{1, \cdots, K\}$ indexing the grid cells. We uniformly initialize the vertices on the 2D image plane, and define DefGrid as a neural network h that predicts the relative offset for each vertex:

$$\{\Delta x_i, \Delta y_i\}_{i=1}^{n} = h(\mathbf{v}, I). \tag{1}$$

We discuss the choice of h in Sect. 4. The deformed vertices are thus:

$$v_i = [x_i + \Delta x_i, y_i + \Delta y_i]^T, \quad i = 1, \ldots, n. \tag{2}$$

3.2 Training of DefGrid

We now discuss training of the grid deforming network h using a variety of unsupervised loss functions. We want all our losses to be differentiable with respect to vertex positions to allow for the gradient to be backpropagated analytically.

Differentiable Variance: As the grid deforms (its vertices move), the grid cells will cover different pixel regions in the image. Our first loss aims to minimize the variance of pixel features in each grid cell. Each pixel p_i has a feature vector $\mathbf{f}_i$, which in our case is chosen to contain RGB values. When supervision is available in the form of segmentation masks, we can optionally append a one hot vector representing the class of the mask. Pixel's position in the image is denoted with $p_i = [p_i^x, p_i^y]^T$, $i \in \{1, \cdots, N\}$, with N indicating the total number of pixels in the image. Variance of a cell C_k is defined as:

$$V_k = \sum\nolimits_{p_i \in S_k} ||\mathbf{f}_i - \overline{\mathbf{f}}_k||_2^2, \tag{3}$$

where S_k denotes the set of pixels inside C_k, and $\overline{\mathbf{f}}_k$ is the mean feature of C_k: $\overline{\mathbf{f}}_k = \frac{\sum_{p_i \in S_k} \mathbf{f}_i}{\sum_{p_i \in S_k} 1}$. Note that this definition of variance is not naturally differentiable with respect to the vertex positions. We thus reformulate the variance function by softly assigning every pixel p_i to each grid cell C_k with a probability $P_{i \to k}(\mathbf{v})$:

$$P_{i \to k}(\mathbf{v}) = \frac{\exp(\text{SignDis}(p_i, C_k)/\delta)}{\sum_{j=1}^{K} \exp(\text{SignDis}(p_i, C_j)/\delta)}, \tag{4}$$

$$\text{SignDis}(p_i, C_k) = \begin{cases} -\text{Dis}(p_i, C_k), & \text{if } p_i \text{ is outside } C_k, \\ \text{Dis}(p_i, C_k), & \text{if } p_i \text{ is inside } C_k. \end{cases} \tag{5}$$

$$\text{Dis}(p_i, C_k) = \min(D(p_i, v_{a_k} v_{b_k}), D(p_i, v_{b_k} v_{c_k}), D(p_i, v_{c_k} v_{a_k})), \tag{6}$$

where $D(p_i, v_i v_j)$ is the L1 distance between a pixel and a line segment $v_i v_j$, and δ is a hyperparameter to control the slackness. We use $P_{i \to k}(\mathbf{v})$ to indicate that the probability of assignment depends on the grid's vertex positions, and is in our case a differentiable function. Intuitively, if the pixel is very close or inside a cell, then $P_{i\to k}(\mathbf{v})$ is close to 1, and close to 0 otherwise. To check whether the pixel is inside a cell, we calculate the barycentric weight of the pixel with respect to three vertices of the cell. If all the barycentric weights are between 0 and 1, then the pixel is inside, otherwise it falls outside the triangle.

We now re-define the cell's variance as follows:

$$\tilde{V}_k(\mathbf{v}) = \sum_{i=1}^{N} P_{i\to k}(\mathbf{v}) \cdot ||\mathbf{f}_i - \overline{\mathbf{f}}_k||_2^2, \tag{7}$$

which is therefore a differentiable function of grid's vertex positions. Our variance-based loss function aims to minimize the sum of variances across all grid cells:

$$L_{\text{var}}(\mathbf{v}) = \sum_{k=1}^{K} \tilde{V}_k(\mathbf{v}) \tag{8}$$

Differentiable Reconstruction: Inspired by SSN [24], we further differentiably reconstruct an image using the deformed grid by taking into account the probability of assignments: $P_{i\to k}(\mathbf{v})$. Intuitively, we represent each cell using its mean feature $\overline{\mathbf{f}}_k$, and "paste" it back into the image plane according to the positions of the cell's deformed vertices. Specifically, we reconstruct each pixel in an image by softly gathering information from each grid cell using $P_{i\to k}(\mathbf{v})$:

$$\hat{\mathbf{f}}_i(\mathbf{v}) = \textstyle\sum_{k=1}^{K} P_{i\to k}(\mathbf{v}) \cdot \overline{\mathbf{f}}_k, \tag{9}$$

The reconstruction loss is the distance between the reconstructed pixel feature and original pixel feature:

$$L_{\text{recons}}(\mathbf{v}) = \textstyle\sum_{i=1}^{N} ||\hat{\mathbf{f}}_i(\mathbf{v}) - \mathbf{f}_i||_1. \tag{10}$$

We experimentally found that L1 distance works better than L2.

Regularization: To regularize the shape of the grid and prevent self-intersections, we introduce two regularizers. We employ an **Area balancing** loss function that encourages the areas of the cells to be similar, and thus, avoids self-intersections by minimizing the variance of the areas:

$$L_{\text{area}}(\mathbf{v}) = \sum_{j=1}^{K} ||a_k(\mathbf{v}) - \overline{a}(\mathbf{v})||_2^2, \tag{11}$$

where $\overline{a}$ is the mean area and a_k is the area of cell C_k. We also utilize **Laplacian regularization** following works on 3D mesh prediction [8,38]. In particular, this loss encourages the neighboring vertices to move along similar directions with respect to the center vertex:

$$L_{\text{lap}}(\mathbf{v}) = \sum_{i=1}^{n} ||\Delta_i - \frac{1}{||\mathcal{N}(i)||} \sum_{j\in\mathcal{N}(i)} \Delta_j||_2^2, \tag{12}$$

where $\Delta_i = [\Delta x_i, \Delta y_i]^T$ is the predicted offset of vertex v_i and $\mathcal{N}(i)$ is the set of neighboring vertices of vertex v_i.

The final loss to train our network h is a weighted sum of all the above terms:

$$L_{\text{def}} = L_{\text{var}} + \lambda_{\text{recons}} L_{\text{recons}} + \lambda_{\text{area}} L_{\text{area}} + \lambda_{\text{lap}} L_{\text{lap}}, \tag{13}$$

where $\lambda_{\text{recons}}, \lambda_{\text{area}}, \lambda_{\text{lap}}$ are hyperparameters that balance different terms.

4 Applications

Our DefGrid supports many computer vision tasks that are done on fixed image grids today. We discuss three possible use cases in this section. DefGrid can be inserted as a plug-and-play module at several levels of processing. By inserting it at the input level we utilize DefGrid as a learnable geometric downsampling layer to replace standard pooling methods. We showcase its effectiveness through an application to semantic segmentation in Sect. 4.1. We further show an application to object mask annotation in Sect. 4.2 where we propose a model that reasons on the boundary-aligned grid output by a deep DefGrid to produce object polygons. Lastly, we showcase DefGrid as a standalone module for unsupervised image partitioning in Sect. 4.3.

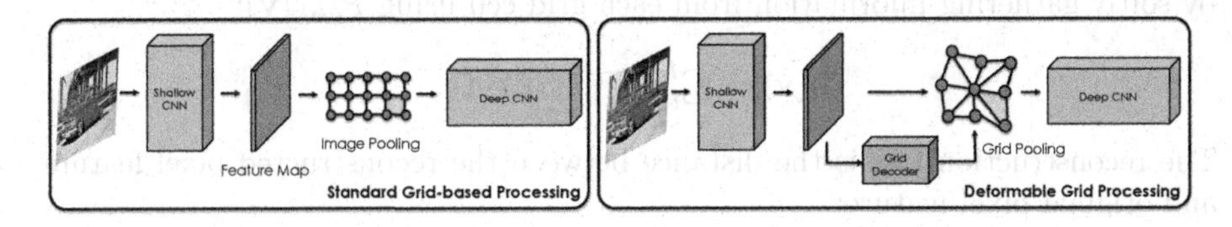

Fig. 3. Feature pooling in fixed grid versus DefGrid. DefGrid can easily be used in existing deep CNNs and perform learnable downsampling.

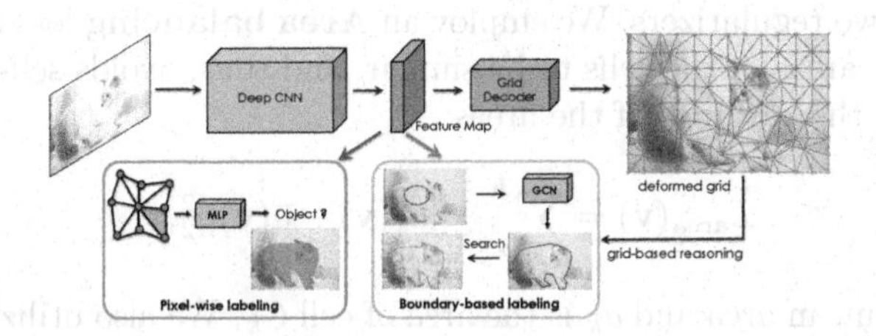

Fig. 4. Object mask annotation by reasoning on a DefGrid's boundary-aligned grid. We support both pixel-wise labeling and curve-based tracing.

4.1 Learnable Geometric Downsampling

Semantic segmentation of complex scenes typically requires high resolution images as input and thus produces high resolution feature maps which are computationally intensive. Existing deep CNNs often take downsampled images as input and use feature pooling and bottleneck structures to relieve the memory usage [18,19,42]. We argue that downsampling the features with our DefGrid can preserve finer geometric information. Given an arbitrary deep CNN architecture, we propose to insert a DefGrid using a shallow CNN encoder to predict a deformed grid. The predicted boundary-preserving grid can be used for geometry-aware feature pooling. Specifically, to represent each cell we can apply mean or max pooling by averaging or selecting maximum feature values within each triangular cell. Due to the regular grid topology, these features can be directly passed to a standard CNN. Note that the grid pooling operation warps the original feature map from the image coordinates to grid coordinates. Thus the final output (predicted semantic segmentation) is pasted back into the image plane by checking in which grid's cell the pixel lies. The full pipeline is end-to-end differentiable. We can jointly train the model in a multi-task manner with a cross-entropy loss for the semantic segmentation branch and grid deformation loss in Eq. 13. The DefGrid module is lightweight and thus bears minimal computational overhead. The architecture is illustrated in Fig. 3.

4.2 Object Mask Annotation

Object mask annotation is the problem of outlining a foreground object given a user-provided bounding box [3,7,27,29,39]. Two dominant approaches have been proposed to tackle this task. The first approach utilizes a deep neural network to predict a pixel-wise mask [26,29,39]. The second approach tries to outline the boundary with a polygon/spline [3,7,14,20,27]. Our DefGrid supports both approaches, and improves upon them via a polygonal grid-based reasoning (Fig. 4).

Boundary-Based Segmentation: We formulate the boundary-based segmentation as a minimal energy path searching problem. We search for a closed path along the grid's edges that has minimal Distance Transform energy[1]:

$$Q = \arg\min_{Q \in \mathcal{Q}} \sum_{i=1}^{M} DT(v_{Q_i}, v_{Q_{(i+1)\%M}}), \tag{14}$$

where $\mathcal{Q}$ denotes the set of all possible paths on our grid, and M is the length of path Q. We first predict a distance transform energy map for an object using a deep network trained with the L2 loss. We then compute the energy in each grid vertex via bilinear sampling. We obtain the energy for each grid edge by averaging the energy values for the points along the line defined by two vertices. Note that directly searching on the grid may result in many local minima. We

[1] The DT energy for each pixel is its distance to the nearest boundary.

employ Curve-GCN [27] to predict 40 seed points and snap each of these points to the grid vertex that has the minimal energy among its top-k closest vertices. Then for each neighboring seed points pair, we use Dijkstra algorithm to find the minimal energy path between them. We provide algorithm details in the appendix. Our approach improves over Curve-GCN in two aspects: 1) it better aligns with image boundaries as it explicitly reasons on our boundary-aligned grid, 2) since we search for a minimal energy path between neighboring points output by Curve-GCN, our approach can handle objects with more complex boundaries that cannot be represented with only 40 points.

Pixel-Wise Segmentation: Rather than producing a pixel-wise mask, we instead predict the class label for each grid cell. Specifically, we first use a deep neural network to obtain a feature map from the image. Then, for every grid cell, we average pool the feature of all pixels that are inside the cell, and use a MLP network to predict the class label for each cell. The model is trained with the cross-entropy loss. As the grid boundary aligns well with the object boundary, pooling the feature inside the grid is more efficient and effective for learning.

4.3 Unsupervised Image Partitioning

We can already view our deformed triangular cells as "superpixels", trained with unsupervised loss functions. We can go further and cluster cells by using the affinity between them. In particular, we view the deformed grid as an undirected weighted graph where each grid cell is a node and an edge connects two nodes if they share an edge in the grid. The weight for each edge is the affinity between two cells, which can be calculated using RGB values of pixels inside the cells.

Different clustering techniques can be used and exploring all is out of scope for this paper. To show the effectiveness of DefGrid as an unsupervised image partitioning method, we here utilize simple greedy agglomerative clustering. We average the affinity to represent a new node after merging. Clustering stops when we reach the desired number of superpixels or the affinity is lower than a threshold. Note that our superpixels are, by design, polygons. Note that supervised loss functions are naturally supported in our framework, however we do not explore them in this paper.

Table 1. Learnable downsampling on cityscapes Semantic Segm benchmark.

Downsampling Ratio	1/4			1/8			1/16			1/32		
Metric	mIoU	F (4px)	F (16px)	mIoU	F (4px)	F (16px)	mIoU	F (4px)	F (16px)	mIoU	F (4px)	F (16px)
Strided convolution	65.76	60.59	73.97	59.18	53.80	70.14	51.02	47.33	60.41	41.03	44.12	51.05
Max pooling	66.32	60.92	74.18	59.83	54.22	70.71	52.93	49.00	63.59	42.44	45.23	52.47
Average Pooling	66.53	61.09	74.39	59.45	0.5361	70.67	52.01	47.58	61.33	43.54	45.37	53.39
Grid max pooling	67.87	64.37	75.10	64.75	61.02	72.95	55.87	**53.98**	**66.62**	47.20	**48.74**	**60.73**
Grid average pooling	**67.91**	**64.99**	**75.43**	**65.36**	**61.12**	**73.03**	**56.94**	53.77	66.38	**48.30**	48.67	60.62

5 Experiments

We extensively evaluate DefGrid on downstream tasks. We first show application to learnable downsampling for semantic segmentation. We then evaluate on the object annotation task with boundary-based and pixel-wise methods. We finally show the effectiveness of DefGrid for unsupervised image partitioning.

5.1 Learnable Geometric Downsampling

To verify the effectiveness of our DefGrid as an effective downsampling method, we compare it with (fixed) image grid feature pooling methods as baselines, namely max/average pooling and stride convolution, on the Cityscapes [9] semantic segmentation benchmark. The baseline methods perform max/average pooling, or stride convolution on the shallow feature map, while our grid pooling methods apply the max/average pooling on deformed triangle cells. We compare our grid pooling with baselines when the height and width of the feature map is downsampled to 1/4, 1/8, 1/16 and 1/32 of the original image size. We use a modified ResNet50 [18] which is more lightweight than SOTA models [35].

Evaluation Metrics: Following [26,27,39], we evaluate the performance using mean Intersection-over-Union (mIoU), and the boundary F score, with 4 and 16 pixels threshold on the full image. All metrics are averaged across all classes.

Results: Performances (mIoU and boundary F scores) are reported in Table 1. Our DefGrid pooling methods consistently outperform the baselines, especially on the boundary score. We benefit from the edge-aligned property of the DefGrid coordinates. At 1/8 with 1/4 downsampling ratios, the baseline performance drops significantly due to missing the tiny instances, while our DefGrid pooling methods cope with this issue more gracefully. We also outperform baselines when the downsampling ratio is small, showing an efficient usage of limited spatial capacity. We visualize qualitative results for the predicted grids in Fig. 5. Our DefGrid better aligns with boundaries and thus what the downstream network "sees" is more informative than the fixed uniform grid.

Fig. 5. Illustration of learnable downsampling: We show DefGrid and its reconstructed image (left), comparing it to FixedGrid (right). [Please zoom in]

Table 2. Boundary-based object annotation on Cityscapes-Multicomp.

Model	Bicycle	Bus	Person	Train	Truck	Mcycle	Car	Rider	mIoU	F1	F2
Curve-GCN [27]	75.40	86.02	79.87	82.89	86.44	75.69	90.21	76.61	81.64	59.45	75.43
Pixel-wise	74.95	86.19	80.35	81.10	86.10	75.82	89.78	77.14	81.43	60.25	74.49
FixedGrid	75.10	85.73	79.84	84.35	86.02	75.97	89.76	76.56	81.67	59.01	74.77
DefGrid	**75.46**	**86.29**	**80.40**	**84.91**	**86.58**	**76.13**	**90.42**	**77.20**	**82.17**	**61.94**	**77.04**

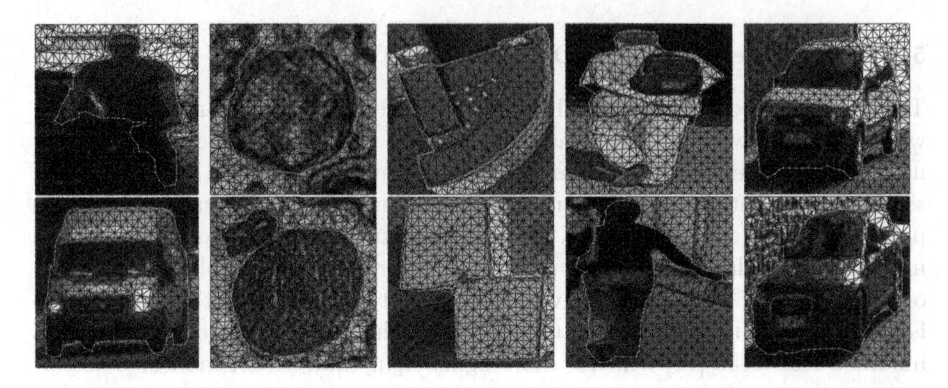

Fig. 6. Deformed Grid: We show examples of predicted grids both on in-domain (Cityscapes) and cross-domain images (Medical, Rooftop, ADE, KITTI). Orange line is the obtained minimal energy path along the grid's edges.

5.2 Object Annotation

Dataset: Following [3,7,26,27,39], we train and test both of our instance segmentation models on the Cityscapes dataset [9]. We assume the bounding box of each object is provided by the annotator and the task is to trace the boundary of the object. We evaluate under two different settings, depending on the model. To compare with pixel-wise methods, we follow the setting proposed in DELSE [39], where an image is first cropped with a 10 pixel expansion around the ground truth bounding box, and resized to the size of 224×224. This setting is referred to as Cityscapes-Stretch. Our boundary-based annotation model builds on top of Curve-GCN which predicts a polygon that is topologically equivalent to a sphere. To compare with the baseline, we thus assume that the annotator creates a box around each connected component of the object individually. We process each box in the same way as above, and evaluate performance on all component boxes. We refer to this setting as Cityscapes-Multicomp.

Boundary-Based Object Annotation

Network Architecture: We use the image encoder from DELSE [39], and further add three branches to predict grid deformation, Curve-GCN points and Distance Transform energy map. For each branch, we first separately apply one 3×3 conv filter to the feature map, followed by batch normalization [21] and ReLU activation. For grid deformation, we extract the feature for each vertex

Table 3. Cross-domain results for boundary-based methods. DefGrid significantly outperforms baselines, particularly evident in F-scores.

Method	Metrics	KITTI	ADE	Rooftop	Card.MR	ssTEM
Curve-GCN [27]	mIoU	87.43	76.71	81.11	86.18	68.97
	F1	64.90	39.37	30.99	62.73	44.74
	F2	78.40	52.86	45.08	78.22	59.85
Pixel-wise	mIoU	86.99	78.23	80.81	88.00	69.37
	F1	62.73	42.88	28.44	62.63	46.07
	F2	77.17	56.15	42.01	77.94	59.77
DefGrid	mIoU	**88.05**	**78.54**	**83.10**	**89.01**	**71.82**
	F1	**66.70**	**43.54**	**34.39**	**65.31**	**50.31**
	F2	**80.23**	**57.20**	**49.79**	**81.92**	**65.14**

with bilinear interpolation, and use a GCN to predict the offset for each vertex. For predicting spoints, we follow Curve-GCN [27]. For the DT energy, we apply two 3×3 conv filters with batch normalization and ReLU activation.

Baselines: For Curve-GCN [27], we compare with Spline-GCN, as it gives better performance than Polygon-GCN in the original paper [27]. We use the official codebase but instead use our image encoder to get the feature map (with negligible performance gap) for a fair comparison. We also compare with pixel-wise methods, where we add two conv filters after the image encoder, which is similar to DELSE [39] and DEXTR [29] but without extreme points. We further compare to our version of the model where the grid is fixed, referred to as FixedGrid.

Results: Table 2 reports results. Our method outperforms baselines in all metrics. Performance gains are particularly significant in terms of F-scores, even when compared with pixel-wise methods. Since network details are the same across the methods, these results signify the importance of reasoning on the grid. Compared with FixedGrid at different grid sizes in Fig. 7, DefGrid achieves superior performance. We attribute the performance gains due to better alignment with boundaries and more flexibility in tracing a longer contour. We show qualitative results in Fig. 6 and 8.

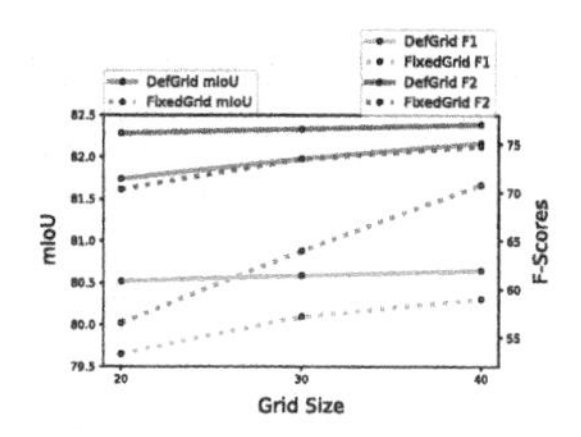

Fig. 7. Object annotation for Cityscapes: FixedGrid vs DefGrid.

Cross Domain Results: Following [27], we evaluate models' ability in generalizing to other datasets out of the box. Quantitative and qualitative results are reported in Table 3 and Fig. 6, Fig. 8, respectively. We outperform all baselines on all datasets, in terms of all metrics, and the performance gains are particularly evident for the F-scores. Qualitatively, in all cross-domain examples the predicted grid's edges align well with real object boundaries (without any finetuning), demonstrating superior generalization capability for DefGrid.

Table 4. Interactive annotation results on both in-domain (first two lines) and cross domain datasets. Our DefGrid starts with a higher automatic performance and keeps this gap across (simulated) annotation rounds.

		Round 0 (Automatic)			Round 1			Round 2			Round 3			Round 4			Round 5		
Dataset	Model	mIoU	F1	F2	mIoU	F1	F2	mIoU	F1	F2	mIoU	F1	F2	mIoU	F1	F2	mIoU	F1	F2
CityScapes [9]	CurveGCN [27]	80.74	58.01	73.67	83.04	60.47	76.35	84.58	64.42	79.42	85.56	67.41	81.40	86.27	69.87	82.94	86.79	71.91	84.12
	DefGrid	**81.98**	**61.43**	**76.56**	**84.48**	**66.59**	**80.96**	**85.73**	**69.98**	**83.41**	**86.41**	**72.05**	**84.71**	**86.81**	**73.60**	**85.51**	**87.12**	**74.84**	**86.20**
ADE [43]	CurveGCN [27]	76.71	39.37	52.86	79.26	42.43	56.62	81.09	46.04	60.51	82.47	49.61	64.00	83.44	52.56	66.61	84.38	55.42	69.00
	DefGrid	**78.54**	**43.54**	**57.20**	**80.67**	**48.57**	**62.76**	**82.27**	**52.33**	**66.45**	**83.43**	**55.60**	**69.37**	**84.23**	**58.22**	**71.79**	**84.69**	**60.03**	**73.29**
KITTI [15]	CurveGCN [27]	87.43	64.90	78.40	89.41	69.20	82.84	90.50	73.11	85.84	91.22	75.94	87.73	91.67	78.16	89.20	**92.00**	**79.90**	90.21
	DefGrid	**88.05**	**66.70**	**80.23**	**89.94**	**70.80**	**84.16**	**90.90**	**74.65**	**87.12**	**91.45**	**77.08**	**88.81**	**91.70**	**78.67**	**89.78**	91.87	79.72	**90.33**
Rooftop [34]	CurveGCN [27]	81.11	30.99	45.08	83.94	34.00	49.24	85.53	37.42	53.39	86.72	41.39	57.85	87.68	45.29	61.88	88.28	48.96	65.60
	DefGrid	**83.10**	**34.39**	**49.79**	**85.37**	**39.60**	**56.04**	**86.77**	**43.74**	**60.73**	**87.53**	**47.30**	**64.51**	**88.19**	**49.95**	**67.50**	**88.55**	**52.20**	**69.66**
Card.MR [33]	CurveGCN [27]	86.18	62.73	78.22	89.22	68.26	84.91	90.42	73.10	89.16	91.25	76.94	91.94	92.00	80.67	94.45	92.45	83.45	95.99
	DefGrid	**89.01**	**65.31**	**81.92**	**91.01**	**72.28**	**88.83**	**91.90**	**77.64**	**92.74**	**92.55**	**81.03**	**94.89**	**92.96**	**83.24**	**95.98**	**93.31**	**85.47**	**96.96**
ssTEM [16]	CurveGCN [27]	68.97	44.74	59.85	71.66	47.81	63.21	74.75	52.78	68.82	76.57	56.88	72.99	78.06	60.67	76.50	79.22	63.84	79.17
	DefGrid	**71.82**	**50.31**	**65.14**	**73.51**	**55.43**	**70.44**	**75.25**	**60.64**	**75.69**	**77.28**	**65.18**	**79.72**	**78.46**	**68.44**	**82.24**	**78.89**	**70.59**	**83.62**

Table 5. Pixel-based methods on Cityscapes-Stretch. Note that PolyTransform [26] uses 512 × 512 resolution, while other methods use 224 × 224 resolution.

Model	Bicycle	Bus	Person	Train	Truck	Mcycle	Car	Rider	mIoU	F1	F2
DELSE* [39]	74.32	**88.85**	80.14	**80.35**	86.05	74.10	86.35	76.74	80.86	60.29	74.40
PolyTransform [26]	74.22	88.78	80.73	77.91	86.45	**74.42**	86.82	**77.85**	80.90	62.33	76.55
OurBack + SLIC [1]	73.88	85.47	79.80	77.97	86.32	72.62	87.85	76.14	80.01	57.95	72.17
DefGrid	**74.82**	87.09	**80.87**	**81.05**	**87.52**	73.44	**89.19**	77.36	**81.42**	**63.38**	**76.89**

Interactive Instance Annotation: We follow Curve-GCN [27] and also report performance for the interactive setting in which an annotator corrects errors by moving the predicted polygon vertices. We follow the original setting but restrict our reasoning on the deformed grid. Results for different simulated rounds of annotation are reported in Table 4 with evident performance gains.

Fig. 8. Qualitative results: Note that the method employs ground-truth boxes. The first row is from Cityscapes. In the bottom two rows, from left to right: Medical, KITTI, Rooftop, ADE.

Pixel-Wise Object Instance Annotation

Network Architecture: To predict the class label for each deformed grid's cell, we first average the feature of all pixels inside each cell, and use a 4-layer MLP to predict the probability of foreground/background. For the hyperparameters and architecture details, we refer to the appendix.

Results: Table 5 provides quantitative results. We show qualitative results in the appendix. Predicting the (binary) class label over the deformed grid's cells achieves higher performance than carefully designed pixel-wise baselines, demonstrating the effectiveness of reasoning on our deformed grid.

5.3 Unsupervised Image Partitioning

Dataset: Following SSN [24], we train the model on 200 training images in the BSDS500 [4] and evaluate on 200 test images. Details are provided in appendix.

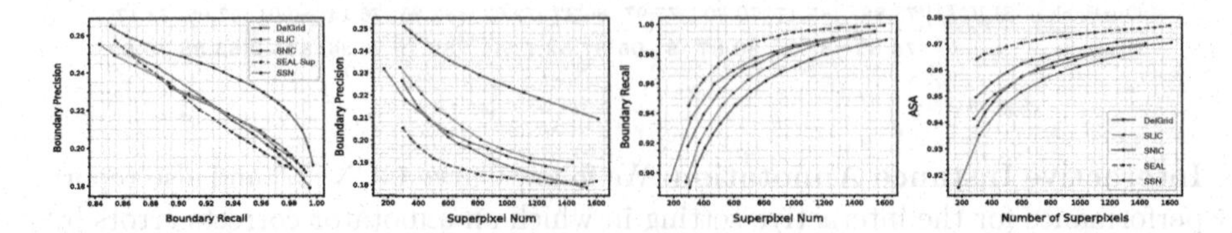

Fig. 9. Unsupervised image partitioning. From left to right: BP-BR, BP, BR and ASA. We use dotted line to represent supervised method.

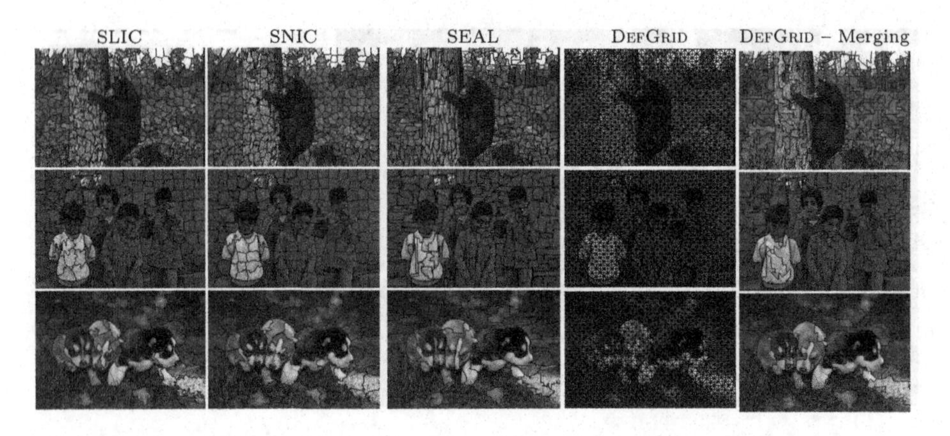

Fig. 10. Unsupervised image partitioning. We compare the DefGrid and results after clustering with existing superpixel baselines. [Please zoom in]

Evaluation Metric: Following the SSN and SEAL [24,36], we use Achievable Segmentation Accuracy (ASA), Boundary Precision (BP) and Boundary Recall (BR) to evaluate the performance of superpixels. For BP and BR, we set the tolerance to be 3 pixels. The evaluation scripts are from SEAL[2].

Baselines: We compare our method with both traditional superpixel methods, SLIC [1], SNIC [2], and deep learning based method SSN [24], SEAL [36]. Note that SEAL not only utilizes ground-truth annotation for training, but also is trained on validation set. We use the official codebase and trained model provided by the authors. We perform all comparisons at different numbers of superpixels.

---

[2] https://github.com/wctu/SEAL.

Results: Quantitative and qualitative results are presented in Fig. 9, Fig. 10, respectively. The deformed grid aligns well with object boundary and outperforms all unsupervised baselines in terms of BP and BP-BR curve, with comparable performance with counterparts in terms of ASA and BR. Our intuition is that, since the edge between two vertices in our grid is constrained to be a straight line, while the ground truth annotation is labelled pixel-by-pixel, our grid sacrifices a little boundary recall while achieving higher boundary precision. It is worth to note that our method outperforms the supervised method SEAL [36]. This reflects the fact that an appearance feature provides a useful signal for training our DefGrid , and our method effectively utilizes this signal.

6 Conclusion

In this paper, we proposed to deform a regular grid to better align with image boundaries as a more efficient way to process images. Our DefGrid is a neural network that predicts offsets for vertices in a grid to perform the alignment, and can be trained entirely with unsupervised losses. We showcase our approach in several downstream tasks with significant performance gains. Our method produces accurate superpixel segmentations, is significantly more precise in outlining object boundaries particularly on out of domain datasets, and leads to a large improvements for semantic segmentation. We hope that our DefGrid benefits other computer vision tasks when combined with deep networks.

Acknowledgments. This work was supported by NSERC. SF acknowledges the Canada CIFAR AI Chair award at the Vector Institute. We thank Frank Shen, Wenzheng Chen and Huan Ling for helpful discussions, and also thank the anonymous reviewers for valuable comments.

References

1. Achanta, R., Shaji, A., Smith, K., Lucchi, A., Fua, P., Süsstrunk, S.: Slic superpixels. Technical report (2010)
2. Achanta, R., Susstrunk, S.: Superpixels and polygons using simple non-iterative clustering. In: CVPR, pp. 4651–4660 (2017)
3. Acuna, D., Ling, H., Kar, A., Fidler, S.: Efficient interactive annotation of segmentation datasets with polygon-RNN++. In: CVPR (2018)
4. Arbelaez, P., Maire, M., Fowlkes, C., Malik, J.: Contour detection and hierarchical image segmentation. IEEE Trans. Pattern Anal. Mach. Intell. **33**(5), 898–916 (2010)
5. den Bergh, M.V., Roig, G., Boix, X., Manen, S., Gool, L.V.: Online video superpixels for temporal window objectness. In: ICCV (2013)
6. Bódis-Szomorú, A., Riemenschneider, H., Van Gool, L.: Superpixel meshes for fast edge-preserving surface reconstruction. In: CVPR, pp. 2011–2020 (2015)
7. Castrejon, L., Kundu, K., Urtasun, R., Fidler, S.: Annotating object instances with a polygon-RNN. In: CVPR (2017)
8. Chen, W., et al.: Learning to predict 3D objects with an interpolation-based differentiable renderer. In: Advances in Neural Information Processing Systems (2019)

9. Cordts, M., et al.: The cityscapes dataset for semantic urban scene understanding. In: Proceedings of the IEEE Conference on Computer Vision and Pattern Recognition, pp. 3213–3223 (2016)
10. Dai, J., et al.: Deformable convolutional networks. In: Proceedings of the IEEE International Conference on Computer Vision, pp. 764–773 (2017)
11. Duan, L., Lafarge, F.: Image partitioning into convex polygons. In: CVPR, pp. 3119–3127 (2015)
12. Felzenszwalb, P.F., Huttenlocher, D.P.: Efficient graph-based image segmentation. Int. J. Comput. Vis. **59**(2), 167–181 (2004)
13. Gadde, R., Jampani, V., Kiefel, M., Kappler, D., Gehler, P.V.: Superpixel convolutional networks using bilateral inceptions. In: Leibe, B., Matas, J., Sebe, N., Welling, M. (eds.) ECCV 2016. LNCS, vol. 9905, pp. 597–613. Springer, Cham (2016). https://doi.org/10.1007/978-3-319-46448-0_36
14. Gao, J., Tang, C., Ganapathi-Subramanian, V., Huang, J., Su, H., Guibas, L.J.: DeepSpline: data-driven reconstruction of parametric curves and surfaces. arXiv preprint arXiv:1901.03781 (2019)
15. Geiger, A., Lenz, P., Urtasun, R.: Are we ready for autonomous driving? The KITTI vision benchmark suite. In: CVPR (2012)
16. Gerhard, S., Funke, J., Martel, J., Cardona, A., Fetter, R.: Segmented anisotropic ssTEM dataset of neural tissue (2013). https://doi.org/10.6084/m9.figshare.856713.v1. https://figshare.com/articles/Segmented_anisotropic_ssTEM_dataset_of_neural_tissue/856713
17. Grundmann, M., Kwatra, V., Han, M., Essa, I.: Efficient hierarchical graph-based video segmentation. In: CVPR (2010)
18. He, K., Zhang, X., Ren, S., Sun, J.: Deep residual learning for image recognition. In: Proceedings of the IEEE Conference on Computer Vision and Pattern Recognition, pp. 770–778 (2016)
19. Huang, G., Liu, Z., Van Der Maaten, L., Weinberger, K.Q.: Densely connected convolutional networks. In: Proceedings of the IEEE Conference on Computer Vision and Pattern Recognition, pp. 4700–4708 (2017)
20. Huang, J., et al.: DeepPrimitive: image decomposition by layered primitive detection. Comput. Vis. Media **4**(4), 385–397 (2018)
21. Ioffe, S., Szegedy, C.: Batch normalization: accelerating deep network training by reducing internal covariate shift. In: ICML, pp. 448–456 (2015)
22. Jacobson, A., Baran, I., Popović, J., Sorkine, O.: Bounded biharmonic weights for real-time deformation. SIGGRAPH **30**(4), 78:1–78:8 (2011)
23. Jaderberg, M., Simonyan, K., Zisserman, A., et al.: Spatial transformer networks. In: Advances in Neural Information Processing Systems, pp. 2017–2025 (2015)
24. Jampani, V., Sun, D., Liu, M.Y., Yang, M.H., Kautz, J.: Superpixel samping networks. In: ECCV (2018)
25. Levinshtein, A., Stere, A., Kutulakos, K., Fleet, D., Dickinson, S., Siddiqi, K.: TurboPixels: fast superpixels using geometric flows. PAMI **31**(12), 2290–2297 (2009)
26. Liang, J., Homayounfar, N., Ma, W.C., Xiong, Y., Hu, R., Urtasun, R.: PolyTransform: deep polygon transformer for instance segmentation. arXiv preprint arXiv:1912.02801 (2019)
27. Ling, H., Gao, J., Kar, A., Chen, W., Fidler, S.: Fast interactive object annotation with curve-GCN. In: CVPR, pp. 5257–5266 (2019)
28. Liu, M.Y., Tuzel, O., Ramalingam, S., Chellappa, R.: Entropy rate superpixel segmentation. In: CVPR 2011, pp. 2097–2104. IEEE (2011)
29. Maninis, K.K., Caelles, S., Pont-Tuset, J., Van Gool, L.: Deep extreme cut: from extreme points to object segmentation. In: CVPR (2018)

30. Moore, A., Prince, S., Warrell, J., Mohammed, U., Jones, G.: Superpixel lattices. In: CVPR (2008)
31. Recasens, A., Kellnhofer, P., Stent, S., Matusik, W., Torralba, A.: Learning to zoom: a saliency-based sampling layer for neural networks. In: Proceedings of the European Conference on Computer Vision (ECCV), pp. 51–66 (2018)
32. Shi, J., Malik, J.: Normalized cuts and image segmentation. Departmental Papers (CIS), p. 107 (2000)
33. Suinesiaputra, A., et al.: A collaborative resource to build consensus for automated left ventricular segmentation of cardiac MR images. Med. Image Anal. **18**(1), 50–62 (2014)
34. Sun, X., Christoudias, C.M., Fua, P.: Free-shape polygonal object localization. In: Fleet, D., Pajdla, T., Schiele, B., Tuytelaars, T. (eds.) ECCV 2014. LNCS, vol. 8694, pp. 317–332. Springer, Cham (2014). https://doi.org/10.1007/978-3-319-10599-4_21
35. Takikawa, T., Acuna, D., Jampani, V., Fidler, S.: Gated-SCNN: gated shape CNNs for semantic segmentation. ArXiv **abs/1907.05740** (2019)
36. Tu, W.C., et al.: Learning superpixels with segmentation-aware affinity loss. In: Proceedings of the IEEE Conference on Computer Vision and Pattern Recognition, pp. 568–576 (2018)
37. Vincent, L., Soille, P.: Watersheds in digital spaces: An efficient algorithm based on immersion simulations. PAMI **13**(6), 583–598 (1991)
38. Wang, N., Zhang, Y., Li, Z., Fu, Y., Liu, W., Jiang, Y.-G.: Pixel2Mesh: generating 3D mesh models from single RGB images. In: Ferrari, V., Hebert, M., Sminchisescu, C., Weiss, Y. (eds.) ECCV 2018. LNCS, vol. 11215, pp. 55–71. Springer, Cham (2018). https://doi.org/10.1007/978-3-030-01252-6_4
39. Wang, Z., Acuna, D., Ling, H., Kar, A., Fidler, S.: Object instance annotation with deep extreme level set evolution. In: CVPR (2019)
40. Yamaguchi, K., McAllester, D., Urtasun, R.: Efficient joint segmentation, occlusion labeling, stereo and flow estimation. In: Fleet, D., Pajdla, T., Schiele, B., Tuytelaars, T. (eds.) ECCV 2014. LNCS, vol. 8693, pp. 756–771. Springer, Cham (2014). https://doi.org/10.1007/978-3-319-10602-1_49
41. Yao, J., Boben, M., Fidler, S., Urtasun, R.: Real-time coarse-to-fine topologically preserving segmentation. In: CVPR (2015)
42. Zhao, H., Shi, J., Qi, X., Wang, X., Jia, J.: Pyramid scene parsing network. In: CVPR (2017)
43. Zhou, B., Zhao, H., Puig, X., Fidler, S., Barriuso, A., Torralba, A.: Scene parsing through ADE20K dataset. In: CVPR (2017)

Soft Expert Reward Learning for Vision-and-Language Navigation

Hu Wang, Qi Wu, and Chunhua Shen(✉)

The University of Adelaide, Adelaide, Australia
{hu.wang,qi.wu01,chunhua.shen}@adelaide.edu.au

Abstract. Vision-and-Language Navigation (VLN) requires an agent to find a specified spot in an unseen environment by following natural language instructions. Dominant methods based on supervised learning clone expert's behaviours and thus perform better on seen environments, while showing restricted performance on unseen ones. Reinforcement Learning (RL) based models show better generalisation ability but have issues as well, requiring large amount of manual reward engineering is one of which. In this paper, we introduce a Soft Expert Reward Learning (SERL) model to overcome the reward engineering designing and generalisation problems of the VLN task. Our proposed method consists of two complementary components: Soft Expert Distillation (SED) module encourages agents to behave like an expert as much as possible, but in a soft fashion; Self Perceiving (SP) module targets at pushing the agent towards the final destination as fast as possible. Empirically, we evaluate our model on the VLN seen, unseen and test splits and the model outperforms the state-of-the-art methods on most of the evaluation metrics.

Keywords: Soft expert distillation · Self perceiving reward · Vision-and-language navigation

1 Introduction

Vision-and-Language Navigation (VLN) tasks [2] define a comprehensive problem: an embodied agent is placed at a spot in a photo-realistic house and the agent is called to navigate to a specific spot based on given natural language instructions. Rising research interests have been put into the VLN since multi-modal data are involved. One of the biggest challenges for this task is to ask an agent to perform appropriate actions in an unseen environment. This in turn requires the agent to learn human behaviours to understand and explore the scene, instead of memorising it.

Current VLN models [2,4,7,9,10] rely much on behavioural cloning (BC) that treats expert behaviours as strong supervision signals. By doing this, it enables the agents to gain better performance on seen scenarios, however the agents meet trouble on unseen environments due to the error accumulation. As stated in [14], teacher forcing models suffer from distribution shift issues because of the greediness of imitating demonstrated expert actions.

© Springer Nature Switzerland AG 2020
A. Vedaldi et al. (Eds.): ECCV 2020, LNCS 12354, pp. 126–141, 2020.
https://doi.org/10.1007/978-3-030-58545-7_8

Some other works [16,19], instead, adopt reinforcement learning (RL) along with supervised learning methods intending to overcome the error accumulation issue caused by hard behavioural cloning. However, the reward engineering in RL suffers issues: the reward functions designed at one environment/task may not generalise well to other scenarios; in many practical and complicated tasks, it is hard to define concrete reward functions as game scores. What is more, a hand-crafted reward is defined to target at a certain functionality, it thus inevitably incurs lacking comprehensive considering of the system dynamics. The designing of a reward function requires careful manual tuning and it also suffers generalisation problem due to environment-oriented reward designing, which may affect the model performance while inference.

In this paper, we propose a Soft Expert Reward Learning (SERL) model to address above issues. Our proposed method consists of two orthogonal parts: the Soft Expert Distillation (SED) module that portrays the expert data distribution by distilling knowledge from a random projection space and a Self Perceiving (SP) module that encourages agents to reach the goal as soon as possible. For the SED module, intuitively, a higher reward should be assigned to an agent who takes an action "close" to its expert. To measure the similarity continuously, a density function was adopted to reflect this process in a soft manner rather than leveraging behaviour cloning directly. This density function is implemented to calculate the similarity between observation-action pairs of the expert and the agent in a randomly projected space, by doing which it transforms the expert behaviour into a soft reward signal for the reinforcement learning branch. For the Self Perceiving (SP) module, our model first predicts the schedule to the target location and then utilises the predicted schedule information as an additional reward. As a result, the agent can perceive its current schedule and use it to further pushing itself forward to the goal.

The two newly designed reward modules work complementarily: the Soft Expert Distillation (SED) reward encourages agents to behave as an expert, but the soften behaviour-imitation process makes it more robust; Self Perceiving (SP) module targets at pushing the agents towards the final destination by introducing the current schedule information as another intrinsic reward signal. In summary, this paper makes the following three main contributions.

- We propose a Soft Expert Distillation (SED) formulation, which is very simple yet offers a highly effective reward signal for obtaining expressive navigational ability. The SED reward encourages the agent to have a better alignment with its expert in a soft manner.
- We introduce another complementary reward signal with aforementioned SED reward termed as Self Perceiving reward that can help the agent use the current schedule information to push itself to reach the destination as soon as possible.
- As a result, we show our instantiated model termed as SERL that enables better performance than current state-of-the-art competing methods in both validation unseen and test unseen set of VLN Room-to-Room dataset [2].

2 Related Work

2.1 Vision-and-Language Navigation

In order to gain promising performance on Vision-and-Language (VLN) [2] task, numerous methods have been proposed, as listed in Table 1. Many existing works adopt supervised learning and behaviour cloning based methods. Seq2seq [2] model is the most naive baseline that utilises an LSTM-based sequence-to-sequence architecture with attention mechanism to predict the next action. Speaker-Follower [4] model designs a language model ("speaker") to learn the relationship between visual and language information, as well as a policy network ("follower") to take actions based on multi-modal inputs. It uses "speaker" to synthesise new instructions for data augmentation and help the policy network to select routes. [7] claims its proposed FAST model is able to balance local and global signals while exploring an unobserved environment. It enables the agent act greedily but allows the agent backtrack if necessary according to global signals. [9] proposes a visual-language co-grounding framework named as self-monitoring model to better fuse the instructions and visual inputs. Building upon self-monitoring model, [10] provides a strategy for the agent to retrieve and re-choose paths based on monitored progress.

Reinforcement learning [8,12,15] is another paradigm for parameter optimisation. Wang *et al.* [19] propose a novel Reinforced Cross-modal Matching (RCM) via reinforcement learning to enforce cross-modal matching locally and globally along with imitation learning. In RCM model, an extrinsic reward measuring the reduced distance toward the target location after taking actions, as well as an intrinsic cross-modal matching reward between trajectories and instructions, are proposed. Most recently, [16] introduces a novel environment dropout to drop features channel-wisely targeting at feature maps inconsistency issue through combining behaviour cloning and reinforcement learning.

Table 1. Performance Evaluation across different methods.

Methods	Behaviour cloning	Reinforcement learning	Reward engineering	Reward learning
Random [2]				
Seq2seq [2]	✓			
Speaker-Follower [4]	✓			
FAST [7]	✓			
Reinforced Cross-Modal [19]	✓	✓	✓	
Self-Monitoring [9]	✓			
Regretful Agent [10]	✓			
EnvDrop [16]	✓	✓	✓	
SERL (Ours)	✓	✓	✓	✓

However, these approaches require either exact imitation of the expert demonstrations or careful reward designing. Behaviour cloning techniques unfortunately lead to error accumulation and further result in catastrophic failure while the agent is exploring unknown environments. Moreover, reward engineering requires careful manual tuning, which motivates us to propose SERL model to learn reward functions from the expert distribution directly.

2.2 Reward Learning

Reward engineering is commonly used to design reward functions for reinforcement learning algorithms. In conventional reinforcement learning tasks, such as playing Atari games [3], rewards are individually shaped by each game simulators. However, reward engineering has obvious drawbacks—the reward functions are designed targeting at different environments which is not generic. There are some methods have been proposed to solve this problem. Recently, Inverse reinforcement learning (IRL) [13] framework is proposed to extract reward functions from expert behaviours by updating both of the reward functions and the policy networks. Random Expert Distillation (RED) [18] proposed an expert policy support estimation method to distil rewards from given expert trajectories. Generative Adversarial Imitation Learning (GAIL) [6] is also a recently proposed model which tries to bypass the reward function and learn experts behaviour directly with generative adversarial networks.

Comparing with the IRL and GAIL models, our proposed Soft Expert Distillation module learns expert demonstration data distribution directly by comparing the output similarity between a randomised network and a distillation network, rather than utilising iterative model updating and generative adversarial networks. The RED model designs state and action in relatively small spaces for the Mujoco environment [17] and its driving task; while we design our SED module in fundamentally different state and action spaces for navigation in photo-realistic Matterport3D environments. We are the first to introduce soft expert reward learning framework into Vision-and-Language task.

3 Soft Expert Reward Learning Model

3.1 Overview and Problem Definition

Vision-and-Language Navigation task requires an agent placed at a unknown photo-realistic house to understand multi-modal data comprehensively, so that the agent can navigate to the specified location. The multi-modal data includes natural image data and natural language instructions. More specifically, after an agent is spawn, at each time step t the observation of the agent consists of 36 images of panoramic views, denoted as $V_t = \{v_{t,1}, v_{t,2}, ..., v_{t,36}\}$. The navigable views $N_t = \{n_{t,1}, n_{t,2}, ..., n_{t,k}, n_{t,k+1}\}$ are given as well, where k denotes the maximum number of navigable viewpoints and $n_{t,k+1}$ represents "stay" action. A m words length instruction is given which is denoted as $X = \{x_1, x_2, ..., x_m\}$.

Based on the visual and language information, actions at each time a_t will be selected and eventually a trajectory $\tau = \{a_1, a_2, ..., a_T\}$ is formed. The objective of VLN task is to find the optimal action a_t^* at each step to quickly reach the target location, while keep the trajectory τ as short as possible. Since Vision-and-Language Navigation task is a sequential decision problem, it can be modelled as a Markov Decision Process (MDP), which is noted as a four-element-tuple $(\mathcal{S}, \mathcal{A}, \mathcal{P}, \mathcal{R})$. $\mathcal{S}$ and $\mathcal{A}$ represent state and action sets relatively. $\mathcal{P}$ is the environment dynamics and it can be presented in the form $\mathcal{P}(s, s') = P(s'|s, a)$. $\mathcal{R}$ is the reward function.

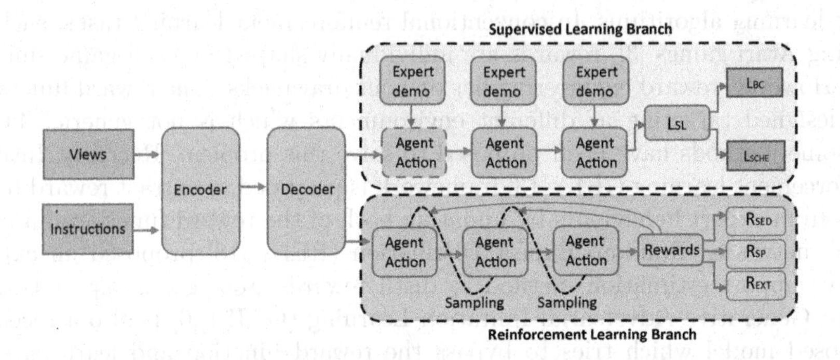

Fig. 1. The proposed Soft Expert Reward Learning (SERL) framework. After getting the visual features and language features through the encoder, they are fed into the decoder to obtain the selected action a_t for time step t. The training process of SERL is divided into two parts: a supervised learning branch and a reinforcement learning branch. We introduce two novel rewards (marked with yellow stars in the figure): Soft Expert Distillation (SED) reward and Self Perceiving (SP) reward. (Color figure online)

In this paper we introduce a Soft Expert Reward Learning model to distil reward function directly from expert demonstrations and soften the process of behaviour cloning to alleviate the drawbacks from error accumulation. The structure of our model is illustrated through Fig. 1. We follow a standard Encoder-Decoder paradigm. The encoder plays the role as a multi-modal data feature extractor to fetch the features from both visual images and language instructions. The decoder is a LSTM (long short-term memory) network with attention mechanism to predict actions according to the abovementioned two branches: the supervised learning branch helps the agent imitate the expert demonstration and perceive the current schedule to the target location; the reinforcement learning branch optimises the outputted action probability distribution from reinforcement learning aspects. The key difference of our proposed SERL model with previous models is that we proposed two novel intrinsic reward signals: Soft Expert Distillation reward R_{SED} encourages the agent to align with expert actions but in a soft fashion and Self Perceiving reward R_{SP} motivates the agent to reach the goal as fast as possible with predicted schedule information. In the following sections, we will first introduce the Encoder-Decoder structure and then introduce the two reward functions.

3.2 Encoder-Decoder Structure

Encoder-Decoder structure (as shown in Fig. 2) is adopted as the main structure of our method. Natural image data and natural language instructions are inputted to an encoder to extract corresponding features maps. Following the paper [9,16], we extract ResNet [5] features of the navigable views concatenated with the orientation as the visual features $VisF_t$. We then use a Bi-Directional Long Short-Term Memory (Bi-LSTM) to pull out language features $LangF_t$. The multi-modal features are fed into a decoder to output the next action probability vectors later on.

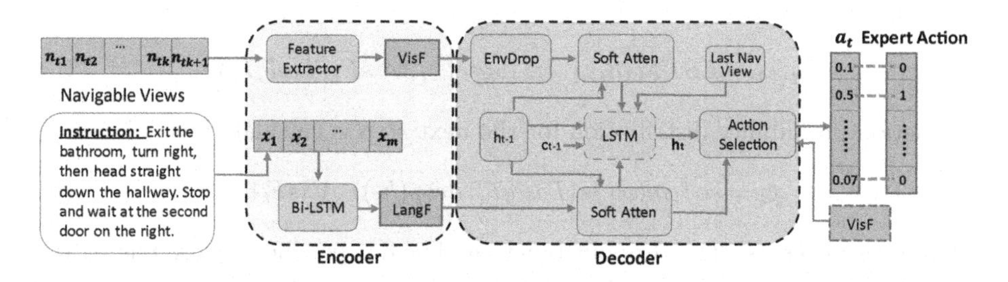

Fig. 2. Encoder-Decoder Structure of Soft Expert Reward Learning (SERL) framework. After fetching the visual and language features from the encoder, the multi-modal features are fed into the decoder to obtain cross-modal attentions. Finally, actions will be chosen according to the attentive features.

Encoder: On the encoder side, after pre-extracting ResNet features of different views, the feature maps of each navigable view $n_{t,i}$ is attached with an orientation tag $(\cos\gamma_{t,i}, \sin\gamma_{t,i}, \cos\varphi_{t,i}, \sin\varphi_{t,i})$ to form the visual feature $VisF_t$:

$$VisF_t = concat(resnet(n_{t,i}), (\cos\gamma_{t,i}, \sin\gamma_{t,i}, \cos\varphi_{t,i}, \sin\varphi_{t,i})), \tag{1}$$

where $concat(.)$ is a concatenation function.

For the language perspective, after each word of the instruction is tokenised into a vector, the token vectors are fed into a Bi-LSTM network to extract the language features $LangF_t$. As Eq. 2, formally we have

$$LangF_t = \{x'_1, x'_2, ..., x'_m\} = Bi\text{-}LSTM(\{x_1, x_2, ..., x_m\}), \tag{2}$$

where x'_i is the corresponding i-th encoded word tokenised by Bi-LSTM.

Decoder: On the decoder side, after the visual feature $VisF_t$ and language features $LangF_t$ are formed, along with the last cross-modal hidden state h_{t-1}, they are fed into soft attention layers to fetch the attentive visual and language features. Following the work [16], the environment dropout is used on $VisF_t$ before feeding into soft attention layer to obtain feature-wise dropout for consistency in different views. Formally,

$$\widetilde{VisF}_t = Soft\text{-}Atten(EnvDrop(VisF_t), h_{t-1}), \tag{3}$$

$$\widetilde{LangF}_t = Soft\text{-}Atten(LangF_t, h_{t-1}). \tag{4}$$

Together with previous navigated view pre^v_{t-1}, last cross-modal hidden state h_{t-1}, cell state c_{t-1}, attentive visual and language features are fed into a LSTM layer to form the cross-modal hidden state h_t and cell state c_t at step t. This step is critical for the model to fuse the visual and language multi-modal signals to choose the action.

$$h_t, c_t = LSTM(h_{t-1}, c_{t-1}, pre^v_{t-1}, \widetilde{VisF}_t, \widetilde{LangF}_t). \tag{5}$$

The action probability distribution for the next step is calculated as:

$$p_t = softmax(fc(\widetilde{LangF}_t, drop(h_t)) \cdot VisF_t), \tag{6}$$

where $drop(.)$ represents a dropout function. The dot product $\cdot$ is used hereafter for matrix multiplication operation.

The decoder is connected to two branches: supervised learning branch and reinforcement learning branch. These two branches optimise the outputted action probability distribution from two different learning paradigms. In this case, the total loss function is:

$$L = L_{SL} + L_{RL}. \tag{7}$$

SL Branch: In the supervised learning branch, the cross-entropy loss between the predicted action logits and expert actions one-hot vector is calculated to force the agent to mimic its teacher's behaviours. This loss is termed as behaviour cloning loss L_{BC}. Following the work [9], besides the behaviour cloning loss, another loss to predict current schedule towards the goal is adopted. This loss is named as schedule loss L_{SCHE} working as an additional supervisory signal. Formally, the loss function for the supervised learning branch is:

$$L_{SL} = L_{BC} + L_{SCHE}. \tag{8}$$

where the behaviour cloning loss L_{BC} can be presented detailedly:

$$L_{BC} = -\sum_{i}^{n} y^{act}_{t,i} log(p_{t,i}), \tag{9}$$

where p_t and y^{act}_t are predicted action logits and expert actions one-hot vector at step t respectively.

To calculate the L_{SCHE}, the model ought to predict distance improvement ratio in advance at each step as its current schedule information. Then, L2 distance between predicted schedule and the genuine schedule is chosen as the loss function. Formally,

$$L_{SCHE} = (y_t^{sche} - V_t^{sche})^2, \tag{10}$$

where V_t^{sche} represents the predicted schedule which will be described in detail in the subsequent section and y_t^{sche} is the corresponding true schedule value.

RL Branch: As the reinforcement learning branch shown in Fig. 1, we adopt actor-critic algorithm [11] as our reinforcement learning method. For the reinforcement learning branch, the training loss L_{RL} can be formally represented as:

$$L_{RL} = \underbrace{\sum_t -log(p_t) * (\overline{R_t} - v(h_t))}_{actor\ loss} + \underbrace{\sum_t (\overline{R_t} - v(h_t))^2}_{critic\ loss}, \tag{11}$$

where $v(.)$ is the value function of critic. $\overline{R_t}$ represents the discounted reward for time step t and it can be formulated as:

$$\overline{R_t} = \overline{R_{t+1}} * \gamma + R_t, \tag{12}$$

in which the γ is the discount factor. The reward R_t is made up of three parts: an extrinsic reward R_{EXT} and another two complementary and newly proposed reward functions—Soft Expert Distillation (SED) reward R_{SED} and Self Perceiving reward R_{SP}. The total reward function thus can be formalised as:

$$R_t = \alpha R_{SED} + \beta R_{SP} + R_{EXT}, \tag{13}$$

where (1) SED reward R_{SED}, an automatically learnt reward function through aligning agent's behaviours to the provided expert demonstrations. (2) SP reward R_{SP}, a reward function comes from predicted schedule to encourage the agent to reach the goal as soon as possible. (3) The extrinsic reward R_{EXT} assigns the agent a positive reward, if the agent stops within three-meter from target or the agent reduces the distance to the goal; otherwise, a negative reward will be returned. α, β are the trade-off factors of SED reward and SP reward respectively. The details of individual proposed reward function will be revealed in the following sections.

3.3 Soft Expert Distillation

Inspired by the work [18], we propose to learn the reward function from inputted expert demonstration in Vision-and-Language Navigation task. We train a neural network to predict the output of a random-initialised but frozen network to distil the expert knowledge. The Soft Expert Distillation networks structure is shown in Fig. 3. The key intuition behind this is: given a certain amount of random projection information, the representation learner is required to fit the structure of these given data points in the random projection space to achieve a similar projected distribution. The learning function is expected to predict relatively better where more expert data lays. In this case, a strong density function is formed. It models the likelihood of the agent performing a similar action with its expert in a situation through distillation. A higher prediction distance, which results in a low SED reward in turn, will be assigned to unexpected observation-action pairs that differs from given expert demonstrations. Thus, a higher reward will be assigned to an agent who takes an action similar with its expert. This encapsulation of density function gives us another view of learning expert demonstrations directly other than [13] and [6].

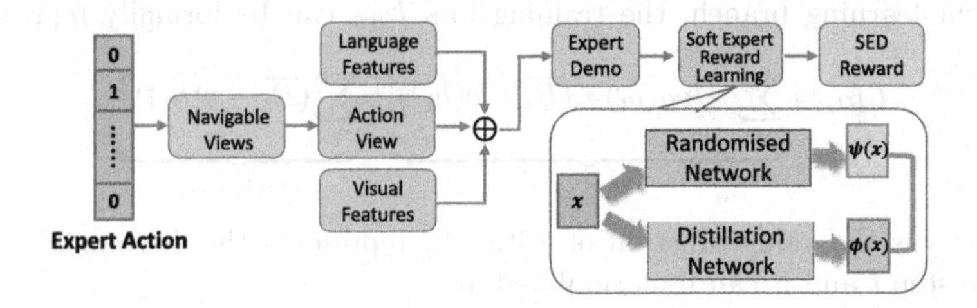

Fig. 3. The Soft Expert Distillation networks structure. Given an expert demonstrated data point $x \in \mathbb{R}^N$, it is fed into a weight-fixed randomly initialised neural network $\psi(\mathbf{x})$; simultaneously, the data point x is inputted into a distillation network $\phi(\mathbf{x};\theta)$ with different structure but same output dimensions with the parameters θ.

Precisely, for a given expert demonstrated data point $x \in \mathbb{R}^N$, we rst feed it into a weight-fixed and random-initialised neural network $\psi(\mathbf{x})$; at the same time the data point x is inputted into a distillation network $\phi(\mathbf{x};\theta)$ with different structure but same output dimensions. The data is projected to a M-dimensional new space by a representation learner $\phi : \mathbb{R}^N \mapsto \mathbb{R}^M$ with the parameters θ. We emphasise here, the function capacity of network ϕ is less than network ψ, by doing which can prevent overfitting. As we adopt L2 distance as our loss function, then we formulate the subsequent step as a prediction task and define a loss function as:

$$L_{sed}(\mathbf{x}) = (\phi(\mathbf{x};\theta), \psi(\mathbf{x}))^2 , \qquad (14)$$

Empirically, both of ψ and ϕ are implemented by multi-layer perceptrons. $\psi : \mathbb{R}^N \mapsto \mathbb{R}^M$ plays the role of a random data mapping function to project points into a randomly projected space. By doing so, this loss offers a simple yet powerful supervisory signal for the distillation network to learn semantic-rich feature representations from given expert data processed by the random projection function ψ.

In order to distil the expert behaviour distribution, the data points are consist of expert's visual observation, language instructions and actions. The equation is formally shown as:

$$L_{sed}^t = (\phi(\{VisF_t, LangF_t, a_t\};\theta) - \psi(\{VisF_t, LangF_t, a_t\}))^2 . \qquad (15)$$

The SED module preserves semantic-rich information w.r.t. distribution of expert demonstration for the representation learner. So the module is an ideal density function to measure the similarity of an agent's behaviour with the expert demonstration. Differ from the behaviour cloning process, it is formed in a soft manner. The SED intrinsic reward function is formally presented as:

$$R_{SED} = \begin{cases} +2, if Dis_t^{sed} <= thresh \\ -2, if Dis_t^{sed} > thresh \end{cases} \tag{16}$$

The L2 distance between $\phi(\{VisF_t, LangF_t, a_t\}; \theta)$ and $\psi(\{VisF_t, LangF_t, a_t\})$ is denoted as Dis_t^{sed}. Intuitively, if Dis_t^{sed} is less than the threshold, it represents the current behaviour of the agent is similar with the expert distribution where a positive reward should be awarded; otherwise, a negative reward will be returned. In contrast, behaviour cloning based models encourage the agent to copy expert demonstrations exactly; while our proposed soft expert distillation module learns the demonstrated behaviour in a soft manner by depicting the distribution of expert behaviours. In the case, the agent can retain the expert knowledge but will not suffer from the error accumulation problem. Thus, it increases the robustness of the model across various VLN environments.

3.4 Self Perceiving Reward

To perceive the schedule information towards the goal is crucial for the agent to complete the VLN task. A self perceiving module is designed to predict distance improvement ratio at each step as current schedule information of the agent. In order to utilise the information more adequately, we take one more step ahead by making use of this schedule information as another intrinsic reward—self perceiving reward. Formally, the self perceiving reward is calculated from:

$$C_{attn} = softmax(fc(h_{t-1}) \cdot LangF_t), \tag{17}$$

$$R_{SP} = V_t^{sche} = \sigma(fc(drop(\tanh(c_t) \odot \sigma(fc(h_{t-1}, \widetilde{VisF}_t))), C_{attn})), \tag{18}$$

where C_{attn} represents the language attention over different vocabularies within the instruction sentence. $\odot$ is the element-wise Hadamard product. Intuitively, the Self Perceiving reward indicates the predicted schedule information toward the destination. The more distance improvement ratio of the current action archived, the higher reward ought to be assigned. Moreover, this reward offers more information of distance change than raw distances. The more self perceiving reward the agent collected, the closer the agent believes to reach the target location.

4 Experiments

Following previous works [2,4,7,9,10,16,19], we evaluate our model on the Room-to-Room (R2R) dataset [2] for VLN task. Furthermore, we test our method on the VLN test server[1] [20] to validate the proposed Soft Expert Reward Learning Model. Ablation study is further conveyed to examine the contribution of each individual component of the model. The experimental results show the effectiveness of the proposed model.

4.1 Experimental Setup

Evaluation Metrics. Currently, a variety of metrics are used to evaluate VLN models. We adopt the following metrics: Navigation Error (NE) is to measure the shortest path distance between the stopping position and the goal; Success Rate (SR) quantifies the rate of success if the agent can stop within three meters from the target; Oracle Success Rate (OSR) is the success percentage if the agent can stop at the closest point along its trajectory; the Success rate weighted by Path Length (SPL) [1] is also adopted to indicate the weighted SR.

Implementation Detail. Following [4,16], we utilise the ResNet-152 model pre-trained on ImageNet to extract CNN features as visual inputs. Empirically, we set the M equal to 128, and set both of the reward trade-off factors α and β to 0.1. In Soft Expert Distillation networks, the randomised network is made up of two hidden linear layers with 512 and 256 neurons respectively; the distillation network has one hidden linear layers with 256 neurons. Between every two linear layers, both of the randomised network and the distillation network adopt leaky-relu as their activation function. To prevent overfitting, we early-stopped the training process of models according to the performance on the validation set. The Soft Expert Distillation module is not jointly trained with the rest of the model. This decoupling prevents performance unstableness during training and increase the robustness of the model.

4.2 Overall Performance

In this section, we convey the evaluation experiments on three individual sets, validation seen, validation unseen and test set, shown in Table 2, to compare the effectiveness of our proposed soft expert reward learning model with other models. The comparison is split into two groups: models trained on non-augmented data and augmented data. Within twelve indicators of validation set and test set, we achieve ten best results on the non-augmented group and nine best results on the augmented group, which reveals the effectiveness of SERL model. More specifically, for the non-augmented group, on validation unseen set, our SERL model reduces the navigation error by 7%, increase the success rate by 4% and SPL by 2%. Our method also receives remarkable results on test unseen set. Similarly for the augmented group, on validation unseen set, it is clear that our model is the best performer. SERL model reduces the navigation error by 5% and gets 0.56 successful rate. Our model also increases 10% for the oracle successful rate and gets 0.48 SPL respectively compared to the second-best model.

[1] The VLN leaderboard address is https://evalai.cloudcv.org/web/challenges/challenge-page/97/leaderboard/270.

Table 2. Performance Evaluation across different methods. The first place of each column is bolded. All of the results are reported on models without beam search, except FAST [7] model using a beam-search style strategy. The ↑ means that the higher the better; vice versa. The * sign represents data augmentation.

Methods	Val seen				Val unseen				Test unseen			
	NE ↓	SR ↑	OSR ↑	SPL ↑	NE ↓	SR ↑	OSR ↑	SPL ↑	NE ↓	SR ↑	OSR ↑	SPL ↑
Random [2]	9.45	0.16	0.21	–	9.23	0.16	0.22	–	9.77	0.13	0.18	–
Seq2seq [2]	6.01	0.39	0.53	–	7.81	0.22	0.28	–	7.85	0.20	0.27	0.18
Self-Monitoring [9]	3.72	0.63	**0.75**	0.56	5.98	0.44	0.58	0.30	–	–	–	–
Regretful-Agent [10]	3.69	0.65	0.72	**0.59**	5.36	0.48	**0.61**	0.37	–	–	–	–
EnvDrop [16]	4.71	0.55	–	0.53	5.49	0.47	–	0.43	–	–	–	–
SERL (Ours)	**3.67**	**0.66**	0.71	0.58	**4.97**	**0.50**	0.59	**0.44**	**5.70**	**0.51**	**0.57**	**0.47**
Speaker-Follower* [4]	3.36	0.66	0.74	–	6.62	0.36	0.45	–	6.62	0.35	–	0.28
RCM* [19]	3.37	0.67	0.77	–	5.88	0.43	0.52	–	6.12	0.43	0.50	0.38
FAST* [7]	–	–	–	–	4.97	**0.56**	–	0.43	**5.14**	**0.54**	–	0.41
Self-Monitoring* [9]	3.22	0.67	**0.78**	0.58	5.52	0.45	0.56	0.32	5.67	0.48	0.59	0.35
Regretful-Agent* [10]	3.23	**0.69**	0.77	0.63	5.32	0.50	0.59	0.41	5.69	0.48	0.56	0.40
EnvDrop* [16]	3.99	0.62	–	0.59	5.22	0.52	–	**0.48**	5.23	0.51	0.59	0.47
EnvDrop-Our-Impl*	3.77	0.66	0.72	0.62	5.49	0.49	0.56	0.45	–	–	–	–
SERL* (Ours)	**3.20**	**0.69**	0.75	**0.64**	**4.74**	**0.56**	**0.65**	**0.48**	5.63	0.53	**0.61**	**0.49**

On the test unseen set, our SERL model can achieve performance better than, or comparably well to, the other competing methods in Table 2. When compared to the second-best model, the model increases 3% for the oracle successful rate and 4% SPL respectively. The FAST [7] model applies a beam-search style strategy, thus it is expected to produce better successful rate (SR) but it leads to a relatively worse SPL.

4.3 Ablation Study

Ablation Study of Different Components Performance. This section examines the contribution of each component of SERL model. Different components are added to the baseline model. The ablation results are represented as Table 3. The results are shown on validation seen and unseen sets and the models are trained with the same data augmentation strategy. In the first column, SED represents our proposed soft expert distillation module, while SP is the self perceiving module. BS represents beam search setting. We check different components in the second column to examine each variant. Row model #1 shows the performance of the environment dropout methods that we implemented. From the table we can clearly find that when comparing to row #1, excluding the beam search setting on the validation unseen set, the model with SED module alone (method #2) achieves higher SR by 6% and increases SPL score from 0.45 to 0.48; the model with SP module alone (#3) receives better success rate as 0.53 from 0.49 and better SPL score as 0.46 from 0.45. This is because the SED module encourages the agent to have better alignment with expert trajectories, but in a soft way; the SP module pushes the agent to find the target location as fast as possible. The full SERL model (method #4) combines the advantages of individual module and it achieves 0.56 of successful rate and 0.48 of SPL, which outperforms other variants.

Table 3. Ablation study of different components in SERL model. We evaluate the results on validation seen set and validation unseen set. The best result are bolded.

Models	SED	SP	BS	Val seen				Val unseen			
				NE ↓	SR ↑	OSR ↑	SPL ↑	NE ↓	SR ↑	OSR ↑	SPL ↑
1				3.77	0.66	0.72	0.62	5.49	0.49	0.56	0.45
2	✓			3.67	0.66	0.74	0.63	5.10	0.52	0.58	**0.48**
3		✓		3.19	0.67	0.72	0.61	4.93	0.53	0.61	0.46
4	✓	✓		3.20	0.69	0.75	**0.64**	4.74	0.56	0.65	**0.48**
5	✓	✓	✓	**2.47**	**0.77**	**0.99**	0.02	**3.01**	**0.71**	**0.99**	0.02

Additionally, beam search is another popular Vision-and-Language Navigation setting. In the beam search setting, the agents are given the chance to choose the trajectories with the highest success rate. In this case, it can further boost the success rate of our SERL model (method #5) to 0.77 on validation seen set and 0.71 on validation seen set. Moreover, SERL model receives 0.70 in successful rate on the test unseen set with beam search.

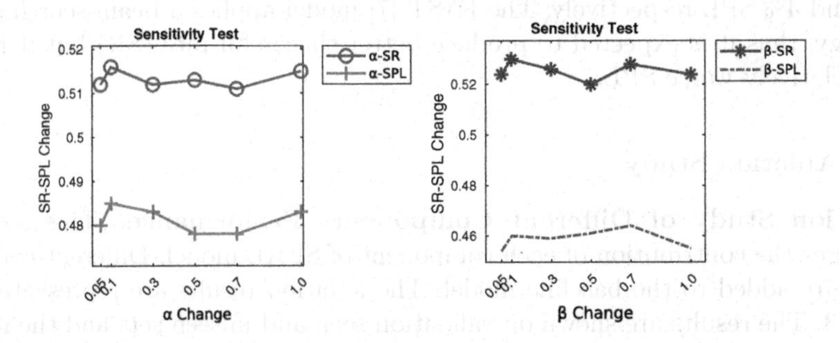

Fig. 4. The sensitivity test of our Soft Expert Reward Learning (SERL) model. The figures show the SR and SPL performance of the model on validation unseen set with different α and β values.

Sensitivity Test. This section presents the performances of SERL model with different α and β weights to trade-off the proposed individual intrinsic reward. Figure 4 shows the sensitivity test results, which is evaluated in SR and SPL on validation unseen set. It is clear that SERL generally performs stably w.r.t. the use of different α and β weights. This demonstrates the general stability of our SERL method by setting different hyper-parameters. In general, $\alpha = \beta = 0.1$ is recommended for SERL to achieve effective visual and language navigation performance.

Step 1 : continue down the stairs. you'll see a big tile on the floor , turn right and go into the first doorway on the right . there will be a big mirror in the room . you'll stop and wait just inside this room .

Step 2 : continue down the stairs. you'll see a big tile on the floor , turn right and go into the first doorway on the right . there will be a big mirror in the room . you'll stop and wait just inside this room .

Step 3 : continue down the stairs. you'll see a big tile on the floor , turn right and go into the first doorway on the right . there will be a big mirror in the room . you'll stop and wait just inside this room .

Step 4 : continue down the stairs. you'll see a big tile on the floor , turn right and go into the first doorway on the right . there will be a big mirror in the room . you'll stop and wait just inside this room .

Step 5 : continue down the stairs. you'll see a big tile on the floor , turn right and go into the first doorway on the right . there will be a big mirror in the room . you'll stop and wait just inside this room .

Step 6 : continue down the stairs. you'll see a big tile on the floor , turn right and go into the first doorway on the right . there will be a big mirror in the room . you'll stop and wait just inside this room .

(a) Our Baseline

Step 1 : continue down the stairs. you'll see a big tile on the floor , turn right and go into the first doorway on the right . there will be a big mirror in the room . you'll stop and wait just inside this room .

Step 2 : continue down the stairs. you'll see a big tile on the floor , turn right and go into the first doorway on the right . there will be a big mirror in the room . you'll stop and wait just inside this room .

Step 3 : continue down the stairs. you'll see a big tile on the floor , turn right and go into the first doorway on the right . there will be a big mirror in the room . you'll stop and wait just inside this room .

Step 4 : continue down the stairs. you'll see a big tile on the floor , turn right and go into the first doorway on the right . there will be a big mirror in the room . you'll stop and wait just inside this room .

Step 5 : continue down the stairs. you'll see a big tile on the floor , turn right and go into the first doorway on the right . there will be a big mirror in the room . you'll stop and wait just inside this room .

Step 6 : continue down the stairs. you'll see a big tile on the floor , turn right and go into the first doorway on the right . there will be a big mirror in the room . you'll stop and wait just inside this room .

(b) SERL Model

Fig. 5. The visualisation of our proposed Soft Expert Reward Learning (SERL) model. The figure shows the comparison between SERL model and the baseline model. The yellow colours in the sentence represents the attention maps over the instruction. The depth of the colours indicates the strength of the attention. The darker the colours, the more attention is put on the specific vocabularies. The check mark means the agents take a same action as the expert; the cross mark represents the opposite.

4.4 Visualisation

Figure 5 shows the actions taken by our baseline agents and proposed SERL agent, respectively. The attention maps over the instruction at each step are also illustrated in the figure. On the left column of the figure, the agent is trained by behaviour cloning solely and it performs correctly at the first three steps. But the agent takes a wrong action at the fourth step and it results in failure navigation in the next three steps. This is because subtle errors will be accumulated at each step by just copy expert demonstrations in the training phase. However, our SERL model can attend over the instruction in a better way and it does not encounter the error accumulation problem in the case.

5 Conclusions

In this paper, we propose a Soft Expert Reward Learning (SERL) model to address the behaviour cloning error accumulation and the reinforcement learning reward engineering issues for VLN task. From the experimental results, we show that our SERL model gains better performance generally than current state-of-the-art methods in both validation unseen and test unseen set on VLN Room-to-Room dataset. The ablation study shows that our proposed the Soft Expert Distillation (SED) module and the Self Perceiving (SP) module are complementary to each other. Moreover, the visualisation experiments further verify the SERL model can overcome the error accumulation problem. In the future, we will further investigate more reward learning methods on VLN task.

References

1. Anderson, P., et al.: On evaluation of embodied navigation agents. arXiv preprint arXiv:1807.06757 (2018)
2. Anderson, P., et al.: Vision-and-language navigation: Interpreting visually-grounded navigation instructions in real environments. In: Proceedings of the IEEE Conference on Computer Vision and Pattern Recognition, pp. 3674–3683 (2018)
3. Brockman, G., et al.: Openai gym. arXiv preprint arXiv:1606.01540 (2016)
4. Fried, D., et al.: Speaker-follower models for vision-and-language navigation. In: Advances in Neural Information Processing Systems, pp. 3314–3325 (2018)
5. He, K., Zhang, X., Ren, S., Sun, J.: Deep residual learning for image recognition. In: Proceedings of the IEEE Conference on Computer Vision and Pattern Recognition, pp. 770–778 (2016)
6. Ho, J., Ermon, S.: Generative adversarial imitation learning. In: Advances in Neural Information Processing Systems, pp. 4565–4573 (2016)
7. Ke, L., et al.: Tactical rewind: self-correction via backtracking in vision-and-language navigation. In: Proceedings of the IEEE Conference on Computer Vision and Pattern Recognition, pp. 6741–6749 (2019)
8. Lillicrap, T.P., et al.: Continuous control with deep reinforcement learning. arXiv preprint arXiv:1509.02971 (2015)
9. Ma, C.Y., et al.: Self-monitoring navigation agent via auxiliary progress estimation. arXiv preprint arXiv:1901.03035 (2019)

10. Ma, C.Y., Wu, Z., AlRegib, G., Xiong, C., Kira, Z.: The regretful agent: heuristic-aided navigation through progress estimation. In: Proceedings of the IEEE Conference on Computer Vision and Pattern Recognition, pp. 6732–6740 (2019)
11. Mnih, V., et al.: Asynchronous methods for deep reinforcement learning. In: International Conference on Machine Learning, pp. 1928–1937 (2016)
12. Mnih, V., et al.: Playing atari with deep reinforcement learning. arXiv preprint arXiv:1312.5602 (2013)
13. Ng, A.Y., Russell, S.J., et al.: Algorithms for inverse reinforcement learning. In: ICML, vol. 1, pp. 663–670 (2000)
14. Reddy, S., Dragan, A.D., Levine, S.: SQIL: imitation learning via regularized behavioral cloning. arXiv preprint arXiv:1905.11108 (2019)
15. Schulman, J., Wolski, F., Dhariwal, P., Radford, A., Klimov, O.: Proximal policy optimization algorithms. arXiv preprint arXiv:1707.06347 (2017)
16. Tan, H., Yu, L., Bansal, M.: Learning to navigate unseen environments: back translation with environmental dropout. arXiv preprint arXiv:1904.04195 (2019)
17. Todorov, E., Erez, T., Tassa, Y.: Mujoco: a physics engine for model-based control. In: 2012 IEEE/RSJ International Conference on Intelligent Robots and Systems, pp. 5026–5033. IEEE (2012)
18. Wang, R., Ciliberto, C., Amadori, P., Demiris, Y.: Random expert distillation: imitation learning via expert policy support estimation. arXiv preprint arXiv:1905.06750 (2019)
19. Wang, X., et al.: Reinforced cross-modal matching and self-supervised imitation learning for vision-language navigation. In: Proceedings of the IEEE Conference on Computer Vision and Pattern Recognition, pp. 6629–6638 (2019)
20. Yadav, D., et al.: EvalAI: towards better evaluation systems for AI agents (2019)

Part-Aware Prototype Network for Few-Shot Semantic Segmentation

Yongfei Liu[1], Xiangyi Zhang[1], Songyang Zhang[1], and Xuming He[1,2(✉)]

[1] School of Information Science and Technology, ShanghaiTech University, Shanghai, China
{liuyf3,zhangxy9,zhangsy1,hexm}@shanghaitech.edu.cn
[2] Shanghai Engineering Research Center of Intelligent Vision and Imaging, Shanghai, China

Abstract. Few-shot semantic segmentation aims to learn to segment new object classes with only a few annotated examples, which has a wide range of real-world applications. Most existing methods either focus on the restrictive setting of one-way few-shot segmentation or suffer from incomplete coverage of object regions. In this paper, we propose a novel few-shot semantic segmentation framework based on the prototype representation. Our key idea is to decompose the holistic class representation into a set of part-aware prototypes, capable of capturing diverse and fine-grained object features. In addition, we propose to leverage unlabeled data to enrich our part-aware prototypes, resulting in better modeling of intra-class variations of semantic objects. We develop a novel graph neural network model to generate and enhance the proposed part-aware prototypes based on labeled and unlabeled images. Extensive experimental evaluations on two benchmarks show that our method outperforms the prior art with a sizable margin (Code is available at: https://github.com/Xiangyi1996/PPNet-PyTorch).

1 Introduction

Semantic segmentation is a core task in modern computer vision with many potential applications ranging from autonomous navigation [4] to medical image understanding [6]. A particular challenge in deploying segmentation algorithms in real-world applications is to adapt to novel object classes efficiently in dynamic environments. Despite the remarkable success achieved by deep convolutional networks in semantic segmentation [2,5,19,38], a notorious disadvantage of those supervised approaches is that they typically require thousands of pixel-wise

Y. Liu and X. Zhang—Contributed equally to the work. This work was supported by Shanghai NSF Grant (No. 18ZR1425100).

Electronic supplementary material The online version of this chapter (https://doi.org/10.1007/978-3-030-58545-7_9) contains supplementary material, which is available to authorized users.

© Springer Nature Switzerland AG 2020
A. Vedaldi et al. (Eds.): ECCV 2020, LNCS 12354, pp. 142–158, 2020.
https://doi.org/10.1007/978-3-030-58545-7_9

labeled images, which are very costly to obtain. While much effort has been made to alleviate such burden on data annotation, such as weak supervision [15], most of them still rely on collecting large-sized datasets.

A promising strategy, inspired by human visual recognition [25], is to enable the algorithm to learn to segment a new object class with only a few annotated examples. Such a learning task, termed as *few-shot semantic segmentation*, has attracted much attention recently [3,21,22,33]. Most of those initial attempts adopt the metric-based meta-learning framework [32], in which they first match learned features from support and query images, and then decode the matching scores into final segmentation.

However, the existing matching-based methods often suffer from several drawbacks due to the challenging nature of semantic segmentation. First, some prior works [35–37] solely focus on the task of one-way few-shot segmentation. Their approaches employ dense pair-wise feature matching and specific decoding networks to generate segmentation, and hence it is non-trivial or computationally expensive to generalize to the multi-way setting. Second, other prototype-based methods [7,27,33] typically use a holistic representation for each semantic class, which is difficult to cope with diverse appearance in objects with different parts, poses or subcategories. More importantly, all those methods represent a semantic class based on a small support set, which is restrictive for capturing rich and fine-grained feature variations required for segmentation tasks.

In this work, we propose a novel prototype-based few-shot learning framework of semantic segmentation to tackle the aforementioned limitations. Our main idea is to enrich the prototype representations of semantic classes in two directions. First, we decompose the commonly used holistic prototype representation into a small set of part-aware prototypes, which is capable of capturing diverse and fine-grained object features and yields better spatial coverage in semantic object regions. Moreover, inspired by the prior work in image classification [1,24], we incorporate a set of unlabeled images into our support set so that our part-aware prototypes can be learned from both labeled and unlabeled data source. This enables us to go beyond the restricted small support set and to better model the intra-class variation in object features. We refer to this new problem setting as semi-supervised few-shot semantic segmentation. Based on our new prototypes, we also design a simple and yet flexible matching strategy, which can be applied to either one-way or multi-way setting.

Specifically, we develop a deep neural network for the task of few-shot semantic segmentation, which consists of three main modules: an embedding network, a prototypes generation network and a part-aware mask generation network. Given a few-shot segmentation task, our embedding network module first computes a 2D conv feature map for each image. Taking as input all the feature maps, the prototype generation module extracts a set of part-aware representations of semantic classes from both labeled and unlabeled support images. To achieve this, we first cluster object features into a set of prototype candidates and then use a graph attention network to refine those prototypes using all the support data. Finally, the part-aware mask generation network fuses the score maps generated by matching the part-aware prototypes to the query image and predicts the semantic segments. We train our deep network using the meta-learning strategy with an augmented loss [34] that exploits the original semantic classes for efficient network learning.

We conduct extensive experiments evaluation on the PASCAL-5^i [3,37] and COCO-20^i dataset [20,33] to validate our few-shot semantic segmentation strategy. The results show that our part-aware prototype learning outperforms the state of the art with a large margin. We also include the detailed ablation study in order to provide a better understanding of our method.

The main contribution of this work can be summarized as the following:

- We develop a flexible prototype-based method for few-shot semantic segmentation, achieving superior performances in one-way and multi-way setting.
- We propose a part-aware prototype representation for semantic classes, capable of encoding fine-grained object features for better segmentation.
- To better capture the intra-class variation, we leverage unlabeled data for semi-supervised prototype learning with a graph attention network.

2 Related Work

2.1 Few-Shot Classification

Few-shot learning aims to learn a new concept representation from only a few annotated examples. Most of existing works can be categorized into metric-learning based [28,32,34], optimization-learning based [8,23], and graph neural network [9,18] based methods. Our work is inspired by the metric-learning based methods. In particular, Oriol et al. [32] propose to encode an input into an embedded feature and to perform a weighted nearest neighbor matching for classification. The prototypical network [28] aims to learn a metric space in which an input is classified according to its distance to class prototypes. Our work is in line with the prototypical network, but we adopt this idea in more challenging segmentation tasks, enjoying a simple design and yet high performance.

There have been several recent attempts aiming to improve the few-shot learning by incorporating a set of unlabeled data, referred to as semi-supervised few-shot learning [1,16,24]. Ren et al. [24] first try to leverage unlabeled data to refine the prototypes by Soft K-means. Ayyad et al. [1] introduced a consistency loss both in local and global for utilizing unlabeled data effectively. These methods are initially proposed for solving semi-supervised problems in few-shot classification regime and hence it is non-trivial to extend them to few-shot segmentation directly. We are the first to leverage unlabeled data in the challenging few-shot segmentation task for capturing the large intra-class variations.

2.2 Few-Shot Semantic Segmentation

Few-shot semantic segmentation aims to segment semantic objects in an image with only a few annotated examples, and attracted much attention recently.

The existing works can be largely grouped into two types: parametric matching-based methods [3,20,35–37] and prototype-based methods [27,33]. A recent exception, MetaSegNet [29], adopts the optimization-based few-shot learning strategy and formulates few-shot segmentation as a pixel classification problem.

In the parametric-matching based methods, Shaban et al. [3] first develop a weight imprinting mechanism to generate the classification weight for few-shot segmentation. Zhang et al. [36] propose to concatenate the holistic objects representation with query features in each spatial position and introduce a dense comparison module to estimate their prediction.The subsequent method, proposed by Zhang et al. [35], attends to foreground features for each query feature with a graph attention mechanism. These methods however mainly focus on the restrictive one-way few-shot setting and it is computationally expensive to generalize them to the multi-way setting.

Prototype-based methods conduct pixel-wise matching on query images with holistic prototypes of semantic classes. Wang et al. [33] propose to learn class-specific prototype representation by introducing the prototypes alignment regularization between support and query images. Siam et al. [27] adopt a novel multi-resolution prototype imprinting scheme for few-shot segmentation. All these prototype-based methods are limited by their holistic representations. To tackle this issue, we propose to decompose object representation into a small set of part-level features for modeling diverse object features at a fine-grained level.

2.3 Graph Neural Networks

Our work is related to learning deep networks on graph-structured data. The Graph Neural Networks are first proposed in [10,26] which learn a feature representation via a recurrent message passing process. Graph convolutional networks are a natural generalization of convolutional neural networks to non-Euclidean graphs. Kipf et al. [14] introduce learning polynomials of the graph laplacian instead of computing eigenvectors to alleviate the computational bottleneck, and validated its effectiveness on semi-supervised learning. Velic et al. [31] incorporate the attention mechanism into the graph neural network to augment node representation with their contextual information. Garcia et al. [9] firstly introduce the graph neural network into the few-shot image classification. By contrast, our work employ graph neural network to learn a set of prototypes for the task of semantic segmentation.

3 Problem Setting

We consider the problem of few-shot semantic segmentation, which aims to learn to segment semantic objects from only a few annotated training images per class. To this end, we adopt a meta-learning strategy [3,32] that builds a meta learner $\mathcal{M}$ to solve a family of few-shot semantic segmentation tasks $\mathcal{T} = \{T\}$ sampled from an underlying task distribution $P_{\mathcal{T}}$.

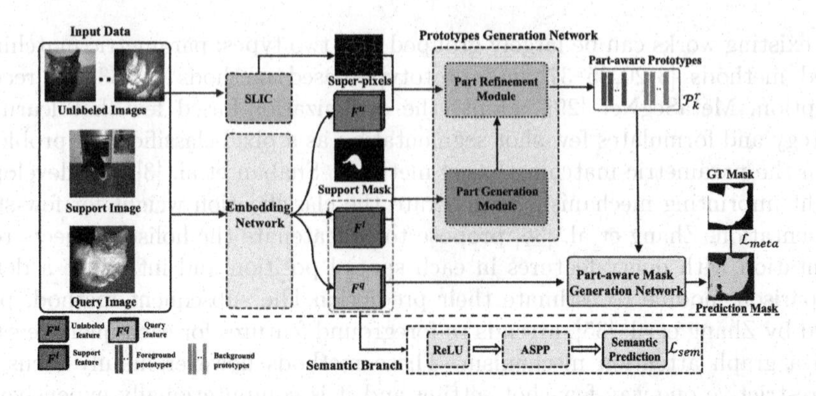

Fig. 1. Model Overview: For each task T, **Embedding Network** first aims to prepare convolutional feature maps of support, unlabeled and query images. **Prototypes Generation Network** then generates a set of part-aware prototypes by taking support and unlabeled image features as input. It consists of two submodules: **Part Generation Module** and **Part Refinement Module** (see below for details). Finally, the **Part-aware Mask Generation Network** performs segmentation on query features based on a set of part-aware prototypes. In addition, **Semantic Branch** generates mask predictions over the global semantic class space $\mathcal{C}^{tr}$.

Formally, each few-shot segmentation task T (also called an episode) is composed of a set of support data $\mathcal{S}$ with ground-truth masks and a set of query images $\mathcal{Q}$. In our semi-supervised few-shot semantic segmentation setting, the support data $\mathcal{S} = \{\mathcal{S}^l, \mathcal{S}^u\}$ where $\mathcal{S}^l$ and $\mathcal{S}^u$ are the annotated image-label pairs and unlabeled images, respectively. More specifically, for the C-way K-shot setting, the annotated support data consists of K image-label pairs from each class, denoted as $\mathcal{S}^l = \{(\mathbf{I}^l_{c,k}, \mathbf{Y}^l_{c,k})\}^{c\in\mathcal{C}_T}_{k=1,...,K}$, where $\{\mathbf{Y}^l_{c,k}\}$ are pixel-wise annotations, $\mathcal{C}_T$ is the subset of class sets for the task T and $|\mathcal{C}_T| = C$. The unlabeled support images $\mathcal{S}^u = \{\mathbf{I}^u_i\}^{N_u}_{i=1}$ are randomly sampled from the semantic class set $\mathcal{C}$ with their class labels removed during training and inference. Similarly, the query set $\mathcal{Q} = \{(\mathbf{I}^q_j, \mathbf{Y}^q_j)\}^{N_q}_{j=1}$, contains N_q images from the class set $\mathcal{C}_T$ whose ground-truth annotations $\{\mathbf{Y}^q_j\}$ are provided during training but *unknown* in test.

The meta learner $\mathcal{M}$ aims to learn a functional mapping from the support set $\mathcal{S}$ and a query image $\mathbf{I}^q$ to its segmentation $\mathbf{Y}^q$ for all the tasks. To achieve this, we construct a training set of segmentation tasks $\mathcal{D}^{tr} = \{(\mathcal{S}_n, \mathcal{Q}_n)\}^{|\mathcal{D}^{tr}|}_{n=1}$ with a class set $\mathcal{C}^{tr}$, and train the meta learner episodically on the tasks in $\mathcal{D}^{tr}$. After the meta-training, the model $\mathcal{M}$ encodes the knowledge on how to perform segmentation on different semantic classes across tasks. We finally evaluate the learned model in a test set of tasks $\mathcal{D}^{te} = \{(\mathcal{S}_m, \mathcal{Q}_m)\}^{|\mathcal{D}^{te}|}_{m=1}$ whose class set $\mathcal{C}^{te}$ is non-overlapped with $\mathcal{C}^{tr}$.

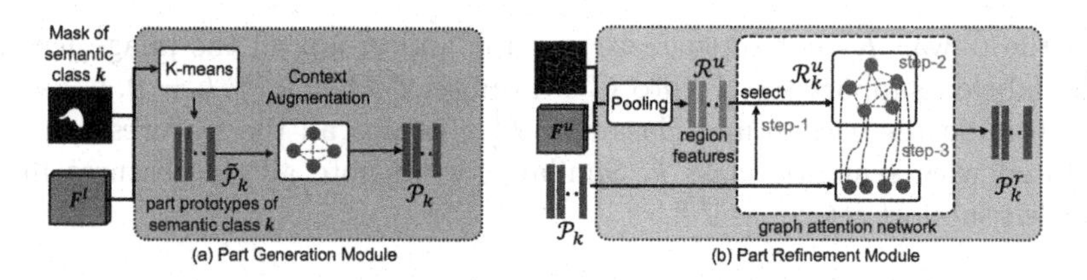

Fig. 2. Part Generation Module aims to generate the initial part-aware prototypes on support images and further incorporate with their global context of the same semantic class. **Part Refinement Module** further improves part-aware prototypes representation with unlabeled images features by a graph attention network.

4 Our Approach

In this work, we adopt a prototype-based few-shot learning framework to build a meta learner $\mathcal{M}$ for semantic segmentation. The main idea of our method is to capture the intra-class variation and fine-grained features of semantic classes by a new prototype representation. Specifically, we propose to decompose the commonly-used holistic representations of support objects into a set of part-aware prototypes for each class, and additionally utilize unlabeled data to enrich their representations.

To this end, we develop a deep graph network, as our meta learner, to encode such a new representation and to segment the query images. Our network consists of three main networks: an *embedding network* that computes convolutional feature maps for the images within a task (in Sect. 4.1); a *prototypes generation network* that extracts a set of part-aware prototypes from the labeled and unlabeled support images (in Sect. 4.2); and *a part-aware mask generation network* that generates the final semantic segmentation of the query images (in Sect. 4.3).

To train our meta model, we adopt a hybrid loss and introduce an auxiliary *semantic branch* that exploits the original semantic classes for efficient learning (in Sect. 4.4). We refer to our deep model as the **Part-aware Prototype Network (PPNet)**. An overview of our framework is illustrated in Fig. 1 and we will introduce the model details in the remaining of this section (Fig. 2).

4.1 Embedding Network

Given a task (or episode), the first module of our PPNet is an embedding network that extracts the convolutional feature maps of all images in the task. Following prior work [35,36], we adopt ResNet [12] as our embedding network, and introduce the dilated convolution to enlarge the receptive field and preserve more spatial details. Formally, we denote the embedding network as f_{em}, and compute the feature maps of images in a task T as $\mathbf{F} = f_{\text{em}}(\mathbf{I})$, $\forall \mathbf{I} \in \mathcal{S} \cup \mathcal{Q}$. Here $\mathbf{F} \in \mathbb{R}^{H_f \times W_f \times n_{ch}}$, n_{ch} is the number of feature channels, and (H_f, W_f) is the height and width of the feature map. We also resize the annotation mask $\mathbf{Y}$ into the same spatial size as the feature map, denoted as $\mathbf{M} \in \mathbb{R}^{H_f \times W_f}$.

In the C-way K-shot setting, we reshape and group all the image features in the labeled support set $\mathcal{S}^l$ into $C+1$ subsets: $\mathcal{F}^l = \{\mathcal{F}^l_k, k = 0, 1, \cdots, C\}$, where 0 indicates background class and $\mathcal{F}^l_k$ contains all the features $\mathbf{f} \in \mathbb{R}^{n_{ch}}$ annotated with semantic class k. Similarly, we denote all the features in the unlabeled support set $\mathcal{S}^u$ as $\mathcal{F}^u$.

4.2 Prototypes Generation Network

Our second module, the prototypes generation network, aims to generate a set of discriminative part-aware prototypes for each class. For notation clarity, here we focus on a single semantic class k. The module takes the image feature set $\mathcal{F}^l_k$ and $\mathcal{F}^u$ as input, and outputs the set of prototypes $\mathcal{P}_k$.

To this end, we introduce a graph neural network defined on the features, which computes the prototypes in two steps according to the different nature of the labeled and unlabeled support sets. Specifically, the network first extracts part-aware prototypes directly from the labeled support data $\mathcal{F}^l_k$ and then refines their representation by making use of the unlabeled data $\mathcal{F}^u$. As a result, the prototypes generation network consists of two submodules: a *Part Generation Module* and a *Part Refinement Module*, which are described in detail as following.

Part Generation with Labeled Data. We first introduce the part generation module, which builds a set of part-aware prototypes from the labeled support set in order to capture fine-grained part-level variation in object regions.

Specifically, we denote the number of prototypes per class as N_p and the prototype set $\mathcal{P}_k = \{\mathbf{p}_i\}_{i=1}^{N_p}$, $\mathbf{p}_i \in \mathbb{R}^{n_{ch}}$. To define our prototypes, we first compute a data partition $\mathcal{G} = \{G_1, G_2, \cdots, G_{N_p}\}$ on the feature set $\mathcal{F}^l_k$ using the K-means clustering and then generate an initial set of prototypes $\tilde{\mathcal{P}}_k = \{\tilde{\mathbf{p}}_i\}_{i=1}^{N_p}$ with an average pooling layer as follows,

$$\tilde{\mathbf{p}}_i = \frac{1}{|G_i|} \sum_{j \in G_i} \mathbf{f}_j, \quad \mathbf{f}_j \in \mathcal{F}^l_k \tag{1}$$

We further incorporate a global context of the semantic class into the part-aware prototypes by augmenting each initial prototype with a context vector, which is estimated from other prototypes in the same class based on the attention mechanism [30]:

$$\mathbf{p}_i = \tilde{\mathbf{p}}_i + \lambda_p \sum_{j=1 \wedge j \neq i}^{N_p} \mu_{ij} \tilde{\mathbf{p}}_j, \quad \mu_{ij} = \frac{d(\tilde{\mathbf{p}}_i, \tilde{\mathbf{p}}_j)}{\sum_{j \neq i} d(\tilde{\mathbf{p}}_i, \tilde{\mathbf{p}}_j)} \tag{2}$$

where λ_p is a scaling parameter and μ_{ij} is the attention weight calculated with a similarity metric d, e.g., cosine similarity.

Part Refinement with Unlabeled Data: The second submodule, the part refinement module, aims to capture the intra-class variation of each semantic class by enriching the prototypes on additional unlabeled support images. However, exploiting the unlabeled data is challenging due to the fact that the set of unannotated image features $\mathcal{F}^u$ is much more noisy and in general has a much larger volume than the labeled set $\mathcal{F}^l_k$.

We tackle the above two problems by a grouping and pruning process, which yields a smaller and more relevant set of features $\mathcal{R}^u_k$ for class k. Based on $\mathcal{R}^u_k$, we then design a graph attention network to smooth the unlabeled features and to refine the part-aware prototypes by aggregating those features. Concretely, our refinement process includes the following three steps:

Step-1: *Relevant feature generation.* We first compute a region-level feature representation of unlabeled images by utilizing the idea of superpixel generation. Concretely, we apply SLIC [13] to all the unlabeled images and generate a set of groupings on $\mathcal{F}^u$. Denoting the groupings as $\mathcal{R} = \{R_1, R_2, \cdots, R_{N_r}\}$, we use the average pooling to produce a pool of region-level features $\mathcal{R}^u = \{\mathbf{r}_i\}_{i=1}^{N_r}$. We then select a set of relevant features for class k as follows:

$$\mathcal{R}^u_k = \{\mathbf{r}_j : \mathbf{r}_j \in \mathcal{R}^u \wedge \exists \mathbf{p}_i \in \mathcal{P}_k, d(\mathbf{p}_i, \mathbf{r}_j) > \sigma\} \tag{3}$$

where $d(\cdot,\cdot)$ is a cosine similarity function between prototype and feature, and σ is a threshold that determines the relevance of the features.

Step-2: *Unlabeled feature augmentation.* With the selected unlabeled features, the second step aims to enhance those region-level representations by incorporating contextual information in the unlabeled feature set. This allows us to encode both local and global cues of a semantic class.

Specifically, we build a fully-connected graph on the feature set $\mathcal{R}^u_k$ and use the following message passing function to compute the update $\tilde{\mathcal{R}}^u_k = \{\tilde{\mathbf{r}}_i\}_{i=1}^{|\tilde{\mathcal{R}}^u_k|}$:

$$\tilde{\mathbf{r}}_i = \mathbf{r}_i + h\left(\frac{1}{Z^u_i} \sum_{j=1 \wedge j \neq i}^{|\mathcal{R}^u_k|} d(\mathbf{r}_i, \mathbf{r}_j)\mathbf{W}\mathbf{r}_j\right) \tag{4}$$

where $\tilde{\mathbf{r}}_i$ represents the updated representation at node i, h is an element-wise activate function (e.g., ReLU). d is a similarity function encoding the relations between two feature vectors $\mathbf{r}_i$ and $\mathbf{r}_j$, and Z^u_i is a normalization factor for node i. $\mathbf{W} \in \mathbb{R}^{n_{ch} \times n_{ch}}$ is the weight matrix defining a linear mapping to encode the message from node j.

Step-3: *Part-aware prototype refinement.* Given the augmented unlabeled features, we refine the original part-aware prototypes with an attention strategy similar to the labeled one. We use the part-aware prototypes $\mathcal{P}_k$ as attention query to choose similar unlabeled features in $\tilde{\mathcal{R}}^u_k$ and aggregate them into $\mathcal{P}_k$:

$$\mathbf{p}^r_i = \mathbf{p}_i + \lambda_r \sum_{j=1}^{|\tilde{\mathcal{R}}^u_k|} \phi_{ij}\tilde{\mathbf{r}}_j, \quad \phi_{ij} = \frac{d(\mathbf{p}_i, \tilde{\mathbf{r}}_j)}{\sum_j d(\mathbf{p}_i, \tilde{\mathbf{r}}_j)} \tag{5}$$

where λ_r is a scaling parameter and ϕ_{ij} is the attention weight. The final refined prototype set for class k is denoted as $\mathcal{P}^r_k = \{\mathbf{p}^r_1, \mathbf{p}^r_2, \cdots, \mathbf{p}^r_{N_p}\}$.

4.3 Part-Aware Mask Generation Network

Given the part-aware prototypes $\{\mathcal{P}_k^r\}_{k=0}^C$ of every semantic class and background in each task, we introduce a simple and yet flexible matching strategy to generate the semantic mask prediction on a query image $\mathbf{I}^q$. We denote its conv feature map as $\mathbf{F}^q$ and its feature column at location (m,n) as $\mathbf{f}_{m,n}^q$.

Specifically, we first generate a similarity score map for each part-aware prototype performing the *prototype-pixel* matching as follows,

$$\mathbf{S}_{k,j}(m,n) = d(\mathbf{f}_{m,n}^q, \mathbf{p}_j^r), \quad \mathbf{p}_j^r \in \mathcal{P}_k^r, \quad \mathbf{S}_{k,j} \in \mathbb{R}^{H_f \times W_f} \tag{6}$$

where d is the cosine similarity function and $\mathbf{S}_{k,j}(m,n)$ is the score at location (m,n). We then fuse together all the score maps from the class k by max-pooling and generate the output segmentation scores by concatenating score maps from all the classes:

$$\mathbf{S}_k^q = \text{MaxPool}(\{\mathbf{S}_{k,j}\}_{j=1}^{N_p}), \quad \hat{\mathbf{Y}}^q = \bigoplus\{\mathbf{S}_k^q\}_{k=0}^C \tag{7}$$

where $\bigoplus$ indicates concatenation. To generate the final segmentation, we upsample the score output $\hat{\mathbf{Y}}^q$ by bilinear interpolation and choose the class label with the highest score at each pixel location.

4.4 Model Training with Semantic Regularization

To estimate the parameters of proposed model, we train our PPNet in the meta-learning framework. Specifically, we adopt the standard cross-entropy loss to train our entire network on all the tasks in the training set $\mathcal{D}^{tr}$. Inspired by [33], we compute the cross-entropy loss on both support and query images. The loss for each task can be written as:

$$\mathcal{L}_{meta} = \mathcal{L}_{CE}(\hat{\mathbf{Y}}^q, \mathbf{Y}^q) + \mathcal{L}_{CE}(\hat{\mathbf{Y}}^l, \mathbf{Y}^l) \tag{8}$$

where $\mathcal{L}_{CE}$ is the cross-entropy function, $\mathbf{Y}^l, \hat{\mathbf{Y}}^l$ are the ground-truth and prediction mask for support image. We note that while our full model is not strictly differentiable w.r.t the embedding network thanks to the prototype clustering and candidate region selection, we are able to compute an approximate gradient by fixing the clustering and selection outcomes. This approximation works well empirically largely due to a well pre-trained embedding network.

In order to learn better visual representation for few-shot segmentation, we introduce another **semantic branch** [34] for computing a semantic loss defined on the global semantic class space C^{tr} (in contrast to C classes in individual tasks). To achieve this, we augment the network with an Atrous Spatial Pyramid Pooling module (ASPP) decoder to predict mask scores $\hat{\mathbf{Y}}_{sem}^q, \hat{\mathbf{Y}}_{sem}^l$ of support and query image respectively in the global class space C^{tr}, and compute the semantic loss as below,

$$\mathcal{L}_{sem} = \mathcal{L}_{CE}(\hat{\mathbf{Y}}_{sem}^q, \mathbf{Y}_{sem}^q) + \mathcal{L}_{CE}(\hat{\mathbf{Y}}_{sem}^l, \mathbf{Y}_{sem}^l) \tag{9}$$

Here $\mathbf{Y}^{q}_{sem}, \mathbf{Y}^{l}_{sem}$ are ground-truth masks defined over shared class space C^{tr}. The overall training loss for each task is:

$$\mathcal{L}_{full} = \mathcal{L}_{meta} + \beta \mathcal{L}_{sem} \tag{10}$$

where β is hyper-parameter to balance the weight of task loss and semantic loss.

5 Experiments

We evaluate our method on the task of few-shot semantic segmentation by conducting a set of experiments on two datasets, including PASCAL-5^i [3] and COCO-20^i [20,33]. In each dataset, we compare our approach with the state-of-the-art methods in terms of prediction accuracy.

Below we first introduce the implementation details and experimental configuration in Sect. 5.1. Then we present our experimental analysis on PASCAL-5^i dataset in Sect. 5.2, followed by our results on the COCO-20^i dataset in Sect. 5.3. We report comparisons of quantitative results and analysis on each dataset. Finally, we conduct a series of ablation studies to evaluate the importance of each component of the model in Sect. 5.4.

5.1 Experimental Configuration

Network Details: We adopt ResNet [12], initialized with weights pre-trained on ILSVRC [25], as feature extractor to compute the convolutional feature maps. In last two res-blocks, the strides of max-pooling are set as 1 and dilated convolutions are taken with dilation rate 2, 4 respectively. The last ReLU operation is removed for generating the prototypes. The input images are resized into a fixed resolution [417, 417] and horizontal random flipping is used for data augmentation. For the part-aware prototypes network, the typical hyper-parameter of the parts is $N_p = 5$. In the part refinement module, we first generate $N_r = 100$ candidate regions on unlabeled data, and select the relevant regions for each semantic class by setting similarity threshold σ as 0. In addition, λ_p in Eq. 2 and λ_r in Eq. 5 are set to 0.8 and 0.2 respectively, which control the proportion of parts and unlabeled information passed.

Training Setting: For the meta-training phase, the model is trained with the SGD optimizer, initial learning rate 5e−4, weight decay 1e−4 and momentum 0.9. We train 24k iterations in total, and decay the learning rate 10 times in 10k, 20k iteration respectively. The weight β of semantic loss $\mathcal{L}_{sem}$ is set as 0.5. At the testing phase, we average the mean-IoU of 5-runs [33] with different random seeds in each fold with each run containing 1000 tasks.

Baseline and Evaluation Metrics: We adopt ResNet-50 [12] as feature extractor in PANet [33] to be our baseline model, denoted as PANet*. Following previous works [3,22,33,35,37], mean-IoU and binary-IoU are adopted for model evaluation. Mean-IoU measures the averaged Intersection-over-Union (IoU) of

Table 1. Mean-IoU of 1-way on PASCAL-5^i. $*$ denotes the results implemented by ourselves. MS denotes the model evaluated with multi-scale inputs. [35,36]. Red numbers denote the averaged mean-IoU over 4 folds.

Methods	MS	Backbone	1-shot					5-shot					#params
			fold-1	fold-2	fold-3	fold-4	mean	fold-1	fold-2	fold-3	fold-4	mean	
OSLSM [3]	x	VGG16	33.60	55.30	40.90	33.50	40.80	35.90	58.10	42.70	39.10	43.90	272.6M
co-FCN [22]	x	VGG16	31.67	50.60	44.90	32.40	41.10	37.50	50.00	44.10	33.90	41.40	34.20M
SG-one [37]	x	VGG16	40.20	58.40	48.40	38.40	46.30	41.90	58.60	48.60	39.40	47.10	19.00M
AMP [27]	x	VGG16	36.80	51.60	46.90	36.00	42.80	44.60	58.00	53.30	42.10	49.50	15.8M
PANet [33]	x	VGG16	42.30	58.00	51.10	41.20	48.10	51.80	64.60	59.80	46.50	55.70	14.7M
PANet* [33]	x	RN50	44.03	57.52	50.84	44.03	49.10	55.31	67.22	61.28	53.21	59.26	23.5M
PGNet* [35]	x	RN50	53.10	63.60	47.60	47.70	53.00	56.30	66.10	48.00	53.20	55.90	32.5M
FWB[20]	x	RN101	51.30	64.49	56.71	**52.24**	56.19	54.84	67.38	62.16	55.30	59.92	43.0M
CANet [36]	✓	RN50	52.50	65.90	51.30	51.90	55.40	55.50	67.80	51.90	53.20	57.10	36.35M
PGNet [35]	✓	RN50	**56.00**	**66.90**	50.60	50.40	56.00	57.70	68.70	52.90	54.60	58.50	32.5M
Ours(w/o $\mathcal{S}^u$)	x	RN50	47.83	58.75	53.80	45.63	51.50	58.39	67.83	64.88	56.73	61.96	23.5M
our	x	RN50	48.58	60.58	55.71	46.47	52.84	**58.85**	**68.28**	**66.77**	**57.98**	62.97	31.5M
Ours	x	RN101	52.71	62.82	**57.38**	47.74	55.16	**60.25**	**70.00**	**69.41**	**60.72**	65.10	50.5M

Table 2. Mean-IoU of 2-way on PSACAL-5^i. $*$ denotes our implementation. Red numbers denote the averaged mean-IoU over 4 folds.

Methods	Backbone	1-shot					5-shot				
		fold-1	fold-2	fold-3	fold-4	mean	fold-1	fold-2	fold-3	fold-4	mean
MetaSegNet [29]	RN9	-	-	-	-	-	41.9	41.75	46.31	43.63	43.30
PANet[33]	VGG16	-	-	-	-	45.1	-	-	-	-	53.10
PANet*[33]	RN50	42.82	56.28	48.72	45.53	48.33	54.65	64.80	57.61	54.94	58.00
Ours(**w/o** $\mathcal{S}^u$)	RN50	45.63	58.00	51.65	45.69	50.24	55.34	66.38	63.79	56.85	60.59
Ours	RN50	**47.36**	**58.34**	**52.71**	**48.18**	51.65	**55.54**	**67.26**	**64.36**	**58.02**	61.30

all the classes. Binary-IoU[1] is calculated by treating all object classes as the foreground and averaging the IoU of foreground and background. In our experiments, we mainly focus on mean-IoU metrics for evaluation since it reflects the model generalization ability more precisely.

5.2 Experiments on PASCAL-5^i

Dataset: The PASCAL-5^i is introduced in [3,37], which is created from PASCAL VOC 2012 dataset with SBD [11] augmentation. Specifically, the 20 classes in PASCAL VOC is split into 4-folds evenly, each containing 5 categories. Models are trained on three folds and evaluated on the rest using cross-validation.

Quantitative Results: We compare the performance of our PPNet with the previous state-of-the-art methods. The detail results of 1-way setting are reported in Table 1. With the ResNet-50 as the embedding network, our model achieves **61.96**% mean-IoU in 5-shot setting, which outperforms the state-of-the-art method with a sizable margin of **2.71**%. The performance can be further improved to **52.84**%(1-shot) and **62.97**%(5-shot) by refining the part prototypes with the unlabeled data. Compared with the PANet [33], our method achieves the considerable improvement in both 1-shot and 5-shot setting, which

[1] We report Binary-IoU in supplementary material for a clear comparison with the previous works.

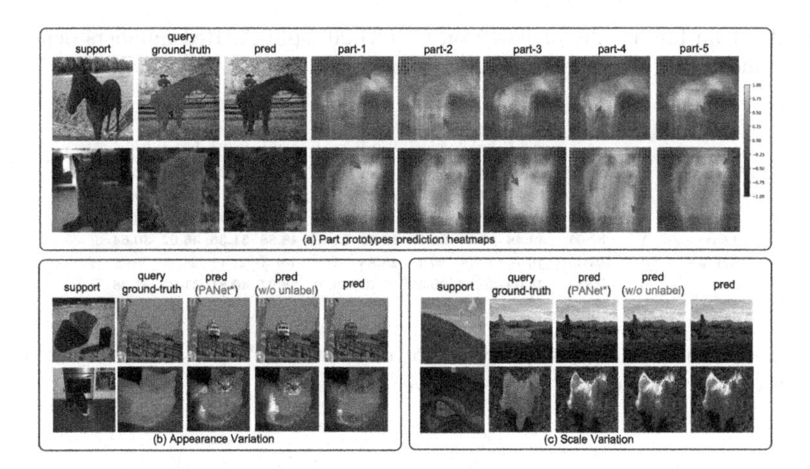

Fig. 3. Qualitative Visualization of 1-way 1-shot setting on PASCAL-5^i. (a) demonstrates the part-aware prototypes response heatmaps. The bright region denotes a high similarity between prototypes and query images. (b) and (c) show the capabilities of our model in coping with appearance and scale variation by utilizing unlabeled data. Red masks denote prediction results of our models. Blue and green masks denote the ground-truth of support and query images. See suppl. for more visualization results. (Color figure online)

demonstrates the part-aware prototypes can provide a superior representation than the holistic prototype. Moreover, our method achieves the state-of-the-art performance at **65.10**% in 5-shot setting relied on the ResNet-101 backbone[2].

To investigate the effectiveness of our method on multi-way setting, a 2-way experiment is conducted and the results are reported in Table 2. Our method can outperform the previous works both in with/without unlabeled data with a large margin. Our model can achieve **51.65**% and **61.30**% for 1-shot and 5-shot setting, which has **3.32**% and **3.30**% performance gain compared with PANet*, respectively. The quantitative results indicate our PPNet can achieve the state-of-the-art in a more challenging setting.

Visualization Analysis: To better understand our part-aware prototype framework, we visualize the responding region of our part prototypes and the prediction results in Fig. 3. The response heatmaps are presented in the column 4–8 of the (a). For example, given a support image (horse or cat), the part prototypes are corresponding to different body parts, which is capable of modeling one semantic class at a fine-grained level.

[2] We note that our 1-shot performance is affected by the limited representation power of the prototypes learned from a single support image while prior methods [35,36] employ a complex Convnet decoder to exploit additional spatial smoothness prior.

Table 3. Mean-IoU results of 1-way on COCO-20^i split-A. Red numbers denote the averaged mean-IoU over 4 folds. * is our implementation

Methods	Split	Backbone	1-shot					5-shot				
			fold-1	fold-2	fold-3	fold-4	mean	fold-1	fold-2	fold-3	fold-4	mean
PANet [33]	A	VGG16	28.70	21.20	19.10	14.80	20.90	39.43	28.30	28.20	22.70	29.70
PANet* [33]	A	RN50	31.50	22.58	21.50	16.20	22.95	45.85	29.15	30.59	29.59	33.80
Ours(w/o $\mathcal{S}^u$)	A	RN50	34.53	25.44	24.33	18.57	25.71	48.30	30.90	35.65	30.20	36.24
Ours	A	RN50	**36.48**	**26.53**	**25.99**	**19.65**	27.16	**48.88**	**31.36**	**36.02**	**30.64**	36.73
FWB [20]	B	RN101	16.98	17.78	20.96	**28.85**	21.19	19.13	21.46	23.39	30.08	23.05
Ours	B	RN50	**28.09**	**30.84**	**29.49**	27.70	29.03	**38.97**	**40.81**	**37.07**	**37.28**	38.53

Table 4. Ablation Studies of 1-way 1-shot on COCO-20^i split-A in every fold. Red numbers denote the averaged mean-IoU over 4 folds.

Model	PAP	SEM	UD	1-shot				
				fold-1	fold-2	fold-3	fold-4	mean
Baseline (PANet*)	-	-	-	31.50	22.58	21.50	16.20	22.95
	✓(w/o context)	-	-	32.05	23.09	21.33	16.94	23.35
	✓	-	-	34.45	24.37	23.46	17.79	25.02
	✓	✓	-	34.53	25.44	24.33	18.57	25.71
	✓	✓	✓(w/o *step-2*)	36.02	24.45	25.82	19.07	26.34
Ours	✓	✓	✓	**36.48**	**26.53**	**25.99**	**19.65**	27.16

Moreover, our model can cope with the large appearance and scale variation between support and query images, which is illustrated in (b) and (c). Compared with the PANet*, our method can enhance the modeling capability with the part prototypes, and has a significant improvement on the segmentation prediction by utilizing the unlabeled images to better model the intra-class variations.

5.3 Experiments on COCO-20^i

Dataset: COCO-20^i [20,33] is another more challenging benchmark built from MSCOCO [17]. Similar to PASCAl-5^i, MS-COCO dataset is divided into 4-folds with 20 categories in each fold. There are two splits of MSCOCO: we refer the data partition in [33] as *split-A* while the split in [20] as *split-B*. We mainly focus on *split-A* and also report the performance on *split-B*. Models are trained on three folds and evaluated on the rest with a cross-validation strategy.

Quantitative Results: We report the performance of our method on this more challenging benchmark in Table 3. Compared with the recent works [20,33], our method can achieve the state-of-the-art performance in different splits that use the same type of embedding networks by a sizable margin. Compared with the baseline PANet*, the same performance improvement trends are shown in both setting. In split-B [20], our model is superior to FWB [20] nearly in every fold, except for fold-4 in 1-shot, achieving 29.03% in 1-shot and 38.53% in 5-shot.

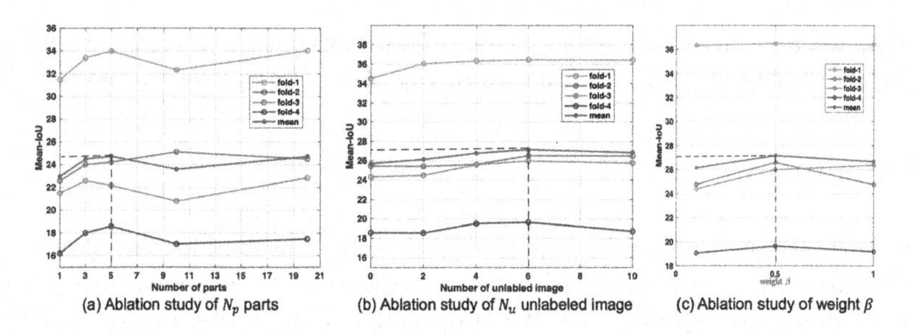

Fig. 4. Ablation studies of N_p parts, N_u unlabeled data and weight β for semantic loss on COCO-20^i split-A 1-way 1-shot. (Color figure online)

5.4 Ablation Study

In this subsection, we conduct several experiments to evaluate the effectiveness of our model components on COCO-20^i *split*-A 1-way 1-shot setting.

Part-Aware Prototypes (PAP): As in Table 4, by decomposing the holistic object representation [33,36,37] into a small set of part-level representations, the averaged mean-IoU is improved from 22.95% to **23.35%**. We further demonstrate the effectiveness of the global context used for augmenting part prototype (in Eq. 2). The performance can achieve continuous improvement to 25.02%, which suggests that global semantic is important for part-level representations.

Semantic Branch (SEM): We also conduct experiments to validate the semantic branch [34]. It is evident that the semantic branch is able to improve the convergence and the final performance significantly, which indicates that the full PPNet exploits the semantic information efficiently.

Unlabel Data (UD): We also investigate the graph attention network for the exploitation of the unlabeled data. As discussed in the method, we propose to utilize the graph attention network to refine the part prototypes. We compare the performance of the full PPNet with the PPNet without the GNN module used in *step-2*. The performance demonstrates the effectiveness of the GNN, and our full PPNet can achieve 27.16% in terms of averaged mean-IoU over 4 folds.

Hyper-Parameters N_p, N_u and β: We conduct several ablation studies to explore the influence of the hyper-parameters of our PPNet. We first investigate the part number N_p on 'baseline+PAP' model and plot the performance curve in Fig. 4(a). In our experiments, the highest performance is achieved when N_p is 5 and 20 (red line) over 4 folds, and we set $N_p = 5$ for computation efficiency. In our semi-supervised few-shot segmentation task, we also investigate the influence of the unlabeled image number N_u. In Fig. 4(b), we can achieve the highest averaged mean-IoU over 4 folds (red line) with our full PPNet when $N_u = 6$. In addition, we also investigate the weight β for semantic loss $\mathcal{L}_{sem}$ in our final model during training stage. As shown in Fig. 4(c), the optimal value is $\beta = 0.5$.

6 Conclusion

In this work, we presented a flexible prototype-based method for few-shot semantic segmentation. Our approach is able to capture diverse appearances of each semantic class. To achieve this, we proposed a part-aware prototypes representation to encode the fine-grained object features. In addition, we leveraged unlabeled data to capture the intra-class variations of the prototypes, where we introduce the first framework of semi-supervised few-shot semantic segmentation. We developed a novel graph neural network model to generate and enhance the part-aware prototypes based on support images with and without pixel-wise annotations. We evaluated our method on several few-shot segmentation benchmarks, in which our approach outperforms the prior works with a large margin, achieving the state-of-the-art performance.

References

1. Ayyad, A., Navab, N., Elhoseiny, M., Albarqouni, S.: Semi-supervised few-shot learning with local and global consistency. arXiv preprint arXiv (2019)
2. Badrinarayanan, V., Kendall, A., Cipolla, R.: Segnet: a deep convolutional encoder-decoder architecture for image segmentation. IEEE Trans. Pattern Anal. Mach. Intell. **39**(12), 2481–2495 (2017)
3. Boots, Z.L.I.E.B., Shaban, A., Bansal, S.: One-shot learning for semantic segmentation. In: British Machine Vision Conference (BMVC) (2017)
4. Brabandere, B.D., Neven, D., Gool, L.V.: Semantic instance segmentation for autonomous driving. In: Proceedings of the IEEE Conference on Computer Vision and Pattern Recognition Workshops (CVPR) (2017)
5. Chen, L.C., Papandreou, G., Schroff, F., Adam, H.: Rethinking atrous convolution for semantic image segmentation. arXiv preprint arXiv (2017)
6. Chung, Y.A., Weng, W.H.: Learning deep representations of medical images using siamese CNNs with application to content-based image retrieval. In: NIPS Machine Learning for Health Workshop (NIPS Workshop) (2017)
7. Dong, N., Xing, E.: Few-shot semantic segmentation with prototype learning. In: British Machine Vision Conference (BMVC) (2018)
8. Finn, C., Abbeel, P., Levine, S.: Model-agnostic meta-learning for fast adaptation of deep networks. In: Proceedings of the 34th International Conference on Machine Learning (ICML) (2017)
9. Garcia, V., Bruna, J.: Few-shot learning with graph neural networks. arXiv preprint arXiv (2017)
10. Gori, M., Monfardini, G., Scarselli, F.: A new model for learning in graph domains. In: Proceedings 2005 IEEE International Joint Conference on Neural Networks. IEEE (2005)
11. Hariharan, B., Arbeláez, P., Bourdev, L., Maji, S., Malik, J.: Semantic contours from inverse detectors. In: 2011 International Conference on Computer Vision (ICCV) (2011)

12. He, K., Zhang, X., Ren, S., Sun, J.: Deep residual learning for image recognition. In: Proceedings of the IEEE Conference on Computer Vision and Pattern Recognition (CVPR) (2016)
13. Kim, A.: Fast slic. https://github.com/Algy/fast-slic
14. Kipf, T.N., Welling, M.: Semi-supervised classification with graph convolutional networks. arXiv preprint arXiv:1609.02907 (2016)
15. Kipf, T.N., Welling, M.: Semi-supervised classification with graph convolutional networks. In: International Conference on Learning Representations (ICLR) (2017)
16. Li, X., et al.: Learning to self-train for semi-supervised few-shot classification. In: Advances in Neural Information Processing Systems (NIPS) (2019)
17. Lin, T.-Y., et al.: Microsoft COCO: common objects in context. In: Fleet, D., Pajdla, T., Schiele, B., Tuytelaars, T. (eds.) ECCV 2014. LNCS, vol. 8693, pp. 740–755. Springer, Cham (2014). https://doi.org/10.1007/978-3-319-10602-1_48
18. Liu, Y., et al.: Learning to propagate labels: transductive propagation network for few-shot learning. arXiv preprint arXiv (2018)
19. Long, J., Shelhamer, E., Darrell, T.: Fully convolutional networks for semantic segmentation. In: Proceedings of the IEEE Conference on Computer Vision and Pattern Recognition (CVPR) (2015)
20. Nguyen, K., Todorovic, S.: Feature weighting and boosting for few-shot segmentation. In: Proceedings of the IEEE International Conference on Computer Vision (ICCV) (2019)
21. Rakelly, K., Shelhamer, E., Darrell, T., Efros, A.A., Levine, S.: Few-shot segmentation propagation with guided networks. arXiv preprint (2018)
22. Rakelly, K., Shelhamer, E., Darrell, T., Efros, A., Levine, S.: Conditional networks for few-shot semantic segmentation (2018)
23. Ravi, S., Larochelle, H.: Optimization as a model for few-shot learning (2016)
24. Ren, M., et al.: Meta-learning for semi-supervised few-shot classification. arXiv preprint arXiv (2018)
25. Russakovsky, O., et al.: Imagenet large scale visual recognition challenge. International Journal of Computer Vision (2015)
26. Scarselli, F., Gori, M., Tsoi, A.C., Hagenbuchner, M., Monfardini, G.: The graph neural network model. IEEE Trans. Neural Netw. **20**(1), 61–80 (2008)
27. Siam, M., Oreshkin, B.: Adaptive masked weight imprinting for few-shot segmentation. arXiv preprint arXiv (2019)
28. Snell, J., Swersky, K., Zemel, R.: Prototypical networks for few-shot learning. In: Advances in Neural Information Processing Systems (NIPS) (2017)
29. Tian, P., Wu, Z., Qi, L., Wang, L., Shi, Y., Gao, Y.: Differentiable meta-learning model for few-shot semantic segmentation. arXiv preprint arXiv (2019)
30. Vaswani, A., et al.: Attention is all you need. In: Advances in Neural Information Processing Systems (NIPS) (2017)
31. Veličković, P., Cucurull, G., Casanova, A., Romero, A., Lio, P., Bengio, Y.: Graph attention networks. arXiv preprint arXiv:1710.10903 (2017)
32. Vinyals, O., Blundell, C., Lillicrap, T., Wierstra, D., et al.: Matching networks for one shot learning. In: Advances in Neural Information Processing Systems (NIPS) (2016)
33. Wang, K., Liew, J.H., Zou, Y., Zhou, D., Feng, J.: Panet: few-shot image semantic segmentation with prototype alignment. arXiv preprint arXiv (2019)
34. Yan, S., Zhang, S., He, X., et al.: A dual attention network with semantic embedding for few-shot learning. In: Proceedings of the AAAI Conference on Artificial Intelligence (AAAI) (2019)

35. Zhang, C., Lin, G., Liu, F., Guo, J., Wu, Q., Yao, R.: Pyramid graph networks with connection attentions for region-based one-shot semantic segmentation. In: Proceedings of the IEEE International Conference on Computer Vision (ICCV) (2019)
36. Zhang, C., Lin, G., Liu, F., Yao, R., Shen, C.: Canet: class-agnostic segmentation networks with iterative refinement and attentive few-shot learning. In: Proceedings of the IEEE Conference on Computer Vision and Pattern Recognition (CVPR) (2019)
37. Zhang, X., Wei, Y., Yang, Y., Huang, T.: SG-one: similarity guidance network for one-shot semantic segmentation. arXiv preprint arXiv (2018)
38. Zhao, H., Shi, J., Qi, X., Wang, X., Jia, J.: Pyramid scene parsing network. In: The IEEE Conference on Computer Vision and Pattern Recognition (CVPR) (2017)

Learning from Extrinsic and Intrinsic Supervisions for Domain Generalization

Shujun Wang[1], Lequan Yu[2(✉)], Caizi Li[3], Chi-Wing Fu[1,3], and Pheng-Ann Heng[1,3]

[1] The Chinese University of Hong Kong, Shatin, Hong Kong
{sjwang,cwfu,pheng}@cse.cuhk.edu.hk
[2] Stanford University, Stanford, USA
lequany@stanford.edu
[3] Guangdong Provincial Key Laboratory of Computer Vision and Virtual Reality Technology, Shenzhen Institutes of Advanced Technology, Chinese Academy of Sciences, Shenzhen, China
cz.li@siat.ac.cn

Abstract. The generalization capability of neural networks across domains is crucial for real-world applications. We argue that a generalized object recognition system should well understand the relationships among different images and also the images themselves at the same time. To this end, we present a new domain generalization framework (called EISNet) that learns how to generalize across domains *simultaneously* from *extrinsic relationship supervision* and *intrinsic self-supervision* for images from multi-source domains. To be specific, we formulate our framework with feature embedding using a multi-task learning paradigm. Besides conducting the common supervised recognition task, we seamlessly integrate a momentum metric learning task and a self-supervised auxiliary task to collectively integrate the extrinsic and intrinsic supervisions. Also, we develop an effective momentum metric learning scheme with the K-hard negative mining to boost the network generalization ability. We demonstrate the effectiveness of our approach on two standard object recognition benchmarks VLCS and PACS, and show that our EISNet achieves state-of-the-art performance.

Keywords: Domain generalization · Unsupervised learning · Metric learning · Self-supervision

1 Introduction

The rise of deep neural networks has achieved promising results in various computer vision tasks. Most of these achievements are based on supervised learning,

Electronic supplementary material The online version of this chapter (https://doi.org/10.1007/978-3-030-58545-7_10) contains supplementary material, which is available to authorized users.

© Springer Nature Switzerland AG 2020
A. Vedaldi et al. (Eds.): ECCV 2020, LNCS 12354, pp. 159–176, 2020.
https://doi.org/10.1007/978-3-030-58545-7_10

which assumes that the models are trained and tested on the samples drawn from the same distribution or domain. However, in many real-world scenarios, the training and test samples are often acquired under different criteria. Therefore, the trained network may perform poorly on "unseen" test data with domain discrepancy from the training data. To address this limitation, researchers have studied how to alleviate the performance degradation of a trained network among different domains. For instance, by utilizing labeled (or unlabeled) target domain samples, various domain adaptation methods have been proposed to minimize the domain discrepancy by aligning the source and target domain distributions [11,18,23,35,41,42].

Although these domain adaptation methods can achieve better performance on the target domain, there exists an indispensable demand to pre-collect and access target domain data during the network training. Moreover, it needs to re-train the network to adapt to every new target domain. However, in real-world applications, it is often the case that adequate target domain data is not available during the training process [28,49]. For example, it is difficult for an automated driving system to know which domain (*e.g.*, city, weather) the self-driving car will be used. Therefore, it has a broad interest in studying how to learn a generalizable network that can be directly applied to new "unseen" target domains. Recently, the community develops *domain generalization* methods to improve the model generalization ability on unseen target domains by utilizing the multiple source domains.

Most existing domain generalization methods attempt to extract the shared domain-invariant semantic features among multiple source domains [8,26–28,31]. For example, Li *et al.* [28] extend an adversarial auto-encoder by imposing the Maximum Mean Discrepancy (MMD) measure to align the distributions among different domains. Since there is no specific prior information from target domains during the training, some works have investigated the effectiveness of increasing the diversity of the inputs by creating synthetic samples to improve the generalization ability of networks [49,50]. For instance, Yue *et al.* [49] propose a domain randomization method with Generative Adversarial Networks (GANs) to learn a model with high generalizability. Meta-learning has also been introduced to address the domain generalization problem via an episodic training [8,27]. Very recently, Carlucci *et al.* [2] introduce a self-supervision task by predicting relative positions of image patches to constrain the semantic feature learning for domain generalization. This shows that the self-supervised task can discover invariance in images with different patch orders and thus improve the network generalization. Such self-supervision task only considers the regularization within images but does not explore the valuable relationship among images across different domains to further enhance the discriminability and transferability of semantic features.

The generalization of deep neural networks relies crucially on the ability to learn and adapt knowledge across various domains. We argue that a generalized object recognition system should well understand the relationships among different objects and the objects themselves at the same time. Particularly, on

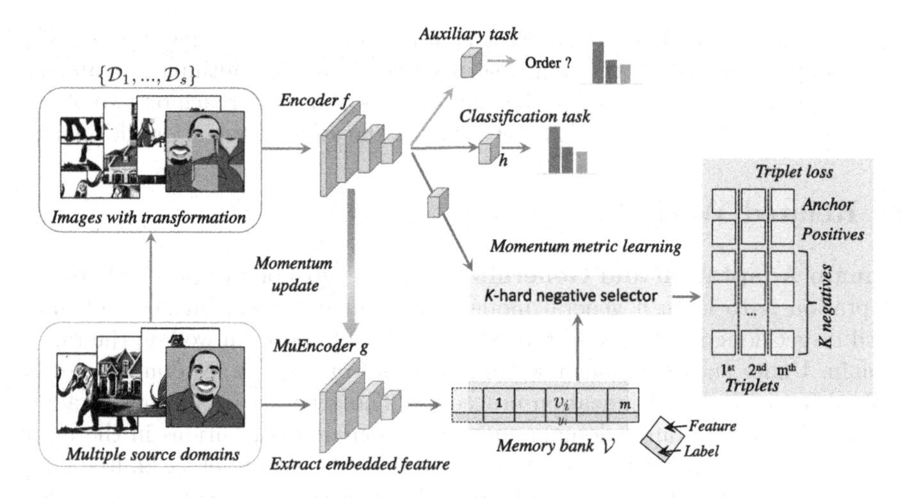

Fig. 1. The framework of the proposed EISNet for domain generalization. We train a feature Encoder f for discriminative and transferable feature extraction and a classifier for object recognition. Two complementary tasks, a momentum metric learning task and a self-supervised auxiliary task, are introduced to prompt general feature learning. We maintain a momentum updated Encoder (MuEncoder) to generate momentum updated embeddings stored in a large memory bank. Also, we design a K-hard negative selector to locate the informative hard triplets from the memory bank to calculate the triplet loss. The auxiliary self-supervised task predicts the order of patches within an image.

the one hand, exploring the relationship among different objects (*i.e.*, *extrinsic supervision*) guides the network to extract domain-independent yet category-specific representation, facilitating decision-boundary learning. On the other hand, exploring context or shape constraint within a single image (*i.e.*, *intrinsic supervision*) introduces necessary regularization for network training, broadening the network understanding of the object.

To this end, we present a new framework called EISNet that learns how to generalize across domains by simultaneously incorporating *extrinsic supervision* and *intrinsic supervision* for images from multi-source domains. We formulate our framework as a multi-task learning paradigm for general feature learning, as shown in Fig. 1. Besides conducting the common supervised recognition task, we seamlessly integrate a momentum metric learning task and a self-supervised auxiliary task into our framework to utilize the *extrinsic* and *intrinsic* supervisions, respectively. Specifically, we develop an effective momentum metric learning scheme with the K-hard negative selector to encourage the network to explore the image relationship and enhance the discriminability learning. The K-hard negative selector is able to filter the informative hard triplets, while the momentum updated encoder guarantees the consistency of embedded features stored in the memory bank, which stabilizes the training process. We then introduce a jigsaw puzzle solving task to learn the spatial relationship of images parts. The three kinds of tasks share the same feature encoder and are optimized in an end-to-end manner. We demonstrate the effectiveness of our approach on two object recognition benchmarks. Our EISNet achieves the state-of-the-art performance.

2 Related Work

Domain Adaptation and Generalization. The goal of unsupervised domain adaptation is to learn a general model with source domain images and unlabeled target domain images, so that the model could perform well on the target domain. Under such a problem setting, images from the target domain can be utilized to guide the optimization procedure. The general idea of domain adaptation is to align the source domain and target domain distributions in the input level [3,19], semantic feature level [9,34], or output space level [5,39,40,43,45]. Most methods adopt Generative Adversarial Networks and achieve better performance on the target domain data. However, training domain adaptation models need to access unlabeled target domain data, making it impractical for some real-world applications.

Domain generalization is an active research area in recent years. Its goal is to train a neural network on multiple source domains and produce a trained model that can be applied directly to unseen target domain. Since there is no specific prior guidance from the target domain during the training procedure, some domain generalization methods proposed to generate synthetic images derived from the given multiple source domains to increase the diversity of the input images, so that the network could learn from a larger data space [49,50]. Another promising direction is to extract domain-invariant features over multiple source domains [12,26–28,31]. For example, Li *et al.* [25] developed a low-rank parameterized CNN model for domain generalization and proposed the domain generalization benchmark dataset PACS. Motiian *et al.* [31] presented a unified framework by exploiting the Siamese architecture to learn a discriminative space. A novel framework based on adversarial autoencoders was presented by Li *et al.* [28] to learn a generalized latent feature representation across domains. Recently, meta-learning-based episodic training was designed to tackle domain generalization problems [8,27]. Li *et al.* [27] developed an episodic training procedure to expose the network to domain shift that characterizes a novel domain at runtime to improve the robustness of the network. Our work is most related to [2], which introduced self-supervision signals to regularize the semantic feature learning. However, besides the self-supervision signals within a single image, we further exploit the extrinsic relationship among image samples across different domains to improve the feature compactness.

Metric Learning. Our work is also related to metric learning, which aims to learn a metric to minimize the intra-class distances and maximize the inter-class variations [14,46]. With the development of deep learning, distance metric also benefits the feature embedding learning for better discrimination [17,44]. Recently, the metric learning strategies have attracted a lot of attention on face verification and recognition [36], fine-grained object recognition [44], image retrieval [48], and so on. Different from previous applications, in this work, we adopt the conventional triplet loss with more informative negative selection and momentum feature extraction for domain generalization.

Self-supervision. Self-supervision is a recent paradigm for unsupervised learning. The idea is to design annotation-free (*i.e.*, self-supervised) tasks for feature learning to facilitate the main task learning. Annotation-free tasks can be predictions of the image colors [24], relative locations of patches from the same image [2,32], image inpainting [33], and image rotation [13]. Typically, self-supervised tasks are used as network pre-train to learn general image features. Recently, it is trained as an auxiliary task to promote the mainstream task by sharing semantic features [4]. In this paper, we inherit the advantage of self-supervision to boost the network generalization ability.

3 Method

We aim to learn a model that can perform well on "unseen" target domain by utilizing multiple source domains. Formally, we consider a set of S source domains $\{\mathcal{D}_1, ..., \mathcal{D}_s\}$, with the j-th domain $\mathcal{D}_j$ having N_j sample-label pairs $\{(x_i^j, y_i^j)\}_{i=1}^{N_j}$, where x_i^j is the i-th sample in $\mathcal{D}_j$ and $y_i^j \in \{1, 2, ..., C\}$ is the corresponding label. In this work, we consider the object recognition task and aim to learn an Encoder $f_\theta : \mathcal{X} \rightarrow \mathcal{Z}$ mapping an input sample x_i into the feature embedding space $f_\theta(x_i) \in \mathcal{Z}$, where θ denotes the parameters of Encoder f_θ. We assume that Encoder f_θ could extract discriminative and transferable features, so that the task network (*e.g.*, classifier) $h_\psi : \mathcal{Z} \rightarrow \mathbb{R}^C$ can be prompted on the unseen target domain.

The overall framework of the proposed EISNet is illustrated in Fig. 1. We adopt the classical classification loss, *i.e.*, Cross-Entropy, to minimize the objective $\mathcal{L}_c(h_\psi(f_\theta(x)), y)$ that measures the difference between the ground truth y and the network prediction $\hat{y} = h_\psi(f_\theta(x))$. To avoid performance degradation on unseen target domain, we introduce two additional complementary supervisions to our framework. One is an extrinsic supervision with momentum metric learning, and the other is an intrinsic supervision with a self-supervised auxiliary task. The momentum metric learning is employed by a triplet loss with a K-hard negative selector on the momentum updated embeddings stored in a large memory bank. We implement a self-supervised auxiliary task by predicting the order of patches within an image. All these tasks adopt a shared encoder f and are seamlessly integrated into an end-to-end learning framework. Below, we introduce the extrinsic supervision and intrinsic self-supervision in detail.

3.1 Extrinsic Supervision with Momentum Metric Learning

For the domain generalization problem, it is necessary to ensure the features of samples with the same label close to each other, while the features of different class samples being far apart. Otherwise, the predictions on the unseen

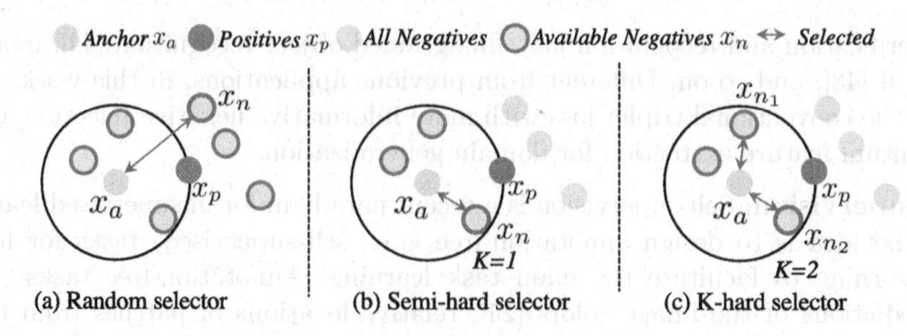

Fig. 2. The schematic diagram of triplet negative sample selectors. We draw a circle with the anchor (x_a) as the center and the distance between the anchor (x_a) and positive (x_p) as the radius. We ignore the relaxation margin here and set K as 2 for illustration. The selected negatives (x_n) are shown with arrows.

target domain may suffer from ambiguous decision boundaries and performance degradation [8,20]. This is well aligned with the philosophy of metric learning. Therefore, we design a momentum metric-learning scheme to encourage the network to learn such domain-independent yet class-specific features by considering the mutual relation among samples across domains. Specifically, we propose a novel K-hard negative selector for triple loss to improve the training effectiveness by selecting informative triplets in the memory bank, and a momentum updated encoder to guarantee the representation consistency among the embeddings stored in the memory bank.

***K*-Hard Negative Selector for Triplet Loss.** The triplet loss is widely used to learn feature embedding based on the relative similarity of the sampled pairs. The goal of the original triplet loss is to assign close distance to pairs of similar samples (*i.e.*, positive pair) and long distance to pairs of dissimilar samples (*i.e.*, negative pair). For example, we can extract the feature representation v_i of each image x_i from multi-source domains with the feature Encoder f_θ. Then by fixing an anchor sample x_a, we choose a corresponding positive sample x_p with the same class label as x_a, and a random negative sample x_n with different class label from x_a to form a triplet $\mathcal{T} = \{(x_a, x_p, x_n)|y_a = y_p,\ y_a \neq y_n\}$. Accordingly, the objective of the original triplet loss is formulated as

$$\mathcal{L}_\mathcal{T} = [d(x_a, x_p)^2 - d(x_a, x_n)^2 + \text{margin}]_+, \tag{1}$$

where $[\cdot]_+ = max(0, \cdot)$, $d(x_i, x_j)$ represents the distance between the samples, and the margin is a standard relaxation coefficient. In general, we use the Euclidean distance to measure distances between the embedded features. Then, the distance between samples x_i and x_j is defined as

$$d(x_i, x_j) = \sqrt{\| v_i - v_j \|^2} = \sqrt{\| f_\theta(x_i) - f_\theta(x_j) \|^2}. \tag{2}$$

The negative sample selection process in the original triplet loss is shown in Fig. 2(a). Since the selected negative sample may already obey the triplet constraint, the training with the original triplet loss selector may not be efficient. To avoid useless training, inspired by [36,37], we propose a novel K-hard negative online selector, which extends the triplet with K negatives that violate the triplet constraint within a certain of margin. Specifically, given a sampled anchor, we randomly choose one positive sample with the same class label as the anchor, and select K hard negative samples $x_{n_i}, i = \{1, 2, ..., K\}$ following

$$\mathcal{T}' = \{(x_a, x_p, x_{n_i}) | y_a = y_p, y_a \neq y_{n_i}, d(x_a, x_{n_i})^2 < d(x_a, x_p)^2 + \text{margin}\}. \quad (3)$$

In an extreme case, the number of hard negative samples may be zero, then we random select negative samples without the distance constraint. Therefore, the objective of the proposed triple loss with K-hard negative selector can be represented as

$$\mathcal{L}_{\mathcal{T}'} = \frac{1}{K} \sum_{i=1}^{K} [d(x_a, x_p)^2 - d(x_a, x_{n_i})^2 + \text{margin}]_+. \quad (4)$$

We illustrate the triplet selection process of semi-hard selector ($K = 1$) and K-hard selector ($K = 2$) in Fig. 2 (b) (c) for a better understanding. Compared with the original triplet loss, our proposed triple loss equipped with the K-hard negative selector considers more informative hard negatives for each anchor, thus facilitating the feature encoder to learn more discriminative features.

Efficient Learning with Memory Bank. The way to select informative triplet pairs has a large influence on the feature embedding. Good features can be learned from a large sample pool that includes a rich set of negative samples [15]. However, selecting K-hard triplets from the whole sample pool is not efficient. To increase the diversity of selected triplet pairs while reducing the computation burden, we maintain memory bank $\mathcal{V}$ to store the feature representation v_i of historical samples [47] with a size of m. Instead of calculating the embedded features of all the images at each iteration, we utilize the stored features to select the K-hard triplet samples. Note that we also keep the class label y_i along with representation v_i in the memory bank to filter the negatives, as shown in Fig. 1. During the network training, we dynamically update the memory bank by discarding the oldest items and feeding the new batch of embedded features, where the memory bank acts as a queue.

Momentum Updated Encoder. With the memory bank, we can improve the efficiency of triplet sample selection. However, the representation consistency between the current samples and historical samples in the memory bank is reduced due to the rapidly-changed encoder [15]. Therefore, instead of utilizing the same feature encoder to extract the representation of current samples and historical samples, we adopt a new **M**omentum **u**pdated **E**ncoder (MuEncoder) to generate feature representation for the samples in the memory bank. Formally, we denote the parameters of Encoder and MuEncoder as θ_f and θ_g, respectively. The Encoder parameter θ_f is optimized by a back-propagation of the loss function, while the MuEncoder parameter θ_g is updated as a moving average of Encoder parameters θ_f following

$$\theta_g = \delta * \theta_g + (1 - \delta) * \theta_f, \quad \text{where } \delta \in [0, 1), \tag{5}$$

where δ is a momentum coefficient to control the update degree of MuEncoder. Since the MuEncoder evolves more smoothly than Encoder, the update of different features in the memory bank is not rapid, thereby easing the triplet loss update. This is confirmed by the experimental results. In our preliminary experiments, we found that a large momentum coefficient δ by slowly updating θ_g could generate better results than rapid updating, which indicates that a slow update of MuEncoder is able to guarantee the representation consistency.

3.2 Intrinsic Supervision with Self-supervised Auxiliary Task

To broaden the network understanding of the data, we propose to utilize the intrinsic supervision within a single image to impose a regularization into the feature embedding by adding auxiliary self-supervised tasks on all the source domain images. A similar idea has been adopted in domain adaptation and Generative Adversarial Networks training [4,38]. The auxiliary self-supervised task is able to exploit the intrinsic semantic information within a single image to provide informative feature representations for the main task.

There are plenty of works focusing on designing auxiliary self-supervised tasks, such as rotation degree prediction and relative location prediction of two patches in one image [7,13,21]. Here, we employ the recently-proposed *solving jigsaw puzzles* [2,32] as our auxiliary task. However, most of the self-supervised tasks focusing on high-level semantic feature learning can be incorporated into our framework. Specifically, we first divide an image into nine (3×3) patches, and shuffle these patches within the 30 different combinations following [2]. As pointed by [2], the model achieves the highest performance when the class number is set as 30 and the order prediction performance decreases when the task becomes more difficult with more orders. A new auxiliary task branch h_a follows the extracted feature representation f_θ to predict the ordering of the patches. A Cross-Entropy loss is applied to tackle this order classification task:

$$\mathcal{L}_a = -\frac{1}{N * 31} \sum_{i=1}^{N} \sum_{c_a=0}^{30} y^a_{i,c_a} * \log(p^a_{i,c_a}), \tag{6}$$

where y^a and p^a are the ground-truth order and predicted order from the auxiliary task branch, respectively. We use $c_a = 0$ to represent the original images without patch shuffle, leading to a total of 31 classes.

Overall, we formulate the whole framework as a multi-task learning paradigm. The total objective function to train the network is represented as

$$\mathcal{L} = \alpha * \mathcal{L}_c + \beta * \mathcal{L}_{T'} + \gamma * \mathcal{L}_a, \tag{7}$$

where α, β, and γ are hyper-parameters to balance the weights of the basic classification supervision, extrinsic relationship supervision, and intrinsic self-supervision, respectively.

4 Experiments

4.1 Datasets

We evaluate our method on two public domain generalization benchmark datasets: **VLCS** and **PACS**. **VLCS** [10] is a classic domain generalization benchmark for image classification, which includes five object categories from four domains (PASCAL VOC 2007, LabelMe, Caltech, and Sun datasets). **PACS** [25] is a recent domain generalization benchmark for object recognition with larger domain discrepancy. It consists of seven object categories from four domains (Photo, Art Paintings, Cartoon, and Sketches datasets) and the domain discrepancy among different datasets is more severe than VLCS, making it more challenging.

4.2 Network Architecture and Implementation Details

Our framework is flexible and one can use different network backbones as the feature Encoder. We utilized a fully-connected layer with 31-dimensional output as the self-supervised auxiliary classification layer following the setting in [2] for a fair comparison. To enable the momentum metric learning, we further employed a fully-connected layer with 128 output channels following the Encoder part and added an L2 normalization layer to normalize the feature representation v of each sample. The MuEncoder has the same network architecture as the Encoder, and the weight of MuEncoder was initialized with the same weight as Encoder. We followed the previous works in the literature [1,2,8,26] and employed the leave-one-domain-out cross-validation strategy to produce the experiment results, i.e., we take turns to choose each domain for testing, and train a network model with the remaining three domains.

We implemented our framework with the PyTorch library on one NVIDIA TITAN Xp GPU. Our framework was optimized with the SGD optimizer. We totally trained 100 epochs, and the batch size was 128. The learning rate was set as 0.001 and decreased to 0.0001 after 80 epochs. We empirically set the margin of the triplet loss as 2. We also adopted the same on-the-fly data augmentation as JiGen [2], which includes random cropping, horizontal flipping, and jitter.

4.3 Results on VLCS Dataset

We followed the same experiment setting in previous work [2] to train and evaluate our method. The extrinsic metric learning and intrinsic self-supervised learning was developed upon the "FC7" features of AlexNet [22] pretrained on ImageNet [6]. We set the size of the memory bank as 1024 and the number of

Table 1. Domain generalization results on **VLCS** dataset with object recognition accuracy (%) using **AlexNet** backbone. The top results are highlighted in **bold**.

Target	Within domain	D-MTAE [12]	CIDDG [29]	CCSA [31]	DBADG [25]	MMD-AAE [28]	MLDG [26]	Epi-FCR [27]	JiGen [2]	MASF [8]	EISNet (Ours)
PASCAL	82.07	63.90	64.38	67.10	69.99	67.70	67.7	67.1	**70.62**	69.14	69.83 ± 0.48
LabelMe	74.32	60.13	63.06	62.10	63.49	62.60	61.3	64.3	60.90	**64.90**	63.49 ± 0.82
Caltech	100.0	89.05	88.83	92.30	93.63	94.40	94.4	94.1	96.93	94.78	**97.33** ± 0.36
Sun	77.33	61.33	62.10	59.10	61.32	64.40	65.9	65.9	64.30	67.64	**68.02** ± 0.81
Average	83.43	68.60	69.59	70.15	72.11	72.28	72.3	72.9	73.19	74.11	**74.67**

Table 2. Domain generalization results on **PACS** dataset with object recognition accuracy (%) using **AlexNet** backbone. The top results are highlighted in **bold**.

Target	Within domain	D-MTAE [12]	CIDDG [29]	DBADG [25]	MLDG [26]	Epi-FCR [27]	MetaReg [1]	JiGen [2]	MASF [8]	EISNet (Ours)
Photo	97.80	91.12	78.65	89.50	88.00	86.1	91.07	89.00	90.68	**91.20** ± 0.00
Art painting	90.36	60.27	62.70	62.86	66.23	64.7	69.82	67.63	70.35	**70.38** ± 0.37
Cartoon	93.31	58.65	69.73	66.97	66.88	72.3	70.35	71.71	**72.46**	71.59 ± 1.32
Sketch	93.88	47.68	64.45	57.51	58.96	65.0	59.26	65.18	67.33	**70.25** ± 1.36
Average	93.84	64.48	68.88	69.21	70.01	72.0	72.62	73.38	75.21	**75.86**

negatives K in the triplet loss Eq. (4) as 256. The hyper-parameters α, β, and γ in total objective function Eq. (7) were set as 1, 0.1, and 0.05, respectively. For our results, we report the average performance and standard deviation over three independent runs.

We compare our method with other nine previous state-of-the-art methods. **D-MTAE** [12] utilized the multi-task auto-encoders to learn robust features across domains. **CIDDG** [29] was a conditional invariant adversarial network that learns the domain-invariant representations under distribution constraints. **CCSA** [31] exploited a Siamese network to learn a discriminative embedding subspace with distribution distances and similarities. **DBADG** [25] developed a low-rank parametrized CNN model for domain generalization. **MMD-AAE** [28] aligned the distribution through an adversarial auto-encoder by Maximum Mean Discrepancy. **MLDG** [26] was a meta-learning method by simulating train/test domain shift during training. **Epi-FCR** [27] was an episodic training method. **JiGen** [2] solved a jigsaw puzzle auxiliary task based on self-supervision. **MASF** [8] employed a meta-learning based strategy with two complementary losses for encoder regularization. Moreover, we include the **Within domain** performance of all the datasets as a comparison to reveal the performance drop due to domain discrepancy. We trained **Within domain** using a supervised way with training and test images from the same domain.

The comparison results with the above methods are shown in Table 1. It is observed that our EISNet achieves the best performance on both Caltech and Sun datasets and comparable results on PASCAL VOC and LabelMe datasets. Overall, EISNet achieves an average accuracy of 74.67% over four domains, outperforming the previous state-of-the-art method **MASF** [8]. Our method also

Table 3. Domain generalization results on **PACS** dataset with object recognition accuracy (%) using **ResNet** backbones. The top results are highlighted in **bold**.

Target	ResNet-18			ResNet-50		
	DeepAll	MASF [8]	EISNet (Ours)	DeepAll	MASF [8]	EISNet (Ours)
Photo	94.25	94.99	**95.93** $\pm$ 0.06	94.83	95.01	**97.11** $\pm$ 0.40
Art painting	77.38	80.29	**81.89** $\pm$ 0.88	81.47	82.89	**86.64** $\pm$ 1.41
Cartoon	75.65	**77.17**	76.44 $\pm$ 0.31	78.61	80.49	**81.53** $\pm$ 0.64
Sketch	69.64	71.69	**74.33** $\pm$ 1.37	69.69	72.29	**78.07** $\pm$ 1.43
Average	79.23	81.04	**82.15**	81.15	82.67	**85.84**

outperforms **JiGen** [2] on three domains and achieves comparable results on the remaining PASCAL VOC domain, demonstrating that utilizing extrinsic relationship supervision can further improve the network generalization ability.

4.4 Results on PACS Dataset

To show the effectiveness of our framework under different network backbones on PACS dataset, we evaluate our method with three different backbones: AlexNet, ResNet-18, and ResNet-50 [16]. The size of memory bank was set as 1024 and K in the triplet loss Eq. (4) was set as 256. The hyper-parameters in total objective function Eq. (7) were set as 1, 0.5, and 0.7 for α, β, and γ, respectively. For our results, we also report the average performance and standard deviation over three independent runs.

Table 2 summarizes the experimental results developed with AlexNet backbone. We compare our methods with eight other methods that achieved previous best results on this benchmark dataset. **MetaReg** [1] utilized a novel classifier regularization in the meta-learning framework. As we can observe from Table 2, by simultaneously utilizing momentum metric learning and intrinsic self-supervision for images across different source domains, our method achieves the best performance on three datasets. Across all domains, our method achieves an average accuracy of 75.86%, setting a new state-of-the-art performance.

We also compare our method with baseline method (DeepAll) and the state-of-the-art method **MASF** [8] using ResNet-18 and ResNet-50 backbones in Table 3. In the ResNet-50 experiment, we reduce the batch size to 64 to fit the limited GPU memory. The DeepAll method is trained with all the source domains without any specific network design. As shown in Table 3, our method consistently outperforms **MASF** about 1.11% and 3.17% on average accuracy with ResNet-18 and ResNet-50 backbone, respectively. This indicates that our designed framework is very general and can be migrated to different network backbones. Note that the improvement over **MASF** is more obvious with a deeper network backbone, showing that our proposed algorithm is more beneficial for domain generalization with deeper feature extractors.

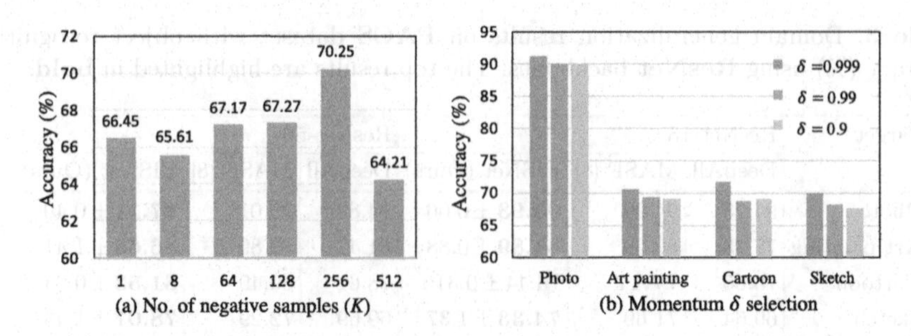

Fig. 3. The performance of our method under different number of negative samples K and momentum update coefficient δ.

Table 4. Ablation study on key components of our method with the **PACS** dataset (%). The top results are highlighted in **bold**.

Extrinsic	Intrinsic	Photo	Art painting	Cartoon	Sketch	Average
–	–	94.85	81.47	78.61	69.69	81.15
✓	–	97.06	81.97	80.70	76.81	84.14
–	✓	97.02	85.17	76.35	76.97	83.88
✓	✓	**97.11**	**86.64**	**81.53**	**78.07**	**85.84**

Table 5. Comparison of our proposed K-hard negative selector with original random selector and semi-hard negative selector.

Selector	Random	Semi-hard	K-hard
Accuracy (%)	65.38	68.08	70.25

4.5 Analysis of Our Method

We conduct extensive analysis of our method. Firstly, we investigate the effectiveness of extrinsic and intrinsic supervision using ResNet-50 backbone on **PACS** dataset, and the experimental results are illustrated in Table 4. The **Extrinsic** supervision indicates that the momentum metric learning is used, while **Intrinsic** supervision denotes that the auxiliary self-supervision loss is optimized. The method without these two supervisions is the baseline model, which is the same with DeepAll results in Table 3. From the results in Table 4, we observe that each supervision plays an important role in our framework. Specifically, equipping the extrinsic supervision into the baseline model yields about 2.99% average accuracy improvement. Meanwhile, we also achieve 2.73% average accuracy improvement over the baseline model by incorporating intrinsic self-supervision of the images. By combing extrinsic and intrinsic supervision, performance is further improved across all settings, indicating these two supervisions are complementary.

We then analyze five key components in our framework, that is a) the number of different negative samples K in momentum metric learning, b) the effective-

Table 6. Comparison among different memory bank size.

Memory bank size m	1024	512	256	128
No. of negatives K	256	128	64	32
Accuracy (%)	70.25	68.80	68.24	67.93

ness of momentum update coefficient δ, c) the effectiveness of hard negative selector, d) the size of memory bank m, and e) time cost. All below comparison experiments are implemented with AlexNet backbone on the PACS benchmark.

a. The number of negative samples K is a key parameter of our designed K-hard negative selector in momentum metric learning. We investigate the network performance under different options. We select six K values at different magnitudes, which are 1, 8, 64, 128, 256, and 512. The Sketch dataset results are shown in Fig. 3 (a), We can observe that a large number of negative samples would lead to better results in general and the network generates the best result with $K = 256$. However, the performance drops drastically if we set $K = 512$, demonstrating that too large K will produce a burden on the metric distance calculation and make the network difficult to learn.
b. The momentum update coefficient δ is important to control the feature consistency among different batches of embedded features in the memory bank. We show the accuracy with different momentum coefficient δ in Fig. 3 (b). It is observed that the network performs well when δ is relatively large, *i.e.*, 0.999. A small coefficient would degrade the network performance, suggesting that a slow updating MuEncoder is beneficial to the feature consistency.
c. To validate the effectiveness of K-hard negative selector in our proposed metric learning, we compare our proposed K-hard negative selector with original random triplet selector and semi-hard negative selector. The Sketch dataset results are shown in Table 5. Equipped with semi-hard negative selector, the accuracy improves 2.70%. By selecting more negative pairs from the memory bank, we obtain the accuracy of 70.25%, demonstrating the effectiveness of the proposed K-hard negative selector.
d. The size of memory bank m can be adjusted according to different tasks. Here, we show the results of four different settings with the number of negatives changing as well in Table 6. In general, our method is able to generate better results with a large memory bank size and negative samples. However, a too large memory bank will increase the burden to calculate the pair-wise distance in triplet loss. Therefore, we need to balance the accuracy and computation burden.
e. Apart from the performance improvement over other methods, our method has much lower computation cost. Under the same server setting (one TITAN XP GPU) and AlexNet backbone, our method only takes 1.5 h to train the network on PACS dataset, while the total training time of the state-of-the-art MASF is about 17 h. Therefore, our method could save more than 91% time cost on training phase.

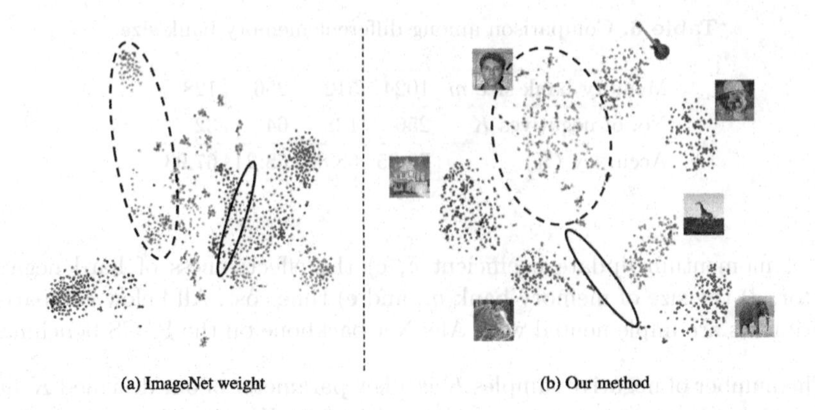

Fig. 4. t-SNE visualization on one target domain to show the discrimination of the network. (a) is the feature embedding extracted from the IMAGENET pre-trained network. (b) shows the feature embedding distributions extracted from our EISNet.

We also employ t-SNE [30] to analyze the feature level discrimination of our method and the visualization results are shown in Fig. 4. Compared with the feature extracted from the ImageNet pre-trained network, the distance between different class clusters in our method becomes evident, indicating that equipped with our proposed extrinsic and intrinsic supervision, the model is able to learn more discriminative features among different object categories regardless domains.

5 Conclusions

We have presented a multi-task learning paradigm to learn how to generalize across domains for domain generalization. The main idea is to learn a feature embedding simultaneously from the extrinsic relationship of different images and the intrinsic self-supervised constraint within the single image. We design an effective and efficient momentum metric learning module to facilitate compact feature learning. Extensive experimental results on two public benchmark datasets demonstrate that our proposed method is able to learn discriminative yet transferable feature, which lead to state-of-the-art performance for domain generalization. Moreover, our proposed framework is flexible and can be migrated to various network backbones.

Acknowledgments. We thank anonymous reviewers for the comments and suggestions. The work described in this paper was supported in parts by the following grants: Key-Area Research and Development Program of Guangdong Province, China (2020B010165004), Hong Kong Innovation and Technology Fund (Project No. ITS/426/17FP and ITS/311/18FP), and National Natural Science Foundation of China with Project No. U1813204.

References

1. Balaji, Y., Sankaranarayanan, S., Chellappa, R.: MetaReg: towards domain generalization using meta-regularization. In: Advances in Neural Information Processing Systems, pp. 998–1008 (2018)
2. Carlucci, F.M., D'Innocente, A., Bucci, S., Caputo, B., Tommasi, T.: Domain generalization by solving jigsaw puzzles. In: Proceedings of the IEEE Conference on Computer Vision and Pattern Recognition, pp. 2229–2238 (2019)
3. Chen, C., Dou, Q., Chen, H., Qin, J., Heng, P.A.: Synergistic image and feature adaptation: towards cross-modality domain adaptation for medical image segmentation. In: Proceedings of the Thirty-Third Conference on Artificial Intelligence (AAAI), pp. 865–872 (2019)
4. Chen, T., Zhai, X., Ritter, M., Lucic, M., Houlsby, N.: Self-supervised GANs via auxiliary rotation loss. In: Proceedings of the IEEE Conference on Computer Vision and Pattern Recognition, pp. 12154–12163 (2019)
5. Chen, Y., Li, W., Van Gool, L.: ROAD: reality oriented adaptation for semantic segmentation of urban scenes. In: Proceedings of the IEEE Conference on Computer Vision and Pattern Recognition, pp. 7892–7901 (2018)
6. Deng, J., Dong, W., Socher, R., Li, L.J., Li, K., Fei-Fei, L.: ImageNet: a large-scale hierarchical image database. In: Proceedings of the IEEE Conference on Computer Vision and Pattern Recognition, pp. 248–255 (2009)
7. Doersch, C., Gupta, A., Efros, A.A.: Unsupervised visual representation learning by context prediction. In: Proceedings of the IEEE International Conference on Computer Vision, pp. 1422–1430 (2015)
8. Dou, Q., de Castro, D.C., Kamnitsas, K., Glocker, B.: Domain generalization via model-agnostic learning of semantic features. In: Advances in Neural Information Processing Systems, pp. 6447–6458 (2019)
9. Dou, Q., Ouyang, C., Chen, C., Chen, H., Heng, P.A.: Unsupervised cross-modality domain adaptation of convnets for biomedical image segmentations with adversarial loss. In: Proceedings of the 27th International Joint Conference on Artificial Intelligence, pp. 691–697 (2018)
10. Fang, C., Xu, Y., Rockmore, D.N.: Unbiased metric learning: On the utilization of multiple datasets and web images for softening bias. In: Proceedings of the IEEE International Conference on Computer Vision, pp. 1657–1664 (2013)
11. Ganin, Y., et al.: Domain-adversarial training of neural networks. J. Mach. Learn. Res. **17**(1), 2030–2096 (2016)
12. Ghifary, M., Bastiaan Kleijn, W., Zhang, M., Balduzzi, D.: Domain generalization for object recognition with multi-task autoencoders. In: Proceedings of the IEEE International Conference on Computer Vision, pp. 2551–2559 (2015)
13. Gidaris, S., Singh, P., Komodakis, N.: Unsupervised representation learning by predicting image rotations. arXiv preprint arXiv:1803.07728 (2018)
14. Hadsell, R., Chopra, S., LeCun, Y.: Dimensionality reduction by learning an invariant mapping. In: Proceedings of the IEEE Conference on Computer Vision and Pattern Recognition, vol. 2, pp. 1735–1742. IEEE (2006)
15. He, K., Fan, H., Wu, Y., Xie, S., Girshick, R.: Momentum contrast for unsupervised visual representation learning. In: Proceedings of the IEEE Conference on Computer Vision and Pattern Recognition (2020)

16. He, K., Zhang, X., Ren, S., Sun, J.: Deep residual learning for image recognition. In: Proceedings of the IEEE Conference on Computer Vision and Pattern Recognition (2016)
17. Hoffer, E., Ailon, N.: Deep metric learning using triplet network. In: Feragen, A., Pelillo, M., Loog, M. (eds.) SIMBAD 2015. LNCS, vol. 9370, pp. 84–92. Springer, Cham (2015). https://doi.org/10.1007/978-3-319-24261-3_7
18. Hoffman, J., et al.: CyCADA: cycle-consistent adversarial domain adaptation. arXiv preprint arXiv:1711.03213 (2017)
19. Hong, W., Wang, Z., Yang, M., Yuan, J.: Conditional generative adversarial network for structured domain adaptation. In: Proceedings of the IEEE Conference on Computer Vision and Pattern Recognition, pp. 1335–1344 (2018)
20. Kamnitsas, K., et al.: Semi-supervised learning via compact latent space clustering. arXiv preprint arXiv:1806.02679 (2018)
21. Kolesnikov, A., Zhai, X., Beyer, L.: Revisiting self-supervised visual representation learning. In: Proceedings of the IEEE Conference on Computer Vision and Pattern Recognition, pp. 1920–1929 (2019)
22. Krizhevsky, A., Sutskever, I., Hinton, G.E.: ImageNet classification with deep convolutional neural networks. Advances in Neural Information Processing Systems 25, pp. 1097–1105 (2012)
23. Kumar, A., Saha, A., Daume, H.: Co-regularization based semi-supervised domain adaptation. In: Advances in Neural Information Processing Systems, pp. 478–486 (2010)
24. Larsson, G., Maire, M., Shakhnarovich, G.: Learning representations for automatic colorization. In: Leibe, B., Matas, J., Sebe, N., Welling, M. (eds.) ECCV 2016. LNCS, vol. 9908, pp. 577–593. Springer, Cham (2016). https://doi.org/10.1007/978-3-319-46493-0_35
25. Li, D., Yang, Y., Song, Y.Z., Hospedales, T.M.: Deeper, broader and artier domain generalization. In: Proceedings of the IEEE International Conference on Computer Vision, pp. 5542–5550 (2017)
26. Li, D., Yang, Y., Song, Y.Z., Hospedales, T.M.: Learning to generalize: meta-learning for domain generalization. In: Thirty-Second AAAI Conference on Artificial Intelligence (2018)
27. Li, D., Zhang, J., Yang, Y., Liu, C., Song, Y.Z., Hospedales, T.M.: Episodic training for domain generalization. In: Proceedings of the IEEE International Conference on Computer Vision, pp. 1446–1455 (2019)
28. Li, H., Jialin Pan, S., Wang, S., Kot, A.C.: Domain generalization with adversarial feature learning. In: Proceedings of the IEEE Conference on Computer Vision and Pattern Recognition, pp. 5400–5409 (2018)
29. Li, Y., et al.: Deep domain generalization via conditional invariant adversarial networks. In: Proceedings of the European Conference on Computer Vision (ECCV), pp. 624–639 (2018)
30. Maaten, L.V.D., Hinton, G.: Visualizing data using t-SNE. J. Mach. Learn. Res. **9**(Nov), 2579–2605 (2008)
31. Motiian, S., Piccirilli, M., Adjeroh, D.A., Doretto, G.: Unified deep supervised domain adaptation and generalization. In: Proceedings of the IEEE International Conference on Computer Vision, pp. 5715–5725 (2017)
32. Noroozi, M., Favaro, P.: Unsupervised learning of visual representations by solving jigsaw puzzles. In: Leibe, B., Matas, J., Sebe, N., Welling, M. (eds.) ECCV 2016. LNCS, vol. 9910, pp. 69–84. Springer, Cham (2016). https://doi.org/10.1007/978-3-319-46466-4_5

33. Pathak, D., Krahenbuhl, P., Donahue, J., Darrell, T., Efros, A.A.: Context encoders: feature learning by inpainting. In: Proceedings of the IEEE Conference on Computer Vision and Pattern Recognition, pp. 2536–2544 (2016)
34. Ren, J., Hacihaliloglu, I., Singer, E.A., Foran, D.J., Qi, X.: Adversarial domain adaptation for classification of prostate histopathology whole-slide images. In: Frangi, A.F., Schnabel, J.A., Davatzikos, C., Alberola-López, C., Fichtinger, G. (eds.) MICCAI 2018. LNCS, vol. 11071, pp. 201–209. Springer, Cham (2018). https://doi.org/10.1007/978-3-030-00934-2_23
35. Saito, K., Watanabe, K., Ushiku, Y., Harada, T.: Maximum classifier discrepancy for unsupervised domain adaptation. In: Proceedings of the IEEE Conference on Computer Vision and Pattern Recognition, pp. 3723–3732 (2018)
36. Schroff, F., Kalenichenko, D., Philbin, J.: Facenet: a unified embedding for face recognition and clustering. In: Proceedings of the IEEE Conference on Computer Vision and Pattern Recognition, pp. 815–823 (2015)
37. Sohn, K.: Improved deep metric learning with multi-class N-pair loss objective. In: Advances in Neural Information Processing Systems, pp. 1857–1865 (2016)
38. Sun, Y., Tzeng, E., Darrell, T., Efros, A.A.: Unsupervised domain adaptation through self-supervision. arXiv preprint arXiv:1909.11825 (2019)
39. Tsai, Y.H., Hung, W.C., Schulter, S., Sohn, K., Yang, M.H., Chandraker, M.: Learning to adapt structured output space for semantic segmentation. In: Proceedings of the IEEE Conference on Computer Vision and Pattern Recognition, pp. 7472–7481 (2018)
40. Tsai, Y.H., Sohn, K., Schulter, S., Chandraker, M.: Domain adaptation for structured output via discriminative patch representations. In: Proceedings of the IEEE International Conference on Computer Vision, pp. 1456–1465 (2019)
41. Tzeng, E., Hoffman, J., Darrell, T., Saenko, K.: Simultaneous deep transfer across domains and tasks. In: Proceedings of the IEEE International Conference on Computer Vision, pp. 4068–4076 (2015)
42. Tzeng, E., Hoffman, J., Saenko, K., Darrell, T.: Adversarial discriminative domain adaptation. In: Proceedings of the IEEE Conference on Computer Vision and Pattern Recognition, pp. 7167–7176 (2017)
43. Vu, T.H., Jain, H., Bucher, M., Cord, M., Pérez, P.: ADVENT: adversarial entropy minimization for domain adaptation in semantic segmentation. In: Proceedings of the IEEE Conference on Computer Vision and Pattern Recognition, pp. 2517–2526 (2019)
44. Wang, J., et al.: Learning fine-grained image similarity with deep ranking. In: Proceedings of the IEEE Conference on Computer Vision and Pattern Recognition, pp. 1386–1393 (2014)
45. Wang, S., Yu, L., Yang, X., Fu, C.W., Heng, P.A.: Patch-based output space adversarial learning for joint optic disc and cup segmentation. IEEE Trans. Med. Imaging **38**(11), 2485–2495 (2019)
46. Weinberger, K.Q., Saul, L.K.: Distance metric learning for large margin nearest neighbor classification. J. Mach. Learn. Res. **10**(Feb), 207–244 (2009)
47. Wu, Z., Xiong, Y., Yu, S.X., Lin, D.: Unsupervised feature learning via non-parametric instance discrimination. In: Proceedings of the IEEE Conference on Computer Vision and Pattern Recognition, pp. 3733–3742 (2018)
48. Yuan, Y., Yang, K., Zhang, C.: Hard-aware deeply cascaded embedding. In: Proceedings of the IEEE International Conference on Computer Vision, pp. 814–823 (2017)

49. Yue, X., Zhang, Y., Zhao, S., Sangiovanni-Vincentelli, A., Keutzer, K., Gong, B.: Domain randomization and pyramid consistency: simulation-to-real generalization without accessing target domain data. In: Proceedings of the IEEE International Conference on Computer Vision, pp. 2100–2110 (2019)
50. Zakharov, S., Kehl, W., Ilic, S.: DeceptionNet: network-driven domain randomization. In: Proceedings of the IEEE International Conference on Computer Vision, pp. 532–541 (2019)

Joint Learning of Social Groups, Individuals Action and Sub-group Activities in Videos

Mahsa Ehsanpour[1,3(✉)], Alireza Abedin[1], Fatemeh Saleh[2,3], Javen Shi[1], Ian Reid[1,3], and Hamid Rezatofighi[1]

[1] The University of Adelaide, Adelaide, Australia
mahsa.ehsanpour@adelaide.edu.au
[2] Australian National University, Canberra, Australia
[3] Australian Centre for Robotic Vision, Brisbane, Australia

Abstract. The state-of-the art solutions for human activity understanding from a video stream formulate the task as a spatio-temporal problem which requires joint localization of all individuals in the scene and classification of their actions or group activity over time. Who is interacting with whom, e.g. not everyone in a queue is interacting with each other, is often not predicted. There are scenarios where people are best to be split into sub-groups, which we call social groups, and each social group may be engaged in a different social activity. In this paper, we solve the problem of simultaneously grouping people by their social interactions, predicting their individual actions and the social activity of each social group, which we call the social task. Our main contributions are: i) we propose an end-to-end trainable framework for the social task; ii) our proposed method also sets the state-of-the-art results on two widely adopted benchmarks for the traditional group activity recognition task (assuming individuals of the scene form a single group and predicting a single group activity label for the scene); iii) we introduce new annotations on an existing group activity dataset, re-purposing it for the social task. The data and code for our method is publicly available (https://github.com/mahsaep/Social-human-activity-understanding-and-grouping).

Keywords: Collective behaviour recognition · Social grouping · Video understanding

1 Introduction

Recognising individuals' action and group activity from video is a widely studied problem in computer vision. This is crucial for surveillance systems, autonomous driving cars and robot navigation in environments where humans are present

Electronic supplementary material The online version of this chapter (https://doi.org/10.1007/978-3-030-58545-7_11) contains supplementary material, which is available to authorized users.

© Springer Nature Switzerland AG 2020
A. Vedaldi et al. (Eds.): ECCV 2020, LNCS 12354, pp. 177–195, 2020.
https://doi.org/10.1007/978-3-030-58545-7_11

Fig. 1. Examples of our annotated data representing social groups of people in the scene and their common activity within their group. Realistically, individuals in the scene may perform their own actions, but they may also belong to a social group with a mutual activity, *e.g., walking together*. In this figure, social groups and their corresponding common activities have been color-coded.

[5,12,34]. In the last decade, most effort from the community has been dedicated to extract reliable spatio-temporal representations from video sequences. Dominantly, this was investigated in a simplified scenario where each video clip was trimmed to involve a single action, hence a classification problem [6,35,53,54]. Recently, the task of human activity understanding has been extended to more challenging and realistic scenarios, where the video is untrimmed and may include multiple individuals performing different actions [22,23,28,56]. In parallel, there are works focusing on predicting a group activity label to represent a collective behaviour of all the actors in the scene [11,14,37]. There are also independent works aiming at only grouping individuals in the scene based on their interactions [7,20,25,47,57] or joint inferring of groups, events and human roles in aerial videos [49] by utilizing hand-crafted features.

In a real scenario, a scene may contain several people, individuals may perform their own actions while they might be connected to a social group. In other words, a real scene generally comprises several groups of people with potentially different social connections, *e.g.* contribution toward a common activity or goal (Fig. 1). To this end, in our work we focus on the problem of "*Who is with whom and what they are doing together?*". Although the existing works mentioned in the previous paragraphs tackle some elements, we propose a holistic approach that considers the multi-task nature of the problem, where these tasks are not completely independent, and which can benefit each other. Understanding of this scene-wide social context would be conducive for many video understanding applications, *e.g.* anomalous behaviour detection in the crowd from a surveillance footage or navigation of an autonomous robot or car through a crowd.

To tackle this real-world problem, we propose an end-to-end trainable framework which takes a video sequence as input and learns to predict *a*) each individual's action; *b*) their social connections and groups; and, *c*) a social activity for each predicted social group in the scene. We first introduce our framework for a relevant conventional problem, *i.e.* group activity recognition [11,14,37]. We propose an architecture design that incorporates: *i*) *I3D backbone* [23] as a state-of-the art feature extractor to encode spatio-temporal representation of individuals in a video clip, *ii*) *Self-attention module* [59] to refine individuals' feature representations, and *iii*) *Graph attention module* [60] to directly model the interactions and connections among the individuals. Our framework outperforms the state-of-the-art on two widely adopted group activity recognition datasets. We then introduce our extended framework that can elegantly handle social grouping and social activity recognition for each group. We also augment an existing group activity dataset with enriched social group and social activity annotations for the social task. Our main contributions are:

1. We propose an end-to-end framework for the group activity recognition task through integration of I3D backbone, self-attention module and graph attention module in a well-justified architecture design. Our pipeline also outperforms existing solutions and sets a new state-of-the-art for the group activity recognition task on two widely adopted benchmarks.
2. We show that by including a graph edge loss in the proposed group activity recognition pipeline, we obtain an end-to-end trainable solution to the multi-task problem of simultaneously grouping people, recognizing individuals' action and social activity of each social group (social task).
3. We introduce new annotations, *i.e.* social groups and social activity label for each sub-group on a widely used group activity dataset to serve as a new benchmark for the social task.

2 Related Work

Action Recognition. Video understanding is one of the main computer vision problems widely studied over the past decades. Deep convolutional neural networks (CNNs) have shown promising performance in action recognition on short trimmed video clips. A popular approach in this line involves adoption of two-stream networks with 2D kernels to exploit the spatial and temporal information [18,19,53,61]. Recurrent neural networks (RNNs) have also been utilized to capture the temporal dependencies of visual features [15,41]. Unlike these approaches, [30,58] focused on CNNs with 3D kernels to extract features from a sequence of dense RGB frames. Recently, [6] proposed I3D, a convolutional architecture that is based on inflating 2D kernels pretrained on ImageNet into 3D ones. By pretraining on large-scale video datasets such as Kinetics [32], I3D outperforms the two-stream 2D CNNs on video recognition benchmark datasets.

Spatio-Temporal Action Detection. Temporal action detection methods aim to recognize humans' actions and their corresponding start and end times in untrimmed videos [51,52,67,69]. By introducing spatio-temporal action annotation for each subject, *e.g.* as in AVA [23], spatio-temporal action detection received considerable attention [17,21,22,38,56,64]. In particular, [56] models the spatio-temporal relations to capture the interactions between human, objects, and context that are crucial to infer human actions. More recently, action transformer network [22] has been proposed to localize humans and recognize their actions by considering the relation between actors and context.

Group Activity Recognition. The aforementioned methods mostly focus on predicting individuals' actions in the scene. However, there are works focusing on group activity recognition where the aim is to predict a single group activity label for the whole scene. The early approaches typically extracted hand-crafted features and applied probabilistic graphical models [2,8,9,11,36,37,49] or AND-OR grammar models [1,49] for group activity recognition. In recent years, deep learning approaches especially RNNs achieve impressive performance largely due to their ability of both learning informative representations and modelling temporal dependencies in the sequential data [4,13,14,26,27,40,44,45,50,62]. For instance, [27] uses a two-stage LSTM model to learn a temporal representation of person-level actions and pools individuals' features to generate a scene-level representation. In [45], attention mechanism is utilized in RNNs to identify the key individuals responsible for the group activity. Later, authors of [48] extended these works by utilizing an energy layer for obtaining more reliable predictions in presence of uncertain visual inputs. Following these pipelines, [26] introduces a relational layer that can learn compact relational representations for each person. The method proposed in [4] is able to jointly localize multiple people and classify the actions of each individual as well as their collective activity. In order to consider the spatial relation between the individuals in the scene, an attentive semantic RNN has been proposed in [44] for understanding group activities. Recently, the Graph Convolutional Network (GCN) has been used in [65] to learn the interactions in an Actor Relation Graph to simultaneously capture the appearance and position relation between actors for group activity recognition. Similarly [3] proposed a CNN model encoding spatial relations as an intermediate activity-based representation to be used for recognizing group activities. There are also a number of attempts to simultaneously track multiple people and estimate their collective activities in multi-stage frameworks [8,39]. Although these approaches try to recognize the interactions between pairs of people by utilizing hand-crafted features, they are not capable of inferring social groups.

Despite the progress made towards action recognition, detection and group activity recognition [55], what still remains a challenge is simultaneously understanding of social groups and their corresponding social activity. Some existing works aim at only detecting groups in the scene [7,20] by relying on hand-crafted rules *e.g.* face orientations, which are only applicable to very specific tasks such as conversational group detection [25,43,47,57] and by utilizing small-scale datasets. In contrast to previous solutions which are task dependent and require domain expert knowledge to carefully design hand-crafted rules, we extend the concept of grouping to more general types of interactions. To this end, we propose an end-to-end trainable framework for video data to jointly predict individuals' action as well as their social groups and social activity of each predicted group.

3 Social Activity Recognition

Social activity recognition seeks to answer the question of "*Who is with whom and what they are doing together?*". Traditional group activity recognition can be seen as a simplified case of social activity recognition where all the individuals are assumed to form a single group in the scene and a single group activity label is predicted for the whole scene. For the ease of conveying our ideas, we present our framework first in the simpler setting of group activity recognition task, and then show how to augment it for the social task.

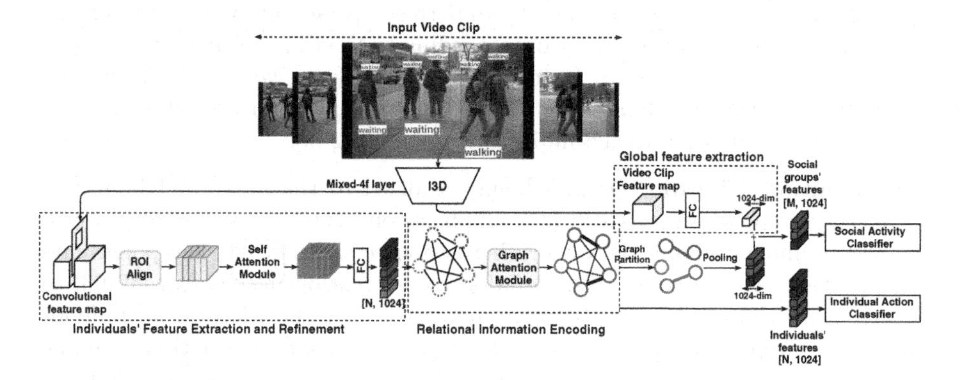

Fig. 2. Our network architecture for the social task. The set of aligned individuals' features are initially refined by the self-attention module. Projected feature maps are then fed into the GAT module to encode relational information between individuals. During training, feature representation of each social group is pooled from the feature maps of its members according to the ground-truth social connections. At test time, we adopt a graph partitioning algorithm on the inter-node attention coefficients provided by GAT and accordingly infer the social groups and social activity of each sub-group. M and N refer to the number of social groups and number of individuals respectively.

3.1 Group Activity Recognition Framework

A group activity recognition framework should be capable of: *a*) generating a holistic and enriched spatio-temporal representation from the entire video clip; *b*) extracting fine-detailed spatial features from bounding box of each person to accurately predict the individual actions; and *c*) learning an aggregated representation from all individuals for precise realization of their collective activities. Illustrated in Fig. 2, we carefully design effective components in our framework to achieve the above desirable properties. We elaborate the components as follows.

I3D Backbone. We use the Inflated 3D ConvNet (I3D) [6] (based on Inception architecture [29]) as the backbone to capture the spatio-temporal context of an input video clip. In I3D, ImageNet pre-trained convolutional kernels are expanded into 3D, allowing it to seamlessly learn effective spatio-temporal representations. Motivated by the promising performance of the pre-trained I3D models in a wide range of action classification benchmarks, we exploit the feature representations offered by this backbone at multiple resolutions. More specifically, we use the deep spatio-temporal feature maps extracted from the final convolutional layer as a rich semantic representation describing the entire video clip. These deeper features provide low-resolution yet high-level representations that encode a summary of the video. In addition, accurate recognition of individuals' action rely on finer details which are often absent in very deep coarse representations. To extract fine spatio-temporal representations for the individuals, we use the higher resolution feature maps from the intermediate Mixed-4f layer of I3D. As depicted in Fig. 2, from this representation we extract the temporally-centered feature map corresponding to the centre frame of the input video clip. Given the bounding boxes in the centre frame, we conduct ROIAlign [24] to project the coordinates on the frame's feature map and slice out the corresponding features for each individual's bounding box.

Self-attention Module. Despite being localized to the individual bounding boxes, these representations still lack emphasis on visual clues that play a crucial role in understanding the underlying actions *e.g.* a person's key-points and body posture. To overcome this, we adopt self-attention mechanism [59,63] to directly learn the interactions between any two feature positions of an individual's feature representation and accordingly leverage this information to refine each individual's feature map. In our framework the self-attention module functions as a non-local operation and computes the response at each position by attending to all positions in an individual's feature map. The output of the self-attention module contextualizes the input bounding box feature map with visual clues and thus, enriches the individual's representation by highlighting the most informative features. As substantiated by ablation studies in Sect. 5.1, capturing such fine details significantly contribute to the recognition performance.

Graph Attention Module. Uncovering subtle interactions among individuals present in a multi-person scenario is fundamental to the problem of group activity recognition; each person individually performs an action and the set of inter-connected actions together result in the underlying global activity context. As such, this problem can elegantly be modeled by a graph, where the nodes represent refined individuals' feature map and the edges represent the interactions among individuals. We adopt the recently proposed Graph Attention Networks (GATs) [60] to directly learn the underlying interactions and seamlessly capture the global activity context. GATs flexibly allow learning attention weights between nodes through parameterized operations based on a self-attention strategy and have successfully demonstrated state-of-the-art results by outperforming existing counterparts [33]. GATs compute attention coefficients for every possible pair of nodes, which can be represented in an adjacency matrix $\hat{O}^{\alpha}$.

Training. In our framework, the GAT module consumes the individuals' feature map obtained from the self-attention component, encodes inter-node relations, and generates an updated representation for each individual. We acquire the group representation by max-pooling the enriched individuals' feature map and adding back a linear projection of the holistic video's features obtained from the I3D backbone. A classifier is then applied on this representation to generate group activity scores denoted by $\hat{O}^{\mathrm{G}}$. Similarly, another classifier is applied on the individuals' representation to govern the individual action scores denoted by

$\hat{O}^{\mathrm{I}}_n$. The associated operations provide a fully differentiable mapping from the input video clip to the output predictions, allowing the framework to be trained in an end-to-end fashion by minimizing the following objective function,

$$\mathcal{L} = \mathcal{L}_{\mathrm{gp}}(O^{\mathrm{G}}, \hat{O}^{\mathrm{G}}) + \lambda \sum_{n} \mathcal{L}_{\mathrm{ind}}(O^{\mathrm{I}}_n, \hat{O}^{\mathrm{I}}_n), \tag{1}$$

where $\mathcal{L}_{\mathrm{gp}}$ and $\mathcal{L}_{\mathrm{ind}}$ respectively denote the cross-entropy loss for group activity and individual action classification. Here, O^{G} and O^{I}_n represent the ground-truth group activities and individual actions, n identifies the individual and λ is the balancing coefficient for the loss functions.

3.2 Social Activity Recognition Framework

In a real-world multi-person scene, a set of social groups each with different number of members and social activity labels often exist. We refer to this challenging problem as the *social activity recognition* task. In this section, we propose a novel yet simple modification to our group activity recognition framework that naturally allows understanding of social groups and their corresponding social activities in a multi-person scenario. The *I3D backbone* and the *self-attention module* remain exactly the same as elucidated in Sect. 3.1. We explain the required modifications for the *graph attention module* as follows.

Training. Previously, the GAT's inter-node attention coefficients were updated with the supervision signal provided by the classification loss terms in Eq. 1. To satisfy the requirements of the new problem, *i.e.* to generate social groups and their corresponding social activity label, we augment the training objective with a graph edge loss $\mathcal{L}_c$ that incentivizes the GAT's self-attention strategy to converge to the individuals' social connections

$$\mathcal{L} = \sum_{s} \mathcal{L}_{\mathrm{sgp}}(O^{\mathrm{SG}}_s, \hat{O}^{\mathrm{SG}}_s) + \lambda_1 \sum_{n} \mathcal{L}_{\mathrm{ind}}(O^{\mathrm{I}}_n, \hat{O}^{\mathrm{I}}_n) + \lambda_2 \mathcal{L}_{\mathrm{c}}(O^{\alpha}, \hat{O}^{\alpha}), \tag{2}$$

where, $\mathcal{L}_c$ is the binary cross-entropy loss to reduce the discrepancy between GAT's learned adjacency matrix $\hat{O}^{\alpha}$ and the ground-truth social group connections O^{α}. Further, $\mathcal{L}_{\mathrm{sgp}}$ and $\mathcal{L}_{\mathrm{ind}}$ respectively denote the cross-entropy loss for social activity of each social group and individual actions classification. Notably, given the ground-truth social groupings during training, we achieve the representation for each social group by max-pooling its corresponding nodes' feature-map and adding back a linear projection of the video features obtained from the I3D backbone (similar to learning group activity representations in our simplified group activity framework). A classifier is then applied on top to generate the social activity scores $\hat{O}^{\mathrm{SG}}_s$. At inference time however, we require a method to infer the social groups in order to compute the corresponding social representations. To this end, we propose to utilize graph spectral clustering [68] on the learned attention coefficients by GAT, $\hat{O}^{\alpha}$ and achieve a set of disjoint partitions representing the social groups. In the above formulation s is the social group identifier and (λ_1, λ_2) are the loss balancing coefficients.

4 Datasets

We evaluate our group activity recognition framework on two widely adopted benchmarks: Volleyball dataset and Collective Activity dataset (CAD). We also perform evaluation of our social activity recognition framework on our provided social task dataset by augmenting the CAD with social groups and social activity label for each group annotations.

4.1 Group Activity Recognition Datasets

Volleyball Dataset [27] contains 4830 videos from 55 volleyball games partitioned into 3493 clips for training and 1337 clips for testing. Each video has a group activity label from the following activities: *right set, right spike, right pass, right win-point, left set, left spike, left pass, left win-point*. The centered frame of each video is annotated with players' bounding boxes and their individual action including *waiting, setting, digging, failing, spiking, blocking, jumping, moving, standing*.

Collective Activity Dataset (CAD) [10] has 44 video sequences captured from indoor and street scenes. In each video, actors' bounding box, their actions and a single group activity label are annotated for the key frames (*i.e.* 1 frame out of every 10 frames). Individual actions include *crossing, waiting, queuing, walking, talking, N/A*. The group activity label associated with each key frame is assigned according to the majority of individuals' actions in the scene. We adopt the same train/test splits as previous works [3,44,65].

4.2 Social Activity Recognition Dataset

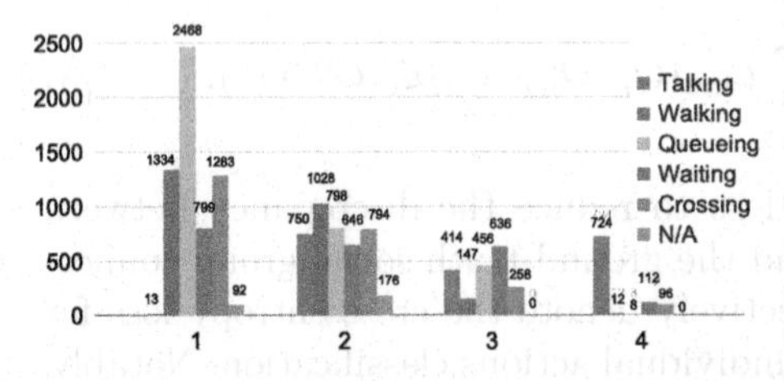

Fig. 3. The histogram of social activities for varying social group sizes (1 to 4 people per social group).

In order to solve the problem of social activity recognition, a video dataset containing scenes with different social groups of people, each performing a social activity is required. Thus, we decided to utilize CAD which is widely used in the group activity recognition task and its properties suit well our problem. Other video action datasets [23,31,32] could not be used in this problem since they mostly consist of scenes with only one social group or a number of sigleton groups. We provide enriched annotations on CAD for solving the social task, which we call *Social-CAD*. As such, for each key frame, we maintain the exact same bounding box coordinates and individual action annotations as the original dataset. However, rather than having a single group activity label for the scene, we annotate different social groups and their corresponding social activity labels. Since there may only exist a subtle indicator in the entire video sequence, *e.g.* an eye contact or a hand shake, suggesting a social connection between the individuals, determining *"who is with whom"* can be a challenging and subjective task. Therefore, to generate the social group annotations as reliable as possible: *a*) we first annotate the trajectory of each person by linking his/her bounding boxes over the entire video, *b*) Given the trajectories, we asked three annotators to independently divide the tracks into different sub-groups according to their social interactions, and *c*) we adopted majority voting to confirm the final social groups. Similar to the CAD annotation, the social activity label for each social group is defined by the dominant action of its members. We use the same train/test splits in Social-CAD as in CAD. Detailed activity distributions of Social-CAD are given in Fig. 3.

5 Experimental Results

To evaluate our proposed framework, we first show that our group activity recognition pipeline outperforms the state-of-arts in both individual action and group activity recognition tasks on two widely adopted benchmarks. Then we evaluate the performance of our social activity recognition framework on Social-CAD (Figs. 4 and 5).

5.1 Group Activity Recognition

Implementation Details. In our model, we use an I3D backbone which is initialized with Kintetics-400 [32] pre-trained model. We utilize ROI-Align with crop size of 5×5 on extracted feature-map from Mixed-4f layer of I3D. We perform self-attention on each individuals' feature map with query, key and value being different linear projections of individuals' feature map with output sizes being $1/8, 1/8, 1$ of the input size. We then learn a 1024-dim feature map for each individual features obtained from self-attention module. Aligned individuals' feature maps are fed into our single-layer, multi-head GAT module with 8 heads

Fig. 4. Visual results of our method on Volleyball dataset for the individual/group activity recognition task (better viewed in color). The bounding boxes around players are produced by our detection-based approach. The numbers above the boxes denote the predicted action IDs. The ground-truth action ID for each player is indicated in red when the predicted action ID is wrong. The label on top of each key frame shows the predicted group activity. Please refer to the dataset section to map IDs to their corresponding actions.

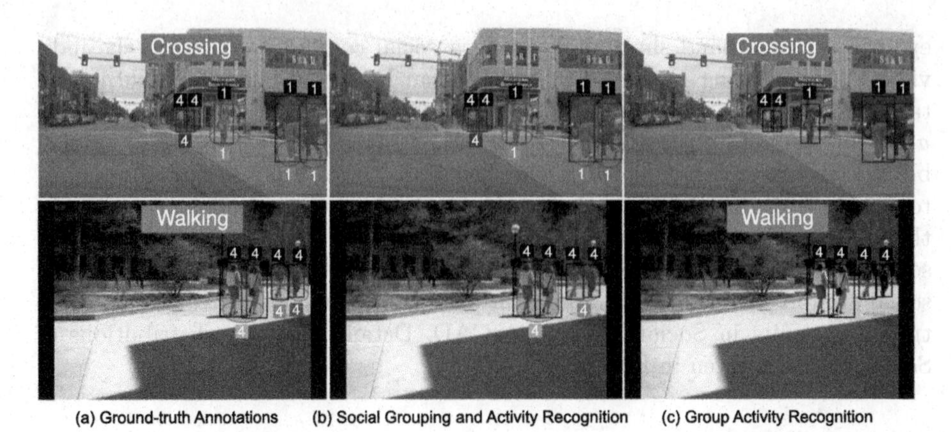

(a) Ground-truth Annotations (b) Social Grouping and Activity Recognition (c) Group Activity Recognition

Fig. 5. Visual results of our method on CAD for both social and group activity recognition tasks. Column(a) shows the ground-truth annotation for both tasks. Column(b) represents our prediction for the social task. Column(c) is our predictions for the group task. Note that social groups are denoted by a colored cylinder with their social group labels underneath. The numbers on top of bounding boxes denote the individual action IDs and the label tag above each key frame is the group activity label for the whole scene. 1 and 4 refer to crossing and walking activities respectively.

and input-dim, hidden-dim, output-dim being 1024 and droupout probability of 0.5 and $\alpha = 0.2$ [60]. We utilize ADAM optimizer with $\beta_1 = 0.9$, $\beta_2 = 0.999$, $\epsilon = 10^{-8}$ following [65]. We train the network in two stages. First, we train the network without the graph attention module. Then we fine-tune the network end-to-end including GAT. For the Volleyball dataset, we train the network in 200 epochs with a mini-batch size of 3 and a learning rate ranging from 10^{-4} to 10^{-6} and $\lambda_1 = 8$. For the CAD, we train the network in 150 epochs with a mini-batch size of 4 and a learning rate ranging from 10^{-5} to 10^{-6} and $\lambda_1 = 10$. Input video clips to the model are 17 frames long, with the annotated key frame in the centre. At test time, we perform our experiments based on two widely used settings in group activity recognition literature namely groundtruth-based and Detection-based settings. In the groundtruth-based setting, ground-truth bounding boxes of individuals are given to the model to infer the individual action for each box and the group activity for the whole scene. In the detection-based setting, we fine-tune a Faster-RCNN [46] on both datasets and utilize the predicted boxes for inferring the individuals' action and group activity.

Evaluation. In order to evaluate the performance of our model for the group activity recognition task, we adopt the commonly used metric, *i.e.* average accuracy, reported in all previous works [3,26,44,65]. To report the performance of individuals' action in GT-based setting, similar to [65], we used the average accuracy as the measure. In the case of Detection-based setting, average accuracy for evaluating the individuals action is not a valid measure due to the presence of false and missing detections. Instead, we report the commonly used measure for object detection, *i.e.* mean average precision (mAP) [16].

Table 1. Ablation study of our method for group activity recognition. w/o SA: without self-attention module. w/o GAT: without graph attention module.

	Volleyball	Collective activity
	Group (Individual) **Acc.% (Acc.%)**	Group (Individual) **Acc.% (Acc.%)**
Ours[group-only]	91.0 (-)	84.6 (-)
Ours[w/o SA- w/o GAT]	92.0 (82.0)	88.2 (73.4)
Ours[w/o GAT]	92.5 (83.2)	88.3 (75.3)
Ours[final]	**93.1(83.3)**	**89.4 (78.3)**

Ablation Study. We justify the choice of each component in our framework with detailed ablation studies. The results on volleyball and CAD are shown in Table 1. As the simplest baseline denoted by Ours[group-only], we use a Kinetics-400 pre-trained I3D backbone and fine-tune it by utilizing the input videos' feature representation obtained from the last layer of I3D and using a cross-entropy loss to learn the group activity without considering individuals' action. It is worth mentioning that surprisingly, our simplest baseline already outperforms many existing frameworks on group activity recognition (see the group accuracy in Table 2). This shows the importance of extracting spatio-temporal features simultaneously using 3D models as well as taking into account the whole video clip rather than solely focusing on individuals and their relations. To consider the effect of jointly training the model on group activity and individual action tasks, we add a new cross-entropy loss to our simplest baseline for training the individuals' action. As Ours[w/o SA- w/o GAT] experiment shows, training both tasks jointly helps improve the group activity recognition performance. In Ours[w/o GAT], we add the self-attention module performing on each individual's feature-map in order to highlight the most important features and improve the individual action recognition performance. As shown in Ours[w/o GAT], utilizing self-attention module improves the individual action accuracy by 1.2% on volleyball dataset and by 2.1% on CAD which also contributes to a slight improvement in the group activity accuracy on both datasets. Finally, we add the GAT module to capture the interactions among individuals which is essential in recognizing group activity. As shown in the Ours[final] experiment, utilizing GAT increases the group activity accuracy by 0.6% on volleyball and by 1.1% on collective activity dataset. GAT also improves the individual action performance on both volleyball and collective active datasets by 0.1% and 3% respectively. The higher boost in individual action performance on CAD compared to the one in the volleyball dataset shows the effectiveness of GAT in highlighting social sub-groups and updating individual feature representations accordingly as it is benefiting from a self attention strategy between nodes.

Comparison with the State-of-the-Arts. We compare our results on Volleyball and CAD with the state-of-the-art methods in Table 2, using group activity accuracy and individual action accuracy as the evaluation metrics. The

Table 2. Comparison with the state-of-the-arts on Volleyball dataset and CAD for group activity recognition.

	Volleyball Group (Individual) Acc.% (Acc.%)	Collective activity Group (Individual) Acc.% (Acc.%)
HDTM [27]	81.9 (-)	81.5 (-)
CERN [48]	83.3 (-)	87.2 (-)
StagNet [44]	89.3 (-)	89.1 (-)
HRN [26]	89.5 (-)	- (-)
SSU [4]	90.6 (81.8)	- (-)
CRM [3]	93.0 (-)	85.8 (-)
ARG† [65]	92.5 (83.0)	88.1 (77.3)
Ours	**93.1 (83.3)**	**89.4 (78.3)**
	Acc.% (mAP%)	Acc.% (mAP%)
StagNet(Det) [44]	87.6 (-)	87.9 (-)
SSU(Det) [4]	86.2 (-)	- (-)
ARG(Det)† [65]	91.5 (39.8)	86.1 (49.6)
Ours(Det)	**93.0 (41.8)**	**89.4 (55.9)**

Table 3. The mean per class group activity accuracies (MPCA) and per class group activity accuracies of our model compared to the existing models on CAD. M, W, Q and T stand for Moving, Waiting, Queuing and Talking respectively. Note that Crossing and Walking are merged as Moving.

Method	M	W	Q	T	MPCA
HDTM [27]	95.9	66.4	96.8	99.5	89.7
SBGAR [40]	90.08	81.4	99.2	84.6	89.0
CRM [3]	91.7	86.3	100.0	98.91	94.2
ARG† [65]	**100.0**	76.0	100.0	100.0	94.0
Ours	98.0	**91.0**	**100.0**	**100.0**	**97.2**

top section of the table demonstrates the performance of the approaches in the groundtruth-based setting, where ground-truth bounding box of each person is used for prediction of the individual action as well as group activity of the whole scene. However, in the detection-based settings (indicated by (Det) in Table 2), a Faster-RCNN is fine-tuned on both datasets and predicted bounding boxes for individuals are used during inference (which is more realistic in practice). In group activity recognition using predicted bounding boxes, our model has the least performance drop compared to other methods. In Table 2, ARG† is the result that we obtained by running [65]'s released code with the same setting mentioned for each dataset, and the reproduced results perfectly matched the reported results on volleyball dataset. However, we could not reproduce their reported results on CAD. Having their source code available, in order to have a fair comparison with our framework, we also reported their performance on individuals' action on CAD datasets using both groundtruth-based and detection-based settings (not reported in the original paper). Our framework outperforms all existing methods in all settings on both datasets. We observe that a common wrong prediction in all the existing methods on CAD is the confusion between *crossing* and *walking* in the previous setting. *crossing* and *walking* are essentially same activities being performed at different locations. Thus, we merge these two classes into a single *moving* class and report the Mean Per Class Accuracy (MPCA) in Table 3 as in [3]. ARG† outperforms our method in one class *moving*, and our model performs the best in all other 3 classes and the overall metric MPCA.

5.2 Social Activity Recognition

Implementation Details. Our model is trained end-to-end for 200 epochs with a mini-batch size of 4 and a learning rate of 10^{-5} and $\lambda_1 = 5$ and $\lambda_2 = 2$. Other hyper-parameters have the same values as in the group activity recognition experiments on CAD. For graph partitioning at test time, we used graph spectral clustering technique [42,68].

Evaluation. Similar to the group activity recognition task, we perform experiments in two groundtruth-based and detection-based settings. For the social task, we evaluate three sub-tasks: 1) *social grouping*, 2) *social activity recognition of each group* and 3) *individuals' action recognition*. In the GT-based setting, for (1) we calculate the *membership accuracy* as the accuracy of predicting each person's assignment to a social group (including singleton groups). This accuracy is also known as unsupervised clustering accuracy [66]. For (2), we evaluate if both the membership and the social activity label of a person are jointly correct. If so, we consider this instance as a true positive, otherwise, it is assumed a false positive. The final measure, *i.e. social accuracy* is attained as a ratio between the number of true positives and the number of predictions. For (3), we evaluate if the individual's action label is correctly predicted and report *individual action accuracy*. In the detection-based setting, mAP is reported in order to evaluate predicted sub-groups, social activity of each group and individuals' action. For this experiment, bounding boxes with N/A groundtruth are excluded.

Results and Comparison. The performance of our method on social task is reported in Table 4, using membership accuracy, social activity accuracy and individual action accuracy in GT-based setting and mAP for each sub-task in detection-based setting as evaluation metrics. We consider three scenarios of baselines for evaluating this task:

1) Single group setting: forcing all individuals form a single social group, and then assessing algorithms' performance. ARG[group] [65] and ours[group] essentially are the approaches in Table 2, but are evaluated in membership and social activity metrics. GT[group] uses ground-truth activity labels serving as the upper bound performance for group activity recognition frameworks;

2) Individuals setting: forcing each individual as a unique social group, *e.g.*, if there are 10 people in the scene, they will be considered as 10 social groups. GT[individuals] uses ground-truth action labels serving as the upper bound performance for group activity recognition frameworks;

3) Social group setting: Partitioning individuals into multiple social groups. Our first approach uses group activity recognition pipeline in training and uses graph spectral clustering technique to find social groups at inference, named as Ours[cluster] (third part of Table 4). This produces better performance compared to the other baselines, but it is outperformed by our final framework denoted by Ours[learn2cluster], where we learn representations via the the additional graph edge loss.

Table 4. Social activity recognition results. In each column we report accuracy and mAP for the groundtruth-based and detection-based settings respectively.

	Membership GT(Det) Acc.%, (mAP%)	Social activity GT(Det) Acc.%, (mAP%)	Individual action GT(Det) Acc.%, (mAP%)
ARG[group] [65]	54.4(49.0)	47.2(34.8)	78.4(62.6)
Ours [group]	54.4(49.0)	47.7(35.6)	79.5(64.2)
GT[group]	54.4(-)	51.6(-)	-(-)
ARG[individuals] [65]	62.4(52.4)	49.0(41.1)	78.4(62.6)
Ours[individuals]	62.4(52.4)	49.5(41.8)	79.5(64.2)
GT[individuals]	62.4(-)	54.9(-)	-(-)
Ours[cluster]	78.2(68.2)	52.2(46.4)	79.5(64.2)
Ours[learn2cluster]	**83.0(74.9)**	**69.0(51.3)**	**83.3(66.6)**

Discussion. As mentioned in Sect. 4.2, there might exist only a single frame with a subtle gesture in the entire video, demonstrating a social connection between people in the scene. Therefore as future direction, incorporating individuals' track and skeleton pose might disambiguate some challenging cases for social grouping. Moreover, the performance of our proposed social grouping framework heavily relies on the performance of the graph spectral clustering technique, which is not part of the learning pipeline. Improving this step by substituting it with a more reliable graph clustering approach, or making it a part of learning pipeline can potentially ameliorate the final results.

6 Conclusion

In this paper, we propose the novel social task which requires jointly predicting of individuals' action, grouping them into social groups based on their interactions and predicting the social activity of each social group. To tackle this problem, we first considered addressing the simpler task of group activity recognition, where all the individuals are assumed to form a single group and a single group activity label is predicted for the scene. As such, we proposed a novel deep framework incorporating well-justified choice of state-of-the-art modules such as I3D backbone, self-attention and graph attention network. We demonstrate that our proposed framework achieves state-of-the-art results on two widely adopted datasets for the group activity recognition task. Next, we introduced our social task dataset by providing additional annotations and re-purposing an existing group activity dataset. We discussed how our framework can readily be extended to handle social grouping and social activity recognition of groups through incorporation of a graph partitioning loss and a graph partitioning algorithm *i.e.* graph spectral clustering. In the future, we aim to use the social activity context for development of better forecasting models, *e.g.* the task of social trajectory prediction, or social navigation system for an autonomous mobile robot.

References

1. Amer, M.R., Xie, D., Zhao, M., Todorovic, S., Zhu, S.C.: Cost-sensitive top-down/bottom-up inference for multiscale activity recognition. In: Proceedings of the European Conference on Computer Vision, pp. 187–200 (2012)
2. Amer, M.R., Lei, P., Todorovic, S.: Hirf: hierarchical random field for collective activity recognition in videos. In: Proceedings of the European Conference on Computer Vision, pp. 572–585 (2014)
3. Azar, S.M., Atigh, M.G., Nickabadi, A., Alahi, A.: Convolutional relational machine for group activity recognition. In: Proceedings of the IEEE Conference on Computer Vision and Pattern Recognition, pp. 7892–7901 (2019)
4. Bagautdinov, T., Alahi, A., Fleuret, F., Fua, P., Savarese, S.: Social scene understanding: end-to-end multi-person action localization and collective activity recognition. In: Proceedings of the IEEE Conference on Computer Vision and Pattern Recognition, pp. 4315–4324 (2017)
5. Caba Heilbron, F., Escorcia, V., Ghanem, B., Carlos Niebles, J.: Activitynet: a large-scale video benchmark for human activity understanding. In: Proceedings of the IEEE Conference on Computer Vision and Pattern Recognition, pp. 961–970 (2015)
6. Carreira, J., Zisserman, A.: Quo vadis, action recognition? a new model and the kinetics dataset. In: Proceedings of the IEEE Conference on Computer Vision and Pattern Recognition, pp. 6299–6308 (2017)
7. Choi, W., Chao, Y.W., Pantofaru, C., Savarese, S.: Discovering groups of people in images. In: Proceedings of the European Conference on Computer Vision, pp. 417–433 (2014)
8. Choi, W., Savarese, S.: A unified framework for multi-target tracking and collective activity recognition. In: Proceedings of the European Conference on Computer Vision, pp. 215–230 (2012)

9. Choi, W., Savarese, S.: Understanding collective activities of people from videos. IEEE Trans. Pattern Anal. Mach. Intell. **36**(6), 1242–1257 (2013)
10. Choi, W., Shahid, K., Savarese, S.: What are they doing?: collective activity classification using spatio-temporal relationship among people. In: Proceedings of the IEEE 12th International Conference on Computer Vision Workshops, pp. 1282–1289 (2009)
11. Choi, W., Shahid, K., Savarese, S.: Learning context for collective activity recognition. In: Proceedings of the IEEE Conference on Computer Vision and Pattern Recognition, pp. 3273–3280 (2011)
12. Collins, R.T., Lipton, A.J., Kanade, T.: Introduction to the special section on video surveillance. IEEE Trans. Pattern Anal. Mach. Intell. **22**(8), 745–746 (2000)
13. Deng, Z., Vahdat, A., Hu, H., Mori, G.: Structure inference machines: recurrent neural networks for analyzing relations in group activity recognition. In: Proceedings of the IEEE Conference on Computer Vision and Pattern Recognition, pp. 4772–4781 (2016)
14. Deng, Z., et al.: Deep structured models for group activity recognition. arXiv preprint arXiv:1506.04191 (2015)
15. Donahue, J., et al.: Long-term recurrent convolutional networks for visual recognition and description. In: Proceedings of the IEEE Conference on Computer Vision and Pattern Recognition, pp. 2625–2634 (2015)
16. Everingham, M., Van Gool, L., Williams, C.K., Winn, J., Zisserman, A.: The pascal visual object classes (VOC) challenge. Int. J. Comput. Vision **88**(2), 303–338 (2010)
17. Feichtenhofer, C., Fan, H., Malik, J., He, K.: Slowfast networks for video recognition. In: Proceedings of the IEEE International Conference on Computer Vision, pp. 6202–6211 (2019)
18. Feichtenhofer, C., Pinz, A., Wildes, R.: Spatiotemporal residual networks for video action recognition. In: Advances in Neural Information Processing Systems, pp. 3468–3476 (2016)
19. Feichtenhofer, C., Pinz, A., Zisserman, A.: Convolutional two-stream network fusion for video action recognition. In: Proceedings of the IEEE Conference on Computer Vision and Pattern Recognition, pp. 1933–1941 (2016)
20. Ge, W., Collins, R.T., Ruback, R.B.: Vision-based analysis of small groups in pedestrian crowds. IEEE Trans. Pattern Anal. Mach. Intell. **34**(5), 1003–1016 (2012)
21. Girdhar, R., Carreira, J., Doersch, C., Zisserman, A.: A better baseline for ava. arXiv preprint arXiv:1807.10066 (2018)
22. Girdhar, R., Carreira, J., Doersch, C., Zisserman, A.: Video action transformer network. In: Proceedings of the IEEE Conference on Computer Vision and Pattern Recognition, pp. 244–253 (2019)
23. Gu, C., et al.: Ava: a video dataset of spatio-temporally localized atomic visual actions. In: Proceedings of the IEEE Conference on Computer Vision and Pattern Recognition, pp. 6047–6056 (2018)
24. He, K., Gkioxari, G., Dollár, P., Girshick, R.: Mask R-CNN. In: Proceedings of the IEEE International Conference on Computer Vision, pp. 2961–2969 (2017)
25. Hung, H., Kröse, B.: Detecting f-formations as dominant sets. In: Proceedings of the 13th International Conference on Multimodal Interfaces, pp. 231–238 (2011)
26. Ibrahim, M.S., Mori, G.: Hierarchical relational networks for group activity recognition and retrieval. In: Proceedings of the European Conference on Computer Vision, pp. 721–736 (2018)

27. Ibrahim, M.S., Muralidharan, S., Deng, Z., Vahdat, A., Mori, G.: A hierarchical deep temporal model for group activity recognition. In: Proceedings of the IEEE Conference on Computer Vision and Pattern Recognition, pp. 1971–1980 (2016)
28. The multiview extended video with activities (MEVA) dataset. https://mevadata.org/
29. Ioffe, S., Szegedy, C.: Batch normalization: Accelerating deep network training by reducing internal covariate shift. arXiv preprint arXiv:1502.03167 (2015)
30. Ji, S., Xu, W., Yang, M., Yu, K.: 3D convolutional neural networks for human action recognition. IEEE Trans. Pattern Anal. Mach. Intell. **35**(1), 221–231 (2012)
31. Joo, H., et al.: Panoptic studio: a massively multiview system for social interaction capture. IEEE Trans. Pattern Anal. Mach. Intell. **41**(1), 190–204 (2017)
32. Kay, W., et al.: The kinetics human action video dataset (2017)
33. Kipf, T.N., Welling, M.: Semi-supervised classification with graph convolutional networks. arXiv preprint arXiv:1609.02907 (2016)
34. Kruse, T., Pandey, A.K., Alami, R., Kirsch, A.: Human-aware robot navigation: a survey. Robot. Auton. Syst. **61**(12), 1726–1743 (2013)
35. Kuehne, H., Jhuang, H., Garrote, E., Poggio, T., Serre, T.: HMDB: a large video database for human motion recognition. In: Proceedings of the IEEE International Conference on Computer Vision, pp. 2556–2563 (2011)
36. Lan, T., Sigal, L., Mori, G.: Social roles in hierarchical models for human activity recognition. In: Proceedings of the IEEE Conference on Computer Vision and Pattern Recognition, pp. 1354–1361 (2012)
37. Lan, T., Wang, Y., Yang, W., Robinovitch, S.N., Mori, G.: Discriminative latent models for recognizing contextual group activities. IEEE Trans. Pattern Anal. Mach. Intell. **34**(8), 1549–1562 (2011)
38. Li, D., Qiu, Z., Dai, Q., Yao, T., Mei, T.: Recurrent tubelet proposal and recognition networks for action detection. In: Proceedings of the European Conference on Computer Vision, pp. 303–318 (2018)
39. Li, W., Chang, M.C., Lyu, S.: Who did what at where and when: simultaneous multi-person tracking and activity recognition. arXiv preprint arXiv:1807.01253 (2018)
40. Li, X., Choo Chuah, M.: SBGAR: semantics based group activity recognition. In: Proceedings of the IEEE International Conference on Computer Vision, pp. 2876–2885 (2017)
41. Li, Z., Gavrilyuk, K., Gavves, E., Jain, M., Snoek, C.G.: VideoLSTM convolves, attends and flows for action recognition. Comput. Vis. Image Underst. **166**, 41–50 (2018)
42. Ng, A.Y., Jordan, M.I., Weiss, Y.: On spectral clustering: analysis and an algorithm. In: Advances in Neural Information Processing Systems, pp. 849–856 (2002)
43. Patron-Perez, A., Marszalek, M., Zisserman, A., Reid, I.D.: High five: recognising human interactions in TV shows. In: BMVC, vol. 1, p. 33 (2010)
44. Qi, M., Qin, J., Li, A., Wang, Y., Luo, J., Van Gool, L.: stagNet: an attentive semantic RNN for group activity recognition. In: Proceedings of the European Conference on Computer Vision, pp. 101–117 (2018)
45. Ramanathan, V., Huang, J., Abu-El-Haija, S., Gorban, A., Murphy, K., Fei-Fei, L.: Detecting events and key actors in multi-person videos. In: Proceedings of the IEEE Conference on Computer Vision and Pattern Recognition, pp. 3043–3053 (2016)
46. Ren, S., He, K., Girshick, R., Sun, J.: Faster R-CNN: towards real-time object detection with region proposal networks. In: Advances in Neural Information Processing Systems, pp. 91–99 (2015)

47. Setti, F., Lanz, O., Ferrario, R., Murino, V., Cristani, M.: Multi-scale f-formation discovery for group detection. In: Proceedings of the IEEE International Conference on Image Processing, pp. 3547–3551 (2013)
48. Shu, T., Todorovic, S., Zhu, S.C.: CERN: confidence-energy recurrent network for group activity recognition. In: Proceedings of the IEEE Conference on Computer Vision and Pattern Recognition, pp. 5523–5531 (2017)
49. Shu, T., Xie, D., Rothrock, B., Todorovic, S., Chun Zhu, S.: Joint inference of groups, events and human roles in aerial videos. In: Proceedings of the IEEE Conference on Computer Vision and Pattern Recognition, pp. 4576–4584 (2015)
50. Shu, X., Tang, J., Qi, G., Liu, W., Yang, J.: Hierarchical long short-term concurrent memory for human interaction recognition. IEEE Trans. Pattern Anal. Mach. Intell. (2019)
51. Sigurdsson, G.A., Divvala, S., Farhadi, A., Gupta, A.: Asynchronous temporal fields for action recognition. In: Proceedings of the IEEE Conference on Computer Vision and Pattern Recognition, pp. 585–594 (2017)
52. Sigurdsson, G.A., Varol, G., Wang, X., Farhadi, A., Laptev, I., Gupta, A.: Hollywood in homes: crowdsourcing data collection for activity understanding. In: Proceedings of the European Conference on Computer Vision, pp. 510–526 (2016)
53. Simonyan, K., Zisserman, A.: Two-stream convolutional networks for action recognition in videos. In: Advances in Neural Information Processing Systems, pp. 568–576 (2014)
54. Soomro, K., Zamir, A.R., Shah, M.: UCF101: a dataset of 101 human actions classes from videos in the wild. arXiv preprint arXiv:1212.0402 (2012)
55. Stergiou, A., Poppe, R.: Analyzing human-human interactions: a survey. Comput. Vis. Image Underst. **188**, 102799 (2019)
56. Sun, C., Shrivastava, A., Vondrick, C., Murphy, K., Sukthankar, R., Schmid, C.: Actor-centric relation network. In: Proceedings of the European Conference on Computer Vision, pp. 318–334 (2018)
57. Swofford, M., Peruzzi, J.C., Vázquez, M., Martín-Martín, R., Savarese, S.: DANTE: deep affinity network for clustering conversational interactants. arXiv preprint arXiv:1907.12910 (2019)
58. Tran, D., Bourdev, L., Fergus, R., Torresani, L., Paluri, M.: Learning spatiotemporal features with 3D convolutional networks. In: Proceedings of the IEEE International Conference on Computer Vision, pp. 4489–4497 (2015)
59. Vaswani, A., et al.: Attention is all you need (2017)
60. Veličković, P., Cucurull, G., Casanova, A., Romero, A., Liò, P., Bengio, Y.: Graph attention networks (2017)
61. Wang, L., et al.: Temporal segment networks: towards good practices for deep action recognition. In: Proceedings of the Proceedings of the European Conference on Computer Vision, pp. 20–36 (2016)
62. Wang, M., Ni, B., Yang, X.: Recurrent modeling of interaction context for collective activity recognition. In: Proceedings of the IEEE Conference on Computer Vision and Pattern Recognition, pp. 3048–3056 (2017)
63. Wang, X., Girshick, R., Gupta, A., He, K.: Non-local neural networks. In: Proceedings of the IEEE Conference on Computer Vision and Pattern Recognition, pp. 7794–7803 (2018)
64. Wu, C.Y., Feichtenhofer, C., Fan, H., He, K., Krahenbuhl, P., Girshick, R.: Long-term feature banks for detailed video understanding. In: Proceedings of the IEEE Conference on Computer Vision and Pattern Recognition, pp. 284–293 (2019)

65. Wu, J., Wang, L., Wang, L., Guo, J., Wu, G.: Learning actor relation graphs for group activity recognition. In: Proceedings of the IEEE Conference on Computer Vision and Pattern Recognition, pp. 9964–9974 (2019)
66. Xie, J., Girshick, R., Farhadi, A.: Unsupervised deep embedding for clustering analysis. In: International Conference on Machine Learning, pp. 478–487 (2016)
67. Xu, H., Das, A., Saenko, K.: R-C3D: region convolutional 3D network for temporal activity detection. In: Proceedings of the IEEE International Conference on Computer Vision, pp. 5783–5792 (2017)
68. Zelnik-Manor, L., Perona, P.: Self-tuning spectral clustering. In: Advances in Neural Information Processing Systems, pp. 1601–1608 (2005)
69. Zhou, B., Andonian, A., Oliva, A., Torralba, A.: Temporal relational reasoning in videos. In: Proceedings of the European Conference on Computer Vision, pp. 803–818 (2018)

Whole-Body Human Pose Estimation in the Wild

Sheng Jin[1,2], Lumin Xu[2,3], Jin Xu[2], Can Wang[2], Wentao Liu[2(✉)], Chen Qian[2], Wanli Ouyang[4], and Ping Luo[1]

[1] The University of Hong Kong, Pok Fu Lam, Hong Kong
pluo@cs.hku.hk
[2] SenseTime Research, Beijing, China
{jinsheng,xulumin,wangcan,liuwentao,qianchen}@sensetime.com
[3] The Chinese University of Hong Kong, Shatin, Hong Kong
[4] The University of Sydney, Sydney, Australia
wanli.ouyang@sydney.edu.au

Abstract. This paper investigates the task of 2D human whole-body pose estimation, which aims to localize dense landmarks on the entire human body including face, hands, body, and feet. As existing datasets do not have whole-body annotations, previous methods have to assemble different deep models trained independently on different datasets of the human face, hand, and body, struggling with dataset biases and large model complexity. To fill in this blank, we introduce COCO-WholeBody which extends COCO dataset with whole-body annotations. To our best knowledge, it is the first benchmark that has manual annotations on the entire human body, including 133 dense landmarks with 68 on the face, 42 on hands and 23 on the body and feet. A single-network model, named ZoomNet, is devised to take into account the hierarchical structure of the full human body to solve the scale variation of different body parts of the same person. ZoomNet is able to significantly outperform existing methods on the proposed COCO-WholeBody dataset. Extensive experiments show that COCO-WholeBody not only can be used to train deep models from scratch for whole-body pose estimation but also can serve as a powerful pre-training dataset for many different tasks such as facial landmark detection and hand keypoint estimation. The dataset is publicly available at https://github.com/jin-s13/COCO-WholeBody.

Keywords: Whole-body human pose estimation · Facial landmark detection · Hand keypoint estimation

1 Introduction

Human pose estimation has significant progress in the past few years. Recently, a more challenging task called *whole-body* pose estimation is proposed and attracts

Electronic supplementary material The online version of this chapter (https://doi.org/10.1007/978-3-030-58545-7_12) contains supplementary material, which is available to authorized users.

© Springer Nature Switzerland AG 2020
A. Vedaldi et al. (Eds.): ECCV 2020, LNCS 12354, pp. 196–214, 2020.
https://doi.org/10.1007/978-3-030-58545-7_12

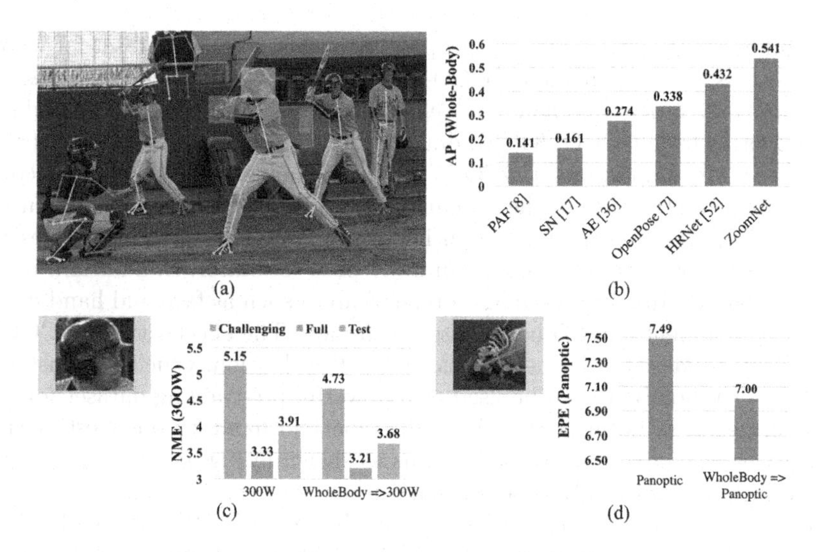

Fig. 1. The proposed COCO-WholeBody dataset provides manual annotations of dense landmarks on the entire human body including body, face, hands, and feet. (a) visualizes an image as an example. The whole-body human pose estimation is challenging because different body parts have different variations such as scale. (b) shows that ZoomNet significantly outperforms the prior arts on this challenging task. (c) and (d) show that existing facial/hand landmark estimation algorithms can be improved by pretraining on COCO-WholeBody.

much attention. As shown in Fig. 1a, whole-body pose estimation aims at localizing keypoints of body, face, hand, and foot simultaneously. This task is important for the development of downstream applications, such as virtual reality, augmented reality, human mesh recovery, and action recognition.

In recent years, deep neural networks (DNNs) become popular for keypoint estimation. However, to our knowledge, existing datasets of human pose estimation do not have manual annotations of the entire human body. Therefore, previous works trained their models separately on different datasets of face, hand and human body. For example, OpenPose [8] combines multiple DNNs trained independently on different datasets, including one DNN for body pose estimation on COCO [30], one DNN for face keypoint detection by combining many datasets (*i.e.* Multi-PIE [14], FRGC [43] and i-bug [48]), and another DNN for hand keypoint detection on Panoptic [50]. These methods may have several drawbacks. First, the data size of the current in-the-wild datasets of 2D hand keypoints is limited. Most approaches of hand pose estimation have to use lab-recorded [13,55] or synthetic datasets [33,34,49], hampering the performance of the existing methods in real-world scenarios. Second, the variations such as illumination, pose and scales in the existing human face [4,25,28,32,48,67], hand [13,13,33,55], and body datasets [2,3,30,60] are different, inevitably introducing dataset biases to the learned deep networks, thus hindering the development of algorithms to comprehensively consider the task as a whole.

To address the above issues, we propose a novel large-scale dataset for whole-body pose estimation, named COCO-WholeBody, which fully annotates the bounding boxes of face and hand, as well as the keypoints of face, hand, and foot for the images from COCO [30]. To our knowledge, this is the first dataset that has whole-body annotations. COCO-WholeBody enables us to take into account the hierarchical structure of the human body and the correlations between different body parts to estimate the entire body pose. Therefore, it enables the development of a more reliable human body pose estimator. In addition, it will also stimulate productive research on related areas such as face and hand detection, face alignment and 2D hand pose estimation. The effectiveness of COCO-WholeBody is validated by using cross-dataset evaluation, which demonstrates that COCO-WholeBody can be used as a powerful pre-training dataset for various tasks, such as facial landmark localization and hand keypoint estimation. We overview the cross-dataset evaluations as shown in Fig. 1c, d.

The task of whole-body pose estimation has not been fully exploited in the literature because of missing a representative benchmark. Previous works [7,17] are mainly the bottom-up approaches, which simultaneously detect the keypoints for all persons in an image at once. They are generally efficient, however, they might suffer from scale variance of persons, causing inferior performance for small persons. Recent works [52,63] found that the top-down alternatives would have higher accuracy, because top-down methods normalize the human instances to roughly the same scale and are less sensitive to the scale variance of different human instances. However, to our knowledge, there is no existing top-down approach for whole-body pose estimation. With COCO-WholeBody, we are able to fill in this blank by designing a top-down whole-body pose estimator. However, predicting all the keypoints for whole-body pose estimation will lead to inferior performance, because the scales of human body, face and hand are different. For example, human body pose estimation requires a large receptive field to handle occlusion and complex poses, while face and hand keypoint estimation requires higher image resolution for accurate localization. If all the keypoints are treated equally and directly predicted at once, the performance is suboptimal.

To solve this technical problem, we propose ZoomNet to effectively handle the scale variance in whole-body pose estimation. ZoomNet follows the top-down paradigm. Given a human bounding box of each person, ZoomNet first localizes the easy-to-detect body keypoints and estimates the rough position of hands and face. Then it zooms in to focus on the hand/face areas and predicts keypoints using features with higher resolution for accurate localization. Unlike previous approaches [7] which usually assemble multiple networks, ZoomNet has a single network that is end-to-end trainable. It unifies five network heads including the human body pose estimator, hand and face detectors, and hand and face pose estimators into a single network with shared low-level features. Extensive experiments show that ZoomNet outperforms the state-of-the-arts [7,17] by a large margin, *i.e.* 0.541 vs 0.338 [7] for whole-body mAP on COCO-WholeBody.

Our major contributions can be summarized as follows. (1) We propose the first benchmark dataset for whole-body human pose estimation, termed

Table 1. Overview of some popular public datasets for 2D keypoint estimation in RGB images. Kpt stands for keypoints, and #Kpt means the annotated number. "Wild" denotes whether the dataset is collected in-the-wild. * means head box.

DataSet	Images	#Kpt	Wild	Body box	Hand box	Face box	Body Kpt	Hand Kpt	Face Kpt	Total instances
MPII [3]	25K	16	✓	✓		*	✓			40K
MPII-TRB [10]	25K	40	✓	✓		*	✓			40K
CrowdPose [29]	20K	14	✓	✓			✓			80K
PoseTrack [2]	23K	15	✓	✓			✓			150K
AI Challenger [60]	300K	14	✓	✓			✓			700K
COCO [30]	200K	17	✓	✓		*	✓			250K
OneHand10K [59]	10K	21	✓		✓			✓		–
SynthHand [34]	63K	21			✓			✓		–
RHD [68]	41K	21			✓			✓		–
FreiHand [69]	130K	21						✓		–
MHP [13]	80K	21			✓			✓		–
GANerated [33]	330K	21						✓		–
Panoptic [50]	15K	21			✓			✓		–
WFLW [61]	10K	98	✓			✓			✓	–
AFLW [25]	25K	19	✓			✓			✓	–
COFW [5]	1852	29	✓			✓			✓	–
300W [48]	3837	68	✓			✓			✓	–
COCO-WholeBody	200K	133	✓	✓	✓	✓	✓	✓	✓	250K

COCO-WholeBody, which encourages more exploration of this task. To evaluate the effectiveness of COCO-WholeBody, we extensively examine the performance of several representative approaches on this dataset. Also, the generalization ability of COCO-WholeBody is validated by cross-dataset evaluations, showing that COCO-WholeBody can serve as a powerful pre-training dataset for many tasks, such as facial landmark localization and hand keypoint estimation. **(2)** We propose a top-down single-network model, ZoomNet to solve the scale variance of different body parts in a single person. Extensive experiments show that the proposed method significantly outperforms previous state-of-the-arts.

2 Related Work

2.1 2D Keypoint Localization Dataset

As shown in Table 1, there are many datasets separately annotated for localizing the keypoints of body [2,3,11,30,60], hand [13,33,50,55,65] or face [4,25,28,32,48,67]. These datasets are briefly discussed in this section.

Body Pose Dataset. There have been several body pose datasets [2,3,10,29,30,60]. *COCO* [30] is one of the most popular, which offers 17-keypoint annotations in uncontrolled conditions. Our COCO-WholeBody is an extension of COCO, with densely annotated 133 face/hand/foot keypoints. The task of whole-body pose estimation is more challenging, due to 1) higher localization accuracy required for face/hands and 2) scale variance between body and face/hands.

Hand Keypoint Dataset. Most existing 2D RGB-based hand keypoint datasets are either synthetic [33,68] or captured in the lab environment [13,50,69]. For example, *Panoptic* [50] is a well-known hand pose estimation dataset, recorded in the CMU's Panoptic studio with multiview dome settings. However, it is limited to a controlled laboratory environment with a simple background. OneHand10K [59] is a recent in-the-wild 2d hand pose dataset. However, the size is still limited. Our COCO-WholeBody is complementary to these RGB-based hand keypoint datasets. It contains about 100 K 21-keypoint labeled hands and hand boxes that are captured in unconstrained environment. To the best of our knowledge, it is the largest in-the-wild dataset for 2D RGB-based hand keypoint estimation. It is very challenging, due to occlusion, hand-hand interaction, hand-object interaction, motion blur, and small scales.

Face Keypoint Dataset. Face keypoint datasets [5,25,48,61] play a crucial role for the development of facial landmark detection a.k.a. face alignment. Among them, *300W* [48] is the most popular. It is a combination of LFPW [4], AFW [67], HELEN [28], XM2VTS [32] with 68 landmarks annotated for each face image. Our proposed COCO-WholeBody follows the same annotation settings as 300 W and 68 keypoints for each face are annotated. Compared to 300W, COCO-WholeBody is much larger and is more challenging as it contains more blurry and small-scale facial images (see Fig. 5a).

DensePose Dataset. Our work is also related to DensePose [1] which provides a dense 3D surface-based representation for human shape. However, since the keypoints in DensePose are uniformly sampled, they lack specific joint articulation information and details of face/hands are missing.

2.2 Keypoints Localization Method

Body Pose Estimation. Recent multi-person body pose estimation approaches can be divided into bottom-up and top-down approaches. Bottom-up approaches [8,19–23,36,38,40,44] first detect all the keypoints of every person in images and then group them into individuals. Top-down methods [9,12,16, 31,37,41,52,63] first detect the bounding boxes and then predict the human body keypoints in each box. By resizing and cropping, top-down approaches normalize the poses to approximately the same scale. Therefore, they are more robust to human-level scale variance and recent state-of-the-arts are obtained by top-down approaches. However, direct usage of the top-down methods for whole-body pose estimation will encounter the problem of scale variance of different body parts (body vs face/hand). To tackle this problem, we propose ZoomNet, a single-network top-down approach that zooms in to the hand/face regions and predicts the hand/face keypoints using higher image resolution for accurate localization.

Face/Hand/Foot Keypoint Localization. Previous works mostly treat the tasks of face/hand/foot keypoint localization independently and solve by different models. For facial keypoint localization, cascaded networks [6,54,57,64] and

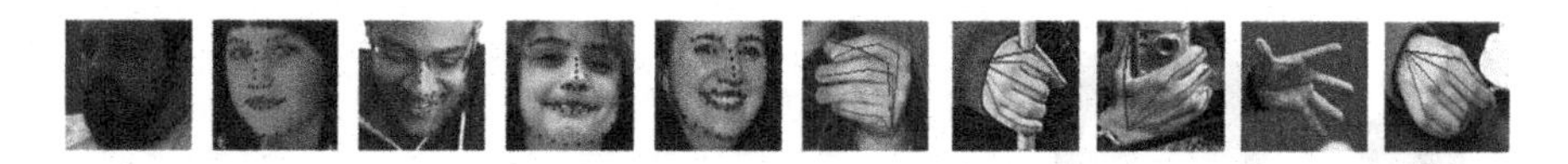

Fig. 2. Annotation examples for face/hand keypoints in COCO-WholeBody.

multi-task learning [56,66] are widely adopted. For hand keypoint estimation, most work rely on auxiliary information such as depth information [39,49,51] or multi-view [15,35] information. For foot keypoint estimation, Cao *et al.* [7] proposed a generic bottom-up method. In this paper, we propose ZoomNet to solve the tasks of face/hand/foot keypoint localization as a whole. It takes into account the inherent hierarchical structure of the full human body to solve the scale variation of different parts in the same person.

Whole-Body Pose Estimation. Whole-body pose estimation has not been well studied in the literature due to the lack of a representative benchmark. OpenPose [7,8,50] applies multiple models (body keypoint estimator) to handle different kinds of keypoints. It first detects body and foot keypoints, and estimates the hand and face position. Then it applies extra models for face and hand pose estimation. Since OpenPose relies on multiple networks, it is hard to train and suffers from increased runtime and computational complexity. Unlike OpenPose, our proposed ZoomNet is a "single-network" method as it integrates five previously separated models (human body pose estimator, hand/face detectors, and hand/face pose estimators) into a single network with shared low-level features. Recently, Hidalgo *et al.* proposes an elegant method SN [17] for bottom-up whole-body keypoint estimation. SN is based on PAF [8] which predicts the keypoint heatmaps for detection and part affinity maps for grouping. Since there exists no such dataset with whole-body annotations, they used a set of different datasets and carefully designed the sampling rules to train the model. However, bottom-up approaches cannot handle scale variation problem well and would have difficulty in detecting face and hand keypoints accurately. In comparison, our ZoomNet is a top-down approach that well handles the extreme scale variance problem. Recent works [24,46,62] also explore the task of monocular 3D whole-body capture. Romero *et al.* proposes a generative 3D model [46] to express body and hands. Xiang *et al.* introduces a 3D deformable human model [62] to reconstruct whole-body pose and Joo *et al.* presents Adam [24] which encompasses the expressive power for body, hands, and facial expression. Their methods still rely on OpenPose [7] to localize 2d body keypoints in images.

3 COCO-WholeBody Dataset

COCO-WholeBody is the first large-scale dataset with the whole-body pose annotation available, to the best of our knowledge. In this section, we will describe the annotation protocols and some informative statistics.

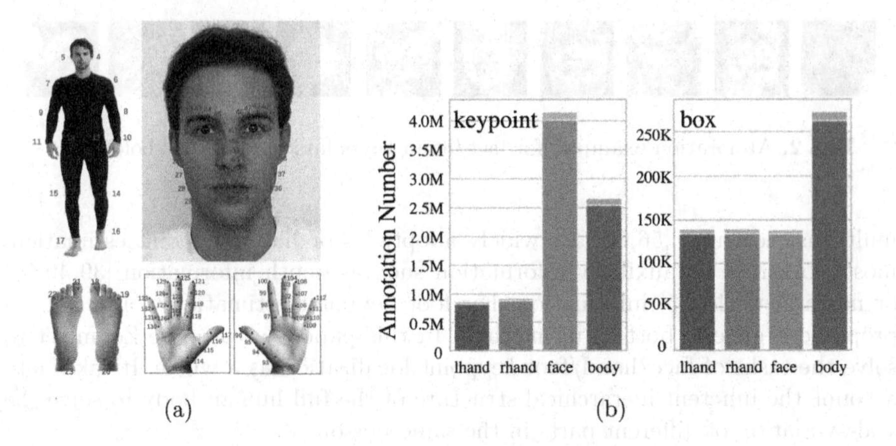

Fig. 3. (a) COCO-WholeBody annotation for 133 keypoints. (b)Statistics of COCO-WholeBody. The number of annotated keypoints and boxes of left hand (lhand), right hand (rhand), face and body are reported.

3.1 Data Annotation

We annotate the face, hand and foot keypoints on the whole train/val set of COCO [30] dataset and form the whole-body annotations with the original body keypoint labels together (see Fig. 2). For each person, we annotate 4 types of bounding boxes (person box, face box, left-hand box, and right-hand box) and 133 keypoints (17 for body, 6 for feet, 68 for face and 42 for hands). The face/hand box is defined as the minimal bounding rectangle of the keypoints. The keypoint annotations are illustrated in Fig. 3a. The face/hand boxes are labeled as *valid*, only if the face/hand images are clear enough for keypoint labeling. Invalid boxes may be blurry or severely occluded. We only label keypoints for *valid* boxes. Manual annotation for whole-body poses in an uncontrolled environment, especially for massive and dense hand and face keypoints, requires trained experts and enormous workload. As a rough estimate, the manual labeling cost of a professional annotator is up to: 10 min/face, 1.5 min/hand, and 10 sec/box. To speed up the annotation process, we follow the semi-automatic methodology to use a set of pre-trained models (for face and hand separately) to pre-annotate and then conduct manual correction. Foot keypoints are directly manually labeled, since its labeling cost is relatively low. Specifically, the annotation process contains the following steps:

1. For each individual person, we manually label the face box, the left-hand box, and the right-hand box. The validity of the boxes is also labeled.
2. Quality control. The annotation quality of the boxes is guaranteed through the strict quality inspection performed by another group of the annotators.
3. For each valid face/hand box, we use pre-trained face/hand keypoint detectors to produce pseudo keypoint labels. We use a combination of the publicly available datasets to train a robust face keypoint detector and a hand keypoint detector based on HRNetV2 [52].

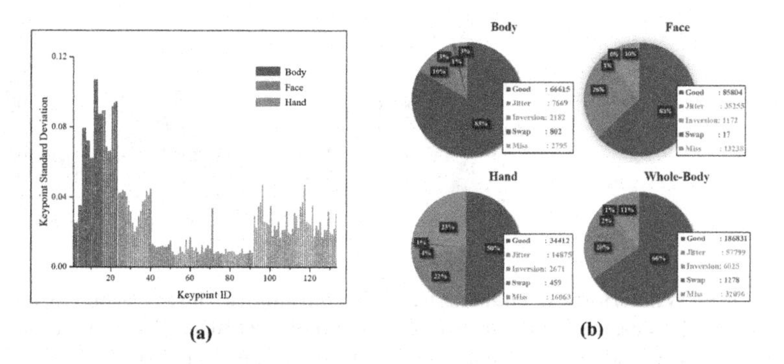

Fig. 4. (a) The normalized standard deviation of manual annotation for each keypoint. Body keypoints have larger manual annotation variance than face and hand keypoints. (b) An example of error diagnosis results of ZoomNet for whole-body pose estimation: jitter, inversion, swap and missing.

4. Manual correction of pseudo labels and further quality control. About 28% of the hand keypoints and 6% of the face keypoints are labeled as invalid and manually corrected by human annotators. By using the semi-automatic annotation, we saw about 89% reduction in the time required for annotation.

To measure the annotation quality, we also had 3 annotators to label the same batch of 500 images for face/hand/foot keypoints. The standard deviation of the human annotation is calculated for each keypoint (see Fig. 4a), which is used to calculate the normalized factor of whole-body keypoint for evaluation. For "body keypoints", we directly use the standard deviation reported in COCO [30].

3.2 Evaluation Protocol and Evaluation Metrics

The evaluation protocol of whole-body pose estimation follows the current practices in the literature [30,60]. All algorithms are trained on COCO-WholeBody training set and evaluated on COCO-WholeBody validation set. We use mean Average Precision (mAP) and Average Recall (mAR) for evaluation, where Object Keypoint Similarity (OKS) is used to measure the similarity between the prediction and the ground truth poses. Invalid boxes and keypoints are masked out during both training and evaluation, thus not affecting the results. The ignored regions are masked out, and only visible keypoints are considered during evaluation. As shown in Fig. 4b, we also develop a tool for deeper performance analysis based on [47] which will be provided to facilitate offline evaluation.

3.3 Dataset Statistics

Dataset Size. COCO-WholeBody is a large-scale dataset with keypoint and bounding box annotations. The number of annotated keypoints as well as boxes of left hand (lhand), right hand (rhand), face and body are shown in Fig. 3b.

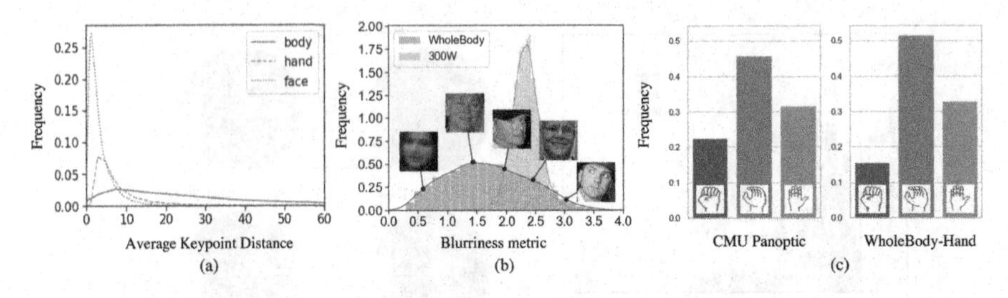

Fig. 5. COCO-WholeBody is challenging as it contains (a) large "scale variance" of body/face/hand, measured by the average keypoint distance, (b) more blurry face images than 300 W and (c) more complex hand poses than Panoptic.

About 130K face and left/right hand boxes are labeled, resulting in more than 800K hand keyponits and 4M face keypoints in total.

Scale Difference. Distribution of the average keypoint distance of different parts in WholeBody Dataset is summarized in Fig. 5a. We calculate the distance between keypoint pairs in the tree-structured skeleton. Hand/face have obviously much smaller scales than body. The various scale distribution makes it challenging to localize keypoints of different human parts simultaneously.

Facial Image "Blurriness". Face image "blurriness" is a key factor for facial landmark localization. We choose a variation of the Laplacian method [42] to measure it. Specifically, an image is first converted into a grayscale image and resized into 112 × 112. The log10 of the Laplacian of the converted image is used as the "blurriness" measurement (the higher the better). The distribution of the blurriness is shown in Fig. 5b. We find that most facial images fall in the interval between 1 and 3 and are clear enough for accurate keypoint localization. Compared with 300W [48], WholeBody has a larger variance of blurriness and contains more challenging images (blurriness < 1).

Gesture Variances for Hands. We first normalize the 2D hand poses by rotating and scaling and then cluster them into three main categories: "fist", "palm" and "others". Unlike most previous hand datasets that are collected in constrained environments, our WholeBody-Hand is collected in-the-wild. Compared with Panoptic [13], WholeBody-Hand is more challenging as it contains a larger proportion of hand images grasping or holding objects.

Overall, COCO-WholeBody is a large-scale dataset with great diversity, which will not only promote researches on the whole-body pose estimation but also contribute to other related areas, such as face and hand keypoint estimation.

4 ZoomNet: Whole-Body Pose Estimation

In this section, we will introduce our whole-body pose estimation pipeline. Given an RGB image, we follow [52,63] to use an off-the-shelf FasterRCNN [45] human detector to generate human body candidates. For each human body candidate,

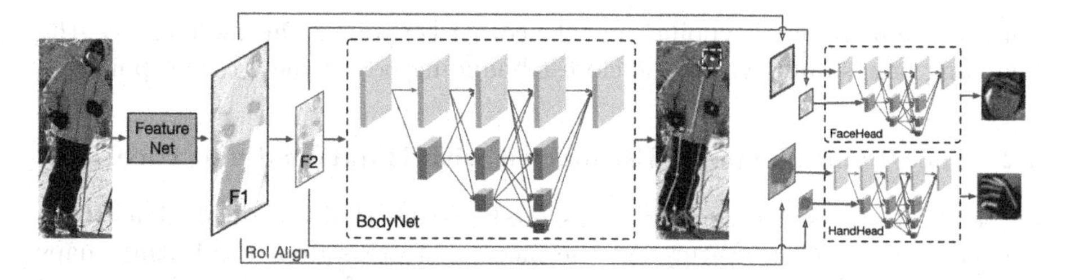

Fig. 6. ZoomNet is a single-network model, which consists of FeatureNet, BodyNet and Face/HandHead. FeatureNet extracts low-level shared features for BodyNet and Face/HandHead. BodyNet predicts body/foot keypoints and the approximate regions of face/hands, while Face/HandHead zooms in to these regions and predict face/hand keypoints with features of higher resolution.

ZoomNet localizes the whole-body keypoints. As shown in Fig. 6, ZoomNet predicts body/foot keypoints and face/hand keypoints successively in a single network, consisting of the following submodules:

FeatureNet: the input image is processed by FeatureNet to extract shared features ($F1$ and $F2$). It consists of two convolutional layers, each of which downsamples the corresponding input to 1/2 resolution, and a bottleneck block for effective feature learning. The input image size is 384×288 and the output feature map sizes for $F1$ and $F2$ are 192×144 and 96×72, respectively.

BodyNet: using the features extracted from FeatureNet, BodyNet predicts body/foot keypoints and face/hand bounding boxes at the same time. Each bounding box is represented by four corner points and one center point. In total, 38 keypoints are generated for each person simultaneously. BodyNet is a multi-resolution network with 38 output channels.

HandHead and FaceHead: Using face and hand bounding boxes predicted by BodyNet, we crop the features in the corresponding areas from F1 and F2. The features from F1 are resized to 64×64 and those from F2 are resized to 32×32. Then HandHead and FaceHead are applied to predict the heatmaps of face/hand keypoints with the output resolution of 64×64 in parallel.

ZoomNet can be based on any state-of-the-art network architecture. In our implementation, we choose HRNet-W32 [52] as the backbone of BodyNet and HRNetV2p-W18 [53] as the backbone of FaceHead/HandHead. Please refer to Supplementary for more implementation details.

4.1 Localizing Body Keypoints and Face/hand Boxes with BodyNet

Our face/hand box localization is inspired by CornerNet [27], which represents the object with keypoint pairs and designs a one-stage keypoint-based detector. In our case, each person has three types of bounding boxes to predict: the face box, the left-hand box, and the right-hand box. Four corner points and one center point are used to represent a box. We use 2D confidence heatmaps to encode both the human body keypoints and the corner keypoints. During inference, the bounding box is obtained by the closest bounding box of the 4 corner points.

4.2 Face/hand Keypoint Estimation with HandHead and FaceHead

Given the face/hand bounding boxes predicted by BodyNet, RoIAlign [16] is applied to extract the features of the face/hand areas from the feature maps $F1$ and $F2$ of FeatureNet. The corresponding visual features are cropped and up-scaled to a higher resolution. With the extracted features, HandHead and FaceHead are used for face and hand keypoint estimation. HandHead and FaceHead use the same network architecture (HRNet-W18). The features extracted by RoIAlign are processed by the HandHead and FaceHead separately. In this way, we are able to preserve the high-resolution for the hand/face regions, and larger receptive fields for body keypoint estimation at the same time.

5 Experiments

5.1 Evaluation on COCO-WholeBody Dataset

To the best of our knowledge, there are only two existing approaches that target at the 2D whole-body pose estimation task, *i.e.* OpenPose [7] and SN [17]. To extensively evaluate the performance of the existing methods on the proposed COCO-WholeBody Dataset, we also build upon the existing multi-person human body pose estimation approaches, including both bottom-up (*i.e.* Partial Affinity Field (PAF) [8] and Associate Embedding (AE) [36]) and top-down methods (*i.e.* HRNet [52]), and adapt them to the more challenging whole-body pose estimation task using official codes (see Supplementary for more details). For fair comparisons, we retrain all methods on COCO-WholeBody and evaluate their performance with single-scale testing as shown in Table 2. We show that our proposed ZoomNet outperforms them by a large margin.

Among these methods, SN [17], PAF [8], AE [36] and HRNet [52] follow a one-stage paradigm and predict all the keypoints simultaneously. Interestingly, we find that in the task of whole-body pose estimation, directly learning to predict all 133 keypoints simultaneously, including body, face, hand keypoints, may harm the original body keypoint estimation accuracy. In Table 2, "-body" means that we only train the model on the original COCO-body keypoint (17 keypoints). We compare the body keypoint estimation results of the model learning the whole-body keypoints versus the model learning the body keypoints only. We observe considerable accuracy decrease by comparing PAF vs PAF-body (−14.3% mAP and -14.2% mAR), AE vs AE-body (−17.7% mAP and −17.0% mAR) and HRNet vs HRNet-body (−9.9% mAP and −10.0% mAR). In comparison, our proposed ZoomNet uses a two-stage framework, which decouples the body keypoint estimation and face/hand keypoint estimation. The accuracy of body keypoint estimation is less affected (−1.5% mAP and −0.7% mAR).

HRNet [52] can be viewed as the *one-stage* alternative of ZoomNet, since they share the same network backbone (HRNet-W32). ZoomNet significantly outperforms HRNet by 10.9% mAP and 13.8% mAR, demonstrating the effectiveness of the "zoom-in" design for solving the scale variation.

OpenPose [7] is a *multi-model* approach, where the hand/face model and the body model are not jointly trained, leading to sub-optimal results. In addition, the hand/face boxes of OpenPose are roughly estimated by hand-crafted rules from the estimated body keypoints. Therefore, the accuracy of the hand/face boxes is limited, which will hinder hand/face pose estimation.

Table 2. Whole-body pose estimation results on COCO-WholeBody dataset. For fair comparisons, results are obtained using single-scale testing.

Method	Body		Foot		Face		Hand		Whole-body	
	AP	AR	AP	AR	AP	AR	AP	AR	AP	AR
OpenPose [7]	0.563	0.612	0.532	0.645	0.482	0.626	0.198	0.342	0.338	0.449
SN [17]	0.280	0.336	0.121	0.277	0.382	0.440	0.138	0.336	0.161	0.209
PAF [8]	0.266	0.328	0.100	0.257	0.309	0.362	0.133	0.321	0.141	0.185
PAF-body [8]	0.409	0.470	–	–	–	–	–	–	–	–
AE [36]	0.405	0.464	0.077	0.160	0.477	0.580	0.341	0.435	0.274	0.350
AE-body [36]	0.582	0.634	–	–	–	–	–	–	–	–
HRNet [52]	0.659	0.709	0.314	0.424	0.523	0.582	0.300	0.363	0.432	0.520
HRNet-body [52]	**0.758**	**0.809**	–	–	–	–	–	–	–	–
ZoomNet	0.743	0.802	**0.798**	**0.869**	**0.623**	**0.701**	**0.401**	**0.498**	**0.541**	**0.658**

Model Complexity Analysis. The model complexity of ZoomNet is 27.36G Flops. By contrast, the model complexity of OpenPose [7] is 451.09G Flops in total (137.52G for BodyNet, 106.77G for FaceNet and $103.40 \times 2 = 206.80$G for HandNet), and that of SN [17] is 272.30G Flops. We also report the average runtime cost on COCO-WholeBody on one GTX-1080 GPU. SN is about 215.5 ms/image, while ZoomNet is about 174.7 ms/image on average (including a Faster RCNN human detector which takes about 106 ms/image).

5.2 Cross-Dataset Evaluation

In this section, we show that the proposed COCO-WholeBody is complementary to other separately labeled benchmarks by evaluating its generalization ability.

WholeBody-Face (WBF) Dataset. We build WholeBody-Face (WBF) by extracting cropped face images/annotations from COCO-WholeBody. We conduct experiments on 300W [48] benchmark. We follow the common settings [53] to train models on 3,148 training images, validate on the "common" set and evaluate on the "challenging", "full" and "test" sets. We use the normalized mean error (NME) for evaluation and inter-ocular distance as normalization. The results are shown in Table 3a. HR-Ours is our implementation of HRNetV2-W18 [53] (HR). *HR-Ours is obtained by training HR on WBF only and directly testing on 300W, which already outperforms RCN [18]. After finetuning on 300W, it gets significantly better performance on "challenging" (4.73 vs 5.15), "full" (3.21 vs 3.33) and "test" (3.68 vs 3.91) than the prior arts.

Table 3. (a) Facial landmark localization (NME) on 300W: "common" (for val), "challenging", "full" and "test". * means only training on WBF. ↓ means lower is better. (b) Cross-dataset evaluation results of HR. Different training and testing settings are evaluated on two datasets: WBH and Panoptic (Pano.) [50]

	extra.	comm. ↓	chall. ↓	full ↓	test ↓
RCN [18]	-	4.67	8.44	5.41	-
DAN [26]	-	3.19	5.24	3.59	4.30
DCFE [58]	w/3D	**2.76**	5.22	3.24	3.88
LAB [61]	w/B	2.98	5.19	3.49	-
HR [53]	-	2.87	5.15	3.32	3.85
*HR-Ours	-	4.61	7.50	5.17	5.66
HR-Ours	-	2.89	5.15	3.33	3.91
HR-Ours	WBF	2.84	**4.73**	**3.21**	**3.68**

(a)

#	Train-set	Test-set	EPE ↓	NME ↓
1	Pano.	Pano.	7.49	0.68
2	WBH ⇒ Pano.	Pano.	**7.00**	**0.63**
3	WBH	WBH	2.76	6.66
4	Pano. ⇒ WBH	WBH	**2.70**	**6.49**

(b)

WholeBody-Hand (WBH) Dataset. For hand pose estimation, we experiment with HRNetV2-W18 (HR) on CMU Panoptic [50] (Pano.), which is a standard benchmark for hand keypoint localization. We randomly split Pano [50] by a rule of 70%-30% for training and validation. We report both the end-point-error (EPE) and the normalized mean error (NME) for evaluation. In NME, the hand bounding box is used as normalization. As shown in Table 3b, we analyze the generalization ability of WholeBody-Hand (WBH) by comparing the (1) HR trained on Pano, (2) HR pretrained on WBH and then finetuned on Pano, (3) HR trained on WBH, and (4) HR pretrained on Pano. and then finetuned on WBH. Comparing #1 and #2, we observe that pretraining on WBH brings about 6.5% improvement (from 7.49 to 7.00) in EPE on Pano. Comparing #1 and #3, we find that WBH vs Pano. is (6.66 vs 0.68) NME and (2.76 vs 7.49) EPE, when training/testing on its own dataset. This implies that the proposed WBH is much more challenging and that hand scales in WBH are smaller.

5.3 Analysis

Effect of the Bounding Box Accuracy on the Keypoint Estimation. We experiment by replacing our predicted face/hand bounding boxes with the ground-truth bounding boxes and re-run our FaceHead/HandHead of ZoomNet to obtain the final face/hand keypoint detection result. As shown in Table 4a, using ground truth bounding boxes (Oracle) significantly improves the mAP of face/hand/whole-body by 19.6%, 8.4% and 23.6% respectively.

Table 4. Effect of bounding box accuracy on keypoint estimation, where Oracle means using gt boxes. (b) Effect of person scales on whole-body pose estimation.

Method	face		hand		whole-body	
	AP	AR	AP	AR	AP	AR
Oracle	0.819	0.854	0.485	0.578	0.777	0.856
Ours	0.623	0.701	0.401	0.498	0.541	0.658

(a)

Method	mAP		mAR	
	medium	large	medium	large
PAF [8]	0.100	0.220	0.113	0.284
SN [17]	0.117	0.252	0.132	0.315
AE [36]	0.190	0.401	0.241	0.499
OpenPose [7]	0.398	0.302	0.425	0.484
HRNet [52]	0.471	0.410	0.538	0.497
Ours	**0.594**	**0.519**	**0.677**	**0.635**

(b)

Effect of the Person Scale on Whole-Body Pose Estimation. As shown in Table 4b, we investigate the effect of person scales. Interestingly, for bottom-up whole-body methods (PAF, SN and AE), the mAP for medium scale is worse than that of large scale, since they are more sensitive to the scale variance and are difficult in detecting smaller-scale people. For top-down approaches such as HRNet and ZoomNet, mAP for medium scale is better, since larger-scale person requires relatively more accurate keypoint localization. For ZoomNet, the gap between the medium and large person scale is about 7.5% mAP and 4.2% mAR.

Effect of Blurriness and Poses on Facial Landmark Detection. In Table 5, we evaluate the performance on different levels of image blurriness and facial poses (yaw angles) on WBF. The model is significantly affected by image blur (2.51 vs 19.13), while more robust to different face poses (9.02 vs 13.77).

Effect of Hand Poses on Hand Keypoint Estimation. As shown in Table 5, we evaluate the performance on different hand poses (fist, palm or others) on WBH (NME). We show that estimating the poses of "palm" or "others" (with various gestures) is more challenging than that of "fist" (with similar patterns).

Table 5. *left:* Effect of blurriness/poses on facial landmark detection (NME) on WholeBody-Face (WBF). *right:* Effect of hand poses on hand keypoint estimation (NME) on WholeBody-Hand (WBH).

WBF (NME ↓)										WBH (NME ↓)			
Blurriness					Yaw angles					Pose			
< 1	1–2	2–3	> 3	ALL	$< 15^\circ$	$15^\circ - 30^\circ$	$30^\circ - 45^\circ$	$> 45^\circ$	ALL	Fist	Palm	Others	ALL
19.13	10.85	4.91	2.51	10.17	9.02	10.56	12.10	13.77	10.17	6.09	7.10	6.33	6.66

6 Conclusion

In this paper, we proposed the first large-scale benchmark for whole-body human pose estimation. We extensively evaluate the performance of the existing approaches on our proposed COCO-WholeBody Dataset. Cross-dataset evaluation also demonstrates the generalization ability of the proposed dataset. Moreover, to solve the problem of extreme scale difference among body parts, ZoomNet is proposed to pay more attention to the hard-to-detect face/hand keypoints. Experiments show that ZoomNet significantly outperforms the prior arts.

Acknowledgments. This work is partially supported by the SenseTime Donation for Research, HKU Seed Fund for Basic Research, Startup Fund, General Research Fund No.27208720, the Australian Research Council Grant DP200103223 and Australian Medical Research Future Fund MRFAI000085.

References

1. Alp Güler, R., Neverova, N., Kokkinos, I.: DensePose: dense human pose estimation in the wild. In: Proceedings of the IEEE Conference on Computer Vision and Pattern Recognition (CVPR) (2018)
2. Andriluka, M., et al.: PoseTrack: a benchmark for human pose estimation and tracking. In: Proceedings of the IEEE Conference on Computer Vision and Pattern Recognition (CVPR) (2018)
3. Andriluka, M., Pishchulin, L., Gehler, P., Schiele, B.: 2D human pose estimation: new benchmark and state of the art analysis. In: Proceedings of the IEEE Conference on Computer Vision and Pattern Recognition (CVPR) (2014)
4. Belhumeur, P.N., Jacobs, D.W., Kriegman, D.J., Kumar, N.: Localizing parts of faces using a consensus of exemplars. IEEE Trans. Pattern Anal. Mach. Intell. **35**, 2930–2940 (2013)
5. Burgos-Artizzu, X.P., Perona, P., Dollár, P.: Robust face landmark estimation under occlusion. In: Proceedings of the 2013 IEEE International Conference on Computer Vision (2013)
6. Cao, X., Wei, Y., Wen, F., Sun, J.: Face alignment by explicit shape regression. Int. J. Comput. Vis. **107**, 177–190 (2014)
7. Cao, Z., Hidalgo, G., Simon, T., Wei, S.E., Sheikh, Y.: OpenPose: real-time multi-person 2d pose estimation using part affinity fields. arXiv preprint arXiv:1812.08008 (2018)
8. Cao, Z., Simon, T., Wei, S.E., Sheikh, Y.: Realtime multi-person 2D pose estimation using part affinity fields. In: Proceedings of the IEEE Conference on Computer Vision and Pattern Recognition (CVPR) (2017)
9. Chen, Y., Wang, Z., Peng, Y., Zhang, Z., Yu, G., Sun, J.: Cascaded pyramid network for multi-person pose estimation. In: Proceedings of the IEEE Conference on Computer Vision and Pattern Recognition (CVPR) (2018)
10. Duan, H., Lin, K.Y., Jin, S., Liu, W., Qian, C., Ouyang, W.: TRB: a novel triplet representation for understanding 2D human body. In: Proceedings of the IEEE International Conference on Computer Vision, pp. 9479–9488 (2019)
11. Eichner, M., Ferrari, V.: We are family: joint pose estimation of multiple persons. In: Daniilidis, K., Maragos, P., Paragios, N. (eds.) ECCV 2010. LNCS, vol. 6311, pp. 228–242. Springer, Heidelberg (2010). https://doi.org/10.1007/978-3-642-15549-9_17

12. Fang, H.S., Xie, S., Tai, Y.W., Lu, C.: RMPE: regional multi-person pose estimation. In: Proceedings of the IEEE Conference on Computer Vision and Pattern Recognition (CVPR) (2017)
13. Gomez-Donoso, F., Orts-Escolano, S., Cazorla, M.: Large-scale multiview 3D hand pose dataset. arXiv preprint arXiv:1707.03742 (2017)
14. Gross, R., Matthews, I., Cohn, J., Kanade, T., Baker, S.: Multi-pie. In: Image and Vision Computing (2010)
15. Guan, H., Chang, J.S., Chen, L., Feris, R.S., Turk, M.: Multi-view appearance-based 3D hand pose estimation. In: IEEE Conference on Computer Vision and Pattern Recognition Workshop (2006)
16. He, K., Gkioxari, G., Dollár, P., Girshick, R.: Mask R-CNN. arXiv preprint arXiv:1703.06870 (2017)
17. Hidalgo, G., et al.: Single-network whole-body pose estimation. In: Proceedings of the IEEE Conference on Computer Vision and Pattern Recognition (CVPR) (2019)
18. Honari, S., Yosinski, J., Vincent, P., Pal, C.: Recombinator networks: learning coarse-to-fine feature aggregation. In: Proceedings of the IEEE Conference on Computer Vision and Pattern Recognition (CVPR) (2016)
19. Insafutdinov, E., et al.: ArtTrack: articulated multi-person tracking in the wild. In: Proceedings of the IEEE Conference on Computer Vision and Pattern Recognition (CVPR) (2017)
20. Insafutdinov, E., Pishchulin, L., Andres, B., Andriluka, M., Schiele, B.: DeeperCut: a deeper, stronger, and faster multi-person pose estimation model. In: Leibe, B., Matas, J., Sebe, N., Welling, M. (eds.) ECCV 2016. LNCS, vol. 9910, pp. 34–50. Springer, Cham (2016). https://doi.org/10.1007/978-3-319-46466-4_3
21. Iqbal, U., Milan, A., Gall, J.: Pose-track: joint multi-person pose estimation and tracking. arXiv preprint arXiv:1611.07727 (2016)
22. Jin, S., Liu, W., Ouyang, W., Qian, C.: Multi-person articulated tracking with spatial and temporal embeddings. In: Proceedings of the IEEE Conference on Computer Vision and Pattern Recognition (CVPR) (2019)
23. Jin, S., et al.: Towards multi-person pose tracking: bottom-up and top-down methods. In: IEEE International Conference on Computer Vision Workshop (2017)
24. Joo, H., Simon, T., Sheikh, Y.: Total capture: a 3D deformation model for tracking faces, hands, and bodies. In: Proceedings of the IEEE Conference on Computer Vision and Pattern Recognition (CVPR) (2018)
25. Koestinger, M., Wohlhart, P., Roth, P.M., Bischof, H.: Annotated facial landmarks in the wild: a large-scale, real-world database for facial landmark localization. In: IEEE International Conference on Computer Vision Workshop (2011)
26. Kowalski, M., Naruniec, J., Trzcinski, T.: Deep alignment network: a convolutional neural network for robust face alignment. In: IEEE Conference on Computer Vision and Pattern Recognition Workshop (2017)
27. Law, H., Deng, J.: CornerNet: detecting objects as paired keypoints. In: Proceedings of the European Conference on Computer Vision (ECCV) (2018)
28. Le, V., Brandt, J., Lin, Z., Bourdev, L., Huang, T.S.: Interactive facial feature localization. In: Fitzgibbon, A., Lazebnik, S., Perona, P., Sato, Y., Schmid, C. (eds.) ECCV 2012. LNCS, vol. 7574, pp. 679–692. Springer, Heidelberg (2012). https://doi.org/10.1007/978-3-642-33712-3_49
29. Li, J., Wang, C., Zhu, H., Mao, Y., Fang, H.S., Lu, C.: CrowdPose: efficient crowded scenes pose estimation and a new benchmark. In: Proceedings of the IEEE Conference on Computer Vision and Pattern Recognition, pp. 10863–10872 (2019)

30. Lin, T.-Y., et al.: Microsoft COCO: Common objects in context. In: Fleet, D., Pajdla, T., Schiele, B., Tuytelaars, T. (eds.) ECCV 2014. LNCS, vol. 8693, pp. 740–755. Springer, Cham (2014). https://doi.org/10.1007/978-3-319-10602-1_48
31. Liu, W., Chen, J., Li, C., Qian, C., Chu, X., Hu, X.: A cascaded inception of inception network with attention modulated feature fusion for human pose estimation. In: The Thirty-Second AAAI Conference on Artificial Intelligence (2018)
32. Messer, K., Matas, J., Kittler, J., Luettin, J., Maitre, G.: XM2VTSDB: the extended M2VTS database. In: Second International Conference on Audio and Video-Based Biometric Person Authentication (1999)
33. Mueller, F., et al.: Ganerated hands for real-time 3D hand tracking from monocular RGB. In: Proceedings of the IEEE Conference on Computer Vision and Pattern Recognition (CVPR) (2018)
34. Mueller, F., Mehta, D., Sotnychenko, O., Sridhar, S., Casas, D., Theobalt, C.: Real-time hand tracking under occlusion from an egocentric RGB-D sensor. In: Proceedings of International Conference on Computer Vision (ICCV) (2017)
35. Neverova, N., Wolf, C., Taylor, G.W., Nebout, F.: Multi-scale deep learning for gesture detection and localization. In: Agapito, L., Bronstein, M.M., Rother, C. (eds.) ECCV 2014. LNCS, vol. 8925, pp. 474–490. Springer, Cham (2015). https://doi.org/10.1007/978-3-319-16178-5_33
36. Newell, A., Huang, Z., Deng, J.: Associative embedding: end-to-end learning for joint detection and grouping. In: Advances in Neural Information Processing Systems (2017)
37. Newell, A., Yang, K., Deng, J.: Stacked hourglass networks for human pose estimation. In: Leibe, B., Matas, J., Sebe, N., Welling, M. (eds.) ECCV 2016. LNCS, vol. 9912, pp. 483–499. Springer, Cham (2016). https://doi.org/10.1007/978-3-319-46484-8_29
38. Nie, X., Feng, J., Xing, J., Yan, S.: Generative partition networks for multi-person pose estimation. arXiv preprint arXiv:1705.07422 (2017)
39. Oikonomidis, I., Kyriazis, N., Argyros, A.A.: Tracking the articulated motion of two strongly interacting hands. In: IEEE Conference on Computer Vision and Pattern Recognition (2012)
40. Papandreou, G., Zhu, T., Chen, L.C., Gidaris, S., Tompson, J., Murphy, K.: PersonLab: person pose estimation and instance segmentation with a bottom-up, part-based, geometric embedding model. arXiv preprint arXiv:1803.08225 (2018)
41. Papandreou, G., et al.: Towards accurate multi-person pose estimation in the wild. arXiv preprint arXiv:1701.01779 (2017)
42. Pech-Pacheco, J. L., Cristóbal, G., Chamorro-Martinez, J., Fernández-Valdivia, J.: Diatom autofocusing in brightfield microscopy: a comparative study. In: ICPR (2000)
43. Phillips, P.J., et al.: Overview of the face recognition grand challenge. In: Proceedings of the IEEE Conference on Computer Vision and Pattern Recognition (CVPR) (2005)
44. Pishchulin, L., et al.: DeepCut: joint subset partition and labeling for multi person pose estimation. In: Proceedings of the IEEE Conference on Computer Vision and Pattern Recognition (CVPR) (2016)
45. Ren, S., He, K., Girshick, R., Sun, J.: Faster R-CNN: towards real-time object detection with region proposal networks. In: Advances in Neural Information Processing Systems (NIPS) (2015)
46. Romero, J., Tzionas, D., Black, M.J.: Embodied hands: modeling and capturing hands and bodies together. ACM Trans. Graph. (ToG) **36**, 245 (2017)

47. Ronchi, M.R., Perona, P.: Benchmarking and error diagnosis in multi-instance pose estimation. In: Proceedings of International Conference on Computer Vision (ICCV) (2017)
48. Sagonas, C., Tzimiropoulos, G., Zafeiriou, S., Pantic, M.: 300 faces in-the-wild challenge: the first facial landmark localization challenge. In: IEEE International Conference on Computer Vision Workshop (2013)
49. Sharp, T., et al.: Accurate, robust, and flexible real-time hand tracking. In: Proceedings of the 33rd Annual ACM Conference on Human Factors in Computing Systems (2015)
50. Simon, T., Joo, H., Matthews, I., Sheikh, Y.: Hand keypoint detection in single images using multiview bootstrapping. In: Proceedings of the IEEE Conference on Computer Vision and Pattern Recognition (CVPR) (2017)
51. Sridhar, S., Mueller, F., Oulasvirta, A., Theobalt, C.: Fast and robust hand tracking using detection-guided optimization. In: Proceedings of the IEEE Conference on Computer Vision and Pattern Recognition (CVPR) (2015)
52. Sun, K., Xiao, B., Liu, D., Wang, J.: Deep high-resolution representation learning for human pose estimation. arXiv preprint arXiv:1902.09212 (2019)
53. Sun, K., et al.: High-resolution representations for labeling pixels and regions. arXiv preprint arXiv:1904.04514 (2019)
54. Sun, Y., Wang, X., Tang, X.: Deep convolutional network cascade for facial point detection. In: Proceedings of the IEEE Conference on Computer Vision and Pattern Recognition (CVPR) (2013)
55. Tompson, J., Stein, M., Lecun, Y., Perlin, K.: Real-time continuous pose recovery of human hands using convolutional networks. ACM Trans. Graph. (ToG) **33**, 1–10 (2014)
56. Trigeorgis, G., Snape, P., Nicolaou, M.A., Antonakos, E., Zafeiriou, S.: Mnemonic descent method: a recurrent process applied for end-to-end face alignment. In: Proceedings of the IEEE Conference on Computer Vision and Pattern Recognition (CVPR) (2016)
57. Tzimiropoulos, G.: Project-out cascaded regression with an application to face alignment. In: Proceedings of the IEEE Conference on Computer Vision and Pattern Recognition (CVPR) (2015)
58. Valle, R., Buenaposada, J.M., Valdes, A., Baumela, L.: A deeply-initialized coarse-to-fine ensemble of regression trees for face alignment. In: Proceedings of the European Conference on Computer Vision (ECCV) (2018)
59. Wang, Y., Peng, C., Liu, Y.: Mask-pose cascaded cnn for 2D hand pose estimation from single color image. IEEE Trans. Circ. Syst. Video Technol. **29**, 3258–3268 (2018)
60. Wu, J., et al.: Ai challenger: a large-scale dataset for going deeper in image understanding. arXiv preprint arXiv:1711.06475 (2017)
61. Wu, W., Qian, C., Yang, S., Wang, Q., Cai, Y., Zhou, Q.: Look at boundary: a boundary-aware face alignment algorithm. In: Proceedings of the IEEE Conference on Computer Vision and Pattern Recognition (CVPR) (2018)
62. Xiang, D., Joo, H., Sheikh, Y.: Monocular total capture: posing face, body, and hands in the wild. In: Proceedings of the IEEE Conference on Computer Vision and Pattern Recognition (CVPR) (2019)
63. Xiao, B., Wu, H., Wei, Y.: Simple baselines for human pose estimation and tracking. In: Proceedings of the European Conference on Computer Vision (ECCV) (2018)
64. Xiong, X., De la Torre, F.: Supervised descent method and its applications to face alignment. In: Proceedings of the IEEE Conference on Computer Vision and Pattern Recognition (CVPR) (2013)

65. Yuan, S., Ye, Q., Stenger, B., Jain, S., Kim, T.K.: BigHand2. 2m benchmark: hand pose dataset and state of the art analysis. In: Proceedings of the IEEE Conference on Computer Vision and Pattern Recognition (CVPR) (2017)
66. Zhang, Z., Luo, P., Loy, C.C., Tang, X.: Learning deep representation for face alignment with auxiliary attributes. IEEE Trans. Pattern Anal. Mach. Intell. **38**, 918–930 (2015)
67. Zhu, X., Ramanan, D.: Face detection, pose estimation, and landmark localization in the wild. In: Proceedings of the IEEE Conference on Computer Vision and Pattern Recognition (CVPR) (2012)
68. Zimmermann, C., Brox, T.: Learning to estimate 3D hand pose from single RGB images. arXiv preprint arXiv: 1705.01389 (2017)
69. Zimmermann, C., Ceylan, D., Yang, J., Russell, B., Argus, M., Brox, T.: FreiHand: a dataset for markerless capture of hand pose and shape from single RGB images. In: Proceedings of International Conference on Computer Vision (ICCV) (2019)

Relative Pose Estimation of Calibrated Cameras with Known SE(3) Invariants

Bo Li[1](✉), Evgeniy Martyushev[2], and Gim Hee Lee[1]

[1] National University of Singapore, Singapore, Singapore
prclibo@gmail.com, gimhee.lee@nus.edu.sg
[2] South Ural State University, Chelyabinsk 454080, Russia
martiushevev@susu.ru

Abstract. The SE(3) invariants of a pose include its rotation angle and screw translation. In this paper, we present a complete comprehensive study of the relative pose estimation problem for a calibrated camera constrained by known SE(3) invariant, which involves 5 minimal problems in total. These problems reduces the minimal number of point pairs for relative pose estimation and improves the estimation efficiency and robustness. The SE(3) invariant constraints can come from extra sensor measurements or motion assumption. Unlike conventional relative pose estimation with extra constraints, no extrinsic calibration is required to transform the constraints to the camera frame. This advantage comes from the invariance of SE(3) invariants cross different coordinate systems on a rigid body and makes the solvers more convenient and flexible in practical applications. In addition to the concept of relative pose estimation constrained by SE(3) invariants, we also present a comprehensive study of existing polynomial formulations for relative pose estimation and discover their relationship. Different formulations are carefully chosen for each proposed problems to achieve best efficiency. Experiments on synthetic and real data shows performance improvement compared to conventional relative pose estimation methods. Our source code is available at: http://github.com/prclibo/relative_pose.

1 Introduction

Minimal relative pose solver of a camera is a fundamental component in modern 3D vision applications including robot localization and mapping, augmented reality, autonomous driving, 3D modeling, etc. Well-known solvers include the 7-point algorithm [11] and the 5-point algorithm [30,39]. It is generally admitted that an n-point solver with smaller n performs more robustly, has less degenerate configurations and requires less iterations when integrated in a RANSAC framework.

As the first contribution of this paper, we show that two measurements – rotation angle and screw translation – can be respectively integrated into relative pose solvers to reduce the number n of minimal points. Typical scenarios for these

Electronic supplementary material The online version of this chapter (https://doi.org/10.1007/978-3-030-58545-7_13) contains supplementary material, which is available to authorized users.

© Springer Nature Switzerland AG 2020
A. Vedaldi et al. (Eds.): ECCV 2020, LNCS 12354, pp. 215–231, 2020.
https://doi.org/10.1007/978-3-030-58545-7_13

measurements include a robot equipped with a camera and an IMU, and a robot with planar motion. These measurements are referred as SE(3) invariants as they stay invariant cross different coordinate systems on a rigid body. Consequently, the proposed methods do not require known extrinsic pose of the camera with respect to the IMU or the motion plane, which is an important advantage over previous relative pose estimation methods [4,8,13,20,23,35,36]. This advantage make the proposed methods more flexible and convenient. For example, when estimating visual odometry to hand-eye calibrate the camera-IMU extrinsics, the proposed methods improve trajectory estimation even though the extrinsics are unavailable. For robot systems subjected to long term operation, the proposed methods avoid re-calibration. All the 5 minimal problems introduced by different combination of SE(3) invariants as constraints are comprehensively studied in this paper.

The second contribution is a comprehensive study of all existing polynomial formulations for relative pose solvers. We show pros and cons of each formulation under different minimal problem settings and reveal connections between the formulations. For each proposed relative pose problem with SE(3) invariants, we evaluate these formulations and propose solvers with the best efficiency.

2 Related Works

A fundamental matrix for a pair of pinhole cameras has 7 DoF and can be estimated minimally from 7 and linearly from 8 point correspondences [10,11,25]. The estimation can be reduced to the 6-point algorithm [12,16,40] when all camera intrinsics except a common focal length are calibrated. In the case where the focal length is also calibrated, the 5-point algorithm [30,39] is naturally introduced.

Beyond 5-point solution, extra constraints can be exploited. With known gravity direction measured by IMU or by the knowledge of motion plane, [4,8,13,35] obtain two rotation angles and hence reduce the minimal number of point pairs to 3. Camera extrinsics are required to be calibrated for these methods to transform the two angles to the camera frame. In [36], it is assumed that the camera follows the Ackermann motion. In this case only one point is needed for pose estimation. This method requires the camera to be specifically mounted. [23] proposes the 4-point algorithm given a known rotation angle measurement from other sensors. This is the first work on integrating SE(3) invariants in relative pose estimation. The known rotation angle can also be used for the camera self-calibration as demonstrated in [27]. As mentioned in the previous section, extrinsic calibration is not required to use SE(3) invariants.

Minimal problems are usually formulated in terms of multivariate polynomial systems and a plenty of methods have been proposed to solve these systems. Some of these methods make use of the Gröbner basis computation [15,16,27,28,39]. The roots are then derived from the eigenvectors of the so-called action matrix constructed from the Gröbner basis [5]. Apart from the action matrix, alternative matrix decomposition methods are also proposed, e.g. PolyEig [17] and QuEst [33]. To avoid significant computational cost of matrix decomposition, the hidden variable approach has been used in several solvers [12,30]. This approach reduces the problem to finding real roots of a univariate polynomial.

3 Preliminaries

3.1 Notation

We preferably use α, β, ... for scalars, a, b, ... for column 3-vectors, and A, B, ... for matrices. For a matrix A, the transpose is $A^\top$, the determinant is $\det A$, and the trace is $\operatorname{tr} A$. For two 3-vectors a and b the cross product is $a \times b$. For a vector a, the entries are a_i, the notation $[a]_\times$ stands for the skew-symmetric matrix such that $[a]_\times b = a \times b$ for any vector b. We use I for the identity matrix and $\| \cdot \|$ for the Frobenius norm. A notation f^*_* is used to refer a polynomial.

A rotation matrix R can be represented by a unit quaternion $[\sigma \; u^\top]$ as follows

$$R = 2(uu^\top - \sigma[u]_\times) + (\sigma^2 - \|u\|^2)I, \tag{1}$$

$$\mathsf{f}^\sigma := \|u\|^2 + \sigma^2 - 1 = 0. \tag{2}$$

With θ as the rotation angle, we have

$$\operatorname{tr} R = 4\sigma^2 - 1 = 2\cos\theta + 1, \tag{3}$$

Another way to represent a rotation matrix R comes from the Cayley transform if and only if it is not a rotation through an angle $\pi + 2\pi k$.

$$R = (I - [v]_\times)(I + [v]_\times)^{-1}, \tag{4}$$

where $v = u/\sigma$ is a 3-vector.

The special Euclidean group SE(3) consists of all orientation-preserving rigid motions of 3-dimensional Euclidean space. Any element $H \in \mathrm{SE}(3)$ can be represented by a 4×4 matrix of the form

$$H = \begin{bmatrix} R & t \\ 0^\top & 1 \end{bmatrix}, \tag{5}$$

where $R \in \mathrm{SO}(3)$ and $t \in \mathbb{R}^3$ are the rotational and translational parts of H respectively. In the sequel, saying about elements of group SE(3) we always imply 4×4 matrices of type (5).

3.2 Epipolar Constraint

Let $P' = [R' \; t']$ and $P'' = [R'' \; t'']$, where $R', R'' \in \mathrm{SO}(3)$ and $t', t'' \in \mathbb{R}^3$, be calibrated camera matrices. Let q'_i and q''_i be the corresponding images of a 3D point Q_i. Then the epipolar constraint reads

$$\mathsf{f}_i := q''^\top_i (R[t']_\times - [t'']_\times R) q'_i = 0, \tag{6}$$

where i counts the point pairs and $R = R''R'^{\top}$ is called the relative rotation matrix. We notice that Eq. (6) can be rewritten in form

$$q_i''^{\top}[t]_\times R q_i' = 0, \tag{7}$$

where $t = Rt' - t''$ is called the relative translation. Matrix $E = [t]_\times R$ is well known in the computer vision community as an essential matrix.

3.3 SE(3) Invariants

Given an element $H \in \mathrm{SE}(3)$, with its rotational part R represented by (1). We denote the unit rotation axis of R by $r = \frac{u}{\|u\|}$, then the value

$$\delta = r^{\top} t \tag{8}$$

is the *screw translation* of H. In this paper, we are specifically interested in the case of $\delta = 0$, which is also equivalent to

$$\mathrm{f}^0 := u^{\top} t = 0. \tag{9}$$

Consider a robot with planar motion. Its rotation axis r must be the normal vector of the motion plane, and its translation vector t must lie on the motion plane. Therefore, it is obvious that the condition of zero screw translation ($\delta = 0$) holds for any planar motion regardless of the camera orientation with respect to the ground plane direction.

Figure 1 illustrates the definition of θ and δ. We refer them as the SE(3) invariants, i.e. scalar values invariant under the conjugation by an SE(3) element. In robotics, this conjugation is known as the hand eye transformation. The difference between SE(3) invariants and an easily mixed-up concept bi-invariant metrics can be found from [3].

Theorem 1 (SE(3) **Invariants**). *For a transform $H \in \mathrm{SE}(3)$, its rotation angle θ and screw translation δ are invariant under the hand eye transformation $H' = X^{-1}HX$ with $X \in \mathrm{SE}(3)$.*

Proof. Let the rotational and translational parts of H, H', X be R, R', R_X and t, t', t_X respectively. Then we have

$$H' = \begin{bmatrix} R_X^{\top} & -R_X^{\top} t_X \\ 0^{\top} & 1 \end{bmatrix} \begin{bmatrix} R & t \\ 0^{\top} & 1 \end{bmatrix} \begin{bmatrix} R_X & t_X \\ 0^{\top} & 1 \end{bmatrix} = \begin{bmatrix} R_X^{\top} R R_X & R_X^{\top}(R t_X - t_X + t) \\ 0^{\top} & 1 \end{bmatrix}, \tag{10}$$

that is $R' = R_X^{\top} R R_X$ and $t' = R_X^{\top}(R t_X - t_X + t)$. The invariance of θ follows from Eq. (3), since $\mathrm{tr}(R') = \mathrm{tr}(R_X^{\top} R R_X) = \mathrm{tr}(R)$.

Furthermore, let r and r' be the unit rotation axes of R and R', respectively. It is clear that $r = Rr$ and $r' = R'r' = R_X^{\top} R R_X r'$ (Lemma II [43]). Hence, axes r and r' are related by $r' = R_X^{\top} r$. Substituting this into the definition of δ' yields

$$\delta' = r'^{\top} t' = r^{\top} R_X R_X^{\top}(R t_X - t_X + t) = r^{\top} R t_X - r^{\top} t_X + r^{\top} t = r^{\top} t = \delta. \tag{11}$$

□

Theorem 1 is a well-known result in robotics [2,3]. Different proof for the rotation part can also be found from [2,3,38].

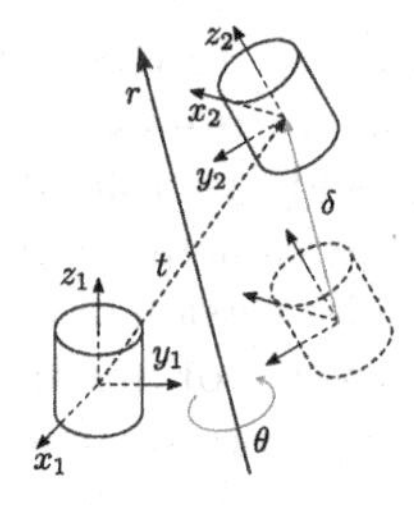

Fig. 1. A rigid motion in SE(3) can always be decomposed as a rotation around an axis r with angle θ, and a screw translation δ along r. θ and δ remain consistent for different parts of the rigid body, regardless of the part offset and the local coordinate system.

4 Minimal Problem Formulations

The relative pose estimation problem aims to solve for the relative rotation R and relative translation t given several image point pairs. It is well known that if no additional constraints are used, the relative pose can be estimated minimally from 5 point pairs [12,30,39]. With known SE(3) invariants (rotation angle θ and screw translation δ), the number of point pairs required for a minimal solution is reduced. Table 1 summarizes the minimal relative pose estimation problems that can be formulated for different combinations of image data and SE(3) invariants.

Table 1. Relative pose problems with SE(3) invariants. RA: Relative angle (θ) ST0: Zero screw translation ($\delta = 0$) ST1: Non-zero screw translation ($\delta \neq 0$) n: Number of points for minimal cases

Problem	SE(3) Inv	n	DoF	Constraints
5P [12,30,39]	–	5	5	$\mathfrak{f}_1, \mathfrak{f}_2, \mathfrak{f}_3, \mathfrak{f}_4, \mathfrak{f}_5$
4P-RA [23,28]	θ	4	5	$\mathfrak{f}_1, \mathfrak{f}_2, \mathfrak{f}_3, \mathfrak{f}_4, \mathfrak{f}^\sigma$
4P-ST0	$\delta = 0$	4	5	$\mathfrak{f}_1, \mathfrak{f}_2, \mathfrak{f}_3, \mathfrak{f}_4, \mathfrak{f}^0$
3P-RA-ST0	$\theta, \delta = 0$	3	5	$\mathfrak{f}_1, \mathfrak{f}_2, \mathfrak{f}_3, \mathfrak{f}^\sigma, \mathfrak{f}^0$
5P-ST1	$\delta \neq 0$	5	6	$\mathfrak{f}_1, \mathfrak{f}_2, \mathfrak{f}_3, \mathfrak{f}_4, \mathfrak{f}_5$
4P-RA-ST1	$\theta, \delta \neq 0$	4	6	$\mathfrak{f}_1, \mathfrak{f}_2, \mathfrak{f}_3, \mathfrak{f}_4, \mathfrak{f}^\sigma$

Remark 1. In 5P, 4P-RA, 4P-ST0 and 3P-RA-ST0, each of SE(3) invariants $\mathfrak{f}^\sigma$ and $\mathfrak{f}^0$ can replace one of point correspondences $\mathfrak{f}_i$ as constraints and the relative pose can be estimated up to an ambiguous scale (5 DoF). In 5P-ST1 and 4P-RA-ST1, condition (8) with $\delta \neq 0$ can not be used to replace point correspondences as the essential matrix does not change regardless of the value of a non-zero δ. Instead, δ can be used to determine the length of translation, and hence the overall scale is observable (6 DoF). 5P-ST1 and 4P-RA-ST1 are therefore equivalent to problems 5P and 4P-RA, respectively.

Remark 2 (Beyond Pinhole Cameras). In this paper, we only focus on relative pose estimation for a pinhole camera. However, it is worth mentioning the difference between pinhole camera and generalized camera models under known SE(3) invariants. For a generalized camera model [32,41], the relative translation length is observable. The vanilla version of relative pose estimation problem requires 6 points to fully recover the 6 DoF relative pose. When the screw translation is known, regardless of being zero or non-zero, 5 points are required to fully recover the 6 DoF relative pose. The case of known relative rotation angle for generalized cameras was covered in [28].

5 Solution Formulations for Relative Pose Estimation

Various solutions have been studied in the past decades on relative estimation problems, i.e. to solve (6) or (7) under different constraints. We provide a comprehensive summary for these formulations in this section and also discover how SE(3) invariants can be denoted for each formulation. All mentioned previous formulations are also listed in Table 2.

Table 2. Representative works of polynomial formulations for relative pose estimation.

Problem	Form	#R	#t	Templ	#S
PC 5P [15]	SIR3	9 (R)	3	66×197	20
PC 5P [6]	SIR6	4 (u, σ)	3	40×56	35
PC 5P [14]	Direct	4 (u, σ)	3	NR	80
PC+θ [23]	SIR3	3 (u)	3	270×290	20
PC+θ [28]	SIR2	3 (u)	2	16×36	20
PC+g (IMU) [13]	SIR3	1 (yaw)	3	CF	4
PC+g (pl) [4]	Direct	1 (yaw)	1	CF	4
PC+Ack [37]	Direct	1 (yaw)	0	CF	1
GC 6P [39]	SIR2	3 (v)	2	60×120[1]	64
GC+θ [28]	SIR2	3 (u)	2	37×81	44
GC+g (IMU) [20]	SIR3	1 (yaw)	3	CF	8
GC+g (pl) [22]	SIR3	1 (yaw)	2	CF	6
GC+Ack [21]	Direct	1 (yaw)	1	CF	2

Problem	Form	Templ	#S
PC 5P [30,39]	NullE	10×20	10
PC−fc [16,40]	NullE	31×46	15
PC+θ−fc−pp [27]	NullE	19×32[2]	6
PC+g (IMU) [8]	NullE	6×10	4

* #R, #t: Number of parameters
* #S: Number of solutions
* +/−: With a constraint or an unknown
* PC/GC: Pinhole camera / generalized camera
* NR: Not reported
* CF: Closed form solution w/o template matrix
* g: Vertical direction
* Ack: Ackermann motion model
* fc/pp: Focal length / principle point
[1] Insufficient for full Gröbner basis generation
[2] The largest of cascaded templates reported

5.1 Solutions by Decomposing E

Directly Solving R and t. As an intuitive start, it is possible to directly solve (6), (7) by considering R and t as polynomial unknowns. In [23], R is parameterized by angle-axis to integrate rotation angle and t is constrained by $\|t\|^2 = 1$ to remove scale ambiguity. In theory R can be also parameterized by a quaternion with constraint of f^σ or by 9 matrix elements with constraint of

$R^\top R = I$. However, these formulations involve $6 \sim 12$ unknowns (including 3 for translation) and require quite complicated polynomial elimination, which makes high computational burden in real-time applications. A simpler specialization is when the vertical direction is known from IMU [13] or known ground plane [4], R can be parameterized by a yaw angle rotation. With Ackermann motion assumed, the parameters can even be further reduced [21,37].

Since each f_i is linear in t, cf. Eq. (7), several formulations have been proposed in previous works to eliminate translation variables and solve rotation parameters only for simplicity.

SIR3: Solving Isolated Rotation by a 3×3 Determinant. We can isolate unknown translation by rewriting the epipolar constraints (7) for n point pairs in the form

$$G\, t = 0, \tag{12}$$

where G is a matrix of size $n \times 3$. Elements of the i-th row of matrix G are polynomials in unknown rotation parameters and known q_i', q_i''. It follows from Eq. (12) that all 3×3 minors of G must vanish for a valid translation. Thus we obtain new polynomial constraints of the total degree 6 on the rotation parameters only. This solution formulation will be further referred to as *SIR3*. If the rotation matrix is represented by (1) (resp. by (4)), then the formulation is denoted by SIR3+u (resp. SIR3+v). When rotation is constrained to have only one unknown, SIR3 generates a closed form univariate polynomial [13,20]. However, with more unknowns in rotation, it is still not satisfactory as leads to large matrix templates for the Gröbner basis computation. For example, in [15] SIR3 is used to solve 5P by a reduction on a 66×197 matrix. In [23], the SIR3+u formulation of 4P-RA involves even larger template matrix and has to be solved by numerical search.

The known rotation angle can be easily integrated into SIR3+u. If the motion is planar, i.e. the screw translation is zero, polynomial f^0 from (9) can be written as a new row $u^\top$ of matrix G.

SIR6: Solving Isolated Rotation by a 6×6 Determinant. Let Q_i be the i-th 3D point so that

$$\lambda_i q_i' = P' Q_i, \quad \mu_i q_i'' = P'' Q_i, \tag{13}$$

where λ_i and μ_i are some scalars. The relative pose R and t satisfies $\lambda_i q_i'' = \mu_i R q_i' + t$. Consider this equation for the i, j, k-th points and subtract the i-th equation over the j-th and k-th respectively to eliminate t. We can obtain two 3D linear equations, forming a 6×6 matrix M_{ijk}.

$$M_{ijk} [\lambda_i \; \mu_i \; \lambda_j \; \mu_j \; \lambda_k \; \mu_k]^\top = 0. \tag{14}$$

Similar to SIR3, the determinant of M_{ijk} must vanish, resulting in polynomials on the rotation parameters only. SIR6 is proposed by [6] as an alternative solution to 5P. Symbolic computation reveals the relationship between SIR6 and SIR3:

Theorem 2 *Consider a 3×3 submatrix G_{ijk} of matrix G whose three rows correspond to the i-th, j-th, and k-th point pairs. We have $\det G_{ijk} = \det M_{ijk}$, up to a sign.*

SIR2: Solving Isolated Rotation by a 2×2 Determinant. Using the rigid motion ambiguity of the world coordinate frame, we set $Q_i = [0\ 0\ 0\ 1]^\top$ in (13) for a certain i. This yields

$$t' = \lambda_i q_i',\ t'' = \mu_i q_i''. \tag{15}$$

Substituting t' and t'' into Eq. (6) for a j-th pair with $j \neq i$, we convert Eq. (6) into $F_{ij} \left[\lambda_i\ \mu_i\right]^\top = 0$. Construct F_{ik} from a k-th point pair and stack with F_{ij} as F_{ijk}, we have

$$F_{ijk} \left[\lambda_i\ \mu_i\right]^\top = 0. \tag{16}$$

where matrix F_{ijk} is of size 2×2. F_{ijk} must have zero determinant. This leads to degree 4 polynomial equations in the rotation parameters. The proposed solution formulation will be further referred to as *SIR2*. The two versions of SIR2 corresponding to the quaternion (1) and Cayley (4) parametrizations are denoted by SIR2+u and SIR2+v respectively. The SIR2+v form was earlier used in [41] for solving the relative pose problem for generalized cameras. In [28], SIR2+u was used to solve 4P-RA. Symbolic computation also reveals the connection between SIR2 and SIR3 and explains why SIR2 is a simpler formulation compared to SIR3/SIR6.

Theorem 3 *Consider G_{ijk} in Theorem 2. Under SIR2/SIR3+u, we have:*

$$\det G_{ijk} = (\|u\|^2 + \sigma^2)\cdot \det F_{ijk}, \tag{17}$$

up to a sign. This equation also holds if replacing $(\|u\|^2 + \sigma^2)$ with $(\|v\|^2 + 1)$ under the SIR2/SIR3+v.

Remark 3 According to Eq. (17), since equation $\|v\|^2+1 = 0$ has infinite number of complex solutions, 5P in SIR3+v is not zero-dimensional over $\mathbb{C}$.

The known rotation angle can be easily integrated into SIR2+u. If the motion is planar, i.e. $\delta = 0$, then polynomial f^0 can add a row $[-u^\top q_i'\ u^\top q_i'']$ to F_{ijk}.

5.2 Solutions by Constraining E

NullE: Solving Essential Matrix Represented by NullSpace Bases. Instead of direct solving for R and t, a more classical approach to the relative pose problem is solving first for the essential matrix E which is a mixed form of rotation and translation parameters. Unknown E is parameterized by $\sum_{i=1}^{9-n} \gamma_i E^{(i)}$, where matrices $E^{(i)}$ form the nullspace basis of the underdetermined linear system $\{\mathrm{f}_i \mid i = 1,\dots,n\}$, γ_i are new unknowns which are usually scaled so that $\gamma_1 = 1$. Traditional 5P solvers [12,30,39] use the following constraints to form a polynomial system on γ_i:

$$\det E = 0, \tag{18}$$

$$2EE^\top E - \mathrm{tr}(EE^\top)E = 0. \tag{19}$$

In addition, [8] found that known vertical can be denoted as constraints on E. Known SE(3) invariants also can be formulated as constraints on E as follows.

Theorem 4 ([27]). *Let $E = [t]_\times R$ be an essential matrix and* $\operatorname{tr} R = \tau$*. Then E fulfills the following equation*

$$\frac{1}{2}(\tau^2 - 1)\operatorname{tr}(EE^\top) + (\tau + 1)\operatorname{tr}(E^2) - \tau \operatorname{tr}^2 E = 0. \tag{20}$$

Theorem 5. *Let $E = [t]_\times R$ be an essential matrix and R be a rotation through an angle θ around a vector $r \neq 0$. If $\delta = r^\top t = 0$, then*

$$\operatorname{tr} E = 0. \tag{21}$$

Conversely, if $\operatorname{tr} E = 0$*, then either $\delta = 0$ or $\theta = \pi k$ for a certain integer k.*

Proof. We utilize Proposition 2.20 from [29] that $\operatorname{tr} E = -2\sin\theta(r^\top t)$ and the statement follows.

Using Eqs. (18)–(21), problems 4P-RA, 4P-ST0, and 3P-RA-ST0 can be formulated in terms of matrix E. Table 3 summarizes these formulations with reference template matrix size generated by automatic polynomial solver generators [16,24].

Remark 4. As is well known, the 5P problem in the NullE formulation has 10 solutions. Each essential matrix corresponds to a twisted pair of rotations [11] and each rotation, being represented by a unit quaternion, doubles due to the sign ambiguity. Therefore, as shown in Table 3, the 5P problem in SIR2/SIR3+u has 40 solutions while in SIR2/SIR3+v has only 20 solutions. The 4P-RA problem in the NullE formulation has 20 solutions, corresponding to 40 rotations. For each pair of rotations, there is a unique one whose rotation angle equals known θ. Similarly for 4P-ST0 in NullE, each pair of rotations corresponding to an essential matrix contains a unique valid rotation.

Remark 5. The 3P-RA-ST0 problem has 12 solutions in SIR2+u. However, in NullE the system consisting of Eqs. (18)–(21) has 20 solutions. The obtained contradiction indicates that there must exist additional polynomial constraints on essential matrix E. Using the implicitization algorithm [5], we found that the entries of E additionally satisfy 7 cubic equations. We provide them in the supplementary material. The above polynomial system complemented with the new 7 cubics has 12 solutions (NullEx in Table 3). However, due to the lack of geometric interpretability for the additional cubics, in this paper we use a hand-crafted solver with slightly larger template matrix.

6 Minimal Relative Pose Solvers with SE(3) Constraints

The goodness of different solver formulations is reflected by the size of the matrix template for Gröbner basis computation since it directly affects both the speed and numerical accuracy of a minimal solver. Different formulated solvers are reported in Table 3. We compared our proposed formulation with [16], the most widely used generator the past years, and [24], a wrapper of a newer generator [19].

Table 3. Comparison of different solution formulations for the minimal relative pose problems with SE(3) invariants. #V: Number of variables D: Highest degree #S: Number of solutions AG: Generator from [16] GAPS: Generator from [24] 1: Mirrored roots u merged by [18] 2: The largest of cascaded templates reported

Problem	Form	#V	D	#S	AG	GAPS	Proposed
5P	SIR3 + u	4	6	40	146×186	116×200[1]	–
5P	SIR3 + v	3	6	∞	–	–	–
5P	SIR2 + u	4	4	40	90×130	60×80[1]	–
5P	SIR2 + v	3	4	20	31×51	36×56	–
5P	NullE	3	3	10	10×20	10×20	–
4P-RA	SIR2 + u	3	4	20	26×46	36×56	16×36 [28]
4P-RA	NullE	4	3	20	34×54[2]	50×70	–
4P-ST0	SIR2 + u	3	4	20	62×82[2]	38×65[1]	–
4P-ST0	SIR2 + v	3	4	10	25×35	27×35	–
4P-ST0	NullE	3	3	10	10×20	10×20	10×20
3P-RA-ST0	SIR2 + u	3	4	12	23×35[2]	28×35	13×25
3P-RA-ST0	NullE	5	3	20	34×54	50×70	–
3P-RA-ST0	NullEx	5	3	12	22×35[2]	53×65	–

6.1 5P, 4P-RA, 5P-ST1 and 4P-RA-ST1

NullE is the most widely used polynomial formulation for 5P, with a template matrix of 10×20. Template matrix of 4P-RA was recently reduced from 270×290 to 16×36 using SIR2-u [28]. 5P-ST1 can be solved by a 5P solver and multiply the unit translation solution t by $\frac{\delta}{r^\top t}$. 4P-RA-ST1 can be solved by a 4P-RA solver in the same way.

6.2 4P-ST0

The NullE formulation for 4P-ST0 produces the smallest template of size 10×20. Note that in NullE of 4P-ST0, Eq. (21) replaces an epipolar constraint of 5P and they are both linear on E. Therefore, 4P-ST0 can be simply solved by a NullE 5P solver by replacing the coefficients of one epipolar constraint.

6.3 3P-RA-ST0

For problem 3P-RA-ST0, the SIR2+u formulation is preferable as it leads to the smallest 13×25 matrix template. The algorithm is summarized as follows.

Three image point pairs are first used to form a 2×2 matrix F_{123}, see Subsect. 5.1. We set $u^\top = [\alpha\ \beta\ \gamma]$. Then our system consists of the following polynomial equations:

- 10 equations of $m \cdot \mathrm{f}^{\sigma} = 0$ for m being every monomial with degree up to 2;
- 1 equation $\det F_{123} = 0$;
- 12 equations of $m \cdot \det F'_{ij} = 0$, with $i \neq j$, $m \in \{\alpha, \beta, \gamma, 1\}$ and

$$F'_{ij} = \begin{bmatrix} F_{ij} \\ -u^\top q'_i \; u^\top q''_i \end{bmatrix}.$$

In matrix form the system can be written as $Ax = 0_{23\times1}$, where A is the 23×35 coefficient matrix whose i-th row consists of coefficients of the i-th polynomial, x is a monomial vector. Matrix A is exactly the template produced by the Automatic Generator, see Table 3. However, the template's size can be further reduced if we take into account the special structure of matrix A. Namely, if the first 10 monomials in x are

$$\alpha^4,\ \alpha^3\beta,\ \alpha^2\beta^2,\ \alpha^3\gamma,\ \alpha^2\beta\gamma,\ \alpha^2\gamma^2,\ \alpha^3,\ \alpha^2\beta,\ \alpha^2\gamma,\ \alpha^2, \tag{22}$$

then matrix A has the following block form $A = \begin{bmatrix} U & V \\ W & X \end{bmatrix}$, where U is an upper-triangular 10×10 matrix with 1's on its main diagonal. We conclude that matrix A is equivalent to $\begin{bmatrix} U & V \\ 0_{13\times10} & B \end{bmatrix}$, where matrix $B = X - WU^{-1}V$ is our final template of size 13×25. Matrix B contains all necessary data for deriving solutions either by constructing an action matrix or by forming the 12-th degree univariate polynomial in accordance with the hidden variable method. We provided more details on the 3P-RA-ST0 solver in the supplementary material. Readers can refer to [7,28] for more usage of the above simplification.

Handling Degeneracy

Condition (8) becomes degenerate when the rotation matrix is close to I. In this case the rotation axis r is ill-posed and vector u becomes arbitrarily small. Enforcing condition (9) in this case might lead to a large deviation in the direction of translation. Nevertheless, this degenerate case can be easily covered by fitting relative pose to a translation-only motion (2P-TO). The skew-symmetric essential matrix $[t]_\times$ can be easily estimated from two image feature pairs. In this paper, we estimate 4P-ST0 and 3P-RA-ST0 together with 2P-TO and accept the results with more inliers.

7 Experiments

7.1 Implementation Details

All algorithms compared in experiments are implemented by C++. The hidden variable method is used to derive solutions of polynomial systems. Roots of univariate polynomials are found using Sturm sequences. We implement 4P-RA [28], 4P-ST0, and 3P-RA-ST0. The C++ 5P solver from [12] is used, which is regarded as the state-of-the-art fast implementation. Runtime statistics on an i5-4288U is listed in Table 4.

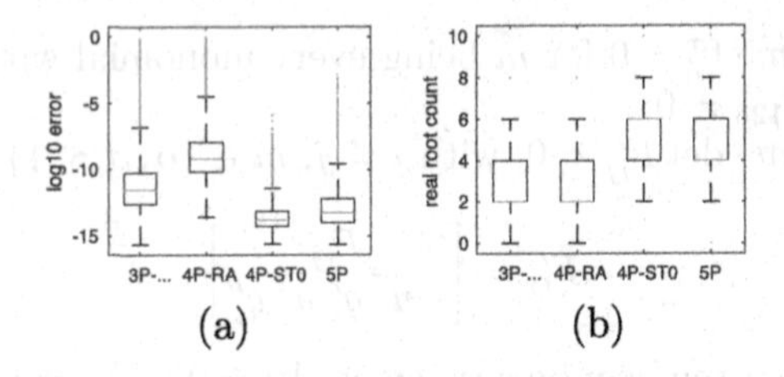

Fig. 2. (a) Numerical accuracy comparison of the solvers; (b) Statistics on the number of real roots for each solver

Table 4. Average runtime comparison of the solvers

Minimal solver	3P-RA-ST0	4P-RA	4P-ST0	5P
Average time	28 µs	34 µs	26 µs	25 µs

7.2 Synthetic Data

Synthetic data are generated to illustrate the algorithm performance. Synthetic image features are generated from a 60° field of view with focal length in 500px. We test algorithm performance under Gaussian image noise whose std ranges in 0–1px. Synthetic data is generated for forward motion and sideway motion. Rotation angle of a pose pair is randomly generated from Gaussian with std of 5°. The rotation angle measurement is disturbed by Gaussian noise (derived from the widely used Brownian process model for IMU noise) with std ranging in 0–1°. To test the performance of 4P-ST0 and 3P-RA-ST0 under non-perfectly planar motion, we first produces unit translation with zero component on rotation axis. Then the translation is disturbed along the rotation axis with Gaussian noise whose std ranges in 0–5%.

The numerical accuracy of each algorithm is compared and listed in Fig. 2(a). The numerical error is measured by the value $\min_i \|R_i - \bar{R}\|$, where i counts all real solutions and $\bar{R}$ is the ground truth relative rotation matrix. The number of real roots is also counted for each algorithm and listed in Fig. 2(b). We observe that 4P-ST0 in NullE formulation generally has more real roots compared to 5P. The number of real roots also affects the computational efficiency in some RANSAC frameworks like OpenCV where each real solution must be verified by computing the reprojection error over all image feature pairs.

In the experiments on both synthetic data and real data, the error of rotation is measured by the rotation angle between the estimated and groundtruth rotation. The error of translation is measure by the angle between the unit groundtruth translation and the estimated translation. Forward motion and sideway motion are experimented separately. The mean estimation error of 4P-ST0 and 5P against image ray disturb is shown in Fig. 3. The estimation is executed on 100 image feature pairs with 30% outliers under RANSAC. Green curves from bottom to top represent 4P-ST0 estimation with different screw translation disturbance along rotation axis $\{0\%, 1.66\%, 3.33\%, 5\%\}$. As it is mentioned

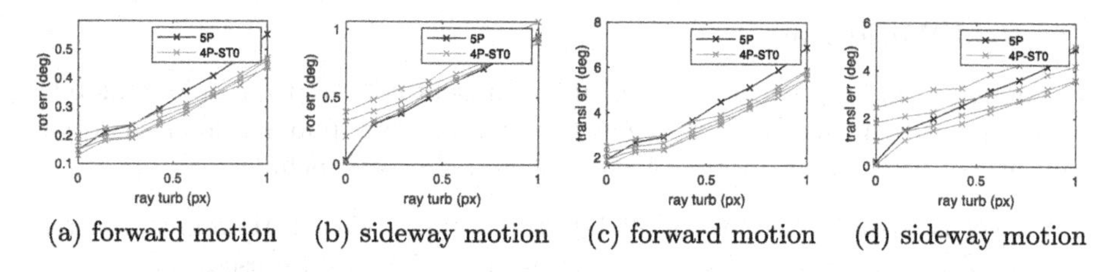

(a) forward motion (b) sideway motion (c) forward motion (d) sideway motion

Fig. 3. Estimation error plot of 4P-ST0 and 5P on synthetic data: (a, b) rotation errors; (c, d) translation errors

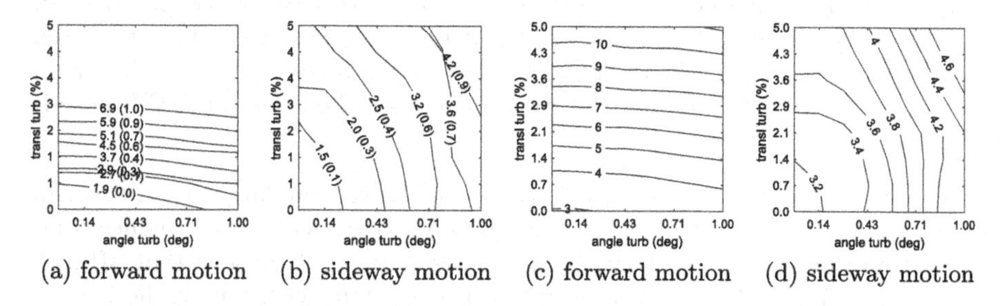

(a) forward motion (b) sideway motion (c) forward motion (d) sideway motion

Fig. 4. Estimation error comparison of 3P-RA-ST0 and 5P: (a, b) contour curves $\xi(\epsilon)$ denote that under image feature noise std ϵ px, when the error of rotation angle and screw translation is at the bottom left of the curve, 3P-RA-ST0 outperforms 5P with translation error no more than ξ; (c, d) translation error contour curve over different rotation angle and screw translation disturbance, with image feature noise std fixed as 1px

in many previous works, the rotation error is generally small for different solvers. Regarding translation error, the advantage of 4P-ST0 over 5P is more significant for forward motion, which is considered as the more common and difficult case than sideway motion.

The performance of 3P-RA-ST0 is affected by rotation angle error, screw translation error and also image feature error. To simplify our visualization, in Fig. 4(a) and Fig. 4(b), we fix the image feature error at different level and consider the translation error surface against the rotation angle and screw translation errors. The intersection contour of this surface with the error surface of 5P is plotted. The bottom-left area of each curve denotes the error level under which 3P-RA-ST0 can outperform 5P. We only compare the translation error here as the rotation error is generally similar for different methods. Compared to 4P-ST0, we find that 3P-RA-ST0 is more sensitive to screw translation error in forward motion. For example, with perfect rotation angle and image feature noise std as 1px in forward motion, 3P-RA-ST0 outperforms 5P only when screw translation error is less than 3% (Fig. 4(a)), while 4P-ST0 outperforms 5P even when screw translation error is 5% (Fig. 3(c)).

7.3 Real-World Data

We compare our approaches on multiple datasets collected on indoor mobile robots or outdoor autonomous vehicles, which are two popular modern robot applications with planar motion. Experimented datasets include:

- RawSeeds-Bicocca [1]: Indoor mobile robot data, with IMU and odometer available for rotation angle. Front camera (FC) images are used.
- TUM-RGBD-SLAM [42], Robot@Home [34]: Indoor mobile robot data, without available angle measurement. Images from the front RGBD camera (FC) are used. The left RGBD camera (LC) of Robot@Home is also experimented.
- KITTI [9], UMich [31], RobotCar [26]: Autonomous vehicle data, with fused GPS/INS data available. Front camera (FC) images are used. The left camera (LC) of UMich is also used. Rotation angle from the fused GPS/INS pose is used.

Consecutive image pairs with translational movement larger than 0.1m in indoor data and 1m in outdoor data are used in experiments. Performance comparison on indoor data is shown in Appendix Table 1. It is seen that 4P-ST0 outperforms 5P in almost all cases, which is consistent with the synthetic data results. With IMU data on RawSeeds, 3P-RA-ST0 further improves the estimation. Results with odometry angle have no improvement, implying the accuracy of odometry angle might be low in this dataset.

However, the performance varies on autonomous driving data in Appendix Table 2. We observe that 4P-ST0 outperforms 5P for broad road or highway environments. For urban narrow road environments, 5P has the better accuracy. The difference in the environments corresponds to different screw translation disturbance on a planar motion assumption as roads are less planar and vehicles might tilt more on urban road. With further analysis, we found that a portion of relative poses have screw translation of more than 20% ($r^\top t/\|t\| > 0.2$) in urban autonomous driving scenarios, which explains the poor performance of 4P-ST0.

We also note that 3P-RA-ST0 performs better on UMich left camera than 4P-ST0, corresponding to the observation in Fig. 4 that 3P-RA-ST0 is less sensitive to screw translation error for sideway motion.

8 Conclusions

In this paper, we show that known SE(3) invariants can be used to constrain the minimal relative pose estimation problem. Compared to existing relative pose problems with contraints, the proposed methods are more flexible and convenient since extrinsics are not required to transform SE(3) invariant to the camera frame. We also comprehensively revise and relate to each other existing formulations of the relative pose problem. The discovered relationship provides a deeper understanding to these previous methods. This knowledge help formulate the most efficient solvers for the proposed relative pose problem with SE(3) constraints. A series of experiments on synthetic and real datasets show practicality of the proposed solvers in robotic perception especially for indoor robots.

Acknowledgements. The work is supported in part by the Singapore MOE Tier 1 grant R-252-000-A65-114 and the Act 211 Government of the Russian Federation, contract No. 02.A03.21.0011.

References

1. Bonarini, A., Burgard, W., Fontana, G., Matteucci, M., Sorrenti, D.G., Tardos, J.D.: Rawseeds: robotics advancement through web-publishing of sensorial and elaborated extensive data sets. In: proceedings of IROS, vol. 6 (2006)
2. Chen, H.: A screw motion approach to uniqueness analysis of head-eye geometry. In: Proceedings. 1991 IEEE Computer Society Conference on Computer Vision and Pattern Recognition, pp. 145–151 (1991). https://doi.org/10.1109/CVPR.1991.139677
3. Chirikjian, G.S.: Partial bi-invariance of SE(3) metrics 1. J. Comput. Inf. Sci. Eng. **15**(1), 011008 (2014). https://doi.org/10.1115/1.4028941
4. Choi, S., Kim, J.H.: Fast and reliable minimal relative pose estimation under planar motion. Image Vis. Comput. **69**, 103–112 (2018)
5. Cox, D., Little, J., O'Shea, D.: Ideals, Varieties, and Algorithms, vol. 3. Springer, New York (2007). https://doi.org/10.1007/978-0-387-35651-8
6. Fathian, K., Jin, J., Wee, S.G., Lee, D.H., Kim, Y.G., Gans, N.R.: Camera relative pose estimation for visual servoing using quaternions. Robot. Autonom. Syst. **107**, 45–62 (2018). https://doi.org/10.1016/j.robot.2018.05.014
7. Faugère, J.C., Lachartre, S.: Parallel Gaussian elimination for gröbner bases computations in finite fields **68**(2), 5 (2007). https://doi.org/10.1145/1837210.1837225
8. Fraundorfer, F., Tanskanen, P., Pollefeys, M.: A minimal case solution to the calibrated relative pose problem for the case of two known orientation angles. In: Daniilidis, K., Maragos, P., Paragios, N. (eds.) ECCV 2010. LNCS, vol. 6314, pp. 269–282. Springer, Heidelberg (2010). https://doi.org/10.1007/978-3-642-15561-1_20
9. Geiger, A., Lenz, P., Urtasun, R.: Are we ready for autonomous driving? The kitti vision benchmark suite. In: Conference on Computer Vision and Pattern Recognition (CVPR) (2012)
10. Hartley, R.: In defence of the 8-point algorithm. In: Fifth International Conference on Computer Vision, 1995. Proceedings, pp. 1064–1070. IEEE (1995)
11. Hartley, R., Zisserman, A.: Multiple View Geometry in Computer Vision. Cambridge University Press, Cambridge (2003)
12. Hartley, R.I., Li, H.: An efficient hidden variable approach to minimal-case camera motion estimation. IEEE Trans. Pattern Anal. Mach. Intell. **34**(12), 2303–2314 (2012)
13. Kalantari, M., Hashemi, A., Jung, F., Guédon, J.P.: A new solution to the relative orientation problem using only 3 points and the vertical direction. J. Math. Imaging Vision **39**(3), 259–268 (2011)
14. Kalantari, M., Jung, F., Guedon, J.-P., Paparoditis, N.: The five points pose problem: a new and accurate solution adapted to any geometric configuration. In: Wada, T., Huang, F., Lin, S. (eds.) PSIVT 2009. LNCS, vol. 5414, pp. 215–226. Springer, Heidelberg (2009). https://doi.org/10.1007/978-3-540-92957-4_19
15. Kneip, L., Siegwart, R., Pollefeys, M.: Finding the exact rotation between two images independently of the translation. In: Fitzgibbon, A., Lazebnik, S., Perona, P., Sato, Y., Schmid, C. (eds.) ECCV 2012. LNCS, vol. 7577, pp. 696–709. Springer, Heidelberg (2012). https://doi.org/10.1007/978-3-642-33783-3_50

16. Kukelova, Z., Bujnak, M., Pajdla, T.: Automatic generator of minimal problem solvers. In: Forsyth, D., Torr, P., Zisserman, A. (eds.) ECCV 2008. LNCS, vol. 5304, pp. 302–315. Springer, Heidelberg (2008). https://doi.org/10.1007/978-3-540-88690-7_23
17. Kukelova, Z., Bujnak, M., Pajdla, T.: Polynomial eigenvalue solutions to the 5-PT and 6-PT relative pose problems. In: British Machine Vision Conference, vol. 2 (2008)
18. Larsson, V., Åström, K.: Uncovering symmetries in polynomial systems. In: Leibe, B., Matas, J., Sebe, N., Welling, M. (eds.) ECCV 2016. LNCS, vol. 9907, pp. 252–267. Springer, Cham (2016). https://doi.org/10.1007/978-3-319-46487-9_16
19. Larsson, V., Astrom, K., Oskarsson, M.: Efficient solvers for minimal problems by syzygy-based reduction. In: Proceedings of the IEEE Conference on Computer Vision and Pattern Recognition, pp. 820–829 (2017)
20. Lee, G., Pollefeys, M., Fraundorfer, F.: Relative pose estimation for a multi-camera system with known vertical direction. In: Proceedings of the IEEE Conference on Computer Vision and Pattern Recognition, pp. 540–547. IEEE (2014)
21. Lee, G.H., Faundorfer, F., Pollefeys, M.: Motion estimation for self-driving cars with a generalized camera. In: Proceedings of the IEEE Conference on Computer Vision and Pattern Recognition, pp. 2746–2753 (2013)
22. Lee, G.H., Fraundorfer, F., Pollefeys, M.: Structureless pose-graph loop-closure with a multi-camera system on a self-driving car. In: 2013 IEEE/RSJ International Conference on Intelligent Robots and Systems, pp. 564–571. IEEE (2013)
23. Li, B., Heng, L., Lee, G., Pollefeys, M.: A 4-point algorithm for relative pose estimation of a calibrated camera with a known relative rotation angle. In: IEEE/RSJ International Conference on Intelligent Robots and Systems, pp. 1595–1601. IEEE (2013)
24. Li, B., Larsson, V.: Gaps: Generator for automatic polynomial solvers. arXiv preprint arXiv:2004.11765 (2020)
25. Longuet-Higgins, H.C.: A computer algorithm for reconstructing a scene from two projections. Nature **293**(5828), 133 (1981)
26. Maddern, W., Pascoe, G., Linegar, C., Newman, P.: 1 Year, 1000km: The Oxford RobotCar Dataset. Int. J. Robot. Res.(IJRR) **36**(1), 3–15 (2017). https://doi.org/10.1177/0278364916679498
27. Martyushev, E.: Self-calibration of cameras with euclidean image plane in case of two views and known relative rotation angle. In: Ferrari, V., Hebert, M., Sminchisescu, C., Weiss, Y. (eds.) ECCV 2018. LNCS, vol. 11208, pp. 435–449. Springer, Cham (2018). https://doi.org/10.1007/978-3-030-01225-0_26
28. Martyushev, E., Li, B.: Efficient relative pose estimation for cameras and generalized cameras in case of known relative rotation angle. J. Math. Imaging Vision **62**(8), 1076–1086 (2020). https://doi.org/10.1007/s10851-020-00958-5
29. Maybank, S.: Theory of reconstruction from image motion, vol. 28. Springer, Heidelberg (2012). https://doi.org/10.1007/978-3-642-77557-4
30. Nistér, D.: An efficient solution to the five-point relative pose problem. IEEE Trans. Pattern Anal. Mach. Intell. **26**(6), 756–770 (2004)
31. Pandey, G., McBride, J.R., Eustice, R.M.: Ford campus vision and lidar data set. Int. J. Robot. Res. **30**(13), 1543–1552 (2011)
32. Pless, R.: Using many cameras as one. In: CVPR (2). pp. 587–593. IEEE Computer Society (2003)
33. Ramirez-paredes, J.P., Doucette, E.A., Curtis, J.W., Gans, N.R.: QuEst : a quaternion-based approach for camera. IEEE Robot. Automat. Lett. **3**(2), 857–864 (2018)

34. Ruiz-Sarmiento, J.R., Galindo, C., González-Jiménez, J.: Robot@home, a robotic dataset for semantic mapping of home environments. International Journal of Robotics Research (2017)
35. Saurer, O., Vasseur, P., Boutteau, R., Demonceaux, C., Pollefeys, M., Fraundorfer, F.: Homography based egomotion estimation with a common direction. IEEE Trans. Pattern Anal. Mach. Intell. **39**(2), 327–341 (2017). https://doi.org/10.1109/TPAMI.2016.2545663
36. Scaramuzza, D.: 1-point-ransac structure from motion for vehicle-mounted cameras by exploiting non-holonomic constraints. Int. J. Comput. Vision **95**(1), 74–85 (2011)
37. Scaramuzza, D., Fraundorfer, F.: Tutorial: visual odometry. IEEE Robot. Autom. Mag. **18**, 80–92 (2011)
38. Shiu, Y.C., Ahmad, S.: Calibration of wrist-mounted robotic sensors by solving homogeneous transform equations of the form ax= xb. IEEE Trans. Robot. Autom. **5**(1), 16–29 (1989)
39. Stewénius, H., Engels, C., Nistér, D.: Recent developments on direct relative orientation. ISPRS J. Photogramm. Remote Sensing **60**(4), 284–294 (2006)
40. Stewénius, H., Nistér, D., Kahl, F., Schaffalitzky, F.: A minimal solution for relative pose with unknown focal length. Image Vis. Comput. **26**(7), 871–877 (2008)
41. Stewénius, H., Nistér, D., Oskarsson, M., Åström, K.: Solutions to minimal generalized relative pose problems. In: Workshop on Omnidirectional Vision (ICCV) (2005)
42. Sturm, J., Engelhard, N., Endres, F., Burgard, W., Cremers, D.: A benchmark for the evaluation of RGB-D slam systems. In: Proceedings of the International Conference on Intelligent Robot Systems (IROS) (2012)
43. Tsai, R.Y., Lenz, R.K.: Real time versatile robotics hand/eye calibration using 3d machine vision. In: Proceedings. 1988 IEEE International Conference on Robotics and Automation, pp. 554–561. IEEE (1988)

Sequential Convolution and Runge-Kutta Residual Architecture for Image Compressed Sensing

Runkai Zheng[1], Yinqi Zhang[1], Daolang Huang[1], and Qingliang Chen[1,2,3](✉)

[1] Department of Computer Science, Jinan University, Guangzhou 510632, China
tpchen@jnu.edu.cn
[2] Guangzhou Xuanyuan Research Institute Company, Ltd., Guangzhou 510006, China
[3] Guangdong E-Tong Software Co., Ltd., Guangzhou 510520, China

Abstract. In recent years, Deep Neural Networks (DNN) have empowered Compressed Sensing (CS) substantially and have achieved high reconstruction quality and speed far exceeding traditional CS methods. However, there are still lots of issues to be further explored before it can be practical enough. There are mainly two challenging problems in CS, one is to achieve efficient data sampling, and the other is to reconstruct images with high-quality. To address the two challenges, this paper proposes a novel Runge-Kutta Convolutional Compressed Sensing Network (RK-CCSNet). In the sensing stage, RK-CCSNet applies Sequential Convolutional Module (SCM) to gradually compact measurements through a series of convolution filters. In the reconstruction stage, RK-CCSNet establishes a novel Learned Runge-Kutta Block (LRKB) based on the famous Runge-Kutta methods, reformulating the process of image reconstruction as a discrete dynamical system. Finally, the implementation of RK-CCSNet achieves state-of-the-art performance on influential benchmarks with respect to prestigious baselines, and all the codes are available at https://github.com/rkteddy/RK-CCSNet.

Keywords: Compressed sensing · Convolutional sensing · Runge-Kutta methods

1 Introduction

Compressed Sensing (CS) [5] is a prominent technique that combines sensing and compression together at the hardware level, and can ensure high-fidelity

This research is supported by National Natural Science Foundation of China grant No. 61772232.

Electronic supplementary material The online version of this chapter (https://doi.org/10.1007/978-3-030-58545-7_14) contains supplementary material, which is available to authorized users.

© Springer Nature Switzerland AG 2020
A. Vedaldi et al. (Eds.): ECCV 2020, LNCS 12354, pp. 232–248, 2020.
https://doi.org/10.1007/978-3-030-58545-7_14

Fig. 1. This figure shows the test results of the proposed RK-CCSNet on BSDS100 [1] in different sampling ratios. Note that almost perfect visual effect is achieved when the sampling ratio is 6.2500%, which implies that our model is capable of reconstructing high-quality images even in low sampling ratios

reconstruction from limited observations received. In the CS framework, signals are acquired by linear projection, which is proved to have the ability to preserve most of the features in a few measurements if the sensing matrices satisfy the *Restricted Isometry Property (RIP)* [3]. Compared with Nyquist's theory, this method uses the sparse nature of the signal to restore the almost perfect original one from a much smaller number of measurements, leading to large reduction in the cost of sensing, storing and transmitting. Several applications such as Single Pixel Camera (SPC) [7], Hyperspectral Compressive Imaging (HCI) [2], Compressive Spectral Imaging System [9], High-Speed Video Camera [13], and CS Magnetic Resonance Imaging (MRI) system [21] have been introduced and implemented. Taking SPC as an example, it uses only a number of single-pixel signals in each shot, and merely a few shots are integrated to reconstruct the original image in the receiving end. Therefore, before decompressing images, the amount of signals needed is much smaller and thus is conducive to long-distance transmission (Fig. 1).

Over the years, a great deal of CS algorithms have been proposed such as Orthogonal Matching Pursuit (OMP) [31], Basis Pursuit (BP) [4] and Total Variance minimization by Augmented Lagrangian and ALternating direction ALgorithms (TVAL3) [19]. For instance, Zhang et al. [34] proposed a Group Sparse Representation (GSR) method to enhance both image sparseness and non-local self-similarity. But the common weaknesses of them are that they all demand high computational overhead and perform poorly at low sampling ratios (especially when the measurement is lower than 10%). With the rapid development of deep learning, researchers were inspired to use new end-to-end models to develop algorithms in CS, called Deep Compressed Sensing (DCS). These algorithms do not use the prior knowledge of any signal, but are fed a large number of training data for neural networks instead. The linear sensing module and reconstruction module form an Auto-Encoder structure [25]. Through end-to-end training, both sensing module and reconstruction module can be jointly optimized. Pure data-driven optimization learns how to make the best of the data structure to speed up the reconstruction process.

There are two challenges in DCS: the linear encoding and the non-linear reconstruction, respectively. For the former one, traditional CS algorithms usually apply hand-designed models according to the nature of the data. However, general DCS models treat the sparse transformation with a fully connected layer, which contains no priors and thus is hard to learn a concise embedding. Since convolution can be an efficient prior that can well describe the structural features of images, and can be easily combined with DCS, we replace the fully connected layer with continuous convolutions (for linear observations, there are no activation after every convolutional layer), named Sequential Convolutional Module (SCM). For the latter one, the main approaches are to develop a powerful reconstruction module with elaborate structures. According to the recent studies on the relationship between ODE [26] and ResNet [11], the conventional residual architecture with simple skip connections can be seen as an approximation of the forward Euler method [26], a simple numerical method. Accordingly, we introduce a novel architecture called Learned Runge-Kutta Block (LRKB) originating from Runge-Kutta methods [26], the higher-order numerical schemes than the forward Euler method.

The main contributions of the paper are summarized as follows:

1. We propose a SCM for image CS, which applies local connectivity priors during the sensing stage. SCM is empirically proved to have the ability to preserve spatial features and thus avoid block artifacts and high frequency noise in the final reconstruction.
2. We further develop a novel LRKB to achieve higher reconstruction quality, by reformulating the process of image reconstruction as a discrete dynamical system. Hence we can adopt highly efficient algorithms from ODE such as Runge-Kutta methods [26], which can offer higher order of accuracy for numerical solutions.
3. An end-to-end Runge-Kutta Convolutional Compressed Sensing Network (RK-CCSNet) is introduced to encapsulate the two modules above, resulting in a novel end-to-end structure. And the implementation of RK-CCSNet are extensively evaluated on influential benchmarks, achieving state-of-the-art performance with respect to prestigious baselines.

The paper is structured as follows. We will present preliminaries in the next section, followed by the section to detail the proposed RK-CCSNet, and then comes the section for empirical and comparative studies on different benchmarks compared with influential models. And we will conclude the paper in the last section.

2 Preliminaries

2.1 Compressed Sensing

CS [5] is a signal acquisition and manipulation paradigm consisting of sensing and compressing simultaneously, which leads to significant reduction in computational cost. Given a high-dimensional signal $\mathbf{x} \in \mathbb{R}^N$, the compressive measurement $\mathbf{y} \in \mathbb{R}^M$ about $\mathbf{x}$ can be obtained by $\mathbf{y} = \mathbf{\Phi x}$, where $\mathbf{\Phi} \in \mathbb{R}^{M \times N}$ $(M \ll N)$ denotes the sensing matrix. The aim of CS is to reconstruct the original signal $\mathbf{x}$ from a much lower dimensional measurement $\mathbf{y}$.

2.2 Data-Driven Methods for Image Compressed Sensing

Inspired by the great success of DNN in representation learning, Mousavi et al. [24] designed a new measurement and signal reconstruction framework. Stacked Denoising Autoencoders (SDA) is used as an self-supervised feature learner in the reconstruction network to obtain the statistical correlation between different elements of signals and improve the performance of signal reconstruction. And Kulkarni et al. introduced ReconNet [17], which takes image reconstruction as a task similar to super-resolution, with Convolutional Neural Networks (CNN) to carry out pixel-wise mapping. Later, Mousavi et al. [22] argued that the real-world data is not completely sparse on a fixed basis, and moreover the traditional reconstruction algorithms take a lot of time to converge. And they proposed DeepInverse that utilizes Fully Convolutional Networks (FCN) [27] to recover the original image, which is able to learn a structured representation from training data. And Yao et al. [33] presented DR2Net that applied residual architecture to further improve the reconstruction quality. And Xu et al. [32] used multiple stages of reconstructive adversarial networks through Laplacian pyramid architecture to achieve high-quality image reconstruction. And Shi et al. put forward CSNet [30] and CSNet+ [29] to further improve the reconstruction quality. Most recently, Shi et al. [28] tried to solve the problem that different models should be trained in different sampling ratios by introducing a Scalable Convolutional Neural Network (SCSNet). Parallel convolutions were applied in sensing stage [23] to avoid block-based sensing for better adaptability to different signals like Fourier signals. However, these methods do not make any assumptions about the data (i.e., the natural images), which is very essential to obtain low dimensional embeddings for a specific type of data. And recently it was proposed to use convolution as measurement matrix in [6], in which there is only one convolutional layer, not enough to capture the hierarchical structures.

2.3 Residual Neural Network

The Residual Neural Network (ResNet) was first presented in [11], which introduced the *identity skip connection* that allows data to flow directly to subsequent layers, bypassing residual layers. Generally, a residual block can be written as: $y_{n+1} = y_n + F(y_n)$. Skip connection brings shortcut into neural networks, which propagates the gradients in a more efficient way, making it possible to build a much deeper neural network without gradient vanishing, and thus can obtain impressive performance in many image tasks. ResNet and its variants [15] have been widely used in different applications besides computer vision.

2.4 ResNet and ODEs

Taking x as the time variable, a first-order dynamical system has the form [18]: $y'(x) = F(x, y(x))$ and $y'(x) = y' = \frac{dy}{dx}$ where y is a dependent variable of the changing system state. This ODE describes the process of a system change, in which the rate of change is a function of current time x and system state y. When the initial value satisfies: $y(x_0) = y_0$, this is called the Initial Value Problem (IVP) [18]. Euler method [18] is a first-order numerical method for IVP, including forward Euler method, backward Euler method and improved Euler method. Forward Euler method approximates the system change by truncating Taylor series and integral as: $y_{n+1} = y_n + hF(x_n, y_n)$ and $h = x_{n+1} - x_n$, which has the similar form to a basic block of ResNet. Over the past few years, this link between residual connection and ODEs has been widely discussed by some literature [8,20]. It leads to a novel perspective that the neural network can be reformulated as a discrete sequence of a time-dependent dynamical system, providing good theoretical guidance for the design of neural network architectures. And conventional residual architectures have been used in many DCS models and have gained substantial effects [28,29,33].

As forward Euler method is just the first-order numerical solution of ODEs, we can naturally think of building a more accurate neural network with higher-order numerical approaches such as Runge-Kutta methods [26]. This motivates us to build a residual architecture with LRKB, to achieve higher precision for image reconstruction.

3 The Proposed Model

3.1 Sequential Convolutional Module

Conventional sensing modules consist of a single fully connected layer to replace the sensing matrix which projects the original image into a measurement of much

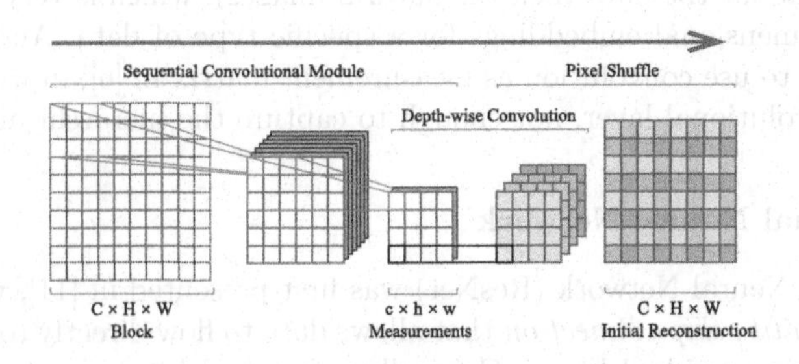

Fig. 2. Sequential Convolutional Module (SCM)

lower dimension linearly. Here, instead of standard sensing strategies, we propose the SCM, which is also a valid linear operation for CS because convolution can be represented by matrix-matrix multiplication. For a given single channel image $\mathcal{I} \in \mathbb{R}^{1\times H\times W}$, the convolution operations squeeze the image into the shape of $c^2 \times \frac{H}{cr} \times \frac{W}{cr}$, where r^2 is the compression ratio and c^2 is the hyperparameter, both of which depend on the configuration of convolution filters. Then a depth-wise convolution layer expanding the feature channels follows and the shape becomes $c^2r^2 \times \frac{H}{cr} \times \frac{W}{cr}$. Finally, the pixel-shuffle layer will rearrange the elements of $c^2r^2 \times \frac{H}{cr} \times \frac{W}{cr}$ tensor to form a $1 \times H \times W$ tensor, illustrated in Fig. 2.

SCM senses the original image by gradually compacting the image size through a sequence of filters. Compared with conventional sensing strategies, which sense the image block by block through a single shared weight matrix multiplication, our method has the advantage to preserve the spatial features thanks to the sparse local connectivity nature of convolution operations. Moreover, continuous convolution can effectively capture the hierarchical structures in the image. And it can be seen in the following section for experimental studies that SCM is justified to have the ability to eliminate noises introduced by long distance high-frequency component in the block and avoid block artifacts.

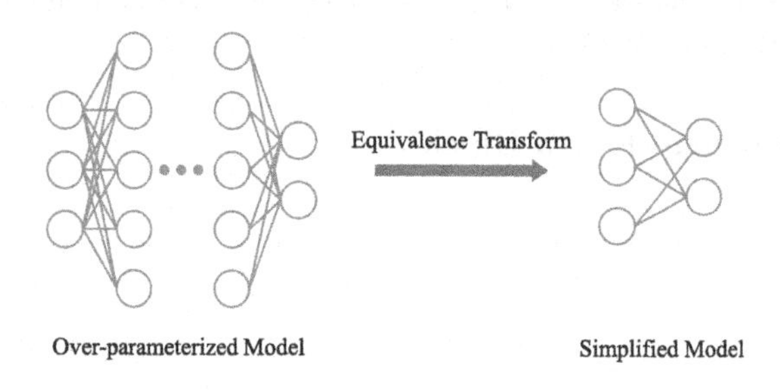

Fig. 3. Simplifying linear over-parameterized model.

The number of feature channels of intermediate layers of SCM during training can be relatively large, as long as the final output shape can meet the required measurements. Since there are no activation functions and no biases, no matter how wide the SCM is, these linear combinations can be finally squeezed into one matrix multiplication as shown in Fig. 3.

To be more specific, we take the feed forward network as an example. Assume that the network input is an n_0-elements vector x, the lth layer contains n_l hidden cells and the ith hidden cell in l^{th} layer is denoted as $h_{l,i}$, the weight in l^{th} layer connecting $h_{l-1,i}$ and $h_{l,j}$ is $w_{l,i,j}$. Then the feed forward network can be modeled by $h_{l,i} = \sum_{j=0}^{n_l} w_{l,i,j}h_{l,j}$.

Every subsequent hidden cell can be represented as a linear combination of x as follows:

$$
\begin{aligned}
h_{l,i} &= \sum_{j_l=0}^{n_l} w_{l,i,j} \sum_{j_{l-1}=0}^{n_{l-1}} w_{l-1,i,j} \cdots \sum_{j_0=0}^{n_0} w_{0,i,j} x_j \\
&= \sum_{j_l=0}^{n_l} \sum_{j_{l-1}=0}^{n_{l-1}} \cdots \sum_{j_0=0}^{n_0} w_{0,i,j} w_{1,i,j} \cdots w_{l,i,j} x_j \\
&= \sum_{j=0}^{n_0} W_{i,j} x_j .
\end{aligned}
$$

This indicates that the final output of the network y can also be represented as a linear combination of the input x. Thus we can utilize the learning ability of an over-parameterized model to converge to a better optimal point. However, wider model is more likely to cause unstable gradient problems during training. So it is a trade-off to choose a proper width.

Note that SCM still pertains to block-based sensing, but it applies local connectivity priors during training time. It is the same in deployment as block-based methods since all the convolution kernels can be transformed into one matrix during test time. So, SCM just changes the training behavior, leading to better performance on natural images.

3.2 Learned Runge-Kutta Block

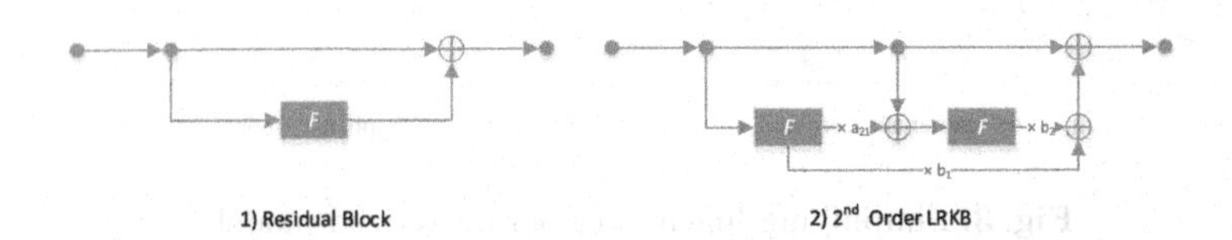

Fig. 4. Comparison of a residual block and a learned Runge-Kutta block.

The gradual reconstruction of the image can be reformulated as a dynamical system, where the initial condition is the measurements and the ideal termination condition is the original image. In such a dynamical system, each CNN block is a state transition, training data are fed to learn the mapping from low dimension measurements to the original image. Thus, we are able to see the residual block with a single skip connection as a forward Euler method [26], which is just a first-order scheme. So by mimicking higher order numerical methods, we can expect higher accuracy. Hence, we consider Runge-Kutta methods, which is a family of high-precision single step algorithms for numerical solution of ODE, to build a novel residual architecture with better performance.

Specifically, second order Runge-Kutta method takes the following form:

$$
y_{n+1} = y_n + b_1 K_1 + b_2 K_2, \tag{1}
$$

$$K_1 = hF(x_n, y_n), \tag{2}$$

$$K_2 = hF(x_n + c_2 h, y_n + a_{21} K_1), \tag{3}$$

where a_{21}, b_1, b_2 and c_2 are the coefficients. To specify the exact values of the coefficients, we expand K_2 at (x_n, y_n) according to Taylor's formula:

$$\begin{aligned} & hF(x_n + c_2 h, y_n + a_{21} K_1) \\ = \;& h\left[F(x_n, y_n) + c_2 h F'_x + a_{21} K_1 F'_y + O(h^2)\right] \\ = \;& h\left[F(x_n, y_n) + c_2 h F'_x + a_{21} h F F'_y + O(h^3)\right], \end{aligned} \tag{4}$$

where F denotes $F(x_n, y_n)$ and F'_x, F'_y are the partial derivatives of F with respect to x and y, respectively. Then we get:

$$\begin{aligned} y_{n+1} &= y_n + (b_1 K_1 + b_2 K_2) \\ &= y(x_n) + b_1 h F(x_n, y_n) + b_2 h\left[F(x_n, y_n) + c_2 h F'_x + a_{21} h F F'_y\right] + O(h^3) \\ &= y(x_n) + (b_1 + b_2) h F(x_n, y_n) + c_2 b_2 h^2 F'_x + a_{21} b_2 h^2 F F'_y + O(h^3). \end{aligned} \tag{5}$$

And we expand $y(x_{n+1})$ at x_n:

$$\begin{aligned} y(x_{n+1}) &= y(x_n) + h y'(x_n) + \frac{h^2}{2!} y''(x_n) + O(h^3) \\ &= y(x_n) + h F(x_n, y_n) + \frac{h^2}{2!}\left[F'_x + F F'_y\right] + O(h^3). \end{aligned} \tag{6}$$

Let $y(x_{n+1}) = y_{n+1}$, we get:

$$b_2 + b_1 = 1, \; b_2 c_2 = \frac{1}{2}, \; b_2 a_{21} = \frac{1}{2}, \tag{7}$$

which is an under-determined system of equation, all methods satisfying the above forms are collectively referred to as Second-Order Runge-Kutta Method. As we can see, b_2 can be the only free variable, and can be jointly optimized during training time.

Actually the neural network can be trained to predict the auxiliary variable of $\alpha = \log(-\log(b_2))$, to avoid division by zero. Also, when regressing the unconstrained value of α, b_2 is resolved to the value between 0 and 1. Hence we can have: $b_2 = e^{-e^{\alpha}}$, $b_1 = 1 - e^{-e^{\alpha}}$, and $a_{21} = \frac{e^{e^{\alpha}}}{2}$.

Regarding each non-linear state transition function F as an independent CNN block, we build a residual block as shown in Fig. 4, where the state transition functions F is illustrated in Fig. 5. Moreover, we use PReLU [10] as the activation function and adopt pre-activation structure [12], where the two convolution filters share the same weights.

3.3 The Overall Structure

The overall structure of our model is an end-to-end auto-encoder structure as shown in Fig. 6, where the encoder is a sequence of sub-sampling convolutional

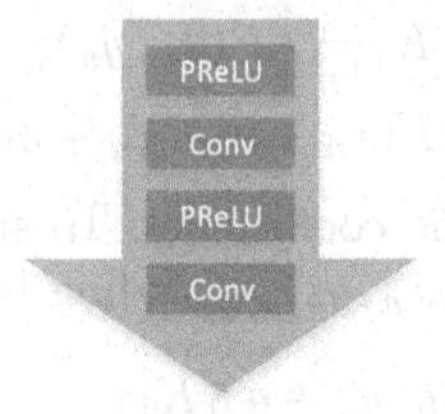

Fig. 5. The state transition function F.

sensing filters without activation functions, producing measurements. A followed depth-wise convolution layer expands the feature channels and the resulting feature map is to be rearranged to match the original size by a pixel shuffle layer, whose product is called initial reconstruction. Then the output of encoder is to be fed to the subsequent reconstruction network consisting of a head, body and tail. The head first converts the initial reconstruction to image features by convolution block, followed by a ReLU function. Afterwards the feature maps are further processed by the body consisting of several LRKBs. Then the tail will turn the resulting feature maps back to the final reconstructed image.

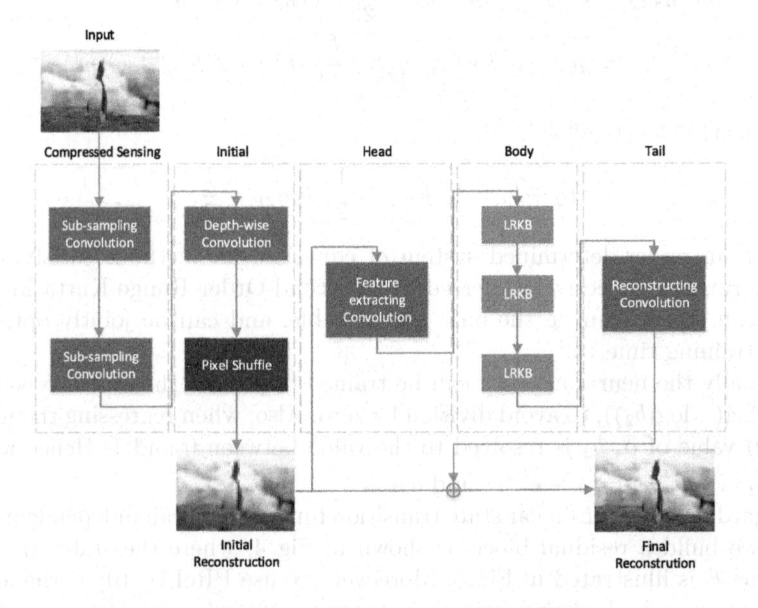

Fig. 6. The overall structure of RK-CCSNet.

4 Experimental Studies

4.1 Weights Initialization

Because of the introduction of sequential convolution filters without activation functions, it was observed that the gradient is unstable during training process. By comparative empirical studies, we have identified the source of the problem in weights initialization.

To be more specific, in each convolution step, we define $\mathcal{X} \in \mathbb{R}^{C\times H\times W}$ to be the input matrix convoluted with a filter $\mathcal{F} \in \mathbb{R}^{O\times C\times H_f\times W_f}$, and $\mathcal{Y} \in \mathbb{R}^{O\times H\times W}$ to be the output matrix, then each element $\mathcal{Y}_{ij}$ in the output matrix $\mathcal{Y}$ is defined as:

$$\mathcal{Y}_{k,i,j} = \sum_{n=0}^{C}\sum_{h=0}^{H_f}\sum_{w=0}^{W_f} \mathcal{X}_{n,i+h-\frac{(H_f-1)}{2},j+w-\frac{(W_f-1)}{2}} \mathcal{F}_{k,n,h,w} \tag{8}$$

where we simply assume that stride equals 1 with the same padding strategy, and the height and width of the filter to be odd number, which can be extended to more general situations. As we can see, each output element is the summation of CH_fW_f products of $\mathcal{X}$ and $\mathcal{F}$. We assume that $\mathcal{X}$ is normalized such that $\mathcal{X} \sim N(0,1)$ and we initialize the weight matrix of $\mathcal{F}$ with normal distribution without considering the shape of filter, let's say $\mathcal{F} \sim N(0,1)$, then after convolution we will get $\mathcal{Y} \sim N(0, \sqrt{CH_fW_f})$. If CH_fW_f is greater than 1 (which is surely the case), after a sequence of convolution steps, the elements in the resulting matrix will grow dramatically, leading to gradient explosion. The similar situation will cause gradient vanishing when we initialize the weights with normal distribution of which standard deviation is too small. To address this problem, we simply initialize the weight matrix with scaled normal distribution as:

$$\mathcal{F} \sim N(0, \frac{1}{\sqrt{CH_fW_f}}). \tag{9}$$

To illustrate the effect of our initialization method, we build a toy example which took a tensor of shape $(8, 64, 96, 96)$ as the input, and is sequentially convoluted by 10 filters of shape $(64, 64, 3, 3)$ with stride of 1 and some padding strategy to keep the shape of the input tensor and the output tensor remain unchanged. We initialize the weights with three different distribution: $\mathcal{F}_1 \sim N(0,1)$, $\mathcal{F}_2 \sim N(0, 0.01)$ and $\mathcal{F}_3 \sim N(0, \frac{1}{\sqrt{CH_fW_f}})$. And Fig. 7 shows the changing standard deviation of the output tensor in each convolution stage.

The figure clearly shows that the general weights initialization method is not suitable for continuous convolution operations without activation functions, which will lead to either gradient explosion or gradient vanishing. And this example verifies that the scaled version of $\mathcal{F}$ remains very stable and thus can have better performance and generalization.

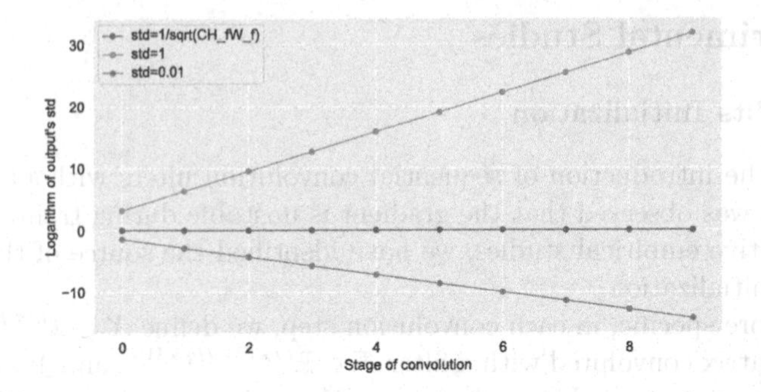

Fig. 7. The comparison of weights initialization with different standard deviations.

4.2 Datasets and Implementation Details

To compare with state-of-the-art deep learning based models, we trained the models on the training set and test set of BSDS500[1], with 400 images for training and 100 images for testing (BSDS100). As the original images are either 321×481 or 481×321, we randomly crop the images into patches of 96×96 and randomly flip horizontally for data augmentation. In addition, we also compare our method with TVAL3 [19] and GSR [34] on Set5 and Set14[2], which contains 5 images and 14 images, respectively. Because those images are not shape consistent, we resize them into (256, 256) for evaluation. All the images are first converted to YCbCr color space and only the Y channel is used as the input of all the models. We use Adam optimizer [16] for training and set the exponential decay rates to 0.9 and 0.999 for the first and second moment estimate. The batch size is set to 4 and both CSNet+ [29] and RK-CCSNet were trained for 200 epochs at all, with the initial learning rate of $1e-3$ and decay of 0.25 at 60, 90, 120, 150 and 180 epochs respectively. The sampling ratio for testing was set from 1/64 to 1/2, i.e., 1.5625%, 3.1250%, 6.2500%, 12.5000%, 25.0000%, and 50.0000%. PSNR (Peak Signal-to-Noise Ratio) and SSIM (Structural SIMilarity) [14] are chosen as the evaluation metrics throughout our experiments.

4.3 Experimental Results

Table 1 presents the test results of CSNet+ [29] and RK-CCSNet on BSDS100 with the corresponding PSNR and SSIM, and the best results are marked in bold font. It can be seen that our model exhibits significantly better performance compared with CSNet+ across all sampling ratios. In average, our model gains 2.14% and 1.72% improvements in PSNR and SSIM, respectively.

Further experimental results of our model on Set5 and Set14 compared with TVAL3, GSR and CSNet+ are provided in Table 2. Our model outperforms all

---

[1] https://www2.eecs.berkeley.edu/Research/Projects/CS/vision/grouping/resources.html#bsds500.

[2] http://vllab.ucmerced.edu/wlai24/LapSRN/.

Table 1. Comparisons of CSNet+ and RK-CCSNet on BSDS100

		CSNet+		RK-CCSNet (our)	
Data	Ratio	PSNR	SSIM	PSNR	SSIM
BSDS100	1.56%	25.01	0.6904	**25.56**	**0.7055**
	3.12%	26.55	0.7413	**26.99**	**0.7564**
	6.25%	28.14	0.7977	**28.60**	**0.8133**
	12.5%	30.11	0.8602	**30.56**	**0.8759**
	25.0%	32.81	0.9206	**33.43**	**0.9335**
	50.0%	36.62	0.9659	**37.92**	**0.9766**
Average		29.87	0.8294	**30.51**	**0.8437**

Table 2. Comparisons of different CS algorithms on Set5 and Set14

		TVAL3		GSR		CSNet+		RK-CCSNet (our)	
Data	Ratio	PSNR	SSIM	PSNR	SSIM	PSNR	SSIM	PSNR	SSIM
Set5	1.5625%	19.00	0.4844	21.39	0.5815	25.02	0.6888	**25.63**	**0.7186**
	3.1250%	19.89	0.5415	23.70	0.6822	27.42	0.7778	**28.03**	**0.8142**
	6.2500%	22.03	0.6175	27.59	0.8163	30.11	0.8605	**30.91**	**0.8867**
	12.5000%	23.75	0.7365	31.61	0.9016	33.57	0.9250	**35.05**	**0.9461**
	25.0000%	27.39	0.8522	36.32	0.9510	37.94	0.9665	**39.29**	**0.9758**
	50.0000%	33.11	0.9430	42.18	0.9908	42.70	0.9856	**44.72**	**0.9913**
Set14	1.5625%	16.79	0.3993	18.93	0.4399	23.13	0.5768	**23.32**	**0.5933**
	3.1250%	18.40	0.4514	20.26	0.5184	25.03	0.6660	**25.42**	**0.6968**
	6.2500%	19.65	0.5287	23.59	0.6526	27.25	0.7651	**27.48**	**0.7897**
	12.5000%	21.03	0.6379	28.08	0.7915	30.16	0.8630	**30.93**	**0.8880**
	25.0000%	22.69	0.7731	31.82	0.8939	33.92	0.9354	**35.03**	**0.9505**
	50.0000%	26.61	0.9004	37.47	0.9619	38.67	0.9756	**40.66**	**0.9848**
Average		22.53	0.6555	28.57	0.7650	31.24	0.8322	**32.21**	**0.8530**

the other ones across all different datasets, exhibiting excellent generalization and achieving state-of-the-art results. We selected some representative images to demonstrate visual comparisons of each model, in Fig. 8 and 9. It can be seen that neither TVAL3 nor GSR can reconstruct meaningful features from sample images in extremely low ratios. CSNet+ can roughly restore the original image but results in serious blocking artifacts, while RK-CCSNet produces much smoother boundary between blocks. Moreover, RK-CCSNet has less luminance loss compared with CSNet+. When the sampling ratio comes to 12.5%, these models all perform well. However, the one reconstructed by TVAL3 has a lot of noises. GSR does a bit better in visual but brings about distortions. CSNet's reconstruction performs poorly in details of the image. We also found that in the case of block by block sensing, if the block contains high-frequency components, the noise will be distributed across all parts of the reconstructed block, causing the whole reconstructed block less smooth. And this difference between blocks exacerbates blocking artifacts. However, RK-CCSNet with SCM as the sensing module, will not lead to this phenomenon, which is thus significantly better in the reconstruction of high frequency details of the image. All in all, RK-CCSNet has the highest reconstruction quality among all the models.

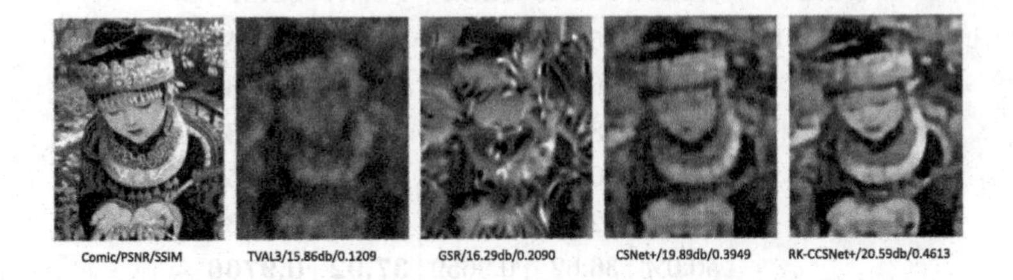

Fig. 8. Visual comparisons of the reconstructed image in sampling ratio of 1.5625%

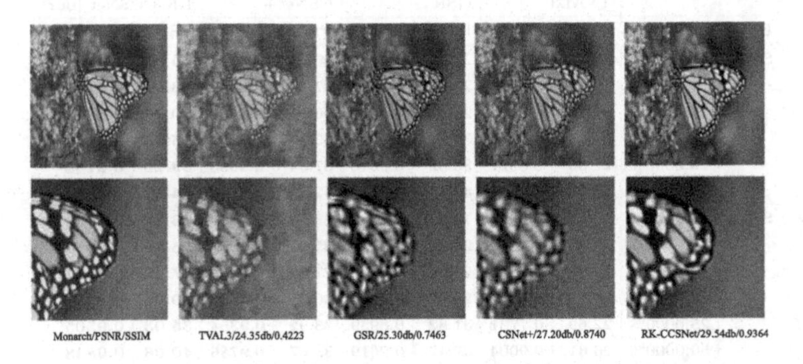

Fig. 9. Visual comparisons of the reconstructed image in sampling ratio of 12.5000%

4.4 Ablation Studies

Ablation studies are further carried out to justify the efficacy of the two modules proposed in our model. In general, we divide a CS model into two submodules: sensing module and reconstruction module. We replace different modules of CSNet+ to form different models. To be more specific, the models compared are listed as follows: Baseline (CSNet+), Baseline with SCM, Baseline with LRKB, and RK-CCSNet. The experimental results are shown in Table 3. It can be seen that both proposed modules can lead to appreciable improvements over the baseline model. LRKB has more non-linear reconstruction strength for the global structure of the image when the observation rate is limited, since LRKB is from ODE theory with higher order of accuracy for numerical analysis than the one in the baseline with a conventional residual architecture. SCM can restore more details and eliminate most noises when the observation rate is sufficient. And SCM's power of preserving spatial features grows with the increasing of sampling ratios, because the larger spatial shape of the measurement will contain more spatial information, while the standard fully connected layer cannot capture spatial features well. Moreover, SCM's local sensing strategy can also avoid introducing noises.

Table 3. Ablation results on BSDS100

		1.5625%		3.1250%		6.2500%		12.5000%		25.0000%		50.0000%	
SCM	LRKB	PSNR	SSIM	PSNR	SSIM	PSNR	SSIM	PSNR	SSIM	PSNR	SSIM	PSNR	SSIM
		25.01	0.6904	26.55	0.7413	28.14	0.7977	30.11	0.8602	32.81	0.9206	36.62	0.9659
✓		25.31	0.7014	26.64	0.7488	28.25	0.8075	30.26	0.8710	33.05	0.9294	37.59	0.9753
	✓	25.49	0.7010	26.91	0.7499	28.48	0.8055	30.11	0.8584	32.43	0.9138	36.91	0.9670
✓	✓	**25.56**	**0.7055**	**26.99**	**0.7564**	**28.60**	**0.8133**	**30.56**	**0.8759**	**33.43**	**0.9335**	**37.92**	**0.9766**

Fig. 10. Visual comparisons of reconstructed image in the sampling ratio of 12.5%

To further present the effect of SCM and LRKB, we compare the visual quality of the reconstructed image by three different models in Fig. 10. It can be seen that our proposed SCM can eliminate most noises inside the block caused by high frequency components as mentioned above, and alleviate luminance loss. Combined with LRKB's powerful reconstruction strength, our model can restore images to a higher level.

5 Conclusion

In this paper, we have proposed a sensing module and a reconstruction module respectively to enhance DCS frameworks. In the sensing stage, the proposed SCM applies continuous convolution operations to replace the conventional single matrix multiplication to preserve spatial features. In the reconstruction module of the proposed LRKB, we reformulate the forward process of ResNet as a discrete dynamical system and introduce a novel residual architecture inspired by Runge-Kutta methods, which can lead to much more precise reconstructions. Furthermore, we have introduced an end-to-end RK-CCSNet to encapsulate the two modules above. The implementation of RK-CCSNet has outperformed other prestigious baselines when extensively evaluated on influential benchmarks. In addition, ablation studies are also carried out that have justified the efficacy of the two modules individually.

References

1. Arbelaez, P., Maire, M., Fowlkes, C.C., Malik, J.: Contour detection and hierarchical image segmentation. IEEE Trans. Pattern Anal. Mach. Intell. **33**(5), 898–916 (2011)
2. August, Y., Vachman, C., Rivenson, Y., Stern, A.: Compressive hyperspectral imaging by random separable projections in both the spatial and the spectral domains. Appl. Opt. **52**(10), D46–D54 (2013)
3. Candès, E.J.: The restricted isometry property and its implications for compressed sensing. C.R. Math. **346**(9–10), 589–592 (2008)
4. Chen, S.S., Donoho, D.L., Saunders, M.A.: Atomic decomposition by basis pursuit. SIAM Rev. **43**(1), 129–159 (2001)
5. Donoho, D.L., et al.: Compressed sensing. IEEE Trans. Inf. Theory **52**(4), 1289–1306 (2006)
6. Du, J., Xie, X., Wang, C., Shi, G., Xu, X., Wang, Y.: Fully convolutional measurement network for compressive sensing image reconstruction. Neurocomputing **328**, 105–112 (2019)
7. Duarte, M.F., et al.: Single-pixel imaging via compressive sampling. IEEE Signal Process. Mag. **25**(2), 83–91 (2008)
8. Weinan, E.: A proposal on machine learning via dynamical systems. Commun. Math. Stat. **5**(1), 1–11 (2017). https://doi.org/10.1007/s40304-017-0103-z
9. Gehm, M., John, R., Brady, D., Willett, R., Schulz, T.: Single-shot compressive spectral imaging with a dual-disperser architecture. Opt. Express **15**(21), 14013–14027 (2007)
10. He, K., Zhang, X., Ren, S., Sun, J.: Delving deep into rectifiers: surpassing human-level performance on ImageNet classification. In: 2015 IEEE International Conference on Computer Vision, ICCV 2015, Santiago, Chile, 7–13 December 2015, pp. 1026–1034. IEEE Computer Society (2015)
11. He, K., Zhang, X., Ren, S., Sun, J.: Deep residual learning for image recognition. In: 2016 IEEE Conference on Computer Vision and Pattern Recognition, CVPR 2016, Las Vegas, NV, USA, 27–30 June 2016, pp. 770–778. IEEE Computer Society (2016)
12. He, K., Zhang, X., Ren, S., Sun, J.: Identity mappings in deep residual networks. In: Leibe, B., Matas, J., Sebe, N., Welling, M. (eds.) ECCV 2016. LNCS, vol. 9908, pp. 630–645. Springer, Cham (2016). https://doi.org/10.1007/978-3-319-46493-0_38
13. Hitomi, Y., Gu, J., Gupta, M., Mitsunaga, T., Nayar, S.K.: Video from a single coded exposure photograph using a learned over-complete dictionary. In: Metaxas, D.N., Quan, L., Sanfeliu, A., Gool, L.V. (eds.) IEEE International Conference on Computer Vision, ICCV 2011, Barcelona, Spain, 6–13 November 2011, pp. 287–294. IEEE Computer Society (2011)

14. Horé, A., Ziou, D.: Image quality metrics: PSNR vs. SSIM. In: 20th International Conference on Pattern Recognition, ICPR 2010, Istanbul, Turkey, 23–26 August 2010, pp. 2366–2369 (2010)
15. Huang, G., Liu, Z., van der Maaten, L., Weinberger, K.Q.: Densely connected convolutional networks. In: 2017 IEEE Conference on Computer Vision and Pattern Recognition, CVPR 2017, Honolulu, HI, USA, 21–26 July 2017, pp. 2261–2269. IEEE Computer Society (2017)
16. Kingma, D.P., Ba, J.: Adam: a method for stochastic optimization. In: 3rd International Conference on Learning Representations, ICLR 2015, San Diego, CA, USA, 7–9 May 2015, Conference Track Proceedings (2015)
17. Kulkarni, K., Lohit, S., Turaga, P.K., Kerviche, R., Ashok, A.: ReconNet: non-iterative reconstruction of images from compressively sensed measurements. In: 2016 IEEE Conference on Computer Vision and Pattern Recognition, CVPR 2016, Las Vegas, NV, USA, 27–30 June 2016, pp. 449–458. IEEE Computer Society (2016)
18. Lambert, J.D.: Numerical Methods for Ordinary Differential Systems: The Initial Value Problem. Wiley, New York (1991)
19. Li, C.: An efficient algorithm for total variation regularization with applications to the single pixel camera and compressive sensing. Master's thesis, Rice University (2010)
20. Lu, Y., Zhong, A., Li, Q., Dong, B.: Beyond finite layer neural networks: bridging deep architectures and numerical differential equations. In: Dy, J.G., Krause, A. (eds.) Proceedings of the 35th International Conference on Machine Learning, ICML 2018, Stockholmsmässan, Stockholm, Sweden, 10–15 July 2018. Proceedings of Machine Learning Research, vol. 80, pp. 3282–3291. PMLR (2018)
21. Lustig, M., Donoho, D.L., Santos, J.M., Pauly, J.M.: Compressed sensing MRI. IEEE Signal Process. Mag. **25**(2), 72 (2008)
22. Mousavi, A., Baraniuk, R.G.: Learning to invert: signal recovery via deep convolutional networks. In: 2017 IEEE International Conference on Acoustics, Speech and Signal Processing, ICASSP 2017, New Orleans, LA, USA, 5–9 March 2017, pp. 2272–2276. IEEE (2017)
23. Mousavi, A., Dasarathy, G., Baraniuk, R.G.: A data-driven and distributed approach to sparse signal representation and recovery. In: International Conference on Learning Representations (2018)
24. Mousavi, A., Patel, A.B., Baraniuk, R.G.: A deep learning approach to structured signal recovery. In: 53rd Annual Allerton Conference on Communication, Control, and Computing, Allerton 2015, Allerton Park & Retreat Center, Monticello, IL, USA, 29 September–2 October 2015, pp. 1336–1343. IEEE (2015)
25. Rumelhart, D.E., Hinton, G.E., Williams, R.J.: Learning internal representations by error propagation, pp. 318–362. MIT Press, Cambridge (1986)
26. Sauer, T.: Numerical Analysis, 2nd edn. Addison-Wesley Publishing Company, USA (2011)
27. Shelhamer, E., Long, J., Darrell, T.: Fully convolutional networks for semantic segmentation. IEEE Trans. Pattern Anal. Mach. Intell. **39**(4), 640–651 (2017)
28. Shi, W., Jiang, F., Liu, S., Zhao, D.: Scalable convolutional neural network for image compressed sensing. In: IEEE Conference on Computer Vision and Pattern Recognition, CVPR 2019, Long Beach, CA, USA, 16–20 June 2019, pp. 12290–12299. Computer Vision Foundation/IEEE (2019)
29. Shi, W., Jiang, F., Liu, S., Zhao, D.: Image compressed sensing using convolutional neural network. IEEE Trans. Image Process. **29**, 375–388 (2020)

30. Shi, W., Jiang, F., Zhang, S., Zhao, D.: Deep networks for compressed image sensing. In: 2017 IEEE International Conference on Multimedia and Expo, ICME 2017, Hong Kong, China, 10–14 July 2017, pp. 877–882. IEEE Computer Society (2017)
31. Tropp, J.A., Gilbert, A.C.: Signal recovery from random measurements via orthogonal matching pursuit. IEEE Trans. Inf. Theory **53**(12), 4655–4666 (2007)
32. Xu, K., Zhang, Z., Ren, F.: LAPRAN: a scalable laplacian pyramid reconstructive adversarial network for flexible compressive sensing reconstruction. In: Ferrari, V., Hebert, M., Sminchisescu, C., Weiss, Y. (eds.) ECCV 2018. LNCS, vol. 11214, pp. 491–507. Springer, Cham (2018). https://doi.org/10.1007/978-3-030-01249-6_30
33. Yao, H., Dai, F., Zhang, D., Ma, Y., Zhang, S., Zhang, Y.: DR2-net: deep residual reconstruction network for image compressive sensing. Neurocomputing **359**, 483–493 (2017)
34. Zhang, J., Zhao, D., Jiang, F., Gao, W.: Structural group sparse representation for image compressive sensing recovery. In: Bilgin, A., Marcellin, M.W., Serra-Sagristà, J., Storer, J.A. (eds.) 2013 Data Compression Conference, DCC 2013, Snowbird, UT, USA, 20–22 March 2013, pp. 331–340. IEEE (2013)

Deep Hough Transform for Semantic Line Detection

Qi Han, Kai Zhao, Jun Xu, and Ming-Ming Cheng(✉)

TKLNDST, CS, Nankai University, Tianjin, China
{qhan,kz}@mail.nankai.edu.cn, {csjunxu,cmm}@nankai.edu.cn

Abstract. In this paper, we put forward a simple yet effective method to detect meaningful straight lines, a.k.a. semantic lines, in given scenes. Prior methods take line detection as a special case of object detection, while neglect the inherent characteristics of lines, leading to less efficient and suboptimal results. We propose a one-shot end-to-end framework by incorporating the classical Hough transform into deeply learned representations. By parameterizing lines with slopes and biases, we perform Hough transform to translate deep representations to the parametric space and then directly detect lines in the parametric space. More concretely, we aggregate features along candidate lines on the feature map plane and then assign the aggregated features to corresponding locations in the parametric domain. Consequently, the problem of detecting semantic lines in the spatial domain is transformed to spotting individual points in the parametric domain, making the post-processing steps, *i.e.* non-maximal suppression, more efficient. Furthermore, our method makes it easy to extract contextual line features, that are critical to accurate line detection. Experimental results on a public dataset demonstrate the advantages of our method over state-of-the-arts. Codes are available at https://mmcheng.net/dhtline/.

Keywords: Straight line detection · Hough transform · CNN

1 Introduction

We investigate an interesting problem of detecting meaningful straight lines in natural scenes. This kind of line structure which outlines the conceptual structure of images is referred to as 'semantic line' in a recent study [29]. The organization of such line structure is an early yet important step in the transformation of the visual signal into useful intermediate concepts for visual interpretation [5]. As demonstrated in Fig. 1, semantic lines belong to a special kind of line structure,

Q. Han and K. Zhao—Equal contribution.

Electronic supplementary material The online version of this chapter (https://doi.org/10.1007/978-3-030-58545-7_15) contains supplementary material, which is available to authorized users.

© Springer Nature Switzerland AG 2020
A. Vedaldi et al. (Eds.): ECCV 2020, LNCS 12354, pp. 249–265, 2020.
https://doi.org/10.1007/978-3-030-58545-7_15

Fig. 1. Example pictures from [31] reveals that semantic lines may help in photographic composition. (a): a photo taken with arbitrary pose. (b): a photo fits the golden ratio principle [7,27] which abtained by the method described in [31] using so-called 'prominent lines' and salient objects [13,14,18] in the image. (c): Our detection result is clean and comprises only few meaningful lines that are potentially helpful in photographic composition. (d): Line detection result by the classical line detection algorithms often focus on fine detailed straight edges.

which outlines the global structure of the image. Identifying such semantic lines is of crucial importance for applications such as photographic composition [31] and artistic creation [27].

The research of detecting line structures (straight lines and line segments) dates back to the very early stage of computer vision. Originally described in [22], the Hough transform (HT) is invented to detect straight lines from bubble chamber photographs. The core idea of the Hough transform is translating the problem of pattern detection in point samples to detecting peaks in the parametric space. This idea is quickly extended [11] to the computer vision community for digital image analysis, and generalized by [3] to detect complex shapes. In the case of line detection, Hough transform collects line evidence from a given edge map and then votes the evidence into the parametric space, thus converting the global line detection problem into a peak response detection problem. Classical Hough transform based methods [16,26,35,44] usually detect continuous straight edges while neglecting the semantics in line structures. Moreover, these methods are very sensitive to light changes and occlusion. Consequently, the results are noisy [2] and often contain irrelevant lines, as shown in Fig. 1(d). In this paper, our mission is to only detect clean, meaningful and outstanding lines, as shown in Fig. 1(c), which is helpful to photographic composition.

Since the success of Convolutional Neural Networks (CNNs) in vast computer vision applications, several recent studies take line detection as a special case of object detection, and adopt existing CNN-based object detectors, *e.g.* faster R-CNN [38] and CornerNet [28], for line detection. Lee *et al.* [29] make several modifications to the faster R-CNN [38] framework for semantic line detection. In their method, proposals are in the form of lines instead of bounding boxes, and features are aggregated along straight lines instead of rectangular areas. Zhang *et al.* [46] adopt the idea from CornetNet, which identifies object locations by detecting a pair of key points, *e.g.* the top-left and bottom-right corners. [46] detects line segments by localizing two corresponding endpoints. Limited by the ROI pooling and non-maximal suppression of lines, both [29] and [46] are less efficient in terms of running time. Moreover, ROI pooling [19] aggregates features along a single line, while many recent studies reveal that richer context information is critical to many tasks [17,21], *e.g.* video classification [42], edge detection [33], salient object detection [4,15], and semantic segmentation [23]. In Table 3, we experimentally verify that only aggregating features along a single line leads to suboptimal results.

In this paper, we propose to incorporate CNNs with Hough transform for straight line detection in natural images. We firstly extract pixel-wise representations with a CNN-based encoder, and then perform Hough transform on the deep representations to convert representations from feature space into parametric space. Then the global line detection problem is converted into simply detecting peak response in the transformed features, making the problem simpler. For example, the time-consuming non-maximal suppression (NMS) is simply calculating the centroids of connected areas in the parametric space, making our method very efficient that can detect lines in real-time. Moreover, in the detection stage, we use several convolutional layers on top of the transformed features to aggregate context-aware features of nearby lines. Consequently, the final decision is made upon not only features of a single line, but also information of lines nearby.

In addition to the proposed method, we introduce a principled metric to assess the agreement of a detected line w.r.t its corresponding ground-truth line. Although [29] has proposed an evaluation metric that uses intersection areas to measure the similarity between a pair of lines, this measurement may lead to ambiguous and misleading results. The contributions are summarized below:

- We propose an end-to-end framework for incorporating the feature learning capacity of CNN with Hough transform, resulting in an efficient real-time solution for semantic line detection.
- We introduce a principled metric which measures the similarity between two lines. Compared with previous IoU based metric [29], our metric has straightforward interpretation without ambiguity in implementation, as detailed in Sect. 4.
- Evaluation results on an open benchmark demonstrate that our method outperforms prior arts with a significant margin.

2 Related Work

The research of line detection in digital images dates back to the very early stage of computer vision. Since the majority of line detection methods are based on the Hough transform [11], we first brief the Hough transform, and then summarize several early methods for line detection using Hough transform. Finally, we describe two recently proposed CNN-based methods for line/segments detection from natural images.

Hough Based Line Detectors. Hough transform (HT) is originally devised by Hough [22] to detect straight lines from bubble chamber photographs.

The algorithm is then extended [11] and generalized [3] to localize arbitrary shapes, *e.g.* ellipses and circles, from digital images. Traditional line detectors start by edge detection in an image, typically with the Canny [6] and Sobel [40] operators. Then the next step is to apply the Hough transform and finally detect lines by picking peak response in the transformed space. HT collects edge response alone a line and accumulates them to a single point in the parametric space.

There are many variants of Hough transform (HT) trying to remedy different shortcomings of the original algorithm. The original HT maps each image point to all points in the parameter space, resulting in a many-to-many voting scheme. Consequently, the original HT presents high computational cost, especially when dealing with large-size images. Kiryati *et al.* [26] try to accelerate HT by proposing the 'probabilistic Hough transform' that randomly picks sample points from a line. Princen *et al.* [35] and Yacoub and Jolion [44] partition the input image into hierarchical image patches, and then apply HT independently to these patches. Fernandes *et al.* [16] use an oriented elliptical-Gaussian kernel to cast votes for only a few lines in the parameter space. Illingworth *et al.* [24] use a 'coarse to fine' accumulation and search strategy to identify significant peaks in the Hough parametric spaces. [1] approaches line detection within a regularized framework, to suppress the effect of noise and clutter corresponding to image features which are not linear. It's worth noting that a clean input edge map is critical to these HT-based detectors.

Line Segments Detection. Despite its robustness and parallelism, Hough transform cannot directly be used for line segments detection because the outputs of HT are infinite long lines. In addition to Hough transform, many other studies have been developed to detect line segments. Burns *et al.* [5] use the edge orientation as the guide for line extraction. The main advantage is that the orientation of the gradients can help to discover low-contrast lines. Etemadi *et al.* [12] establish a chain from the given edge map, and then extract line segments and orientations by walking over these chains. Chan *et al.* [8] use quantized edge orientation to search and merge short line segments.

CNN-Based Line Detectors. There are two CNN-based line (segment) detectors that are closely related to our method. Lee *et al.* [29] regard line detection as a special case of object detection, and adopt the faster R-CNN [38] framework for line detection. Given an input image and predefined line proposals, they first extract spatial feature maps with an encoder network, and then extract line-wise feature vectors by uniformly sampling and pooling along line proposals on the feature maps. A classification network and a regression network are applied to the extracted feature vectors to identify positive lines and adjust the line positions. Zhang *et al.* [46] adopt the CornerNet [28] framework to extract line segments as a pair of key points. Both of the methods as mentioned above extract line-wise feature vectors by aggregating deep features solely along each line, leading to inadequate context information. Besides, there are many works [36,37] using the conception of Hough voting in 3D object detection.

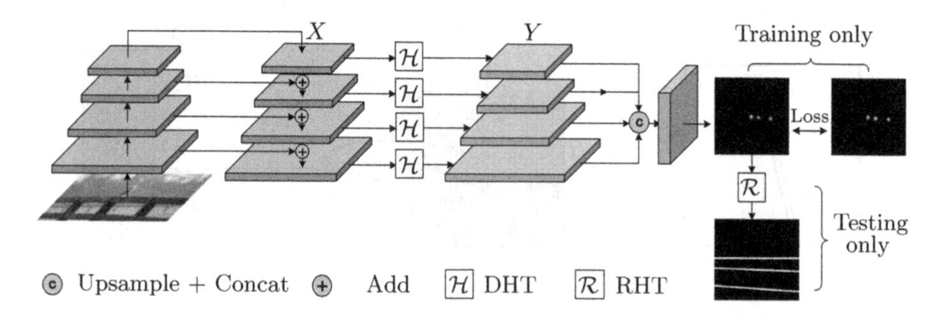

Fig. 2. Pipeline of our proposed method. DHT is short for the proposed Deep Hough Transform, and RHT represents the Reverse Hough Transform.

3 Deep Hough Transform for Line Detection

Our method comprises the following four major components: 1) a CNN encoder that extracts pixel-wise deep representations; 2) the deep Hough transform (DHT) that converts the spatial representations to a parametric space; 3) the line detector that is responsible to detect lines in the parametric space, and 4) a reverse Hough transform (RHT) that converts the detected lines back to image space. All these components are unified in a framework that performs forward inference and backward training in an end-to-end manner.

3.1 Line Parameterization and Reverse

In the 2D case, all straight lines can be parameterized with two parameters: an orientation parameter and a distance parameter. As shown in Fig. 3(a), given a 2D image $I_{W \times H}$ where H and W are the spatial size, we set the origin to the center of the image. Then a line l can be parameterized with r_l and $\theta_l \in [0, \pi)$, representing the distance between l and the origin, and the angle between l and the x-axis, respectively. Obviously $\forall\ l \in I, r_l \in [-\sqrt{W^2 + H^2}/2, \sqrt{W^2 + H^2}/2]$.

Given any line l from I, we can parameterize it with the above formulations, and also we can perform a reverse mapping to translate any (r, θ) pair to a line instance. Formally, we define the line parameterization and reverse as:

$$\begin{aligned} r_l, \theta_l &= P(l), \\ l &= P^{-1}(r_l, \theta_l). \end{aligned} \tag{1}$$

Obviously, both P and P^{-1} are bijective functions. In practice, r and θ are quantized to discrete bins to be processed by computer programs. Suppose the quantization interval for r and θ are Δr and $\Delta\theta$, respectively. Then the quantization can be formulated as below:

$$\hat{r}_l = \left\lceil \frac{r_l}{\Delta r} \right\rceil,\ \hat{\theta}_l = \left\lceil \frac{\theta_l}{\Delta\theta} \right\rceil, \tag{2}$$

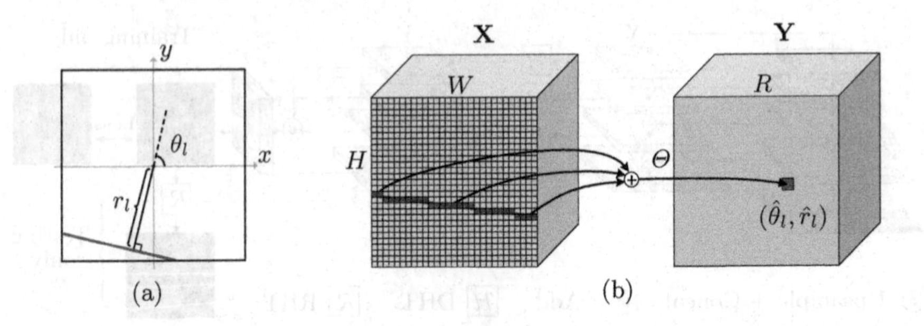

Fig. 3. (a): A line is parameterized by r_l and θ_l; (b): Features along a line in the feature space (left) are accumulated to a point $(\hat{r}_l, \hat{\theta}_l)$ in the parametric space (right).

where $\hat{r}_l$ and $\hat{\theta}_l$ are the quantized line parameters. The number of quantization levels, donoted with Θ and R, are:

$$\Theta = \frac{\pi}{\Delta\theta}, \ R = \frac{\sqrt{W^2 + H^2}}{\Delta r}, \tag{3}$$

as shown in Fig. 3(b).

3.2 Feature Transformation with Deep Hough Transform

Deep Hough Transform. Given an input image I, we first extract deep CNN features $\mathbf{X} \in \mathbb{R}^{C \times H \times W}$ with the encoder network, where C indicates the number of channels and H and W are the spatial size. Afterward, the deep Hough transform (DHT) takes $\mathbf{X}$ as input and produces the transformed features, $\mathbf{Y} \in \mathbb{R}^{C \times \Theta \times R}$. The size of transformed features, Θ, R, are determined by the quantization intervals, as described in Eq. (3).

As shown in Fig. 3(b), given a line $l \in \mathbf{X}$ in the feature space, we accumulate features of all pixels along l, to $(\hat{\theta}_l, \hat{r}_l)$ in the parametric space Y:

$$\mathbf{Y}(\hat{\theta}_l, \hat{r}_l) = \sum_{i \in l} \mathbf{X}(i), \tag{4}$$

where i is the positional index. $\hat{\theta}_l$ and $\hat{r}_l$ are determined by the parameters of line l, according to Eq. (1), and then quantized into discrete grids according to Eq. (2).

The DHT is applied to all unique lines in an image. These lines are obtained by connecting an arbitrary pair of pixels on the edges of an image, and then excluding the duplicated lines. It is worth noting that DHT is order-agnostic in both the feature space and the parametric space, making it highly parallelizable.

Multi-scale DHT with FPN. Our proposed DHT could be easily applied to arbitrary spatial features. We use an FPN network [30] as our encoder, which helps to extract multi-scale feature representations. Specifically, the FPN outputs 4 feature maps X_1, X_2, X_3, X_4 and their respective resolutions are (100, 100), (50, 50), (25, 25), (25, 25). Then each feature map is transformed by a DHT module independently, as shown in Fig. 2. Since these feature maps are in different resolutions, the transformed features Y_1, Y_2, Y_3, Y_4 also have different sizes, because we use the same quantization interval in all stages (see Eq. (3) for details). To fuse transformed features together, we interpolate Y_2, Y_3, Y_4 to the size of Y_1, and then fuse them by concatenation.

3.3 Line Detection in the Parametric Space

Context-Aware Line Detector. After the deep Hough transform (DHT), features are translated to the parametric space where grid location (θ, r) corresponds to features along an entire line $l = P^{-1}(\theta, r)$ in the feature space. An important reason to transform the features to the parametric space is that the line structures could be more compactly represented. As shown in Fig. 4, lines nearby a specific line l are translated to surrounding points near (θ_l, r_l). Consequently, features of nearby lines can be efficiently aggregated using convolutional layers in the parametric space.

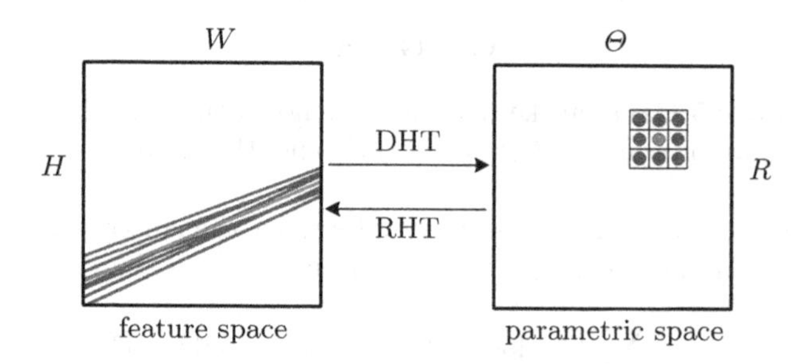

Fig. 4. Illustration of the proposed context-aware feature aggregation. Features of nearby lines in the feature space (left) are translated into neighbor points in the parametric space (right). In the parametric space, a simple 3×3 convolutional operation can easily capture contextual information for the central line (orange). Best viewed in color. (Color figure online)

In each stage of the FPN, we use two 3×3 convolutional layers to aggregate contextual line features. Then we interpolate features to match the resolution of features from various stages, and concatenate the interpolated features together. Finally, a 1×1 convolutional layer is applied to the concatenated feature maps to produce the pointwise predictions.

Loss Function. Since the prediction is directly produced in the parametric space, we calculate the loss in the same space as well. For a training image I, the ground-truth lines are first converted into the parametric space with the standard Hough

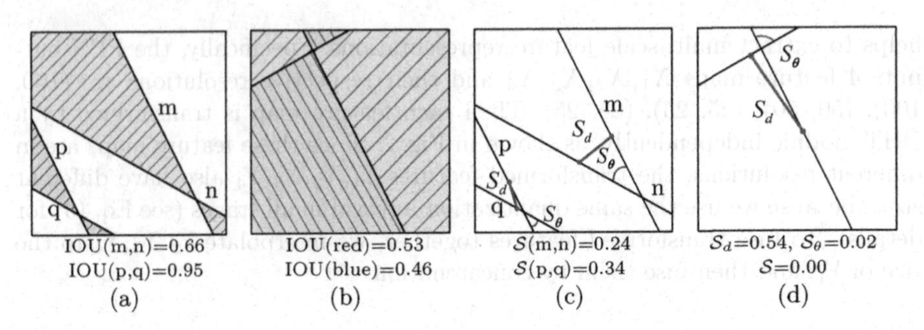

Fig. 5. (a): Two pairs of lines with similar relative position could have very different IOU scores. (b): Even humans cannot determine which area (blue or red) should be considered as the intersection in the IOU-based metric [29]. (c) and (d): Our proposed metric considers both Euclidean distance and angular distance between a pair of lines, resulting in consistent and reasonable scores. Best viewed in color. (Color figure online)

transform. Then to help converging faster, we smooth and expand the ground-truth with a Gaussian kernel. Similar tricks have been used in many other tasks like crowed counting [10,32] and road segmentation [41]. Formally, let $\mathbf{G}$ be the binary ground-truth map in the parametric space, $\mathbf{G}_{i,j} = 1$ indicates there is a line located at i, j in the parametric space. The expanded ground-truth map is

$$\hat{\mathbf{G}} = \mathbf{G} \circledast K,$$

where K is a 5×5 Gaussian kernel and $\circledast$ denotes the convolution operation. An example pair of smoothed ground-truth and the predicted map is shown in Fig. 2.

Finally, we compute the cross-entropy between the smoothed ground-truth and the predicted map in the parametric space:

$$\mathcal{L} = -\sum_i (\hat{\mathbf{G}}_i \cdot \log(\mathbf{P}_i) + (1 - \hat{\mathbf{G}}_i) \cdot \log(1 - \mathbf{P}_i)) \tag{5}$$

3.4 Reverse Mapping

Our detector produces predictions in the parametric space representing the probability of the existence of lines. The predicted map is then binarized with a threshold (*e.g.* 0.01). Then we find each connected area and calculate respective centroids. These centroids are regarded as the parameters of detected lines. At last, all lines are mapped back to the image space with $P^{-1}(\cdot)$, as formulated in Eq. (1). We refer to the "mapping back" step as "Reverse Mapping of Hough Transform (RHT)", as shown in Fig. 2.

4 The Proposed Evaluation Metric

In this section, we elaborate on the proposed evaluation metric that measures the agreement, or alternatively, the similarity between the two lines in an image.

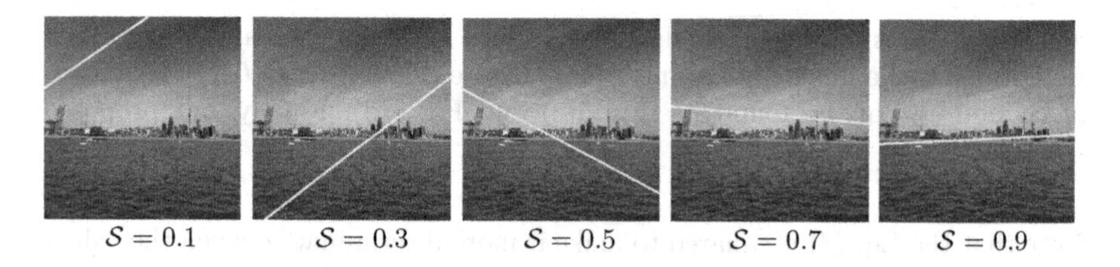

Fig. 6. Example lines with various EA-scores.

Firstly, we review several widely used metrics in the computer vision community and then explain why these existing metrics are not proper for our task. Finally, we introduce our newly proposed metric, which measures the agreement between two lines considering both Euclidean distance and angular distance.

4.1 Review of Existing Metrics

The intersection over union (IOU) is widely used in object detection, semantic segmentation and many other tasks to measure the agreement between detected bounding boxes (segments) w.r.t the ground-truth. Lee *et al.* [29] adopt the original IOU into line detection, and propose the line-based IOU to evaluate the quality of detected lines. Concretely, the similarity between the two lines is measured by the intersection areas of lines divided by the image area. Take Fig. 5(a) as an example, the similarity between line m and n is $\text{IOU}(m, n) = area(red)/area(I)$.

However, we argue that this IOU-based metric is improper and may lead to unreasonable or ambiguous results under specific circumstances. As illustrated in Fig. 5(a), two pairs of lines (m, n, and p, q) with similar structure could have very different IOU scores. In Fig. 5(b), even humans cannot determine which areas (red or blue) should be used as intersection areas in line based IOU. To remedy the aforementioned deficiencies, we elaborately design a new metric that measures the similarity of two lines.

4.2 The Proposed Metric

We propose a simple yet reasonable metric to assess the similarity between a pair of lines. Our metric $\mathcal{S}$, named **EA-score**, considers both **E**uclidean distance and **A**ngular distance between a pair of lines. Let l_i, l_j be a pair of lines to be measured, the angular distance $\mathcal{S}_\theta$ is defined according to the angle between two lines:

$$\mathcal{S}_\theta = 1 - \frac{\theta(l_i, l_j)}{\pi/2}, \tag{6}$$

where $\theta(l_i, l_j)$ is the angle between l_i and l_j. The Euclidean distance is defined as:

$$\mathcal{S}_d = 1 - D(l_i, l_j), \tag{7}$$

where $D(l_i, l_j)$ is the Euclidean distance between midpoints of l_i and l_j. Note that we normalize the image into a unit square before calculating $D(l_i, l_j)$. Examples of $\mathcal{S}_d$ and $\mathcal{S}_\theta$ can be found in Fig. 5(c) and Fig. 5(d). Finally, our proposed EA-score is:

$$\mathcal{S} = (\mathcal{S}_\theta \cdot \mathcal{S}_d)^2. \tag{8}$$

Note that the Eq. (8) is squared to make it more discriminative when the values are high. Several example line pairs and corresponding EA-scores are demonstrated in Fig. 6.

5 Experiments

In this section, we introduce the implementation details of our system, and report experimental results compared with existing methods.

5.1 Implementation Details

Our system is implemented with the PyTorch [34] framework. Since the proposed deep Hough transform (DHT) is highly parallelizable, we implement DHT with native CUDA programming, and all other parts are implemented based on PyTorch's Python API. We use a single RTX 2080 Ti GPU for all experiments.

Network Architectures. We use two representative network architectures, ResNet50 [20] and VGGNet16 [39], as our backbone and the FPN [30] to extract multi-scale deep representations. For the ResNet network, following the common practice in previous works [9,47], the dilated convolution [45] is used in the last layer to increase the resolution of feature maps.

Hyper-Parameters. The size of the Gaussian kernel used in Sect. 3.3 is 5×5. All images are resized to (400, 400) and then wrapped into a mini-batch of 8. We train all models for 30 epochs using the Adam optimizer [25] without weight decay. The learning rate and momentum are set to 2×10^{-4} and 0.9, respectively. The quantization intervals $\Delta\theta, \Delta r$ will be detailed in Sect. 5.3 and Eq. (12).

Datasets and Data Augmentation. Lee *et al.* [29] construct a dataset named SEL, which is, to the best of our knowledge, the only dataset for semantic line detection. The SEL dataset is composed of 1715 images, 1541 images for training and 174 for testing. There are 1.63 lines per image on average, and each image contains 1 line at least, and 6 lines at most. Following the setup in [29], we use only left-right flip data augmentation in all our experiments.

5.2 Evaluation Protocol

Given the metric in Eq. (8), we evaluate the detection results in terms of precision, recall, and F-measure.

For a pair of predicted and ground-truth line $(\hat{l}, l)$, we first calculate the similarity $\mathcal{S}(\hat{l}, l)$ as depicted in Eq. (8). $\hat{l}$ is identified as positive only if $\mathcal{S}(\hat{l}, l) > \epsilon$, where ϵ is a threshold. We calculate the precision and recall as:

$$Precision = \frac{\sum_{\hat{l} \in \mathcal{P}} \mathbb{1}(\mathcal{S}(\hat{l}, l) \geq \epsilon)}{||\mathcal{P}||}, \tag{9}$$

$$Recall = \frac{\sum_{l \in \mathcal{G}} \mathbb{1}(\mathcal{S}(l, \hat{l}) \geq \epsilon)}{||\mathcal{G}||}. \tag{10}$$

$\mathcal{P}$ and $\mathcal{G}$ are sets of predicted and ground-truth lines, respectively, and $|| \cdot ||$ denotes the number of elements in a set. $\mathbb{1}(\cdot)$ is the indicator function evaluating to 1 only if the condition is true. In Eq. (9), given a predicted line $\hat{l}$, l is the nearest ground-truth in the same image. Whereas in Eq. (10), $\hat{l}$ is the nearest prediction given a ground-truth line l in the same image. Accordingly, the F-measure is:

$$F\text{-}measure = \frac{2 \cdot precision \cdot recall}{precision + recall} \tag{11}$$

We apply a series thresholds, *i.e.* $\epsilon = 0.01, 0.02, ..., 0.99$, to predictions. Accordingly, we derive a series of precision, recall and F-measure scores. Finally, we evaluate the performance in terms of average precision, recall and F-measure.

5.3 Grid Search for Quantization Interval

The quantization intervals $\Delta\theta$ and Δr in Eq. (2) are important factors to the performance and running efficiency. Larger intervals lead to fewer quantization levels, *i.e.* Θ and R, and the model will be faster. With smaller intervals, there will be more quantization levels, and the computational overhead is heavier. To achieve a balance between performance and efficiency, we perform a grid search to find proper intervals that are computationally efficient and functionally effective.

We first fix the angular quantization interval to $\Delta\theta = \pi/100$ and then search for different distance quantization intervals Δr. According to the results in Fig. 7(a), with fixed angular interval $\Delta\theta$, the performance first increases with the decrease of Δr, and then gets saturated nearly after $\Delta r = \sqrt{2}$.

Afterward, we fix $\Delta r = \sqrt{2}$ and try different $\Delta\theta$. The results in Fig. 7(b) demonstrate that, with the decrease of $\Delta\theta$, the performance first increases until reaching the peak, and then slightly fall down. Hence, the peak value $\Delta\theta = \pi/100$ is a proper choice for angular quantization.

In summary, we use $\Delta\theta = \pi/100$ and $\Delta r = \sqrt{2}$ in quantization, and corresponding quantization levels are:

$$\Theta = 100, \ R = \sqrt{\frac{W^2 + H^2}{2}}, \tag{12}$$

where H, W are the size of feature maps to be transformed in DHT.

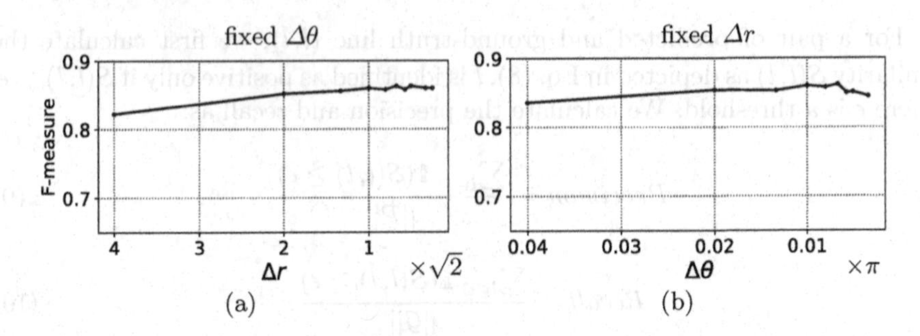

Fig. 7. (a): Performance under different distance quantization intervals Δr with a fixed angular quantization interval $\Delta\theta = \pi/100$. Larger Δr indicates smaller quantization levels R. (b): Performance under different angular quantization intervals $\Delta\theta$ with a fixed distance quantization interval $\Delta r = \sqrt{2}$.

5.4 Comparisons

Quantitative Comparison with Previous Arts. We compare our proposed method with the SLNet [29] and the classical Hough line detection [11] with HED [43] as the edge detector. Note that we train the HED edge detector on the SEL [29] training set using the line annotations as edge ground-truth.

The results in Table 1 illustrates that our method, with either VGG16 or ResNet50 as backbone, consistently outperforms SLNet and HT+HED with a considerable margin. In addition to Table 1, we plot the F-measure *v.s.* threshold and the precision *v.s.* recall curves. Figure 8 reveals that our method achieves higher F-measure than others under a wide range of thresholds.

Table 1. Quantitative comparisions across different methods. Our method significantly outperforms other competitors in terms of average F-measure.

Method	Precision	Recall	F-measure	FPS
SLNet-iter1 [29]	0.747	**0.862**	0.799	2.67
SLNet-iter3 [29]	0.793	0.845	0.817	1.92
SLNet-iter5 [29]	0.798	0.842	0.819	–
SLNet-iter10 [29]	0.814	0.831	0.822	1.10
HED [43] + HT [11]	0.839	0.812	0.825	6.46
Ours (VGG16)	0.844	0.834	0.839	30.01
Ours (ResNet50)	**0.899**	0.824	**0.860**	**49.99**

Runtime Efficiency. In this section, we benchmark the runtime of different methods including SLNet [29] with various iteration steps, classical Hough transform and our proposed method.

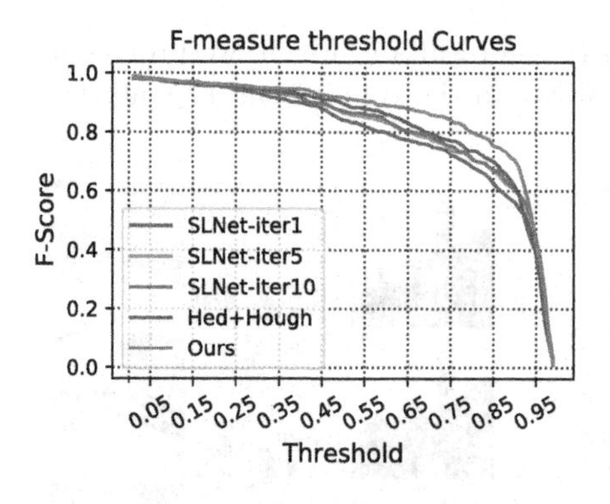

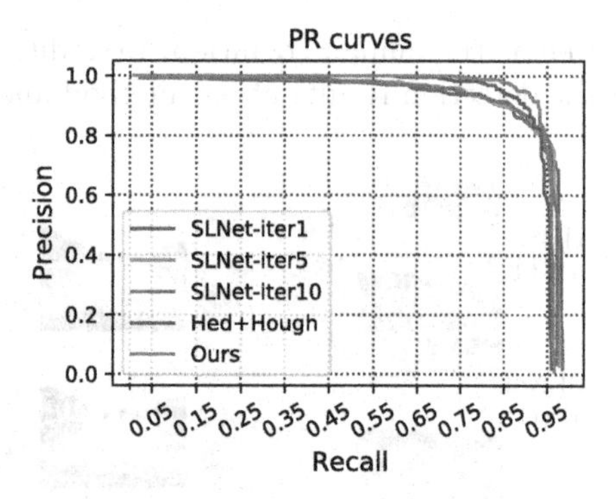

Fig. 8. Left: F-measure under various thresholds. Right: The precision-recall curve. Out method outperforms SLNet [29] and classical Hough transform [11] with a considerable margin. Moreover, even with 10 rounds of location refinement, SLNet still presents inferior performance.

Both SLNet [29] and HT require edge detection, *e.g.* HED [43], as a pre-processing step. The non-maximal suppression (NMS) in SLNet requires edge maps as guidance, and the classical Hough transform takes an edge map as input. Moreover, SLNet uses a refining network to enhance the results iteratively, therefore, the inference speed is related to the iteration steps. In contrast, our method produces results with a single forward pass, and the NMS is as simple as computing the centroids of each connected area in the parametric space.

Table 2. Quantitative speed comparisons. Our method is much faster than the other two competitors in forward pass and post-processing, and our method doesn't require any extra-process *e.g.* edge detection. Consequently, our method can run at 49 frames per second, which is remarkably higher than the other two methods.

Method	Network forward	NMS	Edge	Total
SLNet-iter1 [29]	0.354 s	0.079 s	0.014 s	0.447 s
SLNet-iter3 [29]	0.437 s	0.071 s	0.014 s	0.522 s
SLNet-iter10 [29]	0.827 s	0.068 s	0.014 s	0.909 s
HED [43] + HT [11]	0.014 s	0.117 s	0.024 s	0.155 s
Ours (VGG16)	0.03 s	0.003 s	0	0.033 s
Ours (ResNet50)	0.017 s	0.003 s	0	**0.020 s**

Results in Table 2 illustrate that our method is significantly faster than all other competitors with a very considerable margin. Even with only 1 iteration step, SLNet is still slower than our method.

Qualitative Comparisions. Here we give several example results of our proposed method along with SNLet and HED+HT. As shown in Fig. 9, compared with other methods, our results are more compatitable with the ground-truth as well as the human cognition. In addition to the results in Fig. 9, we provide all the detection results of our method and SLNet in the supplementary material.

Fig. 9. Example detection results by different methods. Compared to SLNet [29] and classical Hough transform [11], our results are more consistent with the ground-truth.

Table 3. Ablation study for each component. MS indicates DHTs with multi-scale features and CTX means context-aware aggregation as described in Sect. 3.2 and 3.3.

DHT	MS	CTX	F-measure
✓			0.845
✓	✓		0.852
✓		✓	0.847
✓	✓	✓	0.860

5.5 Ablation Study

In this section, we ablate each of the components in our method. Specifically, they are: (a) the Deep Hough transform (DHT) module detailed in Sect. 3.2; (b) the multi-scale (MS) DHT architecture described in Sect. 3.2; (c) the context-aware (CTX) line detector proposed in Sect. 3.3. Experimental results are shown in Table 3.

We first construct a baseline model with plain ResNet50 and DHT module. Note that the baseline model achieves 0.845 average F-measure, which has already surpassed the SLNet competitor.

Then we verify the effectiveness of the multi-scale (MS) strategy and context-aware line detector (CTX), individually. We separately append MS and CTX to the baseline model and then evaluate their performance, respectively. Results in Table 3 indicate that both MS and CTX can improve the performance of the baseline model.

At last, we combine all the components together to form our final full method, which achieves the best performance among all other combinations. Experimental results in this section clearly demonstrate that each component of our proposed method contributes to the success of our method.

6 Conclusions

In this paper, we proposed a simple yet effective method for semantic line detection in natural images. By incorporating the strong learning ability of CNNs into classical Hough transform, our method is able to capture complex textures and rich contextual semantics of lines. A new evaluation metric was proposed for line structures, considering both Euclidean distance and angular distance. Both quantitative and qualitative results revealed that our method significantly outperforms previous arts in terms of both detection quality and speed.

Acknowledgments. This research was supported by Major Project for New Generation of AI under Grant No. 2018AAA0100400, NSFC (61922046), Tianjin Natural Science Foundation (18ZXZNGX00110), and the Fundamental Research Funds for the Central Universities (Nankai University: 63201169).

References

1. Aggarwal, N., Karl, W.C.: Line detection in images through regularized hough transform. IEEE Trans. Image Process. **15**(3), 582–591 (2006)
2. Akinlar, C., Topal, C.: Edlines: a real-time line segment detector with a false detection control. Pattern Recogn. Lett. **32**(13), 1633–1642 (2011)
3. Ballard, D.: Generating the hough transform to detect arbitary shapes. Pattern Recogn. **13**(2) (1981)
4. Borji, A., Cheng, M.M., Hou, Q., Jiang, H., Li, J.: Salient object detection: a survey. Comput. Vis. Media **5**(2), 117–150 (2019). https://doi.org/10.1007/s41095-019-0149-9
5. Burns, J.B., Hanson, A.R., Riseman, E.M.: Extracting straight lines. IEEE Trans. Pattern Anal. Mach. Intell. **PAMI–8**(4), 425–455 (1986)
6. Canny, J.: A computational approach to edge detection. IEEE Trans. Pattern Anal. Mach. Intell. **PAMI–8**(6), 679–698 (1986)
7. Caplin, S.: Art and Design in Photoshop. Elsevier/Focal (2008)
8. Chan, T., Yip, R.K.: Line detection algorithm. In: Proceedings of 13th International Conference on Pattern Recognition, vol. 2, pp. 126–130. IEEE (1996)
9. Chen, L.C., Zhu, Y., Papandreou, G., Schroff, F., Adam, H.: Encoder-decoder with atrous separable convolution for semantic image segmentation. In: Proceedings of the European Conference on Computer Vision (ECCV), pp. 801–818 (2018)

10. Cheng, Z.Q., Li, J.X., Dai, Q., Wu, X., Hauptmann, A.G.: Learning spatial awareness to improve crowd counting. In: Proceedings of the IEEE International Conference on Computer Vision, pp. 6152–6161 (2019)
11. Duda, R.O., Hart, P.E.: Use of the hough transformation to detect lines and curves in pictures. Technical report, Sri International Menlo Park Ca Artificial Intelligence Center (1971)
12. Etemadi, A.: Robust segmentation of edge data. In: 1992 International Conference on Image Processing and its Applications, pp. 311–314. IET (1992)
13. Fan, D.P., Lin, Z., Zhang, Z., Zhu, M., Cheng, M.M.: Rethinking RGB-D salient object detection: models, datasets, and large-scale benchmarks. IEEE TNNLS (2020)
14. Fan, D.-P., Zhai, Y., Borji, A., Yang, J., Shao, L.: BBS-Net: RGB-D salient object detection with a bifurcated backbone strategy network. In: Vedaldi, A., Bischof, H., Brox, T., Frahm, J.-M. (eds.) ECCV 2020. LNCS, vol. 12357, pp. 275–292. Springer, Cham (2020). https://doi.org/10.1007/978-3-030-58610-2_17
15. Fan, R., Cheng, M.M., Hou, Q., Mu, T.J., Wang, J., Hu, S.M.: S4Net: single stage salient-instance segmentation. Comput. Vis. Media **6**(2), 191–204 (2020). https://doi.org/10.1007/s41095-020-0173-9
16. Fernandes, L.A., Oliveira, M.M.: Real-time line detection through an improved hough transform voting scheme. Pattern Recogn. **41**(1), 299–314 (2008)
17. Gao, S.H., Cheng, M.M., Zhao, K., Zhang, X.Y., Yang, M.H., Torr, P.: Res2Net: a new multi-scale backbone architecture. IEEE Trans. Pattern Anal. Mach. Intell. 1 (2020)
18. Gao, S.H., Tan, Y.Q., Cheng, M.M., Lu, C., Chen, Y., Yan, S.: Highly efficient salient object detection with 100k parameters. In: European Conference on Computer Vision (ECCV) (2020)
19. Girshick, R.: Fast R-CNN. In: Proceedings of the IEEE International Conference on Computer Vision, pp. 1440–1448 (2015)
20. He, K., Zhang, X., Ren, S., Sun, J.: Deep residual learning for image recognition. In: Proceedings of the IEEE Conference on Computer Vision and Pattern Recognition, pp. 770–778 (2016)
21. Hou, Q., Cheng, M.M., Hu, X., Borji, A., Tu, Z., Torr, P.: Deeply supervised salient object detection with short connections. IEEE TPAMI **41**(4), 815–828 (2019). https://doi.org/10.1109/TPAMI.2018.2815688
22. Hough, P.V.: Method and means for recognizing complex patterns. US Patent 3,069,654 (1962)
23. Huang, Z., Wang, X., Huang, L., Huang, C., Wei, Y., Liu, W.: CCNet: criss-cross attention for semantic segmentation. In: Proceedings of the IEEE International Conference on Computer Vision, pp. 603–612 (2019)
24. Illingworth, J., Kittler, J.: The adaptive hough transform. IEEE Trans. Pattern Anal. Mach. Intell. **PAMI–9**(5), 690–698 (1987)
25. Kingma, D.P., Ba, J.: Adam: a method for stochastic optimization. arXiv preprint arXiv:1412.6980 (2014)
26. Kiryati, N., Eldar, Y., Bruckstein, A.M.: A probabilistic hough transform. Pattern Recogn. **24**(4), 303–316 (1991)
27. Krages, B.: Photography: The Art of Composition. Simon and Schuster, New York (2012)
28. Law, H., Deng, J.: CornerNet: detecting objects as paired keypoints. In: Proceedings of the European Conference on Computer Vision (ECCV), pp. 734–750 (2018)

29. Lee, J.T., Kim, H.U., Lee, C., Kim, C.S.: Semantic line detection and its applications. In: Proceedings of the IEEE International Conference on Computer Vision, pp. 3229–3237 (2017)
30. Lin, T.Y., Dollár, P., Girshick, R., He, K., Hariharan, B., Belongie, S.: Feature pyramid networks for object detection. In: Proceedings of the IEEE Conference on Computer Vision and Pattern Recognition, pp. 2117–2125 (2017)
31. Liu, L., Chen, R., Wolf, L., Cohen-Or, D.: Optimizing photo composition. Comput. Graph. Forum **29**(2), 469–478 (2010)
32. Liu, W., Salzmann, M., Fua, P.: Context-aware crowd counting. In: Proceedings of the IEEE Conference on Computer Vision and Pattern Recognition, pp. 5099–5108 (2019)
33. Liu, Y., et al.: Richer convolutional features for edge detection. IEEE Trans. Pattern Anal. Mach. Intell. **41**(8), 1939–1946 (2019). https://doi.org/10.1109/TPAMI.2018.2878849
34. Paszke, A., et al.: PyTorch: an imperative style, high-performance deep learning library. In: Advances in Neural Information Processing Systems, pp. 8024–8035 (2019)
35. Princen, J., Illingworth, J., Kittler, J.: A hierarchical approach to line extraction based on the hough transform. Comput. Vis. Graph. Image Process. **52**(1), 57–77 (1990)
36. Qi, C.R., Chen, X., Litany, O., Guibas, L.J.: ImvoteNet: boosting 3D object detection in point clouds with image votes. In: Proceedings of the IEEE/CVF Conference on Computer Vision and Pattern Recognition, pp. 4404–4413 (2020)
37. Qi, C.R., Litany, O., He, K., Guibas, L.J.: Deep hough voting for 3D object detection in point clouds. In: Proceedings of the IEEE International Conference on Computer Vision, pp. 9277–9286 (2019)
38. Ren, S., He, K., Girshick, R., Sun, J.: Faster R-CNN: towards real-time object detection with region proposal networks. In: Advances in Neural Information Processing Systems, pp. 91–99 (2015)
39. Simonyan, K., Zisserman, A.: Very deep convolutional networks for large-scale image recognition. In: ICLR (2015)
40. Sobel, I.: An isotropic 3 $\times$ 3 image gradient operator. Presentation at Stanford A.I. Project 1968, February 2014
41. Tan, Y.Q., Gao, S., Li, X.Y., Cheng, M.M., Ren, B.: Vecroad: Point-based iterative graph exploration for road graphs extraction. In: IEEE CVPR (2020)
42. Wang, X., Girshick, R., Gupta, A., He, K.: Non-local neural networks. In: Proceedings of the IEEE Conference on Computer Vision and Pattern Recognition, pp. 7794–7803 (2018)
43. Xie, S., Tu, Z.: Holistically-nested edge detection. In: Proceedings of the IEEE International Conference on Computer Vision, pp. 1395–1403 (2015)
44. Yacoub, S.B., Jolion, J.M.: Hierarchical line extraction. IEE Proc.-Vis. Image Signal Process. **142**(1), 7–14 (1995)
45. Yu, F., Koltun, V.: Multi-scale context aggregation by dilated convolutions. arXiv preprint arXiv:1511.07122 (2015)
46. Zhang, Z., et al.: PPGnet: learning point-pair graph for line segment detection. In: IEEE Conference on Computer Vision and Pattern Recognition, CVPR 2019, Long Beach, CA, USA, 16–20 June 2019 (2019)
47. Zhao, H., Shi, J., Qi, X., Wang, X., Jia, J.: Pyramid scene parsing network. In: Proceedings of the IEEE Conference on Computer Vision and Pattern Recognition, pp. 2881–2890 (2017)

Structured Landmark Detection via Topology-Adapting Deep Graph Learning

Weijian Li[1,2(✉)], Yuhang Lu[1,3], Kang Zheng[1], Haofu Liao[2], Chihung Lin[4], Jiebo Luo[2], Chi-Tung Cheng[4], Jing Xiao[5], Le Lu[1], Chang-Fu Kuo[4], and Shun Miao[1]

[1] PAII. Inc., Bethesda, MD, USA
[2] Department of Computer Science, University of Rochester, Rochester, NY, USA
[3] Department of Computer Science and Engineering, University of South Carolina, Columbia, SC, USA
[4] Chang Gung Memorial Hospital, Linkou, Taiwan
[5] Ping An Technology, Shenzhen, China

Abstract. Image landmark detection aims to automatically identify the locations of predefined fiducial points. Despite recent success in this field, higher-ordered structural modeling to capture implicit or explicit relationships among anatomical landmarks has not been adequately exploited. In this work, we present a new topology-adapting deep graph learning approach for accurate anatomical facial and medical (e.g., hand, pelvis) landmark detection. The proposed method constructs graph signals leveraging both local image features and global shape features. The adaptive graph topology naturally explores and lands on task-specific structures which are learned end-to-end with two Graph Convolutional Networks (GCNs). Extensive experiments are conducted on three public facial image datasets (*WFLW, 300W, and COFW-68*) as well as three real-world X-ray medical datasets (*Cephalometric (public), Hand and Pelvis*). Quantitative results comparing with the previous state-of-the-art approaches across all studied datasets indicating the superior performance in both robustness and accuracy. Qualitative visualizations of the learned graph topologies demonstrate a physically plausible connectivity laying behind the landmarks.

Keywords: Landmark detection · GCN · Adaptive topology

1 Introduction

Image landmark detection has been a fundamental step for many high-level computer vision tasks to extract and distill important visual contents, such as image registration [22], pose estimation [4], identity recognition [75] and image super-resolution [5]. Robust and accurate landmark localization becomes a vital component determining the success of the downstream tasks.

Electronic supplementary material The online version of this chapter (https://doi.org/10.1007/978-3-030-58545-7_16) contains supplementary material, which is available to authorized users.

© Springer Nature Switzerland AG 2020
A. Vedaldi et al. (Eds.): ECCV 2020, LNCS 12354, pp. 266–283, 2020.
https://doi.org/10.1007/978-3-030-58545-7_16

Recently, heatmap regression based methods [48,53,60,72] have achieved encouraging performance on landmark detection. They model landmark locations as heatmaps and train deep neural networks to regress the heatmaps. Despite popularity and success, they usually suffer from a major drawback of lacking a global representation for the structure/shape, which provides high-level and reliable cues in individual anatomical landmark localization. As a result, heatmap-based methods could make substantial errors when being exposed to large appearance variations such as occlusions.

In contrast, coordinate regression based methods [35,52,67,69] have an innate potential to incorporate structural knowledge since the landmark coordinates are directly expressed. Most existing methods initialize landmark coordinates using mean or canonical shapes, which indirectly inject weak structural knowledge [52]. While the exploitation of the structural knowledge in existing methods has still been insufficient as well as further exploitation of the structural knowledge considering the underlying relationships between the landmarks. Effective means for information exchange among landmarks to facilitate landmark detection are also important but have yet to be explored. Due to these limitations, the performance of the latest coordinate-based methods [61] falls behind the heatmap-based ones [56].

In this work, we introduce a new topology-adapting deep graph learning approach for landmark detection, termed *Deep Adaptive Graph (DAG)*. We model the landmarks as a graph and employ global-to-local cascaded Graph Convolutional Networks (GCNs) to move the landmarks towards the targets in multiple steps. Graph signals of the landmarks are built by combining local image features and graph shape features. Two GCNs operate in a cascaded manner, with the first GCN estimating a global transformation of the landmarks and the second GCN estimating local offsets to further adjust the landmark coordinates. The graph topology, represented by the connectivity weights between landmarks, are learned during the training phase.

By modeling landmarks as a graph and processing it with GCNs, our method is able to effectively exploit the structural knowledge and allow rich information exchange among landmarks for accurate coordinate estimation. The graph topology learned for landmark detection task is capable of revealing reasonable landmark relationships for the given task. It also reduces the need for manually defining landmark relations (or grouping), making our method to be easily adopted for different tasks. By incorporating shape features into graph signal in addition to the local image feature, our model can learn and exploit the landmark shape prior to achieve high robustness against large appearance variations (e.g., occlusions). In summary, our main contributions are four-fold:

1. By representing the landmarks as a graph and detecting them using GCNs, our method effectively exploits the structural knowledge for landmark coordinate regression, closes the performance gap between coordinate- and heatmap-based landmark detection methods.

2. Our method automatically reveals physically meaningful relationships among landmarks, leading to a task-agnostic solution for exploiting structural knowledge via step-wise graph transformations.
3. Our model combines both visual contextual information and spatial positional information into the graph signal, allowing structural shape prior to be learned and exploited.
4. Comprehensive quantitative evaluations and qualitative visualizations on six datasets across both facial and medical image domains demonstrate the consistent state-of-the-art performance and general applicability of our method.

2 Related Work

A large number of studies have been reported in this domain including the classic Active Shape Models [11,12,36], Active Appearance Models [10,32,45], Constraind Local Models [3,13,30,44], and more recently the deep learning based models which can be further categorized into heatmap or regression based models.

Heatmap Based Landmark Detection: These methods [9,37,38,48,50,58] generate localized predictions of likelihood heatmaps for each landmark and achieve encouraging performances. A preliminary work by Wei *et al.* [58] introduce a Convolutional Pose Machine (CPM) which models the long-range dependency with a multistage network. Newell *et al.* [37] propose a Stacked Hourglass model leveraging the repeated bottom-up and top-down structure and intermediate supervision. Tang *et al.* [50] investigate a stacked U-Net structure with dense connections. Lately, Sun *et al.* [48] present a deep model named High-Resolution Network (HRNet18) which extracts feature maps in a joint deep and high resolution manner via conducting multi-scale fusions across multiple branches under different resolutions. Based on these models, other methods also integrate additional supervision cues such as the object structure constraints [62,76], the variety of image, and object styles [17,41] to solve specific tasks.

Coordinate Based Landmark Detection: Another common approach directly locates landmark coordinates from input images [33,35,47,49,51,52,73]. Most of these methods consist of multiple steps to progressively update predictions based on visual signals, widely known as Cascaded-Regression. Toshev *et al.* [51] and Sun *et al.* [49] adopt cascaded Convolutional Neural Networks (CNNs) to predict landmark coordinates. Trigeorgis *et al.* [52] model the cascaded regression process using a Recurrent Neural Network (RNN) based deep structure. Lv *et al.* [35] propose a two-stage regression model with global and local reinitializations. From different perspectives, Zhu *et al.* [73] investigate the methods of optimal initialization by searching the object shape space; Valle *et al.* [53] present a combined model with a tree structured regressor to infer landmark locations based on heatmap prediction results; Wu *et al.* [61] leverage uniqueness and discriminative characteristics across datasets to assist landmark detection.

Landmark Detection with Graphs: The structure of landmarks can be naturally modeled as a graph considering the landmark locations and landmark to landmark relationships [44,65,66,71,73]. Zhou *et al.* [71] propose a Graph-Matching method which obtains landmark locations by selecting the set of landmark candidates that would best fit the shape constraints learned from the examplars. Yu *et al.* [66] describe a two-stage deformable shape model to first extract a coarse optimum by maximizing a local alignment likelihood in the region of interest then refine the results by maximizing an energy function under shape constraints. Later, Yu *et al.* [65] present a hierarchical model to extract semantic features by constructing intermediate graphs from bottom-up node clustering and top-down graph deconvolution operations, leveraging the graph layout information. Zou *et al.* [76] introduce a landmark structure construction method with covering set algorithm. While their method is based on heatmap detection results, we would like to directly regress landmark locations from raw input image to avoid potential errors incurred from heatmap detections.

Recently, Ling *et al.* [31] propose a fast object annotation framework, where contour vertices are regressed using GCN to perform segmentation, indicating the benefit of position prediction with iterative message exchanges. In their task, each point is considered with the same semantics towards coarse anonymous matching which is not appropriate for precise targeted localization tasks like landmark detection. Adaptively learning graph connectivities instead of employing a fixed graph structure based on prior knowledge should be explored to improve the model's generalizability to different tasks.

3 Method

Our method adopts the cascaded-regression framework, where given the input image and initial landmarks (from the mean shape), the predicted landmark coordinates are updated in multiple steps. Yet differently, we feature the cascaded-regression framework with a graph representation of the landmarks, denoted by $G = (V, E, F)$, where $V = \{\mathbf{v}_i\}$ denotes the landmarks, $E = \{e_{ij}\}$ denotes the learned connectivity between landmarks and $F = \{\mathbf{f}_i\}$ denotes graph signals capturing appearance and shape information. The graph is processed by cascaded GCNs to progressively update landmark coordinates. An overview of our method is shown in Fig. 1. Details of the cascaded GCNs, graph signal and learned connectivity are presented in Sect. 3.1, Sect. 3.2 and Sect. 3.3, respectively. The training scheme of our method can be found in Sect. 3.4.

3.1 Cascaded GCNs

Given a graph representation of landmarks $G = (V, E, F)$, two-stage cascaded GCN modules are employed to progressively update the landmark coordinates. The first stage, *GCN-global*, estimates a global transformation to coarsely move the landmarks to the targets. The second stage, *GCN-local*, estimates local landmark coordinate offsets to iteratively move the landmarks toward the targets. Both modules employ the same GCN architecture (weights not shared) and the same learnable graph connectivity.

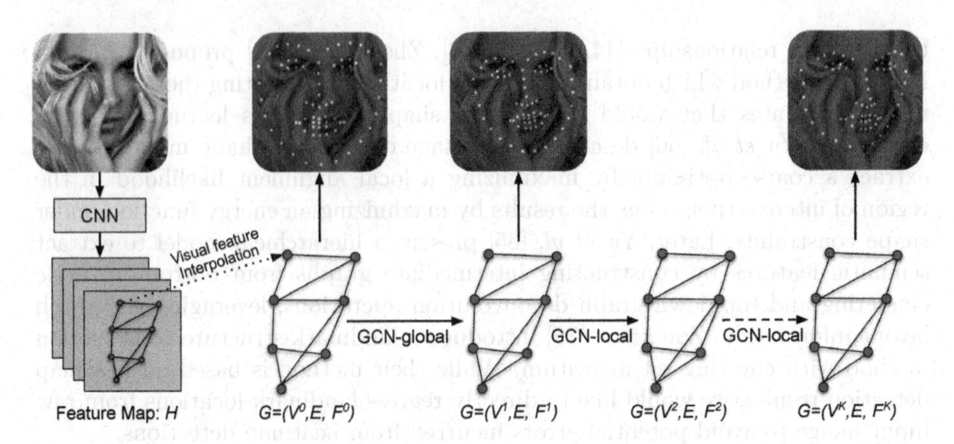

Fig. 1. Overview of the proposed Deep Adaptive Graph (DAG). Initial graph is initialized with the mean value computed from training data. We first deform the landmark graph through a perspective transformation predicted by GCN-global and then precisely shift the graph by GCN-local through iterations. The visual features and shape features are re-interpolated from feature map and re-calculated after each GCN module, respectively.

Graph Convolution: Given a graph connectivity E and a graph feature F, the k-th graph convolution operation updates the i-th node feature $\mathbf{f}_k^j$ by aggregating all node features weighted by the connectivity:

$$\mathbf{f}_{k+1}^i = \mathbf{W}_1\mathbf{f}_k^i + \sum_j e_{ij}\mathbf{W}_2\mathbf{f}_k^j \tag{1}$$

where $\mathbf{W}_1$ and $\mathbf{W}_2$ are learnable weight matrices. The graph convolutions can be seen as the mechanism of information collection among the neighborhoods. The connectivity E serves as pathways for information flow from one landmark to another.

Global Transformation GCN: Previous work [26,35] learn an affine transformation with a deep neural network by predicting a two by three affine transformation matrix which deforms the image to the satisfied posture. Inspired by this work, we employ a GCN on the initial landmarks to coarsely move them to the targets. Considering our graph is more flexible that does not have to maintain the parallelism and respective ratios among the edges, we model the global transformation using a perspective transformation [16]. A perspective transformation can be parameterized by 9 scalars $M = [a,b,c,d,e,f,g,h,i]^T \in \mathbb{R}^{9\times 1}$ with the operation written as:

$$\begin{bmatrix} x' \\ y' \\ 1 \end{bmatrix} \cong \begin{bmatrix} rx' \\ ry' \\ r \end{bmatrix} = \begin{pmatrix} a & b & c \\ d & e & f \\ g & h & i \end{pmatrix} \begin{bmatrix} x \\ y \\ 1 \end{bmatrix} \tag{2}$$

Given a target image, we initialize landmark locations V^0 using the mean shape of landmarks in the training set, and placed it at the center of the image.

The graph is processed by the GCN-global to estimate a perspective transformation to bring the initial structure closer to the target.

Specifically, a graph isomorphism network (GIN) [64] is employed to process the graph features $\{\mathbf{f}_k^i\}$ produced by the GCN to output a 9-dimensional vector representing the perspective transformation:

$$\mathbf{f}^G = \text{MLP}\left(\text{CONCAT}\left(\text{READOUT}\left(\{\mathbf{f}_k^i | i \in G\}\right) | k = 0, 1, \ldots, K\right)\right), \quad (3)$$

where the READOUT operator sums the features from all the nodes in the graph G. The transformation matrix M is obtained by transforming and reshaping $\mathbf{f}^G$ into a 3 by 3 matrix. We then apply this transformation matrix on the initial landmark node coordinates to obtain the aligned landmark coordinates:

$$V^1 = \{\mathbf{v}_i^1\} = \{\mathbf{M}\mathbf{v}_i^0\} \quad (4)$$

Local Refinement GCN: Given the transformed landmarks, we employ GCN-local to further shift the graph in a cascaded manner. GCN-local employs the same architecture as GCN-global, with a difference that the last layer produces a 2-dimensional vector for each landmark, representing the coordinate offset of the landmark. The updated landmark coordinates can be written as:

$$\mathbf{v}_i^{t+1} = \mathbf{v}_i^t + \Delta\mathbf{v}_i^t, \quad (5)$$

where $\Delta\mathbf{v}_i^t = (\Delta x_i^t, \Delta y_i^t)$ is the output of the GCN-local at the t-th step. In all our experiments, we perform $T = 3$ iterations of the GCN-local. Note that the graph signal is re-calculated after each GCN-local iteration.

3.2 Graph Signal with Appearance and Shape Information

We formulate a graph signal F as a set of node features $\mathbf{f}_i$, each associated with a landmark $\mathbf{v}_i$. The graph signal contains a *visual feature* to encode local image appearance and a *shape feature* to encode the global landmark shape.

Visual Feature: Specifically, given a feature map H with D channels produced by a backbone CNN, visual features, denoted by $\mathbf{p}_i \in R^D$, are extracted by interpolating H at the landmark coordinates $\mathbf{v}_i$. The interpolation is performed via a differentiable bi-linear interpolation [26]. In this way, visual feature of each landmark is collected from the feature map, encoding the appearance of its neighborhood.

Shape Feature: While the visual feature encodes the appearance in a neighborhood of the landmark, it does not explicitly encode the global shape of the landmarks. To incorporate this structural information into the graph signal, for each landmark, we compute its displacement vectors to all other landmarks, denoted as $\mathbf{q}_i = \{\mathbf{v}_j - \mathbf{v}_i\}_{j \neq i} \in R^{2\times(N-1)}$, where N is the number of landmarks. Such shape feature allows structural information of the landmarks to be exploited to facilitate landmark detection. For example, when the mouth of a face is occluded, the coordinates of the mouth landmarks can be inferred from the eyes and nose. Wrong landmark detection results that violate the shape prior can also be avoided when the shape is explicitly captured in the graph signal.

The graph signal F is then constructed for each landmark by concatenating the visual feature $\mathbf{p}_i$ and the shape feature $\mathbf{q}_i$ (flattened), resulting in a feature vector $\mathbf{f}_i \in R^{D+2(N-1)}$.

3.3 Landmark Graph with Learnable Connectivity

The graph connectivity determines the relationship between each pair of landmarks in the graph and serves as the information exchange channel in GCN. In most existing applications of GCN [31,40,54,59,70], the graph connectivity is given based on the prior knowledge of the task. In our landmark detection application, it is non-trivial to manually define the optimal underlying graph connectivity for the learning task. Therefore, relying on hand-crafted graph connectivity would introduce a subjective element into the model, which could lead to sub-optimal performance. To address this limitation, we learn task-specific graph connectivities during the training phase in an end-to-end manner. The connectivity weight e_{ij} behaves as information propagation gate in graph convolutions (Eqn. 1). We treat the connectivity $\{e_{ij}\}$, represented as an adjacency matrix, as a learnable parameter that is trained with the network during the training phase. In this way, the task-specific optimal graph connectivity is obtained by optimizing the performance of the target landmark detection task, allowing our method to be applied to different landmark detection tasks without manual intervention.

Graph connectivity learning has been studied before by the research community. One notable example is Graph Attention Networks [54], which employs a self-attention mechanism to adaptively generate connectivity weights during the model inference. We conjugate that in structured landmark detection problems, the underlying relationship between the landmarks remains the same for a given task, instead of varying across individual images. Therefore, we share the same connectivity across images on the same task, and directly optimize the connectivity weights during the training phase.

3.4 Training

GCN-global: Since the perspective transformation estimated by GCN-global has limited degree of freedom, directly penalizing the distance between the predicted and the ground truth landmarks will lead to unstable optimization behavior. As the goal of GCN-global is to coarsely locate the landmarks, we propose to use a margin loss on the L_1 distance, written as:

$$\mathcal{L}_{global} = \left[\left(\frac{1}{N} \sum_{i \in N} \sum_{x,y} |\mathbf{v}_i^1 - \mathbf{v}_i| \right) - m \right]_+ \quad (6)$$

where $[u]_+ := max(0, u)$. $\mathbf{v}_i^1 = (x_i^1, y_i^1)$ and $\mathbf{v}_i = (x_i, y_i)$ denote the predicted and ground truth landmark coordinates for the i-th landmark. m is a hyper-parameter representing a margin which controls how well we want the alignment to be. Following this procedure, we aim to obtain a high robustness of the coarse landmark detection, while forgive small errors.

GCN-local: To learn a precise localization, we directly employ L1 loss on all predicted landmark coordinates after the GCN-local, written as:

$$\mathcal{L}_{local} = \frac{1}{N} \sum_{i \in N} \sum_{x,y} |\mathbf{v}_i^T - \mathbf{v}_i| \tag{7}$$

where $\mathbf{v}_i^T$ is the T-th step (the last step) coordinate predictions, and $\mathbf{v}_i$ is the ground truth coordinate for the i-th landmark.

The overall loss to train DAG is a combination of the above two losses:

$$\mathcal{L} = \lambda_1 \mathcal{L}_{global} + \lambda_2 \mathcal{L}_{local} \tag{8}$$

where λ_k is the weight parameter for each loss.

4 Experiments

4.1 Datasets

We conduct evaluations on three public facial image and three medical image datasets:

WFLW [60] dataset contains 7,500 facial images for training and 2,500 facial images for testing. The testing set is further divided into 6 subsets focusing on particular challenges in the images namely large pose set, expression set, illumination set, makeup set, occlusion set, and blur set. 98 manually labeled landmarks are provided for each image.

300W [43] dataset consists of 5 facial datasets namely LFPW, AFW, HELEN, XM2VTS and IBUG. They are split into a training set with 3,148 images, and a testing set with 689 images where 554 images are from LFPW and HELEN, 135 from IBUG. Each image is labeled with 68 landmarks.

COFW [6] dataset contains 1,345 facial images for training and 507 for testing, under different occlusion conditions. Each image is originally labeled with 29 landmarks and re-annotated with 68 landmarks [21]. We follow previous studies [41,60] to conduct inferences on the re-annotated COFW-68 dataset to test our model's cross-dataset performance which is trained on 300W dataset.

Cephalometric X-ray [55] is a public dataset originally for a challenge in IEEE ISBI-2015. It contains 400 X-ray Cephalometric images with resolution of $1,935 \times 2,400$, 150 images are used as training set, the rest 150 images and 100 images are used as validation and test sets. Each cephalometric image contains 19 landmarks. In this paper, we only focus on the landmark detection task.

Hand X-ray [34] is a real-world medical dataset collected by a hospital. The X-ray images are taken with different hand poses with resolutions in $1,500s \times 2,000s$. In total, 471 images are randomly split into a training set (80%, $N = 378$) and a testing set (20%, $N = 93$). 30 landmarks are manually labeled for each image.

Pelvic X-ray [8,57] another real-world medical dataset collected by the same hospital. Images are taken over patient's pelvic bone with resolutions in $2,500s \times 2,000s$. The challenges in this dataset is the high structural and appearance variation, caused by bone fractures and metal prosthesis. In total, 1,000 imagesare randomly splited into a training set (80%, N = 800) and a testing set (20%, N = 200). 16 landmarks are manually labeled for each image.

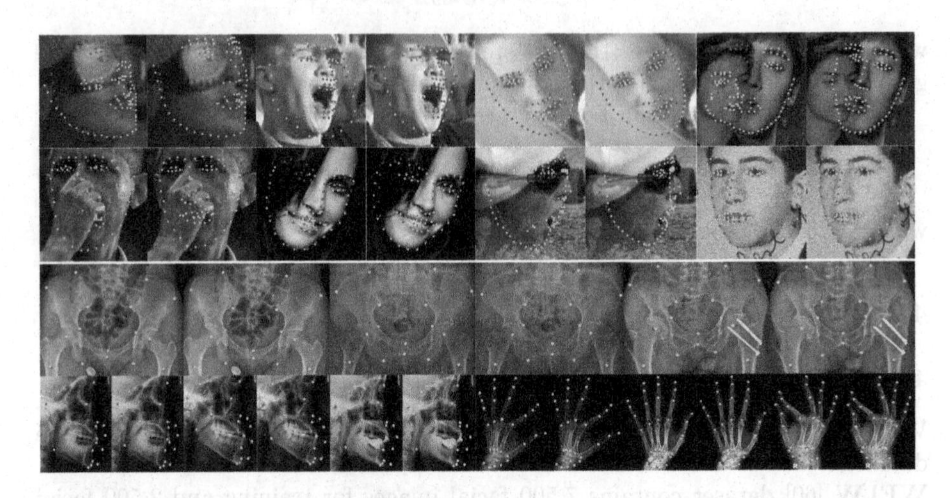

Fig. 2. Visualization of landmark detection results. Pairs of reseults are displayed side by side. For each pair, **Left image:** detection result from a SOTA method [48]. **Right image:** result produced by our method. **Green dot**: predicted landmark location. **Red dot**: groundtruth landmark location.

4.2 Experiment Settings

Evaluation Metrics: We evaluate the proposed method following two sets of metrics. For the facial image datasets, we employ the widely adopted Normalized Mean Error (NME), Area Under the Curve (AUC), Failure Rate for a maximum error of 0.1 (FR@0.1) and Cumulative Errors Distribution (CED) curve (supplementary material). To compare with previous methods, we conduct both "inter-ocular" (outer-eye-corner-distance) and "inter-pupil" (eye-center-distance) normalizations on the detected landmark coordinates.

For the Cephalometric X-ray images, we follow the original evaluation protocol to compare two sets of metrics: Mean Radial Error (MRE) which computes the average of Euclidean Distances of predicted coordinates and ground truth coordinates of all the landmarks; the corresponding Successful Detection Rate (SDR) under 2 mm, 2.5 mm, 3 mm and 4 mm. For the Hand and Pelvic X-rays, we compute MRE, Hausdorff Distance (HD) and Standard Deviations (STD). Recall that Hausdorff Distance measures the maximum value of the minimum distances between two sets of points. In our case, we aim to evaluate the error upper-bound for the detected landmarks.

Table 1. Evaluation on the WFLW dataset (98 Landmarks). *: focus on loss function. #: focus on data augmentation.

Metric	Method	Test	Pose	Expression	Illumination	Make-up	Occlusion	Blur
Mean Error %	CFSS [73]	9.07	21.36	10.09	8.30	8.74	11.76	9.96
	DVLN [61]	6.08	11.54	6.78	5.73	5.98	7.33	6.88
	LAB [60]	5.27	10.24	5.51	5.23	5.15	6.79	6.32
	SAN [17] #	5.22	10.39	5.71	5.19	5.49	6.83	5.80
	WING [20] *	5.11	8.75	5.36	4.93	5.41	6.37	5.81
	HRNet18 [48]	4.60	7.94	4.85	4.55	4.29	5.44	5.42
	STYLE [41] #	4.39	8.42	4.68	4.24	4.37	5.60	4.86
	AWING [56] *	4.36	7.38	4.58	4.32	4.27	5.19	4.96
	Ours	**4.21**	**7.36**	**4.49**	**4.12**	**4.05**	**4.98**	**4.82**
Failure Rate @0.1	CFSS [73]	20.56	66.26	23.25	17.34	21.84	32.88	23.67
	DVLN [61]	19.84	46.93	11.15	7.31	11.65	16.30	13.71
	LAB [60]	7.56	28.83	6.37	6.73	7.77	13.72	10.74
	SAN [17] #	6.32	27.91	7.01	4.87	6.31	11.28	6.60
	WING [20] *	6.00	22.70	4.78	4.30	7.77	12.50	7.76
	HRNet18 [48]	4.64	23.01	3.50	4.72	2.43	8.29	6.34
	STYLE [41] #	4.08	18.10	4.46	2.72	4.37	7.74	4.40
	AWING [56] *	**2.84**	**13.50**	**2.23**	**2.58**	2.91	5.98	**3.75**
	Ours	3.04	15.95	2.86	2.72	**1.45**	**5.29**	4.01
AUC @0.1	CFSS [73]	0.3659	0.0632	0.3157	0.3854	0.3691	0.2688	0.3037
	DVLN [61]	0.4551	0.1474	0.3889	0.4743	0.4494	0.3794	0.3973
	HRNet18 [48]	0.5237	0.2506	0.5102	0.5326	0.5445	0.4585	0.4515
	LAB [60]	0.5323	0.2345	0.4951	0.5433	0.5394	0.4490	0.4630
	SAN [17] #	0.5355	0.2355	0.4620	0.5552	0.5222	0.4560	0.4932
	WING [20] *	0.5504	0.3100	0.4959	0.5408	0.5582	0.4885	0.4932
	AWING [56] *	0.5719	0.3120	0.5149	0.5777	0.5715	0.5022	0.5120
	STYLE [41] #	**0.5913**	0.3109	0.5490	**0.6089**	0.5812	0.5164	**0.5513**
	Ours	0.5893	**0.3150**	**0.5663**	0.5953	**0.6038**	**0.5235**	0.5329

Implementation Details: Following previous studies, we crop and resize facial images into 256×256 based on the provided bounding boxes. We follow [9] to resize the Cephalometric X-rays to 640×800. For the Hand and Pelvic X-rays, we resize each image into 512×512 preserving the original height and width ratio by padding zero values to the empty regions. The proposed model is implemented in PyTorch and is experimented on a single NVIDIA Titan V GPU. We choose $\lambda_1 = \lambda_2 = 1$ for different parts in the overall loss function. HRNet18 [48] pretrained on ImageNet is used as our backbone network to extract visual feature maps for its parallel multi-resolution fusion mechanism and deep network design which fits our need for both high resolution and semantic feature representation. The last output after fusion is extracted as feature map of dimension $H \in R^{256\times64\times64}$. We employ 4 residual GCN blocks [29,31] in GCN-global and GCN-local and perform 3 iterations of GCN-local. Adjacency matrix values are initialized to 1/N so that the total weight for each node is 1 to avoid message explosion.

Table 2. Evaluation on 300W Common set, Challenge set and Fullset.

Inter-Pupil Normalization				
Method	Year	Comm.	Challenge	Full.
CFAN [68]	2014	5.50	16.78	7.69
ESR [7]	2014	5.28	17.00	7.58
SDM [63]	2013	5.57	15.40	7.52
3DDFA [74]	2016	6.15	10.59	7.01
LBF [42]	2014	4.95	11.98	6.32
CFSS [73]	2015	4.73	9.98	5.76
SeqMT [24]	2018	4.84	9.93	5.74
TCDCN [69]	2015	4.80	8.60	5.54
RCN [25]	2016	4.67	8.44	5.41
TSR [35]	2017	4.36	7.56	4.99
DVLN [61]	2017	3.94	7.62	4.66
HG-HSLE [76]	2019	3.94	7.24	4.59
DCFE [53]	2018	3.83	7.54	4.55
STYLE [41] #	2019	3.98	7.21	4.54
AWING [56] *	2019	3.77	**6.52**	4.31
LAB [60]	2018	3.42	6.98	4.12
WING [20] *	2018	**3.27**	7.18	**4.04**
Ours	2020	3.64	6.88	4.27
Inter-Ocular Normalization				
Method	Year	Comm.	Challenge	Full.
PCD-CNN [28]	2018	3.67	7.62	4.44
ODN [72]	2019	3.56	6.67	4.17
CPM+SBR [18]	2018	3.28	7.58	4.10
SAN [17] #	2018	3.34	6.60	3.98
STYLE [41] #	2019	3.21	6.49	3.86
LAB [60]	2018	2.98	5.19	3.49
HRNet18 [48]	2019	2.91	5.11	3.34
HG-HSLE [76]	2019	2.85	5.03	3.28
LUVLi [27]	2020	2.76	5.16	3.23
AWING [56] *	2019	2.72	**4.52**	3.07
Ours	2020	**2.62**	4.77	**3.04**

Table 3. Evaluation on 300W and COFW-68 testsets with the model trained on 300W training set.

300W			
Method	Year	AUC@0.1	FR@0.1
Deng *et al.* [14]	2016	0.4752	5.50
Fan *et al.* [19]	2016	0.4802	14.83
DensReg+DSM [1]	2017	0.5219	3.67
JMFA [15]	2019	0.5485	1.00
LAB [60]	2018	0.5885	0.83
HRNet18 [48]	2019	0.6041	0.66
AWING [56] *	2019	**0.6440**	0.33
Ours	2020	0.6361	**0.33**
COFW-68			
Method	Year	Mean Error %	FR@0.1
CFSS [73]	2015	6.28	9.07
HRNet18 [48]	2019	5.06	3.35
LAB [60]	2018	4.62	2.17
STYLE [41] #	2019	4.43	2.82
Ours	2020	**4.22**	**0.39**

Table 4. Evaluations on the hand X-ray and pelvic X-ray images.

Hand X-ray Dataset				
Method	Year	MRE (pix)	Hausdorff	STD
HRNet18 [48]	2019	12.79	26.36	6.07
Chen *et al.* [9]	2019	7.14	18.71	14.43
Payer *et al.* [39]	2019	6.11	16.55	4.01
Ours	2020	**5.57**	**14.83**	**3.63**
Pelvic X-ray Dataset				
Method	Year	MRE (pix)	Hausdorff	STD
HRNet18 [48]	2019	24.77	71.31	19.98
Payer *et al.* [39]	2019	20.96	68.19	21.93
Chen *et al.* [9]	2019	20.10	59.92	20.14
Ours	2020	**18.39**	**56.72**	**17.67**

4.3 Comparison with the SOTA Methods

WFLW: WFLW is a comprehensive public facial landmark detection dataset focusing on multi-discipline and difficult detection scenarios. Summary of results is shown in Table 1. Following previous works, three evaluation metrics are computed: Mean Error, FR@0.1 and AUC@0.1. Our model achieves *4.21%* mean error which outperforms all the strong state-of-the-art methods including AWING [56] which adopts a new adaptive loss function, SAN [17] and STYLE [41] which leverage additional generated images for training. The most significant improvements lie in *Make-up* and *Occlusion* subsets, where only partial landmarks are visible. Our model is able to accurately infer those hard cases based on the visible landmarks due to the benefit of preserving and leveraging graph structural knowledge. This can be further illustrated by examining the visualization results for the occlusion scenarios in Fig. 2.

300W: There are two evaluation protocols, namely inter-pupil and inter-ocular normalizations. In this paper, we conduct experiments under both settings on

Table 5. Evaluation on the public Cephalometric dataset.

Model	Year	Validation set					Test set				
		MRE	2mm	2.5mm	3mm	4mm	MRE	2mm	2.5mm	3mm	4mm
Arik *et al.* [2]	2017	-	75.37	80.91	84.32	88.25	-	67.68	74.16	79.11	84.63
HRNet18 [48]	2019	1.59	78.11	86.81	90.88	96.74	1.84	69.89	78.95	85.16	92.32
Payer *et al.* [39]	2019	1.34	81.47	89.36	93.15	97.01	1.65	69.94	78.84	85.74	93.89
Chen *et al.* [9]	2019	1.17	86.67	92.67	95.54	**98.53**	1.48	75.05	82.84	**88.53**	**95.05**
Ours	-	**1.04**	**88.49**	**93.12**	**95.72**	98.42	**1.43**	**76.57**	**83.68**	88.21	94.31

the detection results in order to comprehensively evaluate with the other state-of-the-arts. As can be seen from Table 2, our model achieves competitive results in both evaluation settings comparing to the previous best models, STYLE [41], LAB [60] and AWING [56] which are all heatmap-based. Comparing to the latest coordinate-based model ODN [72] and DVLN [61], our method achieves improvements in large margins (*27%* and *8%* respectively) which sets a remarkable milestone for coordinate-based models, closing the gap between coordinate- and heatmap-based methods.

COFW-68 and 300W testset: To verify the robustness and generalizability of our model, we conduct inference on images from COFW-68 and 300W testset using the model trained on 300W training set and validated on 300W fullset. Results summarized in Table 3 indicating our model's superior performance over most of the other state-of-the-art methods in both datasets. In particular for the COFW-68 dataset, the Mean Error and FR@0.1 are significantly improved (*5%* and *86%*) comparing to the previous best model, STYLE [41], demonstrating a strong cross-dataset generalizability of our method.

Cephalometric X-rays: We further applied our model on a public Cephalometric X-ray dataset and compare with HRNet18 [48] and three domain specific state-of-the-art models on this dataset, Arik *et al.* [2], Payer et al. [39] and Chen *et al.* [9]. As is shown in Table 5, our model significantly outperforms Arik *et al.*, HRNet18 [48] and Payer et al. [39] in all metrics. Comparing to Chen *et al.* [9], we also achieve improved overall accuracy evaluated under MRE. A closer look at the error distribution reveals that our model is able to achieve more precise localization under smaller error ranges, i.e., 2mm and 2.5mm.

Hand and Pelvic X-rays: As shown in Table 4, our model achieves susbstantial performance improvements comparing to the HRNet18 [48], Payer et al. [39] and Chen *et al.* [9] on both the Hand and Pelvic X-ray datasets. On Hand X-ray, where the bone structure can vary in different shapes depending on the hand pose, our method still achieves largely reduced Hausdorff distance as well as its standard deviation, reveling DAG's ability in capturing landmark relationships under various situations toward robust landmark detection.

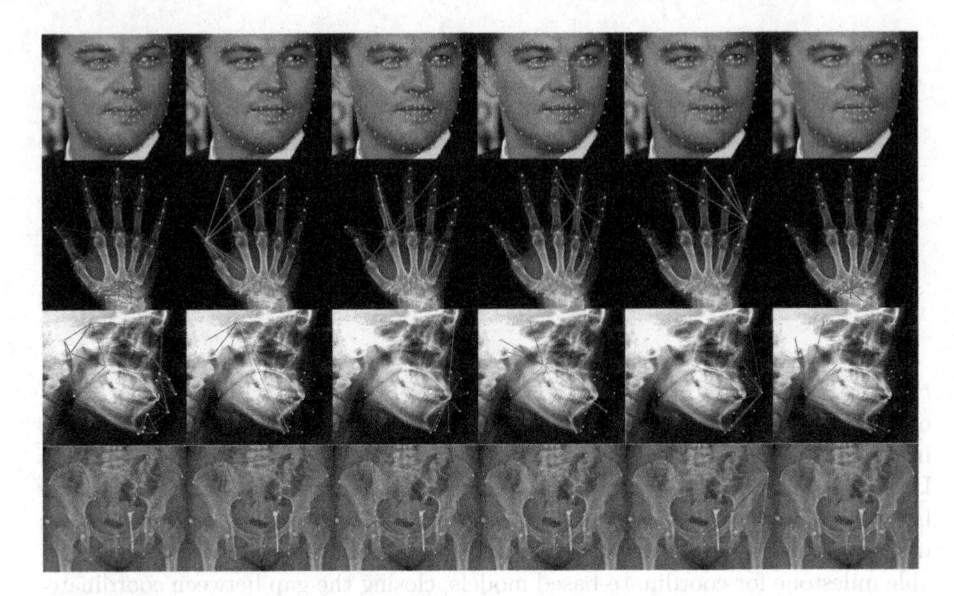

Fig. 3. Graph structure visualization. Red lines: edges. Green dots: landmarks. Deeper red means higher edge weights. [**Leftmost column**]: the constructed graphs (3 highest weighted edges for each landmark). [**Right 5 columns**]: for the 5 landmarks, the most related neighbors (10 highest weighted edges).

4.4 Graph Structure Visualization

To better understand learning outcomes, we look into the visualization on the learned graph structure. As shown in Fig. 3, the learned structures in different domains are meaningful indicating strong connections between 1) spatially close landmarks, and 2) remote but related landmarks that move coherently, e.g. symmetrical body parts. We believe the mechanism behind our algorithm is relying on these locations to provide reliable inductions when it makes movement predictions, such as similar movements by neighbors, or fixed spatial relationships by the symmetrical body parts (e.g., eyes, pelvis). With the learnable graph connectivity, we are able to capture the underlying landmarks relationships for different objects.

4.5 Ablation Studies

In this section, we examine the performance of the proposed methods by conducting ablation studies on the 300W fullset. We analyze: 1) the overall effect of using the proposed DAG to regress landmark coordinates, 2) the individual effect of learning the graph connectivity, 3) the individual effect of incorporating shape feature into the graph signal. More ablation studies can be found in the y material.

Overall effect of the proposed DAG: We analyze the effect of using DAG to regress landmark coordinates in comparison with two baselines, namely

Table 6. Ablation studies on the effectiveness of the proposed method DAG.

	VGG16	ResNet50	StackedHG4	HRNet18
Global feature	4.66	4.33	4.31	4.30
Local feature	4.42	4.10	3.96	3.72
Proposed DAG	**3.66**	**3.65**	**3.07**	**3.04**

Table 7. Ablation study on graph connectivity and shape feature.

	w.o Shape Feature	w. Shape Feature
Self	3.31	3.16
Uniform	3.16	3.12
Learned	**3.08**	**3.04**

1) *Global feature*: The last feature map of the backbone network is global average pooled to produce a feature vector, which connects to a fully connected layer to regress landmark coordinates. This approach is similar to previous coordinate regression based methods, e.g. [61,69]. 2) *Local feature*: The feature vectors are interpolated at each landmark's initial location on the last feature map of the backbone CNN. Then each landmark's feature vector is connected to a fully connected layer to regress the landmark's coordinate. To decouple the effect of the backbone strength, each experiment is conducted on four popular landmark detection backbone networks, namely VGG16 [46], ResNet50 [23], StackedHourGlass4 [37], HRNet18 [48]. Results are listed in Table 6. By comparing different regression methods with the same backbone (columnwise), DAG achieves the best results indicating the proposed framework's strong localization ability. By comparing DAG's results under different backbones (last row), we observe DAG's consistent performance boost demonstrating its effectiveness and promising generalizability.

Individual effect of learning graph connectivity: We study three kinds of graph connectivity schemes, namely 1) *Self*-connectivity: The landmarks only connect to themselves and no other landmarks. 2) *Uniform* connectivity: The landmarks connects to all other landmarks using the same edge weight. 3) *Learned* connectivity: learned edge weights as proposed. As summarized in Table 7, regardless of using shape feature or not, using uniform connectivity performs results in better performance than self-connectivity, demonstrating the importance of allowing information exchange on the graph. The learned connectivity performance the best, further demonstrating that learned edge weights further improve the effectiveness of information exchange on the graph. **Individual effect of incorporating shape feature:** We analyze the effect of incorporating the shape feature using self, uniformed and learned connectivities, respectively. As shown in Table 7, on all three types of connectivities, incorporating the proposed shape feature into graph signal results in improved performance especially for self-connective graphs, where the shape feature adds the missing global structure information.

5 Conclusion

In this paper, we introduce a robust and accurate landmark detection model named Deep Adaptive Graph (DAG). The proposed model deploys an initial landmark graph, and then deforms and progressively updates the graph by learning the adjacency matrix. Graph convolution operations follow the strong structural prior to enable effective local information exchange as well as global structural constraints for each step's movements. The superior performances on three public facial image datasets and three X-ray datasets prove both the effectiveness and generalizability of the proposed method in multiple domains.

Acknowledgments. This work is supported in part by NSF through award IIS-1722847, NIH through the Morris K. Udall Center of Excellence in Parkinson's Disease Research. The main work was done when Weijian Li was a research intern at PAII Inc.

References

1. Alp Guler, R., Trigeorgis, G., Antonakos, E., Snape, P., Zafeiriou, S., Kokkinos, I.: Densereg: Fully convolutional dense shape regression in-the-wild. In: CVPR. pp. 6799–6808 (2017)
2. Arik, S.Ö., Ibragimov, B., Xing, L.: Fully automated quantitative cephalometry using convolutional neural networks. J. Med. Imag. **4**(1), 014501 (2017)
3. Asthana, A., Zafeiriou, S., Cheng, S., Pantic, M.: Robust discriminative response map fitting with constrained local models. In: CVPR. pp. 3444–3451 (2013)
4. Bulat, A., Tzimiropoulos, G.: Binarized convolutional landmark localizers for human pose estimation and face alignment with limited resources. In: ICCV. pp. 3706–3714 (2017)
5. Bulat, A., Tzimiropoulos, G.: Super-fan: Integrated facial landmark localization and super-resolution of real-world low resolution faces in arbitrary poses with gans. In: CVPR. pp. 109–117 (2018)
6. Burgos-Artizzu, X.P., Perona, P., Dollár, P.: Robust face landmark estimation under occlusion. In: CVPR. pp. 1513–1520 (2013)
7. Cao, X., Wei, Y., Wen, F., Sun, J.: Face alignment by explicit shape regression. IJCV **107**(2), 177–190 (2014)
8. Chen, H., et al.: Anatomy-aware siamese network: Exploiting semantic asymmetry for accurate pelvic fracture detection in x-ray images (2020)
9. Chen, R., Ma, Y., Chen, N., Lee, D., Wang, W.: Cephalometric landmark detection by attentive feature pyramid fusion and regression-voting. In: Shen, D., Liu, T., Peters, T.M., Staib, L.H., Essert, C., Zhou, S., Yap, P.-T., Khan, A. (eds.) MICCAI 2019. LNCS, vol. 11766, pp. 873–881. Springer, Cham (2019). https://doi.org/10.1007/978-3-030-32248-9_97
10. Cootes, T.F., Edwards, G.J., Taylor, C.J.: Active appearance models. TPAMI **6**, 681–685 (2001)
11. Cootes, T.F., Taylor, C.J.: Active shape models-'smart snakes'. In: BMVC, pp. 266–275. Springer (1992)

12. Cootes, T.F., Taylor, C.J., Cooper, D.H., Graham, J.: Active shape models-their training and application. Comput. Vis. Image Underst. **61**(1), 38–59 (1995)
13. Cristinacce, D., Cootes, T.F.: Feature detection and tracking with constrained local models. In: BMVC. vol. 1, p. 3. Citeseer (2006)
14. Deng, J., Liu, Q., Yang, J., Tao, D.: M3 csr: Multi-view, multi-scale and multi-component cascade shape regression. Image Vision Comput. **47**, 19–26 (2016)
15. Deng, J., Trigeorgis, G., Zhou, Y., Zafeiriou, S.: Joint multi-view face alignment in the wild. TIP **28**(7), 3636–3648 (2019)
16. DeTone, D., Malisiewicz, T., Rabinovich, A.: Deep image homography estimation. arXiv preprint arXiv:1606.03798 (2016)
17. Dong, X., Yan, Y., Ouyang, W., Yang, Y.: Style aggregated network for facial landmark detection. In: CVPR. pp. 379–388 (2018)
18. Dong, X., Yu, S.I., Weng, X., Wei, S.E., Yang, Y., Sheikh, Y.: Supervision-by-registration: An unsupervised approach to improve the precision of facial landmark detectors. In: CVPR. pp. 360–368 (2018)
19. Fan, H., Zhou, E.: Approaching human level facial landmark localization by deep learning. Image Vision Comput. **47**, 27–35 (2016)
20. Feng, Z.H., Kittler, J., Awais, M., Huber, P., Wu, X.J.: Wing loss for robust facial landmark localisation with convolutional neural networks. In: CVPR. pp. 2235–2245 (2018)
21. Ghiasi, G., Fowlkes, C.C.: Occlusion coherence: Detecting and localizing occluded faces. arXiv preprint arXiv:1506.08347 (2015)
22. Han, D., Gao, Y., Wu, G., Yap, P.T., Shen, D.: Robust anatomical landmark detection with application to mr brain image registration. Comput. Med. Imag. Graph. **46**, 277–290 (2015)
23. He, K., Zhang, X., Ren, S., Sun, J.: Deep residual learning for image recognition. In: CVPR. pp. 770–778 (2016)
24. Honari, S., Molchanov, P., Tyree, S., Vincent, P., Pal, C., Kautz, J.: Improving landmark localization with semi-supervised learning. In: CVPR. pp. 1546–1555 (2018)
25. Honari, S., Yosinski, J., Vincent, P., Pal, C.: Recombinator networks: Learning coarse-to-fine feature aggregation. In: CVPR. pp. 5743–5752 (2016)
26. Jaderberg, M., Simonyan, K., Zisserman, A., et al.: Spatial transformer networks. In: NeurIPS. pp. 2017–2025 (2015)
27. Kumar, A., et al.: Luvli face alignment: Estimating landmarks' location, uncertainty, and visibility likelihood. In: CVPR. pp. 8236–8246 (2020)
28. Kumar, A., Chellappa, R.: Disentangling 3d pose in a dendritic cnn for unconstrained 2d face alignment. In: CVPR. pp. 430–439 (2018)
29. Li, G., Müller, M., Thabet, A., Ghanem, B.: Can gcns go as deep as cnns? In: CVPR (2019)
30. Lindner, C., Bromiley, P.A., Ionita, M.C., Cootes, T.F.: Robust and accurate shape model matching using random forest regression-voting. TPAMI **37**(9), 1862–1874 (2014)
31. Ling, H., Gao, J., Kar, A., Chen, W., Fidler, S.: Fast interactive object annotation with curve-gcn. In: CVPR. pp. 5257–5266 (2019)
32. Liu, X.: Generic face alignment using boosted appearance model. In: CVPR. pp. 1–8. IEEE (2007)
33. Liu, Z., Yan, S., Luo, P., Wang, X., Tang, X.: Fashion landmark detection in the wild. In: ECCV. pp. 229–245. Springer (2016)
34. Lu, Y., et al.: Learning to segment anatomical structures accurately from one exemplar. arXiv preprint arXiv:2007.03052 (2020)

35. Lv, J., Shao, X., Xing, J., Cheng, C., Zhou, X.: A deep regression architecture with two-stage re-initialization for high performance facial landmark detection. In: CVPR. pp. 3317–3326 (2017)
36. Milborrow, S., Nicolls, F.: Locating facial features with an extended active shape model. In: ECCV. pp. 504–513. Springer (2008)
37. Newell, A., Yang, K., Deng, J.: Stacked hourglass networks for human pose estimation. In: ECCV. pp. 483–499. Springer (2016)
38. Payer, C., Štern, D., Bischof, H., Urschler, M.: Regressing heatmaps for multiple landmark localization using cnns. In: MICCAI. pp. 230–238. Springer (2016)
39. Payer, C., Štern, D., Bischof, H., Urschler, M.: Integrating spatial configuration into heatmap regression based CNNs for landmark localization. MIA **54**, 207–219 (2019). https://doi.org/10.1016/j.media.2019.03.007
40. Qi, M., Li, W., Yang, Z., Wang, Y., Luo, J.: Attentive relational networks for mapping images to scene graphs. In: CVPR. pp. 3957–3966 (2019)
41. Qian, S., Sun, K., Wu, W., Qian, C., Jia, J.: Aggregation via separation: Boosting facial landmark detector with semi-supervised style translation. In: ICCV. pp. 10153–10163 (2019)
42. Ren, S., Cao, X., Wei, Y., Sun, J.: Face alignment at 3000 fps via regressing local binary features. In: CVPR. pp. 1685–1692 (2014)
43. Sagonas, C., Tzimiropoulos, G., Zafeiriou, S., Pantic, M.: 300 faces in-the-wild challenge: The first facial landmark localization challenge. In: CVPRW. pp. 397–403 (2013)
44. Saragih, J.M., Lucey, S., Cohn, J.F.: Face alignment through subspace constrained mean-shifts. In: ICCV. pp. 1034–1041. IEEE (2009)
45. Sauer, P., Cootes, T.F., Taylor, C.J.: Accurate regression procedures for active appearance models. In: BMVC. pp. 1–11 (2011)
46. Simonyan, K., Zisserman, A.: Very deep convolutional networks for large-scale image recognition. arXiv preprint arXiv:1409.1556 (2014)
47. Su, J., Wang, Z., Liao, C., Ling, H.: Efficient and accurate face alignment by global regression and cascaded local refinement. In: CVPRW (2019)
48. Sun, K., Xiao, B., Liu, D., Wang, J.: Deep high-resolution representation learning for human pose estimation. In: CVPR. pp. 5693–5703 (2019)
49. Sun, Y., Wang, X., Tang, X.: Deep convolutional network cascade for facial point detection. In: CVPR. pp. 3476–3483 (2013)
50. Tang, Z., Peng, X., Geng, S., Wu, L., Zhang, S., Metaxas, D.: Quantized densely connected u-nets for efficient landmark localization. In: ECCV. pp. 339–354 (2018)
51. Toshev, A., Szegedy, C.: Deeppose: Human pose estimation via deep neural networks. In: CVPR. pp. 1653–1660 (2014)
52. Trigeorgis, G., Snape, P., Nicolaou, M.A., Antonakos, E., Zafeiriou, S.: Mnemonic descent method: A recurrent process applied for end-to-end face alignment. In: CVPR. pp. 4177–4187 (2016)
53. Valle, R., Buenaposada, J.M., Valdés, A., Baumela, L.: A deeply-initialized coarse-to-fine ensemble of regression trees for face alignment. In: ECCV. pp. 585–601 (2018)
54. Veličković, P., Cucurull, G., Casanova, A., Romero, A., Lio, P., Bengio, Y.: Graph attention networks. arXiv preprint arXiv:1710.10903 (2017)
55. Wang, C.W., Huang, C.T., Lee, J.H., Li, C.H., Chang, S.W., Siao, M.J., Lai, T.M., Ibragimov, B., Vrtovec, T., Ronneberger, O., et al.: A benchmark for comparison of dental radiography analysis algorithms. MIA **31**, 63–76 (2016)
56. Wang, X., Bo, L., Fuxin, L.: Adaptive wing loss for robust face alignment via heatmap regression. In: ICCV. pp. 6971–6981 (2019)

57. Wang, Y., Lu, L., Cheng, C.T., Jin, D., Harrison, A.P., Xiao, J., Liao, C.H., Miao, S.: Weakly supervised universal fracture detection in pelvic x-rays. In: Shen, D., Liu, T., Peters, T.M., Staib, L.H., Essert, C., Zhou, S., Yap, P.T., Khan, A. (eds.) MICCAI, pp. 459–467. Springer International Publishing, Cham (2019)
58. Wei, S.E., Ramakrishna, V., Kanade, T., Sheikh, Y.: Convolutional pose machines. In: CVPR. pp. 4724–4732 (2016)
59. Wu, S., Tang, Y., Zhu, Y., Wang, L., Xie, X., Tan, T.: Session-based recommendation with graph neural networks. AAAI. **33**, 346–353 (2019)
60. Wu, W., Qian, C., Yang, S., Wang, Q., Cai, Y., Zhou, Q.: Look at boundary: A boundary-aware face alignment algorithm. In: CVPR. pp. 2129–2138 (2018)
61. Wu, W., Yang, S.: Leveraging intra and inter-dataset variations for robust face alignment. In: CVPRW. pp. 150–159 (2017)
62. Wu, Y., Ji, Q.: Facial landmark detection: a literature survey. IJCV **127**(2), 115–142 (2019)
63. Xiong, X., De la Torre, F.: Supervised descent method and its applications to face alignment. In: CVPR. pp. 532–539 (2013)
64. Xu, K., Hu, W., Leskovec, J., Jegelka, S.: How powerful are graph neural networks? arXiv preprint arXiv:1810.00826 (2018)
65. Yu, W., Liang, X., Gong, K., Jiang, C., Xiao, N., Lin, L.: Layout-graph reasoning for fashion landmark detection. In: CVPR. pp. 2937–2945 (2019)
66. Yu, X., Huang, J., Zhang, S., Metaxas, D.N.: Face landmark fitting via optimized part mixtures and cascaded deformable model. TPAMI **38**(11), 2212–2226 (2015)
67. Yu, X., Zhou, F., Chandraker, M.: Deep deformation network for object landmark localization. In: Leibe, B., Matas, J., Sebe, N., Welling, M. (eds.) ECCV 2016. LNCS, vol. 9909, pp. 52–70. Springer, Cham (2016). https://doi.org/10.1007/978-3-319-46454-1_4
68. Zhang, J., Shan, S., Kan, M., Chen, X.: Coarse-to-fine auto-encoder networks (CFAN) for real-time face alignment. In: Fleet, D., Pajdla, T., Schiele, B., Tuytelaars, T. (eds.) ECCV 2014. LNCS, vol. 8690, pp. 1–16. Springer, Cham (2014). https://doi.org/10.1007/978-3-319-10605-2_1
69. Zhang, Z., Luo, P., Loy, C.C., Tang, X.: Learning deep representation for face alignment with auxiliary attributes. TPAMI **38**(5), 918–930 (2015)
70. Zhao, L., Peng, X., Tian, Y., Kapadia, M., Metaxas, D.N.: Semantic graph convolutional networks for 3d human pose regression. In: CVPR. pp. 3425–3435 (2019)
71. Zhou, F., Brandt, J., Lin, Z.: Exemplar-based graph matching for robust facial landmark localization. In: ICCV. pp. 1025–1032 (2013)
72. Zhu, M., Shi, D., Zheng, M., Sadiq, M.: Robust facial landmark detection via occlusion-adaptive deep networks. In: CVPR. pp. 3486–3496 (2019)
73. Zhu, S., Li, C., Change Loy, C., Tang, X.: Face alignment by coarse-to-fine shape searching. In: CVPR. pp. 4998–5006 (2015)
74. Zhu, X., Lei, Z., Liu, X., Shi, H., Li, S.Z.: Face alignment across large poses: A 3d solution. In: CVPR. pp. 146–155 (2016)
75. Zhu, Z., Luo, P., Wang, X., Tang, X.: Deep learning identity-preserving face space. In: ICCV. pp. 113–120 (2013)
76. Zou, X., Zhong, S., Yan, L., Zhao, X., Zhou, J., Wu, Y.: Learning robust facial landmark detection via hierarchical structured ensemble. In: ICCV (2019)

3D Human Shape and Pose from a Single Low-Resolution Image with Self-Supervised Learning

Xiangyu Xu[1(✉)], Hao Chen[2], Francesc Moreno-Noguer[3], László A. Jeni[1], and Fernando De la Torre[1,4]

[1] Robotics Institute, Carnegie Mellon University, Pittsburgh, USA
[2] Electrical and Computer Engineering, Carnegie Mellon University, Pittsburgh, USA
[3] Institut de Robòtica i Informàtica Industrial (CSIC-UPC), Barcelona, Spain
[4] Facebook Reality Labs (Oculus), Pittsburgh, USA

Abstract. 3D human shape and pose estimation from monocular images has been an active area of research in computer vision, having a substantial impact on the development of new applications, from activity recognition to creating virtual avatars. Existing deep learning methods for 3D human shape and pose estimation rely on relatively high-resolution input images; however, high-resolution visual content is not always available in several practical scenarios such as video surveillance and sports broadcasting. Low-resolution images in real scenarios can vary in a wide range of sizes, and a model trained in one resolution does not typically degrade gracefully across resolutions. Two common approaches to solve the problem of low-resolution input are applying super-resolution techniques to the input images which may result in visual artifacts, or simply training one model for each resolution, which is impractical in many realistic applications.

To address the above issues, this paper proposes a novel algorithm called RSC-Net, which consists of a Resolution-aware network, a Self-supervision loss, and a Contrastive learning scheme. The proposed network is able to learn the 3D body shape and pose across different resolutions with a single model. The self-supervision loss encourages scale-consistency of the output, and the contrastive learning scheme enforces scale-consistency of the deep features. We show that both these new training losses provide robustness when learning 3D shape and pose in a weakly-supervised manner. Extensive experiments demonstrate that the RSC-Net can achieve consistently better results than the state-of-the-art methods for challenging low-resolution images.

Keywords: 3d human shape and pose · Low-resolution · Neural network · Self-supervised learning · Contrastive learning.

Electronic supplementary material The online version of this chapter (https://doi.org/10.1007/978-3-030-58545-7_17) contains supplementary material, which is available to authorized users.

© Springer Nature Switzerland AG 2020
A. Vedaldi et al. (Eds.): ECCV 2020, LNCS 12354, pp. 284–300, 2020.
https://doi.org/10.1007/978-3-030-58545-7_17

1 Introduction

3D human shape and pose estimation from 2D images is of great interest to the computer vision and graphics community. Whereas significant progress has been made in this field, it is often assumed that the input image is high-resolution and contains sufficient information for reconstructing the 3D human geometry in detail [1,2,6,21,22,24,25,34,40–42,52]. However, this assumption does not always hold in practice, since lots of images in real scenes have low resolutions, such as surveillance cameras and sports videos [35,36,38,46–48]. As a result, existing algorithms designed for high-resolution images are prone to fail when applied to low-resolution inputs as shown in Fig. 1. In this paper, we study the relatively unexplored problem of estimating 3D human shape and pose from low-resolution images.

There are two major challenges of this low-resolution 3D estimation problem. First, the resolutions of the input images in real scenarios vary in a wide range, and a network trained for one specific resolution does not always work well for another. One might consider overcoming this problem by simply training different models, one for each image resolution. However, this is impractical in terms of memory and training computation. Alternatively, one could super-resolve the images to a sufficiently large resolution, but the super-resolution step often results in visual artifacts, which leads to poor 3D estimation. To address this issue, we propose a resolution-aware deep neural network for 3D human shape and pose estimation that is robust to different image resolutions. Our network builds upon two main components: a feature extractor shared across different resolutions and a set of resolution-dependent parameters to adaptively integrate the different-level features.

Fig. 1. 3D human shape and pose estimation from a low-resolution image captured from a real surveillance video. SOTA method [25] that works well for high-resolution images performs poorly at low-resolution ones.

Another challenge we encounter is due to the fact that high-quality 3D annotations are hard to obtain, especially for in-the-wild data, and only a small portion of the training images have 3D ground truth labels [21,25], which complicates the training process. Whereas most training images have 2D keypoint labels, they are usually not sufficient for predicting the 3D outputs due to the inherent ambiguities in the 2D-to-3D mapping. This problem is further accentuated in our task, as the low-resolution 3D estimation is not well constrained and has a large solution space due to limited pixel observations. Therefore, directly training low-resolution models with incomplete information typically does not achieve good results. Inspired by the self-supervised learning [26,44], we propose a directional self-supervision loss to remedy the above issue. Specifically, we enforce the consistency across the outputs of the same input image with different resolutions, such that the results of the higher-resolution images can act as guidance for lower-resolution input. This strategy significantly improves the 3D estimation results.

In addition to enforcing output consistency, we also devise an approach to enforce consistency of the feature representations across different resolutions. Nevertheless, we find that the commonly used mean squared error is not effective in measuring discrepancies between high-dimensional feature vectors. Instead, we adapt the contrastive learning [7,14,39] which aims to maximize the mutual information across the feature representations at different resolutions, and encourages the network to produce better features for the low-resolution input.

To summarize, we make the following contributions in this work. First, we study the relatively unexplored problem of 3D human shape and pose estimation from low-resolution images and present a simple yet effective solution for it, called RSC-Net, which is based on a novel resolution-aware network that can handle arbitrary-resolution input with one single model. Second, we propose a self-supervision loss to address the issue of weak supervision. Furthermore, we introduce contrastive learning which effectively enforces the feature consistency across different resolutions. Extensive experiments demonstrate that the proposed method outperforms the state-of-the-art algorithms on challenging low-resolution inputs and achieves robust performance for high-quality 3D human shape and pose estimation.

2 Related Work

We first review the state-of-the-art methods for 3D human shape and pose estimation and then discuss the low-resolution image recognition algorithms.

3D human shape and pose estimation. Recent years have witnessed significant progress in the field of 3D human shape and pose estimation from a single image [1–3,6,9,21,22,24,25,34,40–42,49,50,52]. Existing methods for this task can be broadly categorized into two classes. The first kind of approaches generally splits the 3D human estimation process into two stages: first transforming the input image into new representations, such as human 2D keypoints [1,2,6,9,34,40], human silhouettes [2,34,40], body part segmentations [1], UV mappings [3], and optical flow [9], and then regressing the 3D human parameters [29] from the transformed outputs of the last stage either with iterative optimization [2,6] or neural networks [1,9,34,40]. As these methods map the original input images into simpler representation forms which are generally sparse and can be easily rendered, they can exploit a large amount of synthetic data for training where there are sufficient high-quality 3D labels. However, these two-stage systems are error-prone, as the errors from early stage may be accumulated or even deteriorated [21]. In addition, the intermediate results may throw away valuable information in the image such as context. More importantly, the task of the first stage, *i.e.*, to estimate the intermediate representations, is usually difficult for low-resolution images, and thereby, the aforementioned two-stage models are not suitable to solve our problem of low-resolution 3D human shape and pose estimation.

Without relying on new representations, the second kind of approaches can directly regress the 3D parameters from the input image [21,22,24,25,41,42,50], where most methods are based on deep neural networks. While being concise and not requiring the estimation of intermediate results, these methods usually suffer from the problem of weak supervision due to a lack of high-quality 3D ground truth. Most existing works focus on this problem and have developed different techniques to solve it. As a typical example, Kanazawa *et al.* [21] include a generative adversarial network (GAN) [11] to constrain the solution space using the prior learned from 3D human data. However, we find the GAN-based algorithm less effective for low-resolution input images where substantially fewer pixels are available. Kolotouros [25] *et al.* integrate the optimization-based method [6] into the training process of the deep network to more effectively exploit the 2D keypoints. While achieving good improvements over [21] on high-resolution images, [25] cannot be easily applied to low-resolution input, as the low-resolution network cannot provide good initial results to start the optimization loop. In addition, it significantly increases the training time. On the other hand, temporal information has also been exploited to enforce temporal consistency of the 3D estimation results, which however requires high-resolution video input [22,24,50]. Different from the above methods, we propose a 3D human shape and pose estimation algorithm using a single low-resolution image as input. We propose self-supervision loss and contrastive feature loss which effectively remedy the problem of insufficient 3D supervision.

Low-resolution image recognition. While there is no prior work for low-resolution 3D human shape and pose estimation, there are some related approaches to process low-resolution inputs for other image recognition tasks, such as 2D body pose estimation [35], face recognition [8,10,48], image classification [46], image retrieval [37,43], and object detection [12,27]. Most of these methods address the low-resolution issue by enhancing the degraded input, in either the image space [8,12,46] or the feature space [10,27,37,43]. One typical image-space method [12] applies a super-resolution network which is trained to improve both the image quality (*i.e.*, per-pixel similarity such as PSNR) and the object detection performance. However, the loss functions for higher PSNR and better recognition performance do not always agree with each other, which may lead to inferior solutions. Moreover, the super-resolution model may bring unpleasant artifacts, resulting in domain gap between the super-resolved and

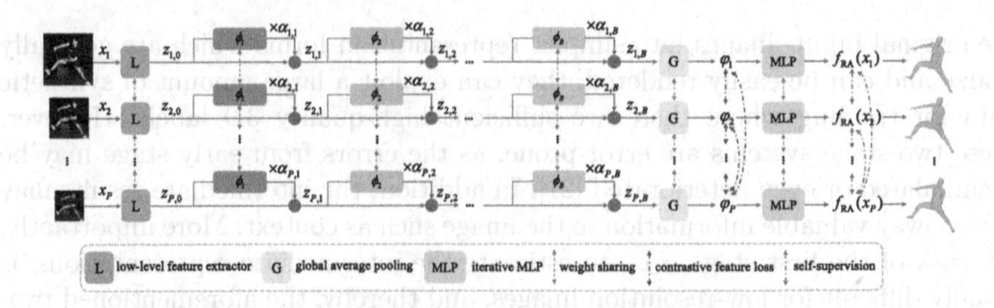

Fig. 2. Overview of the proposed algorithm. The resolution-aware network f_{RA} is trained with a combination of the basic loss (omitted in the figure for simplicity), self-supervision loss and contrastive feature loss. The modules with the same colors are shared across different resolutions, while the matrix α is resolution-dependent. Note that we resize the different resolution inputs $\{x_i\}$ to 224×224 with bicubic interpolation before feeding them into the network.

real high-resolution images. Unlike the image enhancement based approaches, the feature enhancement based methods [10,27,37,43] are not distracted by the image quality loss and thus can better focus on improving the recognition performance. As a representative example, Ge *et al.* [10] use mean squared error (MSE) to enforce the similarity between the features of low-resolution and high-resolution images, which achieves good results for face recognition. Different from the above approaches, Neumann *et al.* [35] propose a novel method for low-resolution 2D body pose estimation by predicting a probability map with Gaussian Mixture Model, which, however, cannot be easily extended to 3D human shape and pose estimation. In this work, we apply the feature enhancement strategy to low-resolution 3D human shape and pose estimation. Instead of using MSE for measuring feature similarity, we introduce the contrastive learning [39] which can more effectively maximize the mutual information across the features of different resolutions. In addition, we handle different-resolution input with a resolution-aware neural network.

3 Algorithm

We study the problem of 3D human shape and pose estimation for a low-resolution image x. Instead of training different networks for each specific resolution, we propose a resolution-aware neural network f_{RA} which can handle the complex inputs with different resolutions. We first introduce the 3D human representation model and the baseline network for 3D human estimation with a single 2D image. Then we describe the proposed resolution-aware model as well as the self-supervision loss and the contrastive learning strategy for training the network. An overview of our method is shown in Fig. 2.

3.1 3D Human Representation

We represent the 3D human body using the Skinned Multi-Person Linear (SMPL) model [29]. The SMPL is a parametric model which describes the body shape and pose with two sets of parameters β and θ, respectively. The body shape is represented by a basis in a low-dimensional shape space learned from a training set of 3D human scans, and the parameters $\beta \in \mathbb{R}^{10}$ are coefficients of the basis vectors. The body pose is defined by a skeleton rig with $K = 24$ joints including the body root, and the pose parameters $\theta \in \mathbb{R}^{3K}$ are the axis-angle representations of the relative rotation between different body parts as well as the global rotation of the body root. With β and θ, we can obtain the 3D body mesh: $M = f_{\text{SMPL}}(\beta, \theta)$, where $M \in \mathbb{R}^{N \times 3}$ is a triangulated surface with $N = 6890$ vertices.

Similar to the prior works [21,25], we can predict the 3D locations of the body joints X with the body mesh using a pretrained mapping matrix $W \in \mathbb{R}^{K \times N}$:

$$X \in \mathbb{R}^{K \times 3} = WM. \tag{1}$$

With the 3D human joints, we use a perspective camera model to project the body joints from 3D to 2D. Assuming the camera parameters are $\delta \in \mathbb{R}^3$ which define the 3D translation of the camera, the 2D keypoints can be formulated as:

$$J \in \mathbb{R}^{K \times 2} = f_{\text{project}}(X, \delta), \tag{2}$$

where f_{project} is the perspective projection function [13].

3.2 Resolution-Aware 3D Human Estimation

Baseline network. Similar to the existing methods [21,25], we use the deep convolutional neural network (CNN) for 3D human estimation, where the ResNet-50 [15] is employed to extract features from the input image. The building block of the ResNet (*i.e.*, ResBlock [16]) can be formulated as:

$$z_k = z_{k-1} + \phi_k(z_{k-1}), \tag{3}$$

where z_k is the output features of the k-th ResBlock, and ϕ_k represents the nonlinear function used to learn the feature residuals, which is modeled by several convolutional layers with ReLU activation [33]. The ResNet stacks B ResBlocks together, and the final output can be written as:

$$z_B = z_0 + \sum_{k=1}^{B} \phi_k(z_{k-1}), \tag{4}$$

where z_0 is the low-level features extracted from the input image x with convolutional layers, and z_B is a combination of different level residual maps from all the ResBlocks. Note that we do not explicitly consider the downsampling ResBlocks in (4) for clarity. With the output features of the ResNet, we can use global average pooling to obtain a feature vector φ and employ an iterative MLP for regressing the 3D parameters β, θ, δ similar to [21,25].

Resolution-aware network. The baseline network is originally designed for high-resolution images with input size 224 × 224 pixels, whereas the image resolutions for human in real scenarios can be much lower and vary in a wide range. A straightforward way to deal with these low-resolution inputs is to train different networks for all possible resolutions and choose the suitable one for each test image. However, this is impractical for real applications.

To solve this problem, we propose a resolution-aware network, and the main idea is that the different-resolution images with the same contents are largely similar as shown in Fig. 2 and can share most parts of the feature extractor. And only a small amount of parameters are needed to be resolution-dependent to account for the characteristics of different image resolutions. Towards this end, instead of directly combining the different level features as in (4), we learn a matrix α to adaptively fuse the residual maps from the ResBlocks for each input resolution as shown in Fig. 2, such that different resolutions can have suitable features for 3D estimation. Specifically, we formulate the output of the proposed resolution-aware network as:

$$z_{i,B} = z_{i,0} + \sum_{k=1}^{B} \alpha_{i,k} \phi_k(z_{i,k-1}), \quad i = 1, 2, \ldots, R, \tag{5}$$

where i is the index for different image resolutions, and larger i indicates smaller image. $i = 1$ corresponds to the original high-resolution input. $\alpha \in \mathbb{R}^{R \times B}$, where R denotes the number of all the image resolutions considered in this work. $z_{i,k}$ and $\alpha_{i,k}$ respectively represent the output and the fusion weight of the k-th ResBlock for the i-th input resolution. According to (5), the original ResBlock in (3) is modified as: $z_{i,k} = z_{i,k-1} + \alpha_{i,k}\phi_k(z_{i,k-1})$. Note that we use a slightly different notation here compared with (3) and (4) which do not have the index i for image resolution, as the baseline network is not resolution-aware and applies the same operations to different resolution inputs.

Note that for training the above network, each high-resolution image in the training dataset needs to conduct the downsampling operation for $R - 1$ times, such that each row of parameters in α have their corresponding training data. Whereas the original training datasets [4,18,28,31,32] are already quite large for the diversity of the training images, it will be further augmented by $R-1$ times, which significantly increases the computational burden of the training process. To remedy the training issues and reduce the parameters in α, we divide all the R resolutions into P ranges and only learn one set of parameters for each range. We design the first resolution range to only have the original high-resolution image, and for the other ranges, we randomly sample a resolution in each range during each training iteration. The training images with different resolutions can be denoted as $\{x_i, i = 1, 2, \ldots, P\}$ where the smaller images $x_2, x_3, \ldots, x_P$ are synthesized from the same high-resolution image x_1 with bicubic interpolation. With this strategy, the training set can be much smaller without losing diversity, and we can have a lower-dimensional matrix $\alpha \in \mathbb{R}^{P \times B}$, where the number of parameters can be reduced from RB to PB. During inference, we first decide the resolution range of the input image and then choose the suitable row of parameters in α for usage in the network.

Progressive training. Directly using different resolution images for training all at once can lead to difficulties in optimizing the proposed model since the network needs to handle inputs with complex resolution properties simultaneously. Instead, we train the proposed network in a progressive manner, where the higher-resolution images are easier to handle and thus first processed in training, and more challenging ones with lower resolutions are subsequently added. In this way, we alleviate the difficulty of the training process and the proposed model can evolve progressively.

Basic loss function. Similar to the previous algorithms [21,25], the basic loss of our network is a combination of 3D and 2D losses. Suppose the output of the proposed network for input image x_i is $[\hat{\beta}_i, \hat{\theta}_i, \hat{\delta}_i] = f_{\text{RA}}(x_i)$ where i is the resolution index, and $X_g, J_g, \beta_{\text{g}}, \theta_{\text{g}}$ are the ground truth 3D and 2D keypoints and SMPL parameters. The basic loss function can be written as:

$$L_{\text{b}} = \sum_i \|[\hat{\beta}_i, \hat{\theta}_i] - [\beta_{\text{g}}, \theta_{\text{g}}]\|_2^2 + \lambda_1 \|\hat{X}_i - X_{\text{g}}\|_2^2 + \lambda_2 \|\hat{J}_i - J_{\text{g}}\|_2^2, \tag{6}$$

where $\hat{X}_i$ and $\hat{J}_i$ are estimated with (1) and (2), respectively. λ_1 and λ_2 are hyper-parameters for balancing different terms. Note that while all the training images have 2D keypoint labels J_g in (6), only a limited portion of them have 3D ground truth X_g, β_g, θ_g. For the training images without 3D labels, we simply omit the first two terms in (6) similar to [21,22,25].

3.3 Self-Supervision

The 3D human shape and pose estimation is a weakly-supervised problem as only a small part of the training data has 3D labels, and it is especially the case for in-the-wild images where accurate 3D annotations cannot be easily captured. This issue gets even worse for the low-resolution images, as the 3D estimation is not well constrained by limited pixel observations, which requires strong supervision signal during training to find a good solution.

To remedy this problem, we propose a self-supervision loss to assist the basic loss for training the resolution-aware network f_{RA}. This new loss term is inspired by the self-supervised learning algorithm [26] which improves the training by minimizing the MSE between the network predictions under different input augmentation conditions. For our problem, we naturally have the same input with different data augmentations, *i.e.*, the different-resolution images synthesized from the same high-resolution image. Thus, the self-supervision loss can be formulated by enforcing the consistency across the outputs of different image resolutions:

$$\sum_{i,j} \|f_{\text{RA}}(x_i) - f_{\text{RA}}(x_j)\|_2^2. \tag{7}$$

However, a major difference between our work and the original self-supervision method [26] is that we are generally more confident in the predictions of the higher-resolution images while [26] treats the results under different input augmentations equally. To exploit this prior knowledge, we improve the loss in (7) and propose a directional self-supervision loss:

$$L_{\text{s}} = \sum_{i,j} w_{i,j} \| \bar{f}_{\text{RA}}(x_i) - f_{\text{RA}}(x_j) \|_2^2, \\ w_{i,j} = \mathbb{1}(j - i > 0) \cdot (j - i), \tag{8}$$

where $w_{i,j}$ is the loss weight for an image pair (x_i, x_j), and it is nonzero only when x_i has higher-resolution than x_j. $\bar{f}_{\text{RA}}$ represents a fixed network, and the gradients are not back-propagated through it such that the lower-resolution image x_j is encouraged to have similar output to higher-resolution x_i but not vice versa. In addition, since higher-resolution results usually provide higher-quality guidance during training, we give a larger weight to larger resolution difference by the term $(j-i)$ in $w_{i,j}$. Note that we use all the resolutions that are higher than x_j as supervision in (8) instead of only using the highest resolution x_1, as the results of x_j and x_1 can differ from each other significantly for a large j, and the results of the resolutions between x_j and x_1 can act as soft targets during training. In [17], Hinton *et al.* show the effectiveness of the "dark knowledge" in soft targets, and similarly for low-resolution 3D human shape and pose estimation, we also find that it is important to provide the challenging input a hierarchical supervision signal such that the learning targets are not too difficult for the network to follow.

3.4 Contrastive Learning

While the self-supervision loss enforces the consistency of the network outputs across different image resolutions, we can further improve the model training by encouraging the consistency of the final feature representation φ encoded by the network, such that features of lower-resolution images are closer to those of higher-resolution ones. Similar to (8), we have the feature consistency loss:

$$L_{\text{f}} = \sum_{i,j} w_{i,j} g(\bar{\varphi}_i, \varphi_j), \tag{9}$$

where φ_i is the feature vector of the i-th resolution input image x_i, and $\bar{\varphi}$ denotes a fixed feature extractor without gradient back-propagation. $w_{i,j}$ is identical to that in (8). The function g is used to measure the distance between two feature vectors, and a straightforward choice is the MSE as in (8). However, the extracted features φ usually have very high dimensions, and the MSE loss is not effective in modeling correlations of the complex structures in high-dimensional representations, due to the fact that it can be decomposed element-wisely, *i.e.*, assuming independence between elements in the feature vectors [39,45]. Moreover, the unimodal losses such as MSE can be easily affected by the noise or insignificant structures in the features, while a better loss function should exploit more global structures [39].

Towards this end, we propose a contrastive feature loss similar to [7,14,39,45] to maximize the mutual information across the feature representations of different resolutions. The main idea behind our contrastive loss is to encourage the feature representation to be close for the same image with different resolutions but far for different images. Mathematically, the contrastive function can be written as:

$$g(\bar{\varphi}_i, \varphi_j) = -\log \frac{\exp(s(\bar{\varphi}_i, \varphi_j)/\tau)}{\exp(s(\bar{\varphi}_i, \varphi_j)/\tau) + \sum_{q \in \mathcal{Q}} \exp(s(q, \varphi_j)/\tau)}, \tag{10}$$

where s represents the cosine similarity function, and τ is a temperature hyper-parameter. φ_i, φ_j are the features of the same input with different resolutions. $\mathcal{Q}$ is a queue of data samples, which is constructed and progressively updated during training, and $\varphi_i, \varphi_j \notin \mathcal{Q}$. We use a method similar to [14] to update the queue, *i.e.*, after each iteration, the current mini-batch is enqueued, and the oldest mini-batch in the queue is removed. Supposing the size of the queue is $|\mathcal{Q}|$, the contrastive loss is essentially a $(|\mathcal{Q}| + 1)$-way softmax-based classifier which classifies different resolutions (φ_i, φ_j) as a positive pair while different contents (q, φ_j) as a negative pair. As the feature extractor of the higher resolution image does not have gradients in (10), the proposed loss function enforces the network to generate higher-quality features for the low-resolution input image.

Our final loss is a combination of the basic loss, self-supervision loss, and contrastive feature loss: $L_\mathrm{b} + \lambda_\mathrm{s} L_\mathrm{s} + \lambda_\mathrm{f} L_\mathrm{f}$, where λ_s and λ_f are hyper-parameters.

4 Experiments

We first describe the implementation details of the proposed RSC-Net. Then we compare our results with the state-of-the-art 3D human estimation approaches for different image resolutions. We also perform a comprehensive ablation study to demonstrate the effect of our contributions.

4.1 Implementation Details

We train our model and the baselines using a combination of 2D and 3D datasets similar to previous works [21,25]. For the 3D datasets, we use Human3.6M [18] and MPI-INF-3DHP [32] with ground truth of 3D keypoints, 2D keypoints, and SMPL parameters. These datasets are mostly captured in constrained environments, and models trained on them do not generalize well to diverse images in real world. For better performance on in-the-wild data, we also use the 2D datasets including LSP [19], LSP-Extended [20], MPII [4], and MS COCO [28], which only have 2D keypoint labels. We crop the human regions from the images and resize them to 224×224. Images with significant occlusions or small human are discarded from the dataset. We consider human image resolutions ranging from 224 to 24. As introduced in Sect. 3.2, we split all the resolutions into $P = 5$

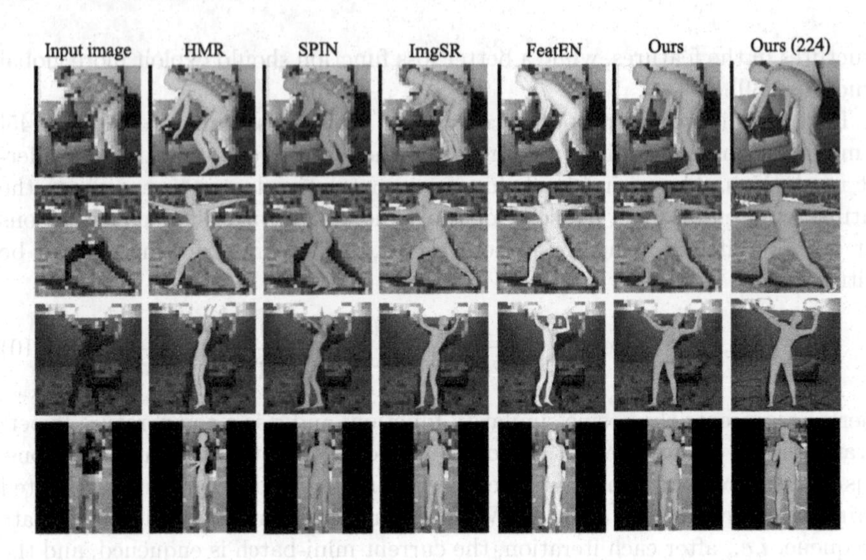

Fig. 3. Visual comparisons with the state-of-the-art methods on challenging low-resolution input. The input image has a resolution of 32 × 32. The results of high-resolution images are also included as a reference. All the baselines are trained with the same training data as our method.

Table 1. Quantitative evaluations against the state-of-the-arts on 3DPW [31].

Methods	MPJPE				MPJPE-PA			
	176	96	52	32	176	96	52	32
HMR	117.86	118.91	125.95	142.29	70.28	70.89	73.64	79.73
SPIN	112.72	113.60	120.71	137.61	69.20	69.40	72.21	78.44
ImgSR	116.47	117.74	127.78	146.58	66.62	67.48	72.34	81.07
FeaEN	107.97	109.42	119.08	143.51	61.37	62.13	66.62	77.21
Ours	**96.36**	**97.36**	**103.49**	**117.12**	**58.98**	**59.34**	**61.81**	**67.59**

ranges: $\{224, (224, 128], (128, 64], (64, 40], (40, 24]\}$, where the first range corresponds to the original high-resolution image x_1. We obtain the lower-resolution images by downsampling the high-resolution images and resize them back to 224 with bicubic interpolation. During training, we apply data augmentations to the images including Gaussian noise, color jitters, rotation, and random flipping. For the loss functions, we set $\lambda_1 = 5$, $\lambda_2 = 5$, $\lambda_s = 0.1$, and $\lambda_f = 0.1$. For contrastive learning, we set the size of the queue as 8192 and $\tau = 0.1$ in (10) similar to [7]. As in [24], we initialize the baseline networks and our model with the parameters of [25]. We use the Adam algorithm [23] to optimize the network with a learning rate 5e-5. Similar to [24], we conduct evaluations on a large in-the-wild dataset 3DPW [31] with 3D joint ground truth to demonstrate the strength of our model in an in-the-wild setting. We also provide results for constrained indoor images using the MPI-INF-3DHP dataset [32]. Following [21,24,25], we compute the procrustes aligned mean per joint position error (MPJPE-PA) and mean per joint position error (MPJPE) for measuring the 3D keypoint accuracy. To evaluate the performance of different image resolutions, we report results for the middle point of each resolution range, *i.e.*, 176, 96, 52, and 32.

4.2 Comparison to State-of-the-Art Methods

We compare against the state-of-the-art 3D human shape and pose estimation methods HMR [21] and SPIN [25] by fine-tuning them on different resolution images with the same training settings as our model. Since no previous approach has focused on the problem of low-resolution 3D human shape and pose estimation, we adapt the low-resolution image recognition algorithms to our task as new baselines, including both image super-resolution based [12] and feature enhancement based [43]. For the image super-resolution based method (denoted as ImgSR), we first use a state-of-the-art network RDN [51] to super-resolve the low-resolution image, and the output is then fed into SPIN [25] for regressing the SMPL parameters. Similar to [12], the network is trained to improve both the perceptual image quality and the 3D human shape and pose estimation accuracy. For feature enhancement (denoted as FeaEN), we apply the strategy in [43] which uses a GAN loss to enhance the discriminative ability of the low-resolution features for better image retrieval performance. Nevertheless, we find the WGAN [5] used in the original work [43] does not work well in our experiments, and we instead use the LSGAN [30] combined with the basic loss (6) to train a stronger baseline network.

Table 2. Quantitative evaluations against the state-of-the-arts on MPI-INF-3DHP [32].

Methods	MPJPE				MPJPE-PA			
	176	96	52	32	176	96	52	32
HMR	114.89	113.27	114.82	133.25	74.77	74.45	76.35	85.30
SPIN	108.46	108.25	113.36	127.27	71.19	71.53	74.76	83.38
ImgSR	107.98	107.56	112.14	125.91	72.13	72.76	75.64	83.52
FeaEN	110.40	109.91	113.09	124.99	71.49	71.52	73.92	81.80
Ours	**103.36**	**103.39**	**106.04**	**115.80**	**70.01**	**70.27**	**72.56**	**78.68**

As shown in Table 1 and 2, the proposed method compares favorably against the baseline approaches on both 3DPW and MPI-INF-3DHP datasets for all the image resolutions. Note that we achieve significant improvement over the baselines on the 3DPW dataset as shown in Table 1, which demonstrates the effectiveness of the proposed method on the challenging in-the-wild images. We also provide a qualitative comparison against the baseline models in Fig. 3, where the proposed method generates higher-quality 3D human estimation results on the challenging low-resolution input.

4.3 Ablation Study

We provide an ablation study using the 3DPW dataset in Fig. 4 and Table 3 to evaluate the proposed resolution-aware network, self-supervision loss, and contrastive feature loss. We first compare the proposed resolution-aware network with the baseline model ResNet50 [15,21]. As shown by "RA" and "Ba" in Table 3, our network can obtain slightly better results than the baseline network with the basic loss (6) as loss function. Further, we can achieve a more significant improvement over the baseline when adding the self-supervision loss (8) for training, *i.e.*, "RA+SS" *vs.* "Ba+SS", which further demonstrates the effectiveness of the resolution-aware structure.

Table 3. Ablation study of the proposed method. Ba: baseline network with basic loss function, RA: resolution-aware network with basic loss function, SS: self-supervision loss, MS: MSE feature loss, CD: cosine distance feature loss, CL: contrastive learning feature loss.

Methods	MPJPE				MPJPE-PA			
	176	96	52	32	176	96	52	32
Ba	112.26	115.18	124.88	143.63	65.04	66.41	71.12	79.43
Ba+SS	107.51	109.58	116.54	128.88	62.32	63.27	66.78	72.49
RA	111.55	112.18	118.70	135.29	64.53	68.88	68.01	75.49
RA+SS	102.56	104.18	110.17	124.23	60.17	60.84	63.71	69.87
RA+SS+MS	105.96	106.15	111.33	124.85	60.90	61.76	64.55	70.40
RA+SS+CD	104.95	105.96	111.41	125.08	61.29	61.91	64.30	70.17
RA+SS+CL	**96.36**	**97.36**	**103.49**	**117.12**	**58.98**	**59.34**	**61.81**	**67.59**

Second, we use the self-supervision loss in (8) to exploit the consistency of the outputs of the same input image with different resolutions. By comparing "RA+SS" against "RA" in Table 3, we show that the self-supervision loss is important for addressing the weak supervision issue of 3D human pose and shape estimation and thus effectively improves the results. The comparison between "Ba+SS" and "Ba" also leads to similar conclusions.

In addition, we propose to enforce the consistency of the features across different image resolutions. However, a normally-used MSE loss does not work well as show in "RA+SS+MS" of Table 3, which is mainly due to that the unimodal losses are not effective in modeling the correlations between high-dimensional vectors and can be easily affected by noise and insignificant structures in the embedded features [39]. In contrast, the proposed contrastive feature loss can more effectively improve the feature representations by maximizing the mutual information across the features of different resolutions, and achieve better results as in "RA+SS+CL" of Table 3. Note that we adopt the cosine similarity in the contrastive feature loss (10) similar to prior methods [14,39,45]. Alternatively, one may only use the cosine distance function for measuring the distance of two features instead of using the whole contrastive loss (10). Nevertheless, this strategy does not work well as shown by "RA+SS+CD" in Table 3, which demonstrates the effectiveness of the proposed algorithm.

Analysis of training strategies. We also provide a detailed analysis of the alternative training strategies of our model. First, as described in Sect. 3.2, we train our model as well as the baselines in a progressive manner to deal with the challenging multi-resolution input. As shown in the first row of Table 4 (*i.e.*, "w/o PT"), directly training the model for all image resolutions without the progressive strategy leads to degraded results.

Second, the original self-supervision loss (7) treats the images under different augmentations equally, while we are generally more confident in the high-resolution predictions. Therefore, we propose a directional self-supervision loss in (8) to exploit this prior knowledge. As shown in the second row of Table 4 (*i.e.*, "w/ SS-o"), using the original self-supervision loss (7) is not able to achieve high-quality results, as the network can minimize (7) by simply degrading the high-resolution predictions without improving the results of low resolution. In addition, we provide hierarchical supervision for low-resolution images in (8) which can act as soft targets during training. As shown in Table 4, only using the highest-resolution predictions as guidance (*i.e.*, "w/ SS-h") cannot produce as good results as the proposed approach (*i.e.*, "full model").

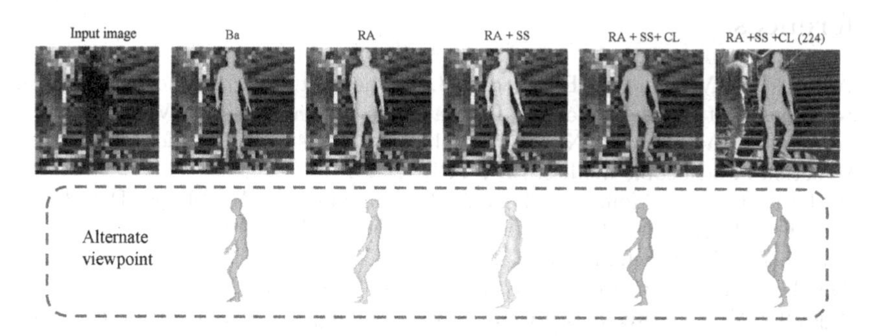

Fig. 4. Visual example which shows the effectiveness of the resolution-aware network, the self-supervision loss, and the contrastive learning feature loss.

Table 4. Analysis of the alternative training strategies. PT: Progressive Training, SS-o: original self-supervision loss, SS-h: only using the highest-resolution for supervision.

Methods	MPJPE				MPJPE-PA			
	176	96	52	32	176	96	52	32
w/o PT	105.11	106.60	113.41	127.05	61.46	62.22	65.47	71.30
w/ SS-o	143.31	142.32	145.61	156.25	77.75	77.51	79.06	82.97
w/ SS-h	104.16	105.24	109.94	122.01	62.46	62.73	64.47	68.89
full model	**96.36**	**97.36**	**103.49**	**117.12**	**58.98**	**59.34**	**61.81**	**67.59**

5 Conclusion

In this work, we study the challenging problem of low-resolution 3D human shape and pose estimation and present an effective solution, the RSC-Net. We propose a resolution-aware neural network which can deal with different resolution images with a single model. For training the network, we propose a directional self-supervision loss which can exploit the output consistency across different resolutions to remedy the issue of lacking high-quality 3D labels. In addition, we introduce a contrastive feature loss which is more effective than MSE for measuring high-dimensional vectors and helps learn better feature representations. Our method performs favorably against the state-of-the-art methods on different resolution images and achieves high-quality results for low-resolution 3D human shape and pose estimation.

References

1. Alldieck, T., Magnor, M., Bhatnagar, B.L., Theobalt, C., Pons-Moll, G.: Learning to reconstruct people in clothing from a single rgb camera. In: CVPR (2019)
2. Alldieck, T., Magnor, M., Xu, W., Theobalt, C., Pons-Moll, G.: Video based reconstruction of 3d people models. In: CVPR (2018)
3. Alldieck, T., Pons-Moll, G., Theobalt, C., Magnor, M.: Tex2shape: Detailed full human body geometry from a single image. In: ICCV (2019)
4. Andriluka, M., Pishchulin, L., Gehler, P., Schiele, B.: 2d human pose estimation: New benchmark and state of the art analysis. In: CVPR (2014)
5. Arjovsky, M., Chintala, S., Bottou, L.: Wasserstein generative adversarial networks. In: ICML (2017)
6. Bogo, F., Kanazawa, A., Lassner, C., Gehler, P., Romero, J., Black, M.J.: Keep it smpl: Automatic estimation of 3d human pose and shape from a single image. In: ECCV (2016)
7. Chen, T., Kornblith, S., Norouzi, M., Hinton, G.: A simple framework for contrastive learning of visual representations. In: ICML (2020)
8. Cheng, Z., Zhu, X., Gong, S.: Low-resolution face recognition. In: ACCV (2018)
9. Doersch, C., Zisserman, A.: Sim2real transfer learning for 3d human pose estimation: motion to the rescue. In: NeurIPS (2019)
10. Ge, S., Zhao, S., Li, C., Li, J.: Low-resolution face recognition in the wild via selective knowledge distillation. TIP **28**(4), 2051–2062 (2018)

11. Goodfellow, I., et al.: Generative adversarial nets. In: NIPS (2014)
12. Haris, M., Shakhnarovich, G., Ukita, N.: Task-driven super resolution: Object detection in low-resolution images. arXiv:1803.11316 (2018)
13. Hartley, R., Zisserman, A.: Multiple view geometry in computer vision. Cambridge University Press (2003)
14. He, K., Fan, H., Wu, Y., Xie, S., Girshick, R.: Momentum contrast for unsupervised visual representation learning. In: CVPR (2020)
15. He, K., Zhang, X., Ren, S., Sun, J.: Deep residual learning for image recognition. In: CVPR (2016)
16. He, K., Zhang, X., Ren, S., Sun, J.: Identity mappings in deep residual networks. In: ECCV (2016)
17. Hinton, G., Vinyals, O., Dean, J.: Distilling the knowledge in a neural network. arXiv:1503.02531 (2015)
18. Ionescu, C., Papava, D., Olaru, V., Sminchisescu, C.: Human3.6m: Large scale datasets and predictive methods for 3d human sensing in natural environments. TPAMI **36**(7), 1325–1339 (2013)
19. Johnson, S., Everingham, M.: Clustered pose and nonlinear appearance models for human pose estimation. In: BMVC (2010)
20. Johnson, S., Everingham, M.: Learning effective human pose estimation from inaccurate annotation. In: CVPR (2011)
21. Kanazawa, A., Black, M.J., Jacobs, D.W., Malik, J.: End-to-end recovery of human shape and pose. In: CVPR (2018)
22. Kanazawa, A., Zhang, J.Y., Felsen, P., Malik, J.: Learning 3d human dynamics from video. In: CVPR (2019)
23. Kingma, D., Ba, J.: Adam: A method for stochastic optimization. In: ICLR (2014)
24. Kocabas, M., Athanasiou, N., Black, M.J.: Vibe: Video inference for human body pose and shape estimation. In: CVPR (2020)
25. Kolotouros, N., Pavlakos, G., Black, M.J., Daniilidis, K.: Learning to reconstruct 3d human pose and shape via model-fitting in the loop. In: ICCV (2019)
26. Laine, S., Aila, T.: Temporal ensembling for semi-supervised learning. In: ICLR (2017)
27. Li, J., Liang, X., Wei, Y., Xu, T., Feng, J., Yan, S.: Perceptual generative adversarial networks for small object detection. In: CVPR (2017)
28. Lin, T.Y., et al.: Microsoft coco: Common objects in context. In: ECCV (2014)
29. Loper, M., Mahmood, N., Romero, J., Pons-Moll, G., Black, M.J.: Smpl: A skinned multi-person linear model. ACM Trans. Graph. **34**(6), 248 (2015)
30. Mao, X., Li, Q., Xie, H., Lau, R.Y., Wang, Z., Paul Smolley, S.: Least squares generative adversarial networks. In: ICCV (2017)
31. von Marcard, T., Henschel, R., Black, M.J., Rosenhahn, B., Pons-Moll, G.: Recovering accurate 3d human pose in the wild using imus and a moving camera. In: ECCV (2018)
32. Mehta, D., et al.: Monocular 3d human pose estimation in the wild using improved cnn supervision. In: 3DV (2017)
33. Nair, V., Hinton, G.E.: Rectified linear units improve restricted boltzmann machines. In: ICML (2010)
34. Natsume, R., et al.: Siclope: Silhouette-based clothed people. In: CVPR (2019)
35. Neumann, L., Vedaldi, A.: Tiny people pose. In: ACCV (2018)
36. Nishibori, K., Takahashi, T., Deguchi, D., Ide, I., Murase, H.: Exemplar-based human body super-resolution for surveillance camera systems. In: International Conference on Computer Vision Theory and Applications (VISAPP) (2014)

37. Noh, J., Bae, W., Lee, W., Seo, J., Kim, G.: Better to follow, follow to be better: Towards precise supervision of feature super-resolution for small object detection. In: ICCV (2019)
38. Oh, S., et al.: A large-scale benchmark dataset for event recognition in surveillance video. In: CVPR (2011)
39. Oord, A.v.d., Li, Y., Vinyals, O.: Representation learning with contrastive predictive coding. arXiv:1807.03748 (2018)
40. Pavlakos, G., Zhu, L., Zhou, X., Daniilidis, K.: Learning to estimate 3d human pose and shape from a single color image. In: CVPR (2018)
41. Pumarola, A., Sanchez-Riera, J., Choi, G., Sanfeliu, A., Moreno-Noguer, F.: 3dpeople: Modeling the geometry of dressed humans. In: ICCV (2019)
42. Saito, S., Huang, Z., Natsume, R., Morishima, S., Kanazawa, A., Li, H.: Pifu: Pixel-aligned implicit function for high-resolution clothed human digitization. In: ICCV (2019)
43. Tan, W., Yan, B., Bare, B.: Feature super-resolution: Make machine see more clearly. In: CVPR (2018)
44. Tarvainen, A., Valpola, H.: Mean teachers are better role models: Weight-averaged consistency targets improve semi-supervised deep learning results. In: NIPS (2017)
45. Tian, Y., Krishnan, D., Isola, P.: Contrastive multiview coding. arXiv preprint arXiv:1906.05849 (2019)
46. Wang, Z., Chang, S., Yang, Y., Liu, D., Huang, T.S.: Studying very low resolution recognition using deep networks. In: CVPR (2016)
47. Xu, X., Ma, Y., Sun, W.: Towards real scene super-resolution with raw images. In: CVPR (2019)
48. Xu, X., Sun, D., Pan, J., Zhang, Y., Pfister, H., Yang, M.H.: Learning to super-resolve blurry face and text images. In: ICCV (2017)
49. Zanfir, A., Marinoiu, E., Sminchisescu, C.: Monocular 3d pose and shape estimation of multiple people in natural scenes-the importance of multiple scene constraints. In: CVPR (2018)
50. Zhang, J.Y., Felsen, P., Kanazawa, A., Malik, J.: Predicting 3d human dynamics from video. In: ICCV (2019)
51. Zhang, Y., Tian, Y., Kong, Y., Zhong, B., Fu, Y.: Residual dense network for image super-resolution. In: CVPR (2018)
52. Zheng, Z., Yu, T., Wei, Y., Dai, Q., Liu, Y.: Deephuman: 3d human reconstruction from a single image. In: ICCV (2019)

Learning to Balance Specificity and Invariance for In and Out of Domain Generalization

Prithvijit Chattopadhyay[1(✉)], Yogesh Balaji[2], and Judy Hoffman[1]

[1] Georgia Institute of Technology, Atlanta, Georgia
{prithvijit3,judy}@gatech.edu
[2] University of Maryland, Maryland, USA
yogesh@cs.umd.com

Abstract. We introduce **D**omain-specific **M**asks for **G**eneralization, a model for improving both in-domain and out-of-domain generalization performance. For domain generalization, the goal is to learn from a set of source domains to produce a single model that will best generalize to an unseen target domain. As such, many prior approaches focus on learning representations which persist across all source domains with the assumption that these domain agnostic representations will generalize well. However, often individual domains contain characteristics which are unique and when leveraged can significantly aid in-domain recognition performance. To produce a model which best generalizes to both seen and unseen domains, we propose learning domain specific masks. The masks are encouraged to learn a balance of domain-invariant and domain-specific features, thus enabling a model which can benefit from the predictive power of specialized features while retaining the universal applicability of domain-invariant features. We demonstrate competitive performance compared to naive baselines and state-of-the-art methods on both PACS and DomainNet (Our code is available at https://github.com/prithv1/DMG).

Keywords: Distribution shift · Domain generalization

1 Introduction

The success of deep learning has propelled computer vision systems from purely academic endeavours to key components of real-world products. This deployment into unconstrained domains has forced researchers to focus attention beyond a closed-world supervised learning paradigm, where learned models are only evaluated on held-out in-domain test data, and instead produce models capable of generalizing to diverse test time data distributions.

This problem has been formally studied and progress measured in the *domain generalization* literature [15,34]. Most prior work in domain generalization focuses

Electronic supplementary material The online version of this chapter (https://doi.org/10.1007/978-3-030-58545-7_18) contains supplementary material, which is available to authorized users.

© Springer Nature Switzerland AG 2020
A. Vedaldi et al. (Eds.): ECCV 2020, LNCS 12354, pp. 301–318, 2020.
https://doi.org/10.1007/978-3-030-58545-7_18

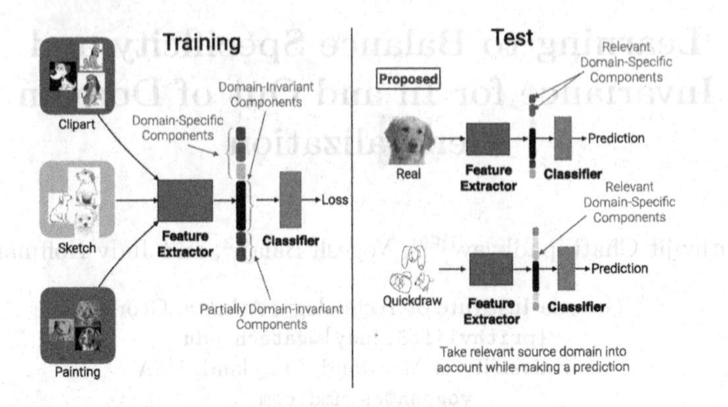

Fig. 1. Balancing specificity and invariance. At training time, we optimize for a combination of domain-specific (shown in blue, yellow, red) and domain invariant (shown in black) learned representations. Partially invariant representations are indicated as color combinations (i.e. blue + yellow = green). At test-time, these learned representations that capture a balance of domain-specificity and invariance allow the classifier to make a better prediction for given test-instance by leveraging domain-specific features from the most similar source domains. (color figure online)

on learning a model which generalizes to unseen domains by either directly optimizing for domain invariance [34] or designing regularizers that induce such a bias [3], the idea being that features which are present across multiple training distributions are more likely to persist in the novel distributions. However, in practice, as the number of training time data sources increases it becomes ever more likely that at least some of the data encountered at test time will be very similar to one or more source domains. In such a situation, ignoring features specific to only a domain or two may artificially limit the efficacy of the final model. However, leveraging a balance between "*invariance*" – features that are shared across domains – and "*specificity*" – features which are specific to individual domains – might actually aid the model in making a better prediction.

It is important to note that the similarity of data encountered at test-time to a source domain can be understood clearly only in the context of the other available source domains. Consider the example in Fig. 1, where a classifier trained on *clipart*, *sketch* and *painting* encounters an instance from a novel domain *quickdraw* at test-time. Due to the severe domain-shift involved, leveraging the relative similarity of the test-instance to samples from *sketch* might result in a better prediction compared to a setting where the model relies solely on invariant characteristics across domains. However, manually crafting such a balance or creating an explicit separation between domain-specificity and invariance [20] is not scalable as the number and diversity of the source distributions available during training increases.

In this paper, we propose DMG: **D**omain-specific **M**asks for **G**eneralization, an algorithm for automatically learning to balance between domain-invariant and domain-specific features producing a single model capable of simultaneously achieving strong performance across multiple distinct domains.

At a high-level, we cast this problem of *balanced* feature selection as one of learning distribution-specific binary masks over features of a shared deep convolutional network (CNN). Specifically, for a given layer in the CNN, we associate domain-specific mask parameters for each neuron which decide whether to turn that neuron *on* or *off* during a forward pass. We learn these masks end-to-end via backpropagation along with the network parameters. To promote discriminative features and strong end-task performance, we simultaneously minimize the standard classification error and, to encourage domain-specificity in the selected features, we penalize for overlap amongst masks from different source domains. Importantly, our approach uses straightforward optimization across all pooled source data without any need for multi-stage training or meta-learning. At test-time we average the predictions obtained by applying all the individual source domain masks thus making a prediction that is informed by both characteristics which are shared across the source domains and are specific to individual domains. Based on our experiments, we find that not only does our modeling choice result in at par or improved performance compared to other complex alternatives that explicitly model domain-shift during training, but also allows us to explicitly characterize activations specific to individual source domains. Compared to prior work, we find that our approach is much more scalable and is faster to train as training time is essentially equivalent to the same as training a vanilla aggregate baseline which pools data from multiple source domains and trains a single deep network.

Additionally, we note that efforts towards domain generalization in the computer vision literature have focused primarily on measuring novel domain performance at test time. Since it is likely that in a realistic scenario the model might also encounter data from the source distributions at test-time, it is equally important to *retain* strong performance on the source distributions in addition to improved generalization to novel domains. Thus, given that measuring continued holistic progress in domain generalization requires benchmarking proposed solutions in terms of both in and out-of-domain generalization performance, we also report in-domain generalization performance on the large DomainNet [35] benchmark proposed for domain adaptation. Concretely, we make the following contributions.

- We introduce an approach, DMG: **D**omain-specific **M**asks for **G**eneralization, that learns models capable of balancing specificity and invariance over multiple distinct domains. We demonstrate that despite our relatively simple approach, DMG achieves competitive out-of-domain performance on the commonly used PACS [25] benchmark and on the challenging DomainNet [35] dataset. In addition, we demonstrate that our model can be used as a drop-in replacement for an aggregate model when evaluated on in-domain test samples, or can be trivially converted into a high performing domain-specific model given a known test time domain label.
- We verify that our model does indeed lead to the emergence of domain specificity and show that our test time performance is stable across a variety of allowed domain overlap settings. Though not the focus of this paper, this domain specificity may be a helpful tool towards model interpretability.

2 Related Work

Domain Adaptation. Significant progress has been made in the problem of unsupervised domain adaptation where given access to a labeled source and an unlabeled target dataset, the task is to improve performance on the target domain. One popular line of approaches include learning a domain invariant representation by minimizing the distributional shift between source and target feature distributions using an adversarial loss [14,44], or MMD-based loss [28–30]. While these approaches perform alignment in the feature space, pixel-level alignment is performed using cross-domain generative models such as GANs in [6]. A combination of feature-level and pixel-level aligment is explored in [18,40]. In addition, several regularization strategies have also been proven to be effective for domain adaptation such as dropout regularization [38], classifier discrepancy [39], self-ensembling [13], etc. Most existing domain adaptation methods consider the setting where the source and the target datasets contain one domain each. In multi-source domain adaptation, the source dataset consists of a mixture of multiple domains where domain alignment is performed using an adversarial interplay involving a k-way domain discriminator in [47], and multi-domain moment matching in [35].

Domain Generalization. Similar to the multi-source domain adaptation problem, domain generalization considers multiple domains in the input data distribution. However, no access to the target distribution (including the unlabeled target) is assumed during training. This makes domain generalization a much harder problem than multi-source adaptation. One common approach to the problem involves decomposing a model into domain-specific and domain-invariant components, and using the domain-invariant component to make predictions at test time [16,20]. Recently, the use of meta-learning for domain generalization has gained much attention. [26] extends the MAML framework of [12] for domain generalization by learning parameters that adapt quickly to target domains. In [3], a regularization function is estimated using meta-learning, which when used with multi-domain training results in a robust minima with improved domain generalization. Use of data augmentation techniques for domain generalization is explored in [46]. Recently, a novel variant of empirical risk minimization framework, called Invariant Risk Minimization (IRM) has been proposed in [1,2] to make machine learning models invariant to spurious correlations in data when training across multiple sources.

Disentangled Representations. The goal of learning disentangled representations is to be able to disentangle learned features into multiple factors of variations, each factor representing a semantically meaningful concept. The problem has primarily been studied in the unsupervised setting. Typical approaches involve training a generative model such as a GAN or VAE while imposing constraints in the latent space using KL-divergence [8,21] or mutual information [9]. In the context of domain adaptation, disentangling features into domain-specific and domain-independent factors have been proposed in [7,36]. The domain-independent factors are then used to obtain predictions in the target domain.

Our approach performs a similar implicit disentanglement, where domain-specific and domain-invariant factors are mined using a masking operation.

Dropout, Pruning, Sparsification and Attention. Our approach to learn domain-specific masks is similar to the techniques adopted in the network pruning and sparsification literature. Relevant to our work are approaches that directly learn a pruning strategy during training [41,43,45]. [41] involves learning masks over parameters under a sparsity constraint to discover small sub-networks. In addition to model compression, pruning strategies have also been used in multi-task and continual learning. In [42], catastrophic forgetting is prevented while learning tasks (and subsequently attending over them) in a sequential manner. In [31], a binary mask corresponding to individual tasks are learnt for a fixed backbone network. The resulting task-specific network is obtained by applying the learnt masks on the backbone network. In [32], weights of a network are iteratively pruned to free up packets of neurons. The free neurons are in-turn updated to learn new tasks without forgetting. A similar approach is proposed in [5] for multi-domain learning where domain-specific networks are constructed by masking convolution filters under a budget on new parameters being introduced for each domain. Similarly, several approaches building on top of Dropout [43] have also been proposed for domain adaptation. In [38], a pair of sub-networks are sampled from dropout that give maximal classifier discrepancy. Feature network is trained to minimize this discrepancy, thus making it insensitive to perturbations in classifier weights. An efficient implementation of this idea using adversarial dropout is proposed in [24]. In [48], saliency supervision is used to develop explainable models for domain generalization. While DMG is akin to attention being used as learned masks for subset selection [31,32], our focus is on implicitly learning to disentangle domain-specific and invariant feature components for multi-source domain generalization.

3 Approach

Our motivation to ensure a balance between specificity and invariance is to aid prediction in situations where an instance at test-time might benefit from some of the domain-specific components captured by the domain-specific masks. In what follows, we first describe the problem setup, ground associated notations and then describe our proposed approach, DMG.

3.1 Problem Setup

Domain generalization involves training a model on data, denoted as $\mathcal{X}$, sampled from p source distributions that generalizes well to q unknown target distributions which lack training data. Without loss of generality we focus on the classification case, where the goal is to learn a model which maps inputs to the desired output label, $M : \mathcal{X} \rightarrow \mathcal{Y}$. Let $\{\mathcal{D}_i\}_{i=1}^{p+q}$ denote the $p+q$ distributions with same support $\mathcal{X} \times \mathcal{Y}$. Let $D_i = \{(x_j^{(i)}, y_j^{(i)})\}_{i=1}^{|D_i|}$ refer to the dataset sampled

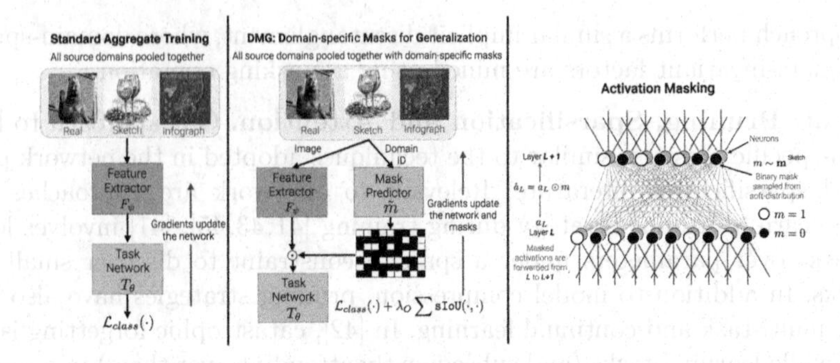

Fig. 2. Illustration of our approach (DMG): We introduce domain-specific activation masks for learning a balance between domain-specific and domain-agnostic features. [Left] Our training pipeline involves incorporating domain-specific masks in the vanilla aggregate training process. [Middle] For an image belonging to *sketch*, we sample a binary mask from the corresponding mask parameters, which is then applied to the neurons of the task-network. [Right] Post feature extraction, an elementwise product of the obtained binary masks is performed with the neurons of the task network layer (L) to obtain the *effective* activations being passed on to the next layer ($L+1$). The mask and network parameters are learned end-to-end based on the standard cross-entropy coupled with the `sIoU` loss penalizing mask overlap among the source domains.

from the ith distribution, i.e., $D_i \sim \mathcal{D}_i$. We operate in the setting where all the distributions share the same label space and distributional variations exist only in the input data (space $\mathcal{X}$). We are interested in learning a parametric model $M_\Theta : \mathcal{X} \rightarrow \mathcal{Y}$, that we can decompose into a feature extractor (F_ψ) and a task-network (T_θ) i.e., $M_\Theta(x) = (T_\theta \circ F_\psi)(x)$, where Θ, ψ, θ denote the parameters of the complete, feature and the task networks respectively. For the remaining subsections, we refer to the set of source domains as D_S and index individual source domains by d. We learn domain specific masks only on the neurons present in the task network.

3.2 Activation or Feature Selection via Domain-Specific Masks

Our goal is to learn representations which capture a balance of domain specific components (useful for predictive performance on a specific domain) and domain invariant components (useful in general for the discriminative task). Capturing information contained in multiple source distributions in such a manner allows us to make better predictions by automatically relying more on characteristics of a specific source domain in situations where an instance observed at test-time is relatively similar to one of the sources. We cast this problem of disentangling domain-specific and domain-invariant feature components as that of learning binary masks on the neurons of the task network specific to individual source domains. More specifically, for each of the p source distributions, we initialize masks $\mathbf{m}^d$ over neurons (or activations) of the task-network T_θ. Our masks can be viewed as layer-wise gates which decide which neurons to turn *on* or *off* during a forward pass through the network.

Given k neurons at some layer L of T_θ, we introduce parameters $\tilde{\mathbf{m}}^d \in \mathbb{R}^k$ for each of the source distributions $d \in D_S$. During training, for instances x_i^d from domain d, we first form mask probabilities $\mathbf{m}^d$ via a sigmoid operation as $\mathbf{m}^d = \sigma(\tilde{\mathbf{m}}^d)$. Then, the binary masks m_i^d are sampled from a bernoulli distribution given by the mask probabilities. i.e., $m_i^d \sim \mathbf{m}^d$, with $m_i^d \in \{0,1\}^k$. Upon sampling masks for individual neurons, the effective activations which are passed on to the next layer $L+1$ are $\hat{a}_L = a_L \odot m_i^d$, i.e., an elementwise product of the obtained activations and the sampled binary masks (see Fig. 2, right). During training, we sample such binary masks corresponding to the source domain of the input instance, thereby making feedforward predictions by only using domain-specific masks. Under this setup, the prediction made by the entire network M_Θ for an instance $x_i^d \in d$ can be expressed as $\hat{y}_i = M_\Theta : (x_i^d; m_i^d)$ where m_i^d denotes the sampled mask (for domain d) being applied to all neurons in the task-network T_θ. Note that, akin to dropout [43], these domain-specific masks identify *domain-specific* sub-networks – for an instance x_i^d, the sampled binary mask m_i^d identifies a specific "thinner" subnetwork.

We learn the mask-parameters in addition to the parameters of the network during training. However, note that the mask-parameters $\tilde{\mathbf{m}}^d$ cannot be updated directly using back-propagation as the sampled binary mask is discrete. We approximate gradients through sampled discrete masks using the straight-through estimator [4], i.e., we use a discretized m_i^d during a forward pass but use the continuous version $\mathbf{m}^d$ during the backward pass by approximating $\nabla_{m_i^d}\mathcal{L} \approx \nabla_{\mathbf{m}^d}\mathcal{L}$. Even though the hard sampling step is non-differentiable, gradients with respect to m_i^d serve as a noisy estimator of $\nabla_{\mathbf{m}^d}\mathcal{L}$.

Incentivizing Domain-Specificity. To ensure the masks capture neurons that are specific to individual source domains, we need to encourage specificity in the masks while maximizing predictive performance on the source set of distributions. To incentivize domain-specificity, we introduce an additional *soft*-overlap loss that ensures masks associated with each of the source distributions overlap minimally. To quantify overlap we compute the Jaccard Similarity Coefficient [19] (also known as IoU score) among pairs of source domain masks. However, as IoU is non-differentiable it is not possible to directly optimize for the same using gradient descent. Therefore, inspired by prior work [37], we minimize the following *soft*-overlap loss for every pair of source domain masks $\{\mathbf{m}^{d_i}, \mathbf{m}^{d_j}\}$ at a layer L as,

$$\texttt{sIoU}(\mathbf{m}^{d_i}, \mathbf{m}^{d_j}) = \frac{\mathbf{m}^{d_i} \cdot \mathbf{m}^{d_j}}{\sum_k (\mathbf{m}^{d_i} + \mathbf{m}^{d_j} - \mathbf{m}^{d_i} \odot \mathbf{m}^{d_j})} \tag{1}$$

where $\mathbf{m}^{d_i} \cdot \mathbf{m}^{d_j}$ approximates the intersection for the pair of source domain masks as the inner product of the mask distributions, $\odot$ denotes the elementwise product and k denotes the number of neurons in layer L. During training `sIoU` ensures predictions for instances from different source domains are made using different sub-networks (as identified by the domain-specific binary masks).

To summarize, for a set of source domains D_S the overall objective we optimize during training ensures – (1) good predictive performance on the discriminative task at hand and (2) minimal overlap among source-domain masks,

$$\mathcal{L}(\theta, \psi, \tilde{\mathbf{m}}^{d_1}, .., \tilde{\mathbf{m}}^{d_{|D_S|}}) = \sum_{d \in D_S} \sum_{x_i^d \in d} \mathcal{L}_{\texttt{class}}(\theta, \psi, m_i^d)$$
$$+ \lambda_O \sum_{L \in T_\theta} \sum_{(d_i, d_j) \in D_S} \texttt{sIoU}(\mathbf{m}^{d_i}, \mathbf{m}^{d_j}) \quad (2)$$

where $m_i^d \sim \mathbf{m}_i^d$ for every instance x_i^d of the source domain d and $\mathcal{L}_{\texttt{class}}(\cdot)$ denotes the standard cross entropy loss. Figure 2 summarizes our training pipeline in context of a standard aggregation method where a CNN is trained jointly on data pooled from all the source domains.

Prediction at Test-Time. To obtain a prediction at test-time, we follow a soft-scaling scheme similar to Dropout [43]. Recall that sampling from domain-specific *soft*-masks essentially amounts to sampling a "thinned" sub-network from the orginal task-network. However, since it is intractable to obtain predictions from all such possible (exponential) domain-specific sub-networks, we follow a simple averaging scheme that ensures that the *expected* output under the distribution induced by the masks is the same as the actual output at test-time. Specifically, we scale every neuron by the associated domain-specific *soft*-mask $\mathbf{m}^d$ instead of turning neurons *on* or *off* based on a discrete mask $m \sim \mathbf{m}^d$ and average the predictions obtained by applying $\mathbf{m}^d$ for all the source domains to the task network.[1]

4 Experiments

4.1 Experimental Settings

Datasets and Metrics. We conduct domain generalization (DG) experiments on the following datasets:

PACS [25] – PACS is a recently proposed benchmark for domain generalization which consists of only 9991 images of 7 classes, distributed across 4 domains - *photo*, *art-painting*, *cartoon* and *sketch*. Following standard practice, we conduct 4 sets of experiments – treating one domain as the unseen target and the rest as the source set of domains. The authors of [25] provide specified `train` and `val` splits for each domain to ensure fair comparison and treat the entirety of `train` + `val` as the `test`-split of the target domain. We use the same splits for our experiments. As such, the proposed splits do not include an in-domain `test`-split, thereby limiting us from computing in-domain performance in addition to measuring out-of-domain generalization.

[1] We experimented with learning a domain-classifier on source domains to use the predicted probabilities as weights for test-time averaging. We observed insignificant difference in out-of-domain performance but significantly worse in-domain performance, though we believe this may be dataset-specific.

Table 1. Out of Domain Accuracy (%) on DomainNet (λ_O = 0.1) ∓We were unable to optimize the MetaReg [3] objective with Adam [22] as the optimizer and therefore, we also include comparisons with Aggregate and MetaReg trained with SGD.

	Method	C	I	P	Q	R	S	Overall
AlexNet	Aggregate	47.17	10.15	31.82	11.75	44.35	26.33	28.60
	Aggregate-SGD∓	42.30	12.42	31.45	9.52	42.76	29.34	27.97
	Multi-Headed	45.96	10.56	31.07	12.05	43.56	25.93	28.19
	MetaReg [3]∓	42.86	**12.68**	32.47	9.37	43.43	29.87	28.45
	DMG (Ours)	**50.06**	12.23	**34.44**	**13.07**	**46.98**	**30.13**	**31.15**
ResNet-18	Aggregate	57.15	17.69	43.21	13.87	54.91	39.41	37.71
	Aggregate-SGD∓	56.56	18.44	**45.30**	12.47	57.90	38.83	38.25
	Multi-Headed	55.46	17.51	40.85	11.19	52.92	38.65	36.10
	MetaReg [3]∓	53.68	**21.06**	45.29	10.63	**58.47**	**42.31**	38.57
	DMG (Ours)	**60.07**	18.76	44.53	**14.16**	54.72	41.73	**39.00**
ResNet-50	Aggregate	62.18	19.94	45.47	13.81	57.45	44.36	40.54
	Aggregate-SGD∓	64.04	23.63	**51.04**	13.11	64.45	47.75	**44.00**
	Multi-Headed	61.74	21.25	46.80	13.89	58.47	45.43	41.27
	MetaReg [3]∓	59.77	**25.58**	50.19	11.52	**64.56**	**50.09**	43.62
	DMG (Ours)	**65.24**	22.15	50.03	**15.68**	59.63	49.02	43.63

DomainNet [35] – DomainNet is a recently proposed large-scale dataset for domain adaptation which consists of ~0.6 million images of 345 classes distributed across 6 domains – *real*, *clipart*, *sketch*, *painting*, *quickdraw* and *infograph*. DomainNet surpasses all prior datasets for domain adaptation significantly in terms of size and diversity. The authors of [35] recently released annotated `train` and `test` splits for all the 6 domains. We divide the `train` split from [35] randomly in a 90-10% proportion to obtain `train` and `val` splits for our experiments. Similar to PACS, we conduct 6 sets of leave-one-out experiments. We report out-of-domain performance as the accuracy on the `test` split of the unseen domain. For in-domain performance, we report accuracy averaged over all the source domain `test` splits.

Models. We experiment with ImageNet [10] pretrained AlexNet [23], ResNet-18 [17] and ResNet-50 [17] backbone architectures. For AlexNet, we apply domain-specific masks on the input activations of the last three fully-connected layers – our task network T_θ – and turn dropout [43] *off* while learning the domain-specific masks. For ResNet-18 and 50, we apply domain specific masks on the input activations of the last residual block and the first fully connected layer.[2]

Baselines and Points of Comparison. We compare DMG with two simple baselines (treating dropout [43] as usual if present in the backbone CNN) – (1)

[2] Specifically, for ResNet, the domain-specific masks are trained to *drop* or *keep* specific channels in the input activations as opposed to every spatial feature in every channel in order to reduce complexity in terms of the number of mask parameters to be learnt.

Table 2. Out of Domain Accuracy (%) on PACS (λ_O = 0.1) *We include the aggregate baseline both as reported in [27] as well as our own implementation (indicated as Aggregate*).

	Method	A	C	P	S	Overall
AlexNet	Aggregate [27]	63.40	66.10	88.50	56.60	68.70
	Aggregate*	56.20	70.69	86.29	60.32	68.38
	Multi-Headed	61.67	67.88	82.93	59.38	67.97
	DSN [7]	61.10	66.50	83.30	58.60	67.40
	Fusion [33]	64.10	66.80	90.20	60.10	70.30
	MLDG [26]	66.20	66.90	88.00	59.00	70.00
	MetaReg [3]	63.50	69.50	87.40	59.10	69.90
	CrossGrad [46]	61.00	67.20	87.60	55.90	67.90
	Epi-FCR [27]	64.70	72.30	86.10	65.00	72.00
	MASF [11]	**70.35**	**72.46**	**90.68**	67.33	**75.21**
	DMG (Ours)	64.65	69.88	87.31	**71.42**	73.32
ResNet-18	Aggregate [27]	77.60	73.90	94.40	74.30	79.10
	Aggregate*	72.61	78.46	93.17	65.20	77.36
	Multi-Headed	78.76	72.10	94.31	71.77	79.24
	MLDG [26]	79.50	77.30	94.30	71.50	80.70
	MetaReg [3]	79.50	75.40	94.30	72.20	80.40
	CrossGrad [46]	78.70	73.30	94.00	65.10	77.80
	Epi-FCR [27]	**82.10**	77.00	93.90	73.00	**81.50**
	MASF [11]	80.29	77.17	**94.99**	71.68	81.03
	DMG (Ours)	76.90	**80.38**	93.35	**75.21**	81.46
ResNet-50	Aggregate*	75.49	**80.67**	93.05	64.29	78.38
	Multi-Headed	75.15	76.37	**95.27**	75.26	80.51
	MASF [11]	**82.89**	80.49	95.01	72.29	82.67
	DMG (Ours)	82.57	78.11	94.49	**78.32**	**83.37**

Aggregate - the CNN backbone trained jointly on data accumulated from all the source domains and (2) Multi-Headed - the CNN backbone with different classifier heads corresponding to each of the source domains (at test-time we average predictions from all the classifier heads). Note, this baseline has more parameters than our model due to the repeated classification heads. In addition to the above baselines, we also compare with the recently proposed domain generalization approaches (cited in Tables 1, 2 and 3). Please refer to the supplementary material for implementation details.

Table 3. In Domain Accuracy (%) on DomainNet (λ_O = 0.1). For the case where inputs have known domain (KD) label, we can use the corresponding learning mask (DMG-KD) to achieve the strongest performance without requiring additional models or parameters. Column headers identify the target domains in the corresponding multi-source shifts. ∓We were unable to optimize the MetaReg [3] objective with Adam [22] as the optimizer and therefore, we also include comparisons with Aggregate and MetaReg trained with SGD.

	Method	C	I	P	Q	R	S	Overall
AlexNet	Aggregate	48.56	57.24	51.38	49.60	47.48	50.72	50.83
	Aggregate-SGD∓	48.14	54.93	50.55	48.33	47.57	49.98	49.92
	Multi-Headed	48.16	56.73	51.31	49.75	47.65	50.82	50.74
	MetaReg [3]∓	48.87	56.06	51.23	49.60	48.66	50.12	50.76
	DMG (Ours)	49.63	58.47	52.88	51.33	49.07	52.42	52.30
	DMG-KD (Ours)	**51.91**	**61.01**	**54.93**	**53.84**	**51.08**	**54.47**	**54.54**
ResNet-18	Aggregate	56.58	65.27	59.29	59.15	55.47	58.84	59.10
	Aggregate-SGD∓	55.32	63.63	57.40	57.98	53.99	57.37	57.62
	Multi-Headed	47.79	56.80	50.85	54.86	46.92	49.50	51.12
	MetaReg [3]∓	56.25	63.07	57.74	58.73	55.40	58.04	58.21
	DMG (Ours)	57.39	65.73	58.87	59.66	55.95	58.63	59.37
	DMG-KD (Ours)	**58.61**	**66.98**	**59.86**	**60.98**	**57.24**	**59.84**	**60.59**
ResNet-50	Aggregate	61.68	69.73	63.90	63.88	60.29	63.62	63.85
	Aggregate-SGD∓	61.64	69.36	63.65	64.08	60.52	63.82	63.85
	Multi-Headed	53.77	62.09	56.54	60.32	51.38	55.10	56.53
	MetaReg [3]∓	61.86	68.80	63.23	64.75	60.59	63.21	63.74
	DMG (Ours)	61.78	69.49	63.93	64.09	59.92	63.50	63.79
	DMG-KD (Ours)	**63.16**	**70.79**	**65.03**	**65.67**	**61.30**	**64.86**	**65.14**

4.2 Results

We report results on both PACS (out-of-domain) and DomainNet (in-domain and out-of-domain). For DomainNet, we use C, I, P, Q, R, S to denote the domains – *clipart*, *infograph*, *painting*, *quickdraw*, *real* and *sketch* respectively. On PACS, we use A, C, P and S to denote the domains – *art-painting*, *cartoon*, *photo* and *sketch* respectively. We summarize the observed trends below:

Out-of-Domain Generalization. Tables 1 and 2[3] summarize out of domain generalization results on the DomainNet and PACS datasets, respectively.

DomainNet - On DomainNet, we observe that DMG beats the naive aggregate baseline, the multi-headed baseline and MetaReg [3] using AlexNet as the backbone architecture in terms of overall performance – with an improvement of 2.7% over MetaReg [3] and 2.6% over the Aggregate baseline. Interestingly, this corresponds to an almost 2.89% improvement on the I, P, Q, R, S→C and a 2.63% improvement on the C, I, P, Q, S→R shifts (see Table 1, AlexNet set of

[3] For more comparisons to prior work, please refer to the supplementary material.

Table 4. Domain-Specialized Masks ($\lambda_O = 0.1$). We show how optimizing for `sIoU` leads to masks which are specialized for the individual source domains in terms of predictive performance. We consider two multi-source shifts I, P, Q, R, S→C [top-half] and C, I, P, R, S→Q [bottom-half] on DomainNet [35] with the AlexNet as the backbone architecture and find that using corresponding source domain masks leads to significantly improved in-domain performance.

		Source					Target
	Chosen Mask	I	P	Q	R	S	C
AlexNet	$\mathbf{m}^{\text{Infograph}}$	**23.84**	45.56	59.13	62.43	46.70	46.91
	$\mathbf{m}^{\text{Painting}}$	19.88	**52.41**	59.00	60.36	45.75	46.87
	$\mathbf{m}^{\text{Quickdraw}}$	21.72	48.47	**62.52**	65.32	48.69	50.33
	$\mathbf{m}^{\text{Real}}$	18.42	43.48	58.80	**68.62**	44.81	47.69
	$\mathbf{m}^{\text{Sketch}}$	19.45	45.41	57.64	61.78	**52.16**	48.36
	Combined	22.28	49.55	60.45	66.14	49.72	50.06
	Chosen Mask	C	I	P	R	S	Q
AlexNet	$\mathbf{m}^{\text{Clipart}}$	**66.70**	21.36	46.60	64.35	49.70	13.37
	$\mathbf{m}^{\text{Infograph}}$	60.71	**24.95**	47.06	63.78	49.36	12.58
	$\mathbf{m}^{\text{Painting}}$	59.21	20.59	**53.21**	60.67	48.14	12.01
	$\mathbf{m}^{\text{Real}}$	59.62	19.41	43.82	**69.82**	47.22	11.31
	$\mathbf{m}^{\text{Sketch}}$	60.97	20.29	45.69	62.40	**54.51**	13.08
	Combined	64.13	23.21	50.05	67.03	52.24	13.07

rows). Using ResNet-18 as the backbone architecture, we observe that DMG is competitive with MetaReg [3] (improvement margin of 0.43%) accompanied by improvements on the I, P, Q, R, S→C and C, I, P, R, S→Q shifts. We observe similar trends using ResNet-50, where DMG is competitive with the best performing Aggregate-SGD$^{\mp}$ baseline.

PACS - To compare DMG with prior work in the Domain Generalization literature, we also report results on the more commonly used PACS [25] benchmark in Table 2. We find that in terms of overall performance, DMG with AlexNet as the backbone architecture outperforms baselines and prior approaches including MetaReg [3][4] – which learns regularizers by modeling domain-shifts within the source set of distributions, MLDG [26] – which learns robust network parameters using meta-learning and Epi-FCR [27] – a recently proposed episodic scheme to learn network parameters robust to domain-shift, and performs competitively with MASF [11] – which introduces complementary losses to explicitly regularize the semantic structure of the feature space via a model-agnostic episodic learning procedure. Notice that this improvement also comes with a 4.09% improvement over MASF [11] on the A, C, P→S shift. Using ResNet-18 and ResNet-50 as the backbone architectures, we observe that DMG leads to comparable and improved overall performance, with margins of 0.04% and 0.7% for ResNet-18 and ResNet-50, respectively. For ResNet-18, this is accompanied with a 0.91% and 1.92% improvement on the A, C, P→S and A, P, S→C shifts. Similarily for ResNet-50, we observe a 3.06% improvement on the A, C, P→S shift.

[4] We report the performance for MetaReg [3] from [27] as the official PACS `train-val` data split changed post MetaReg [3] publication.

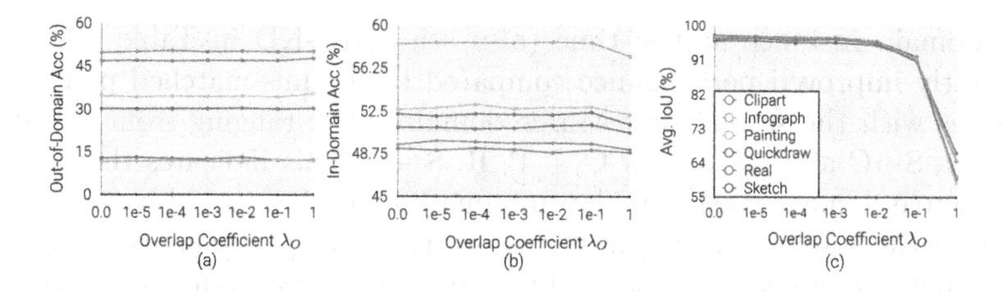

Fig. 3. Sensitivity to λ_O. DMG is relatively insensitive to the setting of the hyper-parameter λ_O as measured by out-of-domain accuracy (a), in-domain accuracy (b), and average IoU score measured among pairs of source domain masks (c). The legends in (c) indicate the target domain in the corresponding multi-source shift. AlexNet is the backbone CNN.

Due to its increased size, both in terms of number of images and number of categories, DomainNet proves to be a more challenging benchmark than PACS. Likely due to this difficulty, we find that performance on some of the hardest shifts (with Quickdraw and Infograph as the target domain) is significantly low (<25% for Quickdraw). Furthermore, DMG and prior domain generalization approaches perform comparably to naive baselines (ex. Aggregate) on these shifts, indicating that there is significant room for improvement.

In-Domain Generalization. For each of the domain-shifts in Table 1, we further report in-domain generalization performance on DomainNet in Table 3. For in-domain evaluation, we present both our standard approach as well as a version which assumes knowledge of the domain corresponding to each test instance. For the latter, we report the performance of DMG using only the mask corresponding to the known domain (KD) label and refer to this as DMG-KD. Notably, for this case where a test instance is drawn from one of the source domains, DMG-KD provides significant performance improvement over the baselines (see Table 3). Compared to DMG, we observe that DMG-KD results in a consistent improvement of ~1–2%. This alludes to the fact that the learnt domain-specific masks are indeed specialized for individual source domains.

5 Analysis

Domain Specialization. We demonstrate that as an outcome of DMG, using masks corresponding to the source domain at hand leads to siginificantly improved in-domain performance compared to a mismatched domain-mask pair, indicating the emergence of domain-specialized masks. In Table 4, we report results on the I, P, Q, R, S→C (easy) and C, I, P, R, S→Q (hard) shifts using AlexNet as the backbone CNN. We report both in and out-of-domain performance using each of the source domain masks and compare it with the setting when predictions from all the source domain masks are averaged. The cells highlighted in gray represent in-domain accuracies when masks are paired with the corresponding source domain. Clearly, using the mask corresponding to the source domain instance at test-time (also see DMG-KD in Table 3) leads to significantly improved performance compared to the mis-matched pairs – with differences with the second best source domain mask ranging from ∼2–4% for I, P, Q, R, S→C and ∼3–6% for C, I, P, R, S→Q. This indicates that not only do the source domain masks overlap minimally, but they are also "specialized" for each of the source domains in terms of predictive performance. We further observe that averaging predictions obtained from all the source domain masks leads to performance that is relatively closer to the DMG-KD setting compared to a mismatched mask-domain pair (but still falls behind by ∼2–3%). We note that certain source domain masks do lead to out-of-domain accuracies which are close (within 1%) to the combined setting – $\mathbf{m}^{\text{Quickdraw}}$ for the I, P, Q, R, S→C shift and $\mathbf{m}^{\text{Clipart}}$, $\mathbf{m}^{\text{Infograph}}$, $\mathbf{m}^{\text{Sketch}}$ for the C, I, P, R, S→Q shift. This highlights the motivation at the heart of our approach – how leveraging characteristics specific to individual source domains in addition to the invariant ones are useful for generalization.

Sensitivity to λ_O. A key component of our approach is the *soft*-IoU loss which encourages domain specificity by minimizing overlapping features across domains. During optimization, we require setting of a loss balancing hyper-parameter, λ_O. Here, we explore the sensitivity of our model to λ_O by sweeping from 0 to 1 in logarithmic increments. Figure 3 shows the final in and out-of-domain accuracies (Fig. 3 (b) and (a)) and overlap (Fig. 3 (c)) measured as the IoU [19] among pairs of *discrete* source domain masks obtained by thresholding the soft-mask values per-domain at 0.5, i.e., $m = \mathbf{1}_{\mathbf{m}^d > 0.5}$ for domain d. We observe that both in and out-of-domain generalization performance is robust to the choice of λ_O, with only minor variations and a slight drop in in-domain performance at extreme values of λ_O (0.1 and 1). In Fig. 3 (c), we observe that initially average pairwise IoU measures stay stable till $\lambda_O = 10^{-3}$ but drop at high values of $\lambda_O = 0.1$ and 1 (as low as <60% for some shifts)– indicating an increase in the "domain specificity" of the masks involved. Note that low IoU at high-values of λ_O is accompanied only by a minor drop in in-domain performance and almost no-drop in out-of-domain performance! It is crucial to note here that although there is an expected trade-off between specificity and generalization performance this trade-off does not result in large fluctuations for DMG. Please refer to the supplementary document for more analysis of DMG.

6 Conclusion

To summarize, we propose DMG: **D**omain-specific **M**asks for **G**eneralization, a method for multi-source domain learning which balances domain-specific and domain-invariant feature representations to produce a single strong model capable of effective domain generalization. We learn this balance by introducing domain-specific masks over neurons and optimizing such masks so as to minimize cross-domain feature overlap. Thus, our model, DMG, benefits from the predictive power of features specific to individual domains while retaining the generalization capapbilities of components shared across the source domains.

DMG achieves competitive out-of-domain performance on the commonly used PACS dataset and competitive in and out-of-domain performance on the challenging DomainNet dataset. Although beyond the scope of this paper, encouraging a blend of domain specificity and invariance may be useful not only in the context of generalization performance but also in terms of model interpretability.

Acknowledgements. We thank Viraj Prabhu, Daniel Bolya, Harsh Agrawal and Ramprasaath Selvaraju for fruitful discussions and feedback. This work was partially supported by DARPA award FA8750-19-1-0504.

References

1. Ahuja, K., Shanmugam, K., Varshney, K., Dhurandhar, A.: Invariant risk minimization games. arXiv preprint arXiv:2002.04692 (2020)
2. Arjovsky, M., Bottou, L., Gulrajani, I., Lopez-Paz, D.: Invariant risk minimization. arXiv preprint arXiv:1907.02893 (2019)
3. Balaji, Y., Sankaranarayanan, S., Chellappa, R.: Metareg: towards domain generalization using meta-regularization. In: Advances in Neural Information Processing Systems. pp. 998–1008 (2018)
4. Bengio, Y., Léonard, N., Courville, A.: Estimating or propagating gradients through stochastic neurons for conditional computation. arXiv preprint arXiv:1308.3432 (2013)
5. Berriel, R., et al.: Budget-aware adapters for multi-domain learning. In: Proceedings of the IEEE International Conference on Computer Vision. pp. 382–391 (2019)
6. Bousmalis, K., Silberman, N., Dohan, D., Erhan, D., Krishnan, D.: Unsupervised pixel-level domain adaptation with generative adversarial networks. In: The IEEE Conference on Computer Vision and Pattern Recognition (CVPR) (2017)
7. Bousmalis, K., Trigeorgis, G., Silberman, N., Krishnan, D., Erhan, D.: Domain separation networks. In: Advances in neural information processing systems. pp. 343–351 (2016)
8. Burgess, C.P., et al.: Pre: Understanding disentangling in β-vae. arXiv preprint arXiv:1804.03599 (2018)
9. Chen, X., Duan, Y., Houthooft, R., Schulman, J., Sutskever, I., Abbeel, P.: Infogan: interpretable representation learning by information maximizing generative adversarial nets. In: Advances in Neural Information Processing Systems. pp. 2172–2180 (2016)
10. Deng, J., Dong, W., Socher, R., Li, L.J., Li, K., Fei-Fei, L.: Imagenet: a large-scale hierarchical image database. In: 2009 IEEE Conference on Computer Vision and Pattern Recognition. pp. 248–255. IEEE (2009)
11. Dou, Q., de Castro, D.C., Kamnitsas, K., Glocker, B.: Domain generalization via model-agnostic learning of semantic features. In: Advances in Neural Information Processing Systems. pp. 6447–6458 (2019)
12. Finn, C., Abbeel, P., Levine, S.: Model-agnostic meta-learning for fast adaptation of deep networks. In: Proceedings of the 34th International Conference on Machine Learning. vol. 70, pp. 1126–1135. JMLR. org (2017)
13. French, G., Mackiewicz, M., Fisher, M.: Self-ensembling for visual domain adaptation. In: International Conference on Learning Representations (2018), https://openreview.net/forum?id=rkpoTaxA-

14. Ganin, Y., Ustinova, E., Ajakan, H., Germain, P., Larochelle, H., Laviolette, F., Marchand, M., Lempitsky, V.: Domain-adversarial training of neural networks. J. Mach. Learn. Res. **17**(1), 2030–2096 (2016)
15. Ghifary, M., Bastiaan Kleijn, W., Zhang, M., Balduzzi, D.: Domain generalization for object recognition with multi-task autoencoders. In: Proceedings of the IEEE international conference on computer vision. pp. 2551–2559 (2015)
16. Ghifary, M., Kleijn, W.B., Zhang, M., Balduzzi, D.: Domain generalization for object recognition with multi-task autoencoders. In: 2015 IEEE International Conference on Computer Vision, ICCV 2015, Santiago, Chile, December 7–13 (2015)
17. He, K., Zhang, X., Ren, S., Sun, J.: Deep residual learning for image recognition. In: Proceedings of the IEEE Conference on Computer Vision and Pattern Recognition. pp. 770–778 (2016)
18. Hoffman, J., et al.: Cycada: cycle-consistent adversarial domain adaptation. In: Proceedings of the 35th International Conference on Machine Learning, ICML 2018, Stockholmsmässan, Stockholm, Sweden, July 10–15, 2018 pp. 1994–2003 (2018)
19. Jaccard, P.: Etude de la distribution florale dans une portion des alpes et du jura. Bulletin de la Societe Vaudoise des Sciences Naturelles **37**, 547–579 (1901). https://doi.org/10.5169/seals-266450
20. Khosla, A., Zhou, T., Malisiewicz, T., Efros, A.A., Torralba, A.: Undoing the damage of dataset bias. In: Fitzgibbon, A., Lazebnik, S., Perona, P., Sato, Y., Schmid, C. (eds.) ECCV 2012. LNCS, vol. 7572, pp. 158–171. Springer, Heidelberg (2012). https://doi.org/10.1007/978-3-642-33718-5_12
21. Kim, H., Mnih, A.: Disentangling by factorising. In: Dy, J., Krause, A. (eds.) Proceedings of the 35th International Conference on Machine Learning. Proceedings of Machine Learning Research, vol. 80, pp. 2649–2658. PMLR, Stockholmsmässan, Stockholm Sweden (2018)
22. Kingma, D.P., Ba, J.: Adam: A method for stochastic optimization. arXiv preprint arXiv:1412.6980 (2014)
23. Krizhevsky, A., Sutskever, I., Hinton, G.E.: Imagenet classification with deep convolutional neural networks. In: Advances in neural information processing systems. pp. 1097–1105 (2012)
24. Lee, S., Kim, D., Kim, N., Jeong, S.G.: Drop to adapt: Learning discriminative features for unsupervised domain adaptation. In: Proceedings of the IEEE International Conference on Computer Vision. pp. 91–100 (2019)
25. Li, D., Yang, Y., Song, Y.Z., Hospedales, T.: Deeper, broader and artier domain generalization. In: International Conference on Computer Vision (2017)
26. Li, D., Yang, Y., Song, Y.Z., Hospedales, T.M.: Learning to generalize: Meta-learning for domain generalization. In: Thirty-Second AAAI Conference on Artificial Intelligence (2018)
27. Li, D., Zhang, J., Yang, Y., Liu, C., Song, Y.Z., Hospedales, T.M.: Episodic training for domain generalization. In: Proceedings of the IEEE International Conference on Computer Vision. pp. 1446–1455 (2019)
28. Long, M., Cao, Y., Wang, J., Jordan, M.I.: Learning transferable features with deep adaptation networks. In: Proceedings of the 32nd International Conference on Machine Learning. pp. 97–105 (2015)
29. Long, M., Wang, J., Jordan, M.I.: Unsupervised domain adaptation with residual transfer networks. CoRR abs/1602.04433 (2016)

30. Long, M., Zhu, H., Wang, J., Jordan, M.I.: Deep transfer learning with joint adaptation networks. In: Precup, D., Teh, Y.W. (eds.) Proceedings of the 34th International Conference on Machine Learning, ICML 2017, Sydney, NSW, Australia, 6–11 August 2017. Proceedings of Machine Learning Research, vol. 70, pp. 2208–2217. PMLR (2017)
31. Mallya, A., Davis, D., Lazebnik, S.: Piggyback: Adapting a single network to multiple tasks by learning to mask weights. In: Proceedings of the European Conference on Computer Vision (ECCV). pp. 67–82 (2018)
32. Mallya, A., Lazebnik, S.: Packnet: Adding multiple tasks to a single network by iterative pruning. In: Proceedings of the IEEE Conference on Computer Vision and Pattern Recognition. pp. 7765–7773 (2018)
33. Mancini, M., Bulò, S.R., Caputo, B., Ricci, E.: Best sources forward: domain generalization through source-specific nets. In: 2018 25th IEEE International Conference on Image Processing (ICIP). pp. 1353–1357. IEEE (2018)
34. Muandet, K., Balduzzi, D., Schölkopf, B.: Domain generalization via invariant feature representation. In: International Conference on Machine Learning. pp. 10–18 (2013)
35. Peng, X., Bai, Q., Xia, X., Huang, Z., Saenko, K., Wang, B.: Moment matching for multi-source domain adaptation. In: Proceedings of the IEEE International Conference on Computer Vision. pp. 1406–1415 (2019)
36. Peng, X., Huang, Z., Sun, X., Saenko, K.: Domain agnostic learning with disentangled representations. In: ICML (2019)
37. Rahman, M.A., Wang, Y.: Optimizing intersection-over-union in deep neural networks for image segmentation. In: Bebis, G., et al. (eds.) ISVC 2016. LNCS, vol. 10072, pp. 234–244. Springer, Cham (2016). https://doi.org/10.1007/978-3-319-50835-1_22
38. Saito, K., Ushiku, Y., Harada, T., Saenko, K.: Adversarial dropout regularization. In: International Conference on Learning Representations (2018)
39. Saito, K., Watanabe, K., Ushiku, Y., Harada, T.: Maximum classifier discrepancy for unsupervised domain adaptation. arXiv preprint arXiv:1712.02560 (2017)
40. Sankaranarayanan, S., Balaji, Y., Castillo, C.D., Chellappa, R.: Generate to adapt: Aligning domains using generative adversarial networks. In: The IEEE Conference on Computer Vision and Pattern Recognition (CVPR) (2018)
41. Savarese, P., Silva, H., Maire, M.: Winning the lottery with continuous sparsification. arXiv preprint arXiv:1912.04427 (2019)
42. Serra, J., Suris, D., Miron, M., Karatzoglou, A.: Overcoming catastrophic forgetting with hard attention to the task. In: International Conference on Machine Learning. pp. 4548–4557 (2018)
43. Srivastava, N., Hinton, G., Krizhevsky, A., Sutskever, I., Salakhutdinov, R.: Dropout: a simple way to prevent neural networks from overfitting. J. Mach. Learn. Res. **15**(1), 1929–1958 (2014)
44. Tzeng, E., Hoffman, J., Saenko, K., Darrell, T.: Adversarial discriminative domain adaptation. In: Proceedings of the IEEE Conference on Computer Vision and Pattern Recognition. pp. 7167–7176 (2017)
45. Venkatesh, B., Thiagarajan, J.J., Thopalli, K., Sattigeri, P.: Calibrate and prune: Improving reliability of lottery tickets through prediction calibration. arXiv preprint arXiv:2002.03875 (2020)
46. Volpi, R., Namkoong, H., Sener, O., Duchi, J.C., Murino, V., Savarese, S.: Generalizing to unseen domains via adversarial data augmentation. In: Advances in Neural Information Processing Systems. pp. 5334–5344 (2018)

47. Xu, R., Chen, Z., Zuo, W., Yan, J., Lin, L.: Deep cocktail network: multi-source unsupervised domain adaptation with category shift. In: Proceedings of the IEEE Conference on Computer Vision and Pattern Recognition. pp. 3964–3973 (2018)
48. Zunino, A., et al.: Explainable deep classification models for domain generalization. arXiv preprint arXiv:2003.06498 (2020)

Contrastive Learning for Unpaired Image-to-Image Translation

Taesung Park[1], Alexei A. Efros[1(✉)], Richard Zhang[2], and Jun-Yan Zhu[2]

[1] University of California, Berkeley, USA
efros@eecs.berkeley.edu
[2] Adobe Research, San Jose, USA

Abstract. In image-to-image translation, each patch in the output should reflect the *content* of the corresponding patch in the input, independent of domain. We propose a straightforward method for doing so – maximizing mutual information between the two, using a framework based on contrastive learning. The method encourages two elements (corresponding patches) to map to a similar point in a learned feature space, relative to other elements (other patches) in the dataset, referred to as negatives. We explore several critical design choices for making contrastive learning effective in the image synthesis setting. Notably, we use a multilayer, patch-based approach, rather than operate on entire images. Furthermore, we draw negatives from *within* the input image itself, rather than from the rest of the dataset. We demonstrate that our framework enables one-sided translation in the unpaired image-to-image translation setting, while improving quality and reducing training time. In addition, our method can even be extended to the training setting where each "domain" is only a single image.

Keywords: Contrastive learning · Noise contrastive estimation · Mutual information · Image generation

1 Introduction

Consider the image-to-image translation problem in Fig. 1. We wish for the output to take on the *appearance* of the target domain (a zebra), while retaining the structure, or *content*, of the specific input horse. This is, fundamentally, a disentanglement problem: separating the content, which needs to be preserved across domains, from appearance, which must change. Typically, target appearance is enforced using an adversarial loss [21,31], while content is preserved using cycle-consistency [37,81,89]. However, cycle-consistency assumes that the relationship between the two domains is a bijection, which is often too restrictive. In this paper, we propose an alternative, rather straightforward way of maintaining correspondence in content but not appearance – by maximizing the mutual information between corresponding input and output patches.

Electronic supplementary material The online version of this chapter (https://doi.org/10.1007/978-3-030-58545-7_19) contains supplementary material, which is available to authorized users.

© Springer Nature Switzerland AG 2020
A. Vedaldi et al. (Eds.): ECCV 2020, LNCS 12354, pp. 319–345, 2020.
https://doi.org/10.1007/978-3-030-58545-7_19

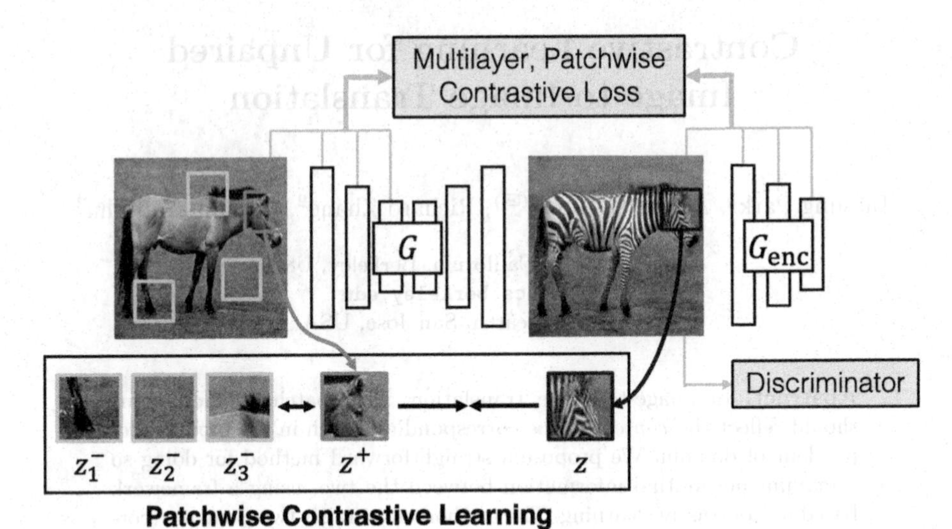

Fig. 1. Patchwise Contrastive Learning for one-sided translation. A generated **outputpatch** should appear closer to its correspondinginputpatch, in comparison to other randompatches. We use a multilayer, patchwise contrastive loss, which maximizes *mutual information* between corresponding input and output patches. This enables one-sided translation in the unpaired setting.

In a successful result, given a specific patch on the output, for example, the generated zebra forehead highlighted in blue, one should have a good idea that it came from the horse forehead, and not the other parts of the horse or the background vegetation. We achieve this by using a type of contrastive loss function, InfoNCE loss [57], which aims to learn an embedding or an encoder that *associates* corresponding patches to each other, while *disassociating* them from others. To do so, the encoder learns to pay attention to the commonalities between the two domains, such as object parts and shapes, while being invariant to the differences, such as the textures of the animals. The two networks, the generator and encoder, conspire together to generate an image such that patches can be easily traceable to the input.

Contrastive learning has been an effective tool in unsupervised visual representation learning [9,24,57,80]. In this work, we demonstrate its effectiveness in a conditional image synthesis setting and systematically study several key factors to make it successful. We find it pertinent to use it on a multilayer, patchwise fashion. In addition, we find that drawing negatives *internally* from within the input image, rather than externally from other images in the dataset, forces the patches to better preserve the content of the input. Our method requires neither memory bank [24,80] nor specialized architectures [3,25].

Extensive experiments show that our faster, lighter model outperforms both prior one-sided translation methods [4,18] and state-of-the-art models that rely on several auxiliary networks and multiple loss functions. Furthermore, since our contrastive representation is formulated within the same image, our method can even be trained on single images. Our code and models are available at GitHub.

2 Related Work

Image translation and cycle-consistency. Paired image-to-image translation [31] maps an image from input to output domain using an adversarial loss [21], in conjunction with a reconstruction loss between the result and target. In unpaired translation settings, corresponding examples from domains are not available. In such cases, *cycle-consistency* has become the de facto method for enforcing correspondence [37,81,89], which learns an inverse mapping from the output domain back to the input and checks if the input can be reconstructed. Alternatively, UNIT [44] and MUNIT [30] propose to learn a shared intermediate "content" latent space. Recent works further enable multiple domains and multi-modal synthesis [1,10,41,45,90] and improve the quality of results [20,43,72,79,88]. In all of the above examples, cycle-consistency is used, often in multiple aspects, between (a) two image domains [37,81,89] (b) image to latent [10,30,41,44,90], or (c) latent to image [30,90]. While effective, the underlying bijective assumption behind cycle-consistency is sometimes too restrictive. Perfect reconstruction is difficult to achieve, especially when images from one domain have additional information compared to the other domain.

Relationship preservation. An interesting alternative approach is to encourage relationships present in the input be analogously reflected in the output. For example, perceptually similar patches *within* an input image should be similar in the output [88], output and input images share similar content regarding a pre-defined distance [5,68,71], vector arithmetic between input images is preserved using a margin-based triplet loss [3], distances *between* input images should be consistent in output images [4], the network should be equivariant to geometric transformations [18]. Among them, TraVeLGAN [3], DistanceGAN [4] and GcGAN [18] enable one-way translation and bypass cycle-consistency. However, they rely on relationship between entire images, or often with predefined distance functions. Here we seek to replace cycle-consistency by instead learning a cross-domain similarity function *between input and output patches* through information maximization, without relying on a pre-specified distance.

Emergent perceptual similarity in deep network embeddings. Defining a "perceptual" distance function between high-dimensional signals, e.g., images, has been a longstanding problem in computer vision and image processing. The majority of image translation work mentioned uses a per-pixel reconstruction metric, such as ℓ_1. Such metrics do not reflect human perceptual preferences and can lead to blurry results. Recently, the deep learning community has found that the VGG classification network [69] trained on ImageNet dataset [14] can be re-purposed as a "perceptual loss" [16,19,34,52,75,87], which can be used in paired image translation tasks [8,59,77], and was known to outperform traditional metrics such as SSIM [78] and FSIM [84] on human perceptual tests [87]. In particular, the Contextual Loss [52] boosts the perceptual quality of pretrained VGG features, validated by human perceptual judgments [51]. In these cases, the frozen network weights cannot adapt to the data on hand. Furthermore, the frozen similarity function may not be appropriate when comparing data *across* two domains, depending on the pairing. By posing our constraint via mutual information, our method makes use of negative samples from the data, allowing the cross-domain similarity function to adapt to the particular input and output domains, and bypass using a pre-defined similarity function.

Contrastive representation learning. Traditional unsupervised learning has sought to learn a compressed code which can effectively reconstruct the input [27]. Data imputation – holding one subset of raw data to predict from another – has emerged as a more effective family of pretext tasks, including denoisin [76], context prediction [15,60], colorization [40,85], cross-channel encoding [86], frame prediction [46,55], and multi-sensory prediction [56,58]. However, such methods suffer from the same issue as before—the need for a pre-specified, hand-designed loss function to measure predictive performance.

Recently, a family of methods based on *maximizing mutual information* has emerged to bypass the above issue [9,24,25,28,47,54,57,73,80]. These methods make use of noise contrastive estimation [23], learning an embedding where associated signals are brought together, in *contrast* to other samples in the dataset (note that similar ideas go back to classic work on metric learning with Siamese nets [12]). Associated signals can be an image with itself [17,24,49,67,80], an image with its downstream representation [28,47], neighboring patches within an image [25,33,57], or multiple views of the input image [73], and most successfully, an image with a set of transformed versions of itself [9,54]. The design choices of the InfoNCE loss, such as the number of negatives and how to sample them, hyperparameter settings, and data augmentations all play a critical role and need to be carefully studied. We are the first to use InfoNCE loss for the conditional image synthesis tasks. As such, we draw on these important insights, and find additional pertinent factors, unique to image synthesis.

3 Methods

We wish to translate images from input domain $\mathcal{X} \subset \mathbb{R}^{H\times W\times C}$ to appear like an image from the output domain $\mathcal{Y} \subset \mathbb{R}^{H\times W\times 3}$. We are given a dataset of unpaired instances $X = \{\boldsymbol{x} \in \mathcal{X}\}, Y = \{\boldsymbol{y} \in \mathcal{Y}\}$. Our method can operate even when X and Y only contain a single image each.

Our method only requires learning the mapping in one direction and avoids using inverse auxiliary generators and discriminators. This can largely simplify the training procedure and reduce training time. We break up our generator function G into two components, an encoder G_{enc} followed by a decoder G_{dec}, which are applied sequentially to produce output image $\hat{\boldsymbol{y}} = G(\boldsymbol{z}) = G_{\text{dec}}(G_{\text{enc}}(\boldsymbol{x}))$.

Adversarial loss. We use an adversarial loss [21], to encourage the output to be visually similar to images from the target domain, as follows:

$$\mathcal{L}_{\text{GAN}}(G, D, X, Y) = \mathbb{E}_{\boldsymbol{y}\sim Y} \log D(\boldsymbol{y}) + \mathbb{E}_{\boldsymbol{x}\sim X} \log(1 - D(G(\boldsymbol{x}))). \tag{1}$$

Mutual information maximization. We use a noise contrastive estimation framework [57] to maximize mutual information between input and output.

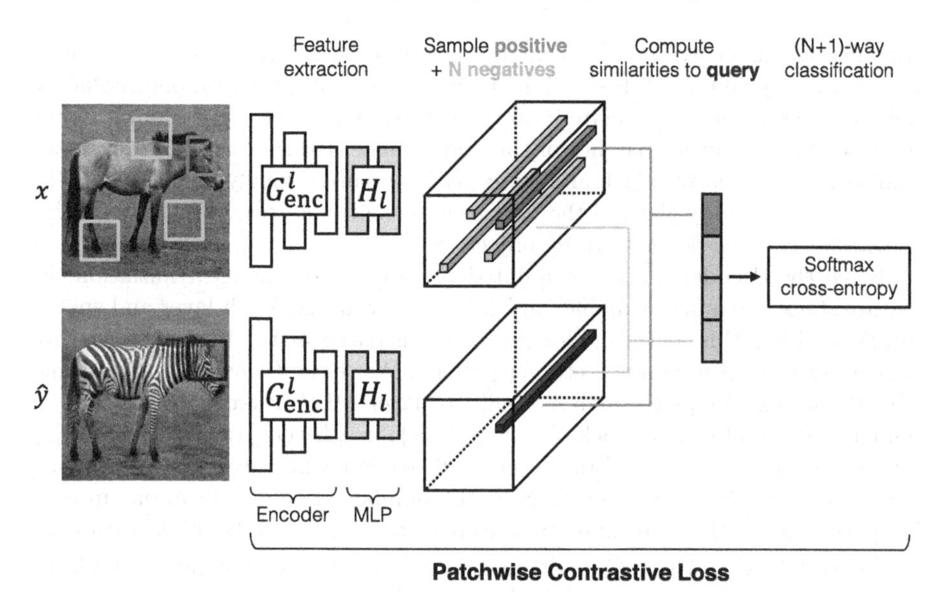

Fig. 2. Patchwise Contrastive Loss. Both images, $\boldsymbol{x}$ and $\hat{\boldsymbol{y}}$, are encoded into feature tensor. We sample a **query** patch from the output $\hat{\boldsymbol{y}}$ and compare it to the input patch at the same location. We set up an (N+1)-way classification problem, where N negative patches are sampled from the same input image at different locations. We reuse the encoder part G_{enc} of our generator and add a two-layer MLP network. This network learns to project both the input and output patch to a shared embedding space.

The idea of contrastive learning is to associate two signals, a "query" and its "positive" example, in contrast to other points within the dataset, referred to as "negatives". The query, positive, and N negatives are mapped to K-dimensional vectors $\boldsymbol{v}, \boldsymbol{v}^+ \in \mathbb{R}^K$ and $\boldsymbol{v}^- \in \mathbb{R}^{N \times K}$, respectively. $\boldsymbol{v}^-{}_n \in \mathbb{R}^K$ denotes the n-th negative. We normalize vectors onto a unit sphere to prevent the space from collapsing or expanding. An $(N+1)$–way classification problem is set up, where the distances between the query and other examples are scaled by a temperature $\tau = 0.07$ and passed as logits [24,80]. The cross-entropy loss is calculated, representing the probability of the positive example being selected over negatives.

$$\ell(\boldsymbol{v}, \boldsymbol{v}^+, \boldsymbol{v}^-) = -\log\left[\frac{\exp(\boldsymbol{v}\cdot\boldsymbol{v}^+/\tau)}{\exp(\boldsymbol{v}\cdot\boldsymbol{v}^+/\tau) + \sum_{n=1}^{N}\exp(\boldsymbol{v}\cdot\boldsymbol{v}^-{}_n/\tau)}\right]. \quad (2)$$

Our goal is to associate the input and output data. In our context, query refers to an output. positive and negatives are corresponding and noncorresponding input. Below, we explore several important design choices, including how to map the images into vectors and how to sample the negatives.

Multilayer, patchwise contrastive learning. In the unsupervised learning setting, contrastive learning has been used both on an image and patch level [3,25]. For our application, we note that not only should the whole images share content, but also corresponding patches between the input and output images. For example, given a patch showing the legs of an output zebra, one should be able to more strongly associate it to the corresponding legs of the input horse, more so than the other patches of the horse image. Even at the pixel level, the colors of a zebra body (black and white) can be more strongly associated to the color of a horse body than to the background shades of grass. Thus, we employ a *multilayer, patch-based* learning objective.

Since the encoder G_{enc} is computed to produce the image translation, its feature stack is readily available, and we take advantage. Each layer and spatial location within this feature stack represents a patch of the input image, with deeper layers corresponding to bigger patches. We select L layers of interest and pass the feature maps through a small two-layer MLP network H_l, as used in SimCLR [9], producing a stack of features $\{\boldsymbol{z}_l\}_L = \{H_l(G^l_{\text{enc}}(\boldsymbol{x}))\}_L$, where G^l_{enc} represents the output of the l-th chosen layer. We index into layers $l \in \{1, 2, ..., L\}$ and denote $s \in \{1, ..., S_l\}$, where S_l is the number of spatial locations in each layer. We refer to the corresponding feature as $\boldsymbol{z}^s_l \in \mathbb{R}^{C_l}$ and the other features as $\boldsymbol{z}^{S\setminus s}_l \in \mathbb{R}^{(S_l-1)\times C_l}$, where C_l is the number of channels at each layer. Similarly, we encode the output image $\hat{\boldsymbol{y}}$ into $\{\hat{\boldsymbol{z}}_l\}_L = \{H_l(G^l_{\text{enc}}(G(\boldsymbol{x})))\}_L$.

We aim to match corresponding input-output patches at a specific location. We can leverage the other patches *within* the input as negatives. For example, a zebra leg should be more closely associated with an input horse leg than the other patches of the same input, such as other horse parts or the background sky and vegetation. We name it as the *PatchNCE* loss, as illustrated in Fig. 2. Appendix C.3 provides pseudocode.

$$\mathcal{L}_{\text{PatchNCE}}(G, H, X) = \mathbb{E}_{\boldsymbol{x}\sim X} \sum_{l=1}^{L} \sum_{s=1}^{S_l} \ell(\hat{\boldsymbol{z}}^s_l, \boldsymbol{z}^s_l, \boldsymbol{z}^{S\setminus s}_l). \tag{3}$$

Alternatively, we can also leverage image patches from the rest of the dataset. We encode a random negative image from the dataset $\tilde{\boldsymbol{x}}$ into $\{\tilde{\boldsymbol{z}}_l\}_L$, and use the following *external* NCE loss. In this variant, we maintain a large, consistent dictionary of negatives using an auxiliary moving-averaged encoder, following MoCo [24]. MoCo enables negatives to be sampled from a longer history, and performs more effective than end-to-end updates [25,57] and memory bank [80].

$$\mathcal{L}_{\text{external}}(G, H, X) = \mathbb{E}_{\boldsymbol{x}\sim X, \tilde{z}\sim Z^-} \sum_{l=1}^{L} \sum_{s=1}^{S_l} \ell(\hat{\boldsymbol{z}}^s_l, \boldsymbol{z}^s_l, \tilde{\boldsymbol{z}}_l), \tag{4}$$

where dataset negatives $\tilde{\boldsymbol{z}}_l$ are sampled from an external dictionary Z^- from the source domain, whose data are computed using a moving-averaged encoder $\hat{H}_l$ and moving-averaged MLP $\hat{H}$. We refer our readers to the original work for more details [24].

In Sect. 4.1, we show that our encoder G_{enc} learns to capture domain-invariant concepts, such as animal body, grass, and sky for horse $\rightarrow$ zebra, while our decoder G_{dec} learns to synthesize domain-specific features such as zebra stripes.

Interestingly, through systematic evaluations, we find that using internal patches only outperforms using external patches. We hypothesize that by using internal statistics, our encoder does not need to model large intra-class variation such as white horse vs. brown horse, which is not necessary for generating output zebras. Single image internal statistics has been proven effective in many vision tasks such as segmentation [32], super-resolution, and denoising [66,91].

Final objective. Our final objective is as follows. The generated image should be realistic, while patches in the input and output images should share correspondence. Figure 1 illustrates our minimax learning objective. Additionally, we may utilize PatchNCE loss $\mathcal{L}_{\text{PatchNCE}}(G, H, Y)$ on images from domain $\mathcal{Y}$ to prevent the generator from making unnecessary changes. This loss is essentially a learnable, domain-specific version of the identity loss, commonly used by previous unpaired translation methods [71,89].

$$\mathcal{L}_{\text{GAN}}(G, D, X, Y) + \lambda_X \mathcal{L}_{\text{PatchNCE}}(G, H, X) + \lambda_Y \mathcal{L}_{\text{PatchNCE}}(G, H, Y). \quad (5)$$

We choose $\lambda_X = 1$ when we jointly train with the identity loss $\lambda_Y = 1$, and choose a larger value $\lambda_X = 10$ without the identity loss ($\lambda_Y = 0$) to compensate for the absence of the regularizer. We find that the former configuration, named *Contrastive Unpaired Translation (CUT)* hereafter, achieves superior performance to existing methods, whereas the latter, named *FastCUT*, can be thought as a faster and lighter version of CycleGAN. Our model is relatively simple compared to recent methods that often use 5–10 losses and hyper-parameters.

Discussion. Li et al. [42] has shown that cycle-consistency loss is the upper bound of conditional entropy $\text{H}(X|Y)$ (and $\text{H}(Y|X)$). Therefore, minimizing cycle-consistency loss encourages the output $\hat{\boldsymbol{y}}$ to be more dependent on input $\boldsymbol{x}$. This is related to our objective of maximizing the mutual information $\text{I}(X, Y)$, as $\text{I}(X, Y) = \text{H}(X) - \text{H}(X|Y)$. As entropy $\text{H}(X)$ is a constant and independent of the generator G, maximizing mutual information is equivalent to minimizing the conditional entropy. Notably, using contrastive learning, we can achieve a similar goal without introducing inverse mapping networks and additional discriminators. In the unconditional modeling scenario, InfoGAN [7] shows that simple losses (e.g., L2 or cross-entropy) can serve as a lower bound for maximizing mutual information between an image and a low-dimensional code. In our setting, we maximize the mutual information between two high-dimensional image spaces, where simple losses are no longer effective. Liang et al. [43] proposes an adversarial loss based on Siamese networks that encourages the output to be closer to the target domain than to its source domain. The above method still builds on cycle-consistency and two-way translations. Different from the above work, we use contrastive learning to enforce content consistency, rather than to improve the adversarial loss itself. To measure the similarity between two distributions, the Contextual Loss [52] used softmax over cosine disntances of features extracted from pre-trained networks. In contrast, we learn the encoder with the NCE loss to associate the input and output patches at the same location.

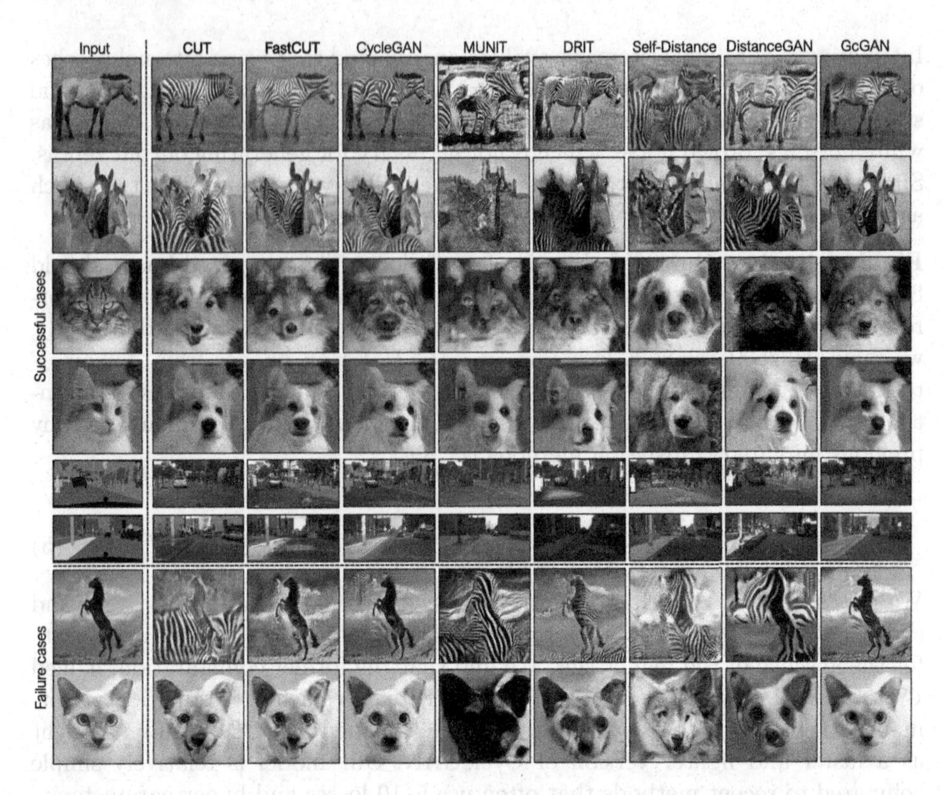

Fig. 3. Results. We compare our methods (CUT and FastCUT) with existing methods on the horse→zebra, cat→dog, and Cityscapes datasets. CycleGAN [89], MUNIT [44], and DRIT [41], are two-sided methods, while SelfDistance, DistanceGAN [4], and GcGAN [18] are one-sided. We show successful cases above the dotted lines. Our full version CUT is able to add the zebra texture to the horse bodies. Our fast variant FastCUT can also generate competitive results at the least computational cost of training. The final rows show failure cases. In the first, we are unable to identify the unfamiliar pose of the horse and instead add texture to the background. In the second, the method hallucinates a tongue.

4 Experiments

We test across several datasets. We first show that our method improves upon baselines in unpaired image translation. We then show that our method can extend to *single-image* training. Full results are available at our website.

Training details. We follow the setting of CycleGAN [89], except that the ℓ_1 cycle-consistency loss is replaced with our contrastive loss. In detail, we used LSGAN [50] and Resnet-based generator [34] with PatchGAN [31]. We define our encoder as the first half of the generator, and accordingly extract our multilayer features from five evenly distributed points of the encoder. For single image translation, we use a StyelGAN2-based generator [36]. To embed the encoder's

Table 1. Comparison with baselines We compare our methods across datasets on common evaluation metrics. CUT denotes our model trained with the identity loss ($\lambda_X = \lambda_Y = 1$), and FastCUT without it ($\lambda_X = 10, \lambda_Y = 0$). We show FID, a measure of image quality [26] (lower is better). For Cityscapes, we show the semantic segmentation scores (mAP, pixAcc, classAcc) to assess the discovered correspondence (higher is better for all metrics). Based on quantitative measures, CUT produces higher quality and more accurate generations with light footprint in terms of training speed (seconds per sample) and GPU memory usage. Our variant FastCUT also produces competitive results with even lighter computation cost of training.

Method	Cityscapes				Cat→Dog	Horse→Zebra		
	mAP↑	pixAcc↑	classAcc↑	FID↓	FID↓	FID↓	sec/iter↓	Mem(GB)↓
CycleGAN [89]	20.4	55.9	25.4	76.3	85.9	77.2	0.40	4.81
MUNIT [44]	16.9	56.5	22.5	91.4	104.4	133.8	0.39	3.84
DRIT [41]	17.0	58.7	22.2	155.3	123.4	140.0	0.70	4.85
Distance [4]	8.4	42.2	12.6	81.8	155.3	72.0	**0.15**	2.72
SelfDistance [4]	15.3	56.9	20.6	78.8	144.4	80.8	0.16	2.72
GCGAN [18]	21.2	63.2	26.6	105.2	96.6	86.7	0.26	2.67
CUT	**24.7**	**68.8**	**30.7**	**56.4**	**76.2**	**45.5**	0.24	3.33
FastCUT	19.1	59.9	24.3	68.8	94.0	73.4	**0.15**	**2.25**

features, we apply a two-layer MLP with 256 units at each layer. We normalize the vector by its L2 norm. See Appendix C.1 for more training details.

4.1 Unpaired Image Translation

Datasets We conduct experiments on the following datasets.

- *Cat→Dog* contains 5,000 training and 500 val images from AFHQ Dataset [11].
- *Horse→Zebra* contains 2,403 training and 260 zebra images from ImageNet [14] and was introduced in CycleGAN [89].
- *Cityscapes* [13] contains street scenes from German cities, with 2,975 training and 500 validation images. We train models at 256 × 256 resolution. Unlike previous datasets listed, this does have corresponding labels. We can leverage this to measure how well our unpaired algorithm discovers correspondences.

Evaluation protocol. We adopt the evaluation protocols from [26,89], aimed at assessing *visual quality* and *discovered correspondence.* For the first, we utilize the widely-used Fréchet Inception Distance (FID) metric, which empirically estimates the distribution of real and generated images in a deep network space and computes the divergence between them. Intuitively, if the generated images are realistic, they should have similar summary statistics as real images, in any feature space. For *Cityscapes* specifically, we have ground truth of paired label maps. If accurate correspondences are discovered, the algorithm should generate images that are recognizable as the correct class. Using an off-the-shelf network to test "semantic interpretability" of image translation results has been commonly used [31,85]. We use the pretrained semantic segmentation network

Method	Training settings					Testing datasets		
						Horse→Zebra	Cityscapes	
	Id	Negs	Layers	Int	Ext	FID↓	FID↓	mAP↑
CUT (default)	✓	255	All	✓	✗	45.5	**56.4**	**24.7**
no id	✗	255	All	✓	✗	39.3	68.5	22.0
no id, 15 neg	✗	15	All	✓	✗	44.1	59.7	23.1
no id, 15 neg, last	✗	15	Last	✓	✗	**38.1**	114.1	16.0
last	✓	255	Last	✓	✗	441.7	141.1	14.9
int and ext	✓	255	All	✓	✓	56.4	64.4	20.0
ext only	✓	255	All	✗	✓	53.0	110.3	16.5
ext only, last	✓	255	Last	✗	✓	60.1	389.1	5.6

Fig. 4. Ablations. The PatchNCE loss is trained with negatives from each layer output of the same (internal) image, with the identity preservation regularization. *(Left)* We try removing the identity loss [**Id**], using less negatives [**Negs**], using only the last layer of the encoder [**Layers**], and varying where patches are sampled, internal [**Int**] vs external [**Ext**]. *(Right)* We plot the FIDs on horse→zebra and Cityscapes dataset. Removing the identity loss (`no id`) and reducing negatives (`no id, 15 neg`) still perform strongly. In fact, our variant FastCUT does not use the identity loss. However, reducing number of layers (`last`) or using external patches (`ext`) hurts performance.

DRN [83]. We train the DRN at 256 × 128 resolution, and compute mean average precision (mAP), pixel-wise accuracy (pixAcc), and average class accuracy (classAcc). See Appendix C.2 for more evaluation details.

Comparison to baselines. In Table 1, we show quantitative measures of our and Fig. 3, we compare our method to baselines. We present two settings of our method in Eqn. 5: CUT with the identity loss ($\lambda_X = \lambda_Y = 1$), and FastCUT without it ($\lambda_X = 10, \lambda_Y = 0$). On image quality metrics across datasets, our methods outperform baselines. We show qualitative results in Fig. 3 and additional results in Appendix A. In addition, our Cityscapes semantic segmentation scores are higher, suggesting that our method is able to find correspondences between output and input.

Speed and memory. Since our model is one-sided, our method is memory-efficient and fast. For example, our method with the identity loss was 40% faster and 31% more memory-efficient than CycleGAN at training time, using the same architectures as CycleGAN (Table 1). Furthermore, our faster variant FastCUT is 63% faster and 53% lighter, while achieving superior metrics to CycleGAN. Table 1 contains the speed and memory usage of each method measured on NVIDIA GTX 1080Ti, and shows that FastCUT achieves competitive FIDs and segmentation scores with a lower time and memory requirement. Therefore, our method can serves as a practical, lighter alternative in scenarios, when an image translation model is jointly trained with other components [29,62].

4.2 Ablation Study and Analysis

We find that in the image synthesis setting, similarly to the unsupervised learning setting [9,24,25], implementation choices for contrastive loss are important.

Fig. 5. Qualitative ablation results of our full method (CUT) are shown: *without* the identity loss $\mathcal{L}_{\text{PatchNCE}}(G, H, Y)$ on domain Y (`no id`), using only one layer of the encoder (`last layer only`), and using external instead of internal negatives (`external only`). The ablations cause noticeable drop in quality, including repeated building or vegetation textures when using only external negatives or the last layer output.

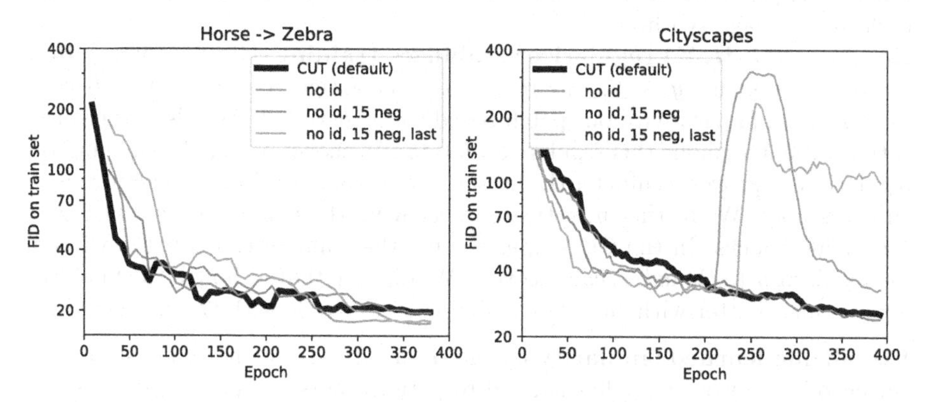

Fig. 6. Identity loss $\mathcal{L}_{\text{PatchNCE}}(G, H, Y)$ on domain Y adds stability. This regularizer encourages an image from the output domain y to be unchanged by the generator. Using it (shown in **bold, black** curves), we observe better stability in comparison to other variants. On the left, our variant without the regularizer, `no id`, achieves better FID. However, we see higher variance in the training curve. On the right, training without the regularizer can lead to collapse.

Here, try various settings and ablations of our method, summarized in Fig. 4. By default, we use the ResNet-based generator used in CycleGAN [89], with patchNCE using (a) negatives sampled from the input image, (b) multiple layers of the encoder, and (c) a PatchNCE loss $\mathcal{L}_{\text{PatchNCE}}(G, H, Y)$ on domain Y. In Fig. 4, we show results using several variants and ablations, taken after training for 400 epochs. We show qualitative examples in Fig. 5.

Internal negatives are more effective than external. By default, we sample negatives from *within* the same image (internal negatives). We also try adding negatives from other images, using a momentum encoder [24]. However, the external negatives, either as addition (`int and ext`) or replacement of internal negatives (`ext only`), hurts performance. In Fig. 5, we see a loss of quality, such as repeated texture in the Cityscapes dataset, indicating that sampling negatives from the same image serves as a stronger signal for preserving content.

Importance of using multiple layers of encoder. Our method uses multiple layers of the encoder, every four layers from pixels to the 16th layer. This is consistent with the standard use of ℓ_1+VGG loss, which uses layers from the pixel level up to a deep convolutional layer. On the other hand, many contrastive learning-based unsupervised learning papers map the whole image into a single representation. To emulate this, we try only using the last layer of the encoder (`last`), and try a variant using external negatives only (`ext only, last`). Performance is drastically reduced in both cases. In unsupervised representation learning, the input images are fixed. For our application, the loss is being used as a signal for synthesizing an image. As such, this indicates that the dense supervision provided by using multiple layers of the encoder is important when performing image synthesis.

$\mathcal{L}_{\text{PatchNCE}}(G, H, Y)$ regularizer stabilizes training. Given an image from the output domain $\boldsymbol{y} \in \mathcal{Y}$, this regularizer encourages the generator to leave the image unchanged with our patch-based contrastive loss. We also experiment with a variant without this regularizer, `no id`. As shown in Fig. 4, removing the regularizer improves results for the horse→zebra task, but decreases performance on Cityscapes. We further investigate by showing the training curves in Fig. 6, across 400 epochs. In the Cityscapes results, the training can collapse without the regularizer (although it can recover). We observe that although the final FID is sometimes better without, the training is more stable with the regularizer.

Visualizing learned similarity by encoder G_{enc} To further understand why our encoder network G_{enc} has learned to perform horse→ zebra task, we study the output space of the 1st residual block for both horse and zebra features. As shown in Fig. 7. Given an input and output image, we compute the distance between a query patch's feature vector $\boldsymbol{v}$ (highlighted as red or blue dot) to feature vectors $\boldsymbol{v}^-$ of all the patches in the input using $\exp(\boldsymbol{v} \cdot \boldsymbol{v}^-/\tau)$ (Eqn. 2). Additionally, we perform a PCA dimension reduction on feature vectors from both horse and zebra patches. In (d) and (e), we show the top three principal components, which looks similar before and after translation. This indicates that our encoder is able to bring the corresponding patches from two domains into a similar location in the feature embedding space.

Additional applications. Figure 8 shows additional results: Parisian street → Burano's brightly painted houses and Russian Blue cat → Grumpy cat.

4.3 High-Resolution Single Image Translation

Finally, we conduct experiments in the single image setting, where both the source and target domain only have one image each. Here, we transfer a Claude Monet's painting to a natural photograph. Recent methods [64, 65] have explored

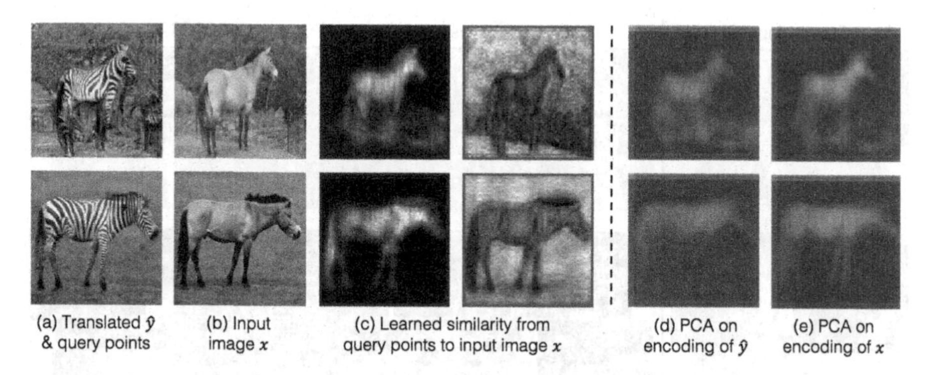

Fig. 7. Visualizing the learned similarity by G_{enc}. Given query points (blue or red) on an output image (a) and input (b), we visualize the learned similarity to patches on the input image by computing $\exp(\boldsymbol{v} \cdot \boldsymbol{v}^{-}/\tau)$ in (c). Here $\boldsymbol{v}$ is the query patch in the output and $\boldsymbol{v}^{-}$ denotes patches from the input. This suggests that our encoder may learn cross-domain correspondences implicitly. In (d) and (e), we visualize the top 3 PCA components of the shared embedding.

Fig. 8. Additional applications on Parisian street → Burano's colored houses and Russian Blue cat → Grumpy cat.

training unconditional models on a single image. Bearing the additional challenge of respecting the structure of the input image, conditional image synthesis using only one image has not been explored by previous image-to-image translation methods. Our painting → photo task is also different from neural style transfer [19,34] (photo → painting) and photo style transfer [48,82] (photo → photo).

Since the whole image (at HD resolution) cannot fit on a commercial GPU, at each iteration we train on 16 random crops of size 128 × 128. We also randomly scale the image to prevent overfitting. Furthermore, we observe that limiting the receptive field of the discriminator is important for preserving the structure of the input image, as otherwise the GAN loss will force the output image to be identical to the target image. Therefore, the crops are further split into 64 × 64 patches before passed to the discriminator. Lastly, we find that using gradient penalty [35,53] stabilizes optimization. We call this variant SinCUT.

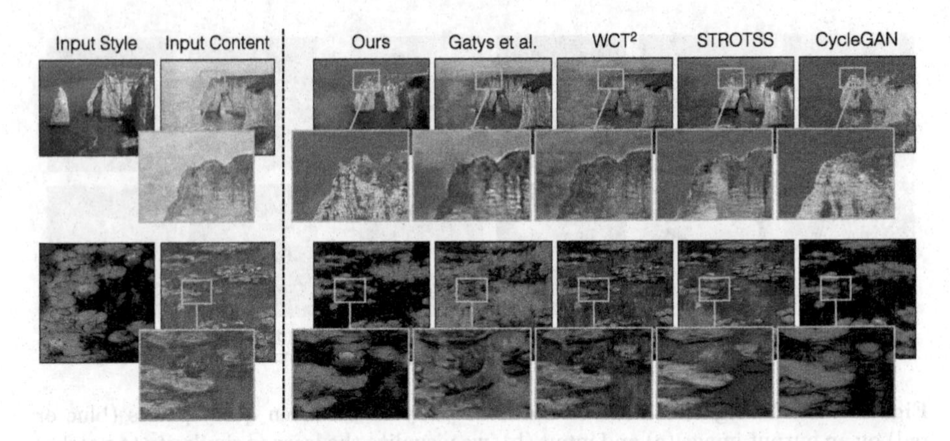

Fig. 9. High-res painting to photo translation. We transfer Claude Monet's paintings to reference natural photographs. The training only requires a single image from each domain. We compare our results (SinCUT) to recent style and photo transfer methods including Gatys et al. [19], WCT^2 [82], STROTSS [39], and patch-based CycleGAN [89]. Our method generates can reproduce the texture of the reference photo while retaining structure of input painting. Our generation is at 1k $\sim$ 1.5k resolution.

Figure 9 shows a qualitative comparison between our results and baseline methods including two neural style transfer methods (Gatys et al. [19] and STROTSS [39]), one leading photo style transfer method WCT^2 [82], and a CycleGAN baseline [89] that uses the ℓ_1 cycle-consistency loss instead of our contrastive loss at the patch level. The input paintings are high-res, ranging from 1k to 1.5k. Appendix B includes additional examples. We observe that Gatys et al. [19] fails to synthesize realistic textures. Existing photo style transfer methods such as WCT^2 can only modify the color of the input image. Our method SinCUT outperforms CycleGAN and is comparable to a leading style transfer method [39], which is based on optimal transport and self-similarity. Interestingly, our method is not originally designed for this application. This result suggests the intriguing connection between image-to-image translation and neural style transfer.

5 Conclusion

We propose a straightforward method for encouraging content preservation in unpaired image translation problems – by maximizing the mutual information between input and output with contrastive learning. The objective learns an embedding to bringing together corresponding patches in input and output, while pushing away noncorresponding "negative" patches. We study several important design choices. Interestingly, drawing negatives from *within* the image itself, rather than other images, provides a stronger signal. Our method *learns a cross-domain similarity function* and is the first image translation algorithm, to our knowledge, to not use any pre-defined similarity function (such as ℓ_1 or perceptual loss). As our method does not rely on cycle-consistency, it can enable one-sided image translation, with better quality than established baselines. In addition, our method can be used for *single-image* unpaired translation.

Acknowledgments. We thank Allan Jabri and Phillip Isola for helpful discussion and feedback. Taesung Park is supported by a Samsung Scholarship and an Adobe Research Fellowship, and some of this work was done as an Adobe Research intern. This work was partially supported by NSF grant IIS-1633310, grant from SAP, and gifts from Berkeley DeepDrive and Adobe.

Appendix A Additional Image-to-Image Results

We first show additional, randomly selected results on datasets used in our main paper. We then show results on additional datasets.

A.1 Additional Comparisons

In Fig. 10, we show additional, randomly selected results for Horse→Zebra and Cat→Dog. This is an extension of Fig. 3 in the main paper. We compare to baseline methods CycleGAN [89], MUNIT [30], DRIT [41], Self-Distance and DistanceGAN [4], and GcGAN [18].

B.2 Additional Datasets

In Fig. 11 and 12, we show additional datasets, compared against baseline method CycleGAN [89]. Our method provides better or comparable results, demonstrating its flexibility across a variety of datasets.

- *Apple→Orange* contains 996 apple and 1,020 orange images from ImageNet and was introduced in CycleGAN [89].
- *Yosemite Summer→Winter* contains 1,273 summer and 854 winter images of Yosemite scraped using the FlickAPI was introduced in CycleGAN [89].
- *GTA→Cityscapes* GTA contains 24,966 images [63] and Cityscapes [13] contains 19,998 images of street scenes from German cities. The task was originally used in CyCADA [29].

Appendix B Additional Single Image Translation Results

We show additional results in Fig. 13 and Fig. 14, and describe training details below.

Training details. At each iteration, the input image is randomly scaled to a width between 384 to 1024, and we randomly sample 16 crops of size 128 × 128. To avoid overfitting, we divide crops into 64 × 64 tiles before passing them to the

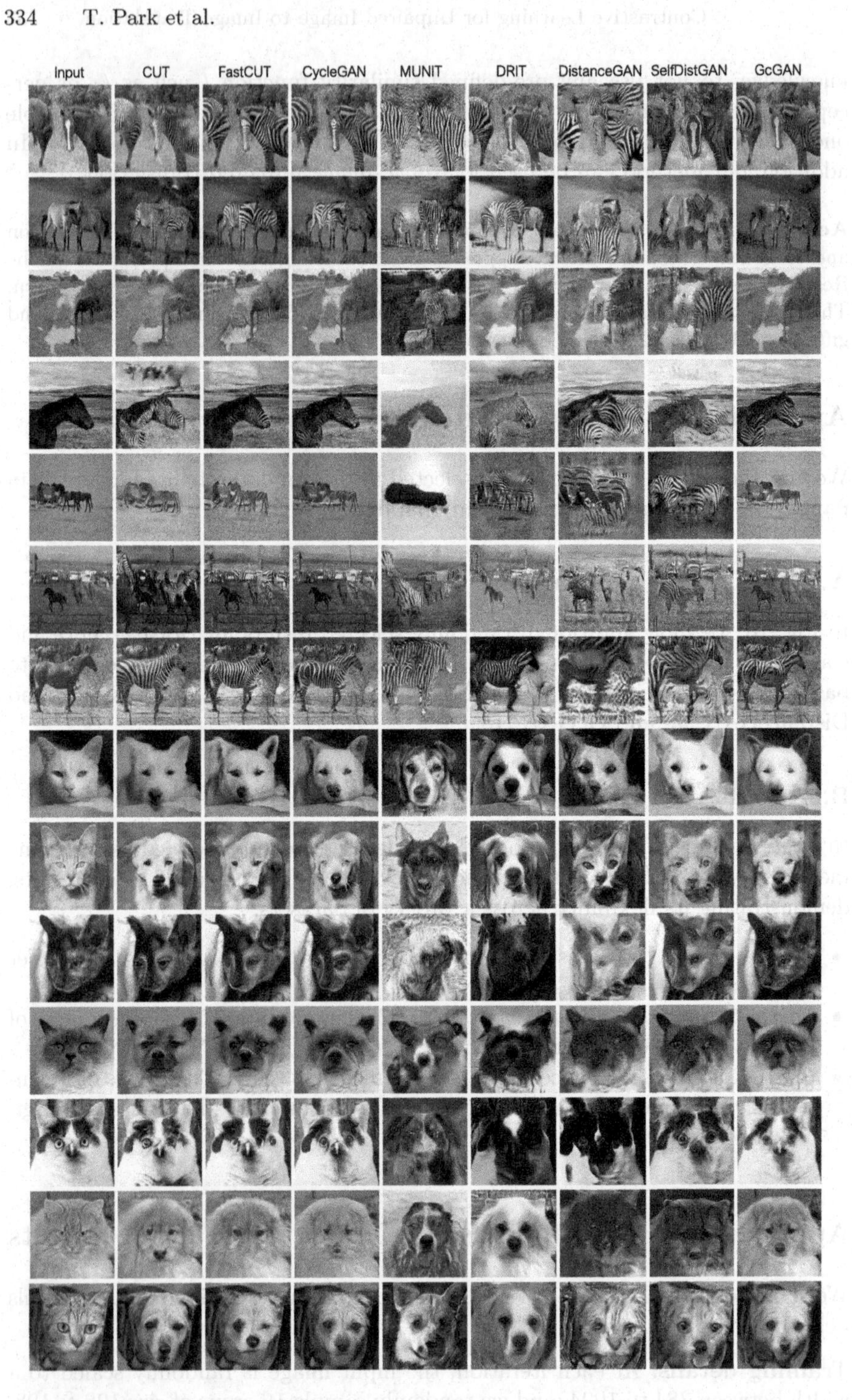

Fig. 10. Randomly selected Horse→Zebra and Cat→Dog results. This is an extension of Fig. 3 in the main paper.

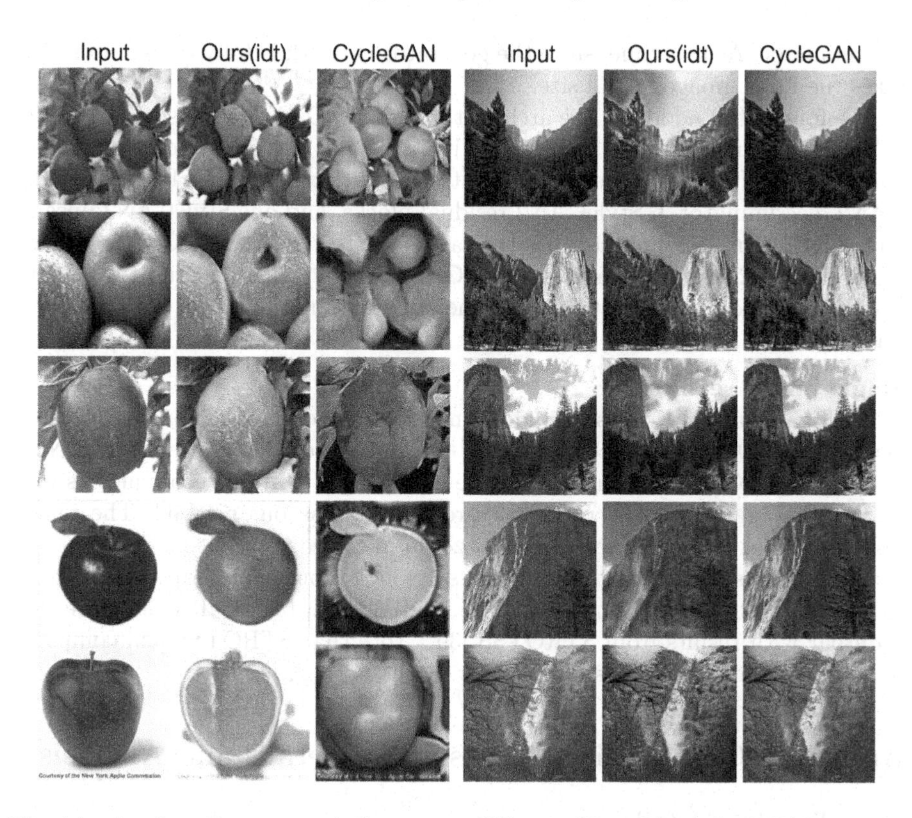

Fig. 11. **Apple→Orange** and **Summer→Winter Yosemite**. CycleGAN models were downloaded from the authors' public code repository. Apple→Orange shows that CycleGAN may suffer from color flipping issue.

Fig. 12. **GTA→Cityscapes** results at 1024 × 512 resolution. The model was trained on 512 × 512 crops.

discriminator. At test time, since the generator network is fully convolutional, it takes the input image at full size.

We found that adopting the architecture of StyleGAN2 [36] instead of CycleGAN slightly improves the output quality, although the difference is marginal. Our StyleGAN2-based generator consists of one downsampling block of StyleGAN2 discriminator, 6 StyleGAN2 residual blocks, and one StyleGAN2 upsampling block. Our discriminator has the same architecture as StyleGAN2. Following StyleGAN2, we use non-saturating GAN loss [61] with R1 gradient penalty [53]. Since we do not use style code, the style modulation layer of StyleGAN2 was removed.

Single image results.
In Fig. 13 and 14, we show additional comparison results for our method, Gatys et al. [19], STROTSS [39], WCT^2 [82], and CycleGAN baseline [89]. Note that the CycleGAN baseline adopts the same augmentation techniques as well as the same generator/discriminator architectures as our method. The image resolution is at 1–2 Megapixels. Please zoom in to see more visual details.

Both figures demonstrate that our results look more photorealistic compared to CycleGAN baseline, Gatys et al. [19], and WCT^2. The quality of our results is on par with results from STROTSS [39]. Note that STROTSS [39] compares to and outperforms recent style transfer methods (e.g., [22,52]).

Appendix C Unpaired Translation Details and Analysis

C.1 Training Details

To show the effect of the proposed patch-based contrastive loss, we intentionally match the architecture and hyperparameter settings of CycleGAN, except the loss function. This includes the ResNet-based generator [34] with 9 residual blocks, PatchGAN discriminator [31], Least Square GAN loss [50], batch size of 1, and Adam optimizer [38] with learning rate 0.002.

Our full model CUT is trained up to 400 epochs, while the fast variant FastCUT is trained up to 200 epochs, following CycleGAN. Moreover, inspired by GcGAN [18], FastCUT is trained with flip-equivariance augmentation, where the input image to the generator is horizontally flipped, and the output features are flipped back before computing the PatchNCE loss. Our encoder G_{enc} is the first half of the CycleGAN generator [89]. In order to calculate our multi-layer, patch-based contrastive loss, we extract features from 5 layers, which are RGB pixels, the first and second downsampling convolution, and the first and the fifth residual block. The layers we use correspond to receptive fields of sizes 1×1, 9×9, 15×15, 35×35, and 99×99. For each layer's features, we sample 256 random locations, and apply 2-layer MLP to acquire 256-dim final features. For our baseline model that uses MoCo-style memory bank [24], we follow the setting of MoCo, and used momentum value 0.999 with temperature 0.07. The size of the memory bank is 16384 per layer, and we enqueue 256 patches per image per iteration.

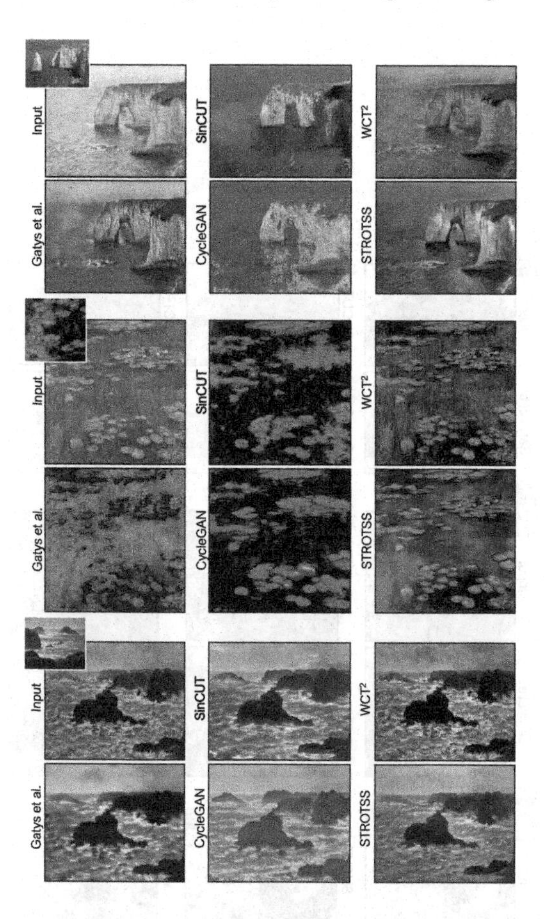

Fig. 13. High-res painting to photo translation (I). We transfer Monet's paintings to reference natural photos shown as insets at top-left corners. The training only requires a single image from each domain. We compare our results to recent style and photo transfer methods including Gatys et al. [19], WCT^2 [82], STROTSS [39], and our modified patch-based CycleGAN [89]. Our method can reproduce the texture of the reference photos while retaining structure of the input paintings. Our results are at 1k $\sim$ 1.5k resolution.

C.2 Evaluation Details

We list the details of our evaluation protocol.

Fréchet Inception Distance (FID [26]) throughout this paper is computed by resizing the images to 299-by-299 using bilinear sampling of PyTorch framework, and then taking the activations of the last average pooling layer of a pretrained Inception V3 [70] using the weights provided by the TensorFlow framework. We use the default setting of https://github.com/mseitzer/pytorch-fid. All test set images are used for evaluation, unless noted otherwise.

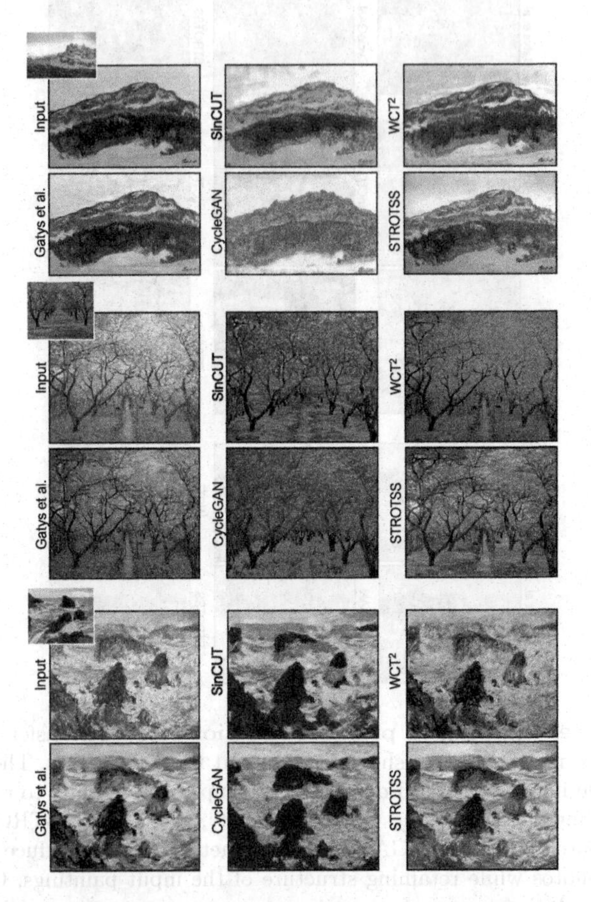

Fig. 14. High-res painting to photo translation (II). We transfer Monet's paintings to reference natural photos shown as insets at top-left corners. The training only requires a single image from each domain. We compare our results to recent style and photo transfer methods including Gatys et al. [19], WCT^2 [82], STROTSS [39], and our modified patch-based CycleGAN [89]. Our method can reproduce the texture of the reference photos while retaining structure of the input paintings. Our results are at 1k $\sim$ 1.5k resolution.

Semantic segmentation metrics on the Cityscapes dataset are computed as follows. First, we trained a semantic segmentation network using the DRN-D-22 [83] architecture. We used the recommended setting from https://github.com/fyu/drn, with batch size 32 and learning rate 0.01, for 250 epochs at 256 × 128 resolution. The output images of the 500 validation labels are resized to 256 × 128 using bicubic downsampling, passed to the trained DRN network, and compared against the ground truth labels downsampled to the same size using nearest-neighbor sampling.

C.3 Pseudocode

Here we provide the pseudo-code of PatchNCE loss in the PyTorch style. Our code and models are available at our GitHub repo.

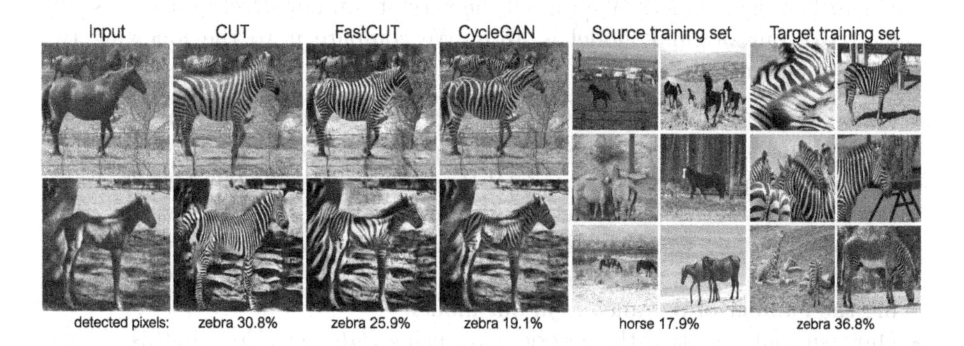

Fig. 15. Distribution matching. We measure the percentage of pixels belonging to the horse/zebra bodies, using a pre-trained semantic segmentation model. We find a distribution mismatch between sizes of horses and zebras images – zebras usually appear larger (36.8% vs. 17.9%). Our full method CUT has the flexibility to enlarge the horses, as a means of better matching of the training statistics than CycleGAN [89]. Our faster variant FastCUT, trained with a higher PatchNCE loss ($\lambda_X = 10$) and flip-equivariance augmentation, behaves more conservatively like CycleGAN.

C.4 Distribution Matching

In Fig. 15, we show an interesting phenomenon of our method, caused by the training set imbalance of the horse→zebra set. We use an off-the-shelf DeepLab model [7] trained on COCO-Stuff [6], to measure the percentage of pixels that belong to horses and zebras[1]. The training set exhibits dataset bias [74]. On average, zebras appear in more close-up pictures than horses and take up about twice the number of pixels (37% vs 18%). To perfectly satisfy the discriminator, a translation model should attempt to match the statistics of the training set.

[1] Pretrained model from https://github.com/kazuto1011/deeplab-pytorch.

Our method allows the flexibility for the horses to change the size, and the percentage of output zebra pixels (31%) better matches the training distribution (37%) than the CycleGAN baseline (19%). On the other hand, our fast variant *FastCUT* uses a larger weight ($\lambda_X = 10$) on the Patch NCE loss and flip-equivariance augmentation, and hence behaves more conservatively and more similar to CycleGAN. The strong distribution matching capacity has pros and cons. For certain applications, it can create introduce undesired changes (e.g., zebra patterns on the background for horse→zebra). On the other hand, it can enable dramatic geometric changes for applications such as Cat→Dog.

C.5 Additional Ablation Studies

In the paper, we mainly discussed the impact of loss functions and the number of patches on the final performance. Here we present additional ablation studies on more subtle design choices. We run all the variants on horse2zebra datasets [89]. The FID of our original model is **46.6**. We compare it to the following two variants of our model:

- Ours without weight sharing for the encoder G_{enc} and MLP projection network H: for this variant, when computing features $\{z_l\}_L = \{H_l(G^l_{\text{enc}}(\boldsymbol{x}))\}_L$, we use two separate encoders and MLP networks for embedding input images (e.g., horse) and the generated images (e.g., zebras) to feature space. They do not share any weights. The FID of this variant is **50.5**, worse than our method. This shows that weight sharing helps stabilize training while reducing the number of parameters in our model.
- Ours without updating the decoder G_{dec} using *PatchNCE* loss: in this variant, we exclude the gradient propagation of the decoder G_{dec} regarding *PatchNCE* loss $\mathcal{L}_{\text{PatchNCE}}$. In other words, the decoder G_{dec} only gets updated through the adversarial loss $\mathcal{L}_{\text{GAN}}$. The FID of this variant is **444.2**, and the results contain severe artifacts. This shows that our $\mathcal{L}_{\text{PatchNCE}}$ not only helps learn the encoder G_{enc}, as done in previous unsupervised feature learning methods [24], but also learns a better decoder G_{dec} together with the GAN loss. Intuitively, if the generated result has many artifacts and is far from realistic, it would be difficult for the encoder to find correspondences between the input and output, producing a large *PatchNCE* loss.

References

1. Almahairi, A., Rajeswar, S., Sordoni, A., Bachman, P., Courville, A.: Augmented cyclegan: Learning many-to-many mappings from unpaired data. In: International Conference on Machine Learning (ICML) (2018)
2. Amodio, M., Krishnaswamy, S.: Travelgan: Image-to-image translation by transformation vector learning. In: Proceedings of the IEEE Conference on Computer Vision and Pattern Recognition. pp. 8983–8992 (2019)
3. Bachman, P., Hjelm, R.D., Buchwalter, W.: Learning representations by maximizing mutual information across views. In: Advances in Neural Information Processing Systems (NeurIPS) (2019)

4. Benaim, S., Wolf, L.: One-sided unsupervised domain mapping. In: Advances in Neural Information Processing Systems (NeurIPS) (2017)
5. Bousmalis, K., Silberman, N., Dohan, D., Erhan, D., Krishnan, D.: Unsupervised pixel-level domain adaptation with generative adversarial networks. In: IEEE Conference on Computer Vision and Pattern Recognition (CVPR) (2017)
6. Caesar, H., Uijlings, J., Ferrari, V.: Coco-stuff: Thing and stuff classes in context. In: IEEE Conference on Computer Vision and Pattern Recognition (CVPR) (2018)
7. Chen, L.C., Papandreou, G., Kokkinos, I., Murphy, K., Yuille, A.L.: Deeplab: Semantic image segmentation with deep convolutional nets, atrous convolution, and fully connected crfs. IEEE Trans. Pattern Anal. Mach. Intell. (TPAMI) **40**(4), 834–848 (2018)
8. Chen, Q., Koltun, V.: Photographic image synthesis with cascaded refinement networks. In: IEEE International Conference on Computer Vision (ICCV) (2017)
9. Chen, T., Kornblith, S., Norouzi, M., Hinton, G.: A simple framework for contrastive learning of visual representations. In: International Conference on Machine Learning (ICML) (2020)
10. Choi, Y., Choi, M., Kim, M., Ha, J.W., Kim, S., Choo, J.: Stargan: Unified generative adversarial networks for multi-domain image-to-image translation. In: IEEE Conference on Computer Vision and Pattern Recognition (CVPR) (2018)
11. Choi, Y., Uh, Y., Yoo, J., Ha, J.W.: Stargan v2: Diverse image synthesis for multiple domains. In: IEEE Conference on Computer Vision and Pattern Recognition (CVPR) (2020)
12. Chopra, S., Hadsell, R., LeCun, Y.: Learning a similarity metric discriminatively, with application to face verification. In: IEEE Conference on Computer Vision and Pattern Recognition (CVPR) (2005)
13. Cordts, M., et al.: The cityscapes dataset for semantic urban scene understanding. In: IEEE Conference on Computer Vision and Pattern Recognition (CVPR) (2016)
14. Deng, J., Dong, W., Socher, R., Li, L.J., Li, K., Fei-Fei, L.: ImageNet: A Large-Scale Hierarchical Image Database. In: IEEE Conference on Computer Vision and Pattern Recognition (CVPR) (2009)
15. Doersch, C., Gupta, A., Efros, A.A.: Unsupervised visual representation learning by context prediction. In: IEEE International Conference on Computer Vision (ICCV) (2015)
16. Dosovitskiy, A., Brox, T.: Generating images with perceptual similarity metrics based on deep networks. In: Advances in Neural Information Processing Systems (2016)
17. Dosovitskiy, A., Fischer, P., Springenberg, J.T., Riedmiller, M., Brox, T.: Discriminative unsupervised feature learning with exemplar convolutional neural networks. IEEE Trans. Pattern Anal. Mach. Intell. (TPAMI) **38**(9), 1734–1747 (2015)
18. Fu, H., Gong, M., Wang, C., Batmanghelich, K., Zhang, K., Tao, D.: Geometry-consistent generative adversarial networks for one-sided unsupervised domain mapping. In: IEEE Conference on Computer Vision and Pattern Recognition (CVPR) (2019)
19. Gatys, L.A., Ecker, A.S., Bethge, M.: Image style transfer using convolutional neural networks. In: IEEE Conference on Computer Vision and Pattern Recognition (CVPR) (2016)
20. Gokaslan, A., Ramanujan, V., Ritchie, D., In Kim, K., Tompkin, J.: Improving shape deformation in unsupervised image-to-image translation. In: European Conference on Computer Vision (ECCV) (2018)
21. Goodfellow, I., et al.: Generative adversarial nets. In: Advances in Neural Information Processing Systems (2014)

22. Gu, S., Chen, C., Liao, J., Yuan, L.: Arbitrary style transfer with deep feature reshuffle. In: IEEE Conference on Computer Vision and Pattern Recognition (CVPR) (2018)
23. Gutmann, M., Hyvärinen, A.: Noise-contrastive estimation: A new estimation principle for unnormalized statistical models. In: International Conference on Artificial Intelligence and Statistics (AISTATS) (2010)
24. He, K., Fan, H., Wu, Y., Xie, S., Girshick, R.: Momentum contrast for unsupervised visual representation learning. In: IEEE Conference on Computer Vision and Pattern Recognition (CVPR) (2020)
25. Hénaff, O.J., Razavi, A., Doersch, C., Eslami, S., Oord, A.v.d.: Data-efficient image recognition with contrastive predictive coding. In: IEEE Conference on Computer Vision and Pattern Recognition (CVPR) (2019)
26. Heusel, M., Ramsauer, H., Unterthiner, T., Nessler, B., Hochreiter, S.: GANs trained by a two time-scale update rule converge to a local Nash equilibrium. In: Advances in Neural Information Processing Systems (2017)
27. Hinton, G.E., Salakhutdinov, R.R.: Reducing the dimensionality of data with neural networks. Science **313**(5786), 504–507 (2006)
28. Hjelm, R.D., et al.: Learning deep representations by mutual information estimation and maximization. arXiv preprint arXiv:1808.06670 (2018)
29. Hoffman, J., et al.: Cycada: Cycle-consistent adversarial domain adaptation. In: International Conference on Machine Learning (ICML) (2018)
30. Huang, X., Liu, M.Y., Belongie, S., Kautz, J.: Multimodal unsupervised image-to-image translation. European Conference on Computer Vision (ECCV) (2018)
31. Isola, P., Zhu, J.Y., Zhou, T., Efros, A.A.: Image-to-image translation with conditional adversarial networks. In: IEEE Conference on Computer Vision and Pattern Recognition (CVPR) (2017)
32. Isola, P., Zoran, D., Krishnan, D., Adelson, E.H.: Crisp boundary detection using pointwise mutual information. In: European Conference on Computer Vision (ECCV) (2014)
33. Isola, P., Zoran, D., Krishnan, D., Adelson, E.H.: Learning visual groups from co-occurrences in space and time. arXiv preprint arXiv:1511.06811 (2015)
34. Johnson, J., Alahi, A., Fei-Fei, L.: Perceptual losses for real-time style transfer and super-resolution. In: European Conference on Computer Vision (ECCV) (2016)
35. Karras, T., Laine, S., Aila, T.: A style-based generator architecture for generative adversarial networks. In: IEEE Conference on Computer Vision and Pattern Recognition (CVPR) (2019)
36. Karras, T., Laine, S., Aittala, M., Hellsten, J., Lehtinen, J., Aila, T.: Analyzing and improving the image quality of stylegan. In: IEEE Conference on Computer Vision and Pattern Recognition (CVPR) (2020)
37. Kim, T., Cha, M., Kim, H., Lee, J., Kim, J.: Learning to discover cross-domain relations with generative adversarial networks. In: International Conference on Machine Learning (ICML) (2017)
38. Kingma, D.P., Ba, J.: Adam: A method for stochastic optimization. In: International Conference on Learning Representations (ICLR) (2015)
39. Kolkin, N., Salavon, J., Shakhnarovich, G.: Style transfer by relaxed optimal transport and self-similarity. In: IEEE Conference on Computer Vision and Pattern Recognition (CVPR) (2019)
40. Larsson, G., Maire, M., Shakhnarovich, G.: Colorization as a proxy task for visual understanding. In: IEEE Conference on Computer Vision and Pattern Recognition (CVPR). pp. 6874–6883 (2017)

41. Lee, H.Y., Tseng, H.Y., Huang, J.B., Singh, M.K., Yang, M.H.: Diverse image-to-image translation via disentangled representation. In: European Conference on Computer Vision (ECCV) (2018)
42. Li, C., et al.: Alice: Towards understanding adversarial learning for joint distribution matching. In: Advances in Neural Information Processing Systems (2017)
43. Liang, X., Zhang, H., Lin, L., Xing, E.: Generative semantic manipulation with mask-contrasting gan. In: European Conference on Computer Vision (ECCV) (2018)
44. Liu, M.Y., Breuel, T., Kautz, J.: Unsupervised image-to-image translation networks. In: Advances in Neural Information Processing Systems (2017)
45. Liu, M.Y., et al.: Few-shot unsupervised image-to-image translation. In: IEEE International Conference on Computer Vision (ICCV) (2019)
46. Lotter, W., Kreiman, G., Cox, D.: Deep predictive coding networks for video prediction and unsupervised learning. arXiv preprint arXiv:1605.08104 (2016)
47. Löwe, S., O'Connor, P., Veeling, B.: Putting an end to end-to-end: Gradient-isolated learning of representations. In: Advances in Neural Information Processing Systems (NeurIPS) (2019)
48. Luan, F., Paris, S., Shechtman, E., Bala, K.: Deep photo style transfer. In: IEEE Conference on Computer Vision and Pattern Recognition (CVPR) (2017)
49. Malisiewicz, T., Gupta, A., Efros, A.A.: Ensemble of Exemplar-SVMs for object detection and beyond. In: IEEE International Conference on Computer Vision (ICCV) (2011)
50. Mao, X., Li, Q., Xie, H., Lau, Y.R., Wang, Z., Smolley, S.P.: Least squares generative adversarial networks. In: IEEE International Conference on Computer Vision (ICCV) (2017)
51. Mechrez, R., Talmi, I., Shama, F., Zelnik-Manor, L.: Maintaining natural image statistics with the contextual loss. In: Asian Conference on Computer Vision (ACCV) (2018)
52. Mechrez, R., Talmi, I., Zelnik-Manor, L.: The contextual loss for image transformation with non-aligned data. In: European Conference on Computer Vision (ECCV) (2018)
53. Mescheder, L., Geiger, A., Nowozin, S.: Which training methods for gans do actually converge? In: International Conference on Machine Learning (ICML) (2018)
54. Misra, I., van der Maaten, L.: Self-supervised learning of pretext-invariant representations. arXiv preprint arXiv:1912.01991 (2019)
55. Misra, I., Zitnick, C.L., Hebert, M.: Shuffle and learn: unsupervised learning using temporal order verification. In: Leibe, B., Matas, J., Sebe, N., Welling, M. (eds.) ECCV 2016. LNCS, vol. 9905, pp. 527–544. Springer, Cham (2016). https://doi.org/10.1007/978-3-319-46448-0_32
56. Ngiam, J., Khosla, A., Kim, M., Nam, J., Lee, H., Ng, A.Y.: Multimodal deep learning. In: International Conference on Machine Learning (ICML) (2011)
57. Oord, A.v.d., Li, Y., Vinyals, O.: Representation learning with contrastive predictive coding. arXiv preprint arXiv:1807.03748 (2018)
58. Owens, A., Wu, J., McDermott, J.H., Freeman, W.T., Torralba, A.: Ambient sound provides supervision for visual learning. In: European Conference on Computer Vision (ECCV) (2016)
59. Park, T., Liu, M.Y., Wang, T.C., Zhu, J.Y.: Semantic image synthesis with spatially-adaptive normalization. In: IEEE Conference on Computer Vision and Pattern Recognition (CVPR) (2019)

60. Pathak, D., Krahenbuhl, P., Donahue, J., Darrell, T., Efros, A.A.: Context encoders: Feature learning by inpainting. In: Proceedings of the IEEE conference on computer vision and pattern recognition. pp. 2536–2544 (2016)
61. Radford, A., Metz, L., Chintala, S.: Unsupervised representation learning with deep convolutional generative adversarial networks. In: International Conference on Learning Representations (ICLR) (2016)
62. Rao, K., Harris, C., Irpan, A., Levine, S., Ibarz, J., Khansari, M.: Rl-cyclegan: Reinforcement learning aware simulation-to-real. In: IEEE Conference on Computer Vision and Pattern Recognition (CVPR) (2020)
63. Richter, S.R., Vineet, V., Roth, S., Koltun, V.: Playing for data: Ground truth from computer games. In: European Conference on Computer Vision (ECCV) (2016)
64. Shaham, T.R., Dekel, T., Michaeli, T.: Singan: Learning a generative model from a single natural image. In: IEEE International Conference on Computer Vision (ICCV) (2019)
65. Shocher, A., Bagon, S., Isola, P., Irani, M.: Ingan: Capturing and remapping the" dna" of a natural image. In: IEEE International Conference on Computer Vision (ICCV) (2019)
66. Shocher, A., Cohen, N., Irani, M.: "zero-shot" super-resolution using deep internal learning. In: IEEE Conference on Computer Vision and Pattern Recognition (CVPR) (2018)
67. Shrivastava, A., Malisiewicz, T., Gupta, A., Efros, A.A.: Data-driven visual similarity for cross-domain image matching. ACM Transactions on Graphics (SIGGRAPH Asia) 30(6) (2011)
68. Shrivastava, A., Pfister, T., Tuzel, O., Susskind, J., Wang, W., Webb, R.: Learning from simulated and unsupervised images through adversarial training. In: IEEE Conference on Computer Vision and Pattern Recognition (CVPR) (2017)
69. Simonyan, K., Zisserman, A.: Very deep convolutional networks for large-scale image recognition. In: International Conference on Learning Representations (ICLR) (2015)
70. Szegedy, C., Vanhoucke, V., Ioffe, S., Shlens, J., Wojna, Z.: Rethinking the inception architecture for computer vision. In: IEEE Conference on Computer Vision and Pattern Recognition (CVPR) (2016)
71. Taigman, Y., Polyak, A., Wolf, L.: Unsupervised cross-domain image generation. In: International Conference on Learning Representations (ICLR) (2017)
72. Tang, H., Xu, D., Sebe, N., Yan, Y.: Attention-guided generative adversarial networks for unsupervised image-to-image translation. In: International Joint Conference on Neural Networks (IJCNN) (2019)
73. Tian, Y., Krishnan, D., Isola, P.: Contrastive multiview coding. arXiv preprint arXiv:1906.05849 (2019)
74. Torralba, A., Efros, A.A.: Unbiased look at dataset bias. In: IEEE Conference on Computer Vision and Pattern Recognition (CVPR) (2011)
75. Ulyanov, D., Vedaldi, A., Lempitsky, V.: Improved texture networks: Maximizing quality and diversity in feed-forward stylization and texture synthesis. In: IEEE Conference on Computer Vision and Pattern Recognition (CVPR) (2017)
76. Vincent, P., Larochelle, H., Bengio, Y., Manzagol, P.A.: Extracting and composing robust features with denoising autoencoders. In: International Conference on Machine Learning (ICML) (2008)
77. Wang, T.C., Liu, M.Y., Zhu, J.Y., Tao, A., Kautz, J., Catanzaro, B.: High-resolution image synthesis and semantic manipulation with conditional gans. In: IEEE Conference on Computer Vision and Pattern Recognition (CVPR) (2018)

78. Wang, Z., Bovik, A.C., Sheikh, H.R., Simoncelli, E.P.: Image quality assessment: from error visibility to structural similarity. IEEE Trans. Image Process. **13**(4), 600–612 (2004)
79. Wu, W., Cao, K., Li, C., Qian, C., Loy, C.C.: Transgaga: Geometry-aware unsupervised image-to-image translation. In: IEEE Conference on Computer Vision and Pattern Recognition (CVPR) (2019)
80. Wu, Z., Xiong, Y., Yu, S.X., Lin, D.: Unsupervised feature learning via non-parametric instance discrimination. In: IEEE Conference on Computer Vision and Pattern Recognition (CVPR) (2018)
81. Yi, Z., Zhang, H., Tan, P., Gong, M.: Dualgan: Unsupervised dual learning for image-to-image translation. In: IEEE International Conference on Computer Vision (ICCV) (2017)
82. Yoo, J., Uh, Y., Chun, S., Kang, B., Ha, J.W.: Photorealistic style transfer via wavelet transforms. In: IEEE International Conference on Computer Vision (ICCV) (2019)
83. Yu, F., Koltun, V., Funkhouser, T.: Dilated residual networks. In: IEEE Conference on Computer Vision and Pattern Recognition (CVPR) (2017)
84. Zhang, L., Zhang, L., Mou, X., Zhang, D.: Fsim: A feature similarity index for image quality assessment. IEEE Trans. Image Process. **20**(8), 2378–2386 (2011)
85. Zhang, R., Isola, P., Efros, A.A.: Colorful image colorization. In: European Conference on Computer Vision (ECCV) (2016)
86. Zhang, R., Isola, P., Efros, A.A.: Split-brain autoencoders: Unsupervised learning by cross-channel prediction. In: IEEE Conference on Computer Vision and Pattern Recognition (CVPR) (2017)
87. Zhang, R., Isola, P., Efros, A.A., Shechtman, E., Wang, O.: The unreasonable effectiveness of deep features as a perceptual metric. In: IEEE Conference on Computer Vision and Pattern Recognition (CVPR). pp. 586–595 (2018)
88. Zhang, R., Pfister, T., Li, J.: Harmonic unpaired image-to-image translation. In: International Conference on Learning Representations (ICLR) (2019)
89. Zhu, J.Y., Park, T., Isola, P., Efros, A.A.: Unpaired image-to-image translation using cycle-consistent adversarial networks. In: IEEE International Conference on Computer Vision (ICCV) (2017)
90. Zhu, J.Y., et al.: Toward multimodal image-to-image translation. In: Advances in Neural Information Processing Systems (2017)
91. Zontak, M., Irani, M.: Internal statistics of a single natural image. In: IEEE Conference on Computer Vision and Pattern Recognition (CVPR) (2011)

DLow: Diversifying Latent Flows for Diverse Human Motion Prediction

Ye Yuan(✉) and Kris Kitani

Robotics Institute, Carnegie Mellon University, Pittsburgh, USA
{yyuan2,kkitani}@cs.cmu.edu

Abstract. Deep generative models are often used for human motion prediction as they are able to model multi-modal data distributions and characterize diverse human behavior. While much care has been taken into designing and learning deep generative models, how to efficiently produce diverse samples from a deep generative model *after* it has been trained is still an under-explored problem. To obtain samples from a pretrained generative model, most existing generative human motion prediction methods draw a set of independent Gaussian latent codes and convert them to motion samples. Clearly, this random sampling strategy is not guaranteed to produce diverse samples for two reasons: (1) The independent sampling cannot force the samples to be diverse; (2) The sampling is based solely on likelihood which may only produce samples that correspond to the major modes of the data distribution. To address these problems, we propose a novel sampling method, Diversifying Latent Flows (DLow), to produce a diverse set of samples from a pretrained deep generative model. Unlike random (independent) sampling, the proposed DLow sampling method samples a single random variable and then maps it with a set of learnable mapping functions to a set of correlated latent codes. The correlated latent codes are then decoded into a set of correlated samples. During training, DLow uses a diversity-promoting prior over samples as an objective to optimize the latent mappings to improve sample diversity. The design of the prior is highly flexible and can be customized to generate diverse motions with common features (e.g., similar leg motion but diverse upper-body motion). Our experiments demonstrate that DLow outperforms state-of-the-art baseline methods in terms of sample diversity and accuracy (Code: https://github.com/Khrylx/DLow. Video: https://youtu.be/64OEdSadb00).

Keywords: Generative models · Diversity · Human motion forecasting

1 Introduction

Human motion prediction, i.e., predicting the future 3D poses of a person based on past poses, is an important problem in computer vision and has many

Electronic supplementary material The online version of this chapter (https://doi.org/10.1007/978-3-030-58545-7_20) contains supplementary material, which is available to authorized users.

© Springer Nature Switzerland AG 2020
A. Vedaldi et al. (Eds.): ECCV 2020, LNCS 12354, pp. 346–364, 2020.
https://doi.org/10.1007/978-3-030-58545-7_20

Fig. 1. In the latent space of a conditional variational autoencoder (CVAE), samples (stars) from our method DLow are able to cover more modes (colored ellipses) than the CVAE samples. In the motion space, DLow generates a diverse set of future human motions while the CVAE only produces perturbations of the motion of the major mode.

useful applications in autonomous driving [53], human robot interaction [37] and healthcare [65]. It is a challenging problem because the future motion of a person is potentially diverse and multi-modal due to the complex nature of human behavior. For many safety-critical applications, it is important to predict a diverse set of human motions instead of just the most likely one. For examples, an autonomous vehicle should be aware that a nearby pedestrian can suddenly cross the road even though the pedestrian will most likely remain in place. This diversity requirement calls for a generative approach that can fully characterize the multi-modal distribution of future human motion.

Deep generative models, e.g., variational autoencoders (VAEs) [36], are effective tools to model multi-modal data distributions. Most existing work [3,6, 40,44,59,66,69] using deep generative models for human motion prediction is focused on the design of the generative model to allow it to effectively learn the data distribution. After the generative model is learned, little attention has been paid to the sampling method used to produce *motion samples* (predicted future motions) from the *pretrained* generative model (weights kept fixed). Most of prior work predicts a set of motions by randomly sampling a set of latent codes from the latent prior and decoding them with the generator into motion samples. We argue that such a sampling strategy is not guaranteed to produce a diverse set of samples for two reasons: (1) The samples are independently drawn, which makes it difficult to enforce diversity; (2) The samples are drawn based on likelihood only, which means many samples may concentrate around the major modes (which have more observed data) of the data distribution and fail to cover the minor modes (as shown in Fig. 1 (Bottom)). The poor sample efficiency of random sampling means that one needs to draw a large number of samples in order to cover all the modes which is computationally expensive and can lead to high latency, making it unsuitable for real-time applications such as autonomous driving and virtual reality. This prompts us to address an overlooked aspect of diverse human motion prediction—the sampling strategy.

We propose a novel sampling method, Diversifying Latent Flows (DLow), to obtain a diverse set of samples from a pretrained deep generative model. For this work, we use a conditional variational autoencoder (CVAE) as our pretrained generative model but other generative models can also be used with our approach. DLow is inspired by the two previously mentioned problems with random (independent) sampling. To tackle problem (1) where sample independence limits model diversity, we introduce a new random variable and a set of learnable deterministic mapping functions to correlate the motion samples. We first transform the random variable with the mapping functions to generate a set of correlated latent codes which are then decoded into motion samples using the generator. As all motion samples are generated from a common random factor, this formulation allows us to model the joint sample distribution and offers us the opportunity to impose diversity on the samples by optimizing the parameters of the mapping functions. To address problem (2) where likelihood-based sampling limits diversity, we introduce a diversity-promoting prior (loss function) on the samples during the training of DLow. The prior follows an energy-based formulation using an energy function based on pairwise sample distance. We optimize the mapping functions during training to minimize the cross entropy between the joint sample distribution and diversity-promoting prior to increase sample diversity. To strike a balance between diversity and likelihood, we add a KL term to the optimization to enhance the likelihood of each sample. The relative weights between the prior term and the KL term represent the trade-off between the diversity and likelihood of the generated motion samples. Furthermore, our approach is highly flexible in that by designing different forms of the diversity-promoting prior we can impose a variety of structures on the samples besides diversity. For example, we can design the prior to ask the motion samples to cover the ground truth better to achieve higher sample accuracy. Additionally, other designs of the prior can enable new applications, such as controllable motion prediction, where we generate diverse motion samples that share some common features (e.g., similar leg motion but diverse upper-body motion).

The contributions of this work are the following: (1) We propose a novel perspective for addressing sample diversity in deep generative models—designing sampling methods for a *pretrained* generative model. (2) We propose a principled sampling method, DLow, which formulates diversity sampling as a constrained optimization problem over a set of learnable mapping functions using a diversity-promoting prior on the samples and KL constraints on the latent codes, which allows us to balance between sample diversity and likelihood. (3) Our approach allows for flexible design of the diversity-promoting prior to obtain more accurate samples or enable new applications such as controllable motion prediction. (4) We demonstrate through human motion prediction experiments that our approach outperforms state-of-the-art baseline methods in terms of sample diversity and accuracy.

2 Related Work

Human Motion Prediction. Most previous work takes a deterministic approach to modeling human motion and regress a single future motion from past 3D poses [1,9,13,16,17,21,33,43,49,50,55,67] or video frames [10,73,75]. While these approaches are able to predict the most likely future motion, they fail to model the multi-modal nature of human motion, which is essential for safety-critical applications. More related to our work, stochastic human motion prediction methods start to gain popularity with the development of deep generative models. These methods [3,6,40,44,59,66,69,74] often build upon popular generative models such as conditional generative adversarial networks (CGANs; [20]) or conditional variational autoencoders (CVAEs; [36]). The aforementioned methods differ in the design of their generative models, but at test time they follow the same sampling strategy—randomly and independently sampling trajectories from the pretrained generative model without considering the correlation between samples. In this work, we propose a principled sampling method that can produce a diverse set of samples, thus improving sample efficiency compared to the random sampling typically used in prior work.

Diverse Inference. Producing a diverse set of solutions has been investigated in numerous problems in computer vision and machine learning. A branch of these diversity-driven methods stems from the M-Best MAP problem [52,60], including diverse M-Best solutions [7] and multiple choice learning [27,42]. Alternatively, submodular function maximization has been applied to select a diverse subset of garments from fashion images [30]. Another type of methods [5,18,19,31, 38,68,72] seeks diversity using determinantal point processes (DPPs; [39,48]) which are efficient probabilistic models that can measure the global diversity and quality within a set. Similarly, Fisher information [58] has been used for diverse feature [22] and data [62] selection. Diversity has also been a key aspect in generative modeling. A vast body of work has tried to alleviate the mode collapse problem in GANs [4,11,12,15,24,45,63,70] and the posterior collapse problem in VAEs [8,28,35,46,64,76]. Normalizing flows [56] have also been used to promote diversity in trajectory forecasting [23,57]. This line of work aims to improve the diversity of the data distribution learned by deep generative models. We address diversity from a different angle by improving the strategy for producing samples from a pretrained deep generative model.

3 Diversifying Latent Flows (DLow)

For many existing methods on generative vision tasks such as multi-modal human motion prediction, the primary focus is to learn a good generative model that can capture the multi-modal distribution of the data. In contrast, once the generative model is learned, little attention has been paid to devising sampling strategies for producing diverse samples from the *pretrained* generative model.

In this section, we will introduce our method, Diversifying Latent Flows (DLow), as a principled way for drawing a diverse and likely set of samples from a pretrained generative model (weights kept fixed). To provide the proper context, we will first start with a brief review of deep generative models and how traditional methods produce samples from a pretrained generative model.

Background: Deep Generative Models. Let $\mathbf{x} \in \mathcal{X}$ denote data (e.g., human motion) drawn from a data distribution $p(\mathbf{x}|\mathbf{c})$ where $\mathbf{c}$ is some conditional information (e.g., past motion). One can reparameterize the data distribution by introducing a latent variable $\mathbf{z} \in \mathcal{Z}$ such that $p(\mathbf{x}|\mathbf{c}) = \int_{\mathbf{z}} p(\mathbf{x}|\mathbf{z},\mathbf{c})p(\mathbf{z})d\mathbf{z}$, where $p(\mathbf{z})$ is a Gaussian prior distribution. Deep generative models learn $p(\mathbf{x}|\mathbf{c})$ by modeling the conditional distribution $p(\mathbf{x}|\mathbf{z},\mathbf{c})$, and the generative process can be described as sampling $\mathbf{z}$ and mapping them to data samples $\mathbf{x}$ using a deterministic *generator* function $G_\theta : \mathcal{Z} \to \mathcal{X}$ as

$$\mathbf{z} \sim p(\mathbf{z}), \tag{1}$$

$$\mathbf{x} = G_\theta(\mathbf{z}, \mathbf{c}), \tag{2}$$

where the generator G_θ is instantiated as a deep neural network parametrized by θ. This generative process produces samples from the implicit sample distribution $p_\theta(\mathbf{x}|\mathbf{c})$ of the generative model, and the goal of generative modeling is to learn a generator G_θ such that $p_\theta(\mathbf{x}|\mathbf{c}) \approx p(\mathbf{x}|\mathbf{c})$. There are various approaches for learning the generator function G_θ, which yield different types of deep generative models such as variational autoencoders (VAEs; [36]), normalizing flows (NFs; [56]), and generative adversarial networks (GANs; [20]). Note that even though the discussion in this work is focused on conditional generative models, our method can be readily applied to the unconditional case.

Random Sampling. Once the generator function G_θ is learned, traditional approaches produce samples from the learned data distribution $p_\theta(\mathbf{x}|\mathbf{c})$ by first randomly sampling a set of latent codes $Z = \{\mathbf{z}_1, \dots, \mathbf{z}_K\}$ from the latent prior $p(\mathbf{z})$ (Eq. (1)) and decode Z with the generator G_θ into a set of data samples $X = \{\mathbf{x}_1, \dots, \mathbf{x}_K\}$ (Eq. (2)). We argue that such a sampling strategy may result in a less diverse sample set for two reasons: (1) Independent sampling cannot model the repulsion between samples within a diverse set; (2) The sampling is only based on the data likelihood and many samples can concentrate around a small number of modes that have more training data. As a result, random sampling can lead to low sample efficiency because many samples are similar to one another and fail to cover other modes in the data distribution.

DLow Sampling. To address the above issues with the random sampling approach, we propose an alternative sampling method, Diversifying Latent Flows (DLow), that can generate a diverse and likely set of samples from a pretrained deep generative model. Again, we stress that the weights of the generative model are kept fixed for DLow. We later apply DLow to the task of human motion prediction in Sect. 4 to demonstrate DLow's ability to improve sample diversity.

Instead of sampling each latent code $\mathbf{z}_k \in Z$ independently according to $p(\mathbf{z})$, we introduce a random variable $\boldsymbol{\epsilon}$ and conditionally generate the latent codes Z

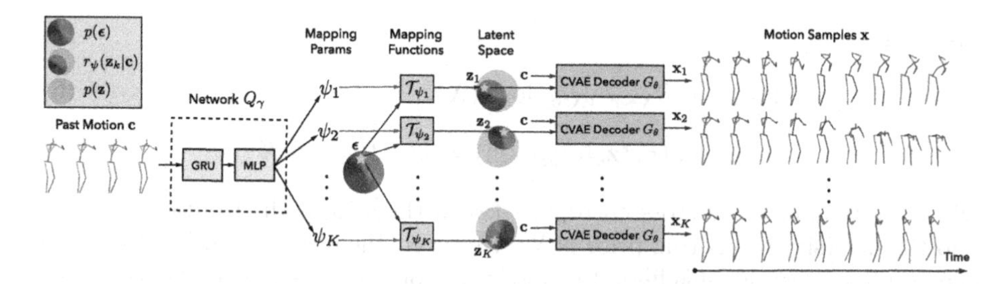

Fig. 2. Overview of our DLow framework applied to diverse human motion prediction. The network Q_γ takes past motion $\mathbf{c}$ as input and outputs the parameters of the mapping functions $\mathcal{T}_{\psi_1}, \dots, \mathcal{T}_{\psi_K}$. Each mapping $\mathcal{T}_{\psi_k}$ transforms the random variable $\boldsymbol{\epsilon}$ to a different latent code $\mathbf{z}_k$ and also warps the density $p(\boldsymbol{\epsilon})$ to the latent code density $r_\psi(\mathbf{z}_k|\mathbf{c})$. Each latent code $\mathbf{z}_k$ is decoded by the CVAE decoder into a motion sample $\mathbf{x}_k$.

and data samples X as follows:

$$\boldsymbol{\epsilon} \sim p(\boldsymbol{\epsilon})\,, \tag{3}$$

$$\mathbf{z}_k = \mathcal{T}_{\psi_k}(\boldsymbol{\epsilon})\,, \qquad 1 \le k \le K\,, \tag{4}$$

$$\mathbf{x}_k = G_\theta(\mathbf{z}_k, \mathbf{c})\,, \qquad 1 \le k \le K\,, \tag{5}$$

where $p(\boldsymbol{\epsilon})$ is a Gaussian distribution, $\mathcal{T}_{\psi_1}, \dots, \mathcal{T}_{\psi_K}$ are latent mapping functions with parameters $\psi = \{\psi_1, \dots, \psi_K\}$, and each $\mathcal{T}_{\psi_k}$ maps $\boldsymbol{\epsilon}$ to a different latent code $\mathbf{z}_k$. The above generative process defines a joint distribution $r_\psi(X, Z|\mathbf{c}) = p_\theta(X|Z, \mathbf{c})r_\psi(Z|\mathbf{c})$ over the samples X and latent codes Z, where $p_\theta(X|Z, \mathbf{c})$ is the conditional distribution induced by the generator $G_\theta(\mathbf{z}, \mathbf{c})$. Notice that in our setup, $r_\psi(X, Z|\mathbf{c})$ depends only on ψ as the generator parameters θ are learned in advance and are kept fixed. The data samples X can be viewed as a sample from the joint sample distribution $r_\psi(X|\mathbf{c}) = \int r_\psi(X, Z|\mathbf{c})dZ$ and the latent codes Z can be regarded as a sample from the joint latent distribution $r_\psi(Z|\mathbf{c})$ induced by warping $p(\boldsymbol{\epsilon})$ through $\mathcal{T}_{\psi_1}, \dots, \mathcal{T}_{\psi_K}$. If we further marginalize out all variables except for $\mathbf{x}_k$ from $r_\psi(X|\mathbf{c})$, we obtain the marginal sample distribution $r_\psi(\mathbf{x}_k|\mathbf{c})$ from which each sample $\mathbf{x}_k$ is drawn. Similarly, each latent code $\mathbf{z}_k \in Z$ can be viewed as a latent sample from the marginal latent distribution $r_\psi(\mathbf{z}_k|\mathbf{c})$.

The above distribution reparametrizations are illustrated in Fig. 2. We can see that all latent codes Z and data samples X are correlated as they are uniquely determined by $\boldsymbol{\epsilon}$, and by sampling $\boldsymbol{\epsilon}$ one can easily produce Z and X from the joint latent distribution $r_\psi(Z|\mathbf{c})$ and joint sample distribution $r_\psi(X|\mathbf{c})$. Because $r_\psi(Z|\mathbf{c})$ and $r_\psi(X|\mathbf{c})$ are controlled by the latent mapping functions $\mathcal{T}_{\psi_1}, \dots, \mathcal{T}_{\psi_K}$, we can impose structural constraints on $r_\psi(Z|\mathbf{c})$ and $r_\psi(X|\mathbf{c})$ by optimizing the parameters ψ of the latent mapping functions.

To encourage the diversity of samples X, we introduce a diversity-promoting prior $p(X)$ (specific form defined later) and formulate a constrained optimization problem:

$$\min_{\psi} \quad -\mathbb{E}_{X\sim r_\psi(X|\mathbf{c})}[\log p(X)], \tag{6}$$

$$\text{s.t.} \quad \mathrm{KL}(r_\psi(\mathbf{z}_k|\mathbf{c})\|p(\mathbf{z}_k)) = 0, \quad 1 \le k \le K, \tag{7}$$

where we minimize the cross entropy between the sample distribution $r_\psi(X|\mathbf{c})$ and the diversity-promoting prior $p(X)$. However, the objective in Eq. (6) alone can result in very low-likelihood samples $\mathbf{x}_k$ corresponding to latent codes $\mathbf{z}_k$ that are far away from the Gaussian prior $p(\mathbf{z}_k)$. To ensure that each sample $\mathbf{x}_k$ also has high likelihood under the generative model $p_\theta(\mathbf{x}|\mathbf{c})$, we add constraints in Eq. (7) on the KL divergence between $r_\psi(\mathbf{z}_k|\mathbf{c})$ and the Gaussian prior $p(\mathbf{z}_k)$ (same as $p(\mathbf{z})$) to make $r_\psi(\mathbf{z}_k|\mathbf{c}) = p(\mathbf{z}_k)$ and thus $r_\psi(\mathbf{x}_k|\mathbf{c}) = p_\theta(\mathbf{x}_k|\mathbf{c})$ where $r_\psi(\mathbf{x}_k|\mathbf{c}) = \int p_\theta(\mathbf{x}_k|\mathbf{z}_k,\mathbf{c}) r_\psi(\mathbf{z}_k|\mathbf{c}) d\mathbf{z}_k$ and $p_\theta(\mathbf{x}_k|\mathbf{c}) = \int p_\theta(\mathbf{x}_k|\mathbf{z}_k,\mathbf{c}) p(\mathbf{z}_k) d\mathbf{z}_k$. To optimize this constrained objective, we soften the constraints with the Lagrangian function:

$$\min_{\psi} -\mathbb{E}_{X\sim r_\psi(X|\mathbf{c})}[\log p(X)] + \beta \sum_{k=1}^{K} \mathrm{KL}(r_\psi(\mathbf{z}_k|\mathbf{c})\|p(\mathbf{z}_k)), \tag{8}$$

where we use the same Lagrangian multiplier β for all constraints. Despite having similar form, the above objective is very *different* from the objective function of β-VAE [29] in many ways: (1) our goal is to learn a diverse sampling distribution $r_\psi(X|\mathbf{c})$ for a pretrained generative model rather than learning the generative model itself; (2) The first part in our objective is a diversifying term instead of a reconstruction term; (3) Our objective function applies to most deep generative models, not just VAEs. In this objective, the softening of the hard KL constraints allows for the trade-off between the diversity and likelihood of the samples X. For small β, $r_\psi(\mathbf{z}_k|\mathbf{c})$ is allowed to deviate from $p(\mathbf{z}_k)$ so that $r_\psi(\mathbf{z}_1|\mathbf{c}), \ldots, r_\psi(\mathbf{z}_K|\mathbf{c})$ can potentially attend to different regions in the latent space as shown in Fig. 2 (latent space) to further improve sample diversity. For large β, the objective will focus on minimizing the KL term so that $r_\psi(\mathbf{z}_k|\mathbf{c}) \approx p(\mathbf{z}_k)$ and $r_\psi(\mathbf{x}_k|\mathbf{c}) \approx p_\theta(\mathbf{x}_k|\mathbf{c})$, and thus the sample $\mathbf{x}_k$ will have high likelihood under $p_\theta(\mathbf{x}_k|\mathbf{c})$.

The overall DLow objective is defined as:

$$L_{\mathrm{DLow}} = L_{\mathrm{prior}} + \beta L_{\mathrm{KL}}, \tag{9}$$

where L_{prior} and L_{KL} are the first and second term in Eq. (8) respectively. In the following, we will discuss in detail how we design the latent mapping functions $\mathcal{T}_{\psi_1}, \ldots, \mathcal{T}_{\psi_K}$ and the diversity-promoting prior $p(X)$.

Latent Mapping Functions. Each latent mapping $\mathcal{T}_{\psi_k}$ transforms the Gaussian distribution $p(\epsilon)$ to the marginal latent distribution $r_\psi(\mathbf{z}_k|\mathbf{c})$ for latent code $\mathbf{z}_k$ where $\mathcal{T}_{\psi_k}$ is also conditioned on $\mathbf{c}$. As $r_\psi(\mathbf{z}_k|\mathbf{c})$ should stay close to the Gaussian latent prior $p(\mathbf{z}_k)$, it would be ideal if the mapping $\mathcal{T}_{\psi_k}$ makes $r_\psi(\mathbf{z}_k|\mathbf{c})$ also a Gaussian. Thus, we design $\mathcal{T}_{\psi_k}$ to be an invertible affine transformation:

$$\mathcal{T}_{\psi_k}(\epsilon) = \mathbf{A}_k(\mathbf{c})\epsilon + \mathbf{b}_k(\mathbf{c}), \tag{10}$$

where the mapping parameters $\psi_k = \{\mathbf{A}_k(\mathbf{c}), \mathbf{b}_k(\mathbf{c})\}$, $\mathbf{A}_k \in \mathbb{R}^{n_z \times n_z}$ is a nonsingular matrix, $\mathbf{b}_k \in \mathbb{R}^{n_z}$ is a vector, and n_z is the number of dimensions for $\mathbf{z}_k$ and $\boldsymbol{\epsilon}$. As shown in Fig. 2, we use a K-head network $Q_\gamma(\mathbf{c})$ to output $\psi_1, \ldots, \psi_K$, and the parameters γ of the network $Q_\gamma(\mathbf{c})$ are the parameters to be optimized with the DLow objective in Eq. (9).

Under the invertible affine transformation $\mathcal{T}_{\psi_k}$, $r_\psi(\mathbf{z}_k|\mathbf{c})$ becomes a Gaussian distribution $\mathcal{N}(\mathbf{b}_k, \mathbf{A}_k\mathbf{A}_k^T)$. This allows us to compute the KL divergence terms in L_{KL} analytically:

$$\mathrm{KL}(r_\psi(\mathbf{z}_k|\mathbf{c})\|p(\mathbf{z}_k)) = \frac{1}{2}\left(\mathrm{tr}\left(\mathbf{A}_k\mathbf{A}_k^T\right) + \mathbf{b}_k^T\mathbf{b}_k - n_z - \log\det\left(\mathbf{A}_k\mathbf{A}_k^T\right)\right). \quad (11)$$

The KL divergence is minimized when $r_\psi(\mathbf{z}_k|\mathbf{c}) = p(\mathbf{z}_k)$ which implies that $\mathbf{A}_k\mathbf{A}_k^T = \mathbf{I}$ and $\mathbf{b}_k = \mathbf{0}$. Geometrically, this means that $\mathbf{A}_k$ is in the orthogonal group $O(n_z)$, which includes all rotations and reflections in an n_z-dimensional space. This means any mapping $\mathcal{T}_{\psi_k}$ that is a rotation or reflection operation will minimize the KL divergence. As mentioned before, there is a trade-off between diversity and likelihood in Eq. (9). To improve sample diversity (minimize L_{prior}) without compromising likelihood (KL divergence), we can optimize $\mathcal{T}_{\psi_1}, \ldots, \mathcal{T}_{\psi_K}$ to be different rotations or reflections to map $\boldsymbol{\epsilon}$ to different feasible points $\mathbf{z}_1, \ldots, \mathbf{z}_k$ in the latent space. This geometric understanding sheds light on the mapping space admitted by the hard KL constraints. In practice, we use soft KL constraints in the DLow objective to further enlarge the feasible mapping space which allows us to achieve lower L_{prior} and better sample diversity.

Diversity-Promoting Prior. In the DLow objective, a diversity-promoting prior $p(X)$ on the joint sample distribution is used to guide the optimization of the latent mapping functions $\mathcal{T}_{\psi_1}, \ldots, \mathcal{T}_{\psi_K}$. With an energy-based formulation, the prior $p(X)$ can be defined using an energy function $E(X)$:

$$p(X) = \exp(-E(X))/\mathcal{S}\,, \quad (12)$$

where $\mathcal{S}$ is a normalizing constant. Dropping the constant $\mathcal{S}$, the first term in Eq. (8) can be rewritten as

$$L_{\mathrm{prior}} = \mathbb{E}_{X\sim r_\psi(X|\mathbf{c})}[E(X)]\,. \quad (13)$$

To promote sample diversity of X, we design an energy function $E := E_d$ based on a pairwise distance metric $\mathcal{D}$:

$$E_d(X) = \frac{1}{K(K-1)}\sum_{i=1}^{K}\sum_{j\neq i}^{K}\exp\left(-\frac{\mathcal{D}^2(\mathbf{x}_i, \mathbf{x}_j)}{\sigma_d}\right), \quad (14)$$

where we use the Euclidean distance for $\mathcal{D}$ and an RBF kernel with scale σ_d. Minimizing L_{prior} moves the samples towards a lower-energy (diverse) configuration. L_{prior} can be evaluated efficiently with the reparametrization trick [36].

Up to this point, we have described the proposed sampling method, DLow, for generating a diverse set of samples from a pretrained generative model $p_\theta(\mathbf{x}|\mathbf{c})$.

By introducing a common random variable ϵ, DLow allows us to generate correlated samples X. Moreover, by introducing learnable mapping functions $\mathcal{T}_{\psi_k}$, we can model the joint sample distribution $r_\psi(X|\mathbf{c})$ and impose structural constraints, such as diversity, on the sample set X which cannot be modeled by random sampling from the generative model.

4 Diverse Human Motion Prediction

Equipped with a method to generate diverse samples from a pretrained deep generative model, we now turn our attention to the task of diverse human motion prediction. Suppose the pose of a person is a V-dimensional vector consisting of 3D joint positions, we use $\mathbf{c} \in \mathbb{R}^{H \times V}$ to denote the past motion of H time steps and $\mathbf{x} \in \mathbb{R}^{T \times V}$ to denote the future motion over a future time horizon of T. Given a past motion $\mathbf{c}$, the goal of diverse human motion prediction is to generate a diverse set of future motions $X = \{\mathbf{x}_1, \dots, \mathbf{x}_K\}$.

To capture the multi-modal distribution of the future trajectory $\mathbf{x}$, we take a generative approach and use a conditional variational autoencoder (CVAE) to learn the future trajectory distribution $p_\theta(\mathbf{x}|\mathbf{c})$. Here we use the CVAE for its stability over other popular approaches such as CGANs, but other suitable deep generative models could also be used. The CVAE uses a variational lower bound [34] as a surrogate for the intractable true data log-likelihood:

$$\mathcal{L}(\mathbf{x}; \theta, \phi) = \mathbb{E}_{q_\phi(\mathbf{z}|\mathbf{x},\mathbf{c})}\left[\log p_\theta(\mathbf{x}|\mathbf{z}, \mathbf{c})\right] - \mathrm{KL}\left(q_\phi(\mathbf{z}|\mathbf{x}, \mathbf{c}) \| p(\mathbf{z})\right), \tag{15}$$

where $q_\phi(\mathbf{z}|\mathbf{x}, \mathbf{c})$ is an ϕ-parametrized approximate posterior distribution. We use multivariate Gaussians for the prior, posterior (encoder distribution) and likelihood (decoder distribution): $p(\mathbf{z}) = \mathcal{N}(\mathbf{0}, \mathbf{I})$, $q_\phi(\mathbf{z}|\mathbf{x}, \mathbf{c}) = \mathcal{N}(\boldsymbol{\mu}, \mathrm{Diag}(\boldsymbol{\sigma}^2))$, and $p_\theta(\mathbf{x}|\mathbf{z}, \mathbf{c}) = \mathcal{N}(\tilde{\mathbf{x}}, \alpha\mathbf{I})$ where α is a hyperparameter. Both the encoder and decoder are implemented as recurrent neural networks (RNNs) (network architectures given in the supplementary materials). The encoder network F_ϕ outputs the parameters of the posterior distribution: $(\boldsymbol{\mu}, \boldsymbol{\sigma}) = F_\phi(\mathbf{x}, \mathbf{c})$; the decoder network G_θ outputs the reconstructed future trajectory $\tilde{\mathbf{x}} = G_\theta(\mathbf{z}, \mathbf{c})$. The CVAE is learned via jointly optimizing the encoder and decoder with Eq. (15).

4.1 Diversity Sampling with DLow

Once the CVAE is learned, we follow the DLow framework proposed in Sect. 3 to optimize the network Q_γ and learn the latent mapping functions $\mathcal{T}_{\psi_1}, \dots, \mathcal{T}_{\psi_K}$. Before doing this, to fully leverage the DLow framework, we will look at one of DLow's key feature, i.e., the design of the diversity-promoting prior $p(X)$ in L_{prior} can be flexibly changed by modifying the underlying energy function $E(X)$. This allows us to impose various structural constraints besides diversity on the sample set X. Below, we will provide two examples of such prior designs that (1) improve sample accuracy or (2) enable new applications such as controllable motion prediction.

Reconstruction Energy. To ensure that the sample set X is both diverse and accurate, i.e., the ground truth future motion $\hat{\mathbf{x}}$ is close to one of the samples in X, we can modify the prior's energy function E in Eq. (12) by adding a reconstruction term E_r:

$$E(X) = E_d(X) + \lambda_r E_r(X)\,, \tag{16}$$

$$E_r(X) = \min_k \mathcal{D}^2(\mathbf{x}_k, \hat{\mathbf{x}})\,, \tag{17}$$

where λ_r is a weighting factor and we use Euclidean distance as the distance metric $\mathcal{D}$. As DLow produces a correlated set of samples X instead of independent samples, the network Q_γ can learn to distribute samples in a way that are both diverse and accurate, covering the ground truth better. We use this prior design for our main experiments.

Controllable Motion Prediction. Another possible design of the diversity-promoting prior $p(X)$ is one that promotes diversity in a certain subspace of the sample space. In the context of human motion prediction, we may want certain body parts to move similarly but other parts to move differently. For example, we may want leg motion to be similar but upper-body motion to be diverse across motion samples. We call this task controllable motion prediction, i.e., finding a set of diverse samples that share some common features, which can allow users or down-stream systems to explore variations of a certain type of samples.

Formally, we divide the human joints into two sets, J_s and J_d, and ask samples in X to have similar motions for joints J_s but diverse motions for joints J_d. We can slice a motion sample $\mathbf{x}_k$ into two parts: $\mathbf{x}_k = \left(\mathbf{x}_k^s, \mathbf{x}_k^d\right)$ where $\mathbf{x}_k^s$ and $\mathbf{x}_k^d$ correspond to J_s and J_d respectively. Similarly, we can slice the sample set X into two sets: $X_s = \{\mathbf{x}_1^s, \ldots, \mathbf{x}_K^s\}$ and $X_d = \{\mathbf{x}_1^d, \ldots, \mathbf{x}_K^d\}$. We then define a new energy function E for the prior $p(X)$:

$$E(X) = E_d(X_d) + \lambda_s E_s(X_s) + \lambda_r E_r(X)\,, \tag{18}$$

$$E_s(X_s) = \frac{1}{K(K-1)} \sum_{i=1}^{K} \sum_{j \neq i}^{K} \mathcal{D}^2(\mathbf{x}_i^s, \mathbf{x}_j^s)\,, \tag{19}$$

where we add another energy term E_s weighted by λ_s to minimize the motion distance between samples for joints J_s, and we only compute the diversity-promoting term E_d using motions of joints J_d. After optimizing Q_γ using the DLow objective with the new energy E, we can produce diverse samples X that have similar motions for joints J_s.

Furthermore, we may also want to use a reference motion sample $\mathbf{x}_{\text{ref}}$ to provide the desired features. To achieve this, we can treat $\mathbf{x}_{\text{ref}}$ as the first sample $\mathbf{x}_1$ in X. We first find its corresponding latent code $\mathbf{z}_1 := \mathbf{z}_{\text{ref}}$ using the CVAE encoder: $\mathbf{z}_{\text{ref}} = F_\phi^{\boldsymbol{\mu}}(\mathbf{x}_{\text{ref}}, \mathbf{c})$. We can then find the common variable $\boldsymbol{\epsilon}_{\text{ref}}$ for generating X using the inverse mapping $\mathcal{T}_{\psi_1}^{-1}$:

$$\boldsymbol{\epsilon}_{\text{ref}} = \mathcal{T}_{\psi_1}^{-1}(\mathbf{z}_{\text{ref}}) = \mathbf{A}_1^{-1}(\mathbf{z}_{\text{ref}} - \mathbf{b}_1)\,. \tag{20}$$

With $\boldsymbol{\epsilon}_{\text{ref}}$ known, we can generate X that includes $\mathbf{x}_{\text{ref}}$. In practice, we force $\mathcal{T}_{\psi_1}$ to be an identity mapping to enforce $r_\psi(\mathbf{z}_1|\mathbf{c}) = p(\mathbf{z}_1)$ so that $r_\psi(\mathbf{z}_1|\mathbf{c})$ covers the posterior distribution of $\mathbf{z}_{\text{ref}}$. Otherwise, if $\mathbf{z}_{\text{ref}}$ lies outside of the high density region of $r_\psi(\mathbf{z}_1|\mathbf{c})$, it may lead to low-likelihood $\boldsymbol{\epsilon}_{\text{ref}}$ after the inverse mapping.

5 Experiments

Datasets. We perform evaluation on two public motion capture datasets: Human3.6M [32] and HumanEva-I [61]. Human3.6M is a large-scale dataset with 11 subjects (7 with ground truth) and 3.6 million video frames in total. Each subject performs 15 actions and the human motion is recorded 50 Hz. Following previous work [47,51,54,71], we adopt a 17-joint skeleton and train on five subjects (S1, S5, S6, S7, S8) and test on two subjects (S9 and S11). HumanEva-I is a relatively small dataset, containing only three subjects recorded 60 Hz. We adopt a 15-joint skeleton [54] and use the same train/test split provided in the dataset. By using both a large dataset with more variation in motion and a small dataset with less variation, we can better evaluate the generalization of our method to different types of data. For Human3.6M, we predict future motion for 2 s based on observed motion of 0.5 s. For HumanEva-I, we forecast future motion for 1 s given observed motion of 0.25 s.

Baselines. To fully evaluate our method, we consider three types of baselines: (1) Deterministic motion prediction methods, including **ERD** [16] and **acLSTM** [43]; (2) Stochastic motion prediction methods, including CVAE based methods, **Pose-Knows** [66] and **MT-VAE** [69], as well as a CGAN based method, **HP-GAN** [6]; (3) Diversity-promoting methods for generative models, including **Best-of-Many** [8], **GMVAE** [14], **DeLiGAN** [26], and **DSF** [72].

Metrics. We use the following metrics to measure both sample *diversity* and *accuracy*. (1) **Average Pairwise Distance (APD)**: average $L2$ distance between all pairs of motion samples to measure diversity within samples, which is computed as $\frac{1}{K(K-1)}\sum_{i=1}^{K}\sum_{j\neq i}^{K}\|\mathbf{x}_i - \mathbf{x}_j\|$. (2) **Average Displacement Error (ADE)**: average $L2$ distance over all time steps between the ground truth motion $\hat{\mathbf{x}}$ and the closest sample, which is computed as $\frac{1}{T}\min_{\mathbf{x}\in X}\|\hat{\mathbf{x}}-\mathbf{x}\|$. (3) **Final Displacement Error (FDE)**: $L2$ distance between the final ground truth pose $\mathbf{x}^T$ and the closest sample's final pose, which is computed as $\min_{\mathbf{x}\in X}\|\hat{\mathbf{x}}^T - \mathbf{x}^T\|$. (4) **Multi-Modal ADE (MMADE)**: the multi-modal version of ADE that obtains multi-modal ground truth future motions by grouping similar past motions. (5) **Multi-Modal FDE (MMFDE)**: the multi-modal version of FDE.

In these metrics, APD has been used to measure sample diversity [3]. ADE and FDE are common metrics for evaluating sample accuracy in trajectory forecasting literature [2,25,41]. MMADE and MMFDE [72] are metrics used to measure a method's ability to produce multi-modal predictions.

Table 1. Quantitative results on Human3.6M and HumanEva-I.

Method	Human3.6M [32]					HumanEva-I [61]				
	APD ↑	ADE ↓	FDE ↓	MMADE ↓	MMFDE ↓	APD ↑	ADE ↓	FDE ↓	MMADE ↓	MMFDE ↓
DLow (Ours)	**11.741**	**0.425**	**0.518**	**0.495**	**0.531**	**4.855**	**0.251**	**0.268**	**0.362**	**0.339**
ERD [16]	0	0.722	0.969	0.776	0.995	0	0.382	0.461	0.521	0.595
acLSTM [43]	0	0.789	1.126	0.849	1.139	0	0.429	0.541	0.530	0.608
Pose-Knows [66]	6.723	0.461	0.560	0.522	0.569	2.308	0.269	0.296	0.384	0.375
MT-VAE [69]	0.403	0.457	0.595	0.716	0.883	0.021	0.345	0.403	0.518	0.577
HP-GAN [6]	7.214	0.858	0.867	0.847	0.858	1.139	0.772	0.749	0.776	0.769
Best-of-Many [8]	6.265	0.448	0.533	0.514	0.544	2.846	0.271	0.279	0.373	0.351
GMVAE [14]	6.769	0.461	0.555	0.524	0.566	2.443	0.305	0.345	0.408	0.410
DeLiGAN [26]	6.509	0.483	0.534	0.520	0.545	2.177	0.306	0.322	0.385	0.371
DSF [72]	9.330	0.493	0.592	0.550	0.599	4.538	0.273	0.290	0.364	0.340

5.1 Quantitative Results

We summarize the quantitative results on Human3.6M and HumanEva-I in Table 1. The metrics are computed with the sample set size $K = 50$. For both datasets, we can see that our method, DLow, outperforms all baselines in terms of both sample diversity (APD) and accuracy (ADE, FDE) as well as covering multi-modal ground truth (MMADE, MMFDE). Deterministic methods like ERD [16] and acLSTM [43] do not perform well because they only predict one future trajectory which can lead to mode averaging. Methods like MT-VAE [69] produce trajectories samples that lack diversity so they fail to cover the multi-modal ground-truth (indicated by high MMADE and MMFDE) despite having decently low ADE and FDE. We would also like to point out the closest competitor DSF [72] can only generate one deterministic set of samples, while our method can produce multiple diverse sets by sampling $\boldsymbol{\epsilon}$. We also show how each metric changes against various K in the supplementary materials.

Table 2. Ablation study on Human3.6M and HumanEva-I.

Energy		Human3.6M [32]					HumanEva-I [61]				
E_d	E_r	APD ↑	ADE ↓	FDE ↓	MMADE ↓	MMFDE ↓	APD ↑	ADE ↓	FDE ↓	MMADE ↓	MMFDE ↓
✓	✓	11.741	**0.425**	**0.518**	**0.495**	**0.531**	4.855	**0.251**	**0.268**	**0.362**	**0.339**
✓	✗	**13.091**	0.546	0.663	0.599	0.669	**4.927**	0.263	0.281	0.368	0.347
✗	✓	6.844	0.432	0.525	0.500	0.539	2.355	0.252	0.277	0.376	0.366
✗	✗	6.383	0.520	0.629	0.577	0.638	2.247	0.281	0.317	0.395	0.393

Ablation Study. We further perform an ablation study (Table 2) to analyze the effects of the two energy terms E_d and E_r in Eq. (16). First, without the reconstruction term E_r, the DLow variant is able to achieve higher diversity (APD) at the cost of sample accuracy (ADE, FDE, MMADE, MMFDE). This is expected because the network only optimizes the diversity term E_d and focuses solely on diversity. Second, for the variant without E_d, both sample diversity and accuracy decrease. It is intuitive to see why the diversity (APD) decreases. To see why the sample accuracy (ADE, FDE, MMADE, MMFDE) also decreases, we should consider the fact that a more diverse set of samples have a better chance at covering the ground truth. Finally, when we remove both E_d and E_r (i.e., only optimize L_{KL}), the results are the worst, which is expected.

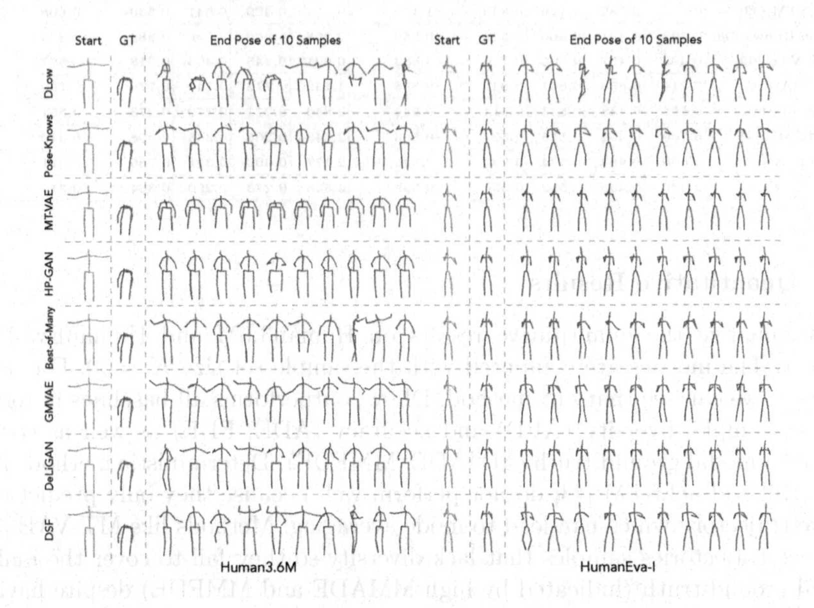

Fig. 3. Qualitative Results on Human3.6M and HumanEva-I.

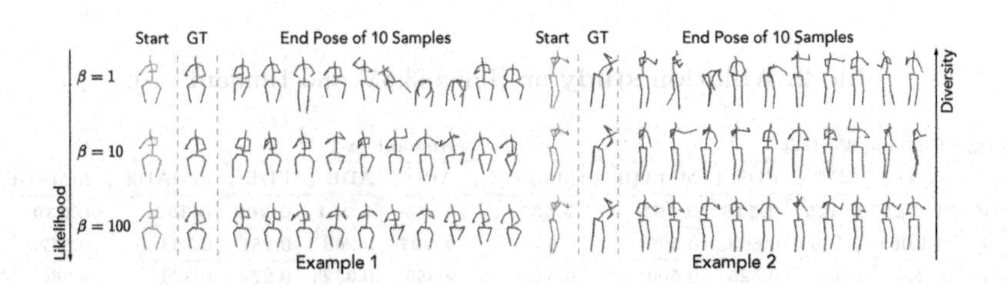

Fig. 4. Varying β in DLow allows us to balance between diversity and likelihood.

5.2 Qualitative Results

To visually evaluate the diversity and accuracy of each method, we present a qualitative comparison in Fig. 3 where we render the start pose, the end pose of the ground truth future motion, and the end pose of 10 motion samples. Note that we do not model the global translation of the person, which is why some sitting motions appear to be floating. For Human3.6M, we can see that our method DLow can predict a wide array of future motions, including standing, sitting, bending, crouching, and turning, which cover the ground truth bending motion. In contrast, the baseline methods mostly produce perturbations of a single motion—standing. For HumanEva-I, we can see that DLow produces interesting variations of the fighting motion, while the baselines produce almost identical future motions.

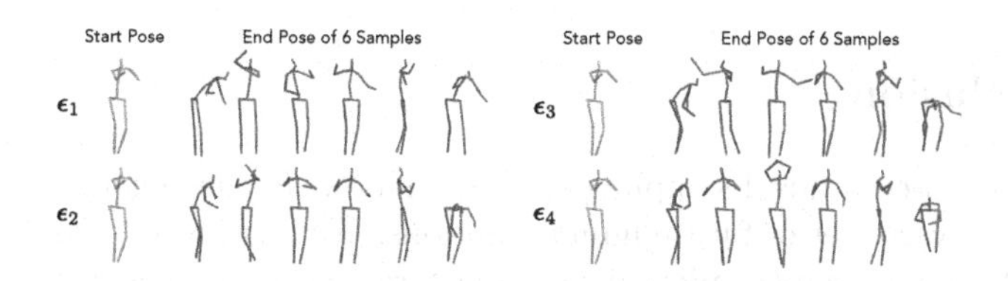

Fig. 5. Effect of varying ϵ on motion samples.

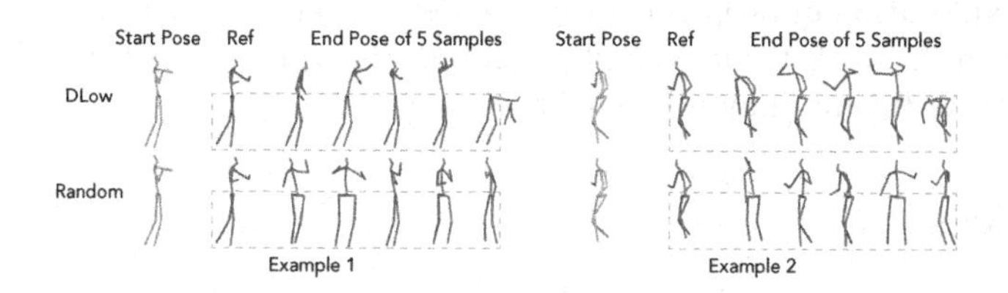

Fig. 6. Controllable Motion Prediction. DLow enables samples to have more similar leg motion to the reference.

Diversity vs. Likelihood. As discussed in the approach section, the β in Eq. (8) represents the trade-off between sample diversity and likelihood. To verify this, we trained three DLow models with different β (1, 10, 100) and visualize the motion samples generated by each model in Fig. 4. We can see that a larger β leads to less diverse samples which correspond to the major mode of the generator distribution, while a smaller β can produce more diverse motion samples covering other plausible yet less likely future motions.

Effect of Varying ϵ. A key difference between our method and DSF [72] is that we can generate multiple diverse sets of samples while DSF can only produce a fixed diverse set. To demonstrate this, we show in Fig. 5 how the motion samples of DLow change with different ϵ. By comparing the four sets of motion samples, one can conclude that changing ϵ varies each set of samples but preserves the main structure of each motion.

Controllable Motion Prediction. As highlighted before, the flexible design of the diversity-promoting prior enables a new application, controllable motion prediction, where we predict diverse motions that share some common features. We showcase this application by conducting an experiment using the energy function defined in Eq. (18). The network is trained so that the leg motion of the motion samples is similar while the upper-body motion is diverse. The results are shown in Fig. 6. We can see that given a reference motion, our method can generate diverse upper-body motion and preserve similar leg motion, while random samples from the CVAE cannot enforce similar leg motion. Please refer to the supplementary materials for more results.

6 Conclusion

We have proposed a novel sampling strategy, DLow, for deep generative models to obtain a diverse set of future human motions. We introduced learnable latent mapping functions which allowed us to generate a set of correlated samples, whose diversity can be optimized by a diversity-promoting prior. Experiments demonstrated superior performance in generating diverse motion samples. Moreover, we showed that the flexible design of the diversity-promoting prior further enables new applications, such as controllable human motion prediction. We hope that our exploration of deep generative models through the lens of diversity will encourage more work towards understanding the complex nature of modeling and predicting future human behavior.

References

1. Aksan, E., Kaufmann, M., Hilliges, O.: Structured prediction helps 3D human motion modelling. In: Proceedings of the IEEE International Conference on Computer Vision, pp. 7144–7153 (2019)
2. Alahi, A., Goel, K., Ramanathan, V., Robicquet, A., Fei-Fei, L., Savarese, S.: Social LSTM: human trajectory prediction in crowded spaces. In: Proceedings of the IEEE Conference on Computer Vision and Pattern Recognition, pp. 961–971 (2016)
3. Aliakbarian, S., Saleh, F.S., Salzmann, M., Petersson, L., Gould, S.: A stochastic conditioning scheme for diverse human motion prediction. In: Proceedings of the IEEE/CVF Conference on Computer Vision and Pattern Recognition, pp. 5223–5232 (2020)
4. Arjovsky, M., Chintala, S., Bottou, L.: Wasserstein GAN. arXiv preprint arXiv:1701.07875 (2017)
5. Azadi, S., Feng, J., Darrell, T.: Learning detection with diverse proposals. In: Proceedings of the IEEE Conference on Computer Vision and Pattern Recognition, pp. 7149–7157 (2017)
6. Barsoum, E., Kender, J., Liu, Z.: HP-GAN: probabilistic 3D human motion prediction via GAN. In: Proceedings of the IEEE Conference on Computer Vision and Pattern Recognition Workshops, pp. 1418–1427 (2018)
7. Batra, D., Yadollahpour, P., Guzman-Rivera, A., Shakhnarovich, G.: Diverse M-best solutions in Markov random fields. In: Fitzgibbon, A., Lazebnik, S., Perona, P., Sato, Y., Schmid, C. (eds.) ECCV 2012. LNCS, vol. 7576, pp. 1–16. Springer, Heidelberg (2012). https://doi.org/10.1007/978-3-642-33715-4_1

8. Bhattacharyya, A., Schiele, B., Fritz, M.: Accurate and diverse sampling of sequences based on a "best of many" sample objective. In: Proceedings of the IEEE Conference on Computer Vision and Pattern Recognition, pp. 8485–8493 (2018)
9. Butepage, J., Black, M.J., Kragic, D., Kjellstrom, H.: Deep representation learning for human motion prediction and classification. In: Proceedings of the IEEE Conference on Computer Vision and Pattern Recognition, pp. 6158–6166 (2017)
10. Chao, Y.W., Yang, J., Price, B., Cohen, S., Deng, J.: Forecasting human dynamics from static images. In: Proceedings of the IEEE Conference on Computer Vision and Pattern Recognition, pp. 548–556 (2017)
11. Che, T., Li, Y., Jacob, A.P., Bengio, Y., Li, W.: Mode regularized generative adversarial networks. arXiv preprint arXiv:1612.02136 (2016)
12. Chen, X., Duan, Y., Houthooft, R., Schulman, J., Sutskever, I., Abbeel, P.: InfoGAN: interpretable representation learning by information maximizing generative adversarial nets. In: Advances in Neural Information Processing Systems, pp. 2172–2180 (2016)
13. Chiu, H.k., Adeli, E., Wang, B., Huang, D.A., Niebles, J.C.: Action-agnostic human pose forecasting. In: 2019 IEEE Winter Conference on Applications of Computer Vision (WACV), pp. 1423–1432. IEEE (2019)
14. Dilokthanakul, N., et al.: Deep unsupervised clustering with Gaussian mixture variational autoencoders. arXiv preprint arXiv:1611.02648 (2016)
15. Elfeki, M., Couprie, C., Riviere, M., Elhoseiny, M.: GDPP: learning diverse generations using determinantal point process. arXiv preprint arXiv:1812.00068 (2018)
16. Fragkiadaki, K., Levine, S., Felsen, P., Malik, J.: Recurrent network models for human dynamics. In: Proceedings of the IEEE International Conference on Computer Vision, pp. 4346–4354 (2015)
17. Ghosh, P., Song, J., Aksan, E., Hilliges, O.: Learning human motion models for long-term predictions. In: 2017 International Conference on 3D Vision (3DV), pp. 458–466. IEEE (2017)
18. Gillenwater, J.A., Kulesza, A., Fox, E., Taskar, B.: Expectation-maximization for learning determinantal point processes. In: Advances in Neural Information Processing Systems, pp. 3149–3157 (2014)
19. Gong, B., Chao, W.L., Grauman, K., Sha, F.: Diverse sequential subset selection for supervised video summarization. In: Advances in Neural Information Processing Systems, pp. 2069–2077 (2014)
20. Goodfellow, I., et al.: Generative adversarial nets. In: Advances in Neural Information Processing Systems, pp. 2672–2680 (2014)
21. Gopalakrishnan, A., Mali, A., Kifer, D., Giles, L., Ororbia, A.G.: A neural temporal model for human motion prediction. In: Proceedings of the IEEE Conference on Computer Vision and Pattern Recognition, pp. 12116–12125 (2019)
22. Gu, Q., Li, Z., Han, J.: Generalized fisher score for feature selection. arXiv preprint arXiv:1202.3725 (2012)
23. Guan, J., Yuan, Y., Kitani, K.M., Rhinehart, N.: Generative hybrid representations for activity forecasting with no-regret learning. In: Proceedings of the IEEE Conference on Computer Vision and Pattern Recognition (2020)
24. Gulrajani, I., Ahmed, F., Arjovsky, M., Dumoulin, V., Courville, A.C.: Improved training of wasserstein gans. In: Advances in Neural Information Processing Systems, pp. 5767–5777 (2017)

25. Gupta, A., Johnson, J., Fei-Fei, L., Savarese, S., Alahi, A.: Social GAN: socially acceptable trajectories with generative adversarial networks. In: Proceedings of the IEEE Conference on Computer Vision and Pattern Recognition, pp. 2255–2264 (2018)
26. Gurumurthy, S., Kiran Sarvadevabhatla, R., Venkatesh Babu, R.: DeliGAN: generative adversarial networks for diverse and limited data. In: Proceedings of the IEEE Conference on Computer Vision and Pattern Recognition, pp. 166–174 (2017)
27. Guzman-Rivera, A., Batra, D., Kohli, P.: Multiple choice learning: learning to produce multiple structured outputs. In: Advances in Neural Information Processing Systems, pp. 1799–1807 (2012)
28. He, J., Spokoyny, D., Neubig, G., Berg-Kirkpatrick, T.: Lagging inference networks and posterior collapse in variational autoencoders. arXiv preprint arXiv:1901.05534 (2019)
29. Higgins, I., et al.: beta-VAE: learning basic visual concepts with a constrained variational framework. ICLR **2**(5), 6 (2017)
30. Hsiao, W.L., Grauman, K.: Creating capsule wardrobes from fashion images. In: Proceedings of the IEEE Conference on Computer Vision and Pattern Recognition, pp. 7161–7170 (2018)
31. Huang, D.A., Ma, M., Ma, W.C., Kitani, K.M.: How do we use our hands? discovering a diverse set of common grasps. In: Proceedings of the IEEE Conference on Computer Vision and Pattern Recognition, pp. 666–675 (2015)
32. Ionescu, C., Papava, D., Olaru, V., Sminchisescu, C.: Human3.6M: large scale datasets and predictive methods for 3D human sensing in natural environments. IEEE Trans. Pattern Anal. Mach. Intell. **36**(7), 1325–1339 (2013)
33. Jain, A., Zamir, A.R., Savarese, S., Saxena, A.: Structural-RNN: deep learning on spatio-temporal graphs. In: Proceedings of the IEEE Conference on Computer Vision and Pattern Recognition, pp. 5308–5317 (2016)
34. Jordan, M.I., Ghahramani, Z., Jaakkola, T.S., Saul, L.K.: An introduction to variational methods for graphical models. Mach. Learn. **37**(2), 183–233 (1999)
35. Kim, Y., Wiseman, S., Miller, A.C., Sontag, D., Rush, A.M.: Semi-amortized variational autoencoders. arXiv preprint arXiv:1802.02550 (2018)
36. Kingma, D.P., Welling, M.: Auto-encoding variational bayes. arXiv preprint arXiv:1312.6114 (2013)
37. Koppula, H.S., Saxena, A.: Anticipating human activities for reactive robotic response. In: IROS, Tokyo, p. 2071 (2013)
38. Kulesza, A., Taskar, B.: k-dpps: Fixed-size determinantal point processes. In: Proceedings of the 28th International Conference on Machine Learning (ICML 2011), pp. 1193–1200 (2011)
39. Kulesza, A., Taskar, B., et al.: Determinantal point processes for machine learning. Found. Trends® Mach. Learn. **5**(2–3), 123–286 (2012)
40. Kundu, J.N., Gor, M., Babu, R.V.: BIHMP-GAN: bidirectional 3D human motion prediction GAN. In: Proceedings of the AAAI Conference on Artificial Intelligence, vol. 33, pp. 8553–8560 (2019)
41. Lee, N., Choi, W., Vernaza, P., Choy, C.B., Torr, P.H., Chandraker, M.: Desire: distant future prediction in dynamic scenes with interacting agents. In: Proceedings of the IEEE Conference on Computer Vision and Pattern Recognition, pp. 336–345 (2017)
42. Lee, S., Prakash, S.P.S., Cogswell, M., Ranjan, V., Crandall, D., Batra, D.: Stochastic multiple choice learning for training diverse deep ensembles. In: Advances in Neural Information Processing Systems, pp. 2119–2127 (2016)

43. Li, Z., Zhou, Y., Xiao, S., He, C., Huang, Z., Li, H.: Auto-conditioned recurrent networks for extended complex human motion synthesis. arXiv preprint arXiv:1707.05363 (2017)
44. Lin, X., Amer, M.R.: Human motion modeling using DVGANs. arXiv preprint arXiv:1804.10652 (2018)
45. Lin, Z., Khetan, A., Fanti, G., Oh, S.: PACGAN: the power of two samples in generative adversarial networks. In: Advances in Neural Information Processing Systems, pp. 1498–1507 (2018)
46. Liu, X., Gao, J., Celikyilmaz, A., Carin, L., et al.: Cyclical annealing schedule: a simple approach to mitigating KL vanishing. arXiv preprint arXiv:1903.10145 (2019)
47. Luvizon, D.C., Picard, D., Tabia, H.: 2D/3D pose estimation and action recognition using multitask deep learning. In: Proceedings of the IEEE Conference on Computer Vision and Pattern Recognition, pp. 5137–5146 (2018)
48. Macchi, O.: The coincidence approach to stochastic point processes. Adv. Appl. Probab. **7**(1), 83–122 (1975)
49. Mao, W., Liu, M., Salzmann, M., Li, H.: Learning trajectory dependencies for human motion prediction. In: Proceedings of the IEEE International Conference on Computer Vision, pp. 9489–9497 (2019)
50. Martinez, J., Black, M.J., Romero, J.: On human motion prediction using recurrent neural networks. In: Proceedings of the IEEE Conference on Computer Vision and Pattern Recognition, pp. 2891–2900 (2017)
51. Martinez, J., Hossain, R., Romero, J., Little, J.J.: A simple yet effective baseline for 3D human pose estimation. In: Proceedings of the IEEE International Conference on Computer Vision, pp. 2640–2649 (2017)
52. Nilsson, D.: An efficient algorithm for finding the m most probable configurations in probabilistic expert systems. Stat. Comput. **8**(2), 159–173 (1998)
53. Paden, B., Čáp, M., Yong, S.Z., Yershov, D., Frazzoli, E.: A survey of motion planning and control techniques for self-driving urban vehicles. IEEE Trans. Intell. Veh. **1**(1), 33–55 (2016)
54. Pavllo, D., Feichtenhofer, C., Grangier, D., Auli, M.: 3D human pose estimation in video with temporal convolutions and semi-supervised training. In: Proceedings of the IEEE Conference on Computer Vision and Pattern Recognition, pp. 7753–7762 (2019)
55. Pavllo, D., Grangier, D., Auli, M.: QuaterNet: a quaternion-based recurrent model for human motion. arXiv preprint arXiv:1805.06485 (2018)
56. Rezende, D.J., Mohamed, S.: Variational inference with normalizing flows. arXiv preprint arXiv:1505.05770 (2015)
57. Rhinehart, N., Kitani, K.M., Vernaza, P.: R2P2: a reparameterized pushforward policy for diverse, precise generative path forecasting. In: Ferrari, V., Hebert, M., Sminchisescu, C., Weiss, Y. (eds.) ECCV 2018. LNCS, vol. 11217, pp. 794–811. Springer, Cham (2018). https://doi.org/10.1007/978-3-030-01261-8_47
58. Rissanen, J.J.: Fisher information and stochastic complexity. IEEE Trans. Inf. Theory **42**(1), 40–47 (1996)
59. Ruiz, A.H., Gall, J., Moreno-Noguer, F.: Human motion prediction via spatio-temporal inpainting. arXiv preprint arXiv:1812.05478 (2018)
60. Seroussi, B., Golmard, J.L.: An algorithm directly finding the k most probable configurations in Bayesian networks. Int. J. Approx. Reason. **11**(3), 205–233 (1994)
61. Sigal, L., Balan, A.O., Black, M.J.: HUMANEVA: synchronized video and motion capture dataset and baseline algorithm for evaluation of articulated human motion. Int. J. Comput. Vis. **87**(1–2), 4 (2010)

62. Sourati, J., Akcakaya, M., Erdogmus, D., Leen, T.K., Dy, J.G.: A probabilistic active learning algorithm based on fisher information ratio. IEEE Trans. Pattern Anal. Mach. Intell. **40**(8), 2023–2029 (2017)
63. Srivastava, A., Valkov, L., Russell, C., Gutmann, M.U., Sutton, C.: VeeGAN: reducing mode collapse in GANs using implicit variational learning. In: Advances in Neural Information Processing Systems, pp. 3308–3318 (2017)
64. Tolstikhin, I., Bousquet, O., Gelly, S., Schoelkopf, B.: Wasserstein auto-encoders. arXiv preprint arXiv 1711, 01558 (2017)
65. Troje, N.F.: Decomposing biological motion: a framework for analysis and synthesis of human gait patterns. J. Vis. **2**(5), 2 (2002)
66. Walker, J., Marino, K., Gupta, A., Hebert, M.: The pose knows: video forecasting by generating pose futures. In: Proceedings of the IEEE International Conference on Computer Vision, pp. 3332–3341 (2017)
67. Wang, B., Adeli, E., Chiu, H.k., Huang, D.A., Niebles, J.C.: Imitation learning for human pose prediction. In: Proceedings of the IEEE International Conference on Computer Vision, pp. 7124–7133 (2019)
68. Weng, X., Yuan, Y., Kitani, K.: Joint 3d tracking and forecasting with graph neural network and diversity sampling. arXiv:2003.07847 (2020)
69. Yan, X., et al.: MT-VAE: learning motion transformations to generate multimodal human dynamics. In: Ferrari, V., Hebert, M., Sminchisescu, C., Weiss, Y. (eds.) ECCV 2018. LNCS, vol. 11209, pp. 276–293. Springer, Cham (2018). https://doi.org/10.1007/978-3-030-01228-1_17
70. Yang, D., Hong, S., Jang, Y., Zhao, T., Lee, H.: Diversity-sensitive conditional generative adversarial networks. arXiv preprint arXiv:1901.09024 (2019)
71. Yang, W., Ouyang, W., Wang, X., Ren, J., Li, H., Wang, X.: 3D human pose estimation in the wild by adversarial learning. In: Proceedings of the IEEE Conference on Computer Vision and Pattern Recognition, pp. 5255–5264 (2018)
72. Yuan, Y., Kitani, K.: Diverse trajectory forecasting with determinantal point processes. arXiv preprint arXiv:1907.04967 (2019)
73. Yuan, Y., Kitani, K.: Ego-pose estimation and forecasting as real-time PD control. In: Proceedings of the IEEE International Conference on Computer Vision, pp. 10082–10092 (2019)
74. Yuan, Y., Kitani, K.: Residual force control for agile human behavior imitation and extended motion synthesis. arXiv preprint arXiv:2006.07364 (2020)
75. Zhang, J.Y., Felsen, P., Kanazawa, A., Malik, J.: Predicting 3D human dynamics from video. In: Proceedings of the IEEE International Conference on Computer Vision, pp. 7114–7123 (2019)
76. Zhao, S., Song, J., Ermon, S.: InfoVAE: information maximizing variational autoencoders. arXiv preprint arXiv:1706.02262 (2017)

GRNet: Gridding Residual Network for Dense Point Cloud Completion

Haozhe Xie[1,2,3], Hongxun Yao[1,2](✉), Shangchen Zhou[4], Jiageng Mao[5], Shengping Zhang[2,6], and Wenxiu Sun[7]

[1] State Key Laboratory of Robotics and System, Harbin Institute of Technology, Harbin, China
[2] Faculty of Computing, Harbin Institute of Technology, Harbin, China
[3] SenseTime Research, Shenzhen, China
[4] Nanyang Technological University, Singapore, Singapore
[5] The Chinese University of Hong Kong, Hong Kong SAR, China
[6] Peng Cheng Laboratory, Shenzhen, China
[7] SenseTime Research, Hong Kong SAR, China
`h.yao@hit.edu.cn`
`https://haozhexie.com/project/grnet`

Abstract. Estimating the complete 3D point cloud from an incomplete one is a key problem in many vision and robotics applications. Mainstream methods (*e.g.,* PCN and TopNet) use Multi-layer Perceptrons (MLPs) to directly process point clouds, which may cause the loss of details because the structural and context of point clouds are not fully considered. To solve this problem, we introduce 3D grids as intermediate representations to regularize unordered point clouds and propose a novel Gridding Residual Network (GRNet) for point cloud completion. In particular, we devise two novel differentiable layers, named *Gridding* and *Gridding Reverse*, to convert between point clouds and 3D grids without losing structural information. We also present the differentiable *Cubic Feature Sampling* layer to extract features of neighboring points, which preserves context information. In addition, we design a new loss function, namely *Gridding Loss*, to calculate the L1 distance between the 3D grids of the predicted and ground truth point clouds, which is helpful to recover details. Experimental results indicate that the proposed *GRNet* performs favorably against state-of-the-art methods on the ShapeNet, Completion3D, and KITTI benchmarks.

Keywords: Point cloud completion · Gridding · Cubic feature sampling

1 Introduction

With the rapid development of 3D acquisition technologies, 3D sensors (*e.g.*, LiDARs) are becoming increasingly available and affordable. As a commonly

Electronic supplementary material The online version of this chapter (https://doi.org/10.1007/978-3-030-58545-7_21) contains supplementary material, which is available to authorized users.

© Springer Nature Switzerland AG 2020
A. Vedaldi et al. (Eds.): ECCV 2020, LNCS 12354, pp. 365–381, 2020.
https://doi.org/10.1007/978-3-030-58545-7_21

used format, point clouds are the preferred representation for describing the 3D shape of an object. Complete 3D shapes are required in many applications, including semantic segmentation and SLAM [2]. However, due to limited sensor resolution and occlusion, highly sparse and incomplete point clouds can be acquired, which causes loss in geometric and semantic information. Consequently, recovering the complete point clouds from partial observations, named point cloud completion, is very important for practical applications.

In the recent few years, convolutional neural networks (CNNs) have been applied to 2D images and 3D voxels. Since the convolution can not be directly applied to point clouds due to their irregularity and unorderedness, most of the existing methods [3,7,26,29,34,35,40] voxelize the point cloud into binary voxels, where 3D convolutional neural networks can be applied. However, the voxelization operation leads to an irreversible loss of geometric information. Other approaches [27,38,51] use the Multi-Layer Perceptrons (MLPs) to process point clouds directly. However, these approaches use max pooling to aggregate information across points in a global or hierarchical manner, which do not fully consider the connectivity across points and the context of neighboring points. More recently, several attempts [41,42] have been made to incorporate graph convolutional networks (GCN) [14] to build local graphs in the neighborhood of each point in the point cloud. However, constructing the graph relies on the K-nearest neighbor (KNN) algorithm, which is sensitive to the point cloud density [39].

Several attempts in point cloud segmentation have been made to capture spatial relationships in point clouds through more general convolution operations. SPLATNet [36] and InterpConv [28] perform convolution on high-dimensional lattices and 3D cubes interpolated from neighboring points, respectively. However, both of them are based on a strong assumption that the 3D coordinates of the output points are the same as the input points and thus can not be used for 3D point completion.

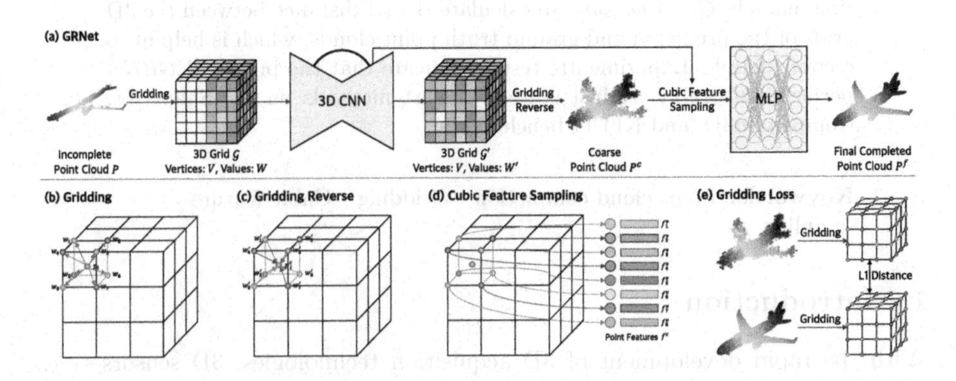

Fig. 1. Overview of the proposed **(a)** GRNet, **(b)** Gridding, **(c)** Gridding Reverse, **(d)** Cubic Feature Sampling, and **(e)** Gridding Loss.

To address the issues mentioned above, we introduce 3D grids as intermediate representations to regularize unordered point clouds, which explicitly preserves the structural and context of point clouds. Consequently, we propose a novel Gridding Residual Network (*GRNet*) for point cloud completion, as shown in Fig. 1. Besides 3D CNN and MLP, we devise three differentiable layers: *Gridding*, *Gridding Reverse*, and *Cubic Feature Sampling*. In *Gridding*, for each point of the point cloud, eight vertices of the 3D grid cell that the point lies in are first weighted using an interpolation function that explicitly measures the geometric relations of the point cloud. Then, a 3D convolutional neural network (3D CNN) with skip connections is adopted to learn context-aware and spatially-aware features, which allows the network to complete missing parts of the incomplete point cloud. Next, *Gridding Reverse* converts the output 3D grid to a coarse point cloud by replacing each 3D grid cell with a new point whose coordinate is the weighted sum of the eight vertices of the 3D grid cell. The following *Cubic Feature Sampling* extracts features for each point in the coarse point cloud by concatenating the features of the corresponding eight vertices of the 3D grid cell that the point lies in. The coarse point cloud and the features are forwarded to an MLP to obtain the final completed point cloud.

Existing methods adopt Chamfer Distance in PSGN [4] as the loss function to train the neural networks. This loss function penalizes the prediction deviating from the ground-truth. However, there is no guarantee that the predicted point clouds follow the geometric layout of objects, and the networks tend to output a mean shape that minimizes the distance [11,48]. Some recent works [11,12, 18,21,48] attempt to solve the unorderness while preserving fine-grained details by projecting the 3D point cloud to an image, which is then supervised by the corresponding ground truth masks. However, the projection requires extrinsic camera parameters, which are challenging to estimate in most scenarios [31]. To solve the unorderedness of point clouds, we propose *Gridding Loss*, which calculates the L1 distance between the generated points and ground truth by representing them in regular 3D grids with the proposed *Gridding* layer.

The contributions can be summarized as follows:

- We innovatively introduce 3D grids as intermediate representations to regularize unordered point clouds, which explicitly preserve the structural and context of point clouds.
- We propose a novel Gridding Residual Network (*GRNet*) for point cloud completion. We design three differentiable layers: *Gridding*, *Gridding Reverse*, and *Cubic Feature Sampling*, as well as a new *Gridding Loss*.
- Extensive experiments are conducted on the ShapeNet, Completion3D, and KITTI benchmarks, which indicate that the proposed *GRNet* performs favorably against state-of-the-art methods.

2 Related Work

According to the network architecture used in point cloud completion and reconstruction, existing networks can be roughly categorized into MLP-based, graph-based, and convolution-based networks.

MLP-Based Networks. Pioneered by PointNet [32], several works use MLP for point cloud processing [1,22] and reconstruction [27,51] because of its simplicity and strong representation ability. These methods model each point independently using several Multi-layer Perceptrons and then aggregate a global feature using a symmetric function (*e.g.*, Max Pooling). However, the geometric relationships among 3D points are not fully considered. PointNet++ [33] and TopNet [38] incorporate a hierarchical architecture to consider the geometric structure. To relief the structure loss caused by MLP, AtlasNet [6] and MSN [23] recover the complete point cloud of an object by estimating a collection of parametric surface elements.

Graph-Based Networks. By considering each point in a point cloud as a vertex of a graph, graph-based networks generate directed edges for the graph based on the neighbors of each point. In these methods, convolution is usually operated on spatial neighbors, and pooling is used to produce a new coarse graph by aggregating information from each point's neighbors. Compared with MLP-based methods, graph-based networks take local geometric structures into account. In DGCNN [42], a graph is constructed in the feature space and dynamically updated after each layer of the network. Further, LDGCNN [52] removes the transformation network and link the hierarchical features from different layers in DGCNN to improve its performance and reduce the model size. Inspired by DGCNN, Hassani and Haley [8] introduce the multi-scale graph-based network to learn point and shape features for self-supervised classification and reconstruction. DCG [41] also follows DGCNN to encode additional local connection into a feature vector and progressively evolves from coarse to fine point clouds.

Convolution-Based Networks. Early works [3,7,17] usually apply 3D convolutional neural networks (CNNs) build upon the volumetric representation of 3D point clouds. However, converting point clouds into 3D volumes introduces a quantization effect that discards some details of the data [43] and is not suitable for representing fine-grained information. To the best of our knowledge, no work directly applies CNNs on irregular point clouds for shape completion. In point cloud understanding, several works [10,15,16,20,28] develop CNNs operating on discrete 3D grids that are transformed from point clouds. Hua *et al.* [10] define convolutional kernels on regular 3D grids, where the points are assigned with the same weights when falling into the same grid. PointCNN [20] achieves permutation invariance through a χ-conv transformation. Besides CNNs on discrete space, several methods [9,24–26,36,39,44,49] define convolutional kernels on continuous space. Thomas *et al.* [39] propose both rigid and deformable kernel point convolution (KPConv) operators for 3D point clouds using a set of learnable kernel points. Compared with graph-based networks, convolution-based networks are more efficient and robust to point cloud density [28].

3 Gridding Residual Network

3.1 Overview

The proposed GRNet aims to recover the complete point cloud from an incomplete one in a coarse-to-fine fashion. It consists of five components, including *Gridding* (Sect. 3.2), 3D Convolutional Neural Network (Sect. 3.3), *Gridding Reverse* (Sect. 3.4), *Cubic Feature Sampling* (Sect. 3.5), and Multi-layer Perceptron (Sect. 3.6), as shown in Fig. 1. Given an incomplete point cloud P as input, *Gridding* is first used to obtain a 3D grid $\mathcal{G} = <V, W>$, where V and W are the vertex set and value set of $\mathcal{G}$, respectively. Then, W is fed to a 3D CNN, whose output is W'. Next, *Gridding Reverse* produces a coarse point cloud P^c from the 3D grid $\mathcal{G}' = <V, W'>$. Subsequently, *Cubic Feature Sampling* generates features F^c for the coarse point cloud P^c. Finally, MLP takes the coarse point cloud P^c and the corresponding features F^c as input to produce the final completed point cloud P^f.

3.2 Gridding

2D and 3D convolutions have been developed to process regularly arranged data such as images and voxel grids. However, it is challenging to directly apply standard 2D and 3D convolutions to unordered and irregular point clouds. Several methods [3,7,17,26] convert point clouds into 3D voxels and then apply 3D convolutions to them. However, the voxelization process leads to an irreversible loss of geometric information. Recent methods [38,51] adopt Multi-layer Perceptrons (MLPs) to directly operate on point clouds and aggregate information across points with max pooling. However, MLP-based methods may lose local context information because the connectivity and layouts of points are not fully considered. Recent studies also indicate that simply applying MLPs to point clouds cannot always work in practice [28,49].

In this paper, we introduce 3D grids as intermediate representations to regularize point clouds and further propose a differentiable *Gridding* layer, which converts an unordered and irregular point cloud $P = \{p_i\}_{i=1}^{n}$ into a regular 3D grid $\mathcal{G} = <V, W>$ while preserving spatial layouts of the point cloud, where $p_i \in \mathbb{R}^3$, $V = \{v_i\}_{i=1}^{N^3}$, $W = \{w_i\}_{i=1}^{N^3}$, $v_i \in \{(-\frac{N}{2}, -\frac{N}{2}, -\frac{N}{2}), \dots, (\frac{N}{2}-1, \frac{N}{2}-1, \frac{N}{2}-1)\}$, $w_i \in \mathbb{R}$, n is the number of points in P, and N is the resolution of the 3D grid $\mathcal{G}$. As shown in Fig. 1 (b), we define a cell as a cubic consisting of eight vertices. For each vertex $v_i = (x_i^v, y_i^v, z_i^v)$ of the 3D grid cell $\mathcal{G}$, we define the neighboring points $\mathcal{N}(v_i)$ as points that lie in the adjacent 8 cells of this vertex. The point $p = (x, y, z) \in \mathcal{N}(v_i)$ is defined as a neighboring point of vertex v_i by satisfying $p \in P$, $x_i^v - 1 < x < x_i^v + 1$, $y_i^v - 1 < y < y_i^v + 1$, and $z_i^v - 1 < z < z_i^v + 1$, respectively. In standard voxelization, value w_i at the vertex v_i is computed as

$$w_i = \begin{cases} 0 & \forall p \notin \mathcal{N}(v_i) \\ 1 & \exists p \in \mathcal{N}(v_i) \end{cases} \tag{1}$$

However, this voxelization process introduces a quantization effect that discards some details of an object. In addition, voxelization is not differentiable and thus can not be applied to point cloud reconstruction. As illustrated in Fig. 1 (b), given a vertex v_i and its neighboring points $p \in \mathcal{N}(v_i)$, the proposed *Gridding* layer computes the corresponding value w_i of this vertex v_i as

$$w_i = \sum_{p \in \mathcal{N}(v_i)} \frac{w(v_i, p)}{|\mathcal{N}(v_i)|} \tag{2}$$

where $|\mathcal{N}(v_i)|$ is the number of neighboring points of v_i. Specially, we define $w_i = 0$ if $|\mathcal{N}(v_i)| = 0$. The interpolation function $w(v_i, p)$ is defined as

$$w(v_i, p) = (1 - |x_i^v - x|)(1 - |y_i^v - y|)(1 - |z_i^v - z|) \tag{3}$$

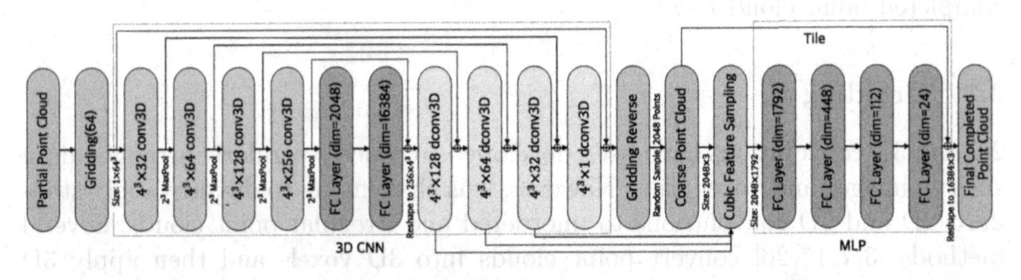

Fig. 2. The network architecture of GRNet. ⊕ denotes the sum operation. Tile creates a new tensor of size 16384 × 3 by replicating the "Coarse Point Cloud" 8 times.

3.3 3D Convolutional Neural Network

The 3D Convolutional Neural Network (3D CNN) with skip connections aims to complete the missing parts of the incomplete point cloud. It follows the idea of a 3D encoder-decoder with U-net connections [46,47]. Given W as input, the 3D CNN can be formulated as

$$W' = \text{3DCNN}(W) \tag{4}$$

where $W' = \{w_i'\}_{i=1}^{N^3}$ and $w_i' \in \mathbb{R}$.

As shown in Fig. 2, the encoder of the 3D CNN has four 3D convolutional layers, each of which has a bank of 4^3 filters with padding of 2, followed by batch normalization, leaky ReLU activation, and a max pooling layer with a kernel size of 2^3. The numbers of output channels of convolutional layers are 32, 64, 128, 256, respectively. The encoder is finally followed by two fully connected layers with dimensions of 2048 and 16384. The decoder consists of four transposed convolutional layers, each of which has a bank of 4^3 filters with padding of 2 and stride of 1, followed by a batch normalization layer and a ReLU activation.

3.4 Gridding Reverse

As illustrated in Fig. 1 (c), we propose *Gridding Reverse* to generate the coarse point cloud $P^c = \{p_i^c\}_{i=1}^m$ from the 3D grid $\mathcal{G}' = <V, W'>$, where $p_i^c \in \mathbb{R}^3$ and m is the number of points in the coarse point cloud P^c. Let $\Theta^i = \{\theta_j^i\}_{j=1}^8$ be the index set of vertices of the $i-$th 3D grid cell. *Gridding Reverse* generates one point coordinate p_i^c for this grid cell by a weighted combination of eight vertices coordinates $\{v_\theta | \theta \in \Theta^i\}$ and the corresponding values $\{w'_\theta | \theta \in \Theta^i\}$ in this cell, which is computed as

$$p_i^c = \frac{\sum_{\theta \in \Theta^i} w'_\theta v_\theta}{\sum_{\theta \in \Theta^i} w'_\theta} \tag{5}$$

Specially, we ignore the point p_i^c for this cell if $\sum_{\theta \in \Theta^i} w'_\theta = 0$.

3.5 Cubic Feature Sampling

MLP-based methods (*e.g.*, PCN) are unable to take the context of neighboring points into account due to no local spatial connectivity across points. These methods use max-pooling to aggregate information globally, which may lose local context information.

To overcome this issue, we present *Cubic Feature Sampling* to aggregate features $F^c = \{f^c\}_{i=1}^m$ for the coarse point cloud P^c, which is helpful for the following MLP to recover the details of point clouds, as shown in Fig. 1 (d). Let $\mathcal{F} = \{f_1^v, f_2^v, \dots, f_{t^3}^v\}$ be the feature map of 3D CNN, where $f_i^v \in \mathbb{R}^c$ and t^3 is the size of the feature map. For a point p_i^c of the coarse point cloud P^c, its features f_i^c are computed as

$$f_i^c = [f_{\theta_1^i}^v, f_{\theta_2^i}^v, \dots, f_{\theta_8^i}^v] \tag{6}$$

where $[\cdot]$ is the concatenation operation. $\{f_{\theta_j^i}^v\}_{j=1}^8$ denotes the features of eight vertices of the i-th 3D gird cell where p_i^c lies in.

In *GRNet*, *Cubic Feature Sampling* extracts the point features from feature maps generated by the first three transposed convolutional layers in 3D CNN. To reduce the redundancy of these features and generate a fixed number of points, we randomly sample $2,048$ points from the coarse point cloud P^c. Consequently, it produces a feature map of size 2048×1792.

3.6 Multi-layer Perceptron

The Multi-layer Perceptron (MLP) is used to recover the details from the coarse point cloud by learning residual offsets between the coordinates of points in the coarse and final completed point cloud. It takes the coarse point cloud P^c and the corresponding features F^c as input, and outputs the final completed point cloud $P^f = \{p_i^f\}_{i=1}^k$ as

$$P^f = \text{MLP}(F^c) + \text{Tile}(P^c, r) \tag{7}$$

where $p_i^f \in \mathbb{R}^3$ and k is the number of points in the final completed point cloud P^f. Tile creates a new tensor of size $rm \times 3$ by replicating P^c r times.

In *GRNet*, r is set to 8. The MLP consists of four fully connected layers with dimensions of 1792, 448, 112, and 24, respectively. The output of MLP is reshaped to 16384×3, which corresponds to the offsets of the coordinates of $16,384$ points.

3.7 Gridding Loss

Existing methods adopt Chamfer Distance [4] as the loss function to train the neural networks. This loss function penalizes the prediction deviating from the ground-truth. However, it can not guarantee that the predicted points follow the geometric layout of the object. Therefore the networks tend to output a mean shape that minimizes the distance, which causes the loss of the object's details [11,48].

Due to the unorderedness of point clouds, it is difficult to directly apply binary cross-entropy like voxels or L1/L2 loss like images. With the proposed *Gridding*, we can convert unordered point clouds into regular 3D grids (Fig. 1 (e)). Therefore, we design a new loss function based on *Gridding*, namely *Gridding Loss*, which is defined as the L1 distance between value sets of the two 3D grids. Let $\mathcal{G}_{pred} = <V^{pred}, W^{pred}>$ and $\mathcal{G}_{gt} = <V^{gt}, W^{gt}>$ be the 3D grids obtained by *Gridding* the predicted and ground truth point clouds, respectively, where $W^{pred} \in \mathbb{R}^{N_G^3}$, $W^{gt} \in \mathbb{R}^{N_G^3}$, and N_G is the resolution of the two 3D grids. The *Gridding Loss* can be defined as

$$\mathcal{L}_{Gridding}(W^{pred}, W^{gt}) = \frac{1}{N^3} \sum ||W^{pred} - W^{gt}|| \tag{8}$$

4 Experiments

4.1 Datasets

ShapeNet. The ShapeNet dataset [45] for point cloud completion is derived from PCN [51], which consists of 30,974 3D models from 8 categories. The ground truth point clouds containing 16,384 points are uniformly sampled on mesh surfaces. The partial point clouds are generated by back-projecting 2.5D depth maps into 3D. For a fair comparison, we use the same train/val/test splits as PCN.

Completion3D. The Completion3D benchmark [38] is composed of 28,974 and 800 samples for training and validation, respectively. Different from the ShapeNet dataset generated by PCN, there are only 2,048 points in the ground truth point clouds.

KITTI. The KITTI dataset [5] is composed of a sequence of real-world Velodyne LiDAR scans, also derived from PCN [51]. For each frame, the car objects are extracted according to the 3D bounding boxes, which results in 2,401 partial point clouds. The partial point clouds in KITTI are highly sparse and do not have complete point clouds as ground truth.

4.2 Evaluation Metrics

Let $\mathcal{T} = \{(x_i, y_i, z_i)\}_{i=1}^{n_\mathcal{T}}$ be the ground truth and $\mathcal{R} = \{(x_i, y_i, z_i)\}_{i=1}^{n_\mathcal{R}}$ be a reconstructed point set being evaluated, where $n_\mathcal{T}$ and $n_\mathcal{R}$ are the numbers of points of $\mathcal{T}$ and $\mathcal{R}$, respectively. In our experiments, we use both Chamfer Distance and F-Score as quantitative evaluation metrics.

Chamfer Distance. Follow PSGN [4] and TopNet [38], the distance between $\mathcal{T}$ and $\mathcal{R}$ are defined as

$$\text{CD} = \frac{1}{n_\mathcal{T}} \sum_{t \in \mathcal{T}} \min_{r \in \mathcal{R}} ||t - r||_2^2 + \frac{1}{n_\mathcal{R}} \sum_{r \in \mathcal{R}} \min_{t \in \mathcal{T}} ||t - r||_2^2 \tag{9}$$

F-Score. As pointed out in [37], Chamfer Distance may sometimes be misleading. As suggested in [37], we take F-Score as an extra metric to evaluate the performance of point completion results, which can be defined as following

$$\text{F-Score}(d) = \frac{2P(d)R(d)}{P(d) + R(d)} \tag{10}$$

where $P(d)$ and $R(d)$ denote the precision and recall for a distance threshold d, respectively.

$$P(d) = \frac{1}{n_\mathcal{R}} \sum_{r \in \mathcal{R}} \left[\min_{t \in \mathcal{T}} ||t - r|| < d \right] \tag{11}$$

$$R(d) = \frac{1}{n_\mathcal{T}} \sum_{t \in \mathcal{T}} \left[\min_{r \in \mathcal{R}} ||t - r|| < d \right] \tag{12}$$

4.3 Implementation Details

We implement our network using PyTorch [30] and CUDA[1]. All models are optimized with an Adam optimizer [13] with $\beta_1 = 0.9$ and $\beta_2 = 0.999$. We train the network with a batch size of 32 on two NVIDIA TITAN Xp GPUs. The initial learning rate is set to 1e−4 and decayed by 2 after 50 epochs. The optimization is set to stop after 150 epochs.

4.4 Shape Completion on ShapeNet

To compare the performance of *GRNet* with other state-of-the-art methods, we conduct experiments on the ShapeNet dataset. **AtlasNet** [6] generates a point cloud with a set of parametric surface elements. To compare with other methods fairly, we sample 16,384 points from the generated primitive surface elements. **PCN** [51] completes the partial point cloud with a stacked version of PointNet [32], which directly outputs the coordinates of 16,384 points. **FoldingNet** [50] is a baseline method adopted in PCN [51], which deforms a 128×128

[1] The source code is available at https://github.com/hzxie/GRNet.

Table 1. Point completion results on ShapeNet compared using Chamfer Distance (CD) with L2 norm computed on 16,384 points and multiplied by 10^4. The best results are highlighted in bold.

Methods	Airplane	Cabinet	Car	Chair	Lamp	Sofa	Table	Watercraft	Overall
AtlasNet [6]	1.753	5.101	3.237	5.226	6.342	5.990	4.359	4.177	4.523
PCN [51]	**1.400**	4.450	**2.445**	4.838	6.238	5.129	3.569	4.062	4.016
FoldingNet [50]	3.151	7.943	4.676	9.225	9.234	8.895	6.691	7.325	7.142
TopNet [38]	2.152	5.623	3.513	6.346	7.502	6.949	4.784	4.359	5.154
MSN [23]	1.543	7.249	4.711	4.539	6.479	5.894	3.797	3.853	4.758
GRNet	1.531	**3.620**	2.752	**2.945**	**2.649**	**3.613**	**2.552**	**2.122**	**2.723**

Table 2. Point completion results on ShapeNet compared using F-Score@1%. Note that the F-Score@1% is computed on 16,384 points. The best results are highlighted in bold.

Methods	Airplane	Cabinet	Car	Chair	Lamp	Sofa	Table	Watercraft	Overall
AtlasNet [6]	0.845	0.552	0.630	0.552	0.565	0.500	0.660	0.624	0.616
PCN [51]	0.881	**0.651**	**0.725**	0.625	0.638	0.581	0.765	0.697	0.695
FoldingNet [50]	0.642	0.237	0.382	0.236	0.219	0.197	0.361	0.299	0.322
TopNet [38]	0.771	0.404	0.544	0.413	0.408	0.350	0.572	0.560	0.503
MSN [23]	**0.885**	0.644	0.665	0.657	0.699	0.604	**0.782**	0.708	0.705
GRNet	0.843	0.618	0.682	**0.673**	**0.761**	**0.605**	0.751	**0.750**	**0.708**

2D grid into 3D point cloud. **TopNet** [38] incorporates a decoder following a hierarchical rooted tree structure to consider the topology of point clouds. Due to the scalable architecture of TopNet, it can easily generate 16,384 points by setting the number of nodes and the size of feature embedding. A very recent method **MSN** [23] generates dense point cloud containing 8,192 points in a coarse-to-fine fashion. To generate 16,384 points, we combine the generated points of 2 times forward propagation.

Quantitative results in Tables 2 and 1 indicate that *GRNet* outperforms all competitive methods in terms of Chamfer Distance and F-Score@1%. Figure 3 shows the qualitative results for point completion on ShapeNet, which indicates that the proposed method recovers better details of objects (*e.g.*, chairs and lamps) than the other methods.

4.5 Shape Completion on Completion3D

Using the model with the lowest Chamfer Distance (CD) on the validation set, we recover the complete point clouds for 1,184 objects in the Completion3D testing set. Then, random subsampling is applied to the generated point clouds to obtain 2,048 points for benchmark evaluation. According to the online leaderboard[2], as shown in Table 3, the overall CD for the proposed *GRNet* is 10.64,

[2] https://completion3d.stanford.edu/results.

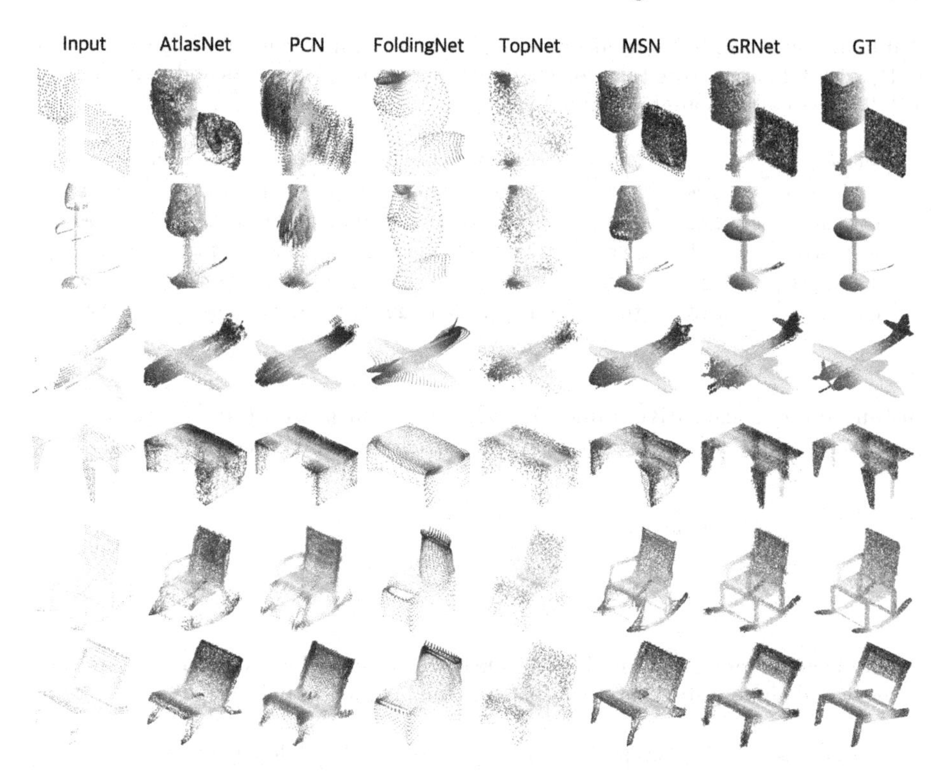

Fig. 3. Qualitative completion results on the ShapeNet testing set. GT stands for the ground truth of the 3D object.

which remarkably outperforms all state-of-the-art methods and ranks first on this benchmark.

4.6 Shape Completion on KITTI

To evaluate the performance of the proposed method on real-world LiDAR scans, we test *GRNet* on the KITTI dataset for completing sparse point clouds of cars. Unlike ShapeNet generated by back-projected from 2.5D images, point clouds from LiDAR scans can be highly sparse, which are much sparser than those in ShapeNet.

We fine-tuned all competitive methods on ShapeNetCars (the cars from ShapeNet) except PCN that directly uses released output for evaluation. During testing, each point cloud is transformed into the bounding box's coordinates and transformed back to the world frame after completion. The models trained specifically on cars are able to incorporate prior knowledge of the object class.

Since there are no complete ground truth point clouds for KITTI, we use Consistency and Uniformity to evaluate the performance of all competitive methods. Consistency in PCN [51] is the average CD between the output of the same car

Table 3. Point completion results on Completion3D compared using Chamfer Distance (CD) with L2 norm. Note that the CD is computed on 2,048 points and multiplied by 10^4. The best results are highlighted in bold.

Methods	Airplane	Cabinet	Car	Chair	Lamp	Sofa	Table	Watercraft	Overall
AtlasNet [6]	10.36	23.40	13.40	24.16	20.24	20.82	17.52	11.62	17.77
FoldingNet [50]	12.83	23.01	14.88	25.69	21.79	21.31	20.71	11.51	19.07
PCN [51]	9.79	22.70	12.43	25.14	22.72	20.26	20.27	11.73	18.22
TopNet [38]	7.32	18.77	12.88	19.82	14.60	16.29	14.89	8.82	14.25
GRNet	**6.13**	**16.90**	**8.27**	**12.23**	**10.22**	**14.93**	**10.08**	**5.86**	**10.64**

instance in n_f consecutive frames. Let $\mathcal{R}^j_{t_i}$ be the output for the j-th car instance at time t_i. The Consistency for the j-th car can be calculated as

$$\text{Consistency} = \frac{1}{n_f - 1} \sum_{i=2}^{n_f} \text{CD}(\mathcal{R}^j_{t_{i-1}}, \mathcal{R}^j_{t_i}) \tag{13}$$

Table 4. Point completion results on LiDAR scans from KITTI compared using Consistency and Uniformity. The best results are highlighted in bold.

Methods	Consistency $\times 10^{-3}$	Uniformity for different p				
		0.4%	0.6%	0.8%	1.0%	1.2%
AtlasNet [6]	0.700	1.146	1.005	0.874	0.761	0.686
PCN [51]	1.557	3.662	5.812	7.710	9.331	10.823
FoldingNet [50]	1.053	1.245	1.303	1.262	1.162	1.063
TopNet [38]	0.568	1.353	1.326	1.219	1.073	0.950
MSN [23]	1.951	0.822	0.675	0.523	0.462	0.383
GRNet	**0.313**	**0.632**	**0.572**	**0.489**	**0.410**	**0.352**

Following PU-GAN [19], we adopt Uniformity to evaluate the distribution uniformity of the completed point clouds, which can be formulated as

$$\text{Uniformity}(p) = \frac{1}{M} \sum_{i=1}^{M} \text{U}_{\text{imbalance}}(S_i)\text{U}_{\text{clutter}}(S_i) \tag{14}$$

where $S_i (i = 1, 2, \ldots, M)$ is a point subset cropped from a patch of the output $\mathcal{R}$ using the farthest sampling and ball query of radius $\sqrt{p}$. The term $\text{U}_{\text{imbalance}}$ and $\text{U}_{\text{clutter}}$ account for the global and local distribution uniformity, respectively.

$$\text{U}_{\text{imbalance}}(S_i) = \frac{(|S_i| - \hat{n})^2}{\hat{n}} \tag{15}$$

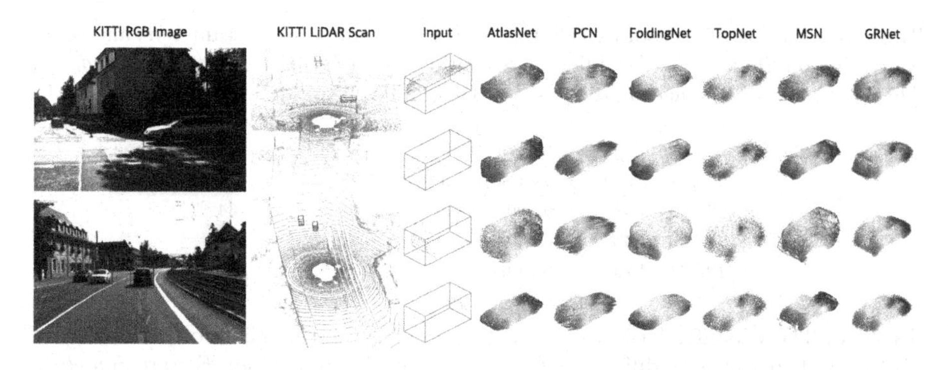

Fig. 4. Qualitative completion results on the LiDAR scans from KITTI. The incomplete input point cloud is extracted and normalized from the scene according to its 3D bounding box.

where $\hat{n} = p|\mathcal{R}|$ is the expected number of points in S_i.

$$\mathrm{U}_{\mathrm{clutter}}(S_i) = \frac{1}{|S_i|} \sum_{j=1}^{|S_i|} \frac{(d_{i,j} - \hat{d})^2}{\hat{d}} \tag{16}$$

where $d_{i,j}$ represents the distance to the nearest neighbor for the j-th point in S_i, and $\hat{d}$ is roughly $\sqrt{\frac{2\pi p}{|S_i|\sqrt{3}}}$ if S_i has a uniform distribution [19].

Table 4 shows the completion results for cars in the LiDAR scans from the KITTI dataset. Experimental results indicate that *GRNet* outperforms other competitive methods in terms of Consistency and Uniformity. Benefited from *Gridding* and *Gridding Reverse*, *GRNet* is more sensitive to the spatial structure of the input points, which leads to better consistency between the two consecutive frames. As shown in Fig. 4, the cars are barely recognizable due to incompleteness of the input data. In contrast, the completed point clouds provide more geometric information. In addition, the qualitative results also demonstrate the proposed method generates more reasonable shape completion.

4.7 Ablation Study

The performance improvement of *GRNet* should be attributed to three key components, including *Gridding*, *Cubic Feature Sampling*, and *Gridding Loss*. To demonstrate the effectiveness of each component in the proposed method, we evaluate the performance with different parameters.

Gridding. Table 5 shows the results of different resolutions of 3D grids generated by *Gridding*. The F-Score of final completed point clouds increases with the 3D grids' resolutions. However, the numbers of parameters and the backward time also increases. To archive a balance between effect and efficiency, we choose the resolution of size 64^3 for *Gridding* in *GRNet*.

Table 5. The Chamfer Distance (CD), F-Score@1%, numbers of parameters, and backward time on ShapeNet with different resolutions of 3D grids generated by *Gridding*. The backward time is measured on an NVIDIA TITAN Xp GPU with batch size of 1.

Resolutions	CD ($\times 10^{-4}$)		F-Score@1%		# Parameters	Backward time
	Coarse	Complete	Coarse	Complete	(M)	(ms)
32^3	23.339	5.943	0.329	0.549	69.54	64
64^3	**11.259**	**2.723**	0.340	0.708	76.70	100
128^3	12.383	2.732	**0.366**	**0.712**	76.77	302

Table 6. The Chamfer Distance (CD), F-Score@1%, and numbers of parameters of MLPs on ShapeNet with different features maps feeding into *Cubic Feature Sampling*. The backward time is measured on an NVIDIA TITAN Xp GPU with batch size of 1.

The size of feature maps			CD	F-Score	# Parameters	Backward time
128×8^3	64×16^3	32×32^3	($\times 10^{-4}$)	@1%	(M)	(ms)
			11.375	0.343	0	72
		✓	2.922	0.640	0.11	80
	✓	✓	2.805	0.686	0.96	88
✓	✓	✓	**2.723**	**0.708**	4.07	100

Table 7. The Chamfer Distance (CD) and F-Score@1% on ShapeNet with different resolutions of 3D grids generated by *Gridding Loss*. The backward time is measured on an NVIDIA TITAN Xp GPU with batch size of 1.

Resolutions	CD ($\times 10^{-4}$)		F-Score@1%		Backward time (ms)
	Coarse	Complete	Coarse	Complete	
Not Used	11.259	4.460	0.340	0.624	86
64^3	10.275	3.427	0.364	0.672	92
128^3	**9.324**	**2.723**	**0.386**	**0.708**	100

Cubic Feature Sampling. To quantitatively evaluate the effect of *Cubic Feature Sampling*, we compare the performance without *Cubic Feature Sampling* and with different feature maps fed into it. The experimental results presented in Table 6 indicate that *Cubic Feature Sampling* improves the point cloud completion results significantly. In addition, with more feature maps are fed, the completion quality becomes better without a significant increase in the numbers of parameters and backward time.

Gridding Loss. We further validate the effects of *Gridding Loss*, as shown in Table 7. There is a decrease in terms of both CD and F-Score when removing *Gridding Loss*. When increasing the resolution of 3D grids from 64^3 to 128^3, there are 25.9% and 5.4% improvements in CD and F-Score, respectively.

5 Conclusion

In this paper, we study how to recover the complete 3D point cloud from an incomplete one. The main motivation of this work is to enable the convolutions on 3D point clouds while preserving their structural and context information. To this aim, we introduce 3D grids as intermediate representations to regularize unordered point clouds. We then propose a novel Gridding Residual Network (GRNet) for point cloud completion, which contains three novel differentiable layers: *Gridding*, *Gridding Reverse*, and *Cubic Feature Sampling*, as well as a new *Gridding Loss*. Extensive comparisons are conducted on the ShapeNet, Completion3D, and KITTI benchmarks, which indicate that the proposed *GRNet* performs favorably against state-of-the-art methods.

Acknowledgements. This work is supported by the National Natural Science Foundation of China (Nos. 61772158, 61702136 and 61872112), National Key Research and Development Program of China (Nos. 2018YFC0806802 and 2018YFC0832105), and Self-Planned Task (No. SKLRS202002D) of State Key Laboratory of Robotics and System (HIT).

References

1. Achlioptas, P., Diamanti, O., Mitliagkas, I., Guibas, L.J.: Learning representations and generative models for 3D point clouds. In: ICML 2018 (2018)
2. Cadena, C., et al.: Past, present, and future of simultaneous localization and mapping: toward the robust-perception age. IEEE Trans. Rob. **32**(6), 1309–1332 (2016)
3. Dai, A., Qi, C.R., Nießner, M.: Shape completion using 3D-encoder-predictor CNNs and shape synthesis. In: CVPR 2017 (2017)
4. Fan, H., Su, H., Guibas, L.J.: A point set generation network for 3D object reconstruction from a single image. In: CVPR 2017 (2017)
5. Geiger, A., Lenz, P., Stiller, C., Urtasun, R.: Vision meets robotics: the KITTI dataset. Int. J. Robot. Res. (IJRR) **32**(11), 1231–1237 (2013)
6. Groueix, T., Fisher, M., Kim, V.G., Russell, B.C., Aubry, M.: A papier-mâché approach to learning 3D surface generation. In: CVPR 2018 (2018)
7. Han, X., Li, Z., Huang, H., Kalogerakis, E., Yu, Y.: High-resolution shape completion using deep neural networks for global structure and local geometry inference. In: ICCV 2017 (2017)
8. Hassani, K., Haley, M.: Unsupervised multi-task feature learning on point clouds. In: ICCV 2019 (2019)
9. Hermosilla, P., Ritschel, T., Vázquez, P., Vinacua, A., Ropinski, T.: Monte Carlo convolution for learning on non-uniformly sampled point clouds. ACM Trans. Graph. **37**(6), 235:1–235:12 (2018)
10. Hua, B., Tran, M., Yeung, S.: Pointwise convolutional neural networks. In: CVPR 2018 (2018)
11. Jiang, L., Shi, S., Qi, X., Jia, J.: GAL: geometric adversarial loss for single-view 3D-object reconstruction. In: Ferrari, V., Hebert, M., Sminchisescu, C., Weiss, Y. (eds.) ECCV 2018. LNCS, vol. 11212, pp. 820–834. Springer, Cham (2018). https://doi.org/10.1007/978-3-030-01237-3_49

12. Kar, A., Häne, C., Malik, J.: Learning a multi-view stereo machine. In: NIPS 2017 (2017)
13. Kingma, D.P., Ba, J.: Adam: a method for stochastic optimization. In: ICLR 2015 (2015)
14. Kipf, T.N., Welling, M.: Semi-supervised classification with graph convolutional networks. In: ICLR 2017 (2017)
15. Lan, S., Yu, R., Yu, G., Davis, L.S.: Modeling local geometric structure of 3D point clouds using Geo-CNN. In: CVPR 2019 (2019)
16. Lei, H., Akhtar, N., Mian, A.: Octree guided CNN with spherical kernels for 3D point clouds. In: CVPR 2019 (2019)
17. Li, D., Shao, T., Wu, H., Zhou, K.: Shape completion from a single RGBD image. IEEE Trans. Visual Comput. Graphics **23**(7), 1809–1822 (2017)
18. Li, K., Pham, T., Zhan, H., Reid, I.: Efficient dense point cloud object reconstruction using deformation vector fields. In: Ferrari, V., Hebert, M., Sminchisescu, C., Weiss, Y. (eds.) ECCV 2018. LNCS, vol. 11216, pp. 508–524. Springer, Cham (2018). https://doi.org/10.1007/978-3-030-01258-8_31
19. Li, R., Li, X., Fu, C., Cohen-Or, D., Heng, P.: PU-GAN: a point cloud upsampling adversarial network. In: ICCV 2019 (2019)
20. Li, Y., Bu, R., Sun, M., Wu, W., Di, X., Chen, B.: PointCNN: convolution on x-transformed points. In: NeurIPS 2018 (2018)
21. Lin, C., Kong, C., Lucey, S.: Learning efficient point cloud generation for dense 3D object reconstruction. In: AAAI 2018 (2018)
22. Lin, H., Xiao, Z., Tan, Y., Chao, H., Ding, S.: Justlookup: One millisecond deep feature extraction for point clouds by lookup tables. In: ICME 2019 (2019)
23. Liu, M., Sheng, L., Yang, S., Shao, J., Hu, S.M.: Morphing and sampling network for dense point cloud completion. In: AAAI 2020 (2020)
24. Liu, Y., Fan, B., Meng, G., Lu, J., Xiang, S., Pan, C.: DensePoint: learning densely contextual representation for efficient point cloud processing. In: ICCV 2019 (2019)
25. Liu, Y., Fan, B., Xiang, S., Pan, C.: Relation-shape convolutional neural network for point cloud analysis. In: CVPR 2019 (2019)
26. Liu, Z., Tang, H., Lin, Y., Han, S.: Point-voxel CNN for efficient 3D deep learning. In: NeurIPS 2019 (2019)
27. Mandikal, P., Radhakrishnan, V.B.: Dense 3D point cloud reconstruction using a deep pyramid network. In: WACV 2019 (2019)
28. Mao, J., Wang, X., Li, H.: Interpolated convolutional networks for 3D point cloud understanding. In: ICCV 2019 (2019)
29. Nguyen, D.T., Hua, B., Tran, M., Pham, Q., Yeung, S.: A field model for repairing 3D shapes. In: CVPR 2016 (2016)
30. Paszke, A., et al.: PyTorch: an imperative style, high-performance deep learning library. In: NeurIPS 2019 (2019)
31. Peng, S., Liu, Y., Huang, Q., Zhou, X., Bao, H.: PVNet: pixel-wise voting network for 6DoF pose estimation. In: CVPR 2019 (2019)
32. Qi, C.R., Su, H., Mo, K., Guibas, L.J.: PointNet: deep learning on point sets for 3D classification and segmentation. In: CVPR 2017 (2017)
33. Qi, C.R., Yi, L., Su, H., Guibas, L.J.: PointNet++: deep hierarchical feature learning on point sets in a metric space. In: NIPS 2017 (2017)
34. Sharma, A., Grau, O., Fritz, M.: VConv-DAE: deep volumetric shape learning without object labels. In: Hua, G., Jégou, H. (eds.) ECCV 2016. LNCS, vol. 9915, pp. 236–250. Springer, Cham (2016). https://doi.org/10.1007/978-3-319-49409-8_20
35. Stutz, D., Geiger, A.: Learning 3D shape completion from laser scan data with weak supervision. In: CVPR 2018 (2018)

36. Su, H., et al.: SPLATNet: sparse lattice networks for point cloud processing. In: CVPR 2018 (2018)
37. Tatarchenko, M., Richter, S.R., Ranftl, R., Li, Z., Koltun, V., Brox, T.: What do single-view 3D reconstruction networks learn? In: CVPR 2019 (2019)
38. Tchapmi, L.P., Kosaraju, V., Rezatofighi, H., Reid, I.D., Savarese, S.: TopNet: structural point cloud decoder. In: CVPR 2019 (2019)
39. Thomas, H., Qi, C.R., Deschaud, J., Marcotegui, B., Goulette, F., Guibas, L.J.: KPConv: flexible and deformable convolution for point clouds. In: ICCV 2019 (2019)
40. Varley, J., DeChant, C., Richardson, A., Ruales, J., Allen, P.K.: Shape completion enabled robotic grasping. In: IROS 2017 (2017)
41. Wang, K., Chen, K., Jia, K.: Deep cascade generation on point sets. In: IJCAI 2019 (2019)
42. Wang, Y., Sun, Y., Liu, Z., Sarma, S.E., Bronstein, M.M., Solomon, J.M.: Dynamic graph CNN for learning on point clouds. ACM Trans. Graph. **38**(5), 146:1–146:12 (2019)
43. Wang, Z., Lu, F.: VoxSegNet: volumetric CNNs for semantic part segmentation of 3D shapes. IEEE Trans. Vis. Comput. Graph. (2019). https://doi.org/10.1109/TVCG.2019.2896310
44. Wu, W., Qi, Z., Li, F.: PointConv: deep convolutional networks on 3D point clouds. In: CVPR 2019 (2019)
45. Wu, Z., et al.: 3D ShapeNets: a deep representation for volumetric shapes. In: CVPR 2015 (2015)
46. Xie, H., Yao, H., Sun, X., Zhou, S., Zhang, S.: Pix2Vox: context-aware 3D reconstruction from single and multi-view images. In: ICCV 2019 (2019)
47. Xie, H., Yao, H., Zhang, S., Zhou, S., Sun, W.: Pix2Vox++: multi-scale context-aware 3D object reconstruction from single and multiple images. Int. J. Comput. Vision **128**(12), 2919–2935 (2020). https://doi.org/10.1007/s11263-020-01347-6
48. Xu, Q., Wang, W., Ceylan, D., Mech, R., Neumann, U.: DISN: deep implicit surface network for high-quality single-view 3D reconstruction. In: NeurIPS 2019 (2019)
49. Xu, Y., Fan, T., Xu, M., Zeng, L., Qiao, Yu.: SpiderCNN: deep learning on point sets with parameterized convolutional filters. In: Ferrari, V., Hebert, M., Sminchisescu, C., Weiss, Y. (eds.) ECCV 2018. LNCS, vol. 11212, pp. 90–105. Springer, Cham (2018). https://doi.org/10.1007/978-3-030-01237-3_6
50. Yang, Y., Feng, C., Shen, Y., Tian, D.: FoldingNet: point cloud auto-encoder via deep grid deformation. In: CVPR 2018 (2018)
51. Yuan, W., Khot, T., Held, D., Mertz, C., Hebert, M.: PCN: point completion network. In: 3DV 2018 (2018)
52. Zhang, K., Hao, M., Wang, J., de Silva, C.W., Fu, C.: Linked dynamic graph CNN: learning on point cloud via linking hierarchical features. arXiv:1904.10014 (2019)

Gait Lateral Network: Learning Discriminative and Compact Representations for Gait Recognition

Saihui Hou[1,3], Chunshui Cao[3], Xu Liu[2,3], and Yongzhen Huang[1,3](✉)

[1] Institute of Automation, Chinese Academy of Sciences, Beijing, China
yzhuang@nlpr.ia.ac.cn
[2] Beijing University of Technology, Beijing, China
[3] WATRIX AI, Beijing, China

Abstract. Gait recognition aims at identifying different people by the walking patterns, which can be conducted at a long distance without the cooperation of subjects. A key challenge for gait recognition is to learn representations from the silhouettes that are invariant to the factors such as clothing, carrying conditions and camera viewpoints. Besides being discriminative for identification, the gait representations should also be compact for storage to keep millions of subjects registered in the gallery. In this work, we propose a novel network named Gait Lateral Network (GLN) which can learn both *discriminative* and *compact* representations from the silhouettes for gait recognition. Specifically, GLN leverages the inherent feature pyramid in deep convolutional neural networks to enhance the gait representations. The silhouette-level and set-level features extracted by different stages are merged with the lateral connections in a top-down manner. Besides, GLN is equipped with a *Compact Block* which can significantly reduce the dimension of the gait representations without hindering the accuracy. Extensive experiments on CASIA-B and OUMVLP show that GLN can achieve state-of-the-art performance using the 256-dimensional representations. Under the most challenging condition of walking in different clothes on CASIA-B, our method improves the rank-1 accuracy by 6.45%.

Keywords: Gait recognition · Lateral connections · Discriminative representations · Compact representations

1 Introduction

Gait recognition aims at identifying different people using videos recording the walking patterns [38]. Compared to other biometrics such as face [33], fingerprint [27] and iris [39], human gait can be obtained at a long distance without the cooperation of subjects, which contributes to its broad applications in crime prevention, forensic identification and social security [4,18]. However, gait recognition suffers from a lot of variations such as clothing, carrying conditions and

© Springer Nature Switzerland AG 2020
A. Vedaldi et al. (Eds.): ECCV 2020, LNCS 12354, pp. 382–398, 2020.
https://doi.org/10.1007/978-3-030-58545-7_22

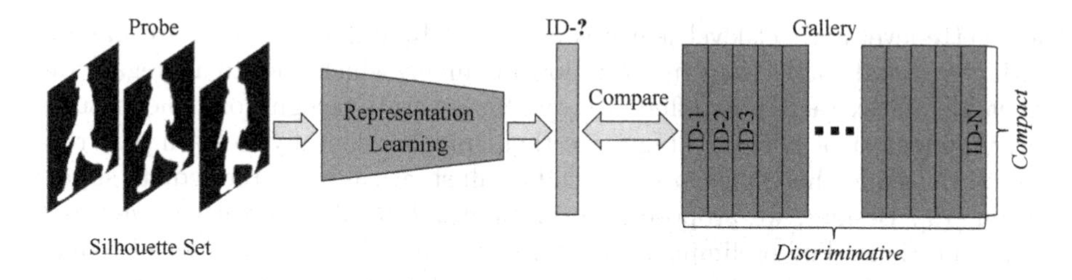

Fig. 1. Illustration of the silhouette-based gait recognition. The learned representations should be *discriminative* to identify different people, and should also be *compact* for the convenience of storage

camera viewpoints [36,42]. A key challenge is to learn representations from the silhouettes of gait sequences that are invariant to the factors mentioned above.

To address the issue, various methods have been proposed which can be roughly divided into three categories. The first category [8,10,35,41] aggregates the silhouettes of a complete gait sequence into an image (or template) for recognition, *e.g.* Gait Energy Image [8]. Despite the simplicity, the temporal and fine-grained spatial information is inevitably lost in the pre-processing. The second [20,40] regards the silhouettes of a gait sequence as a video. For example, in [40], a 3D-CNN [14] is adopted to extract the spatial and temporal information while the model is relatively hard to train. The third [6] is recently proposed and treats the silhouettes of a gait sequence as a unordered set, which is robust to the number of the silhouettes and achieves significant improvements. However, the dimension of the representations learned by [6] reaches up to 15872 that is much higher than those for face recognition (*e.g.* 180 [33]) or person re-identification (*e.g.* 2048 [25]).

In this work, we deal with gait recognition with the aims of learning both *discriminative* and *compact* representations from the silhouettes for gait recognition. We propose a novel network named Gait Lateral Network (denoted as GLN) where the silhouettes of each gait sequence are regarded as an unordered set. As illustrated in Fig. 1, besides being discriminative to identify different people, the learned representation for each silhouette set should also be as compact as possible, which would otherwise incur a heavy storage burden to keep millions of subjects registered in the gallery. It is noteworthy that, the dimension of the representations learned by GLN is fixed to 256 which is reduced by nearly two orders of magnitude compared to [6] and the performance for all walking conditions are improved simultaneously.

Specifically, we propose to leverage the inherent feature pyramid in deep CNNs to learn *discriminative* gait representations. The features extracted by different layers capture various visual details of the input [43]. We notice that the silhouettes for different subjects only have subtle differences in many cases, which makes it vital to explore the shallow features encoding the local spatial structural information for gait recognition. Particularly, we modify the network of [6] as the backbone and explicitly divide the layers into three stages. The silhouette-level and set-level features extracted by different stages are merged with the lateral connections in a top-down manner, which tries to aggregate the visual details extracted by different layers for accurate recognition. The features after refinement of different stages are then split horizontally to learn part representations and the triplet loss is added at all stages as the intermediate supervision [19]. Besides, we propose a novel *Compact Block* to learn *compact* gait representations. The preliminary study reveals that there exists a lot of redundancy in the high-dimensional representations learned by HPM [6,7] which is widely adopted for part representation learning. The proposed *Compact Block* can distill the knowledge of high-dimensional gait representations into compact ones without hindering the accuracy. Its architecture is simple but non-trivial which can be seamlessly integrated with the backbone and trained in an end-to-end manner. We regard the high-dimensional representations as an ensemble of low-dimensional ones and utilize *Dropout* to select a small subset, which is then mapped into a compact space by *Fully Connected Layer*.

In summary, our contributions of this work lie in three folds: (1) We propose to leverage the inherent feature pyramid in deep CNNs to enhance the gait representations for accurate recognition. The silhouette-level and set-level features extracted by different stages are merged with the lateral connections in a top-down manner. (2) We propose a *Compact Block* which can significantly reduce the dimension of the gait representations without hindering the accuracy. (3) The resulting GLN can learn both *discriminative* and *compact* representations from the silhouettes for gait recognition. The experiments on CASIA-B [42] and OUMVLP [36] show that GLN can achieve state-of-the-art performance for all walking conditions using the 256-dimensional representations. In particular, under the most challenging condition of walking in different clothes on CASIA-B, the rank-1 accuracy achieved by GLN exceeds GaitSet [6] with the 15872-dimensional representations by 6.45%.

2 Related Work

Motion-Based Gait Recognition. These methods including [1,3,16] attempt to model the human body structures and then extract motion features for gait recognition, which have the advantage of being robust to clothing and carrying conditions. Nevertheless, they usually fail on low-resolution videos where it is difficult to estimate the body parameters accurately.

Appearance-Based Gait Recognition. These methods including [8,17,26, 37] directly learn features from the gait sequences without explicitly modeling the body structures, which suit for the low-resolution conditions and thus attract increasing attention [44,46]. The silhouettes are usually taken as the input and a key challenge is to learn representations from the silhouettes that are robust to the factors such as clothing, carrying conditions and camera viewpoints [36,42]. The silhouette-based gait recognition can be roughly divided into three categories where the silhouettes of a complete gait sequence are respectively regarded as an image [8,10,35,41], a video [20,40] or an unordered image set [6].

Deep learning that innovates the field of computer vision is also widely used for gait recognition. Specifically, a comprehensive study on deep convolutional neural networks for gait recognition is conducted in [41]. An auto-encoder framework is proposed by [46] to explicitly disentangle the appearance and pose features in the representation learning. JUCNet [44] integrates the cross-gait and unique-gait supervision with a tailored quintuplet loss. DiGGAN [12] takes advantage of a Conditional GAN [28] to learn the view-invariant gait features. GaitSet [6] treats the silhouettes of each gait sequence as an unordered set and splits the features horizontally to learn part representations for gait recognition, which achieves significant improvements and holds the best performance across different datasets. However, the dimension of the final representations learned by [6] is too high, *i.e.* 15872.

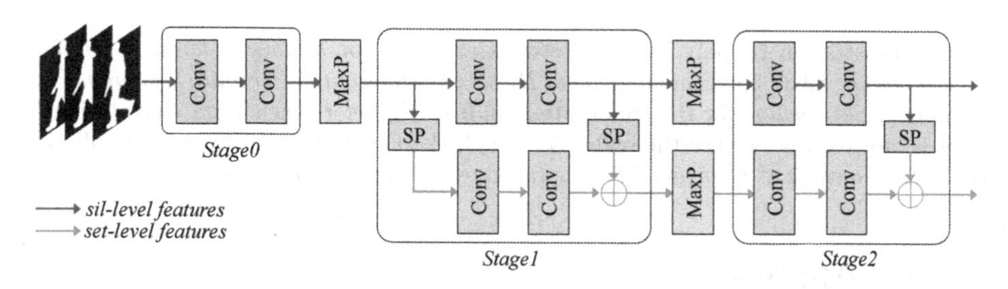

Fig. 2. Illustration of the division for the backbone, *Sil-level* for *Silhouette-level*, *MaxP* for *Max Pooling*, *SP* for *Set Pooling*. The *silhouette-level* features are extracted from each silhouette separately while the *set-level* features are extracted from all silhouettes. *Set Pooling* is a function to aggregate the features in a silhouette set

Inherent Feature Pyramid. The inherent feature pyramid in deep convolutional neural networks has been exploited in many visual tasks. For example, FCN [24] utilizes the features of different layers to progressively refine the predictions for semantic segmentation. Hypercolumns [9] proposes an efficient computation strategy to aggregate the features of different layers for object segmentation and localization. SSD [23] detects the objects using the features of different layers separately without fusing features or scores.

The top-down manner to merge the features of different stages in GLN is inspired by FPN [21] for object detection. However, our approach differs from FPN in three aspects. First, there are two branches in the last two stages of GLN as shown in Fig. 2 and the lateral connections in GLN are utilized to merge the silhouette-level and set-level features simultaneously. Second, the training labels for different stages in FPN are assigned according to the receptive fields, while the supervision signals for different stages in GLN are the same. Third, FPN shares the parameters in the heads following different stages, while the subsequent layers for different stages in GLN have independent parameters.

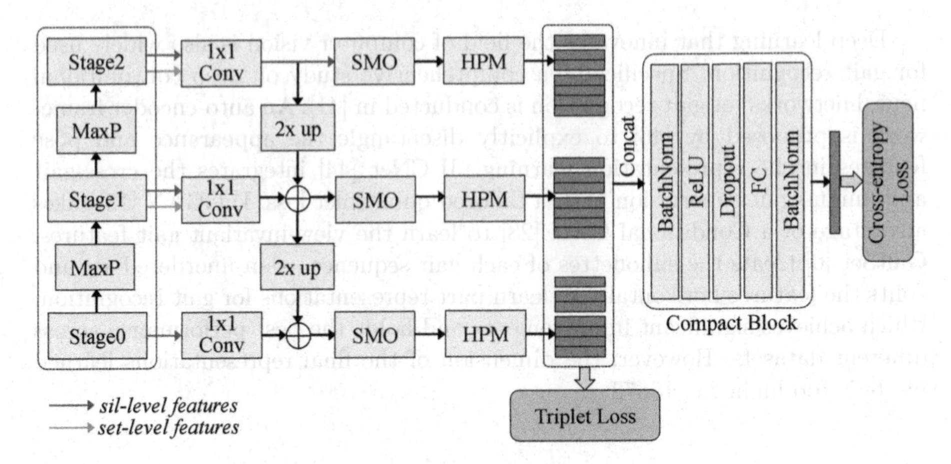

Fig. 3. Illustration of Gait Lateral Network, *Sil-level* for *Silhouette-level, MaxP* for *Max Pooling, SMO* for *Smooth Layer, HPM* for *Horizontal Pyramid Mapping*. For simplicity, *Set Pooling* for the silhouette-level features before each 1×1 convolutional layer is omitted and we use the scales $S = \{1, 2, 4\}$ to split the features horizontally in HPM. The output of *Compact Block* is taken as the final representation

3 Our Approach

In this work, we propose a novel network named Gait Lateral Network (GLN) which can learn both *discriminative* and *compact* representations from the silhouettes for gait recognition. The silhouettes of a complete gait sequence are regarded as an unordered set. The network structure is illustrated in Fig. 3. The silhouette-level and set-level features extracted by different stages in the backbone are merged with the lateral connections in a top-down manner, which aims to enhance the gait representations for accurate recognition. And we propose a *Compact Block* which can significantly reduce the dimension of the gait representations without hindering the accuracy. In what follows, we will first elaborate the lateral connections in GLN. Then we will introduce the composition of *Compact Block*. Finally, we will describe the corresponding training strategy for GLN.

3.1 Lateral Connections

In GLN, we propose to leverage the inherent feature pyramid in deep convolutional neural networks to learn *discriminative* gait representations. The features extracted by different layers in the backbone are aggregated to enhance the gait representations.

Specifically, we modify the network of [6] as the backbone where the order of the second *Set Pooling* and *Max Pooling* is switched. As shown in Fig. 2, we explicitly divide the layers in the backbone into three stages. The first stage is comprised of two convolutional layers which transform the silhouettes into the internal features. The second and third stages consist of two branches that learn the silhouette-level and set-level features respectively. *Set Pooling* is a function to aggregate the features in a silhouette set, which should be permutation invariant to the order of the silhouettes and is implemented by *Max Pooling* for simplicity. Note that, different from *Max Pooling* between different stages operating along the spatial dimensions (height and width), *Max Pooling* for *Set Pooling* operates along the set dimension[1]. The backbone extracts the silhouette-level features as well as the set-level features in a *bottom-up* way. The features extracted by the three stages are respectively denoted as $\{C_0, C_1, C_2\}$, which have the strides of $\{1, 2, 4\}$ with respect to the input silhouettes.

The features of different stages in the backbone capture various visual details of the silhouettes [43], and we propose to merge the features extracted by different stages with the lateral connections in a *top-down* manner. The strategy is illustrated in Fig. 3. Specifically, first, at the last two stages, we adopt *Set Pooling* to deal with the silhouette-level features and concatenate the output with the set-level features along the channel dimension. And at the first stage, only the silhouette-level features are available which are also processed by *Set Pooling*. Then at each stage, a 1×1 convolutional layer is taken to rearrange the features and adjust the channel dimension. Next, starting from the features generated at the last stage, we upsample the spatial dimensions (height and width) by a factor of 2 and add to the features generated at the previous stage (which have the same channel dimensions after the 1×1 convolutional layers) by element-wise addition. This process is iterated until the features generated at all stages are merged. Finally, a smooth layer is appended after each stage to alleviate the aliasing effect caused by upsampling and semantic gaps between different stages. The output of the smooth layers are denoted as $\{F_0, F_1, F_2\}$ for the three stages, corresponding to $\{C_0, C_1, C_2\}$, which respectively have the same spatial dimensions.

It is worth noting that, the output of the 1×1 convolutional layers as well as the smooth layers have the same channel dimensions, which are fixed to 256 through our experiments. Every smooth layer is implemented by a 3×3 convolutional layer. Besides, there are no non-linear activation functions involved in the lateral connections and we use the nearest neighbor upsampling between different stages.

3.2 Compact Block

In this section, we will elaborate the composition of *Compact Block* which is proposed to learn *compact* gait representations. Before that, we first review *Horizontal Pyramid Mapping* (HPM) [6] as the background which results in the high representation dimension.

HPM is equivalent to *Horizontal Pyramid Pooling* (HPP) [7] for person Re-ID, which is adopted by GLN to learn part representations for gait recognition.

[1] The silhouette-level features have the shape of [*batch, set, channel, height, width*] where the *set* dimension denotes the number of the silhouettes in an unordered set. And the set-level features have the shape of [*batch, channel, height, width*].

Despite the effectiveness, the representations obtained by HPM hold a very high dimension, *e.g.* 15872 [6]. Specifically, HPM first splits the features horizontally using multiple scales S, *e.g.* $S = \{1, 2, 4, 8, 16\}$. For each scale $s \in S$, the features F are sliced into s bins horizontally and equally. Then, the global max and average pooling are taken to generate the features $G_{s,t}$ for each bin:

$$G_{s,t} = \text{MaxPool}\left(F_{s,t}\right) + \text{AvgPool}\left(F_{s,t}\right) \tag{1}$$

where $s \in S$ and $t \in \{1, \cdots, s\}$. Finally, a fully connected layer is applied to $G_{s,t}$ and the output is denoted as $\widehat{G}_{s,t}$. In the training phase, the loss is added to the features of each part. And in the test phase, the features of all parts are concatenated as the final representations. As a result, the dimension of the final representations is proportional to the sum of scales (*e.g.* $sum(S) = 31$ with $S = \{1, 2, 4, 8, 16\}$) and the feature dimension of each part (*e.g.* 256), which is infeasible for the real-world applications. By delving into the formulation of HPM, we observe that the representations across different scales encode some duplicate information. For example, the part representations with the indexes $(s, t) = (4, 1)$ and $(s, t) = \{(8, 1), (8, 2)\}$ correspond to the same regions in the input silhouettes. Thus we conjecture that there exists a lot of redundancy in the high-dimensional representations obtained by HPM.

To tackle the issue, we propose a *Compact Block* with the aims of distilling the knowledge of high-dimensional representations into compact ones without hindering the accuracy. As shown in Fig. 3, *Compact Block* has a plain structure which is composed of *Batch Normalization (BN-I)* [13], *ReLU* [15], *Dropout* [32], *Fully Connected Layer (FC)* and another *Batch Normalization (BN-II)*. The block is simple yet effective, and here we provide the design principles for each layer:

(1) *BN-I* is adopted to normalize the concatenated features obtained by HPM, which helps stabilize the training processing.
(2) *ReLU* is introduced as the activation function to increase the non-linearity for *Compact Block*.
(3) *Dropout* is the key of *Compact Block*. As mentioned above, the representations obtained by HPM can be regarded as an ensemble of low-dimensional ones. Here we take advantage of *Dropout* to select a small subset from each high-dimensional representation.
(4) *FC* is used to map the small subset from *Dropout* into a more discriminative space. The output of *FC* determines the dimension of the final representations which is set to 256 through our experiments.
(5) *BN-II* is introduced for the convenience of optimizing the cross-entropy loss inspired by [22,25], where each subject in the training set is treated as a separate class.

In summary, *Compact Block* significantly reduces the gait representations to a fixed dimension (*e.g.* 256 in our experiments), which is seamlessly integrated with the backbone and trained in an end-to-end manner. It is worth noting that, we adopt the implementation of *Dropout* that is available in PyTorch [29]. It only works in the training phase, and at inference time the final representations can be treated as an ensemble of multiple reductions.

3.3 Training Strategy

The training strategy for GLN consists of two steps: *Lateral Pretraining* and *Global Training*. As shown in Fig. 3, there are two types of losses involved in the training, *i.e.* triplet loss and cross-entropy loss. The triplet loss is deployed after HPM as the intermediate supervision [19], while the cross-entropy loss is added at the end of GLN to learn the global representations.

First, in order to obtain a reasonable initialization for the lateral connections, we propose *Lateral Pretraining* supervised by the triplet loss only. Specifically, the *batch all* version of triplet loss [11] is added to the features of each part obtained by HPM at all stages. Formally:

$$L_{tp} = \frac{1}{N_{tp_+}} \overbrace{\sum_{s \in S} \sum_{t=1}^{s}}^{bins} \overbrace{\sum_{i=1}^{P} \sum_{j=1}^{K}}^{anchors} \overbrace{\sum_{\substack{a=1 \\ a \neq j}}^{K}}^{pos.} \overbrace{\sum_{\substack{b=1 \\ b \neq i}}^{P} \sum_{c=1}^{K}}^{negative} \left[m + d_{s,t,i,j,i,a}^{s,t,i,j,b,c} \right]_+ \tag{2}$$

$$d_{s,t,i,j,i,a}^{s,t,i,j,b,c} = dist(f(sil_{i,j}^{s,t}), f(sil_{i,a}^{s,t})) - dist(f(sil_{i,j}^{s,t}), f(sil_{b,c}^{s,t}))$$

where N_{tp_+} is the number of triplets resulting in the non-zero loss terms over a mini-batch, S is the multiple scales for HPM, (P, K) are the number of subjects and the number of sequences for each subject in a mini-batch, m is the margin threshold, f denotes the feature extraction, sil denotes the silhouette set, $dist$ measures the similarity between two features, *e.g.* euclidean distance. Note that, in *Lateral Pretraining*, we do not decrease the learning rate to prevent overfitting [47].

Then, *Global Training* is conducted to train the whole network with the sum of triplet loss and cross-entropy loss. For cross-entropy loss, each subject in the training set is treated as a separate class and the label smooth technique [34] is adopted. Formally:

$$L_{ce} = -\frac{1}{P \times K} \sum_{i=1}^{P} \sum_{j=1}^{K} \sum_{n=1}^{N} q_n^{ij} \log p_n^{ij} \tag{3}$$

where N is the number of all subjects in the training set, p is the probabilities belonging to each subject, q encodes the identity information which is computed as follows (taking the y-th subject as an example):

$$q_n^{ij} = \begin{cases} 1 - \frac{N-1}{N}\epsilon & \text{if } n = y \\ \frac{\epsilon}{N} & \text{otherwise} \end{cases} \tag{4}$$

where ϵ is a small constant to encourage the model to be less confident on the training set. In our experiments, ϵ is set to 0.1. The total loss for *Global Training* is computed as:

$$L = L_{tp} + L_{ce} \tag{5}$$

It is worth noting that, in the training phase, another *Fully Connected Layer* is introduced after *Compact Block* to compute the probabilities for each subject, which, however, is deprecated at inference time. The output of *Compact Block* is taken as the final representation for each silhouette set to match the probe and gallery.

Table 1. The dataset statistics. *NM* for *normal walking*, *BG* for *walking with bags*, *CL* for *walking in different clothes*

Dataset	Subjects		Walking conditions			Views
	Train	Test	NM	BG	CL	
CASIA-B	74	50	6	2	2	11
OUMVLP	5153	5154	2	–	–	14

4 Experiment

4.1 Settings

Datasets. The experiments are conducted on two popular gait datasets: CASIA-B [42] and OUMVLP [36]. The dataset statistics are shown in Table 1.

CASIA-B. It is a typical gait dataset that consists of 124 subjects. The walking conditions contain normal walking (NM, 6 variants per subject), walking with bags (BG, 2 variants per subject) and walking in different clothes (CL, 2 variants per subject). The 11 views for each walking condition are uniformly distributed in $[0°, 180°]$ at an interval of 18°. In total, there are $(6+2+2) \times 11 = 110$ sequences for each subject. There is no partition for training and test provided in this dataset. In our experiments, we take the first 74 subjects as the training set and the rest 50 as the test set. For evaluation, we regard the first 4 variants of normal walking (NM) for each subject as the gallery with the rest as the probe. The probe can be further divided into three subsets according to the walking conditions, *i.e.* NM, BG, CL.

OUMVLP. It is the largest gait dataset in public which consists of 10307 subjects. However, only the sequences of normal walking (NM, 2 variants per subject) are available for each subject. The 14 views are uniformly distributed between $[0°, 90°]$ and $[180°, 270°]$ at an interval of 15°. In total, there are $2 \times 14 = 28$ sequences for each subject. According to the provided partition, we take 5153 subjects as the training set with the rest 5154 as the test set. For evaluation, the first variant of normal walking (NM) for each subject is treated as the gallery with the rest as the probe.

Implementation Details. All models are implemented with PyTorch [29]. The silhouettes in both datasets are pre-processed using the methods in [35]. The number of subjects and the sequences for each subject in a mini-batch as well as the input size of each silhouette, are set to $(8, 16, 128 \times 88)$ for CASIA-B and $(32, 16, 64 \times 44)$ for OUMVLP. In the training phase, we randomly select 30 silhouettes for each gait sequence. For evaluation, all silhouettes of a gait sequence are taken to obtain the final representation.

The convolutional channels in the three stages shown in Fig. 2 are set to $(32, 64, 128)$ for CASIA-B and $(64, 128, 256)$ for OUMVLP. In the lateral connections shown in Fig. 3, the output dimensions of the 1×1 convolutional layers and the smooth layers are all set to 256. We use the multiple scales $S = \{1, 2, 4, 8, 16\}$ to split the features horizontally at all stages and the feature dimension of each part obtained by HPM is set to 256. For *Compact Block*, an aggressive dropping ratio 0.9 is adopted for *Dropout* and the output dimension is set to 256.

We adopt SGD with momentum [30] as the optimizer. The initial learning rate is set to 0.1 which is not decreased in *Lateral Pretraining*. While in *Global Training*, the learning rate is scaled to its 1/10 three times until convergence. The step size is set to 10000 iterations for CAISA-B and 50000 iterations for OUMVLP. We use the momentum 0.9 and the weight decay 5e−4 for the optimization. The margin threshold m for L_{tp} in Eq. 2 is set to 0.2. Besides, the warmup strategy [25] is adopted at the start of training.

Baselines. GaitSet [6] holds the best performance for the silhouette-based gait recognition and is taken as an important baseline in our experiments. It proposes to treat the silhouettes of a gait sequence as an unordered set and splits the features horizontally to learn part representations for gait recognition, which outperforms the previous works [31,41] by a large margin. It is worth mentioning that, we reproduce the results for GaitSet by ourselves which are a little higher than those reported in [6]. Besides, for a comprehensive study, we also re-implement GEINet [31] which is a representative method taking Gait Energy Image [8] as the input. It customizes a network for gait recognition and treats each subject as a separate class in the training. The features before the softmax layer are taken to match the probe and gallery for evaluation. Finally, to enable a more fair comparison on CASIA-B, we implement an improved version of GaitSet (denoted as GaitSet-L) where the input size of each silhouette is enlarged from 64×44 to 128×88.

4.2 Performance Comparison

CASIA-B. Table 2 shows the performance comparison on CASIA-B. The dimensions of the final representations learned by different methods are also compared. The probe sequences are divided into three subsets, *i.e.* NM, BG, CL, which are respectively evaluated. The accuracy for each probe view is averaged on all gallery views excluding the identical-view cases.

From the results in Table 2, we observe that GaitSet and GaitSet-L outperform GEINet by a large margin, which, however, generate the gait representations with a very high dimension (*i.e.* 15872). The comparisons between GaitSet-L and GaitSet indicate that enlarging the input size is beneficial to gait recognition especially for walking with bags (BG) and walking in different clothes (CL), although the consumption of GPU memory is simultaneously increased. Particularly, compared to GaitSet and GaitSet-L, GLN reduces the representation dimension by nearly two orders of magnitude (15872 → 256) and achieves state-of-the-art performance under all walking conditions (NM-96.88%, BG-94.04%, CL-77.50%). Under the most challenging condition of walking in different clothes (CL), GLN exceeds GaitSet by 6.45% with the representation dimension significantly reduced to 256. The improvements under the other two walking conditions compared to GaitSet are also impressive, *i.e.* +1.67% for normal walking (NM) and +5.96% for walking with bags (BG). Besides, we notice that, though the average performance is inferior to GLN, GaitSet-L achieves the best performance in some probe views (*e.g.* 126°) for walking in different clothes (CL). This phenomenon needs further exploration.

Table 2. The rank-1 accuracy (%) on CAISA-B across different views excluding the identical-view cases, *DIM* for *Dimension.* For evaluation, the first 4 variants of normal walking (NM) for each subject are taken as the gallery. The probe sequences are divided into three subsets according to the walking conditions, *i.e.* NM, BG and CL

Probe	Method	DIM	Probe view											Average
			0°	18°	36°	54°	72°	90°	108°	126°	134°	162°	180°	
NM	GEINet [31]	1024	40.20	38.90	42.90	45.60	51.20	42.00	53.50	57.60	57.80	51.80	47.70	48.11
	GaitSet [6]	15872	**93.40**	98.10	98.50	97.80	92.60	90.90	94.20	97.30	98.40	97.00	89.10	95.21
	GaitSet-L	15872	91.40	98.50	98.80	97.20	94.80	92.90	95.40	97.90	98.80	96.50	89.10	95.57
	GLN (ours)	**256**	93.20	**99.30**	**99.50**	**98.70**	**96.10**	**95.60**	**97.20**	**98.10**	**99.30**	**98.60**	**90.10**	**96.88**
BG	GEINet [31]	1024	34.20	29.29	31.21	35.20	35.20	27.60	35.90	43.50	45.00	38.99	36.80	35.72
	GaitSet [6]	15872	85.90	92.12	93.94	90.41	86.40	78.70	85.00	91.60	93.10	91.01	80.70	88.08
	GaitSet-L	15872	89.00	95.25	95.56	93.98	89.70	86.70	89.70	94.30	95.40	92.73	84.40	91.52
	GLN (ours)	**256**	**91.10**	**97.68**	**97.78**	**95.20**	**92.50**	**91.20**	**92.40**	**96.00**	**97.50**	**94.95**	**88.10**	**94.04**
CL	GEINet [31]	1024	19.90	20.30	22.50	23.50	26.70	21.30	27.40	28.20	24.20	22.50	21.60	23.46
	GaitSet [6]	15872	63.70	75.60	80.70	77.50	69.10	67.80	69.70	74.60	76.10	71.10	55.70	71.05
	GaitSet-L	15872	66.30	79.40	84.50	80.70	74.60	73.20	74.10	**80.30**	**79.70**	72.30	62.90	75.27
	GLN (ours)	**256**	**70.60**	**82.40**	**85.20**	**82.70**	**79.20**	**76.40**	**76.20**	78.90	77.90	**78.70**	**64.30**	**77.50**

OUMVLP. Table 3 displays the performance comparison on OUMVLP where GLN also achieves state-of-the-art performance with the 256-dimensional representations. The input size of each silhouette on this dataset is set to 64 × 44 due to the limits of GPU memory and thus the performance of GaitSet-L is not available. It is worth noting that, the dimension of the representations learned by GEINet is doubled to 2048 for this large-scale dataset, and the reproduced results for GEINet is much higher than those reported in [6]. In spite of the high representation dimension, GaitS et al. so holds the best performance on this large-scale dataset before this work and outperforms GEINet by a large margin. According to the results shown in Table 3, we observe that GLN improves the rank-1 accuracy by 2.13% compared to GaitSet and the representation dimension is significantly reduced to 256.

Besides, we notice that the gait data for some subjects in OUMVLP is incomplete. As a result, for some probe sequences, there are not the corresponding sequences in the gallery. Thus we further conduct the evaluation ignoring the probe sequences which have no corresponding ones in the gallery. As shown in the last three rows of Table 3, GLN finally achieves the rank-1 accuracy of 95.57% with the 256-dimensional representations, which exceeds GaitSet with the 15872-dimensional representations by 2.32% on this large-scale dataset.

4.3 Ablation Study

In this section we provide the ablation study to further analyze GLN. The experiments are conducted on CASIA-B using the settings described in Sect. 4.1.

Table 3. The rank-1 accuracy (%) on OUMVLP across different views excluding the identical-view cases, *DIM* for *Dimension*. For evaluation, the first variant of normal walking (NM) for each subject is taken as the gallery with the rest as the probe. The last three rows show the results ignoring the probe sequences which have no corresponding ones in the gallery

Method	DIM	Probe view														Average
		0°	15°	30°	45°	60°	75°	90°	180°	195°	210°	225°	240°	255°	270°	
GEINet [31]	2048	23.20	38.09	47.95	51.81	47.53	48.09	43.75	27.25	37.89	46.78	49.85	45.94	45.65	40.96	42.48
GaitSet [6]	15872	79.33	87.59	89.96	90.09	87.96	88.74	87.69	81.82	86.46	88.95	89.17	87.16	87.60	86.15	87.05
GLN(ours)	**256**	**83.81**	**90.00**	**91.02**	**91.21**	**90.25**	**89.99**	**89.43**	**85.28**	**89.09**	**90.47**	**90.59**	**89.60**	**89.31**	**88.47**	**89.18**
GEINet [31]	2048	24.91	40.65	51.55	55.13	49.81	51.05	46.37	29.17	40.67	50.53	53.27	48.39	48.64	43.49	45.26
GaitSet [6]	15872	84.50	93.27	96.72	96.58	93.48	95.28	94.15	87.04	92.50	96.00	95.96	92.99	94.34	92.69	93.25
GLN(ours)	**256**	**89.28**	**95.84**	**97.87**	**97.82**	**96.01**	**96.68**	**96.07**	**90.71**	**95.34**	**97.66**	**97.54**	**95.69**	**96.24**	**95.27**	**95.57**

Lateral Connections. In GLN, the silhouette-level and set-level features extracted by different stages are merged with the lateral connections in a top-down manner. Here we separately evaluate the effect of lateral connections. Specifically, the network shown in Fig. 3 is trained without *Compact Block* until convergence. HPM is applied to the features generated by the lateral connections at the three stages. And the features of all parts are concatenated as the final representations where the dimension reaches up to 23808. According to the results shown in Table 4, the lateral connections can improve the performance under all walking conditions especially for walking in different clothes (CL).

Label Smooth. As stated in Sect. 3.3, the label smooth is adopted to prevent overfitting on the subjects in the training set. Here we conduct the experiment ignoring the label smooth and the standard cross-entropy loss [15] is computed for GLN. As shown in the last two rows of Table 4, the label smooth is beneficial to gait recognition under all walking conditions. Besides, the experimental results in Table 4 indicate that *Compact Block* can simultaneously reduce the dimension of the representations and improve the performance especially for normal walking (NM) and walking with bags (BG). The performance for walking in different clothes (CL) is comparable before and after reduction.

Training Strategy. The training strategy for GLN consists of two steps: *Lateral Pretraining* and *Global Training*. We have also tried to train the whole network globally from scratch, however, the performance is inferior under all walking conditions on CASIA-B (NM-96.48%, BG-93.07%, CL-77.07%). The comparison indicates that it is necessary to pretrain the lateral connections.

Table 4. The ablation study for lateral connections and label smooth, *DIM* for *Dimension*, *CBlock* for *Compact Block*. The results are reported on CASIA-B

Method	DIM	Label Smooth	NM	BG	CL
GaitSet [6]	15872	–	95.21	88.08	71.05
GaitSet-L	15872	–	95.57	91.52	75.27
GLN (without *CBlock*)	23808	–	95.58	91.98	77.22
GLN (with *CBlock*)	256	×	96.48	94.03	77.03
GLN (with *CBlock*)	256	√	**96.88**	**94.04**	**77.50**

Output Dimensions. The dimension of the final representations learned by GLN is empirically set to 256 through our experiments. Here we provide the experimental results with different output dimensions including 128 and 512. As shown in Table 5, the performance for normal walking (NM) and walking with bags (BG) are comparable with the three dimensions. The dimension 256 achieves the best performance for walking in different clothes (CL) which is the most challenging and occurs frequently in the real-world applications.

Variants of Compact Block. As shown in Table 5, we conduct the experiments in comparison to some variants of *Compact Block*. We use the same settings as described in Sect. 4.1 except that the structure of *Compact Block* is replaced by the variants shown in Table 5. And we have also tried some classical methods for dimension reduction such as Principal Components Analysis (PCA, NM-95.47%, BG-91.90%, CL-76.99%) and Linear Discriminant Analysis (LDA, NM-87.97%, BG-81.85%, CL-63.19%). The performance comparisons indicate that *Compact Block* can be treated as a reasonable choice to reduce the dimension of the gait representations.

Time Statistics. Here we provide the running time comparison on CASIA-B between GaitSet-L and GLN in the training (GaitSet-L: 0.96s per iteration v.s. GLN: 1.01s per iteration) and test (GaitSet-L: 0.021s per sequence v.s. GLN: 0.022s per sequence). Though the running time of GLN is marginally increased compared to GaitSet-L, our method can reduce the representation dimension by nearly two orders of magnitude ($15872 \rightarrow 256$) and the performance for all walking conditions are improved simultaneously.

Table 5. The ablation study for output dimensions and variants of *Compact Block*, *DIM* for *Dimension*. The results are reported on CASIA-B

DIM	Variants of *compact block*	NM	BG	CL
512	*BN+ReLU+Dropout+FC+BN*	96.98	94.09	77.10
256	*BN+ReLU+Dropout+FC+BN*	96.88	94.04	**77.50**
128	*BN+ReLU+Dropout+FC+BN*	**97.07**	**94.10**	76.84
256	*BN+FC+BN*	94.58	90.81	70.05
256	*BN+ReLU+FC+BN*	95.29	90.74	71.48
256	*BN+Dropout+FC+BN*	96.37	93.83	75.23
256	*BN+ReLU+Dropout+FC+BN*	**96.88**	**94.04**	**77.50**

Comparison to More Baselines. As stated in Sect. 4.1, GaitSet [6] holds state-of-the-art performance for the silhouettes-based gait recognition before this work and GEINet [31] is a representative method taking Gait Energy Image [8] as input, which are more related to our work and compared thoroughly in our experiments. Here we provide more silhouette-based methods for comparison such as CNN-LB [41] (NM-89.9%, BG-72.4%, CL-54.0%) and J-CNN [45] (NM-91.2%, BG-75.0%, 54.0%). Besides, we notice that there are some methods taking other types of input for gait recognition such as GaitNet [46] (RGB frames, NM-92.3%, BG-88.9%, CL-62.3%), GaitMotion [2] (optical flow, NM-97.5%, BG-83.6%, CL-48.8%), SM-Prod [5] (gray images and optical flow, NM-99.8%, BG-96.1%, CL-67.0%). Though some methods [2,5] report a little higher performance for NM, the optical flow needs a lot of computation cost and the performance for the challenging CL is much inferior to our method (CL-77.50%). The results here are all reported on CASIA-B.

5 Conclusion

In this work, we propose a novel network named Gait Lateral Network (GLN) which can learn both *discriminative* and *compact* representations from the silhouettes for gait recognition. Specifically, the inherent feature pyramid in deep convolutional networks is leveraged to learn *discriminative* gait representations. The silhouette-level and set-level features extracted by different stages in the backbone are merged with the lateral connections in a top-down manner, which enhances the gait representations by aggregating more visual details. And we propose a *Compact Block* to learn *compact* gait representations, which can significantly reduce the dimension of the gait representations without hindering the accuracy. Extensive experiments on CASIA-B and OUMVLP demonstrate that GLN achieves state-of-the-art performance under all walking conditions using the 256-dimensional representations.

Acknowledgments. We are grateful to Prof. Dongbin Zhao for his support to this work.

References

1. Ariyanto, G., Nixon, M.S.: Model-based 3D gait biometrics. In: International Joint Conference on Biometrics, pp. 1–7 (2011)
2. Bashir, K., Xiang, T., Gong, S., Mary, Q.: Gait representation using flow fields. In: BMVC, pp. 1–11 (2009)
3. Bodor, R., Drenner, A., Fehr, D., Masoud, O., Papanikolopoulos, N.: View-independent human motion classification using image-based reconstruction. Image Vis. Comput. **27**(8), 1194–1206 (2009)
4. Bouchrika, I., Goffredo, M., Carter, J., Nixon, M.: On using gait in forensic biometrics. J. Forensic Sci. **56**(4), 882–889 (2011)
5. Castro, F.M., Marín-Jiménez, M.J., Guil, N., de la Blanca, N.P.: Multimodal feature fusion for CNN-based gait recognition: an empirical comparison. Neural Comput. Appl. **32**, 14173–14193 (2020). https://doi.org/10.1007/s00521-020-04811-z
6. Chao, H., He, Y., Zhang, J., Feng, J.: Gaitset: regarding gait as a set for cross-view gait recognition. In: AAAI, vol. 33, pp. 8126–8133 (2019)
7. Fu, Y., et al.: Horizontal pyramid matching for person re-identification. In: AAAI, vol. 33, pp. 8295–8302 (2019)
8. Han, J., Bhanu, B.: Individual recognition using gait energy image. TPAMI **28**(2), 316–322 (2005)
9. Hariharan, B., Arbeláez, P., Girshick, R., Malik, J.: Hypercolumns for object segmentation and fine-grained localization. In: CVPR, pp. 447–456 (2015)
10. He, Y., Zhang, J., Shan, H., Wang, L.: Multi-task GANs for view-specific feature learning in gait recognition. IEEE Trans. Inf. Forensics Secur. **14**(1), 102–113 (2018)
11. Hermans, A., Beyer, L., Leibe, B.: In defense of the triplet loss for person re-identification. arXiv preprint arXiv:1703.07737 (2017)
12. Hu, B., Gao, Y., Guan, Y., Long, Y., Lane, N., Ploetz, T.: Robust cross-view gait identification with evidence: a discriminant gait GAN (DIGGAN) approach on 10000 people. arXiv preprint arXiv:1811.10493 (2018)
13. Ioffe, S., Szegedy, C.: Batch normalization: accelerating deep network training by reducing internal covariate shift. ICML **37**, 448–456 (2015)
14. Ji, S., Xu, W., Yang, M., Yu, K.: 3D convolutional neural networks for human action recognition. TPAMI **35**(1), 221–231 (2012)
15. Krizhevsky, A., Sutskever, I., Hinton, G.E.: ImageNet classification with deep convolutional neural networks. In: NeurIPS, pp. 1097–1105 (2012)

16. Kusakunniran, W., Wu, Q., Li, H., Zhang, J.: Multiple views gait recognition using view transformation model based on optimized gait energy image. In: ICCV Workshops, pp. 1058–1064 (2009)
17. Kusakunniran, W., Wu, Q., Zhang, J., Ma, Y., Li, H.: A new view-invariant feature for cross-view gait recognition. IEEE Trans. Inf. Forensics Secur. **8**(10), 1642–1653 (2013)
18. Larsen, P.K., Simonsen, E.B., Lynnerup, N.: Gait analysis in forensic medicine. J. Forensic Sci. **53**(5), 1149–1153 (2008)
19. Lee, C.Y., Xie, S., Gallagher, P.W., Zhang, Z., Tu, Z.: Deeply-supervised nets. ArXiv abs/1409.5185 (2014)
20. Liao, R., Cao, C., Garcia, E.B., Yu, S., Huang, Y.: Pose-based temporal-spatial network (PTSN) for gait recognition with carrying and clothing variations. In: Zhou, J., et al. (eds.) CCBR 2017. LNCS, vol. 10568, pp. 474–483. Springer, Cham (2017). https://doi.org/10.1007/978-3-319-69923-3_51
21. Lin, T.Y., Dollár, P., Girshick, R., He, K., Hariharan, B., Belongie, S.: Feature pyramid networks for object detection. In: CVPR, pp. 2117–2125 (2017)
22. Liu, C.T., Wu, C.W., Wang, Y.C.F., Chien, S.Y.: Spatially and temporally efficient non-local attention network for video-based person re-identification. arXiv preprint arXiv:1908.01683 (2019)
23. Liu, W., et al.: SSD: single shot multibox detector. In: Leibe, B., Matas, J., Sebe, N., Welling, M. (eds.) ECCV 2016. LNCS, vol. 9905, pp. 21–37. Springer, Cham (2016). https://doi.org/10.1007/978-3-319-46448-0_2
24. Long, J., Shelhamer, E., Darrell, T.: Fully convolutional networks for semantic segmentation. In: CVPR, pp. 3431–3440 (2015)
25. Luo, H., Gu, Y., Liao, X., Lai, S., Jiang, W.: Bag of tricks and a strong baseline for deep person re-identification. In: CVPR Workshops (2019)
26. Makihara, Y., Sagawa, R., Mukaigawa, Y., Echigo, T., Yagi, Y.: Gait recognition using a view transformation model in the frequency domain. In: Leonardis, A., Bischof, H., Pinz, A. (eds.) ECCV 2006. LNCS, vol. 3953, pp. 151–163. Springer, Heidelberg (2006). https://doi.org/10.1007/11744078_12
27. Maltoni, D., Maio, D., Jain, A.K., Prabhakar, S.: Handbook of Fingerprint Recognition. Springer, London (2009). https://doi.org/10.1007/978-1-84882-254-2
28. Mirza, M., Osindero, S.: Conditional generative adversarial nets. ArXiv abs/1411.1784 (2014)
29. Paszke, A., et al.: Pytorch: an imperative style, high-performance deep learning library. In: NeurIPS, pp. 8024–8035 (2019)
30. Ruder, S.: An overview of gradient descent optimization algorithms. arXiv preprint arXiv:1609.04747 (2016)
31. Shiraga, K., Makihara, Y., Muramatsu, D., Echigo, T., Yagi, Y.: GeiNet: view-invariant gait recognition using a convolutional neural network. In: International Conference on Biometrics, pp. 1–8 (2016)
32. Srivastava, N., Hinton, G.E., Krizhevsky, A., Sutskever, I., Salakhutdinov, R.: Dropout: a simple way to prevent neural networks from overfitting. J. Mach. Learn. Res. **15**, 1929–1958 (2014)
33. Sun, Y., Chen, Y., Wang, X., Tang, X.: Deep learning face representation by joint identification-verification. In: NeurIPS,pp. 1988–1996 (2014)
34. Szegedy, C., Vanhoucke, V., Ioffe, S., Shlens, J., Wojna, Z.: Rethinking the inception architecture for computer vision. In: CVPR, pp. 2818–2826 (2016)
35. Takemura, N., Makihara, Y., Muramatsu, D., Echigo, T., Yagi, Y.: On input/output architectures for convolutional neural network-based cross-view gait recognition. IEEE Trans. Circuits Syst. Video Technol. (2017)

36. Takemura, N., Makihara, Y., Muramatsu, D., Echigo, T., Yagi, Y.: Multi-view large population gait dataset and its performance evaluation for cross-view gait recognition. IPSJ Trans. Comput. Vis. Appl. **10**(1), 1–14 (2018). https://doi.org/10.1186/s41074-018-0039-6
37. Wang, C., Zhang, J., Wang, L., Pu, J., Yuan, X.: Human identification using temporal information preserving gait template. TPAMI **34**(11), 2164–2176 (2011)
38. Wang, L., Tan, T., Ning, H., Hu, W.: Silhouette analysis-based gait recognition for human identification. TPAMI **25**(12), 1505–1518 (2003)
39. Wildes, R.P.: Iris recognition: an emerging biometric technology. Proc. IEEE **85**(9), 1348–1363 (1997)
40. Wolf, T., Babaee, M., Rigoll, G.: Multi-view gait recognition using 3D convolutional neural networks. In: ICIP, pp. 4165–4169 (2016)
41. Wu, Z., Huang, Y., Wang, L., Wang, X., Tan, T.: A comprehensive study on cross-view gait based human identification with deep CNNs. TPAMI **39**(2), 209–226 (2016)
42. Yu, S., Tan, D., Tan, T.: A framework for evaluating the effect of view angle, clothing and carrying condition on gait recognition. In: International Conference on Pattern Recognition, vol. 4, pp. 441–444 (2006)
43. Zeiler, M.D., Fergus, R.: Visualizing and understanding convolutional networks. In: Fleet, D., Pajdla, T., Schiele, B., Tuytelaars, T. (eds.) ECCV 2014. LNCS, vol. 8689, pp. 818–833. Springer, Cham (2014). https://doi.org/10.1007/978-3-319-10590-1_53
44. Zhang, K., Luo, W., Ma, L., Liu, W., Li, H.: Learning joint gait representation via quintuplet loss minimization. In: CVPR, pp. 4700–4709 (2019)
45. Zhang, Y., Huang, Y., Wang, L., Yu, S.: A comprehensive study on gait biometrics using a joint CNN-based method. Pattern Recogn. **93**, 228–236 (2019)
46. Zhang, Z., et al.: Gait recognition via disentangled representation learning. In: CVPR, pp. 4710–4719 (2019)
47. Zhu, W., Hu, J., Sun, G., Cao, X., Qiao, Y.: A key volume mining deep framework for action recognition. In: CVPR, pp. 1991–1999 (2016)

Blind Face Restoration via Deep Multi-scale Component Dictionaries

Xiaoming Li[1,4,6], Chaofeng Chen[2,4], Shangchen Zhou[3], Xianhui Lin[4], Wangmeng Zuo[1,5](✉), and Lei Zhang[4,6]

[1] Faculty of Computing, Harbin Institute of Technology, Harbin, China
{csxmli,wmzuo}@hit.edu.cn
[2] Department of Computer Science, The University of Hong Kong, Hong Kong, China
cfchen@cs.hku.hk
[3] School of Computer Science and Engineering, Nanyang Technological University, Singapore, UK
shangchenzhou@gmail.com
[4] DAMO Academy, Alibaba Group, Hangzhou, China
xianhui.lxh@alibaba-inc.com
[5] Peng Cheng Lab, Shenzhen, China
[6] Department of Computing, The Hong Kong Polytechnic University, Hong Kong, China
cslzhang@comp.polyu.edu.hk

Abstract. Recent reference-based face restoration methods have received considerable attention due to their great capability in recovering high-frequency details on real low-quality images. However, most of these methods require a high-quality reference image of the same identity, making them only applicable in limited scenes. To address this issue, this paper suggests a deep face dictionary network (termed as DFDNet) to guide the restoration process of degraded observations. To begin with, we use K-means to generate deep dictionaries for perceptually significant face components (*i.e.*, left/right eyes, nose and mouth) from high-quality images. Next, with the degraded input, we match and select the most similar component features from their corresponding dictionaries and transfer the high-quality details to the input via the proposed dictionary feature transfer (DFT) block. In particular, component AdaIN is leveraged to eliminate the style diversity between the input and dictionary features (*e.g.*, illumination), and a confidence score is proposed to adaptively fuse the dictionary feature to the input. Finally, multi-scale dictionaries are adopted in a progressive manner to enable the coarse-to-fine restoration. Experiments show that our proposed method can achieve plausible performance in both quantitative and qualitative evaluation, and more importantly, can generate realistic and promising results on real degraded images without requiring an identity-belonging reference. The source code and models are available at https://github.com/csxmli2016/DFDNet.

Electronic supplementary material The online version of this chapter (https://doi.org/10.1007/978-3-030-58545-7_23) contains supplementary material, which is available to authorized users.

© Springer Nature Switzerland AG 2020
A. Vedaldi et al. (Eds.): ECCV 2020, LNCS 12354, pp. 399–415, 2020.
https://doi.org/10.1007/978-3-030-58545-7_23

Keywords: Face hallucination · Deep face dictionary · Guided image restoration · Convolutional neural networks

1 Introduction

Blind face restoration (or face hallucination) aims at recovering realistic details from real low-quality (LQ) image to its high-quality (HQ) one, without knowing the degradation types or parameters. Compared with single image restoration tasks, *e.g.*, image super-resolution [9,36,46], denoising [42,43], and deblurring [22,23], blind image restoration suffers from more challenges, yet is of great practical value in restoring real LQ images.

Recently, benefited from the carefully designed architecture and the incorporation of related priors in deep neural convolutional networks, the restoration results tend to be more plausible and acceptable. Though great achievements have been made, the real LQ images usually contain complex and diverse distributions that are impractical to synthesize, making the blind restoration problem intractable. To solve this issue, reference-based methods [7,26,35,47] have been suggested by using reference prior in image restoration task to improve the process of network learning and alleviate the dependency of network on degraded input. Among these methods, GFRNet [26] and GWAINet [7] adopt a frontal HQ image as reference to guide the restoration of degraded observation. However, these two methods suffer from two drawbacks. 1) They have to obtain a frontal HQ reference which is from the same identity with LQ image. 2) The differences of poses and expressions between the reference and degraded input will affect the reconstruction performance. These two requirements limit their applicative ability to some specific scenarios (*e.g.*, old film restoration or phone album that supports identity group).

In this paper, we present a DFDNet by building deep face dictionaries to address the aforementioned difficulties. We note that the four face components (*i.e.*, left/right eyes, nose and mouth) are similar among different people. Thus, in this work, we off-line build face component dictionaries by adopting K-means on large amounts of HQ face images. This manner can obtain more accurate component reference without requiring the corresponding identity-belonging HQ images, which makes the proposed model applicable in most face restoration scenes. To be specific, we firstly use pre-trained VggFace [3] to extract the multi-scale features of HQ face images in different feature scale (*e.g.*, output of different convolutional layers). Secondly, we adopt RoIAlign [14] to crop their component features based on the facial landmarks. K-means is then applied on these features to generate the K clusters for each component on different feature levels. After that, component adaptive instance normalization (CAdaIN) is proposed to norm the corresponding dictionary feature which helps to eliminate the effect of style diversity (*i.e.*, illumination or skin color). Finally, with the degraded input, we match and select the dictionary component clusters which have the smallest feature distance to guide the following restoration process in an adaptive and progressive manner. A confidence score is predicted to balance the input component feature and the selected dictionary feature. In addition, we use multi-scale dictionaries to guide the restoration progressively which further improves the performance. Compared with the former reference-based methods (*i.e.*, GFRNet [26] and GWAINet [7]), which have only one HQ reference, our DFDNet has more component candidates to be selected as a reference, thus making our model achieve superior performance.

Extensive experiments are conducted to evaluate the performance of our proposed DFDNet. The quantitative and qualitative results show the benefits of deep multi-scale face dictionaries brought in our method. Moreover, DFDNet can also generate plausible and promising results on real LQ images. Without requiring identity-belonging HQ reference, our method is flexible and practical in most face restoration applications.

To sum up, the main contributions of this work are:

- We use deep component dictionaries as reference candidates to guide the degraded face restoration. The proposed DFDNet can generalize to face images without requiring the identity-belonging HQ reference, which is more applicative and efficient than those reference-based methods.
- We suggest a DFT block by utilizing CAdaIN to eliminate the distribution diversity between the input and dictionary clusters for better dictionary feature transfer, and we also propose a confidence score to adaptively fuse the dictionary feature to the input with different degradation level.
- We adopt a progressive manner for training DFDNet by incorporating the component dictionaries in different feature scales. This can make our DFDNet learn coarse-to-fine details.
- Our proposed DFDNet can achieve promising performance on both synthetic and real degraded images, showing its potential in real applications.

2 Related Work

In this section, we discuss recent works about single image and reference-based image restoration methods which are closely related to our work.

2.1 Single Image Restoration

Along with the benefits brought by deep CNNs, single image restoration has achieved great success in many tasks, *e.g.*, image super-resolution [9,19,24,44,46], denoising [13,38,42,43], deblurring [22,29,41], and compression artifact removal [8,10,12]. Due to the specific facial structure, there are also several well-developed methods for face hallucination [2,4–6,15,37,39,40,48]. Among these methods, Huang *et al.* [15] suggest to ultra-resolve a very low resolution face image by using the neural networks to predict the wavelet coefficients of HQ images. Cao *et al.* [2] propose reinforcement learning to discover the attended regions and then enhance them with a learnable local network. To better recover the structure details, there are also some methods that incorporate the image prior knowledge in the restoring process. Wang *et al.* [35] propose to use semantic segmentation probability maps as class prior to recover class-aware textures on natural image super-resolution task. It firstly takes the LR images through a segmentation network to generate the class probability maps. And then these maps and LQ features are fused together by spatial feature transformation. As for face images, Shen *et al.* [33] propose to learn a global semantic face prior as input to impose local structure on the output. Similarly, Xu *et al.* [39] use a multi-tasks model to predict the facial components heatmaps and use them for incorporating structure information. Chen *et al.* [4] learn the facial geometry prior (*i.e.*, landmarks heatmaps and parsing maps) and take them to recover the high-resolution results. Yu *et al.* [40] develop a facial attribute-embedded network by incorporating face attributes vector in the LR feature space. Kim *et al.* [6] adopt a progressive manner to generate the successive higher resolution output and propose a facial attention loss on landmarks to constrain the structure of reconstruction. However, most of these facial prior knowledge mainly focus on geometry constrains (*i.e.*, landmarks or heatmaps), which may not bring direct facial details for the restoration of LQ image. Thus, most of these single image restoration methods failed to generate plausible and realistic details on real LQ face images because of the ill-posed problem and the limitation of a single image or facial structure prior brought to the learning process of networks.

2.2 Reference-Based Image Restoration

Due to the limitation of single image restoration methods on real-world LQ images, there are some works that use an additional image to guide the restoration process, which can bring the object structure details to the final result. As for natural image restoration, Zhang *et al.* [47] utilize a reference image which has similar content with a LR image and then adopt a global matching scheme to search the similar content patches. These reference feature patches are then used to swap the texture feature of LR images. This method can achieve great visual improvements. However, it is very time and memory consuming in searching similar patches from the global content. Moreover, the requirement of reference further limits its application, because finding a natural image with a similar content for each LR input is also terrible and sometimes it is impossible to obtain these types of image.

Different from natural image, face owns specific structures and share the similar components on different images of the same identity. Based on this observation, two reference-based methods have been developed for face restoration. Li *et al.* [26] and Dogan *et al.* [7] use a fixed frontal HQ reference for each identity to provide identity-aware features to benefit the restoration process. However, we note that face images are usually taken under unconstrained conditions, *e.g.*, different background, poses, expressions, illuminations, *etc.* To solve this problem, they utilize a WarpNet to predict flow field to warp the reference to align with the LQ image. However, the alignment still does not solve all the differences between the reference and input, *i.e.*, mouth close to open. Besides, the warped reference is usually unnatural and may take obvious artifacts to the final reconstruction result. We note that each component between different identity has the similar structure (*i.e.*, teeth, nose, and eyes). It is intuitive to split the whole face into different parts, and generate the representative components for each one. To achieve this goal, we firstly use K-means on HQ images to cluster different component features off-line.

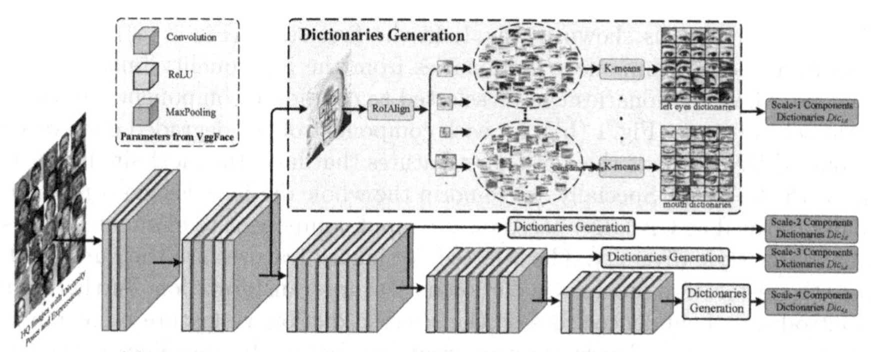

(a) Off-line generation of multi-scale component dictionaries.

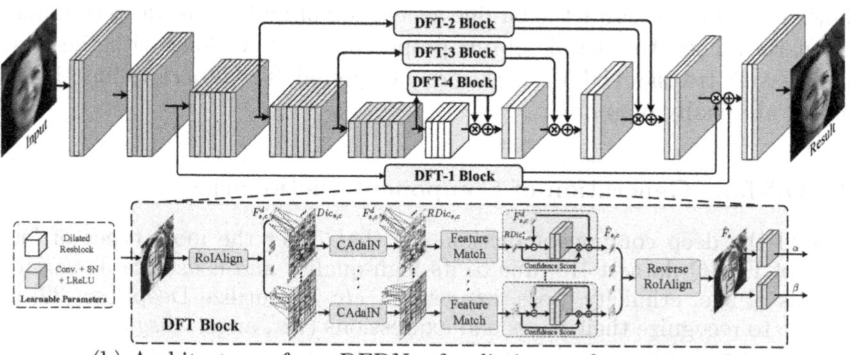

(b) Architecture of our DFDNet for dictionary feature transfer.

Fig. 1. Overview of our proposed method. It mainly contains two parts: (a) the off-line generation of multi-scale component dictionaries from large amounts of high-quality images which have diverse poses and expressions. K-means is adopted to generate K clusters for each component (*i.e.*, left/right eyes, nose and mouth) on different feature scales. (b) The restoration process and dictionary feature transfer (DFT) block that are utilized to provide the reference details in a progressive manner. Here, DFT-i block takes the Scale-i component dictionaries for reference in the same feature level.

Then we match the LQ features from the conducted component dictionaries to select the one with the similar structures to guide the latter restoration. Moreover, with the conducted dictionaries, we do not require an identity-belonging reference anymore, and more component candidates can be selected as reference. It is much more accurate and effective than only one face image in reference-based restoration and can be applied in the unconstrained applications.

3 Proposed Method

Inspired by the former reference-based image restoration methods [7,26,47], this work attempts to overcome the limitation of requiring reference image in face restoration. Given a LQ image I^d, our proposed DFDNet aims to generate plausible and realistic HQ one $\hat{I}^h$ with the conducted component dictionaries.

The whole pipeline is shown in Fig. 1. In the first stage (Fig. 1 (a)), we firstly generate the deep component dictionaries from the high-quality images I^h via k-means. These dictionaries can be selected as candidate component references. In the second stage (Fig. 1 (b)), for each component of the degraded observation I^d, our DFDNet selects the dictionary features that have the most similar structure with the input. Specially, we re-norm the whole dictionaries via component AdaIN (termed as CAdaIN) based on the input component to eliminate the distribution or style diversity. The selected dictionary features are then utilized to guide the restoration process via dictionary feature transformation. Furthermore, we introduce a confidence score on the selected dictionary feature to generalize different degradation levels through weighted feature fusion. The progressive manner from coarse to fine is also beneficial to the restoration process. In the following, we first describe the off-line generation of multi-scale deep component dictionaries. Then the details of our proposed DFDNet along with the dictionaries feature transfer (DFT) blocks are interpreted. The objective functions for training are finally presented.

3.1 Off-Line Generation of Component Dictionaries

To build the deep component dictionaries that cover the most types of faces, we adopt FFHQ dataset [18] due to its high-quality and considerable variation in terms of age, ethnicity, pose, expression, *etc.* We utilize DeepPose [32] and Face++[1] to recognize their poses and expressions (*i.e.*, anger, disgust, fear, happiness, neutral, sadness and surprise), respectively, to balance the distribution of each attribute. Among these 70,000 high-quality images of FFHQ, we select 10,000 ones to build our dictionaries. Given a high-quality image I^h, we first use pre-trained VggFace [3] to extract its features on different scales. With the facial landmarks L^h detected by dlib [20], we utilize RoIAlign [14] to crop and re-sample these four components on each scale to a fixed size. We then adopt K-means [30] to generate K clusters for each component, resulting in our component dictionaries. In particular, for handling 256×256 images, the feature sizes of left/right eyes, nose and mouth on scale-1 are set to 40/40, 25, 55, respectively. The sizes are down-sampled one by one by two times for the following scale-{2, 3, 4}. These dictionary feature can be formulated as:

$$Dic_{s,c} = \mathcal{F}_{Dic}\left(I^h|L^h; \Theta_{Vgg}\right), \tag{1}$$

where $s \in \{1, 2, 3, 4\}$ is the dictionary scale, $c \in$ {left eye, right eye, nose, mouth} is the type of components, and Θ_{Vgg} is the fixed parameters from VggFace.

3.2 Deep Face Dictionary Network

After building the high-quality component dictionFaries, our DFDNet is then proposed to transfer the dictionary features to the degraded input I^d. The proposed

[1] https://www.faceplusplus.com.cn/emotion-recognition/.

DFDNet can be formulated as:

$$\hat{I} = \mathcal{F}(I^d | L^d, Dic; \Theta), \tag{2}$$

where L^d and Dic represent the facial landmarks of I^d and the component dictionaries in Eq. 1, respectively. Θ denotes the learnable parameters of DFDNet.

To guarantee the features of I^d and Dic in the same feature space, we take the pre-trained VggFace model as the encoder of DFDNet, which has the same network architecture and parameters in the dictionary generation network (Fig. 1 (a)). Suppose that the encoder of DFDNet is different from VggFace or trainable in the training phase, it easily generates different features which are inconsistent with the pre-conducted dictionaries. For better transferring the dictionary feature to the input components, we suggest a DFT block and use it in a progressive manner. It mainly contains five parts, *i.e.*, RoIAlign, CAdaIN, Feature Match, Confidence Score and Reverse RoIAlign. As for the encoder features of I^d, we first utilize RoIAlign to generate four component regions. We note that these input components may have different distribution/style with the cluster of conducted dictionaries $Dic_{s,c}$, we here suggest a component adaptive instance norm [16] (CAdaIN) to re-norm each cluster in the dictionaries. The feature match scheme is then utilized to select the cluster with the similar texture. In addition, a confidence score is predicted based on the residual between the selected cluster and the input feature to better provide complementary details on input. The reverse RoIAlign is finally adopted to paste the restored features to the corresponding locations. For better transformation of restored features to the decoder, we modify the UNet [31] and propose to use spatial feature transform (SFT) [35] to transfer the dictionary features to the degraded input.

CAdaIN. We note that face images are usually under unconstrained conditions, *e.g.*, different illuminations, skin color. To eliminate the effect of these diversities between the input components and dictionaries, we adopt component AdaIN (CAdaIN) to re-norm the clusters in component dictionaries for accurate feature matching. AdaIN [16] can remain the structure while translate the content to the desired style. Denote $F^d_{s,c}$ and $Dic^k_{s,c}$ as the *c-th* component features of the input I^d and the *k-th* cluster from the component dictionaries at scale s, respectively. The re-normed dictionaries $RDic_{s,c}$ by CAdaIN is formulated by:

$$RDic^k_{s,c} = \sigma\left(F^d_{s,c}\right)\left(\frac{Dic^k_{s,c} - \mu\left(Dic^k_{s,c}\right)}{\sigma\left(Dic^k_{s,c}\right)}\right) + \mu\left(F^d_{s,c}\right) \tag{3}$$

where s and c are the dictionary scale and the type of components defined in Eq. 1. σ and μ are the mean and standard deviation. The re-normed dictionaries $RDic^k_{s,c}$ has the similar distribution with input components $F^d_{s,c}$, which can not only eliminate the style difference, but also facilitate the feature match scheme.

Feature Match. As for the input component feature $F^d_{s,c}$ and the re-normed dictionaries $RDic_{s,c}$, we adopt inner product to measure the similarity between the $F^d_{s,c}$ and all the clusters in $RDic_{s,c}$. For *k-th* cluster in component dictionary, the similarity is defined as:

$$S^k_{s,c} = \left\langle F^d_{s,c}, RDic^k_{s,c} \right\rangle, \tag{4}$$

The input component feature $F^d_{s,c}$ matches across all the clusters in the re-normed component dictionaries to select the most similar one. $F^d_{s,c}$ has the same size with *k-th* cluster in the corresponding dictionaries, thus this inner product operation can be regarded as a convolutional layer with zero bias and weights of $F^c_{s,d}$ performed over all the clusters. This is very efficient to obtain the dictionaries' similarity scores. Among all the scores $S_{s,c}$, we select the re-normed cluster with the highest similarity as the matched dictionaries, termed as $RDic^*_{s,c}$. This selected component feature $RDic^*_{s,c}$ is then utilized to provide the high-quality details to guide the restoration of the input components in the following section.

Confidence Score. We note that the slight degradation of input (*e.g.*, ×2 super-resolution) relies little on the dictionaries and vice versa. To generalize our DFDNet to different degradation level, we take the residual between $F^d_{s,c}$ and $RDic^*_{s,c}$ as input to predict a confidence score that performs on the selected dictionary feature $RDic^*_{s,c}$. The result is expected to contain the absent high-quality details which can add back to $F^d_{s,c}$. The output of confidence score can be formulated by:

$$\hat{F}_{s,c} = F^d_{s,c} + RDic^*_{s,c} * \mathcal{F}_{Conf}(RDic^*_{s,c} - F^d_{s,c}; \Theta_C), \tag{5}$$

where Θ_C is the learnable parameters of confidence score block $\mathcal{F}_{Conf}$.

Reverse RoIAlign. After all the input components are processed by the former section, here we utilize a reverse operation of RoIAlign by taking $\hat{F}_{s,c}$ and $c \in$ {left/right eyes, nose and mouth} to their original locations of $F^d_{s,c}$. Denote the result of reverse RoIAlign $\hat{F}_s$. This manner can easily keep and translate other features (*e.g.*, background) to the decoder for better restoration.

Inspired by SFT [35], which is proposed to learn a feature modulation function that incorporates some prior condition through affine transformation. The scale α and shift β parameters are learned from the restored features $\hat{F}_s$ with two convolutional layers. The scale-s SFT layer is formulated as:

$$SFT_s = \alpha \odot F^{decoder}_s + \beta, \tag{6}$$

where α and β are both element-wise weights which have the same shape (i.e., height, width, number of channels) with $F^{decoder}_s$. After the progressive DFT block, our DFDNet can gradually learn the fine details for the final result $\hat{I}$.

3.3 Model Objective

The learning objective for training our DFDNet contains two parts, 1) reconstruction loss that constrains the result $\hat{I}$ close to the ground-truth I^h, 2) adversarial loss [11] for recovering realistic details.

Reconstruction Loss. We adopt mean square error (MSE) on both pixel and feature space (perceptual loss [17]). The whole reconstruction loss is defined as,

$$\mathcal{L}_{rec} = \lambda_{l2}\|\hat{I} - I^h\|^2 + \sum_{m=1}^{M} \frac{\lambda_{p,m}}{C_m H_m W_m} \left\|\Psi_m(\hat{I}) - \Psi_m(I^h)\right\|^2 \tag{7}$$

where Ψ_m denotes the $m\text{-}th$ convolution layer of VggFace model Ψ. C, H and W are the channel, height, and width for the $m\text{-}th$ feature. λ_{l2} and $\lambda_{p,m}$ are the trade-off parameters. The first term tends to generate blurry results, while the second one (perceptual loss) is beneficial for improving visual quality for the reconstruction results. The combination of the two terms is common in computer vision tasks and also is effective in the stable training of neural networks. In our experimental settings, we set M equal to 4.

Adversarial Loss. It is widely used to generate realistic details in image restoration tasks. In this work, we adopt multi-scale discriminators [34] at different size of the restoration results. Moreover, for stable training of each discriminator, we adopt SNGAN [28] by incorporating the spectral normalization after each convolution layer. The objective function for training multi-scale discriminators is defined as:

$$\ell_{\text{adv},D_r} = \sum_{r}^{R} \mathbb{E}_{I^h_{\downarrow r} \sim P(I^h_{\downarrow r})} \left[\min\left(0, D_r(I^h_{\downarrow r}) - 1\right)\right] + \mathbb{E}_{\hat{I}_{\downarrow r} \sim P(\hat{I}_{\downarrow r})} \left[\min\left(0, -1 - D_r(\hat{I}_{\downarrow r})\right)\right], \tag{8}$$

where $_{\downarrow r}$ denotes the down-sampling operation with scale factor r and $r \in \{1, 2, 4, 8\}$. Similarly, the loss for training generator $\mathcal{F}$ is defined as:

$$\ell_{\text{adv},G} = -\lambda_{a,r} \sum_{r}^{R} \mathbb{E}_{I^d \sim P(I^d)} \left[D_r\left(\mathcal{F}\left(I^d | L^d, Dic; \Theta\right)_{\downarrow r}\right)\right], \tag{9}$$

where $\lambda_{a,r}$ is the trade-off parameters for each scale discriminator.

To sum up, the full objective function for training our DFDNet can be written as the combination of reconstruction and adversarial loss,

$$\mathcal{L} = \ell_{rec} + \ell_{\text{adv},G}. \tag{10}$$

4 Experiments

Since the performance of reference-based methods are usually superior to other single image or face restoration methods [26], in this paper, we mainly compare our DFDNet with reference-based (*i.e.*, GFRNet [26], GWAINet [7]) and face prior-based methods (*i.e.*, Shen *et al.* [33], Kim *et al.* [6]). We also report the results of single natural image (*i.e.*, RCAN [46], ESRGAN [36]) and face (*i.e.*, WaveletSR [15]) super-resolution methods. Among these methods, Shen *et al.* [33] and Kim *et al.* [6] can only handle 128 × 128 images, while others can restore 256 × 256 images. For fair comparisons, our DFDNet is trained on these two sizes (termed as DFDNet128 and DFDNet256). RCAN [46] and ESRGAN [36] were originally trained on the natural images, thus we retrain them using our training data for further fair comparison (termed as *RCAN and *ESRGAN). WaveletSR [15] was also retrained by using our training data with their released training code (termed as *WaveletSR). Following [26], PSNR, SSIM and LPIPS [45] are reported on the super-resolution task ($\times 4$ and $\times 8$) which also has the random injection of Gaussian noise and blur operation for quantitatively evaluating on the blind restoration task. In terms of qualitative comparison, we demonstrate the comparisons on the synthetic and real-world low-quality images. More visual results including high resolution restoration performance (*i.e.*, 512×512) can be found in our supplemental materials.

4.1 Training Details

As mentioned in Sect. 3.1, we select 10,000 images from FFHQ [18] to build our component dictionaries. We note that GFRNet, GWAINet and WaveletSR adopt VggFace2 [3] as their training data, we also use it for training and validating our DFDNet for fair comparison. To evaluate the generality of our method, we build two test datasets, *i.e.*, 2,000 test images from VggFace2 [3] which are not overlapped with the training data, and another 2,000 images from CelebA [27]. Each of them has a high-quality reference from the same identity for running GFRNet and GWAINet. To synthesize the training data that approximate to the real LQ images, we adopt the same degradation model suggested in GFRNet [26],

$$I^d = \left((I^h \otimes \mathbf{k})_{\downarrow_r} + \mathbf{n}_\sigma\right)_{JPEG_q} \tag{11}$$

where $\mathbf{k}$ denotes two common types of blur kernel, *i.e.*, Gaussian blur with $\varrho \in \{1 : 0.1 : 5\}$ and 32 motion blur kernels from [1,25]. Down-sampler r, Gaussian noise n_σ and JPEG compression quality q are randomly sampled from $\{1 : 0.1 : 8\}$, $\{0 : 1 : 15\}$ and $\{40 : 1 : 80\}$, respectively. The trade-off parameters for training DFDNet are set as follows: $\lambda_{l2} = 100$, $\lambda_{p,1} = 0.5$, $\lambda_{p,2} = 1$, $\lambda_{p,3} = 2$, $\lambda_{p,4} = 4$, $\lambda_{a,1} = 4$, $\lambda_{a,2} = 2$, $\lambda_{a,4} = 1$, $\lambda_{a,8} = 1$. The Adam optimizer [21] is adopted to train our DFDNet with learning rate $lr = 2 \times 10^{-4}$, $\beta_1 = 0.5$ and $\beta_2 = 0.999$. lr is reduced by 2 times when the reconstruction loss on validation set becomes non-decreasing. The whole model including the generation of multi-scale component dictionaries and the training of DFDNet are executed on a server with 128G RAM and 4 Tesla V100. It takes 4 days to train our DFDNet.

4.2 Results on Synthetic Images

Qualitative Evaluation. The quantitative results of these competing methods on super-resolution task are shown in Table 1. We can have the following observations: 1) Compared with all the competing methods, our DFDNet is superior to others by a large margin on two datasets and two super-resolution tasks (*i.e.*, at least 0.4 dB in $\times 4$ and 0.3 dB in $\times 8$ higher than the 2-*nd* best method). 2)

Table 1. Quantitative comparisons on two datasets and two tasks (×4 and ×8).

Methods	VggFace2 [3]						CelebA [27]					
	×4			×8			×4			×8		
	PSNR↑	SSIM↑	LPIPS↓	PSNR↑	SSIM↑	LPIPS↓	PSNR↑	SSIM↑	LPIPS↓	PSNR↑	SSIM↑	LPIPS↓
Shen *et al.* [33]	20.56	.745	.080	18.79	.717	.126	21.04	.751	.079	18.64	.714	.131
Kim *et al.* [6]	–	–	–	20.99	.759	.095	–	–	–	20.72	.749	.104
DFDNet128	25.76	.893	.035	23.42	.841	.071	25.92	.899	.031	23.40	.839	.080
RCAN [46]	24.87	.889	.283	21.36	.819	.295	24.93	.892	.267	21.11	.814	.302
*RCAN	25.32	.896	.247	22.94	.836	.271	25.47	.901	.217	22.84	.831	.283
ESRGAN [36]	24.13	.876	.223	–	–	–	24.31	.878	.210	–	–	–
*ESRGAN	24.91	.891	.194	–	–	–	25.04	.896	.193	–	–	–
WaveletSR [15]	24.30	.878	.236	21.70	.823	.273	24.51	.884	.247	21.42	.820	.279
GFRNet [26]	27.13	.912	.132	23.37	.856	.269	27.32	.915	.124	23.12	.852	.273
GWAINet [7]	–	–	–	23.41	.860	.260	–	–	–	23.38	.859	.270
DFDNet256	**27.54**	**.923**	**.114**	**23.73**	**.872**	**.239**	**27.77**	**.925**	**.103**	**23.69**	**.872**	**.241**

Fig. 2. Visual comparisons of these competing methods on ×4 SR task. Close-up in the right bottom of GFRNet is the required guidance.

Even though the retrained *RCAN and *ESRGAN have achieved great improvements, the performance is still inferior to GFRNet, GWAINet and our DFDNet, mainly due to the lack of high-quality facial references. 3) With the same training data, reference-based methods (*i.e.*, GFRNet [26] and GWAINet [7]) outperform other methods, but are still inferior to our DFDNet, which can be attributed to the incorporation of high-quality component dictionaries and the progressive dictionary feature transfer manner. Given a LQ image, our DFDNet has more candidates to be selected as component reference, resulting in the flexible and effective restoration. 4) Our component dictionaries are conducted on FFHQ [18] and DFDNet is trained on VggFace2 [3], but the performance on CelebA [27] still outperforms other methods, indicating the great generalization of our DFDNet.

Visual Comparisons. Figures. 2 and 3 show the restoration results of these competing methods on ×4 and ×8 super-resolution tasks. Shen *et al.* [33] and Kim *et al.* [6] were proposed to handle face deblur and super-resolution problems. Since they only released their test model, we did not re-implement them with the same training data and degradation model in this paper, resulting in

Fig. 3. Visual comparisons of these competing methods on ×8 SR task. Close-up in the right bottom of GFRNet is the required guidance.

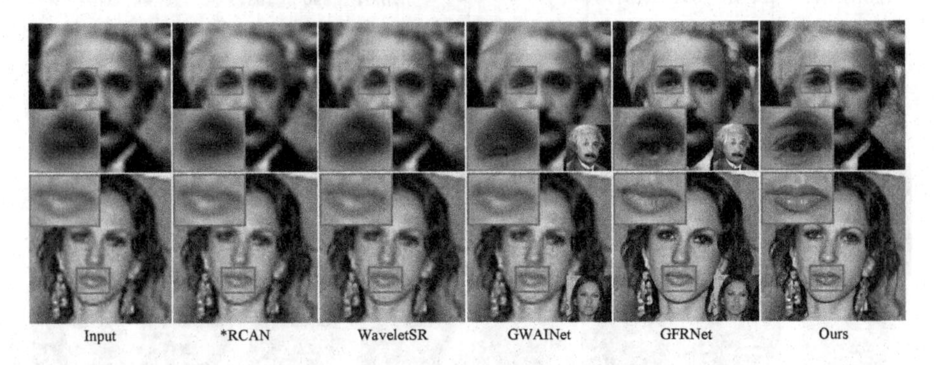

Fig. 4. Visual comparisons of competing methods with top performance on real-world low-quality images. Close-up at the right bottom is the required guidance.

their poor performance. The retrained *RCAN, *ESRGAN and *WaveletSR still limited in generating plausible facial structure, which may be caused by the lack of reasonable guidance for face restoration. In terms of reference-based methods, GFRNet [26] and GWAINet [7] generate plausible structures but fail to restore realistic details. In contrast to these competing methods, our DFDNet can reconstruct promising structure with richer details on these notable face regions (*i.e.*, eyes and mouth). Moreover, even though the degraded input is not frontal, our DFDNet can also have plausible performance (2-*nd* rows in Figs. 2 and 3).

Performance on Real-World Low-Quality Images. Our goal is to restore the real low-quality images without knowing the degradation types and parameters. To evaluate the performance of our DFDNet on blind face restoration, we select the real images from Google Image with face resolution lower than 80 × 80 and each of them has an identity-belonging high-quality reference for running GFRNet [26] and GWAINet [7]. Here we only show the visual results on competing methods with top-5 quantitative performance in Fig. 4. Among these competing methods, only GFRNet [26] is proposed to handle blind face restoration, thus can well generalize to real degraded images. However, its results still contain obvious artifacts due to the inconsistent reference of only one high-quality image. With the incorporation of component dictionaries, our DFDNet

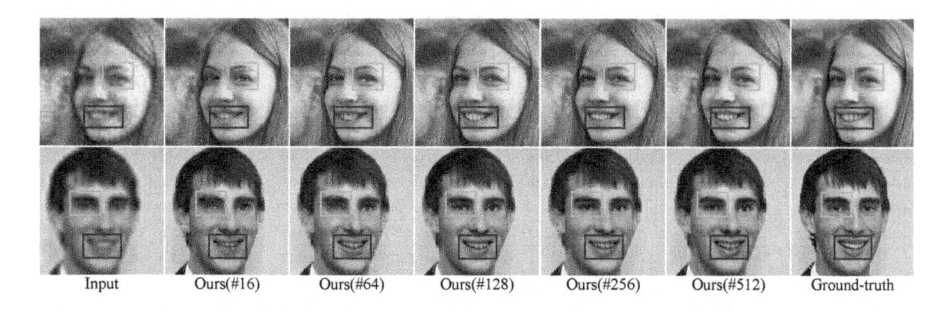

Fig. 5. Restoration results of our DFDNet with different cluster numbers.

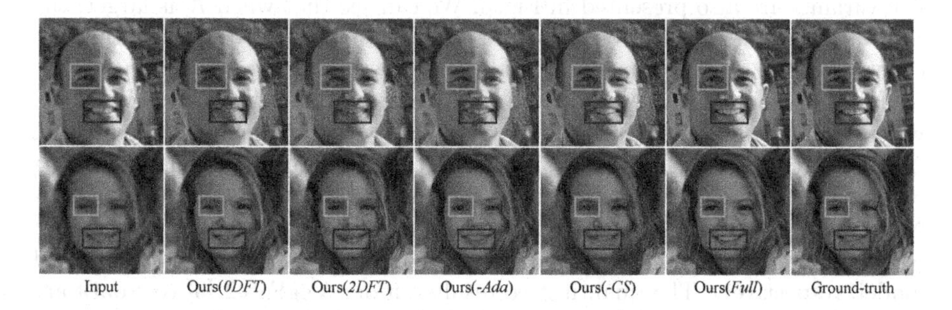

Fig. 6. Restoration results of our DFDNet variants.

can generate plausible and realistic results, especially in the eyes and mouth region, indicating the effectiveness of our DFDNet in handling real degraded observations. Moreover, our DFDNet does not require the identity-belonging reference, showing practical values in wide applications.

4.3 Ablation Study

To evaluate the effectiveness of our proposed DFDNet, we conduct two groups of ablative experiments, *i.e.*, the cluster number K for each component dictionary, and the progressive dictionary feature transfer block (DFT). For the first one, we generate different number of clusters in our component dictionaries. In this paper, we consider the cluster $K \in \{16, 64, 128, 256, 512\}$. For each variant, we retrain our DFDNet256 with the same experimental settings but with different cluster numbers, which are defined as Ours($\#K$). The quantitative results on our VggFace2 test data are shown in Table 2. One can see that Ours(#64) has nearly the same performance with GFRNet [26]. We analyze that because GFRNet [26] adopts alignment between reference and degraded input, making Ours(#16) performs poorer than it. By increasing the cluster numbers, our DFDNet tends to achieve better results. We note that Ours(#256) performs on par with Ours(#512) but has less time-consuming in feature match. Thus, we adopt Ours(#256) as our default model. Visual comparisons between these

Table 2. Comparisons on cluster number.

Methods	×4			×8		
	PSNR↑	SSIM↑	LPIPS↓	PSNR↑	SSIM↑	LPIPS↓
Ours(#16)	26.79	.908	.144	23.21	.839	.257
Ours(#64)	27.15	.914	.126	23.38	.856	.266
Ours(#128)	27.43	.919	.120	23.56	.867	.248
Ours(#256)	27.54	.923	.114	23.73	.872	.239
Ours(#512)	27.55	.923	.110	23.75	.873	.231

Table 3. Comparisons on variants of DFT.

Methods	×4			×8		
	PSNR↑	SSIM↑	LPIPS↓	PSNR↑	SSIM↑	LPIPS↓
Ours(*0DFT*)	25.30	.896	.239	23.06	.839	.253
Ours(*2DFT*)	26.43	.905	.161	23.24	.848	.261
Ours(*-Ada*)	25.47	.897	.190	22.97	.836	.270
Ours(*-CS*)	27.23	.914	.129	23.51	.862	.246
Ours(*Full*)	27.54	.923	.114	23.73	.872	.239

five variants are also presented in Fig. 5. We can see that when K is larger, the restoration results tend to be clear and are much more realistic, indicating the effectiveness of our dictionaries in guiding the restoration process.

For the second one, to evaluate the effectiveness of our progressive DFT block, we consider the following variants: 1) Ours(*Full*): the final model in this paper, 2) Ours(*0DFT*): our DFDNet by removing all the DFT blocks and directly using SFT to transfer the encoder feature to the decoder, 3) Ours(*2DFT*): our DFDNet with two DFT blocks (*i.e.*, DFT-{3,4} block), 4) Ours(*-Ada*) and Ours(*-CS*): by removing the CAdaIN and Confidence Score in all the DFT blocks of final model, respectively. The quantitative results on our VggFace2 [3] test data are reported in Table 3. We can have the following observations. (i) By increasing the number of DFT block, obvious gains (at least 2.2 dB in ×4 and 0.6 dB in ×8) are achieved, indicating the effectiveness of our progressive manner. (ii) The performance is severely degraded when removing the CAdaIN. This may be caused by the inconsistent distribution of degraded feature and dictionaries, resulting in the wrong matched features for restoration. (iii) With the incorporation of confidence score, which can help balance the input and the matched dictionary feature, our DFDNet can also achieve plausible improvements. Figure 6 shows the restoration results of these variants. We can see that compared with Ours(*0DFT*) and Ours(*2DFT*), Ours(*Full*) is much clear and contains rich details. Results of Ours(*-Ada*) are inconsistent with ground-truth (*i.e.*, mouth region in 1-*st* row). By the way, when the degradation is slight (1-*st* row), Ours(*-CS*) which directly swaps the dictionary feature to the degraded image can easily change the original content (mouth region), making the undesired modification of face components.

5 Conclusion

In this paper, we present a blind face restoration model, *i.e.*, DFDNet, to solve the limitation of reference-based methods. To eliminate the dependence of identity-belonging high-quality reference, we firstly suggest traditional K-means on large amount of high-quality images to cluster perceptually significant facial component. For dictionary feature transfer, we then propose a DFT block by addressing the following problems, distribution diversity between degraded input and dictionary feature with proposed component AdaIN, feature match scheme with fast inner product similarity, and generalization to degradation level with the confidence score. Finally, the multi-scale component dictionaries are incorporated in the multiple DFT blocks in a progressive manner, which can make our DFDNet learn the coarse-to-fine details for face restoration. Experiments validate the effectiveness of our DFDNet in handling the synthetic and real-world low-quality images. Moreover, we did not require an identity-belonging reference, showing the practical value in wide scenes in the real-world applications.

Acknowledgments. This work is partially supported by the National Natural Science Foundation of China (NSFC) under Grant No.s 61671182, U19A2073 and Hong Kong RGC RIF grant (R5001-18).

References

1. Boracchi, G., Foi, A.: Modeling the performance of image restoration from motion blur. IEEE Trans. Image Process. **21**(8), 3502–3517 (2012)
2. Cao, Q., Lin, L., Shi, Y., Liang, X., Li, G.: Attention-aware face hallucination via deep reinforcement learning. In: CVPR (2017)
3. Cao, Q., Shen, L., Xie, W., Parkhi, O.M., Zisserman, A.: Vggface2: a dataset for recognising faces across pose and age. In: FG (2018)
4. Chen, Y., Tai, Y., Liu, X., Shen, C., Yang, J.: Fsrnet: end-to-end learning face super-resolution with facial priors. In: CVPR June 2018
5. Chrysos, G.G., Zafeiriou, S.: Deep face deblurring. In: CVPRW (2017)
6. Kim, D.K, Minseon, K.G., Kim, D.S.: Progressive face super-resolution via attention to facial landmark. In: BMVC (2019)
7. Dogan, B., Gu, S., Timofte, R.: Exemplar guided face image super-resolution without facial landmarks. In: CVPRW (2019)
8. Dong, C., Deng, Y., Change Loy, C., Tang, X.: Compression artifacts reduction by a deep convolutional network. In: ICCV (2015)
9. Dong, C., Loy, C.C., He, K., Tang, X.: Learning a deep convolutional network for image super-resolution. In: Fleet, D., Pajdla, T., Schiele, B., Tuytelaars, T. (eds.) ECCV 2014. LNCS, vol. 8692, pp. 184–199. Springer, Cham (2014). https://doi.org/10.1007/978-3-319-10593-2_13
10. Galteri, L., Seidenari, L., Bertini, M., Del Bimbo, A.: Deep generative adversarial compression artifact removal. In: ICCV (2017)
11. Goodfellow, I., et al.: Generative adversarial nets. In: NeurIPS (2014)
12. Guo, J., Chao, H.: One-to-many network for visually pleasing compression artifacts reduction. In: CVPR (2017)
13. Guo, S., Yan, Z., Zhang, K., Zuo, W., Zhang, L.: Toward convolutional blind denoising of real photographs. In: CVPR (2019)
14. He, K., Gkioxari, G., Dollár, P., Girshick, R.: Mask R-CNN. In: ICCV (2017)
15. Huang, H., He, R., Sun, Z., Tan, T.: Wavelet-srnet: a wavelet-based CNN for multi-scale face super resolution. In: ICCV (2017)
16. Huang, X., Belongie, S.: Arbitrary style transfer in real-time with adaptive instance normalization. In: ICCV (2017)
17. Johnson, J., Alahi, A., Fei-Fei, L.: Perceptual losses for real-time style transfer and super-resolution. In: Leibe, B., Matas, J., Sebe, N., Welling, M. (eds.) ECCV 2016. LNCS, vol. 9906, pp. 694–711. Springer, Cham (2016). https://doi.org/10.1007/978-3-319-46475-6_43

18. Karras, T., Laine, S., Aila, T.: A style-based generator architecture for generative adversarial networks. In: CVPR (2019)
19. Kim, J., Kwon Lee, J., Mu Lee, K.: Accurate image super-resolution using very deep convolutional networks. In: CVPR (2016)
20. King, D.E.: Dlib-ml: a machine learning toolkit. J. Mach. Learn. Res. **10**, 1755–1758 (2009)
21. Kingma, D.P., Ba, J.: Adam: a method for stochastic optimization (2014). arXiv preprint arXiv:1412.6980
22. Kupyn, O., Budzan, V., Mykhailych, M., Mishkin, D., Matas, J.: Deblurgan: Blind motion deblurring using conditional adversarial networks. In: CVPR (2018)
23. Kupyn, O., Martyniuk, T., Wu, J., Wang, Z.: Deblurgan-v2: Deblurring (orders-of-magnitude) faster and better. In: ICCV (2019)
24. Ledig, C., et al.: Photo-realistic single image super-resolution using a generative adversarial network. In: CVPR (2017)
25. Levin, A., Weiss, Y., Durand, F., Freeman, W.T.: Understanding and evaluating blind deconvolution algorithms. In: CVPR (2009)
26. Li, X., Liu, M., Ye, Y., Zuo, W., Lin, L., Yang, R.: Learning warped guidance for blind face restoration. In: ECCV (2018)
27. Liu, Z., Luo, P., Wang, X., Tang, X.: Deep learning face attributes in the wild. In: ICCV (2015)
28. Miyato, T., Kataoka, T., Koyama, M., Yoshida, Y.: Spectral normalization for generative adversarial networks. In: ICLR (2018)
29. Nah, S., Hyun Kim, T., Mu Lee, K.: Deep multi-scale convolutional neural network for dynamic scene deblurring. In: CVPR (2017)
30. Pedregosa, F., et al.: Scikit-learn: machine learning in Python. J. Mach. Learn. Res. **12**, 2825–2830 (2011)
31. Ronneberger, O., Fischer, P., Brox, T.: U-net: convolutional networks for biomedical image segmentation. In: Navab, N., Hornegger, J., Wells, W.M., Frangi, A.F. (eds.) MICCAI 2015. LNCS, vol. 9351, pp. 234–241. Springer, Cham (2015). https://doi.org/10.1007/978-3-319-24574-4_28
32. Ruiz, N., Chong, E., Rehg, J.M.: Fine-grained head pose estimation without keypoints. In: CVPRW (2018)
33. Shen, Z., Lai, W.S., Xu, T., Kautz, J., Yang, M.H.: Deep semantic face deblurring. In: CVPR (2018)
34. Wang, T.C., Liu, M.Y., Zhu, J.Y., Tao, A., Kautz, J., Catanzaro, B.: High-resolution image synthesis and semantic manipulation with conditional gans. In: CVPR (2018)
35. Wang, X., Yu, K., Dong, C., Change Loy, C.: Recovering realistic texture in image super-resolution by deep spatial feature transform. In: CVPR (2018)
36. Wang, X., et al.: Esrgan: enhanced super-resolution generative adversarial networks. In: ECCVW (2018)
37. Xu, X., Sun, D., Pan, J., Zhang, Y., Pfister, H., Yang, M.H.: Learning to super-resolve blurry face and text images. In: ICCV (2017)
38. Yang, D., Sun, J.: Bm3D-net: a convolutional neural network for transform-domain collaborative filtering. IEEE Sign. Process. Lett. **25**(1), 55–59 (2017)
39. Yu, X., Fernando, B., Ghanem, B., Porikli, F., Hartley, R.: Face super-resolution guided by facial component heatmaps. In: ECCV (2018)
40. Yu, X., Fernando, B., Hartley, R., Porikli, F.: Super-resolving very low-resolution face images with supplementary attributes. In: CVPR (2018)
41. Zhang, H., Dai, Y., Li, H., Koniusz, P.: Deep stacked hierarchical multi-patch network for image deblurring. In: CVPR (2019)

42. Zhang, K., Zuo, W., Chen, Y., Meng, D., Zhang, L.: Beyond a Gaussian denoiser: residual learning of deep CNN for image denoising. IEEE Trans. Image Process. **26**(7), 3142–3155 (2017)
43. Zhang, K., Zuo, W., Zhang, L.: Ffdnet: toward a fast and flexible solution for CNN-based image denoising. IEEE Trans. Image Process. **27**(9), 4608–4622 (2018)
44. Zhang, K., Zuo, W., Zhang, L.: Deep plug-and-play super-resolution for arbitrary blur kernels. In: CVPR (2019)
45. Zhang, R., Isola, P., Efros, A.A., Shechtman, E., Wang, O.: The unreasonable effectiveness of deep features as a perceptual metric. In: CVPR (2018)
46. Zhang, Y., Li, K., Li, K., Wang, L., Zhong, B., Fu, Y.: Image super-resolution using very deep residual channel attention networks. In: ECCV (2018)
47. Zhang, Z., Wang, Z., Lin, Z., Qi, H.: Image super-resolution by neural texture transfer. In: CVPR (2019)
48. Zhu, S., Liu, S., Loy, C.C., Tang, X.: Deep cascaded bi-network for face hallucination. In: Leibe, B., Matas, J., Sebe, N., Welling, M. (eds.) ECCV 2016. LNCS, vol. 9909, pp. 614–630. Springer, Cham (2016). https://doi.org/10.1007/978-3-319-46454-1_37

Robust Neural Networks Inspired by Strong Stability Preserving Runge-Kutta Methods

Byungjoo Kim[1], Bryce Chudomelka[2], Jinyoung Park[1], Jaewoo Kang[1](✉), Youngjoon Hong[2], and Hyunwoo J. Kim[1](✉)

[1] Department of Computer Science, Korea University, Seoul, Republic of Korea
{byung4329,lpmn678,kangj,hyunwoojkim}@korea.ac.kr

[2] Department of Mathematics and Statistics, San Diego State University, San Diego, CA, USA
{bchudomelka,yhong2}@sdsu.edu

Abstract. Deep neural networks have achieved state-of-the-art performance in a variety of fields. Recent works observe that a class of widely used neural networks can be viewed as the Euler method of numerical discretization. From the numerical discretization perspective, Strong Stability Preserving (SSP) methods are more advanced techniques than the explicit Euler method that produce both accurate and stable solutions. Motivated by the SSP property and a generalized Runge-Kutta method, we proposed Strong Stability Preserving networks (SSP networks) which improve robustness against adversarial attacks. We empirically demonstrate that the proposed networks improve the robustness against adversarial examples without any defensive methods. Further, the SSP networks are complementary with a state-of-the-art adversarial training scheme. Lastly, our experiments show that SSP networks suppress the blow-up of adversarial perturbations. Our results open up a way to study robust architectures of neural networks leveraging rich knowledge from numerical discretization literature.

1 Introduction

Recent progress in deep learning has shown promising results in various research areas, such as computer vision, natural language processing and recommendation systems. In particular, on the ImageNet classification task [18], deep neural networks show state-of-the-art performance, e.g., residual networks (ResNet), which outperform humans in image classification [15]. Despite the success, deep neural networks often suffer from the lack of robustness against adversarial attacks [30]. ResNet, which is a widely used base network, also suffers from adversarial

B. Kim and B. Chudomelka—Equal Contribution.

Electronic supplementary material The online version of this chapter (https://doi.org/10.1007/978-3-030-58545-7_24) contains supplementary material, which is available to authorized users.

© Springer Nature Switzerland AG 2020
A. Vedaldi et al. (Eds.): ECCV 2020, LNCS 12354, pp. 416–432, 2020.
https://doi.org/10.1007/978-3-030-58545-7_24

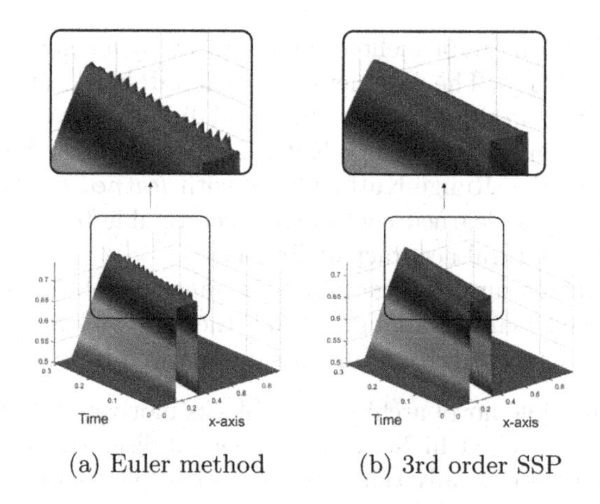

(a) Euler method (b) 3rd order SSP

Fig. 1. We illustrate the difference between a forward Euler discretization and a third-order SSP discretization applied to the inviscid Burgers' solution. After computing numerical solutions, the solutions are filtered through the sigmoid function as an activation function. Evidently, in (a) the Euler scheme, i.e., a *ResBlock*, produces notable numerical errors while the SSP3 discretization in (b) shows a stable numerical approximation. For more details, please see the supplement.

attacks which necessitates a more fundamental understanding of the architecture at hand.

One interesting interpretation of the ResNet architecture is that of the explicit Euler discretization scheme, i.e., $x(t_{k+1}) = x(t_k) + F(x(t_k))$, because it allows one to view neural networks as numerical methods. The explicit Euler method is one of the simplest first-order numerical schemes but often leads to large numerical errors due to its low order. Thus, we would expect that applying advanced numerical discretizations would produce a more accurate numerical solution than the Euler method, such as an explicit high-order Runge-Kutta method. However, an arbitrary explicit high-order Runge-Kutta method can pose a stability problem if the numerical solution becomes unstable [27]. To tackle this issue, [9] and [27] introduce the notion of total variation diminishing (TVD); also called Strong Stability Preserving (SSP) methods. The strong stability preserving approach produces a more accurate solution of the differential equation than the Euler method. We would expect to obtain a more accurate solution of the underlying function with non-smooth initial data (shocks) compared to the Euler method without notable numerical errors, see Fig. 1. This phenomenon is directly related to the problem of adversarial attack and robustness of neural networks [30].

Motivated by the advanced numerical discretization schemes, we propose novel network architectures with the SSP property that address robustness; SSP networks (SSPNets). The use of the SSP property consistently demonstrates that all of our proposed architectures outperform ResNet in terms of robustness. SSP architectural blocks do not increase the amount of model parameters compared to ResNet, can be easily implemented, and realized by a convex combination of existing ResNet modules. The parameters used in SSP blocks are *mathematically derived coefficients* from the advanced numerical discretization methods. In addition, starting from an explicit Runge-Kutta method with the SSP property, we propose novel Adaptive Runge-Kutta blocks with *learned coefficients* obtained by training. With these learned coefficients, we are able to improve robustness while retaining the natural accuracy of ResNet.

The simple architectural change, SSPNets, improve robustness and are complementary with adversarial training, which is the *de facto* state-of-the-art defensive methodology. Our **contributions** are summarized as follows:

- We propose multiple novel architectural blocks motivated by the Strong Stability Preserving explicit higher-order numerical discretization method.
- We demonstrate empirically that these proposed blocks improve the robustness of Strong Stability Preserving networks consistently; against adversarial examples and without any defensive methods.
- We further improve on robustness with a novel adaptive architectural block motivated by a generalized Runge-Kutta method and the SSP property.
- Last but not least, we show that Strong Stability Preserving Networks suppress the blow-up of adversarial perturbations added to inputs.

2 Background and Related Work

2.1 Neural Networks and Differential Equations

Neural networks such as ResNet [15], PolyNet [40] and recurrent neural networks share a common operation represented as $x_{t+1} = x_t + F(x_t; \Theta_t)$. Interestingly, a sequence of the operations (or equivalently the network architectures) can be interpreted as an explicit Euler method for numerical discretization [6,7,20,24, 25]. For instance, ResNet can be written mathematically as

$$\begin{aligned} x_0 &= x, \\ x_{k+1} &= x_k + F(x_k; \Theta_k), \quad k \in \{0, 1, \dots, A-1\}, \\ \hat{y} &= f(x_A), \end{aligned} \tag{1}$$

where A denotes the number of layers in the network.

If we multiply the function F by Δt, i.e., $x_{k+1} = x_k + \Delta t F(x_k; \Theta_k)$, then ResNet can be seen as the explicit Euler numerical scheme discretization with an initial condition, $x(0)$, to solve the initial value problem given as

$$\begin{aligned} x(0) &= x, \\ \frac{dx(t)}{dt} &= F(x(t); \Theta(t)), \\ \hat{y} &= f(x(A)). \end{aligned} \tag{2}$$

The explicit Euler method is the simplest Runge-Kutta method and often suffers from low accuracy because it is a first-order method. In this regard, higher-order numerical methods are natural candidates to obtain a more precise numerical solution, but the higher accuracy from higher-order methods may come with the cost of instability, e.g., poor convergence behaviour on stiff differential equations compared to the first order Euler method [4]. Therefore, it is important to understand the trade-off between accuracy and stability when considering a numerical method.

Recently, some network architectures inspired by the computational similarity between ResNet and Euler discretization have been proposed, e.g., NeuralODE and FFJORD [6,11]. Unlike ResNet, which requires the discretization of observation/emission intervals to be represented by a finite number of hidden layers, NeuralODE and FFJORD use numerical discretization methods in the forward propagation to define continuous-depth and continuous-time latent variable models. These require ODE solvers for training and inference, unlike our implementation of SSP networks. Since we changed only computational graphs and coefficients based on the numerical discretization theory, our methods perform the standard forward/backward propagation in the discrete space as ResNet.

Another approach to design new blocks/layers of neural networks is to make them have operations similar to advanced numerical discretization techniques that possess desirable properties [20,25]. From the partial differential equation perspective, analysis on numerical stability of conventional residual connections lead to the development of new architectures: parabolic/hyperbolic CNNs to achieve better stability as parabolic/hyperbolic PDEs [25]. The models use theoretical assumptions on the function to achieve stability with a positive semidefinite Jacobian of the function resulting in constraints on convolutional kernels; alternatively, our networks do not require such constraints.

2.2 Robust Machine Learning and Adversarial Attacks

Stability and robustness of neural networks have been studied in the context of adversarial attacks after the success of deep learning [2,30,36]. Gradient-based adversarial attacks create adversarial examples solving optimization problems. One example is the maximization of loss against ground truth labels within a small ball, e.g., $\max_\delta \mathcal{L}(h_\theta(x+\delta), y)$, *s.t.* $\|\delta\|_\infty \leq \epsilon$, where h_θ is a model parameterized by θ, x, y are the input (natural sample) and its target label respectively, and $\mathcal{L}$ is a loss function. The simplest procedure to approximate the solution is to use the fast gradient sign method (FGSM) [8]. It can be seen as an optimal solution to a linearized loss function, i.e., $\arg\max_{\|v\|_\infty \leq \alpha} v^T \nabla_\delta \mathcal{L}(h_\theta(x+\delta), y) = \alpha \cdot \text{sign}(\nabla_\delta \mathcal{L}(h_\theta(x+\delta), y))$. Furthermore, the FGSM can be more powerful when it is used with iterative methods such as the projected gradient descent (PGD). PGD has been used in both untargeted and targeted attacks [5,21].

One of the early attempts to defend against adversarial attacks is adversarial training using FGSM, a single-step method [8]. After that, various defensive techniques have been proposed [3,22,26,29,31]. Many of them were defeated by iterative attack methods [5] and Backward Pass Differentiable Approximation [1]. Adversarial training with stronger multi-step attack methods is still promising and shows state-of-the-art performance [21,32,35]. More recently, *provably* robust neural networks have been successfully trained by minimizing the lower bound of risk based on convex duality and convex relaxation [33,34]. Most adversarial training methods above assume that attack methods are known *a priori*, i.e., a *white-box* attack, and generate augmented samples using the attacks. Another defensive technique is to alleviate the effect of perturbation by augmentation and reconstruction [23,37], or denoising [35]. These methods alongside adversarial training achieved comparable robustness. Similarly, in this work we will introduce our approach and evaluate it with adversarial training.

3 Strong Stability Preserving Networks

In this section, we introduce the mathematical framework for the Strong Stability Preserving property and describe how to implement SSP blocks with *mathematically derived coefficients*. Next, we provide a variance analysis to compare high-order Runge-Kutta blocks with residual blocks. Lastly, we introduce adaptive Runge-Kutta blocks with *learnable coefficients* which possess the SSP property.

3.1 Motivation of Strong Stability Preserving Method

Our objective is to solve the *non-autonomous* differential equation given as

$$\frac{\partial u}{\partial t} = L(u(t), t), \quad t \in [t_0, ..., t_N], \tag{3}$$

where t_0, t_N are the initial and terminal time state respectively; a non-autonomous system permits a time varying solution, e.g., the learned function varies as the depth of the network increases. The function L is a linear (or non-linear) function and $u(t_0)$ is given by the initial condition. The objective is to figure out the terminal state of the function u, i.e., $u(t_N)$.

A general high-order Runge-Kutta time discretization for solving the initial value problem (3) introduced in [28] is given as

$$\begin{aligned} u^{(0)} &= u^n, \\ u^{(i)} &= \sum_{k=0}^{i-1} \left(\alpha_{i,k} u^{(k)} + \Delta t \beta_{i,k} L(u^{(k)}) \right), \quad i \in \{1, \cdots, m\}, \\ u^{n+1} &= u^{(m)}, \end{aligned} \tag{4}$$

where $\sum_{k=0}^{i-1} \alpha_{i,k} = 1$ and $\alpha_{i,k} \geq 0$. For example, if $m = 1$, it becomes the first-order Euler method as in Eq. (1) with $\alpha_{1,0} = \beta_{1,0} = 1$.

Shu et al. [27,28] propose a TVD time discretization method that is called the SSP time discretization method; for more discussion on the TVD method, we refer the reader to [12,13]. The procedure of TVD time discretization is to take the high-order method to decrease the local truncation error and maintain the stability under a suitable restriction on the time step. While applying the TVD scheme into the explicit high-order Runge-Kutta methods, there needs the assumption to hold it: *The first-order Euler method in time is strongly stable under a certain (semi) norm when the time step Δt is suitably restricted* [10]. More precisely, if we assume that the forward Euler time discretization is stable under a certain norm, the SSP methods find a higher-order time discretization that maintains strong stability for the same norm; improving accuracy.

Followed by this assumption, for a sufficiently small time step known as Courant-Friedrichs-Lewy (CFL) condition $\Delta t \leq \Delta t_{CFL}$, the total variation semi-norm of the numerical scheme does not increase in time, that is,

$$TV(u^{n+1}) \leq TV(u^n), \tag{5}$$

where the total variation is defined by

$$TV(u^n) := \sum_j |u_{j+1}^n - u_j^n|, \tag{6}$$

where j is the spatial discretization. The explicit high-order Runge-Kutta discretization with the SSP property maintains a higher order accuracy with a modified CFL condition $\Delta t \leq c\Delta t_{CFL}$. In other words, the high-order SSP Runge-Kutta scheme improves accuracy while retaining its stability. This has been theoretically studied by the following Lemma 1.

Lemma 1. *If the forward Euler method is strongly stable under the CFL condition, i.e.* $||u^n + \Delta t L(u^n)|| \leq ||u^n||$, *then the Runge-Kutta method possesses SSP,* $||u^{n+1}|| \leq ||u^n||$, *provided that* $\Delta t \leq c\Delta t_{CFL}$.

We provide a sketch of the proof of Lemma 1 in the supplement. The full proof of the Lemma 1 can be found in [28]. Following this representation, we can figure out the specific coefficients $\alpha_{i,k}$ and $\beta_{i,k}$ in equation (4). In particular, the second and third order nonlinear SSP Runge-Kutta method was studied in [28].

Lemma 2. *An optimal second-order SSP Runge-Kutta method is given by,*

$$\begin{aligned} u^{(1)} &= u^n + \Delta t L(u^n), \\ u^{n+1} &= \frac{1}{2}u^n + \frac{1}{2}u^{(1)} + \frac{1}{2}\Delta t L(u^{(1)}), \end{aligned} \tag{7}$$

with a CFL coefficient $c = 1$. *In addition, an optimal third-order SSP Runge-Kutta method is of the form*

$$\begin{aligned} u^{(1)} &= u^n + \Delta t L(u^n), \\ u^{(2)} &= \frac{3}{4}u^n + \frac{1}{4}u^{(1)} + \frac{1}{4}\Delta t L(u^{(1)}), \\ u^{n+1} &= \frac{1}{3}u^n + \frac{2}{3}u^{(2)} + \frac{2}{3}\Delta t L(u^{(2)}), \end{aligned} \tag{8}$$

with a CFL coefficient $c = 1$.

A sketch of the proof for Lemma 2 can be found in the supplement and for the detailed proof, we refer the reader to [9,10,28].

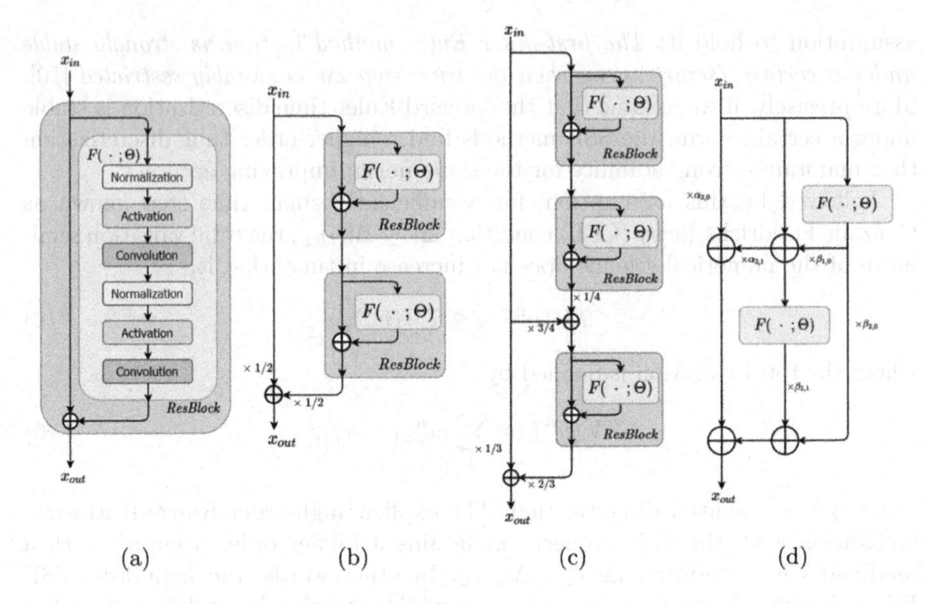

Fig. 2. Network modules with *ResBlock* and SSP blocks. (a): *ResBlock*. (b): *SSP2-block* (c): *SSP3-block*, (d): *ArkBlock*

3.2 Strong Stability Preserving Networks

Next, we show how to incorporate the explicit SSP Runge-Kutta method into neural networks. Equation (7) and (8) can be implemented with standard residual blocks and simple operations, as shown in Fig. 2.

Let *ResBlock* denote a standard residual block written as $ResBlock(x(t_k);\Theta(t_k)) = x(t_k) + F(x(t_k);\Theta(t_k))$, where $\Theta(t_k)$ are the parameters of $ResBlock(\cdot;\Theta(t_k))$. The function F is typically composed of two or three sets of normalization, activation, and convolutional layers, e.g., Fig. 2 and [15,16]. When the numbers of input and output channels differ, we use the expansive residual block *ResBlock-E*; this can be implemented with a 1×1 convolutional filter to expand the number of channels.

Using the standard modules in ResNet (*ResBlock* and *ResBlock-E*), SSPNets can be constructed. First, SSP blocks can be implemented using linear combinations of *ResBlocks*. As the Euler method interpretation of ResNet requires $\Delta t = 1$, we assume $\Delta t = 1$ in Eq. (7), then the *SSP2-block* is given by,

$$\begin{aligned} x(t_{k+\frac{1}{2}}) &= \underbrace{x(t_k) + F(x(t_k);\Theta(t_k))}_{ResBlock(x(t_k);\Theta(t_k))}, \\ x(t_{k+1}) &= \frac{1}{2}x(t_k) + \underbrace{\frac{1}{2}x(t_{k+\frac{1}{2}}) + \frac{1}{2}F\left(x(t_{k+\frac{1}{2}});\Theta(t_k)\right)}_{\frac{1}{2}ResBlock\left(x(t_{k+1/2});\Theta(t_k)\right)}. \end{aligned} \tag{9}$$

Similarly, the third order SSP in Eq. (8) (*SSP3-block*) is written as

$$\begin{aligned} x(t_{k+\frac{1}{3}}) &= \underbrace{x(t_k) + F\left(x(t_k); \Theta(t_k)\right)}_{ResBlock(x(t_k);\Theta(t_k))}, \\ x(t_{k+\frac{2}{3}}) &= \frac{3}{4}x(t_k) + \underbrace{\frac{1}{4}x(t_{k+\frac{1}{3}}) + \frac{1}{4}F\left(x(t_{k+\frac{1}{3}}); \Theta(t_k)\right)}_{\frac{1}{4}ResBlock\left(x\left(t_{k+1/3}\right);\Theta(t_k)\right)}, \\ x(t_{k+1}) &= \frac{1}{3}x(t_k) + \underbrace{\frac{2}{3}x(t_{k+\frac{2}{3}}) + \frac{2}{3}F\left(x(t_{k+\frac{2}{3}}); \Theta(t_k)\right)}_{\frac{2}{3}ResBlock\left(x\left(t_{k+2/3}\right);\Theta(t_k)\right)}. \end{aligned} \tag{10}$$

The SSP block schematic is presented in Fig. 2 and SSP blocks are only used when the number of channels does not change.

The explicit SSP Runge-Kutta methods in Eq. (7) and (8) use the same function L multiple times. Similarly, SSP blocks in Eq. (9) and (10) apply the same *ResBlock* multiple times. Using the same *ResBlock* multiple times can be viewed as parameter sharing, which is a kind of regularization. In other words, without increasing the number of parameters, a SSP block implementation improves the robustness of neural networks by utilizing higher-order schemes.

Midpoint Runge-Kutta Second-Order Methods. For contrast, one may ask whether or not the stability preserving properties are key to the robustness against adversarial perturbation. We address this important question by training another network that utilizes a second-order midpoint Runge-Kutta method (mid-RK2) which does not have the strong stability preserving property [4,10]. Recall that this method is implemented numerically as

$$x(t_{k+1}) = x(t_k) + F\left(x(t_k) + \frac{1}{2}F(x(t_k); \Theta(t_k)); \Theta(t_k)\right), \tag{11}$$

and does not have the SSP property. This network will provide a comparison of numerical discretization methods with regard to stability in attacked accuracy.

Variance Analysis of SSP Networks. We analyze the variance increase of SSP blocks following previous works [14,39], which compare the variance of input and output of functional modules. Next, we show that SSP blocks suppress the variance increase compared to *ResBlock*; as well as comparing the variance of the midpoint Runge-Kutta second-order numerical method for further justification.

Lemma 3. *If* $\mathit{Var}[F(x)] = \mathit{Var}[x]$, $\mathit{Cov}[x, F(y)] = 0$ *then the variance increases by*

$$\begin{aligned} \mathit{Var}[ResBlock(x)] &= 2\,\mathit{Var}[x], & \mathit{Var}[\textit{mid-RK2}(x)] &= \frac{9}{4}\mathit{Var}[x], \\ \mathit{Var}[\textit{SSP2-Block}(x)] &= \frac{7}{4}\mathit{Var}[x], & \mathit{Var}[\textit{SSP3-Block}(x)] &= \frac{29}{18}\mathit{Var}[x]. \end{aligned} \tag{12}$$

The variance of SSP blocks is smaller than that of *ResBlock*. The variance adds to our argument that the SSP property is the reason for improved robustness; for more detailed derivation and proof, see the supplement.

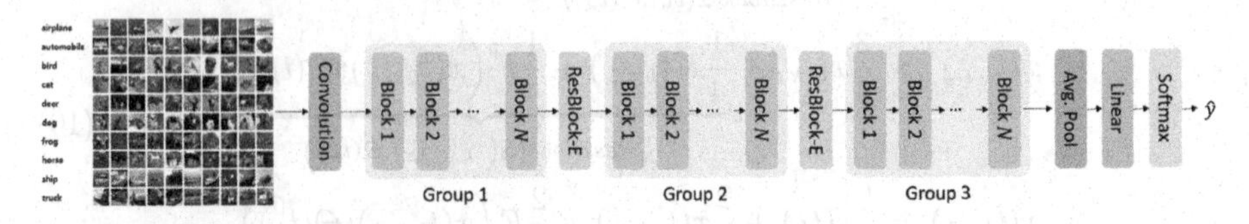

Fig. 3. The overall architecture of neural networks used in experiments. Each group has $N \in \{6, 10\}$ blocks and the block is either *ResBlock*, SSP blocks (2 or 3) or *ArkBlock*. The *ResBlock-E* is inserted between groups to expand the number of channels for all the architectures.

Adaptive SSP Networks. Also, we generalize Eq. (4) with the second-order Adaptive Runge-Kutta block (*ArkBlock*) that has the SSP property by construction. These novel computational blocks slightly increase the number of parameters compared to *ResBlock* but also provide greater robustness and natural accuracy than *SSP2-Block* or *SSP3-Block*. Finally, we explore different computational architectures within each group to retain natural accuracy and further improve robustness.

A naive implementation of Eq.(4) yields 5 additional parameters. We can retain the SSP property in *ArkBlocks* by reducing the number of parameters with Ralston's method [9]. Thus, the number of additional learned parameters per block, when compared with *ResBlock*, is 2 and is defined as

$$\begin{aligned} &\alpha_{1,0} = 1, \qquad \alpha_{2,0} = 1 - \alpha_{2,1}, \\ &\beta_{2,0} = 1 - \frac{1}{2\beta_{1,0}} - \alpha_{2,1}\beta_{1,0}, \qquad \beta_{2,1} = \frac{1}{2\beta_{1,0}}. \end{aligned} \tag{13}$$

We further improve performance by reducing the number of parameters by fixing $\alpha_{2,1}$ and simply learning $\beta_{1,0}$ in each block.

Adaptive SSP networks still maintain the same architecture, as in Fig. 3, but are comprised of blocks that have the form

$$\begin{aligned} u^{(1)} &= u^n + \beta_{1,0} L(u^n), \\ u^{(n+1)} &= \alpha_{2,0} u^n + \beta_{2,0} L(u^n) + \alpha_{2,1} u^{(1)} + \beta_{2,1} L(u^{(1)}). \end{aligned} \tag{14}$$

We implement *ArkBlock*s with,

$$\begin{aligned} x(t_{k+\frac{1}{2}}) &= x(t_k) + \beta_{1,0} F\left(x(t_k); \Theta(t_k)\right), \\ x(t_{k+1}) &= \alpha_{2,0} x(t_k) + \beta_{2,0} F\left(x(t_k); \Theta(t_k)\right) \\ &\quad + \alpha_{2,1} x(t_{k+\frac{1}{2}}) + \beta_{2,1} F\left(x(t_{k+\frac{1}{2}}); \Theta(t_k)\right). \end{aligned} \tag{15}$$

The *ArkBlocks* are inspired by the generalized Runge-Kutta method in (14). However, the numerical scheme in Eq. (14), keeps $\alpha_{2,1}$ and $\beta_{1,0}$ constant in all blocks, while *ArkBlocks* set those parameters as *learnable*; varying in each block. Such an adaptivity based on data and architectures cannot be obtained by mathematically derived coefficients. To our knowledge, this is the first attempt.

Table 1. The accuracy against adversarial attacks with standard training on the MNIST dataset; all models were trained with 6 blocks. Note that PGD_i represents a projected gradient descent attack with i iterations and that all the SSPNets are more robust against adversarial attacks than ResNet.

Model	Clean	FGSM	PGD_{20}	PGD_{30}
ResNet	0.9961	0.7674	0.5799	0.1773
SSP-2	0.9954	0.7984	0.5979	0.1850
SSP-3	0.9960	0.8022	0.6176	0.1930
SSP-adap	0.9946	**0.8586**	**0.7611**	**0.5102**

4 Experiments

We evaluate the robustness of various SSP networks against adversarial examples. MNIST [19] and CIFAR10 [18] are used for evaluation; for results on other datasets, see the supplement. The robustness is measured by the classification accuracy on adversarial examples generated by FGSM [8] and PGD [21].

In this section, we empirically address the following three questions:

- Are deep neural networks with the SSP property more robust than ResNet when the models are trained with or without adversarial training?
- Can we further improve upon adversarial robustness and simultaneously retain the natural accuracy of ResNet?
- Do Strong Stability Preserving networks suppress the perturbation growth during forward propagation?

4.1 Experimental Setup

ResNet and SSP Networks. Each group has N blocks where each block can be either *ResBlock*, *SSP2-block*, *SSP3-block*, or *ArkBlock*, as seen in Fig. 3. Networks are named after the type of blocks: ResNet, SSP-2, SSP-3, and SSP-adap. The blocks in each group have the same number of input/output channels. The convolutional layers in group 1, group 2, and group 3 have 16, 32, 64 channels respectively. The classification layer of our networks consist of an average pooling and softmax layer, in order to calculate the confidence score.

4.2 Evaluation on MNIST with Standard Training

We demonstrate that SSPNets are more robust than ResNet with standard training. Since MNIST has relatively low-resolution images compared to CIFAR10, we used a smaller architecture by skipping group 1 and 2 in Fig. 3.

Experimental Details. We evaluate the models on MNIST. When training the models, samples are augmented by adding random noise δ drawn from a uniform distribution $Uniform(-\epsilon, \epsilon)$. We set the maximum perturbation magnitude $\epsilon = 0.3$ for both training and evaluation. For optimization, Adam [17] is used with learning rate 0.0001 and $(\beta_1, \beta_2) = (0.9, 0.999)$, minibatch size of 128. Models are trained for 100 epochs.

Robustness Comparison. The results in Table 1 show that all four models have high accuracy ($99.5 \sim 99.6\%$) in classifying clean samples. This means that SSP blocks do not lead to a significant loss of accuracy on clean samples. Further, the improvement by SSP compared to ResNet is consistently observed in different settings. SSP-2 improves the robustness by 3% against FGSM and 1% against PGD. SSP-3 shows larger improvement about 4% and 2% against FGSM and PGD. SSP-adap shows the largest improvement about 9% and 33% against FGSM and PGD. It is known that adversarial training on MNIST is sufficiently robust against FGSM and PGD. All models trained by adversarial training achieve $96 \sim 97\%$ on MNIST, which makes it hard to demonstrate the benefit of SSP networks with adversarial training compared to ResNet.

4.3 SSP with Adversarial Training

We analyze the robustness of SSP networks, on the CIFAR10 dataset. Our preliminary experiments show that all the models, e.g., ResNet, SSP-2, SSP-3, and SSP-adap trained without adversarial training are easily fooled by PGD attacks, but more analysis is needed on a more challenging dataset. For this reason, we focus on the adversarial training setting for CIFAR10. Please see supplementary materials for more analysis on SSP networks with adversarial training.

Adversarial Training. Before experimental results, we briefly summarize the adversarial training proposed by [21]. The objective of adversarial training is to minimize the adversarial risk given as,

$$R_{adv}(h_\theta) = \mathbb{E}_{(x,y)\sim D}\big[\max_{\delta \in \Delta} \mathcal{L}(h_\theta(x+\delta), y)\big], \tag{16}$$

where the h_θ is a model parameterized by θ, $\mathcal{L}$ is a loss function, y is the label of corresponding image x, D is a true data distribution, and Δ is a set of small perturbations satisfying $\|\delta\|_p \leq \epsilon$. In our experiments, the ℓ_∞ metric is used, i.e., $p = \infty$. Finding the exact solution to $\max_{\delta \in \Delta} \mathcal{L}(h_\theta(x+\delta), y)$ is intractable, so [21] approximate it with a sample generated by the PGD attack. PGD attack finds the adversarial example given as $x_{i+1} = \Pi(x_i + \alpha \nabla_{x_i} \mathcal{L}(h_\theta(x_i), y))$, where $i \in \{0, 1, \cdots, K-1\}$, K is the number of iterations of PGD attack, Π denotes the projection to a small ball Δ and a valid pixel range. In our experiment, x_0 is initialized with the input image augmented by adding the random perturbation δ_0 sampled from the uniform distribution $Uniform(-\epsilon, \epsilon)$.

To summarize, our adversarial training procedure works as follows: First, randomly perturb the image within the allowed perturbation range ϵ. Next, generate the candidate adversarial example by PGD attack. Finally, take the gradient descent step on a minibatch composed of only candidate adversarial examples. The adversarial training is closely related to the Frank-Wolfe Algorithm and two projections in the original adversarial training can be simplified to one projection to the intersection of two convex sets. The pseudocode of adversarial training and a detailed discussion of implementation are provided in the supplement.

Table 2. CIFAR10 robustness evaluation against adversarial attacks. The column index N indicates the number of blocks in each group of Fig. 3, K indicates the number of PGD iterations during training while PGD_i represents the attack with i iterations during attack. The SSP-adap model indicates an adaptive Runge-Kutta structure. All the SSP networks are more robust against adversarial attack than ResNet. Moreover, SSP-adap maintains the natural accuracy.

N	K	Model	Clean	FGSM	PGD_7	PGD_{12}	PGD_{20}
6	7	ResNet	0.8357	0.5116	0.4389	0.4215	0.4150
6	7	mid-RK2	**0.8407**	0.5156	0.4377	0.4193	0.4129
6	7	SSP-2	0.8257	0.5223	0.4577	0.4426	0.4368
6	7	SSP-3	0.8376	0.5165	0.4478	0.4305	0.4246
6	7	SSP-adap	0.8376	**0.5283**	**0.4640**	**0.4455**	**0.4403**
6	12	ResNet	**0.8010**	0.5304	0.4817	0.4691	0.4650
6	12	mid-RK2	0.7957	0.5326	0.4849	0.4740	0.4693
6	12	SSP-2	0.7899	0.5426	0.5073	0.4983	0.4961
6	12	SSP-3	0.7966	0.5440	**0.5092**	**0.4999**	**0.4976**
6	12	SSP-adap	0.7988	**0.5504**	0.5066	0.4964	0.4943
10	7	ResNet	**0.8516**	0.5225	0.4398	0.4188	0.4111
10	7	mid-RK2	0.8451	0.5146	0.4343	0.4122	0.4045
10	7	SSP-2	0.8437	**0.5373**	0.4714	0.4502	0.4427
10	7	SSP-3	0.8505	0.5350	**0.4719**	**0.4558**	**0.4497**
10	7	SSP-adap	0.8504	0.5308	0.4592	0.4376	0.4310
10	12	ResNet	0.8181	0.5467	0.4957	0.4799	0.4755
10	12	mid-RK2	**0.8198**	0.5522	0.4968	0.4818	0.4775
10	12	SSP-2	0.8144	0.5497	0.5074	0.4957	0.4932
10	12	SSP-3	0.8119	0.5507	0.5032	0.4929	0.4890
10	12	SSP-adap	0.8156	**0.5643**	**0.5166**	**0.5054**	**0.5016**

Experimental Details. We use the Stochastic Gradient Descent method with Nesterov momentum, learning rate of 0.1, weight decay of 0.0005, momentum 0.9, and a minibatch size of 128 samples. All models are trained for 200 epochs and in every 60, 100, 140 epochs, the learning rate decayed with a decaying factor 0.1.

Both adversarial training and robustness evaluation, we set the maximum perturbation range $\epsilon = 8/255$. To evaluate the robustness, we use FGSM [8] and PGD [21]; similar to our MNIST experiments. We set the PGD attack parameters to $\alpha = 2/255$, and the number of iterations $K = 7, 12, 20$ in evaluation.

Robustness Comparison. The experimental results are shown in Table 2. Models are evaluated in four different settings varying both the number of blocks (6 or 10 in column N) for each group in Fig. 3 and the number of iterations in PGD (7 or 12 in column K) to generate adversarial examples during training.

Before discussing about the effectiveness of SSP, we briefly show the relationship among robustness, the amount of model parameters, and the strength of attacks used in adversarial training. As shown in Table 2, the robustness of all the models is improved by stronger attacks during training (e.g., larger K in PGD). The same observation is reported in [32]. For instance, SSP-3 (N = 6, K = 12) shows higher accuracy than SSP-3 (N = 6, K = 7) against all the attacks and especially the improvement is about 7% against PGD with 20 iterations. Also, a bigger model size (e.g., larger N) increases the robustness against adversarial examples. This is closely related to the finding in [21] that increasing the number of channels in hidden layers often improves the robustness. Our experiments show that increasing the model size by adding more layers improves the robustness. For example, when $K = 7$, SSP-3 with $N = 10$ blocks show overall higher accuracy than SSP-3 with $N = 6$ blocks, the gain is about 2%. From the numerical discretization perspective, more blocks can be seen as a finer time discretization that leads to a more accurate numerical solution (or prediction).

All SSPNets, SSP-2, SSP-3 and SSP-adap, consistently outperform ResNet by, roughly, $1 \sim 3.9\%$ when ResNet and SSPNets have the same number of blocks, N, and iterations, K, in adversarial training. Note that we compare SSP-2 (and SSP-3) with ResNet, which has the same amount of parameters, and this is important to assure that the gain is not from an increased the amount of model parameters. Also, SSP-2, SSP-3 and SSP-adap have the same time discretization as ResNet. So, we conclude that the improvement in robustness against adversarial attacks solely comes from the strength of a higher-order numerical discretization. Table 2 shows one more interesting property of SSP networks. Unlike adversarial training and defensive methods that usually cause the label leaking effects [31,38], SSP-2, SSP-3 and SSP-adap (our architectural changes) do not bring any additional loss of accuracy on natural samples.

On the other hand, Table 2 also shows that the mid-RK2 architecture does not outperform ResNet, SSP-2, SSP-3 or SSP-adap even though the mid-RK2 is derived from the second order numerical scheme. This gives credence to the implementation of SSPNets and implies that the robust performance is not a result of arbitrary high-order methods. In addition, SSP-adap achieves comparable natural accuracy as ResNet and improves robustness. Table 2 demonstrates the consistent improvement across various settings. For example, SSP-adap achieves nearly 4% absolute performance improvement for $N = 10, K = 7$. The improvement by SSP networks compared to ResNet and the performance difference between different SSP networks are relatively smaller than Table 1.

Our conjecture is that this is due to the improvement by adversarial training. We believe that the Strong Stability Preserving property imposed by our architectural change allows the SSPNets to improve the robustness against adversarial attacks.

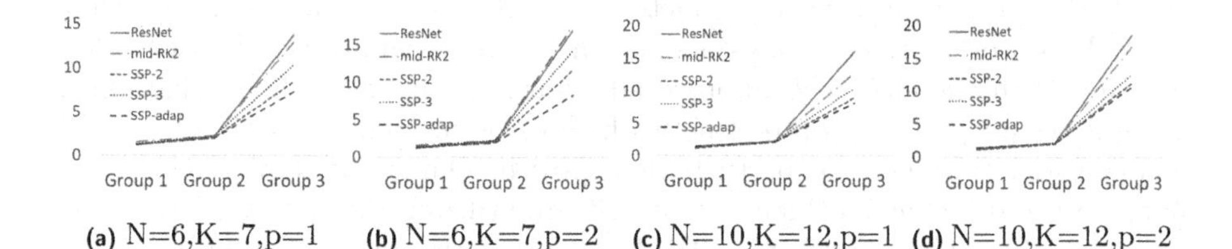

(a) N=6,K=7,p=1 (b) N=6,K=7,p=2 (c) N=10,K=12,p=1 (d) N=10,K=12,p=2

Fig. 4. Perturbation growth ratio in Eq. (17) of clean samples and its adversarial counterparts. As the SSP networks suppress the perturbation growth during forward propagation, SSP-2, SSP-3 and SSP-adap have a lower ratio than ResNet. For full version of this figure, see the supplement.

Perturbation Growth Ratio Comparison. We investigate how the distance between clean samples and adversarial examples evolves through networks by calculating the perturbation growth ratio between input/output of groups given by

$$\texttt{PGR}(f) = \mathbb{E}_{x \sim \mathcal{D}} \left[\mathbb{E}_{x' \sim \mathcal{X}'} \left[\frac{\|f(x) - f(x')\|_p}{\|x - x'\|_p} \right] \right], \quad p \in \{1, 2\} \tag{17}$$

where $f(\cdot)$ is a function of a group, x' is a corrupted sample from x and is an element of the set $\mathcal{X}'$, $\mathcal{X}'$ is a small neighborhood of x, and p defines a type of norm either ℓ_1 (related to TV in Eq. (5)) or ℓ_2 (related to Lemma 3). Since each model has a different scale of feature maps, to compare, the distance needs proper normalization. So, we first measure the distance between a clean sample and its adversarial example before/after each group in Fig. 3. x' is the adversarial example generated by PGD attack with 20 iterations for each model.

Figure 4 presents the perturbation growth ratio when $N = 6, K = 7$ and $N = 10, K = 12$ at each group in the models. Since the adversarial examples change the final predictions, the perturbation growth ratio increases in all the models. However, for SSPNets, the perturbation growth ratio is significantly lower than ResNet. This result supports that the proposed SSP blocks improve robustness of networks against adversarial attacks when compared to ResNet. We also conducted an experiment when x' is corrupted by adding a random perturbation to x and the result is consistent with Fig. 4. For full version of Fig. 4 and more discussion, see the supplement.

5 Conclusion

In this work, we leverage the Strong Stability Preserving property of numerical discretization in order to improve adversarial robustness. Inspired by the Strong Stability Preserving methods, we design a series of SSPNets by applying the same *ResBlock* multiple times with parameters derived from numerical analysis. All of the SSP networks provide robustness against adversarial attacks. In particular, SSPNets with the *ArkBlock* improve adversarial robustness while maintaining natural accuracy. The proposed networks are complementary with adversarial training and suppress the perturbation growth. Our work shows the way to improve the robustness of neural networks by utilizing the theory of advanced numerical discretization schemes. We believe that the intersection of numerical discretization and robust deep learning will provide new opportunities to study robust neural networks.[1]

Acknowledgments. This work was supported by Institute of Information & communications Technology Planning & Evaluation (IITP) grant funded by the Korea government (MSIT) (No. 2019-0-00533), National Supercomputing Center with supercomputing resources including technical support (KSC-2019-CRE-0186), National Research Foundation of Korea (NRF-2020R1A2C3010638), and Simons Foundation Collaboration Grants for Mathematicians.

References

1. Athalye, A., Carlini, N., Wagner, D.: Obfuscated gradients give a false sense of security: Circumventing defenses to adversarial examples. In: ICML, pp. 274–283 (2018)
2. Ben-Tal, A., El Ghaoui, L., Nemirovski, A.: Robust Optimization, vol. 28. Princeton University Press, Princeton (2009)
3. Buckman, J., Roy, A., Raffel, C., Goodfellow, I.: Thermometer encoding: One hot way to resist adversarial examples. In: ICLR (2018)
4. Butcher, J.C.: The Numerical Analysis of Ordinary Differential Equations. Runge Kutta and General Linear Methods. A Wiley-Interscience Publication, John Wiley & Sons Ltd, Chichester (1987)
5. Carlini, N., Wagner, D.: Towards evaluating the robustness of neural networks. In: 2017 IEEE Symposium on Security and Privacy (SP), pp. 39–57 (2017)
6. Chen, T.Q., Rubanova, Y., Bettencourt, J., Duvenaud, D.K.: Neural ordinary differential equations. In: NeurIPS, pp. 6572–6583 (2018)
7. Ciccone, M., Gallieri, M., Masci, J., Osendorfer, C., Gomez, F.: NAIS-net: stable deep networks from non-autonomous differential equations. In: NeurIPS, pp. 3025–3035 (2018)
8. Goodfellow, I., Shlens, J., Szegedy, C.: Explaining and harnessing adversarial examples. In: ICLR (2015)
9. Gottlieb, S., Shu, C.W.: Total variation diminishing Runge-Kutta schemes. Math. Comput. Am. Math. Soc. **67**(221), 73–85 (1998)

[1] The codes are available at https://github.com/matbambbang/sspnet.

10. Gottlieb, S., Shu, C.W., Tadmor, E.: Strong stability-preserving high-order time discretization methods. SIAM Rev. **43**(1), 89–112 (2001)
11. Grathwohl, W., Chen, R.T., Betterncourt, J., Sutskever, I., Duvenaud, D.: FFJORD: Free-form continuous dynamics for scalable reversible generative models (2018). arXiv preprint arXiv:1810.01367
12. Harten, A.: High resolution schemes for hyperbolic conservation laws. J. Comput. Phys. **49**(3), 357–393 (1983)
13. Harten, A., Engquist, B., Osher, S., Chakravarthy, S.R.: Uniformly high-order accurate essentially nonoscillatory schemes. III. J. Comput. Phys. **71**(2), 231–303 (1987)
14. He, K., Zhang, X., Ren, S., Sun, J.: Delving deep into rectifiers: surpassing human-level performance on imagenet classification. In: Proceedings of the IEEE International Conference on Computer Vision, pp. 1026–1034 (2015)
15. He, K., Zhang, X., Ren, S., Sun, J.: Deep residual learning for image recognition. In: CVPR, pp. 770–778 (2016)
16. He, K., Zhang, X., Ren, S., Sun, J.: Identity mappings in deep residual networks. In: Leibe, B., Matas, J., Sebe, N., Welling, M. (eds.) ECCV 2016. LNCS, vol. 9908, pp. 630–645. Springer, Cham (2016). https://doi.org/10.1007/978-3-319-46493-0_38
17. Kingma, D.P., Ba, J.: Adam: a method for stochastic optimization. In: ICLR (2014)
18. Krizhevsky, A., Hinton, G.: Learning multiple layers of features from tiny images. Technical report, Citeseer (2009)
19. LeCun, Y., Cortes, C.: MNIST handwritten digit database (2010). http://yann.lecun.com/exdb/mnist/
20. Lu, Y., Zhong, A., Li, Q., Dong, B.: Beyond finite layer neural networks: bridging deep architectures and numerical differential equations. In: ICML, pp. 5181–5190 (2018)
21. Madry, A., Makelov, A., Schmidt, L., Tsipras, D., Vladu, A.: Towards deep learning models resistant to adversarial attacks. In: ICLR (2018)
22. Papernot, N., McDaniel, P., Wu, X., Jha, S., Swami, A.: Distillation as a defense to adversarial perturbations against deep neural networks. In: 2016 IEEE Symposium on Security and Privacy (SP), pp. 582–597. IEEE (2016)
23. Raff, E., Sylvester, J., Forsyth, S., McLean, M.: Barrage of random transforms for adversarially robust defense. In: CVPR June 2019
24. Rubanova, Y., Chen, R.T., Duvenaud, D.: Latent odes for irregularly-sampled time series (2019). arXiv preprint arXiv:1907.03907
25. Ruthotto, L., Haber, E.: Deep neural networks motivated by partial differential equations (2018). arXiv preprint arXiv:1804.04272
26. Samangouei, P., Kabkab, M., Chellappa, R.: Defense-GAN: protecting classifiers against adversarial attacks using generative models. In: ICLR (2018)
27. Shu, C.W.: Total-variation-diminishing time discretizations. SIAM J. Sci. Stat. Comput. **9**(6), 1073–1084 (1988)
28. Shu, C.W., Osher, S.: Efficient implementation of essentially non-oscillatory shock-capturing schemes. J. Comput. Phys. **77**(2), 439–471 (1988)
29. Song, Y., Kim, T., Nowozin, S., Ermon, S., Kushman, N.: Pixeldefend: Leveraging generative models to understand and defend against adversarial examples. In: ICLR (2018)
30. Szegedy, C., et al.: Intriguing properties of neural networks (2013). arXiv preprint arXiv:1312.6199
31. Tsipras, D., Santurkar, S., Engstrom, L., Turner, A., Madry, A.: Robustness may be at odds with accuracy. In: ICLR (2019)

32. Wang, Y., Ma, X., Bailey, J., Yi, J., Zhou, B., Gu, Q.: On the convergence and robustness of adversarial training. In: ICML, pp. 6586–6595 (2019)
33. Wong, E., Kolter, Z.: Provable defenses against adversarial examples via the convex outer adversarial polytope. In: ICML, pp. 5283–5292 (2018)
34. Wong, E., Schmidt, F., Metzen, J.H., Kolter, J.Z.: Scaling provable adversarial defenses. In: NeurIPS, pp. 8400–8409 (2018)
35. Xie, C., Wu, Y., Maaten, L.v.d., Yuille, A.L., He, K.: Feature denoising for improving adversarial robustness. In: CVPR, pp. 501–509 (2019)
36. Xu, H., Caramanis, C., Mannor, S.: Robustness and regularization of support vector machines. J. Mach. Learn. Res. **10**(7), 1485–1510 (2009)
37. Yang, Y., Zhang, G., Xu, Z., Katabi, D.: Me-net: Towards effective adversarial robustness with matrix estimation. In: ICML, pp. 7025–7034 (2019)
38. Zhang, H., Yu, Y., Jiao, J., Xing, E., Ghaoui, L.E., Jordan, M.: Theoretically principled trade-off between robustness and accuracy. In: ICML, pp. 7472–7482 (2019)
39. Zhang, H., Dauphin, Y.N., Ma, T.: Residual learning without normalization via better initialization. In: International Conference on Learning Representations (2019)
40. Zhang, X., Li, Z., Change Loy, C., Lin, D.: Polynet: a pursuit of structural diversity in very deep networks. In: CVPR, pp. 718–726 (2017)

Inequality-Constrained and Robust 3D Face Model Fitting

Evangelos Sariyanidi[1](✉), Casey J. Zampella[1], Robert T. Schultz[1,2], and Birkan Tunc[1,2]

[1] Center for Autism Research, Children's Hospital of Philadelphia, Philadelphia, USA
{sariyanide,zampellac,schultzrt,tuncb}@chop.edu
[2] University of Pennsylvania, Philadelphia, USA

Abstract. Fitting 3D morphable models (3DMMs) on faces is a well-studied problem, motivated by various industrial and research applications. 3DMMs express a 3D facial shape as a linear sum of basis functions. The resulting shape, however, is a plausible face only when the basis coefficients take values within limited intervals. Methods based on unconstrained optimization address this issue with a weighted ℓ_2 penalty on coefficients; however, determining the weight of this penalty is difficult, and the existence of a single weight that works universally is questionable. We propose a new formulation that does not require the tuning of any weight parameter. Specifically, we formulate 3DMM fitting as an inequality-constrained optimization problem, where the primary constraint is that basis coefficients should not exceed the interval that is learned when the 3DMM is constructed. We employ additional constraints to exploit sparse landmark detectors, by forcing the facial shape to be within the error bounds of a reliable detector. To enable operation "in-the-wild", we use a robust objective function, namely Gradient Correlation. Our approach performs comparably with deep learning (DL) methods on "in-the-wild" data that have inexact ground truth, and better than DL methods on more controlled data with exact ground truth. Since our formulation does not require any learning, it enjoys a versatility that allows it to operate with multiple frames of arbitrary sizes. This study's results encourage further research on 3DMM fitting with inequality-constrained optimization methods, which have been unexplored compared to unconstrained methods.

Keywords: 3D model fitting · 3D face reconstruction · 3D shape

1 Introduction

Estimation of 3D facial shape from 2D data via 3D morphable models (3DMMs), *a.k.a.* face reconstruction, is a fundamental computer vision problem that

Electronic supplementary material The online version of this chapter (https://doi.org/10.1007/978-3-030-58545-7_25) contains supplementary material, which is available to authorized users.

© Springer Nature Switzerland AG 2020
A. Vedaldi et al. (Eds.): ECCV 2020, LNCS 12354, pp. 433–449, 2020.
https://doi.org/10.1007/978-3-030-58545-7_25

attracts great interest due to its various applications [10], such as facial expression synthesis or analysis [18], gaze estimation [38] and facial landmark detection [45].

Most 3DMMs reconstruct facial shape as a linear sum of basis functions that are typically learned via a variant of principal component analysis (PCA) [3,5,7] or other models [10,14]. Then, 3D facial shape reconstruction from 2D data is performed by inferring the basis coefficients in this sum, often using unconstrained pseudo-second-order (PSO) optimization. The magnitude of basis coefficients must not be too large; otherwise, the resulting shape may hardly look like a face (Fig. 1a). More specifically, the coefficients should be within the bounds of their distributions, which are learned when the 3DMM is constructed. For example, if the 3DMM is learned with PCA (*e.g.*, [21]), the coefficients should very rarely exceed ±3 standard deviations (Sect. 2.2). A similar interval can be found for 3DMMs learned with other stochastic approaches (*e.g.*, [14]).

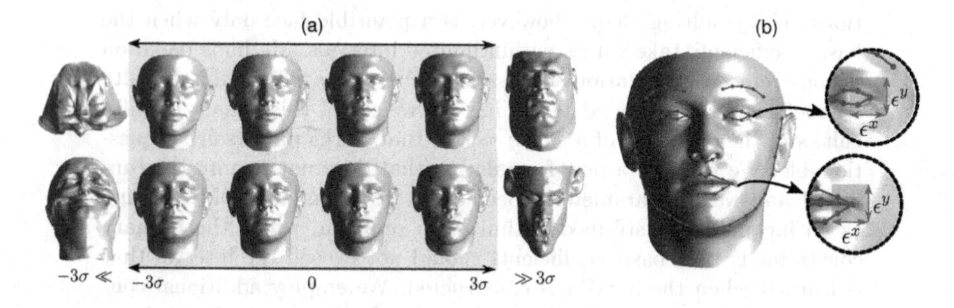

Fig. 1. Illustration of why inequality constraints are useful. (a) Morphable model constraints. Each row shows the effect of a basis function of the Basel'09 model [21] when generating facial shapes. The shapes may become implausible when the basis coefficient is outside a certain interval (in this case, ±3 standard deviations, $[-3\sigma, 3\sigma]$); thus, the optimization algorithm is constrained to this interval. (b) Sparse landmark constraints. The purple dots show the output of a landmark detector, and the red rectangles depict the maximal error for two landmarks (*i.e.*, eye and mouth corner), learned on a large dataset (Sect. 2.2). When a dense facial mesh is fit to this image, the mouth and eye corner of the mesh should remain inside the red rectangles (Color figure online)

Many methods address the above-mentioned issue by adding a weighted ℓ_2 penalty on the coefficients (Sect. 1.1). Unfortunately, the weights of those penalties are not easy to tune, and such ℓ_2 regularization can lead to overly smooth faces that miss personal characteristics, or images that exaggerate those characteristics to minimize reconstruction error (Fig. 2). Using deep learning (DL) is another alternative to 3DMM fitting. However, the DL methods that achieve the best performance "in-the-wild" also tend to produce overly-smooth faces (Fig. 2). Moreover, DL methods may lack versatility, as they rely on a fixed architecture that may not be suitable for working with images of arbitrary size, or with multiple frames of a person when available (Sect. 1.1).

This paper introduces a novel and theoretically compelling alternative to unconstrained PSO optimization, which achieves a robustness on par with DL methods, without sacrificing the versatility of 3DMMs. Specifically, we formulate 3DMM fitting as an inequality-constrained optimization problem, where the primary constraint is that no basis function coefficient should be outside the coefficient interval that is learned while the 3DMM is constructed (Fig. 1a). Thus, we prevent coefficients from taking prohibitively large values, but unlike ℓ_2 regularization, do not require the tuning of any application-specific parameter. Additional inequality constraints are used to exploit (sparse) facial landmark detectors (Fig. 1b). To enable our approach to operate "in-the-wild", we use Gradient Correlation (GC) [34] as the objective function, since GC is robust against illumination variations and occlusion. Finally, our approach can fit a 3DMM to multiple frames of a person. We refer to our formulation as *3DI* (3D estimation via Inequality constraints).

This paper has two technical contributions. First, we propose a novel formulation for 3DMM fitting as an inequality-constrained optimization problem. Second, we show how to use GC as a robust objective function for fitting 3DMMs. Our results on a widely-used "in-the-wild" dataset, AFLW2000-3D [44], are comparable to those of DL methods, even though our method does not require any training at all, and the 3DMM that we use is learned under controlled conditions [14]. Moreover, we demonstrate that our method actually outperforms the state-of-the-art methods on data where exact ground truth is known (improvement between 21% and 39% on the BU-4DFE dataset [41] and a synthesized dataset). The performance improves significantly with multiple frames, highlighting the benefits of a versatile method that can use multiple images.

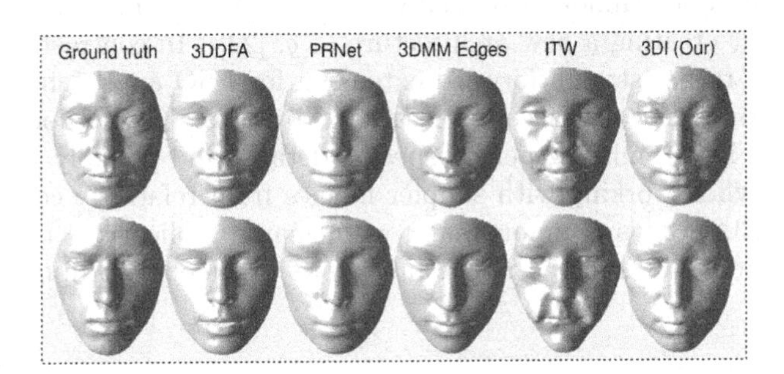

Fig. 2. Illustration of the output of two state-of-the-art deep learning-based methods, PRNet [11] and 3DDFA [45], in comparison to our method (3DI; used with Basel 2009 model [21]), two optimization-based methods with ℓ_2 regularization (3DMM edges [1] and ITW [7]) and ground truth. Deep learning methods are remarkably robust, but tend to produce over-smooth faces, whereas our method is better capable of applying the right amount of detail. More illustrations are provided in Supplementary Material

1.1 Related Work

Many 3DMM fitting methods use unconstrained PSO optimization [10], such as stochastic gradient [3,24], Levenberg-Marquardt [23,27] or Gauss-Newton [6,7,15,17,30,40,46]. These methods usually use various weighted ℓ_2 regularization terms in the cost function [1–4,6,7,12,15,17,24,27,30,39,46]. Unfortunately, the determination of the weight of these terms can be ad hoc [3]. Moreover, it is highly unlikely that there are specific optimal weights that can work on all images under all circumstances. For example, the optimal weight for the ℓ_2 penalty on deviation from landmarks should ideally depend on the error of the landmark detector, which is different for each image. A key novelty of our approach is to replace ℓ_2 regularization strategies with inequality constraints. Our formulation requires only determination of the bounds of the inequalities, which is learned for each 3DMM only once, without tuning w.r.t. a problem-specific metric (Sect. 2.2). Some recent methods can operate "in-the-wild" by using robust representations [1,6,7]. We also use a robust approach to fit morphable models, namely GC, which is robust against illumination variation and has a built-in outlier elimination, rendering it robust against occlusions [34].

Deep learning (DL) techniques are increasingly popular for 3D shape estimation [11,13,20,22,26,28,29,31,43]. Herein, we focus on DL methods that prioritize 3D shape reconstruction, which is the purpose of our study (but, see a recent survey on using DL for other tasks; *e.g.*, texture reconstruction [10]). Two DL methods are particularly robust, namely 3DDFA [45] and PRNet [11]. To our knowledge, PRNet achieves the best performance on one of the most popular "in-the-wild" datasets, AFLW2000-3D [45], as also shown in a recent independent study [43]. Like most DL methods, PRNet and 3DDFA work with a single frame, and cannot be trivially extended to use multiple frames of a person without creating a new architecture (*e.g.* [22]). In contrast, multi-frame operation is a rather straightforward extension for PSO approaches [7] and for our formulation. Also, the faces reconstructed by those DL methods tend to be too smooth (Fig. 2), missing person-specific details, and possibly limiting their performance when working with simpler images from relatively controlled conditions (Sect. 3.2). This is unfortunate, as in many applications the conditions are not "in-the-wild", such as Skype interviews or video recordings for clinical research [36].

2 Inequality-Constrained 3D Model Fitting

We first give background for fitting a 3D model to 2D frames and introduce our notation (Sect. 2.1). We then explain why it is natural to formulate 3DMM fitting as an inequality-constrained optimization problem (Sect. 2.2). Finally, we describe the robust objective function that we use (Sect. 2.3) and how to optimize it subject to inequality constraints (Sect. 2.4).

2.1 Background and Notation

Fitting a 3DMM to a set of 2D frames, $\mathbf{I}_1, \dots, \mathbf{I}_T$, amounts to finding the facial shape, texture, camera view, and illumination coefficients that best reconstruct the frames.

3D Facial Shape. The 3D facial shape is represented with a dense mesh of N points at each frame. Let $\mathbf{p}_t$ be the mesh at frame t, $\mathbf{p}_t := (\mathbf{p}_{t1}^T, \mathbf{p}_{t2}^T, \dots, \mathbf{p}_{tN}^T)^T \in \mathbb{R}^{3N}$, where $\mathbf{p}_{ti}$ is a single point, *i.e.*, $\mathbf{p}_{ti} = (p_{ti}^x, p_{ti}^y, p_{ti}^z)^T$. Then, morphable models represent the facial shape as a linear sum,

$$\mathbf{p}_t = \bar{\mathbf{p}} + \mathbf{A}\boldsymbol{\alpha} + \mathbf{E}\boldsymbol{\varepsilon}_t, \tag{1}$$

where $\bar{\mathbf{p}}$ is the mean face shape, $\mathbf{A} \in \mathbb{R}^{3N \times K_\alpha}$ is the (shape) identity basis of the morphable model, $\boldsymbol{\alpha} \in \mathbb{R}^{K_\alpha}$ is the vector of shape parameters, $\mathbf{E} \in \mathbb{R}^{3N \times K_\varepsilon}$ is the facial expression basis and $\boldsymbol{\varepsilon}_t \in \mathbb{R}^{K_\varepsilon}$ is the vector of expression coefficients. Note that $\boldsymbol{\alpha}$ does not depend on t as facial identity does not change over time, but the expression $\boldsymbol{\varepsilon}_t$ can. The points undergo a camera view transformation; that is, a rotation by a matrix $\mathbf{R}_t \in SO(3)$ and translation by a vector $\boldsymbol{\tau}_t = (\tau_{tx}, \tau_{ty}, \tau_{tz})^T$. The view-transformed points are represented as $\mathbf{v}_{t1}, \mathbf{v}_{t2}, \dots, \mathbf{v}_{tN}$, where

$$\mathbf{v}_{ti} := (v_{ti}^x, v_{ti}^y, v_{ti}^z)^T := \mathbf{R}_t \mathbf{p}_{ti} + \boldsymbol{\tau}_t. \tag{2}$$

The rotation matrix can be represented via quaternion parameters q_0^t, q_1^t, q_2^t and q_3^t [7]. The camera view transformation is represented concisely as a 6-vector $\mathbf{c}_t := (c_{t1}, \dots, c_{t6})^T := (q_1^t, q_2^t, q_3^t, \boldsymbol{\tau}_t^T)^T$; q_0^t is ignored, as it can be determined when q_1^t, q_2^t, q_3^t are known due to the unit-norm constraint of quaternions [7].

3D-to-2D Mapping. The next step towards reconstructing the face image is to project each 3D point $\mathbf{v}_{ti}$ onto the image plane. For a CCD camera, this process is carried out with a perspective transformation [16], and the $2N$-vector containing image points, $\mathbf{x}_t := (x_1^t, y_1^t, \dots, x_N^t, y_N^t)$, is obtained as:

$$x_i^t = \phi_x v_{ti}^x / v_{ti}^z + c_x, \quad y_i^t = \phi_y v_{ti}^y / v_{ti}^z + c_y \tag{3}$$

where ϕ_x and ϕ_y are the parameters of the perspective transformation, and c_x and c_y are the coordinates of the image center.

Texture. To obtain the reconstructed face image, $\hat{\mathbf{I}}_t$, one needs to determine the texture (*i.e.*, the pixel intensity) that will be assigned to each image point. This essentially depends on two factors: the facial texture of the person (*e.g.*, color of skin) and the illumination. A morphable model represents the facial texture of a person, $\hat{\mathbf{I}}_t^f$, as a linear sum, $\hat{\mathbf{I}}_t^f := \bar{\mathbf{t}} + \mathbf{B}\boldsymbol{\beta}$, where $\bar{\mathbf{t}}$ is the mean texture and $\bar{\mathbf{t}} \in \mathbb{R}^N$ for a grayscale image; $\mathbf{B} \in \mathbb{R}^{N \times K_\beta}$ is the texture basis of the morphable model; and $\boldsymbol{\beta} \in \mathbb{R}^{K_\beta}$ is the vector of texture coefficients. Using a simplified version of the Phong illumination model (*i.e.*, we ignore specular reflection [3]), the pixel intensities of the reconstructed image can finally be computed as

$$\hat{\mathbf{I}}_t = \hat{\mathbf{I}}_t^f + \Lambda \hat{\mathbf{I}}_t^f \odot \hat{\mathbf{I}}_t^d, \tag{4}$$

where $\odot$ is element-wise vector production, $\hat{\mathbf{I}}_t^d$ is the diffuse reflection component of the Phong model and Λ is a scalar—the diffuse reflection coefficient. The ith element of $\hat{\mathbf{I}}_t^d$ is $\hat{\mathbf{I}}_t^d[i] := \langle \mathbf{n}_{ti}, \boldsymbol{\lambda}_t - \mathbf{v}_{ti} \rangle$, where $\mathbf{n}_{ti}$ is the unit-norm the surface normal vector of the facial mesh at the ith point, $\langle \cdot, \cdot \rangle$ is the standard inner product on ℓ_2, and $\boldsymbol{\lambda}_t := (\lambda_{tx}, \lambda_{ty}, \lambda_{tz})^T$ is the 3D location of the illumination source. Equation (4) can be extended to use multiple illumination sources [3].

Image Formation. Rendering of reconstructed face image is carried out by filling the pixels whose location is specified in $\mathbf{x}_t$ with the intensity values specified in $\hat{\mathbf{I}}_t$. The true rendering process is slightly more complicated, as it requires rasterization to identify the pixels that will be rendered and Z-buffering to discard occluded pixels. However, for notational simplicity, we will suppose that $\hat{\mathbf{I}}_t$ is the (vectorized) rendered image—that the ith value of $\hat{\mathbf{I}}_t$ contains the pixel intensity at image location (x_i, y_i).

2.2 Inequality Constraints

The facial shape and texture bases of morphable models are learned from a large number of 3D facial scans [3]. Often, a statistical learning approach that underlies the assumption of Normal distribution (*e.g.*, PCA [2,7]) is used, in which case basis coefficients should very rarely exceed ±3 standard distribution [35]. Thus, the distributions learned while constructing the 3DMM can be used to determine hard upper and lower bounds for basis coefficients. Importantly, one can set those bounds *a priori*, in an application-independent manner. Nevertheless, since the basis coefficients are empirical distributions, it is worth visually confirming that the hard bounds generated using the statistics of those distributions do indeed generate plausible-looking faces, as the aim is to have universally-valid bounds.

Let us define $\mathbf{h}^\alpha$, $\mathbf{h}^\beta$ and $\mathbf{h}_t^\varepsilon$ as the constraint functions $\mathbf{h}^\alpha := \boldsymbol{\alpha}$, $\mathbf{h}^\beta := \boldsymbol{\beta}$, and $\mathbf{h}_t^\varepsilon := \boldsymbol{\varepsilon}_t$. Then, the *morphable model constraints* are

$$\boldsymbol{\alpha}^- \preceq \mathbf{h}^\alpha \preceq \boldsymbol{\alpha}^+, \quad \boldsymbol{\varepsilon}^- \preceq \mathbf{h}_t^\varepsilon \preceq \boldsymbol{\varepsilon}^+, \quad \boldsymbol{\beta}^- \preceq \mathbf{h}^\beta \preceq \boldsymbol{\beta}^+, \tag{5}$$

where $\boldsymbol{\alpha}^- \in \mathbb{R}^{K_\alpha}, \boldsymbol{\alpha}^+ \in \mathbb{R}^{K_\alpha}$, $\boldsymbol{\varepsilon}^- \in \mathbb{R}^{K_\varepsilon}, \boldsymbol{\varepsilon}^+ \in \mathbb{R}^{K_\varepsilon}$, $\boldsymbol{\beta}^- \in \mathbb{R}^{K_\beta}$ and $\boldsymbol{\beta}^+ \in \mathbb{R}^{K_\beta}$ are vectors containing the bounds for the morphable model's facial shape, expression and texture coefficients. The symbol $\preceq$ is componentwise inequality [8].

Additional inequality constraints can be used to further improve the fitting via (sparse) 2D landmark detectors, as in Fig. 1b. Suppose that we have a detector that estimates the locations of L landmark points. Let those landmark points on the facial mesh be $\mathbf{x}'_t := (x^t_{i_1}, y^t_{i_1}, \dots, x^t_{i_L}, y^t_{i_L})$, and let $\hat{\mathbf{x}}'_t := (\hat{x}^t_{i_1}, \hat{y}^t_{i_1}, \dots, \hat{x}^t_{i_L}, \hat{y}^t_{i_L})$ be the location of the same landmarks as estimated by the detector. Let us suppose that the maximal error of the landmark detector is measured for each landmark on a very large dataset and encoded in a vector $\boldsymbol{\epsilon}$ as $\boldsymbol{\epsilon} := (\epsilon^x_1, \epsilon^y_1, \dots, \epsilon^x_L, \epsilon^y_L)$ (Fig. 1b). Here, the maximal error of a landmark, $\epsilon^x_j, \epsilon^y_j$, can be defined in the strict sense (*i.e.*, the error on the ith landmark does not exceed $\epsilon^x_i, \epsilon^y_i$ for any image in the dataset) or in a slightly loose sense, such as the error that is valid for 99% of the images. Then, the discrepancy between $\hat{\mathbf{x}}'_t$ and the image location of the same landmarks under the morphable model,

$\mathbf{x}'_t := (x^t_{i_1}, y^t_{i_1}, \ldots, x^t_{i_L}, y^t_{i_L})$ [see (3)], should not exceed ϵ. Thus, the *sparse landmark constraint* can be represented as

$$-s^t_b \epsilon \preceq \mathbf{h}^L_t \preceq s^t_b \epsilon, \tag{6}$$

where $\mathbf{h}^L_t := \mathbf{x}'_t - \hat{\mathbf{x}}'_t$, and s_b is the bounding box size $s^t_b := \sqrt{w^t_{bbox} \times h^t_{bbox}}$, which is used for normalizing the error [9]. The width and height of the box, w^t_{bbox} and h^t_{bbox}, are computed from the landmarks.

2.3 Objective Function

Fitting a 3DMM to an input image $\mathbf{I}_t$ requires a cost function to measure the quality of fit. One may simply use the squared pixel-wise difference between the input and reconstructed image [3], but this would hardly be robust (*e.g.*, against occlusions). We use GC, as it is a robust function due to its outlier elimination property [34]. GC has been used for rigid [32,34] and non-rigid 2D registration [33]; however, to our knowledge, it has not been used for 3DMM fitting. We derive the mathematical expressions needed for the latter in Sect. 2.4.

To compute GC, we need to compute the *magnitude-normalized gradient* of the input image $\mathbf{I}_t$ and the fitted image $\hat{\mathbf{I}}_t$ [34]. Let us denote the magnitude-normalized gradients of $\mathbf{I}_t$ along the x and y axes with the N-dimensional vectors $\mathbf{g}_{tx}$ and $\mathbf{g}_{ty}$. If we approximate the ideal gradient operator with centered difference, then the kth entry of those vectors can be computed as

$$\mathbf{g}_{tx}[k] := (\mathbf{I}_t[k_r] - \mathbf{I}_t[k_l])/h, \quad \mathbf{g}_{ty}[k] := (\mathbf{I}_t[k_b] - \mathbf{I}_t[k_a])/h, \tag{7}$$

where k_a, k_b, k_l, and k_r are the pixels above, below, to the left, and right of the kth pixel, and $h := \sqrt{(\mathbf{I}_t[k_r] - \mathbf{I}_t[k_l])^2 + (\mathbf{I}_t[k_b] - \mathbf{I}_t[k_a])^2}$ is the magnitude. The magnitude-normalized gradients of $\hat{\mathbf{I}}_t$, $\hat{\mathbf{g}}_{tx}$ and $\hat{\mathbf{g}}_{ty}$, are computed similarly. For notational simplicity, we concatenate those gradients and represent them as $\mathbf{g}_t := (\mathbf{g}^T_{tx}, \mathbf{g}^T_{ty})^T$ and $\hat{\mathbf{g}}_t := (\hat{\mathbf{g}}^T_{tx}, \hat{\mathbf{g}}^T_{ty})^T$. The objective function f that we aim to maximize, namely the GC between the input and the fitted frames, is

$$f = \sum_{t=1}^{T} \mathbf{g}^T_t \hat{\mathbf{g}}_t. \tag{8}$$

2.4 Optimization

Inequality-constrained optimization problems are more difficult to solve than unconstrained problems [8]. Algorithms that are standard for 3DMM fitting, such as Gauss-Newton, cannot be used with inequality constraints. Fortunately, there are high-quality solvers for inequality-constrained problems, such as IPOPT [37], that require only the derivative of the objective function f and the Jacobian of inequality constraints. We derive these terms below.

Derivative of Objective Function. The derivative of f in (8) w.r.t. any parameter $\mathbf{y}$ that affects the rendered image $\hat{\mathbf{I}}_t$ is

$$\frac{\partial f}{\partial \mathbf{y}} = \sum_{t=1}^{T} \mathbf{g}_t^T \frac{\partial \hat{\mathbf{g}}_t}{\partial \mathbf{y}} = \sum_{t=1}^{T} \mathbf{g}_t^T \frac{\partial \hat{\mathbf{g}}_t}{\partial \hat{\mathbf{I}}_t} \frac{\partial \hat{\mathbf{I}}_t}{\partial \mathbf{y}}. \tag{9}$$

Since the normalized gradient depends only on neighboring values of the input image [see (7)], $\partial \hat{\mathbf{g}}_t / \partial \hat{\mathbf{I}}_t$ is a sparse $2N \times N$ matrix. This matrix is obtained by horizontally concatenating $\partial \hat{\mathbf{g}}_{tx} / \partial \hat{\mathbf{I}}_t$ and $\partial \hat{\mathbf{g}}_{ty} / \partial \hat{\mathbf{I}}_t$. The entries of the latter matrices are provided in Supplementary Material. To compute the partial derivatives of f for all needed variables, we must compute $\partial \hat{\mathbf{I}}_t / \partial \boldsymbol{\alpha}$, $\partial \hat{\mathbf{I}}_t / \partial \boldsymbol{\beta}$, $\partial \hat{\mathbf{I}}_t / \partial \boldsymbol{\varepsilon}_t$, $\partial \hat{\mathbf{I}}_t / \partial \mathbf{c}_t$ and $\partial \hat{\mathbf{I}}_t / \partial \boldsymbol{\lambda}_t$ and then replace them in turn with $\partial \hat{\mathbf{I}}_t / \partial \mathbf{y}$ in (9). $\partial \hat{\mathbf{I}}_t / \partial \boldsymbol{\beta}$ is rather simple as $\boldsymbol{\beta}$ affects only the $\hat{\mathbf{I}}_t^f$ in (4):

$$\frac{\partial \hat{\mathbf{I}}_t}{\partial \boldsymbol{\beta}} = \frac{\partial \hat{\mathbf{I}}_t^f}{\partial \boldsymbol{\beta}} + \frac{\partial \hat{\mathbf{I}}_t^f}{\partial \boldsymbol{\beta}} \odot (\mathbf{1}_{K_\beta}^T \otimes \Lambda \hat{\mathbf{I}}_t^d) = \mathbf{B} + \mathbf{B} \odot (\mathbf{1}_{K_\beta}^T \otimes \Lambda \hat{\mathbf{I}}_t^d), \tag{10}$$

where $\mathbf{1}_{K_\beta}^T$ is the transpose of the K_β-dimensional column vector whose all entries are 1, and $\otimes$ is the Kronecker product; therefore, $(\mathbf{1}_{K_\beta}^T \otimes \hat{\mathbf{I}}_t^d)$ is an $N \times K_\beta$ matrix whose every column is $\Lambda \hat{\mathbf{I}}_t^d$. The derivative w.r.t. illumination source $\boldsymbol{\lambda}$ is also simple as $\boldsymbol{\lambda}$ has no effect on $\hat{\mathbf{I}}_t^f$:

$$\frac{\partial \hat{\mathbf{I}}_t}{\partial \boldsymbol{\lambda}} = (\mathbf{1}_3^T \otimes \hat{\mathbf{I}}_t^f) \odot \frac{\partial \hat{\mathbf{I}}_t^d}{\partial \boldsymbol{\lambda}} = (\mathbf{1}_3^T \otimes \hat{\mathbf{I}}_t^f) \odot (\mathbf{n}_{t1}, \mathbf{n}_{t2}, \ldots, \mathbf{n}_{tN})^T. \tag{11}$$

The derivatives w.r.t. remaining parameters are obtained as follows:

$$\begin{aligned}
\frac{\partial \hat{\mathbf{I}}_t}{\partial \boldsymbol{\alpha}} &= \frac{\partial \hat{\mathbf{I}}_t^f}{\partial \mathbf{x}_t} \frac{\partial \mathbf{x}_t}{\partial \mathbf{p}_t} \frac{\partial \mathbf{p}_t}{\partial \boldsymbol{\alpha}} \odot \left(\mathbf{1}_{NK_\alpha} + \mathbf{1}_{K_\alpha}^T \otimes \Lambda \hat{\mathbf{I}}_t^d\right) + \left(\mathbf{1}_{K_\alpha}^T \otimes \Lambda \hat{\mathbf{I}}_t^f\right) \odot \frac{\partial \hat{\mathbf{I}}_t^d}{\partial \mathbf{x}_t} \frac{\partial \mathbf{x}_t}{\partial \mathbf{p}_t} \frac{\partial \mathbf{p}_t}{\partial \boldsymbol{\alpha}}, \\
\frac{\partial \hat{\mathbf{I}}_t}{\partial \boldsymbol{\varepsilon}_t} &= \frac{\partial \hat{\mathbf{I}}_t^f}{\partial \mathbf{x}_t} \frac{\partial \mathbf{x}_t}{\partial \mathbf{p}_t} \frac{\partial \mathbf{p}_t}{\partial \boldsymbol{\varepsilon}_t} \odot \left(\mathbf{1}_{NK_\varepsilon} + \mathbf{1}_{K_\varepsilon}^T \otimes \Lambda \hat{\mathbf{I}}_t^d\right) + \left(\mathbf{1}_{K_\varepsilon}^T \otimes \Lambda \hat{\mathbf{I}}_t^f\right) \odot \frac{\partial \hat{\mathbf{I}}_t^d}{\partial \mathbf{x}_t} \frac{\partial \mathbf{x}_t}{\partial \mathbf{p}_t} \frac{\partial \mathbf{p}_t}{\partial \boldsymbol{\varepsilon}_t}, \\
\frac{\partial \hat{\mathbf{I}}_t}{\partial \mathbf{c}_t} &= \frac{\partial \hat{\mathbf{I}}_t^f}{\partial \mathbf{x}_t} \frac{\partial \mathbf{x}_t}{\partial \mathbf{c}_t} \odot \left(\mathbf{1}_{N6} + \mathbf{1}_6^T \otimes \Lambda \hat{\mathbf{I}}_t^d\right) + \left(\mathbf{1}_6^T \otimes \Lambda \hat{\mathbf{I}}_t^f\right) \odot \frac{\partial \hat{\mathbf{I}}_t^d}{\partial \mathbf{c}_t}
\end{aligned} \tag{12}$$

where $\mathbf{1}_N$ is an N-dimensional vector of ones. $\partial \hat{\mathbf{I}}_t^f / \partial \mathbf{x}_t$ is an $N \times 2N$ block-diagonal matrix that contains the (un-normalized) gradient of the image $\hat{\mathbf{I}}_t^f$; specifically, the nth block on its diagonal is a 1×2 matrix comprising the horizontal and vertical gradient of the nth pixel of $\hat{\mathbf{I}}_t^f$. $\partial \mathbf{p} / \partial \boldsymbol{\alpha}$ and $\partial \mathbf{p} / \partial \boldsymbol{\varepsilon}_t$ are respectively $\mathbf{A}$ and $\mathbf{E}$. $\partial \mathbf{x}_t / \partial \mathbf{p}_t$ is a block-diagonal matrix whose nth block is a 2×3 matrix containing the derivative of the nth image point w.r.t. $\mathbf{p}_n$. The remaining terms that are needed to complete the computation of derivatives, namely $\partial \mathbf{x}_t / \partial \mathbf{c}_t$ and $\partial \hat{\mathbf{I}}_t^d / \partial \mathbf{c}_t$, are provided in Supplemental Material.

Jacobian of Constraints. Our problem has $2 + 2T$ constraint functions $\mathbf{h}^\alpha, \mathbf{h}^\beta, \mathbf{h}_1^\varepsilon, \ldots, \mathbf{h}_T^\varepsilon, \mathbf{h}_1^L, \ldots, \mathbf{h}_T^L$ and $2 + 3T$ sets of variables $\boldsymbol{\alpha}$, $\boldsymbol{\beta}$, $\boldsymbol{\varepsilon}_1, \ldots, \boldsymbol{\varepsilon}_T$,

$\mathbf{c}_1, \dots, \mathbf{c}_T$, $\boldsymbol{\lambda}_1, \dots, \boldsymbol{\lambda}_T$. The Jacobian of the constraints is therefore a matrix partitioned into a grid of $(2 + 2T) \times (2 + 3T)$ blocks, where each partition is the partial derivative of one of the afore-listed constraint functions w.r.t. one of the sets of variables. $\mathbf{J}$ is a sparse matrix, as each constraint depends only on a small set of variables. We list all of the non-zero derivatives in this partitioning below. The derivatives for the morphable model constraints are

$$\frac{\partial \mathbf{h}^{\alpha}}{\partial \boldsymbol{\alpha}} = I_{K_\alpha}, \quad \frac{\partial \mathbf{h}_t^{\varepsilon}}{\partial \boldsymbol{\varepsilon}_t} = I_{K_\varepsilon} \quad \frac{\partial \mathbf{h}^{\beta}}{\partial \boldsymbol{\beta}} = I_{K_\beta}, \tag{13}$$

where I_{K_α}, I_{K_ε} and I_{K_β} are identity matrices of size K_α, K_ε and K_β, respectively. The derivatives for the landmark constraints are

$$\frac{\partial \mathbf{h}_t^L}{\partial \boldsymbol{\varepsilon}_t} = \frac{\partial \mathbf{x}_t'}{\partial \mathbf{p}_t'} \frac{\partial \mathbf{p}_t'}{\partial \boldsymbol{\varepsilon}_t}, \quad \frac{\partial \mathbf{h}_t^L}{\partial \mathbf{c}_t} = \frac{\partial \mathbf{x}_t'}{\partial \mathbf{c}_t}, \tag{14}$$

where $\mathbf{p}_t'$ is the $3L$-vector obtained by concatenating the 3D points corresponding to landmarks, $\mathbf{p}_t' := (\mathbf{p}_{ti_1}^T, \dots, \mathbf{p}_{ti_L}^T)^T$. The matrix $\partial \mathbf{x}_t' / \partial \mathbf{p}_t'$ is a block-diagonal matrix obtained similarly to the $\partial \mathbf{x}_t / \partial \mathbf{p}_t$ described for (12); the only difference is that it comprises L blocks –corresponding to L landmarks– and not N blocks. The derivative $\partial \mathbf{p}_t' / \partial \boldsymbol{\alpha}$ is a $3L \times K_\alpha$ matrix containing the rows of $\mathbf{A}$ corresponding to the L landmarks. Similarly, $\partial \mathbf{p}_t' / \partial \boldsymbol{\varepsilon}_t$ contains the rows of $\mathbf{E}$ corresponding to landmarks. The partial derivative $\frac{\partial \mathbf{x}_t'}{\partial \mathbf{c}_t}$ is also similar to the $\frac{\partial \mathbf{x}_t}{\partial \mathbf{c}_t}$ in (12), with the difference that it is computed from L landmark points.

3 Experimental Validation

We validate our method experimentally on three tasks, namely (sparse) 2D facial landmark estimation (*a.k.a.* face alignment), 3D landmark estimation, and dense (3D) facial shape estimation. We show the robustness of the method, as well as its ability to attain high precision, through experiments conducted with "in-the-wild" data in addition to controlled data.

3.1 Experimental Setup

Evaluation metric and datasets. We evaluate performance with the commonly used Normalized Mean Error (NME) for all tasks [9,11,45]. For 2D landmark estimation, the NME of one image is computed by calculating the estimation error for each landmark via ℓ_2 norm, then computing the average of those errors, and finally normalizing this average by dividing it by the bounding box size computed as $\sqrt{w_{bbox} \times h_{bbox}}$, where the bounding box width w_{bbox} and height h_{bbox} are computed from the labeled landmarks. We report performance on the commonly employed $L = 68$ landmark points (Fig. 1b) as well as the $L = 51$ (inner-face) points [25]. When computing NME for 3D landmark estimation, the point-wise error is computed in terms of 3D points, and Z-normalization is applied to all points to resolve the ambiguity along the depth axis. For NME for dense facial shape estimation, the average error is computed from all the points on the dense facial mesh, and normalization is performed by dividing by outer interocular distance [11]. Similarly to previous studies, we use Iterative Closest Point (ICP) prior to computing the dense NME, but only to establish the point correspondence between the ground truth mesh shape and the estimated facial mesh [19,42] (*i.e.*, rigid alignment is not used).

We use three datasets. First, *AFLW2000-3D* [45]— a widely used dataset [9,19,43,45] that contains 2D and 3D landmark annotations. Second, a *Synthesized* dataset that we generated using the Basel'09 3DMM [21]. This dataset has two advantages: It enables us to compute exact ground truth for 2D and 3D landmarks, and to run multi-frame experiments, as the images of the same face from different angles can be generated trivially (examples of images for the Synthesized dataset are in Supplementary Material). We use the Basel'09 model, as its facial mesh is used by many previous methods [1,6,7,45] and the ground truth location of the $L = 68$ landmarks on this mesh are established. Thus, an *exact* comparison between the estimated and true location of 3D points becomes possible. We synthesize 900 images from 100 subjects in 9 poses as in previous works [19,45]; that is, we apply 9 yaw rotations of, -80, -60, ..., $80°$, and a pitch rotation randomly selected from -15, 20 and $25°$. We also apply a random illumination variation. Finally, we use the *BU-4DFE* [41] dataset for dense facial shape estimation. BU-4DFE contains 3D facial data collected from 101 subjects, and allows us to evaluate our method on faces of real subjects from various ethnic and racial backgrounds. This is important for our study as the Basel models that we use for fitting are constructed from European participants; therefore, our methods's ability to generalize to non-European populations must be explicitly tested. Similarly to the Synthesized dataset, we generate 9 images per subject, but also add one of the six basic facial expressions, namely happiness, sadness, anger, surprise, fear and disgust.

Compared Methods and Implementation of 3DI. To validate different aspects of our method, we compare with five very recent state-of-the-art methods. First, we compare with PRNet [11], which, to our knowledge, attains the best performance on AFLW2000-3D, even in an independent study [43]. Second, we compare with 3DDFA [45]. While 3DFFA was outperformed by some recent studies [11,19,43], it now has an updated code[1] that is considerably improved. Third, we compare with one of the most popular landmark estimation methods, 3D-FAN [9]. Fourth, we compare with two robust optimization-based methods that use ℓ_2 regularization, namely 3DMM Edges [1] and ITW [7]. Finally, we compare with a video-based variant of the latter, ITW-V [7].

We implemented our method, 3DI, in MATLAB. We used IPOPT [37] for inequality-constrained optimization, and 3D-FAN [9] for landmark estimation. To learn the maximal landmark detection error (ϵ) as discussed in Sect. 2.2, we synthesized a large face dataset with the Basel'09 model, and computed the error of 3D-FAN for each landmark. We ignored the outliers by not taking into account the 1% of the images with the highest errors. However, we expanded the

[1] https://github.com/cleardusk/3DDFA.

error bounds by 15% to account for the difficulties associated with "in-the-wild" images. We resized each image to 100×100, and applied Gaussian smoothing as suggested for GC [34]. We used the Basel'17 3DMM[14], except for the experiments on the Synthesized dataset, where we used Basel'09 for landmark estimation experiments with the aim of providing exact comparison using the common keypoints with other methods (see *Datasets* above). We applied the coefficient constraints of ± 3 standard deviations for the Basel'09 model. For the Basel'17 model, we used a reduced interval of ± 1.5 standard deviations as the interval of ± 3 generates implausible looking faces with this model (see Supplementary Material). Note that these intervals were determined only through evaluating Basel 3DMMs, and fixed for all experiments. The source code of our method is available on https://github.com/sariyanidi/3DI for research purposes.

3.2 Results

2D and 3D Landmark Estimation. Our method's qualitative 2D landmark estimation and dense 3DMM fitting results are shown in Fig. 3. Overall, Fig. 3 demonstrates that our method operates well "in-the-wild", producing compelling results even in the presence of large illumination or expression variations, or occlusions (more qualitative results are provided in Supplementary Material). This is a remarkable outcome that validates the theoretical appeal of our formulation in practice; to our knowledge, we propose the first method that can generalize to "in-the-wild" data, without requiring a morphable model constructed from uncontrolled images (*e.g.*, [7]) or a deep architecture trained with large amounts of "in-the-wild" data. Figure 4 shows the 2D and 3D landmark estimation performance of all methods via cumulative error distribution (CED) on the AFLW2000-3D dataset, and Table 1 shows the mean NME. The 2D landmark estimation performance of our method on 51 or 68 landmarks (Fig. 4 and Table 1) is very similar to that of PRNet, which, to our knowledge, is the best-performing method on this dataset. Our method's error is slightly higher than other methods on 3D landmark estimation, but one must take those results with a pinch of salt, because the annotations of AFLW2000-3D are controversial [11]. 3D landmark annotations on AFLW are obtained from single frames. Such annotations can hardly be called ground truth, as inferring 3D points from 2D data is an ill-posed problem. We next investigate the performance of the same methods on BU-4DFE and the Synthesized dataset, where true ground truth is available.

Figure 5 shows the 2D and 3D landmark estimation results on the Synthesized dataset. Results for multiframe methods are computed by using 5 randomly selected frames simultaneously; additional results with 3 and 9 frames are reported in Supplementary Material. Our method outperforms all other methods when the performance metric is reliable (*i.e.*, a true ground truth is available). Of note, using our method with multiple frames significantly improves 3D landmark estimation even though it does not improve 2D landmark estimation. This is because 3DMM fitting is an ill-posed problem; even though single-frame estimation is generally capable of finding a good fit for 2D landmarks, the 3D location of the same landmarks is not necessarily as accurate. In particular, the

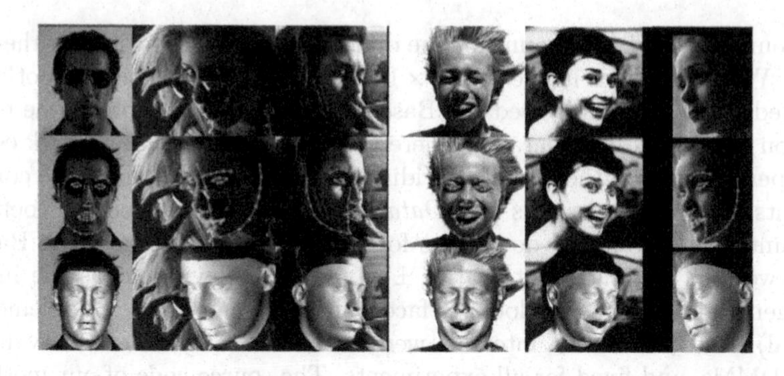

Fig. 3. Qualitative illustration of our method's performance on the AFLW2000-3D dataset. Top row: input images; middle row: 2D landmarks estimated by our method; bottom row: dense 3D shape estimated by our method. It is notable that our method can successfully operate in such uncontrolled conditions, even though we use a 3DMM collected from controlled data, namely Basel 2017 [14]

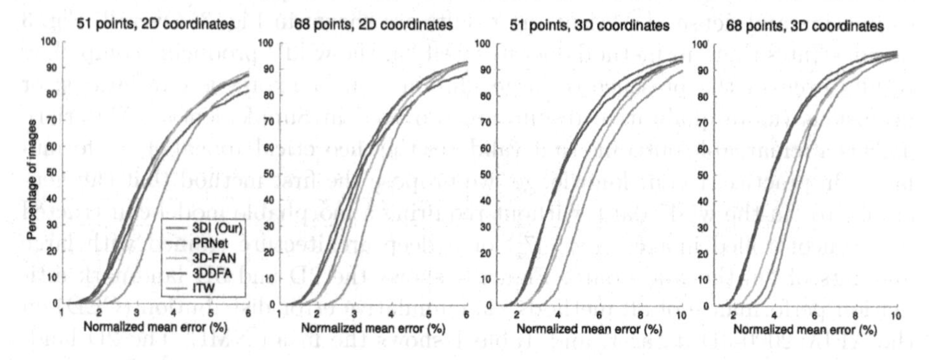

Fig. 4. Cumulative error distribution (CED) of compared methods on the AFLW2000-3D dataset for the tasks of 2D and 3D (sparse) landmark estimation. Performance is reported separately for $L = 68$ and $L = 51$ landmarks

2D landmark estimation of our method on 51 landmarks is ~39% better than the next best method. The 3D estimation is ~21% better than the next best method for single-frame and ~39% better for multi-frame. Note that the morphable model that we used for fitting in the experiments on the Synthesized dataset (*i.e.*, Basel'09) is also the model that was used to generate the images of the dataset. This may be seen as a possible explanation of our method's superiority in this experiment. To alleviate this concern, we re-run experiments using our method with the Basel'17 model and results (see y material) show that our method attains similar results even when we use a different morphable model.

Dense Shape Estimation. Figure 6 and Fig. 7 show the dense shape estimation performance of the compared methods on the Synthesized and BU-4DFE datasets, respectively. Dense estimation is performed on a facial shape mesh with ~23,000 points that cover the facial region and ignore the ear, neck etc.

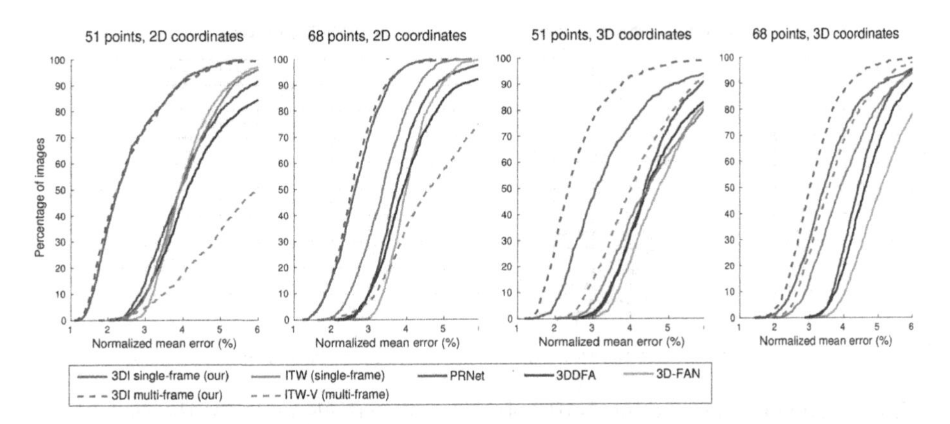

Fig. 5. Cumulative error distribution (CED) of compared methods on the Synthesized dataset for the tasks of 2D and 3D (sparse) landmark estimation. Performance is reported separately for $L = 68$ and $L = 51$ landmarks

Table 1. Mean NME of compared methods on the AFLW2000-3D and Synthesized datasets for the tasks of 2D and 3D landmark estimation for $L = 51$ and $L = 68$ landmarks. Bold and underline indicate best and second best performance, respectively

	AFLW2000-3D dataset				Synthesized dataset			
	2D landmarks		3D landmarks		2D landmarks		3D landmarks	
	$L = 51$	$L = 68$	$L = 51$	$L = 68$	$L = 51$	$L = 68$	$L = 51$	$L = 68$
PRNet	**0.041**	**0.035**	**0.048**	**0.044**	0.041	0.038	0.045	0.045
3DDFA	0.049	0.040	0.050	0.046	0.048	0.042	0.048	0.048
3D-FAN	**0.041**	0.038	0.060	0.053	0.041	0.041	0.050	0.052
ITW	0.047	0.042	0.066	0.060	0.041	0.034	0.048	0.041
ITW-V (multi-frame)	N/A	N/A	N/A	N/A	0.060	0.045	0.043	0.038
3DI Single-frame (our)	0.042	0.038	0.057	0.056	**0.025**	**0.027**	0.034	0.036
3DI Multi-frame (our)	N/A	N/A	N/A	N/A	0.026	**0.027**	**0.026**	**0.032**

Results show that our method outperforms existing methods; in particular, our method's mean NME is ~34% (~21%) lower compared to the next best method on the BU-4DFE (Synthesized) dataset, when we use multiple frames.

Ablation Study. We performed an ablation study to show the effect of our formulation's two critical components, namely the two inequality constraints. Our ablation study is conducted on 2D/3D landmark detection ($L = 68$). When we omit the first constraint, mean NME increases by 17.4% for 2D points and 45% for 3D points on the AFLW2000-3D dataset; and by 16.7% for 2D points and 11.1% for 3D points on the Synthesized dataset. When we omit the second constraint, mean NME increases by 9.5% for 2D points and 12.5% for 3D points on the AFLW2000-3D dataset; and by 19.3% for 2D points and 8.6% for 3D points on the Synthesized dataset. Another component that can be subjected to

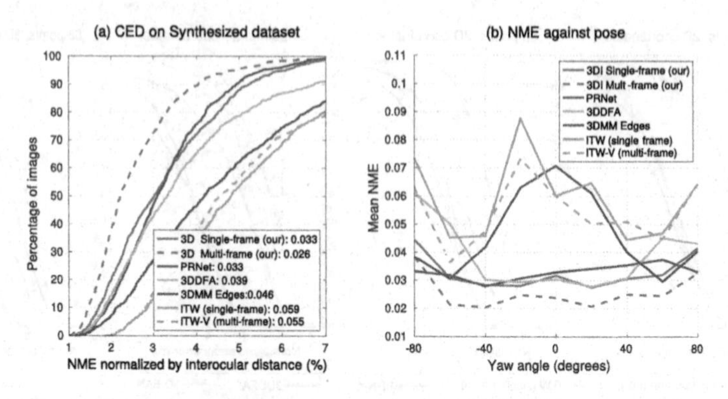

Fig. 6. Normalized mean error (NME) of compared methods on the Synthesized dataset for dense face reconstruction, reported in terms of Cumulative Error Distribution (CED) and NME against pose. Numbers in legend indicate mean NME

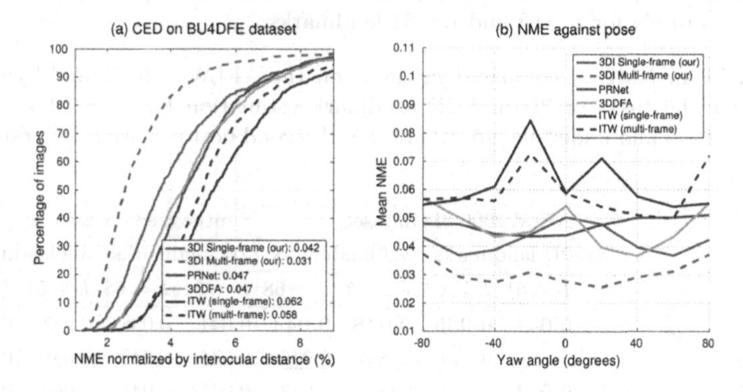

Fig. 7. Normalized mean error (NME) of compared methods on the BU-4DFE dataset for dense face reconstruction, reported in terms of Cumulative Error Distribution (CED) and NME against pose. Numbers in legend indicate mean NME

ablation is GC—it can be replaced by a simpler objective function such as squared pixel-wise difference. However, the latter proved inadequate in uncontrolled conditions with illumination variations and occlusions as was shown in studies of 3DMM fitting [7] or the closely related problem of image alignment [34].

4 Conclusions and Future Work

This paper proposes a new and theoretically compelling formulation to a well-established computer vision problem, namely 3D morphable model (3DMM) fitting. We show that when 3DMM fitting is formulated as an inequality-constrained optimization problem with a robust objective function, the resulting approach performs on par with top-performing deep learning (DL) methods on "in-the-wild" data where ground truth is not exact, and outperforms those methods on more controlled data with exact ground truth. Moreover, this approach enjoys the versatility of standard optimization approaches, as it is capable of working with multiple frames of arbitrary sizes. The results of this paper strongly encourage future research to evaluate the efficiency of existing inequality-constrained minimization algorithms (*e.g.*, the log-barrier method, primal-dual interior-point methods [8]), which, unlike unconstrained methods, remain unexplored in the context of 3DMM fitting and similar problems.

Acknowledgments. This work is partially funded by the Office of the Director, National Institutes of Health (OD) and National Institute of Mental Health (NIMH) of US, under grants R01MH118327, R01MH122599 and R21HD102078.

References

1. Bas, A., Smith, W.A.P., Bolkart, T., Wuhrer, S.: Fitting a 3D morphable model to edges: a comparison between hard and soft correspondences. In: Chen, C.-S., Lu, J., Ma, K.-K. (eds.) ACCV 2016. LNCS, vol. 10117, pp. 377–391. Springer, Cham (2017). https://doi.org/10.1007/978-3-319-54427-4_28
2. Blanz, V., Vetter, T.: A morphable model for the synthesis of 3D faces. In: Proceedings of the Conference on Computer Graphics and Interactive Techniques, pp. 187–194. ACM Press/Addison-Wesley Publishing Co. (1999)
3. Blanz, V., Vetter, T.: Face recognition based on fitting a 3D morphable model. IEEE Trans. Pattern Anal. Mach. Intell. **25**(9), 1063–1074 (2003)
4. Bolkart, T., Wuhrer, S.: 3D faces in motion: fully automatic registration and statistical analysis. Comput. Vis. Image Understand. **131**, 100–115 (2015)
5. Booth, J., Antonakos, E., Ploumpis, S., Trigeorgis, G., Panagakis, Y., Zafeiriou, S.: A 3D morphable model learnt from 10,000 faces. In: Proceedings of the IEEE Conference on Computer Vision and Pattern Recognition, pp. 5464–5473. IEEE (2016)
6. Booth, J., Antonakos, E., Ploumpis, S., Trigeorgis, G., Panagakis, Y., Zafeiriou, S.: 3D face morphable models "in-the-wild". In: Proceedings of the IEEE Conference on Computer Vision and Pattern Recognition, pp. 5464–5473. IEEE (2017)
7. Booth, J., et al.: 3D reconstruction of "in-the-wild" faces in images and videos. IEEE Trans. Pattern Anal. Mach. Intell. **40**(11), 2638–2652 (2018)
8. Boyd, S., Boyd, S.P., Vandenberghe, L.: Convex Optimization. Cambridge University Press, New York (2004)
9. Bulat, A., Tzimiropoulos, G.: How far are we from solving the 2D & 3D face alignment problem? (and a dataset of 230,000 3d facial landmarks). In: Proceedings of the International Conference on Computer Vision. IEEE (2017)
10. Egger, B., et al.: 3D morphable face models-past, present and future. arXiv preprint arXiv:1909.01815 (2019)
11. Feng, Y., Wu, F., Shao, X., Wang, Y., Zhou, X.: Joint 3D face reconstruction and dense alignment with position map regression network. In: Ferrari, V., Hebert, M., Sminchisescu, C., Weiss, Y. (eds.) Computer Vision – ECCV 2018. LNCS, vol. 11218, pp. 557–574. Springer, Cham (2018). https://doi.org/10.1007/978-3-030-01264-9_33
12. Garrido, P., et al.: Reconstruction of personalized 3D face rigs from monocular video. ACM Trans. Graph. **35**(3), 1–15 (2016)

13. Gecer, B., Ploumpis, S., Kotsia, I., Zafeiriou, S.: Ganfit: generative adversarial network fitting for high fidelity 3D face reconstruction. In: Proceedings of the IEEE Conference on Computer Vision and Pattern Recognition, pp. 1155–1164. IEEE (2019)
14. Gerig, T., et al.: Morphable face models - an open framework. In: Proceedings of the IEEE International Conference on Automatic Face Gesture Recognition, pp. 75–82. IEEE (2018)
15. Guo, Y., Cai, J., Jiang, B., Zheng, J., et al.: CNN-based real-time dense face reconstruction with inverse-rendered photo-realistic face images. IEEE Trans. Pattern Anal. Mach. Intell. **41**(6), 1294–1307 (2018)
16. Hartley, R., Zisserman, A.: Multiple View Geometry in Computer Vision. Cambridge University Press, New York (2003)
17. Hernandez, M., Hassner, T., Choi, J., Medioni, G.: Accurate 3D face reconstruction via prior constrained structure from motion. Comput. Graph. **66**, 14–22 (2017)
18. Hu, L., et al.: Avatar digitization from a single image for real-time rendering. ACM Trans. Graph. **36**(6), 1–14 (2017)
19. Jackson, A.S., Bulat, A., Argyriou, V., Tzimiropoulos, G.: Large Pose 3D Face Reconstruction from a Single Image via Direct Volumetric CNN Regression. IEEE (2017)
20. Liu, Y., Jourabloo, A., Ren, W., Liu, X.: Dense face alignment. In: Proceedings of the International Conference on Computer Vision Workshops, pp. 1619–1628. IEEE (2017)
21. Paysan, P., Knothe, R., Amberg, B., Romdhani, S., Vetter, T.: A 3D face model for pose and illumination invariant face recognition. In: Proceedings of IEEE International Conference on Advanced Video and Signal based Surveillance for Security, Safety and Monitoring in Smart Environments, pp. 296–301. IEEE (2009)
22. Piotraschke, M., Blanz, V.: Automated 3D face reconstruction from multiple images using quality measures. In: Proceedings of the IEEE Conference on Computer Vision and Pattern Recognition, pp. 3418–3427. IEEE (2016)
23. Qu, C., Monari, E., Schuchert, T., Beyerer, J.: Adaptive contour fitting for pose-invariant 3D face shape reconstruction. In: Xie, X., Jones, M.W., Tam, G.K.L. (eds.) Proceedings of the British Machine Vision Conference, pp. 87.1-87.12. BMVA Press (2015)
24. Romdhani, S., Vetter, T.: Estimating 3D shape and texture using pixel intensity, edges, specular highlights, texture constraints and a prior. In: Proceedings of the IEEE Conference on Computer Vision and Pattern Recognition, vol. 2, pp. 986–993. IEEE (2005)
25. Sagonas, C., Tzimiropoulos, G., Zafeiriou, S., Pantic, M.: 300 faces in-the-wild challenge: the first facial landmark localization challenge. In: Proceedings of the International Conference on Computer Vision Workshops, pp. 397–403. IEEE (2013)
26. Sela, M., Richardson, E., Kimmel, R.: Unrestricted facial geometry reconstruction using image-to-image translation. In: Proceedings of the International Conference on Computer Vision, pp. 1576–1585. IEEE (2017)
27. Shi, F., Wu, H.T., Tong, X., Chai, J.: Automatic acquisition of high-fidelity facial performances using monocular videos. ACM Trans. Graph. **33**(6), 1–13 (2014)
28. Tewari, A., et al.: FML: face model learning from videos. In: Proceedings of the IEEE Conference on Computer Vision and Pattern Recognition, pp. 10812–10822. IEEE (2019)
29. Tewari, A., et al.: Self-supervised multi-level face model learning for monocular reconstruction at over 250 hz. In: Proceedings of the IEEE Conference on Computer Vision and Pattern Recognition, pp. 2549–2559. IEEE (2018)

30. Thies, J., Zollhofer, M., Stamminger, M., Theobalt, C., Nießner, M.: Face2face: real-time face capture and reenactment of RGB videos. In: Proceedings of the IEEE Conference on Computer Vision and Pattern Recognition, pp. 2387–2395. IEEE (2016)
31. Tran, L., Liu, F., Liu, X.: Towards high-fidelity nonlinear 3D face morphable model. In: Proceedings of the IEEE Conference on Computer Vision and Pattern Recognition, pp. 1126–1135. IEEE (2019)
32. Tzimiropoulos, G., Argyriou, V., Stathaki, T.: Subpixel registration with gradient correlation. IEEE Trans. Image Process. **20**(6), 1761–1767 (2010)
33. Tzimiropoulos, G., Alabort-i-Medina, J., Zafeiriou, S., Pantic, M.: Generic active appearance models revisited. In: Lee, K.M., Matsushita, Y., Rehg, J.M., Hu, Z. (eds.) ACCV 2012. LNCS, vol. 7726, pp. 650–663. Springer, Heidelberg (2013). https://doi.org/10.1007/978-3-642-37431-9_50
34. Tzimiropoulos, G., Zafeiriou, S., Pantic, M.: Robust and efficient parametric face alignment. In: Proceedings of the International Conference on Computer Vision, pp. 1847–1854. IEEE (2011)
35. Upton, G., Cook, I.: A Dictionary of Statistics 3e. Oxford University Press, Oxford (2014)
36. Valstar, M., et al.: Avec 2013: the continuous audio/visual emotion and depression recognition challenge. In: Proceedings of the ACM International Workshop on Audio/visual Emotion Challenge, pp. 3–10. ACM (2013)
37. Wächter, A.: Short tutorial: getting started with ipopt in 90 minutes. In: Naumann, U., Schenk, O., Simon, H.D., Toledo, S. (eds.) Combinatorial Scientific Computing. Schloss Dagstuhl - Leibniz-Zentrum fuer Informatik, Germany (2009)
38. Wang, K., Ji, Q.: Real time eye gaze tracking with 3D deformable eye-face model. In: Proceedings of the International Conference on Computer Vision, pp. 1003–1011. IEEE (2017)
39. Weise, T., Bouaziz, S., Li, H., Pauly, M.: Realtime performance-based facial animation. ACM Trans. Graph. **30**(4), 1–10 (2011)
40. Xue, N., Deng, J., Cheng, S., Panagakis, Y., Zafeiriou, S.: Side information for face completion: a robust PCA approach. IEEE Trans. Pattern Anal. Mach. Intell. **41**(10), 2349–2364 (2019)
41. Zhang, X., et al.: A high-resolution spontaneous 3d dynamic facial expression database. In: Proceedings of the IEEE International Conference and Workshops on Automatic Face and Gesture Recognition, pp. 1–6. IEEE (2013)
42. Zhou, Q.Y., Park, J., Koltun, V.: Open3D: A modern library for 3D data processing. arXiv:1801.09847 (2018)
43. Zhou, Y., Deng, J., Kotsia, I., Zafeiriou, S.: Dense 3D face decoding over 2500fps: joint texture & shape convolutional mesh decoders. In: Proceedings of the IEEE Conference on Computer Vision and Pattern Recognition, pp. 1097–1106 (2019)
44. Zhu, X., Lei, Z., Liu, X., Shi, H., Li, S.Z.: Face alignment across large poses: a 3D solution. In: Proceedings of the IEEE Conference on Computer Vision and Pattern Recognition. IEEE, June 2016
45. Zhu, X., Liu, X., Lei, Z., Li, S.Z.: Face alignment in full pose range: a 3D total solution. IEEE Trans. Pattern Anal. Mach. Intell. **41**(1), 78–92 (2017)
46. Zollhöfer, M., et al.: Real-time non-rigid reconstruction using an RGB-D camera. ACM Trans. Graph. **33**(4), 1–12 (2014)

Gabor Layers Enhance Network Robustness

Juan C. Pérez[1(✉)], Motasem Alfarra[2], Guillaume Jeanneret[1], Adel Bibi[2], Ali Thabet[2], Bernard Ghanem[2], and Pablo Arbeláez[1]

[1] Center for Research and Formation in Artificial Intelligence, Universidad de los Andes, Bogota, Colombia
jc.perez13@uniandes.edu.co
[2] King Abdullah University of Science and Technology (KAUST), Thuwal, Saudi Arabia

Abstract. We revisit the benefits of merging classical vision concepts with deep learning models. In particular, we explore the effect of replacing the first layers of various deep architectures with Gabor layers (*i.e.* convolutional layers with filters that are based on learnable Gabor parameters) on robustness against adversarial attacks. We observe that architectures with Gabor layers gain a consistent boost in robustness over regular models and maintain high generalizing test performance. We then exploit the analytical expression of Gabor filters to derive a compact expression for a Lipschitz constant of such filters, and harness this theoretical result to develop a regularizer we use during training to further enhance network robustness. We conduct extensive experiments with various architectures (LeNet, AlexNet, VGG16, and WideResNet) on several datasets (MNIST, SVHN, CIFAR10 and CIFAR100) and demonstrate large empirical robustness gains. Furthermore, we experimentally show how our regularizer provides consistent robustness improvements.

Keywords: Gabor · Robustness · Adversarial attacks · Regularizer

1 Introduction

Deep neural networks (DNNs) have enabled outstanding performance gains in several fields, from computer vision [16,21] to machine learning [23] and natural language processing [17]. However, despite this success, powerful DNNs are still highly susceptible to small perturbations in their input, known as adversarial attacks [15]. Their accuracy on standard benchmarks can be drastically reduced in the presence of perturbations that are imperceptible to the human eye. Furthermore, the construction of such perturbations is rather undemanding and, in some cases, as simple as performing a single gradient ascent step on a loss function with respect to the image [15].

J. C. Perez, M. Alfarra, G. Jeanneret denotes equal contribution.

Electronic supplementary material The online version of this chapter (https://doi.org/10.1007/978-3-030-58545-7_26) contains supplementary material, which is available to authorized users.

© Springer Nature Switzerland AG 2020
A. Vedaldi et al. (Eds.): ECCV 2020, LNCS 12354, pp. 450–466, 2020.
https://doi.org/10.1007/978-3-030-58545-7_26

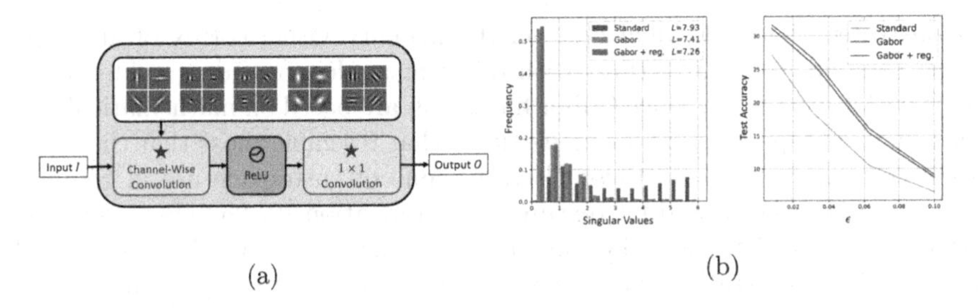

Fig. 1. Gabor layers and their effect on network robustness. (a): Gabor layers convolve each channel of the input with a set of learned Gabor filters. As low-level filters, Gabor filters offer a natural approach to represent local signals. (b): Replacing standard convolutional layers with Gabor layers yields an structured distribution of the filters' singular values, reduces the Lipschitz constant of the filters (L in the legend of the left plot), and improves accuracy under adversarial attacks (right figure). These results are for VGG16 on CIFAR100.

The brittleness of DNNs in the presence of adversarial attacks has spurred interest in the machine learning community, as evidenced by the emerging corpus of recent methods that focus on designing adversarial attacks [6,15,32,49]. This phenomenon is far-reaching and widespread, and is of particular importance in real-world scenarios, *e.g.*, autonomous cars [5,9] and devices for the visually impaired [38]. The risks that this degenerate behavior poses underscore the need for models that are not only accurate, but also robust to adversarial attacks.

Despite the complications that adversarial examples raise in modern computer vision, such inconveniences were not a major concern in the pre-DNN era. Many classical computer vision methods drew inspiration from the animal visual system, and so were designed to extract and use features that were semantically meaningful to humans [25–27,34,37]. As such, these methods were structured, generally comprehensible and, hence, better understood than DNNs. Furthermore, these methods even exhibited rigorous stability properties under robustness analysis [14]. However, mainly due to large performance gaps on several tasks, classical methods were overshadowed by DNNs. It is precisely in the frontier between classical computer vision and DNNs that a stream of works arose to combine tools and insights from both worlds to improve performance. For instance, the works of [46,53] showed that introducing structured layers inspired by the classical compressed sensing literature can outperform pure learning-based DNNs. Moreover, Bai *et al.* [3] achieved large gains in performance in instance segmentation by introducing intuitions from the classical watershed transform into DNNs.

In this paper, and searching for robustness in computer vision, we draw inspiration from biological vision, as the survival of species strongly depends on both the accuracy and robustness of the animal visual system. We note that Marr's and Julesz' work [19,30] argues that the visual cortex initially processes low-level agnostic information, in which the system's input is segmented according to blobs, edges, bars, curves, and boundaries. Furthermore, Hubel and Wiesel [18] demonstrated that individual cells on the primary visual cortex of an animal model respond to wave textures with different angles, providing evidence that supports Marr's theory. Since Gabor filters [13] are based on mathematical functions that are capable of modeling elements that resemble those that the animal visual cortices respond to, these filters became of customary use in computer vision, and have been used for texture characterization [19,25], character recognition [45], edge detection [33], and face recognition [8]. While several works examine their integration into DNNs [1,28,41], none investigate the effect of introducing parameterized Gabor filters into DNNs on the robustness of these networks. Our work fills this gap in the literature, as we provide experimental results demonstrating the significant impact that such architectural change has on improving robustness. Figure 1 shows an overview of our work and results.

Contributions: Our main contributions are two-fold: **(1)** We propose a *parameterized* Gabor-structured convolutional layer as a replacement for early convolutional layers in DNNs. We observe that such layers can have a remarkable impact on robustness. Thereafter, we analyze and derive an analytical expression for a Lipschitz constant of the Gabor filters, and propose a new training regularizer to further boost robustness. **(2)** We empirically validate our claims with a large number of experiments on different architectures (LeNet [24], AlexNet [21], VGG16 [44] and Wide-ResNet [51]) and over several datasets (MNIST [22], SVHN [31], CIFAR10 and CIFAR100 [20]). We show that introducing our proposed Gabor layers in DNNs induces a consistent boost in robustness at negligible cost, while preserving high generalizing test performance. In addition, we experimentally show that our novel regularizer based on the Lipschitz constant we derive can further improve adversarial robustness. For instance, we improve adversarial robustness on certain networks by almost 18% with ℓ_∞ bounded noise of $8/255$. Lastly, we show empirically that combining this architectural change with adversarial training [29,43] can further improve robustness.[1]

2 Related Work

Integrating Gabor Filters with DNNs. Several works attempted to combine Gabor filters and DNNs. For instance, the work of [41] showed that replacing the first convolutional layers in DNNs with Gabor filters speeds up the training procedure, while [28] demonstrated that introducing Gabor layers reduces the parameter count without hurting generalization accuracy. Regarding large scale datasets, Alekseev and Bobe [1] showed that the standard classification accuracy of AlexNet [21] on ImageNet [40] can be attained even when the first convolutional filters are replaced with Gabor filters. Moreover, other works have integrated Gabor filters with DNNs for various applications, *e.g.*, pedestrian detection [35], object recognition [50], hyper-spectral image classification [7], and Chinese optical character recognition [55]. Likewise, in this work, we study the effects of introducing Gabor filters into various DNNs by means of a *Gabor*

[1] Code at https://github.com/BCV-Uniandes/Gabor_Layers_for_Robustness.

layer, a convolution-based layer we propose, in which the convolutional filters are constructed by a parameterized Gabor function with learnable parameters. Furthermore, and based on the well-defined spatial structure of these filters, we study the effect of these layers on robustness, and find encouraging results.

Robust Neural Networks. Recent work demonstrated that DNNs are vulnerable to perturbations in their input. While input perturbations as simple as shifts and translations can cause drastic changes in the output of DNNs [54], the case of carefully-crafted adversarial perturbations has been of particular interest for researchers [47]. This susceptibility to adversarial perturbations incited a stream of research that aimed to develop not only accurate but also robust DNNs. A straightforward approach to this nuisance is the direct augmentation of data corrupted with adversarial examples in the training set [15]. However, the performance of this approach can be computationally limited, since the amount of augmentation needed for a high dimensional input space is computationally prohibitive. Moreover, Papernot *et al.* [36] showed that distilling DNNs into smaller networks can improve robustness. Another approach to robustness is through the functional lens. For instance, Parseval Networks [11] showed that robustness can be achieved by regularizing the Lipschitz constant of each layer in a DNN to be smaller than 1. In this work, along the lines of Parseval Networks [11], and since Gabor filters can be generated by sampling from a continuous Gabor function, we derive an analytical closed form expression for the Lipschitz constant of the filters of the proposed Gabor layer. This derivation allows us to propose well-motivated regularizers that can encourage Lipschitz constant minimization, and then harness such regularizers to improve the robustness of networks with Gabor layers.

Adversarial Training. An orthogonal direction for obtaining robust models is through optimization of a saddle point problem, in which an adversary, whose aim is to maximize the objective, is introduced into the traditional optimization objective. In other words, instead of the typical training scheme, one can minimize the worst adversarial loss over all bounded energy (often measured in ℓ_∞ norm) perturbations around every given input in the training data. This approach is one of the most celebrated for training robust networks, and is now popularly known as adversarial training [29]. However, this training comes at an inconvenient computational cost. To this regard, several works [43,48,52] proposed faster and computationally-cheaper versions of adversarial training capable of achieving similar robustness levels. In this work, we use "free" adversarial training [43] in our experiments to further study Gabor layers and adversarial training as orthogonal approaches to achieve robustness. Our results show how Gabor layers interact positively with adversarial training and hence, can be jointly used for enhancing network robustness.

3 Methodology

As demonstrated by Hubel and Wiesel [18], the first layers of visual processing in the animal brain are responsible for detecting low-level visual information.

Since Gabor filters have the capacity to capture low-level representations, and inspired by the robust properties of the animal visual system, we hypothesize that Gabor filters possess inherent robustness properties that are transferable to other systems, perhaps even DNNs. In this section, we discuss our proposed Gabor layer and its implementation. Then, we derive a Lipschitz constant to the Gabor filter, and design a regularizer, which aims at controlling the robustness properties of the layer by controlling the Lipschitz constant.

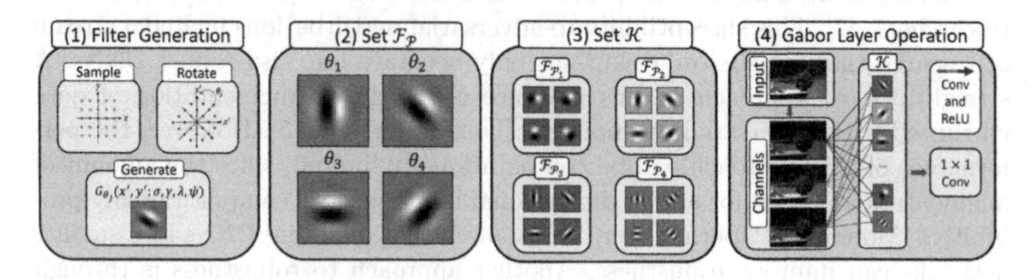

Fig. 2. Gabor layer operations. *(1)* We generate filters by rotating a sampled grid over multiple orientations and then evaluating the Gabor function according to the set of parameters $\mathcal{P}$ to yield the set of filters $\mathcal{F}_\mathcal{P}$ *(2)*. Then, we construct the total set of filters $\mathcal{K}$ by joining multiple sets $\mathcal{F}_{\mathcal{P}_i}$ *(3)*. Finally, the Gabor Layer operation *(4)* separately convolves every filter in $\mathcal{K}$ with every channel from the input, applies ReLU non-linearity, and then applies a 1×1 convolution to the output features to get the desired number of output channels.

3.1 Convolutional Gabor Filter as a Layer

We start by introducing the Gabor functions, defined as follows:

$$\begin{aligned} G_\theta(x',y';\sigma,\gamma,\lambda,\psi) &:= e^{-\,(x'^2+\gamma^2 y'^2)/\sigma^2}\,\cos(\lambda x'+\psi) \\ x' = x\cos\theta - y\sin\theta \quad & y' = x\sin\theta + y\cos\theta. \end{aligned} \tag{1}$$

To construct a discrete Gabor filter, we discretize x and y in Eq. (1) uniformly on a grid, where the number of grid samples determines the filter size. Given a fixed set of parameters $\{\sigma,\gamma,\lambda,\psi\}$, a grid $\{(x_i,y_i)\}_{i=1}^{k^2}$ of size $k \times k$, a rotation angle θ_j and a filter scale α_j, computing Eq. (1) with a scale over the grid yields a single surface $\alpha_j G_{\theta_j}(x',y';\sigma,\gamma,\lambda,\psi) \in \mathbb{R}^{1\times k\times k}$, that we interpret as a filter for a convolutional layer. The learnable parameters [39] for such a function are given by the set $\mathcal{P} = \{\alpha_j,\sigma,\gamma,\lambda,\psi; \forall j = 1,\ldots,r\}$, where the rotations θ_j are restricted to be r angles uniformly sampled from the interval $[0,2\pi]$. Evaluating these functions results in r rotated filters, each with a scale α_j defined by the set $\mathcal{F}_\mathcal{P} = \{\alpha_j G_{\theta_j}\}_{j=1}^r$. In this work, we consider several sets of learnable parameters $\mathcal{P}$, say p of them, thus, the set of all Gabor filters (totaling to rp filters) is given by the set $\mathcal{K} = \{\mathcal{F}_{\mathcal{P}_i}\}_{i=1}^p$. Refer to Fig. 2 for a graphical guide on the construction of $\mathcal{K}$.

3.2 Implementation of the Gabor Layer

Given an input tensor I with m channels, $I \in \mathbb{R}^{m\times h\times w}$, the Gabor layers follow a depth-wise separable convolution-based [10] approach to convolve the Gabor filters in $\mathcal{K}$ with I. The tensor I is first separated into m individual channels. Next, each channel is convolved with each filter in $\mathcal{K}$, and then a ReLU activation is applied to the output. Formally, the Gabor layer with filters in the set $\mathcal{K}$ operating on some input tensor I is presented as $\mathcal{R} = \{\text{ReLU}(I_i \star f_j),\ I_i \in \mathcal{I},\ f_j \in \mathcal{K},\ \forall i,j\}$ where $I_i = I(i,:,:) \in \mathbb{R}^{1\times h\times w}$ and $\star$ denotes the convolution operation. This operation produces $|\mathcal{R}| = mrp$ responses. Finally, the responses are stacked and convolved with a 1×1 filter with n filters. Thus, the final response is of size $n \times h' \times w'$. Figure 2 shows an overview of this operation.

3.3 Regularization

A function $f : \mathbb{R}^n \to \mathbb{R}$ is L-Lipschitz if $\|f(x) - f(y)\| \leq L\|x - y\|\ \forall x, y \in \mathbb{R}^n$, where L is the Lipschitz constant. Studying the Lipschitz constant of a DNN is essential for exploring its robustness properties, since DNNs with a small Lipschitz constant enjoy a better-behaving backpropagated signal of gradients, improved computational stability [42], and an enhanced robustness to adversarial attacks [11]. Therefore, to train networks that are robust, Cisse *et al.* [11] proposed a regularizer that encourages the weights of the networks to be tight frames, which are extensions of orthogonal matrices to non-square matrices. However, the training procedure is nontrivial to implement. Following the intuition of [11], we study the continuity properties of the Gabor layer and derive an expression for a Lipschitz constant as a function of the parameters of the filters, $\mathcal{P}$. This expression allows for direct network regularization on the parameters of the Gabor layer corresponding to the fastest decrease in the Lipschitz constant of the layer. To this end, we present our main theoretical result, which allows us to apply regularization on the Lipschitz constant of the filters of a Gabor layer.

Theorem 1. *Given a Gabor filter $G_\theta(m, n; \sigma, \gamma, \lambda, \psi)$, a Lipschitz constant L of the convolutional layer that has G_θ as its filter, with circular boundary conditions for the convolution, is given by:*

$$L = \left(1 + |X'|e^{-m_*^2/\sigma^2}\right)\left(1 + |Y'|e^{-\gamma^2 n_*^2/\sigma^2}\right),$$

where $X' = X \setminus \{0\}$, $Y' = Y \setminus \{0\}$, $X = \{x_i\}_{i=1}^{k^2}$ and $Y = \{y_i\}_{i=1}^{k^2}$ are sets of sampled values of the rotated (x', y') grid where $\{0\} \in X, Y$, $m_* = \text{argmin}_{x\in X'}|x|$ and $n_* = \text{argmin}_{y\in Y'}|y|$.

Proof. To compute the Lipschitz constant of a convolutional layer, one must compute the largest singular value of the underlying convolutional matrix of the filter. For separable convolutions, this computation is equivalent to the maximum magnitude of the 2D Discrete Fourier Transform (DFT) of the Gabor filter G_θ [4,42]. Thus, the Lipschitz constant of the convolutional layer is given by $L = \max_{u,v}|\text{DFT}(G_\theta(m, n; \sigma, \gamma, \lambda, \psi))|$, where DFT is the 2D DFT over the coordinates m and n in the spatial domain, u and v are the coordinates in the frequency domain, and $|\cdot|$ is the magnitude operator. Note that G_θ can be expressed as a product of two functions that are independent of the sampling sets X and Y as follows:

$$G_\theta(m,n;\sigma,\gamma,\lambda,\psi) := \underbrace{e^{-m^2/\sigma^2}\cos(\lambda m+\psi)}_{f(m;\sigma,\lambda,\psi)}\underbrace{e^{-\gamma^2 n^2/\sigma^2}}_{g(n;\sigma,\gamma)}.$$

Thus, we have

$$\begin{aligned} L &= \max_{u,v}\left|\text{DFT}\left(G_\theta(m,n;\sigma,\gamma,\lambda,\psi)\right)\right| \\ &= \max_{u,v}\Big|\sum_{m\in X} e^{-\omega_m u m} f(m;\sigma,\lambda,\psi)\sum_{n\in Y} e^{-\omega_n v n} g(n;\sigma,\gamma)\Big| \\ &\leq \max_{u,v}\sum_{m\in X}\left|f(m;\sigma,\lambda,\psi)\right|\sum_{n\in Y}\left|g(n;\sigma,\gamma)\right|. \end{aligned}$$

Note that $\omega_m = \frac{j2\pi}{|X|}$, $\omega_n = \frac{j2\pi}{|Y|}$ and $j^2=-1$. The last inequality follows from Cauchy–Schwarz and the fact that $|e^{-\omega_m u m}| = |e^{-\omega_n v n}| = 1$. Note that since $|g(n;\sigma,\gamma)| = g(n;\sigma,\gamma)$, and $|f(m;\sigma,\lambda,\psi)| \leq e^{-m^2/\sigma^2}$ we have that:

$$L \leq \sum_{m\in X} e^{-m^2/\sigma^2}\sum_{n\in Y} e^{-\gamma^2 n^2/\sigma^2} \leq \left(1+|X'|e^{-m_*^2/\sigma^2}\right)\left(1+|Y'|e^{-\gamma^2 n_*^2/\sigma^2}\right).$$

The last inequality follows by construction, since we have $\{0\}\in X,Y$, *i.e.*, the choice of uniform grid contains the 0 element in both X and Y, regardless of the orientation θ, where $m_* = \text{argmin}_{x\in X'}|x|$, and $n_* = \text{argmin}_{y\in Y'}|y|$. □

3.4 Lipschitz Constant Regularization

Theorem 1 provides an explicit expression for a Lipschitz constant of the Gabor filter as a function of its parameters. Note that the expression we derived decreases exponentially fast with σ. In particular, we note that, as σ decreases, G_θ converges to a scaled Dirac-like surface. Hence, this Lipschitz constant is minimized when the filter resembles a Dirac-delta. Therefore, to train DNNs with improved robustness, one can minimize the Lipschitz constant we derived in Theorem 1. Note that the Lipschitz constant of the network can be upper bounded by the product of the Lipschitz constants of individual layers. Thus, decreasing the Lipschitz constant we provide in Theorem 1 can aid in decreasing the overall Lipschitz constant of a DNN, and thereafter enhance the network's robustness. To this end, we propose the following regularized loss:

$$\mathcal{L} = \mathcal{L}_{\text{ce}} + \beta\sum_i \sigma_i^2, \tag{2}$$

where $\mathcal{L}_{\text{ce}}$ is the typical cross-entropy loss and $\beta > 0$ is a trade-off parameter. The loss in Equation (2) can lead to unbounded solutions for σ_i. To alleviate this behavior, we also propose the following loss:

$$\mathcal{L} = \mathcal{L}_{\text{ce}} + \beta \sum_{i} \left(\mu \tanh \sigma_i\right)^2, \tag{3}$$

where μ is a scaling constant for $\tanh \sigma$. In the following section, we present experiments showing the effect of our Gabor layers on network robustness. Specifically, we show the gains obtained from the architectural modification of introducing Gabor layers, the introduction of our proposed regularizer, and the addition of adversarial training to the overall pipeline.

4 Experiments

To demonstrate the benefits and impact on robustness of integrating our Gabor layer to DNNs, we conduct extensive experiments with LeNet [24], AlexNet [21], VGG16 [44], and Wide-ResNet [51] on the MNIST [22], CIFAR10, CIFAR100 [20] and SVHN [31] datasets. In each of the aforementioned networks, we replace up to the first three convolutional layers with Gabor layers, and measure the impact of the Gabor layers in terms of accuracy, robustness, and the distribution of singular values of the layers. Moreover, we perform experiments demonstrating that the robustness of the Gabor-layered networks can be enhanced further by using the regularizer we propose in Eqs. (2) and (3), and even when jointly employing regularization and adversarial training [43].

4.1 Implementation Details

We train all networks with stochastic gradient descent with weight decay of 5×10^{-4}, momentum of 0.9, and batch size of 128. For MNIST, we train the networks for 90 epochs with a starting learning rate of 10^{-2}, which is multiplied by a factor of 10^{-1} at epochs 30 and 60. For SVHN, we train models for 160 epochs with a starting learning rate of 10^{-2} that is multiplied by a factor of 10^{-1} at epochs 80 and 120. For CIFAR10 and CIFAR100, we train the networks for 300 epochs with a starting learning rate of 10^{-2} that is multiplied by a factor of 10^{-1} every 100 epochs.

4.2 Robustness Assessment

Following common practice for robustness evaluation [29,43], we assess the robustness of a DNN by measuring its prediction accuracy when the input is probed with adversarial attacks, which is widely referred to in the literature as "adversarial accuracy". We also measure the "flip rate", which is defined as the percentage of instances of the test set for which the predictions of the network changed when subjected to adversarial attacks.

Formally, if $x \in \mathbb{R}^d$ is some input to a classifier $C : \mathbb{R}^d \to \mathbb{R}^k$, $C(x)$ is the prediction of C at input x. Then, $x^{\text{adv}} = x + \eta$ is an adversarial example if the

Table 1. Test set accuracies on different datasets of various baselines, and their Gabor-layered versions. Gabor-layered architectures can recover the accuracies of their standard counterparts while providing robustness. Δ is the absolute difference between the baselines and the Gabor-layered architectures.

Dataset	Architecture	Baseline	Gabor	Δ
MNIST	LeNet	99.36	99.03	0.33
SVHN	WideResNet	96.62	96.70	0.08
SVHN	VGG16	96.52	96.18	0.34
CIFAR10	VGG16	92.03	91.35	0.68
CIFAR100	AlexNet	46.48	45.15	1.33
CIFAR100	WideResNet	77.68	76.86	0.82
CIFAR100	VGG16	67.54	64.49	3.05

prediction of the classifier has changed, *i.e.* $C(x^{\text{adv}}) \neq C(x)$. Both η and x^{adv} must adhere to constraints, namely: *(i)* the ℓ_p-norm of η must be bounded by some ϵ, *i.e.*, $\|\eta\|_p \leq \epsilon$, and *(ii)* x^{adv} must lie in the space of valid instances X, *i.e.*, $x^{\text{adv}} \in [0,1]^d$. A standard approach to constructing x^{adv} for some input x is by running Projected Gradient Descent (PGD) [29] with x as an initialization for several iterations. For some loss function $\mathcal{L}$, a PGD iteration projects a step of the Fast Gradient Sign Method [15] onto the valid set $\mathcal{S}$, which is defined by the constraints on η and x^{adv}. Formally, one iteration of PGD attack is:

$$x^{k+1} = \prod_{\mathcal{S}} \left(x^k + \delta \operatorname{sign}\left(\nabla_{x^k} \mathcal{L}(x^k, y) \right) \right),$$

where $\prod_{\mathcal{S}}$ is the projection operator onto $\mathcal{S}$ and y is the label. In our experiments, we consider attacks where η is ϵ-ℓ_∞ bounded. For each image, we run PGD for 200 iterations and perform 10 random restarts inside the ϵ-ℓ_∞ ball centered in the image. Following prior art [48], we set $\epsilon \in \{0.1, 0.2, 0.3\}$ for MNIST and $\epsilon \in \{2/255, 8/255, 16/255\}$ for all other datasets. Throughout our experiments, we assess robustness by measuring the test set accuracies under PGD attacks.

4.3 Performance of Gabor-Layered Architectures

The Gabor function in Eq. (1) restricts the space of patterns attainable by the Gabor filters. However, this set of patterns is aligned with what is observed in practice in the early layers of many standard architectures [2,21]. This observation follows the intuition that DNNs learn hierarchical representations, with early layers detecting lines and blobs, and deeper layers learning semantic information [12]. By experimenting with Gabor layers on various DNNs, we find that Gabor-layered DNNs recover close-to-standard, and sometimes better, test-set accuracies on several datasets. In Table 1, we report the test-set accuracies of several dataset-network pairs for standard DNNs and their Gabor-layered counterparts. We show the absolute difference in performance in the last column.

Moreover, in Fig. 3, we provide a visual comparison between the patterns learned by AlexNet in its original implementation [21] and those learned in the Gabor-layered version of AlexNet (trained on CIFAR100). We observe that filters in the Gabor layer converge to filters that are similar to those found in the original implementation of AlexNet, where we observe blob-like structures and oriented edges and bars of various sizes. Note that both sets of filters are, in turn, similar to filter banks traditionally used in computer vision, as those proposed by Leung and Malik [25]. Next, we show that the Gabor-layered networks highlighted in Table 1 not only achieve test set accuracies as high as those of standard DNNs, but also enjoy better robustness properties for free.

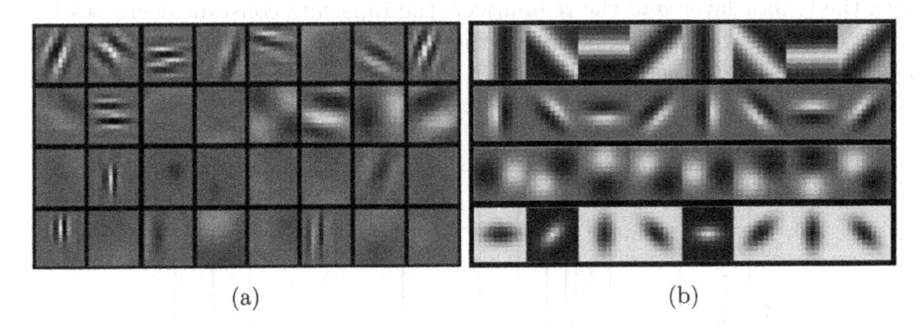

(a) (b)

Fig. 3. Comparison between filters learned by AlexNet and by Gabor-layered AlexNet. (a) Various filters learned in the first convolutional layer of AlexNet in its original implementation [21]. (b) Several filters learned in the Gabor-layered version of AlexNet. Each column in (b) is a different orientation of the same filter, while each row represents a different set of parameters of the Gabor function. Note that standard convolutional layers use multiple-channeled filters, while Gabor layers use single-channeled filters. Compellingly, both sets of filters present blobs and oriented edges and bars of various sizes and intensities.

4.4 Distribution of Singular Values

The Lipschitz constant of a network is an important quantity in the study of the network's robustness properties, since it is a measure of the variation of the network's predictions under input perturbations. Hence, in this work we study the distribution of singular values and, as a consequence, the Lipschitz constant of the filters of the layers in which we introduced Gabor layers instead of regular convolutional layers, in a similar fashion to [11]. In Fig. 4 we report box-plots of the distributions of singular values for the first layer of LeNet trained on MNIST, and the first three layers of VGG16 trained on CIFAR100. Each plot shows the distribution of singular values of the standard architectures (S), Gabor-layered architectures (G), and Gabor-layered architectures trained *with* the regularizers (G+r) proposed in Eqs. (2) and (3).

Figure 4 demonstrates that the singular values of the filters in Gabor layers tend to be concentrated around smaller values, while also being distributed in smaller ranges than those of their standard convolutional counterparts, as shown by the interquartile ranges. Additionally, in most cases, the Lipschitz constant of the filters of Gabor layers, *i.e.* the top notch of each box-plot, is smaller than that of standard convolutional layers.

Moreover, we find that training Gabor-layered networks with the regularizer we introduced in Eqs. (2) and (3) incites further reduction in the singular values of the Gabor filters, as shown in Fig. 4. For instance, the Gabor-layered version of LeNet trained on MNIST has a smaller interquartile range of the singular values, but still suffers from a large Lipschitz constant. However, by jointly introducing both the Gabor layer *and* the regularizer, the Lipschitz constant decreases by a factor of almost 5. This reduction in the Lipschitz constant is consistent in all layers of VGG16 trained on CIFAR100.

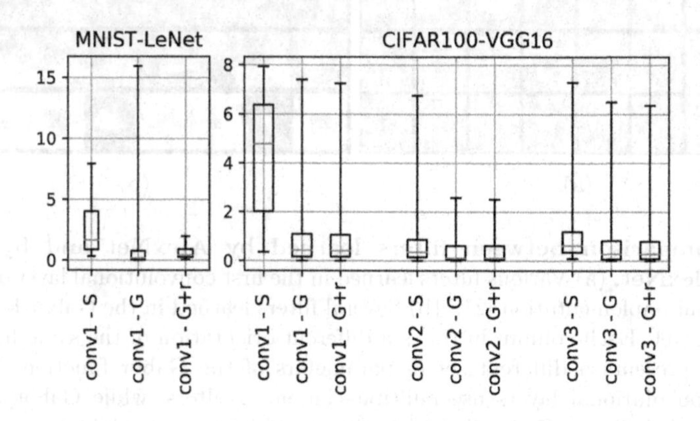

Fig. 4. Box-plot representation of the distribution of singular values in layers of LeNet and VGG16. Left: LeNet on MNIST. Right: VGG16 on CIFAR100. S: Standard; G: Gabor-layered; G+r: Gabor-layered with regularization. The top notch of each box-plot corresponds to the maximum value of the distribution, *i.e.* the Lipschitz constant of the layer.

4.5 Robustness in Gabor-Layered Architectures

After observing significant differences in the distribution of singular values between standard convolutional layers and Gabor layers, we now study the impact on robustness that Gabor layers introduce. We study the robustness properties of different architectures trained on various datasets when Gabor layers are introduced in the first layers of each network. The modifications that we perform on each architecture are:

- **LeNet.** We replace the first layer with a Gabor layer with $p = 2$.
- **AlexNet.** We replace the first layer with a Gabor layer with $p = 7$.

- **WideResNet.** We replace the first layer with a Gabor layer with $p = 4$ for SVHN, and $p = 3$ for CIFAR100.
- **VGG16.** We replace the first three layers with Gabor layers with parameters $p = 3$, $p = 1$ and $p = 3$, respectively.

Note that we fix the number of rotations r to be 8 in all of the experiments, and we leave the ablation on p to the **Appendix**.

Table 2. Adversarial accuracy comparison. We compare Standard (S), Gabor-layered (G), and *regularized* Gabor-layered (G+r) architectures. For each attack strength (ϵ), the highest performance is in **bold**; second-highest is underlined.

ϵ		$2/255$			$8/255$			$16/255$		
Dataset	Network	S	G	G+r	S	G	G+r	S	G	G+r
SVHN	WRN	40.27	49.35	**53.36**	1.02	1.03	**1.13**	1.32	**1.36**	1.19
SVHN	VGG16	57.86	62.88	**64.03**	5.84	14.57	**15.99**	2.33	7.98	**8.88**
CIFAR10	VGG16	34.22	37.60	**38.07**	23.63	30.11	**30.69**	13.88	19.50	**19.93**
CIFAR100	AN	**15.05**	14.77	14.68	4.80	7.71	**7.88**	5.37	6.25	**6.67**
CIFAR100	WRN	4.52	8.06	**9.33**	2.38	2.92	**3.07**	1.65	2.4	**2.59**
CIFAR100	VGG16	27.22	31.12	**31.68**	18.46	25.82	**26.64**	10.49	15.40	**16.06**

Standard Architectures *vs.* Gabor-Layered Architectures. We report the adversarial accuracies on both standard architectures and Gabor-layered architectures, where we refer to each with "S" and "G", respectively. Table 2 presents results on SVHN, CIFAR10 and CIFAR100, and Table 3 presents results on MNIST. We observe that Gabor-layered architectures consistently outperform their standard counterparts across datasets, and can provide up to a 9% boost in adversarial robustness. For instance, with $\epsilon = 8/255$ attacks, introducing Gabor layers into VGG16 boosts adversarial accuracy from 23.63 to 30.11 (6.48% relative increment) and from 5.84 to 14.57 (8.73% relative increment) on CIFAR10 and SVHN (Table 2), respectively. For LeNet on MNIST, under an $\epsilon = 0.2$ attack, introducing Gabor layers can boost adversarial accuracy from 4.39% to 7.94% (80% relative increment). Additionally, we report the flip rates for these experiments and for experiments on ImageNet as well in the **Appendix**. On the flip rates, we conduct a similar analysis, which yields equivalent conclusions: introducing Gabor layers leads to boosts in robustness.

It is worthwhile to note that the increase in robustness we observe in the Gabor-layered networks came *solely* from an architectural change, *i.e.* replacing convolutional layers with Gabor layers, without there being any other modification. Our experimental results demonstrate that: (**1**) Simply introducing Gabor filters, in the form of Gabor layers, as low-level feature extractors in DNNs provides beneficial effects to robustness. (**2**) Such robustness improvements are consistent across datasets and architectures. Inspired by these results, we now investigate the robustness effects of using our proposed regularization approach, as proposed in Eqs. (2) and (3).

Robustness Effects of Introducing Regularization. To better control robustness and to regularize the Lipschitz constant we derived in Theorem 1, we proposed two regularizers for the loss function as per Eqs. (2) and (3). As we noted in Subsect. 4.4 (refer to Fig. 4), Gabor-layered architectures inherently tend to have lower singular values than their standard counterparts. Additionally, upon training the same architectures *with* the proposed regularizer, we observe that Gabor layers tend to enjoy further reduction in the Lipschitz constant. These results suggest that such architectures may have enhanced robustness properties as a consequence.

Table 3. Adversarial accuracy comparison on MNIST. We compare Standard (S), Gabor-layered (G), and *regularized* Gabor-layered (G+r) architectures. For each attack strength (ϵ), the highest performance is in **bold**; second-highest is underlined.

ϵ		0.1			0.2			0.3		
Dataset	Network	S	G	G+r	S	G	G+r	S	G	G+r
MNIST	LeNet	80.04	80.58	**88.42**	4.39	7.94	**22.69**	0.44	**0.78**	0.76

To assess the role of the proposed regularizer on robustness, we train Gabor-layered architectures from scratch following the same parameters from Subsect. 4.1 and include the regularizer. We present the results in Tables 2 and 3, where we refer to these regularized architectures as "G+r". We observe that, in most cases, adding the regularizer improves adversarial accuracy. For instance, for LeNet on MNIST, the regularizer improves adversarial accuracy over the Gabor-layered architecture without any regularization by 8% and 14% with $\epsilon = {}^{2}/_{255}, {}^{8}/_{255}$ attacks, respectively. This improvement is still present in more challenging datasets. For instance, for VGG16 on SVHN under attacks with $\epsilon = {}^{2}/_{255}$ and $\epsilon = {}^{8}/_{255}$, we observe increments of over 1% from the regularized architecture with respect to its non-regularized equivalent. For the rest of the architectures and datasets, we observe modest but, nonetheless, sustained increments in performance. We report the flip rates for these experiments in the **Appendix**. The conclusions obtained from analyzing the flip rates are analogous to what we conclude from the adversarial accuracies: applying regularization on these layers provides minor but consistent improvements in robustness.

It is worthy to note that, although the implementation of the regularizer is trivial and that we perform optimization including the regularizer for the same number of epochs as regular training, our results still show that it is possible to achieve desirable robustness properties. We expect that substantial modifications and optimization heuristics can be applied in the training procedure, with aims at stronger exploitation of the insights we have provided here, and most likely resulting in more significant boosts in robustness.

4.6 Effects of Adversarial Training

Adversarial training [29] has become the standard approach for tackling robustness. In these experiments, we investigate how Gabor layers interact with adversarial training. We study whether the increments in robustness we observed when introducing Gabor layers can still be observed in the regime of adversarially-trained models. To study such interaction, we adversarially train standard, Gabor-layered and regularized Gabor-layered architectures, and then compare their robustness properties. We use the adversarial training described in [43], with 8 mini-batch replays, and $\epsilon = {}^{8}/_{255}$.

In Table 4, we report adversarial accuracies for AlexNet on CIFAR10 and CIFAR100 under $\epsilon = {}^{8}/_{255}$ attacks. Even in the adversarially trained networks-regime, our experiments show that **(1)** Gabor-layered architectures outperform their standard counterparts, and **(2)** regularization of Gabor-layered architectures provides substantial improvements in robustness with respect to their non-regularized equivalents. Such results demonstrate that Gabor layers represent an orthogonal approach towards robustness and, hence, that Gabor layers and adversarial training can be jointly harnessed for enhancing the robustness of DNNs.

The results we present here are empirical evidence that using closed form expressions for filter-generating functions in convolutional layers can be exploited for the purpose of increasing robustness in DNNs. We refer the interested reader to the **Appendix** for the rest of the experimental results.

Table 4. Adversarial accuracy with $\epsilon = {}^{8}/_{255}$. We study the effect of equipping our Gabor-layered and regularized Gabor-layered architectures with adversarial training on the robustness of the network.

			CIFAR10		CIFAR100	
Gabor	Reg.	Adv. training	AlexNet	VGG16	AlexNet	VGG16
		✓	19.24	41.95	13.48	18.49
✓		✓	20.26	42.41	10.94	**19.66**
✓	✓	✓	**22.15**	**44.02**	**13.62**	19.41

5 Conclusions

In this work, we study the effects in robustness of architectural changes in convolutional neural networks. We show that introducing Gabor layers consistently improves the robustness across various neural network architectures and datasets. We also show that the Lipschitz constant of the filters in these Gabor layers tends to be lower than that of traditional filters, which was theoretically and empirically shown to be beneficial to robustness [11]. Furthermore, theoretical analysis allows us to find a closed form expression for a Lipschitz constant of the Gabor filters. We then leverage this expression as a regularizer in the pursuit of enhanced robustness and validate its usefulness experimentally. Finally, we study the interaction between Gabor layers, our regularizer, and adversarial training, and show that the benefits of using Gabor layers are still observed when deep learning models are specifically trained for the purpose of adversarial robustness, showing that Gabor layers can be jointly used with adversarial training for further enhancements in robustness.

Acknowledgments. This work was partially supported by the King Abdullah University of Science and Technology (KAUST) Office of Sponsored Research (OSR) under Award No. OSR-CRG2019-4033.

References

1. Alekseev, A., Bobe, A.: Gabornet: gabor filters with learnable parameters in deep convolutional neural network. In: International Conference on Engineering and Telecommunication (EnT) (2019)
2. Atanov, A., Ashukha, A., Struminsky, K., Vetrov, D., Welling, M.: The deep weight prior. In: International Conference on Learning Representations (ICLR) (2019)
3. Bai, M., Urtasun, R.: Deep watershed transform for instance segmentation. In: IEEE Conference on Computer Vision and Pattern Recognition (CVPR) (2017)
4. Bibi, A., Ghanem, B., Koltun, V., Ranftl, R.: Deep layers as stochastic solvers. In: International Conference on Learning Representations (ICLR) (2019)
5. Cao, Y., et al.: Adversarial sensor attack on lidar-based perception in autonomous driving. In: ACM SIGSAC Conference on Computer and Communications Security (2019)
6. Carlini, N., Wagner, D.: Towards evaluating the robustness of neural networks. In: IEEE Symposium on Security and Privacy (SP) (2017)
7. Chen, Y., Zhu, L., Ghamisi, P., Jia, X., Li, G., Tang, L.: Hyperspectral images classification with gabor filtering and convolutional neural network. IEEE Geosci. Remote Sens. Lett. **14**, 2355–2359 (2017)
8. Liu, C., Wechsler, H.: Independent component analysis of gabor features for face recognition. IEEE Trans. Neural Netw. **14**, 919–928 (2003)
9. Chernikova, A., Oprea, A., Nita-Rotaru, C., Kim, B.: Are self-driving cars secure? evasion attacks against deep neural networks for steering angle prediction. In: IEEE Security and Privacy Workshops (SPW) (2019)
10. Chollet, F.: Xception: deep learning with depthwise separable convolutions. In: IEEE Conference on Computer Vision and Pattern Recognition (CVPR) (2017)
11. Cisse, M., Bojanowski, P., Grave, E., Dauphin, Y., Usunier, N.: Parseval networks: improving robustness to adversarial examples. In: International Conference on Machine Learning (ICML) (2017)
12. Farabet, C., Couprie, C., Najman, L., LeCun, Y.: Learning hierarchical features for scene labeling. IEEE Trans. Pattern Anal. Mach. Intell. **35**(8), 1915–1929 (2012)
13. Gabor, D.: Theory of communication. Part 1: The analysis of information. J. Inst. Electr. Eng. Part III Radio Commun. Eng. **93**, 429–441 (1946)
14. Goldstein, T., Osher, S.: The split Bregman method for L1-regularized problems. SIAM J. Imag. Sci. **2**, 323–343 (2009)
15. Goodfellow, I.J., Shlens, J., Szegedy, C.: Explaining and harnessing adversarial examples. In: International Conference on Learning Representations (ICLR) (2015)

16. He, K., Zhang, X., Ren, S., Sun, J.: Deep residual learning for image recognition. In: IEEE Conference on Computer Vision and Patter Recognition (CVPR) (2016)
17. Hinton, G., et al.: Deep neural networks for acoustic modeling in speech recognition: the shared views of four research groups. IEEE Sig. Process. Mag. **29**, 82–97 (2012)
18. Hubel, D.H., Wiesel, T.N.: Receptive fields of single neurons in the cat's striate cortex. J. Physiol. **148**, 574–591 (1959)
19. Julesz, B.: Textons, the elements of texture perception, and their interactions. Nature **290**, 91–97 (1981)
20. Krizhevsky, A., Hinton, G.: Learning multiple layers of features from tiny images. Technical report, Citeseer (2009)
21. Krizhevsky, A., Sutskever, I., Hinton, G.E.: Imagenet classification with deep convolutional neural networks. In: Neural Information Processing Systems (NeurIPS) (2012)
22. LeCun, Y.: The MNIST database of handwritten digits (1998). http://yann.lecun.com/exdb/mnist/
23. LeCun, Y., Bengio, Y., Hinton, G.: Deep learning. Nature **521**, 436–444 (2015)
24. LeCun, Y., Haffner, P., Bottou, L., Bengio, Y.: Object recognition with gradient-based learning. Shape, Contour and Grouping in Computer Vision. LNCS, vol. 1681, pp. 319–345. Springer, Heidelberg (1999). https://doi.org/10.1007/3-540-46805-6_19
25. Leung, T., Malik, J.: Representing and recognizing the visual appearance of materials using three-dimensional textons. Int. J. Comput. Vis. (IJCV) **43**, 29–44 (2001)
26. Lowe, D.G.: Object recognition from local scale-invariant features. In: International Conference on Computer Vision (ICCV). IEEE (1999)
27. Lowe, D.G.: Distinctive image features from scale-invariant keypoints. Int. J. Comput. Vis. (IJCV) **60**, 91–110 (2004)
28. Luan, S., Chen, C., Zhang, B., Han, J., Liu, J.: Gabor convolutional networks. IEEE Trans. Image Process. **27**, 4357–4366 (2018)
29. Madry, A., Makelov, A., Schmidt, L., Tsipras, D., Vladu, A.: Towards deep learning models resistant to adversarial attacks. In: International Conference on Learning Representations (ICLR) (2018)
30. Marr, D.: Vision: a computational investigation into the human representation and processing of visual information. In: PsycCRITIQUES (1982)
31. Montufar, G., Pascanu, R., Cho, K., Bengio, Y.: On the number of linear regions of deep neural networks. In: Advances in Neural Information Processing Systems (NeurIPS) (2014)
32. Moosavi-Dezfooli, S.M., Fawzi, A., Frossard, P.: Deepfool: a simple and accurate method to fool deep neural networks. In: IEEE Conference on Computer Vision and Pattern Recognition (CVPR) (2016)
33. Namuduri, K.R., Mehrotra, R., Ranganathan, N.: Edge detection models based on Gabor filters. In: International Conference on Pattern Recognition. Conference C: Image, Speech and Signal Analysis (1992)
34. Novak, C.L., Shafer, S.A., et al.: Anatomy of a color histogram. In: IEEE Conference on Computer Vision and Pattern Recognition (CVPR) (1992)
35. Ouyang, W., Wang, X.: Joint deep learning for pedestrian detection. In: IEEE International Conference on Computer Vision (ICCV) (2013)
36. Papernot, N., McDaniel, P., Wu, X., Jha, S., Swami, A.: Distillation as a defense to adversarial perturbations against deep neural networks. In: IEEE Symposium on Security and Privacy (SP) (2016)

37. Perona, P., Malik, J.: Scale-space and edge detection using anisotropic diffusion. IEEE Trans. Pattern Anal. Mach. Intell. **12**, 629–639 (1990)
38. Poggi, M., Mattoccia, S.: A wearable mobility aid for the visually impaired based on embedded 3D vision and deep learning. In: 2016 IEEE Symposium on Computers and Communication (ISCC) (2016)
39. Rumelhart, D.E., Hinton, G.E., Williams, R.J.: Learning representations by back-propagating errors. Nature **323**, 533–536 (1986)
40. Russakovsky, O., et al.: Imagenet large scale visual recognition challenge. In: International Journal of Computer Vision (IJCV) (2015)
41. Sarwar, S.S., Panda, P., Roy, K.: Gabor filter assisted energy efficient fast learning convolutional neural networks. CoRR (2017)
42. Sedghi, H., Gupta, V., Long, P.M.: The singular values of convolutional layers. In: International Conference on Learning Representations (ICLR) (2019)
43. Shafahi, A., et al.: Adversarial training for free! In: Neural Information Processing Systems (NeurIPS) (2019)
44. Simonyan, K., Zisserman, A.: Very deep convolutional networks for large-scale image recognition. In: International Conference on Learning Representations (ICLR) (2015)
45. Su, Y.M., Wang, J.F.: A novel stroke extraction method for Chinese characters using Gabor filters. Pattern Recogn. **36**, 635–647 (2003)
46. Sun, J., Li, H., Xu, Z., et al.: Deep ADMM-net for compressive sensing MRI. In: Neural Information Systems (NeurIPS) (2016)
47. Szegedy, C., et al.: Intriguing properties of neural networks. In: International Conference on Learning Representations (ICLR) (2014)
48. Wong, E., Rice, L., Kolter, J.Z.: Fast is better than free: revisiting adversarial training. In: International Conference on Learning Representations (ICLR) (2020)
49. Xu, K., et al.: Structured adversarial attack: towards general implementation and better interpretability. In: International Conference on Learning Representations (ICLR) (2019)
50. Yao, H., Chuyi, L., Dan, H., Weiyu, Y.: Gabor feature based convolutional neural network for object recognition in natural scene. In: International Conference on Information Science and Control Engineering (ICISCE) (2016)
51. Zagoruyko, S., Komodakis, N.: Wide residual networks. In: British Machine Vision Conference (BMVC) (2016)
52. Zhang, D., Zhang, T., Lu, Y., Zhu, Z., Dong, B.: You only propagate once: accelerating adversarial training via maximal principle. In: Neural Information Processing Systems (NeurIPS) (2019)
53. Zhang, J., Ghanem, B.: Deep learning. In: IEEE Conference on Computer Vision and Patter Recognition (CVPR) (2017)
54. Zhang, R.: Making convolutional networks shift-invariant again. In: International Con-ference on Machine Learning (ICML) (2019)
55. Zhong, Z., Jin, L., Xie, Z.: High performance offline handwritten Chinese character recognition using Googlenet and directional feature maps. In: International Conference on Document Analysis and Recognition (ICDAR) (2015)

Conditional Image Repainting via Semantic Bridge and Piecewise Value Function

Shuchen Weng[1], Wenbo Li[2], Dawei Li[2], Hongxia Jin[2], and Boxin Shi[1,3](✉)

[1] NELVT, Department of Computer Science and Technology, Peking University, Beijing, China
{shuchenweng,shiboxin}@pku.edu.cn
[2] Samsung Research America AI Center, Mountain View, CA, USA
{wenbo.li1,dawei.l,hongxia.jin}@samsung.com
[3] Institute for Artificial Intelligence, Peking University, Beijing, China

Abstract. We study conditional image repainting where a model is trained to generate visual content conditioned on user inputs, and composite the generated content seamlessly onto a user provided image while preserving the semantics of users' inputs. The content generation community has been pursuing to lower the skill barriers. The usage of human language is the rose among thorns for this purpose, because the language is friendly to users but poses great difficulties for the model in associating relevant words with the semantically ambiguous regions. To resolve this issue, we propose a delicate mechanism which bridges the semantic chasm between the language input and the generated visual content. The state-of-the-art image compositing techniques pose a latent ceiling of fidelity for the composited content during the adversarial training process. In this work, we improve the compositing by breaking through the latent ceiling using a novel piecewise value function. We demonstrate on two datasets that the proposed techniques can better assist tackling conditional image repainting compared to the existing ones.

Keywords: Image generation · Semantic · Compositing · Adversarial

1 Introduction

The advanced image editing techniques lower the skill barriers and simplify the required user inputs. For example, FaceApp [1] simplifies the user inputs to just one click for various face editing tasks including the alternations of smile, age and style. The research community also witnessed the great efforts along this direction, *e.g.*, GauGAN [13] is trained to synthesize images collaboratively with users, *i.e.*, users draw contours of objects, and GauGAN fills the object textures.

Electronic supplementary material The online version of this chapter (https://doi.org/10.1007/978-3-030-58545-7_27) contains supplementary material, which is available to authorized users.

© Springer Nature Switzerland AG 2020
A. Vedaldi et al. (Eds.): ECCV 2020, LNCS 12354, pp. 467–482, 2020.
https://doi.org/10.1007/978-3-030-58545-7_27

Aiming to further push the frontier of the practicability of the image editing techniques, we study a practical use case of image editing, *i.e.*, *conditional image repainting*, which is achieved through the collaboration between users and the trained model. By "repainting", we mean that the model is trained to repaint an area of an existing image with some visual content. By "conditional", we mean that the visual content to be repainted is generated by the model conditioned on several user inputs. As such, the conditional image repainting can be formulated into two sub-tasks, *i.e.*, conditional content *generation* and content *compositing*. Figure 1 illustrates our targeted conditional image repainting problem.

Fig. 1. From left to right, we show the input image, input semantic parsing mask (the white indicates the unaltered regions), generated content, and composited image. The input color description is "The grass is green and yellow, the pavement is gray, and the sky is white". (Color figure online)

Conditional content generation refers to visual synthesis tasks conditioned on user inputs. The user inputs cover *three aspects*, *i.e.*, geometry (shape, pose and semantic labels), colors, and gray-scale textures, and can be roughly divided into two categories, *i.e.*, visually-concrete reference (*e.g.*, reference images and semantic parsing masks) and visually-abstract description (*e.g.*, language and latent code). For example, in [13,22], the geometry is provided by users as semantic parsing masks, and the gray-scale textures altogether with colors as a whole is squeezed into a latent code. To enable higher control flexibility, Li *et al.* [12] attempt to use a separate reference image as user input for each aspect. However, it is sometimes cumbersome to find a desirable reference image, or very expensive to modify a reference image if the off-the-shelf one is not satisfying.

This motivates us to study the conditional content generation based on inputs that are more user-friendly. For the geometry, we follow [13,22] to use semantic parsing masks as input which can be easily manipulated by users. The gray-scale textures are highly correlated with the content class to be generated, so we use the latent code as input for the sake of reducing users' burden. A naive solution of providing the color input is to ask users to select different colors from a palette, and associate them with different parts of the geometry. However, this solution may require extensive and precise operations on the user interface, which is impractical on mobile devices with relatively small screens. Natural language is a user-friendly option for summarizing colors and their distributions. In order to enforce the generated content at each region to reflect its relevant words, the generation model needs to associate the word features with the relevant regions on the image plane. The word features correspond to semantically meaningful words, but as an embryo of the generated content, the region features are semantically ambiguous, thus causing a great challenge for the cross-modality association. We name such a challenge as the *semantic chasm*. In order to bridge the chasm, we disambiguate the semantics of region features through the semantic parsing mark which can be considered as a cross-modality mediator. By overlaying a semantic parsing mask on the image plane, regions covered by object masks can be associated with words which are relevant to the object class names. Guided by this philosophy, we propose a delicate and plug-n-play SEmantic-BridgE (SEBE) attention mechanism for assisting using language as the input color condition.

To complete the repainting task, the model needs to adjust the contrast and brightness of the generated visual content according to a user-provide image while preserving the semantics of the user-input conditions, and then composites the adjusted content at the user-indicated location of the user-provided image. The assumption is that the composited content should visually be indistinguishable from the innate content of the provided image. Based on this assumption, state-of-the-art image compositing techniques [3,9] train an adversarial discriminator for segmenting the composited content so as to supervise the compositing model (which plays a similar role to a generator). Such an adversarial training poses a latent ceiling of fidelity for the composited content with a rigid value function by restricting that when training the compositing model, under no circumstances, the discriminator should identify the innate content as the composited one. However, there might be cases being mistakenly penalized where the fidelity of the composited content is high enough to confuse the discriminator. Therefore, we propose a *piecewise value function* for the discriminator which applies the proper penalization opportunistically. In order to pave the way for the piecewise value function, we preprocess the input to the discriminator so as to impede the convergence of the discriminator.

We conduct extensive experiments on CUB-200-2011 dataset [21] and COCO-Stuff dataset [2], and show that the proposed SEBE attention mechanism and piecewise value function are beneficial for conditional image repainting.

2 Related Works

Cross-Modality Attention. Most existing methods [4,15,17,25,26,29] addressed the semantic chasm by attentively estimating the relevance between words and regions. Such an attentive model is trained in a data-driven fashion and the training is partially supervised by a cross-modality retrieval [25] or reconstruction [15] loss. Let $e[i]$ and $h[j]$ denote features of the i-th word and the j-th region, respectively. Their relevance is estimated as an attention weight $\beta[j, i]$:

$$\beta[j,i] = \frac{\exp(s[j,i])}{\sum_{k=1}^{K} \exp(s[j,k])}, \quad s[j,i] = (h[j])^T \phi(e[i]), \tag{1}$$

where $s[j,i]$ is the similarity between $h[j]$ and $e[i]$ which is computed by the dot product, and K is the number of words. $\phi(\cdot)$ is a linear layer for mapping the word features to the domain of the region features. With the estimated attention weights, in most existing methods [4,15,17,25,29], the words' features are aggregated by their relevance to each region on the image plane: $c[j] = \sum_{i=1}^{N} \beta[j,i]\phi(e[i])$, where $c[j]$ is the aggregated word features (or named context feature vector) for the j-th region. For each region, $c[j]$ is concatenated with $h[j]$, so as to enforce the generated content at each region to reflect its relevant words. Despite great improvements achieved, the cross-modality attention estimation is still challenging because of the semantic ambiguity of regions.

Content compositing has been tackled from different angles, *e.g.*, handcrafted feature matching [20,24], fusion of semantic information [19], image reconstruction [6], *etc.* Recent progress in content compositing derives from the compositing models [3,5,9,18] based on the adversarial training. As a concurrent work, Cong *et al.* [5] train a U-Net [16] based model to fuse the composited content and the innate content of the provided image, and also propose a domain verification discriminator for supervising the compositing model. Compared to [3,5,9,18] address the compositing problem in a more parametric fashion, which train a model to infer a set of contrast and brightness affine parameters for adjusting the color tone of the composited content. The compositing model is supervised by a segmentation [3,9] or detection [18] based adversarial discriminator.

Conditional normalization is proposed to alleviate the "condition dilution" problem by performing an affine transformation after each normalization operation. The affine parameters, which are inferred through a network from the input condition, are responsible for modulating the activations either element-by-element [13] or channel-by-channel [10]. The element-wise transformation is tailored for the input condition with spatial dimensions, *e.g.*, parsing mask, while the channel-wise one is much more general and not limited to spatial-explicit condition, and thus should be suitable for our gray-scale texture condition, *i.e.*, Gaussian noise vector.

3 Preliminaries

This section revisits recent techniques for addressing our targeted problems, so as to analyze their limitations in detail, and clarify our motivations technically.

3.1 Object-Driven Attention for Content Generation

To further resolve the semantic chasm challenge, Li *et al.* [11] proposed the object-driven attention in which with the help of objects' names, the cross-modality attention estimation can be converted to the attention estimation within the same modality. The intuition of the object-driven attention (2) is as follows: *(i)* The attention of a region to a word can be estimated by comparing the name embedding of the object that covers this region and the embedding of this word. If the name of an object to be generated in the image matches with one in a sentence, the word embedding of these two names should be similar thus leading to high and reliable attention weight. *(ii)* The descriptive words of an object should reside around the object name in a sentence, so the feature of the object name should contain the information of its descriptive words due to the property of the bi-directional LSTM based text encoder [25]. *(iii)* Given the reliable attention weight and the meaningful feature of the object name, it should be more effective in enforcing the reflection of the relevant words on image regions. Similar to (1), the object-driven attention is formulated as:

$$\beta^{\text{obj}}[t,i] = \frac{\exp(s^{\text{obj}}[t,i])}{\sum_{k=1}^{K}\exp(s^{\text{obj}}[t,k])}, \quad s^{\text{obj}}[t,i] = (\hat{e}_{+}[t])^{T}\hat{e}[i], \tag{2}$$

where $\hat{e}[i]$ and $\hat{e}_{+}[t]$ denote the GloVe embedding [14] of the i-th word and the name of the t-th object, respectively. For the j-th region, if it is covered by the t-th object, its object-driven context feature $c^{\text{obj}}[j]$ can be computed similarly to computing $c[j]$ using the cross-modality attention: $c^{\text{obj}}[j] = \sum_{i=1}^{N}\beta^{\text{obj}}[t,i]\phi(e[i])$. If a region is not covered by any object, its object-driven context feature is set to all-zero. For each region, $c[j]$ and $c^{\text{obj}}[j]$ are concatenated with $h[j]$ to enforce the reflection of the relevant words.

Limitation 1. *Despite the rationality of the object-driven attention, it has three weaknesses that we cannot ignore.* (i) *The dot product is not suitable for computing the similarity in the object-driven attention* (2), *because its output depends on the magnitude of the word embedding vectors. For example, the similarity between any pair of identical word embedding vectors should be high and constant, because their similarity should indicate the "perfect-match". But, the dot product cannot guarantee this.* (ii) *The context feature vectors driven by these two attentions are concatenated with the region features without partiality. When these two attentions are not in consensus for a particular region, it causes extra burdens for the generation model learning to figure how out to use these two attentions in the training stage.* (iii) *The concatenation of two context feature vectors causes large overhead of the runtime memory.*

3.2 Segmentation-Based Adversarial Training for Compositing

We study [3,9] to design our compositing model, which employ the segmentation-based adversarial discriminator for training. During training, the adversarial discriminator D learns to identify the composited foreground content by maximizing the value function V_1:

$$\begin{aligned}\max_{D} V_1(D,G) = \mathbb{E}_{y\sim p_{\text{data}}(y)}\xi\big(\log(\mathbf{1}-D(y|\bar{y}))\big) + \mathbb{E}_{y\sim p_{\text{g}}(y)}\xi\big(\log D(y|\bar{y})\big) + \\ \mathbb{E}_{y\sim p_{\text{data}}(y)}\xi\big(\log(\mathbf{1}-D(\bar{y}|y))\big) + \underline{\mathbb{E}_{y\sim p_{\text{g}}(y)}\xi\big(\log(\mathbf{1}-D(\bar{y}|y))\big)},\end{aligned} \tag{3}$$

where p_{g} is a probability distribution defined by the compositing model G, and *y and $\bar{y}$ represent the foreground and background content, respectively.* $D(y|\bar{y})$

outputs the probability of y being the composited foreground content conditioned on $\bar{y}$. ξ represents the mean-reduction function for pixels inside a content.

D has two directions with the shared parameters: $D(y|\bar{y})$ and $D(\bar{y}|y)$. Given a fixed G, the optimality of $D(y|\bar{y})$ is proved in [7] to be $D_G^*(y|\bar{y}) = \frac{p_{\text{data}}(y)}{p_{\text{data}}(y)+p_{\text{g}}(y)}$. In the supplementary, we prove the optimality of $D(\bar{y}|y)$ to be $D_G^*(\bar{y}|y) = 0$ which has nothing to do with p_{g}, and thus can be regarded as a posterior-collapse state. This is understandable because the amount of real images for training is limited, given enough training steps, $D(\bar{y}|y)$ should be able to memorize the data. At that time, the loss terms related to $D(\bar{y}|y)$ should be invalid.

During the training of G, G minimizes the value function V_2:

$$\min_G V_2(G, D) = \mathbb{E}_{y\sim p_{\text{g}}(y)}\xi\big(\log D(y|\bar{y})\big) + \underline{\mathbb{E}_{y\sim p_{\text{g}}(y)}\xi\big(\log D(\bar{y}|y)\big)}. \tag{4}$$

The *intuition of* $V_2(4)$ is that both the composited foreground content and the innate background content should be identified as the innate content by D, *i.e.*, $D(y|\bar{y}) = 0$ and $D(\bar{y}|y) = 0$, and thus these two contents should be indistinguishable. However, the minimax game shown in $V_1(3)$ and $V_2(4)$ is atypical from the perspective of adversarial training, because the underlined terms in $V_1(3)$ and $V_2(4)$ should be identical for each player in a typical minimax game. The typical value function V_2' can be formed by substituting the underlined term in $V_1(3)$ for that in $V_2(4)$:

$$\min_G V_2'(G, D) = \mathbb{E}_{y\sim p_{\text{g}}(y)}\xi\big(\log D(y|\bar{y})\big) + \underline{\mathbb{E}_{y\sim p_{\text{g}}(y)}\xi\big(\log(1 - D(\bar{y}|y))\big)}. \tag{5}$$

Limitation 2. *Minimizing $V_2'(5)$ pushes G to evolve toward confusing D to identify the innate content as the composited one, i.e., $D(\bar{y}|y) = 1$, which contradicts the intuition of $V_2(4)$. Moreover, as discussed above, it should not be difficult for $D(\bar{y}|y)$ to reach its optimality. Thus, $D(\bar{y}|y)$ could be reliable for most of the training time, so the gradients deriving from the underlined term in $V_2'(5)$ would keep steady no matter how G evolves, thus bringing the potential harm to the training. Considering the high reliability of $D(\bar{y}|y)$, the underlined term in $V_2(4)$ would be kept minimal during the training period no matter how G evolves, so this term makes little sense in supervising G. Here we come to understand that [3, 9] abandon the harm of $V_2'(5)$ and embrace the limitation of $V_2(4)$. The limitation of $V_2(4)$ stems from the convenience for $D(\bar{y}|y)$ in reaching the convergence. Supposing that we can impede the convergence of $D(\bar{y}|y)$, the reliability of $D(\bar{y}|y)$ should be weakened. In this fashion, the aforementioned intuition of $V_2(4)$ may be too strict for the weakened $D(\bar{y}|y)$, because there are higher chances that the fidelity of the composited content is high enough to confuse $D(\bar{y}|y)$ to mistakenly identified the innate content as the composited one. So, G could be excessively penalized by $V_2(4)$ given the weakened $D(\bar{y}|y)$. In other words, $V_2(4)$ poses a latent ceiling of fidelity for the composited content by constraining that the fidelity of the composited content should not be high enough to confuse $D(\bar{y}|y)$.*

4 Conditional Image Repainting

In this work, the conditional image repainting is formulated as a generation-compositing setting. In the generation phase, the content generation model $G^{\rm cg}$ accepts three inputs: *(i)* a semantic parsing mask $x^{\rm g} \in \mathbb{L}^{N_{\rm g} \times H_1 \times W_1}$ for defining the content geometry, where $\mathbb{L} \in \{0, 1\}$, and $N_{\rm g}$, H_1 and W_1 represent the number of object classes, image height and width, respectively; *(ii)* a sentence $x^{\rm c}$ describing the colors and their distributions on the geometry; *(iii)* a Gaussian noise vector $x^{\rm t} \sim \mathcal{N}(0, 1)$ encoding the gray-scale textures. Then, $G^{\rm cg}$ maps these inputs to a visual content $\dot{y}$, concluding the generation phase.

$\dot{y}$ can be composited onto a user-provided image $\bar{y}$ at a user-indicated location. In order to make $\dot{y}$ and $\bar{y}$ more harmonious, the content compositing model $G^{\rm cc}$ infers a set of contrast and brightness affine parameters for $\dot{y}$ to be adjusted.

4.1 Semantic-Bridge Attention for Content Generation

In order to resolve Limitation 1, we propose the SEmantic-BridgE (SEBE) attention mechanism for content generation. The intuition of SEBE attention is the same as that of the object-driven attention in Sect. 3, *i.e.*, bridging the semantic chasm between the word features and the region features through the semantics of the geometry that covers those regions. SEBE achieves improvements over the object-driven attention in three aspects, *i.e.*, attention estimation, attention selection, and computational overhead.

Trustier Attention Estimation. Let $\hat{e}[i]$ and $\hat{e}_+[t]$ denote the GloVe embedding of the i-th word in the input sentence, and the name of the t-th object to be generated on the image plane. For any region, if it is covered by the geometry of the t-th object on the image plane, then its region feature h_j can be represented by $\hat{e}_+[t]$. Given the GloVe embedding of any word $\hat{e}[i]$, the cross-modality attention estimation between words and regions can thus be formulated as the attention estimation of the same modality (viz. the space of GloVe embedding).

The object-driven attention computes the similarity between $\hat{e}[i]$ and $\hat{e}_+[t]$ as their dot product in (2). However, as discussed in Sect. 3, the dot product operation is not suitable for computing the similarity for embedding of the same modality, because its output depends on the magnitude of the word embedding vectors. Therefore, supposing that the j-th region is covered by the t-th object, we formulate the attention estimation of SEBE as

$$\beta^{\rm SEBE}[j,i] = \frac{s^{\rm SEBE}[j,i]+1}{\sum_{k=1}^{K}(s^{\rm SEBE}[j,k]+1)}, \quad s^{\rm SEBE}[j,i] = \frac{(\hat{e}_+[t])^T\hat{e}[i]}{\|\hat{e}_+[t]\|\|\hat{e}[i]\|}, \tag{6}$$

where $s^{\rm SEBE}[j,k] \in [-1,1]$ is the cosine similarity between $\hat{e}_+[t]$ and $\hat{e}[i]$, viz. the dot product operation normalized the magnitude of two vectors. The attention weight $\beta^{\rm SEBE}[j,i] \in [0,1]$ is computed by shifting $s^{\rm SEBE}[j,i]$ to be non-negative and normalizing the shifted value by L1 Norm of attention weights of the t-th object for all words. Thus, if $\hat{e}_+[t]$ and $\hat{e}[i]$ are embedding vectors of the same word (viz. object name), $s^{\rm SEBE}[j,i]$ is able to stay constantly as 1. The coverage relationship between regions and objects is specified in the input semantic parsing mask x^{g}. If a region is not covered by any objects, we set its SEBE attention weight to be zero.

Smarter Attention Selection. As discussed in Sect. 3, the object-driven attention and the cross-modality are used to compute their respective context feature vector for each region, and these two types of context feature vectors are concatenated with the region features for model's further processing. Such an impartial treatment of these two attentions shift the duty of selecting which attention to trust from the input end to model, which causes extra learning burdens for model. Therefore, in SEBE, we address the attention selection at the input end with the philosophy of "loudness is persuasive". For the j-th region, we formulate its context feature vector computation based on two types of attentions as

$$c^{\mathrm{SEBE}}[j] = \sum_{i=1}^{N} \max(\beta^{\mathrm{SEBE}}[j,i], \beta[j,i])\phi(e[i]). \tag{7}$$

If the j-th region is covered by an object, $\beta^{\mathrm{SEBE}}[j,i]$ is computed as in (6), and otherwise is set to zero. $\beta[j,i]$ is a cross-modality attention weight which is computed as in (1). $\phi(\cdot)$ is a linear layer as introduced below (1). Consequently, there is only one context feature vector, *i.e.*, $c^{\mathrm{SEBE}}[j]$, for each region.

Lighter Computational Overhead. Reducing a half of the context feature vectors reduce the runtime memory overhead significantly because the concatenated features are supposed to be fed in a series of residual blocks which need to keep the feature dimensions the same throughout the process.

4.2 Piecewise Value Function for Content Compositing

In order to resolve Limitation 2, we propose a piecewise value function for content compositing. Specifically, we modify V_2(4) by replacing its rigid underlined term with a piecewise term:

$$\mathcal{S}(D(\bar{y}|y)) = \begin{cases} \mathbb{E}_{y\sim p_{\mathrm{g}}(y)}\xi\big(\log D(\bar{y}|y)\big), & \text{if } D(\bar{y}|y) < 0.5 \\ \mathbb{E}_{y\sim p_{\mathrm{g}}(y)}\xi\big(\log(1 - D(\bar{y}|y))\big), & \text{otherwise} \end{cases} \tag{8}$$

The philosophies behind (8) are two-fold: *(i)* when $D(\bar{y}|y) < 0.5$, it retains the intuition of V_2(4) in Sect. 3.2 that the composited content should be indistinguishable from the innate one, *i.e.*, $D(y|\bar{y}) = 0$ and $D(\bar{y}|y) = 0$; *(ii)* otherwise, we consider that $y \sim p_{\mathrm{g}}(y)$ has successfully confused $D(\bar{y}|y)$, so we urge G^{cc} to evolve along the direction of making $D(\bar{y}|y)$ more confused (viz. producing composited content of higher fidelity) by encouraging $D(\bar{y}|y)$ to approximate 1.

Considering the gradient ineffectiveness issue of the underlined term in V_2'(5) as justified in Sect. 3.2, combining these two philosophies in (8) help further improve G^{cc} while circumventing the weakness of V_2'(5). By replacing the underlined term in V_2(4) with (8), we have the piecewise value function V_3 as

$$\min_{G} V_3(G,D) = \mathbb{E}_{y\sim p_{\mathrm{g}}(y)}\xi\big(\log D(y|\bar{y})\big) + \mathcal{S}(D(\bar{y}|y)). \tag{9}$$

$V_1(3)$ and $V_3(9)$ compose a two-player minimax game for G^{cc} and its adversarial discriminator D^{cc}. As discussed in Sect. 3.2, it is convenient for $D^{cc}(y|\bar{y})$ to reach the convergence. Thus, in order to impede the convergence, we propose a novel but delicate strategy to improve D^{cc} based on the design in [9], which will be introduced in the network architecture design Sect. 4.3.

4.3 Network Architecture Design

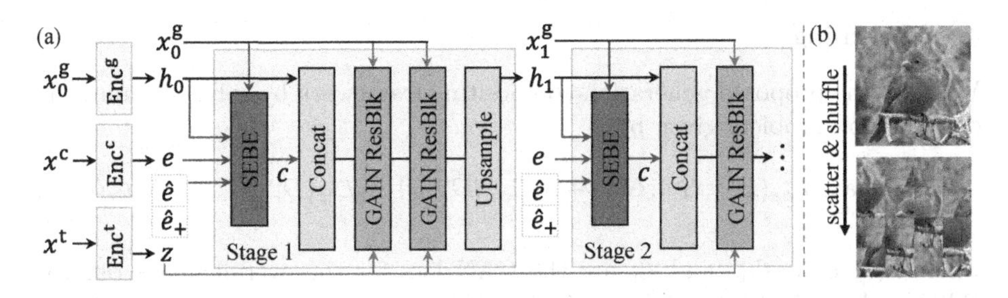

Fig. 2. *(a)* The multistage conditional content generator G^{cg} takes as input three conditions, *i.e.*, geometry x_0^g, color x^c, and gray-scale texture x^t for the first stage. x^c and x^t are encoded by Enc^c and Enc^t to form e and z for condition injection. x_0^g is encoded by Enc^g to form the initial region features h_0. The resolution of x_i^g is doubled with the increment of i, while the specified objects remain the same. $\hat{e}$ and $\hat{e}_+$ denote the GloVe embedding of words in x^c and names of objects specified in x_i^g. Red arrows indicate that regions are associated with their relevant words through SEBE, and the word features e are aggregated by these associations to form the context features c. Green arrows indicate the injection of z under the guidance of x_i^g. *(b)* The input preparation process, "scatter & shuffle", for the compositing discriminator D^{cc}.

Conditional Content Generator. G^{cg} is multistage, in which every stage shares the architecture that stacks two residual blocks as shown in Fig. 2(a). In each block, we employ the Gated Adaptive Instance Normalization (GAIN) [23] for injecting the texture code (*i.e.*, , Gaussian noise), which is proved to overcome the shortcomings of AdaIN in injecting the texture code for the non-rigid geometry. h_{i+1} represents the intermediate features from the previous stage, which can be fed into a conv-tanh block to generate an image $\dot{y}_{i+1}$ (omitted in Fig. 2(a)) The resolution of $\dot{y}_{i+1}$ is doubled with the increment of i. Red arrows indicate the inputs to and output from SEBE, which supplement the introduction in Sect. 4.1.

Compositing Model G^{cc}**.** Its design follows the recently proposed pixel transformation method [3] using a neural network to infer the contrast and brightness transformation parameters given both the composited and the innate contents.

Compositing Discriminator. D^{cc} is not designed following the segmentation-based discriminator in [3], because as discussed in Sect. 3.2, it is not difficult for the segmentation-based discriminator to reach convergence because the discriminator can memorize data after some epochs. This could weaken the effectiveness of the proposed piecewise value function in Sect. 4.2. Therefore, in order to impede the convergence, we exponentially increase the amount of real training images by reorganizing images using a simple "scatter & shuffle" strategy (which is also applied to the composited images) as shown in Fig. 2(b). This makes D^{cc} very hard to go through all reorganized images multiple times to memorize data during training. Then, the real/fake labels are no longer distributed by pixels but by patches. Therefore, we build D^{cc} as a simple CNN for patch-wise classification.

4.4 Learning

We train the proposed generation-compositing framework by solving a minimax optimization problem given by

$$\min_{D^{cg},D^{cc}} \max_{G^{cg},G^{cc}} \mathcal{L}_{cg}(D^{cg},G^{cg},G^{cc}) + \lambda_1 \mathcal{L}_{cm}(G^{cg}) + \lambda_2 \mathcal{L}_{cc}(D^{cc},G^{cc}) + \lambda_3 \mathcal{L}_{r}(G^{cc}), \tag{10}$$

where $\mathcal{L}_{cg}$, $\mathcal{L}_{cm}$, $\mathcal{L}_{cc}$, and $\mathcal{L}_{r}$ are the GAN loss for the overall image quality, DAMSM loss [25] for the color condition, GAN loss and a regularization loss [3] for the compositing performance, respectively. D^{cg} is a set of joint-conditional-unconditional patch discriminators [11] for each stage of G^{cg}. $\mathcal{L}_{cg}$, $\mathcal{L}_{cm}$, and $\mathcal{L}_{r}$ are borrowed from [23]. $\mathcal{L}_{cc}$ is defined by $V_1(3)$ for training D^{cc} and by $V_3(9)$ for training G^{cc}. Let y denote a composited image. Applying D^{cc} to y, we have $\mathbf{p} = D^{cc}(y)$, where $\mathbf{p} = \{p_1, \dots, p_i, \dots, p_N\}$ is a set of probabilities with each indicating how likely a patch belongs to the composited content. We define two index sets for each y, *i.e.*, the composited patch index set $\mathbb{I}^{ci}$ and the innate patch index set $\mathbb{I}^{ii}$. For training D^{cc}, $\mathcal{L}_{cc}$ is a typical classification loss. For training G^{cc}, $\mathcal{L}_{cc}$ is defined as

$$\mathcal{L}_{cc}(G^{cc}, D^{cc}) = -\frac{1}{N}\Big(\sum_{i \in \mathbb{I}^{ci}} \log(1-p_i) + \sum_{i \in \mathbb{I}^{ii}} \log \psi(p_i)\Big), \tag{11}$$

where ψ corresponds to the piecewise term in (8), which is defined as $\psi(p_i) = 1 - p_i$, if $p_i < 0.5$; otherwise, $\psi(p_i) = p_i$.

Based on the experiments on a held-out validation set, we set the hyperparameters in this section as: $\lambda_1 = 20$, $\lambda_2 = 0.03$, and $\lambda_3 = 1.0$.

5 Experiments

Datasets. We use CUB-200-2011 [21] and COCO-Stuff [2] for evaluation. For CUB, we annotate bird images with parsing masks, and follow [25] for data processing. For COCO, we select 9 most common stuff classes to use, including sky, grass, road, clouds, pavement, dirt, sand, bush, and sea. We annotate 10 captions per image, and use 6.2K images for training and 1.4K for test.
Quantitative Evaluation Metrics. Three evaluation metrics are used: *(i)* we use the *Fréchet inception distance* (FID) [8] score to evaluate the general image quality. *(ii)* Following [25], we use R-precision to evaluate whether the generated

Table 1. The quantitative experiments. ↑ (↓) means the higher (lower), the better. The best performances are highlighted in **bold**. The compared baselines are divided into six categories: Rows 1–2 for generation, and Rows 3–4 for compositing.

Category	Methods	CUB-200-2011			COCO-Stuff		
		FID ↓	R-prcn (%) ↑	M-score ↓	FID ↓	R-prcn (%) ↑	M-score ↓
Attn Est	SEBE w/ DotPrdct	12.6	98.72	32.86	19.3	59.14	81.62
	SEBE w/o CrsMod	12.68	98.75	35.68	19.43	58.73	72.11
	CrsMod	12.21	98.7	31.38	19.06	60.48	78.54
Attn Sel	SEBE w/o SAS	12.31	98.91	34.18	20.03	61.13	81.14
Seg	Seg V_2(4)	12.12	98.74	28.1	19.11	60.36	75.16
	Seg V_3(9)	12.25	98.99	33.53	19.25	57.34	76.35
Cls	Cls V_2(4)	12.39	98.81	27.17	19.23	57.4	74.74
	Cls V_2'(5)	12.44	**99.18**	34.95	19	57.65	81.22
	Cls V_3(9) w/o Pwise	12.62	98.99	26.65	18.96	58.53	76.21
Ours	SEBE-GAIN-Cls-V_3(9)	**12.08**	98.94	**24.6**	**18.91**	**65.43**	**67.96**

image is well conditioned on the input color description. More specifically, given a generated image y conditioned on the input sentence x^c and 9 randomly sampled sentences, we rank these 10 sentences by the pre-trained DAMSM model. If the ground truth sentence x^c is ranked the highest, we count this a success retrieval. We perform this retrieval task on all generated images and calculate the percentage of success retrievals as the R-precision score. *(iii)* For measuring the compositing quality, we follow [18] to use the M-score which is the output by a manipulation detection model [28]. The higher M-score, the higher possibility that an image has been manipulated. For each compared method, we randomly pick 500 generated images to calculate the average M-score.

5.1 Content Generation

We evaluate two aspects of our method for content generation, *i.e.*, attention estimation (abbr. Attn Est) and attention selection (abbr. Attn Sel) in Sect. 4.1. For each aspect, we create some baselines either by disabling modules of our model or adapting the existing techniques to our task. The quantitative and qualitative comparison are shown in Table 1 and left side of Fig. 3, respectively. Note that our full-version method outperforms the compared baselines in most metrics on both datasets, which demonstrates the effectiveness of our proposed modules quantitatively, so we focus on analyzing the qualitative results in the following.

Attention Estimation. We create three baselines for this aspect: *(i) SEBE w/ DotPrdct* using the object-driven attention (2) [11] for estimation, and keeping the attention selection (7); *(ii) SEBE w/o CrsMod* by disabling the cross-modality attention in (7); *(iii) CrsMod* using the cross-modality attention estimation (1) [25], and by disabling the SEBE attention in (7). Figure 3 shows that SEBE is effective in controlling the color artifacts such as Column 3 and 5 for *SEBE w/ DotPrdct*, Column 2 and 4 for *SEBE w/o CrsMod*, and Column 4 for *CrsMod*. Please refer to Sect. 2 and Limitation 1 for shortcomings of these baselines.

Attention Selection. We create *SEBE w/o SAS* by disabling the attention selection (7). As justified in Sect. 4.1, this module should be able to shift model's burden in selecting attention to the input end. The texture artifacts in Column 1 and the color artifacts in Column 4 of Fig. 3 are obvious for *SEBE w/o SAS*.

5.2 Content Compositing

For content compositing, we evaluate the influences of discriminator design, *i.e.*, the full-image and segmentation based discriminator [3] (abbr. Seg) vs. the shuffled-patches and classification based one (abbr. Cls), and the influences for different value functions including V_2(4), V_2'(5), V_3(9) w/o the piecewise (abbr. Pwise) term (8), and V_3(9).

Seg vs. Cls. When the Seg and Cls discriminators are evaluated with the same value functions, *i.e.*, V_2(4) and V_3(9), Cls outperforms Seg in terms of M-score on both datasets, which demonstrate the effectiveness of our discriminator design in Sect. 4.3. In the supplementary, we show that the Seg discriminator reaches the convergence much faster than the Cls one. This implies that Seg discriminator and the compositing model reach the Nash equilibrium faster, which prevents further improving the compositing model in the training.

Fig. 3. Qualitative comparison for content generation (left) and compositing (right). Best viewed on the computer, in color and zoomed-in. Input color descriptions: (1) "This bird is black and yellow in color, and has an orange beak." (2) "The bird has a white belly and chest with gray wings and tail and black striped head." (3) "This bird has a white belly and breast, with a long orange hooked bill." (4) "The pavement is brown and gray." (5) "The grass in the picture is brown and green." (Color figure online)

Value Functions. In Table 1, *Cls* $V_3(9)$ outperforms *Cls* $V_2(4)$ significantly in terms of M-score, while this is not the case for the Seg This phenomenon echos our analysis in Limitation 2 that we need to first impede the early convergence of the discriminator before the value function is improved. It also proves the necessity of modifying the discriminator as in Sect. 4.3. In addition, *Cls* $V_2'(5)$ yields much worse M-score than our method, which provides some evidence for our discussions in Limitation 2 about the weakness of $V_2'(5)$. Here we come to know that both $V_2(4)$ and $V_2'(5)$ cannot achieve prominent compositing performance alone. In fact, our proposed $V_3(9)$ implements a mechanism for choosing to apply $V_2(4)$ or $V_2'(5)$ at the right time. To further study the impact of the proposed piecewise term (8), we simply remove it from $V_3(9)$ to see the results. This means that the discriminator only cares about the composited content but ignores the innate one when training the compositing model. From Table 1, we see that the influences are more obvious on COCO-Stuff than on CUB. This might be because that COCO-Stuff is more challenging than CUB.

Fig. 4. Composited images in which the innate content is correctly identified or misidentified by D^{cc} are shown on the left and right, respectively.

Fig. 5. Comparison with [27] for mask based object removal. Masks are placed in the lower-left corner of the edited images, where the gray indicates regions to be filled. From left to right, we show the real image, the result of [27], and our result.

Assumption in Limitation 2 is that the fidelity of the composited content is high enough to confuse $D(\bar{y}|y)$ to mistakenly identified the innate content as the composited one, which is also the motivation for us to improve the value function in Sect. 4.2. Therefore, we select a discriminator in-between the training process, and visualize in Fig. 4 the randomly-sampled composited images in which the innate content is correctly identified or misidentified by our compositing discriminator D^{cc}. We can see that the compositing fidelity of the misidentified images is generally higher than that of the correctly identified images, which provides evidence supporting the assumption.

5.3 Qualitative Study

Object removal is always considered as a task for image inpainting. However, as shown in Fig. 5, the recently proposed [27] cannot handle cases with complicated background well. Surprisingly, our method can successfully remove the objects despite the cost of substituting the generated content for a large portion of content in the original images, *e.g.*, sky and lawn in Fig. 5. In the supplementary, we provide more analyses about this task and the limitations of our work in handling this task, and indicate the future direction of our research.
Iterative image editing in the wild is shown on the right of Fig. 6. From the real image in Column 5 to the final editing results in Column 8, the whole scenes look quite different, which demonstrates the robustness and flexibility of our method.

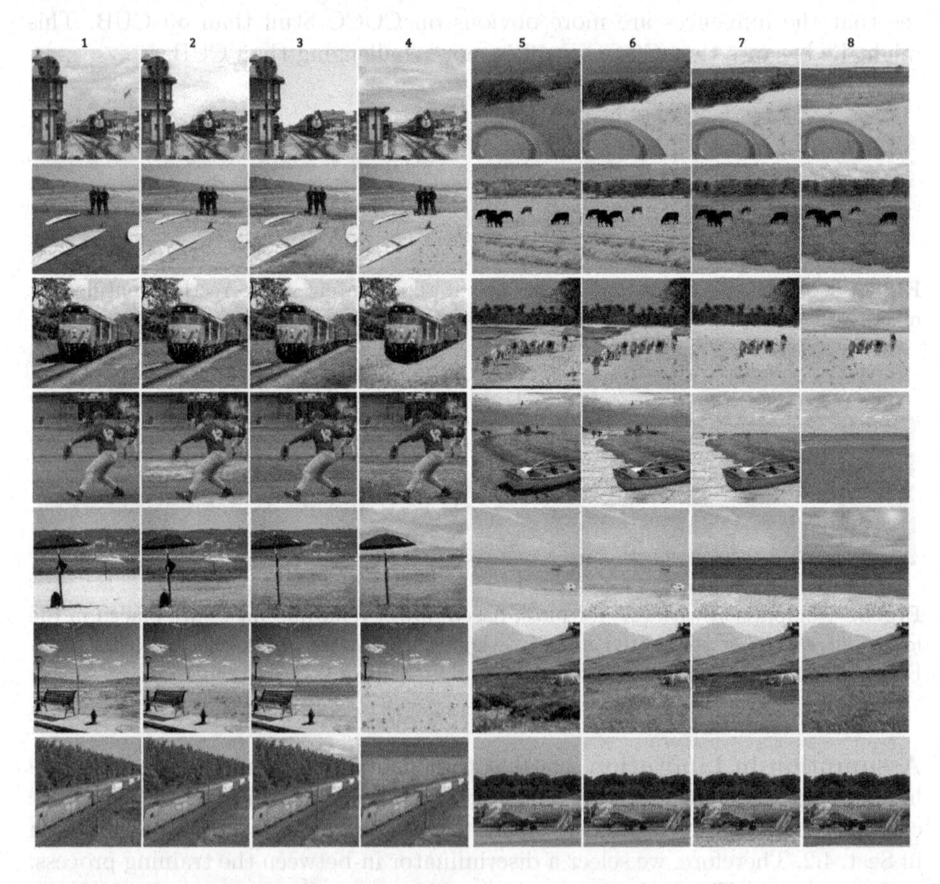

Fig. 6. Image editing in the wild. Column 1 and Column 5 show the real images. Columns 2–4 show the alternation of input conditions. Columns 6–8 show the iterative editing. See the supplementary for detailed input conditions for producing these images.

6 Conclusion

Targeting at a relatively new and practical task, conditional image repainting, we propose two novel and delicate modules for addressing the weaknesses of the existing component technologies, *i.e.*, semantic-bridge attention mechanism for assisting using languages as conditional input, and a piecewise value function to improve the adversarial training of the compositing model. We observe favorable performance with both quantitative and qualitative results, and also explore several interesting potential application scenarios of the proposed techniques.

Acknowledgments. PKU affiliated authors are supported by National Natural Science Foundation of China under Grant No. 61872012, National Key R&D Program of China (2019YFF0302902), and Beijing Academy of Artificial Intelligence (BAAI).

References

1. FaceApp. https://www.faceapp.com/
2. Caesar, H., Uijlings, J.R.R., Ferrari, V.: Coco-stuff: thing and stuff classes in context. In: CVPR (2018)
3. Chen, B., Kae, A.: Toward realistic image compositing with adversarial learning. In: CVPR (2019)
4. Chen, X., Qing, L., He, X., Luo, X., Xu, Y.: FTGAN: a fully-trained generative adversarial networks for text to face generation. CoRR abs/1904.05729 (2019)
5. Cong, W., et al.: Deep image harmonization via domain verification. CoRR abs/1911.13239 (2019)
6. Cun, X., Pun, C.: Improving the harmony of the composite image by spatial-separated attention module. CoRR abs/1907.06406 (2019)
7. Goodfellow, I.J., et al.: Generative adversarial nets. In: NIPS (2014)
8. Heusel, M., Ramsauer, H., Unterthiner, T., Nessler, B., Klambauer, G., Hochreiter, S.: GANs trained by a two time-scale update rule converge to a nash equilibrium. In: NIPS (2017)
9. Huang, H., Xu, S., Cai, J., Liu, W., Hu, S.: Temporally coherent video harmonization using adversarial networks. TIP **29**, 214–224 (2020)
10. Karras, T., Laine, S., Aila, T.: A style-based generator architecture for generative adversarial networks. In: CVPR (2019)
11. Li, W., et al.: Object-driven text-to-image synthesis via adversarial training. In: CVPR (2019)
12. Li, Y., Singh, K.K., Ojha, U., Lee, Y.J.: Mixnmatch: multifactor disentanglement and encoding for conditional image generation. CoRR abs/1911.11758 (2019)
13. Park, T., Liu, M., Wang, T., Zhu, J.: Semantic image synthesis with spatially-adaptive normalization. In: CVPR (2019)
14. Pennington, J., Socher, R., Manning, C.D.: Glove: global vectors for word representation. In: EMNLP (2014)
15. Qiao, T., Zhang, J., Xu, D., Tao, D.: Mirrorgan: learning text-to-image generation by redescription. In: CVPR (2019)
16. Ronneberger, O., Fischer, P., Brox, T.: U-Net: convolutional networks for biomedical image segmentation. In: Navab, N., Hornegger, J., Wells, W.M., Frangi, A.F. (eds.) MICCAI 2015. LNCS, vol. 9351, pp. 234–241. Springer, Cham (2015). https://doi.org/10.1007/978-3-319-24574-4_28

17. Tan, H., Liu, X., Li, X., Zhang, Y., Yin, B.: Semantics-enhanced adversarial nets for text-to-image synthesis. In: ICCV (2019)
18. Tripathi, S., Chandra, S., Agrawal, A., Tyagi, A., Rehg, J.M., Chari, V.: Learning to generate synthetic data via compositing. In: CVPR (2019)
19. Tsai, Y., Shen, X., Lin, Z., Sunkavalli, K., Lu, X., Yang, M.: Deep image harmonization. In: CVPR (2017)
20. Tsai, Y., Shen, X., Lin, Z., Sunkavalli, K., Yang, M.: Sky is not the limit: semantic-aware sky replacement. TOG **35**, 149 (2016)
21. Wah, C., Branson, S., Welinder, P., Perona, P., Belongie, S.: The Caltech-UCSD Birds-200-2011 Dataset. Technical report CNS-TR-2011-001, California Institute of Technology (2011)
22. Wang, T., Liu, M., Zhu, J., Tao, A., Kautz, J., Catanzaro, B.: High-resolution image synthesis and semantic manipulation with conditional gans. In: CVPR (2018)
23. Weng, S., Li, W., Li, D., Jin, H., Shi, B.: Misc: multi-condition injection and spatially-adaptive compositing for conditional person image synthesis. In: CVPR (2020)
24. Wu, H., Zheng, S., Zhang, J., Huang, K.: GP-GAN: towards realistic high-resolution image blending. In: ACM MM (2019)
25. Xu, T., et al.: Attngan: fine-grained text to image generation with attentional generative adversarial networks. In: CVPR (2018)
26. Yin, G., Liu, B., Sheng, L., Yu, N., Wang, X., Shao, J.: Semantics disentangling for text-to-image generation. In: CVPR (2019)
27. Yu, J., Lin, Z., Yang, J., Shen, X., Lu, X., Huang, T.S.: Free-form image inpainting with gated convolution. In: ICCV (2019)
28. Zhou, P., Han, X., Morariu, V.I., Davis, L.S.: Learning rich features for image manipulation detection. In: CVPR (2018)
29. Zhu, M., Pan, P., Chen, W., Yang, Y.: DM-GAN: dynamic memory generative adversarial networks for text-to-image synthesis. In: CVPR (2019)

Learnable Cost Volume Using the Cayley Representation

Taihong Xiao[1(✉)], Jinwei Yuan[2], Deqing Sun[2], Qifei Wang[2], Xin-Yu Zhang[3], Kehan Xu[4], and Ming-Hsuan Yang[1,2]

[1] University of California, Merced, USA
{txiao3,mhyang}@ucmerced.edu
[2] Google Research, Mountain View, USA
{jinwei,deqingsun,qfwang,minghsuan}@google.com
[3] Nankai University, Tianjin, China
xinyuzhang@mail.nankai.edu.cn
[4] Peking University, Beijing, China
yurina@pku.edu.cn

Abstract. Cost volume is an essential component of recent deep models for optical flow estimation and is usually constructed by calculating the inner product between two feature vectors. However, the standard inner product in the commonly-used cost volume may limit the representation capacity of flow models because it neglects the correlation among different channel dimensions and weighs each dimension equally. To address this issue, we propose a *learnable cost volume* (LCV) using an elliptical inner product, which generalizes the standard inner product by a positive definite kernel matrix. To guarantee its positive definiteness, we perform spectral decomposition on the kernel matrix and re-parameterize it via the Cayley representation. The proposed LCV is a lightweight module and can be easily plugged into existing models to replace the vanilla cost volume. Experimental results show that the LCV module not only improves the accuracy of state-of-the-art models on standard benchmarks, but also promotes their robustness against illumination change, noises, and adversarial perturbations of the input signals.

Keywords: Optical flow · Cost volume · Cayley representation · Inner product

1 Introduction

Optical flow estimation is a fundamental computer vision task and has broad applications, such as video interpolation [2], video prediction [21], video segmentation [6,36], and action recognition [22]. Despite the recent progress made by

Electronic supplementary material The online version of this chapter (https://doi.org/10.1007/978-3-030-58545-7_28) contains supplementary material, which is available to authorized users.

© Springer Nature Switzerland AG 2020
A. Vedaldi et al. (Eds.): ECCV 2020, LNCS 12354, pp. 483–499, 2020.
https://doi.org/10.1007/978-3-030-58545-7_28

deep learning models, it is still challenging to accurately estimate optical flow for image sequences with large displacements, textureless regions, motion blur, occlusion, illumination changes, and non-Lambertian reflection.

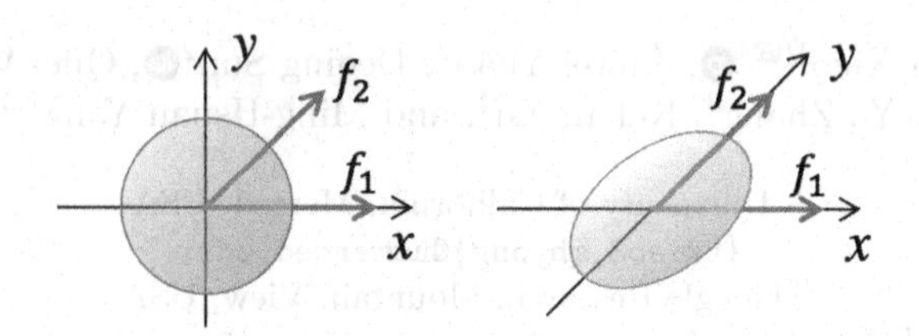

Fig. 1. Standard inner product space v.s. elliptical inner product space.

Most deep optical flow models [12,23,33] adopt the idea of coarse-to-fine processing via feature pyramids and construct *cost volumes* at different levels of the pyramids. The cost volume stores the costs of matching pixels in the source image with their potential matching candidates in the target image. It is typically constructed by calculating the inner product between the convolutional features of one frame and those of the next frame, and then regressed to the estimated optical flow by an estimation sub-network. The accuracy of the estimated optical flow heavily relies on the quality of the constructed cost volume.

While the standard Euclidean inner product is widely used to build the cost volume (a.k.a., vanilla cost volume) for optical flow, we argue that it limits the representation capacity of the flow model for two reasons. First, the correlation among different channel dimensions is not taken into consideration by the standard Euclidean inner product. As shown in Fig. 1, we use a simple 2D example for illustration. Given two feature vectors $\boldsymbol{f}_1$ and $\boldsymbol{f}_2$ with positive correlation in the standard inner product space, we are able to find a proper elliptical inner product space to make these two feature vectors orthogonal to each other, which gives a zero correlation. Therefore, the specific choice of the inner product space influences the values of the matching costs, and thus should be further exploited. Second, each feature dimension contributes equally to the vanilla cost volume, which may give a sub-optimal solution to constructing the cost volume for flow estimation. Ideally, dimensions corresponding to noises and random perturbations should be suppressed, while those containing discriminative signals for flow estimation should be kept or magnified.

To address these limitations, we propose a *learnable cost volume* (LCV) module which accounts for the correlation among different channel dimensions and re-weighs the contribution of each feature channel to the cost volume. The LCV generalizes the Euclidean inner product space to an elliptical inner product space, which is parameterized by a symmetric and positive definite kernel matrix. The spectral decomposition of the kernel matrix gives an orthogonal matrix and a diagonal matrix. The orthogonal matrix linearly transforms the features into a new feature space, which accounts for the correlation among different channel dimensions. The diagonal matrix multiplies each transformed feature by a positive scalar, which weighs each feature dimension differently. From a geometric perspective, the orthogonal matrix rotates the axes and the diagonal matrix stretches the axes so that the feature vectors are represented in a learned elliptical inner product space, which generates more discriminative matching costs for flow estimation.

However, directly learning a kernel matrix in an end-to-end manner cannot guarantee the symmetry and positive definiteness of the kernel matrix, which is required by the definition of inner product. To address this issue, we perform spectral decomposition on the kernel matrix and represent each component via the Cayley transform. Specifically, the special orthogonal matrices that exclude -1 as the eigenvalue can be bijectively mapped into the skew-symmetric matrices, and the diagonal matrices can be similarly represented by the composition of the Cayley transform and the arctangent function. In this way, all parameters of the learnable cost volume can be inferred in an end-to-end fashion without explicitly imposing any constraints.

The proposed learnable cost volume is a general version of the vanilla cost volume, and thus can replace the vanilla cost volume in the existing networks. We finetune the existing architectures equipped with LCV by initializing the kernel matrix as the identity matrix and restoring other parameters from the pre-trained models. Experimental results on the Sintel and KITTI benchmark datasets show that the proposed LCV significantly improves the performance of existing methods in both supervised and unsupervised settings. In addition, we demonstrate that LCV is able to promote the robustness of the existing models against illumination changes, noises, and adversarial attacks.

To summarize, we make the following contributions:

1. We propose a learnable cost volume (LCV) to account for correlations among different feature dimensions and weight each dimension separately.
2. We employ the Cayley representation to re-parameterize the kernel matrix in a way that all parameters can be learned in an end-to-end manner.
3. The proposed LCV can easily replace the vanilla cost volume and improve the accuracy and robustness of the state-of-the-art models.

2 Related Work

Supervised Learning of Optical Flow. Inspired by the success of convolutional neural networks (CNNs) on per-pixel predictions such as semantic segmentation and single-image depth estimation, Dosovitski *et al.* propose FlowNet [8], the first end-to-end deep neural network capable of learning optical flow. FlowNet predicts a dense optical flow map from two consecutive image frames with an encoder-decoder architecture. FlowNet2.0 [15] extends FlowNet by stacking multiple basic FlowNet modules for iterative refinement and its accuracy is fully on par with those of the state-of-the-art methods at the time. Motivated by the idea of coarse-to-fine refinement in traditional optical flow methods, SpyNet [29] introduces a compact spatial pyramid network that warps images at multiple scales to deal with displacements caused by large motions. PWC-Net [33] extracts feature through pyramidal processing and builds a cost volume at each level from the warped and the target features to iteratively refine the estimated flow. VCN [39] improves the cost volume processing by decoupling the 4D convolution into a 2D spatial filter and a 2D winner-take-all (WTA) filter, while still retaining a large receptive field. HD^3 [40] learns a probabilistic matching density distribution at each scale and merges the matching densities at different scales to recover the global matching density.

Unsupervised Learning of Optical Flow. The advantage of unsupervised methods is that it can sidestep the limitations of the synthetic datasets and exploit the large number of training data in the realistic domain. In [17] and [31], the flow guidance comes from warping the target image according to the predicted flow and comparing against the reference image. The photometric loss is adopted to ensure brightness constancy and spatial smoothness. In some work [25,37], occluded regions are excluded from the photometric loss. As pixels occluded in the target image are also absent in the warped one, enforcing matching of the occluded pixels would misguide the training. Wang *et al.* [37] obtain an occlusion mask from the range map inferred from the backward flow, while UnFlow [25] relies on the forward-backward consistency to estimate the occlusion mask. Unlike these two methods that predict the occlusion map in advance with certain heuristic, Back2Future [16] estimates the occlusion and optical flow jointly by introducing a multi-frame formulation and reasoning the occlusion in a more advanced manner. DDFlow [23] performs knowledge distillation by cropping patches from the unlabeled images, which provides flow guidance for the occluded regions. SelFlow [24] hallucinates synthetic occlusions by perturbing super-pixels where the occluded regions are guided by a model pre-trained from non-occluded regions.

Correspondence Matching. Typically, stereo matching algorithms [11,32] involve local correspondence extraction and smoothness regularization, where the smoothness regularization is enforced by energy minimization. Recently, hand-crafted features are replaced by deep features and minimization of the matching cost is substituted by training convolutional neural networks [19,42]. Xu *et al.* [38] construct a 4D cost volume using an adaptation of the semi-global matching, and Yang *et al.* [39] reduce the computation overhead of processing the 4D matching volume by factorizing into two separable filters.

Different from these approaches where the correspondence is represented by a hand-crafted matching cost volume, we propose a learnable cost volume that can capture the correlation among different channels by adapting the features to an elliptical inner product space. Such a correlation is automatically learned by optimizing the kernel matrix using the Cayley representation, which is more flexible and effective in optical flow estimation and can be easily plugged into the existing architectures. To our knowledge, this paper is the first one to use the Cayley representation for learning correspondence in optical flow.

3 Learnable Correlation Volume

3.1 Vanilla Cost Volume

Let $\boldsymbol{F}^1, \boldsymbol{F}^2 \in \mathbb{R}^{c \times h \times w}$ be the convolutional feature of the first frame and the warped feature of the second frame, respectively. The vanilla cost volume is defined as the inner product between the query feature $\boldsymbol{F}^1_{i,j}$ and the potential match candidate $\boldsymbol{F}^2_{k',l'}$, *i.e.*,

$$C(\boldsymbol{F}^1, \boldsymbol{F}^2)_{k,l,i,j} = \boldsymbol{F}^{1\top}_{i,j} \boldsymbol{F}^2_{k',l'}, \tag{1}$$

which maps from the space $\mathbb{R}^{c \times h \times w} \times \mathbb{R}^{c \times h \times w}$ to $\mathbb{R}^{u \times v \times h \times w}$. Here, u and v are usually odd numbers, indicating the displacement ranges in horizontal and vertical directions, (i, j) denotes the spatial location of the feature map $\boldsymbol{F}^1$, and $(k', l') = (i - (u-1)/2 + k, j - (v-1)/2 + l)$ denotes that of $\boldsymbol{F}^2$. For each location (i, j) of the query feature $\boldsymbol{F}^1$, the matching is performed against pixels of $\boldsymbol{F}^2$ within a $u \times v$ search window centered by the location (i, j). Then, the cost volume is either reshaped into $uv \times h \times w$ and post-processed by 2D convolutions [33], or kept as a 4D tensor on which the separable 4D convolutions [39] are applied.

3.2 Learnable Cost Volume

We generalize the standard Euclidean inner product to the elliptical inner product, where the matching cost is computed as follows:

$$C(\boldsymbol{F}^1, \boldsymbol{F}^2)_{k,l,i,j} = \boldsymbol{F}^{1\top}_{i,j} \boldsymbol{W} \boldsymbol{F}^2_{k',l'}. \tag{2}$$

Here, $\boldsymbol{W} \in \mathbb{R}^{c \times c}$ is a learnable kernel matrix that determines the elliptical inner product space, and other notations are the same as those in Eq. (1). According to the definition of inner product, $\boldsymbol{W}$ should be a symmetric and positive definite matrix. By spectral decomposition, we obtain

$$\boldsymbol{W} = \boldsymbol{P}^\top \boldsymbol{\Lambda} \boldsymbol{P}, \tag{3}$$

where $\boldsymbol{P}$ is an orthogonal matrix, and $\boldsymbol{\Lambda}$ is a diagonal matrix with positive entries, *i.e.*, $\boldsymbol{\Lambda} = \mathrm{diag}(\lambda_1, \cdots, \lambda_c)$ with $\lambda_i > 0$, $\forall i \in \{1, \cdots, c\}$. The orthogonal matrix $\boldsymbol{P}$ actually rotates the coordinate axes and the diagonal matrix $\boldsymbol{\Lambda}$ re-weights different dimensions, which directly address the two limitations mentioned in Sect. 1.

3.3 Learning with the Cayley Representation

In the proposed LCV module, the entries of the kernel matrix $\boldsymbol{W}$ are the only learnable parameters. However, the constraints of symmetry and positive-definiteness hinders the gradient-based end-to-end learning of $\boldsymbol{W}$. To address this issue, we propose to optimize $\boldsymbol{P}$ and $\boldsymbol{\Lambda}$ instead of $\boldsymbol{W}$.

One way to optimize $\boldsymbol{P}$ is to employ the Riemann gradient descent on the Stiefel manifold, which is defined as

$$V_k(\mathbb{R}^n) = \{\boldsymbol{A} \in \mathbb{R}^{n \times k} | \boldsymbol{A}^\top \boldsymbol{A} = \boldsymbol{I}_k\}. \tag{4}$$

All orthogonal matrices lie in the Stiefel manifold. Specifically, $\boldsymbol{P} \in V_c(\mathbb{R}^c)$. Therefore, we can apply the Riemann gradient descent on the Stiefel matrix manifold, where the projection and retraction formula [1] are given by

$$\mathcal{P}_{\boldsymbol{X}}(\boldsymbol{Z}) = (\boldsymbol{I} - \boldsymbol{X}\boldsymbol{X}^\top)\boldsymbol{Z} + \boldsymbol{X} \cdot \mathrm{skew}(\boldsymbol{X}^\top \boldsymbol{Z}) \tag{5}$$

$$\mathcal{R}_{\boldsymbol{X}}(\boldsymbol{Z}) = (\boldsymbol{X} + \boldsymbol{Z})(\boldsymbol{I} + \boldsymbol{Z}^\top \boldsymbol{Z})^{-\frac{1}{2}}, \tag{6}$$

where $\mathrm{skew}(\boldsymbol{X}) := (\boldsymbol{X} - \boldsymbol{X}^\top)/2$. However, to perform the Riemann gradient descent, the projection and retraction operations are required in each training step, and the matrix multiplication brings considerable computational overhead.

We can address this issue in a more elegant way using the Cayley Representation [5]. First, we define a set of matrices:

$$\mathrm{SO}^*(n) := \{\boldsymbol{A} \in \mathrm{SO}(n) : -1 \notin \sigma(\boldsymbol{A})\}, \tag{7}$$

where $\sigma(\boldsymbol{A})$ denotes the spectrum, *i.e.*, all eigenvalues, of $\boldsymbol{A}$. $\mathrm{SO}^*(n)$ is a subset of the special orthogonal group $\mathrm{SO}(n)$ and the spectrum of its elements excludes -1. Then, we have the following theorems:

Theorem 1 (Cayley Representation). *Given any matrix $\boldsymbol{P} \in \mathrm{SO}^*(n)$, there exists a unique skew-symmetric matrix $\boldsymbol{S}$,* i.e., *$\boldsymbol{S}^\top = -\boldsymbol{S}$, such that*

$$\boldsymbol{P} = (\boldsymbol{I} - \boldsymbol{S})(\boldsymbol{I} + \boldsymbol{S})^{-1}. \tag{8}$$

Theorem 2. *The set of matrices $\mathrm{SO}^*(n)$ is connected.*

By Theorem 1, we can initialize the matrix $\boldsymbol{P}$ in Eq. (3) as an identity matrix $\boldsymbol{I} \in \mathrm{SO}^*(c)$, and update $\boldsymbol{S}$ so as to update $\boldsymbol{P}$ using gradient-based optimizer. Let $\boldsymbol{P}^*$ be the optimal orthogonal matrix, and we claim that it is possible to reach $\boldsymbol{P}^*$ from initializing as the identity matrix $\boldsymbol{P} = \boldsymbol{I}$. This because $\mathrm{SO}^*(c)$ is a connected set (Theorem 2), so there exists a continuous path joining $\boldsymbol{I} \in \mathrm{SO}^*(c)$ and any $\boldsymbol{P} \in \mathrm{SO}^*(c)$, including $\boldsymbol{P}^*$.

Due to the positive definiteness of $\boldsymbol{W}$, the constraint of the diagonal matrix $\boldsymbol{\Lambda} = \mathrm{diag}(\lambda_1, \cdots, \lambda_c)$ is $\lambda_i > 0$, $\forall i = 1, \ldots, c$. Thus, we map $\mathbb{R}$ to $\mathbb{R}^+$ by applying the composition of the Cayley transform and the arctangent function, *i.e.*,

$$\lambda_i = \frac{\pi + 2\arctan t_i}{\pi - 2\arctan t_i}, \tag{9}$$

where $t_i \in \mathbb{R}$ is free of constraint.

The above re-parameterization trick enables us to update the kernel matrix $\boldsymbol{W}$ in an end-to-end manner using the SGD optimizer or its variants, which alleviates the heavy computation brought by the projection and retraction and makes the training process much easier.

3.4 Interpretation

To better understand the learnable cost volume, we analyze several cases here.
1. $\boldsymbol{W} = \boldsymbol{I}$. This degenerates into the vanilla cost volume, in which the standard Euclidean inner product is adopted.
2. $\boldsymbol{W} = \boldsymbol{\Sigma}^{-1}$. Let $\boldsymbol{\Sigma}$ be the covariance matrix, *i.e.*, Gram matrix, of the convolutional feature, then the learnable cost volume is essentially a whitening transformation. Let $\boldsymbol{Q} = \boldsymbol{\Lambda}^{1/2}\boldsymbol{P}$, and then Eq. (2) can be formulated as

$$C(\boldsymbol{F}^1, \boldsymbol{F}^2)_{k,l,i,j} = \boldsymbol{F}^{1\top}_{i,j}\boldsymbol{P}^\top\boldsymbol{\Lambda}^{1/2}\boldsymbol{\Lambda}^{1/2}\boldsymbol{P}\boldsymbol{F}^2_{k',l'} = (\boldsymbol{Q}\boldsymbol{F}^1_{i,j})^\top(\boldsymbol{Q}\boldsymbol{F}^2_{k',l'}), \tag{10}$$

where $\boldsymbol{Q}\boldsymbol{F}^1_{i,j}$ represents the transformed feature of $\boldsymbol{F}^1_{i,j}$ after PCA [18] whitening. Similarly, letting $\boldsymbol{R} = \boldsymbol{P}^\top\boldsymbol{\Lambda}^{1/2}\boldsymbol{P}$, we can have

$$C(\boldsymbol{F}^1, \boldsymbol{F}^2)_{k,l,i,j} = \boldsymbol{F}^{1\top}_{i,j}\boldsymbol{P}^\top\boldsymbol{\Lambda}^{1/2}\boldsymbol{P}\boldsymbol{P}^\top\boldsymbol{\Lambda}^{1/2}\boldsymbol{P}\boldsymbol{F}^2_{k',l'} = (\boldsymbol{R}\boldsymbol{F}^1_{i,j})^\top(\boldsymbol{R}\boldsymbol{F}^2_{k',l'}), \tag{11}$$

where $\boldsymbol{R}\boldsymbol{F}^1_{i,j}$ is the transformed feature of $\boldsymbol{F}^1_{i,j}$ after ZCA [3] whitening. It has been shown that the high-level styles can be removed with the contextual structures remained by whitening the convolutional features [20].
3. $\boldsymbol{W} = \boldsymbol{P}^\top\boldsymbol{\Lambda}\boldsymbol{P}$. The learnable cost volume shares a similar formula as the whitening process, but $\boldsymbol{W}$ is learned over the whole training dataset rather than statistics of two inputs, thus contains certain holistic information of the entire training dataset. Because it has been verified that the certain holistic characteristics of the underlying image can be captured by the Gram matrix along the channel dimension [9,20]. The learnable cost volume performs as "whitening" features using the common information learned from all frames. Specifically, the orthogonal matrix $\boldsymbol{P}$ re-arranges the information across the channel dimension, while the diagonal matrix $\boldsymbol{\Lambda}$ filters out insignificant signals, making the correlation more robust to the illumination changes and noises. (See Sect. 4.4.)

It should also be pointed out that the whitening matrix $\boldsymbol{R}$ in Eq. (11) could be viewed as a 1×1 conv functioning on the feature, but directly applying a 1×1 conv with learnable parameters on features before computing the standard cost volume cannot replace the proposed learned cost volume. Because $\boldsymbol{R}^\top\boldsymbol{R}$ only gives a positive semi-definite matrix even when $\boldsymbol{R}$ is full-rank, which does not meet the positive definiteness property of an inner product.

3.5 Relation with the Weighted Sum of Squared Difference

The learnable cost volume can be also formulated by re-thinking the simplest matching criterion for comparing two features, *i.e.*, the weighted sum of squared difference (WSSD):

$$\sum_i \lambda_i \left(G_i(\boldsymbol{F}^2) - G_i(\boldsymbol{F}^1)\right)^2, \tag{12}$$

where $G : \mathbb{R}^c \rightarrow \mathbb{R}^c$ denotes a transformation function on the features $\boldsymbol{F}^i \in \mathbb{R}^c$, $i = 1, 2$, and $G_i(\boldsymbol{F})$ indicates the ith element of $G(\boldsymbol{F})$. By the Taylor series expansion, we have

$$\sum_i \lambda_i \left(G_i(\boldsymbol{F}^2) - G_i(\boldsymbol{F}^1)\right)^2 \approx \sum_i \lambda_i \left(\nabla G_i(\boldsymbol{F}^1)^\top \Delta\boldsymbol{F}\right)^2 = \Delta\boldsymbol{F}^\top \boldsymbol{W} \Delta\boldsymbol{F}, \quad (13)$$

where $\Delta\boldsymbol{F}=\boldsymbol{F}^2-\boldsymbol{F}^1$ is the feature difference and $\boldsymbol{W} = \sum_i \lambda_i \nabla G_i(\boldsymbol{F}^1) \nabla G_i(\boldsymbol{F}^1)^\top$ is the auto-correlation matrix. Here, $\boldsymbol{W}$ coincides with the kernel matrix of the proposed LCV module in Eq. (2). When $\lambda_i = 1 (i = 1, \dots, c)$ and G is an identity map, then $\boldsymbol{W} = \boldsymbol{I}$, which corresponds to the vanilla cost volume. If we further expand Eq. (13), we can see the connection with the proposed learnable correlation volume as follows:

$$\begin{aligned} \Delta\boldsymbol{F}^\top \boldsymbol{W} \Delta\boldsymbol{F} &= (\boldsymbol{F}^2 - \boldsymbol{F}^1)^\top \boldsymbol{W} (\boldsymbol{F}^2 - \boldsymbol{F}^1) \\ &= (\boldsymbol{F}^{2\top} \boldsymbol{W} \boldsymbol{F}^2 + \boldsymbol{F}^{1\top} \boldsymbol{W} \boldsymbol{F}^1) - 2\boldsymbol{F}^{1\top} \boldsymbol{W} \boldsymbol{F}^2, \end{aligned} \quad (14)$$

where the last term shares the same formula with the proposed learnable cost volume. This implies that the proposed learnable cost volume is inversely correlated with WSSD. As WSSD measures the discrepancy between two features, the learnable cost volume characterizes a certain kind of similarity between them.

4 Experiments

In this section, we present the experimental results of optical flow estimation in both supervised and unsupervised settings to demonstrate the effectiveness of the proposed learnable cost volume. Also, we carry out ablation studies to show that the LCV module performs favorably against other counterparts. Moreover, we analyze the behavior of LCV and find it beneficial to handling three challenging cases. More results can be found in the supplementary material and the source code and trained models will be made available to the public.

Training Process. It is well-known that the deep optical flow estimation pipeline consists the following stages in the supervised settings [34]: 1) train the model on the FlyingChairs [7] dataset; 2) finetune the model on the FlyingThings3D [28] dataset; and 3) finetune the model on the Sintel [4] and KITTI [26,27] training sets. Besides, there are lots of tricks such as data augmentation and learning rate disruption, making the training process more complicated.

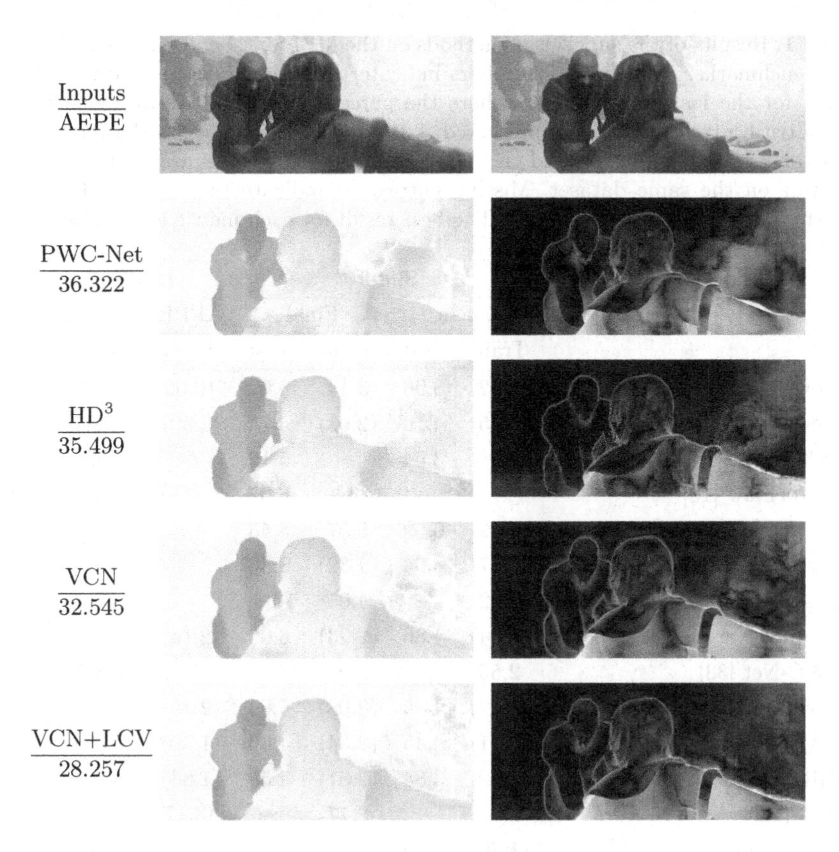

Fig. 2. Visual results on "Ambush 1" from the Sintel test final pass. The number under each method denotes the average end-point error (AEPE). Left: estimated flow; right: error map (increases from black to white).

To avoid the tedious training procedure over multiple datasets, we adopt a more efficient way to train the model equipped with LCV. As mentioned in Sect. 3.4, the vanilla cost volume is a special case of the learnable cost volume when $\boldsymbol{W} = \boldsymbol{I}$, which means that the learnable cost volume is more general and backward compatible with vanilla cost volume. Therefore, we initialize the kernel matrix $\boldsymbol{W}$ as the identity matrix and other parameters are directly restored from the pre-trained models without using LCV. After that, we finetune the model with LCV on the Sintel or KITTI datasets using the same loss function. This training process not only significantly reduces training time but also plays a crucial role in the success under the unsupervised settings. (See Sect. 4.2.) This approach can also be viewed as fixing the kernal matrix as $\boldsymbol{W} = \boldsymbol{I}$ in the first three training stages, and let $\boldsymbol{W}$ be learnable in the final stage.

Table 1. Results of the supervised methods on the MPI Sintel and KITTI 2015 optical flow benchmarks. All reported numbers indicate the average endpoint error (AEPE) except for the last two columns, where the percentage of outliers averaged over all groundtruth pixels (Fl-all) are presented. "-ft" means finetuning on the relative MPI Sintel or KITTI training set and the numbers in the parenthesis are results that train and test on the same dataset. Missing entries (-) indicate that the results are not reported for the respective method. The best result for each metric is printed in bold.

Methods	Sintel				KITTI 2015		
	Clean		Final		AEPE	Fl-all (%)	
	Train	Test	Train	Test	Train	Train	Test
FlowNet2 [15]	2.02	3.96	3.14	6.02	10.06	30.37	–
FlowNet2-ft [15]	(1.45)	4.16	(2.01)	5.74	(2.30)	(8.61)	10.41
DCFlow [38]	–	3.54	–	5.12	–	15.09	14.83
MirrorFlow [13]	–	–	–	6.07	–	9.93	10.29
SpyNet [29]	4.12	6.69	5.57	8.43	–	–	–
SpyNet-ft [29]	(3.17)	6.64	(4.32)	8.36	–	–	35.07
LiteFlowNet [12]	2.52	–	4.05	10.39	–	–	–
LiteFlowNet + ft [12]	(1.64)	4.86	(2.23)	6.09	(2.16)	–	10.24
PWC-Net [33]	2.55	–	3.93	–	10.35	33.67	–
PWC-Net-ft [33]	(2.02)	4.39	(2.08)	5.04	(2.16)	(9.80)	9.60
PWC-Net+-ft [34]	(1.71)	3.45	(2.34)	4.60	(1.50)	(5.30)	7.72
IRR-PWC-ft [14]	(1.92)	3.84	(2.51)	4.58	(1.63)	(5.30)	7.65
HD^3 [40]	3.84	–	8.77	–	13.17	23.99	–
HD^3-ft [40]	(1.70)	4.79	(1.17)	4.67	(1.31)	(4.10)	6.55
VCN [39]	2.21	–	3.62	–	8.36	25.10	8.73
VCN-ft [39]	(1.66)	2.81	(2.24)	4.40	(1.16)	(4.10)	6.30
RAFT [35]	1.09	2.77	1.53	3.61	(1.07)	(3.92)	6.30
RAFT (warm start) [35]	1.10	**2.42**	1.61	3.39	–	–	–
VCN + LCV	(1.62)	2.83	(2.22)	4.20	(1.13)	(3.80)	**6.25**
RAFT + LCV	**(0.94)**	2.75	**(1.31)**	3.55	**(1.06)**	**(3.77)**	6.26
RAFT + LCV (warm start)	(0.99)	2.49	(1.47)	**3.37**	–	–	–

4.1 Supervised Optical Flow Estimation

First, we incorporate the learnable cost volume in the VCN [39] and RAFT [35] framework, and compare them with other existing methods. As shown in Table 1, our method performs favorably against other state-of-the-art methods on the Sintel Clean/Final pass and the KITTI 2015 benchmark.

The proposed LCV module improves the performance of VCN and RAFT by transforming the features of video frames to a whitened space to obtain a clean and robust matching correlation. This could account for the performance improvement on the Sintel Final pass, where the scenarios are much harder.

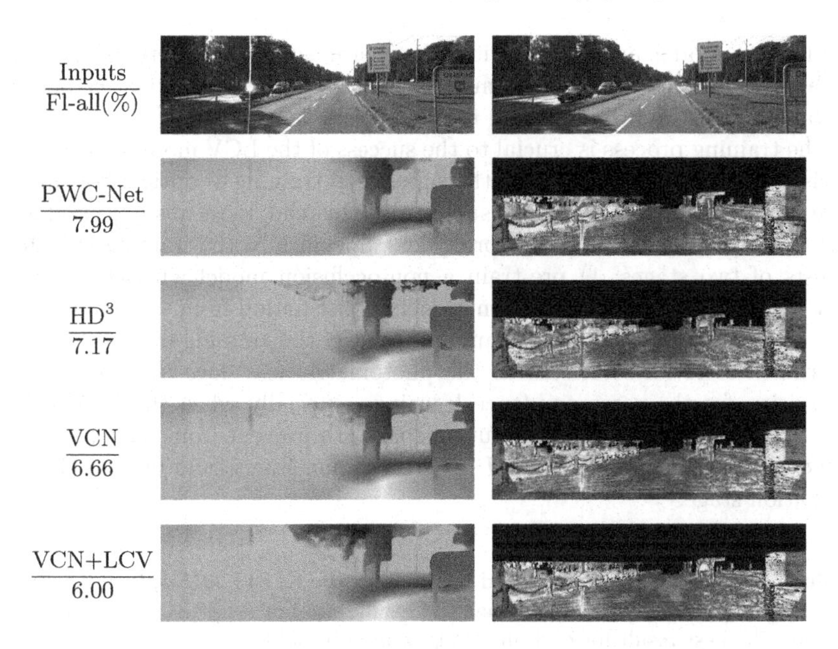

Fig. 3. Visual results on the KITTI 2015 test set. The number under each method name denotes the Fl-all score on the given frames. Left: estimated flow; right: error map (increases from blue to red). (Color figure online)

As shown in Fig. 2, the flow estimation error for the snow background at the right side is smaller than other methods. This is a challenging case because the front person's arm renders occlusion to part of the snow background and the background is nearly all white, providing few clues for matching. However, the LCV module exploits more information from the correlation among different channels, which assists in obtaining the coherent flow estimation in the snow background. The LCV module also has an edge over the vanilla cost volume under the circumstance of light reflection and occlusion. As shown in Fig. 3, the prediction error of our method is smaller around the light reflection region and the rightmost traffic sign.

Although we do not report the model parameters in the table, the proposed LCV module only makes a very slight increase in the model size. The additional parameters come from the kernel matrices $\boldsymbol{W} \in \mathbb{R}^{c \times c}$ at different pyramid levels. Taking VCN+LCV as an example, there are five kernel matrices in total, whose channel dimensions are 64, 64, 128, 128, and 128, respectively. The LCV module only takes up $64^2 \times 2 + 128^2 \times 3 = 57,344$ parameters, which is negligible compared with the entire VCN model of around 6.23M parameters.

4.2 Unsupervised Optical Flow Estimation

We also test the LCV module in unsupervised settings on the KITTI 2015 benchmark. We replace the vanilla cost volume with the LCV module in the

DDFlow [23] model, and compare it with other unsupervised methods. As shown in Table 2, our model outperforms the DDFlow baseline, and even performs favorably against SelFlow [24], an improved version of DDFlow.

The training process is crucial to the success of the LCV module in the unsupervised methods. Different from the supervised training of optical flow models, there is no ground truth for direct supervision. Instead, most unsupervised methods use the photometric loss as a proxy loss. Specifically, the training of DDFlow consists of two stages: 1) pre-train a non-occlusion model with census transform [10], and 2) train an occlusion model by distillation from the non-occlusion model. If we directly follow the same procedure, the training of DDFlow+LCV will run into trivial solutions, as the photometric loss does not give a strong supervision for the correspondence learning, especially when the LCV module increases the dimension of the solution space. To prevent from trivial solutions, we fix the kenrel matrix as $\boldsymbol{W} = \boldsymbol{I}$ in the pre-train stages, and update $\boldsymbol{W}$ in the distillation stage.

Table 2. Results of the unsupervised methods on the KITTI 2015 optical flow benchmark. Missing entries (-) indicate that the results are not reported for the respective method. The best result for each metric is printed in bold.

Methods	KITTI 2015			
	Train	Test		
	AEPE	Fl-bg (%)	Fl-fg (%)	Fl-all (%)
DSTFlow [31]	16.79	–	–	39
GeoNet [41]	10.81	–	–	–
UnFlow [25]	8.88	–	–	28.95
DF-Net [43]	7.45	–	–	22.82
OccAwareFlow [37]	8.88	–	–	31.20
Back2FutureFlow [16]	6.59	22.67	24.27	22.94
SelFlow [24]	**4.84**	**12.68**	21.74	14.19
DDFlow [23]	5.72	13.08	20.40	14.29
DDFlow + LCV (Ours)	5.15	12.98	**19.83**	**14.12**

Table 3. Ablation study of different variants of VCN on the KITTI 2015 dataset.

Methods	VCN	VCN (ct)	VCN ($\boldsymbol{W}$, ct)	VCN ($\boldsymbol{A}$, ct)	VCN ($\boldsymbol{P}$, ct)	VCN(1 × 1 conv)	VCN + LCV
AEPE/Fl-all	3.9/1.144	4.2/1.204	4.1/1.193	3.8/1.136	3.9/1.129	3.9/1.163	3.8/1.132

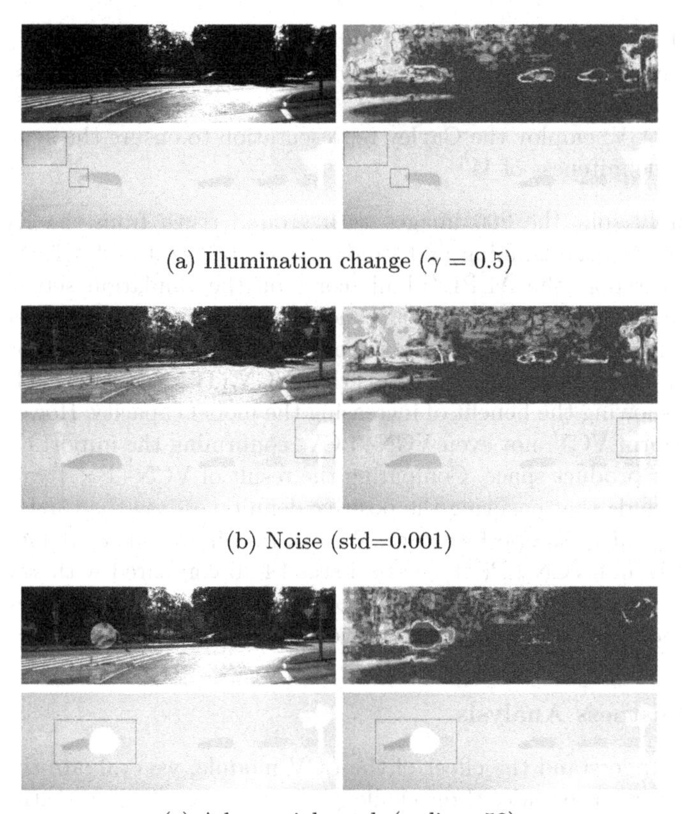

(a) Illumination change ($\gamma = 0.5$)
(b) Noise (std=0.001)
(c) Adversarial patch (radius=50)

Fig. 4. Visual results of three challenging cases, *i.e.*, illumination change, noise, and adversarial patch. Top left: the first input frame; bottom left/right: flow by VCN/VCN+LCV; top right: flow difference between two methods.

4.3 Ablation Study

We evaluate multiple variants of the LCV module based on the VCN baseline:

- VCN: the original VCN baseline.
- VCN (ct): continue training the existing VCN using a small learning rate for more epochs.
- VCN ($\boldsymbol{W}$, ct): remove the symmetry and positive definiteness constraint of $\boldsymbol{W}$, *i.e.*, not using the Cayley representation. We restore the weights from the pre-trained VCN and continue training the model with free $\boldsymbol{W}$.
- VCN ($\boldsymbol{\Lambda}$, ct): fix $\boldsymbol{P}$ to be an identity matrix and make the diagonal matrix $\boldsymbol{\Lambda}$ learnable.
- VCN ($\boldsymbol{P}$, ct): fix $\boldsymbol{\Lambda}$ to be an identity matrix and make the orthogonal matrix $\boldsymbol{P}$ learnable.

- VCN (1 × 1 conv): replace the positive definite $\boldsymbol{W}$ with $\boldsymbol{R}^\top \boldsymbol{R}$, where $\boldsymbol{R}$ is a 1 × 1 conv operating on features with input and output dimensions equal. $\boldsymbol{R}^\top \boldsymbol{R}$ is only a positive semi-definite matrix.
- VCN + LCV: employ the Cayley representation to ensure the symmetry and positive definiteness of $\boldsymbol{W}$.

We randomly split the 200 images with ground truth from the KITTI 2015 training set into the training and validation set by a ratio of 4:1. As shown in Table 3, we report the AEPE/Fl-all scores on the validation set. We observe that continuing training of the VCN model does not bring any benefit, which indicates that the best VCN model is not obtained at the very end of the training. Another interesting observation is that VCN ($\boldsymbol{W}$, ct) performs better than VCN (ct), showing the benefit of increasing the model capacity. However, it does not outperform VCN, not even VCN+LCV, confirming the importance of using a valid inner product space. Comparing the result of VCN (1 × 1 conv), we can further conclude that ensuring the positive definiteness via the Cayley representation is crucial to the performance. We can also find that VCN ($\boldsymbol{\Lambda}$, ct) gets a lower AEPE and VCN ($\boldsymbol{P}$, ct) gets a lower Fl-all compared with vanilla VCN. VCN+LCV combines the advantages of both axis rotation and re-weighting, aiming to address two limitations mentioned in the paper.

4.4 Robustness Analysis

To further understand the effect of the LCV module, we evaluate the flow estimation performance under three challenging cases, *i.e.*, 1) illumination changes: we adjust the illumination of the input frames by changing the value of γ, where $\gamma = 1.0$ is the original image, $\gamma < 1.0$ is for a darker image, and $\gamma > 1.0$ is for a brighter image. 2) adding noises: we adjust the standard deviation to control the noise magnitude. and 3) inserting adversarial patches: we borrow the universal adversarial patch [30] that can perform a black-box attack for all optical flow models, and insert patches of different sizes to the input frames.

Table 4. Results on three challenging cases (numbers: AEPE/Fl-all scores).

(a) Illumination change

γ	0.2	0.3	0.4	0.5	0.7	1.0	2.0	3.0
VCN	16.8/3.240	9.9/1.891	5.9/1.306	3.8/0.995	2.7/0.834	2.5/0.805	2.6/0.819	2.6/0.826
VCN+LCV	17.1/3.232	9.8/1.866	5.9/1.273	3.7/0.967	2.6/0.804	2.4/0.775	2.4/0.790	2.5/0.804

(b) Noise

Standard deviation	0.0001	0.001	0.01	0.1
VCN	2.6/0.816	2.9/0.868	5.0/1.157	19.6/3.213
VCN+LCV	2.4/0.785	2.7/0.838	4.7/1.107	18.9/3.043

(c) Adversarial patch

Patch size	50	100	150	200
VCN	3.5/0.981	5.6/1.419	8.5/2.048	11.9/2.880
VCN+LCV	3.4/0.949	5.5/1.384	8.3/2.004	11.6/2.801

We compare the VCN model and its variant equipped with LCV. Both two models are trained on the KITTI 2015 training set. For qualitative comparison, we perform the above three types of processing on 194 images with the flow groundtruth from the KITTI 2012 as our test set. As shown in Table 4(a), VCN+LCV consistently outperforms the VCN baseline in all three challenging cases. For better illustration, we visualize the effect on an image from KITTI 2015 test set as shown in Fig. 4. It can be seen that the LCV module can help stabilize the flow prediction around the background trees at the top left corner of the frame under the cases of dark illumination and random noise injection. In the third example, the outline of the car body near the patch circle is better preserved by our model. (See the difference map for details.)

5 Conclusions

In this work, we introduce a learnable cost volume (LCV) module for optical flow estimation. The proposed LCV module generalizes the standard Euclidean inner product into an elliptical inner product with a symmetric and positive definite kernel matrix. To keep its symmetry and positive definiteness, we use the Cayley representation to re-parameterize the kernel matrix for end-to-end training. The proposed LCV is a lightweight module and can be easily plugged into any existing networks to replace the vanilla cost volume. Experimental results show that the proposed LCV module improves both the accuracy and the robustness of state-of-the-art optical flow models.

Acknowledgments. This work is supported in part by NSF CAREER Grant 1149783. We also thank Pengpeng Liu and Jingfeng Wu for kind help.

References

1. Absil, P.A., Mahony, R., Sepulchre, R.: Optimization Algorithms on Matrix Manifolds. Princeton University Press, Princeton (2009)
2. Bao, W., Lai, W.S., Ma, C., Zhang, X., Gao, Z., Yang, M.H.: Depth-aware video frame interpolation. In: IEEE Conference on Computer Vision and Pattern Recognition (CVPR) (2019)
3. Brox, T., Malik, J.: Large displacement optical flow: descriptor matching in variational motion estimation. IEEE Trans. Pattern Recogn. Mach. Intell. (PAMI) **33**(3), 500–513 (2010)
4. Butler, D.J., Wulff, J., Stanley, G.B., Black, M.J.: A naturalistic open source movie for optical flow evaluation. In: Fitzgibbon, A., Lazebnik, S., Perona, P., Sato, Y., Schmid, C. (eds.) ECCV 2012. LNCS, vol. 7577, pp. 611–625. Springer, Heidelberg (2012). https://doi.org/10.1007/978-3-642-33783-3_44
5. Cayley, A.: About the algebraic structure of the orthogonal group and the other classical groups in a field of characteristic zero or a prime characteristic. Reine Angewandte Mathematik **32**, 1846 (1846)
6. Cheng, J., Tsai, Y.H., Wang, S., Yang, M.H.: Segflow: joint learning for video object segmentation and optical flow. In: IEEE International Conference on Computer Vision (ICCV) (2017)

7. Dosovitskiy, A., et al.: Flownet: learning optical flow with convolutional networks. In: IEEE International Conference on Computer Vision (ICCV) (2015)
8. Dosovitskiy, A., et al.: Flownet: learning optical flow with convolutional networks. In: IEEE Conference on Computer Vision and Pattern Recognition (CVPR) (2015)
9. Gatys, L.A., Ecker, A.S., Bethge, M.: Image style transfer using convolutional neural networks. In: IEEE Conference on Computer Vision and Pattern Recognition (CVPR) (2016)
10. Hafner, D., Demetz, O., Weickert, J.: Why is the census transform good for robust optic flow computation? In: Kuijper, A., Bredies, K., Pock, T., Bischof, H. (eds.) SSVM 2013. LNCS, vol. 7893, pp. 210–221. Springer, Heidelberg (2013). https://doi.org/10.1007/978-3-642-38267-3_18
11. Horn, B.K., Schunck, B.G.: Determining optical flow. Artif. Intell. **17**(1–3), 185–203 (1981)
12. Hui, T.W., Tang, X., Change Loy, C.: Liteflownet: a lightweight convolutional neural network for optical flow estimation. In: IEEE Conference on Computer Vision and Pattern Recognition (CVPR) (2018)
13. Hur, J., Roth, S.: Mirrorflow: exploiting symmetries in joint optical flow and occlusion estimation. IEEE International Conference on Computer Vision (ICCV), pp. 312–321 (2017)
14. Hur, J., Roth, S.: Iterative residual refinement for joint optical flow and occlusion estimation. In: IEEE Conference on Computer Vision and Pattern Recognition (CVPR) (2019)
15. Ilg, E., Mayer, N., Saikia, T., Keuper, M., Dosovitskiy, A., Brox, T.: Flownet 2.0: evolution of optical flow estimation with deep networks. In: IEEE Conference on Computer Vision and Pattern Recognition (CVPR) (2017)
16. Janai, J., Güney, F., Ranjan, A., Black, M.J., Geiger, A.: Unsupervised learning of multi-frame optical flow with occlusions. In: European Conference on Computer Vision (ECCV) (2018)
17. Yu, J.J., Harley, A.W., Derpanis, K.G.: Back to basics: unsupervised learning of optical flow via brightness constancy and motion smoothness. In: Hua, G., Jégou, H. (eds.) ECCV 2016. LNCS, vol. 9915, pp. 3–10. Springer, Cham (2016). https://doi.org/10.1007/978-3-319-49409-8_1
18. Jolliffe, I.T.: Principal components in regression analysis. In: Principal Component Analysis, pp. 129–155. Springer, New York (1986). https://doi.org/10.1007/978-1-4757-1904-8_8
19. Kendall, A., et al.: End-to-end learning of geometry and context for deep stereo regression. In: IEEE International Conference on Computer Vision (ICCV) (2017)
20. Li, Y., Fang, C., Yang, J., Wang, Z., Lu, X., Yang, M.H.: Universal style transfer via feature transforms. In: Neural Information Processing Systems (NeurIPS) (2017)
21. Li, Y., Fang, C., Yang, J., Wang, Z., Lu, X., Yang, M.H.: Flow-grounded spatial-temporal video prediction from still images. In: European Conference on Computer Vision (ECCV) (2018)
22. Lin, J., Gan, C., Han, S.: Tsm: temporal shift module for efficient video understanding. In: IEEE International Conference on Computer Vision (ICCV) (2019)
23. Liu, P., King, I., Lyu, M.R., Xu, J.: Ddflow: learning optical flow with unlabeled data distillation. In: Association for the Advancement of Artificial Intelligence (AAAI) (2019)
24. Liu, P., Lyu, M.R., King, I., Xu, J.: Selflow: self-supervised learning of optical flow. In: IEEE Conference on Computer Vision and Pattern Recognition (CVPR) (2019)

25. Meister, S., Hur, J., Roth, S.: Unflow: unsupervised learning of optical flow with a bidirectional census loss. In: Association for the Advancement of Artificial Intelligence (AAAI) (2017)
26. Menze, M., Heipke, C., Geiger, A.: Joint 3d estimation of vehicles and scene flow. In: ISPRS Workshop on Image Sequence Analysis (ISA) (2015)
27. Menze, M., Heipke, C., Geiger, A.: Object scene flow. ISPRS J. Photogrammetry Remote Sensing (JPRS) **140**, 60–76 (2018)
28. Mayer, N., et al.: A large dataset to train convolutional networks for disparity, optical flow, and scene flow estimation. In: IEEE Conference on Computer Vision and Pattern Recognition (CVPR) (2016)
29. Ranjan, A., Black, M.J.: Optical flow estimation using a spatial pyramid network. In: IEEE Conference on Computer Vision and Pattern Recognition (CVPR) (2017)
30. Ranjan, A., Janai, J., Geiger, A., Black, M.J.: Attacking optical flow. In: IEEE International Conference on Computer Vision (ICCV) (2019)
31. Ren, Z., Yan, J., Ni, B., Liu, B., Yang, X., Zha, H.: Unsupervised deep learning for optical flow estimation. In: Association for the Advancement of Artificial Intelligence (AAAI) (2017)
32. Scharstein, D., Szeliski, R.: A taxonomy and evaluation of dense two-frame stereo correspondence algorithms. Int. J. Comput. Vis. (IJCV) **47**(1–3), 7–42 (2002). https://doi.org/10.1023/A:1014573219977
33. Sun, D., Yang, X., Liu, M.Y., Kautz, J.: Pwc-net: CNNs for optical flow using pyramid, warping, and cost volume. In: IEEE Conference on Computer Vision and Pattern Recognition (CVPR) (2018)
34. Sun, D., Yang, X., Liu, M.Y., Kautz, J.: Models matter, so does training: an empirical study of CNNs for optical flow estimation. IEEE Trans. Pattern Recogn. Mach. Intell. (PAMI) **42**, 1408–1423 (2019)
35. Teed, Z., Deng, J.: Raft: recurrent all-pairs field transforms for optical flow. arXiv preprint arXiv:2003.12039 (2020)
36. Tsai, Y.H., Yang, M.H., Black, M.J.: Video segmentation via object flow. In: IEEE Conference on Computer Vision and Pattern Recognition (CVPR) (2016)
37. Wang, Y., Yang, Y., Yang, Z., Zhao, L., Xu, W.: Occlusion aware unsupervised learning of optical flow. In: IEEE Conference on Computer Vision and Pattern Recognition (CVPR) (2017)
38. Xu, J., Ranftl, R., Koltun, V.: Accurate optical flow via direct cost volume processing. In: IEEE Conference on Computer Vision and Pattern Recognition (CVPR) (2017)
39. Yang, G., Ramanan, D.: Volumetric correspondence networks for optical flow. In: Neural Information Processing Systems (NeurIPS) (2019)
40. Yin, Z., Darrell, T., Yu, F.: Hierarchical discrete distribution decomposition for match density estimation. In: IEEE Conference on Computer Vision and Pattern Recognition (CVPR), June 2019
41. Yin, Z., Shi, J.: Geonet: unsupervised learning of dense depth, optical flow and camera pose. In: IEEE Conference on Computer Vision and Pattern Recognition (CVPR) (2018)
42. Zbontar, J., LeCun, Y.: Stereo matching by training a convolutional neural network to compare image patches. J. Mach. Learn. Res. (JMLR) **17**, 2287–2318 (2016)
43. Zou, Y., Luo, Z., Huang, J.B.: Df-net: unsupervised joint learning of depth and flow using cross-task consistency. In: European Conference on Computer Vision (ECCV) (2018)

HALO: Hardware-Aware Learning to Optimize

Chaojian Li[1], Tianlong Chen[2], Haoran You[1], Zhangyang Wang[2(✉)], and Yingyan Lin[1(✉)]

[1] Rice University, Houston, TX 77005, USA
{cl114,hy34,yingyan.lin}@rice.edu
[2] The University of Texas at Austin, Austin, TX 78712, USA
{tianlong.chen,atlaswang}@utexas.edu

Abstract. There has been an explosive demand for bringing machine learning (ML) powered intelligence into numerous Internet-of-Things (IoT) devices. However, the effectiveness of such intelligent functionality requires in-situ continuous model adaptation for adapting to new data and environments, while the on-device computing and energy resources are usually extremely constrained. Neither traditional hand-crafted (e.g., SGD, Adagrad, and Adam) nor existing meta optimizers are specifically designed to meet those challenges, as the former requires tedious hyper-parameter tuning while the latter are often costly due to the meta algorithms' own overhead. To this end, we propose *hardware-aware learning to optimize* **(HALO)**, a practical meta optimizer dedicated to resource-efficient on-device adaptation. Our HALO optimizer features the following highlights: (1) *faster adaptation speed* (i.e., taking fewer data or iterations to reach a specified accuracy) by introducing a new regularizer to promote empirical generalization; and (2) *lower per-iteration complexity*, thanks to a stochastic structural sparsity regularizer being enforced. Furthermore, the optimizer itself is designed as a very light-weight RNN and thus incurs negligible overhead. Ablation studies and experiments on five datasets, six optimizees, and two state-of-the-art (SOTA) edge AI devices validate that, while always achieving a better accuracy (↑0.46% - ↑20.28%), HALO can greatly trim down the energy cost (up to ↓60%) in adaptation, quantified using an IoT device or SOTA simulator. Codes and pre-trained models are at https://github.com/RICE-EIC/HALO.

Keywords: On-device learning · Learning to optimize · Meta learning · Efficient training · Internet-of-Things

---

The first two authors Chaojian Li and Tianlong Chen contributed equally.

Electronic supplementary material The online version of this chapter (https://doi.org/10.1007/978-3-030-58545-7_29) contains supplementary material, which is available to authorized users.

© Springer Nature Switzerland AG 2020
A. Vedaldi et al. (Eds.): ECCV 2020, LNCS 12354, pp. 500–518, 2020.
https://doi.org/10.1007/978-3-030-58545-7_29

1 Introduction

The record-breaking success of machine learning (ML) algorithms has fueled an explosive demand for bringing ML-powered intelligent functionality into numerous Internet-of-Things (IoT) devices [37,39]. For practical deployment, many of them (such as autonomous vehicles, drones, mobiles, and wearables) require on-site in-situ learning for enabling them to continuously learn from new data and adapt to new environments [50]. However, the realization of on-device continuous model adaptation remains a bottleneck challenge because powerful performance of ML algorithms often comes at a prohibitive training cost while IoT devices are often extremely resource constrained. To tackle this challenge, existing efficient training techniques such as low-precision and pruning training can largely fall short as they are not designed and optimized for on-device model adaptation. Specifically, in contrast to standard training, on-device adaptation needs to (1) achieve fast model convergence (i.e., reduced training iterations) given that limited data is available or can be stored on IoT devices and (2) be realized with much boosted training energy/time efficiency for possibly wide adoption.

To close the aforementioned gap, we explore from a promising yet unexplored perspective motivated by the observation that neither traditional hand-crafted (e.g., SGD, Adagrad, and Adam) nor existing meta optimizers are dedicated to meet the on-device adaptation challenges. This is because the former requires tedious and manual hyper-parameter tuning, while the latter can be automated, they are often more costly due to the meta algorithms' own overhead. Specifically, we propose, develop, and experimentally validate a *hardware-aware learning to optimize* (**HALO**) framework, targeting to aggressively trim down the energy cost of on-device ML adaptation. This paper makes the following contributions:

- We **for the first time** introduce learning to optimize to a practical and explosively demanded application of resource-efficient, on-device ML adaptation, and demonstrate that it largely outperforms the most competitive SOTA optimizers. The proposed HALO framework is achieved using a Long Short-Term Memory (LSTM) aided with an innovative Jacobian regularizer that is dedicated for faster adaptation.
- To further ensure that the proposed HALO can be practically deployed for model adaptation on numerous resource-limited IoT devices, we next introduce (stochastic) structural sparsity as an extra regularizer for the learning optimizer, so that it can be efficiently implemented on hardware. Thanks to the aforementioned two regularizers, the HALO generated optimizers are enforced to naturally achieve the critical specification of on-device adaptation, i.e., both faster adaptation speed and reduced per-iteration complexity.
- We have evaluated and demonstrated the HALO optimizers on various models, datasets, and experiment settings (including going-wider, going-deeper, going-sparser, and going-lower bits), by exhaustively comparing it with existing off-the-shelf traditional hand-crafted and meta-optimizers. Extensive experiments and ablation studies show that HALO consistently outperforms others, by largely reducing on-device adaptation energy consumption (i.e., the

energy it takes to adapt for achieving the specified accuracy) while always maintaining a better accuracy given the same energy budget.

2 Related Works

Model Adaptation. Model adaptation techniques are commonly exploited to: (1) continuously improve a model's performance in the same domain, as more data is collected; or (2) further tune a model already trained on one domain (source domain) to adapt to a new domain (target domain), assuming the source and target domains to have a certain mismatch (either data distribution or task types) [7]. Many adaptation algorithms have been explored for various ML algorithms, from aligning data distributions [10] to utilizing feature or module transferability [5,16,41], for which [59] provides a comprehensive literature review.

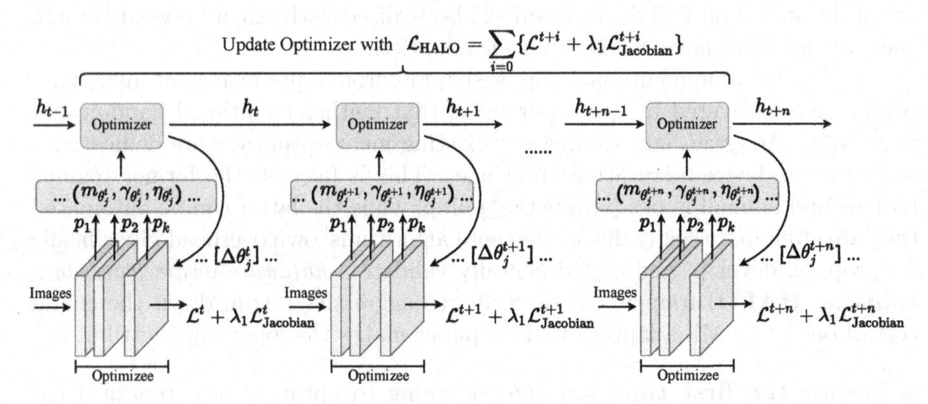

Fig. 1. The overall framework of our proposed hardware-aware learning to optimize (HALO) method. For each time step t, the optimizer will first take the previous hidden vector h_{t-1} and the relative input vector $(m_{\theta_j^t}, \gamma_{\theta_j^t}, \eta_{\theta_j^t})$ which contains the gradient information from the optimizees, and then output a parameter update rule $\Delta\theta_j^t$ for the optimizees. The layers to be updated are selected according to the probability $(p_1, \cdots, p_k)$, i.e., the structural sparsity regularizer. After $(n+1)$ optimization iterations (in our case, $n = 10$), we update the optimizer with the averaged optimizee loss $\mathcal{L}$ and the Jacobian regularizer $\mathcal{L}_{\text{Jacobian}}$.

Adaptation algorithm is the cornerstone for many intelligent edge platforms to perceive and react to the changing new environments (such as drones and outdoor robots) and for wearable devices to personalize their functionality to individual users [29], and so on.

Learning to Optimize. Using machine learning algorithms to design an optimizer is a promising direction towards replacing tedious algorithm crafting and/or hyperparameter tuning. [2] first employs a coordinate-wise LSTM as a learnable optimizer for training neural networks. It takes the gradient of optimizee parameters as inputs and outputs the parameters' update rule. [11] introduces the history of objective values for inputs, and outputs gradients as the actions of reinforcement learning agents. [42] introduces two practical techniques of random scaling and objective convexifying to boost generalization ability. [66] designs a hierarchical RNN architecture, augmenting the inputs with the log gradient magnitudes and the log learning rate; its results remain to be a SOTA among learned optimizers. Lately, [8] combines both point-based and population-based optimization algorithms, and further incorporates the posterior into meta-loss to balance the exploitation-exploration trade-off.

3 The Proposed HALO Framework

In this section, we introduce our HALO framework with two innovations dedicated to the resource-efficient adaptation goal. First, a Jacobian regularizer is designed to boost the *empirical generalization and convergence speed.* Second, we introduce structural sparsity as the desired property to be enforced on the optimizer output, such that the resulting update is more *hardware friendly and energy-efficient.* Both are shown to be experimentally effective in Sect. 4.

3.1 Faster and Better: A Jacobian-Regularized Learned Optimizer

The backbone of HALO follows the classical setting in [66]. We adopt a similar hierarchical RNN as the learned optimizer. Specifically, the hierarchical RNN architecture contains three levels, named "Parameter RNN", "Tensor RNN", and "Global RNN" from the low to high levels. Specifically, the "Parameter RNN" deals with the inputs and outputs update rules for each parameter of the optimizees; the "Tensor RNN" takes as inputs all hidden states from the "Parameter RNN" which processes parameters belonging to the same tensor and returns a bias term to them; and the "Global RNN" takes as inputs all hidden states from the "Tensor RNN" and returns a bias term. The RNN parameters are shared within each level. In this way, the learned optimizer is able to capture the inter-parameter dependencies.

As shown in Fig. 1, following the prior wisdom of learned optimizers [42,66], the inputs of our optimizer are $(m_{\theta_j^t}, \gamma_{\theta_j^t}, \eta_{\theta_j^t})$, corresponding to the scaled averaged gradients $m_{\theta_j^t}$, the relative log gradient magnitudes $\gamma_{\theta_j^t}$, and the relative log learning rate $\eta_{\theta_j^t}$ of layer j's parameter θ_j in iteration t, respectively, more details of which can be found in the supplement. The output is the parameter update $\Delta\theta_j^t$. Our learnable optimizer performs a coordinate-wise update on the parameter θ so that the learned optimizer can scale to training optimizees with any number of parameters: an important **"one-for-all"** feature desired by mobile applications where a number of different models are typically configured to meet different platforms' resource constraints. Between different coordinates, the weights of the optimizer are shared. In HALO, the optimizer is updated by $\mathcal{L}_{\text{HALO}}$, which is the sum of the average optimizee loss $\mathcal{L}$, plus a new Jacobian regularizer term $\mathcal{L}_{\text{Jacobian}}$ (λ_1 is a hyperparameter, more ablation studies are provided in the supplement): $\mathcal{L}_{\text{HALO}} = \mathcal{L} + \lambda_1 \mathcal{L}_{\text{Jacobian}}$.

We next discuss "*what* and *why*" regarding this new regularizer.

Jacobian Regularizer. We propose a powerful regularizer, called *Jacobian regularizer*, that controls the update magnitudes of the optimizee (i.e., the model to be adapted by HALO). Without loss of generalizability, we define our optimizee with k layers as $f(\theta)$, $\theta = (\theta_1, \theta_2, \cdots, \theta_k)$ ($k = 1$ for shallow models). The Jacobian of the optimizee loss $\mathcal{L}$ can be written as $J = \left[\frac{\partial \mathcal{L}}{\partial \theta_1}, \frac{\partial \mathcal{L}}{\partial \theta_2}, \cdots, \frac{\partial \mathcal{L}}{\partial \theta_k}\right]$, and our new regularizer term can be defined as $\mathcal{L}_{\text{Jacobian}} = ||J||_2^2$. ($||\cdot||_2^2$ is a Frobenius norm)

The Jacobian regularizer encourages the optimizee's landscape to be **smoother** and **flatter**. Intuitively, such a landscape facilitates an optimizer to explore **faster** and **more widely** in a neighborhood, which makes it favor our goal of fast adaptation. More formally, recent theories have revealed that optimizing in flat minima leads to more generalizable solutions [21,28]. It is straightforward to see that the larger the components of the Jacobian are, the more unstable the model prediction is with respect to input perturbations. Enforcing $\mathcal{L}_{\text{Jacobian}}$ is therefore a natural way to reduce this instability: it decreases the input-output Jacobian magnitude, potentially reducing the influence of noisy updates during training. That **robustness** is meaningful for practical on-device adaptation whose input samples are often very noisy [6].

Among past works, [51] constrained the Jacobian matrix of the encoder for the regularization of auto-encoders. [22] showed that constraining the Jacobian increases classification margins of neural networks and therefore enhances the model stability. While the above works exploit Jacobian regularizer in classical optimizers, to our best knowledge, we are **the first** to extend this line of ideas into the *learning to optimize* field. Our results demonstrate its effectiveness in improving generalization performance (i.e., adaptation/test accuracy) of the learned optimizer, in addition to the faster empirical convergence speed.

Besides, the analysis in [57] found that a bounded spectral norm of the network's Jacobian matrix is more directly related to the generalization of neural networks. We tested and verified that replacing the Frobenius norm with the spectral norm will yield similar empirical performance and convergence benefits, sometimes the spectral norm being better. However, computing the spectral norm is much more expensive and goes against our goal of resource efficiency: that is why we stay with the Frobenius norm in implementing $\mathcal{L}_{\text{Jacobian}}$.

3.2 More Hardware-Efficient: Stochastic Structural Sparsity

As a learned optimizer, HALO targets faster empirical convergence (e.g., taking fewer iterations to reach a certain accuracy level), which is further boosted by the new Jacobian regularizer. We introduce another regularizer, that enhances the energy efficiency from an orthogonal angle: enforcing structural sparsity on the learned updates (i.e., the outputs of HALO) at each iteration, such that the per-iteration complexity and hence resource costs could be trimmed down.

Structural sparsity is a well-explored regularizer that is typically achieved by weight decay, norm constraints, or various pruning means [65]. In comparison, we choose an extremely cheap "stochastic" way to enforce that. As shown in Fig. 1, for each layer j in the optimizee, we set it to have a probability p_j to be updated by HALO, at each iteration. Correspondingly, only the layers that are updated at the current iteration will back-propagate to update the learned optimizers.

We note that similar ideas of "randomly not updating all layers every time" were previously exploited for training very deep networks [25] and faster dynamic inference [67]. Lately, it was demonstrated to be helpful for energy-efficient training too [64]. We are **the first** to show this heuristic regularizer to work well for learned optimizers in efficient training.

Compared to enforcing filter- and parameter-wise structural sparsity, the purposed layer-wise structural sparsity regularizer is particularly hardware-friendly, as it requires no massive indexing and gathering processing. As our experiments in Sect. 4.2.1 show, this alone can save up to 45.42% training energy per iteration on average while sacrificing little accuracy or convergence speed.

4 Experiments and Analysis

In this section, we present ablation studies and evaluation results of the proposed HALO under five datasets, six optimizees (i.e., the wider one in Fig. 2 (b), the wider and deeper one in Fig. 2 (c), ResNet-18 [20] with quantization and high sparsity, two multilayer perceptrons (MLPs), and a CNN+LSTM [52]), and two SOTA edge AI computing devices.

4.1 Experiment Setup

Here we summarize our experiment details including the datasets and baselines, adaptation/test experiment setting, and evaluation metrics, and details of the optimizer design can be found in the supplement.

Datasets and Baselines. To evaluate the potential of the proposed HALO in handling on-device adaptation under different applications and scenarios, we consider a total of **five datasets**, including (1) MNIST [34], (2) CIFAR-10 [32], (3) Thyroid Disease Prediction (TDP) [12], (4) Gas Sensor Array Drift (GSAD) [62], and (5) Smartphones (SP) [3] (more details on the train/test subset splitting could be found in the supplement). These five diverse sets of datasets can emulate on-device ML applications for tasks of object recognition, healthcare monitoring, environmental monitoring, and activity recognition

For benchmarking, we evaluate HALO's generated optimizers against **five baselines** of SOTA optimizers, including three traditional hand-crafted optimizers (i.e., SGD, Adagrad [13], and Adam [31]) and two meta-optimizers (i.e., the DM-L2O [2] and Hierarchical-L2O [66]).

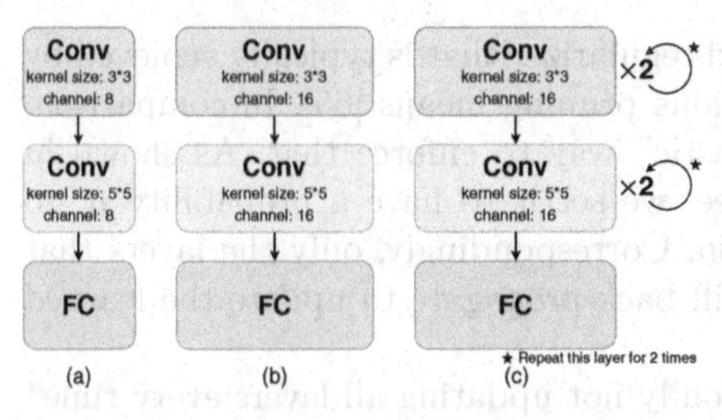

Fig. 2. The convolutional networks adopted for (a) training all the evaluated optimizers, and the optimizee networks including (b) a wider one and (c) a wider and deeper one, as compared to (a).

Table 1. A summary of the splitting details for all the considered datasets.

Dataset	Subset	Domain
MNIST [34]	A	{1, 3, 5, 7, 9}
	B	{0, 2, 4, 6, 8}
CIFAR-10 [32]	A	{plane, bird, deer, frog, ship}
	B	{car, cat, dog, horse, truck}
TDP [12]	A	Female
	B	Male
GSAD [62]	A	{Acetaldehyde, Acetone, Toluene}
	B	{Ethanol, Ethylene, Ammonia}
SP [3]	A	{Walking Upstairs, Sitting, Laying}
	B	{Walking, Walking downstairs, Standing}

Adaptation/Test Setting. To evaluate each optimizer under a given dataset, we split the dataset into two non-overlapping subsets following the prepossessing in [18,23,30], which are termed as A and B, respectively, with samples from different domains and each consisting of non-overlapping classes. Table 1 summarizes the splitting details for all our considered datasets. Following the strategy in real-world deployments [4,19,49], we first pre-train the model on one subset, and then start from the pre-trained model to retrain it on the other subset to see how accurately and efficiently the corresponding optimizee can adapt to the new domain. The same splitting is applied to the test set for accuracy validation in both the pre-train and adaptation training processes. We observe that the accuracy of the optimizee in the TDP dataset, which is trained on the subset B for the male domain and achieves an accuracy of 73.92%, drops to 55.74% when directly applying to the female domain without adaptation, motivating the need of adaptation.

Evaluation Metrics. We evaluate all optimizers in terms of the hardware energy consumption in addition to the optimizees' averaged training loss and adaptation/test accuracy over ten random initialization settings. Specifically, for the full-precision optimizees, we obtain the real-measured energy consumption on two SOTA edge AI computing devices, i.e., NVIDIA TX2 [45] (for more complex CNNs in Fig. 3 - Fig. 6 and Table 4) and Raspberry Pi [61] (for simpler MLPs in Tables 2 - 3); for the quantized optimizees, we adopt a SOTA hardware energy simulator, Bit Fusion [55], to obtain the energy consumption (the one in Fig. 7). The real-device energy measurement setup and energy simulation details are provided in the supplement.

4.2 Ablation Studies of the Proposed HALO

Here we perform ablation studies of HALO's effectiveness (Sect. 4.2.1) and structural sparsity regularizer (Sect. 4.2.2).

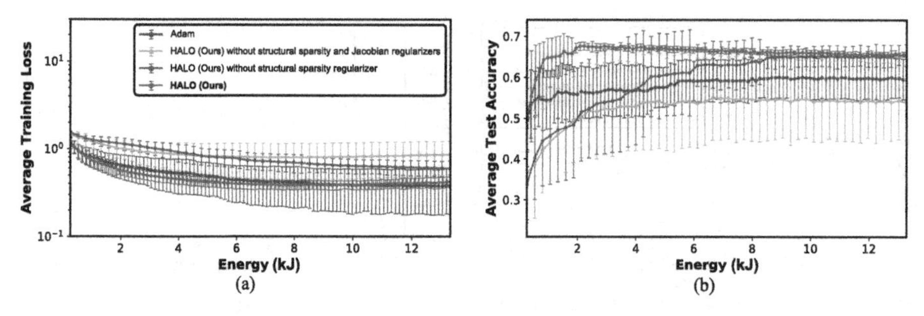

Fig. 3. Ablation studies on **the effectiveness of HALO's regularizers**: (a) The average training loss and (b) adaptation/test accuracy vs. the required energy cost over ten runs, on CIFAR-10-A.

4.2.1 Ablation Studies on the Effectiveness of HALO's Regularizers

For evaluating the effectiveness of HALO's regularizers, we perform a set of experiments using the wider optimizee (see Fig. 2 (b)) and CIFAR-10 dataset. Specifically, the optimizee is evaluated when enforcing the two regularizers of HALO in an incremental manner, with the corresponding optimizer being trained from scratch. Figure 3 shows the average training loss and adaptation/test accuracy versus the corresponding real-device measured energy over ten random initialization settings, from which we can make the following observation:

First, the vanilla HALO without the two regularizers can not surpass the SOTA traditional hand-crafted optimizer, Adam [31], in terms of both the training loss and adaptation/test accuracy after convergence, while at the same time suffering from a larger variance;

Second, after adding the Jacobian regularizer, the corresponding optimizee achieves a better adaptation/test accuracy **after** convergence, which verifies the flatten local minima found by HALO with the help of the Jacobian regularizer is beneficial for generalization ability as introduced in Sect. 3.1, while always performing worse under the same energy budget in the early stage, as compared to the optimizee trained using Adam;

Third, our proposed HALO always leads to a lower training loss (e.g., up to $\downarrow$16.72% lower under the same energy) and higher accuracy (e.g., up to $\uparrow$11.70% higher under the same energy) while having a smaller variance, as compared to all baselines, which seems to align with recent observations that compressing gradient during training can benefit the efficiency without hurting the performance [14,64].

This set of experiments validate that the two regularizers integrated into HALO lead to not only **faster adaptation** with **reduced energy cost**, but also offer a bonus benefit of **improving convergence stability**.

4.2.2 Ablation Studies of HALO's Structural Sparsity Regularizer

Here we present ablation studies on the schedule schemes of HALO's updating probability in the structural sparsity regularizer using the wider optimizee (see

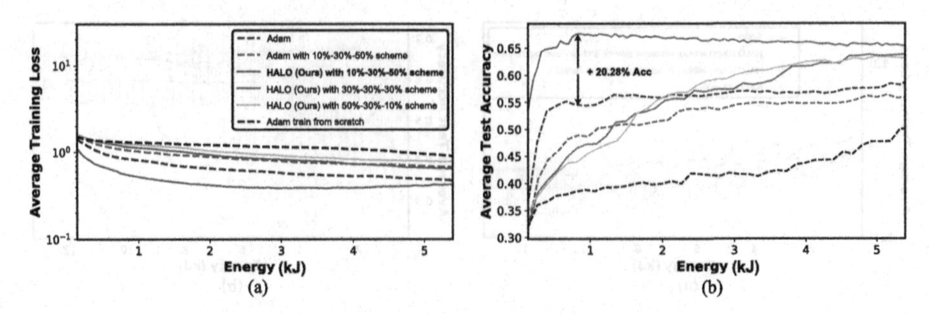

Fig. 4. Ablation studies of HALO's structural sparsity regularizer in terms of the updating probability: (a) The average training loss and (b) adaptation accuracy vs. the required energy cost over ten runs, on CIFAR-10-A.

Fig. 2 (b)) and CIFAR-10 dataset. Specifically, different updating probability can be adopted for the first, second, and third layers of the optimizee.

For example, "10%-30%-50%" means the corresponding updating probability are 10%, 30%, and 50%, respectively. For HALO, we consider three schedule schemes for the updating probability, i.e., "progressively increased", "uniformly equal", and "progressively decreased"; For Adam, we consider the schedule scheme of updating probability under which HALO performs the best (i.e., "progressively increased", more schemes for Adam and other optimizers could be found in the supplement).

The experiment results in Fig. 4 show that: (1) comparing Adam with Adam + structural sparsity regularizer, we find that the hand-crafted optimizer, Adam, does not benefit from this regularizer, as evidenced by the corresponding decreased adaptation/test accuracy; and (2) comparing the three schedule schemes of updating probability for HALO, we observed that the "progressively increased", e.g., "10%-30%-50%", significantly outperforms the other two schedule schemes by offering a higher adaptation/test accuracy under the same energy cost.

Note that the advantage of such a "progressively increased" schedule scheme for HALO is consistently observed under different datasets and models, which seems to coincide with recent findings [1,17,36,63,69] that (1) different stages of DNN training call for different treatments and (2) not all layers are equally important for training convergence.

4.3 HALO Under Different Datasets/Optimizees

4.3.1 HALO on the CIFAR-10 Dataset

In this subsection, we evaluate HALO's performance and generalization capability when being applied to various optimizees, which are (1) wider (Fig. 2 (b)), (2) wider and deeper (Fig. 2 (c)), and (3) wider, deeper, highly sparse and quantized (pruned and quantized ResNet-18 [20]), as compared to the networks used to train the optimizers.

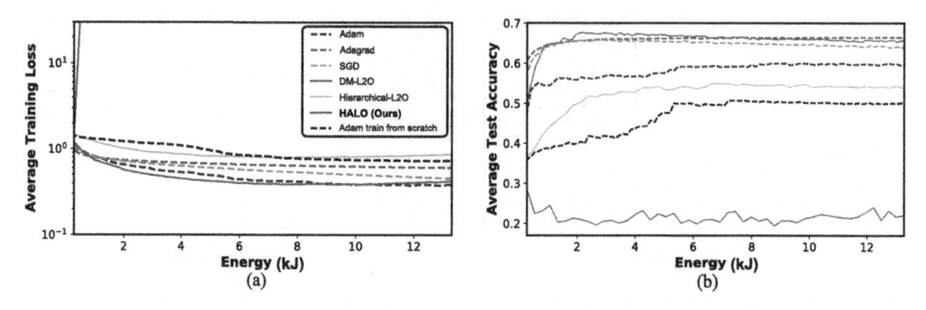

Fig. 5. HALO for the wider optimizee: (a) The average training loss and (b) adaptation/test accuracy vs. the energy cost over ten runs, on CIFAR-10-A.

Experiment settings. For all the aforementioned three optimizees, experiments are performed using the CIFAR-10 dataset. And for all the experiments using the optimizee (3) mentioned above, a compressed ResNet-18 [20] is trained and pruned under a pruning ratio of 70.0%, which leads to a reduction of 43.5% and 61.5% in the computational cost (i.e., FLOPs) and model size over the unpruned one, respectively, while performing quantization-aware training [26] in the optimizee (3) during adaptation. Note that we consider a pruned and quantized optimizee because the current practice often compresses ML models before deploying them into IoT devices [15,46,64].

Experiment results and observations. For each optimizer on the three considered optimizees, we evaluate the optimizees' averaged training loss and adaptation/test accuracy under ten random initialization settings. The corresponding results are plotted in Figs. 5 - 7, from which we can make the following **five observations**:

First, while SOTA learning to optimize works are merely evaluated in terms of the optimizees' training loss [2,42,66], we find that both the training loss and adaptation/test accuracy need to be considered for adaptation tasks, as a lower training loss might not guarantee a higher adaptation/test accuracy. For

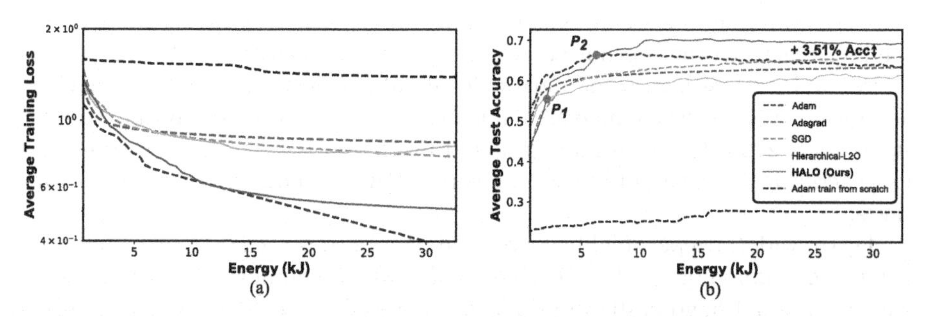

Fig. 6. HALO for the wider and deeper optimizee: (a) The average training loss and (b) adaptation/test accuracy vs. the energy cost over ten runs, on CIFAR-10-A.

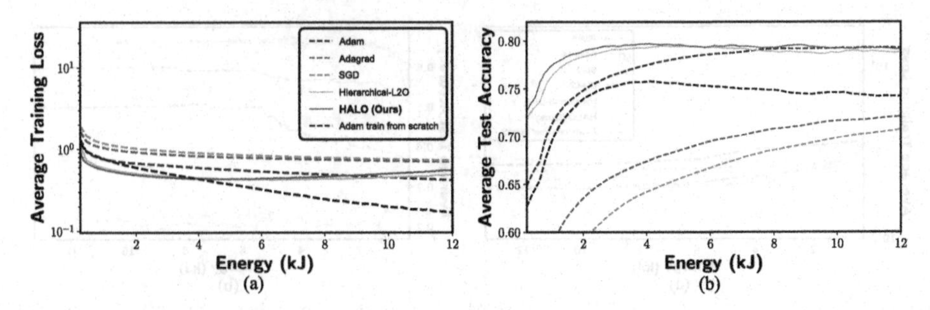

Fig. 7. HALO for the wider, deeper, highly sparse, and quantized optimizee: (a) The average training loss and (b) est accuracy vs. the energy cost over ten runs, on CIFAR-10-A.

example, from Fig. 5 we can see that Adam achieves a smaller training loss but a lower adaptation accuracy, as compared to that of Adagrad.

Second, the HALO optimizer outperforms all other meta-optimizers under all the three optimizees, while (1) the meta-optimizer, DM-L2O [2], fails under the evaluation with the wider optimizee as shown in Fig. 5, which is consistent with observations discussed in prior works of learning to optimize [42,66], thus won't be included in the following experiments, and (2) the meta-optimizer, Hierarchical-L2O [66], always leads to a lower adaptation accuracy (e.g., up to ↓16.88% lower in the wider optimizee as shown in Fig. 5 under the same energy budget) and a larger training loss (e.g., up to ↑65.65% larger in the wider optimizee as shown in Fig. 5 under the same energy budget), as compared to that of HALO.

Third, as shown in Fig. 6, in the early training stage before the cross-points P_1 or P_2, the HALO optimizer does not outperform other optimizers, while in the later stage after the cross-point P_2, HALO significantly surpasses others by a large performance margin (at least ↑3.51% higher adaptation/test accuracy when the energy cost is around 30 kJ). For this interesting and expected phenomenon, we conjecture the possible reasons, which are consistent with empirical observations in [21,28]: (1) with the assistance of the Jacobian regularizer, HALO in the early training stage can explore the training landscape to locate a flatter minima, whereas other optimizers who are not flatness-aware can be easily overfitting and get stuck in some narrow local valleys, leading to results that the adaptation/test accuracy increases quickly at first, and then decays in the later stage; and (2) flatten local minima found by HALO is beneficial for improving its generalization capability [21,28], favoring HALO's large performance advantages.

Forth, wider, deeper, highly sparse and quantized optimizee targets on a difficult (e.g., exploding gradient is a common issue in training quantized network with traditional hand-crafted optimizers as described in [24]) yet practical setting for on-device adaptation [15,46,64]. As shown in Fig. 7, HALO outperforms all traditional hand-crafted optimizers obviously, regardless of the latter being extensively tuned, and show a marginal improvement over the Hierarchical-L2O, indicating the superiority of our purposed HALO (i.e., higher energy efficiency while having one-for-all generalization capability) even in such stringent cases.

Fifth, while the baseline optimizers don't have a fixed performance ranking under various optimizees (e.g., Adam performs better than Adagrad in the wider and deeper optimizee but worse in the wider optimizee, as shown in Fig. 6 and 5), HALO **always** achieves both a comparable or lower training loss (e.g., up to ↓16.72% lower in the wider optimizee as shown in Fig. 5 under the same energy budget) and a comparable or higher adaptation/test accuracy (e.g., up to ↑3.51% in the deeper optimizee as shown in Fig. 6 under the same energy budget) among all considered optimizers (including both traditional hand-crafted and mete-optimizers), indicating **HALO's consistently best capability** in balancing the models' accuracy and on-device adaptation cost in various optimizees. Notably, even in some cases HALO outperforms other optimizers marginally (e.g., HALO and Adadgrad in Fig. 5), the latter ones need massive manual hyperparameters tuning while the former one could be used without any hyper-parameter tuning while having the one-for-all generalization capability.

4.3.2 HALO on the Thyroid Disease Prediction Dataset (TDP)

In this subsection, we evaluate HALO's performance on the TDP dataset, the tasks of which can emulate on-device healthcare monitoring applications, one of the most popular applications on resource-limited IoT devices [33,43,54].

Experiment settings. As introduced in Sect. 4.1, the input data of the TDP dataset is a 26-dimensional vector, we thus adopt a three-layer MLP for the optimizee. Specially, we evaluate HALO in terms of the average training loss and adaptation/test accuracy under the same training iterations, where the adaptation energy budget is set to be no more than 0.5 kJ in Raspberry Pi [61] which is 2% of the most commonly-used Li-Po battery's capacity (27 kJ) [47] adopted by massive IoT devices [56].

Experiment results and observations. Table 2 summarizes the experiment results, from which we can see that (1) HALO outperforms all the baseline optimizers (both traditional hand-crafted and automatically learned ones) in terms of the achieved average adaptation/test accuracy under the same number of iterations; and (2) HALO achieves a higher adaptation/test accuracy (i.e., ↑0.96% - ↑1.00%) while requiring ↓31.8% less adaptation energy as compared to the SOTA learning to optimize optimizer, Hierarchical-L2O, thanks to its reduced energy cost per iteration, indicating the advantage and effectiveness of the proposed HALO for on-device adaptation. Note that although the hand-crafted traditional optimizers require a lower energy cost (e.g., ↓0.32 kJ less energy with a ↓0.31% lower accuracy) over the proposed HALO, they are not applicable for widely adopted on-device adaptation into numerous IoT applications due to their required tedious and manual hyper-parameter tuning.

Table 2. The evaluation results of the trained HALO on the TDP dataset.

Methods	Avg. Test Acc. (%)				Avg. Loss			Energy (kJ)		
	# of Iters.				# of Iters.			# of Iters.		
	1k	2k	3k	Final	1k	2k	3k	1k	2k	3k
Adam	79.41	79.85	79.85	79.85	0.49	0.44	0.43	0.04	0.09	0.13
Adagrad	75.81	75.81	75.82	75.83	0.77	0.73	0.71	0.04	0.09	0.13
SGD	77.08	78.07	78.29	78.29	0.73	0.69	0.67	0.04	0.09	0.13
Hierarchical-L2O	78.42	78.92	79.17	79.19	0.56	0.50	0.48	0.22	0.44	0.66
HALO	**79.42**	**79.88**	**80.16**	**80.28**	0.58	0.53	0.50	0.15	0.30	0.45
Improv. over SOTA L2O	**↑1.00%**	**↑0.96%**	**↑0.99%**	**↑1.09%**	-0.02	-0.03	-0.02	**↓31.8%**	**↓31.8%**	**↓31.8%**

Table 3. The evaluation results of the trained HALO on the GSAD Dataset.

Methods	Avg. Test Acc. (%)				Avg. Loss			Energy (kJ)		
	# of Iters.				# of Iters.			# of Iters.		
	1k	2k	3k	Final	1k	2k	3k	1k	2k	3k
Adam	69.75	69.91	69.91	69.91	0.42	0.38	0.37	0.02	0.04	0.06
Adagrad	59.06	60.31	61.38	62.16	0.93	0.79	0.71	0.02	0.04	0.06
SGD	58.62	59.66	60.56	60.97	0.78	0.69	0.67	0.02	0.04	0.06
Hierarchical-L2O	77.87	79.34	80.12	80.31	0.44	0.34	0.31	0.15	0.29	0.44
HALO	**79.25**	**85.22**	**86.81**	**87.00**	0.46	0.30	0.24	0.06	0.12	0.18
Improv. over SOTA L2O	**↑1.38%**	**↑5.88%**	**↑6.69%**	**↑6.69%**	-0.02	0.04	0.07	**↓60.0%**	**↓58.6%**	**↓59.1%**

4.3.3 HALO on the Gas Sensor Array Drift Dataset (GSAD)

Here we evaluate HALO's performance on the GSAD dataset, the tasks of which aim to classify the individual gas components in the gas mixtures based on the response of various metal oxide collected by the corresponding IoT sensors [27,35].

Experiment settings. The input data of GSAD is a 129-dimensional vector, we thus adopt a three-layer MLP for the optimizee, which is similar to the one in Sect. 4.3.2, except that the first layer's dimension is increased for adapting to the increased input dimension. We also evaluate HALO in terms of the average training loss and adaptation/test accuracy under the same training iterations. The adaptation energy budget of this set of experiments is 0.2 kJ which is smaller than that in Sect. 4.3.2, considering that GSAD's corresponding tasks of environmental monitoring applications often face circumstances with a very limited energy budget for a long time [38,44].

Experiment results and observations. This set of experiment results are shown in Table 3 from which we can see that: (1) HALO outperforms both the traditional hand-crafted and automatically learned baseline optimizers by achieving a higher average adaptation/test accuracy given the same number of iterations; and (2) HALO performs better than the SOTA learning to optimize optimizer, Hierarchical-L2O, as it achieves a higher average adaptation/test accuracy (i.e., ↑1.38% - ↑6.69%) while requiring ↓58.6% - ↓60.0% lower adaptation energy. Similarly, although the hand-crafted optimizers require a lower energy cost (e.g., ↓0.12 kJ less energy consumption at a cost of a ↓16.90% lower accuracy) over the proposed HALO, their required tedious and manual hyper-parameter tuning limits their applicability to on-device adaptation for numerous IoT applications.

4.3.4 HALO on the Smartphones Dataset (SP)

Here we evaluate HALO's performance on the SP dataset, the tasks of which aim to predict human activities based on data collected from smartphones [9,58].

Experiment settings. Considering the sequence data in the SP dataset, we adopt an optimizee with a CNN+LSTM architecture [52], where a CNN of three convolution layers are used for feature extraction followed by LSTMs to support sequence prediction. For this set of experiment, (1) HALO's structural sparsity regularizer is only applied to the optimizee's convolution layers for balancing adaptation cost and accuracy; and (2) we relax the energy budget to 5 kJ in the NVIDIA TX2 [45], which is around 8% of the battery capability of commonly used smartphones (e.g., battery capacity of SAMSUNG Galaxy S20 is around 63 kJ [53]), considering the practicability of smartphone-based applications.

Experiment results and observations. The experiment results of HALO and the baseline optimizers in Table 4 show that (1) HALO again performs the best among all the optimizers including both traditional hand-crafted and automatically learned ones in terms of the achieved average adaptation/test accuracy under the same number of iterations; and (2) HALO requires a lower ↓44.3% - ↓45.1% energy while achieving a higher (i.e., ↑0.96% - ↑1.00%) average adaptation/test accuracy, over the SOTA learning to optimize optimizer, Hierarchical-L2O, thanks to its structural sparsity that enforces reduced energy cost per iteration. Note that while being automated and thus more applicable for a wide adoption into numerous IoT applications, the proposed HALO can achieves higher accuracies than all the hand-crafted optimizers which are limited when it comes to on-device adaptation due to their required tedious and manual hyper-parameter tuning needed for changes in data or application.

Table 4. The evaluation results of the trained HALO on the SP Dataset.

Methods	Avg. Test Acc. (%)				Avg. Loss			Energy (kJ)		
	# of Iters.				# of Iters.			# of Iters.		
	1k	2k	3k	Final	1k	2k	3k	1k	2k	3k
Adam	88.57	92.47	95.59	95.66	6.27E-5	1.79E-5	1.21E-5	1.15	2.92	4.62
Adagrad	69.24	72.31	75.64	76.56	1.17E-2	5.99E-3	4.38E-3	1.15	2.92	4.62
SGD	66.47	68.52	68.52	68.52	2.21E-2	1.15E-2	8.01E-3	1.15	2.92	4.62
Hierarchical-L2O	92.31	96.06	97.08	97.20	2.32E-4	5.38E-5	4.81E-5	1.61	4.08	6.44
HALO	**93.22**	**96.52**	**97.76**	**98.16**	2.49E-3	3.55E-4	3.09E-4	1.17	2.96	4.67
Improv. over SOTA L2O	**↑0.91%**	**↑0.46%**	**↑0.68%**	**↑0.96%**	-2.26E-3	-3.01E-4	-2.61E-4	**↓27.3%**	**↓27.5%**	**↓27.5%**

5 Conclusions

We propose a learning to optimize framework, HALO, a practical meta-optimizer dedicated to resource-efficient on-device adaptation. Specifically, Jacobian and structural sparsity regularizers are integrated in HALO to reduce per-iteration complexity and enforce faster adaptation speed, thus contribute to the adaptation efficiency. Furthermore, we demonstrate that HALO outperforms existing off-the-shelf traditional hand-crafted and meta-optimizers based on extensive experiments on six optimizees, five datasets, and two SOTA edge AI computing devices.

Acknowledgments. The work is supported by the National Science Foundation (NSF) through the Real-Time Machine Learning program (Award number: 1937592, 1937588).

References

1. Achille, A., Rovere, M., Soatto, S.: Critical learning periods in deep networks. In: International Conference on Learning Representations (2019). https://openreview.net/forum?id=BkeStsCcKQ
2. Andrychowicz, M., et al.: Learning to learn by gradient descent by gradient descent. In: Advances in Neural Information Processing Systems, pp. 3981–3989 (2016)
3. Anguita, D., Ghio, A., Oneto, L., Parra, X., Reyes-Ortiz, J.L.: A public domain dataset for human activity recognition using smartphones. In: Esann (2013)
4. Ashqar, B.A., Abu-Naser, S.S.: Identifying images of invasive hydrangea using pre-trained deep convolutional neural networks. Int. J. Acad. Eng. Res. (IJAER) **3**(3), 28–36 (2019)
5. Bengio, Y.: Deep learning of representations for unsupervised and transfer learning. In: Proceedings of ICML Workshop on Unsupervised and Transfer Learning, pp. 17–36 (2012)
6. Bippus, R., Fischer, A., Stahl, V.: Domain adaptation for robust automatic speech recognition in car environments. In: Sixth European Conference on Speech Communication and Technology (1999)
7. Blitzer, J., McDonald, R., Pereira, F.: Domain adaptation with structural correspondence learning. In: Proceedings of the 2006 Conference on Empirical Methods in Natural Language Processing, pp. 120–128 (2006)
8. Cao, Y., Chen, T., Wang, Z., Shen, Y.: Learning to optimize in swarms. In: Advances in Neural Information Processing Systems, vol. 32, pp. 15018–15028. Curran Associates, Inc. (2019). http://papers.nips.cc/paper/9641-learning-to-optimize-in-swarms.pdf
9. Chen, H., Mahfuz, S., Zulkernine, F.: Smart phone based human activity recognition. In: 2019 IEEE International Conference on Bioinformatics and Biomedicine (BIBM), pp. 2525–2532. IEEE (2019)
10. Chen, Q., Liu, Y., Wang, Z., Wassell, I., Chetty, K.: Re-weighted adversarial adaptation network for unsupervised domain adaptation. In: Proceedings of the IEEE Conference on Computer Vision and Pattern Recognition, pp. 7976–7985 (2018)
11. Chen, Y., et al.: Learning to learn without gradient descent by gradient descent. In: Proceedings of the 34th International Conference on Machine Learning, vol. 70, pp. 748–756. JMLR org (2017)

12. Dua, D., Graff, C.: UCI machine learning repository (2017). http://archive.ics.uci.edu/ml
13. Duchi, J., Hazan, E., Singer, Y.: Adaptive subgradient methods for online learning and stochastic optimization. J. Mach. Learn. Res. **12**, 2121–2159 (2011)
14. Elibol, M., Lei, L., Jordan, M.I.: Variance reduction with sparse gradients (2020)
15. Fang, B., Zeng, X., Zhang, M.: Nestdnn: resource-aware multi-tenant on-device deep learning for continuous mobile vision. In: Proceedings of the 24th Annual International Conference on Mobile Computing and Networking, pp. 115–127 (2018)
16. Glorot, X., Bordes, A., Bengio, Y.: Domain adaptation for large-scale sentiment classification: a deep learning approach. In: Proceedings of the 28th International Conference on Machine Learning (2011)
17. Greff, K., Srivastava, R.K., Schmidhuber, J.: Highway and residual networks learn unrolled iterative estimation (2016). arXiv preprint arXiv:1612.07771
18. Grothmann, T., Patt, A.: Adaptive capacity and human cognition: the process of individual adaptation to climate change. Glob. Environ. Change **15**(3), 199–213 (2005)
19. Habibzadeh, M., Jannesari, M., Rezaei, Z., Baharvand, H., Totonchi, M.: Automatic white blood cell classification using pre-trained deep learning models: Resnet and inception. In: Tenth International Conference on Machine Vision (ICMV 2017), vol. 10696, p. 1069612. International Society for Optics and Photonics (2018)
20. He, K., Zhang, X., Ren, S., Sun, J.: Identity mappings in deep residual networks. In: Leibe, B., Matas, J., Sebe, N., Welling, M. (eds.) ECCV 2016. LNCS, vol. 9908, pp. 630–645. Springer, Cham (2016). https://doi.org/10.1007/978-3-319-46493-0_38
21. Hochreiter, S., Schmidhuber, J.: Flat minima. Neural Comput. **9**(1), 1–42 (1997)
22. Hoffman, J., Roberts, D.A., Yaida, S.: Robust learning with Jacobian regularization (2019). arXiv preprint arXiv:1908.02729
23. Hoffman, J., Rodner, E., Donahue, J., Darrell, T., Saenko, K.: Efficient learning of domain-invariant image representations (2013). arXiv preprint arXiv:1301.3224
24. Hou, L., Zhu, J., Kwok, J., Gao, F., Qin, T., Liu, T.Y.: Normalization helps training of quantized LSTM. In: Advances in Neural Information Processing Systems, pp. 7344–7354 (2019)
25. Huang, G., Sun, Y., Liu, Z., Sedra, D., Weinberger, K.Q.: Deep networks with stochastic depth. In: Leibe, B., Matas, J., Sebe, N., Welling, M. (eds.) ECCV 2016. LNCS, vol. 9908, pp. 646–661. Springer, Cham (2016). https://doi.org/10.1007/978-3-319-46493-0_39
26. Jacob, B., et al.: Quantization and training of neural networks for efficient integer-arithmetic-only inference. In: 2018 IEEE/CVF Conference on Computer Vision and Pattern Recognition, June 2018. https://doi.org/10.1109/cvpr.2018.00286, http://dx.doi.org/10.1109/CVPR.2018.00286
27. Keshamoni, K., Hemanth, S.: Smart gas level monitoring, booking and gas leakage detector over IoT. In: 2017 IEEE 7th International Advance Computing Conference (IACC), pp. 330–332. IEEE (2017)
28. Keskar, N.S., Mudigere, D., Nocedal, J., Smelyanskiy, M., Tang, P.T.P.: On large-batch training for deep learning: Generalization gap and sharp minima (2016). arXiv preprint arXiv:1609.04836
29. Kikui, K., Itoh, Y., Yamada, M., Sugiura, Y., Sugimoto, M.: Intra-/inter-user adaptation framework for wearable gesture sensing device. In: Proceedings of the 2018 ACM International Symposium on Wearable Computers, pp. 21–24 (2018)

30. Kikui, K., Itoh, Y., Yamada, M., Sugiura, Y., Sugimoto, M.: Intra-/inter-user adaptation framework for wearable gesture sensing device. In: Proceedings of the 2018 ACM International Symposium on Wearable Computers, New York, NY, USA. ISWC'2018, pp. 21–24, Association for Computing Machinery (2018). https://doi.org/10.1145/3267242.3267256
31. Kingma, D.P., Ba, J.: Adam: a method for stochastic optimization (2014)
32. Krizhevsky, A., et al.: Learning multiple layers of features from tiny images (2009)
33. Lane, N.D., Bhattacharya, S., Mathur, A., Georgiev, P., Forlivesi, C., Kawsar, F.: Squeezing deep learning into mobile and embedded devices. IEEE Pervasive Compu. **16**(3), 82–88 (2017)
34. LeCun, Y.: The mnist database of handwritten digits (1999). http://www.yann.lecun.com/exdb/mnist/
35. Lee, J., et al.: Mems-based no 2 gas sensor using zno nano-rods for low-power IoT application. J. Korean Phys. Soc. **70**(10), 924–928 (2017)
36. Li, Y., Wei, C., Ma, T.: Towards explaining the regularization effect of initial large learning rate in training neural networks (2019). arXiv preprint arXiv:1907.04595
37. Lin, Y., Sakr, C., Kim, Y., Shanbhag, N.: Predictivenet: an energy-efficient convolutional neural network via zero prediction. In: 2017 IEEE International Symposium on Circuits and Systems (ISCAS), pp. 1–4 (2017)
38. Liu, C.H., Fan, J., Branch, J.W., Leung, K.K.: Toward qoi and energy-efficiency in internet-of-things sensory environments. IEEE Trans. Emerg. Top. Comput. **2**(4), 473–487 (2014). https://doi.org/10.1109/TETC.2014.2364915
39. Liu, S., Lin, Y., Zhou, Z., Nan, K., Liu, H., Du, J.: On-demand deep model compression for mobile devices: a usage-driven model selection framework. In: Proceedings of the 16th Annual International Conference on Mobile Systems, Applications, and Services, pp. 389–400. ACM (2018)
40. Liu, Z., Sun, M., Zhou, T., Huang, G., Darrell, T.: Rethinking the value of network pruning. In: International Conference on Learning Representations (2019). https://openreview.net/forum?id=rJlnB3C5Ym
41. Long, M., Zhu, H., Wang, J., Jordan, M.I.: Deep transfer learning with joint adaptation networks. In: Proceedings of the 34th International Conference on Machine Learning, vol. 70, pp. 2208–2217. ICML'2017, JMLR org (2017)
42. Lv, K., Jiang, S., Li, J.: Learning gradient descent: Better generalization and longer horizons. In: Proceedings of the 34th International Conference on Machine Learning, vol. 70, pp. 2247–2255. JMLR org (2017)
43. Miotto, R., Wang, F., Wang, S., Jiang, X., Dudley, J.T.: Deep learning for healthcare: review, opportunities and challenges. Briefings Bioinf. **19**(6), 1236–1246 (2018)
44. Moreno, M., Úbeda, B., Skarmeta, A.F., Zamora, M.A.: How can we tackle energy efficiency in iot basedsmart buildings? Sensors **14**(6), 9582–9614 (2014)
45. NVIDIA Inc.: NVIDIA Jetson TX2. https://www.nvidia.com/en-us/autonomous-machines/embedded-systems/jetson-tx2/, Accessed 01 Sep 2019
46. Park, K., Yi, Y.: Bpnet: branch-pruned conditional neural network for systematic time-accuracy tradeoff in dnn inference: work-in-progress. In: Proceedings of the International Conference on Hardware/Software Codesign and System Synthesis Companion, pp. 1–2 (2019)
47. PARTICLE Inc.: PARTICLE LP103450. https://store.particle.io/products/li-po-battery, Accessed 29 Feb 2020
48. Patrick Mochel and Mike Murphy.: sysfs-The filesystem for exporting kernel objects. https://www.kernel.org/doc/Documentation/filesystems/sysfs.txt, Accessed 21 Nov 2019

49. Peters, M., Ruder, S., Smith, N.A.: To tune or not to tune? adapting pretrained representations to diverse tasks (2019). arXiv preprint arXiv:1903.05987
50. Petrolo, R., Lin, Y., Knightly, E.: Astro: autonomous, sensing, and tetherless networked drones. In: Proceedings of the 4th ACM Workshop on Micro Aerial Vehicle Networks, Systems, and Applications, New York, NY, USA. DroNet'2018, pp. 1–6. Association for Computing Machinery (2018). https://doi.org/10.1145/3213526.3213527
51. Rifai, S., Vincent, P., Muller, X., Glorot, X., Bengio, Y.: Contractive auto-encoders: explicit invariance during feature extraction (2011)
52. Sainath, T.N., Vinyals, O., Senior, A., Sak, H.: Convolutional, long short-term memory, fully connected deep neural networks. In: 2015 IEEE International Conference on Acoustics, Speech and Signal Processing (ICASSP), pp. 4580–4584, April 2015. https://doi.org/10.1109/ICASSP.2015.7178838
53. SAMSUNG Inc.: SAMSUNG Galaxy S20. https://www.samsung.com/global/galaxy/galaxy-s20/specs/, Accessed 29 Feb 2020
54. Sannino, G., Pietro, G.D.: A deep learning approach for ECG-based heartbeat classification for arrhythmia detection. Future Gener. Comput. Syst. **86**, 446–455 (2018). https://doi.org/10.1016/j.future.2018.03.057, http://www.sciencedirect.com/science/article/pii/S0167739X17324548
55. Sharma, H., et al.: Bit fusion: bit-level dynamically composable architecture for accelerating deep neural network. In: 2018 ACM/IEEE 45th Annual International Symposium on Computer Architecture (ISCA), June 2018. https://doi.org/10.1109/isca.2018.00069, http://dx.doi.org/10.1109/ISCA.2018.00069
56. Singh, K.J., Kapoor, D.S.: Create your own internet of things: A survey of iot platforms. IEEE Consum. Electron. Mag. **6**(2), 57–68 (2017). https://doi.org/10.1109/MCE.2016.2640718
57. Sokolić, J., Giryes, R., Sapiro, G., Rodrigues, M.R.: Robust large margin deep neural networks. IEEE Trans. Signal Process. **65**(16), 4265–4280 (2017)
58. Subasi, A., Radhwan, M., Kurdi, R., Khateeb, K.: IoT based mobile healthcare system for human activity recognition. In: 2018 15th Learning and Technology Conference (L&T), pp. 29–34. IEEE (2018)
59. Tan, C., Sun, F., Kong, T., Zhang, W., Yang, C., Liu, C.: A survey on deep transfer learning. In: Kůrková, V., Manolopoulos, Y., Hammer, B., Iliadis, L., Maglogiannis, I. (eds.) ICANN 2018. LNCS, vol. 11141, pp. 270–279. Springer, Cham (2018). https://doi.org/10.1007/978-3-030-01424-7_27
60. Texas Instruments Inc.: INA3221 Triple-Channel, High-Side Measurement, Shunt and Bus Voltage Monitor. http://www.ti.com/product/INA3221, Accessed 21 Nov 2019
61. Upton, E., Halfacree, G.: Raspberry Pi User Guide. John Wiley & Sons, Hoboken (2014)
62. Vergara, A., Vembu, S., Ayhan, T., Ryan, M.A., Homer, M.L., Huerta, R.: Chemical gas sensor drift compensation using classifier ensembles. Sensors Actuators B: Chem. **166**, 320–329 (2012)
63. Wang, X., Yu, F., Dou, Z.Y., Darrell, T., Gonzalez, J.E.: Skipnet: Learning dynamic routing in convolutional networks. In: Proceedings of the European Conference on Computer Vision (ECCV), pp. 409–424 (2018)
64. Wang, Y., et al.: E2-train: training state-of-the-art CNNs with over 80% energy savings. In: Advances in Neural Information Processing Systems, pp. 5139–5151 (2019)

65. Wen, W., Wu, C., Wang, Y., Chen, Y., Li, H.: Learning structured sparsity in deep neural networks. In: Advances in Neural Information Processing Systems, pp. 2074–2082 (2016)
66. Wichrowska, O., et al.: Learned optimizers that scale and generalize. In: Proceedings of the 34th International Conference on Machine Learning, vol. 70, pp. 3751–3760. JMLR org (2017)
67. Wu, Z., et al.: Blockdrop: Dynamic inference paths in residual networks. In: Proceedings of the IEEE Conference on Computer Vision and Pattern Recognition, pp. 8817–8826 (2018)
68. You, H., et al.: Drawing early-bird tickets: towards more efficient training of deep networks (2019)
69. Zhang, C., Bengio, S., Singer, Y.: Are all layers created equal? CoRR abs/1902.01996 (2019)

Structured3D: A Large Photo-Realistic Dataset for Structured 3D Modeling

Jia Zheng[1,2](✉), Junfei Zhang[1], Jing Li[2], Rui Tang[1], Shenghua Gao[2,3], and Zihan Zhou[4]

[1] KooLab, Kujiale.com, Hangzhou, China
[2] ShanghaiTech University, Shanghai, China
zhengjia@shanghaitech.edu.cn
[3] Shanghai Engineering Research Center of Intelligent Vision and Imaging, Shanghai, China
[4] The Pennsylvania State University, State College, USA
https://structured3d-dataset.org

Abstract. Recently, there has been growing interest in developing learning-based methods to detect and utilize salient semi-global or global structures, such as junctions, lines, planes, cuboids, smooth surfaces, and all types of symmetries, for 3D scene modeling and understanding. However, the ground truth annotations are often obtained via human labor, which is particularly challenging and inefficient for such tasks due to the large number of 3D structure instances (*e.g.*, line segments) and other factors such as viewpoints and occlusions. In this paper, we present a new synthetic dataset, Structured3D, with the aim of providing large-scale photo-realistic images with rich 3D structure annotations for a wide spectrum of structured 3D modeling tasks. We take advantage of the availability of professional interior designs and automatically extract 3D structures from them. We generate high-quality images with an industry-leading rendering engine. We use our synthetic dataset in combination with real images to train deep networks for room layout estimation and demonstrate improved performance on benchmark datasets.

Keywords: Dataset · 3D structure · Photo-realistic rendering

1 Introduction

Inferring 3D information from 2D sensory data such as images and videos has long been a central research topic in computer vision. Conventional approach

J. Zheng and J. Zhang—Equal contribution.

The work was partially done when Jia Zheng interned at KooLab, Kujiale.com.

Electronic supplementary material The online version of this chapter (https://doi.org/10.1007/978-3-030-58545-7_30) contains supplementary material, which is available to authorized users.

© Springer Nature Switzerland AG 2020
A. Vedaldi et al. (Eds.): ECCV 2020, LNCS 12354, pp. 519–535, 2020.
https://doi.org/10.1007/978-3-030-58545-7_30

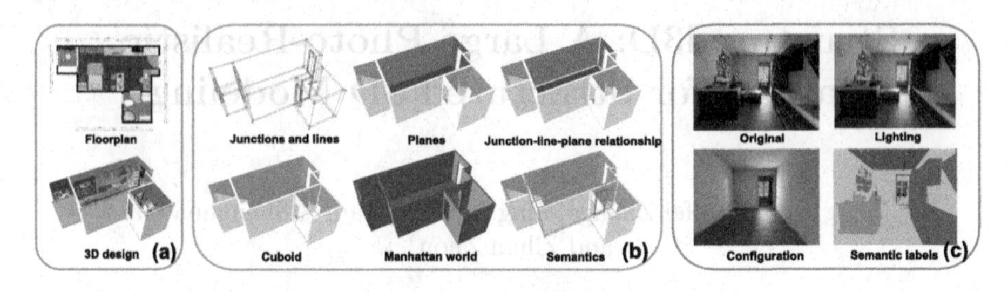

Fig. 1. The Structured3D dataset. From a large collection of house designs **(a)** created by professional designers, we automatically extract a variety of ground truth 3D structure annotations **(b)** and generate photo-realistic 2D images **(c)**.

to building 3D models typically relies on detecting, matching, and triangulating local image features (*e.g.*, patches, superpixels, edges, and SIFT features). Although significant progress has been made over the past decades, these methods still suffer from some fundamental problems. In particular, local feature detection is sensitive to a large number of factors such as scene appearance (*e.g.*, textureless areas and repetitive patterns), lighting conditions, and occlusions. Further, the noisy, point cloud-based 3D model often fails to meet the increasing demand for high-level 3D understanding in real-world applications.

When perceiving 3D scenes, humans are remarkably effective in using salient global structures such as lines, contours, planes, smooth surfaces, symmetries, and repetitive patterns. Thus, if a reconstruction algorithm can take advantage of such global information, it is natural to expect the algorithm to obtain more accurate results. Traditionally, however, it has been computationally challenging to reliably detect such global structures from noisy local image features. Recently, deep learning-based methods have shown promising results in detecting various forms of structure directly from the images, including lines [12,40], planes [16, 19,35,36], cuboids [10], floorplans [17,18], room layouts [14,26,41], abstracted 3D shapes [28,32], and smooth surfaces [11].

With the fast development of deep learning methods comes the need for large amounts of accurately annotated data. In order to train the proposed neural networks, most prior work collects their own sets of images and manually label the structure of interest in them. Such a strategy has several shortcomings. *First*, due to the tedious process of manually labeling and verifying all the structure instances (*e.g.*, line segments) in each image, existing datasets typically have limited sizes and scene diversity. And the annotations may also contain errors. *Second*, since each study primarily focuses on one type of structure, none of these datasets has multiple types of structure labeled. As a result, existing methods are unable to exploit relations between different types of structure (*e.g.*, lines and planes) as humans do for effective, efficient, and robust 3D reconstruction.

In this paper, we present a large synthetic dataset with rich annotations of 3D structure *and* photo-realistic 2D renderings of indoor man-made environments (Fig. 1). At the core of our dataset design is a unified representation of 3D

Table 1. An overview of datasets with structure annotations. †: The actual numbers are not explicitly given and hard to estimate, because these datasets contain images from Internet (LSUN Room Layout, PanoContext), or multiple sources (LayoutNet). *: Dataset is unavailable online at the time of publication.

Datasets	#Scenes	#Rooms	#Frames	Annotated structure
PlaneRCNN [16]	–	–	100,000	Planes
Wireframe [12]	–	–	5,462	Wireframe (2D)
SceneCity 3D [40]	230	–	23,000	Wireframe (3D)
SUN Primitive [34]	–	–	785	Cuboids, other primitives
LSUN Room Layout [39]	–	n/a†	5,394	Cuboid layout
PanoContext [37]	–	n/a†	500 (pano)	Cuboid layout
LayoutNet [41]	–	n/a†	1,071 (pano)	Cuboid layout
MatterportLayout* [42]	–	n/a†	2,295 (RGB-D pano)	Manhattan layout
Raster-to-Vector [17]	870	–	–	Floorplan
Structured3D	3,500	21,835	196,515	"primitive + relationship"

structure which enables us to efficiently capture multiple types of 3D structure in the scene. Specifically, the proposed representation considers any structure as *relationship* among *geometric primitives*. For example, a "wireframe" structure encodes the incidence and intersection relationship between line segments, whereas a "cuboid" structure encodes the rotational and reflective symmetry relationship among its planar faces. With our "primitive + relationship" representation, one can easily derive the ground truth annotations for a wide variety of semi-global and global structures (*e.g.*, lines, wireframes, planes, regular shapes, floorplans, and room layouts), and also exploit their relations in future data-driven approaches (*e.g.*, the wireframe formed by intersecting planar surfaces in the scene).

To create a large-scale dataset with the aim of facilitating research on data-driven methods for structured 3D scene understanding, we leverage the availability of professional interior designs and millions of production-level 3D object models – all coming with fine geometric details and high-resolution textures (Fig. 1(a)). We first use computer programs to automatically extract information about 3D structure from the original house design files. As shown in Fig. 1(b), our dataset contains rich annotations of 3D room structure including a variety of geometric primitives and relationships. To further generate photo-realistic 2D images (Fig. 1(c)), we utilize industry-leading rendering engines to model the lighting conditions. Currently, our dataset consists of more than 196k images of 21,835 rooms in 3,500 scenes (*i.e.*, houses).

To showcase the usefulness and uniqueness of the proposed Structured3D dataset, we train deep networks for room layout estimation on a subset of the dataset. We show that the models trained on both synthetic and real data outperform the models trained on real data only. Further, following the spirit of [8,27], we show how multi-modal annotations in our dataset can benefit domain adaptation tasks.

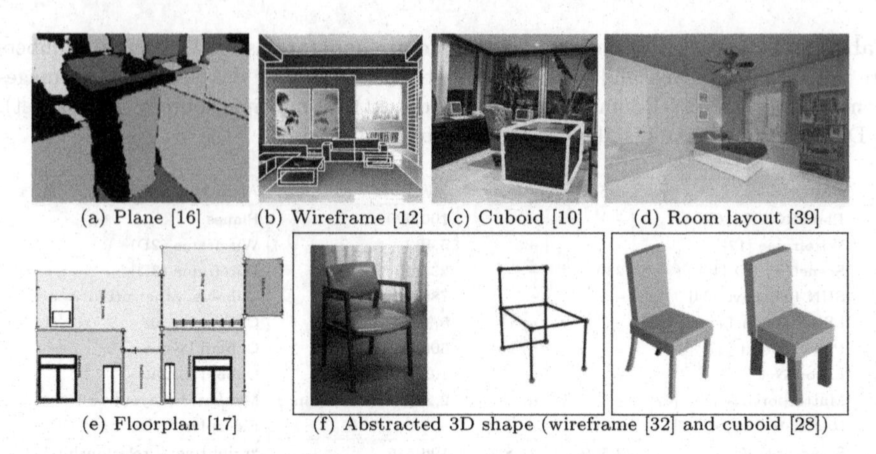

(a) Plane [16] (b) Wireframe [12] (c) Cuboid [10] (d) Room layout [39]
(e) Floorplan [17] (f) Abstracted 3D shape (wireframe [32] and cuboid [28])

Fig. 2. Example annotations of structure in existing datasets. The reference number indicates the paper from which the illustration is originally from.

In summary, the **main contributions** of this paper are:

- We create the Structured3D dataset, which contains rich ground truth 3D structure annotations of 21,835 rooms in 3,500 scenes, and more than 196k photo-realistic 2D renderings of the rooms.
- We introduce a unified "primitive + relationship" representation. This representation enables us to efficiently capture a wide variety of semi-global or global 3D structures and their mutual relationships.
- We verify the usefulness of our dataset by using it to train deep networks for room layout estimation and demonstrating improved performance on public benchmarks.

2 Related Work

Datasets. Table 1 summarizes existing datasets for structured 3D scene modeling. Additionally, [28,32] provide datasets with structured representations of single objects. We show example annotations in these datasets in Fig. 2. Note that ground truth annotations in most datasets are manually labeled. This is one main reason why all these datasets have limited size, *i.e.*, contain no more than a few thousand images. One exception is [16], which employs a multi-model fitting algorithm to automatically extract planes from 3D scans in the ScanNet dataset [9]. But such algorithms are sensitive to data noises and outliers, thus introduce errors in the annotations (Fig. 2(a)). Similar to our work, SceneCity 3D [40] also contains synthetic images with ground truth automatically extracted from CAD models. But the number of scenes is limited to 230. Further, none of these datasets has more than one type of structure labeled, although different types of structure often have strong relations among them. For example, from the wireframe in Fig. 2(b) humans can easily identify other types of structure such as planes and cuboids. Our new dataset sets to bridge the gap between what is needed to train machine learning models to achieve human-level holistic 3D scene understanding and what is being offered by existing datasets.

Note that our dataset is very different from other popular large-scale 3D datasets, such as NYU v2 [23], SUN RGB-D [24], 2D-3D-S [3,4], ScanNet [9], and Matterport3D [6], in which the ground truth 3D information is stored in the format of point clouds or meshes. These datasets lack ground truth annotations of semi-global or global structures. While it is theoretically possible to extract 3D structure by applying structure detection algorithms to the point clouds or meshes (*e.g.*, extracting planes from ScanNet as did in [16]), the detection results are often noisy and even contain errors. In addition, for some types of structure like wireframes and room layouts, how to reliably detect them from raw sensor data remains an active research topic in computer vision.

In recent years, synthetic datasets have played an important role in the successful training of deep neural networks. Notable examples for indoor scene understanding include SUNCG [25], SceneNet RGB-D [20], and InteriorNet [15]. These datasets exceed real datasets in terms of scene diversity and frame numbers. But just like their real counterparts, these datasets lack ground truth structure annotations. Another issue with some synthetic datasets is the degree of realism in both the 3D models and the 2D renderings. [38] shows that physically-based rendering could boost the performance of various indoor scene understanding tasks. To ensure the quality of our dataset, we make use of 3D room models created by professional designers and the state-of-the-art industrial rendering engines. Table 2 summarizes the differences of 3D scene datasets.

Room Layout Estimation. Room layout estimation aims to reconstruct the enclosing structure of the indoor scene, consisting of walls, floor, and ceiling. Existing public datasets (*e.g.*, PanoContext [37] and LayoutNet [41]) assume a simple box-shaped layout. PanoContext [37] collects about 500 panoramas from the SUN360 dataset [33], LayoutNet [41] extends the layout annotations to include panoramas from 2D-3D-S [3]. Recently, MatterportLayout [42] collects 2,295 RGB-D panoramas from Matterport3D [6] and extends annotations to Manhattan layout. We note that all room layout in these real datasets is manually labeled by the human. Since the room structure may be occluded by furniture and other objects, the "ground truth" inferred by humans may not be consistent with the actual layout. In our dataset, all ground truth 3D annotations are automatically extracted from the original house design files.

3 A Unified Representation of 3D Structure

The main goal of our dataset is to provide rich annotations of ground truth 3D structure. A naive way to do so is generating and storing different types of 3D annotations in the same format as existing works, like wireframes as in [12], planes as in [16], floorplans as in [17], and so on. But this leads to a lot of redundancy. For example, planes in man-made environments are often bounded

Table 2. Comparison of 3D scene datasets. †: Meshes are obtained by 3D reconstruction algorithm. Notations for applications: O (object detection), U (scene understanding), S (image synthesis), M (structured 3D modeling).

Datasets	Scene design type	3D annotation	2D rendering	Applications
NYU v2 [23]	Real	Raw RGB-D	Real images	O U
SUN RGB-D [24]	Real	Raw RGB-D	Real images	O U
2D-3D-S [3,4]	Real	Mesh†	Real images	O U
ScanNet [9]	Real	Mesh†	Real images	O U
Matterport3D [6]	Real	Mesh†	Real images	O U
SUNCG [25]	Amateur	Mesh	n/a	O U
SceneNet RGB-D [20]	Random	Mesh	Photo-realistic	O U
InteriorNet [15]	Professional	n/a	Photo-realistic	O U S
Structured3D	Professional	3D structures	Photo-realistic	O U S M

by a number of line segments, which are part of the wireframe. Even worse, by representing wireframes and planes separately, the relationships between them are lost. In this paper, we present a unified representation in order to minimize redundancy while preserving mutual relationships. We show how the most common types of structure studied in the literature (*e.g.*, planes, cuboids, wireframes, room layouts, and floorplans) can be derived from our representation.

Our representation of the structure is largely inspired by the early work of Witkin and Tenenbaum [31], which characterizes structure as *"a shape, pattern, or configuration that replicates or continues with little or no change over an interval of space and time"*. Accordingly, to describe any structure, we need to specify: (i) what pattern is continuing or replicating (*e.g.*, a patch, an edge, or a texture descriptor), and (ii) the domain of its replication or continuation. In this paper, we call the former **primitives** and the latter **relationships**.

3.1 The "Primitive + Relationship" Representation

We now show how to describe a man-made environment using a unified representation. For ease of exposition, we assume all objects in the scene can be modeled by piece-wise planar surfaces. But our representation can be easily extended to more general surfaces. An illustration of our representation is shown in Fig. 3.

Primitives. Generally, a man-made scene has the following geometric primitives:

- **Planes P**: We model the scene as a collection of planes $\mathbf{P} = \{p_1, p_2, \ldots\}$. Each plane is described by its parameters $p = \{\mathbf{n}, d\}$, where $\mathbf{n}$ and d denote the surface normal and the distance to the origin, respectively.
- **Lines L**: When two planes intersect in the 3D space, a line is created. We use $\mathbf{L} = \{l_1, l_2, \ldots\}$ to represent the set of all 3D lines in the scene.
- **Junction Points X**: When two lines meet in the 3D space, a junction point is formed. We use $\mathbf{X} = \{x_1, x_2, \ldots\}$ to represent the set of all junction points.

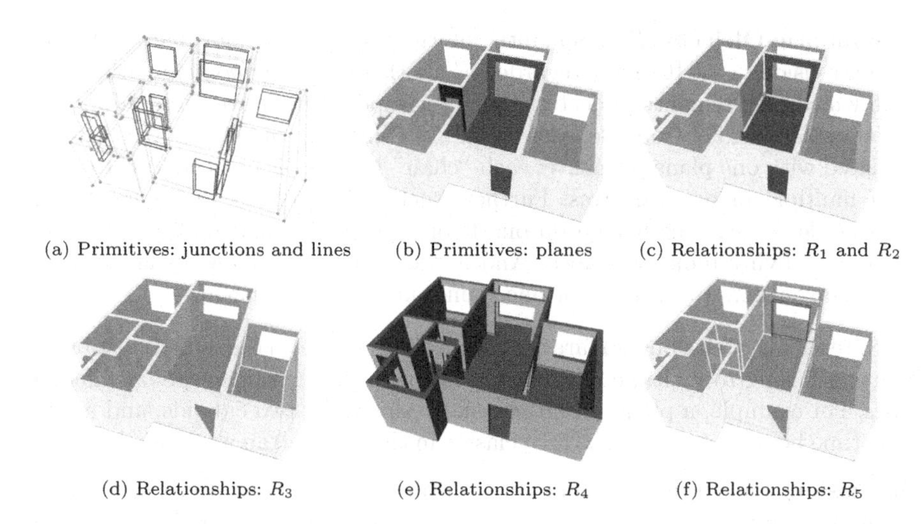

(a) Primitives: junctions and lines (b) Primitives: planes (c) Relationships: R_1 and R_2

(d) Relationships: R_3 (e) Relationships: R_4 (f) Relationships: R_5

Fig. 3. The ground truth 3D structure annotations in our dataset are represented by primitives and relationships. **(a)**: Junctions and lines. **(b)**: Planes. We highlight the planes in a single room. **(c)**: Plane-line and line-junction relationships. We highlight a junction, the three lines intersecting at the junction, and the planes intersecting at each of the lines. **(d)**: Cuboids. We highlight one cuboid instance. **(e)**: Manhattan world. We use different colors to denote planes aligned with different directions. **(f)**: Semantic objects. We highlight a "room", a "balcony", and the "door" connecting them.

Relationships. Next, we define some common types of relationships between the geometric primitives:

- **Plane-Line Relationships** (R_1): We use a matrix W_1 to record all incidence and intersection relationships between planes in $\mathbf{P}$ and lines in $\mathbf{L}$. Specifically, the ij-th entry of W_1 is 1 if l_i is on p_j, and 0 otherwise. Note that two planes are intersected at some line if and only if the corresponding entry in $W_1^T W_1$ is nonzero.
- **Line-Point Relationships** (R_2): Similarly, we use a matrix W_2 to record all incidence and intersection relationships between lines in $\mathbf{L}$ and points in $\mathbf{X}$. Specifically, the mn-th entry of W_2 is 1 if x_m is on l_n, and 0 otherwise. Note that two lines are intersected at some junction if and only if the corresponding entry in $W_2^T W_2$ is nonzero.
- **Cuboids** (R_3): A cuboid is a special arrangement of plane primitives with rotational and reflection symmetry along x-, y- and z-axes. The corresponding symmetry group is the dihedral group D_{2h}.
- **Manhattan World** (R_4): This is a special type of 3D structure commonly used for indoor and outdoor scene modeling. It can be viewed as a *grouping* relationship, in which all the plane primitives can be grouped into three classes, $\mathbf{P}_1$, $\mathbf{P}_2$, and $\mathbf{P}_3$, $\mathbf{P} = \bigcup_{i=1}^{3} \mathbf{P}_i$. Further, each class is represented by a single normal vector $\mathbf{n}_i$, such that $\mathbf{n}_i^T \mathbf{n}_j = 0, i \neq j$.

- **Semantic Objects** (R_5): Semantic information is critical for many 3D computer vision tasks. It can be regarded as another type of *grouping* relationship, in which each semantic object instance corresponds to one or more primitives defined above. For example, each "wall", "ceiling", or "floor" instance is associated with one plane primitive; each "chair" instance is associated with a set of multiple plane primitives. Further, such a grouping is hierarchical. For example, we can further group one floor, one ceiling, and multiple walls to form a "living room" instance. And a "door" or a "window" is an opening which connects two rooms (or one room and the outer space).

Note that the relationships are not mutually exclusive, in the sense that a primitive can belong to multiple relationship instances of the same type or different types. For example, a plane primitive can be shared by two cuboids, and at the same time belong to one of the three classes in the Manhattan world model.

Discussion. The primitives and relationships we discussed above are just a few most common examples. They are by no means exhaustive. For example, our representation can be easily extended to include other primitives such as parametric surfaces. And besides cuboids, there are many other types of regular or symmetric shapes in man-made environments, where type corresponds to a different symmetry group.

Our representation of 3D structures is also related to the graph representations in semantic scene understanding [2,13,30]. As these graphs focus on semantics, geometry is represented in simplified manners by (i) 6D object poses and (ii) coarse, discrete spatial relations such as "supported by", "front", "back", and "adjacent". In contrast, our representation focuses on modeling the scene geometry using fine-grained primitives (*i.e.*, junctions, lines, and planes) and relationships (in terms of topology and regularities). Thus, it is highly complementary to the scene graphs in prior work. Intuitively, it can be used for geometric analysis and synthesis tasks, in a similar way as scene graphs are used for semantic scene understanding.

3.2 Relation to Existing Models

Given our representation which contains primitives $\mathcal{P} = \{\mathbf{P}, \mathbf{L}, \mathbf{X}\}$ and relationships $\mathcal{R} = \{R_1, R_2, \ldots\}$, we show how several types of 3D structure commonly studied in the literature can be derived from it. We again refer readers to Fig. 2 for illustrations of these structures.

Planes: A large volume of studies in the literature model the scene as a collection of 3D planes, where each plane is represented by its parameters and boundary. To generate such a model, we simply use the plane primitives $\mathbf{P}$. For each $p \in \mathbf{P}$, we further obtain its boundary by using matrix W_1 in R_1 to find all the lines in $\mathbf{L}$ that form an incidence relationship with p.

Wireframes: A wireframe consists of lines $\mathbf{L}$ and junction points $\mathbf{P}$, and their incidence and intersection relationships (R_2).

Cuboids: This model is same as R_3.

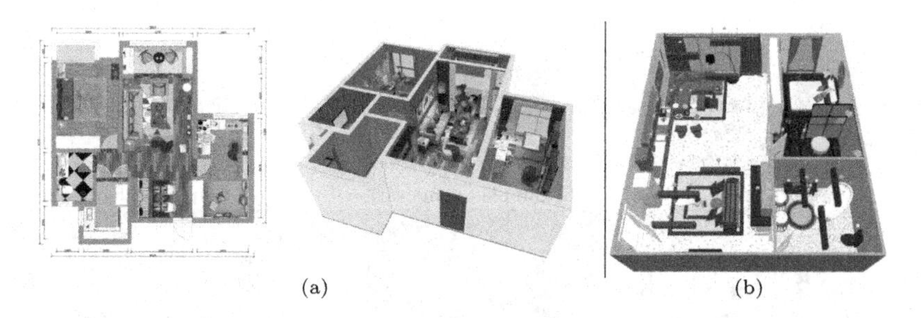

Fig. 4. Comparison of 3D house designs. **(a)**: The 3D models in our database are created by professional designers using high-quality furniture models from world-leading manufacturers. Most designs are being used in real-world production. **(b)**: The 3D models in SUNCG dataset [25] are created using Planner 5D [1], an online tool for amateur interior design.

Manhattan Layouts: A Manhattan room layout model includes a "room" as defined in R_5 which also satisfies the Manhattan world assumption (R_4).

Floorplans: A floorplan is a 2D vector representation that consists of a set of line segments and semantic labels (*e.g.*, room types). To obtain such a vector representation, we can identify all lines in $\mathbf{L}$ and junction points in $\mathbf{X}$ which lie on a "floor" (as defined in R_5). To further obtain the semantic room labels, we can project all "rooms", "doors", and "windows" (as defined in R_5) to this floor.

Abstracted 3D shapes: In addition to room structures, our representation can also be applied to individual 3D object models to create abstractions in the form of wireframes or cuboids, as described above.

4 The Structured3D Dataset

Our unified representation enables us to encode a rich set of geometric primitives and relationships for structured 3D modeling. With this representation, our ultimate goal is to build a dataset that can be used to train machines to achieve the human-level understanding of the 3D environment.

As a first step towards this goal, in this section, we describe our ongoing effort to create a large-scale dataset of indoor scenes which include (i) ground truth 3D structure annotations of the scene and (ii) realistic 2D renderings of the scene. Note that in this work we focus on extracting ground truth annotations on the room structure only. We plan to extend our dataset to include 3D structure annotations of individual furniture models in the future.

In the following, we describe our general procedure to create the dataset. We refer readers to the supplementary materials for additional details, including dataset statistics and example annotations.

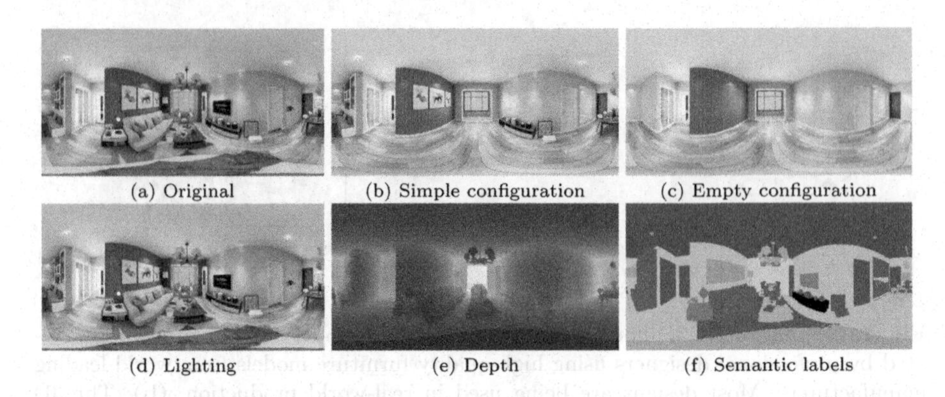

(a) Original (b) Simple configuration (c) Empty configuration

(d) Lighting (e) Depth (f) Semantic labels

Fig. 5. Examples of our rendered panoramic images.

4.1 Extraction of Structured 3D Models

To extract a "primitive + relationship" scene representation, we utilize a large database of house designs hand-crafted by professional designers. An example design is shown in Fig. 4(a). All information of the design is stored in an industry-standard format in the database so that specifications about the geometry (*e.g.*, the precise size of each wall), textures and materials, and functions (*e.g.*, which room the wall belongs to) of all objects can be easily retrieved.

From the database, we have selected 3,500 house designs with 21,835 rooms. We created a computer program to automatically extract all the geometric primitives associated with the room structure, which consists of the ceiling, floor, walls, and openings (doors and windows). Given the precise measurements and associated information of these entities, it is straightforward to generate all planes, lines, and junctions, as well as their relationships (R_1 and R_2).

Since the measurements are highly accurate and noise-free, other types of relationship such a Manhattan world (R_3) and cuboids (R_4) can also be easily obtained by clustering the primitives, followed by a geometric verification process. Finally, to include semantic information (R_5) into our representation, we map the relevant labels provided by the professional designers to the geometric primitives in our representation. Figure 3 shows examples of the extracted geometric primitives and relationships.

4.2 Photo-Realistic 2D Rendering

To ensure the quality of our 2D renderings, our rendering engine is developed in collaboration with a company specialized in interior design rendering. Our engine uses a well-known ray-tracing method [21], a Monte Carlo approach to approximating realistic Global Illumination (GI), for RGB rendering. The other ground truth images are obtained by a customized path-tracer renderer on top of Intel Embree [29], an open-source collection of ray-tracing kernels for x86 CPUs.

Fig. 6. Photo-realistic rendering vs. real-world decoration. The first and third columns are rendered images.

Each room is manually created by professional designers with over one million CAD models of furniture from world-leading manufacturers. These high-resolution furniture models are measured in real-world dimensions and being used in real production. A default lighting setup is also provided. Figure 4 compares the 3D models in our database with those in SUNCG [25], which are created using Planner 5D [1], an online tool for amateur interior design.

At the time of rendering, a panoramic or pin-hole camera is placed at random locations not occupied by objects in the room. We use 512×1024 resolution for panoramas and 720×1280 for perspective images. Figure 5 shows example panoramas rendered by our engine. For each room, we generate different configurations (full, simple, and empty) by removing some or all the furniture. We also modify the lighting setup to generate images with different temperatures. For each image, our dataset al.so includes the depth map and semantic mask. Figure 6 illustrates the degree of photo-realism of our dataset, where we compare the rendered images with photos of real decoration guided by the design.

4.3 Use Cases

Due to the unique characteristics of our dataset, we envision it contributing to computer vision research in terms of both methodology and applications.

Methodology. As our dataset contains multiple types of 3D structure annotations as well as ground truth labels (*e.g.*, semantic maps, depth maps, and 3D object bounding boxes), it enables researchers to design novel multi-modal or multi-task approaches for a variety of vision tasks. As an example, we show in Sect. 5 that, by leveraging multi-modal annotations in our dataset, we can boost the performance of existing room layout estimation methods in the domain adaptation framework.

Applications. Our dataset also facilitates research on a number of problems and applications. For example, as shown in Table 1, all publicly available datasets for room layout estimation are limited to simple cuboid rooms. Our dataset is the first to provide the general (non-cuboid) room layout annotations. As another example, existing datasets for floorplan reconstruction [7,18] contain about 100–150 scenes, whereas our dataset includes 3,500 scenes.

Table 3. Room layout statistics. †: MatterportLayout is the only other dataset with non-cuboid layout annotations, but is unavailable at the time of publication.

#Corners	4	5	6	7	8	9	10+	Total
MatterportLayout†	1211	0	501	0	309	0	274	2295
Structured3D	13743	52	3727	30	1575	17	2691	21835

Another major line of research that would benefit from our dataset is image synthesis. With a photo-realistic rendering engine, we are able to generate images given *any* scene configurations and viewpoints. These images may be used as ground truth for tasks including image inpainting (*e.g.*, completing an image when certain furniture is removed) and novel view synthesis.

Finally, we would like to emphasize the potential of our dataset in terms of extension capabilities. As we mentioned before, the unified representation enables us to include many other types of structure in the dataset. As for 2D rendering, depending on the application, we can easily simulate different effects such as lighting conditions, fisheye and novel camera designs, motion blur, and imaging noise. Furthermore, the dataset may be extended to include videos for applications such as visual SLAM [5].

5 Experiments

5.1 Experiment Setup

To demonstrate the benefits of our dataset, we use it to train deep neural networks for room layout estimation, an important task in structured 3D modeling.

Real Dataset. We use the same dataset as LayoutNet [41]. The dataset consists of images from PanoContext [37] and 2D-3D-S [3], including 818 training images, 79 validation images, and 166 test images. Note that both datasets only provide cuboid layout annotations.

Our Structured3D Dataset. In this experiment, we use a subset of panoramas with the original lighting and full configuration. Each panorama corresponds to a different room in our dataset. We show statistics of different room layouts in our dataset in Table 3. Since the current real dataset only contains cuboid layout annotations (*i.e.*, 4 corners), we choose 12k panoramic images with the cuboid layout in our dataset. We split the images into 10k for training, 1k for validation, and 1k for testing.

Evaluation Metrics. Following [26,41], we adopt three standard metrics: (i) 3D IoU: intersection over union between predicted 3D layout and the ground truth, (ii) Corner Error (CE): normalized ℓ_2 distance between predicted corner and ground truth, and (iii) Pixel Error (PE): pixel-wise error between predicted plane classes and ground truth.

Table 4. Quantitative evaluation under different training schemes. The best and the second best results are boldfaced and underlined, respectively.

Methods	Config.	PanoContext			2D-3D-S		
		3D IoU (%) ↑	CE (%) ↓	PE (%) ↓	3D IoU (%) ↑	CE (%) ↓	PE (%) ↓
LayoutNet [41,42]	s	75.64	1.31	4.10	57.18	2.28	7.55
	r	84.15	0.64	<u>1.80</u>	83.39	0.74	2.39
	s + r	**84.96**	**0.61**	**1.75**	<u>83.66</u>	<u>0.71</u>	<u>2.31</u>
	s → r	<u>84.77</u>	<u>0.63</u>	1.89	**84.04**	**0.66**	**2.08**
HorizonNet [26]	s	75.89	1.13	3.15	67.66	1.18	3.94
	r	83.42	0.73	2.09	84.33	0.64	2.04
	s + r	<u>84.45</u>	<u>0.70</u>	<u>1.89</u>	<u>84.36</u>	**0.59**	<u>1.90</u>
	s → r	**85.27**	**0.66**	**1.86**	**86.01**	<u>0.61</u>	**1.84**

Baselines. We choose two recent CNN-based approaches, LayoutNet [41,42][1] and HorizonNet [26][2], based on their performance and source code availability. LayoutNet uses a CNN to predict a corner probability map and a boundary map from the panorama and vanishing lines, then optimizes the layout parameters based on network predictions. HorizonNet represents room layout as three 1D vectors, *i.e.*, boundary positions of floor-wall, and ceiling-wall, and the existence of wall-wall boundary. It trains CNNs to directly predict the three 1D vectors. In this paper, we follow the default training setting of the respective methods. For specific training procedures, please refer to the supplementary materials.

5.2 Experiment Results

Augmenting Real Datasets. In this experiment, we train LayoutNet and HorizonNet in four different manners: (i) training only on our synthetic dataset ("**s**"), (ii) training only on the real dataset ("**r**"), (iii) training on the synthetic and real dataset with Balanced Gradient Contribution (BGC) [22] ("**s + r**"), and (iv) pre-training on our synthetic dataset, then fine-tuning on the real dataset ("**s → r**"). We adopt the training set of LayoutNet as the real dataset in this experiment. The results are shown in Table 4. As one can see, augmenting real datasets with our synthetic data boosts the performance of both networks. We refer readers to supplementary materials for more qualitative results.

Performance vs. Synthetic Data Size. We further study the relationship between the number of synthetic images used in pre-training and the accuracy on the real dataset. We sample 1k, 5k and 10k synthetic images for pre-training, then fine-tune the model on the real dataset. The results are shown in Table 5. As expected, using more synthetic data generally improves the performance.

Domain Adaptation. Domain adaptation techniques (*e.g.*, [27]) have been shown to be effective in bridging the performance gap when directly applying

---

[1] https://github.com/zouchuhang/LayoutNetv2.

[2] https://github.com/sunset1995/HorizonNet.

Table 5. Quantitative evaluation using varying synthetic data size in pre-training. The best and the second best results are boldfaced and underlined, respectively.

Methods	Synthetic Data Size	PanoContext 3D IoU (%) ↑	PanoContext CE (%) ↓	PanoContext PE (%) ↓	2D-3D-S 3D IoU (%) ↑	2D-3D-S CE (%) ↓	2D-3D-S PE (%) ↓
LayoutNet [41,42]	1k	83.81	<u>0.66</u>	1.99	83.57	0.72	2.31
	5k	<u>84.47</u>	0.67	<u>1.97</u>	**84.55**	<u>0.69</u>	<u>2.21</u>
	10k	**84.77**	**0.63**	**1.89**	<u>84.04</u>	**0.66**	**2.08**
HorizonNet [26]	1k	83.77	0.74	2.11	85.19	<u>0.63</u>	2.01
	5k	<u>84.13</u>	<u>0.73</u>	<u>2.07</u>	**86.35**	**0.61**	<u>1.87</u>
	10k	**85.27**	**0.66**	**1.86**	<u>86.01</u>	**0.61**	**1.84**

Table 6. Domain adaptation results. NA: non-adaptive baseline. +DA: align layout estimation output. +Depth: align both layout estimation and depth outputs. Real: train in the target domain.

Methods	PanoContext 3D IoU (%) ↑	PanoContext CE (%) ↓	PanoContext PE (%) ↓	2D-3D-S 3D IoU (%) ↑	2D-3D-S CE (%) ↓	2D-3D-S PE (%) ↓
NA	75.64	1.31	4.10	57.18	2.28	7.55
+DA	76.91	1.19	3.64	70.08	1.36	4.66
+Depth	78.34	1.03	2.99	72.99	1.24	3.60
Real	81.76	0.95	2.58	81.82	0.96	3.13

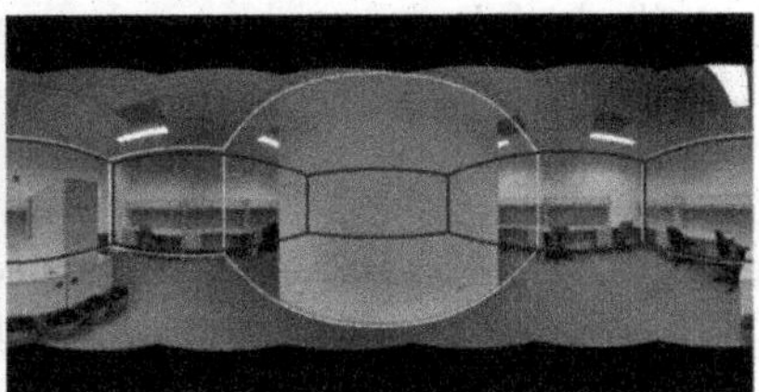

Fig. 7. Limitation of real datasets. **Left:** PanoContext dataset. **Right:** 2D-3D-S dataset. Blue lines are ground truth layout and green lines are predictions (Color figure online).

models learned on synthetic data to real environments. In this experiment, we do not assume access to ground truth layout labels in the real dataset. We adopt LayoutNet as the task network and use PanoContext and 2D-3D-S separately. We apply a discriminator network to align the output features of the LayoutNet for two domains. Inspired by [8], we further leverage multi-modal annotations in our dataset by adding another decoder branch to the LayoutNet for depth prediction. We concatenate the boundary, corner, and depth predictions as the input of the discriminator network. The results are shown in the Table 6. By incorporating additional information, *i.e.*, depth map, we further boost the performance on both datasets. This illustrates the advantage of including multiple types of ground truth in our dataset.

Limitation of Real Datasets. Due to human errors, the annotation in real datasets is not always consistent with the actual room layout. In the left image of Fig. 7, the room is a non-cuboid layout, but the ground truth layout is labeled as cuboid shape. In the right image, the front wall is not labeled as ground truth. These examples illustrate the limitation of using real datasets as benchmarks. We avoid such errors in our dataset by automatically generating ground truth from the original design files.

6 Conclusion

In this paper, we present Structured3D, a large synthetic dataset with rich ground truth 3D structure annotations of 21,835 rooms and more than 196k photo-realistic 2D renderings. Among many potential use cases of our dataset, we further demonstrate its benefit in augmenting real data and facilitating domain adaptation for the room layout estimation task.

We view this work as an important and exciting step towards building intelligent machines which can achieve human-level holistic 3D scene understanding. In the future, we will continue to add more 3D structure annotations of the scenes and objects to the dataset, and explore novel ways to use the dataset to advance techniques for structured 3D modeling and understanding.

Acknowledgments. We would like to thank Kujiale.com for providing the database of house designs and the rendering engine. We especially thank Qing Ye and Qi Wu from Kujiale.com for the help on the data rendering. This work was partially supported by the National Key R&D Program of China (#2018AAA0100704) and the National Science Foundation of China (#61932020). Zihan Zhou was supported by NSF award #1815491.

References

1. Planner 5D. https://planner5d.com
2. Armeni, I., et al.: 3D scene graph: a structure for unified semantics, 3D space, and camera. In: ICCV, pp. 5664–5673 (2019)
3. Armeni, I., Sax, A., Zamir, A.R., Savarese, S.: Joint 2D–3D-semantic data for indoor scene understanding. CoRR abs/1702.01105 (2017)
4. Armeni, I., et al.: 3D semantic parsing of large-scale indoor spaces. In: CVPR, pp. 1534–1543 (2016)
5. Cadena, C., et al.: Past, present, and future of simultaneous localization and mapping: toward the robust-perception age. IEEE Trans. Robot. **32**(6), 1309–1332 (2016)
6. Chang, A.X., et al.: Matterport3D: learning from RGB-D data in indoor environments. In: 3DV, pp. 667–676 (2017)
7. Chen, J., Liu, C., Wu, J., Furukawa, Y.: Floor-SP: inverse CAD for floorplans by sequential room-wise shortest path. In: ICCV, pp. 2661–2670 (2019)
8. Chen, Y., Li, W., Chen, X., Van Gool, L.: Learning semantic segmentation from synthetic data: a geometrically guided input-output adaptation approach. In: CVPR, pp. 1841–1850 (2019)

9. Dai, A., Chang, A.X., Savva, M., Halber, M., Funkhouser, T., Nießner, M.: ScanNet: richly-annotated 3D reconstructions of indoor scenes. In: CVPR, pp. 5828–5839 (2017)
10. Dwibedi, D., Malisiewicz, T., Badrinarayanan, V., Rabinovich, A.: Deep cuboid detection: Beyond 2D bounding boxes. CoRR abs/1611.10010 (2016)
11. Groueix, T., Fisher, M., Kim, V.G., Russell, B., Aubry, M.: A papier-mâché approach to learning 3D surface generation. In: CVPR, pp. 216–224 (2018)
12. Huang, K., Wang, Y., Zhou, Z., Ding, T., Gao, S., Ma, Y.: Learning to parse wireframes in images of man-made environments. In: CVPR, pp. 626–635 (2018)
13. Huang, S., Qi, S., Zhu, Y., Xiao, Y., Xu, Y., Zhu, S.-C.: Holistic 3D scene parsing and reconstruction from a single RGB image. In: Ferrari, V., Hebert, M., Sminchisescu, C., Weiss, Y. (eds.) ECCV 2018. LNCS, vol. 11211, pp. 194–211. Springer, Cham (2018). https://doi.org/10.1007/978-3-030-01234-2_12
14. Lee, C., Badrinarayanan, V., Malisiewicz, T., Rabinovich, A.: RoomNet: end-to-end room layout estimation. In: ICCV, pp. 4875–4884 (2017)
15. Li, W., et al.: InteriorNet: mega-scale multi-sensor photo-realistic indoor scenes dataset. In: BMVC, p. 77 (2018)
16. Liu, C., Kim, K., Gu, J., Furukawa, Y., Kautz, J.: Planercnn: 3D plane detection and reconstruction from a single image. In: CVPR. pp. 4450–4459 (2019)
17. Liu, C., Wu, J., Kohli, P., Furukawa, Y.: Raster-to-vector: revisiting floorplan transformation. In: ICCV. pp. 2214–2222 (2017)
18. Liu, C., Wu, J., Furukawa, Y.: Floornet: a unified framework for floorplan reconstruction from 3D scans. In: ECCV, pp. 203–219 (2018)
19. Liu, C., Yang, J., Ceylan, D., Yumer, E., Furukawa, Y.: Planenet: piece-wise planar reconstruction from a single RGB image. In: CVPR, pp. 2579–2588 (2018)
20. McCormac, J., Handa, A., Leutenegger, S., Davison, A.J.: Scenenet RGB-D: can 5m synthetic images beat generic imagenet pre-training on indoor segmentation? In: ICCV, pp. 2697–2706 (2017)
21. Purcell, T.J., Buck, I., Mark, W.R., Hanrahan, P.: Ray tracing on programmable graphics hardware. ACM Trans. Graph. **21**(3), 703–712 (2002)
22. Ros, G., Stent, S., Alcantarilla, P.F., Watanabe, T.: Training constrained deconvolutional networks for road scene semantic segmentation. CoRR abs/1604.01545 (2016)
23. Silberman, N., Hoiem, D., Kohli, P., Fergus, R.: Indoor segmentation and support inference from RGBD images. In: Fitzgibbon, A., Lazebnik, S., Perona, P., Sato, Y., Schmid, C. (eds.) ECCV 2012. LNCS, vol. 7576, pp. 746–760. Springer, Heidelberg (2012). https://doi.org/10.1007/978-3-642-33715-4_54
24. Song, S., Lichtenberg, S.P., Xiao, J.: SUN RGB-D: a RGB-D scene understanding benchmark suite. In: CVPR, pp. 567–576 (2015)
25. Song, S., Yu, F., Zeng, A., Chang, A.X., Savva, M., Funkhouser, T.A.: Semantic scene completion from a single depth image. In: CVPR, pp. 1746–1754 (2017)
26. Sun, C., Hsiao, C.W., Sun, M., Chen, H.T.: Horizonnet: Learning room layout with 1D representation and pano stretch data augmentation. In: CVPR, pp. 1047–1056 (2019)
27. Tsai, Y.H., Hung, W.C., Schulter, S., Sohn, K., Yang, M.H., Chandraker, M.: Learning to adapt structured output space for semantic segmentation. In: CVPR. pp. 7472–7481 (2018)
28. Tulsiani, S., Su, H., Guibas, L.J., Efros, A.A., Malik, J.: Learning shape abstractions by assembling volumetric primitives. In: CVPR. pp. 2635–2643 (2017)
29. Wald, I., Woop, S., Benthin, C., Johnson, G.S., Ernst, M.: Embree: a kernel framework for efficient CPU ray tracing. ACM Trans. Graph. **33**(4), 143:1–143:8 (2014)

30. Wang, K., Lin, Y.A., Weissmann, B., Savva, M., Chang, A.X., Ritchie, D.: Planit: planning and instantiating indoor scenes with relation graph and spatial prior networks. ACM Trans. Graph. **38**(4), 1–15 (2019)
31. Witkin, A.P., Tenenbaum, J.M.: On the role of structure in vision. In: Beck, J., Hope, B., Rosenfeld, A. (eds.) Human and Machine Vision, pp. 481–543. Academic Press, Cambridge (1983)
32. Wu, J., Xue, T., Lim, J.J., Tian, Y., Tenenbaum, J.B., Torralba, A., Freeman, W.T.: 3D interpreter networks for viewer-centered wireframe modeling. IJCV **126**(9), 1009–1026 (2018). https://doi.org/10.1007/s11263-018-1074-6
33. Xiao, J., Ehinger, K.A., Oliva, A., Torralba, A.: Recognizing scene viewpoint using panoramic place representation. In: CVPR, pp. 2695–2702 (2012)
34. Xiao, J., Russell, B., Torralba, A.: Localizing 3D cuboids in single-view images. In: NeurIPS, pp. 746–754 (2012)
35. Yang, F., Zhou, Z.: Recovering 3D planes from a single image via convolutional neural networks. In: ECCV, pp. 87–103 (2018)
36. Yu, Z., Zheng, J., Lian, D., Zhou, Z., Gao, S.: Single-image piece-wise planar 3D reconstruction via associative embedding. In: CVPR. pp. 1029–1037 (2019)
37. Zhang, Y., Song, S., Tan, P., Xiao, J.: Panocontext: a whole-room 3D context model for panoramic scene understanding. In: ECCV, pp. 668–686 (2014)
38. Zhang, Y., et al.: Physically-based rendering for indoor scene understanding using convolutional neural networks. In: CVPR, pp. 5287–5295 (2017)
39. Zhang, Y., Yu, F., Song, S., Xu, P., Seff, A., Xiao, J.: Large-scale scene understanding challenge: room layout estimation (2016)
40. Zhou, Y., et al.: Learning to reconstruct 3D manhattan wireframes from a single image. In: ICCV, pp. 7698–7707 (2019)
41. Zou, C., Colburn, A., Shan, Q., Hoiem, D.: Layoutnet: reconstructing the 3D room layout from a single RGB image. In: CVPR, pp. 2051–2059 (2018)
42. Zou, C., et al.: 3D manhattan room layout reconstruction from a single 360 image. CoRR abs/1910.04099 (2019)

BroadFace: Looking at Tens of Thousands of People at once for Face Recognition

Yonghyun Kim[1(✉)], Wonpyo Park[2], and Jongju Shin[1]

[1] Kakao Enterprise, Seongnam, Korea
{aiden.d,isaac.giant}@kakaoenterprise.com
[2] Kakao Corp., Seongnam, Korea
tony.nn@kakaocorp.com

Abstract. The datasets of face recognition contain an enormous number of identities and instances. However, conventional methods have difficulty in reflecting the entire distribution of the datasets because a mini-batch of small size contains only a small portion of all identities. To overcome this difficulty, we propose a novel method called BroadFace, which is a learning process to consider a massive set of identities, comprehensively. In BroadFace, a linear classifier learns optimal decision boundaries among identities from a large number of embedding vectors accumulated over past iterations. By referring more instances at once, the optimality of the classifier is naturally increased on the entire datasets. Thus, the encoder is also globally optimized by referring the weight matrix of the classifier. Moreover, we propose a novel compensation method to increase the number of referenced instances in the training stage. BroadFace can be easily applied on many existing methods to accelerate a learning process and obtain a significant improvement in accuracy without extra computational burden at inference stage. We perform extensive ablation studies and experiments on various datasets to show the effectiveness of BroadFace, and also empirically prove the validity of our compensation method. BroadFace achieves the *state-of-the-art* results with significant improvements on nine datasets in 1:1 face verification and 1:N face identification tasks, and is also effective in image retrieval.

Keywords: Face recognition · Large mini-batch learning · Image retrieval

1 Introduction

Face recognition is a key technique for many applications of biometric authentication such as electronic payment, lock screen of smartphones, and video surveillance. The main tasks of face recognition are categorized into face verification and face identification. In face verification, a pair of faces are compared to verify whether their identities are the same or different. In face identification, the identity of a given face is determined by comparing it to a pre-registered gallery

Y. Kim and W. Park—Equal contribution.

© Springer Nature Switzerland AG 2020
A. Vedaldi et al. (Eds.): ECCV 2020, LNCS 12354, pp. 536–552, 2020.
https://doi.org/10.1007/978-3-030-58545-7_31

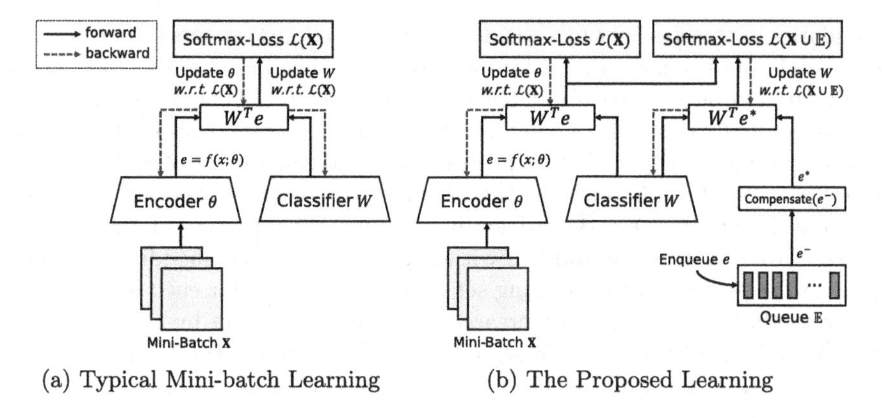

(a) Typical Mini-batch Learning (b) The Proposed Learning

Fig. 1. (a) In typical mini-batch learning, the parameter θ of encoder f and the parameter W of linear classifier are optimized on a small mini-batch $\mathbf{X}$. (b) In the proposed method, the parameter of the encoder is optimized on a small mini-batch, but the parameter of the classifier is optimized on both the mini-batch and the large queue $\mathbb{E}$ that contains embedding vectors e^- of past iterations.

of identities. Many researches [1,3,4,19,27,34,38,44,47] on face recognition have been conducted for decades. The recent adoption [6,7,22,32,37,39–41] of Convolutional Neural Networks (CNNs) has dramatically increased recognition accuracy. However, many difficulties of face recognition still remain to be solved.

Most previous studies focus on improving the discriminative power of an embedding space, because face recognition models are evaluated on independent datasets that include unseen identities. The mainstream of recent studies [6,22,39,40] is to introduce a new objective function to maximize inter-class discriminability and intra-class compactness; they try to consider all identities by referring an identity-representative vector, which is the weight vector of the last fully-connected layer for identity classification.

However, conventional methods still have difficulty in covering a massive set of identities at once, because these methods use a small mini-batch (Fig. 1a) much less than the number of identities due to memory constraints. Inspecting tens of thousands of identities with the mini-batch of the small size requires numerous iterations, and this complicates the task of learning optimal decision boundaries in an embedding space while considering all of the identities, comprehensively. Increasing the size of the mini-batch may alleviate some of the problem, but in general, this solution is impractical because of memory constraints; it also does not guarantee improved accuracy [9,12,18,48].

We propose a novel method, called *BroadFace*, which is a learning process to consider a massive set of identities, comprehensively (Fig. 1b). BroadFace has a large queue to keep a massive number of embedding vectors accumulated over past iterations. Our learning process increases the optimality of decision boundaries of the classifier by considering the embedding vectors of both a given mini-batch and the large queue for each iteration. The parameters of the model are updated iteratively, so after a few iterations the error of enqueued embedding vectors gradually increases. Therefore, we introduce a compensation method that reduces the expected error between the current and enqueued embedding vectors by referencing the difference of the identity-representative vectors of current and past iterations. Our BroadFace has several advantages: (1) the identity-representative vectors are updated with a large number of embedding vectors to increase the portion of the training set that is considered for each iteration, (2) the optimality of the model is increased on the entire dataset by referring to the globally well-optimized identity-representative vectors, (3) the learning process is accelerated. We summarize the contributions as follows:

- We propose a new way that allows an embedding space to distinguish numerous identities in a broad perspective by learning identity-representative vectors from a massive number of instances.
- We perform extensive ablation studies on its behaviors, and experiments on various datasets to show the effectiveness of the proposed method, and to empirically prove the validity of our compensation method.
- BroadFace can be easily applied on many existing face recognition methods to obtain a significant improvement. Moreover, during inference time, it does not require any extra computational burden.

2 Related Works

Recent studies of face recognition tend to introduce a new objective function that learns an embedding space by exploiting an identity-representative vector. NormFace [39] reveal that optimization using cosine similarity between identity-representative vectors and embedding vectors is more effective than optimization using the inner product. To increase the discriminative abilities of learned features, SphereFace [22], CosFace [40] and ArcFace [6] adopted different kinds of margin into the embedding space. Furthermore, some works adopted an additional loss function to regulate the identity-representative vectors. RegularFace [52] minimized a cosine similarity between identity-representative vectors, and UniformFace [8] equalized distances between all the cluster centers. However, those methods can suffer from an enormous number of identities and instances because they are based on a mini-batch learning. Our BroadFace overcomes the limitation of a mini-batch learning, and, it can be easily applied on those face recognition methods.

In terms of preserving knowledge of model on previously visited data, the continual learning [13,20] shares the similar concept with BroadFace. However, BroadFace is different from continual learning, as BroadFace preserves knowledge of previous data from the same dataset while continual learning preserves knowledge of previous data from different datasets.

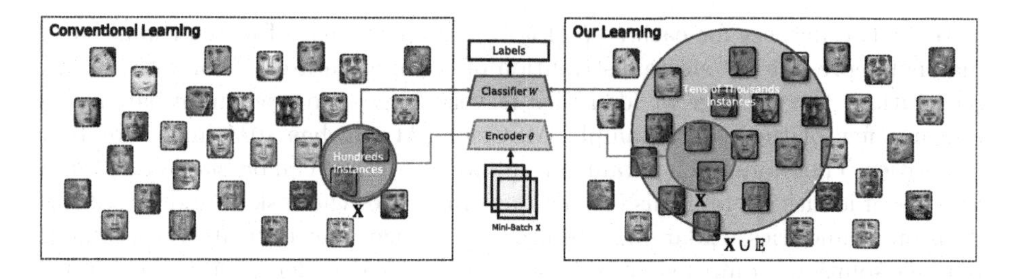

Fig. 2. Our learning refers more instances to learn the classifier in training stage.

3 Proposed Method

We describe the widely-adopted learning scheme in face recognition, and then illustrate the proposed BroadFace in detail.

3.1 Typical Learning

Learning of Face Recognition. In general, a face recognition network is divided into two parts: (1) an encoder network that extracts an embedding vector from a given image and (2) a linear classifier that maps an embedding vector into probabilities of identities. An evaluation is performed by comparing embedding vectors on images of unseen identities, so the classifier is discarded at the inference stage. Here, f is the encoder network that extracts a D-dimensional embedding vector e from a given image x: $e = f(x;\theta)$ with a model parameter θ. The linear classifier performs classification for C identities from an embedding vector e with a weight matrix W of $C \times D$-dimensions. For a mini-batch $\mathbf{X}$, the objective function such as a variant of angular softmax losses [6,22,39] is used to optimize the encoder and the classifier:

$$\mathcal{L}(\mathbf{X}) = \frac{1}{|\mathbf{X}|} \sum_{i \in \mathbf{X}} l(e_i, y_i), \tag{1}$$

$$l(e_i, y_i) = -\log \frac{\exp(\hat{W}_{y_i}^T \hat{e}_i)}{\sum_{j=1}^{C} \exp(\hat{W}_j^T \hat{e}_j)}, \tag{2}$$

where y_i is an labeled identity of x_i and $\hat{\cdot}$ indicates that a given vector is L_2 normalized (*e.g.*, $\|\hat{e}\|_2 = 1$). In Eq. 2, $\hat{W}_{y_i}$ acts as an representative instance of the given identity y_i that maximizes the cosine similarity with an embedding vector e_i. Thus, $\hat{W}_y$ can be regarded as the identity-representative vector, which is the expectation of instances belong to y:

$$\hat{W}_y = E_x \left[\hat{e}_i | y_i = y \right]. \tag{3}$$

Limitations of Mini-batch. The parameters of the model are updated in an iterative process that considers a mini-batch that contains only a small portion of the entire dataset for each step (Fig. 2). However, use of a small mini-batch may not represent the entire distribution of training datasets. Moreover, in face recognition, the number of identities is very large and each mini-batch only contains few of them; for example, MSCeleb-1M [10] has 10M images of 100k celebrities. Therefore, each parameter update of a model can be biased on a small number of identities, and this restriction complicates the task of finding optimal decision boundaries. Enlarging the mini-batch size may mitigate the problem, but this solution requires heavy computation of the encoder, proportional to the batch-size.

3.2 BroadFace

We introduce BroadFace, which is a simple yet effective way to cover a large number of instances and identities. BroadFace learns globally well-optimized identity-representative vectors from a massive number of embedding vectors (Fig. 2). For example, on a single Nvidia V100 GPU, the size of a mini-batch for ResNet-100 is at most 256, whereas BroadFace can utilize more than 8k instances at once. The following describes each step.

(1) Queuing Past Embedding Vectors. BroadFace has two queues of predefined size: $\mathbb{E}$ stores embedding vectors; $\mathbb{W}$ stores identity-representative vectors from the past. For each iteration, after model update, embedding vectors of a given mini-batch $\{e_i\}_{i \in \mathbf{X}}$ are enqueued to $\mathbb{E}$, and corresponding identity-representative vectors $\{W_{y_i}\}_{i \in \mathbf{X}}$ of each instance are enqueued to $\mathbb{W}$. By referring to the past embedding vectors in the queue to compute the loss $\mathcal{L}(\mathbf{X} \cup \mathbb{E})$, the network increase the number of instances and identities explored at each update.

(2) Compensating Past Embedding Vectors. As the model parameter θ of the encoder is updated over iterations, past embedding vectors $e^- \in \mathbb{E}$ conflict with the embedding space of the current parameter (Fig. 3a); $\epsilon = e - e^-$ where θ^- is the past parameter of the encoder and $e^- = f(x; \theta^-)$. The magnitude of the error ϵ is relatively small when few iterations have been passed from the past. However, the error is gradually accumulated over iterations and the error hinders appropriate training. We introduce a compensation function $\rho(y)$ for each identity to reduce the errors as an additive model; $e_i^* = e_i^- + \rho(y)$ where e_i^* is a compensated past embedding vector to a current embedding vector (Fig. 3b). The compensation function should minimize an expected squared error J between the current embedding vectors and the compensated past embedding vectors that belong to y:

$$\begin{aligned} \text{minimize } J(\rho(y)) &= E_x\left[\left(e_i^* - e_i\right)^2 \middle| y_i = y\right], \\ &= E_x\left[\left(e_i^- + \rho(y) - e_i\right)^2 \middle| y_i = y\right]. \end{aligned} \tag{4}$$

The partial derivative of J with respect to $\rho(y)$ is:

$$\frac{\partial J}{\partial \rho(y)} = E_x\left[2\left(e_i^- + \rho(y) - e_i\right) \middle| y_i = y\right]. \tag{5}$$

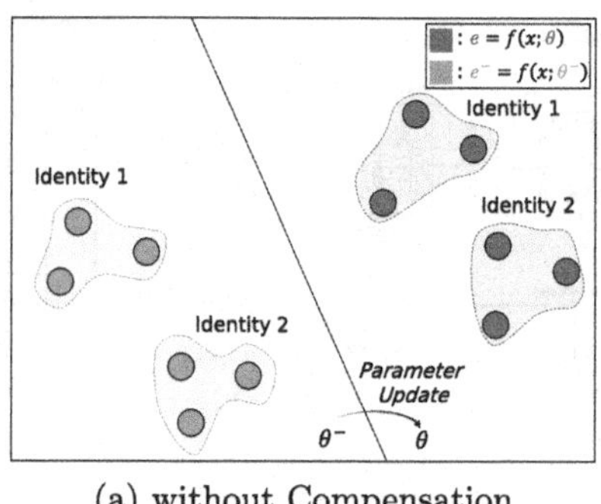

(a) without Compensation

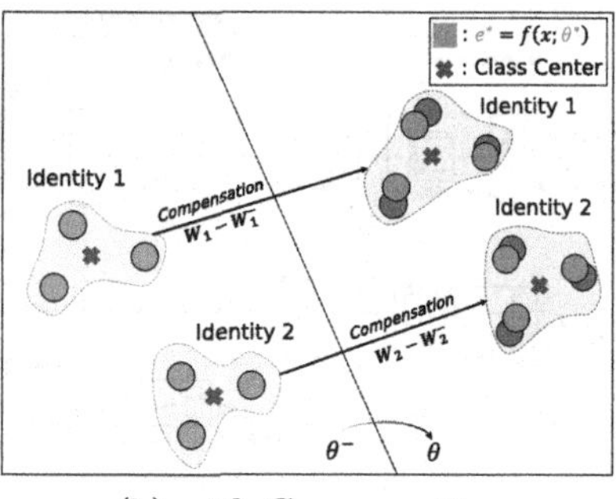

(b) with Compensation

Fig. 3. (a) enqueued embedding vectors (gray circles) at the past are further away from the embedding vectors (blue circles) at the current iteration due to the parameter update and this indicates significant errors. (b) the compensated embedding vectors (orange circles) closely approach to the embedding vectors at the current iteration, by considering the difference between the identity-representative vectors (class centers) at the past and the current iteration. (Color figure online)

Thus, the optimal compensation function is the difference between the expectations of current embedding vectors and past embedding vectors:

$$\begin{aligned} \rho(y) &= E_x\left[e_i \middle| y_i = y\right] - E_x\left[e_i^- \middle| y_i = y\right], \\ &\approx \lambda(W_y - W_y^-), \end{aligned} \tag{6}$$

where $W_y^- \in \mathbb{W}$ is an identity-representative vector, which is enqueued to the queue during the same iteration as $e_i^- \in \mathbb{E}$. As explained in Eq. 3, the identity-representative vector and the expected embedding vector point in the same direction when the vectors are projected onto a hyper-sphere, but the vectors are different in scale. Thus, we deploy a simple normalization term per each instance to adjust these scales: $\lambda = \left\|e_i^-\right\| / \left\|W_{y_i}^-\right\|$. Then the compensated embedding vector e^* is computed as:

$$e_i^* = e_i^- + \frac{\left\|e_i^-\right\|}{\left\|W_{y_i}^-\right\|}(W_{y_i} - W_{y_i}^-). \tag{7}$$

In empirical studies, the compensation function reduces the error significantly.

(3) Learning from Numerous Embedding Vectors. By executing the preceding two steps, BroadFace generates additional large-scale embedding vectors from the past. In our method, the encoder is trained on a mini-batch as before and the classifier is trained on both a mini-batch and the additional embedding vectors. The objective functions for the encoder and the classifier are defined as:

$$\mathcal{L}_{\text{encoder}}(\mathbf{X}) = \frac{1}{|\mathbf{X}|}\left\{\sum_{i \in \mathbf{X}} l(e_i)\right\}, \tag{8}$$

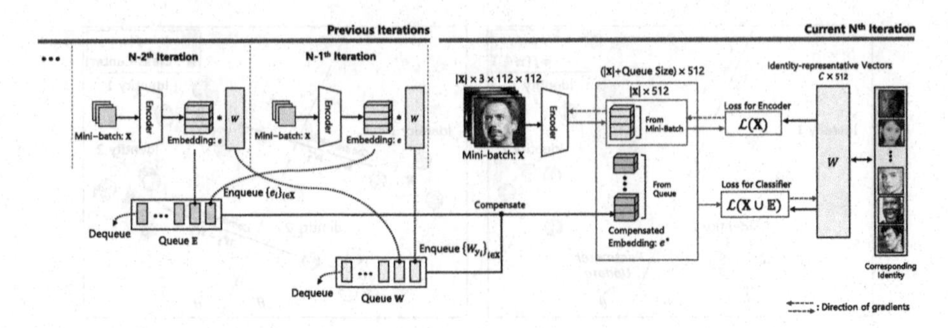

Fig. 4. Learning process of the proposed method. BroadFace deploys large queues to store embedding vectors and their corresponding identity-representative vectors per iteration. The embedding vectors of the past instances stored in the queues are used to compute loss for identity-representative vectors. BroadFace effectively learns from tens of thousands of instances for each iteration.

$$\mathcal{L}_{\text{classifier}}(\mathbf{X} \cup \mathbb{E}) = \frac{1}{|\mathbf{X} \cup \mathbb{E}|} \left\{ \sum_{i \in \mathbf{X}} l(e_i) + \sum_{j \in \mathbb{E}} l(e_j^*) \right\}. \tag{9}$$

The parameter θ of the encoder is updated *w.r.t.* $\mathcal{L}_{\text{encoder}}(\mathbf{X})$ while the parameter of the classifier W is updated *w.r.t.* $\mathcal{L}_{\text{classifier}}(\mathbf{X} \cup \mathbb{E})$. The large number of embedding vectors in the queue helps to learn highly precise identity-representative vectors that show reduced bias on a mini-batch and increased optimality on the entire dataset. The precise identity-representative vectors can accelerate the learning procedure. Moreover, our method can be easily implemented by adding several queues in the learning process (Fig. 4) and significantly improves accuracy in face recognition without any computational cost at inference stage.

3.3 Discussion

Effectiveness of Compensation. We show that the compensation method is empirically effective. After a small number of iterations, the error of the enqueued embedding vectors is also small and the compensation method is not necessary. However, after a large number of iterations, the error increases and the compensation method becomes necessary to keep a large number of embedding vectors (Fig. 5a). A large accumulated error may degrade the training process of the network (Fig. 7a and Fig. 7b). We illustrate how the compensation function reduces the difference between past and current embedding vectors in 2-dimensional space by t-SNE [24] (Fig. 5b). The past embedding vectors approach to current embedding vectors after applying compensation. This shows that the proposed compensation function works properly in practice.

Memory Efficiency. We compare BroadFace with enlarging the size of a mini-batch in terms of memory consumption (Fig. 6). A naïve mini-batch learning requires a huge amount of memory to forward and backward the entire network.

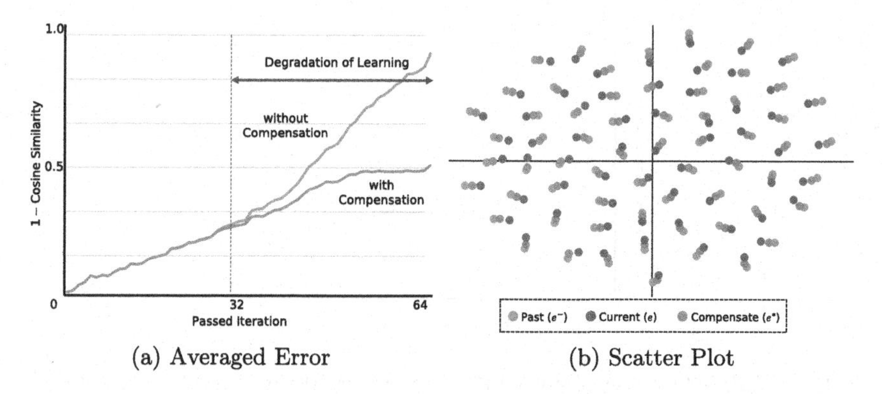

(a) Averaged Error (b) Scatter Plot

Fig. 5. (a) The average of the cosine errors between the embedding vectors at the current iteration and the past iteration with and without compensation. The errors are computed with randomly sampled instances over 64 iterations. (b) the scatter plot of the past (before 64 iterations), current and compensated embedding vectors for 64 instances.

The maximum size of a mini-batch is about 240 instances when a model based on ResNet-100 is trained on NVidia V100 of 32 GB. However, BroadFace requires only a matrix multiplication between the embedding vectors in $\mathbb{E}$ and the weight matrix W (Eq. 2). The marginal computational cost of BroadFace enables a classifier to learn decision boundaries from a massive set of instances, *e.g.*, 8192 instances for a single GPU. Note that, enlarging the size of a mini-batch to 8192 requires about 952 GB of memory which is infeasibly large for a single GPU.

4 Experiments

4.1 Implementation Details

Experimental Setting. As pre-processing, we normalize a face image to 112×112 by warping a face-region using five facial points from two eyes, nose and two corners of mouth [6,22,40]. A backbone network is ResNet-100 [11] that is used in the recent works [6,15]. After the `res5c` layer of ResNet-100, a block of batch normalization, fully-connected and batch normalization layers is deployed to compute a 512-dimensional embedding vector. The computed embedding vectors and the weight vectors of the linear classifier are L_2-normalized and trained by the ArcFace [6]. Our model is trained on 4 synchronized NVidia V100 GPUs and a mini-batch of 128 images is assigned for each GPU. The queue of BroadFace stores up to 8,192 embedding vectors accumulated over 64 iterations for each GPU, thus the total size of the queues is 32,768 for 4 GPUs. To avoid abrupt changes in the embedding space, the network of BroadFace is trained from the pre-trained network that is trained by the softmax based loss [6]. We adopted stochastic gradient descent (SGD) optimizer, and a learning rate is set to $5 \cdot 10^{-3}$

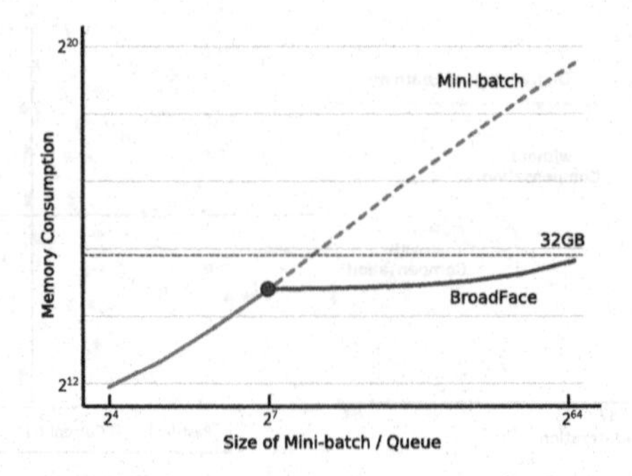

Fig. 6. Illustration on memory consumption of a conventional mini-batch learning (blue line) and the proposed BroadFace (red line) depending on the size of a mini-batch. The blue dotted line indicates memory consumption that is estimated for a large size of mini-batch by linear regression. (Color figure online)

for the first 50k, $5 \cdot 10^{-4}$ for the 20k, and $5 \cdot 10^{-5}$ for the 10k with a weight decay of $5 \cdot 10^{-4}$ and a momentum of 0.9.

Datasets. All the models are trained on MSCeleb-1M [10], which is composed of about 10M images for 100k identities. We use the refined version [6], which contains 3.8M images for 85k identities by removing the noisy labels of MSCeleb-1M. For the test, we perform evaluations on the following various datasets:

- Labeled Faces in the Wild (LFW) [14] contains 13k images of faces that are collected from web for 5,749 different individuals. Cross-Age LFW (CALFW) [54] provides pairs with age variation, and Cross-Pose (CPLFW) [53] provides pairs with pose variation from the images of LFW.
- YouTube Faces (YTF) [43] contains 3,425 videos of 1,595 different people.
- MegaFace [17] contains more than 1M images from 690k identities to evaluate recognition-accuracy with enormous distractors.
- Celebrities in Frontal-Profile (CFP) [33] contains 500 subjects; each subject has 10 frontal and 4 profile images.
- AgeDB-30 [26], which contains 12,240 images of 440 identities with age variations, is suitable to evaluate the sensitivity of a given method in age variation.
- IARPA Janus Benchmark (IJB) [25,42], which is designed to evaluate unconstrained face recognition systems, is one of the most challenging datasets in public. IJB-B [42] is composed of 67k face images, 7k face videos and 10k non-face images. IJB-C [25], which adds additional new subjects with increased occlusion and diversity of geographic origin to IJB-B, is composed of 138k face images, 11k face videos and 10k non-face images.

4.2 Evaluations on Face Recognition

We conduct experiments on the described various datasets to show the effectiveness of the proposed method.

Table 1. Verification accuracy (%) on LFW and YTF.

Method	LFW	YTF	Method	LFW	YTF
DeepID [36]	99.47	93.2	DeepFace [37]	97.35	91.4
VGGFace [28]	98.95	97.3	FaceNet [32]	99.64	95.1
CenterLoss [41]	99.28	94.9	RangeLoss [51]	99.52	93.7
MarginalLoss [7]	99.48	95.9	SphereFace [22]	99.42	95.0
RegularFace [52]	99.61	96.7	CosFace [40]	99.81	97.6
UniformFace [8]	99.80	97.7	AFRN [15]	99.85	97.1
ArcFace [6]	99.83	97.7	BroadFace	**99.85**	**98.0**

Table 2. Verification accuracy (%) on CALFW, CPLFW, CFP-FP and AgeDB-30.

Method	CALFW	CPLFW	CFP-FP	AgeDB-30
CenterLoss [41]	85.48	77.48	–	–
SphereFace [22]	90.30	81.40	–	–
VGGFace2 [2]	90.57	84.00	–	–
CosFace [40]	95.76	92.28	98.12	98.11
ArcFace [6]	95.45	92.08	98.27	98.28
BroadFace	**96.20**	**93.17**	**98.63**	**98.38**

Table 3. Identification and verification evaluation on MegaFace [17]. Ident indicates rank-1 identification accuracy (%) and Verif indicates a true accept rate (%) at a false accept rate of 1e−6.

Method	MF-Large		MF-Large-Refined [6]	
	Ident	Verif	Ident	Verif
RegularFace [52]	75.61	91.13	–	–
UniformFace [8]	79.98	95.36	–	–
SphereFace [22]	–	–	97.91	97.91
AdaptiveFace [21]	–	–	95.02	95.61
CosFace [40]	80.56	96.56	97.91	97.91
ArcFace [6]	81.03	96.98	98.35	98.49
BroadFace	**81.33**	**97.56**	**98.70**	**98.95**

LFW and YTF are widely used to evaluate verification performance under the unrestricted environments. LFW, which contains pairs of images, evaluates a model by comparing two embedding vectors of a given pair. YTF contains videos that are sets of images; from the shortest clip of 48 frames to the longest clip of 6,070 frames. To compare a pair of videos, YTF compares a pair of video-representative embedding vectors that are averaged embedding vectors of images

Table 4. Verification evaluation with a True Accept Rate at a certain False Accept Rate (TAR@FAR) from 1e−4 to 1e−6 on IJB-B and IJB-C. † denotes BroadFace trained by CosFace [40].

Method	IJB-B			IJB-C		
	FAR = 1e−6	FAR = 1e−5	FAR = 1e−4	FAR = 1e−6	FAR = 1e−5	FAR = 1e−4
VGGFace2 [2]	–	0.671	0.800	–	0.747	0.840
CenterFace [41]	–	–	–	–	0.781	0.853
ComparatorNet [46]	–	–	0.849	–	–	0.885
PRN [16]	–	0.721	0.845	–	–	–
AFRN [15]	–	0.771	0.885	–	0.884	0.931
CosFace [40]	0.3649	0.8811	0.9480	0.8591	0.9410	0.9637
BroadFace†	0.4092	0.8997	**0.9497**	0.8596	**0.9459**	**0.9638**
ArcFace [6]	0.3828	0.8933	0.9425	0.8906	0.9394	0.9603
BroadFace	**0.4653**	**0.9081**	0.9461	**0.9041**	0.9411	0.9603

collected from each video. Even though both datasets are highly-saturated in accuracy, our BroadFace outperforms other recent methods (Table 1).

CALFW, CPLFW, CFP-FP and AgeDB-30 are also widely used to verify that methods are robust to pose and age variation. CALFW and AgeDB-30 have multiple instances for same identity of different ages and CPLFW and CFP-FP have multiple instances for same identity of different poses (frontal and profile faces). BroadFace shows better verification-accuracy on all datasets (Table 2).

MegaFace is designed to evaluate both face identification and verification tasks under difficulty caused by a huge number of distractors. We evaluate our BroadFace on Megaface Challenge 1 where the training dataset is more than 0.5 million images. BroadFace outperforms the other top-ranked face recognition models for both face identification and verification tasks (Table 3). On the refined MegaFace [6], where noisy labels are removed, BroadFace also surpasses the other models.

IJB-B and IJB-C are the most challenging datasets to evaluate unconstrained face recognition. We report BroadFace with CosFace [40] and BroadFace with ArcFace [6] in verification task without any augmentations such as horizontal flipping in test time. Our BroadFace shows significant improvements on all FAR criteria (Table 4). In IJB-B [42], BroadFace improves 8.25% points on FAR = 1e−6, 1.48% points on FAR = 1e−5 and 0.36% points on FAR = 1e−4 comparing to the results of ArcFace [6].

4.3 Evaluations on Image Retrieval

Both face recognition and image retrieval have the same goal that learn an optimal embedding space to compare a given pair of items such as face, clothes or industrial products. To show that BroadFace is widely applicable on other applications, we compare BroadFace with recently proposed metric learning methods for image retrieval.

Table 5. Recall@K comparison with state-of-the-art methods. For fair comparison, we divide methods according to the dimension (Dim.) of an embedding vector. The numbers under the datasets refer to recall at K.

Methods	Dim.	In-Shop				SOP			
		1	10	20	30	1	10	10^2	10^3
Margin [45]	128	–	–	–	–	72.7	86.2	93.8	98.0
MIC+Margin [29]	128	88.2	97.0	–	–	77.2	89.4	95.6	–
DC [31]	128	85.7	95.5	96.9	97.5	75.9	88.4	94.9	98.1
ArcFace [6]	128	84.1	94.9	96.2	96.9	73.3	86.4	93.2	97.1
BroadFace	128	**89.8**	**97.4**	**98.1**	**98.4**	**79.7**	**90.7**	**95.7**	**98.4**
TML [49]	512	–	–	–	–	78.0	**91.2**	**96.7**	**99.0**
NSM [50]	512	88.6	**97.5**	**98.4**	**98.8**	78.2	90.6	96.2	–
ArcFace [6]	512	87.3	96.3	97.3	97.9	76.9	89.1	95.0	98.2
BroadFace	512	**90.1**	97.4	98.1	98.4	**80.2**	91.0	95.9	98.4

Experimental Settings. We use ResNet-50 [11] that is pre-trained on ILSVRC 2012-CLS [30] as a backbone network. We use ArcFace [6] as a baseline objective function and set the size of the queue to 32k for BroadFace. We follow the standard input augmentation and evaluation protocol [35]. We evaluate on two large datasets with a large number of classes similar to face-recognition: In-Shop Clothes Retrieval (In-Shop) [23] and Stanford Online Products (SOP) [35].

In-Shop and SOP are the standard datasets in image retrieval. In-Shop contains 11,735 classes of clothes. For training, the first 3,997 classes with 25,882 images are used, and the remaining 7,970 classes with 26,830 images are split into query set gallery set for evaluation. SOP contains 22,634 classes of industrial products. For training, the first 11,318 classes with 59,551 images are used and the remaining 11,316 classes with 60,499 images are used for evaluation. Our baseline models that are trained with ArcFace [6] underperform comparing to the other state-of-the-art methods. BroadFace significantly improves the recall of the baseline models and the improved model even outperforms the other methods (Table 5).

4.4 Analysis of BroadFace

Size of Queue. BroadFace has only one hyper-parameter, the size of the queue, to determine the maximum number of embedding vectors accumulated over past iterations. Using the single parameter makes our method easy to tune and the parameter plays a very important role in determining recognition-accuracy. As the size of the queue grows, the performance increases steadily (Table 6a). Especially, without our compensation method, accuracy degradation occurs when the size of the queue is significantly large. However, our compensation method alleviates the degradation by correcting the enqueued embedding vectors. We show

Table 6. Effects of BroadFace varying the size of the queue and the type of the backbone network on IJB-B dataset in face recognition.

(a) Total Size of Queue

Size of queue	TAR			
	FAR = 1e−6		FAR = 1e−5	
	Without compensation	With compensation	Without compensation	With compensation
0 (Baseline)	0.3828	0.3828	0.8933	0.8933
2048 (512 × 4 GPUs)	0.4310	0.4255	0.9061	0.9077
8192 (2048 × 4 GPUs)	0.4346	0.4394	0.9071	0.9085
32768 (8192 × 4 GPUs)	0.4259	0.4653	0.9078	0.9081

(b) Backbone Network

	Dim.	TAR				GFlops
		FAR = 1e−6		FAR = 1e−5		
		ArcFace	BroadFace	ArcFace	BroadFace	
MobileFaceNet [5]	128	0.3552	0.3665	0.8456	0.8458	0.9G
ResNet-18 [11]	128	0.3678	0.3808	0.8588	0.8638	5.2G
ResNet-34 [11]	512	0.3981	0.4325	0.8798	0.8828	8.9G
ResNet-100 [11]	512	0.3828	0.4653	0.8933	0.9081	24.1G

another experiment on the enormous size of the queue from 0 to 32,000 in image retrieval (Fig. 7a). With the proposed compensation, the recall is consistently improved as the size of the queue is increased. However, without the proposed compensation, the recall of the models degrades when the size of the queue is more than 16k.

Generalization Ability. Our BroadFace is generally applicable to any objective functions and any backbone networks. We apply BroadFace to two widely used objective functions of CosFace [40] and ArcFace [6]. For both CosFace and ArcFace, BroadFace increases recognition-accuracy (Table 4). We also apply BroadFace to several backbone networks such as MobileFaceNet [5], ResNet-18 [11] and ResNet-34 [11]. We set the dimensions of embedding vector to 128 for light backbone networks such as MobileFaceNet and ResNet-18, and 512 for heavy backbone networks such as ResNet-34 and ResNet-100. BroadFace is significantly effective for all backbone networks (Table 6b). In particular, ResNet-34 trained with BroadFace achieves comparable performance to ResNet-100 trained only with ArcFace, even though ResNet-34 has much less GFlops.

Learning Acceleration. Our BroadFace accelerates the learning process of both face recognition and image retrieval. In face recognition, many iterations are still needed to overcome a small gap of performance among the methods

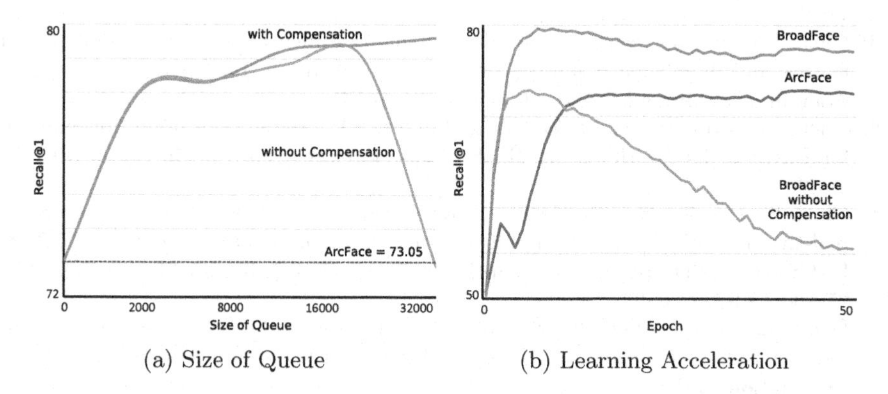

(a) Size of Queue

(b) Learning Acceleration

Fig. 7. (a) The recall depending on the size of the queue in BroadFace with and without our compensation function; the red line indicates the recall of ArcFace (baseline) on the test set. (b) the learning curve for the test set when the size of the queue is 32k; ArcFace reaches the highest recall at the 45th epoch, our BroadFace reaches the highest recall at the 10th epoch, and the learning process collapses without our compensation function.

on the highly-saturated datasets. Thus, we experiment the acceleration of the learning process in image retrieval to clearly show the effectiveness (Fig. 7b). Our BroadFace reaches peak performance much faster and higher than the baseline model. Without our compensation method, the model gradually collapses.

5 Conclusion

We introduce a new way called BroadFace that allows an embedding space to distinguish numerous identities in a broad perspective by increasing the optimality of constructed identity-representative vectors. BroadFace is significantly effective for face recognition and image retrieval where their datasets consist of numerous identities and instances. BroadFace can be easily applied on many existing face recognition methods to obtain a significant improvement without any extra computational cost in the inference stage.

Acknowledgments. We would like to thank AI R&D team of Kakao Enterprise for the helpful discussion. In particular, we would like to thank Yunmo Park who designed the visual materials.

References

1. Ahonen, T., Hadid, A., Pietikäinen, M.: Face recognition with local binary patterns. In: Pajdla, T., Matas, J. (eds.) ECCV 2004. LNCS, vol. 3021, pp. 469–481. Springer, Heidelberg (2004). https://doi.org/10.1007/978-3-540-24670-1_36
2. Cao, Q., Shen, L., Xie, W., Parkhi, O.M., Zisserman, A.: VGGFace2: a dataset for recognising faces across pose and age. In: International Conference on Automatic Face and Gesture Recognition (2018)

3. Chen, D., Cao, X., Wen, F., Sun, J.: Blessing of dimensionality: high-dimensional feature and its efficient compression for face verification. In: IEEE Conference on Computer Vision and Pattern Recognition (2013)
4. Chen, D., Cao, X., Wipf, D., Wen, F., Sun, J.: An efficient joint formulation for Bayesian face verification. IEEE Trans. Pattern Anal. Mach. Intell. **39**, 32–46 (2017)
5. Chen, S., Liu, Y., Gao, X., Han, Z.: MobileFaceNets: efficient CNNs for accurate real-time face verification on mobile devices. In: Zhou, J., et al. (eds.) CCBR 2018. LNCS, vol. 10996, pp. 428–438. Springer, Cham (2018). https://doi.org/10.1007/978-3-319-97909-0_46
6. Deng, J., Guo, J., Xue, N., Zafeiriou, S.: ArcFace: additive angular margin loss for deep face recognition. In: IEEE Conference on Computer Vision and Pattern Recognition (2019)
7. Deng, J., Zhou, Y., Zafeiriou, S.: Marginal loss for deep face recognition. In: IEEE Conference on Computer Vision and Pattern Recognition Workshops (2017)
8. Duan, Y., Lu, J., Zhou, J.: UniformFace: learning deep equidistributed representation for face recognition. In: IEEE Conference on Computer Vision and Pattern Recognition (2019)
9. Goyal, P., et al.: Accurate, large minibatch SGD: training ImageNet in 1 hour. arXiv preprint arXiv:1706.02677 (2017)
10. Guo, Y., Zhang, L., Hu, Y., He, X., Gao, J.: MS-Celeb-1M: a dataset and benchmark for large-scale face recognition. In: Leibe, B., Matas, J., Sebe, N., Welling, M. (eds.) ECCV 2016. LNCS, vol. 9907, pp. 87–102. Springer, Cham (2016). https://doi.org/10.1007/978-3-319-46487-9_6
11. He, K., Zhang, X., Ren, S., Sun, J.: Deep residual learning for image recognition. In: IEEE Conference on Computer Vision and Pattern Recognition (2016)
12. Hoffer, E., Hubara, I., Soudry, D.: Train longer, generalize better: closing the generalization gap in large batch training of neural networks. In: Advances in Neural Information Processing Systems (2017)
13. Hou, S., Pan, X., Loy, C.C., Wang, Z., Lin, D.: Learning a unified classifier incrementally via rebalancing. In: Proceedings of the IEEE Conference on Computer Vision and Pattern Recognition, pp. 831–839 (2019)
14. Huang, G.B., Ramesh, M., Berg, T., Learned-Miller, E.: Labeled faces in the wild: a database for studying face recognition in unconstrained environments. Technical report, University of Massachusetts, Amherst (2007)
15. Kang, B.N., Kim, Y., Jun, B., Kim, D.: Attentional feature-pair relation networks for accurate face recognition. In: IEEE International Conference on Computer Vision (2019)
16. Kang, B.-N., Kim, Y., Kim, D.: Pairwise relational networks for face recognition. In: Ferrari, V., Hebert, M., Sminchisescu, C., Weiss, Y. (eds.) ECCV 2018. LNCS, vol. 11206, pp. 646–663. Springer, Cham (2018). https://doi.org/10.1007/978-3-030-01216-8_39
17. Kemelmacher-Shlizerman, I., Seitz, S.M., Miller, D., Brossard, E.: The MegaFace benchmark: 1 million faces for recognition at scale. In: IEEE Conference on Computer Vision and Pattern Recognition (2016)
18. Keskar, N.S., Mudigere, D., Nocedal, J., Smelyanskiy, M., Tang, P.T.P.: On large-batch training for deep learning: Generalization gap and sharp minima. In: International Conference on Learning Representations (2017)
19. Kumar, N., Berg, A.C., Belhumeur, P.N., Nayar, S.K.: Attribute and simile classifiers for face verification. In: IEEE International Conference on Computer Vision (2009)

20. Li, Z., Hoiem, D.: Learning without forgetting. IEEE Trans. Pattern Anal. Mach. Intell. **40**(12), 2935–2947 (2017)
21. Liu, H., Zhu, X., Lei, Z., Li, S.Z.: AdaptiveFace: adaptive margin and sampling for face recognition. In: IEEE Conference on Computer Vision and Pattern Recognition (2019)
22. Liu, W., Wen, Y., Yu, Z., Li, M., Raj, B., Song, L.: SphereFace: deep hypersphere embedding for face recognition. In: IEEE Conference on Computer Vision and Pattern Recognition (2017)
23. Liu, Z., Luo, P., Qiu, S., Wang, X., Tang, X.: DeepFashion: powering robust clothes recognition and retrieval with rich annotations. In: IEEE Conference on Computer Vision and Pattern Recognition (2016)
24. Maaten, L.V.D., Hinton, G.: Visualizing data using t-SNE. J. Mach. Learn. Res. **9**, 2579–2605 (2008)
25. Maze, B., et al.: Iarpa janus benchmark - c: face dataset and protocol. In: International Conference on Biometrics (2018)
26. Moschoglou, S., Papaioannou, A., Sagonas, C., Deng, J., Kotsia, I., Zafeiriou, S.: AgeDB: the first manually collected, in-the-wild age database. In: IEEE Conference on Computer Vision and Pattern Recognition Workshops (2017)
27. Nguyen, H.V., Bai, L.: Cosine similarity metric learning for face verification. In: Kimmel, R., Klette, R., Sugimoto, A. (eds.) ACCV 2010. LNCS, vol. 6493, pp. 709–720. Springer, Heidelberg (2011). https://doi.org/10.1007/978-3-642-19309-5_55
28. Parkhi, O.M., Vedaldi, A., Zisserman, A.: Deep face recognition. In: British Machine Vision Conference (2015)
29. Roth, K., Brattoli, B., Ommer, B.: Mic: Mining interclass characteristics for improved metric learning. In: IEEE International Conference on Computer Vision (2019)
30. Russakovsky, O., et al.: ImageNet Large Scale Visual Recognition Challenge. Int. J. Comput. Vis. **115**, 211–252 (2015). https://doi.org/10.1007/s11263-015-0816-y
31. Sanakoyeu, A., Tschernezki, V., Buchler, U., Ommer, B.: Divide and conquer the embedding space for metric learning. In: IEEE Conference on Computer Vision and Pattern Recognition (2019)
32. Schroff, F., Kalenichenko, D., Philbin, J.: FaceNet: a unified embedding for face recognition and clustering. In: IEEE Conference on Computer Vision and Pattern Recognition (2015)
33. Sengupta, S., Chen, J., Castillo, C., Patel, V.M., Chellappa, R., Jacobs, D.W.: Frontal to profile face verification in the wild. In: IEEE Winter Conference on Applications of Computer Vision (2016)
34. Simonyan, K., Parkhi, O., Vedaldi, A., Zisserman, A.: Fisher vector faces in the wild. In: British Machine Vision Conference (2013)
35. Song, H.O., Xiang, Y., Jegelka, S., Savarese, S.: Deep metric learning via lifted structured feature embedding. In: IEEE Conference on Computer Vision and Pattern Recognition (2016)
36. Sun, Y., Chen, Y., Wang, X., Tang, X.: Deep learning face representation by joint identification-verification. In: Advances in Neural Information Processing Systems (2014)
37. Taigman, Y., Yang, M., Ranzato, M., Wolf, L.: Deepface: Closing the gap to human-level performance in face verification. In: IEEE Conference on Computer Vision and Pattern Recognition (2014)
38. Turk, M.A., Pentland, A.P.: Face recognition using eigenfaces. In: IEEE Conference on Computer Vision and Pattern Recognition (1991)

39. Wang, F., Xiang, X., Cheng, J., Yuille, A.L.: NormFace: L2 hypersphere embedding for face verification. In: ACM International Conference on Multimedia (2017)
40. Wang, H., Wang, Y., Zhou, Z., Ji, X., Gong, D., Zhou, J., Li, Z., Liu, W.: CosFace: large margin cosine loss for deep face recognition. In: IEEE Conference on Computer Vision and Pattern Recognition (2018)
41. Wen, Y., Zhang, K., Li, Z., Qiao, Yu.: A discriminative feature learning approach for deep face recognition. In: Leibe, B., Matas, J., Sebe, N., Welling, M. (eds.) ECCV 2016. LNCS, vol. 9911, pp. 499–515. Springer, Cham (2016). https://doi.org/10.1007/978-3-319-46478-7_31
42. Whitelam, C., et al.: Iarpa janus benchmark-b face dataset. In: IEEE Conference on Computer Vision and Pattern Recognition Workshops (2017)
43. Wolf, L., Hassner, T., Maoz, I.: Face recognition in unconstrained videos with matched background similarity. In: IEEE Conference on Computer Vision and Pattern Recognition (2011)
44. Wolf, L., Hassner, T., Taigman, Y.: Descriptor based methods in the wild. In: European Conference on Computer Vision Workshops (2008)
45. Wu, C.Y., Manmatha, R., Smola, A.J., Krahenbuhl, P.: Sampling matters in deep embedding learning. In: IEEE International Conference on Computer Vision (2017)
46. Xie, W., Shen, L., Zisserman, A.: Pairwise relational networks for face recognition. In: European Conference on Computer Vision (2018)
47. Yin, Q., Tang, X., Sun, J.: An associate-predict model for face recognition. In: IEEE Conference on Computer Vision and Pattern Recognition (2011)
48. You, Y., Gitman, I., Ginsburg, B.: Scaling SGD batch size to 32k for ImageNet training. arXiv preprint arXiv:1708.03888 (2017)
49. Yu, B., Tao, D.: Deep metric learning with Tuplet margin loss. In: IEEE International Conference on Computer Vision (2019)
50. Zhai, A., Wu, H.Y.: Classification is a strong baseline for deep metric learning. arXiv preprint arXiv:1811.12649 (2018)
51. Zhang, X., Fang, Z., Wen, Y., Li, Z., Qiao, Y.: Range loss for deep face recognition with long-tailed training data. In: IEEE International Conference on Computer Vision (2017)
52. Zhao, K., Xu, J., Cheng, M.M.: RegularFace: Deep face recognition via exclusive regularization. In: IEEE Conference on Computer Vision and Pattern Recognition (2019)
53. Zheng, T., Deng, W.: Cross-pose LFW: a database for studying cross-pose face recognition in unconstrained environments. Technical report, Beijing University of Posts and Telecommunications (2018)
54. Zheng, T., Deng, W., Hu, J.: Cross-age LFW: A database for studying cross-age face recognition in unconstrained environments. arXiv preprint arXiv:1708.08197 (2017)

Interpretable Visual Reasoning via Probabilistic Formulation Under Natural Supervision

Xinzhe Han[1,2], Shuhui Wang[2](✉), Chi Su[3], Weigang Zhang[4], Qingming Huang[1,2,5], and Qi Tian[6]

[1] University of Chinese Academy of Sciences, Beijing, China
hanxinzhe17@mails.ucas.ac.cn, qmhuang@ucas.ac.cn

[2] Key Laboratory of Intelligent Information Processing, Institute of Computing Technology, CAS, Beijing, China
wangshuhui@ict.ac.cn

[3] Kingsoft Cloud, Beijing, China
suchi@kingsoft.com

[4] Harbin Institute of Technology, Weihai, China
wgzhang@hit.edu.cn

[5] Peng Cheng Laboratory, Shenzhen, China

[6] Shenzhen University, Shenzhen, China
wywqtian@gmail.com

Abstract. Visual reasoning is crucial for visual question answering (VQA). However, without labelled programs, implicit reasoning under natural supervision is still quite challenging and previous models are hard to interpret. In this paper, we rethink implicit reasoning process in VQA, and propose a new formulation which maximizes the log-likelihood of joint distribution for the observed question and predicted answer. Accordingly, we derive a Temporal Reasoning Network (TRN) framework which models the implicit reasoning process as sequential planning in latent space. Our model is interpretable on both model design in probabilist and reasoning process via visualization. We experimentally demonstrate that TRN can support implicit reasoning across various datasets. The experimental results of our model are competitive to existing implicit reasoning models and surpass baseline by a large margin on complicated reasoning tasks without extra computation cost in forward stage.

Keywords: Visual Question Answering · Implicit reasoning · Temporal Reasoning Network · Explanable machine learning

1 Introduction

Recent advances in deep learning allow us to investigate emerging research themes lying at the intersection between vision and language. Visual Question

Electronic supplementary material The online version of this chapter (https://doi.org/10.1007/978-3-030-58545-7_32) contains supplementary material, which is available to authorized users.

© Springer Nature Switzerland AG 2020
A. Vedaldi et al. (Eds.): ECCV 2020, LNCS 12354, pp. 553–570, 2020.
https://doi.org/10.1007/978-3-030-58545-7_32

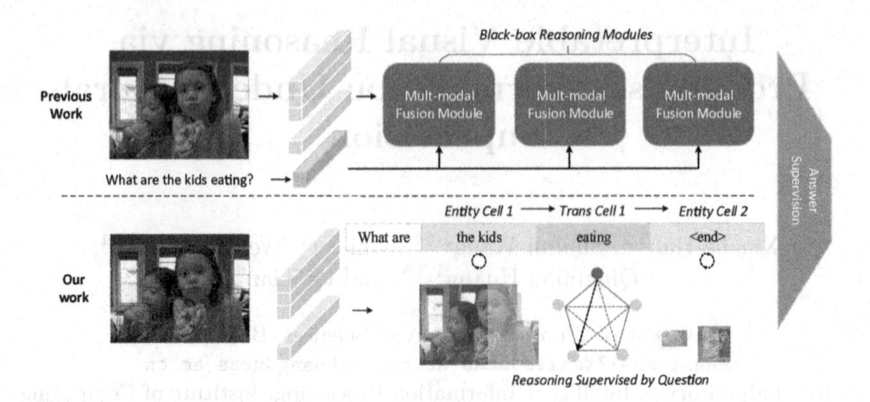

Fig. 1. Comparison between our work and previous work. We regard the question as a hint for reasoning programs, and formulate a Bayesian probabilistic framework for visual reasoning under natural supervision. The highlights in images are sampled from inferred Concrete distribution latent states, which can indicate the most critical image part that should be attended to at each time step

Answering (VQA) [3] is a representative task that aims to get an open-ended answer given an image and a natural language question. Since VQA requires high-level understanding of images and the associated questions, visual reasoning is required to provide primitives for deriving a good answer and make VQA model more interpretable for better human understanding [2,23,26,27,31,36,48,55,58].

Existing study towards interpretable visual reasoning can be divided into two groups. One group of work are conducted on synthetic datasets, *e.g.*, CLEVR [26], making use of external knowledge. Explicit "functional programs" are adopted by modular networks [2,20,21] and neural symbolism [49,53,58] to generated questions and nearly perfect answers are achieved. These explicit reasoning methods are quite interpretable, but they largely rely on strong assumptions like having labelled programs on questions or labelled entity relations. In another research direction, to solve real-world problem, implicit reasoning via stacked attention modules [8,15,45,48] or graph reasoning [22,36,43,49,52] establish specially designed "black box" neural architectures. Changing in attention maps is thought to be implicit reasoning evidence, and is somewhat a kind of "side effect" of answer classifier optimization. Taking a comparison between implicit and explicit reasoning, it can be found that existing real-world reasoning methods only maximize the likelihood of predicted answers without an understandable reasoning procedure. The multi-hoop attention, obtained by back-propagation, identify components that support a predicted answer, still lacks the ability to explain the reasoning process that achieves a specific answer.

It is known that Bayesian models fall below the ceiling of interpretable machine learning, since they convey a clear representation of the relationships between features and targets. For example, Deep Kalman Filters (DKFs) [38] and Deep Variational Bayes Filters (DVBFs) [30] endow deep models with interpretability inherent to probabilistic graphical models, to make them much more explainable as "Bayesian hybrid transparent" [4]. To unveil the black box of implicit reasoning, we view VQA task from a probabilistic perspective and reformulate a new general Bayesian interpretation for visual reasoning under real-world setting. We pay primary attention to the basic visual reasoning problem formulation and expect to enhance the interpretability of implicit reasoning methods.

Specifically, we reinvestigate the question generation process on synthetic datasets [24,26], which infers that questions clearly convey reasoning programs and vice versa. Considering that no program labeling is available on real-world data, we assume that there is a set of discrete latent states lying behind input question words, and use them to sequentially describe which part of the image should be attended to at each time step. These latent states act similarly as labelled programs in explicit reasoning models. Based on the latent states, the questions can be regarded as implicit supervision for underlying reasoning programs, as shown in Fig. 1. Instead of only maximizing answer likelihood in previous works, we reformulate an alternative probabilistic interpretation formulation, which maximizes the log-likelihood of joint distribution for the observed question and predicted answer. In this way, the answer predictor and latent states indicating reasoning evidence can be directly optimized simultaneously.

By decomposing the probabilistic formulation, we show that an interpretable reasoning process should have three basic modules, *i.e.*, *State Transition, State Inference* and *Generative Reconstruction.* We also show that recent developments in implicit reasoning can be steadily explained by our probabilistic framework, from one-step State Inference (one-stage fusion [1,13,16,32]) to multi-step State Inference (stacked attention [31,39]) and then to multi-step State Inference with dependent transitions (relation-based methods [14,22,36,52,54]). As a practitioner of the probabilistic framework, we derive a latent sequence model parametrized by a VAE-based neural network, named Temporal Reasoning Network (TRN). We integrate TRN module into existing representative models, such as the one-stage VQA model UpDn [1] and stack-attention model BAN [31]. The injection of TRN on existing models can be regarded as a regularization term in the training stage, and it can be removed in the testing stage without extra computational cost. The results demonstrate that compared to the baseline models, the enhanced models with TRN achieve improved performance and better interpretability without using extra fusion strategies.

It is worth noting that both architecture and loss function in our work are naturally derived from the basic probabilistic formulation and corresponding graphical model. Every term in TRN is conceptually and mathematically interpretable, which further guarantees the interpretability of the whole model. With *Generative Reconstruction* module and latent state sampling, we can also visualize the reasoning process along with question words, which demonstrates that answer prediction procedure of our model is also interpretable. Major contributions are three folds:

- We formulate a new probabilistic interpretation for visual reasoning in the real-world VQA task under natural supervision.
- Following the new probabilistic framework, we propose a sequential latent state model TRN, which is interpretable on both model design and answer prediction.
- TRN can well collaborate with existing models. It can help shallow models like UpDn achieves comparable result compared to state-of-the-art implicit

reasoning methods on VQA v2, CLEVR, and CLEVR-Human datasets, and enhance the explanation to existing black-box reasoning models like BAN. Code is available at https://github.com/GeraldHan/TRN.

2 Related Work

2.1 Visual Question Answering and Reasoning

The task of visual question answering is to infer the answer based on the input question and image. Primary methods for VQA mainly focus on better attention mechanisms [14,31,39,56,57] and multi-modal fusion strategies [6,13,16,32,59]. Recent research efforts towards VQA have changed from multi-modal matching to visual reasoning. Existing methods of visual reasoning can be categorized into explicit reasoning and implicit reasoning.

Explicit reasoning are mainly conducted on CLEVR [26] with compositional reasoning procedure. Andreas *et al.* [2] first propose Neural Modular Networks (NMN), which explicitly decompose the reasoning procedure into a sequence of sub-tasks handled by specialized modules. Consequent studies improve this work by proposing better layout policy [20,21], or designing more specific modules [27,42]. Similar to our work, Vedantam *et al.* [53] propose Probabilistic NMN, which provides a probabilistic formulation of NMN and requires a smaller number of teaching examples for layouts. Different from [53], our work focuses on more natural supervision. A more explicit symbolic reasoning method over the object-level structural scene representation is introduced in [41,49,58], which divides the perception and reasoning into two specific stages. Nevertheless, "expert layouts" are needed to supervise the layout policy to get compositional behaviour and good accuracy, which limits their performance on real-world datasets and human asked questions.

Implicit reasoning is extensively studied on real-world datasets like VQA [3]. Bottom-up [1] with object and attribute features extracted by Faster R-CNN [46] is a common baseline. A widely-used approach is to perform reasoning by sequential interactions between image representations and question embeddings [14,15,23,31,57]. Another research line focus on relation reasoning, which can be conducted on a fully-connected graph of objects [8,48]. To better model the interactions between multiple objects, labelled relation or question-conditioned graph representations for images are adopted, then a GCN [34] is used to implicitly infer the interactive representation of objects [22,36,43,52,54]. Implicit reasoning is suitable for real-world setting but much less interpretable.

We build a bridge between implicit and explicit reasoning by introducing latent states behind question words. Thus, the implicit reasoning can be performed with comparable interpretability to explicit reasoning under natural supervision.

2.2 Hybrid Transparent with Bayesian Interpretation

The ideas behind variational auto-encoders (VAEs) [33,47] have enabled complex latent dynamical systems like SVAE [29], non-linear SSMs [9,11,12,28] or

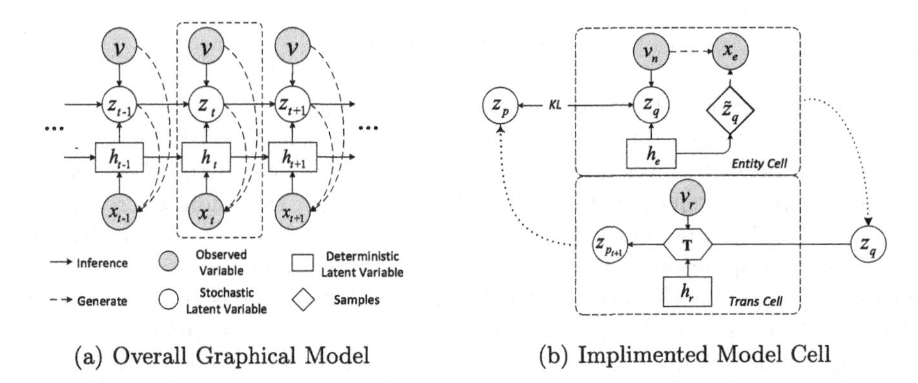

(a) Overall Graphical Model

(b) Implimented Model Cell

Fig. 2. Temporal Reasoning Model. (a) is the overall stochastic graphical model for reasoning process, which constructs a latent state z_t underlying each question item x_t conditioned on the given image $\mathbf{v}$ (image is constructed as a graph in practice), simulating labelled programs in explicit models. (b) is the implementation detail for a single reasoning step (the dash block in (a)). Entity Cell models the intra-block reasoning, which infers $q(z_t|h_t, \mathbf{v})$ and $p(x_t|z_t, \mathbf{v})$ based on node features $\mathbf{v}_e$ and entity phrase embedding h_e. Trans Cell infers the prior $p(z_t|z_{t-1}, \mathbf{v})$ for the next time block with a transition function $\mathbf{T}$

parametrized deep Markov models [35,51]. These methods combine Bayesian probabilistic graphical models in the embedding space with neural networks to enhance the interpretability of deep models. More similar to our work, model-based reinforcement learning [5,7,10,18] planning in latent space typically assumes access to the low-dimensional states of the environment and plan the dynamics directly in continuous state space, which can be more efficient compared to Bellman backups of traditional reinforcement learning. Recent works TD-VAE [17] and PlaNet [19] further extend to time-sequential planning to solve more complex problems with sequential state observations.

To look inside the implicit reasoning procedure, we formulate a latent sequence model generating implicit policies along with sequential question word inputs. In this way, we construct a deep model with Bayesian hybrid transparent [4] and explicitly model the reasoning procedure without extra-label efforts.

3 Method

3.1 Model Definition

Let $\mathbf{v}$ be image representations and $\mathbf{x}$ is a question comprised of a sequence of L words. The goal of VQA is to predict the best answer $\hat{a} \in \mathcal{A}$, where $\mathcal{A}$ is the answer set. As common practice in the VQA literature, answer prediction can be defined as a classification problem:

$$\hat{a} = \arg\max_{a \in \mathcal{A}} p_\gamma(a|\mathbf{v}, \mathbf{x}) \tag{1}$$

where p_γ denotes the trained classification model.

However, this kind of optimization does not explicitly model the reasoning process. In order to investigate the reasoning process behind the classifier, we assume there is a sequence of state variable $\mathbf{z} = \{z_t\}_{i=1}^T$ that indicates reasoning procedure underlying question $\mathbf{x}$. This time-dependent latent state z_t is decided by the current question item and conditioned on image representations. Fig 2(a) gives a detailed illustration of the whole graphical model for the above reasoning process. According to the graphical model, we treat image representation $\mathbf{v}$ as a global condition, question words $\mathbf{x}$ as sequential observations and assume a general form of fully Bayesian state space model. Our probabilistic formulation specifies the joint distribution $p(a, \mathbf{x}, \mathbf{z}|\mathbf{v})$, and our goal is to find an approximation for model evidence $p(a, \mathbf{x}|\mathbf{v})$ with respect to the posterior distribution $p(\mathbf{z}|\mathbf{x}, \mathbf{v})$. The log marginal evidence probability can be decomposed as

$$\log p(a, \mathbf{x}|\mathbf{v}) = \log p(\mathbf{x}|\mathbf{v}) + \log p(a|\mathbf{x}, \mathbf{v}) \tag{2}$$

Thus, the above model can be divided into two separate parts. The latter part is the answer classifier, similar to traditional VQA models. The former part models the **Temporal Reasoning** process optimized by maximum data likelihood of observed question $\mathbf{x}$, where the underlying latent states $\mathbf{z}$ can be optimized as latent variables via variational inference. Explicitly modelling reasoning process with such a fully probabilistic formulation and injecting it into existing methods are the main contributions of this work. The learning diagram will be detailed in Sect. 3.2.

3.2 Learning

Temporal Reasoning is to optimize log-likelihood of the question $\log p(\mathbf{x}|\mathbf{v})$. Following Bayesian rule, it can be auto-regressively decomposed as $\log p(\mathbf{x}|\mathbf{v}) = \sum_t \log p(x_t|x_{1:t-1}, \mathbf{v})$. For a given time step t, we decompose the log marginal probability with respect to the variational posterior $q(z_t, z_{t-1}|x_t, \mathbf{v}, x_{1:t-1})$ as

$$\log p(x_t|x_{1:t-1}, \mathbf{v}) = KL(q(z_t, z_{t-1})\|p(z_t, z_{t-1})) + \mathcal{L}_t \tag{3}$$

where $q(z_t, z_{t-1})$ is the short form of $q(z_t, z_{t-1}|\mathbf{v}, x_{1:t})$ which is a inference posterior distribution for latent states z_t, while $p(z_t, z_{t-1})$ is the abbreviation for $p(z_t, z_{t-1}|\mathbf{v}, x_{1:t-1})$ which is a corresponding generative prior distribution.[1] $\mathcal{L}_t$ is the evidence lower bound (ELBO) of data likelihood at time step t, which is

$$\begin{aligned} \mathcal{L}_t = \mathbb{E}_{(z_{t-1}, z_t) \sim q(z_t, z_{t-1})} [&\log p(x_t|z_t, z_{t-1}, x_{1:t-1}, \mathbf{v}) \\ &+ \log p(z_t, z_{t-1}|\mathbf{v}, x_{1:t-1}) - \log q(z_t, z_{t-1}|\mathbf{v}, x_{1:t})] \end{aligned} \tag{4}$$

[1] Following equations will use the same shorten expressions for convenience. Both distributions are parametrized as neural networks in our work. p indicates generate distributions, while q refers to inference distributions.

Considering the Markov assumption underlying the graphical model in Fig. 2(a), we can simplify $p(x_t|z_t, z_{t-1}, x_{1:t-1}, \mathbf{v}) = p(x_t|z_t, \mathbf{v})$. Moreover, following the Bayes rule, we can decompose $q(z_t, z_{t-1}|\mathbf{v}, x_{1:t})$ and $p(z_t, z_{t-1}|\mathbf{v}, x_{1:t-1})$ as

$$\begin{aligned} p(z_t, z_{t-1}|\mathbf{v}, x_{1:t-1}) &= p(z_{t-1}|x_{1:t-1}, \mathbf{v})p(z_t|z_{t-1}, \mathbf{v}) \\ q(z_t, z_{t-1}|\mathbf{v}, x_{1:t}) &= q(z_t|x_{1:t}, \mathbf{v})q(z_{t-1}|z_t, x_{1:t}, \mathbf{v}) \end{aligned} \tag{5}$$

Similar to [19], to simplify the model, joint representation $p(x_{1:t})$ is deterministically encoded with a Recurrent Neural Networks (RNNs) as

$$h_t = \mathrm{RNN}(x_t, h_{t-1}) \tag{6}$$

where $h_t \in \mathbb{R}^{d_q}$ is a deterministic variable encoding all history observations. Therefore, Eq. 4 can be decomposed as:

$$\begin{aligned} \mathcal{L}_t = \mathbb{E}_{\substack{z_t \sim q(z_t) \\ z_{t-1} \sim q(z_{t-1})}} \Big[&\log p(x_t|z_t, \mathbf{v}) + \log p(z_{t-1}|h_{t-1}, \mathbf{v}) + \log p(z_t|z_{t-1}, \mathbf{v}) \\ &- \log q(z_t|h_t, \mathbf{v}) - \log q(z_{t-1}|z_t, h_t, \mathbf{v}) \Big] \end{aligned} \tag{7}$$

where $q(z_t)$ and $q(z_{t-1})$ abbreviate $q(z_t|h_t, \mathbf{v})$ and $q(z_{t-1}|z_t, h_t, \mathbf{v})$ respectively.

Answer Classification is similar to existing methods. After Temporal Reasoning, the final state b_T is fed into a Multi-layer Perception (MLP) to predict the answer. Thus, the likelihood of answer can be approximated by a deterministic classifier:

$$\log p(a|\mathbf{x}, \mathbf{v}) \triangleq \mathcal{U} = f(a|b_T) \tag{8}$$

The training objective is to maximize the lower bound of joint data likelihood:

$$\arg\max_{\theta,\phi,\nu} \left[\mathcal{U}(f;\nu) + \sum_{t=1}^{T} \mathcal{L}_t(p, q; \theta, \phi) \right] \tag{9}$$

where the Answer Classifier is parametrized as f_ν, generative distribution and inference distribution in Temporal Reasoning are parametrized as p_θ and q_ϕ respectively.

As shown above, all terms in the loss function are completely derived by variational inference applied to Eq. 2 under a basic latent state assumption. Therefore, all parts of our framework can be mathematically explained from probabilistic perspective, which achieves our interpretability on model design.

3.3 Intuitive Explanation

Equation 7 derived from the graphical model is parametrized as four different modules. This section will provide a more intuitive explanation behind mathematical derivation. It can further reveal the interpretability of our method.

Generative Reconstruction $\log p(x_t|z_t, \mathbf{v})$ indicates that currently observed word x_t can be reconstructed from corresponding latent state z_t and global condition $\mathbf{v}$. We measure it by Binary Cross Entropy (BCE) between input x_t and reconstructed $\tilde{x}_t$. This term performs external supervision via the input questions.

State Transition $\log p(z_t|z_{t-1}, \mathbf{v})$ predicts a prior distribution for z_t based on former state z_{t-1}. It can be regarded as forward transition in latent space under Markovian assumption. It can guarantee the time-dependency in reasoning process.

State Inference $\log q(z_t|h_t, \mathbf{v})$ indicates that the posterior of latent state distribution $q(z_t)$ depends on history observations h_t and visual features $\mathbf{v}$. Since $p(z_{t-1}|h_{t-1}, \mathbf{v})$ has a consistent dependency with the posterior $q(z_{t-1}|h_{t-1}, \mathbf{v})$ at former time step $t-1$, we approximate generative distribution $p(z_{t-1}|h_{t-1}, \mathbf{v}) = q_\phi(z_{t-1}|h_{t-1}, \mathbf{v})$ without loss of information.

Backward Transition $\log q(z_{t-1}|z_t, h_t, \mathbf{v})$ indicates the former state z_{t-1} can be re-inferenced from the current state and observations. This term has a similar facility with State Inference but is hard to model, so we ignore this term in practice to simplify our model.

Comparing with recent proposed VQA methods, this probabilistic formulation can help explain recent developments in implicit reasoning. The attention mechanism can be viewed as a type of *State Inference* in latent space. From one-stage attention/fusion model [1,16] to stacked attention methods [31,39], the performance is largely improved due to the introduction of *multi-step State Inference*. Recent proposed relation-based methods [14,22,36,52,54] further strengthen the dependences between stacked modules, that can collaborate with the function of *State Transition* in our formulation. Moreover, stronger fusion strategies can help establish a more informative latent space. From this perspective, implicit reasoning is indeed sequentially supervised by both question and answer but has not been explicitly modelled before.

3.4 Parametrization and Implementation

Following the instruction of probabilistic formulation, we implement the temporal reasoning process as a VAE based latent sequence model, named Temporal Reasoning Network (TRN). It should be stressed that TRN is not a fixed network. We implement it as complementary modules upon existing baseline models based on the proposed graphical model.

Latent State distribution. In order to reveal the reasoning procedure underlying question words and fairly compare with attention-based baselines, we assume latent states following Concrete distribution [40]

$$q(z_t) = \mathcal{C}(\pi_t, \tau) \tag{10}$$

where $\pi_t \in \mathbb{R}^K$ indicates the decision evidence at time step t and τ is the super-parameter for temperature. In practice, we use Exponential Concrete distribution for more stable calculation of logarithm probability. We sample $\tilde{z}_t$ from $q(z_t)$ using Gumbel Softmax trick [25] for gradient back-propagation. The distribution function and calculation of log-probability will be provided in supplementary material.

Feature Parametrization. To better implement the reasoning process, we reformulate the image representation as a graph, where $\mathbf{v}_n \in \mathbb{R}^{K \times d_n}$ are node features indicating K objects, and $\mathbf{v}_r \in \mathbb{R}^{K\times K\times d_r}$ denote edge features of relations between nodes. Moreover, regarding every question word as a state is time-consuming and hard to ground in $\mathbf{v}$. We extract noun phrases from the input question by open-sourced spaCy as *entity phrases*, while the phrase between noun chunks as *transition phrases*. Parsing question by phrases can obtain more semantic information and save computational cost. Following [37], this phrase representation is encoded by Bi-directional Gated Recurrent Unit (BiGRU) to capture the context information.

$$\begin{gathered}(\overrightarrow{w_{1:L}}; \overleftarrow{w_{1:L}}) = \text{Bi-GRU}(\mathbf{x}) \\ h_e^t = [\overrightarrow{w_{e_t}}; \overleftarrow{w_{s_t}}],\ h_r^t = [\overrightarrow{w_{s_{t+1}}}; \overleftarrow{w_{e_t}}],\ h_e^T = [\overrightarrow{w_L}; \overleftarrow{w_0}]\end{gathered} \tag{11}$$

where e_t and s_t are start and end location of the t-th entity phrase, $h_e^t \in \mathbb{R}^{d_q}$ and $h_r^t \in \mathbb{R}^{d_q}$ denote entity embeddings and transition embeddings at time step t. We treat the global GRU output $[\overrightarrow{w_L}; \overleftarrow{w_0}]$ as the last entity embedding h_e^T. Since the number of entity phrases are much smaller than that of the question words, parsing by phrase not only largely save computational cost but obtain more semantic information that can be grounded in the image as well.

Implementation Details. We implement TRN as an injected module to both classical one stage method Bottom-up Top-down Attention (UpDn) [1] and widely used implicit reasoning method Bilinear Attention Network (BAN) [31]. A single Temporal Reasoning cell is divided into two cells, *i.e.*, *Entity Cell* and *Trans Cell.* As shown in Fig. 2(b), Entity Cell models the State Inference $q(z_t|h_t, \mathbf{v})$ and Generative Reconstruction term $\log p(x_t|z_t, \mathbf{v})$. Trans Cell infers State Transition term $p(z_t|z_{t-1}, \mathbf{v})$ for the next time block, which is modelled as one-step Markovian transition on the graph. More implement details can be found in Sect. 3.4 and Algorithm 1 in supplementary material.

It can be seen that the only connection between each reasoning block is just the K-L Divergence of $p(z_t|z_{t-1}, \mathbf{v})$ and $q(z_t|h_t, \mathbf{v})$. For a fair comparison with baseline, no more extra fusion strategies are used in TRN. Therefore, the whole TRN can be regarded as a regularization term in the training stage and can be removed in test stage, which would not bring extra computational cost compared to original methods.

4 Experiments

4.1 Datasets

VQA 2.0 is a commonly used VQA dataset composed of real-world images from MSCOCO with the same train/validation/test splits. Following previous works, we take the answers that appeared more than 9 times in the training set as candidate answers, which produces 3129 answer candidates.

CLEVR is a synthetic dataset, consisting of visual scenes with simple geometric shapes with complicated relational questions like "*What size is the cylinder that*

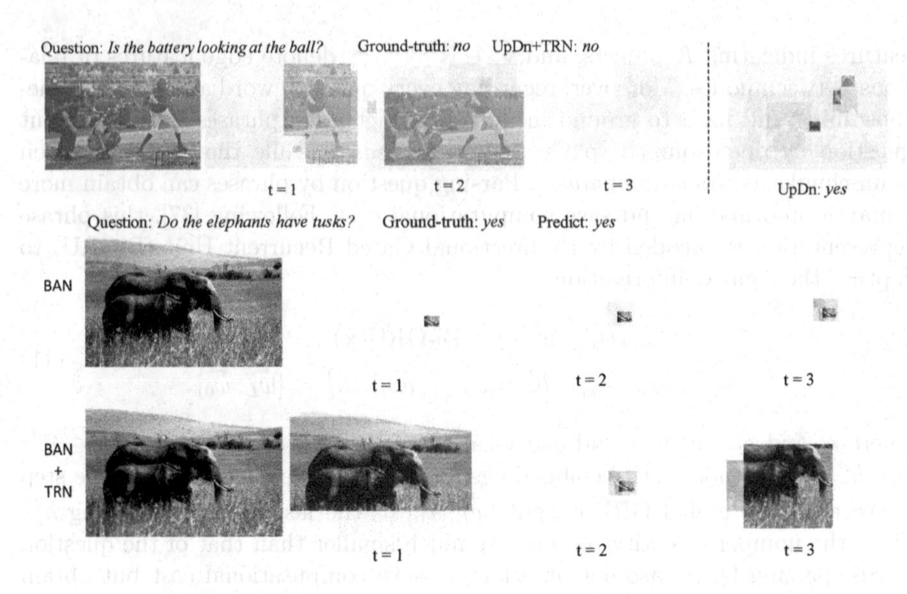

Fig. 3. Examples from VQA v2.0 val, which visualize the change of latent states across TRN Blocks. The highlighted regions are the most important areas that tend to be sampled. The upper example is from UpDn+TRN, the model first finds the batter and then searches the area that he is looking at (no balls). In practice, our model can directly output $t = 3$ and the answer *no*. UpDn gives a wrong answer and attends at the bat. The lower example is from BAN+TRN. The model first finds the elephants and tusks, then finds elephants with tusks. Without dependence between stacked modules, the attention maps of BAN almost remain unchanged. Overlooking the elephants and directly locating tusks may lead to over-fitting on dataset bias and be prone to fail in queries like "tigers with tusks". More examples can be found in the Supplementary

is left of the brown metal thing that is left of the big sphere?". It is the most commonly used dataset for visual reasoning that requires the model's long-chain reasoning ability. Since this work focuses on VQA under natural supervision, we only use question-answer pairs annotation in CLEVR to evaluate our reasoning ability in complicated questions.

CLEVR Human dataset consists of human-generated questions for CLEVR images, which can test the model generalization for real-world questions, since all the questions are generated from programs in the original dataset.

4.2 Evaluation on Real-World Datasets

Experiment Settings. We use the object proposal feature provided by [1]. The node features $\mathbf{v}_n \in \mathbb{R}^{36\times 2048}$ consists of 36 objects features, each object feature v_i is a local visual feature vector $o_i \in \mathbb{R}^{2048}$ extracted by Faster R-CNN [46]. The edge features $\mathbf{v}_r \in \mathbb{R}^{36\times 36\times 1024}$ are concatenation of corresponding node feature transforming to 1024-dim. For question $\mathbf{x}$, each word x_t is first

Table 1. Model accuracy on the VQA v2.0 benchmark (open-ended setting on the test-dev and test-std split). Methods with * are reimplemented by ourselves. Bold colour highlights the best performance, and italic numbers are the second place

Method	Test-dev				Test-std
	Y/N	Num.	Other	All	
MUTAN [6]	82.88	44.54	56.50	66.01	66.38
MuRel [8]	*84.77*	49.84	57.85	68.03	68.41
Dyna Tree [52]	84.28	47.78	59.11	68.19	68.49
DFAF [14]	**86.09**	**53.53**	**60.49**	**70.22**	**70.34**
QCG [43]	82.91	47.13	56.22	56.45	66.17
RAMEN [50]	–	–	–	–	65.96
UpDn* [1]	82.64	45.51	57.21	65.82	65.91
BAN-3* [31]	84.68	*50.71*	58.56	*68.43*	68.47
UpDn + TRN	83.83	45.61	57.44	67.00	67.21
BAN-3 + TRN	84.59	50.23	*58.64*	68.38	*68.76*

initialized by 300-dim GloVe word embeddings [44], then fed into a Bi-GRU. The final representation h_e^t and h_r^t are 1024-dim phrase embeddings with context information. The number of Temporal Reasoning Blocks is set as 3.

Comparison with Baseline Models. Since real-world dataset VQA v2.0 does not contain too many questions that need reasoning, our TRN only sightly improves the performance compared to baselines , as shown in Table 1. The primary function of TRN in VQA v2.0 is to obtain an interpretable reasoning process.

As shown in Fig. 3, one-stage models like UpDn does not perform well on questions considering multiple objects, our model can deal with this drawback and improve the performance on questions that require reasoning Despite integrate multi-step attention, the stacked attentions in BAN are almost independent. Although attending to the right objects, it cannot provide explainable evidence for reasoning. With the help of TRN, the visualization of sampling in latent spaces from BAN+TRN are closer to human understanding.

Comparison with Other Methods. We compare with two recently proposed reasoning models with multi-step attention and fusion strategy (MUTAN [6], MuRel [8]), three models focus on relation reasoning (QCG [43], Dyna Tree [52], DFAF [14]) and RAMEN [50] that claims to work on both real-world and synthetic datasets. All methods are trained with both train and validation split without model ensemble. Our method achieves comparable performance and BAN-3+TRN is the second place in VQA v2.0 test-std split.

Among these methods, DFAF uses at most 100 region proposals and achieves the best single model performance[2]. MuRel is a variant of BAN, and the

[2] Our reproduction with 36 proposals only gets 67.69% accuracy on test-std.

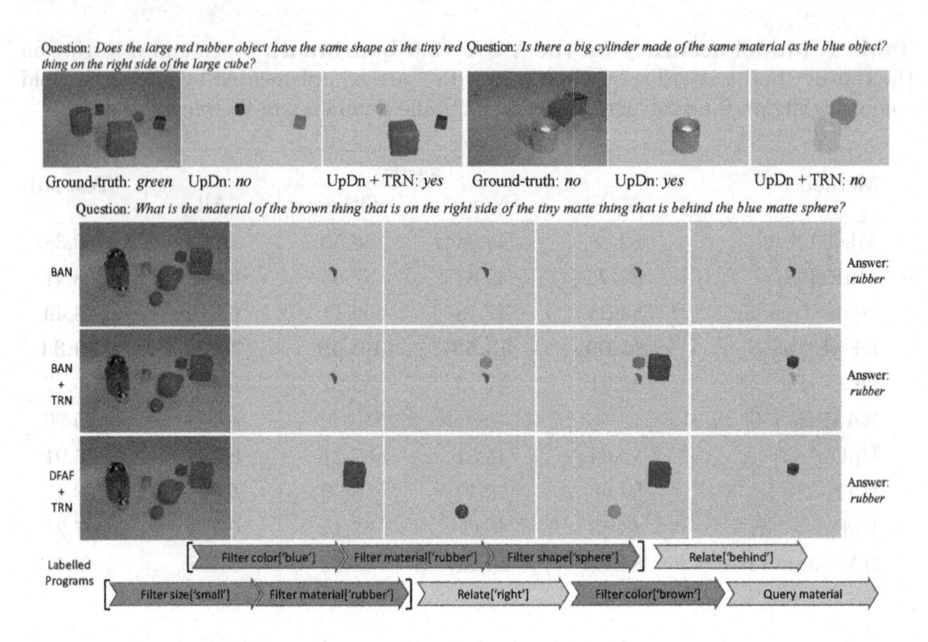

Fig. 4. Examples from our TRN on validation split of CLEVR and CLEVR-Humans. The upper row shows the last latent state between UpDn and UpDn+TRN. The latter two examples are reasoning process visualization of BAN, BAN+TRN, and DFAF+TRN. The depth of colour indicates sample value from latent space. Though outputting the same answers, BAN provides the right answer but wrong evidence, while our model can catch up with the right piece of evidences that is much closer to labelled programs. DFAF+TRN can better locate informative objects, but the time dependence in reasoning process is relatively worse due to strong fusion strategy

reasoning process can also be visualized. The major module DynamicIntraMAF in DFAF and Pairwise module proposed in MuRel play similarily as the State Transition term in TRN. Our probabilistic formulation can help enhance their interpretability with Bayesian transparency.

4.3 Evaluation on Synthetic Datasets

Experiment Settings. For CLEVR dataset, The node representations $\mathbf{v}_n \in \mathbb{R}^{15\times 18}$ include at most 15 proposal features, each object feature v_i is an 18-dim output of object attribute extractor provided in NS-VQA [58], which refers to the shape, colour, material, and 3-dim coordinate position of the proposal object. Since this representation contains enough semantics and the relationships between objects are simple, the edge feature $v_r^{i,j} \in \mathbb{R}^{18}$ are just defined as $v_n^i - v_n^j$. Question $\mathbf{x}$ is randomly initialized, and we get final phrase representations $h_t \in \mathbb{R}^{600}$ after the same operations with real-world setting. The temperature of Concrete distribution is 2.0. The maximum entity number is set to be 5 in experiments. For CLEVR, we train on CLEVR train split and test on validation split. For CLEVR-Humans, we first pre-train our model on CLEVR train split, and then fine-tune on CLEVR-Humans train split.

Table 2. Model accuracy comparison on CLEVR and CLEVR-Humans. Methods with * is reproduced by ourselves. Bold colour highlights the best performance, and italic numbers are the second place

Method	CLEVR	CLEVR-Humans
Film [45]	97.6%	**75.9%**
RN [48]	95.5%	57.6%
MAC [23]	**98.0%**	50.2%
RAMEN [50]	96.9%	57.8%
LCGN [22]	*97.9%*	–
UpDn*[1]	78.1%	56.6%
BAN-5* [45]	83.1%	60.5%
DFAF-5* [14]	95.5%	63.2%
UpDn + TRN	87.7%	69.5%
BAN-5 + TRN	85.2%	65.4%
DFAF-5 + TRN	96.7%	*72.9%*

Table 3. Ablation study for UpDn+TRN on CLEVR validation set. SI, ST and GR stand for *State Inference*, *State Transition* and *Generative Reconstruction*, respectively. Components in TRN are indivisible. On CLEVR val, UpDn+TRN degrades to original UpDn without State Inference or State Transition. UpDn+TRN without Generative Reconstruction is similar with BAN, which can improve to 82.3% accuracy but poor visualization

SI	ST	GR	CLEVR val
✓		✓	78.1%
	✓	✓	77.6%
✓	✓		84.3%
✓	✓	✓	87.7%

Comparison with Baseline Models. Most questions in CLEVR deal with multiple objects and require long-time reasoning ability. As shown in Table 2, TRN can surpass one-stage baseline UpDn by a large margin on both CLEVR and CLEVR-Humans. With help of the additional reasoning process in TRN, many failure cases in UpDn can be corrected. The visualizations of the last latent state in Fig. 4 further demonstrate the effectiveness of our method. Moreover, TRN can be removed in the test stage, which means this improvement does not require extra computational cost.

Visualization of the reasoning process reveals that BAN does not really understand the question. As shown in Fig. 4, although giving correct answers, BAN cannot provide understandable reasoning process[3], and sometimes even attends to wrong evidences. On the contrary, BAN+TRN can offer more explainable reasoning as visualized which is much closer to labelled programs and human understanding. The visualization is much more apparent than on VQA v2.0 that TRN not only grounds noun phrases in image, but illustrates time-dependent reasoning process. Moreover, the final results of BAN+TRN are worse than UpDn+TRN. We speculate that the major reason may be the high-level features we use for node representation. Since BAN mainly focuses on multi-modal fusion, this 18-dim vector may be too abstracted for multi-modal fusion.

Comparison with Other Methods. We compare our model with four previous works (Film [45], RN [48], MAC [23], LCGN [22]) that do not use any

[3] The object attention is Softmax of the sum of $\mathcal{A}$ along question dimension.

functional program information. Recent proposed method LCGN does not report their performance on CLEVR-Humans.

Due to long-time fusion strategies, most of these methods are better than our model on CLEVR. To verify this, we conduct TRN on DFAF [14], which adopts inter-intra attention fusion strategy across stacked modules. The implementation details can be found in the Supplementary. As shown in Table 2, DFAF-5+TRN achieves 96.7% accuracy, which is comparable to state-of-the-art implicit reasoning models specially designed for CLEVR. This indicates that stronger fusion strategies can help establish more informative latent space, which would be a crucial complement for complicated reasoning process. However, visualization of DFAF+TRN shown in Fig. 4 is not as understandable as BAN+TRN, because complex fusion has overly strong fitting ability that can integrate information without considering time dependences. Moreover, we achieve better performance on CLEVR-Humans. Despite not using labelled programs, models with over-parametrized fusion largely rely on the fixed grammar of input questions, resulting in over-fitting.

4.4 Discussion

Ablation Study. As shown in Table 3, components in TRN are indivisible and derived from an integrated process with question supervision. Moreover, we set the number of blocks for VQA/CLEVR as 3/5 because the number of entity phrases in most questions is no more than 2/4 (with 1 global embedding). A more detailed ablation study is provided in Sect. 4.1 in the Supplementary.

Failure Cases. Failure cases are provided in Sect. 6 in the Supplementary. TRN performs poorly on counting problems. This may be a result of the intrinsic weakness of Concrete distribution, which tends to sample one-hot vector and ignore items that are relatively unimportant. This shortcoming also exists in attention mechanism. It can still be improved by choosing better latent state distributions or specially designed modules. Another shortcoming is in dealing with adverbial problems. Reasoning phrase-by-phrase may fail to catch relationships between entities that are distant in the question. Although stronger fusion strategies can help catch up with enough information, it may break the chain of reasoning process due to overly strong fitting ability. How to model the long-time dependences is a common challenge for latent sequence models. Modelling an interpretable fusion strategy may be the future work that needs to be done.

5 Conclusion

Reasoning in VQA under natural supervision is a very challenging task due to super asymmetric information. In this work, we analyse real-world VQA task from a new perspective, and propose a new probabilistic formulation that can explicitly model the reasoning process without extra program labeling. Experiments on both real-world and synthetic datasets demonstrate our model's effectiveness and interpretability. We hope such a probabilistic formulation can provide guidance on further advancements in problems with insufficient natural supervision or other tasks that need multi-step programming. In future work, we will devote our efforts to learning interpretable models for complicated vision-language tasks by combining knowledges.

Acknowledgments. This work was supported in part by the National Key R&D Program of China under Grant 2018AAA0102003, in part by National Natural Science Foundation of China: 61672497, 61620106009, 61836002, 61931008 and U1636214, and in part by Key Research Program of Frontier Sciences, CAS: QYZDJ-SSW-SYS013. Authors are grateful to Kingsoft Cloud for support of free GPU cloud computing resource and Yuecong Min for fruitful discussion.

References

1. Anderson, P., He, X., Buehler, C., Teney, D., Johnson, M., Gould, S., Zhang, L.: Bottom-up and top-down attention for image captioning and visual question answering. In: Proceedings of the IEEE Conference on Computer Vision and Pattern Recognition, pp. 6077–6086 (2018)
2. Andreas, J., Rohrbach, M., Darrell, T., Klein, D.: Neural module networks. In: Proceedings of the IEEE Conference on Computer Vision and Pattern Recognition, pp. 39–48 (2016)
3. Antol, S., Agrawal, A., Lu, J., Mitchell, M., Batra, D., Lawrence Zitnick, C., Parikh, D.: VQA: visual question answering. In: Proceedings of the IEEE International Conference on Computer Vision, pp. 2425–2433 (2015)
4. Arrieta, A.B., et al.: Explainable artificial intelligence (XAI): concepts, taxonomies, opportunities and challenges toward responsible AI. Inf. Fus. **58**, 82–115 (2020)
5. Banijamali, E., Shu, R., Ghavamzadeh, M., Bui, H., Ghodsi, A.: Robust locally-linear controllable embedding. arXiv preprint arXiv:1710.05373 (2017)
6. Ben-Younes, H., Cadene, R., Cord, M., Thome, N.: Mutan: multimodal tucker fusion for visual question answering. In: Proceedings of the IEEE International Conference on Computer Vision, pp. 2612–2620 (2017)
7. Buesing, L., et al.: Learning and querying fast generative models for reinforcement learning. arXiv preprint arXiv:1802.03006 (2018)
8. Cadene, R., Ben-Younes, H., Cord, M., Thome, N.: Murel: multimodal relational reasoning for visual question answering. In: Proceedings of the IEEE Conference on Computer Vision and Pattern Recognition, pp. 1989–1998 (2019)
9. Chen, T.Q., Rubanova, Y., Bettencourt, J., Duvenaud, D.K.: Neural ordinary differential equations. In: Advances in Neural Information Processing Systems, pp. 6571–6583 (2018)
10. Chua, K., Calandra, R., McAllister, R., Levine, S.: Deep reinforcement learning in a handful of trials using probabilistic dynamics models. In: Advances in Neural Information Processing Systems, pp. 4754–4765 (2018)
11. Chung, J., Kastner, K., Dinh, L., Goel, K., Courville, A.C., Bengio, Y.: A recurrent latent variable model for sequential data. In: Advances in Neural Information Processing Systems, pp. 2980–2988 (2015)
12. Doerr, A., Daniel, C., Schiegg, M., Nguyen-Tuong, D., Schaal, S., Toussaint, M., Trimpe, S.: Probabilistic recurrent state-space models. arXiv preprint arXiv:1801.10395 (2018)
13. Fukui, A., Park, D.H., Yang, D., Rohrbach, A., Darrell, T., Rohrbach, M.: Multimodal compact bilinear pooling for visual question answering and visual grounding. arXiv preprint arXiv:1606.01847 (2016)

14. Gao, P., Jiang, Z., You, H., Lu, P., Hoi, S.C., Wang, X., Li, H.: Dynamic fusion with intra-and inter-modality attention flow for visual question answering. In: Proceedings of the IEEE Conference on Computer Vision and Pattern Recognition, pp. 6639–6648 (2019)
15. Gao, P., You, H., Zhang, Z., Wang, X., Li, H.: Multi-modality latent interaction network for visual question answering. arXiv preprint arXiv:1908.04289 (2019)
16. Gao, Y., Beijbom, O., Zhang, N., Darrell, T.: Compact bilinear pooling. In: Proceedings of the IEEE Conference on Computer Vision and Pattern Recognition, pp. 317–326 (2016)
17. Gregor, K., Papamakarios, G., Besse, F., Buesing, L., Weber, T.: Temporal difference variational auto-encoder. arXiv preprint arXiv:1806.03107 (2018)
18. Ha, D., Schmidhuber, J.: World models. arXiv preprint arXiv:1803.10122 (2018)
19. Hafner, D., Lillicrap, T., Fischer, I., Villegas, R., Ha, D., Lee, H., Davidson, J.: Learning latent dynamics for planning from pixels. arXiv preprint arXiv:1811.04551 (2018)
20. Hu, R., Andreas, J., Darrell, T., Saenko, K.: Explainable neural computation via stack neural module networks. In: Ferrari, V., Hebert, M., Sminchisescu, C., Weiss, Y. (eds.) ECCV 2018. LNCS, vol. 11211, pp. 55–71. Springer, Cham (2018). https://doi.org/10.1007/978-3-030-01234-2_4
21. Hu, R., Andreas, J., Rohrbach, M., Darrell, T., Saenko, K.: Learning to reason: end-to-end module networks for visual question answering. In: Proceedings of the IEEE International Conference on Computer Vision (ICCV) (2017)
22. Hu, R., Rohrbach, A., Darrell, T., Saenko, K.: Language-conditioned graph networks for relational reasoning. arXiv preprint arXiv:1905.04405 (2019)
23. Hudson, D.A., Manning, C.D.: Compositional attention networks for machine reasoning. In: ICLR (2018)
24. Hudson, D.A., Manning, C.D.: GQA: a new dataset for real-world visual reasoning and compositional question answering. In: Proceedings of the IEEE Conference on Computer Vision and Pattern Recognition, pp. 6700–6709 (2019)
25. Jang, E., Gu, S., Poole, B.: Categorical reparameterization with gumbel-softmax. arXiv preprint arXiv:1611.01144 (2016)
26. Johnson, J., Hariharan, B., van der Maaten, L., Fei-Fei, L., Lawrence Zitnick, C., Girshick, R.: CLEVR: a diagnostic dataset for compositional language and elementary visual reasoning. In: Proceedings of the IEEE Conference on Computer Vision and Pattern Recognition, pp. 2901–2910 (2017)
27. Johnson, J., et al.: Inferring and executing programs for visual reasoning. In: Proceedings of the IEEE International Conference on Computer Vision, pp. 2989–2998 (2017)
28. Johnson, M., Duvenaud, D.K., Wiltschko, A., Adams, R.P., Datta, S.R.: Composing graphical models with neural networks for structured representations and fast inference. In: Advances in Neural Information Processing Systems, pp. 2946–2954 (2016)
29. Johnson, M.J., Duvenaud, D.K., Wiltschko, A., Adams, R.P., Datta, S.R.: Composing graphical models with neural networks for structured representations and fast inference. In: Advances in Neural Information Processing Systems, pp. 2946–2954 (2016)
30. Karl, M., Soelch, M., Bayer, J., Van der Smagt, P.: Deep variational Bayes filters: Unsupervised learning of state space models from raw data. arXiv preprint arXiv:1605.06432 (2016)
31. Kim, J.H., Jun, J., Zhang, B.T.: Bilinear attention networks. In: Advances in Neural Information Processing Systems, pp. 1564–1574 (2018)

32. Kim, J.H., On, K.W., Lim, W., Kim, J., Ha, J.W., Zhang, B.T.: Hadamard product for low-rank bilinear pooling. arXiv preprint arXiv:1610.04325 (2016)
33. Kingma, D.P., Welling, M.: Auto-encoding variational Bayes. arXiv preprint arXiv:1312.6114 (2013)
34. Kipf, T.N., Welling, M.: Semi-supervised classification with graph convolutional networks. arXiv preprint arXiv:1609.02907 (2016)
35. Krishnan, R.G., Shalit, U., Sontag, D.: Structured inference networks for nonlinear state space models. In: Thirty-First AAAI Conference on Artificial Intelligence (2017)
36. Li, L., Gan, Z., Cheng, Y., Liu, J.: Relation-aware graph attention network for visual question answering. arXiv preprint arXiv:1903.12314 (2019)
37. Liu, J., Hockenmaier, J.: Phrase grounding by soft-label chain conditional random field. arXiv preprint arXiv:1909.00301 (2019)
38. Lu, G., Ouyang, W., Xu, D., Zhang, X., Gao, Z., Sun, M.-T.: Deep Kalman filtering network for video compression artifact reduction. In: Ferrari, V., Hebert, M., Sminchisescu, C., Weiss, Y. (eds.) Computer Vision – ECCV 2018. LNCS, vol. 11218, pp. 591–608. Springer, Cham (2018). https://doi.org/10.1007/978-3-030-01264-9_35
39. Lu, J., Yang, J., Batra, D., Parikh, D.: Hierarchical question-image co-attention for visual question answering. In: Advances In Neural Information Processing Systems, pp. 289–297 (2016)
40. Maddison, C.J., Mnih, A., Teh, Y.W.: The concrete distribution: a continuous relaxation of discrete random variables. arXiv preprint arXiv:1611.00712 (2016)
41. Mao, J., Gan, C., Kohli, P., Tenenbaum, J.B., Wu, J.: The neuro-symbolic concept learner: interpreting scenes, words, and sentences from natural supervision. arXiv preprint arXiv:1904.12584 (2019)
42. Mascharka, D., Tran, P., Soklaski, R., Majumdar, A.: Transparency by design: closing the gap between performance and interpretability in visual reasoning. In: Proceedings of the IEEE Conference on Computer Vision and Pattern Recognition, pp. 4942–4950 (2018)
43. Norcliffe-Brown, W., Vafeias, S., Parisot, S.: Learning conditioned graph structures for interpretable visual question answering. In: Advances in Neural Information Processing Systems, pp. 8334–8343 (2018)
44. Pennington, J., Socher, R., Manning, C.: Glove: global vectors for word representation. In: Proceedings of the 2014 Conference on Empirical Methods in Natural Language Processing (EMNLP), pp. 1532–1543 (2014)
45. Perez, E., Strub, F., De Vries, H., Dumoulin, V., Courville, A.: Film: visual reasoning with a general conditioning layer. In: Thirty-Second AAAI Conference on Artificial Intelligence (2018)
46. Ren, S., He, K., Girshick, R., Sun, J.: Faster R-CNN: towards real-time object detection with region proposal networks. In: Advances in Neural Information Processing Systems, pp. 91–99 (2015)
47. Rezende, D.J., Mohamed, S., Wierstra, D.: Stochastic backpropagation and approximate inference in deep generative models. arXiv preprint arXiv:1401.4082 (2014)
48. Santoro, A., Raposo, D., Barrett, D.G., Malinowski, M., Pascanu, R., Battaglia, P., Lillicrap, T.: A simple neural network module for relational reasoning. In: Advances in Neural Information Processing Systems, pp. 4967–4976 (2017)
49. Shi, J., Zhang, H., Li, J.: Explainable and explicit visual reasoning over scene graphs. In: Proceedings of the IEEE Conference on Computer Vision and Pattern Recognition, pp. 8376–8384 (2019)

50. Shrestha, R., Kafle, K., Kanan, C.: Answer them all! toward universal visual question answering models. In: Proceedings of the IEEE Conference on Computer Vision and Pattern Recognition, pp. 10472–10481 (2019)
51. Tan, Z.X., Soh, H., Ong, D.C.: Factorized inference in deep Markov models for incomplete multimodal time series. arXiv preprint arXiv:1905.13570 (2019)
52. Tang, K., Zhang, H., Wu, B., Luo, W., Liu, W.: Learning to compose dynamic tree structures for visual contexts. In: The IEEE Conference on Computer Vision and Pattern Recognition (CVPR), June 2019
53. Vedantam, R., Desai, K., Lee, S., Rohrbach, M., Batra, D., Parikh, D.: Probabilistic neural-symbolic models for interpretable visual question answering. arXiv preprint arXiv:1902.07864 (2019)
54. Wang, P., Wu, Q., Cao, J., Shen, C., Gao, L., Hengel, A.V.D.: Neighbourhood watch: referring expression comprehension via language-guided graph attention networks. In: Proceedings of the IEEE Conference on Computer Vision and Pattern Recognition, pp. 1960–1968 (2019)
55. Wei, K., Yang, M., Wang, H., Deng, C., Liu, X.: Adversarial fine-grained composition learning for unseen attribute-object recognition. In: Proceedings of the IEEE International Conference on Computer Vision, pp. 3741–3749 (2019)
56. Xu, K., Ba, J., Kiros, R., Cho, K., Courville, A., Salakhudinov, R., Zemel, R., Bengio, Y.: Show, attend and tell: Neural image caption generation with visual attention. In: International Conference on Machine Learning, pp. 2048–2057 (2015)
57. Yang, Z., He, X., Gao, J., Deng, L., Smola, A.: Stacked attention networks for image question answering. In: Proceedings of the IEEE Conference on Computer Vision and Pattern Recognition, pp. 21–29 (2016)
58. Yi, K., Wu, J., Gan, C., Torralba, A., Kohli, P., Tenenbaum, J.: Neural-symbolic VQA: disentangling reasoning from vision and language understanding. In: Advances in Neural Information Processing Systems, pp. 1031–1042 (2018)
59. Yu, Z., Yu, J., Xiang, C., Fan, J., Tao, D.: Beyond bilinear: generalized multimodal factorized high-order pooling for visual question answering. IEEE Trans. Neural Netw. Learn. Syst. **29**(12), 5947–5959 (2018)

Domain Adaptive Semantic Segmentation Using Weak Labels

Sujoy Paul[1(✉)], Yi-Hsuan Tsai[2], Samuel Schulter[2], Amit K. Roy-Chowdhury[1], and Manmohan Chandraker[2,3]

[1] UC Riverside, Riverside, USA
spaul003@ucr.edu
[2] NEC Labs America, Princeton, USA
[3] UC San Diego, San Diego, USA

Abstract. Learning semantic segmentation models requires a huge amount of pixel-wise labeling. However, labeled data may only be available abundantly in a domain different from the desired target domain, which only has minimal or no annotations. In this work, we propose a novel framework for domain adaptation in semantic segmentation with image-level weak labels in the target domain. The weak labels may be obtained based on a model prediction for unsupervised domain adaptation (UDA), or from a human annotator in a new weakly-supervised domain adaptation (WDA) paradigm for semantic segmentation. Using weak labels is both practical and useful, since (i) collecting image-level target annotations is comparably cheap in WDA and incurs no cost in UDA, and (ii) it opens the opportunity for category-wise domain alignment. Our framework uses weak labels to enable the interplay between feature alignment and pseudo-labeling, improving both in the process of domain adaptation. Specifically, we develop a weak-label classification module to enforce the network to attend to certain categories, and then use such training signals to guide the proposed category-wise alignment method. In experiments, we show considerable improvements with respect to the existing state-of-the-arts in UDA and present a new benchmark in the WDA setting. Project page is at http://www.nec-labs.com/~mas/WeakSegDA.

1 Introduction

Unsupervised domain adaptation (UDA) methods for semantic segmentation have been developed to tackle the issue of domain gap. Existing methods aim to adapt a model learned on the source domain with pixel-wise ground truth annotations, e.g., from a simulator which requires the least annotation efforts, to the target domain that does not have any form of annotations. These UDA methods in the literature for semantic segmentation are developed mainly using two mechanisms: pseudo label self-training and distribution alignment between the source and target domains. For the first mechanism, pixel-wise pseudo labels are

Electronic supplementary material The online version of this chapter (https://doi.org/10.1007/978-3-030-58545-7_33) contains supplementary material, which is available to authorized users.

© Springer Nature Switzerland AG 2020
A. Vedaldi et al. (Eds.): ECCV 2020, LNCS 12354, pp. 571–587, 2020.
https://doi.org/10.1007/978-3-030-58545-7_33

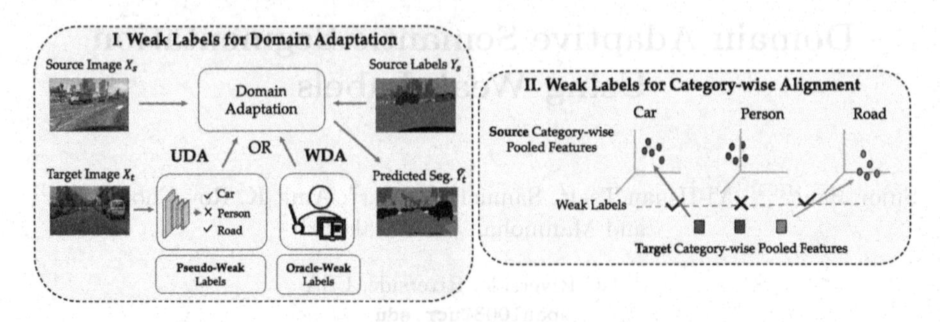

Fig. 1. Our work introduces two key ideas to adapt semantic segmentation models across domains. I: Using image-level weak annotations for domain adaptation, either estimated, i.e., pseudo-weak labels (Unsupervised Domain Adaptation, UDA) or acquired from a human oracle (Weakly-supervised Domain Adaptation (WDA). II: We utilize weak labels to improve the category-wise feature alignment between the source and target domains. ✓/✗ depicts weak labels, i.e., the categories present/absent in an image.

generated via strategies such as confidence scores [23,29] or self-paced learning [60], but such pseudo-labels are specific to the target domain, and do not consider alignment between domains. For the second mechanism, numerous spaces could be considered to operate the alignment procedure, such as pixel [21,35], feature [22,57], output [9,49], and patch [50] spaces. However, alignment performed by these methods are agnostic to the category, which may be problematic as the domain gap may vary across categories.

To alleviate the issue of lacking annotations in the target domain, we propose a concept of utilizing *weak labels* on the domain adaptation task for semantic segmentation, in the form of image- or point-level annotations in the target domain. Such weak labels can be used for category-wise alignment between the source and target domain, and also to enforce constraints on the categories present in an image. It is important to note that our weak labels could be estimated from the model prediction in the UDA setting, or provided by the human oracle in the weakly-supervised domain adaptation (WDA) paradigm (see left of Fig. 1). We are the first to introduce the WDA setting for semantic segmentation with image-level weak-labels, which is practically useful as collecting such annotations is much easier than pixel-wise annotations on the target domain. Benefiting from the concept of weak labels introduced in this paper, we aim to utilize such weak labels to act as an enabler for the interplay between the alignment and pseudo labeling procedures, as they are much less noisy compared to pixel-wise pseudo labels. Specifically, we use weak labels to perform both 1) image-level classification to identify the presence/absence of categories in an image as a regularization, and 2) category-wise domain alignment using such categorical labels. For the image-level classification task, weak labels help our model obtain a better pixel-wise attention map per category. Then, we utilize the category-wise attention maps as the guidance to further pool category-wise features for proposed domain alignment procedure (right of Fig. 1). Note that, although weak labels have been used in domain adaptation for object detection [24], our motivation is different from theirs. More specifically, [24] uses the weak labels to choose pseudo labels for self-training, while we formulate a general framework to learn from weak labels with different forms, i.e., UDA and WDA (image-level or point supervision), as well as to improve feature alignment across domains using weak labels. We conduct experiments on the road scene segmentation problem from GTA5 [39]/SYNTHIA [40] to Cityscapes [11]. We perform extensive experiments to verify the usefulness of each component in the proposed framework, and show that our approach performs favorably against state-of-the-art algorithms for UDA. In addition, we show that our proposed method can be used for WDA and present its experimental results as a new benchmark. For the WDA setting, we also show that our method can incorporate various types of weak labels, such as image-level or point supervision. The **main contributions** of our work are: 1) we propose a concept of using weak labels to help domain adaptation for semantic segmentation; 2) we utilize weak labels to improve category-wise alignment for better feature space adaptation; and 3) we demonstrate that our method is applicable to both UDA and WDA settings.

2 Related Work

In this section, we discuss the literature of unsupervised domain adaptation (UDA) for image classification and semantic segmentation. In addition, we also discuss weakly-supervised methods for semantic segmentation.

UDA for Image Classification. The UDA task for image classification has been developed via aligning distributions across source and target domains. To this end, hand-crafted features [15,18] and deep features [16,51] have been considered to minimize the domain discrepancy and learn domain-invariant features. To further enhance the alignment procedure, maximum mean discrepancy [32] and adversarial learning [17,52] based approaches have been proposed. Recently, several algorithms focus on improving deep models [13,28,33,41], combining distance metric learning [44,45], utilizing pixel-level adaptation [3,47], or incorporating active learning [46].

UDA for Semantic Segmentation. Existing UDA methods in literature for semantic segmentation can be categorized primarily into to two groups: domain alignment and pseudo-label self-training. For domain alignment, numerous algorithms focus on aligning distributions in the pixel [4,10,21,35,56,58], feature [7,22,57], and output [9,49] spaces. For pseudo-label re-training, current methods [30,43,60] aim to generate pixel-wise pseudo labels on the target images, which is utilized to finetune the segmentation model trained on the source domain.

To achieve better performance, recent works [14,29,50,54] attempt to combine the above two mechanisms. AdvEnt [54] adopts adversarial alignment and self-training in the entropy space, while BDL [29] combines output space and pixel-level adaptation with pseudo-label self-training in an iterative updating scheme. Moreover, Tsai et al. [50] propose a patch-level alignment method and show that their approach is complementary to existing modules such as output space adaptation and pseudo-label self-training. Similarly, Du et al. [14] integrate category-wise adversarial alignment with pixel-wise pseudo-labels, which may be noisy, leading to incorrect alignment. In addition, [14] needs to progressively change a ratio for selecting pseudo-labels, and the final performance is sensitive to this chosen parameter.

Compared to the above-mentioned approaches, we propose to exploit weak labels by learning an image classification task, while improving domain alignment through category-wise attention maps. Furthermore, we show that our approach can be utilized even in the case where oracle-weak labels are available on the target domain, in which case the performance will be further improved.

Weakly-Supervised Semantic Segmentation. In this paper, since we are specifically interested in how weak labels can help domain adaptation, we also discuss the literature for weakly-supervised semantic segmentation, which has been tackled through different types of weak labels, such as image-level [1,5,27,37,38], video-level [8,48,59], bounding box [12,25,36], scribble [31,53], and point [2] supervisions. Under this setting, these methods train the model using ground truth weak labels and perform testing in the same domain, which does not require domain adaptation. In contrast, we use a source domain with pixel-wise ground truth labels, but in the target domain, we consider pseudo-weak labels (UDA) or oracle-weak labels (WDA). As a result, we note that performance of weakly-supervised semantic segmentation methods which do not utilize any source domain, is usually much lower than the domain adaptation setting adopted in this paper, e.g., the mean IoU on Cityscapes is only 24.9% as shown in [42].

3 Domain Adaptation with Weak Labels

In this section, we first introduce the problem and then describe details of the proposed framework - the image-level classification module and category-wise alignment method using weak labels. Finally, we present our method of obtaining the weak labels for the UDA and WDA settings.

3.1 Problem Definition

In the source domain, we have images and pixel-wise labels denoted as $\mathcal{I}_s = \{X_s^i, Y_s^i\}_{i=1}^{N_s}$. Whereas, our target dataset contains images and only image-level labels as $\mathcal{I}_t = \{X_t^i, y_t^i\}_{i=1}^{N_t}$. Note that $X_s, X_t \in \mathbb{R}^{H\times W\times 3}$, $Y_s \in \mathbb{B}^{H\times W\times C}$ with pixel-wise one-hot vectors, $y_t \in \mathbb{B}^C$ is a multi-hot vector representing the categories present in the image and C is the number of categories, same for both the source and target datasets. Such image-level labels y_t are often termed as weak labels. We can either estimate them, in which case we call them pseudo-weak labels (Unsupervised Domain Adptation, UDA) or acquire them from a human

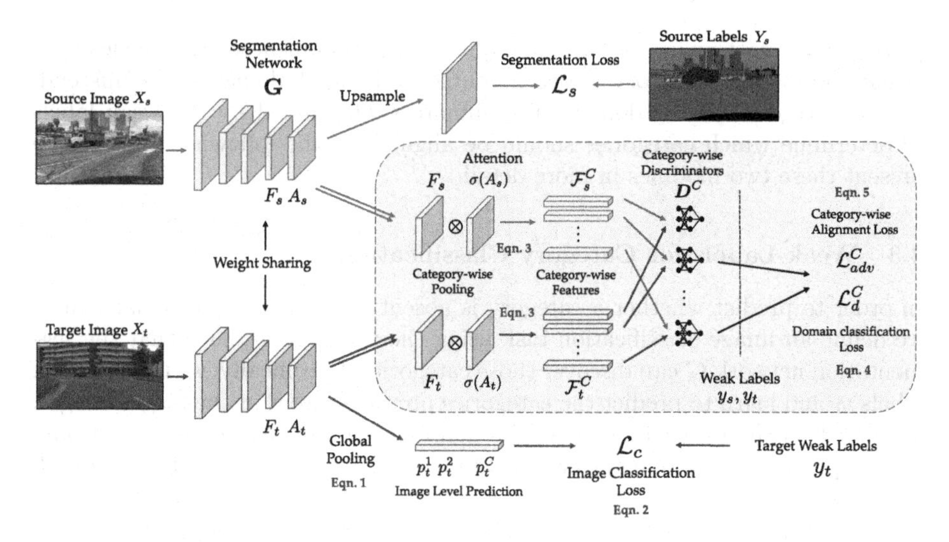

Fig. 2. The proposed architecture consists of the segmentation network $\boldsymbol{G}$ and the weak label module. We compute the pixel-wise segmentation loss $\mathcal{L}_s$ for the source images and image classification loss $\mathcal{L}_c$ using the weak labels y_t for the target images. Note that the weak labels can be estimated as pseudo-weak labels or provided by a human oracle. We then use the output prediction A, convert it to an attention map $\sigma(A)$ and pool category-wise features $\mathcal{F}^C$. Next, these features are aligned between source and target domains using the category-wise alignment loss $\mathcal{L}_{adv}^C$ guided by the category-wise discriminators $\boldsymbol{D}^C$ learned via the domain classification loss $\mathcal{L}_d^C$.

oracle that is called oracle-weak labels (Weakly-supervised Domain Adaptation, WDA). We will further discuss details of acquiring weak labels in Sect. 3.6. Given such data, the problem is to adapt a segmentation model $\boldsymbol{G}$ learned on the source dataset $\mathcal{I}_s$ to the target dataset $\mathcal{I}_t$.

3.2 Algorithm Overview

Figure 2 presents an overview of our proposed method. We first pass both the source and target images through the segmentation network G and obtain their features $F_s, F_t \in \mathbb{R}^{H' \times W' \times 2048}$, segmentation predictions $A_s, A_t \in \mathbb{R}^{H' \times W' \times C}$, and the up-sampled pixel-wise predictions $O_s, O_t \in \mathbb{R}^{H \times W \times C}$. Note that $H'(< H), W'(< W)$ are the downsampled spatial dimensions of the image after passing through the segmentation network. As a baseline, we use the source pixel-wise annotations to learn $\boldsymbol{G}$, while aligning the output space distribution O_s and O_t, following [49].

In addition to having pixel-wise labels on the source data, we also have image-level weak labels on the target data. As discussed before, such weak labels can be either estimated (UDA) or acquired from an oracle (WDA). We then utilize these weak labels to update the segmentation network $\boldsymbol{G}$ in two different ways. First, we introduce a module which learns to predict the categories that are present in a target image. Second, we formulate a mechanism to align the features of each individual category between source and target domains. To this end, we use category-specific domain discriminators $\boldsymbol{D}^c$ guided by the weak labels to determine which categories should be aligned. In the following sections, we present these two modules in more detail.

3.3 Weak Labels for Category Classification

In order to predict whether a category is absent/present in a particular image, we define an image classification task using the weak labels, such that the segmentation network $\boldsymbol{G}$ can discover those categories. Specifically, we use the weak labels y_t and learn to predict the categories present/absent in the target images. We first feed the target images X_t through $\boldsymbol{G}$ to obtain the predictions A_t and then apply a global pooling layer to obtain a single vector of predictions for each category:

$$p_t^c = \sigma_s \left[\frac{1}{k} \log \frac{1}{H'W'} \sum_{h',w'} \exp kA_t^{(h',w',c)} \right], \tag{1}$$

where σ_s is the sigmoid function such that p_t represents the probability that a particular category appears in an image. Note that (1) is a smooth approximation of the `max` function. The higher the value of k, the better it approximates to `max`. We set $k = 1$ as we do not want the network to focus only on the maximum value of the prediction, which may be noisy, but also on other predictions that may have high values. Using p_t and the weak labels y_t, we can compute the category-wise binary cross-entropy loss:

$$\mathcal{L}_c(X_t; \boldsymbol{G}) = \sum_{c=1}^{C} -y_t^c \log(p_t^c) - (1 - y_t^c) \log(1 - p_t^c). \tag{2}$$

This is shown at the bottom stream of Fig. 2. This loss function $\mathcal{L}_c$ helps to identify the categories which are absent/present in a particular image and enforces the segmentation network $\boldsymbol{G}$ to pay attention to those objects/stuff that are partially identified when the source model is used directly on the target images.

3.4 Weak Labels for Feature Alignment

The classification loss using weak labels introduced in (2) regularizes the network focusing on certain categories. However, distribution alignment across the source and target domains is not considered yet. As discussed in the previous section, methods in literature either align feature space [22] or output space [49] across domains. However, such alignment is agnostic to the category, so it may align features of categories that are not present in certain images. Moreover, features belonging to different categories may have different domain gaps. Thereby, performing category-wise alignment could be beneficial but has not been widely studied in UDA for semantic segmentation. Although an existing work [14] attempts to align category-wise features, it utilizes pixel-wise pseudo labels, which may be noisy, and performs alignment in a high-dimensional feature space, which is not only difficult to optimize but also requires more computations (more discussions are provided in the experimental section).

To alleviate all the above issues, we use image-level weak labels to perform category-wise alignment in the feature space. Specifically, we obtain the category-wise features for each image via an attention map, i.e., segmentation prediction, guided by our classification module using weak labels, and then align these features between the source and target domains. We next discuss the category-wise feature pooling mechanism followed by the adversarial alignment technique.

Category-Wise Feature Pooling. Given the last layer features F and the segmentation prediction A, we obtain the category-wise features by using the prediction as an attention over the features. Specifically, we obtain the category-wise feature $\mathcal{F}^c$ as a 2048-dimensional vector for the cth category as follows:

$$\mathcal{F}^c = \sum_{h',w'} \sigma(A)^{(h',w',c)} F^{(h',w')}, \tag{3}$$

where $\sigma(A)$ is a tensor of dimension $H' \times W' \times C$, with each channel along the category dimension representing the category-wise attention obtained by the softmax operation σ over the spatial dimensions. As a result, $\sigma(A)^{(h',w',c)}$ is a scalar and $F^{(h',w')}$ is a 2048-dimensional vector, while $\mathcal{F}^c$ is the summed feature of $F^{(h',w')}$ weighted by $\sigma(A)^{(h',w',c)}$ over the spatial map $H' \times W'$. Note that we drop the subscripts s, t for source and target, as we employ the same operation to obtain the category-wise features for both domains. We next present the mechanism to align these features across domains. Note that we will use $\mathcal{F}^c$ to denote the pooled feature for the cth category and $\mathcal{F}^C$ to denote the set of pooled features for all the categories. Category-wise feature pooling is shown in the middle of Fig. 2.

Category-Wise Feature Alignment. To learn the segmentation network $\boldsymbol{G}$ such that the source and target category-wise features are aligned, we use an adversarial loss while using category-specific discriminators $\boldsymbol{D}^C = \{\boldsymbol{D}^c\}_{c=1}^C$. The reason of using category-specific discriminators is to ensure that the feature distribution for each category could be aligned independently, which avoids the noisy distribution modeling from a mixture of categories. In practice, we train C distinct category-specific discriminators to distinguish between category-wise features drawn from the source and target images. The loss function to train the discriminators $\boldsymbol{D}^C$ is as follows:

$$\mathcal{L}_d^C(\mathcal{F}_s^C, \mathcal{F}_t^C; \boldsymbol{D}^C) = \sum_{c=1}^{C} -y_s^c \log \boldsymbol{D}^c(\mathcal{F}_s^c) - y_t^c \log\left(1 - \boldsymbol{D}^c(\mathcal{F}_t^c)\right). \tag{4}$$

Note that, while training the discriminators, we only compute the loss for those categories which are present in the particular image via the weak labels

$y_s, y_t \in \mathbb{B}^C$ that indicate whether a category occurs in an image or not. Then, the adversarial loss for the target images to train the segmentation network $\boldsymbol{G}$ can be expressed as follows:

$$\mathcal{L}_{adv}^{C}(\mathcal{F}_t^C; \boldsymbol{G}, \boldsymbol{D}^C) = \sum_{c=1}^{C} -y_t^c \log \boldsymbol{D}^c(\mathcal{F}_t^c). \tag{5}$$

Similarly, we use the target weak labels y_t to align only those categories present in the target image. By minimizing $\mathcal{L}_{adv}^{C}$, the segmentation network tries to fool the discriminator by maximizing the probability of the target category-wise feature being considered as drawn from the source distribution. These loss functions in (4) and (5) are obtained in the right of the middle box in Fig. 2.

3.5 Network Optimization

Discriminator Training. We learn a set of C distinct discriminators for each category c. We use the source and target images to train the discriminators, which learn to distinguish between the category-wise features drawn from either the source or the target domain. The optimization problem to train the discriminator can be expressed as: $\min_{\boldsymbol{D}^C} \mathcal{L}_d^C(\mathcal{F}_s^C, \mathcal{F}_t^C)$. Note that each discriminator is trained only with features pooled specific to that particular category. Therefore, given an image, we only update those discriminators corresponding to those categories which are present in the image and ignore the rest.

Segmentation Network Training. We train the segmentation network with the pixel-wise cross-entropy loss $\mathcal{L}_s$ on the source images, image classification loss $\mathcal{L}_c$ and adversarial loss $\mathcal{L}_{adv}^{C}$ on the target images. We combine these loss functions to learn $\mathbf{G}$ as follows:

$$\min_{G} \mathcal{L}_s(X_s) + \lambda_c \mathcal{L}_c(X_t) + \lambda_d \mathcal{L}_{adv}^{C}(\mathcal{F}_t^C). \tag{6}$$

We follow the standard GAN training procedure [19] to alternatively update $\boldsymbol{G}$ and $\boldsymbol{D}^C$. Note that, computing $\mathcal{L}_{adv}^{C}$ involves the category-wise discriminators $\boldsymbol{D}^C$. Therefore, we fix $\boldsymbol{D}^C$ and backpropagate gradients only for the segmentation network $\boldsymbol{G}$.

3.6 Acquiring Weak Labels

In the above sections, we have proposed a mechanism to utilize image-level weak labels of the target images and adapt the segmentation model between source and target domains. In this section, we explain two methods to obtain such image-level weak labels.

Pseudo-Weak Labels (UDA). One way of obtaining weak labels is to directly estimate them using the data we have, i.e., source images/labels and target images, which is the unsupervised domain adaptation (UDA) setting. In this work, we utilize the baseline model [49] to adapt a model learned from the source to the target domain, and then obtain the weak labels of the target images as follows:

$$y_t^c = \begin{cases} 1, & \text{if } p_t^c > T, \\ 0, & \text{otherwise} \end{cases} \tag{7}$$

where p_t^c is the probability for category c as computed in (1) and T is a threshold, which we set to 0.2 in all the experiments unless specified otherwise. In practice, we compute the weak labels online during training and avoid any additional inference step. Specifically, we forward a target image, obtain the weak labels using (7), and then compute the loss functions in (6). As the weak labels obtained in this manner do not require human supervision, adaptation using such labels is unsupervised.

Oracle-Weak Labels (WDA). In this form, we obtain the weak labels by querying a human oracle to provide a list of the categories that occur in the target image. As we use supervision from an oracle on the target images, we refer to this as weakly-supervised domain adaptation (WDA). It is worth mentioning that the WDA setting could be practically useful, as collecting such human annotated weak labels is much easier than pixel-wise annotations. Also, there has not been any prior research involving this setting for domain adaptation.

To show that our method can use different forms of oracle-weak labels, we further introduce the point supervision as in [2], which only increases effort by a small amount compared to the image-level supervision. In this scenario, we randomly obtain one pixel coordinate of each category that belongs in the image, i.e., the set of tuples $\{(h^c, w^c, c) | \forall y_t^c = 1\}$. For an image, we compute the loss as follows: $\mathcal{L}_{point} = -\sum_{\forall y_t^c=1} y_t^c \log(O_t^{(h^c, w^c, c)})$, where $O_t \in \mathbb{R}^{H \times W \times C}$ is the output prediction of target after pixel-wise softmax.

4 Experimental Results

In this section, we perform an evaluation of our domain adaptation framework for semantic segmentation. We present the results for using both pseudo-weak labels, i.e., unsupervised domain adaptation (UDA) and human oracle-weak labels, i.e., weakly-supervised domain adaptation (WDA) and compare it with existing state-of-the-art methods. We also perform ablation studies to analyse the benefit of using pseudo/oracle-weak labels via our proposed weak-label classification module and category-wise alignment.

Datasets and Metric. We evaluate our domain adaptation method under the Sim-to-Real case with two different source-target scenarios. First, we adapt from GTA5 [39] to the Cityscapes dataset [11]. Second, we use SYNTHIA [40] as the source and Cityscapes as the target, which has a larger domain gap than the former case. For all experiments, we use the Intersection-over-Union (IoU) ratio as the metric. For SYNTHIA→Cityscapes, following the literature [54], we report the performance averaged over 16 categories (listed in Table 2) and 13 categories (removing wall, fence and pole), which we denote as mIoU*.

Network Architectures. For the segmentation network G, to have a fair comparison with works in literature, we use the DeepLab-v2 framework [6] with the ResNet-101 [20] architecture. We extract features F_s, F_t before the Atrous Spatial Pyramid Pooling (ASPP) layer. For the category-wise discriminators $D^C = \{D^c\}_{c=1}^C$, we use C separate networks, where each consists of three fully-connected layers, having number of nodes $\{2048, 2048, 1\}$ with ReLU activation.

Training Details. We implement our framework using PyTorch on a single Titan X GPU with 12G memory for all our experiments. We use the SGD method to optimize the segmentation network and the Adam optimizer [26] to train the discriminators. We set the initial learning rates to be 2.5×10^{-4} and 1×10^{-4} for the segmentation network and discriminators, with polynomial decay of power 0.9 [6]. As a common practice in weakly-supervised semantic segmentation [1], we use Dropout of 0.1 and 0.3 for oracle-weak labels and pseudo-weak labels respectively, on the spatial predictions before computing the loss $\mathcal{L}_c$. We choose λ_c to be 0.2 for oracle-weak labels and use a smaller $\lambda_c = 0.01$ for pseudo-weak labels to account for its inaccurate prediction. For the weight on the category-wise adversarial loss $\mathcal{L}_{adv}^C$, we set $\lambda_{adv} = 0.001$. For experiments using pseudo weak labels, to avoid noisy pseudo weak label prediction in the early training stage, we first train the segmentation baseline network using [49] for 60 K iterations. Then, we include the proposed weak-label classification and alignment procedure, and train the entire framework.

4.1 Comparison with State-of-the-art Methods

Unsupervised Domain Adaptation (UDA). We compare our method with existing state-of-the-art UDA methods in Table 1 for GTA5→Cityscapes and in Table 2 for SYNTHIA→Cityscapes. Recent methods [4,21,29,50] show that adapting images from source to target on the pixel level and then adding those translated source images in training enhances the performance. We follow this practice in the final model via adding these adapted images to the source dataset, as their pixel-wise annotations do not change after adaptation. Thus adaptation using weak labels aligns the features not only between the original source and target images, but also between the translated source images and the target images. We show that our method is also complementary to pixel-level adaptation.

Discussions. In terms of applied techniques, e.g, pseudo-label re-training and domain alignment, the closest comparisons to our method are DISE [4], BDL [29], and Patch Space alignment [50]. We show that our method performs favorably against these approaches on both benchmarks. This can be attributed to our introduced concept of using weak labels, in which our UDA model explores pseudo-weak image-level labels, instead of using pixel-level pseudo-labels [29,50] that may be noisy and degrade the performance. In addition, these methods do not perform domain alignment guided by such pseudo labels, whereas we use weak labels to enable our category-wise alignment procedure.

Table 1. Results of adapting GTA5 to Cityscapes. The top group is for UDA, while the bottom group presents our method's performance using the oracle-weak labels for WDA that use either image-level or point supervision.

Method	GTA5 → Cityscapes																			
	Road	Sidewalk	Building	Wall	Fence	Pole	Light	Sign	Veg	Terrain	Sky	Person	Rider	Car	Truck	Bus	Train	Mbike	Bike	mIoU
No Adapt	75.8	16.8	77.2	12.5	21.0	25.5	30.1	20.1	81.3	24.6	70.3	53.8	26.4	49.9	17.2	25.9	6.5	25.3	36.0	36.6
Road [9]	76.3	36.1	69.6	28.6	22.4	28.6	29.3	14.8	82.3	35.3	72.9	54.4	17.8	78.9	27.7	30.3	4.0	24.9	12.6	39.4
AdaptOutput [49]	86.5	25.9	79.8	22.1	20.0	23.6	33.1	21.8	81.8	25.9	75.9	57.3	26.2	76.3	29.8	32.1	7.2	29.5	32.5	41.4
AdvEnt [54]	89.4	33.1	81.0	26.6	26.8	27.2	33.5	24.7	83.9	36.7	78.8	58.7	30.5	84.8	**38.5**	44.5	1.7	31.6	32.4	45.5
CLAN [34]	87.0	27.1	79.6	27.3	23.3	28.3	35.5	24.2	83.6	27.4	74.2	58.6	28.0	76.2	33.1	36.7	6.7	**31.9**	31.4	43.2
SWD [28]	**92.0**	46.4	82.4	24.8	24.0	**35.1**	33.4	34.2	83.6	30.4	80.9	56.9	21.9	82.0	24.4	28.7	6.1	25.0	33.6	44.5
SSF-DAN [14]	90.3	38.9	81.7	24.8	22.9	30.5	37.0	21.2	**84.8**	38.8	76.9	58.8	30.7	**85.7**	30.6	38.1	5.9	28.3	36.9	45.4
DISE [4]	91.5	47.5	82.5	31.3	25.6	33.0	33.7	25.8	82.7	28.8	82.7	**62.4**	**30.8**	85.2	27.7	34.5	6.4	25.2	24.4	45.4
BDL [29]	91.4	47.9	**84.2**	**32.4**	26.0	31.8	37.3	33.0	83.3	**39.2**	79.2	57.7	25.6	81.3	36.3	39.7	2.6	31.3	33.5	47.2
AdaptPatch [50]	92.3	**51.9**	82.1	29.2	25.1	24.5	33.8	33.0	82.4	32.8	82.2	58.6	27.2	84.3	33.4	**46.3**	2.2	29.5	32.3	46.5
Ours (UDA)	91.6	47.4	84.0	30.4	**28.3**	31.4	**37.4**	**35.4**	83.9	38.3	**83.9**	61.2	28.2	83.7	28.8	41.3	**8.8**	24.7	**46.4**	**48.2**
Ours (WDA: Image)	89.5	54.1	83.2	31.7	34.2	37.1	43.2	39.1	85.1	39.6	85.9	61.3	34.1	82.3	42.3	51.9	34.4	33.1	45.4	53.0
Ours (WDA: Point)	94.0	62.7	86.3	36.5	32.8	38.4	44.9	51.0	86.1	43.4	87.7	66.4	36.5	87.9	44.1	58.8	23.2	35.6	55.9	56.4

The only prior work that adopts category-wise feature alignment is SSF-DAN [14]. However, our method is different from theirs in three aspects: 1) We introduce the weak-label classification module to take advantage of image-level weak labels that enables an efficient feature alignment process and the novel WDA setting; 2) Our unified framework can be applied for both UDA and WDA settings with various types of supervisions; 3) Due to the introduced weak-label module, our category-wise feature alignment is operated in the pooled feature space in (3) guided by an attention map, rather than in a much higher-dimensional spatial space as in [14] that uses pixel-wise pseudo-labels. This essentially improves the training efficiency compared to [14], which requires a GPU with 16 GB memory as their discriminator needs much more computation time (>20×) and GPU memory (>8×) compared to our combined output space and category-wise discriminators. Also, the discriminators in [14] require 130 GFLOPS, whereas our discriminators require a total of only 0.5 GFLOPS.

4.2 Weakly-Supervised Domain Adaptation (WDA)

Image-level Supervision. We present the results of our method when using oracle-weak labels (obtained from the ground truth of the training set) in the last rows of Table 1 and 2. To the best of our knowledge, we are the first to work on WDA, i.e., using human oracle-weak labels on domain adaptation for semantic segmentation, and there are no other methods to compare against in the literature. From the results, it is interesting to note that the major boost in performance using WDA compared to UDA occurs for categories such as truck, bus, train, and motorbike for both cases using GTA5 and SYNTHIA as the source domain. One reason is that those categories are most underrepresented in both the source and the target datasets. Thus, they are not predicted in most of the target images, but using the oracle-weak labels helps to identify them better.

Table 2. Results of adapting SYNTHIA to Cityscapes. The top group is for UDA, while the bottom group presents the WDA setting using oracle-weak labels. mIoU and mIoU* are averaged over 16 and 13 categories.

Method	SYNTHIA → Cityscapes																	
	Road	Sidewalk	Building	Wall	Fence	Pole	Light	Sign	Veg	Sky	Person	Rider	Car	Bus	Mbike	Bike	mIoU	mIoU*
No Adapt	55.6	23.8	74.6	9.2	0.2	24.4	6.1	12.1	74.8	79.0	55.3	19.1	39.6	23.3	13.7	25.0	33.5	38.6
AdaptOutput [49]	79.2	37.2	78.8	10.5	0.3	25.1	9.9	10.5	78.2	80.5	53.5	19.6	67.0	29.5	21.6	31.3	39.5	45.9
AdvEnt [54]	85.6	42.2	79.7	8.7	0.4	25.9	5.4	8.1	80.4	84.1	57.9	23.8	73.3	36.4	14.2	33.0	41.2	48.0
CLAN [34]	81.3	37.0	80.1	–	–	–	16.1	13.7	78.2	81.5	53.4	21.2	73.0	32.9	22.6	30.7	–	47.8
SWD [28]	82.4	33.2	**82.5**	–	–	–	**22.6**	**19.7**	**83.7**	78.8	44.0	17.9	75.4	30.2	14.4	39.9	–	48.1
DADA [55]	89.2	44.8	81.4	6.8	0.3	26.2	8.6	11.1	81.8	84.0	54.7	19.3	79.7	**40.7**	14.0	38.8	42.6	49.8
SSF-DAN [14]	84.6	41.7	80.8	–	–	–	11.5	14.7	80.8	**85.3**	57.5	21.6	82.0	36.0	19.3	34.5	–	50.0
DISE [4]	91.7	53.5	77.1	2.5	0.2	**27.1**	6.2	7.6	78.4	81.2	55.8	19.2	**82.3**	30.3	17.1	34.3	41.5	48.8
AdaptPatch [50]	82.4	38.0	78.6	8.7	**0.6**	26.0	3.9	11.1	75.5	84.6	53.5	21.6	71.4	32.6	19.3	31.7	40.0	46.5
Ours (UDA)	**92.0**	**53.5**	80.9	**11.4**	0.4	21.8	3.8	6.0	81.6	84.4	**60.8**	**24.4**	80.5	39.0	**26.0**	**41.7**	**44.3**	**51.9**
Ours (WDA: Image)	92.3	51.9	81.9	21.1	1.1	26.6	22.0	24.8	81.7	87.0	63.1	33.3	83.6	50.7	33.5	54.7	50.6	58.5
Ours (WDA: Point)	94.9	63.2	85.0	27.3	24.2	34.9	37.3	50.8	84.4	88.2	60.6	36.3	86.4	43.2	36.5	61.3	57.2	63.7

Table 3. Ablation of the proposed loss functions for GTA5→Cityscapes.

GTA5 → Cityscapes				
	Supervision	$\mathcal{L}_c$	$\mathcal{L}_{adv}^C$	mIoU
UDA	No Adapt.			36.6
	Baseline [49]			41.4
	Pseudo-Weak	✓		44.2
		✓	✓	**45.6**
WDA	Oracle-Weak	✓		50.8
		✓	✓	**52.1**

Table 4. Ablation of the proposed loss functions for SYNTHIA→Cityscapes.

SYNTHIA → Cityscapes					
	Supervision	$\mathcal{L}_c$	$\mathcal{L}_{adv}^C$	mIoU	mIoU*
UDA	No Adapt.			33.5	38.6
	Baseline [49]			39.5	45.9
	Pseudo-Weak	✓		41.7	49.0
		✓	✓	**42.7**	**49.9**
WDA	Oracle-Weak	✓		47.8	56.0
		✓	✓	**49.2**	**57.2**

Point Supervision. We introduce another interesting setting of point supervision as in [2], which adds only a slight increase of annotation time compared to the image-level supervision. We follow [2] and randomly sample one pixel per category in each target image as the supervision. Note that, all the details and the modules are the same during training in this setting. In Table 1 and 2, the results show that using point supervision improves performance (3.4–6.6%) on both benchmarks compared to the image-level supervision. This shows that our method is a general framework that can be applied to the conventional UDA setting as well as the WDA setting using either image-level or point supervision, while all the settings achieve consistent performance gains.

Figure 3b shows a comparison of annotation time v.s. performance for various levels of supervision. With low annotation cost in WDA cases, our model bridges the gap in performance between UDA and full supervision ones (more results are shown in the supplementary material). Note that, other forms of weak labels such as object count and density can also be effective.

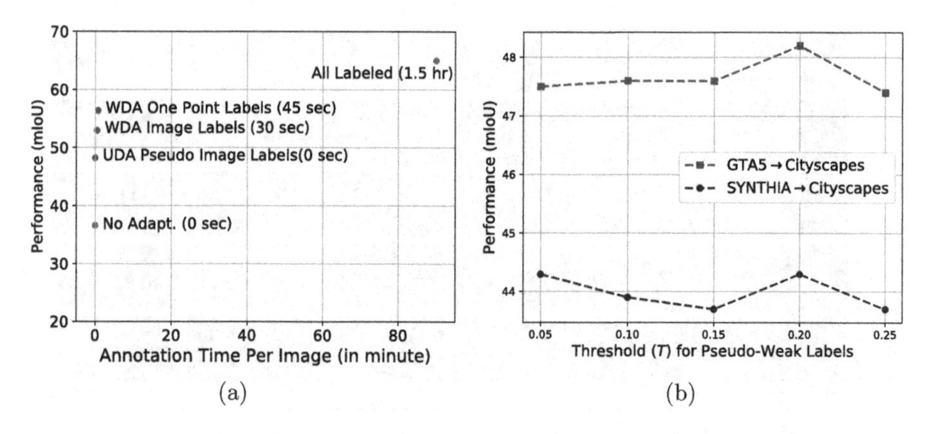

Fig. 3. (a) Performance comparison on GTA5→Cityscapes with different levels of supervision on target images: no target labels ("No Adapt." and "UDA"), weak image labels (30 s), one point labels (45 s), and fully-supervised setting with all pixels labeled ("All Labeled") that takes 1.5 h per image according to [11]. (b) Performance of our method on GTA5→Cityscapes with variations in the threshold, i.e., T in (7), for obtaining the pseudo-weak labels.

4.3 Ablation Study

Effect of Weak Labels. We show results for using both pseudo-weak labels as well as human oracle-weak labels. Table 3 and 4 present the results for different combinations of the modules used in our framework without pixel-level adaptation [21]. It is interesting to note that on GTA5→Cityscapes, even when using pseudo-weak labels, our method obtains a 4.2% boost in performance (41.4 → 45.6), as well as a 3–4% boost for SYNTHIA→Cityscapes. In addition, as expected, using oracle-weak labels performs better than pseudo-weak labels by 6.5% on GTA5→Cityscapes and 6.5–7.3% on SYNTHIA→Cityscapes. It it also interesting to note that using the category-wise alignment consistently improves the performance for all the cases, i.e., different types of weak labels and for different datasets.

Effect of Pseudo-Weak Label Threshold. We use a threshold T in (7) to convert the image-level prediction probability to a multi-hot vector denoting the pseudo-weak labels that indicates absence/presence of the categories. Note that the threshold is on a probability between 0 and 1. We then study the effect of T by varying it and plot the performance in Fig. 3a on GTA5→Cityscapes. The figure shows that our model generally works well with T in a range of 0.05 to 0.25. However, when we make T larger than 0.3, the performance starts to drop significantly, as in this case, the recall of the pseudo-weak labels would be very low compared with the oracle-weak labels (i.e., ground truths), which makes the segmentation network fail to predict most categories.

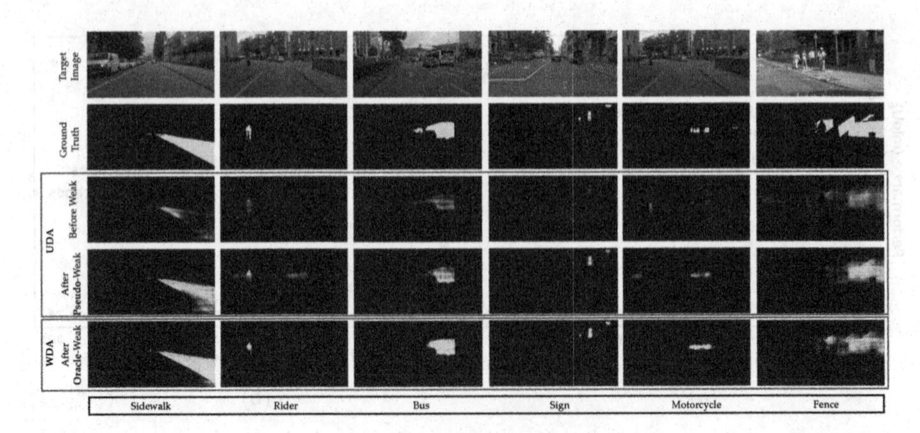

Fig. 4. Visualizations of category-wise segmentation prediction probability before and after using the pseudo-weak labels on GTA5→Cityscapes. Before adaptation, the network only highlights the areas partially with low probability, while using the pseudo-weak labels helps the adapted model obtain much better segments, and is closer to the model using oracle-weak labels.

Output Space Visualization. We present some visualizations of the segmentation prediction probability for each category in Fig. 4. Before using any weak labels (third row), the probabilities may be low, even though there is a category present in that image. However, based on these initial predictions, our model can estimate the categories and then enforce their presence/absence explicitly in the proposed classification loss and alignment loss. The fourth row in Fig. 4 shows that such pseudo-weak labels help the network discover object/stuff regions towards better segmentation. For example, the fourth and fifth column shows that, although the original prediction probabilities are quite low, results using pseudo-weak labels are estimated correctly. Moreover, the last row shows that the predictions can be further improved when we have oracle-weak labels.

5 Conclusions

In this paper, we use weak labels to improve domain adaptation for semantic segmentation in both the UDA and WDA settings, with the latter being a novel setting. Specifically, we design an image-level classification module using weak labels, enforcing the network to pay attention to categories that are present in the image. With such a guidance from weak labels, we further utilize a category-wise alignment method to improve adversarial alignment in the feature space. Based on these two mechanisms, our formulation generalizes both to pseudo-weak and oracle-weak labels. We conduct extensive ablation studies to validate our approach against state-of-the-art UDA approaches.

Acknowledgement. This work was a part of Sujoy Paul's internship at NEC Labs America. This work was also partially funded by NSF grant 1724341.

References

1. Ahn, J., Kwak, S.: Learning pixel-level semantic affinity with image-level supervision for weakly supervised semantic segmentation. In: CVPR (2018)
2. Bearman, A., Russakovsky, O., Ferrari, V., Fei-Fei, L.: What's the point: semantic segmentation with point supervision. In: Leibe, B., Matas, J., Sebe, N., Welling, M. (eds.) ECCV 2016. LNCS, vol. 9911, pp. 549–565. Springer, Cham (2016). https://doi.org/10.1007/978-3-319-46478-7_34
3. Bousmalis, K., Silberman, N., Dohan, D., Erhan, D., Krishnan, D.: Unsupervised pixel-level domain adaptation with generative adversarial networks. In: CVPR (2017)
4. Chang, W.L., Wang, H.P., Peng, W.H., Chiu, W.C.: All about structure: adapting structural information across domains for boosting semantic segmentation. In: CVPR (2019)
5. Chang, Y.T., Wang, Q., Hung, W.C., Piramuthu, R., Tsai, Y.H., Yang, M.H.: Weakly-supervised semantic segmentation via sub-category exploration. In: CVPR (2020)
6. Chen, L.C., Papandreou, G., Kokkinos, I., Murphy, K., Yuille, A.L.: DeepLab: semantic image segmentation with deep convolutional nets, atrous convolution, and fully connected CRFs. CoRR abs/1606.00915 (2016)
7. Chen, Y.H., Chen, W.Y., Chen, Y.T., Tsai, B.C., Wang, Y.C.F., Sun, M.: No more discrimination: cross city adaptation of road scene segmenters. In: ICCV (2017)
8. Chen, Y.W., Tsai, Y.H., Lin, Y.Y., Yang, M.H.: VOSTR: video object segmentation via transferable representations. Int. J. Comput. Vis. (2020)
9. Chen, Y., Li, W., Gool, L.V.: ROAD: reality oriented adaptation for semantic segmentation of urban scenes. In: CVPR (2018)
10. Choi, J., Kim, T., Kim, C.: Self-ensembling with GAN-based data augmentation for domain adaptation in semantic segmentation. In: ICCV (2019)
11. Cordts, M., et al.: The cityscapes dataset for semantic urban scene understanding. In: CVPR (2016)
12. Dai, J., He, K., Sun, J.: BoxSup: exploiting bounding boxes to supervise convolutional networks for semantic segmentation. In: ICCV (2015)
13. Dai, S., Sohn, K., Tsai, Y.H., Carin, L., Chandraker, M.: Adaptation across extreme variations using unlabeled domain bridges. arXiv preprint arXiv:1906.02238 (2019)
14. Du, L., et al.: SSF-DAN: separated semantic feature based domain adaptation network for semantic segmentation. In: ICCV (2019)
15. Fernando, B., Habrard, A., Sebban, M., Tuytelaars, T.: Unsupervised visual domain adaptation using subspace alignment. In: ICCV (2013)
16. Ganin, Y., Lempitsky, V.: Unsupervised domain adaptation by backpropagation. In: ICML (2015)
17. Ganin, Y., et al.: Domain-adversarial training of neural networks. In: JMLR (2016)
18. Gong, B., Shi, Y., Sha, F., Grauman, K.: Geodesic flow kernel for unsupervised domain adaptation. In: CVPR (2012)
19. Goodfellow, I.J., et al.: Generative adversarial nets. In: NIPS (2014)
20. He, K., Zhang, X., Ren, S., Sun, J.: Deep residual learning for image recognition. In: CVPR (2016)
21. Hoffman, J., et al.: CYCADA: cycle-consistent adversarial domain adaptation. In: ICML (2018)

22. Hoffman, J., Wang, D., Yu, F., Darrell, T.: FCNs in the wild: pixel-level adversarial and constraint-based adaptation. CoRR abs/1612.02649 (2016)
23. Hung, W.C., Tsai, Y.H., Liou, Y.T., Lin, Y.Y., Yang, M.H.: Adversarial learning for semi-supervised semantic segmentation. In: BMVC (2018)
24. Inoue, N., Furuta, R., Yamasaki, T., Aizawa, K.: Cross-domain weakly-supervised object detection through progressive domain adaptation. In: Proceedings of the IEEE Conference on Computer Vision and Pattern Recognition, pp. 5001–5009 (2018)
25. Khoreva, A., Benenson, R., Hosang, J., Hein, M., Schiele, B.: Simple does it: weakly supervised instance and semantic segmentation. In: CVPR (2017)
26. Kingma, D.P., Ba, J.: Adam: a method for stochastic optimization. In: ICLR (2015)
27. Kolesnikov, A., Lampert, C.H.: Seed, expand and constrain: three principles for weakly-supervised image segmentation. In: Leibe, B., Matas, J., Sebe, N., Welling, M. (eds.) ECCV 2016. LNCS, vol. 9908, pp. 695–711. Springer, Cham (2016). https://doi.org/10.1007/978-3-319-46493-0_42
28. Lee, C.Y., Batra, T., Baig, M.H., Ulbricht, D.: Sliced Wasserstein discrepancy for unsupervised domain adaptation. In: CVPR (2019)
29. Li, Y., Yuan, L., Vasconcelos, N.: Bidirectional learning for domain adaptation of semantic segmentation. In: CVPR (2019)
30. Lian, Q., Lv, F., Duan, L., Gong, B.: Constructing self-motivated pyramid curriculums for cross-domain semantic segmentation: a non-adversarial approach. In: ICCV (2019)
31. Lin, D., Dai, J., Jia, J., He, K., Sun, J.: ScribbleSup: scribble-supervised convolutional networks for semantic segmentation. In: CVPR (2016)
32. Long, M., Cao, Y., Wang, J., Jordan, M.: Learning transferable features with deep adaptation networks. In: ICML (2015)
33. Long, M., Zhu, H., Wang, J., Jordan, M.I.: Unsupervised domain adaptation with residual transfer networks. In: NIPS (2016)
34. Luo, Y., Zheng, L., Guan, T., Yu, J., Yang, Y.: Taking a closer look at domain shift: category-level adversaries for semantics consistent domain adaptation. In: CVPR (2019)
35. Murez, Z., Kolouri, S., Kriegman, D., Ramamoorthi, R., Kim, K.: Image to image translation for domain adaptation. In: CVPR (2018)
36. Papandreou, G., Chen, L.C., Murphy, K., Yuille, A.L.: Weakly-and semi-supervised learning of a DCNN for semantic image segmentation. In: ICCV (2015)
37. Pathak, D., Krahenbuhl, P., Darrell, T.: Constrained convolutional neural networks for weakly supervised segmentation. In: ICCV (2015)
38. Pinheiro, P.O., Collobert, R.: From image-level to pixel-level labeling with convolutional networks. In: CVPR (2015)
39. Richter, S.R., Vineet, V., Roth, S., Koltun, V.: Playing for data: ground truth from computer games. In: Leibe, B., Matas, J., Sebe, N., Welling, M. (eds.) ECCV 2016. LNCS, vol. 9906, pp. 102–118. Springer, Cham (2016). https://doi.org/10.1007/978-3-319-46475-6_7
40. Ros, G., Sellart, L., Materzynska, J., Vazquez, D., Lopez, A.M.: The SYNTHIA dataset: a large collection of synthetic images for semantic segmentation of urban scenes. In: CVPR (2016)
41. Saito, K., Watanabe, K., Ushiku, Y., Harada, T.: Maximum classifier discrepancy for unsupervised domain adaptation. In: CVPR (2018)
42. Saleh, F.S., Aliakbarian, M.S., Salzmann, M., Petersson, L., Alvarez, J.M.: Bringing background into the foreground: making all classes equal in weakly-supervised video semantic segmentation. In: ICCV (2017)

43. Saleh, F.S., Aliakbarian, M.S., Salzmann, M., Petersson, L., Alvarez, J.M.: Effective use of synthetic data for urban scene semantic segmentation. In: Ferrari, V., Hebert, M., Sminchisescu, C., Weiss, Y. (eds.) ECCV 2018. LNCS, vol. 11206, pp. 86–103. Springer, Cham (2018). https://doi.org/10.1007/978-3-030-01216-8_6
44. Sohn, K., Liu, S., Zhong, G., Yu, X., Yang, M.H., Chandraker, M.: Unsupervised domain adaptation for face recognition in unlabeled videos. In: ICCV (2017)
45. Sohn, K., Shang, W., Yu, X., Chandraker, M.: Unsupervised domain adaptation for distance metric learning. In: ICLR (2019)
46. Su, J.C., Tsai, Y.H., Sohn, K., Liu, B., Maji, S., Chandraker, M.: Active adversarial domain adaptation. In: WACV (2020)
47. Tran, L., Sohn, K., Yu, X., Liu, X., Chandraker, M.: Gotta adapt 'Em all: joint pixel and feature-level domain adaptation for recognition in the wild. In: CVPR (2019)
48. Tsai, Y.-H., Zhong, G., Yang, M.-H.: Semantic co-segmentation in videos. In: Leibe, B., Matas, J., Sebe, N., Welling, M. (eds.) ECCV 2016. LNCS, vol. 9908, pp. 760–775. Springer, Cham (2016). https://doi.org/10.1007/978-3-319-46493-0_46
49. Tsai, Y.H., Hung, W.C., Schulter, S., Sohn, K., Yang, M.H., Chandraker, M.: Learning to adapt structured output space for semantic segmentation. In: CVPR (2018)
50. Tsai, Y.H., Sohn, K., Schulter, S., Chandraker, M.: Domain adaptation for structured output via discriminative patch representations. In: ICCV (2019)
51. Tzeng, E., Hoffman, J., Darrell, T., Saenko, K.: Simultaneous deep transfer across domains and tasks. In: ICCV (2015)
52. Tzeng, E., Hoffman, J., Saenko, K., Darrell, T.: Adversarial discriminative domain adaptation. In: CVPR (2017)
53. Vernaza, P., Chandraker, M.: Learning random-walk label propagation for weakly-supervised semantic segmentation. In: CVPR (2017)
54. Vu, T.H., Jain, H., Bucher, M., Cord, M., Pérez, P.: ADVENT: adversarial entropy minimization for domain adaptation in semantic segmentation. In: CVPR (2019)
55. Vu, T.H., Jain, H., Bucher, M., Cord, M., Pérez, P.: DADA: depth-aware domain adaptation in semantic segmentation. In: ICCV (2019)
56. Wu, Z., et al.: DCAN: dual channel-wise alignment networks for unsupervised scene adaptation. In: ECCV (2018)
57. Zhang, Y., David, P., Gong, B.: Curriculum domain adaptation for semantic segmentation of urban scenes. In: ICCV (2017)
58. Zhang, Y., Qiu, Z., Yao, T., Liu, D., Mei, T.: Fully convolutional adaptation networks for semantic segmentation. In: CVPR (2018)
59. Zhong, G., Tsai, Y.-H., Yang, M.-H.: Weakly-supervised video scene co-parsing. In: Lai, S.-H., Lepetit, V., Nishino, K., Sato, Y. (eds.) ACCV 2016. LNCS, vol. 10111, pp. 20–36. Springer, Cham (2017). https://doi.org/10.1007/978-3-319-54181-5_2
60. Zou, Y., Yu, Z., Kumar, B.V.K.V., Wang, J.: Domain adaptation for semantic segmentation via class-balanced self-training. In: ECCV (2018)

Knowledge Distillation Meets Self-supervision

Guodong Xu[1(✉)], Ziwei Liu[1], Xiaoxiao Li[1], and Chen Change Loy[2]

[1] The Chinese University of Hong Kong, Shatin, Hong Kong
{xg018,zwliu,lx015}@ie.cuhk.edu.hk

[2] Nanyang Technological University, Singapore, Singapore
ccloy@ntu.edu.sg

Abstract. Knowledge distillation, which involves extracting the "dark knowledge" from a teacher network to guide the learning of a student network, has emerged as an important technique for model compression and transfer learning. Unlike previous works that exploit architecture-specific cues such as activation and attention for distillation, here we wish to explore a more general and model-agnostic approach for extracting "richer dark knowledge" from the pre-trained teacher model. We show that the seemingly different self-supervision task can serve as a simple yet powerful solution. For example, when performing contrastive learning between transformed entities, the noisy predictions of the teacher network reflect its intrinsic composition of semantic and pose information. By exploiting the similarity between those self-supervision signals as an auxiliary task, one can effectively transfer the hidden information from the teacher to the student. In this paper, we discuss practical ways to exploit those noisy self-supervision signals with selective transfer for distillation. We further show that self-supervision signals improve conventional distillation with substantial gains under few-shot and noisy-label scenarios. Given the richer knowledge mined from self-supervision, our knowledge distillation approach achieves state-of-the-art performance on standard benchmarks, i.e., CIFAR100 and ImageNet, under both similar-architecture and cross-architecture settings. The advantage is even more pronounced under the cross-architecture setting, where our method outperforms the state of the art by an average of 2.3% in accuracy rate on CIFAR100 across six different teacher-student pairs. The code and models are available at: https://github.com/xuguodong03/SSKD.

1 Introduction

The seminal paper by Hinton et al. [15] show that the knowledge from a large ensemble of models can be distilled and transferred to a student network. Specifically, one can raise the temperature of the final softmax to produce soft targets of the teacher for guiding the training of the student. The guidance is achieved

Electronic supplementary material The online version of this chapter (https://doi.org/10.1007/978-3-030-58545-7_34) contains supplementary material, which is available to authorized users.

© Springer Nature Switzerland AG 2020
A. Vedaldi et al. (Eds.): ECCV 2020, LNCS 12354, pp. 588–604, 2020.
https://doi.org/10.1007/978-3-030-58545-7_34

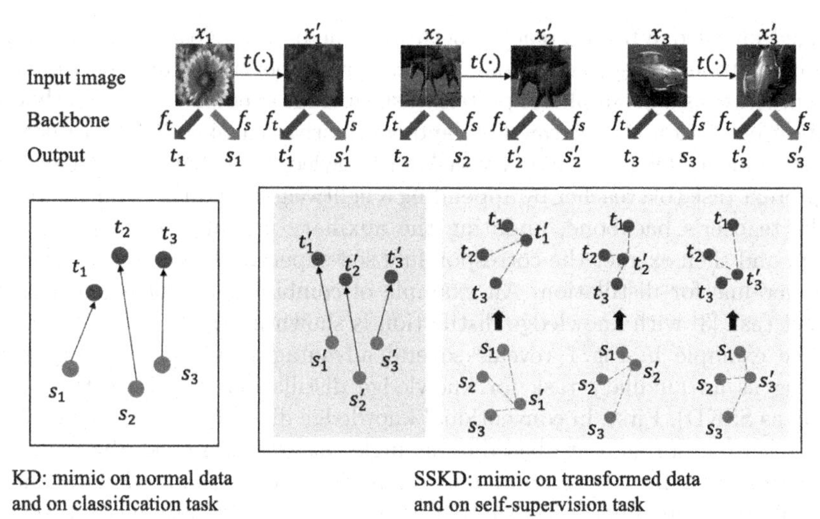

Fig. 1. Difference between conventional KD [15] and SSKD. We extend the mimicking on normal data and on a single classification task to the mimicking on transformed data and with an additional self-supervision pretext task. The teacher's self-supervision predictions contain rich structured knowledge that can facilitate more rounded knowledge distillation on the student. In this example, contrastive learning on transformed images serves as the self-supervision pretext task. It constructs a single positive pair and several negative pairs through image transformations $t(\cdot)$, and then encourages the network to recognize the positive pair. The backbone of the teacher and student are represented as f_t and f_s, respectively, while the corresponding output is given as t and s with subscript representing the index

by minimizing the Kullback-Leibler (KL) divergence between teacher and student outputs. An interesting and inspiring observation is that despite the teacher model assigns probabilities to incorrect classes, the relative probabilities of incorrect answers are exceptionally informative about generalization of the trained model. The hidden knowledge encapsulated in these secondary probabilities is sometimes known as "dark knowledge".

In this work, we are fascinated on how one could extract richer "dark knowledge" from neural networks. Existing studies focus on what types of intermediate representations of teacher networks should student mimic. These representations include feature map [36,37], attention map [44], gram matrix [42], and feature distribution statistics [16]. While the intermediate representations of the network could provide more fine-grained information, a common characteristic shared by these medium of knowledge is that they are all derived from a single task (typically the original classification task). The knowledge is highly task-specific, and hence, such knowledge may only reflect a single facet of the complete knowledge encapsulated in a cumbersome network. To mine for richer "dark knowledge", we need an auxiliary task apart from the original classification task, so as to extract richer information that is complementary to the classification knowledge.

In this study, we show that a seemingly different learning scheme – self-supervised learning, when treated as an auxiliary task, can help gaining more rounded knowledge from a teacher network. The original goal of self-supervised learning is to learn representations with natural supervisions derived from data via a pretext task. Examples of pretext tasks include exemplar-based method [8], rotation prediction [10], jigsaw [29], and contrastive learning [3,26]. To use self-supervised learning as an auxiliary task for knowledge distillation, one can apply the pretext task to a teacher by appending a lightweight auxiliary branch/module to the teacher's backbone, updating the auxiliary module with the backbone frozen, and then extract the corresponding self-supervised signals from the auxiliary module for distillation. An example of combining a contrastive learning pretext task [3] with knowledge distillation is shown in Fig. 1.

The example in Fig. 1 reveals several advantages of using self-supervised learning as an auxiliary task for knowledge distillation (we name the combination as SSKD). First, in conventional knowledge distillation, a student mimics a teacher from normal data based on a single classification task. SSKD extends the notion to a broader extent, i.e., mimicking on transformed data and on an additional self-supervision pretext task. This enables the student to capture richer structured knowledge from the self-supervision predictions of teacher, which cannot be sufficiently captured by a single task. We show that such structured knowledge not only improves the overall distillation performance, but also regularizes the student to generalize better on few-shot and noisy-label scenarios.

Another advantage of SSKD is that it is model-agnostic. Previous knowledge distillation methods suffer from degraded performance under cross-architecture settings, for the knowledge they transfer is architecture-specific. For example, when transfer the feature of ResNet50 [12] to ShuffleNet [49], student may have trouble in mimicking due to the architecture gap. In contrast, SSKD transfers only the last layer's outputs, hence allowing a flexible solution space for the student model to search for intermediate features that best suit its own architecture.

Contributions: We propose a novel framework called SSKD that leverages self-supervised tasks to facilitate extraction of richer knowledge from teacher network to student network. To our knowledge, this is the first work that defines the knowledge through self-supervised tasks. We carefully investigate the influence of different self-supervised pretext tasks and the impact of noisy self-supervised predictions to the performance of knowledge distillation. We show that SSKD greatly boosts the generalizability of student networks and offers significant advantages under few-shot and noisy-label scenarios. Extensive experiments on two standard benchmarks, CIFAR100 [22] and ImageNet [5], demonstrate the effectiveness of SSKD over other state-of-the-art methods.

2 Related Work

Knowledge Distillation. Knowledge distillation trains a smaller network using the supervision signals from both ground truth labels and a larger network. Hinton et al. [15] propose to match the outputs of classifiers of two models by minimizing the KL-divergence of the category distribution. Besides the final layer logits, teacher network also distills compact feature representations from its backbone.

FitNets [37] proposes to mimic the intermediate feature maps of teacher network. AT [44] uses attention transfer to teach student which region is the key for classification. FSP [42] distills the second order statistics (Gram matrix) between different layers. AB [14] forces student to learn the binarized values of pre-activation map. IRG [24] explores transferring the similarity between samples. KDSVD [18] calls its method as self-supervised knowledge distillation. Nevertheless, the study regards the teacher's correlation maps of feature singular vectors as self-supervised labels. The label is obtained from the teacher rather than a self-supervised pretext task. Thus, their notion of self-supervised learning differ from the conventional one. Our work, to our knowledge, is the first study that investigates defining the knowledge via self-supervised pretext tasks. CRD [40] also combines self-supervision (SS) with distillation. The difference is the purpose of SS and how contrastive task is performed. In CRD, contrastive learning is performed across teacher and student networks to maximize the mutual information between two networks. In SSKD, contrastive task serves as a way to define knowledge. It is performed separately in two networks and then matched together through KL-divergence, which is very different from CRD.

Self-supervised Learning. Self-supervision methods design various pretext tasks whose labels can be derived from the data itself. In the process of solving these tasks, the network learn useful representations. Based on pretext tasks, SS methods can be grouped into several categories, including construction-based methods such as inpainting [34] and colorization [48], prediction-based methods [6,8,10,20,27,29,30,45,47], cluster-based methods [2,46], generation-based methods [7,9,11] and contrastive-based methods [3,13,26,31,39]. Exemplar [8] applies heavy transformation to each training image and treat all the images generated from the same image as a separate category. Jigsaw puzzle [29] splits the image into several non-overlapping patches and forces the network to recognise the shuffled order. Jigsaw++ [30] also involves SS and KD. But it utilizes knowledge transfer to boost the self-supervision performance, which solves an inverse problem of SSKD. Rotation [20] feeds the network with rotated images and forces it to recognise the rotation angle. SimCLR [3] applies augmentation to training samples and requires the network to match original image and transformed image through contrastive loss. Considering the excellent performance obtained by SimCLR [3], we adopt it as our main pretext task in SSKD. However, SSKD is not limited to using only contrastive learning, many other pretext tasks [8,20,29] can also serve the purpose. We investigate their usefulness in Sect. 4.1.

3 Methodology

This section is divided into three main sections. We start with a brief review of knowledge distillation and self-supervision in Sect. 3.1. For self-supervision, we discuss contrastive prediction as our desired pretext task, although SSKD is not limited to contrastive prediction. Sect. 3.2 specifies the training process of teacher and student model. Finally, we discuss the influence of noisy self-supervised predictions and ways to handle the noise in Sect. 3.3.

3.1 Preliminaries

Knowledge Distillation. Hinton et al. [15] suggest that the soft targets predicted by a well-optimized teacher model can provide extra information, comparing to one-hot hard labels. The relatively high probabilities assigned to wrong categories encode semantic similarities between different categories. Forcing a student to mimic teacher's prediction causes the student to learn this secondary information that cannot be expressed by hard labels alone. To obtain the soft targets, temperature scaling is introduced in [15] to soften the peaky distribution:

$$p^i(x;\tau) = \text{Softmax}(s(x);\tau) = \frac{e^{s_i(x)/\tau}}{\sum_k e^{s_k(x)/\tau}}, \tag{1}$$

where x is the data sample, i is the category index, $s_i(x)$ is the score logit that x obtains on category i, and τ is the temperature. The knowledge distillation loss L_{kd} measured by KL-divergence is:

$$L_{kd} = -\tau^2 \sum_{x \sim \mathcal{D}_x} \sum_{i=1}^{C} p_t^i(x;\tau) \log(p_s^i(x;\tau)), \tag{2}$$

where t and s denote teacher and student models, respectively, C is the total number of classes, $\mathcal{D}_x$ indicates the dataset. The complete loss function L of the student model is a linear combination of the standard cross-entropy loss L_{ce} and knowledge distillation loss L_{kd}:

$$L = \lambda_1 L_{ce} + \lambda_2 L_{kd} \tag{3}$$

Contrastive Prediction as Self-supervision Task. Motivated by the success of contrastive prediction methods [3,13,26,31,39] for self-supervised learning, we adopt contrastive prediction as the self-supervision task in our framework. The general goal of contrastive prediction is to maximize agreement between a data point and its transformed version via a contrastive loss in latent space.

Given a mini-batch containing N data points $\{x_i\}_{i=1:N}$, we apply independent transformation $t(\cdot)$ (sampled from the same distribution $\mathcal{T}$) to each data point and obtain $\{\widetilde{x}_i\}_{i=1:N}$. Both x_i and $\widetilde{x}_i$ are fed into the teacher or student networks to extract representations $\phi_i = f(x_i), \widetilde{\phi}_i = f(\widetilde{x}_i)$. We follow Chen et al. [3] and add a projection head on the top of the network. The projection head is a 2-layer multilayer perceptron. It maps the representations into a latent space where the contrastive loss is applied, i.e., $z_i = \text{MLP}(\phi_i), \widetilde{z}_i = \text{MLP}(\widetilde{\phi}_i)$.

We take $(\widetilde{x}_i, x_i)$ as the positive pair and $(\widetilde{x}_i, x_k)_{k \neq i}$ as the negative pair. Given some $\widetilde{x}_i$, the contrastive prediction task is to identify the corresponding x_i from the set $\{x_i\}_{i=1:N}$. To meet the goal, the network should maximize the similarity between positive pairs and minimize the similarity between negative pairs. In this work, we use a cosine similarity. If we organize the similarities between $\{\widetilde{x}_i\}$ and $\{x_i\}$ into matrix form $\mathcal{A}$, then we have:

$$\mathcal{A}_{i,j} = \text{cosine}(\widetilde{z}_i, z_j) = \frac{\text{dot}(\widetilde{z}_i, z_j)}{||\widetilde{z}_i||_2 ||z_j||_2}, \tag{4}$$

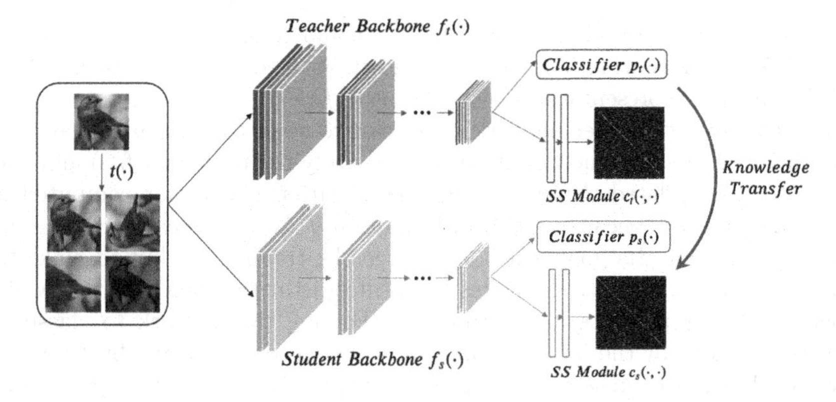

Fig. 2. Training scheme of SSKD. Input images are transformed by designated transformations to prepare data for the self-supervision task. Teacher and student networks both contain three components, i.e., backbone $f(\cdot)$, classifier $p(\cdot)$ and SS module $c(\cdot,\cdot)$. Teacher's training are split into two stages. The first stage trains $f_t(\cdot)$ and $p_t(\cdot)$ with a classification task, and the second stage fine-tunes $c_t(\cdot,\cdot)$ with a self-supervision task. In student's training, we force the student to mimic teacher on both classification output and self-supervision output, besides the standard label loss

where $\mathcal{A}_{i,j}$ represents the similarity between $\widetilde{x}_i$ and x_j. The loss of contrastive prediction is:

$$L = -\sum_i \log\left(\frac{\exp(\text{cosine}(\widetilde{z}_i, z_i)/\tau)}{\sum_k \exp(\text{cosine}(\widetilde{z}_i, z_k)/\tau)}\right) = -\sum_i \log\left(\frac{\exp(\mathcal{A}_{i,i}/\tau)}{\sum_k \exp(\mathcal{A}_{i,k}/\tau)}\right), \tag{5}$$

where τ is another temperature parameter (can be different from τ in Eq. (1)). The loss form is similar to softmax loss and can be understood as maximizing the probability that $\widetilde{z}_i$ and z_i come from a positive pair. In the process of matching $\{\widetilde{x}_i\}$ and $\{x_i\}$, the network learns transformation invariant representations. In SSKD, however, the main goal is not to learn representations invariant to transformations, but to exploit contrastive prediction as an auxiliary task for mining richer knowledge from the teacher model.

3.2 Learning SSKD

The framework of SSKD is shown in Fig. 2. Both teacher and student consist of three components: a backbone $f(\cdot)$ to extract representations, a classifier $p(\cdot)$ for the main task and a self-supervised (SS) module for specific self-supervision task. In this work, contrastive prediction is selected as the SS task, so the SS module $c_t(\cdot,\cdot)$ and $c_s(\cdot,\cdot)$ consist of a 2-layer MLP and a similarity computation module. More SS tasks will be compared in the experiments.

Training the Teacher Network. The inputs are normal data $\{x_i\}$ and transformed version $\{\widetilde{x}_i\}$. The transformation $t(\cdot)$ is sampled from a predefined transformation distribution $\mathcal{T}$. In this study, we select four transformations, i.e., color dropping, rotation, cropping followed by resize and color distortion, as depicted in Fig. 2. More transformations can be included. We feed x and $\widetilde{x}$ to the backbone and obtain their representations $\phi = f_t(x), \widetilde{\phi} = f_t(\widetilde{x})$.

The training of teacher network contains two stages. In the first stage, the network is trained with the classification loss. Only the backbone $f_t(\cdot)$ and classifier $p_t(\cdot)$ are updated. Note that the classification loss is not computed on transformed data $\widetilde{x}$ because the transformation $\mathcal{T}$ is much heavier than usual data augmentation. Its goal is not to enlarge the training set but to make the $\widetilde{x}$ visually less similar to x. It makes the contradistinction much harder, which is beneficial to representation learning [3]. Forcing the network to classify $\widetilde{x}$ correctly can destroy the semantic information learned from x and hurt the performance. In the second stage, we fix $f_t(\cdot)$ and $p_t(\cdot)$, and only update parameters in SS module $c_t(\cdot,\cdot)$ using the contrastive prediction loss in Eq. (5).

The two stages of training have distinct roles. The first stage is simply the typical training of a network for classification. The second stage, aims at adapting the SS module to use the features from the existing backbone for contrastive prediction. This allows us to extract knowledge from the SS module for distillation. It is worth pointing out that the second-stage training is highly efficient given the small MLP head, thus it is easy to prepare a teacher network for SSKD.

Training the Student Network. After training the teacher's SS module, we apply softmax (with temperature scale τ) to the teacher's similarity matrix $\mathcal{A}$ (Eq. (4)) along the row dimension leading to a probability matrix $\mathcal{B}^t$, with $\mathcal{B}^t_{i,j}$ representing the probability that $\widetilde{x}_i$ and x_j is a positive pair. Similar operations are applied to the student to obtain $\mathcal{B}^s$. With $\mathcal{B}^t$ and $\mathcal{B}^s$, we can compute the KL-divergence loss between the SS module's output of both teacher and student:

$$L_{ss} = -\tau^2 \sum_{i,j} \mathcal{B}^t_{i,j} \log(\mathcal{B}^s_{i,j}). \tag{6}$$

The transformed data point $\widetilde{x}$ is the side product of contrastive prediction task. Though we do not require the student to classify them correctly, we can encourage the student's classifier output $p_s(f_s(\widetilde{x}))$ to be close to that of teacher's. The loss function is:

$$L_T = -\tau^2 \sum_{\widetilde{x} \sim \mathcal{T}(\mathcal{D}_x)} \sum_{i=1}^{C} p^i_t(\widetilde{x};\tau) \log(p^i_s(\widetilde{x};\tau)). \tag{7}$$

The final loss for student network is the combination of aforementioned terms, i.e., cross entropy loss L_{ce}, L_{kd} in Eq. (2), L_{ss} in Eq. (6), and L_T in Eq. (7):

$$L = \lambda_1 L_{ce} + \lambda_2 L_{kd} + \lambda_3 L_{ss} + \lambda_4 L_T, \tag{8}$$

where the λ_i is the balancing weight.

3.3 Imperfect Self-supervised Predictions

When performing contrastive prediction, a teacher may produce inaccurate predictions, e.g., assigning x_k to the $\widetilde{x}_i$, $i \neq k$. This is very likely since the backbone of the teacher is not fine-tuned together with the SS module for contrastive prediction. Similar to conventional knowledge distillation, those relative probabilities that the teacher assigns to incorrect answers contain rich knowledge of the teacher. Transferring this inaccurate but structured knowledge is the core of our SSKD.

Nevertheless, we empirically found that an extremely incorrect prediction may still mislead the learning of the student. To ameliorate negative impacts of those outliers, we adopt an heuristic approach to perform selective transfer. Specifically, we define the error level of a prediction as the ranking of the corresponding ground truth label in the classification task. Given a transformed sample $\widetilde{x}_i$ and corresponding positive pair index i, we sort the scores that the network assigns to each $\{x_i\}_{i=1:N}$ in a descending order. The rank of x_i represents the error level of the prediction about $\widetilde{x}_i$. The rank of 1 means the prediction is completely correct. A lower rank indicates a higher degree of error. During the training of student, we sort all the $\widetilde{x}$ in a mini-batch in an ascending order according to error levels of the teacher's prediction, and only transfer all the correct predictions and the top-k% ranked incorrect predictions. This strategy suppresses potential noise in teacher's predictions and transfer only beneficial knowledge. We show our experiments in Sect. 4.1.

4 Experiments

The experiments section consists of three parts. We first conduct ablation study to examine the effectiveness of several components of SSKD in Sect. 4.1. Comparison with state-of-the-art methods is conducted in Sect. 4.2. In Sect. 4.3, we further show SSKD's advantages under few-shot and noisy-label scenarios.

Evaluations are conducted on CIFAR100 [22] and ImageNet [5] datasets, both of which are widely used as the benchmarks for knowledge distillation. CIFAR100 consists of $60,000$ 32×32 colour images, with $50,000$ images for training and $10,000$ images for testing. There are 100 classes, each contains 600 images. ImageNet is a large-scale classification dataset, containing $1,281,167$ images for training and $50,000$ images for testing.

4.1 Ablation Study

Effectiveness of Self-supervision Auxiliary Task. The motivation behind SSKD is that teacher's inaccurate self-supervision output encodes rich structured knowledge of teacher network and mimicking this output can benefit student's learning. To examine this hypothesis, we train a student network whose only training signals come from teacher's self-supervision output, i.e., set λ_1,λ_2, λ_3 in Eq. (8) to be 0, and observe whether student can learn good representations.

We first demonstrate the utility by examining the student's feature distribution. We select vgg13 [38] and vgg8 as the teacher and student networks, respectively. The CIFAR100 [22] training split is selected as the training set. After the training, we use the student backbone to extract features (before logits) of

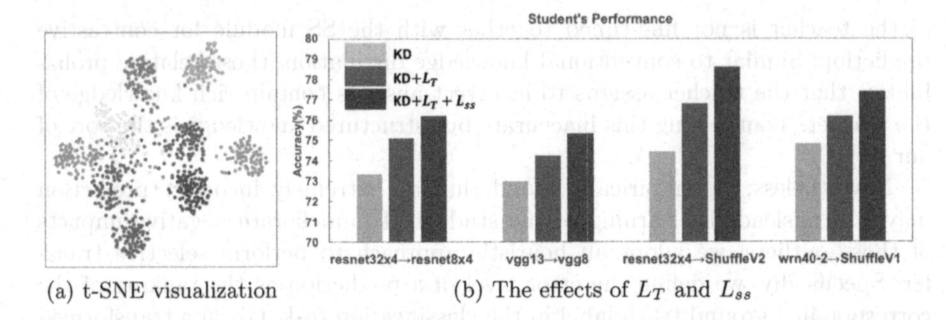

(a) t-SNE visualization (b) The effects of L_T and L_{ss}

Fig. 3. Effectiveness of self-supervision auxiliary task. Mimicking the self-supervision output benefits the feature learning and final classification performance. (a) t-SNE visualization of learned features by mimicking teacher's self-supervision output. Each color represents one category. (b) The consistent improvement across all four tested teacher-student network pairs demonstrates the effectiveness of including self-supervision task as an auxiliary task

CIFAR100 test set. We randomly select 9 categories out of 100 and visualize the features with t-SNE. The results are shown in Fig. 3(a). Though the accuracy of teacher's contrastive prediction is only around 50%, mimicking this inaccurate output still makes student learn highly clustered patterns, showing that teacher's self-supervision output does transfer meaningful structured knowledge.

To test the effectiveness of designed L_T and L_{ss}, we compare three variants of SSKD with CIFAR100 on four teacher-student pairs. The three variants are: 1) conventional KD, 2) KD with additional loss L_T (KD$+L_T$), 3) full SSKD (KD$+L_T+L_{ss}$). The results are shown in Fig. 3(b). On all four different teacher-student pairs, L_T and L_{ss} boost the accuracies by a large margin, showing the effectiveness of our designed components.

Influence of Noisy Self-supervision Predictions. As discussed in Sect. 3.3, removing some extreme outliers are beneficial for SSKD. Some transformed samples with large error levels may play a misleading role. To examine this conjecture, we compare several students that receive different proportions of incorrect predictions from teacher. Specifically, we sort all the transformed $\widetilde{x}$ in a mini-batch according to their error levels in an ascending order. We transfer all the correct predictions. For incorrect predictions, we only transfer top-k% samples with the smallest error levels. A higher k value indicates a higher number of predictions with larger error levels being transferred to student network. Experiments are conducted on CIFAR100 with three teacher-student pairs. The results are shown in Table 1. The general trend shows that incorrect predictions are beneficial ($k = 0$ yields the lowest accuracies). Removing extreme outliers help to give a peak performance between $k = 50$ and $k = 75$ across different achitectures. When comparing with other methods in Sect. 4.2 and 4.3, we fix $k = 75$ for all the teacher-student pairs.

Table 1. Influence of noisy self-supervision predictions to student accuracies(%), when transferring the top-k% smallest error-level samples. As more samples with large error level are transferred, the performances go through a rise-and-fall process. The baseline with $k = 0$ is equivalent to transferring only correct predictions

Teacher-Student pair	$k = 0$	$k = 25$	$k = 50$	$k = 75$	$k = 100$
vgg13 → vgg8	74.19	74.36	74.76	**75.01**	74.77
resnet32 × 4 → ShuffleV2	77.65	77.72	77.96	**78.61**	77.97
wrn40-2 → wrn16-2	75.27	75.34	**75.97**	75.63	75.53

Table 2. Influence of different self-supervision tasks. Self-supervised (SS) performance denotes the linear evaluation accuracy on ImageNet. Student accuracies (vgg13→vgg8) derived from the corresponding SS methods are positively correlated with the performance of the SS method itself. The SS performances are obtained from [3,20,26]

SS method	Exemplar [8]	Jigsaw [29]	Rotation [20]	Contrastive [3]
SS performance	31.5	45.7	48.9	69.3
Student performance	74.57	74.85	75.01	75.48

Influence of Different Self-supervision Tasks. Different pretext tasks in self-supervision would result in different qualities of extracted features. Similarly, distillation with different self-supervision tasks also lead to students with different performances. Here, we examine the influence of SS method's performance on SSKD. We employ the commonly used linear evaluation accuracy as our metric. In particular, each method first trains a network with its own pretext task. A single layer classifier is then trained by using the representations extracted from the fixed backbone. In this way, the classification accuracies represent the quality of SS methods. In Table 2, we compare four widely used self-supervision methods: Exemplar [8], Rotation [20], Jigsaw [29] and Contrastive [3]. We list the linear evaluation accuracies each method obtains on ImageNet with ResNet50 [12] network and also student's accuracies when they are incorporated, respectively, into KD. We find that the performance of SSKD is positively correlated with the corresponding SS method.

4.2 Benchmark

CIFAR100. We compare our method with representative knowledge distillation methods, including: KD [15], FitNet [37], AT [44], SP [41], VID [1], RKD [32], PKT [33], AB [14], FT [19], CRD [40]. ResNet [12], wideResNet [43], vgg [38], ShuffleNet [49] and MobileNet [17] are selected as the network backbones. For all competing methods, we use the implementation of [40]. For a fair comparison, we combine all competing methods with conventional KD [15] (except KD itself).

Table 3. KD between similar architectures. Top-1 accuracy (%) on CIFAR100. **Bold** and underline denote the best and the second best results, respectively. We denote by * methods that we re-run using author-provided code. SSKD obtains the best results on four out of five teacher-student pairs

Teacher	wrn40-2	wrn40-2	resnet56	resnet32×4	vgg13
Student	wrn16-2	wrn40-1	resnet20	resnet8×4	vgg8
Teacher	76.46	76.46	73.44	79.63	75.38
Student	73.64	72.24	69.63	72.51	70.68
KD [15]	74.92	73.54	70.66	73.33	72.98
FitNet [37]	75.75	74.12	71.60	74.31	73.54
AT [44]	75.28	74.45	**71.78**	74.26	73.62
SP [41]	75.34	73.15	71.48	74.74	73.44
VID [1]	74.79	74.20	71.71	74.82	73.96
RKD [32]	75.40	73.87	71.48	74.47	73.72
PKT [33]	76.01	74.40	71.44	74.17	73.37
AB [14]	68.89	75.06	71.49	74.45	74.27
FT [19]	75.15	74.37	71.52	75.02	73.42
CRD* [40]	**76.04**	75.52	71.68	75.90	74.06
Ours	**76.04**	**76.13**	71.49	**76.20**	**75.33**

And we omit "+KD" notation in all the following tables (except for Table 5) and figures for simplicity.[1]

We compare performances on 11 teacher-student pairs to investigate the generalization ability of each method. Following CRD [40], we split these pairs into 2 groups according to whether teacher and student have similar architecture styles. The results are shown in Table 3 and Table 4. In each table, the second partition after the header show the accuracies of the teacher's and student's performance when they are trained individually, while the third partition show the student's performance after knowledge distillation.

For teacher-student pairs with a similar architecture, SSKD performs the best in four out of five pairs (Table 3). The gap between SSKD and the best-performing competing methods is 0.52% (averaged on five pairs). Notably, in all six teacher-student pairs with different architectures, SSKD consistently achieves the best results (Table 4), surpassing the best competing methods by a large margin with an average absolute accuracy difference of 2.14%. Results on cross-architecture pairs clearly demonstrate that our method does not rely on architecture-specific cues. Instead, SSKD distills knowledge only from the outputs of the final layer of teacher model. Such strategy allows a larger solution

[1] For experiments on CIFAR100, since we add the conventional KD with competing methods, the results are slightly better than those reported in CRD [40]. More details on experimental setting are provided in the supplementary material.

Table 4. KD between different architectures. Top-1 accuracy (%) on CIFAR100. **Bold** and underline denote the best and the second best results, respectively. We denote by * methods that we re-run using author-provided code. SSKD consistently obtains the best results on all pairs

Teacher	vgg13	ResNet50	ResNet50	resnet32×4	resnet32×4	wrn40-2
Student	MobileNetV2	MobileNetV2	vgg8	ShuffleV1	ShuffleV2	ShuffleV1
Teacher	75.38	79.10	79.10	79.63	79.63	76.46
Student	65.79	65.79	70.68	70.77	73.12	70.77
KD [15]	67.37	67.35	73.81	74.07	74.45	74.83
FitNet [37]	68.58	68.54	73.84	74.82	75.11	75.55
AT [44]	69.34	69.28	73.45	74.76	75.30	75.61
SP [41]	66.89	68.99	73.86	73.80	75.15	75.56
VID [1]	66.91	68.88	73.75	74.28	75.78	75.36
RKD [32]	68.50	68.46	73.73	74.20	75.74	75.45
PKT [33]	67.89	68.44	73.53	74.06	75.18	75.51
AB [14]	68.86	69.32	74.20	76.24	75.66	76.58
FT [19]	69.19	69.01	73.58	74.31	74.95	75.18
CRD* [40]	68.49	70.32	74.42	75.46	75.72	75.96
Ours	**71.53**	**72.57**	**75.76**	**78.44**	**78.61**	**77.40**

Table 5. Top-1/Top-5 error (%) on ImageNet. Bold and underline denote the best and the second best results, respectively. The competing methods include CC [35], SP [41], Online-KD [23], KD [15], AT [44], and CRD [40]. The results of competing methods are obtained from [40]

	Teacher	Student	CC	SP	Online-KD	KD	AT	CRD	CRD+KD	Ours
Top-1	26.70	30.25	30.04	29.38	29.45	29.34	29.30	28.83	28.62	**28.38**
Top-5	8.58	10.93	10.83	10.20	10.41	10.12	10.00	9.87	9.51	**9.33**

space for student model to search intermediate representations that best suit its own architecture.

ImageNet. Limited by computation resources, we only conduct one teacher-student pair on ImageNet, i.e., ResNet34 as teacher and ResNet18 as student. As shown in Table 5, for both Top-1 and Top-5 error rates, our SSKD obtains the best performances. The results on ImageNet demonstrate the scalability of SSKD to large-scale dataset.

Teacher-Student Similarity. SSKD can extract richer knowledge by mimicking self-supervision output and make student much more similar to teacher than other KD methods. To examine this claim, we analyze the similarity between student and teacher networks using two metrics, i.e., KL-divergence and CKA similarity [21]. Small KL-divergence and large CKA similarity indicate that student is similar to teacher. We use vgg13 and vgg8 as teacher and student, respectively, and use CIFAR100 as the training set. We compute the KL-divergence and CKA similarity between teacher and student on three sets, i.e., test partitions of CIFAR100, STL10 [4] and SVHN [28]. As shown in Table 6, our method achieves

Table 6. Teacher-student similarity. KL-divergence and CKA-similarity [21] between student and teacher networks. **Bold** and underline denote the best and the second best results, respectively. All the models are trained on CIFAR100 training set. ↓ (↑) indicates the smaller (larger) the better. SSKD wins in five out of six comparisons

Dataset	CIFAR100 test set		STL10 test set		SVHN test set	
Metric	KL-div (↓)	CKA-simi (↑)	KL-div (↓)	CKA-simi (↑)	KL-div (↓)	CKA-simi (↑)
KD [15]	6.91	0.7003	16.28	0.8234	15.21	0.6343
SP [41]	6.81	0.6816	16.07	0.8278	14.47	0.6331
VID [1]	6.76	0.6868	16.15	0.8298	12.60	0.6502
FT [19]	6.69	0.6830	15.95	0.8287	12.53	0.6734
RKD [32]	6.68	0.7010	16.14	0.8290	13.78	0.6503
FitNet [37]	6.63	0.6826	15.99	0.8214	16.34	0.6634
AB [14]	6.51	0.6931	15.34	0.8356	11.13	0.6532
AT [44]	6.61	0.6804	16.32	0.8204	15.49	0.6505
PKT [33]	6.73	0.6827	16.17	0.8232	14.08	0.6555
CRD [40]	6.34	0.6878	**14.71**	0.8315	10.85	0.6397
Ours	**6.24**	**0.7419**	14.91	**0.8521**	**10.58**	**0.7382**

the smallest KL-divergence and the largest CKA similarity on CIFAR100 test set. Compared to CIFAR100, STL10 and SVHN have different distributions that have not been seen during training, therefore more difficult to mimic. However, the proposed SSKD still obtains the best results in all the metrics except KL-divergence in STL10. From this similarity analysis, we conclude that SSKD can help student mimic teacher better and get a larger similarity to teacher network.

4.3 Further Analysis

Few-Shot Scenario. In a real-world setting, the number of samples available for training is often limited [25]. To investigate the performance of SSKD under few-shot scenarios, we conduct experiments on subsets of CIFAR100. We randomly sample images of each class to form a new training set. We train student model using newly crafted training set, while maintaining the same test set. Vgg13 and vgg8 are chosen as teacher and student model, respectively. We compare our student's performance with KD [15], AT [44] and CRD [40]. The percentages of reserved samples are 25%, 50%, 75% and 100%. For a fair comparison, we employ the same data for different methods.

The results are shown in Fig. 4(a). In all data proportions, SSKD achieves the best result. As training samples decrease, the superiority of our method becomes more apparent, e.g., ∼7% absolute improvement in accuracy compared to all competing methods when the percentage of reserved samples are 25%. Previous methods mainly focus on learning various intermediate features of teacher or exploring the relations between samples. The excessive mimicking leads to over-fitting on the training set. In SSKD, the transformed images and self-supervision

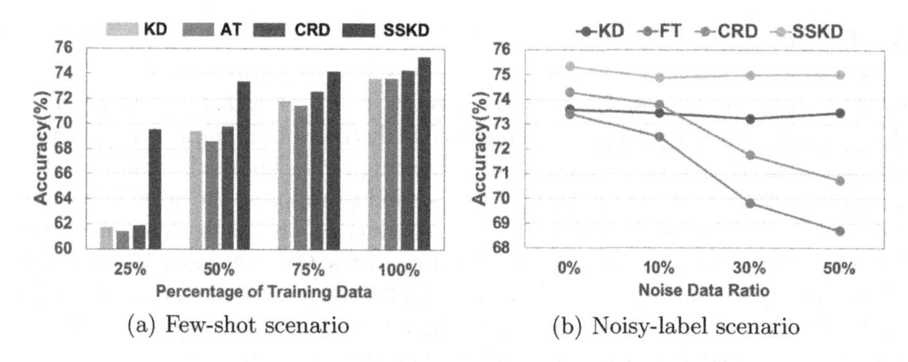

(a) Few-shot scenario (b) Noisy-label scenario

Fig. 4. Accuracies on CIFAR100 test set under few-shot and noisy-label scenarios. (a) Students are trained with subsets of CIFAR100. SSKD achieves the best results in all cases. The superiority is especially striking when only 25% of the training data is available. (b) Students are trained with data with perturbed labels. The accuracies of FT and CRD drop dramatically as noisy labels increase, while SSKD is much more stable and maintains a high performance in all cases

task endow the student model with structured knowledge that provides strong regularization, hence making it generalizes better to test set.

Noisy-Label Scenario. Our SSKD forces student to mimic teacher on both classification task and self-supervision task. The student learns more well rounded knowledge from the teacher model than relying entirely on annotated labels. Such strategy strengthens the ability of student to resist label noise. In this section, we investigate the performance of KD [15], FT [19], CRD [40] and SSKD when trained with noisy label data. We choose vgg13 and vgg8 as the teacher and student models, respectively. We assume the teacher is trained with clean data and will be shared by all students. This assumption does not affect evaluation on robustness of different distillation methods. When training student models, we randomly perturb the labels of certain portions of training data and use the original test data for evaluation. We introduce same disturbances to all methods. Since the loss weight of cross entropy on labels affects how well a model resists label noise, we use the same loss weight for all methods for a fair comparison. We set the percentage of disturbed labels to be 0%, 10%, 30% and 50%. Results are shown in Fig. 4(b). SSKD outperforms competing methods in all noise ratios. As noise data increase, the performance of FT and CRD drop dramatically. KD and SSKD are more stable. Specifically, accuracy of SSKD only drop by a marginal 0.45% when the percentage of noise data increases from 0% to 50%, demonstrating the robustness of SSKD against noisy data labels. We attribute the robustness to the structured knowledge offered by self-supervised tasks.

5 Conclusion

In this work, we proposed a novel framework called SSKD, the first attempt that combines self-supervision with knowledge distillation. It employs contrastive prediction as an auxiliary task to help extracting richer knowledge from teacher network. A selective transfer strategy is designed to suppress the noise in teacher knowledge. We examined our method by conducting thorough experiments on CIFAR100 and ImageNet using various architectures. Our method achieves state-of-the-art performances, demonstrating the effectiveness of our approach. Further analysis showed that our SSKD can make student more similar to teacher and work well under few-shot and noisy-label scenarios.

Acknowledgement. This research was supported by SenseTime-NTU Collaboration Project, Collaborative Research Grant from SenseTime Group (CUHK Agreement No. TS1610626 & No. TS1712093), and NTU NAP.

References

1. Ahn, S., Hu, S.X., Damianou, A., Lawrence, N.D., Dai, Z.: Variational information distillation for knowledge transfer. In: IEEE Conference on Computer Vision and Pattern Recognition (2019)
2. Caron, M., Bojanowski, P., Joulin, A., Douze, M.: Deep clustering for unsupervised learning of visual features. In: The European Conference on Computer Vision (2018)
3. Chen, T., Kornblith, S., Norouzi, M., Hinton, G.: A simple framework for contrastive learning of visual representations. arXiv preprint arXiv:2002.05709 (2020)
4. Coates, A., Ng, A., Lee, H.: An analysis of single-layer networks in unsupervised feature learning. In: Proceedings of the Fourteenth International Conference on Artificial Intelligence and Statistics, vol. 15, pp. 215–223 (2011)
5. Deng, J., Dong, W., Socher, R., Li, L., Kai, L., Fei-Fei, L.: ImageNet: a large-scale hierarchical image database. In: IEEE Conference on Computer Vision and Pattern Recognition, pp. 248–255 (2009)
6. Doersch, C., Gupta, A., Efros, A.A.: Unsupervised visual representation learning by context prediction. In: The IEEE International Conference on Computer Vision (2015)
7. Donahue, J., Simonyan, K.: Large scale adversarial representation learning. In: Advances in Neural Information Processing Systems, pp. 10541–10551 (2019)
8. Dosovitskiy, A., Fischer, P., Springenberg, J.T., Riedmiller, M., Brox, T.: Discriminative unsupervised feature learning with exemplar convolutional neural networks. arXiv preprint arXiv:1406.6909 (2014)
9. Dumoulin, V., et al.: Adversarially learned inference. In: International Conference on Learning Representations (2017)
10. Gidaris, S., Singh, P., Komodakis, N.: Unsupervised representation learning by predicting image rotations. In: International Conference on Learning Representations (2018)
11. Goodfellow, I., et al.: Generative adversarial nets. In: Advances in Neural Information Processing Systems, pp. 2672–2680 (2014)

12. He, K., Zhang, X., Ren, S., Sun, J.: Deep residual learning for image recognition. In: IEEE Conference on Computer Vision and Pattern Recognition, pp. 770–778 (2016)
13. Hénaff, O.J., Razavi, A., Doersch, C., Eslami, S.M.A., van den Oord, A.: Data-efficient image recognition with contrastive predictive coding. arXiv preprint arXiv:1905.09272 (2019)
14. Heo, B., Lee, M., Yun, S., Choi, J.Y.: Knowledge transfer via distillation of activation boundaries formed by hidden neurons. In: AAAI, pp. 3779–3787 (2019)
15. Hinton, G., Vinyals, O., Dean, J.: Distilling the knowledge in a neural network. In: NIPS Deep Learning and Representation Learning Workshop (2015)
16. Hou, Y., Ma, Z., Liu, C., Hui, T.W., Loy, C.C.: Inter-region affinity distillation for road marking segmentation. In: IEEE Conference on Computer Vision and Pattern Recognition (2020)
17. Howard, A.G., et al.: MobileNets: efficient convolutional neural networks for mobile vision applications. arXiv preprint arXiv:1704.04861 (2017)
18. Lee, S.H., Kim, D.H., Song, B.C.: Self-supervised knowledge distillation using singular value decomposition. In: Ferrari, V., Hebert, M., Sminchisescu, C., Weiss, Y. (eds.) ECCV 2018. LNCS, vol. 11210, pp. 339–354. Springer, Cham (2018). https://doi.org/10.1007/978-3-030-01231-1_21
19. Kim, J., Park, S., Kwak, N.: Paraphrasing complex network: network compression via factor transfer. In: Advances in Neural Information Processing Systems, pp. 2760–2769 (2018)
20. Kolesnikov, A., Zhai, X., Beyer, L.: Revisiting self-supervised visual representation learning. In: IEEE Conference on Computer Vision and Pattern Recognition (2019)
21. Kornblith, S., Norouzi, M., Lee, H., Hinton, G.E.: Similarity of neural network representations revisited. In: International Conference on Machine Learning. Proceedings of Machine Learning Research, vol. 97, pp. 3519–3529 (2019)
22. Krizhevsky, A.: Learning multiple layers of features from tiny images. Technical report (2009)
23. Lan, X., Zhu, X., Gong, S.: Knowledge distillation by on-the-fly native ensemble. In: Advances in Neural Information Processing Systems, pp. 7528–7538 (2018)
24. Liu, Y., et al.: Knowledge distillation via instance relationship graph. In: IEEE Conference on Computer Vision and Pattern Recognition (2019)
25. Liu, Z., Miao, Z., Zhan, X., Wang, J., Gong, B., Yu, S.X.: Large-scale long-tailed recognition in an open world. In: IEEE Conference on Computer Vision and Pattern Recognition (2019)
26. Misra, I., van der Maaten, L.: Self-supervised learning of pretext-invariant representations. arXiv preprint arXiv:1912.01991 (2019)
27. Misra, I., Zitnick, C.L., Hebert, M.: Shuffle and learn: unsupervised learning using temporal order verification. In: Leibe, B., Matas, J., Sebe, N., Welling, M. (eds.) ECCV 2016. LNCS, vol. 9905, pp. 527–544. Springer, Cham (2016). https://doi.org/10.1007/978-3-319-46448-0_32
28. Netzer, Y., Wang, T., Coates, A., Bissacco, A., Wu, B., Ng, A.Y.: Reading digits in natural images with unsupervised feature learning. In: NIPS Workshop on Deep Learning and Unsupervised Feature Learning (2011)
29. Noroozi, M., Favaro, P.: Unsupervised learning of visual representations by solving jigsaw puzzles. In: Leibe, B., Matas, J., Sebe, N., Welling, M. (eds.) ECCV 2016. LNCS, vol. 9910, pp. 69–84. Springer, Cham (2016). https://doi.org/10.1007/978-3-319-46466-4_5

30. Noroozi, M., Vinjimoor, A., Favaro, P., Pirsiavash, H.: Boosting self-supervised learning via knowledge transfer. In: IEEE Conference on Computer Vision and Pattern Recognition (2018)
31. van den Oord, A., Li, Y., Vinyals, O.: Representation learning with contrastive predictive coding. arXiv preprint arXiv:1807.03748 (2018)
32. Park, W., Kim, D., Lu, Y., Cho, M.: Relational knowledge distillation. In: IEEE Conference on Computer Vision and Pattern Recognition (2019)
33. Passalis, N., Tefas, A.: Learning deep representations with probabilistic knowledge transfer. In: The European Conference on Computer Vision (2018)
34. Pathak, D., Krähenbühl, P., Donahue, J., Darrell, T., Efros, A.: Context encoders: feature learning by inpainting. In: IEEE Conference on Computer Vision and Pattern Recognition (2016)
35. Peng, B., Jin, X., Liu, J., Li, D., Wu, Y., Liu, Y., Zhou, S., Zhang, Z.: Correlation congruence for knowledge distillation. In: The IEEE International Conference on Computer Vision (2019)
36. Rao, A., et al.: A unified framework for shot type classification based on subject centric lens. In: The European Conference on Computer Vision (2020)
37. Romero, A., Ballas, N., Kahou, S.E., Chassang, A., Gatta, C., Bengio, Y.: FitNets: hints for thin deep nets. arXiv preprint arXiv:1412.6550 (2014)
38. Simonyan, K., Zisserman, A.: Very deep convolutional networks for large-scale image recognition. arXiv preprint arXiv:1409.1556 (2014)
39. Tian, Y., Krishnan, D., Isola, P.: Contrastive multiview coding. arXiv preprint arXiv:1906.05849 (2019)
40. Tian, Y., Krishnan, D., Isola, P.: Contrastive representation distillation. In: International Conference on Learning Representations (2020)
41. Tung, F., Mori, G.: Similarity-preserving knowledge distillation. In: The IEEE International Conference on Computer Vision (2019)
42. Yim, J., Joo, D., Bae, J., Kim, J.: A gift from knowledge distillation: fast optimization, network minimization and transfer learning. In: IEEE Conference on Computer Vision and Pattern Recognition (2017)
43. Zagoruyko, S., Komodakis, N.: Wide residual networks. arXiv preprint arXiv:1605.07146 (2016)
44. Zagoruyko, S., Komodakis, N.: Paying more attention to attention: improving the performance of convolutional neural networks via attention transfer. In: International Conference on Learning Representations (2017)
45. Zhan, X., Pan, X., Liu, Z., Lin, D., Loy, C.C.: Self-supervised learning via conditional motion propagation. In: IEEE Conference on Computer Vision and Pattern Recognition (2019)
46. Zhan, X., Xie, J., Liu, Z., Ong, Y.S., Loy, C.C.: Online deep clustering for unsupervised representation learning. In: IEEE Conference on Computer Vision and Pattern Recognition (2020)
47. Zhang, L., Qi, G.J., Wang, L., Luo, J.: AET vs. AED: unsupervised representation learning by auto-encoding transformations rather than data. In: IEEE Conference on Computer Vision and Pattern Recognition (2019)
48. Zhang, R., Isola, P., Efros, A.A.: Colorful image colorization. In: Leibe, B., Matas, J., Sebe, N., Welling, M. (eds.) ECCV 2016. LNCS, vol. 9907, pp. 649–666. Springer, Cham (2016). https://doi.org/10.1007/978-3-319-46487-9_40
49. Zhang, X., Zhou, X., Lin, M., Sun, J.: ShuffleNet: an extremely efficient convolutional neural network for mobile devices. In: IEEE Conference on Computer Vision and Pattern Recognition (2018)

Efficient Neighbourhood Consensus Networks via Submanifold Sparse Convolutions

Ignacio Rocco[1(✉)], Relja Arandjelović[2], and Josef Sivic[1,3]

[1] WILLOW, Inria, DI -ENS, CNRS, PSL Research University, Paris, France
ignacio.rocco@inria.fr
[2] DeepMind, London, UK
[3] Czech Institute of Informatics, Robotics and Cybernetics, CTU, Prague, Czechia

Abstract. In this work we target the problem of estimating accurately localized correspondences between a pair of images. We adopt the recent Neighbourhood Consensus Networks that have demonstrated promising performance for difficult correspondence problems and propose modifications to overcome their main limitations: large memory consumption, large inference time and poorly localized correspondences. Our proposed modifications can reduce the memory footprint and execution time more than 10×, with equivalent results. This is achieved by *sparsifying* the correlation tensor containing tentative matches, and its subsequent processing with a 4D CNN using submanifold sparse convolutions. localization accuracy is significantly improved by processing the input images in higher resolution, which is possible due to the reduced memory footprint, and by a novel two-stage correspondence relocalization module. The proposed Sparse-NCNet method obtains state-of-the-art results on the HPatches Sequences and InLoc visual localization benchmarks, and competitive results on the Aachen Day-Night benchmark.

Keywords: Image matching · Neighbourhood consensus · Sparse CNN

1 Introduction

Finding correspondences between images depicting the same 3D scene is one of the fundamental tasks in computer vision [24,29,35] with applications in 3D reconstruction [50,51,57], visual localization [15,47,53] or pose estimation [14,18,40]. The predominant approach currently consists of first *detecting* salient local features, by selecting the local extrema of some form of feature selection function, and then *describing* them by some form of feature descriptor [7,28,45]. While hand-crafted features such as Hessian affine detectors [30] with SIFT descriptors [28] achieve impressive performance under strong viewpoint changes and constant illumination [31], their robustness to illumination

Electronic supplementary material The online version of this chapter (https://doi.org/10.1007/978-3-030-58545-7_35) contains supplementary material, which is available to authorized users.

© Springer Nature Switzerland AG 2020
A. Vedaldi et al. (Eds.): ECCV 2020, LNCS 12354, pp. 605–621, 2020.
https://doi.org/10.1007/978-3-030-58545-7_35

(a) Input images (b) Output matches (c) Match confidence

Fig. 1. Correspondence estimation with Sparse-NCNet. Given an input image pair (a), we show the *raw* output correspondences produced by Sparse-NCNet (b) which contain groups of spatially coherent matches. These groups tend to form around highly-confident matches, which are shown in yellow shades (c).

changes is limited [31,63]. More recently, a variety of trainable keypoint detectors [26,27,33,56] and descriptors [5,6,22,32,54,59] have been proposed, with the purpose of obtaining increased robustness over hand-crafted methods. While this approach has achieved some success, extreme illumination changes such as day-to-night matching combined with changes in camera viewpoint remain a challenging open problem [4,13,15]. In particular, all local feature methods, whether hand-crafted or trained, suffer from missing detections under these extreme appearance changes.

In order to overcome this issue, the detection stage can be avoided and, instead, features can be extracted on a dense grid across the image. This approach has been successfully used for both place recognition [1,15,36,55] and image matching [44, 47,57]. However, extracting features densely comes with additional challenges: it is memory intensive and the localization accuracy of the features is limited by the sampling interval of the grid used for the extraction.

In this work we adopt the dense feature extraction approach. In particular, we build on the recent Neighbourhood Consensus Networks (NCNet) [44], that allow for jointly trainable feature extraction, matching, and match-filtering to directly output a strong set of (mostly) correct correspondences. Our proposed approach, Sparse-NCNet, seeks to overcome the limitations of the original NCNet formulation, namely: large memory consumption, high execution time and poorly localized correspondences.

Our contributions are the following. First, we propose the efficient Sparse-NCNet model, which is based on a 4D convolutional neural network operating on a *sparse* correlation tensor, which is obtained by storing only the most promising correspondences, instead of the set of all possible correspondences. Sparse-NCNet processes this sparse correlation tensor with submanifold sparse convolutions [21] and can obtain equivalent results to NCNet while being several times faster (up to 10×) and requiring much less memory (up to 20×) without decrease in performance compared to the original NCNet model. Second, we propose a two-stage relocalization module to improve the localization accuracy of the correspondences output by Sparse-NCNet. Finally, we show that the proposed model significantly outperforms state-of-the-art results on the HPatches Sequences [3] benchmark for image matching with challenging viewpoint and illumination changes and the InLoc [53] benchmark for indoor localization and camera pose estimation. Furthermore, we show our model obtains competitive results on the Aachen Day-Night benchmark [47], which evaluates day-night feature matching for the task of camera localization. An example of the correspondences produced by our method is presented in Fig. 1. **Our code and models are available online** [43].

2 Related Work

In this section, we review the relevant related work.

Matching with Trainable Local Features. Most recent work in trainable local features has focused on learning more robust keypoint *descriptors* [5,6,22,32,54,59]. Initially these descriptors were used in conjunction with classic hand-crafted keypoint detectors, such as DoG [28]. Recently, trainable keypoint *detectors* where also proposed [26,27,33,56], as well as methods providing both *detection and description* [12,13,37,41,58]. From these, some adopt the classic approach of first performing detection on the whole image and then computing descriptors from local image patches, cropped around the detected keypoints [37,58], while the most recent methods compute a joint representation from which both detections and descriptors are computed [12,13,41]. In most cases, local features obtained by these methods are independently matched using nearest-neighbour search with the Euclidean distance [5,6,32,54], although some works have proposed to learn the distance function as well [22,59]. As discussed in the previous section, local features are prone to loss of detections under extreme lighting changes [15]. In order to alleviate this issue, in this work we adopt the usage of densely extracted features, which are described next.

Matching with Densely Extracted Features. Motivated by applications in large-scale visual search, others have found that using densely extracted features provides additional robustness to illumination changes compared to local features extracted at detected keypoints, which suffer from low repeatability under strong illumination changes [55,62]. This approach was also adopted by later work [1,36]. Such densely extracted features used for image retrieval are typically computed on a coarse low resolution grid (*e.g.*40×30). However, such coarse localization of the dense features is not an issue for visual retrieval, as the dense features are not directly matched, but rather aggregated into a single image-level descriptor, which is used for retrieval. Recently, densely extracted features have been also employed directly for 3D computer vision tasks, such as 3D reconstruction [57], indoor localization and camera pose estimation [53], and outdoor localization with night queries [15,47]. In these methods, correspondences are obtained by nearest-neighbour search performed on extracted descriptors, and filtered by the mutual nearest-neighbour criterion [38]. In this work, we build on the NCNet method [44], where the match filtering function is learnt from data. Recent methods for learning to filter matches are discussed next.

Learning to Filter Incorrect Matches. When using both local features extracted at keypoints or densely extracted features, the obtained matches by nearest-neighbour search contain a certain portion of incorrect matches. In the case of local features, a heuristic approach such as Lowe's ratio test [28] can be used to filter these matches. However the ratio threshold value needs to be manually tuned for each method. To avoid this issue, filtering by mutual nearest neighbours can be used instead [13]. Recently, trainable approaches have also been proposed for the task of filtering local feature correspondences [9,34,46,60]. Yi *et al.* [34] propose a neural-network architecture that operates on 4D match coordinates and classifies each correspondence as either correct or incorrect. Brachmann *et al.* [9] propose the Neural-guided RANSAC, which extends the previous method to produce weights instead of classification labels, which are used to guide RANSAC sampling. Zhang *et al.* [60] also extend the work of Yi *et al.*in their proposed Order-Aware Networks, which capture local context by clustering 4D correspondences onto a set of ordered clusters, and global context by processing these clusters with a multi-layer perceptron. Finally, Sarlin *et al.* [46] describe a graph neural network followed by an optimization procedure to estimate correspondences between two set of local features. These methods were specifically designed for filtering local features extracted at keypoint locations and not features extracted on a dense grid. Furthermore, these methods are focused only on learning match filtering, and are decoupled from the problem of learning how to detect and describe the local features.

In this paper we build on the NCNet method [44] for filtering incorrect matches, which was designed for dense features. Furthermore, contrary to the above described methods, our approach performs feature extraction, matching and match filtering in a single pipeline.

Improved Feature Localization. Recent methods for local feature detection and description which use a joint representation [12,13] as well as methods for dense feature extraction [44,57] suffer from poor feature localization, as the features are extracted on a low-resolution grid. Different approaches have been proposed to deal with this issue. The D2-Net method [13] follows the approach used in SIFT [28] for refining the keypoint positions, which consists of locally fitting a quadratic function to the feature detection function around the feature position and solving for the extrema. The Superpoint method [12] uses a CNN decoder that produces a one-hot output for each 8×8 pixel cell of the input image (in case a keypoint is effectively detected in this region), therefore achieving pixel-level accuracy. Others [57] use the intermediate higher resolution features from the CNN to improve the feature localization, by assigning to each pooled feature the position of the feature with highest L2 norm from the preceding higher resolution map (and which participated in the pooling). This process can be repeated up to the input image resolution.

The relocalization approach of NCNet [44] is based on a max-argmax operation on the 4D correlation tensor of exhaustive feature matches. This approach can only increase the resolution of the output matches by a factor of 2. In contrast, we describe a new two-stage relocalization module that builds on the approach used in NCNet, by combining a hard relocalization stage that has similar effects to NCNet's max-argmax operation, with a soft-relocalization stage that obtains sub-feature-grid accuracy via interpolation.

Sparse Convolutional Neural Networks were recently used for the purpose of processing sparse 2D data, such as handwritten characters [20]; 3D data, such as 3D point-clouds [19]; or even 4D data, such as temporal sequences of 3D point clouds [10]. These models have shown great success in 3D point-cloud processing tasks such as semantic segmentation [10,21] and point-cloud registration [11,17]. In this work, we use networks with *submanifold sparse convolutions* [21] for the task of filtering correspondences between images, which can be represented as a sparse set of points in a 4D space of image coordinates. In submanifold sparse convolutions, the active sites remain constant between the input and output of each convolutional layer. As a result, the sparsity level remains fixed and does not change after each convolution operation. To the best of our knowledge this is the first time these models are applied to the task of match filtering.

3 Sparse Neighbourhood Consensus Networks

In this section we detail the proposed Sparse Neighbourhood Consensus Networks. We start with a brief review of Neighbourhood Consensus Networks [44] identifying their main limitations. Next, we describe our approach which overcomes these limitations.

3.1 Review: Neighbourhood Consensus Networks

The Neighbourhood Consensus Network [44] is a method for feature extraction, matching and match filtering. Contrary to most methods, which operate on local features, NCNet operates on dense feature maps $(f^A, f^B) \in \mathbb{R}^{h\times w\times c}$ with c channels, which are extracted over a regular grid of $h \times w$ spatial resolution. These are obtained from the input image pair $(I_A, I_B) \in \mathbb{R}^{H\times W\times 3}$ by a fully convolutional feature extraction network. The resolution $h \times w$ of the extracted dense features is typically 1/8 or 1/16 of the input image resolution $H \times W$, depending on the particular feature extraction network architecture used.

Next, the exhaustive set of all possible matches between the dense feature maps f^A and f^B is computed and stored in a 4D correlation tensor $c^{AB} \in \mathbb{R}^{h\times w\times h\times w}$. Finally, the correspondences in c^{AB} are filtered by a 4D CNN. This network can detect coherent spatial matching patterns and propagate information from the most certain matches to their neighbours, robustly identifying the correct correspondences. This last filtering step is inspired by the neighbourhood consensus procedure [8,48,49,52,61], where a particular match is verified by analyzing the existence of other coherent matches in its spatial neighbourhood in both images.

Despite its promising results, the original formulation of Neighbourhood Consensus Networks has three main drawbacks that limit its practical application: it is (i) memory intensive, (ii) slow, and (iii) matches are poorly localized. These points are discussed in detail next.

High Memory Requirements. The high memory requirements are due to the computation of the correlation tensor $c^{AB} \in \mathbb{R}^{h \times w \times h \times w}$ which stores all matches between the densely extracted image features $(f^A, f^B) \in \mathbb{R}^{h \times w \times c}$. Note that the number of elements in the correlation tensor $(h \times w \times h \times w)$ grows quadratically with respect to the number of features $(h \times w)$ of the dense feature maps (f^A, f^B), therefore limiting the ability to increase the feature resolution. For instance, for dense feature maps of resolution 200×150, the correlation tensor would require by itself 3.4 GB of GPU memory in the standard 32-bit float precision. Furthermore, processing this correlation tensor using the subsequent 4D CNN would require more than 50 GB of GPU memory, which is much more than what is currently available on most standard GPUs. While 16-bit half-float precision could be used to halve these memory requirements, they would still be prohibitively large.

Long Processing Time. In addition, Neighbourhood Consensus Networks are slow as the full dense correlation tensor must be processed. For instance, processing the $100 \times 75 \times 100 \times 75$ correlation tensor containing matches between a pair of dense feature maps of 100×75 resolution takes approximately 10 s on a standard Tesla T4 GPU.

Poor Match Localization. Finally, the high-memory requirements limit the maximum feature map resolution that can be processed, which in turn limits the localization accuracy of the estimated correspondences. For instance, for a pair images with 1600×1200 px resolution, where correspondences are computed using a dense feature map with a resolution of 100×75, the output correspondences are localized within an error of 8 pixels. This can be problematic if correspondences are used for tasks such as pose estimation, where small errors in the localization of correspondences in image-space can yield high camera pose errors in 3D space.

In this paper, we devise strategies to overcome the limitations of the original NCNet method, while keeping its main advantages, such as the usage of dense feature maps which avoids the issue of missing detections, and the processing of multiple matching hypotheses to avoid early matching errors. Our efficient Sparse-NCNet approach is described next.

3.2 Sparse-NCNet: Efficient Neighbourhood Consensus Networks

In this section, we describe the Sparse-NCNet approach in detail. An overview is presented in Fig. 2. Similar to NCNet, the first stage of our proposed method

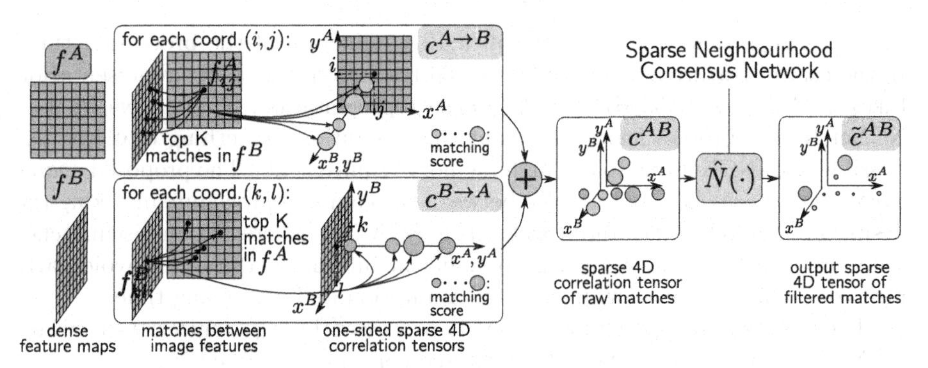

Fig. 2. Overview of Sparse-NCNet. From the dense feature maps f^A and f^B, their top K matches are computed and stored in the one-sided sparse 4D correlation tensors $c^{A\to B}$ and $c^{B\to A}$, which are later combined to obtain the symmetric sparse correlation tensor c^{AB}. The raw matching score values in c^{AB} are processed by the 4D Sparse-NCNet $\hat{N}(\cdot)$ producing the output tensor $\tilde{c}^{AB}$ of filtered matching scores.

consists in dense feature extraction. Given a pair of RGB input images $(I^A, I^B) \in \mathbb{R}^{H\times W\times 3}$, $L2$-normalized dense features $(f^A, f^B) \in \mathbb{R}^{h\times w\times c}$ are extracted via a fully convolutional network $F(\cdot)$:

$$f^A = F(I^A), f^B = F(I^B). \tag{1}$$

Then, these dense features are matched and stored into a *sparse correlation tensor*. Contrary to the original NCNet formulation, where *all* the pairwise matches between the dense features are stored and processed, we propose to keep *only the top K matches* for a given feature, measured by the cosine similarity. In detail, each feature $f^A_{ij:}$ from image A at position (i, j) is matched with its *K nearest-neighbours* in f^B, and vice versa. The one-sided sparse correlation tensor, matching from image A to image B $(A \to B)$ is then described as:

$$c^{A\to B}_{ijkl} = \begin{cases} \langle f^A_{ij:}, f^B_{kl:}\rangle & \text{if } f^B_{kl:} \text{ within K-NN of } f^A_{ij:} \\ 0 & \text{otherwise} \end{cases}. \tag{2}$$

To make the sparse correlation map invariant to the ordering of the input images, we also perform this in the reverse direction $(B \to A)$, and add the two one-sided correlation tensors together to obtain the final (symmetric) *sparse correlation tensor*:

$$c^{AB} = c^{A\to B} + c^{B\to A}. \tag{3}$$

This tensor uses a sparse representation, where only non-zero elements need to be stored. Note that the number of stored elements is, at most, $h \times w \times K \times 2$ which is in practice much less than the $h \times w \times h \times w$ elements of the dense correlation tensor, obtaining great memory savings in both the storage of this tensor and its subsequent processing. For example, for a feature map of size 100×75 and $K = 10$, the sparse representation takes 3.43 MB vs. 215 MB of the dense representation, resulting in a 12× reduction of the processing time. In the case of feature maps with 200 × 150 resolution, the sparse representation takes 13.7 MB vs. 3433 MB for the dense representation. This allows Sparse-NCNet to also process feature maps at this resolution, something that was not possible with NCNet due to the high memory requirements. The proposed *sparse correlation tensor* is a compromise between the common procedure of taking the best scoring match and the approach taken by NCNet, where all pairwise matches are stored. In this way, we can keep sufficient information in order avoid early mistakes, while keeping low memory consumption and processing time.

Then the sparse correlation tensor is processed by a permutation-invariant CNN ($\hat{N}(\cdot)$), to produce the output filtered correlation map $\tilde{c}^{AB}$:

$$\tilde{c}^{AB} = \hat{N}(c^{AB}). \quad (4)$$

The permutation invariant CNN $\hat{N}(\cdot)$ consists of applying the 4D CNN $N(\cdot)$ twice such that the same output matches are obtained regardless of the order of the input images:

$$\hat{N}(c^{AB}) = N(c^{AB}) + \left(N\left((c^{AB})^T\right)\right)^T, \quad (5)$$

where by transposition we mean exchanging the first two dimensions with the last two dimensions, which correspond to the coordinates of the two input images. The 4D CNN $N(\cdot)$ operates on the 4D space of correspondences, and is trained to perform the neighbourhood consensus filtering. Note that while $N(\cdot)$ is a sparse CNN using submanifold sparse convolutions [21], where the active sites between the sparse input and output remain constant, the convolution kernel filters are dense (*i.e.*hypercubic).

While in the original NCNet method, a soft mutual nearest-neighbour operation $M(\cdot)$ is also performed, we have removed it as we noticed its effect was not significant when operating on the sparse correlation tensor. From the output correlation tensor $\tilde{c}^{AB}$, the output matches are computed by applying argmax at each coordinate:

$$\big((i,j),(k,l)\big) \text{ a match if } \begin{cases} (i,j) = \underset{(a,b)}{\text{argmax}}\ \tilde{c}^{AB}_{abkl}, \text{ or} \\ (k,l) = \underset{(c,d)}{\text{argmax}}\ \tilde{c}^{AB}_{ijcd} \end{cases}, \quad (6)$$

where (i,j) is the match coordinate in the sampling grid of f^A, and (k,l) is the match coordinate in the sampling grid of f^B.

3.3 Match Relocalization by Guided Search

While the sparsification of the correlation tensor presented in the previous section allows processing higher resolution feature maps, these are still several times smaller in resolution than the input images. Hence, they are not suitable for applications that require (sub) pixel feature localization such as camera pose estimation or 3D-reconstruction.

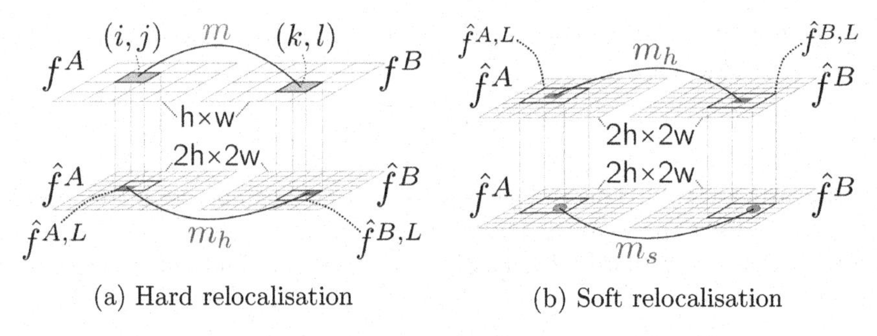

Fig. 3. Two-stage relocalization module. (a) The hard relocalization step allows to increase by 2× the localization accuracy of the matches m outputted by Sparse-NCNet, which are defined on the $h \times w$ feature maps f^A and f^B. This is done by keeping the most similar match m_h between two 2×2 local features $\hat{f}^{A,L}$ and $\hat{f}^{B,L}$, cropped from the $2h \times 2w$ feature maps $\hat{f}^A$ and $\hat{f}^B$. (b) The soft relocalization step then refines the position of these matches in the $2h \times 2w$ grid, by computing sub-feature-grid soft localization displacements based on the softargmax operation.

To address this issue, in this paper we propose a two-stage relocalization module based on the idea of guided search. The intuition is that we search for accurately localized matches on $2h \times 2w$ resolution dense feature maps, guided by the coarse matches output by Sparse-NCNet at $h \times w$ resolution. For this, dense features are first extracted at twice the normal resolution $(\hat{f}^A, \hat{f}^B) \in \mathbb{R}^{2h \times 2w \times c}$, which is done by upsampling the input image by 2× before feeding it into the feature extraction CNN $F(\cdot)$. Note that these higher resolution features are used for relocalization only, *i.e.*they are not used to compute the correlation tensor or processed by the 4D CNN for match-filtering, which would be too expensive. Then, these dense features are downsampled back to the normal $h \times w$ resolution by applying a 2×2 max-pooling operation with a stride of 2, obtaining f^A and f^B. These low resolution features $(f^A, f^B) \in \mathbb{R}^{h \times w \times c}$ are processed by Sparse-NCNet, which outputs matches in the form $m = \big((i,j),(k,l)\big)$, with the coordinates (i,j) and (k,l) indicating the position of the match in f^A and f^B, respectively, as described by (6).

Having obtained the output matches in $h \times w$ resolution, the first step (hard relocalization) consists in finding the best equivalent match in the $2h \times 2w$ resolution grid. This is done by analyzing the matches between two local crops of the high resolution features $\hat{f}^A$ and $\hat{f}^B$, and keeping the highest-scoring one. The second step (soft relocalization) then refines this correspondence further, by obtaining a sub-feature accuracy in the $2h \times 2w$ grid. These two relocalization steps are illustrated in Fig. 3, and are now described in detail.

Hard Relocalization. The first step is hard relocalization, which can improve localization accuracy by 2×. For each match $m = \big((i,j),(k,l)\big)$, the 2× upsampled coordinates $\big((2i,2j),(2k,2l)\big)$ are first computed, and 2×2 local feature crops $\hat{f}^{A,L}, \hat{f}^{B,L} \in \mathbb{R}^{2\times 2\times c}$ are sampled around these coordinates from the high resolution feature maps $\hat{f}^A$ and $\hat{f}^B$:

$$\hat{f}^{A,L} = (\hat{f}^{A}_{ab:})_{\substack{2i\leq a\leq 2i+1,\\ 2j\leq b\leq 2j+1}} \quad (7)$$

and similarly for $\hat{f}^{B,L}$. This is done using a ROI-pooling operation [16]. Finally, exhaustive matches between the local feature crops $\hat{f}^{A,L}$ and $\hat{f}^{B,L}$ are computed, and the output of the hard relocalization module is the displacement associated with the maximal matching score:

$$\Delta m_h = \big((\delta i, \delta j), (\delta k, \delta l)\big) = \operatorname*{argmax}_{(a,b),(c,d)} \langle \hat{f}^{A,L}_{ab:}, \hat{f}^{B,L}_{cd:} \rangle. \quad (8)$$

Then, the final match location from the hard relocalization stage is computed as:

$$m_h = 2m + \Delta m_h = \big((2i + \delta i, 2j + \delta j), (2k + \delta k, 2l + \delta l)\big). \quad (9)$$

Note that the relocalized matches m_h are defined in a $2h \times 2w$ grid, therefore obtaining a $2\times$ increase in localization accuracy with respect to the initial matches m, which are defined in a $h\times w$ grid. Also note that while the implementation is different, the effect of the proposed hard relocalisation is similar to the max-argmax operation used in NCNet [44], while being more memory efficient as it avoids the computation of the a dense correlation tensor in high resolution.

Soft Relocalization. The second step consists of a soft relocalization operation that obtains sub-feature localization accuracy in the $2h \times 2w$ grid of high resolution features $\hat{f}^A$ and $\hat{f}^B$. For this, new 3×3 local feature crops $(\hat{f}^{A,L}, \hat{f}^{B,L}) \in \mathbb{R}^{3\times 3\times c}$ are sampled around the coordinates of the estimated matches m_h from the previous relocalization stage. Note that no upsampling of the coordinates is done in this case, as the matches are already in the $2h \times 2w$ range. Then, soft relocalization displacements are computed by performing the softargmax operation [58] on the matching scores between the central feature of $\hat{f}^{A,L}$ and the whole of $\hat{f}^{B,L}$, and vice versa:

$$\Delta m_s = \big((\delta i, \delta j), (\delta k, \delta l)\big) \text{ where } \begin{cases} (\delta i, \delta j) = \operatorname*{softargmax}_{(a,b)} \langle \hat{f}^{A,L}_{ab:}, \hat{f}^{B,L}_{11:} \rangle \\ (\delta k, \delta l) = \operatorname*{softargmax}_{(c,d)} \langle \hat{f}^{A,L}_{11:}, \hat{f}^{B,L}_{cd:} \rangle \end{cases} \quad (10)$$

The intuition of the softargmax operation is that it computes a weighted average of the candidate positions in the crop where the weights are given by the softmax of the matching scores. The final matches from soft relocalization are obtained by applying the soft displacements to the matches from hard relocalization: $m_s = m_h + \Delta m_s$.

4 Experimental Evaluation

We evaluate the proposed Sparse-NCNet method on three different benchmarks: (i) HPatches Sequences, which evaluates the matching task directly, (ii) InLoc, which targets the problem of indoor 6-dof camera localization and (iii) Aachen Day-Night, which targets the problem of outdoor 6-dof camera localization with challenging day-night illumination changes. We first present the implementation details followed by the results on these three benchmarks. Additional 3D reconstruction results are in the extended version of this work [42].

Implementation Details. We train the Sparse-NCNet model following the training protocol from [44]. We use the IVD dataset with the weakly-supervised mean matching score loss for training [44]. The 4D CNN $N(\cdot)$ has two sparse convolution layers with 3^4 sized kernels, with 16 output channels in the hidden layer. A value of $K = 10$ is used for computing c^{AB} (3). The model is implemented using PyTorch [39], MinkowskiEngine [10] and Faiss [23], and trained for 5 epochs using Adam [25] with a learning rate of 5×10^{-4}. A pretrained ResNet-101 (up to `conv_4_23`) with no strided convolutions in the last block is used as the feature extractor $F(\cdot)$. This feature extraction model is not finetuned as the training dataset is small (3861 image pairs) and that would lead to overfitting and loss of generalization. The softargmax operation in (10) uses a temperature value of 10. In the following experiments, all correspondences are first obtained according to (6), and then only the top-scored correspondences according to the value of $\tilde{c}^{AB}$ are kept (typically between 500–2000).

4.1 HPatches Sequences

The HPatches Sequences [3] benchmark assesses the matching accuracy under strong *viewpoint* and *illumination* variations. We follow the evaluation procedure from [13], where 108 image sequences are employed, each from a different planar scene, and each containing 6 images. The first image from each sequence is matched against the remaining 5 images. The benchmark employs 56 sequences with viewpoint changes, and constant illumination conditions, and 52 sequences with illumination changes and constant viewpoint. The metric used for evaluation is the mean matching accuracy (MMA) [13]. Further details about this metric are provided in the extended version [42].

Ablations. In Fig. 4 we present ablations and a comparison with NCNet. The benefits of sparsification are shown by comparing Sparse-NCNet and NCNet under equal conditions, both without relocalization (methods A1 vs. A2), and with hard relocalization only (methods B1 vs. B2). The results in Fig. 4 show that Sparse-NCNet can obtain significant reductions in processing time and memory consumption, while keeping almost the same matching performance. Furthermore, we show that Sparse-NCNet+hard-relocalization (B1) produces superior results to Sparse-NCNet alone (A1). Finally, we show that using the two-stage relocalization (C1) produces higher matching accuracy than only using hard relocalization (B1), with minimal impact on run time or memory requirements. We have also experimented with replacing our relocalization module with the one from DenseSfM [57]. This resulted in a drop of 11% of the MMA@5px on HPatches, from 87% to 76%, showing the superiority of our approach.

	Method	Feature resolution	Reloc. method	Reloc. resolution	Mean time (s)	Peak VRAM (MB)
A1.	Sparse-NCNet	100 × 75	—	—	0.83	251
A2.	NCNet	100 × 75	—	—	9.81	5763
B1.	Sparse-NCNet	100 × 75	H	200 × 150	1.55	1164
B2.	NCNet	100 × 75	H	200 × 150	10.56	7580
C1.	Sparse-NCNet	100 × 75	H+S	200 × 150	1.56	1164
C2.	Sparse-NCNet	200 × 150	H+S	400 × 300	7.51	2391

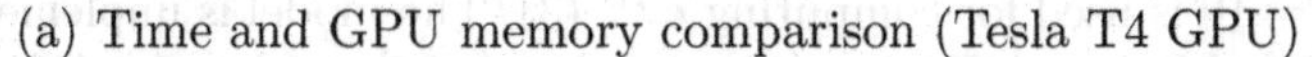
(a) Time and GPU memory comparison (Tesla T4 GPU)

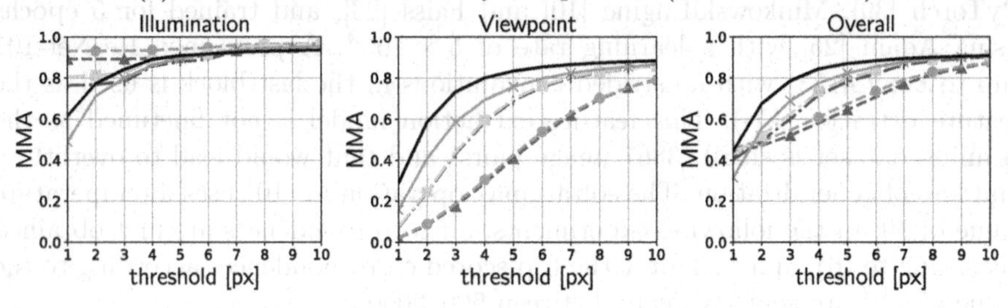

(b) MMA on HPatches Sequences

Fig. 4. Ablations and comparison with NCNet. Sparse-NCNet can obtain equivalent results to NCNet, both without relocalization (*c.f.*A1 vs. A2), and with hard relocalization (H) (*c.f.*B1 vs. B2), while greatly reducing execution time and memory consumption. The proposed two-stage relocalization (H+S) brings an improvement in matching accuracy with a minor increase in execution time (*c.f.*C1 vs. B1). Finally, the reduced memory consumption in Sparse-NCNet allows for processing in higher resolution, which produces the best results, while still being faster and more memory efficient than NCNet (*c.f.*C2 vs. B2).

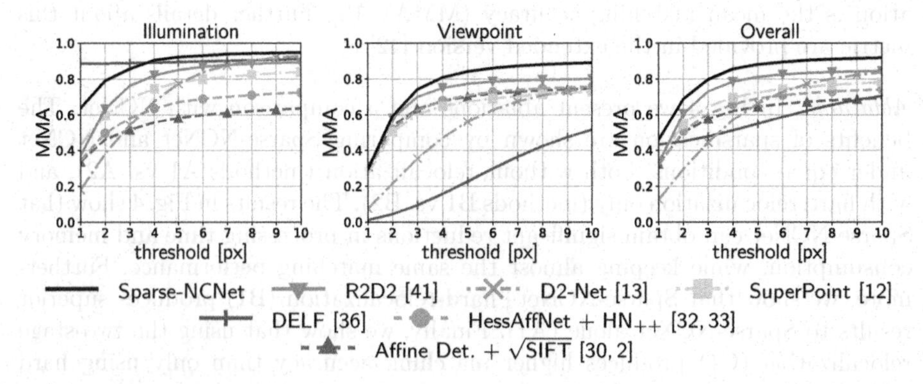

Fig. 5. Sparse-NCNet vs. state-of-the-art on HPatches. The MMA of Sparse-NCNet and several state-of-the-art methods is shown. Sparse-NCNet obtains the best results overall with a large margin over the recent R2D2 method.

Sparse-NCNet vs. State-of-the-art Methods. In addition, we compare the performance of Sparse-NCNet against several methods, including state-of-the-art trainable methods such as SuperPoint [12], D2-Net [13] or R2D2 [41]. The mean-matching accuracy results are presented in Fig. 5. For all other methods, the top 2000 features points where selected from each image, and matched enforcing mutual nearest-neighbours, yielding approximately 1000 correspondences per image pair. For Sparse-NCNet, the top 1000 correspondences where selected for each image pair, for a fair comparison. Sparse-NCNet obtains the best results for the *illumination* sequences for thresholds higher than 4 pixels, and in the *viewpoint* sequences for all threshold values. Sparse-NCNet obtains the best results overall, with a large margin over the state-of-the-art R2D2 method. We believe this could be attributed to the usage of dense descriptors (which avoid the loss of detections) together with an increased matching robustness from performing neighbourhood consensus. Qualitative examples and a comparison with other methods are presented in the extended version of this work [42].

4.2 InLoc Benchmark

The InLoc benchmark [53] targets the problem of indoor localization. It contains a set of *database* images of a building, obtained with a 3D scanner, and a set of *query* images from the same building, captured with a cell-phone several months later. The task is then to obtain the 6-dof camera positions of the query images. We follow the DensePE approach proposed [53] to find the top 10 candidate database images for each query, and employ Sparse-NCNet to obtain matches between them. Then, we follow again the procedure in [53] to obtain the final estimated 6-dof query pose, which consists of running PnP [14] followed by dense pose verification [53].

The results are presented in Fig. 6. First, we observe that Sparse-NCNet with hard relocalization (H) and a resolution of 100×75 obtains equivalent results to NCNet (methods B vs. C), while being almost $7\times$ faster and requiring $6.5\times$ less memory, confirming what was already observed in the HPatches benchmark (*c.f.*B1 vs. B2 in Fig. 4a). Moreover, our proposed Sparse-NCNet method with two-stage relocalization (H+S) in the higher 200×150 resolution (method A) obtains the best results and sets a new state-of-the-art for this benchmark. Recall that it is impossible to use the original NCNet on the higher resolution due to its excessive memory requirements. Qualitative examples are included in the extended version [42].

4.3 Aachen Day-Night

The Aachen Day-Night benchmark [47] targets 6-dof outdoor camera localization under challenging illumination conditions. It contains 98 night-time query images from the city of Aachen, and a shortlist of 20 day-time images for each night-time query. Sparse-NCNet is used to obtain matches between the query and images in the short-list. The resulting matches are then processed by the 3D reconstruction software COLMAP [50] to obtain the estimated query poses.

Table 1. Results on Aachen Day-Night. Sparse-NCNet is able to localize a similar number of queries as R2D2 and D2-Net.

	localized (%)		
Method	0.5m, 2°	1.0m,5°	5.0m,10°
RootSIFT [2,28]	36.7	54.1	72.5
DenseSfM [47]	39.8	60.2	84.7
HessAffNet + HN_{++} [32,33]	39.8	61.2	77.6
DELF [36]	38.8	62.2	85.7
SuperPoint [12]	42.8	57.1	75.5
D2-Net [13]	44.9	66.3	**88.8**
D2-Net (Multi-scale) [13]	44.9	64.3	**88.8**
R2D2 (patch = 16) [41]	44.9	67.3	87.8
R2D2 (patch = 8) [41]	**45.9**	66.3	**88.8**
Sparse-NCNet (H, 200 × 150)	44.9	**68.4**	86.7

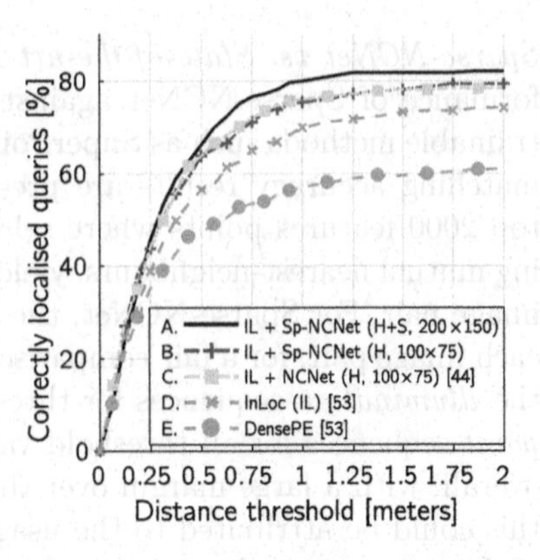

Fig. 6. Results on the InLoc benchmark. Our proposed method (A) obtains state-of-the-art results on this benchmark.

The results are presented in Table 1. Sparse-NCNet presents a similar performance to the state-of-the-art methods D2-Net [13] and R2D2 [41]. Note that the results of these three different methods differ by only a few percent, which represents only 1 or 2 additionally localized queries, from the 98 total night-time queries. The proposed Sparse-NCNet obtains state-of-the-art results for the 1m and 5° threshold, being able to localize 68.4% of the queries (67 out of 98). Qualitative examples are shown in Fig. 1 and in the extended version [42].

5 Conclusion

In this paper we have developed Sparse Neighbourhood Consensus Networks for efficiently estimating correspondences between images. Our approach overcomes the main limitations of the original Neighbourhood Consensus Networks that demonstrated promising results on challenging matching problems, making these models practical and widely applicable. The proposed model jointly performs feature extraction, matching and robust match filtering in a computationally efficient manner, outperforming state-of-the-art results on two challenging matching benchmarks. The entire pipeline is end-to-end trainable, which opens-up the possibility for including additional modules for specific downstream problems such as camera pose estimation or 3D reconstruction.

Acknowledgments. This work was partially supported by the European Regional Development Fund under project IMPACT (reg. no. CZ.02.1.01/0.0/0.0/15 003/0000468), Louis Vuitton ENS Chair on Artificial Intelligence, and the French government under management of Agence Nationale de la Recherche as part of the "Investissements d'avenir" program, reference ANR-19-P3IA-0001 (PRAIRIE 3IA Institute).

References

1. Arandjelović, R., Gronat, P., Torii, A., Pajdla, T., Sivic, J.: NetVLAD: CNN architecture for weakly supervised place recognition. In: CVPR (2016)
2. Arandjelović, R., Zisserman, A.: Three things everyone should know to improve object retrieval. In: Proceedings CVPR, pp. 2911–2918 (2012)
3. Balntas, V., Lenc, K., Vedaldi, A., Mikolajczyk, K.: HPatches: A benchmark and evaluation of handcrafted and learned local descriptors. In: Proceedings CVPR (2017)
4. Balntas, V., Hammarstrand, L., Heijnen, H., Kahl, F., Maddern, W., Mikolajczyk, K., et al.: Workshop in long-term visual localization under changing conditions. In: CVPR (2019). https://www.visuallocalization.net/workshop/cvpr/2019/
5. Balntas, V., Johns, E., Tang, L., Mikolajczyk, K.: PN-Net: Conjoined triple deep network for learning local image descriptors (2016). arXiv preprint arXiv:1601.05030
6. Balntas, V., Riba, E., Ponsa, D., Mikolajczyk, K.: Learning local feature descriptors with triplets and shallow convolutional neural networks. In: Proceedings BMVC (2016)
7. Bay, H., Tuytelaars, T., Van Gool, L.: SURF: speeded up robust features. In: Leonardis, A., Bischof, H., Pinz, Axel (eds.) ECCV 2006. LNCS, vol. 3951, pp. 404–417. Springer, Heidelberg (2006). https://doi.org/10.1007/11744023_32
8. Bian, J., Lin, W.Y., Matsushita, Y., Yeung, S.K., Nguyen, T.D., Cheng, M.M.: GMS: Grid-based motion statistics for fast, ultra-robust feature correspondence. In: Proceedings CVPR (2017)
9. Brachmann, E., Rother, C.: Neural-guided RANSAC: learning where to sample model hypotheses. In: Proceedings of the IEEE International Conference on Computer Vision, pp. 4322–4331 (2019)
10. Choy, C., Gwak, J., Savarese, S.: 4D Spatio-temporal ConvNets: Minkowski convolutional neural networks. In: Proceedings CVPR (2019)
11. Choy, C., Park, J., Koltun, V.: Fully convolutional geometric features. In: Proceedings ICCV (2019)
12. DeTone, D., Malisiewicz, T., Rabinovich, A.: SuperPoint: self-supervised interest point detection and description. In: CVPR Workshops (2018)
13. Dusmanu, M., et al.: D2-Net: a trainable CNN for joint detection and description of local features. In: Proceedings CVPR (2019)
14. Gao, X.S., Hou, X.R., Tang, J., Cheng, H.F.: Complete solution classification for the perspective-three-point problem. IEEE PAMI **25**(8), 930–943 (2003)
15. Germain, H., Bourmaud, G., Lepetit, V.: Sparse-to-dense hypercolumn matching for long-term visual localization. In: 3DV (2019)
16. Girshick, R.: Fast R-CNN. In: Proceedings ICCV (2015)
17. Gojcic, Z., Zhou, C., Wegner, J.D., Guibas, L.J., Birdal, T.: Learning multiview 3D point cloud registration (2020). arXiv preprint arXiv:2001.05119
18. Grabner, A., Roth, P.M., Lepetit, V.: 3D pose estimation and 3D model retrieval for objects in the wild. In: Proceedings CVPR (2018)
19. Graham, B.: Sparse 3D convolutional neural networks (2015). arXiv preprint arXiv:1505.02890
20. Graham, B.: Spatially-sparse convolutional neural networks (2014). arXiv preprint arXiv:1409.6070
21. Graham, B., Engelcke, M., van der Maaten, L.: 3D semantic segmentation with submanifold sparse convolutional networks. In: Proceedings CVPR (2018)

22. Han, X., Leung, T., Jia, Y., Sukthankar, R., Berg, A.C.: MatchNet: unifying feature and metric learning for patch-based matching. In: Proceedings CVPR (2015)
23. Johnson, J., Douze, M., Jégou, H.: Billion-scale similarity search with GPUs (2017). arXiv preprint arXiv:1702.08734
24. Julesz, B.: Towards the automation of binocular depth perception. In: Proceedings IFIP Congress, pp. 439–444 (1962)
25. Kingma, D.P., Ba, J.: Adam: a method for stochastic optimization. In: ICLR (2015)
26. Laguna, A.B., Riba, E., Ponsa, D., Mikolajczyk, K.: Key. Net: keypoint detection by handcrafted and learned CNN filters. In: Proceedings ICCV (2019)
27. Lenc, K., Vedaldi, A.: Learning covariant feature detectors. In: Hua, G., Jégou, He (eds.) ECCV 2016. LNCS, vol. 9915, pp. 100–117. Springer, Cham (2016). https://doi.org/10.1007/978-3-319-49409-8_11
28. Lowe, D.G.: Distinctive image features from scale-invariant keypoints. IJCV **60**(2), 91–110 (2004)
29. Marr, D., Poggio, T.: Cooperative computation of stereo disparity. Science **194**(4262), 283–287 (1976)
30. Mikolajczyk, K., Schmid, C.: An affine invariant interest point detector. In: Heyden, A., Sparr, G., Nielsen, M., Johansen, P. (eds.) ECCV 2002. LNCS, vol. 2350, pp. 128–142. Springer, Heidelberg (2002). https://doi.org/10.1007/3-540-47969-4_9
31. Mikolajczyk, K., et al.: A comparison of affine region detectors. IJCV **65**(1–2), 43–72 (2005)
32. Mishchuk, A., Mishkin, D., Radenovic, F., Matas, J.: Working hard to know your neighbor's margins: local descriptor learning loss. In: NIPS (2017)
33. Mishkin, D., Radenović, F., Matas, J.: Repeatability is not enough: learning discriminative affine regions via discriminability. In: Proceedings ECCV (2018)
34. Moo Yi, K., Trulls, E., Ono, Y., Lepetit, V., Salzmann, M., Fua, P.: Learning to find good correspondences. In: Proceedings of the IEEE Conference on Computer Vision and Pattern Recognition, pp. 2666–2674 (2018)
35. Mori, K.I., Kidode, M., Asada, H.: An iterative prediction and correction method for automatic stereocomparison. Comput. Graph. Image Process. **2**(3–4), 393–401 (1973)
36. Noh, H., Araujo, A., Sim, J., Weyand, T., Han, B.: Large-scale image retrieval with attentive deep local features. In: Proceedings ICCV (2017)
37. Ono, Y., Trulls, E., Fua, P., Yi, K.M.: LF-Net: learning local features from images. In: NIPS (2018)
38. Oron, S., Dekel, T., Xue, T., Freeman, W.T., Avidan, S.: Best-buddies similarity–robust template matching using mutual nearest neighbors. IEEE PAMI **40**(8), 1799–1813 (2017)
39. Paszke, A., et al.: Automatic differentiation in PyTorch (2017)
40. Persson, M., Nordberg, K.: Lambda twist: an accurate fast robust perspective three point (P3P) solver. In: Proceedings ECCV (2018)
41. Revaud, J., Weinzaepfel, P., de Souza, C.R., Humenberger, M.: R2D2: repeatable and reliable detector and descriptor. In: NeurIPS (2019)
42. Rocco, I., Arandjelović, R., Sivic, J.: Efficient neighbourhood consensus networks via submanifold sparse convolutions (2020). https://arxiv.org/abs/2004.10566
43. Rocco, I., Arandjelović, R., Sivic, J.: Sparse neighbouhood consensus networks (2020). https://www.di.ens.fr/willow/research/sparse-ncnet/
44. Rocco, I., Cimpoi, M., Arandjelović, R., Torii, A., Pajdla, T., Sivic, J.: Neighbourhood consensus networks. In: NeurIPS (2018)
45. Rublee, E., Rabaud, V., Konolige, K., Bradski, G.: ORB: An efficient alternative to SIFT or SURF. In: Proceedings ICCV (2011)

46. Sarlin, P.E., DeTone, D., Malisiewicz, T., Rabinovich, A.: Superglue: learning feature matching with graph neural networks (2019). arXiv preprint arXiv:1911.11763
47. Sattler, T., et al.: Benchmarking 6DOF outdoor visual localization in changing conditions. In: Proceedings CVPR (2018)
48. Schaffalitzky, F., Zisserman, A.: Automated scene matching in movies. In: Lew, M.S., Sebe, N., Eakins, J.P. (eds.) CIVR 2002. LNCS, vol. 2383, pp. 186–197. Springer, Heidelberg (2002). https://doi.org/10.1007/3-540-45479-9_20
49. Schmid, C., Mohr, R.: Local grayvalue invariants for image retrieval. IEEE PAMI **19**(5), 530–535 (1997)
50. Schönberger, J.L., Frahm, J.M.: Structure-from-motion revisited. In: Conference on Computer Vision and Pattern Recognition (CVPR) (2016)
51. Schönberger, J.L., Zheng, E., Frahm, J.-M., Pollefeys, M.: Pixelwise view selection for unstructured multi-view stereo. In: Leibe, B., Matas, J., Sebe, N., Welling, M. (eds.) ECCV 2016. LNCS, vol. 9907, pp. 501–518. Springer, Cham (2016). https://doi.org/10.1007/978-3-319-46487-9_31
52. Sivic, J., Zisserman, A.: Video google: a text retrieval approach to object matching in videos. In: Proceedings ICCV (2003)
53. Taira, H., et al.: InLoc: indoor visual localization with dense matching and view synthesis. In: Proceedings CVPR (2018)
54. Tian, Y., Fan, B., Wu, F.: L2-Net: deep learning of discriminative patch descriptor in Euclidean space. In: Proceeding CVPR (2017)
55. Torii, A., Arandjelović, R., Sivic, J., Okutomi, M., Pajdla, T.: 24/7 place recognition by view synthesis. In: CVPR (2015)
56. Verdie, Y., Yi, K., Fua, P., Lepetit, V.: TILDE: a temporally invariant learned detector. In: Proceedings CVPR (2015)
57. Widya, A.R., Torii, A., Okutomi, M.: Structure from motion using dense cnn features with keypoint relocalization. IPSJ Trans. Comput. Vis. Appl. **10**(1), 6 (2018)
58. Yi, K.M., Trulls, E., Lepetit, V., Fua, P.: LIFT: learned invariant feature transform. In: Leibe, B., Matas, J., Sebe, N., Welling, M. (eds.) ECCV 2016. LNCS, vol. 9910, pp. 467–483. Springer, Cham (2016). https://doi.org/10.1007/978-3-319-46466-4_28
59. Zagoruyko, S., Komodakis, N.: Learning to compare image patches via convolutional neural networks. In: Proceedings CVPR (2015)
60. Zhang, J., et al.: Learning two-view correspondences and geometry using order-aware network. In: Proceedings of the IEEE International Conference on Computer Vision, pp. 5845–5854 (2019)
61. Zhang, Z., Deriche, R., Faugeras, O., Luong, Q.T.: A robust technique for matching two uncalibrated images through the recovery of the unknown epipolar geometry. Artif. Intell. **78**(1–2), 87–119 (1995)
62. Zhao, W.L., Jégou, H., Gravier, G.: Oriented pooling for dense and non-dense rotation-invariant features. In: Proceedings BMVC (2013)
63. Zhou, H., Sattler, T., Jacobs, D.W.: Evaluating local features for day-night matching. In: Hua, G., Jégou, H. (eds.) ECCV 2016. LNCS, vol. 9915, pp. 724–736. Springer, Cham (2016). https://doi.org/10.1007/978-3-319-49409-8_60

Reconstructing the Noise Variance Manifold for Image Denoising

Ioannis Marras[1(✉)], Grigorios G. Chrysos[2], Ioannis Alexiou[1], Gregory Slabaugh[1], and Stefanos Zafeiriou[2]

[1] Huawei Noah's Ark, London, UK
{ioannis.marras,ioannis.alexiou,gregory.slabaugh}@huawei.com
[2] Imperial College London, London, UK
{g.chrysos,s.zafeiriou}@imperial.ac.uk

Abstract. Deep Convolutional Neural Networks (CNNs) have been successfully used in many low-level vision problems like image denoising. Although the conditional image generation techniques have led to large improvements in this task, there has been little effort in providing conditional generative adversarial networks (cGANs) with an explicit way of understanding the image noise for object-independent denoising reliable for real-world applications. The task of leveraging structures in the target space is unstable due to the complexity of patterns in natural scenes, so the presence of unnatural artifacts or over-smoothed image areas cannot be avoided. To fill the gap, in this work we introduce the idea of a cGAN which explicitly leverages structure in the image noise variance space. By learning directly a low dimensional manifold of the image noise variance, the generator promotes the removal from the noisy image only that information which spans this manifold. This idea brings many advantages while it can be appended at the end of any denoiser to significantly improve its performance. Based on our experiments, our model substantially outperforms existing state-of-the-art architectures, resulting in denoised images with less over-smoothing and better detail.

1 Introduction

During image acquisition, due to the presence of noise some image corruption is inevitable and can degrade the visual quality considerably. Therefore, noise removal is essential for many digital imaging and computer vision applications [22] and remains an important and active research topic.

Denoising algorithms can be grouped in two categories: learning-based and model-based. Modelling the image prior from a set of noisy and ground-truth image sets is the goal of discriminative learning. The performance of the current learning models is limited by their inadequacy of handling all possible levels of noise in a single model. In this category are methods such as brute force learning

Electronic supplementary material The online version of this chapter (https://doi.org/10.1007/978-3-030-58545-7_36) contains supplementary material, which is available to authorized users.

© Springer Nature Switzerland AG 2020
A. Vedaldi et al. (Eds.): ECCV 2020, LNCS 12354, pp. 622–639, 2020.
https://doi.org/10.1007/978-3-030-58545-7_36

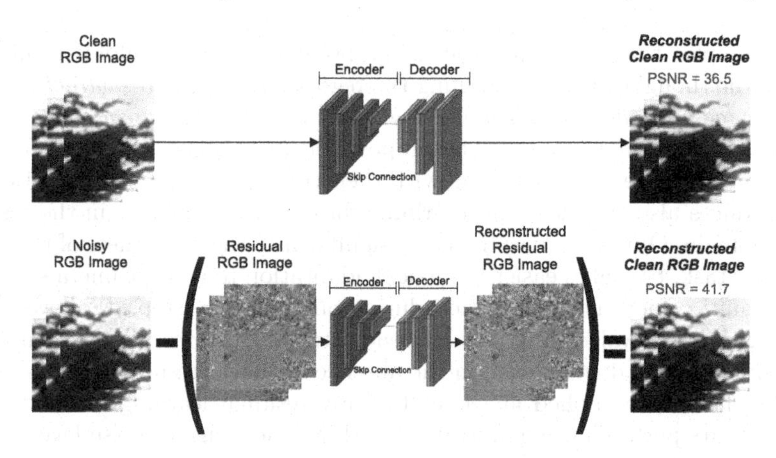

Fig. 1. Motivation of our method: By characterizing directly the image signal dependent noise, the reconstruction of the clean image is much more accurate. Instead of constraining the output of a generator to span the target space, is better to constrain it to remove from the noisy image only that information which spans the manifold of the residual image variance.

like MLP [11], CNNs [58,59] or truncated inference [15]. On the other hand, the model-based algorithms are computationally expensive, and unable to characterize complex image textures. In the this category fall algorithms including external priors [7], Markov random field models [49,52], gradient methods [54,55], non-local self-similarity [33] and sparsity (e.g. MCWNNM [24]).

A denoising algorithm should be efficient, perform denoising using a single model and handle spatially variant noise when the noise standard-deviation is known or unknown. The physics of digital sensors and the steps of an imaging pipeline are well-understood and can be leveraged to generate training data from almost any image using only basic information about the target camera sensor. Recent work has shifted to sophisticated signal-dependent single source noise models [27] that better match the physics of image formation [9,38,43]. Also, adapting a learned denoising algorithm to a new camera sensor may require capturing a new dataset. However, capturing noisy and noise-free image pairs is difficult, requiring long exposures or large bursts of images, and post-processing to combat camera motion and lighting changes.

In this paper, we introduce the idea of a cGAN [44] which directly constrains the image spatially variant noise for image denoising (Fig. 1). In this way, we avoid the direct characterization of the space of clean images, since the complexity of natural image patterns is extremely high. To do so, a combination of supervised (regression) and unsupervised (autoencoder) '*encoder-decoder*' type subnets applies implicit constraints in the residual image (the difference between the noisy observation and the clean image) variance latent subspace. By adopting the idea of residual learning [58] in the regression subnet and using a shared decoder, the unsupervised subnet is explicitly constrained to generate residual image samples that span only the image noise variance manifold. Intuitively, this can be thought of as constraining the regression subnet to *subtract from the noisy image only the residual image that looks like realistic image noise coming from a specific camera sensor*. The proposed idea: a) allows the direct association of one or more camera sensors with their corresponding noise statistics and b) introduces also the idea of a discriminator operating directly in the residual image domain. Our system: a) increases significantly the robustness of the image denoising task, b) makes easier the model adaptation to a new camera sensor, c) allows multi-camera noise reduction during one inference step, d) allows multi-source noise removal during one inference step, e) utilizes all the samples in the residual image domain even in the absence of the corresponding noisy input samples, f) can be applied at the end of any residual learning based denoiser improving its performance and g) deals with a wide range of noise levels.

2 Related Work

2.1 Image Prior Based Methods

Image prior based methods, e.g. NSCR [21], TWSC [56], WNNM [24], can be employed to solve the denoising problem of unknown noise because they do not require training data since they model the image prior over the noisy image directly. The classic BM3D [19] method is based on the idea that natural images usually contain repeated patterns (non-local self-similarity model). In non-local means (NLM) [10], the pixel values are predicted based on their noisy surroundings. Many variants of NLM and BM3D seeking self-similar patches in different transform domains were proposed, e.g. SAPCA [33], NLB [35]. Sparsity is enforced by dictionary-based methods [20] by employing self-similar patches and learning over-complete dictionaries from clean images. In contrast, Noise2Void (N2V) [34] and Noise2Noise (N2N) [37] do not require training noisy image pairs, nor clean target images. N2N attempts to learn a mapping between pairs of independently degraded versions of the same training image. For image patch restoration, maximum likelihood algorithms like Gaussian Mixture Models (GMMs), were employed to learn statistical priors from image patch groups [13,57]. Dictionary learning based and basis-pursuit based algorithms such as KSVD [4], Fields-of-Experts or TNRD [16] operated by finding image representations where sparsity holds or statistical regularities are well-modeled [61]. In [36], an extension of non-local Bayes approach, named NC, was proposed to model the noise of each patch group to be zero-mean correlated and Gaussian distributed. The disadvantage of this category of methods is that external information from possible many other images taken under the same condition with the image to be denoised cannot be used. Furthermore, the generalization capabilities are limited because these methods are defined mostly based on human knowledge.

2.2 Discriminative Deep Learning Methods

In recent years, CNNs have achieved great success in image denoising. The first attempt of employing CNNs for this task was made in [30]. Discriminative deep learning methods are trained offline, extracting information from ground truth annotated training sets before they are applied to test data. In DnCNN [58] and IrCNN [59] networks, stacked convolution, batch normalization and ReLU layers were used to estimate the residual image [26]. By adding symmetric skip connections, an improved encoder-decoder network for image denoising based on residual learning was proposed in [41]. A densely connected denoising network, named Memnet, constructed in [51] to enable memory of the network. A multi-level wavelet CNN (MWCNN) model based on a U-Net architecture used in [40] to incorporate large receptive field for image denoising. By incorporating non-local operations into a recurrent neural network (RNN), a non-local recurrent network (NLRN) for image restoration presented in [39]. A network named N^3Net [47] employed the k-nearest neighbor matching in the denoising network to exploit the non-local property of the image features. A fast and flexible network (FFDNet) which can process images with non-uniform noise corruption proposed in [60]. A residual in the residual structure (RIDNet) used in [6] to ease the flow of low-frequency information and apply feature attention to exploit the channel dependencies. Recently, a blind denoising model for real photographs named CBDNet [25] is composed of two subnetworks: noise estimation and non-blind denoising. A self-guided network (SGN), which adopts a top-down self-guidance architecture to better exploit image multi-scale information presented in [23]. FOCNet network [31] solved a fractional optimal control problem in a multi-scale approach. Although the methods in this category achieved high denoising quality, they cannot work in the absence of paired training data.

2.3 Generative Models

GANs were recently trained to synthesize noise [14], thus pairs of corresponding clean and noisy images were obtained for training CNNs. Any further filtering of the RAW image changes the real noise statistics making that task very difficult [14]. Also, is not realistic to create noisy images by adding random generated noise to the clean images since in real images the noise variance is data dependent. Noise Flow method [1] combined well-established basic parametric noise models (e.g. signal-dependent noise) with the flexibility and expressiveness of normalizing flow architectures to model noise distributions observed from large datasets of real noisy images. However, it is not clear how to quantitatively assess the quality of the generated samples.

3 Our Method

In this section, we introduce our system for the task of image denoising. The goal is to produce a single clean (RGB or RAW) image from a corresponding single noisy (RGB or RAW) image. Firstly, we give a brief overview of the noise signal in real images (Sect. 3.1). Our method falls in the category of conditional image generation methods, thus to make the paper self-contained we briefly describe this category (Sect. 3.2) before introducing our method (Sect. 3.3).

3.1 Image Noise Modeling in Real-World Images

Camera sensors output RAW data in a linear color space where pixel measurements are proportional to the number of photo-electrons collected. The primary sources of noise are shot noise, a Poisson process with variance equal to the signal level, and read noise, an approximately Gaussian process caused by a variety of sensor readout effects. The noise is spatially variant; hence, the assumption that noise is spatially invariant does not hold for real images. The noise is well-modeled by a signal-dependent Gaussian distribution [27]:

$$x_p \sim \mathcal{N}\left(y_p, \sigma_r^2 + \sigma_s y_p\right) \quad (1)$$

where x_p is a noisy measurement of the true intensity y_p at pixel p. The parameters σ_r and σ_s: a) are fixed, given a specific camera sensor, for each sensor gain (ISO) value and varies as ISO changes and b) are different for different camera sensors. Since the noise is structured (not random) a low-dimensional manifold for noise variance exists. A realistic noise model is important aspect in training CNN-based denoising methods for real photographs [6,25].

3.2 Conditional Image Generation

In computer vision, the task of conditional image generation is dominated by approaches similar to a GAN. The GAN consists of a generator and a discriminator module commonly optimized with alternating gradient descent methods. cGAN extend the formulation by providing the generator with additional labels. The generator G takes the form of an encoder-decoder network where the encoder projects the label into a low-dimensional latent subspace and the decoder performs the opposite mapping.

cGAN and its variants like Robust cGAN [17], were applied in the past for the task of object-dependent image denoising. The encoder-decoder generator of Robust cGAN performs a similar regression as its counterpart in cGAN. It accepts a sample from the source domain and maps it to the target domain by using a second CNN in the target domain which promotes more realistic regression outputs. Recently in non GAN-based methods, generators adopting a similar architecture were proposed for object-dependent image denoising. In [29], a two-tailed CNN is employed – for inferring the clean image and the noise separately. The input noisy image is decoupled to the signal and noise in a latent space, while a decoder used to generate the signal and noise in the spatial domain. There are two major drawbacks of all these methods: i) in the absence of skip connections, these methods perform well only in the case of object-dependent image denoising (i.e. face denoising [17,53]). The need of having different models for different objects makes them unsuitable for digital devices with limited resources (e.g. smartphones) where the run-time performance is of importance. ii) the purpose of their unsupervised learning sub-networks, whose (hidden) layers contain representations of the input data, is to be sufficiently powerful for compressing (and decompressing) the data while losing as little information as possible. However, even in the presence of skip connections, this procedure of defining a nonlinear representation which can accurately reconstruct image patterns from a variety of real complex objects/scenes is not realistic. As a result, these methods very often hallucinate complex image structures by introducing severe blurry effects or unusual image patterns/artifacts.

3.3 Image Denoising Based on Noise Variance Manifold Reconstruction

To tackle the problems mentioned in Sect. 3.2, the proposed method introduces the general idea of explicitly constraining the residual image removed by a denoiser to lie in the low-dimensional manifold of the signal dependent image noise variance (Sect. 3.1). Like cGAN, our method consists of a generator and a discriminator. The generator includes two subnets: the first regression (*Reg*) subnet performs regression while the second reconstruction (*Rec*) is an autoencoder in the residual image domain. Both subnets consist of similar encoder-decoder networks, while a backbone network is used prior to the encoder-decoder network of the *Reg subnet*. By sharing the weights of their decoders, the generator adopts the residual learning strategy to remove from the noisy observation that information which spans the image noise variance manifold. A schematic of the proposed generator is illustrated in Fig. 2. Rather than directly outputing the denoised image, the supervised *Reg subnet* is designed to predict the ground-truth residual image $\boldsymbol{v} = \boldsymbol{s} - \boldsymbol{y}$, where $\boldsymbol{s}$ and $\boldsymbol{y}$ stand for the noisy and the clean (ground-truth) image, respectively. Thus, the unsupervised *Rec subnet* works as a conditional auto-encoder in the domain of $\boldsymbol{v}$. The *Rec subnet* during inference is no longer required, therefore the testing complexity remains the same as in standard cGAN. Two Unet style skip connections from the encoder to the decoder used in both subnets improving the learning of the residual between the features corresponding to the image and to the residual image structures. Also, a BEGAN style decoder skip connection [8], which creates a skip connection between the first decoder layer and each successive upsampling layer of the decoder, used to help gradient propagation.

The reconstruction of $\boldsymbol{v}$ is an easier task compared to the reconstruction of $\boldsymbol{y}$ as in cGAN. Thus, we can learn how to turn bad images into good images by only looking at the structure of the residual image. This property: i) makes our cGAN an object-independent image denoiser and ii) helps the denoiser largely avoid image over-smoothing/artifacts, something essential for image denoising.

Noising is a challenging process to be reversed by the few convolutional layers of the encoder in *Reg subnet* especially in an object-independent scenario. This is why a backbone network used to extract complex feature representations, $\boldsymbol{\varphi}$, useful to preserve for later the low and high image frequencies. Different state-of-the-art denoisers could be used as backbone networks. Thus, the proposed idea could be applied at the end of any denoiser constraining its output improving in that way its performance as it is experimentally verified in Sect. 4.

In addition, the *Rec subnet* enables utilization of all the samples in the domain of the residual image even in the absence of the corresponding noisy input samples. In the case of a well defined image noise source, like the one

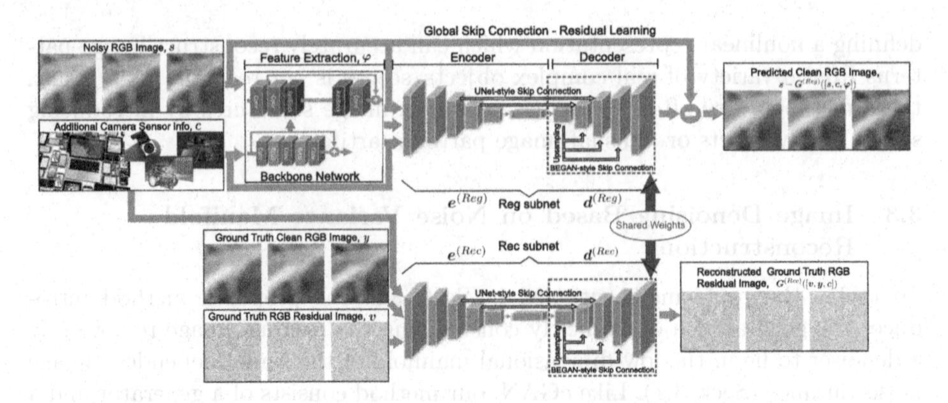

Fig. 2. Schematic of the proposed generator.

described in Sect. 3.1, a huge amount of different residual image realizations (e.g. for different ISOs) could be generated and used to train that subnet.

The adaptation of an existing model to a new camera sensor is an easier task for our method. To do so, only the *Rec subnet* must be retrained from scratch while the *Reg subnet* needs only to be fine-tuned using a small number of paired training samples obtained using the new sensor. Also, our method can remove more than one noise source during one inference step. To do so, a different noise variance manifold for each noise source is obtained, thus a different *Rec subnet* per noise source constrains the denoiser in a sequential manner (Fig. 4(a)).

The task of learning directly the image noise variance manifold can be greatly benefit by any conditional information, $\boldsymbol{c}$, related to the camera sensor. This information varies and it is provided to both subnets. $\boldsymbol{c}$ could contain the two noise parameters σ_r and σ_s (if available) associating in that way a camera sensor with its corresponding noise statistics. In the case of multi-camera noise reduction, $\boldsymbol{c}$ could additionally contain a one hot vector per pixel defining the camera id used to take each picture, thus one or more noise sources are explicitly associated with the corresponding camera sensor. More specifically, *Reg subnet* gets as input $\boldsymbol{c}$ in concatenation (denoted as $[\cdot]$) with $\boldsymbol{s}$ and $\boldsymbol{\varphi}$ and outputs $\boldsymbol{y} - \boldsymbol{G}^{(Reg)}([\boldsymbol{s}, \boldsymbol{c}, \boldsymbol{\varphi}])$, where $\boldsymbol{G}^{(Reg)}([\boldsymbol{s}, \boldsymbol{c}, \boldsymbol{\varphi}])$ is the predicted residual image. The superscript 'Reg' abbreviates modules of the *Reg subnet*. Based on Eq. 1, the noise variance for a pixel p depends, except for the camera sensor-based parameters, on y_p. Thus, the input to *Rec subnet* should be $\boldsymbol{v}$ in concatenation with $\boldsymbol{y}$ and $\boldsymbol{c}$. By giving explicitly $\boldsymbol{y}$ as additional input, *the task of the* Rec subnet *is not to learn the underlying structure of a huge variety of complex image patterns, but to learn how clean image structures are affected by the presence of structured noise.*

The proposed idea deals with a wide range of noise levels in contrast to a standard cGAN or its variants. According to [26], when the original mapping $F(\boldsymbol{s})$ (as in cGAN) is more like an identity mapping, the residual mapping will be much easier to optimize. Note that $\boldsymbol{s}$ is much more like $\boldsymbol{s} - \boldsymbol{G}^{(Reg)}([\boldsymbol{s}, \boldsymbol{c}, \boldsymbol{\varphi}])$

than $\boldsymbol{G}^{(Reg)}([\boldsymbol{s}, \boldsymbol{c}, \boldsymbol{\varphi}])$ (especially when the noise level is low). Thus, $F(\boldsymbol{s})$ would be closer to an identity mapping than $\boldsymbol{G}^{(Reg)}([\boldsymbol{s}, \boldsymbol{c}, \boldsymbol{\varphi}])$, and the residual learning formulation is more suitable for image denoising [58].

In the case of image denoising in the RGB domain, $\boldsymbol{s}$ represents 3-channel image based tensors. Regarding $\boldsymbol{s}$ in the RAW domain, each pixel in a conventional camera (linear Bayer) sensor is covered by a single red, green, or blue color filter, arranged in a 4-channel Bayer pattern (i.e. R-G-G-B). The content loss consists of two terms that compute the per-pixel difference between the predicted clean image, and the clean (ground-truth) image. The two terms are i) the ℓ_1 loss between the ground-truth image and the output of the generator, ii) the ℓ_1 of their gradients; mathematically expressed as:

$$\begin{aligned}\mathcal{L}_c = \lambda_c \cdot \sum_{n=1}^{N} ||(\boldsymbol{s}^{(n)} - \boldsymbol{G}^{(Reg)}([\boldsymbol{s}^{(n)}, \boldsymbol{c}^{(n)}, \boldsymbol{\varphi}^{(n)}])) - \boldsymbol{y}^{(n)}|| \\ + \lambda_{cg} \cdot \sum_{n=1}^{N} ||\nabla(\boldsymbol{s}^{(n)} - \boldsymbol{G}^{(Reg)}([\boldsymbol{s}^{(n)}, \boldsymbol{c}^{(n)}, \boldsymbol{\varphi}^{(n)}])) - \nabla \boldsymbol{y}^{(n)}||,\end{aligned} \tag{2}$$

where $\boldsymbol{G}^{(Reg)}([\boldsymbol{s}^{(n)}, \boldsymbol{c}^{(n)}, \boldsymbol{\varphi}^{(n)}]) = \boldsymbol{d}^{(Reg)}(\boldsymbol{e}^{(Reg)}([\boldsymbol{s}^{(n)}, \boldsymbol{c}^{(n)}, \boldsymbol{\varphi}^{(n)}]))$, N stands for the total number of training samples, $\boldsymbol{e}$ stands for encoder, $\boldsymbol{d}$ stands for decoder and λ_c, $\lambda_{cg} = 0.5 \cdot \lambda_{ae}$ are hyper-parameters to balance the loss terms. The unsupervised *Rec subnet* contributes the following loss term:

$$\mathcal{L}_{Rec} = \sum\nolimits_{n=1}^{N} [f_d(\boldsymbol{v}^{(n)}, \boldsymbol{G}^{(Rec)}([\boldsymbol{v}^{(n)}, \boldsymbol{y}^{(n)}, \boldsymbol{c}^{(n)}]))] \tag{3}$$

where $\boldsymbol{G}^{(Rec)}([\boldsymbol{v}^{(n)}, \boldsymbol{y}^{(n)}, \boldsymbol{c}^{(n)}]) = \boldsymbol{d}^{(Rec)}(\boldsymbol{e}^{(Rec)}([\boldsymbol{v}^{(n)}, \boldsymbol{y}^{(n)}, \boldsymbol{c}^{(n)}]))$ is the *Rec subnet*'s output, f_d is a divergence metric (ℓ_2 loss due to the auto-encoder in the noise domain) and the superscript 'Rec' abbreviates modules of the *Rec subnet.*

Despite sharing the weights of the decoders, the latent representations of the two subnets are forced to span the same space using a latent loss term $\mathcal{L}_{lat}$. This term minimizes the distance between the encoders' outputs, i.e. the two residual noise variance representations are spatially close. The latent loss term is:

$$\mathcal{L}_{lat} = \sum\nolimits_{n=1}^{N} ||\boldsymbol{e}^{(Reg)}([\boldsymbol{s}^{(n)}, \boldsymbol{c}^{(n)}, \boldsymbol{\varphi}^{(n)}]) - \boldsymbol{e}^{(Rec)}([\boldsymbol{v}^{(n)}, \boldsymbol{y}^{(n)}, \boldsymbol{c}^{(n)}])||. \tag{4}$$

As a part of the vanilla cGAN, the feature matching loss [28,50] enables the network to match the data and the model's distribution faster. The intuition is that to match the high-dimensional distribution of the data with *Reg subnet,* their projections in lower-dimensional spaces are encouraged to be similar. The feature matching loss is:

$$\mathcal{L}_f = \sum\nolimits_{n=1}^{N} ||\pi(\boldsymbol{s}^{(n)} - \boldsymbol{G}^{(Reg)}([\boldsymbol{s}^{(n)}, \boldsymbol{c}^{(n)}, \boldsymbol{\varphi}^{(n)}])) - \pi(\boldsymbol{y}^{(n)})||. \tag{5}$$

where $\pi()$ extracts the features from the penultimate layer of the discriminator.

Skip connections enable deeper layers to capture more abstract representations without the need of memorizing all the information. The lower-level representations are propagated directly to the decoder through the shortcut, which makes it harder to train the longer path [48]. The Decov loss term [18] used: a) to penalize the correlations in the representations of one or more layers, b) to implicitly encourage the representations to capture diverse and useful information and c) to maximize the variance captured by the longer path representations. For the jth layer this loss is defined as:

$$\mathcal{L}^j_{decov} = \frac{1}{2}(||\boldsymbol{C}^j||^2_F - ||diag(\boldsymbol{C}^j)||^2_2), \tag{6}$$

where $diag()$ computes the diagonal elements of a matrix and $\boldsymbol{C}^j$ is the covariance matrix of the layer representations. The loss is minimized when the covariance matrix is diagonal, i.e. it imposes a cost to minimize the covariance of hidden units without restricting the diagonal elements that include the variance of the hidden representations.

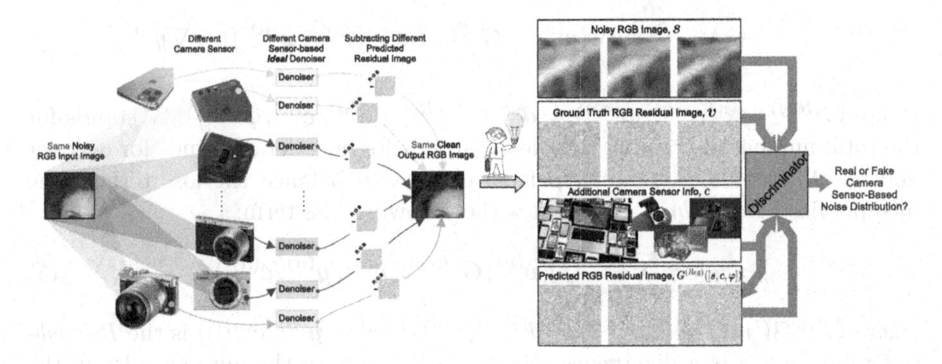

Fig. 3. The proposed discriminator for image denoising operates directly in the residual image domain.

In the case of multi-camera noise reduction, let's assume that the same scene is captured under the same lighting conditions by different camera sensors. Let's also assume that an ideal denoiser per camera sensor exists. In that case, the output of all the denoisers should be the same underlying clean image although the noise statistics of each camera can be very different. This leads to the idea of a discriminator which operates directly in the residual image domain (Fig. 3) thus trying to distinguish between the residual image samples generated by the denoiser and the ground-truth residual image distributions given a specific camera sensor. This is feasible in the proposed method because the *Rec subnet* constraints directly the denoiser to remove only that information which spans the learned noise variance manifold of each camera sensor. The generator samples $\boldsymbol{z}$ from a prior distribution $p_{\boldsymbol{z}}$, e.g. uniform, and tries to model the target distribution p_d; the discriminator D tries to distinguish between the samples generated from the model and the target image noise distributions. More specifically, the discriminator accepts as input $\boldsymbol{G}^{(Reg)}([\boldsymbol{s}, \boldsymbol{c}, \boldsymbol{\varphi}])$ along with $\boldsymbol{v}$, $\boldsymbol{c}$ and $\boldsymbol{s}$, while the standard adversarial loss of cGAN is modified to:

$$\mathcal{L}^{\star}_{adv}(\boldsymbol{G}^{(Reg)}, \boldsymbol{D}) = \mathbb{E}_{\boldsymbol{s},\boldsymbol{v}\sim p_d(\boldsymbol{s},\boldsymbol{v})}[\log \boldsymbol{D}(\boldsymbol{v}|\boldsymbol{s},\boldsymbol{c})] + \mathbb{E}_{\boldsymbol{s}\sim p_d(\boldsymbol{s}),\boldsymbol{z}\sim p_z(\boldsymbol{z})}[\log(1 - \boldsymbol{D}(\boldsymbol{G}^{(Reg)}([\boldsymbol{s},\boldsymbol{c},\boldsymbol{\varphi}])|\boldsymbol{s},\boldsymbol{c}))]. \tag{7}$$

by solving the following min-max problem:

$$\min_{\boldsymbol{w}_G}\max_{\boldsymbol{w}_D} \mathcal{L}^{\star}_{adv}(\boldsymbol{G}^{(Reg)}, \boldsymbol{D}) = \min_{\boldsymbol{w}_G}\max_{\boldsymbol{w}_D} \mathbb{E}_{\boldsymbol{s},\boldsymbol{v}\sim p_d(\boldsymbol{s},\boldsymbol{v})}[\log \boldsymbol{D}(\boldsymbol{v}|\boldsymbol{s},\boldsymbol{c},\boldsymbol{w}_D)] + \mathbb{E}_{\boldsymbol{s}\sim p_d(\boldsymbol{s}),\boldsymbol{z}\sim p_z(\boldsymbol{z})}[\log(1 - \boldsymbol{D}(\boldsymbol{G}^{(Reg)}([\boldsymbol{s},\boldsymbol{c},\boldsymbol{\varphi}]|\boldsymbol{w}_G)|\boldsymbol{s},\boldsymbol{c},\boldsymbol{w}_D))]$$

where $\boldsymbol{w}_G, \boldsymbol{w}_D$ denote the generator's and the discriminator's parameters respectively. The final loss function of our method is:

$$\mathcal{L}_{total} = \mathcal{L}^{\star}_{adv} + \mathcal{L}_c + \lambda_{\pi} \cdot \mathcal{L}_f + \lambda_{ae} \cdot \mathcal{L}_{Rec} + \lambda_l \cdot \mathcal{L}_{lat} + \lambda_d \cdot \sum^{j} \mathcal{L}^{j}_{decov}, \tag{8}$$

where λ_{π}, λ_{ae}, λ_l and λ_d are extra hyper-parameters to balance the loss terms.

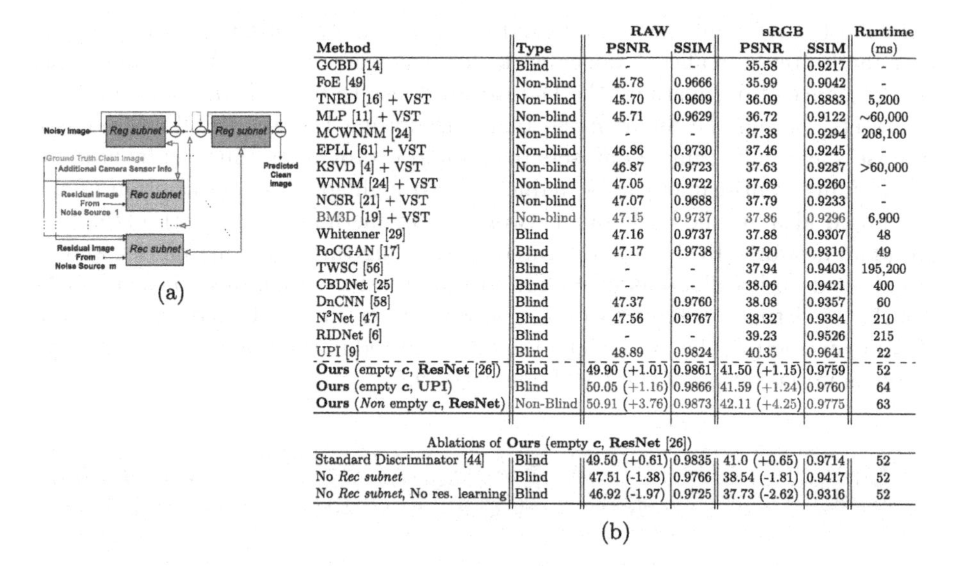

Method	Type	RAW PSNR	RAW SSIM	sRGB PSNR	sRGB SSIM	Runtime (ms)
GCBD [14]	Blind	-	-	35.58	0.9217	-
FoE [49]	Non-blind	45.78	0.9666	35.99	0.9042	-
TNRD [16] + VST	Non-blind	45.70	0.9609	36.09	0.8883	5,200
MLP [11] + VST	Non-blind	45.71	0.9629	36.72	0.9122	~60,000
MCWNNM [24]	Non-blind	-	-	37.38	0.9294	208,100
EPLL [61] + VST	Non-blind	46.86	0.9730	37.46	0.9245	-
KSVD [4] + VST	Non-blind	46.87	0.9723	37.63	0.9287	>60,000
WNNM [24] + VST	Non-blind	47.05	0.9722	37.69	0.9260	-
NCSR [21] + VST	Non-blind	47.07	0.9688	37.79	0.9233	-
BM3D [19] + VST	Non-blind	47.15	0.9737	37.86	0.9296	6,900
Whitenner [29]	Blind	47.16	0.9737	37.88	0.9307	48
RoCGAN [17]	Blind	47.17	0.9738	37.90	0.9310	49
TWSC [56]	Blind	-	-	37.94	0.9403	195,200
CBDNet [25]	Blind	-	-	38.06	0.9421	400
DnCNN [58]	Blind	47.37	0.9760	38.08	0.9357	60
N^{3}Net [47]	Blind	47.56	0.9767	38.32	0.9384	210
RIDNet [6]	Blind	-	-	39.23	0.9526	215
UPI [9]	Blind	48.89	0.9824	40.35	0.9641	22
Ours (empty $\boldsymbol{c}$, **ResNet** [26])	Blind	49.90 (+1.01)	0.9861	41.50 (+1.15)	0.9759	52
Ours (empty $\boldsymbol{c}$, **UPI**)	Blind	50.05 (+1.16)	0.9866	41.59 (+1.24)	0.9760	64
Ours (*Non* empty $\boldsymbol{c}$, **ResNet**)	Non-Blind	50.91 (+3.76)	0.9873	42.11 (+4.25)	0.9775	63
Ablations of **Ours** (empty $\boldsymbol{c}$, **ResNet** [26])						
Standard Discriminator [44]	Blind	49.50 (+0.61)	0.9835	41.0 (+0.65)	0.9714	52
No *Rec subnet*	Blind	47.51 (-1.38)	0.9766	38.54 (-1.81)	0.9417	52
No *Rec subnet*, No res. learning	Blind	46.92 (-1.97)	0.9725	37.73 (-2.62)	0.9316	52

(b)

Fig. 4. (a) In the case of camera multi-source image noise, more than one *Rec subnet* can be employed. Each subnet is responsible for removing noise structure that comes from a specific noise source, and (b) the quantitative results on the DnD benchmark of our method and its ablations. Regarding our method, in parentheses we define the type of denosing plus the used backbone network.

4 Experimental Results

4.1 Training Settings

Synthetic noisy images were combined with real noisy data to improve the generalization ability of our method to real photographs. To generate them, we followed the pipeline in [25]. To do so, we employed BSD500 [42], DIV2K [3], and MIT-Adobe FiveK [12], resulting in 3.5K images while for real noisy images, we extracted cropped patches from SSID [2] and RENOIR [5]. Finally, the data augmentation procedure results in 64×64 image patches. In our '*encoder-decoder*' architecture (same for both subnets) 11 layers were used with an latent space of dimensions MB×2 × 2 × 1024, where MB stands for the mini-batch size. The values of the additional hyper-parameters are $\lambda_{ae} = 2.5 * 10^3$, $\lambda_l = 0.5$ and $\lambda_d = 10^{-6}$. The common hyper-parameters λ_π and λ_c with the vanilla cGAN remain the same. In the beginning, the two subnets were trained separately while afterwards they were jointly trained. For both subnets: the kernel size used was 3×3; Adam [32] was used as the optimizer with default parameters; the learning rate was initially set to 10^{-3} and then halved after 10^6 iterations; ReLU activation used; the network ran for 50 epochs.

4.2 Comparisons on Real-World Images

The most three challenging public datasets that significantly improve upon earlier (and often unrealistic) benchmarks for denoising, were used to evaluate the performance of our method: the Darmstadt Noise Dataset (DnD) [46], the Nam Dataset [45] and the Smartphone Image Denoising Dataset (SIDD) [2]. These datasets are multi-camera datasets (camera id is provided), thus can be used for performing multi-camera noise reduction. To highlight the contribution of the proposed idea, as the backbone network in our method we used: a) a standard residual network (ResNet) [26] created by stacking three building blocks, and b) the best deep learning-based method in the literature according to each benchmark, if existing, excluding the last network layer since this network acts as a feature extractor. The pre-trained weights reported in the literature, if available, used as initialization of the backbone network.

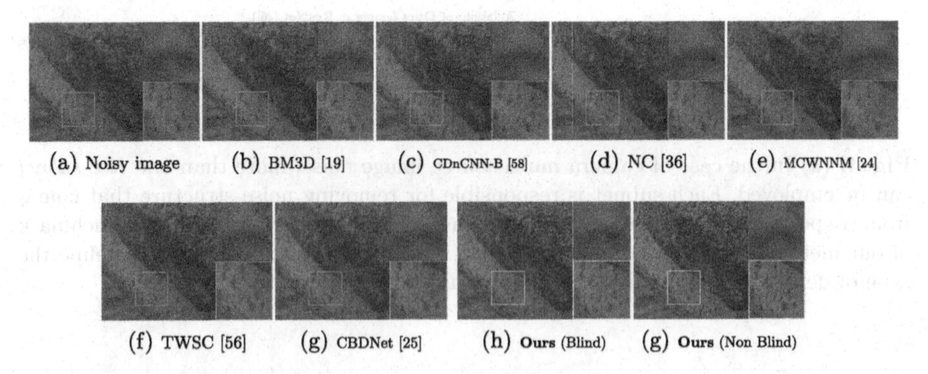

Fig. 5. Example of image denoising of a DnD image. Results of the proposed method shown when a standard ResNet used as backbone network.

Evaluation on DnD: DnD is a novel benchmark dataset which consists of realistic uncompressed photos from 50 scenes taken by 4 different standard consumer cameras of natural "in the wild" scene content. In DnD: the camera metadata have been captured; the noise properties have been carefully calibrated; and the image intensities are presented as RAW unprocessed linear intensities. For each real high-resolution image, the noisy high-ISO image is paired with the corresponding (nearly) noise-free low-ISO ground-truth image.

The evaluation of DnD is separated in two categories: algorithms that use linear Bayer sensor readings or algorithms that use bilinearly demosaiced sRGB images as input. Thus, PSNR and SSIM for each technique are reported for both categories. The quantitative results with respect to prior work of our method and its ablations are shown in Fig. 4(b). For algorithms which have been evaluated with and without a variance stabilizing transformation (VST), the version which performs better is reported. The evaluation of algorithms that only operate on sRGB inputs is also reported. The proposed idea was tested for both categories. A blind and a non-blind version of our method had been tested for each category based on the info that $\boldsymbol{c}$ represents. The blind version uses no extra conditional information along with the noisy input image (empty $\boldsymbol{c}$). As described in Sect. 3.3, in the non-blind version, $\boldsymbol{c}$ could contain information regarding the camera noise model and/or the camera id. As a backbone network for blind image denoising, two variants used: a) the standard ResNet and b) the best method in the literature named UPI [9]. In the case of RAW image domain, the first variant produced significantly higher PSNR (+1.01dB) and SSIM than UPI, while the second one impressively boosted the performance of UPI by 1.16dB. In the case of sRGB image domain, the first variant produced significantly higher PSNR (+1.15dB) and SSIM than UPI, while the second one impressively boosted the performance of UPI by 1.24dB. As a backbone network for non-blind image denoising, only the standard ResNet used since the best methods in the literature are not deep learning techniques. In the case of image RAW domain, our system produced significantly higher PSNR (+3.76dB) and SSIM compared to the second best method named BM3D [19]+VST. In case of sRGB image domain, the improvement over BM3D+VST was 4.25dB. Also, runtimes (mean over 100 runs) reported in the literature are presented as well in Fig. 4(b). The runtime (excluding data transferring to GPU) of our blind model with standard ResNet as backbone network is 52ms while for the non-blind one is 63ms given as input 512×512 images. Some qualitative results are given in Fig. 5.

Evaluation on Nam: The Nam dataset consists of 11 static scenes captured by 3 consumer cameras. For each scene, 500 JPEG noisy temporal images were captured to compute the temporal nearly noise-free mean image and covariance matrix for each pixel. The quantitative results with respect to prior work are shown in Fig. 7(a). Both the blind and non-blind version of our method were evaluated. As a backbone network for blind image denoising, two variants used: a) the standard ResNet and b) the best method in the literature named CBDNet [25]. The first variant produced significantly higher PSNR (+1.03dB) and SSIM than CBDNet, while the second one impressively boosted the

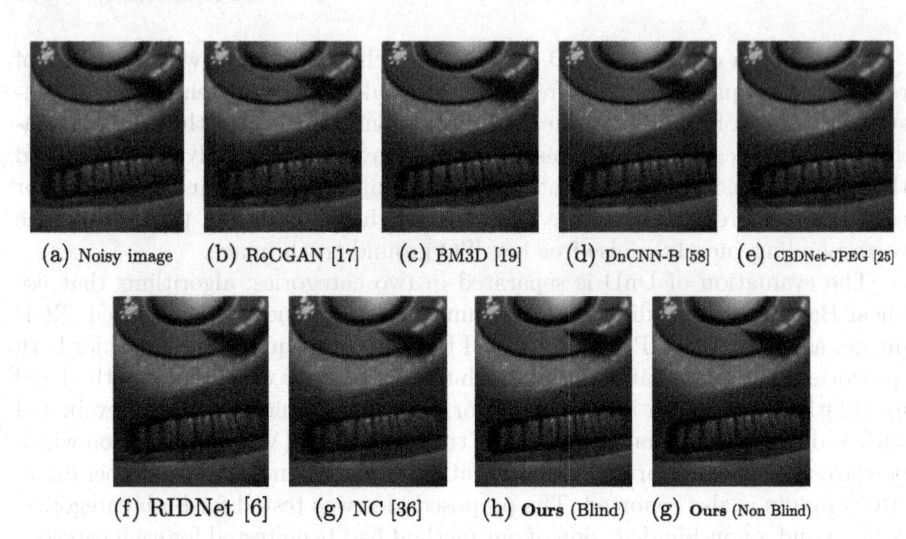

Fig. 6. Example of image denoising of a Nam image. Results of the proposed method shown when a standard ResNet used as backbone network.

performance of CBDNet by 1.18dB. CBDNet-JPEG [25] is a version of CBDNet which specifically deals with the JPEG compression. For fair comparison, we have retrained both variants by adopting this data augmentation technique. In that case, the first variant produced significantly higher PSNR (+0.89dB) and SSIM than CBDNet-JPEG, while the second one impressively boosted the performance of CBDNet-JPEG by 1.07dB. As a backbone network for non-blind image denoising, only the standard ResNet used since the best methods in bibliography are not deep learning techniques. Our system produced significantly higher PSNR (+0.97dB) and SSIM compared to the second best method named WNNM [24]. Some qualitative results are given in Fig. 6.

Evaluation on SIDD: SIDD is real noise dataset with a large number of available test (validation) images. The quantitative results on the SIDD benchmark with respect to prior work are shown in Fig. 7(b). Both the blind and non-blind version of our method were evaluated. As a backbone network for blind image denoising, two variants used: a) the standard ResNet and b) the best method in the literature named RIDNet [6]. The first variant produced significantly higher PSNR (+1.11dB) than RIDNet, while the second one impressively boosted the performance of RIDNet by 1.14dB. As a backbone network for non-blind image denoising, only the standard ResNet was used since the best method in the literature, named BM3D [19], is not a deep learning technique. Our system produced significantly higher PSNR (+8.93dB) compared to BM3D. Some qualitative results are given in Fig. 8.

Since the idea behind our method favours the multi-camera noise reduction task, there is a significant improvement in terms of performance across all datasets. Based on all our experiments, the proposed idea is general and can be appended at the end of existing image denoising methods to significantly improve

Method	**Type**	**PSNR**	**SSIM**
CDnCNN-B [58]	Blind	37.49	0.9272
TWSC [56]	Blind	37.52	0.9292
MCWNNM [24]	Blind	37.91	0.9322
RoCGAN [17]	Blind	38.52	0.9517
Whitenner [29]	Blind	38.62	0.9527
RIDNet [6]	Blind	39.09	0.9591
BM3D [19]	Non-blind	39.84	0.9657
CBDNet [25]	Blind	40.02	0.9687
NC [36]	Blind	40.41	0.9731
WNNM [24]	Non-blind	41.04	0.9768
Ours (empty c, **ResNet** [26])	Blind	41.05 (+1.03)	0.9772
Ours (empty c, CBDNet)	Blind	41.20 (+1.18)	0.9783
Ours (*Non* empty c, **ResNet**)	Non-Blind	42.01 (+0.97)	0.9830
CBDNet-JPEG [25]	Blind	41.31	0.9784
Ours (empty c, **ResNet**)	Blind	42.20 (+0.89)	0.9855
Ours (empty c, CBDNet-JPEG)	Blind	42.38 (+1.07)	0.9867

(a)

Method	**Type**	**PSNR**
DnCNN-B [58]	Blind	26.21
FFDNet [60]	Blind	29.20
CBDNet-JPEG [25]	Blind	30.78
BM3D [19]	Non-blind	30.88
Whitenner [29]	Blind	37.57
RoCGAN [17]	Blind	37.72
RIDNet [6]	Blind	38.71
Ours (empty c, **RIDNet**)	Blind	39.82 (+1.11)
Ours (empty c, **ResNet** [26])	Blind	39.85 (+1.14)
Ours (*Non* empty c, **ResNet**)	Non-Blind	39.81 (+8.93)

(b)

Fig. 7. (a) The quantitative results on the Nam benchmark and (b) the quantitative results on the SIDD benchmark. Regarding our method, in parentheses we define the type of denosing plus the used backbone network.

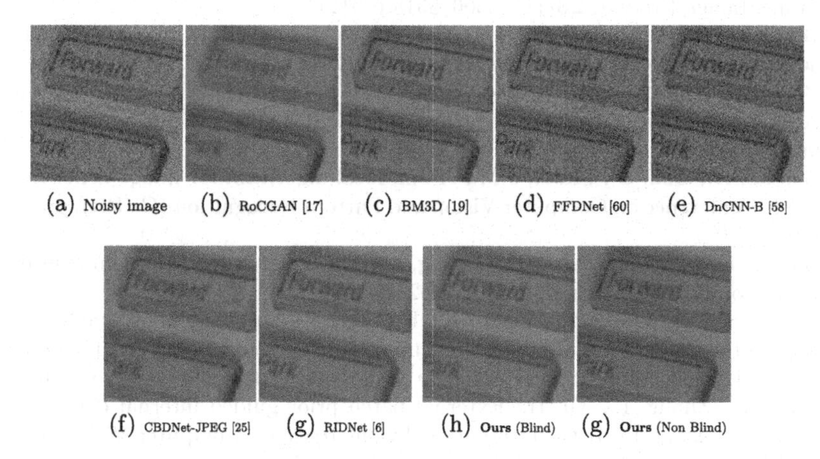

(a) Noisy image (b) RoCGAN [17] (c) BM3D [19] (d) FFDNet [60] (e) DnCNN-B [58]

(f) CBDNet-JPEG [25] (g) RIDNet [6] (h) **Ours** (Blind) (g) **Ours** (Non Blind)

Fig. 8. Example of image denoising of a SIDD image. Results of the proposed method shown when ResNet used as backbone network.

their performance. In addition, the proposed idea better restores the true colors than the competing methods. Also, by directly characterizing the image noise, our method avoids in great degree the image over-smoothing.

5 Conclusions

In this work, we show that is easier to turn noisy images into clean images only by looking at the structure of the residual image. We introduce the idea of a cGAN that explicitly leverages structure in the image noise variance space. By adopting the residual learning, the generator promotes the removal from the noisy image only that information which spans the manifold of the image noise variance. Our method significantly outperforms existing state-of-the-art architectures.

References

1. Abdelhamed, A., Brubaker, M.A., Brown, M.S.: Noise flow: noise modeling with conditional normalizing flows. ArXiv (2019)
2. Abdelhamed, A., Lin, S., Brown, M.S.: A high-quality denoising dataset for smartphone cameras. In: IEEE Conference on Computer Vision and Pattern Recognition (CVPR) (2018)
3. Agustsson, E., Timofte, R.: Ntire 2017 challenge on single image super-resolution: dataset and study, pp. 1122–1131 (2017)
4. Aharon, M., Elad, M., Bruckstein, A.: K-SVD: an algorithm for designing overcomplete dictionaries for sparse representation. IEEE Trans. Signal Process. (TSP) **54**, 4311–4322 (2006)
5. Anaya, J., Barbu, A.: Renoir - a dataset for real low-light noise image reduction. J. Vis. Commun. Image Represent. **51**, 144–154 (2018)
6. Anwar, S., Barnes, N.: Real image denoising with feature attention. In: IEEE International Conference on Computer Vision (ICCV) (2019)
7. Anwar, S., Porikli, F., Huynh, C.P.: Category-specific object image denoising. IEEE Trans. Image Process. **26**(11), 5506–5518 (2017)
8. Berthelot, D., Schumm, T., Metz, L.: Began: Boundary equilibrium generative adversarial networks. ArXiv abs/1703.10717 (2017)
9. Brooks, T., Mildenhall, B., Xue, T., Chen, J., Sharlet, D., Barron, J.T.: Unprocessing images for learned raw denoising. In: IEEE Conference on Computer Vision and Pattern Recognition (CVPR) (2019)
10. Buades, A., Coll, B., Morel, J.M.: A non-local algorithm for image denoising. In: IEEE Conference on Computer Vision and Pattern Recognition (CVPR), pp. 60–65 (2005)
11. Burger, H.C., Schuler, C.J., Harmeling, S.: Image denoising: can plain neural networks compete with bm3d?, pp. 2392–2399 (2012)
12. Bychkovsky, V., Paris, S., Chan, E., Durand, F.: Learning photographic global tonal adjustment with a database of input/output image pairs. In: The Twenty-Fourth IEEE Conference on Computer Vision and Pattern Recognition (2011)
13. Chen, F., Zhang, L., Yu, H.: External patch prior guided internal clustering for image denoising. In: IEEE International Conference on Computer Vision (ICCV), pp. 603–611 (2015)
14. Chen, J., Chen, J., Chao, H., Yang, M.: Image blind denoising with generative adversarial network based noise modeling. In: IEEE Conference on Computer Vision and Pattern Recognition (CVPR) (2018)
15. Chen, Y., Pock, T.: Trainable nonlinear reaction diffusion: a flexible framework for fast and effective image restoration. IEEE Trans. Pattern Anal. Mach. Intell. **39**, 1 (2016)
16. Chen, Y., Yu, W., Pock, T.: On learning optimized reaction diffusion processes for effective image restoration. IEEE Conference on Computer Vision and Pattern Recognition (CVPR) (2015)
17. Chrysos, G.G., Kossaifi, J., Zafeiriou, S.: Robust conditional generative adversarial networks. In: International Conference on Learning Representations (ICLR) (2019)
18. Cogswell, M., Ahmed, F., Girshick, R., Zitnick, L., Batra, D.: Reducing overfitting in deep networks by decorrelating representations. In: International Conference on Learning Representations (ICLR) (2016)
19. Dabov, K., Foi, A., Katkovnik, V., Egiazarian, K.: Image denoising by sparse 3-D transform-domain collaborative filtering. IEEE Trans. Image Process. **16**(8), 2080–2095 (2007)

20. Dong, W., Li, X., Zhang, L., Shi, G.: Sparsity-based image denoising via dictionary learning and structural clustering. In: IEEE Conference on Computer Vision and Pattern Recognition (CVPR), pp. 674–697 (2011)
21. Dong, W., Zhang, L., Shi, G., Li, X.: Nonlocally centralized sparse representation for image restoration. IEEE Trans. Image Process. **22**, 1620–1630 (2013)
22. Gonzalez, R.: Digital Image Processing 2Nd Ed. Prentice-Hall Of India Pvt. Limited (2002). https://books.google.co.uk/books?id=iyJOPgAACAAJ
23. Gu, S., Li, Y., Gool, L.V., Timofte, R.: Self-guided network for fast image denoising. In: IEEE International Conference on Computer Vision (ICCV) (2019)
24. Gu, S., Zhang, L., Zuo, W., Feng, X.: Weighted nuclear norm minimization with application to image denoising. IEEE Conference on Computer Vision and Pattern Recognition, pp. 2862–2869 (2014)
25. Guo, S., Yan, Z., Zhang, K., Zuo, W., Zhang, L.: Toward convolutional blind denoising of real photographs. In: 2019 IEEE Conference on Computer Vision and Pattern Recognition (CVPR) (2019)
26. He, K., Zhang, X., Ren, S., Sun, J.: Deep residual learning for image recognition. In: IEEE Conference on Computer Vision and Pattern Recognition (CVPR), pp. 770–778 (2016)
27. Healey, G., Kondepudy, R.: Radiometric CCD camera calibration and noise estimation. IEEE Trans. Pattern Anal. Mach. Intell. **16**, 267–276 (1994)
28. Isola, P., Zhu, J.Y., Zhou, T., Efros, A.A.: Image-to-image translation with conditional adversarial networks. In: CVPR (2017)
29. Izadi, S., Mirikharaji, Z., Zhao, M., Hamarneh, G.: Whitenner - blind image denoising via noise whiteness priors. In: International Conference on Computer Vision workshop on Visual Recognition for Medical Images (ICCV VRMI) (2019)
30. Jain, V., Sebastian, S.: Natural image denoising with convolutional networks. In: Advances in Neural Information Processing Systems, pp. 769–776 (2009)
31. Jia, X., Liu, S., Feng, X., Zhang, L.: Focnet: a fractional optimal control network for image denoising. In: IEEE Conference on Computer Vision and Pattern Recognition (CVPR) (2019)
32. Kingma, D.P., Ba, J.: Adam: a method for stochastic optimization. arXiv preprint arXiv:1412.6980 (2014)
33. Kostadin, K., Foi, A., Katkovnik, V., Egiazarian, K.: BM3D image denoising with shape-adaptive principal component analysis (2009)
34. Krull, A., Buchholz, T.O., Jug, F.: Noise2void - learning denoising from single noisy images. In: IEEE Conference on Computer Vision and Pattern Recognition (CVPR) (2018)
35. Lebrun, M., Buades, A., Morel, J.M.: A nonlocal bayesian image denoising algorithm. SIAM J. Imaging Sci. **6**, 1665–1688 (2013)
36. Lebrun, M., Colom, M., Morel, J.M.: The noise clinic: a blind image denoising algorithm. Image Process. Line (IPOL) **5**, 1–54 (2015)
37. Lehtinen, J., et al.: Noise2Noise: Learning image restoration without clean data. In: International Conference on Machine Learning (ICML), pp. 2965–2974 (2018)
38. Liu, C., Szeliski, R., Kang, S.B., Zitnick, C.L., Freeman, W.T.: Automatic estimation and removal of noise from a single image. IEEE Trans. Pattern Anal. Mach. Intell. **30**, 299–314 (2008)
39. Liu, D., Wen, B., Fan, Y., Loy, C., Huang, T.S.: Non-local recurrent network for image restoration. In: International Conference on Neural Information Processing Systems (NIPS), pp. 1680–1689 (2018)

40. Liu, P., Zhang, H., Zhang, K., Lin, L., Zuog, W.: Multi-level wavelet-CNN for image restoration. In: IEEE Conference on Computer Vision and Pattern Recognition (CVPR), pp. 773–782 (2018)
41. Mao, X., Shen, C., Yang, Y.B.: Image restoration using very deep convolutional encoder-decoder networks with symmetric skip connections. In: Advances in Neural Information Processing Systems, pp. 2802–2810 (2016)
42. Martin, D., Fowlkes, C., Tal, D., Malik, J.: A database of human segmented natural images and its application to evaluating segmentation algorithms and measuring ecological statistics. In: IEEE International Conference on Computer Vision (ICCV), vol. 2, pp. 416–423 (2001)
43. Mildenhall, B., Barron, J.T., Chen, J., Sharlet, D., Ng, R., Carroll, R.: Burst Denoising With Kernel Prediction Networks. In: The IEEE Conference on Computer Vision and Pattern Recognition (CVPR), June 2018
44. Mirza, M., Osindero, S.: Conditional generative adversarial nets. arXiv preprint arXiv:1411.1784 (2014)
45. Nam, S., Hwang, Y., Matsushita, Y., Kim, S.J.: A holistic approach to cross-channel image noise modeling and its application to image denoising. In: IEEE Conference on Computer Vision and Pattern Recognition (CVPR), pp. 1683–1691 (2016)
46. Plotz, T., Roth, S.: Benchmarking denoising algorithms with real photographs. In: IEEE Conference on Computer Vision and Pattern Recognition (CVPR), pp. 2750–2759 (2017)
47. Plötz, T., Roth, S.: Neural nearest neighbors networks. In: Advances in Neural Information Processing Systems (NeurIPS) (2018)
48. Rasmus, A., Berglund, M., Honkala, M., Valpola, H., Raiko, T.: Semi-supervised learning with ladder networks. In: International Conference on Neural Information Processing Systems (NIPS), pp. 3546–3554 (2015)
49. Roth, S., Black, M.J.: Fields of experts. Int. J. Comput. Vision **82**(2), 205–229 (2009)
50. Salimans, T., Goodfellow, I., Zaremba, W., Cheung, V., Radford, A., Chen, X.: Improved techniques for training gans. In: International Conference on Neural Information Processing Systems (NIPS), pp. 2234–2242 (2016)
51. Tai, Y., Yang, J., Liu, X., Xu, C.: Memnet: A persistent memory network for image restoration. In: IEEE International Conference on Computer Vision (ICCV), pp. 4549–4557 (2017)
52. Tappen, M.F., Liu, C., Adelson, E.H., Freeman, W.T.: Learning gaussian conditional random fields for low-level vision, pp. 1–8, July 2007
53. Tripathi, S., Lipton, Z.C., Nguyen, T.Q.: Correction by projection: denoising images with generative adversarial networks. ArXiv abs/1803.04477 (2018)
54. Weiss, Y., Freeman, W.: What makes a good model of natural images?, pp. 1–8, July 2007
55. Xu, J., Osher, S.: Iterative regularization and nonlinear inverse scale space applied to wavelet-based denoising. IEEE Trans. Image Process. **16**, 534–544 (2007)
56. Xu, J., Zhang, L., Zhang, D.: A trilateral weighted sparse coding scheme for real-world image denoising. In: European Conference on Computer Vision (ECCV), pp. 21–38 (2018)
57. Xu, J., Zhang, L., Zuo, W., Zhang, D., Feng, X.: Patch group based nonlocal self-similarity prior learning for image denoising. In: IEEE International Conference on Computer Vision (ICCV), pp. 244–252 (2015)

58. Zhang, K., Zuo, W., Chen, Y., Meng, D., Zhang, L.: Beyond a Gaussian denoiser: residual learning of deep CNN for image denoising. IEEE Trans. Image Process. **26**(7), 3142–3155 (2017)
59. Zhang, K., Zuo, W., Gu, S., Zhang, L.: Learning deep CNN denoiser prior for image restoration. In: IEEE Conference on Computer Vision and Pattern Recognition (CVPR), pp. 2808–2817 (2017)
60. Zhang, K., Zuo, W., Zhang, L.: FFDnet: toward a fast and flexible solution for CNN-based image denoising. IEEE Trans. Image Process. (TIP) **27**(9), 4608–4622 (2018)
61. Zoran, D., Weiss, Y.: From learning models of natural image patches to whole image restoration. In: IEEE International Conference on Computer Vision (ICCV), pp. 479–486 (2011)

Occlusion-Aware Depth Estimation with Adaptive Normal Constraints

Xiaoxiao Long[1], Lingjie Liu[2], Christian Theobalt[2], and Wenping Wang[1](✉)

[1] The University of Hong Kong, Pok Fu Lam, Hong Kong
{xxlong,wenping}@cs.hku.hk
[2] Max Planck Institute for Informatics, Saarbrücken, Germany
{lliu,theobalt}@mpi-inf.mpg.de

Abstract. We present a new learning-based method for multi-frame depth estimation from a color video, which is a fundamental problem in scene understanding, robot navigation or handheld 3D reconstruction. While recent learning-based methods estimate depth at high accuracy, 3D point clouds exported from their depth maps often fail to preserve important geometric feature (e.g., corners, edges, planes) of man-made scenes. Widely-used pixel-wise depth errors do not specifically penalize inconsistency on these features. These inaccuracies are particularly severe when subsequent depth reconstructions are accumulated in an attempt to scan a full environment with man-made objects with this kind of features. Our depth estimation algorithm therefore introduces a *Combined Normal Map (CNM)* constraint, which is designed to better preserve high-curvature features and global planar regions. In order to further improve the depth estimation accuracy, we introduce a new occlusion-aware strategy that aggregates initial depth predictions from multiple adjacent views into one final depth map and one occlusion probability map for the current reference view. Our method outperforms the state-of-the-art in terms of depth estimation accuracy, and preserves essential geometric features of man-made indoor scenes much better than other algorithms.

Keywords: Multi-view depth estimation · Normal constraint · Occlusion-aware strategy · Deep learning

1 Introduction

Dense multi-view stereo is one of the fundamental problems in computer vision with decades of research, e.g. [4,13,14,31,40,43]. Algorithms vastly differ in their assumptions on input (pairs of images, multi-view images etc.) or employed scene representations (depth maps, point clouds, meshes, patches, volumetric scalar fields etc.) [11].

Electronic supplementary material The online version of this chapter (https://doi.org/10.1007/978-3-030-58545-7_37) contains supplementary material, which is available to authorized users.

© Springer Nature Switzerland AG 2020
A. Vedaldi et al. (Eds.): ECCV 2020, LNCS 12354, pp. 640–657, 2020.
https://doi.org/10.1007/978-3-030-58545-7_37

In this paper, we focus on the specific problem of multi-view depth estimation from a color video from a single moving RGB camera [11]. Algorithms to solve this problem have many important applications in computer vision, graphics and robotics. They empower robots to avoid collisions and plan paths using only one onboard color camera [3]. Such algorithms thus enable on-board depth estimation from mobile phones to properly combine virtual and real scenes in augmented reality in an occlusion-aware way. Purely color-based solutions bear many advantages over RGB-D based approaches [47] since color cameras are ubiquitous, cheap, consume little energy, and work nearly in all scene conditions.

Depth estimation from a video sequence is a challenging problem. Traditional methods [4,13,14,40,43] achieve impressive results but struggle on important scene aspects: texture-less regions, thin structures, shape edges and features, and non-Lambertian surfaces. Recently there are some attempts to employ learning techniques to this problem [16,18,20,36,41]. These methods train an end-to-end network typically with a pixel-wise depth loss function. They lead to significant accuracy improvement in depth estimation compared to non-learning-based approaches. However, most of these works fail to preserve prominent features of 3D shapes, such as corners, sharp edges and planes, because they use only depth for supervision and their loss functions are therefore not built to preserve these structures. This problem is particularly detrimental when reconstructing indoor scenes with man-made objects or regular shapes, as shown in Fig. 4. Another problem is the performance degradation caused by the depth ambiguity in the occluded region, which has been ignored by most of the existing works.

We present a new method for depth estimation with a single moving color camera that is designed to preserve important local features (edges, corners, high curvature features) and planar regions. It takes one video frame as *reference image* and uses some other frames as *source images* to estimate depth in the reference frame. Pairing the reference image with each source image, we first build an initial 3D cost volume from each pair via plane-sweeping warping. Subsequent cost aggregation for each initial cost volume yields an initial depth maps for each image pair (e.g. the reference image and a source image). Then we employ a new occlusion-aware strategy to combine these initial depth maps into one final reference-view depth map, along with an occlusion probability map.

Our first main contribution is a new structure preserving constraint enforced during training. It is inspired by learning-based monocular depth estimation methods using normal constraints for structure preservation. GeoNet [29] enforces a surface normal constraint, but their results have artifacts due to noise in the ground truth depth. Yin et al. [42] propose a global geometric constraint, called *virtual normal*. However, they cannot preserve intrinsic geometric features of real surfaces and local high-curvature features. We therefore propose a new *Combined Normal Map (CNM)* constraint, attached to local features for both local high-curvature regions and global planar regions. For training our network, we use a differentiable least squares module to compute normals directly from the estimated depth and use the *CNM* as ground truth in addition to the standard depth loss. Experiments in Sect. 5.2 show that the use of this novel *CNM* constraint significantly improves the depth estimation accuracy and outperforms those approaches that use only local or global constraints.

Our second contribution is a new neural network that combines depth maps predicted with individual source images into one final reference-view depth map, together with an occlusion probability map. It uses a novel occlusion-aware loss function which assigns higher weights to the non-occluded regions. Importantly, this network is trained without any occlusion ground truth.

We experimentally show that our method significantly outperforms the state-of-the-art multi-view stereo from monocular video, both quantitatively and qualitatively. Furthermore, we show that our depth estimation algorithm, when integrated into a fusion-based handheld 3D scene scanning approach, enables interactive scanning of man-made scenes and objects in much higher shape quality (See Fig. 6).

2 Related Work

Multi-view Stereo. MVS algorithms [4,13,14,40,43] are able to reconstruct 3D models from images under the assumptions of known materials, viewpoints, and lighting conditions. The MVS methods vary significantly according to different scene representations, but the typical workflow is to use hand-crafted photo-consistency metrics to build a cost volume and do the cost aggregation and estimate the 3D geometry from the aggregated cost volume. These methods tend to fail for thin objects, non-diffuse surfaces, or the objects with insufficient features.

Learning-based Depth Estimation. Recently some learning-based methods achieve compelling results in depth estimation. They can be categorized into four groups: i) single-view depth estimation [9,10,22,25]; ii) two-view stereo depth estimation [5,20,26,34,44]; iii) multi-view stereo depth estimation [16,18,41]; iv) depth estimation from a video sequence [23,36]. Single-view depth estimation is an ill-posed problem due to inherent depth ambiguity. On the other hand, two-view and multi-view depth estimation [11] is also challenging due to the difficulties in dense correspondence matching and depth prediction in featureless or specular regions. Some two-view [5,20,26,44] and multi-view [16,18,36,41] depth estimation algorithms have produced promising results in accuracy and speed by integrating cost volume generation, cost volume aggregation, disparity optimization and disparity refinement in an end-to-end network.

Depth estimation from video frames is becoming popular with the high demand in emerging areas, such as AR/VR, robot navigation, autonomous driving and view-dependent realistic rendering. Wang et al. [36] design a real-time multi-view depth estimation network. Liu et al. [23] propose a Bayesian filtering framework to use frame coherence to improve depth estimation. Since these methods enforce depth constraint only, they often fail to preserve important geometric feature (e.g., corners, edges, planes) of man-made scenes.

Surface Normal Constraint. Depth-normal consistency has been explored before for the depth estimation task [8,29,42,46]. Eigen et al. [8] propose a neural network with three decoders to separately predict depth, surface normal and segmentation. Zhang et al. [46] enforce the surface normal constraint for depth completion. Qi et al. [29] incorporate geometric relations between depth and surface normal by introducing the depth-to-normal and normal-to-depth networks. The performance of these methods suffers from the noise in surface normal stemming from ground truth depth. Yin et al. [42] propose a global geometric constraint, called *virtual normal*, which is defined as the normal of a virtual surface plane formed by three randomly sampled non-collinear points with a large distance. The *virtual normal* is unable to faithfully capture the geometric features of real surfaces and the local high-curvature features. Kusupati et al. [21] use the sobel operator to calculate the spatial gradient of depth and enforce the consistency of the spatial gradient and the normal in the pixel coordinate space. However the gradient calculated by the sobel operator is very sensitive to local noise, and causes obvious artifacts in the surface normal calculated from estimated depth (See Fig. 5 for the visual result).

Occlusion Handling for Depth Estimation. Non-learning methods [1,7,19, 27,37,39] use post-processing, such as left-right consistency check [7], to handle the occlusion issue. There are some learning-based methods for handling occlusion as well. Ilg et al. [17] and Wang et al. [35] predict occlusions by training a neural network on a synthetic dataset in a supervised manner. Qiu et al. [30] directly learn an intermediate occlusion mask for a single image; it has no epipolar geometry guarantee and just drives network to learn experienced patterns. We propose an occlusion-aware strategy to jointly refine depth prediction and estimate an occlusion probability map for multi-view depth estimation without any ground truth occlusion labels.

3 Method

In this section we describe the proposed network, as outlined in Fig. 1. Our pipeline takes the frames in a local time window in the video as input. Note that in our setting the video frame rate is 30 fps and we sample video frames with the interval of 10.

For the sake of simplicity in exposition, we suppose that the time window size is 3. We take the middle frame as the reference image I_{ref}, and the two adjacent images as the two source images $\{I_s^1, I_s^2\}$ which are supposed to have sufficient overlap with the reference image. Our goal is to compute an accurate depth map from the view of the reference image.

We first outline our method. With a differentiable homography warping operation [41], all the source images are first warped into a stack of different fronto-parallel planes of the reference camera to form an initial 3D cost volume (see Sect. 3.1). Next, cost aggregation is applied to the initial cost volume to rectify any incorrect cost values and then an initial depth map is extracted from the aggregated cost volume. Besides pixel-wise depth supervision over the initial depth map, we also enforce a novel local and global geometric constraint, namely *Combined Normal Map (CNM)*, for training the network to produce

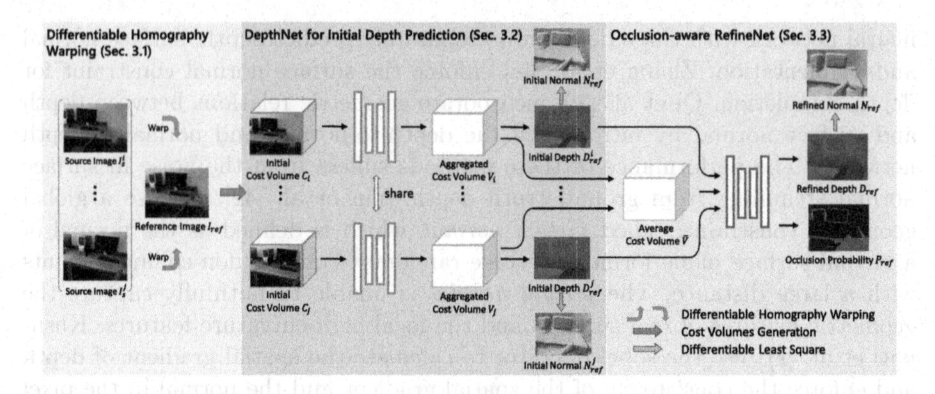

Fig. 1. Overview of our method. The network input consists of one reference image and $n(n = 2, 4, \ldots)$ source images. With homography warping, each source image is combined with the reference image as a pair to generate a cost volume. Then the cost volume is fed into the *DepthNet* to generate an initial depth map, with ℓ_1 depth supervision and the constraint of *Combined Normal Map*. Finally, with the n initial depth maps and the average cost volume of the aggregated cost volumes as input, the *RefineNet* employs an occlusion-aware loss to produce the final reference-view depth map and an occlusion probability map, with again the supervision by the depth and the *CNM* constraints.

superior results (see Sect. 3.2). We then further improve the accuracy of depth estimation by applying a new occlusion-aware strategy to aggregate the depth predictions from different adjacent views into one depth map for the reference view, with an occlusion probability map (see Sect. 3.3). The details of each step will be elaborated in the subsequent sections.

3.1 Differentiable Homography Warping

Consider the reference image I_{ref} and one of its source images, denoted I_s^i. We warp I_s^i into fronto-parallel virtual planes of I_{ref} to form an initial cost volume. Similar to [16,41], we first uniformly sample D depth planes at depth d_n in a range $[d_{min}, d_{max}]$, $n = 1, 2, \ldots, D$. The mapping from the the warped I_s^i to the nth virtual plane of I_{ref} at depth d_n is determined by a planar homography transformation H_n, following the classical plane-sweeping stereo [12].

$$u'_n \sim H_n u, u'_n \sim \mathbf{K}[\mathbf{R}_{s,i}|\mathbf{t}_{s,i}] \begin{bmatrix} (\mathbf{K}^{-1}u)d_n \\ 1 \end{bmatrix}, \tag{1}$$

where u is the homogeneous coordinate of a pixel in reference image, u'_n is projected coordinate of u in source image I_s^i with virtual plane d_n. K denotes the intrinsic parameters of the camera, $\{\mathbf{R}_{s,i}, \mathbf{t}_{s,i}\}$ are the rotation and the translation of the source image I_s^i relative to the reference image I_{ref}.

Next, we measure the visual consistency of the warped I_s^i and I_{ref} at depth d_n and build a cost volume C_i with the size of $W \times H \times D$, where W, H, D are the image width, image height and the number of depth planes, respectively. Unlike previous works [23,41] that use extracted feature maps of an image pair for warping and building a 4D cost volume, here we use the image pair directly to avoid the memory-heavy and time-consuming 3D convolution operation on a 4D cost volume.

3.2 DepthNet for Initial Depth Prediction

Similar to recent learning-based stereo [20] and MVS [18,41] methods, after getting cost volumes $\{C_i\}_{i=1,2}$ from image pairs $\{I_{ref}, I_s^i\}_{i=1,2}$, we first use the neural network *DepthNet* to do cost aggregation for each C_i, which rectifies the incorrect values by aggregating the neighbouring pixel values. Note that we feed C_i stacked with I_s together into the *DepthNet* in order to make use of more detailed context information as [36] does. We denote the aggregated cost volume as V_i. Next, we retrieve the initial depth map D_{ref}^i from the aggregated cost volume V_i using a 2D convolution layer. Note that two initial depth maps, D_{ref}^1 and D_{ref}^2, are generated for the reference view I_{ref}.

We train the network with depth supervision. With only depth supervision, the point cloud converted from the estimated depth does not preserve regular features, such as sharp edges and planar regions. We therefore propose to also enforce the normal constraint for further improvement. Note that, instead of just using local surface normal [29] or just a global virtual normal [42], we use the so called *Combine Normal Map* (CNM) that combines the local surface normal and the global planar structural feature in an adaptive manner.

Pixel-Wise Depth Loss. We use a standard pixel-wise depth map loss as follows,

$$l_{id} = \frac{1}{|Q|} \sum_{q \in Q} \left\| \hat{D}(q) - D_i(q) \right\|_1 \tag{2}$$

where $\boldsymbol{Q}$ is the set of all pixels that are valid in ground truth depth, $\hat{D}(q)$ is the ground truth depth value of pixel q, and $D_i(q)$ is the initial estimated depth value of pixel q.

Combined Normal Map. In order to preserve both local and global structures of scenes, we introduce the *Combined Normal Map (CNM)* as ground truth for the supervision with the normal constraint. To obtain this normal map, we first use PlaneCNN [24] to extract planar regions, such as walls, tables, and floors. Then we apply the local surface normals to non-planar regions, and use the mean of the surface normals in a planar region as the assigned to the region. The visual comparison of the local normal map and the *CNM* can be seen in Fig. 2.

The key insight here is to use local surface normal to capture rich local geometric features in the high-curvature regions and to employ the average normal to filter out noise in the surface normals for the planar regions to preserve global structures. In this way, the *CNM* significantly improves the depth prediction and the recovery of good 3D structures of the scene, compared to using only local or global normal supervision (see the ablation study in Sect. 5.4).

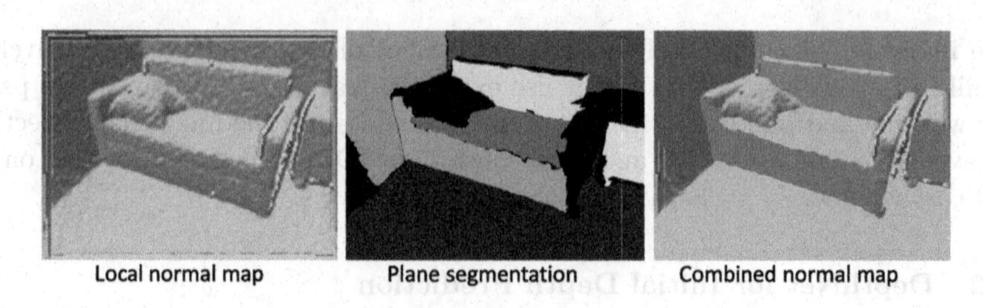

Fig. 2. Visual comparison of local normal map and combined normal map.

Combined Normal Loss. We define the loss on the CNM as follows:

$$l_{in} = -\frac{1}{|Q|} \sum_{q \in Q} \hat{N}(q) \cdot N_i(q), \quad (3)$$

where Q is the set of valid ground truth pixels, $\hat{N}(q)$ is the combined normal of pixel q, and $N_i(q)$ is the surface tangent normal of the 3D point corresponding to the pixel q, both normalized to unit vectors.

To obtain an accurate depth map and preserve geometric features, we combine the Pixel-wise Depth Loss and the Combined Normal Loss together to supervise the network output. The overall loss is:

$$l_i = l_{id} + \lambda l_{in}, \quad (4)$$

where λ is a trade-off parameter, which is set to 1 in all experiments.

3.3 Occlusion-Aware RefineNet

The next step is to combine the initial depth maps D^1_{ref} and D^2_{ref} of the reference image predicted from different image pairs $\{I_{ref}, I^i_s\}_{i=1,2}$ into one final depth map, which is denoted as D_{fin}. We design an occlusion-aware network, namely *RefineNet*, based on an occlusion probability map. Note that the occlusion refers to the region in I_{ref} where it cannot be observed in either I^1_s or I^2_s. In contrast to treating all pixels equally, when calculating the loss we assign the lower weights to the occluded region and the higher weights to the non-occluded region, which shifts the focus of the network to the non-occluded regions, since the depth prediction in the non-occluded regions is more reliable. Furthermore, the occlusion probability map predicted by the network can be used to filter out unreliable depth samples, as shown in Fig. 3, which is useful when the depth maps are fused for 3D reconstruction (see more details in Sect. 5.4).

Here we describe more technical details of this step. We design the *RefineNet* to predict the final depth map and the occlusion probability map from the average cost volume $\bar{V}$ of the two cost volumes $\{V_i\}_{i=1,2}$ and two initial depth maps $\{D^i_{ref}\}_{i=1,2}$. The *RefineNet* has one encoder and two decoders.

The first decoder estimates the occlusion probabilities based on the occlusion information encoded in the average cost volume $\bar{V}$ and the initial depth maps.

Intuitively, for a pixel in the non-occluded region, it has the strongest response (peak) at n^{th} layer with depth d_n of the average cost volume $\bar{V}$, and D^1_{ref} and D^2_{ref} at this pixel have similar depth values. However, for a pixel in the occluded region, it has scattered responses at the depth layers of $\bar{V}$ and has very different values on the initial depth maps at this pixel. The other decoder predicts the refined depth map with the depth constraint and the CNM constraint, as described in Sect. 3.2

To train the *RefineNet*, we design a novel occlusion-aware loss function as follows,

$$L_{refine} = (l_{rd} + \beta \cdot l_{rn}) - \alpha \cdot \frac{1}{|\mathcal{Q}|} \cdot \sum_{q \in \mathcal{Q}} (1 - P(q)), \tag{5}$$

where

$$l_{rd} = \frac{1}{|\mathcal{Q}|} \sum_{q \in \mathcal{Q}} (1 - P(q)) \left\| \hat{D}(q) - D_r(q) \right\|_1 \tag{6}$$

$$l_{rn} = -\frac{1}{|\mathcal{Q}|} \sum_{q \in \mathcal{Q}} (1 - P(q)) \, \hat{N}(q) \cdot N_r(q). \tag{7}$$

Here $D_r(q)$ denotes the refined estimated depth at pixel q, $N_r(q)$ denotes the surface normal of the 3D point corresponding to q, $\hat{D}(q)$ is the ground truth depth of q, $\hat{N}(q)$ is the combined normal of q, and $P(q)$ denotes the occlusion probability value of q ($P(q)$ is high when q is occluded). The weight α is set to 0.2 and β is set to 1 in all experiments.

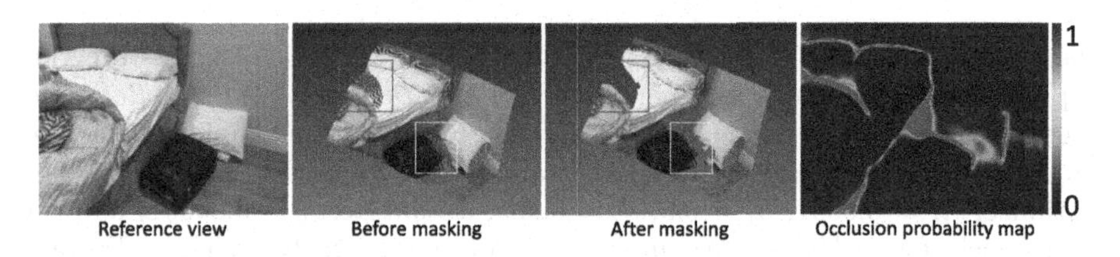

Fig. 3. Efficiency of occlusion probability map. Masked by the occlusion map, the point cloud converted from estimated depth has fewer outlying points.

4 Datasets and Implementation Details

Dataset. We train our network with the ScanNet dataset [6], and evaluate our method on the 7scenes dataset [32] and the SUN3D dataset [38]. ScanNet consists of 1600 different indoor scenes, which are divided into 1000 for training and 600 for testing. ScanNet provides RGB images, ground truth depth maps and camera poses. We generate the *CNM* as described in Sect. 3.2.

Implementation Details. Our training process consists of three stages. First, we train the *DepthNet* using the loss function defined in Eq. 4. Then, we fix

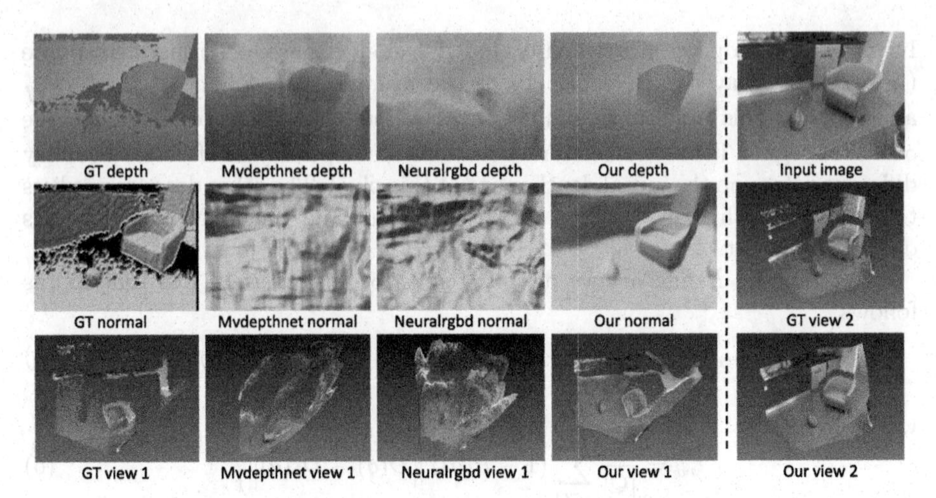

Fig. 4. Visual depth comparison with mvdepthnet [36] and neuralrgbd [23]. Our estimated depth maps preserve shape regularity better than by mvdepthnet and neuralrgbd, for example, in the regions of sofa, cabinet and wall. The result by mvdepthnet (colum 2) shows strong discontinuity of depth value at the lower end of the red wall, which is incorrect. We also show the visualization of normal maps computed from these depth maps to further show the superior quality of our depth estimation in terms of shape regularity. Note that the red region of GT depth (black region of GT normal) is invalid due to scanning failure of the Kinect sensor. More comparisons can be found in the supplementary materials (Color figure online)

the parameters of the *DepthNet* and train the *RefineNet* with the loss function Eq. 5. Next, we finetune the parameters of the *DepthNet* and the *RefineNet* together with the loss terms in Eq. 4 and Eq. 5. For all the training stages, the ground truth depth map and the *CNM* are used as supervision. We use Adam optimizer ($lr = 0.0001$, $\beta_1 = 0.9$, $\beta_2 = 0.999$, $weight\_decay = 0.00001$) and run for 6 epochs for each stage. We implemented our model in Pytorch [28]. Training the network with two GeForce RTX 2080 Ti GPUs, which takes two days for all stages.

5 Experiments

To evaluate our method, we compare our method with state-of-the-arts in three aspects: accuracy of depth estimation, geometric consistency, and video-based 3D reconstruction with TSDF fusion.

5.1 Evaluation Metrics

For depth accuracy, we compute the widely-used statistical metrics defined in [9]: i) accuracy under threshold ($\sigma < 1.25^i$ where $i \in \{1, 2, 3\}$); ii) scale-invariant error (scale.inv); iii) absolute relative error (abs.rel); iv) square relative error (sq.rel); iv) root mean square error (rmse); v) rmse in log space (rmse log).

We evaluate the surface normal accuracy with the following metrics used in the prior works [8,29]: i) the mean of angle error (mean); ii) the median of the angle error (median); iii) rmse; iv) the percentage of pixels with angle error below threshold t where $t \in [11.25°, 22.5°, 30°]$.

Table 1. Comparison of depth estimation over 7-Scenes dataset [32] with the metrics defined in [9].

	$\sigma < 1.25$	$\sigma < 1.25^2$	$\sigma < 1.25^3$	abs.rel	sq.rel	rmse	rmse log	scale.inv
DORN [10]	60.05	87.76	96.33	0.2000	0.1153	0.4591	0.2813	0.2207
GeoNet [29]	55.10	84.46	94.76	0.2574	0.1762	0.5253	0.3952	0.3318
Yin et al. [42]	60.62	88.81	97.44	0.2154	0.1245	0.4500	0.2597	0.1660
DeMoN [34]	31.88	61.02	82.52	0.3888	0.4198	0.8549	0.4771	0.4473
MVSNet-retrain† [41]	64.09	87.73	95.88	0.2339	0.1904	0.5078	0.2611	0.1783
neuralrgbd† [23]	69.26	91.77	96.82	0.1758	0.1123	0.4408	0.2500	0.1899
mvdepthnet [36]	71.97	92.00	97.31	0.1865	0.1163	0.4124	0.2256	0.1691
mvdepthnet-retrain† [36]	71.79	92.56	97.83	0.1925	0.2350	0.4585	0.2301	0.1697
DPSNet [18]	63.65	85.73	94.33	0.2710	0.2752	0.5632	0.2759	0.1856
DPSNet-retrain† [18]	70.96	91.42	97.17	0.1991	0.1420	0.4382	0.2284	0.1685
Kusupati et al.† [21]	73.12	92.33	97.78	0.1801	0.1179	0.4120	0.2184	/
ours†	**76.64**	**94.46**	**98.56**	**0.1612**	**0.0832**	**0.3614**	**0.2049**	**0.1603**

† Trained on ScanNet dataset.

5.2 Comparisons

Depth Prediction. We compare our method with several other depth estimation methods. We categorize them according to their input formats: i) one single image: DORN [10], GeoNet [29] and VNL [42]; ii) two images from monocular camera: DeMoN [34]; iii) multiple unordered images: MVSNet [41], DPSNet [18], kusupati et al. [21]; iv) a video sequence: mvdepthnet [36] and neuralrgbd [23]. The models provided by these methods were trained on different datasets and are evaluated on a separate dataset, 7Scenes dataset [32]. For the multi-view depth estimation methods, neuralrgbd [23] and kusupati et al. [21] are trained on ScanNet, MVSNet [41] is trained on DTU dataset [2], mvdepthnet [36] and DPSNet [18] are trained on mixed datasets (SUN3D [38], TUM RGB-D [33], MVS datasets [34], SceneNN [15] and synthetic dataset Scenes11 [34]). For fair comparison, we further retrained MVSNet [41], mvdepthnet [36] and DPSNet [18] on ScanNet. Table 1 shows that our method outperforms other methods in terms of all evaluation metrics.

Visual Comparison. In Fig. 4, compared to mvdepthnet [36] and neuralrgbd [23], our estimated depth map has less noise, sharper boundaries and spatially consistent depth values, which can be also seen in the surface normal visualization. Furthermore, the 3D point cloud exported from the estimated depth preserves global planar features and local features in the high-curvature regions. More comparison examples are included in the y materials.

Surface Normal Accuracy. To evaluate the accuracy of normal calculated from estimated depth, we choose two single-view depth estimation methods: GeoNet [29] and Yin et al. [42], and one multi-view depth estimation method: Kusupati et al. [21]. GeoNet [29] incorporates surface normal constraint to depth estimation, while Yin et al. [42] propose a global normal constraint, namely *Virtual Normal*, which is defined by three randomly sampled non-collinear points. Kusupati et al. [21] use the sobel operator to calculate depth gradient and enforce the consistency of the depth gradient and the normal in the pixel coordinate space. To demonstrate that the improved performance of our method indeed benefits from the new CNM constraint (rather than entirely due to our multi-view input), we also retrained our network by replacing the Combined Normal Loss with the Virtual Normal loss (denoted as "Ours with VNL" in Table 2) on ScanNet dataset.

As shown in Table 2 and Fig. 5, our method outperforms GeoNet [29], Yin et al. [42] and Kusupati et al. [21] both quantitatively and qualitatively. Compared with our model retrained using VNL, our model with the *CNM* constraint works much better, which preserves local and global features.

Table 2. Evaluation of the calculated surface normal from estimated depth on 7scenes[32] and SUN3D [38]."Ours with VNL" denotes our model retrained using virtual normal loss [42].

Dataset	7scenes						SUN3D					
Method	Error			Accuracy			Error			Accuracy		
	Mean	Median	rmse	11.25°	22.5°	30°	Mean	Median	rmse	11.25°	22.5°	30°
GeoNet [29]	57.04	52.43	65.50	4.396	15.16	23.83	47.39	40.25	57.11	9.311	26.63	37.49
Yin et al. [42]	45.17	37.77	54.77	11.05	29.76	40.78	38.50	30.03	48.90	19.06	40.32	51.17
Ours with VNL	43.08	34.98	53.15	13.84	33.37	43.90	31.68	22.10	42.74	29.50	52.97	62.81
Kusupati et al. [21]	55.54	50.46	64.28	5.774	18.03	26.81	52.86	47.03	62.30	7.507	22.56	32.22
Ours	**36.34**	**27.21**	**46.46**	**20.24**	**43.36**	**54.11**	**29.21**	**20.89**	**38.99**	**30.45**	**55.05**	**65.25**

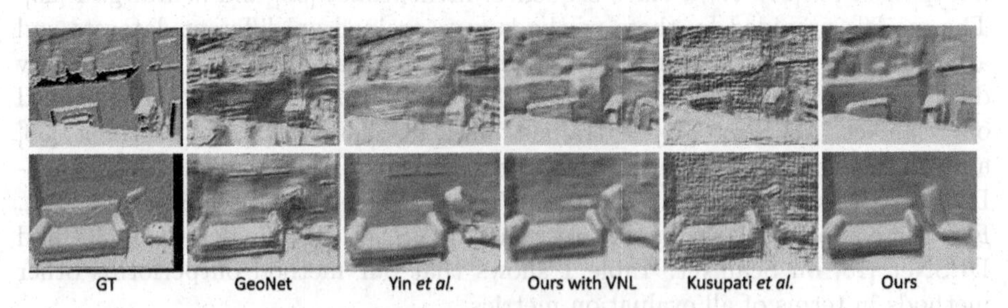

Fig. 5. Visual comparison of surface normal calculated from 3D point cloud exported from estimated depth with GeoNet [29], Yin et al. [42] and Kusupati et al. [21].

5.3 Video Reconstruction

With the high-fidelity depth map and the occlusion probability map obtained by our method, high-quality reconstruction of video even in texture-less environments can be achieved by applying TSDF fusion method [45]. We compare our method with mvdepthnet [36] and neuralrgbd [23]. Note that these three methods are all trained on the ScanNet dataset. As shown in Fig. 6, even for white walls, feature-less sofa and table, our reconstructed result is much better than the other two methods in the aspects of local and global structural recovery, and completion. Also, the color of our reconstruction is closest to the color of the ground truth. This confirms that the individual depth maps have less noise and high-consistency so the colors would not be mixed up in the fusion process.

5.4 Ablation Studies

Compared with prior works, we explicitly drive the depth estimation network to adaptively learn local surface directions and global planar structures. Taking multiple views as inputs, our refinement module jointly refines depth and predicts occlusion probability. In this section, we evaluate the usefulness of each component.

Table 3. Evaluation of the usefulness of *CNM* and the occlusion-aware refinement module. The first four rows show the results without any constraint, with only local normal constraint, with only global normal constraint, and with the *CNM* constraint respectively. The last two rows show the results without/with the occlusion-aware loss.

Components				Estimated depth evaluation						Calculated normal evaluation				
Local	Global	Refine	Occlu.	1.25	1.25^2	abs.rel	sq.rel	rmse	scale.inv	11.25°	22.5°	Mean	Median	rmse
×	×	×	×	71.79	92.56	0.1925	0.2350	0.4585	0.1697	9.877	27.20	46.62	39.91	55.96
✓	×	×	×	71.95	92.45	0.1899	0.1188	0.4060	0.1589	16.97	39.49	37.92	29.60	47.53
×	✓	×	×	66.85	90.17	0.2096	0.1462	0.4570	0.1752	15.39	37.09	39.15	31.33	48.45
✓	✓	×	×	73.17	92.75	0.1812	0.1076	0.3952	0.1654	17.67	40.23	37.65	29.18	47.36
✓	✓	✓	×	74.77	93.22	0.1726	0.0999	0.3877	0.1758	18.61	41.13	37.67	28.77	47.74
✓	✓	✓	✓	**75.80**	**93.79**	**0.1669**	**0.0909**	**0.3731**	**0.1638**	**20.12**	**42.76**	**36.82**	**27.67**	**47.03**

Combined Normal Map. As shown in Table 3, enforced by our *CNM* constraint (local + global), our model achieves better performance in terms of estimated depth and calculated normal from estimated depth, compared to that without geometric constraint, only with local normal constraint, and only with global normal constraint (the plane normal constraint). Figure 7 also shows that with *CNM*, the calculated normal is more consistent in planar regions and the point cloud converted from estimated depth keeps better shape.

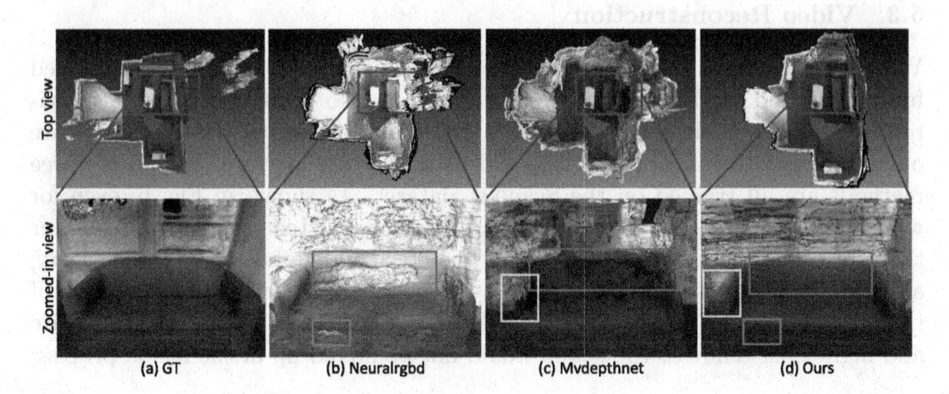

Fig. 6. Comparison with neuralrgbd [23] and mvdepthnet [36] for 3D reconstruction on a scene from ScanNet. (a) With ground truth depth; (b) With estimated depth and confidence map from neuralrgbd; (c) With estimated depth from mvdepthnet; (d) With our estimated depth and occlusion probability map. All reconstructions are done with TSDF fusion [45] using 110 frames uniformly selected from a video of 1100 frames.

Table 4. Effect of the number of source views N and the choice of reference view. We use the model trained with $N = 2$, and test with different number of views over 7-Scenes dataset [32]. The first row shows the result of using the last frame in the local window as the reference frame. The other rows are the results of using the middle frame as the reference frame.

	Estimated depth evaluation					Calculated normal evaluation				
View num	1.25	1.25^2	abs.rel	rmse	scale.inv	$11.25°$	$22.5°$	Mean	Median	rmse
2 views (last)	72.88	92.70	0.1750	0.3910	0.1686	19.14	41.83	37.17	28.27	47.20
2 views	75.80	93.79	0.1669	0.3731	0.1638	20.12	42.76	36.82	27.67	47.03
4 views	76.29	94.28	0.1647	0.3652	0.1608	20.30	43.17	36.57	27.37	46.78
6 views	76.64	94.46	0.1612	0.3614	0.1602	20.37	43.14	36.53	27.37	46.72

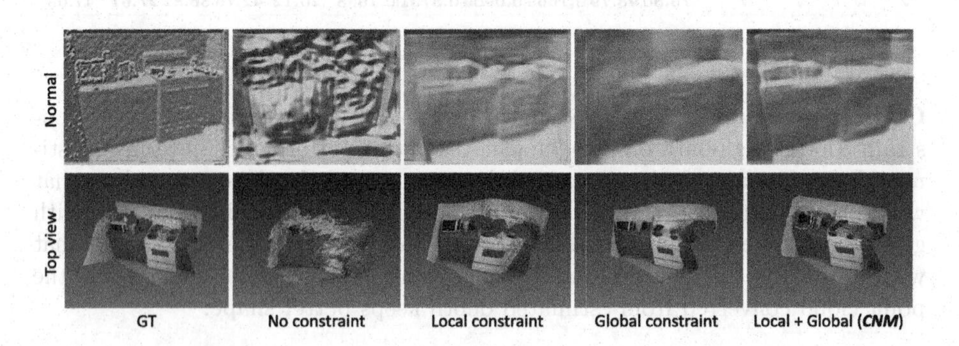

Fig. 7. Effects of using local surface normal and *CNM* as constraints for supervision.

The Number of Source Views for Refinement. Our refinement module can allow any arbitrary even number of source views as input, and generate a refined depth map and an occlusion probability map. In Table 3, with refinement module, the performance of our model has been significantly improved. As shown in Table 4, the quality of refined depth will be improved gradually with the increase of source views. A forward pass with 2/4/6 source views all takes nearly 0.02s on a GeForce RTX 2080 Ti GPU. Furthermore, we evaluate how the performance is affected by the choice of the reference view. We find that the performance of our model will degrade significantly if we use the last frame as the reference view rather than the middle frame in the local time window. This is because the middle frame shares more overlapping areas than the last frame with the other frames in the time window.

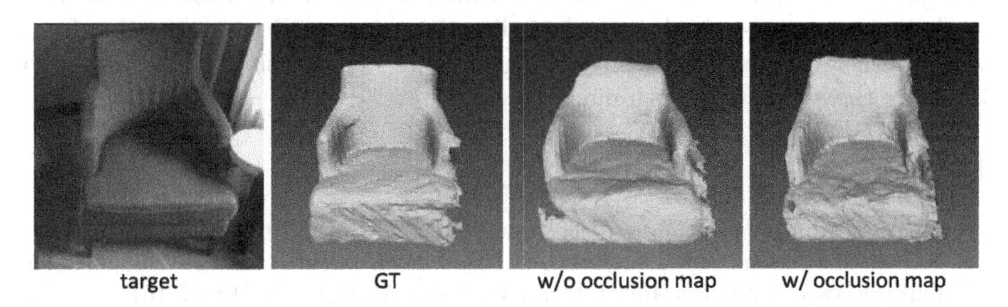

Fig. 8. Effect of occlusion probability map for video reconstruction. The occlusion probability map provides weighting parameters into TSDF fusion system for fusing estimated depth maps, and enables reconstructed model to keep sharp boundaries.

The Usefulness of Occlusion Probability Map. Unlike left-right check as a post-processing operation, our *RefineNet* jointly refines depth and generates an occlusion probability map. As shown in Table 3, our model with the occlusion-aware refinement achieves better results than that with a naive refinement (w/o occlusion map) and without any refinement. The naive refinement treats every pixel equally regardless whether it is on the occluded region or non-occluded region. In contrast, our occlusion-aware refinement penalizes occluded pixels less and pays more attention to the non-occluded region. In Fig. 3, the point cloud converted from estimated depth becomes cleaner and has sharp boundaries by applying the occlusion probability map to the multi-view depth integration process. Moreover unlike binary mask, our occlusion map can be easily used in TSDF fusion system [45] as weights. In Fig. 8, the fused mesh using estimated depth and the occlusion probability map has less artifact than that without occlusion map.

6 Conclusion and Limitations

In this paper, we propose a new method for estimating depth from a video sequence. There are two main contributions. We propose a novel normal-based constraint *CNM*, which is designed to preserve local and global geometric features for depth estimation, and a new occlusion-aware strategy to aggregate multiple depth predictions into one single depth map. Experiments demonstrate that our method outperforms than the state-of-the-art in terms of the accuracy of depth prediction and the recovery of geometric features. Furthermore, the high-fidelity depth prediction and the occlusion detection make the highly-detailed reconstruction with only a commercial RGB camera possible.

Now we discuss the limitations of our work and possible future directions. First, the performance of our method relies on the quality of the *CNM*, which is based on the segmentation of global planar regions. It has been observed that existing plane segmentation methods are not robust for all the scenes. One possible solution is to jointly learn the segmentation labels and the depth prediction. Second, for the task of video-based 3D reconstruction, our work can be extended by designing an end-to-end network to directly generate 3D reconstruction from video, rather than having to invoke the explicit step of using TSDF fusion to integrate the estimated depth maps.

References

1. Learning to find occlusion regions. In: CVPR 2011, pp. 2161–2168. IEEE (2011)
2. Aanæs, H., Jensen, R.R., Vogiatzis, G., Tola, E., Dahl, A.B.: Large-scale data for multiple-view stereopsis. Int. J. Comput. Vision **120**(2), 153–168 (2016)
3. Alvarez, H., Paz, L.M., Sturm, J., Cremers, D.: Collision avoidance for quadrotors with a monocular camera. In: Hsieh, M.A., Khatib, O., Kumar, V. (eds.) Experimental Robotics. STAR, vol. 109, pp. 195–209. Springer, Cham (2016). https://doi.org/10.1007/978-3-319-23778-7_14
4. Bleyer, M., Rhemann, C., Rother, C.: Patchmatch stereo-stereo matching with slanted support windows. BMVC **11**, 1–11 (2011)
5. Chang, J.R., Chen, Y.S.: Pyramid stereo matching network. In: Proceedings of the IEEE Conference on Computer Vision and Pattern Recognition, pp. 5410–5418 (2018)
6. Dai, A., Chang, A.X., Savva, M., Halber, M., Funkhouser, T., Nießner, M.: Scannet: richly-annotated 3d reconstructions of indoor scenes. In: Proceedings of the IEEE Conference on Computer Vision and Pattern Recognition, pp. 5828–5839 (2017)
7. Egnal, G., Wildes, R.P.: Detecting binocular half-occlusions: empirical comparisons of five approaches. IEEE Trans. Pattern Anal. Mach. Intell. **24**(8), 1127–1133 (2002)
8. Eigen, D., Fergus, R.: Predicting depth, surface normals and semantic labels with a common multi-scale convolutional architecture. In: Proceedings of the IEEE international conference on computer vision, pp. 2650–2658 (2015)
9. Eigen, D., Puhrsch, C., Fergus, R.: Depth map prediction from a single image using a multi-scale deep network. In: Advances in neural information processing systems, pp. 2366–2374 (2014)
10. Fu, H., Gong, M., Wang, C., Batmanghelich, K., Tao, D.: Deep ordinal regression network for monocular depth estimation. In: Proceedings of the IEEE Conference on Computer Vision and Pattern Recognition, pp. 2002–2011 (2018)
11. Furukawa, Y., Hernández, C.: Multi-view stereo: a tutorial. Found. Trends.Comput. Graph. Vis. **9**(1-2), 1–148 (2015).https://doi.org/10.1561/0600000052, http://dx.doi.org/10.1561/0600000052

12. Gallup, D., Frahm, J.M., Mordohai, P., Yang, Q., Pollefeys, M.: Real-time plane-sweeping stereo with multiple sweeping directions. In: 2007 IEEE Conference on Computer Vision and Pattern Recognition, pp. 1–8. IEEE (2007)
13. Hosni, A., Bleyer, M., Gelautz, M., Rhemann, C.: Local stereo matching using geodesic support weights. In: 2009 16th IEEE International Conference on Image Processing (ICIP), pp. 2093–2096. IEEE (2009)
14. Hosni, A., Rhemann, C., Bleyer, M., Rother, C., Gelautz, M.: Fast cost-volume filtering for visual correspondence and beyond. IEEE Trans. Pattern Anal. Mach. Intell. **35**(2), 504–511 (2012)
15. Hua, B.S., Pham, Q.H., Nguyen, D.T., Tran, M.K., Yu, L.F., Yeung, S.K.: Scenenn: a scene meshes dataset with annotations. In: 2016 Fourth International Conference on 3D Vision (3DV), pp. 92–101. IEEE (2016)
16. Huang, P.H., Matzen, K., Kopf, J., Ahuja, N., Huang, J.B.: Deepmvs: learning multi-view stereopsis. In: Proceedings of the IEEE Conference on Computer Vision and Pattern Recognition, pp. 2821–2830 (2018)
17. Ilg, E., Saikia, T., Keuper, M., Brox, T.: Occlusions, motion and depth boundaries with a generic network for disparity, optical flow or scene flow estimation. In: Proceedings of the European Conference on Computer Vision (ECCV), pp. 614–630 (2018)
18. Im, S., Jeon, H.G., Lin, S., Kweon, I.S.: Dpsnet: end-to-end deep plane sweep stereo. arXiv preprint arXiv:1905.00538 (2019)
19. Kang, S.B., Szeliski, R., Chai, J.: Handling occlusions in dense multi-view stereo. In: Proceedings of the 2001 IEEE Computer Society Conference on Computer Vision and Pattern Recognition. CVPR 2001, vol. 1, p. I. IEEE (2001)
20. Kendall, A., et al.: End-to-end learning of geometry and context for deep stereo regression. In: The IEEE International Conference on Computer Vision (ICCV), October 2017
21. Kusupati, U., Cheng, S., Chen, R., Su, H.: Normal assisted stereo depth estimation. arXiv preprint arXiv:1911.10444 (2019)
22. Li, Z., Snavely, N.: Megadepth: learning single-view depth prediction from internet photos. In: Proceedings of the IEEE Conference on Computer Vision and Pattern Recognition, pp. 2041–2050 (2018)
23. Liu, C., Gu, J., Kim, K., Narasimhan, S.G., Kautz, J.: Neural RGB (r) d sensing: depth and uncertainty from a video camera. In: Proceedings of the IEEE Conference on Computer Vision and Pattern Recognition, pp. 10986–10995 (2019)
24. Liu, C., Kim, K., Gu, J., Furukawa, Y., Kautz, J.: PlanerCNN: 3d plane detection and reconstruction from a single image. In: The IEEE Conference on Computer Vision and Pattern Recognition (CVPR), June 2019
25. Liu, F., Shen, C., Lin, G.: Deep convolutional neural fields for depth estimation from a single image. In: Proceedings of the IEEE Conference on Computer Vision and Pattern Recognition, pp. 5162–5170 (2015)
26. Luo, W., Schwing, A.G., Urtasun, R.: Efficient deep learning for stereo matching. In: Proceedings of the IEEE Conference on Computer Vision and Pattern Recognition, pp. 5695–5703 (2016)
27. Min, D., Sohn, K.: Cost aggregation and occlusion handling with WLS in stereo matching. IEEE Trans. Image Process. **17**(8), 1431–1442 (2008)
28. Paszke, A., et al.: Automatic differentiation in pytorch (2017)
29. Qi, X., Liao, R., Liu, Z., Urtasun, R., Jia, J.: Geonet: geometric neural network for joint depth and surface normal estimation. In: Proceedings of the IEEE Conference on Computer Vision and Pattern Recognition, pp. 283–291 (2018)

30. Qiu, J., et al.: Deeplidar: deep surface normal guided depth prediction for outdoor scene from sparse lidar data and single color image. In: Proceedings of the IEEE Conference on Computer Vision and Pattern Recognition, pp. 3313–3322 (2019)
31. Seitz, S.M., Curless, B., Diebel, J., Scharstein, D., Szeliski, R.: A comparison and evaluation of multi-view stereo reconstruction algorithms. In: Proceedings of the 2006 IEEE Computer Society Conference on Computer Vision and Pattern Recognition - Volume 1. pp. 519–528. CVPR 2006, IEEE Computer Society, Washington, DC, USA (2006). https://doi.org/10.1109/CVPR.2006.19, http://dx.doi.org/10.1109/CVPR.2006.19
32. Shotton, J., Glocker, B., Zach, C., Izadi, S., Criminisi, A., Fitzgibbon, A.: Scene coordinate regression forests for camera relocalization in RGB-d images. In: Proceedings of the IEEE Conference on Computer Vision and Pattern Recognition, pp. 2930–2937 (2013)
33. Sturm, J., Engelhard, N., Endres, F., Burgard, W., Cremers, D.: A benchmark for the evaluation of RGB-d slam systems. In: 2012 IEEE/RSJ International Conference on Intelligent Robots and Systems, pp. 573–580. IEEE (2012)
34. Ummenhofer, B., et al.: Demon: Depth and motion network for learning monocular stereo. In: Proceedings of the IEEE Conference on Computer Vision and Pattern Recognition, pp. 5038–5047 (2017)
35. Wang, J., Zickler, T.: Local detection of stereo occlusion boundaries. In: Proceedings of the IEEE Conference on Computer Vision and Pattern Recognition, pp. 3818–3827 (2019)
36. Wang, K., Shen, S.: MVdepthnet: real-time multiview depth estimation neural network. In: 2018 International Conference on 3D Vision (3DV), pp. 248–257. IEEE (2018)
37. Xiao, J., Shah, M.: Motion layer extraction in the presence of occlusion using graph cuts. IEEE Trans. Pattern Anal. Mach. Intell. **27**(10), 1644–1659 (2005)
38. Xiao, J., Owens, A., Torralba, A.: Sun3d: a database of big spaces reconstructed using SFM and object labels. In: Proceedings of the IEEE International Conference on Computer Vision, pp. 1625–1632 (2013)
39. Xu, L., Jia, J.: Stereo matching: an outlier confidence approach. In: Forsyth, D., Torr, P., Zisserman, A. (eds.) ECCV 2008. LNCS, vol. 5305, pp. 775–787. Springer, Heidelberg (2008). https://doi.org/10.1007/978-3-540-88693-8_57
40. Yang, Q.: A non-local cost aggregation method for stereo matching. In: 2012 IEEE Conference on Computer Vision and Pattern Recognition, pp. 1402–1409. IEEE (2012)
41. Yao, Y., Luo, Z., Li, S., Fang, T., Quan, L.: MVSnet: Depth inference for unstructured multi-view stereo. In: Proceedings of the European Conference on Computer Vision (ECCV), pp. 767–783 (2018)
42. Yin, W., Liu, Y., Shen, C., Yan, Y.: Enforcing geometric constraints of virtual normal for depth prediction. In: The IEEE International Conference on Computer Vision (ICCV) (2019)
43. Yoon, K.J., Kweon, I.S.: Adaptive support-weight approach for correspondence search. IEEE Trans. Pattern Anal. Mach. Intell. **4**, 650–656 (2006)
44. Zbontar, J., LeCun, Y.: Computing the stereo matching cost with a convolutional neural network. In: Proceedings of the IEEE Conference on Computer Vision and Pattern Recognition, pp. 1592–1599 (2015)
45. Zeng, A., Song, S., Nießner, M., Fisher, M., Xiao, J., Funkhouser, T.: 3dmatch: Learning local geometric descriptors from RGB-D reconstructions. In: CVPR (2017)

46. Zhang, Y., Funkhouser, T.: Deep depth completion of a single RGB-D image. The IEEE Conference on Computer Vision and Pattern Recognition (CVPR) (2018)
47. Zollhöfer, M., Stotko, P., Görlitz, A., Theobalt, C., Nießner, M., Klein, R., Kolb, A.: State of the Art on 3D Reconstruction with RGB-D Cameras. Comput. Graph. Forum (Eurograph. State Art Rep. 2018), **37**(2), 625–652 (2018)

VisualEchoes: Spatial Image Representation Learning Through Echolocation

Ruohan Gao[1,3](✉), Changan Chen[1,3], Ziad Al-Halah[1], Carl Schissler[2], and Kristen Grauman[1,3]

[1] The University of Texas at Austin, Austin, USA
{rhgao,changan,ziad,grauman}@cs.utexas.edu
[2] Facebook Reality Lab, Seattle, USA
carl.schissler@fb.com
[3] Facebook AI Research, Austin, USA

Abstract. Several animal species (e.g., bats, dolphins, and whales) and even visually impaired humans have the remarkable ability to perform echolocation: a biological sonar used to perceive spatial layout and locate objects in the world. We explore the spatial cues contained in echoes and how they can benefit vision tasks that require spatial reasoning. First we capture echo responses in photo-realistic 3D indoor scene environments. Then we propose a novel interaction-based representation learning framework that learns useful *visual* features via echolocation. We show that the learned image features are useful for multiple downstream vision tasks requiring spatial reasoning—monocular depth estimation, surface normal estimation, and visual navigation—with results comparable or even better than heavily supervised pre-training. Our work opens a new path for representation learning for embodied agents, where supervision comes from interacting with the physical world.

1 Introduction

The perceptual and cognitive abilities of embodied agents are inextricably tied to their physical being. We perceive and act in the world by making use of all our senses—especially looking and listening. We see our surroundings to avoid obstacles, listen to the running water tap to navigate to the kitchen, and infer how far away the bus is once we hear it approaching.

By using *two* ears, we perceive spatial sound. Not only can we identify the sound-emitting object (e.g., the revving engine corresponds to a bus), but also we can determine that object's location, based on the time difference between when the sound reaches each ear (Interaural Time Difference, ITD) and the difference in sound level as it enters each ear (Interaural Level Difference, ILD). Critically, even beyond objects, audio is also rich with information about the

Electronic supplementary material The online version of this chapter (https://doi.org/10.1007/978-3-030-58545-7_38) contains supplementary material, which is available to authorized users.

© Springer Nature Switzerland AG 2020
A. Vedaldi et al. (Eds.): ECCV 2020, LNCS 12354, pp. 658–676, 2020.
https://doi.org/10.1007/978-3-030-58545-7_38

environment itself. The sounds we receive are a function of the geometric structure of the space around us and the materials of its major surfaces [5]. In fact, some animals capitalize on these cues by using *echolocation*—actively emitting sounds to perceive the 3D spatial layout of their surroundings [68].

We propose to learn image representations from echoes. Motivated by how animals and blind people obtain spatial information from echo responses, first we explore to what extent the echoes of chirps generated in a scanned 3D environment are predictive of the depth in the scene. Then, we introduce VisualEchoes, a novel image representation learning method based on echolocation. Given a first-person RGB view and an echo audio waveform, our model is trained to predict the correct camera orientation at which the agent would receive those echoes. In this way, the representation is forced to capture the alignment between the sound reflections and the (visually observed) surfaces in the environment. At test time, we observe only pixels—no audio. Our learned VisualEchoes encoder better reveals the 3D spatial cues embedded in the pixels, as we demonstrate in three downstream tasks.

Our approach offers a new way to learn image representations without manual supervision by *interacting* with the environment. In pursuit of this high-level goal there is exciting—though limited—prior work that learns visual features by touching objects [2,59,62,63] or moving in a space [1,27,45]. Unlike mainstream "self-supervised" feature learning work that crafts pretext tasks for large static repositories of human-taken images or video (e.g., colorization [88], jigsaw puzzles [57], audio-visual correspondence [6,50]), in *interaction-based feature learning* an embodied agent[1] performs physical actions in the world that dynamically influence its own first-person observations and possibly the environment itself. Both paths have certain advantages: while conventional self-supervised learning can capitalize on massive static datasets of human-taken photos, interaction-based learning allows an agent to "learn by acting" with rich multi-modal sensing. This has the advantage of learning features adaptable to new environments. Unlike any prior work, we explore feature learning from echoes.

Our contributions are threefold: 1) We explore the spatial cues contained in echoes, analyzing how they inform depth prediction; 2) We propose VisualEchoes, a novel interaction-based feature learning framework that uses echoes to learn an image representation and does not require audio at test time; 3) We successfully validate the learned spatial representation for the fundamental downstream vision tasks of monocular depth prediction, surface normal estimation, and visual navigation, with results comparable to or even outperforming heavily supervised pre-training baselines.

2 Related Work

Auditory Scene Analysis using Echoes: Previous work shows that using echo responses only, one can predict 2D [5] or 3D [14] room geometry and object

[1] Person, robot, or simulated robot.

shape [22]. Additionally, echoes can complement vision, especially when vision-based depth estimates are not reliable, e.g., on transparent windows or featureless walls [49,86]. In dynamic environments, autonomous robots can leverage echoes for obstacle avoidance [77] or mapping and navigation [17] using a bat-like echolocation model. Concurrently with our work, a low-cost audio system called BatVision is used to predict depth maps purely from echo responses [12]. Our work explores a novel direction for auditory scene analysis by employing echoes for spatial visual feature learning, and unlike prior work, the resulting features are applicable in the absence of any audio.

Self-Supervised Image Representation Learning: Self-supervised image feature learning methods leverage structured information within the data itself to generate labels for representation learning [38,69]. To this end, many "pretext" tasks have been explored—for example, predicting the rotation applied to an input image [1,35], discriminating image instances [19], colorizing images [52,88], solving a jigsaw puzzle from image patches [57], predicting unseen views of 3D objects [44], or multi-task learning using synthetic imagery [66]. Temporal information in videos also permits self-supervised tasks, for example, by predicting whether a frame sequence is in the correct order [20,55] or ensuring visual coherence of tracked objects [31,43,82]. Whereas these methods aim to learn features generically useful for recognition, our objective is to learn features generically useful for spatial estimation tasks. Accordingly, our echolocation objective is well-aligned with our target family of spatial tasks (depth, surfaces, navigation), consistent with findings that task similarity is important for positive transfer [87]. Furthermore, unlike any of the above, rather than learn from massive repositories of human-taken photos, the proposed approach learns from interactions with the scene via echolocation.

Feature Learning by Interaction: Limited prior work explores feature learning through interaction. Unlike the self-supervised methods discussed above, this line of work fosters agents that learn from their own observations in the world, which can be critical for adapting to new environments and to realize truly "bottom-up" learning by experience. Existing methods explore touch and motion interactions. In [59], objects are struck with a drumstick to facilitate learning material properties when they sound. In [63], the trajectory of a ball bouncing off surfaces facilitates learning physical scene properties. In [2,62], a robot learns object properties by poking or grasping at objects. In [27], a drone learns not to crash after attempting many crashes. In [1,45], an agent tracks its egomotion in concert with its visual stream to facilitate learning visual categories. In contrast, our idea is to learn visual features by *emitting audio* to acoustically interact with the scene. Our work offers a new perspective on interaction-based feature learning and has the advantages of not disrupting the scene physically and being ubiquitously available, i.e., reaching all surrounding surfaces.

Audio-Visual Learning: Inspiring recent work integrates sound and vision in joint learning frameworks that synthesize sounds for video [59,92], spatialize monaural sounds from video [29,56], separate sound sources [18,24,28,30,58,89], perform cross-modal feature learning [8,60], track audio-visual targets [3,9,26, 34], segment objects with multi-channel audio [42], direct embodied agents to navigate in indoor environments [11,25], recognize actions in videos [32,48], and localize pixels associated with sounds in video frames [7,40,71,75]. None of the prior methods pursues echoes for visual learning. Furthermore, whereas nearly all existing audio-visual methods operate in a passive manner, observing incidental sounds within a video, in our approach the system learns by actively emitting sound—a form of interaction with the physical environment.

Monocular Depth Estimation. To improve monocular depth estimation, recent methods focus on improving neural network architectures [23] or graphical models [53,81,84], employing multi-scale feature fusion and multi-task learning [15,41], leveraging motion cues from successive frames [76], or transfer learning [47]. However, these approaches rely on depth-labeled data that can be expensive to obtain. Hence, recent approaches leverage scenes' spatial and temporal structure to self-supervise depth estimation, by using the camera motion between pairs of images [33,36] or frames [37,46,80,91], or consistency cues between depth and features like surface normals [85] or optical flow [65]. Unlike any of these existing methods, we show that audio in the form of an echo response can be effectively used to recover depth, and we develop a novel feature learning method that benefits a purely visual representation (no audio) at test time.

3 Approach

Our goals are to show that echoes convey spatial information, to learn visual representations by echolocation, and to leverage the learned representations for downstream tasks. In the following, we first describe how we simulate echoes in 3D environments (Sect. 3.1). Then we perform a case study to demonstrate how echoes can benefit monocular depth prediction (Sect. 3.2). Next, we present VisualEchoes, our interaction-based feature learning formulation to learn image representations (Sect. 3.3). Finally, we exploit the learned visual representation for monocular depth, surface normal prediction, and visual navigation (Sect. 3.4).

3.1 Echolocation Simulation

Our echolocation simulation is based on recent work on audio-visual navigation [11], which builds a realistic acoustic simulation on top of the Habitat [70] platform and Replica environments [73]. Habitat [70] is an open-source 3D simulator that supports efficient RGB, depth, and semantic rendering for multiple datasets [10,73,83]. Replica is a dataset of 18 apartment, hotel, office, and room scenes with 3D meshes and high definition range (HDR) textures and renderable reflector information. The platform in [11] simulates acoustics by pre-computing room impulse responses (RIR) between all pairs of possible source and receiver locations, using a form of audio ray-tracing [79]. An RIR is a transfer function between the sound source and the sound microphone, and it is influenced by

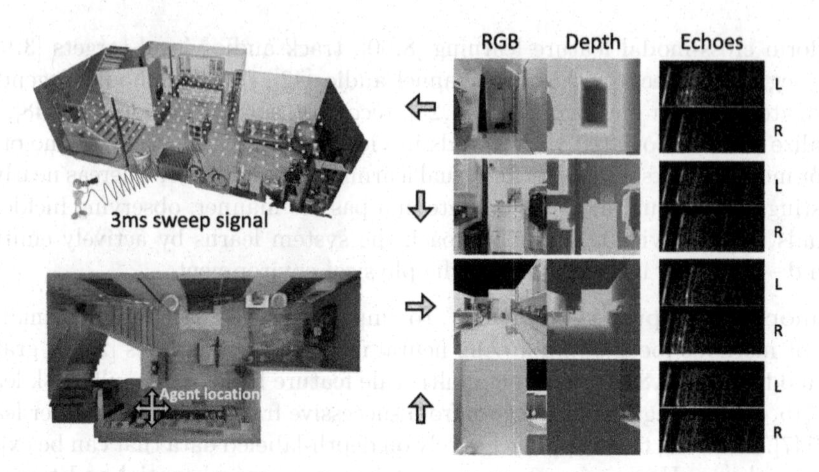

Fig. 1. Echolocation simulation in real-world scanned environments. During training, the agent goes to the densely sampled locations marked with yellow dots. The left bottom figure illustrates the top-down view of one Replica scene where the agent's location is marked. The agent actively emits 3 ms omnidirectional sweep signals to get echo responses from the room. The right column shows the corresponding RGB and depth of the agent's view as well as the echoes received in the left and right ears when the agent faces each of the four directions.

the room geometry, materials, and the sound source location [51]. The sound received at the listener location is computed by convolving the appropriate RIR with the waveform of the source sound.

We use the binaural RIRs for all Replica environments to generate echoes for our approach. As the source audio "chirp" we use a sweep signal from 20 Hz-20 kHz (the human-audible range) within a duration of 3 ms. While technically any emitted sound could provide some echo signal from which to learn, our design (1) intentionally provides the response for a wide range of frequencies and (2) does so in a short period of time to avoid overlap between echoes and direct sounds. We place the source at the *same* location as the receiver and convolve the RIR for this source-receiver pair with the sweep signal. In this way, we compute the echo responses that would be received at the agent's microphone locations. We place the agents at all navigable points on the grid (every 0.5 m [11]) and orient the agent in four cardinal directions (0°, 90°, 180°, 270°) so that the rendered egocentric views (RGB and depth) and echoes capture room geometry from different locations and orientations.

Figure 1 illustrates how we perform echolocation for one scene environment. The agent goes to the densely sampled navigable locations marked with yellow dots and faces four orientations at each location. It actively emits omnidirectional chirp signals and records the echo responses received when facing each direction. Note that the spectrograms of the sounds received at the left (L) and right (R) ears reveal that the agent first receives the direct sound (strong bright curves), and then receives different echoes for the left and right microphones due

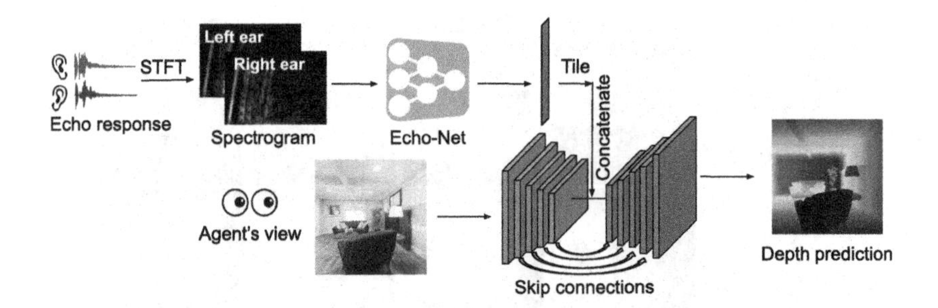

Fig. 2. Our RGB+Echo2Depth network takes the echo responses and the corresponding egocentric RGB view as input, and performs joint audio-visual analysis to predict the depth map for the input image. The injected echo response provides additional cues of the spatial layout of the scene. Note: in later sections we define networks that do not have access to the audio stream at test time.

to ITD, ILD, and pinnae reflections. The subtle difference in the two spectrograms conveys cues about the spatial configuration of the environment, as can be observed in the last column of Fig. 1.

3.2 Case Study: Spatial Cues in Echoes

With the synchronized egocentric views and echo responses in hand, we now conduct a case study to investigate the spatial cues contained in echo responses in these realistic indoor 3D environments. We have two questions: (1) can we directly predict depth maps purely from echoes? and (2) can we use echoes to augment monocular depth estimation from RGB? Answering these questions will inform our ultimate goal of devising a interaction-supervised visual feature learning approach leveraging echoes only at training time (Sect. 3.3). Furthermore, it can shed light on the extent to which low-cost audio sensors can replace depth sensors, which would be especially useful for navigation robots under severe bandwidth or sensing constraints, e.g., nano drones [54,61].

Note that these two goals are orthogonal to that of prior work performing depth prediction from a single view [16,23,41,53,84]. Whereas they focus on developing sophisticated loss functions and architectures, here we explore how an agent *actively interacting with the scene acoustically* may improve its depth predictions. Our findings can thus complement existing monocular depth models.

We devise an RGB+Echo2Depth network (and its simplified variants using only RGB or echo) to test the settings of interest. The RGB+Echo2Depth network predicts a depth map based on the agent's egocentric RGB input and the echo response it receives when it emits a chirp standing at that position and orientation in the 3D environment. The core model is a multi-modal U-Net [67]; see Fig. 2. To directly measure the spatial cues contained in echoes alone, we also test a variant called Echo2Depth. Instead of performing upsampling based on the audio-visual representation, this model drops the RGB input, reshapes the audio feature, and directly upsamples from the audio representation. Similarly,

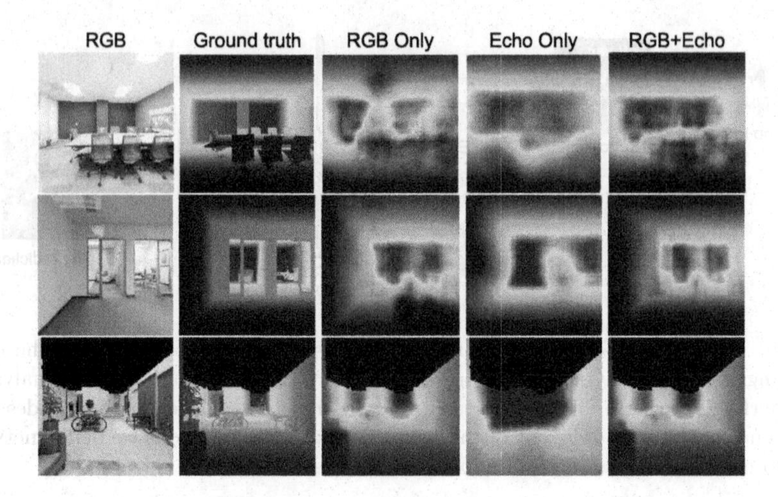

Fig. 3. Qualitative results of our case study on monocular depth estimation in unseen environments using echoes. Together with the quantitative results (Table 1), these examples show that echoes contain useful spatial cues that inform a visual spatial task. For example, in row 1, the RGB+Echo model better infers the depth of the column on the back wall, whereas the RGB-Only model mistakenly infers the strong contours to indicate a much closer surface. The last row shows a typical failure case (see text). See Supp. for more examples.

Table 1. Case study depth prediction results. ↓ lower better, ↑ higher better.

	RMS ↓	REL ↓	log 10 ↓	$\delta < 1.25$ ↑	$\delta < 1.25^2$ ↑	$\delta < 1.25^3$ ↑
AVERAGE	1.070	0.791	0.230	0.235	0.509	0.750
ECHO2DEPTH	0.713	0.347	0.134	0.580	0.772	0.868
RGB2DEPTH	0.374	0.202	0.076	0.749	0.883	0.945
RGB+ECHO2DEPTH	**0.346**	**0.172**	**0.068**	**0.798**	**0.905**	**0.950**

to measure the cues contained in the RGB alone, a variant called RGB2DEPTH drops the echoes and predicts the depth map purely based on the visual features. The RGB2DEPTH model represents existing monocular depth prediction approaches that predict depth from a single RGB image, in the context of the same architecture design as RGB+ECHO2DEPTH to allow apples-to-apples calibration of our findings. We use RGB images of spatial dimension 128×128. See Supp. for network details and loss functions used to train the three models.

Table 1 shows the quantitative results of predicting depth from only echoes, only RGB, or their combination. We evaluate on a heldout set of three Replica environments (comprising 1,464 total views) with standard metrics: root mean squared error (RMS), mean relative error (REL), mean log 10 error (log 10), and thresholded accuracy [16,41]. We can see that depth prediction is possible purely from echoes. Augmenting traditional single-view depth estimation with echoes (bottom row) achieves the best performance by leveraging the additional acoustic spatial cues. Echoes alone are naturally weaker than RGB alone, yet still better than the simple AVERAGE baseline that predicts the average depth values in all training data.

Figure 3 shows qualitative examples. It is clear that echo responses indeed contain cues of the spatial layout; the depth map captures the rough room layout, especially its large surfaces. When combined with RGB, the predictions are more accurate. The last row shows a typical failure case, where the echoes alone cannot capture the depth as well due to far away surfaces with weaker echo signals.

3.3 VisualEchoes Spatial Representation Learning Framework

Having established the scope for inferring depth from echoes, we now present our VISUALECHOES model to leverage echoes for visual representation learning. We stress that our approach assumes audio/echoes are available only during training; at test time, an RGB image alone is the input.

The key insight of our approach is that the echoes and visual input should be consistent. This is because both are functions of the same latent variable—the 3D shape of the environment surrounding the agent's co-located camera and microphones. We implement this idea by training a network to predict their correct association.

In particular, as described in Sect. 3.1, at any position in the scene, we suppose the agent can face four orientations, i.e., at an azimuth angle of $0°$, $90°$, $180°$, and $270°$. When the agent emits the sweep signal (chirp) at a certain position, it will hear different echo responses when it faces each different orientation. If the agent correctly interprets the spatial layout of the current view from *visual* information, it should be able to tell whether that visual input is congruous with the echo response it hears. Furthermore, and more subtly, to the extent the agent implicitly learns about probable views surrounding its current egocentric field of view (e.g., what the view just to its right may look like given the context of what it sees in front of it), it should be able to tell which direction the received echo *would* be congruous with, if not the current view.

We introduce a representation learning network to capture this insight. See Fig. 4. The visual stream takes the agent's current RGB view as input, and the audio stream takes the echo response received from one of the four orientations—not necessarily the one that coincides with the visual stream orientation. The fusion layer fuses the audio and visual information to generate an audio-visual feature of dimension D. A final fully-connected layer is used to make the final prediction among four classes. See Supp. and Sect. 4 for architecture details. The four classes are defined as follows:

$\uparrow$: The echo is received from the same orientation as the agent's current view.
$\rightarrow$: The echo is received from the orientation if the agent turns right by $90°$.
$\downarrow$: The echo is received from the orientation opposite the agent's current view.
$\leftarrow$: The echo is received from the orientation if the agent turns left by $90°$.

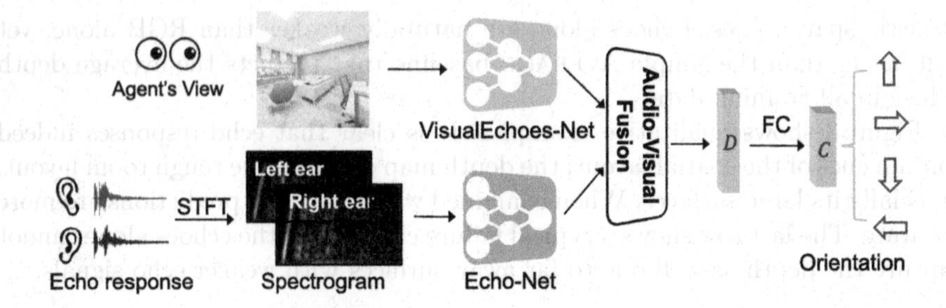

Fig. 4. Our VISUALECHOES network takes the agent's current RGB view as visual input, and the echo responses from one of the four orientations as audio input. The goal is to predict the orientation at which the agent would receive the input echoes based on analyzing the spatial layout in the image. After training with RGB and echoes, the VISUALECHOES-Net is a pre-trained encoder ready to extract spatially enriched features from novel RGB images, as we validate with multiple downstream tasks (cf. Sect. 3.4).

The network is trained with cross-entropy loss. Note that although the emitted source signal is always the same (3 ms *omnidirectional* sweep signal, cf. Sect. 3.1), the agent hears different echoes when facing the four directions because of the shape of the ears and the head shadowing effect modeled in the binaural head-related transfer function (HRTF). Since the classes above are defined relative to the agent's current view, it can only tell the orientation for which it is receiving the echoes if it can correctly interpret the 3D spatial layout within the RGB input. In this way, the agent's aural interaction with the scene enhances spatial feature learning for the visual stream.

The proposed idea generalizes trivially to use more than four discrete orientations—and even arbitrary orientations if we were to use regression rather than classification. The choice of four is simply based on the sound simulations available in existing data [11], though we anticipate it is a good granularity to capture the major directions around the agent. Our training paradigm requires the representation to discern mismatches between the image and echo using echoes generated from the same physical position on the ground plane but different orientations. This is in line with our interactive embodied agent motivation, where an agent can look ahead, then turn and hear echoes from another orientation at the same place in the environment, and learn their (dis)association. In fact, ecological psychologists report that humans can perform more accurate echolocation when moving, supporting the rationale of our design [68,74]. Furthermore, our design ensures the mismatches are "hard" examples useful for learning spatial features because the audio-visual data at offset views will naturally be related to one another (as opposed to views or echoes from an unrelated environment).

3.4 Downstream Tasks for the Learned Spatial Representation

Having introduced our VisualEchoes feature learning framework, next we describe how we repurpose the learned visual representation for three fundamental downstream tasks that require spatial reasoning: monocular depth prediction, surface normal estimation, and visual navigation. For each task, we adopt strong models from the literature and swap in our pre-trained encoder VisualEchoes-Net for the RGB input.

Monocular depth prediction: We explore how our echo-based pre-training can benefit performance for traditional monocular depth prediction. Note that unlike the case study in Sect. 3.2, in this case there are no echo inputs at test time, only RGB. To evaluate the quality of our learned representation, we adopt a strong recent approach for monocular depth prediction [41] consisting of several novel loss functions and a multi-scale network architecture that is based on a backbone network. We pre-train ResNet-50 [39] using VisualEchoes and use it as the backbone for comparison with [41].

Surface normal estimation: We also evaluate the learned spatial representation to predict surface normals from a single image, another fundamental mid-level vision task that requires spatial understanding of the geometry of the surfaces [21]. We adopt the the state-of-the-art pyramid scene parsing network PSPNet architecture [90] for surface normal prediction, again swapping in our pre-trained VisualEchoes network for the RGB feature backbone.

Visual navigation: Finally, we validate on an embodied visual navigation task. In this task, the agent receives a sequence of RGB images as input and a point goal defined by a displacement vector relative to the starting position of the agent [4]. The agent is spawned at random locations and must navigate to the target location quickly and accurately. This entails reasoning about 3D spatial configurations to avoid obstacles and find the shortest path. We adopt a state-of-the-art reinforcement learning-based PointGoal visual navigation model [70]. It consists of a three-layer convolutional network and a fully-connected layer to extract visual feature from the RGB images. We pre-train its visual network using VisualEchoes, then train the full network end to end.

While other architectures are certainly possible for each task, our choices are based on both on the methods' effectiveness in practice, their wide use in the literature, and code availability. Our contribution is feature learning from echoes as a pre-training mechanism for spatial tasks, which is orthogonal to advances on architectures for each individual task. In fact, a key message of our results is that the VisualEchoes-Net encoder boosts multiple spatial tasks, under multiple different architectures, and on multiple datasets.

4 Experiments

We present experiments to validate VisualEchoes for three tasks and three datasets (Replica [73], NYU-V2 [72], and DIODE [78]). The goal is to examine the impact of our features compared to either learning features for that task from scratch or learning features with manual semantic supervision. See Supp. for details of the three datasets.

Implementation Details: All networks are implemented in PyTorch. For the echoes, we use the first 60 ms, which allows most of the room echo responses following the 3 ms chirp to be received. We use an audio sampling rate of 44.1 kHz. STFT is computed using a Hann window of length 64, hop length of 16, and FFT size of 512. The audio-visual fusion layer (see Fig. 4) concatenates the visual and audio feature, and then uses a fully-connected layer to reduce the feature dimension to $D = 128$. See Supp. for details of the network architectures and optimization hyperparameters.

Evaluation Metrics: We report standard metrics for the downstream tasks. 1) *Monocular Depth Prediction:* RMS, REL, and others as defined above, following [16,41]. 2) *Surface Normal Estimation:* mean and median of the angle distance and the percentage of good pixels (i.e., the fraction of pixels with cosine distance to ground-truth less than t) with $t = 11.25°, 22.5°, 30°$, following [21]. 3) *Visual Navigation:* success rate normalized by inverse path length (SPL), the distance to the goal at the end of the episode, and the distance to the goal normalized by the trajectory length, following [4].

4.1 Transferring VisualEchoes Features for RGB2Depth

Having confirmed echoes reveal spatial cues in Sect. 3.2, we now examine the effectiveness of VisualEchoes, our learned representation. Our model achieves 66% test accuracy on the orientation prediction task, while chance performance is only 25%; this shows learning the visual-echo consistency task itself is possible.

First, we use the same RGB2Depth network from our case study in Sect. 3.2 as a testbed to demonstrate the learned spatial features can be successfully transferred to other domains. Instead of randomly initializing the RGB2Depth UNet encoder, we initialize with an encoder 1) pre-trained for our visual-echo consistency task, 2) pre-trained for image classification using ImageNet [13], or 3) pre-trained for scene classification using the MIT Indoor Scene dataset [64]. Throughout, aside from the standard ImageNet pre-training baseline, we also include MIT Indoor Scenes pre-training, in case it strengthens the baseline due to its domain alignment with the indoor scenes in Replica, DIODE, and NYU-2.

Table 2 shows the results on all three datasets: Replica, NYU-V2, and DIODE. The model initialized with our pre-trained VisualEchoes network achieves much better performance compared to the model trained from scratch. Moreover, it even outperforms the supervised model pre-trained on scene classification in some cases. The ImageNet pre-trained model performs much worse; we suspect that the UNet encoder does not have sufficient capacity to handle ImageNet classification, and also the ImageNet domain is much different than indoor scene environments. This result accentuates that task similarity promotes positive transfer [87]: our unsupervised spatial pre-training task is more powerful

Table 2. Depth prediction results on the Replica, NYU-V2, and DIODE datasets. We use the RGB2Depth network from Sect. 3.2 for all models. Our VisualEchoes pre-training transfers well, consistently predicting depth better than the model trained from scratch. Furthermore, it is even competitive with the supervised models, whether they are pre-trained for ImageNet or MIT Indoor Scenes (1M/16K manually labeled images). ↓ lower better, ↑ higher better. (Un)sup = (un)supervised. We boldface the best unsupervised method.

		RMS ↓	REL ↓	log 10 ↓	$\delta < 1.25$ ↑	$\delta < 1.25^2$ ↑	$\delta < 1.25^3$ ↑
Sup	ImageNet Pre-trained	0.356	0.203	0.076	0.748	0.891	0.948
	MIT Indoor Scene Pre-trained	0.334	0.196	0.072	0.770	0.897	0.950
Unsup	Scratch	0.360	0.214	0.078	0.747	0.879	0.940
	VisualEchoes (Ours)	**0.332**	**0.195**	**0.070**	**0.773**	**0.899**	**0.951**

(a) Replica

		RMS ↓	REL ↓	log 10 ↓	$\delta < 1.25$ ↑	$\delta < 1.25^2$ ↑	$\delta < 1.25^3$ ↑
Sup	ImageNet Pre-trained	0.812	0.249	0.102	0.589	0.855	0.955
	MIT Indoor Scene Pre-trained	0.776	0.239	0.098	0.610	0.869	0.959
Unsup	Scratch	0.818	0.252	0.103	0.586	0.853	0.950
	VisualEchoes (Ours)	**0.797**	**0.246**	**0.100**	**0.600**	**0.863**	**0.956**

(b) NYU-V2

		RMS ↓	REL ↓	log 10 ↓	$\delta < 1.25$ ↑	$\delta < 1.25^2$ ↑	$\delta < 1.25^3$ ↑
Sup	ImageNet Pre-trained	2.250	0.453	0.199	0.336	0.591	0.766
	MIT Indoor Scene Pre-trained	2.218	0.424	0.198	0.363	0.632	0.776
Unsup	Scratch	2.352	0.481	0.214	0.321	0.581	0.742
	VisualEchoes (Ours)	**2.223**	**0.430**	**0.198**	**0.340**	**0.610**	**0.769**

(c) DIODE

Table 3. Ablation study on Replica. See Supp. for results on NYU-V2 and Diode.

	RMS ↓	REL ↓	log 10 ↓	$\delta < 1.25$ ↑	$\delta < 1.25^2$ ↑	$\delta < 1.25^3$ ↑
Scratch	0.360	0.214	0.078	0.747	0.879	0.940
SimpleVisualEchoes	0.340	0.198	0.073	0.763	0.892	0.948
BinaryMatching	0.345	0.199	0.074	0.760	0.889	0.944
VisualEchoes (Ours)	**0.332**	**0.195**	**0.070**	**0.773**	**0.899**	**0.951**

for depth inference than a supervised semantic category pre-training task. See Supp. for low-shot experiments varying the amount of training data.

We also perform an ablation study to demonstrate that the design of our spatial representation learning framework is essential and effective. We compare with the following two variants: SimpleVisualEchoes, which simplifies our orientation prediction task to two classes; and BinaryMatching, which mimics prior work [6] that leverages the correspondence between images and audio as supervision by training a network to decide if the echo and RGB are from the same environment. As shown in Table 3, our method performs much better than both baselines. See Supp. for details.

4.2 Evaluating on Downstream Tasks

Next we evaluate the impact of our learned VISUALECHOES representation on all three downstream tasks introduced in Sect. 3.4.

Table 4. Results for three downstream tasks. ↓ lower better, ↑ higher better.

		RMS ↓	REL ↓	log 10 ↓	$\delta < 1.25$ ↑	$\delta < 1.25^2$ ↑	$\delta < 1.25^3$ ↑
Sup	ImageNet Pre-trained [41]	**0.555**	**0.126**	**0.054**	**0.843**	**0.968**	**0.991**
	MIT Indoor Scene Pre-trained	0.711	0.180	0.075	0.730	0.925	0.979
Unsup	Scratch	0.804	0.209	0.086	0.676	0.897	0.967
	VISUALECHOES (Ours)	0.683	0.165	0.069	0.762	0.934	0.981

(a) Depth prediction results on NYU-V2.

		Mean Dist. ↓	Median Dist. ↓	$t < 11.25°$ ↑	$t < 22.5°$ ↑	$t < 30°$ ↑
Sup	ImageNet Pre-trained	26.4	17.1	36.1	59.2	68.5
	MIT Indoor Scene Pre-trained	25.2	17.5	36.5	57.8	67.2
Unsup	Scratch	26.3	16.1	37.9	60.6	69.0
	VISUALECHOES (Ours)	**22.9**	**14.1**	**42.7**	**64.1**	**72.4**

(b) Surface normal estimation results on NYU-V2. The results for the ImageNet Pre-trained baseline and the Scratch baseline are directly quoted from [38].

		SPL ↑	Distance to Goal ↓	Normalized Distance to Goal ↓
Sup	ImageNet Pre-trained	0.833	0.663	0.081
	MIT Indoor Scene Pre-trained	0.798	1.05	0.124
Unsup	Scratch	0.830	0.728	0.096
	VISUALECHOES (Ours)	**0.856**	**0.476**	**0.061**

(c) Visual navigation performance in unseen Replica environments.

Monocular depth prediction: Table 4a shows the results.[2] All methods use the same settings as [41], where they evaluate and report results on NYU-V2. We use the authors' publicly available code[3] and use ResNet-50 as the encoder. See Supp. for details. With this apples-to-apples comparison, the difference in performance can be attributed to whether/how the encoder is pre-trained. Although our VISUALECHOES features are learned from Replica, they transfer reasonably well to NYU-V2, outperforming models trained from scratch by a large margin. This result is important because it shows that despite training with simulated audio, our model generalizes to real-world test images. Our features also compare favorably to supervised models trained with heavy supervision.

Surface normal estimation: Table 4b shows the results. We follow the same setting as [38] and we use the authors' publicly available code.[4] Our model performs much better even compared to the ImageNet-supervised pre-trained model, demonstrating that our interaction-based feature learning framework via echoes makes the learned features more useful for 3D geometric tasks.

[2] We evaluate on NYU-V2, the most widely used dataset for the task of single view depth prediction and surface normal estimation. The authors's code [38,41] is tailored to this dataset.

[3] https://github.com/JunjH/Revisiting_Single_Depth_Estimation.

[4] https://github.com/facebookresearch/fair_self_supervision_benchmark.

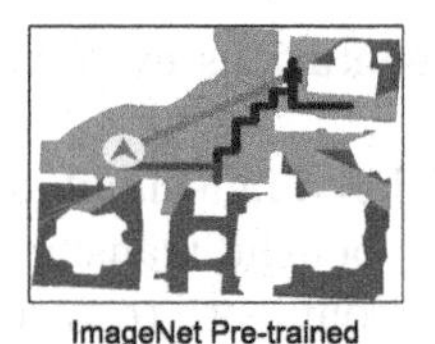

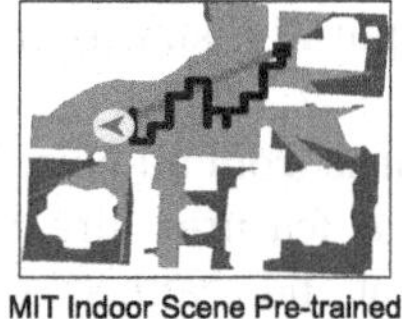

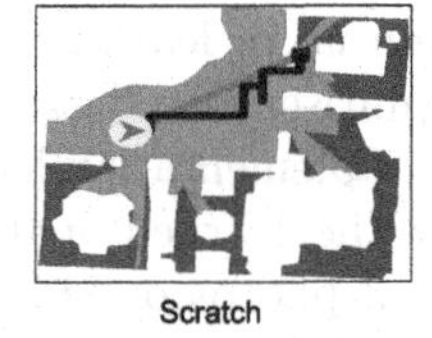

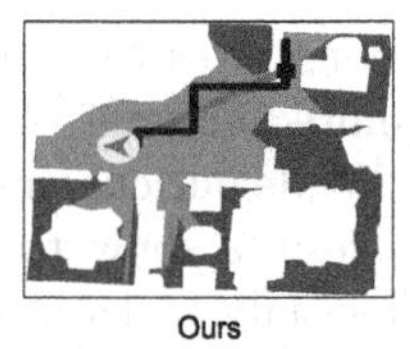

Fig. 5. Qualitative examples of visual navigation trajectories on top-down maps. Blue square and arrow denote agent's starting and ending positions, respectively. The green path indicates the shortest geodesic path to the goal, and the agent's path is in dark blue. Agent path color fades from dark blue to light blue as time goes by. Note, the agent sees a sequence of egocentric views, not the map.

Visual navigation: Table 4c shows the results. By pre-training the visual network, VISUALECHOES equips the embodied agents with a better sense of room geometry and allows them to learn faster (see Supp. for training curves). Notably, the agent also ends much closer to the goal. We suspect it can better gauge the distance because of our VISUALECHOES pre-training. Models pre-trained for classification on MIT Indoor Scenes perform more poorly than Scratch; again, this suggests features useful for recognition may not be optimal for a spatial task like point goal navigation.

This series of results on three tasks consistently shows the promise of our VISUALECHOES features. We see that learning from echoes translates into a strengthened *visual* encoding. Importantly, while it is always an option to train multiple representations entirely from scratch to support each given task, our results are encouraging since they show the *same* fundamental interaction-based pre-training is versatile across multiple tasks.

4.3 Qualitative Results

Figure 5 shows example navigation trajectories on top-down maps. Our visual-echo consistency pre-training task allows the agent to better interpret the room's spatial layout to find the goal more quickly than the baselines. See Supp. for qualitative results on depth estimation and surface normal examples. Initializing with our pre-trained VISUALECHOES network leads to much more accurate depth prediction and surface normal estimates compared to no pre-training, demonstrating the usefulness of the learned spatial features.

5 Conclusions and Future Work

We presented an approach to learn spatial image representations via echolocation. We performed an in-depth study on the spatial cues contained in echoes and how they can inform single-view depth estimation. We showed that the learned spatial features can benefit three downstream vision tasks. Our work opens a new path for interaction-based representation learning for embodied agents and demonstrates the potential of learning spatial visual representations even with a limited amount of multisensory data.

While our current implementation learns from audio rendered in a simulator, the results show that the learned spatial features already benefit transfer to vision-only tasks in real photos outside of the scanned environments (e.g., the NYU-V2 [72] and DIODE [78] images), indicating the realism of what our system learned. Nonetheless, it will be interesting future work to capture the echoes on a real robot. We are also interested in pursuing these ideas within a sequential model, such that the agent could actively decide when to emit chirps and what type of chirps to emit to get the most informative echo responses.

Acknowledgements. UT Austin is supported in part by DARPA Lifelong Learning Machines and ONR PECASE. RG is supported by Google PhD Fellowship and Adobe Research Fellowship.

References

1. Agrawal, P., Carreira, J., Malik, J.: Learning to see by moving. In: ICCV (2015)
2. Agrawal, P., Nair, A.V., Abbeel, P., Malik, J., Levine, S.: Learning to poke by poking: experiential learning of intuitive physics. In: NeurIPS (2016)
3. Alameda-Pineda, X., et al.: Salsa: a novel dataset for multimodal group behavior analysis. IEEE Trans. Pattern Anal. Mach. Intell. **38**(8), 1707–1720 (2015)
4. Anderson, P., et al.: On evaluation of embodied navigation agents (2018). arXiv preprint arXiv:1807.06757
5. Antonacci, F., et al.: Inference of room geometry from acoustic impulse responses. IEEE Trans. Audio Speech Lang. Process. **20**(10), 2683–2695 (2012)
6. Arandjelovic, R., Zisserman, A.: Look, listen and learn. In: ICCV (2017)
7. Arandjelović, R., Zisserman, A.: Objects that sound. In: ECCV (2018)
8. Aytar, Y., Vondrick, C., Torralba, A.: Soundnet: learning sound representations from unlabeled video. In: NeurIPS (2016)
9. Ban, Y., Li, X., Alameda-Pineda, X., Girin, L., Horaud, R.: Accounting for room acoustics in audio-visual multi-speaker tracking. In: ICASSP (2018)
10. Chang, A., et al.: Matterport3D: Learning from RGB-D data in indoor environments. 3DV (2017)
11. Chen, C., et al.: Audio-visual embodied navigation. In: ECCV (2020)
12. Christensen, J., Hornauer, S., Yu, S.: Batvision - learning to see 3D spatial layout with two ears. In: ICRA (2020)
13. Deng, J., Dong, W., Socher, R., Li, L.J., Li, K., Fei-Fei, L.: Imagenet: a large-scale hierarchical image database. In: CVPR (2009)
14. Dokmanić, I., Parhizkar, R., Walther, A., Lu, Y.M., Vetterli, M.: Acousticechoes reveal room shape. Proc. Natl. Acad. Sci. **110**(30), 12186–12191 (2013)
15. Eigen, D., Fergus, R.: Predicting depth, surface normals and semantic labels with a common multi-scale convolutional architecture. In: ICCV (2015)
16. Eigen, D., Puhrsch, C., Fergus, R.: Depth map prediction from a single image using a multi-scale deep network. In: NeurIPS (2014)
17. Eliakim, I., Cohen, Z., Kosa, G., Yovel, Y.: A fully autonomous terrestrialbat-like acoustic robot. PLoS Comput. Biol. **14**(9), e1006406 (2018)
18. Ephrat, A., et al.: Looking to listen at the cocktail party: a speaker-independent audio-visual model for speech separation. In: SIGGRAPH (2018)

19. Feng, Z., Xu, C., Tao, D.: Self-supervised representation learning by rotation feature decoupling. In: CVPR (2019)
20. Fernando, B., Bilen, H., Gavves, E., Gould, S.: Self-supervised video representation learning with odd-one-out networks. In: CVPR (2017)
21. Fouhey, D.F., Gupta, A., Hebert, M.: Data-driven 3D primitives for single image understanding. In: ICCV (2013)
22. Frank, N., Wolf, L., Olshansky, D., Boonman, A., Yovel, Y.: Comparing vision-based to sonar-based 3D reconstruction. ICCP (2020)
23. Fu, H., Gong, M., Wang, C., Batmanghelich, K., Tao, D.: Deep ordinal regression network for monocular depth estimation. In: CVPR (2018)
24. Gan, C., Huang, D., Zhao, H., Tenenbaum, J.B., Torralba, A.: Music gesture for visual sound separation. In: CVPR (2020)
25. Gan, C., Zhang, Y., Wu, J., Gong, B., Tenenbaum, J.B.: Look, listen, and act: towards audio-visual embodied navigation. In: ICRA (2020)
26. Gan, C., Zhao, H., Chen, P., Cox, D., Torralba, A.: Self-supervised moving vehicle tracking with stereo sound. In: ICCV (2019)
27. Gandhi, D., Pinto, L., Gupta, A.: Learning to fly by crashing. In: IROS (2017)
28. Gao, R., Feris, R., Grauman, K.: Learning to separate object sounds by watching unlabeled video. In: ECCV (2018)
29. Gao, R., Grauman, K.: 2.5D visual sound. In: CVPR (2019)
30. Gao, R., Grauman, K.: Co-separating sounds of visual objects. In: ICCV (2019)
31. Gao, R., Jayaraman, D., Grauman, K.: Object-centric representation learning from unlabeled videos. In: Lai, S.-H., Lepetit, V., Nishino, K., Sato, Y. (eds.) ACCV 2016. LNCS, vol. 10115, pp. 248–263. Springer, Cham (2017). https://doi.org/10.1007/978-3-319-54193-8_16
32. Gao, R., Oh, T.H., Grauman, K., Torresani, L.: Listen to look: action recognition by previewing audio. In: CVPR (2020)
33. Garg, R., B.G., V.K., Carneiro, G., Reid, I.: Unsupervised CNN for single view depth estimation: geometry to the rescue. In: Leibe, B., Matas, J., Sebe, N., Welling, M. (eds.) ECCV 2016. LNCS, vol. 9912, pp. 740–756. Springer, Cham (2016). https://doi.org/10.1007/978-3-319-46484-8_45
34. Gebru, I.D., Ba, S., Evangelidis, G., Horaud, R.: Tracking the active speaker based on a joint audio-visual observation model. In: ICCV Workshops (2015)
35. Gidaris, S., Singh, P., Komodakis, N.: Unsupervised representation learning by predicting image rotations. In: ICLR (2018)
36. Godard, C., Mac Aodha, O., Brostow, G.J.: Unsupervised monocular depth estimation with left-right consistency. In: CVPR (2017)
37. Godard, C., Mac Aodha, O., Firman, M., Brostow, G.J.: Digging into self-supervised monocular depth estimation. In: ICCV (2019)
38. Goyal, P., Mahajan, D., Gupta, A., Misra, I.: Scaling and benchmarking self-supervised visual representation learning. In: ICCV (2019)
39. He, K., Zhang, X., Ren, S., Sun, J.: Deep residual learning for image recognition. In: CVPR (2016)
40. Hershey, J.R., Movellan, J.R.: Audio vision: using audio-visual synchrony to locate sounds. In: NeurIPS (2000)
41. Hu, J., Ozay, M., Zhang, Y., Okatani, T.: Revisiting single image depth estimation: toward higher resolution maps with accurate object boundaries. In: WACV (2019)
42. Irie, G., et al.: Seeing through sounds: predicting visual semantic segmentation results from multichannel audio signals. In: ICASSP (2019)
43. Jayaraman, D., Grauman, K.: Slow and steady feature analysis: higher order temporal coherence in video. In: CVPR (2016)

44. Jayaraman, D., Gao, R., Grauman, K.: Shapecodes: self-supervised feature learning by lifting views to viewgrids. In: ECCV (2018)
45. Jayaraman, D., Grauman, K.: Learning image representations equivariant to ego-motion. In: ICCV (2015)
46. Jiang, H., Larsson, G., Maire Greg Shakhnarovich, M., Learned-Miller, E.: Self-supervised relative depth learning for urban scene understanding. In: ECCV (2018)
47. Karsch, K., Liu, C., Kang, S.B.: Depth transfer: depth extraction from video using non-parametric sampling. IEEE Trans. Pattern Anal. Mach. Intell. **36**(11), 2144–2158 (2014)
48. Kazakos, E., Nagrani, A., Zisserman, A., Damen, D.: Epic-fusion: Audio-visual temporal binding for egocentric action recognition. In: ICCV (2019)
49. Kim, H., Remaggi, L., Jackson, P.J., Fazi, F.M., Hilton, A.: 3D room geometry reconstruction using audio-visual sensors. In: 3DV (2017)
50. Korbar, B., Tran, D., Torresani, L.: Co-training of audio and video representations from self-supervised temporal synchronization. In: NeurIPS (2018)
51. Kuttruff, H.: Electroacoustical systems in rooms. Room Acoustics, pp. 267–293. CRC Pres, Boca Raton (2017)
52. Larsson, G., Maire, M., Shakhnarovich, G.: Colorization as a proxy task for visual understanding. In: CVPR (2017)
53. Liu, F., Shen, C., Lin, G.: Deep convolutional neural fields for depth estimation from a single image. In: CVPR (2015)
54. McGuire, K., De Wagter, C., Tuyls, K., Kappen, H., de Croon, G.: Minimal navigation solution for a swarm of tiny flying robots to explore an unknown environment. Sci. Robot. **4**(35), eaaw9710 (2019)
55. Misra, I., Zitnick, C.L., Hebert, M.: Shuffle and learn: unsupervised learning using temporal order verification. In: Leibe, B., Matas, J., Sebe, N., Welling, M. (eds.) ECCV 2016. LNCS, vol. 9905, pp. 527–544. Springer, Cham (2016). https://doi.org/10.1007/978-3-319-46448-0_32
56. Morgado, P., Vasconcelos, N., Langlois, T., Wang, O.: Self-supervised generation of spatial audio for 360° video. In: NeurIPS (2018)
57. Noroozi, M., Favaro, P.: Unsupervised learning of visual representations by solving jigsaw puzzles. In: Leibe, B., Matas, J., Sebe, N., Welling, M. (eds.) ECCV 2016. LNCS, vol. 9910, pp. 69–84. Springer, Cham (2016). https://doi.org/10.1007/978-3-319-46466-4_5
58. Owens, A., Efros, A.A.: Audio-visual scene analysis with self-supervised multisensory features. In: ECCV (2018)
59. Owens, A., Isola, P., McDermott, J., Torralba, A., Adelson, E.H., Freeman, W.T.: Visually indicated sounds. In: CVPR (2016)
60. Owens, A., Wu, J., McDermott, J.H., Freeman, W.T., Torralba, A.: Ambient sound provides supervision for visual learning. In: Leibe, B., Matas, J., Sebe, N., Welling, M. (eds.) ECCV 2016. LNCS, vol. 9905, pp. 801–816. Springer, Cham (2016). https://doi.org/10.1007/978-3-319-46448-0_48
61. Palossi, D., Loquercio, A., Conti, F., Flamand, E., Scaramuzza, D., Benini, L.: A 64-mw DNN-based visual navigation engine for autonomous nano-drones. IEEE Internet Things J. **6**(5), 8357–8371 (2019)
62. Pinto, L., Gupta, A.: Supersizing self-supervision: learning to grasp from 50k tries and 700 robot hours. In: ICRA (2016)
63. Purushwalkam, S., Gupta, A., Kaufman, D.M., Russell, B.: Bounce and learn: modeling scene dynamics with real-world bounces. In: ICLR (2019)
64. Quattoni, A., Torralba, A.: Recognizing indoor scenes. In: CVPR (2009)

65. Ranjan, A., et al.: Competitive collaboration: joint unsupervised learning of depth, camera motion, optical flow and motion segmentation. In: CVPR (2019)
66. Ren, Z., Jae Lee, Y.: Cross-domain self-supervised multi-task feature learning using synthetic imagery. In: CVPR (2018)
67. Ronneberger, O., Fischer, P., Brox, T.: U-net: convolutional networks for biomedical image segmentation. In: Navab, N., Hornegger, J., Wells, W.M., Frangi, A.F. (eds.) MICCAI 2015. LNCS, vol. 9351, pp. 234–241. Springer, Cham (2015). https://doi.org/10.1007/978-3-319-24574-4_28
68. Ronneberger, O., Fischer, P., Brox, T.: U-net: convolutional networks for biomedical image segmentation. In: Navab, N., Hornegger, J., Wells, W.M., Frangi, A.F. (eds.) MICCAI 2015. LNCS, vol. 9351, pp. 234–241. Springer, Cham (2015). https://doi.org/10.1007/978-3-319-24574-4_28
69. de Sa, V.R.: Learning classification with unlabeled data. In: NeurIPS (1994)
70. Savva, M., et al.: Habitat: a platform for embodied AI research. In: ICCV (2019)
71. Senocak, A., Oh, T.H., Kim, J., Yang, M.H., So Kweon, I.: Learning to localize sound source in visual scenes. In: CVPR (2018)
72. Silberman, N., Hoiem, D., Kohli, P., Fergus, R.: Indoor segmentation and support inference from RGBD images. In: Fitzgibbon, A., Lazebnik, S., Perona, P., Sato, Y., Schmid, C. (eds.) ECCV 2012. LNCS, vol. 7576, pp. 746–760. Springer, Heidelberg (2012). https://doi.org/10.1007/978-3-642-33715-4_54
73. Straub, J., et al.: The replica dataset: a digital replica of indoor spaces (2019). arXiv preprint arXiv:1906.05797
74. Stroffregen, T.A., Pittenger, J.B.: Human echolocation as a basic form of perception and action. Ecol. Psychol. **7**(3), 181–216 (1995)
75. Tian, Y., Shi, J., Li, B., Duan, Z., Xu, C.: Audio-visual event localization in unconstrained videos. In: ECCV (2018)
76. Ummenhofer, B., et al.: Demon: Depth and motion network for learning monocular stereo. In: CVPR (2017)
77. Vanderelst, D., Holderied, M.W., Peremans, H.: Sensorimotor model of obstacleavoidance in echolocating bats. PLoS Comput. Biol. **11**(10), e1004484 (2015)
78. Vasiljevic, Iet al.: DIODE: A Dense Indoor and Outdoor DEpth Dataset (2019). arXiv preprint arXiv:1908.00463
79. Veach, E., Guibas, L.: Bidirectional estimators for light transport. In: Photorealistic Rendering Techniques (1995)
80. Vijayanarasimhan, S., Ricco, S., Schmid, C., Sukthankar, R., Fragkiadaki, K.: SFM-net: Learning of structure and motion from video (2017). arXiv preprint arXiv:1704.07804
81. Wang, P., Shen, X., Lin, Z., Cohen, S., Price, B., Yuille, A.L.: Towards unified depth and semantic prediction from a single image. In: CVPR (2015)
82. Wang, X., Gupta, A.: Unsupervised learning of visual representations using videos. In: ICCV (2015)
83. Xia, F., Zamir, A.R., He, Z., Sax, A., Malik, J., Savarese, S.: Gibson env: Realworld perception for embodied agents. In: CVPR (2018)
84. Xu, D., Ricci, E., Ouyang, W., Wang, X., Sebe, N.: Multi-scale continuous CRFS as sequential deep networks for monocular depth estimation. In: CVPR (2017)
85. Yang, Z., Wang, P., Xu, W., Zhao, L., Nevatia, R.: Unsupervised learning of geometry with edge-aware depth-normal consistency. In: AAAI (2018)
86. Ye, M., Zhang, Y., Yang, R., Manocha, D.: 3d reconstruction in the presence of glasses by acoustic and stereo fusion. In: ICCV (2015)
87. Zamir, A.R., Sax, A., Shen, W., Guibas, L.J., Malik, J., Savarese, S.: Taskonomy: Disentangling task transfer learning. In: CVPR (2018)

88. Zhang, R., Isola, P., Efros, A.A.: Colorful image colorization. In: Leibe, B., Matas, J., Sebe, N., Welling, M. (eds.) ECCV 2016. LNCS, vol. 9907, pp. 649–666. Springer, Cham (2016). https://doi.org/10.1007/978-3-319-46487-9_40
89. Zhao, H., Gan, C., Rouditchenko, A., Vondrick, C., McDermott, J., Torralba, A.: The sound of pixels. In: ECCV (2018)
90. Zhao, H., Shi, J., Qi, X., Wang, X., Jia, J.: Pyramid scene parsing network. In: CVPR (2017)
91. Zhou, T., Brown, M., Snavely, N., Lowe, D.G.: Unsupervised learning of depth and ego-motion from video. In: CVPR (2017)
92. Zhou, Y., Wang, Z., Fang, C., Bui, T., Berg, T.L.: Visual to sound: Generating natural sound for videos in the wild. In: CVPR (2018)

Smooth-AP: Smoothing the Path Towards Large-Scale Image Retrieval

Andrew Brown(✉), Weidi Xie, Vicky Kalogeiton, and Andrew Zisserman

Visual Geometry Group, University of Oxford, Oxford, UK
{abrown,weidi,vicky,az}@robots.ox.ac.uk
https://www.robots.ox.ac.uk/~vgg/research/smooth-ap/

Abstract. Optimising a ranking-based metric, such as Average Precision (AP), is notoriously challenging due to the fact that it is non-differentiable, and hence cannot be optimised directly using gradient-descent methods. To this end, we introduce an objective that optimises instead a *smoothed approximation* of AP, coined *Smooth-AP*. Smooth-AP is a plug-and-play objective function that allows for end-to-end training of deep networks with a simple and elegant implementation. We also present an analysis for why directly optimising the ranking based metric of AP offers benefits over other deep metric learning losses.

We apply Smooth-AP to standard retrieval benchmarks: Stanford Online products and VehicleID, and also evaluate on larger-scale datasets: INaturalist for fine-grained category retrieval, and VGGFace2 and IJB-C for face retrieval. In all cases, we improve the performance over the state-of-the-art, especially for larger-scale datasets, thus demonstrating the effectiveness and scalability of Smooth-AP to real-world scenarios.

1 Introduction

Our objective in this paper is to improve the performance of 'query by example', where the task is: given a query image, rank all the instances in a retrieval set according to their relevance to the query. For instance, imagine that you have a photo of a friend or family member, and want to search for all of the images of that person within your large smart-phone image collection; or on a photo licensing site, you want to find all photos of a particular building or object, starting from a single photo. These use cases, where high recall is premium, differ from the 'Google Lens' application of identifying an object from an image, where only one 'hit' (match) is sufficient.

The benchmark metric for retrieval quality is Average Precision (AP) (or its generalized variant, Normalized Discounted Cumulative Gain, which includes non-binary relevance judgements). With the resurgence of deep neural networks, end-to-end training has become the *de facto* choice for solving specific vision

Electronic supplementary material The online version of this chapter (https://doi.org/10.1007/978-3-030-58545-7_39) contains supplementary material, which is available to authorized users.

© Springer Nature Switzerland AG 2020
A. Vedaldi et al. (Eds.): ECCV 2020, LNCS 12354, pp. 677–694, 2020.
https://doi.org/10.1007/978-3-030-58545-7_39

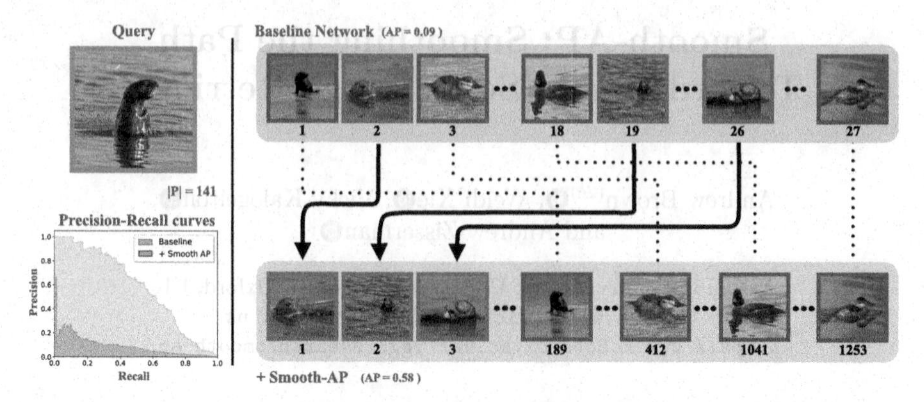

Fig. 1. Ranked retrieval sets before (top) and after (bottom) applying Smooth-AP on a baseline network (*i.e.* ImageNet pre-trained weights) for a given query (pink image). The precision-recall curve is shown on the left. Smooth-AP results in large boost in AP, as it moves positive instances (green) high up the ranks and negative ones (red) low down. $|\mathcal{P}|$ is the number of positive instances in the retrieval set for this query. Images are from the INaturalist dataset. (Color figure online)

tasks with well-defined metrics. However, the core problem with AP and similar metrics is that they include a discrete ranking function that is neither differentiable nor decomposable. Consequently, their direct optimization, *e.g.* with gradient-descent methods, is notoriously difficult.

In this paper, we introduce a novel differentiable AP approximation, *Smooth-AP*, that allows end-to-end training of deep networks for ranking-based tasks (Fig. 1). Smooth-AP is a simple, elegant, and scalable method that takes the form of a plug-and-play objective function by relaxing the Indicator function in the non-differentiable AP with a sigmoid function. To demonstrate its effectiveness, we perform experiments on two commonly used image retrieval benchmarks, Stanford Online Products and VehicleID, where Smooth-AP outperforms all recent AP approximation approaches [5,51] as well as recent deep metric learning methods. We also experiment on three further large-scale retrieval datasets (VGGFace2, IJB-C, INaturalist), which are orders of magnitude larger than the existing retrieval benchmarks. To our knowledge, this is the first work that demonstrates the possibility of training networks for AP on datasets with millions of images for the task of image retrieval. We show large performance gains over all recently proposed AP approximating approaches and, somewhat surprisingly, also outperform strong verification systems [13,34] by a significant margin, reflecting the fact that metric learning approaches are indeed inefficient for training large-scale retrieval systems that are measured by global ranking metrics.

2 Related Work

As an essential component of information retrieval [36], algorithms that optimize rank-based metrics have been the focus of extensive research over the years.

In general, the previous approaches can be split into two lines of research, namely metric learning, and direct approximation of Average Precision.

Image Retrieval. This is one of the most researched topics in the vision community. Several themes have been explored in the literature, for example, one theme is on the speed of retrieval and explores methods of approximate nearest neighbors [12,26,27,29,45,57]. Another theme is on how to obtain a compact image descriptor for retrieval in order to reduce the memory footprint. Descriptors were typically constructed through an aggregation of local features, such as Fisher vectors [44] and VLAD [2,28]. More recently, neural networks have made impressive progress on learning representations for image retrieval [1,3,17,48,66], but common to all is the choice of the loss function used for training; in particular, it should ideally be a loss that will encourage 'good' ranking.

Metric Learning. To avoid the difficulties from directly optimising rank-based metrics, such as Average Precision, there is a great body of work that focuses on metric learning [1,2,4,8,10,11,29,32,40,42,49,61,67,70]. For instance, the contrastive [11] and triplet [70] losses, which consider pairs or triplets of elements, and force all positive instances to be close in the high-dimensional embedding space, while separating negatives by a fixed distance (margin). However, due to the limited rank-positional awareness that a pair/triplet provides, a model is likely to waste capacity on improving the order of positive instances at low (poor) ranks at the expense of those at high ranks, as was pointed out by Burges *et al.* [4]. Of more relevance, the list-wise approaches [4,8,40,42,67] look at many examples from the retrieval set, and have been proven to improve training efficiency and performance. Despite being successful, one drawback of metric learning approaches is that they are mostly driven by minimizing distances, and therefore remain ignorant of the importance of shifting ranking orders – the latter is essential when evaluating with a rank-based metric.

Optimizing Average Precision (AP). The trend of directly optimising the non-differentiable AP has been recently revived in the retrieval community. Sophisticated methods [5,10,15,21–23,38,50,51,58,60,61,73,75] have been developed to overcome the challenge of non-decomposability and non-differentiability in optimizing AP. Methods include: creating a distribution over rankings by treating each relevance score as a Gaussian random variable [60], loss-augmented inference [38], direct loss minimization [23,58], optimizing a smooth and differentiable upper bound of AP [38,39,75], training a LSTM to approximate the discrete ranking step [15], differentiable histogram binning [5,21,22,50,61], error driven update schemes [10], and the very recent black-box optimization [51]. Significant progress on optimizing AP was made by the information retrieval community [4,9,18,33,47,60], but the methods have largely been ignored by the vision community, possibly because they have never been demonstrated on large-scale image retrieval or due to the complexity of the proposed smooth objectives. One of the motivations of this work is to show that with the progress of deep learning research, *e.g.* auto-differentiation, better optimization techniques, large-scale datasets, and fast computation devices, it is possible and in fact very easy to directly optimize a close approximation to AP.

3 Background

In this section, we define the notations used throughout the paper.

Task Definition. Given an input query, the goal of a retrieval system is to rank all instances in a retrieval set $\Omega = \{I_i, i = 0, \cdots, m\}$ based on their relevance to the query. For *each* query instance I_q, the retrieval set is split into the positive $\mathcal{P}_q$ and negative $\mathcal{N}_q$ sets, which are formed by all instances of the same class and of different classes, respectively. Note that there is a different positive and negative set for each query.

Average Precision (AP). AP is one of the standard metrics for information retrieval tasks [36]. It is a single value defined as the area under a Precision-Recall curve. For a query I_q, the predicted relevance scores of all instances in the retrieval set are measured via a chosen metric. In our case, we use the cosine similarity (though the Smooth-AP method is independent of this choice):

$$\mathcal{S}_\Omega = \left\{ s_i = \left\langle \frac{v_q}{\|v_q\|} \cdot \frac{v_i}{\|v_i\|} \right\rangle, i = 0, \cdots, n \right\}, \tag{1}$$

where $\mathcal{S}_\Omega = \mathcal{S}_P \cup \mathcal{S}_N$, and $\mathcal{S}_P = \{s_\zeta, \forall \zeta \in \mathcal{P}_q\}$, $\mathcal{S}_N = \{s_\xi, \forall \xi \in \mathcal{N}_q\}$ are the positive and negative relevance score sets, respectively, v_q refers to the query vector, and v_i to the vectorized retrieval set. The AP of a query I_q can be computed as:

$$AP_q = \frac{1}{|\mathcal{S}_P|} \sum_{i \in \mathcal{S}_P} \frac{\mathcal{R}(i, \mathcal{S}_P)}{\mathcal{R}(i, \mathcal{S}_\Omega)}, \tag{2}$$

where $\mathcal{R}(i, \mathcal{S}_P)$ and $\mathcal{R}(i, \mathcal{S}_\Omega)$ refer to the rankings of the instance i in $\mathcal{P}$ and Ω, respectively. Note that, the rankings referred to in this paper are assumed to be *proper rankings*, meaning no two samples are ranked equally.

Ranking Function ($\mathcal{R}$). Given that AP is a ranking-based method, the key element for direct optimisation is to define the ranking $\mathcal{R}$ of one instance i. Here, we define it in the following way [47]:

$$\mathcal{R}(i, \mathcal{S}) = 1 + \sum_{j \in \mathcal{S}, j \neq i} \mathbb{1}\{(s_i - s_j) < 0\}, \tag{3}$$

where $\mathbb{1}\{\cdot\}$ acts as an Indicator function, and $\mathcal{S}$ any set, *e.g.* Ω. Conveniently, this can be implemented by computing a difference matrix $D \in \mathbb{R}^{m \times m}$:

$$D = \begin{bmatrix} s_1 & \dots & s_m \\ \vdots & \ddots & \vdots \\ s_1 & \dots & s_m \end{bmatrix} - \begin{bmatrix} s_1 & \dots & s_1 \\ \vdots & \ddots & \vdots \\ s_m & \dots & s_m \end{bmatrix} \tag{4}$$

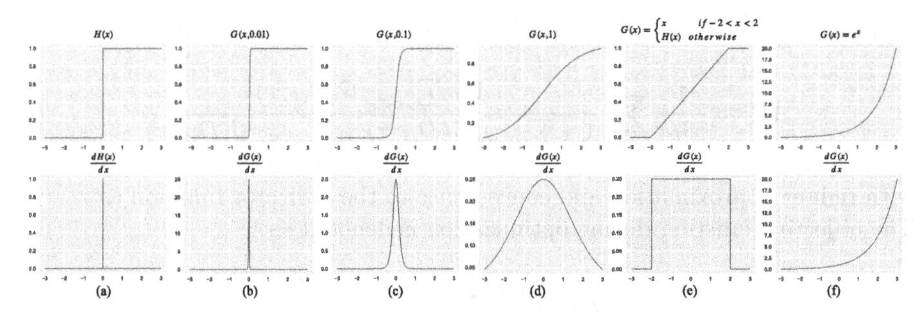

Fig. 2. The possible **different approximations** to the discrete Indicator function. First row: Indicator function (a), three sigmoids with increasing temperatures (b, c, d), linear (e), exponential (f). Second row: their derivatives.

The exact AP for a query instance I_q from Eq. 2 becomes:

$$AP_q = \frac{1}{|\mathcal{S}_P|} \sum_{i \in \mathcal{S}_P} \frac{1 + \sum_{j \in \mathcal{S}_p, j \neq i} \mathbb{1}\{D_{ij} > 0\}}{1 + \sum_{j \in \mathcal{S}_P, j \neq i} \mathbb{1}\{D_{ij} > 0\} + \sum_{j \in \mathcal{S}_N} \mathbb{1}\{D_{ij} > 0\}} \tag{5}$$

Derivatives of Indicator. The particular Indicator function used in computing AP is a Heaviside step function $\mathcal{H}(\cdot)$ [47], with its distributional derivative defined as Dirac delta function:

$$\frac{\mathrm{d}\mathcal{H}(x)}{\mathrm{d}x} = \delta(x),$$

This is either flat everywhere, with zero gradient, or discontinuous, and hence cannot be optimized with gradient based methods (Fig. 2).

4 Approximating Average Precision (AP)

As explained, AP and similar metrics include a discrete ranking function that is neither differentiable nor decomposable. In this section, we first describe *Smooth-AP*, which essentially replaces the discrete indicator function with a sigmoid function, and then we provide an analysis on its relation to other ranking losses, such as triplet loss [24, 70], FastAP [5] and Blackbox AP [51].

4.1 Smoothing AP

To smooth the ranking procedure, which will enable direct optimization of AP, Smooth-AP takes a simple solution which is to replace the Indicator function $\mathbb{1}\{\cdot\}$ by a sigmoid function $\mathcal{G}(\cdot;\tau)$, where the τ refers to the temperature adjusting the sharpness:

$$\mathcal{G}(x;\tau) = \frac{1}{1 + e^{\frac{-x}{\tau}}}. \tag{6}$$

Substituting $\mathcal{G}(\cdot;\tau)$ into Eq. 5, the true AP can be approximated as:

$$AP_q \approx \frac{1}{|\mathcal{S}_P|} \sum_{i \in \mathcal{S}_P} \frac{1 + \sum_{j \in \mathcal{S}_P} \mathcal{G}(D_{ij};\tau)}{1 + \sum_{j \in \mathcal{S}_P} \mathcal{G}(D_{ij};\tau) + \sum_{j \in \mathcal{S}_N} \mathcal{G}(D_{ij};\tau)}$$

with tighter approximation and convergence to the indicator function as $\tau \rightarrow 0$. The objective function during optimization is denoted as:

$$\mathcal{L}_{AP} = \frac{1}{m} \sum_{k=1}^{m} (1 - AP_k) \tag{7}$$

Smoothing Parameter. τ governs the temperature of the sigmoid that replaces the Indicator function $\mathbb{1}\{\cdot\}$. It defines an operating region, where terms of the difference matrix are given a gradient by the Smooth-AP loss. If the terms are mis-ranked, Smooth-AP will attempt to shift them to the correct order. Specifically, a small value of τ results in a small operating region (Fig. 2 (b) – note the small region with gradient seen in the sigmoid derivative), and a tighter approximation of true AP. The strong acceleration in gradient around the zero point (Fig. 2 (b)-(c) second row) is essential to replicating the desired qualities of AP, as it encourages the shifting of instances in the embedding space that result in a change of rank (and hence change in AP), rather than shifting instances by some large distance but not changing the rank. A large value of τ offers a large operating region, however, at the cost of a looser approximation to AP due to its divergence from the indicator function.

Relation to Triplet Loss. Here, we demonstrate that the triplet loss (a popular surrogate loss for ranking) is in fact optimising a distance metric rather than a ranking metric, which is sub-optimal when evaluating using a ranking metric. As shown in Eq. 5, the goal of optimizing AP is equivalent to minimizing all the $\sum_{i \in \mathcal{S}_P, j \in \mathcal{S}_N} \mathbb{1}\{D_{ij} < 0\}$, *i.e.* the violating terms. We term these as such because these terms refer to cases where a negative instance is ranked above a positive instance in terms of relevance to the query, and optimal AP is only acquired when all positive instances are ranked above all negative instances.

For example, consider one query instance with predicted relevance score and ground-truth relevance labels as:

$$\text{Instances ordered by score}: (s_0\ s_4\ s_1\ s_2\ s_5\ s_6\ s_7\ s_3)$$
$$\text{Ground truth labels}: (1\ \ 0\ \ 1\ \ 1\ \ 0\ \ 0\ \ 0\ \ 1)$$

the violating terms are: $\{(s_4 - s_1), (s_4 - s_2), (s_4 - s_3), (s_5 - s_3), (s_6 - s_3), (s_7 - s_3)\}$. An ideal AP loss would actually treat each of the terms unequally, *i.e.* the model would be forced to spend more capacity on shifting orders between s_4 and s_1, rather than s_3 and s_7, as that makes a larger impact on improving the AP.

Another interpretation of these violating cases can also be drawn from the triplet loss perspective. Specifically, if we treat the query instance as an "anchor", with s_j denoting the similarity between the "anchor" and negative instance, and

s_i denoting the similarity between the "anchor" and positive instance. In this example, the triplet loss tries to optimize a margin hinge loss:

$$\begin{aligned}\mathcal{L}_{\text{triplet}} &= \max(s_4 - s_1 + \alpha, 0) + \max(s_4 - s_2 + \alpha, 0)\\ &+ \max(s_4 - s_3 + \alpha, 0) + \max(s_5 - s_3 + \alpha, 0)\\ &+ \max(s_6 - s_3 + \alpha, 0) + \max(s_7 - s_3 + \alpha, 0)\end{aligned}$$

This can be viewed as a differentiable approximation to the goal of optimizing AP where the Indicator function has been replaced with the margin hinge loss, thus solving the gradient problem. Nevertheless, using a triplet loss to approximate AP may suffer from two problems: *First*, all terms are linearly combined and treated equally in $\mathcal{L}_{\text{triplet}}$. Such a surrogate loss may force the model to optimize the terms that have only a small effect on AP, *e.g.* optimizing $s_4 - s_1$ is the same as $s_7 - s_4$ in the triplet loss, however, from an AP perspective, it is important to correct the mis-ordered instances at high rank. *Second*, the linear derivative means that the optimization process is purely based on distance (not ranking orders), which makes it sub-optimal when evaluating AP. For instance, in the triplet loss case, reducing the distance $s_4 - s_1$ from 0.8 to 0.5 is the same as from 0.2 to -0.1. In practise, however, the latter case (shifting orders) will clearly have a much larger impact on the AP computation than the former.

Comparison to Other AP-Optimising Methods. The two key differences between Smooth-AP and the recently introduced FastAP and Blackbox AP, are that Smooth-AP (i) provides a closer approximation to Average Precision, and (ii) is far simpler to implement. Firstly, due to the sigmoid function, Smooth-AP optimises a ranking metric, and so has the same objective as Average Precision. In contrast, FastAP and Blackbox AP linearly interpolate the non-differentiable (piecewise constant) function, which can potentially lead to the same issues as triplet loss, *i.e.* optimizing a distance metric, rather than rankings. Secondly, Smooth-AP simply needs to replace the indicator function in the AP objective with a sigmoid function. While FastAP uses abstractions such as Histogram Binning, and Blackbox AP uses a variant of numerical derivative. These differences are positively affirmed through the improved performance of Smooth-AP over several datasets (Sect. 6.5).

5 Experimental Setup

In this section, we describe the datasets used for evaluation, the test protocols, and the implementation details. The procedure followed here is to take a pre-trained network and fine-tune with Smooth-AP loss. Specifically, ImageNet pretrained networks are used for the object/animal retrieval datasets, and high-performing face-verification models for the face retrieval datasets.

5.1 Datasets

We evaluate the Smooth-AP loss on five datasets containing a wide range of domains and sizes. These include the commonly used retrieval benchmark

Table 1. Datasets used for training and evaluation.

Dataset		# Images	# Classes	# Ims/Class
Object/animal	SOP train	59,551	11,318	5.3
retrieval	SOP test	60,502	11,316	5.3
datasets	VehicleID train	110,178	13,134	8.4
	VehicleID test	40,365	4,800	8.4
	INaturalist train	325,846	5,690	57.3
	INaturalist test	136,093	2,452	55.5
Face retrieval	VGGFace2 train	3.31 M	8,631	363.7
datasets	VGGFace2 test	169,396	500	338.8
	IJB-C	148,824	3,531	42.1

datasets, as well as several additional *large-scale* (>100K images) datasets. Table 1 describes their details.

Stanford Online Product (SOP) [58] was initially collected for investigating the problem of metric learning. It includes 120K images of products that were sold online. We use the same evaluation protocol and train/test split as [67].

VehicleID [64] contains 221,736 images of 26,267 vehicle categories, 13,134 of which are used for training (containing 110,178 images). By following the same test protocol as [64], three test sets of increasing size are used for evaluation (termed small, medium, large), which contain 800 classes (7,332 images), 1600 classes (12,995 images) and 2400 classes (20,038 images) respectively.

INaturalist [62] is a large-scale animal and plant species classification dataset, designed to replicate real-world scenarios through 461,939 images from 8,142 classes. It features many visually similar species, captured in a wide variety of environments. We construct a new image retrieval task from this dataset, by keeping 5,690 classes for training, and 2,452 unseen classes for evaluating image retrieval at test time, according to the same test protocols as existing benchmarks [67]. We will make the train/test splits publicly available.

VGGFace2 [7] is a large-scale face dataset with over 3.31 million images of 9,131 subjects. The images have large variations in pose, age, illumination, ethnicity and profession, *e.g.* actors, athletes, politicians. For training, we use the pre-defined *training set* with 8,631 identities, and for testing we use the *test set* with 500 identities, totalling 169K testing images.

IJB-C [37] is a challenging public benchmark for face recognition, containing images of subjects from both still frames and videos. Each video is treated as a single instance by averaging the CNN-produced vectors for each frame to a single vector. Identities with less than 5 instances (images or videos) are removed.

5.2 Test Protocol

Here, we describe the protocols for evaluating retrieval performance, mean Average Precision (mAP) and Recall@K (R@K). For all datasets, every instance of each class is used in turn as the query I_q, and the retrieval set Ω is formed out of all the remaining instances. We ensure that each class in all datasets contains several images (Table 1), such that if an instance from a class is used as the query, there are plenty of remaining positive instances in the retrieval set. For object/animal retrieval evaluation, we use the *Recall@K* metric in order to compare to existing works. For face retrieval, AP is computed from the resulting output ranking for each query, and the mAP score is computed by averaging the APs across every instance in the dataset, resulting in a single value.

5.3 Implementation Details

Object/Animal Retrieval (SOP, VehicleID, INaturalist). In line with previous works [5,51,53,55,69,71], we use ResNet50 [20] as the backbone architecture, which was pretrained on ImageNet [54]. We replace the final softmax layer with one linear layer (following [5,51], with dimension being set to 512). All images are resized to 256×256. At training time, we use random crops and flips as augmentations, and at test time, a single centre crop of size 224×224 is used. For all experiments we set τ to 0.01 (Sect. 6.4).

Face Retrieval Datasets (VGGFace2, IJB-C). We use two high performing face verification networks: the method from [7] using the SENet-50 architecture [25] and the state-of-the-art ArcFace [13] (using ResNet-50), both trained on the VGGFace2 training set. For SENet-50, we follow [7] and use the same face crops (extended by the recommended amount), resized to 224×224 and we L2-normalize the final 256D embedding. For ArcFace, we generate normalised face crops (112×112) by using the provided face detector [13], and align them with the predicted 5 facial key points, then L2-normalize the final 512D embedding. For both models, we set the batch size to 224 and τ to 0.01 (Sect. 6.4).

Mini-Batch Training. During training, we form each mini-batch by randomly sampling classes such that each represented class has $|\mathcal{P}|$ samples per class. For all experiments, we L2-normalize the embeddings, use cosine similarity to compute the relevance scores between the query and the retrieval set, set $|\mathcal{P}|$ to 4, and use an Adam [31] optimiser with a base learning rate of 10^{-5} with weight decay $4e^{-5}$. We employ the same hard negative mining technique as [5,51] only for the Online Products dataset. Otherwise we use no special sampling strategies.

6 Results

In this section, we first explore the effectiveness of the proposed Smooth-AP by examining the performance of various models on the five retrieval datasets. Specifically, we compare with the recent AP optimization and broader metric

Fig. 3. Qualitative results for the INaturalist dataset using Smooth-AP loss. For each query image (top row), the top 3 instances from the retrieval set are shown ranked from top to bottom. Every retrieved instance shown is a true positive.

learning methods on the standard benchmarks SOP and VehicleID (Sect. 6.1), and then shift to further large-scale experiments, *e.g.* INaturalist for animal/plant retrieval, and IJB-C and VGGFace2 for face retrieval (Sects. 6.2–6.3). Then, we present an ablation study of various hyper-parameters that affect the performance of Smooth-AP: the sigmoid temperature, the size of the positive set, and the batch size (Sect. 6.4). Finally, we discuss various findings and analyze the performance gaps between various models (Sect. 6.5).

Note that, although there has been a rich literature on metric learning methods [14,16,19,30,32,35,40,41,43,46,51,58,59,61,65,67–69,71,72,74] using these image retrieval benchmarks, we only list the very recent state-of-the-art approaches, and try to compare with them as fairly as we can, *e.g.* no model ensemble, and using the same backbone network and image resolution. However, there still remain differences on some small experimental details, such as embedding dimensions, optimizer, and learning rates. Qualitative results for the INaturalist dataset are illustrated in Fig. 3.

6.1 Evaluation on Stanford Online Products (SOP)

We compare with a wide variety of state-of-the-art image retrieval methods, *e.g.* deep metric learning methods [53,55,69,71], and AP approximation methods [5,51]. As shown in Table 2, we observe that Smooth-AP achieves state-of-the-art results on the SOP benchmark. In particular, our best model outperforms the very recent AP approximating methods (Blackbox AP and FastAP) by a 1.5% margin for Recall@1. Furthermore, Smooth-AP performs on par with the concurrent work (Cross-Batch Memory [69]). This is particularly impressive as [69] harnesses memory techniques to sample from many mini-batches simultaneously for each weight update, whereas Smooth-AP only makes use of a single mini-batch on each training iteration.

Table 2. Results on Stanford Online Products. Deep metric learning and recent AP approximating methods are compared to using the ResNet50 architecture. BS: mini-batch size.

	SOP			
Recall@K	1	10	100	1000
Margin [71]	72.7	86.2	93.8	98.0
Divide [55]	75.9	88.4	94.9	98.1
FastAP [5]	76.4	89.0	95.1	98.2
MIC [53]	77.2	89.4	95.6	-
Blackbox AP [51]	78.6	90.5	96.0	98.7
Cont. w/M [69]	**80.6**	**91.6**	96.2	98.7
Smooth-AP BS = 224	79.2	91.0	96.5	98.9
Smooth-AP BS = 384	80.1	91.5	**96.6**	**99.0**

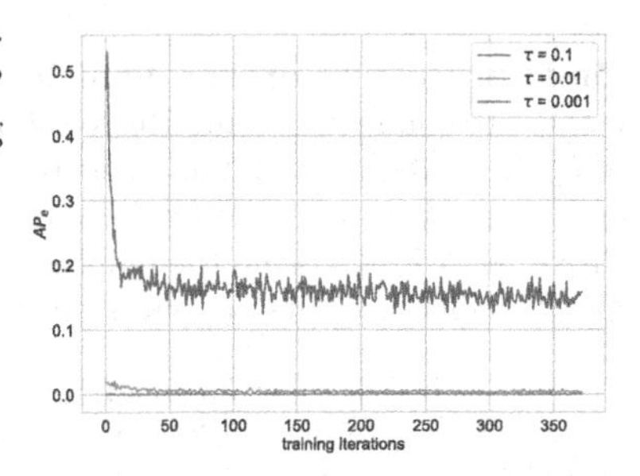

Fig. 4. The AP approximation error, AP_e over one training epoch for Online Products for different values of sigmoid annealing temperature, τ.

Figure 4 provides a quantitative analysis into the effect of sigmoid temperature τ on the tightness of the AP approximation, which can be plotted via the AP approximation error:

$$AP_e = |AP_{pred} - AP| \tag{8}$$

where AP_{pred} is the predicted approximate AP when the sigmoid is used in place of the indicator function in Eq. 5, and AP is the true AP. As expected, a lower value of τ leads to a tighter approximation to Average Precision, shown by the low approximation error.

6.2 Evaluation on VehicleID and INaturalist

In Table 3, we show results on the VehicleID and INaturalist dataset. We observe that Smooth-AP achieves state-of-the-art results on the challenging and large-scale VehicleID dataset. In particular, our model outperforms FastAP by a significant 3% for the Small protocol Recall@1. Furthermore, Smooth-AP exceeds the performance of [69] on 4 of the 6 recall metrics.

As we are the first to report results on INaturalist for *image retrieval*, in addition to Smooth-AP, we re-train state-of-the-art metric learning and AP approximating methods, with the respective official code, *e.g.* Triplet and ProxyNCA [52], FastAP [6], Blackbox AP [63]. As shown in Table 3. Smooth-AP outperforms all methods by $2-5\%$ on Recall@1 for the experiments when the same batch size is used (224). Increasing the batch size to 384 for Smooth-AP leads to a further boost of 1.4% to 66.6 for Recall@1. These results demonstrate that Smooth-AP is particularly suitable for *large-scale* retrieval datasets, thus revealing its scalability to *real-world* retrieval problems. We note here that these large-scale datasets (>100k images) are less influenced by hyper-parameter tuning and so provide ideal test environments to demonstrate improved image retrieval techniques.

Table 3. Results on the VehicleID (left) and INaturalist (right). All experiments are conducted using ResNet50 as backbone. All results for INaturalist are from publicly available official implementations in the PyTorch framework with a batch size of 224. † refers to the recent re-implementation [52] - we make the design choice for Proxy NCA loss to keep the number of proxies equal to the number of training classes. The VehicleID results are obtained with a batch-size of 384.

	VehicleID					
	Small		Medium		Large	
Recall@K	1	5	1	5	1	5
Divide [55]	87.7	92.9	85.7	90.4	82.9	90.2
MIC [53]	86.9	93.4	-	-	82.0	91.0
FastAP [5]	91.9	96.8	90.6	95.9	87.5	95.1
Cont. w/M [69]	94.7	96.8	**93.7**	95.8	**93.0**	95.8
Smooth-AP	**94.9**	**97.6**	93.3	**96.4**	91.9	**96.2**

	INaturalist			
Recall@K	1	4	16	32
Triplet Semi-Hard [71]	58.1	75.5	86.8	90.7
Proxy NCA† [40]	61.6	77.4	87.0	90.6
FastAP [5]	60.6	77.0	87.2	90.6
Blackbox AP [51]	62.9	79.0	88.9	92.1
Smooth-AP BS=224	65.9	80.9	89.8	92.7
Smooth-AP BS=384	**67.2**	**81.8**	**90.3**	**93.1**

Table 4. mAP results on face retrieval datasets. Smooth-AP *consistently* boosts the AP performance for both VGGFace2 and ArcFace, while outperforming other standard metric learning losses (Pairwise contrastive and Triplet).

VGGFace2	VF2 Test	IJB-C
Softmax	0.828	0.726
+Pairwise	0.828	0.728
+Triplet	0.845	0.740
+Smooth-AP	**0.850**	**0.754**

ArcFace	VF2 Test	IJB-C
ArcFace	0.858	0.772
+Pairwise	0.861	0.775
+Triplet	0.880	0.787
+Smooth-AP	**0.902**	**0.803**

6.3 Evaluation on Face Retrieval

Due to impressive results [7,13], face retrieval is considered saturated. Nevertheless, we demonstrate here that Smooth-AP can further boost the face retrieval performance. Specifically, we append Smooth-AP on top of modern methods (VGGFace2 and ArcFace) and evaluate mAP on IJB-C and VGGFace2, *i.e.* one of the largest face recognition datasets.

As shown in Table 4, when appending the Smooth-AP loss, retrieval metrics such as mAP can be significantly improved upon the baseline model for both datasets. This is particularity impressive as both baselines have already shown very strong performance on facial verification and identification tasks, yet Smooth-AP is able to increase mAP by up to 4.4% on VGGFace2 and 3.1% on ArcFace. Moreover, Smooth-AP strongly outperforms both the pairwise [11] and triplet [56] losses, *i.e.* the two most popular surrogates to a ranking loss. As discussed in Sect. 4.1, these surrogates optimise a distance metric rather than a ranking metric, and the results show that the latter is optimal for AP.

Table 5. Ablation study over different parameters: temperature τ, size of positive set during minibatch sampling $|\mathcal{P}|$, and batch size B. Performance is benchmarked on VGGFace2-Test and IJB-C.

τ	mAP	
	VF2	IJB-C
0.1	0.824	0.726
0.01	**0.844**	**0.736**
0.001	0.839	0.733

$|\mathcal{P}| = 4$, $B = 128$

$\|\mathcal{P}\|$	mAP	
	VF2	IJB-C
4	**0.844**	**0.736**
8	0.833	0.734
16	0.824	0.726

$\tau = 0.01$, $B = 128$

$\|\mathcal{B}\|$	mAP	
	VF2	IJB-C
64	0.824	0.726
128	0.844	0.736
256.	**0.853**	**0.754**

$\tau = 0.01$, $|\mathcal{P}| = 4$

6.4 Ablation Study

To investigate the effect of different hyper-parameter settings, *e.g.* the sigmoid temperature τ, the size of the positive set $|\mathcal{P}|$, and batch size B (Table 5), we use VGGFace2 and IJB-C with SE-Net50 [7], as the large-scale datasets are unlikely to lead to overfitting, and therefore provide a fair understanding about these hyper-parameters. Note that we only vary one parameter at a time.

Effect of Sigmoid Temperature τ. As explained in Sect. 4.1, τ governs the smoothing of the sigmoid that is used to approximate the indicator function in the Smooth-AP loss. The ablation shows that a value of 0.01 leads to the best mAP scores, which is the optimal trade-off between AP approximation and a large enough operating region in which to provide gradients. Surprisingly, this value (0.01) corresponds to a small operating region. We conjecture that a tight approximation to true AP is the key, and when partnered with a large enough batch size, enough elements of the difference matrix will lie within the operating region in order to induce sufficient re-ranking gradients. The sigmoid temperature can further be viewed from the margin perspective (inter-class margins are commonly used in metric learning to help generalisation [11,42,70]). Smooth-AP only stops providing gradients to push a positive instance above a negative instance once they are a distance equal to the width of the operating region apart, hence enforcing a margin that equates to roughly 0.1 for this choice of τ.

Effect of Positive Set $|\mathcal{P}|$. In this setting, the positive set represents the *instances* that come from the same class in the mini-batch during training. We observe that a small value (4) results in the highest mAP scores, this is because mini-batches are formed by sampling at the class level, where a low value for $|\mathcal{P}|$ means a larger number of sampled classes and a higher probability of sampling hard-negative instances that violate the correct ranking order. Increasing the number of classes in the batch results in a better batch approximation of the true class distribution, allowing each training iteration to enforce a more optimally structured embedding space.

Effect of Batch Size B. Table 5 shows that large batch sizes result in better mAP, especially for VGGFace2. This is expected, as it again increases the chance of getting hard-negative samples in the batch.

6.5 Further Discussion

There are several important observations in the above results. Smooth-AP outperforms all previous AP approximation approaches, as well as the metric learning techniques (pair, triplet, and list-wise) on *three* image retrieval benchmarks, SOP, VehicleID, Inaturalist, with the performance gap being particularly apparent on the large-scale *INaturalist* dataset. Similarly, when scaled to face datasets containing millions of images, Smooth-AP is able to improve the retrieval metrics for state-of-the-art face verification networks. We hypothesis that these performance gains upon the previous AP approximating methods come from a tighter approximation to AP than other existing approaches, hence demonstrating the effectiveness and scalability of Smooth-AP. Furthermore, many of the properties that deep metric learning losses handcraft into their respective methods (distance-based weighting [58,71], inter-class margins [11,56,58,67], intra-class margins [67]), are naturally built into our AP formulation, and result in improved generalisation capabilities.

7 Conclusions

We introduce *Smooth-AP*, a novel loss that directly optimizes a smoothed approximation of AP. This is in contrast to modern contrastive, triplet, and list-wise deep metric learning losses which act as surrogates to encourage ranking. We show that Smooth-AP outperforms recent AP-optimising methods, as well as the deep metric learning methods, and with a simple and elegant, plug-and-play style method. We provide an analysis for the reasons why Smooth-AP outperforms these other losses, *i.e.* Smooth-AP preserves the goal of AP which is to optimise ranking rather than distances in the embedding space. Moreover, we also show that fine-tuning face-verification networks by appending the Smooth-AP loss can strongly improve the performance. Finally, in an effort to bridge the gap between experimental settings and real-world retrieval scenarios, we provide experiments on several large-scale datasets and show Smooth-AP loss to be considerably more scalable than previous approximations.

Acknowledgements. We are grateful to Tengda Han, Olivia Wiles, Christian Rupprecht, Sagar Vaze, Quentin Pleple and Maya Gulieva for proof-reading, and to Ernesto Coto for the initial motivation for this work. Funding for this research is provided by the EPSRC Programme Grant Seebibyte EP/M013774/1. AB is funded by an EPSRC DTA Studentship.

References

1. Arandjelović, R., Gronat, P., Torii, A., Pajdla, T., Sivic, J.: NetVLAD: CNN architecture for weakly supervised place recognition. In: Proceedings of CVPR (2016)
2. Arandjelovic, R., Zisserman, A.: All about VLAD. In: Proceedings of CVPR (2013)
3. Babenko, A., Slesarev, A., Chigorin, A., Lempitsky, V.: Neural codes for image retrieval. In: Fleet, D., Pajdla, T., Schiele, B., Tuytelaars, T. (eds.) ECCV 2014. LNCS, vol. 8689, pp. 584–599. Springer, Cham (2014). https://doi.org/10.1007/978-3-319-10590-1_38
4. Burges, C.J., Ragno, R., Le, Q.V.: Learning to rank with nonsmooth cost functions. In: NeurIPS (2007)
5. Cakir, F., He, K., Xia, X., Kulis, B., Sclaroff, S.: Deep metric learning to rank. In: Proceedings of CVPR (2019)
6. Cakir, F., He, K., Xia, X., Kulis, B., Sclaroff, S.: Fastap: deep metric learning to rank (2019). https://github.com/kunhe/Deep-Metric-Learning-Baselines
7. Cao, Q., Shen, L., Xie, W., Parkhi, O.M., Zisserman, A.: VGGFace2: a dataset for recognising faces across pose and age. In: Proceedings of International Conference on Automatic Face and Gesture Recognition (2018)
8. Cao, Z., Qin, T., Liu, T.Y., Tsai, M.F., Li, H.: Learning to rank: from pairwise approach to listwise approach. In: Proceedings of ICML (2007)
9. Chapelle, O., Le, Q., Smola, A.: Large margin optimization of ranking measures. In: NeurIPS (2007)
10. Chen, H., Xie, W., Vedaldi, A., Zisserman, A.: Autocorrect: deep inductive alignment of noisy geometric annotations. In: Proceedings of BMVC (2019)
11. Chopra, S., Hadsell, R., LeCun, Y.: Learning a similarity metric discriminatively, with application to face verification. In: Proceedings of CVPR (2005)
12. Chum, O., Mikulik, A., Perdoch, M., Matas, J.: Total recall II: query expansion revisited. In: Proceedings of CVPR (2011)
13. Deng, J., Guo, J., Xue, N., Zafeiriou, S.: Arcface: additive angular margin loss for deep face recognition. In: Proceedings of CVPR (2019)
14. Duan, Y., Zheng, W., Lin, X., Lu, J., Zhou, J.: Deep adversarial metric learning. In: Proceedings of CVPR (2018)
15. Engilberge, M., Chevallier, L., Pérez, P., Cord, M.: Sodeep: a sorting deep net to learn ranking loss surrogates. In: Proceedings of CVPR (2019)
16. Ge, W., Huang, W., Dong, D., Scott, M.R.: Deep metric learning with hierarchical triplet loss. In: Ferrari, V., Hebert, M., Sminchisescu, C., Weiss, Y. (eds.) ECCV 2018. LNCS, vol. 11210, pp. 272–288. Springer, Cham (2018). https://doi.org/10.1007/978-3-030-01231-1_17
17. Gordo, A., Almazán, J., Revaud, J., Larlus, D.: Deep image retrieval: learning global representations for image search. In: Leibe, B., Matas, J., Sebe, N., Welling, M. (eds.) ECCV 2016. LNCS, vol. 9910, pp. 241–257. Springer, Cham (2016). https://doi.org/10.1007/978-3-319-46466-4_15
18. Guiver, J., Snelson, E.: Learning to rank with softrank and gaussian processes. In: Proceedings of the 31st Annual International ACM SIGIR Conference on Research and Development in Information Retrieval (2008)
19. Harwood, B., Kumar, B., Carneiro, G., Reid, I., Drummond, T., et al.: Smart mining for deep metric learning. In: Proceedings of ICCV (2017)
20. He, K., Zhang, X., Ren, S., Sun, J.: Deep residual learning for image recognition. In: Proceedings of CVPR (2016)

21. He, K., Cakir, F., Adel Bargal, S., Sclaroff, S.: Hashing as tie-aware learning to rank. In: Proceedings of CVPR (2018)
22. He, K., Lu, Y., Sclaroff, S.: Local descriptors optimized for average precision. In: Proceedings of CVPR (2018)
23. Henderson, P., Ferrari, V.: End-to-end training of object class detectors for mean average precision. In: Lai, S.-H., Lepetit, V., Nishino, K., Sato, Y. (eds.) ACCV 2016. LNCS, vol. 10115, pp. 198–213. Springer, Cham (2017). https://doi.org/10.1007/978-3-319-54193-8_13
24. Hermans, A., Beyer, L., Leibe, B.: In defense of the triplet loss for person re-identification. arXiv preprint arXiv:1703.07737 (2017)
25. Hu, J., Shen, L., Sun, G.: Squeeze-and-excitation networks. In: Proceedings of CVPR (2018)
26. Jegou, H., Douze, M., Schmid, C.: Hamming embedding and weak geometric consistency for large scale image search. In: Forsyth, D., Torr, P., Zisserman, A. (eds.) ECCV 2008. LNCS, vol. 5302, pp. 304–317. Springer, Heidelberg (2008). https://doi.org/10.1007/978-3-540-88682-2_24
27. Jégou, H., Douze, M., Schmid, C.: Product quantization for nearest neighbor search. In: IEEE PAMI (2011)
28. Jégou, H., Douze, M., Schmid, C., Pérez, P.: Aggregating local descriptors into a compact image representation. In: Proceedings of CVPR (2010)
29. Jégou, H., Perronnin, F., Douze, M., Sánchez, J., P'erez, P., Schmid, C.: Aggregating local image descriptors into compact codes. In: IEEE PAMI (2011)
30. Kim, W., Goyal, B., Chawla, K., Lee, J., Kwon, K.: Attention-based ensemble for deep metric learning. In: Ferrari, V., Hebert, M., Sminchisescu, C., Weiss, Y. (eds.) ECCV 2018. LNCS, vol. 11205, pp. 760–777. Springer, Cham (2018). https://doi.org/10.1007/978-3-030-01246-5_45
31. Kingma, D.P., Ba, J.: Adam: a method for stochastic optimization. CoRR (2014)
32. Law, M.T., Urtasun, R., Zemel, R.S.: Deep spectral clustering learning. In: Proceedings of ICML (2017)
33. Li, K., Huang, Z., Cheng, Y.C., Lee, C.H.: A maximal figure-of-merit learning approach to maximizing mean average precision with deep neural network based classifiers. In: Proceedings of ICASSP (2014)
34. Liu, W., Wen, Y., Yu, Z., Li, M., Raj, B., Song, L.: Sphereface: deep hypersphere embedding for face recognition. In: Proceedings of CVPR (2017)
35. Lu, J., Xu, C., Zhang, W., Duan, L.Y., Mei, T.: Sampling wisely: deep image embedding by top-k precision optimization. In: Proceedings of ICCV (2019)
36. Manning, C.D., Raghavan, P., Schütze, H.: Introduction to Information Retrieval. Cambridge University Press, New York (2008)
37. Maze, B., et al.: IARPA janus benchmark-c: face dataset and protocol. In: 2018 International Conference on Biometrics (ICB) (2018)
38. McFee, B., Lanckriet, G.R.: Metric learning to rank. In: Proceedings of ICML (2010)
39. Mohapatra, P., Rolinek, M., Jawahar, C., Kolmogorov, V., Pawan, K.: Efficient optimization for rank-based loss functions. In: Proceedings of CVPR (2018)
40. Movshovitz-Attias, Y., Toshev, A., Leung, T.K., Ioffe, S., Singh, S.: No fuss distance metric learning using proxies. In: Proceedings of CVPR (2017)
41. Oh Song, H., Jegelka, S., Rathod, V., Murphy, K.: Deep metric learning via facility location. In: Proceedings of CVPR (2017)
42. Oh Song, H., Xiang, Y., Jegelka, S., Savarese, S.: Deep metric learning via lifted structured feature embedding. In: Proceedings of CVPR (2016)

43. Opitz, M., Waltner, G., Possegger, H., Bischof, H.: BIER: boosting independent embeddings robustly. In: Proceedings of ICCV (2017)
44. Perronnin, F., Liu, Y., Sánchez, J., Poirier, H.: Large-scale image retrieval with compressed fisher vectors. In: Proceedings of CVPR (2010)
45. Philbin, J., Chum, O., Isard, M., Sivic, J., Zisserman, A.: Object retrieval with large vocabularies and fast spatial matching. In: Proceedings of CVPR (2007)
46. Qian, Q., Shang, L., Sun, B., Hu, J., Li, H., Jin, R.: Softtriple loss: deep metric learning without triplet sampling. In: Proceedings of ICCV (2019)
47. Qin, T., Liu, T.Y., Li, H.: A general approximation framework for direct optimization of information retrieval measures. Inf. Retrieval **13**, 375–397 (2010)
48. Radenović, F., Tolias, G., Chum, O.: CNN image retrieval learns from BoW: unsupervised fine-tuning with hard examples. In: Leibe, B., Matas, J., Sebe, N., Welling, M. (eds.) ECCV 2016. LNCS, vol. 9905, pp. 3–20. Springer, Cham (2016). https://doi.org/10.1007/978-3-319-46448-0_1
49. Rao, Y., Lin, D., Lu, J., Zhou, J.: Learning globally optimized object detector via policy gradient. In: Proceedings of CVPR (2018)
50. Revaud, J., Almazá, J., Rezende, R.S., de Souza, C.R.: Learning with average precision: training image retrieval with a listwise loss. In: Proceedings of ICCV (2019)
51. Rolínek, M., Musil, V., Paulus, A., Vlastelica, M., Michaelis, C., Martius, G.: Optimizing rank-based metrics with blackbox differentiation. In: Proceedings of CVPR (2020)
52. Roth, K., Brattoli, B.: Deep metric learning baselines (2019). https://github.com/Confusezius/Deep-Metric-Learning-Baselines
53. Roth, K., Brattoli, B., Ommer, B.: Mic: mining interclass characteristics for improved metric learning. In: Proceedings of ICCV (2019)
54. Russakovsky, O., et al.: Imagenet large scale visual recognition challenge. IJCV **115**, 211–252 (2015)
55. Sanakoyeu, A., Tschernezki, V., Buchler, U., Ommer, B.: Divide and conquer the embedding space for metric learning. In: Proceedings of CVPR (2019)
56. Schroff, F., Kalenichenko, D., Philbin, J.: Facenet: a unified embedding for face recognition and clustering. In: Proceedings of CVPR (2015)
57. Sivic, J., Zisserman, A.: Video Google: a text retrieval approach to object matching in videos. In: Proceedings of ICCV (2003)
58. Song, H.O., Xiang, Y., Jegelka, S., Savarese, S.: Deep metric learning via lifted structured feature embedding. In: Proceedings of CVPR (2016)
59. Suh, Y., Han, B., Kim, W., Lee, K.M.: Stochastic class-based hard example mining for deep metric learning. In: Proceedings of CVPR (2019)
60. Taylor, M., Guiver, J., Robertson, S., Minka, T.: Softrank: optimising non-smooth rank metrics. In: WSDM (2008)
61. Ustinova, E., Lempitsky, V.: Learning deep embeddings with histogram loss. In: NeurIPS (2016)
62. Van Horn, G., et al.: The INaturalist species classification and detection dataset. In: Proceedings of CVPR (2018)
63. Vlastelica, M., Paulus, A., Musil, V., Martius, G., Rolínek, M.: Differentiation of blackbox combinatorial solvers (2020). https://github.com/martius-lab/blackbox-backprop
64. Wah, C., Branson, S., Welinder, P., Perona, P., Belongie, S.: The Caltech-UCSD Birds-200-2011 Dataset. Technical report CNS-TR-2011-001, California Institute of Technology (2011)

65. Wang, J., Zhou, F., Wen, S., Liu, X., Lin, Y.: Deep metric learning with angular loss. In: Proceedings of ICCV (2017)
66. Wang, J., et al.: Learning fine-grained image similarity with deep ranking. In: Proceedings of CVPR (2014)
67. Wang, X., Hua, Y., Kodirov, E., Hu, G., Garnier, R., Robertson, N.: Ranked list loss for deep metric learning. In: Proceedings of CVPR (2019)
68. Wang, X., Han, X., Huang, W., Dong, D., Scott, M.R.: Multi-similarity loss with general pair weighting for deep metric learning. In: Proceedings of CVPR (2019)
69. Wang, X., Zhang, H., Huang, W., Scott, M.R.: Cross-batch memory for embedding learning. In: Proceedings of CVPR (2020)
70. Weinberger, K., Blitzer, J., Saul, L.: Distance metric learning for large margin nearest neighbor classification. In: NeurIPS (2006)
71. Wu, C.Y., Manmatha, R., Smola, A.J., Krahenbuhl, P.: Sampling matters in deep embedding learning. In: Proceedings of CVPR (2017)
72. Xuan, H., Souvenir, R., Pless, R.: Deep randomized ensembles for metric learning. In: Ferrari, V., Hebert, M., Sminchisescu, C., Weiss, Y. (eds.) ECCV 2018. LNCS, vol. 11220, pp. 751–762. Springer, Cham (2018). https://doi.org/10.1007/978-3-030-01270-0_44
73. Yang, H.F., Lin, K., Chen, C.S.: Cross-batch reference learning for deep classification and retrieval. In: Proceedings of ACMMM (2016)
74. Yuan, Y., Yang, K., Zhang, C.: Hard-aware deeply cascaded embedding. In: Proceedings of ICCV (2017)
75. Yue, Y., Finley, T., Radlinski, F., Joachims, T.: A support vector method for optimizing average precision. In: SIGIR (2007)

Naive-Student: Leveraging Semi-Supervised Learning in Video Sequences for Urban Scene Segmentation

Liang-Chieh Chen[1(✉)], Raphael Gontijo Lopes[1], Bowen Cheng[2], Maxwell D. Collins[1], Ekin D. Cubuk[1], Barret Zoph[1], Hartwig Adam[1], and Jonathon Shlens[1]

[1] Google Research, Mountain View, USA
lcchen@google.com
[2] UIUC, Champaign, USA

Abstract. Supervised learning in large discriminative models is a mainstay for modern computer vision. Such an approach necessitates investing in large-scale human-annotated datasets for achieving state-of-the-art results. In turn, the efficacy of supervised learning may be limited by the size of the human annotated dataset. This limitation is particularly notable for image segmentation tasks, where the expense of human annotation is especially large, yet large amounts of unlabeled data may exist. In this work, we ask if we may leverage semi-supervised learning in unlabeled video sequences and extra images to improve the performance on urban scene segmentation, simultaneously tackling semantic, instance, and panoptic segmentation. The goal of this work is to avoid the construction of sophisticated, learned architectures specific to label propagation (*e.g.*, patch matching and optical flow). Instead, we simply predict pseudo-labels for the unlabeled data and train subsequent models with both human-annotated and pseudo-labeled data. The procedure is iterated for several times. As a result, our Naive-Student model, trained with such simple yet effective iterative semi-supervised learning, attains state-of-the-art results at all three Cityscapes benchmarks, reaching the performance of 67.8% PQ, 42.6% AP, and 85.2% mIOU on the test set. We view this work as a notable step towards building a simple procedure to harness unlabeled video sequences and extra images to surpass state-of-the-art performance on core computer vision tasks.

Keywords: Semi-supervised learning · Pseudo label · Semantic segmentation · Instance segmentation · Panoptic segmentation

---

Electronic supplementary material The online version of this chapter (https://doi.org/10.1007/978-3-030-58545-7_40) contains supplementary material, which is available to authorized users.

© Springer Nature Switzerland AG 2020
A. Vedaldi et al. (Eds.): ECCV 2020, LNCS 12354, pp. 695–714, 2020.
https://doi.org/10.1007/978-3-030-58545-7_40

1 Introduction

Significant advances in computer vision due to deep learning [10,26,40,66] have been tempered by the fact that these advances have been accrued through supervised learning on large-scale, human-annotated datasets [49,69]. The paradigm of supervised learning requires the expenditure of a large amount of resources to manually label static images – whether through the development of specialized annotation tools [5,8,70], or the amount of human hours for the annotation itself [49,69]. Such an approach does not scale effectively to comprehensively label real-time video frames (but see [2,7,65]). More importantly, supervised training is rather sample-inefficient as many examples are required for good generalization [29,39]. Ideally, one would expect and hope that a training method may be able to learn in a more self-supervised manner particularly on video – much as presumed to occur in human visual learning [41,82].

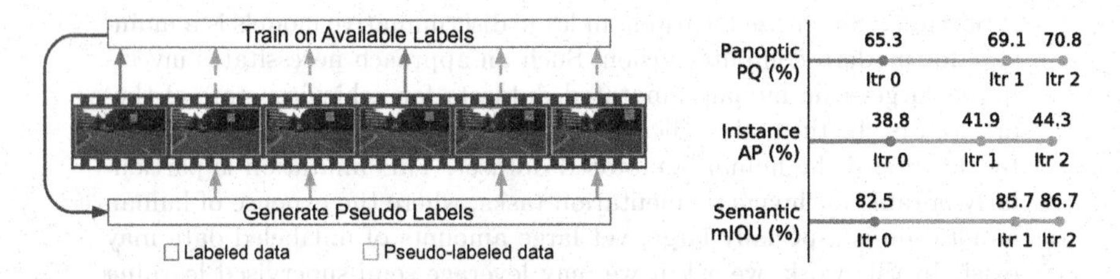

Fig. 1. Naive-Student: semi-supervised learning in video sequences for scene segmentation. We iteratively train on human-annotated frames (1 out of 30 in each Cityscapes video sequence), and generate pseudo-labels for the other unlabeled video frames (left). Segmentation performances (val set) improve at each iteration (right).

The limitations of supervised learning is most pronounced in the task of image segmentation [22]. Human annotation of static images for segmentation is particularly expensive, requiring, for instance, 90 min per image [17] or 22 worker hours per 1,000 mask segmentations [49]. In the case of self-driving cars, the annotation of video is a critical supervised learning problem [24,75], and in turn has fostered an industry of specialized companies for data annotation.

In contrast, recent findings on the benefits of pre-training on ImageNet for segmentation [47] indicate that current segmentation approaches may benefit from large-scale image classification datasets. This direction has been further pursued by [12,74] on an extremely large image classification dataset [30]. Additionally, many segmentation methods [9,92] apply transfer learning by pre-training on augmented segmentation datasets [27,49,55] and then fine-tuning on the target datasets [17,21]. Likewise, other works attempt to exploit label propagation in video to improve segmentation. However, these methods require building specialized modules to propagate labels across video frames [23,54,56,95].

In this work, we leverage both unlabeled video frames and extra unlabeled images to improve the urban scene segmentation evaluated in terms of semantic segmentation, instance segmentation, and panoptic segmentation. Importantly, we do not require any specialized methods for propagating label information across video frames, such as optical flow [23,54,56], patch matching [4,6], or learned motion vector [95]. Instead, we propose to employ a simple iterative semi-supervised learning procedure. At each iteration, the model from the previous iteration generates pseudo-labels for unlabeled video frames (Fig. 1). Specifically, a pseudo-label is generated through a distillation across multiple augmentations applied to each unlabeled video frame. Subsequent iterations of the training procedure train on the original labeled data as well as the newly pseudo-labeled data. Our model, trained with such a simple yet effective method, simultaneously sets new state-of-the-art results on the Cityscapes urban scene segmentation [17], achieving 67.8% PQ, 42.6% AP, and 85.2% mIOU on test set. We hope that such an iterative semi-supervised learning may provide more label-efficient methods for developing a machine learning solution to segmentation.

2 Related Works

Our method is related to both *self-training* [20,25,67,71,88,91], where the predictions of a model on unlabeled data is used to train the model, and *semi-supervised learning* [44,57,64,68,86], where additionally extra human-annotated data is available to guide the training with unlabeled data. In particular, our model is trained with some human-annotated images and abundant pseudo-labeled [3,34,42,72] video sequences.

Semi-supervised learning has been widely applied to several computer vision tasks, including semantic segmentation [19,31,35,57,59,73,80,81,96], object detection [64,68,76], instance segmentation [35,60], panoptic segmentation [45], human pose estimation [58], person re-identification [93], multi-object tracking and segmentation [62,77], and so on. A comprehensive literature survey is beyond the scope of this work, and thus we focus on comparing our proposed method with the most related ones.

Our proposed iterative semi-supervised learning is similar to the work by Papandreou *et al.* [57], STC [80], Simple-Does-It [35], the work by Li *et al.* [45], and Noisy-Student [84]. In particular, our iterative semi-supervised learning is similar to the Expectation-Maximization method by Papandreou *et al.* [57] which alternates between estimating the latent pixel labels (*i.e.*, pseudo labels) and optimizing the network parameters with bounding box or image-level annotations. Similarly, Li *et al.* [45] generate pseudo labels for panoptic segmentation by exploiting both fully-annotated and weakly-annotated images, where bounding boxes for 'thing' classes and image-level tags for 'stuff' classes are provided. However, unlike those two works, we do not exploit any weakly-annotated data. Additionally, we do not sort the images by the annotation difficulty and do not exploit any other assistance, such as saliency maps, as in STC [80]. Simple-Does-It [35] adopts a complicated de-noising procedure to clean the pseudo labels, while we simply use the outputs from a neural network. Finally, following Noisy-Student [84], we employ a stronger Student network in the subsequent iterations, but we do not employ any noisy data augmentation (*i.e.*, RandAugment [18]).

Algorithm 1. Iterative semi-supervised learning for urban scene segmentation.

Labeled data: n pairs of image x_i and corresponding human annotation y_i
Unlabeled data: m images collected from multiple video sequences or extra images with no human annotations $\{\widetilde{x}_1, \widetilde{x}_2, ..., \widetilde{x}_m\}$.
Step 1: Train a Teacher network θ_t (with prediction function f) on the manually labeled images by minimizing the total loss $\mathcal{L}$ for scene segmentation.

$$\theta_t^* = \arg\min_{\theta_t} \frac{1}{n} \sum_{i=1}^{n} \mathcal{L}(y_i, f(x_i, \theta_t))$$

where $\mathcal{L} = \lambda_{\text{sem}} \mathcal{L}_{\text{sem}} + \lambda_{\text{heatmap}} \mathcal{L}_{\text{heatmap}} + \lambda_{\text{offset}} \mathcal{L}_{\text{offset}}$ in our framework.
Step 2: Generate pseudo-labels $\widetilde{y}_i$ for unlabeled images with test-time augmentations (*i.e.*, multi-scale inputs and left-right flips).

$$\widetilde{y}_i = f(Aug(\widetilde{x}_i), \theta_t^*), \forall i = 1, ..., m$$

where $Aug(\cdot)$ is test-time augmentation.
Step 3: Train an *equal or larger* Student network θ_s on pseudo-labeled images $(\widetilde{x}_i, \widetilde{y}_i)$ with the same objective.

$$\theta_s^* = \arg\min_{\theta_s} \frac{1}{m} \sum_{i=1}^{m} \mathcal{L}(\widetilde{y}_i, f(\widetilde{x}_i, \theta_s))$$

Step 4: Fine-tune the Student network θ_s^* from step 3 on the manually labeled image annotations (x_i, y_i) using the same objective.

$$\theta_s^{**} = \arg\min_{\theta_s^*} \frac{1}{n} \sum_{i=1}^{n} \mathcal{L}(y_i, f(x_i, \theta_s^*))$$

Step 5: Return to step 2 but employ the Student network θ_s^{**} as a Teacher until reaching desired number of iterations.

When generating pseudo labels, we employ a simple test-time augmentation, *i.e.*, multi-scale inputs and left-right flips, a common strategy used by segmentation models [12,92], which bears a similarity to Data-Distillation [64]. However, our framework is deployed in an iterative manner, and we exploit unlabeled video sequences for scene segmentation, simultaneously tackling semantic, instance, and panoptic segmentation. Additionally, we do not set a threshold as [64] to remove false positives, avoiding tuning of another hyper-parameter.

Video sequences have also been exploited in semi-supervised learning for semantic segmentation. Human-annotated ground-truth labels of certain frames in a video sequence could be propagated to other unlabeled frames via patch matching [4,6] or optical flow [23,53,54,56,94]. Recently, Zhu *et al.* [95] generate pseudo-labeled video sequences by jointly propagating the image-label pair with learned motion vectors, and demonstrate promising results. Similarly, our method also exploits unlabeled video sequences. However, our method is much simpler since we do not employ any label-propagation modules (*e.g.*, patch matching [4,6], optical flow [23,54,56], or motion vectors [95]) but instead directly generate the pseudo labels for each video frame.

3 Methods

Algorithm 1 gives an overview of our proposed iterative semi-supervised learning for scene segmentation. Suppose two sets of images are given, where one contains human annotations and the other does not. The human-annotated images are exploited to train a Teacher network using the loss function for scene segmentation. Pseudo-labels for those un-annotated images are then generated by the Teacher network with a test-time augmentation function. A Student network is subsequently trained with the pseudo-labeled images using the same loss function for scene segmentation. The Student network is then fine-tuned on human-labeled images before evaluating on the validation set or test set. Finally, one could optionally replace the Teacher network with the Student network and iterate the procedure again. Our method, dubbed *Naive-Student*, is motivated by Noisy-Student [84] where we adopt a stronger Student network in the following iterations, but we do not inject noise (*i.e.*, RandAugment [18]) to the Student. Our algorithm is illustrated in Fig. 2. We elaborate on the details below.

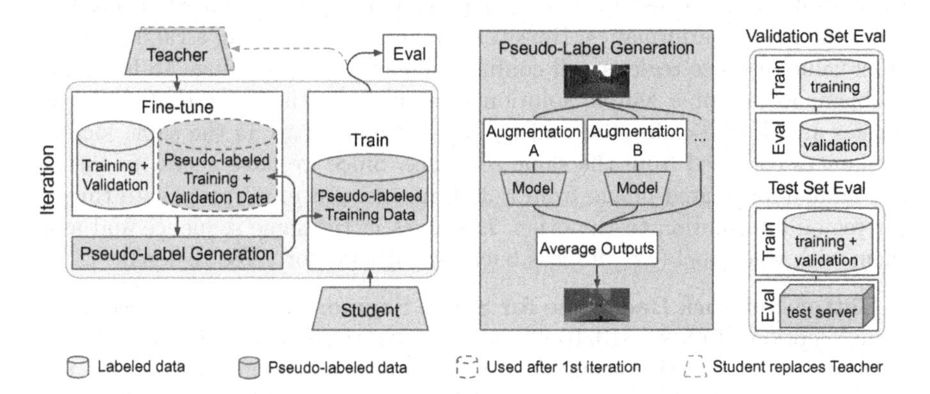

Fig. 2. Overview of our proposed iterative semi-supervised learning for scene segmentation. The Teacher network is trained with all available human-annotated images (and extra pseudo-labeled images after 1st iteration), and then generates pseudo-labels for all the unlabeled images with a simple test-time augmentation (*i.e.*, multi-scale inputs and left-right flips). The Student network is subsequently trained with the pseudo-labeled data, and optionally replaces the Teacher network in following iterations. Before evaluating the validation or test set performance, the Student network is fine-tuned on the human-annotated images. Note that the validation set is only exploited by the Teacher in order to generate high-quality pseudo-labels, and the Student has no access to it. Additionally, the final test set results are evaluated on a fair test server where the annotations are held-out.

The Loss for Scene Segmentation: Our core building block is the state-of-the-art bottom-up panoptic segmentation model, Panoptic-DeepLab [14], which improves the semantic segmentation model DeepLabv3+ [13] by incorporating another class-agnostic instance segmentation prediction. Its instance segmentation prediction involves a simple instance center prediction as well as the offset regression from each pixel to its corresponding center. As a result, the total loss function $\mathcal{L}$ for scene segmentation boils down to three loss functions: softmax cross entropy loss $\mathcal{L}_{sem}$ for semantic segmentation, mean squared error loss $\mathcal{L}_{heatmap}$ for instance center prediction, and L_1 loss $\mathcal{L}_{offset}$ for offset regression. In our algorithm, the Teacher and the Student networks are trained with the same total loss function $\mathcal{L}$.

Pseudo-Label Generation: After training the Teacher network on all human-annotated images (and all pseudo-labeled images after iteration 1), we generate (or update) the pseudo labels for all un-annotated images with a test-time augmentation function $Aug(\cdot)$. We simply use the common test-time augmentations, *i.e.*, multi-scale inputs and left-right flips. We only generate hard pseudo labels (*i.e.*, a one-hot distribution) in order to save disk space when processing large resolution images (*e.g.*, Cityscapes image size is 1024×2048).

Ego-Car Region in Pseudo Labels: Cityscapes images are collected (or recorded) with a driving vehicle. A part of the vehicle, called "ego-car" region, is thus visible in all frames of a video sequence. This region is ignored during evaluating the model performance. However, we find that assigning a random pseudo label value to those regions will confuse models during training. To handle this problem, we adopt a simple solution by exploiting the prior that Cityscapes images are all well-calibrated and the ego-car regions are in the same locations for images collected from the same sequence. Since we have access to the only one human-annotated image from a 30-frame sequence, we propagate this ego-car region information to the other 29 frames in the same sequence and assign them with void label (*i.e.*, no loss back-propagation for those regions).

A Better Network Backbone for Scene Segmentation: The efficient backbone Xception-71 (X-71) [13,16,63] is adopted in the Teacher network at the first iteration in our iterative semi-supervised learning algorithm. In the next iteration, a stronger backbone should be used to generate pseudo labels with a better quality. In this work, we modify the powerful Wide ResNet-38 (WR-38) [83,90] for scene segmentation. In particular, we remove the last residual block B7 in WR-38 [83] and repeat the residual block B6 two more times, resulting in our proposed WR-41. Additionally, we adopt drop path [32,50] (with a constant survival probability 0.8) and multi-grid scheme [12,79] in the last three residual blocks (with unit rate $\{1, 2, 4\}$, same as [12]). As a result, the proposed WR-41 attains better performance than X-71 in the fully supervised setting.

4 Experiments

We conduct experiments on the popular Cityscapes dataset [17], which consists of a large and diverse set of street-view video sequences recorded from 50 cities primarily in Germany. From the video sequences, 5000 images are provided with high-quality pixel-wise annotations in which 2975, 500, and 1525 images are used for training, validation, and test, respectively. Each image is selected from the 20th frame of a 30-frame video snippet. Additionally, another 20000 images are accompanied with coarse annotations. We define each dataset split below.

train-fine: Training set (2,975 images) with fine pixel-wise annotations.

val-fine: Validation set (500 images) with fine pixel-wise annotations.

test-fine: Test set (1,525 images) where the fine pixel-wise annotations are held-out, and the evaluation is performed on a fair test server.

train-extra: Extra 20,000 images with coarse annotations. Our proposed method is not limited to video sequences, and thus we also generate pseudo-labels for this set, instead of using the provided coarse annotations.

train-sequence: The video sequences where the train-fine set is selected from. This set contains $2975 \times 30 = 89,250$ frames.

val-sequence: The video sequences where the val-fine set is selected from. This set contains $500 \times 30 = 15,000$ frames.

Furthermore, one could merge training and validation splits (*e.g.*, trainval-fine is merged from train-fine and val-fine, and similarly for trainval-sequence).

Experimental Setup: We report mean intersection-over-union (mIOU), average precision (AP), and panoptic quality (PQ) to evaluate the semantic, instance, and panoptic segmentation results, respectively.

The state-of-art bottom-up panoptic segmentation model, Panoptic-DeepLab [15], is included in our proposed iterative semi-supervised learning pipeline. Panoptic-DeepLab is a simple framework and simultaneously produces semantic, instance, and panoptic segmentation results without the need to fine-tune on each task. We adopt the same training protocol as [15] when using Panoptic-DeepLab. For example, our models are trained using TensorFlow [1] on 32 TPUs. We use the 'poly' learning rate policy [52] with an initial learning rate of 0.001 for Xception-71 (X-71) backbone [16,63] and 0.0001 for our proposed Wide ResNet-41 (WR-41) [28,83,90], respectively. During training, the batch normalization [33] is fine-tuned, random scale data augmentation and Adam [36] optimizer without weight decay are adopted. On Cityscapes, we employ training crop size equal to 1025×2049 with batch size 32, and 180K training iterations. Similar to other works on panoptic segmentation [37,38,43,61,78,85,87], we re-assign to void label all 'stuff' segments whose areas are smaller than a threshold of 4096. Additionally, we employ multi-scale inference (scales equal to $\{0.5, 0.75, 1, 1.25, 1.5, 1.75, 2\}$ for Cityscapes) and left-right flipped inputs, to further improve the performance for test server evaluation.

4.1 Urban Scene Segmentation Results

In this subsection, we summarize our main results on the Cityscapes dataset.

Cityscapes Val-Fine Set: In our iterative semi-supervised learning framework, at each iteration, all data splits, including Mapillary Vistas [55] and Cityscapes trainval-fine, (also trainval-sequence and train-extra after 1st iteration), are exploited for the Teacher networks in order to generate better pseudo-labels, while the Student networks are always initialized from the Mapillary Vistas pre-trained checkpoint (unless it is specified that it is initialized from previous iterations). In Table 1, we report the validation set results. At iteration 0, we employ the state-of-art Panoptic-DeepLab with Xception-71 (validation set results from [15] are shown in the table for comparison) as the Teacher network to generate pseudo-labels for train-sequence and train-extra splits which are subsequently used to train our Student network using the proposed Wide ResNet-41 as backbone. As a result, at iteration 1, we improve over the Panoptic-DeepLab (X-71) baseline by a margin of 3.8% PQ, 3.1% AP, and 3.2% mIOU. The Student network is then selected as the new Teacher network after fine-tuning on all the available data splits (*i.e.*, trainval-sequence, train-extra, and trainval-fine). At iteration 2, by training with the better quality pseudo-labels, we observe an additional improvement of 1.3% PQ, 1.5% AP, and 0.8% mIOU for the new Student network. Additionally, one could further slightly improve the performance by initializing the Student network from iteration 1, as shown in the last row.

Cityscapes Test-Fine Set: In Table 2, we report our Cityscapes test set results. As shown in the table, our single model simultaneously ranks 1st at all three Cityscapes benchmarks. In particular, for the panoptic segmentation benchmark, our model outperforms Panoptic-DeepLab (X-71) [15] by 2.3% PQ, Li *et al.* [46] by 4.5% PQ, and Seamless [61] by 5.2% PQ. For the instance segmentation benchmark, our model outperforms PolyTransform [48] by 2.5% AP, Panoptic-DeepLab (X-71) [15] by 3.6% AP, and PANet [51] by 6.2% AP. Finally, for the competitive semantic segmentation benchmark, our model outperforms Panoptic-DeepLab (X-71) [15] by 1.0% mIOU, OCR [89] by 1.5% mIOU, and Zhu *et al.* [95] by 1.7% mIOU.

Table 1. Iterative semi-supervised training systematically improves urban scene segmentation results. Results presented on Cityscapes validation data. The Students are pretrained on ImageNet [69] and Mapillary Vistas [55]. Baseline Xception-71 (X-71) at iteration 0 is obtained from [15], while Wide-ResNet-41 (WR-41) is *modified* from [83]. All Teacher networks have been trained on ImageNet, Mapillary Vistas, and Cityscapes. Labeled data is from Cityscapes train-fine set. Pseudo-Labeled data is from Cityscapes train-sequence and train-extra sets. † indicates that model was initialized from the checkpoint in the previous iteration. The text color indicates how the Student is selected as the new Teacher (*e.g.*, the Student X-71 at iteration 0 becomes the Teacher at iteration 1 after fine-tuning).

	Architecture		Training set		Validation set		
Itr	Student	Teacher	Labeled	Pseudo-Labeled	PQ (%)	AP (%)	mIOU (%)
0	X-71	-	✓		65.3	38.8	82.5
1	WR-41	X-71		✓	69.1	41.9	85.7
2	WR-41	WR-41		✓	70.4	43.4	86.5
2	WR-41†	WR-41		✓	70.8	44.3	86.7

Visualization of Generated Pseudo Labels: We observe visually subtle differences between iteration 1 and iteration 2, since both Teachers yield high-quality results. To further look into the minor differences, we zoom-in some generated pseudo-labels in Fig. 3. As shown in the figure, the Teacher at iteration 2 generates slightly better pseudo-labels along the thin and small objects.

Visualization of Segmentation Results: In Fig. 4, we visualize some segmentation results obtained by the Student network on val-fine set.

Table 2. Iterative semi-supervised learning achieves state-of-the-art urban scene segmentation results. Results presented on Cityscapes *test-fine* set. **C:** Cityscapes coarse annotation. **V:** Cityscapes video. **MV:** Mapillary Vistas. Note that we do not exploit the train-extra coarse annotations, but instead we generate pseudo-labels for them.

Model	Extra data	Training method	PQ (%)	AP (%)	mIOU (%)
Naive-Student (ours)	C, V, MV	Iterative semi-supervised	67.8	42.6	85.2
Seamless [61]	MV	Supervised	62.6	-	-
Li *et al.* [46]	COCO	Supervised	63.3	-	-
Panoptic-DeepLab (X-71) [15]	MV	Supervised	65.5	39.0	84.2
PANet [51]	COCO	Supervised	-	36.4	-
PolyTransform [48]	COCO	Supervised	-	40.1	-
Zhu *et al.* [95]	C, V, MV	Semi-supervised	-	-	83.5
OCR [89]	C, MV	Supervised	-	-	83.7

4.2 Ablation Studies

In this subsection, we provide ablation studies on several design choices. Xception-71 is used as the backbone if not specified.

Training Iterations: First, we verify that the performance improvement does not solely result from longer training iterations, but from the extra large pseudo-labeled images. We train Panoptic-DeepLab [15] with 60K iterations on Cityscapes train-fine set and obtain a PQ of 62.9%. We increase the training iterations to 120K iterations, but do not observe any improvement (62.7% PQ) (*i.e.*, performance saturates after 60K iterations). On the other hand, our proposed Naive-Student attains a better performance with 180K iterations (65.3% PQ) when trained with the larger train-sequence set.

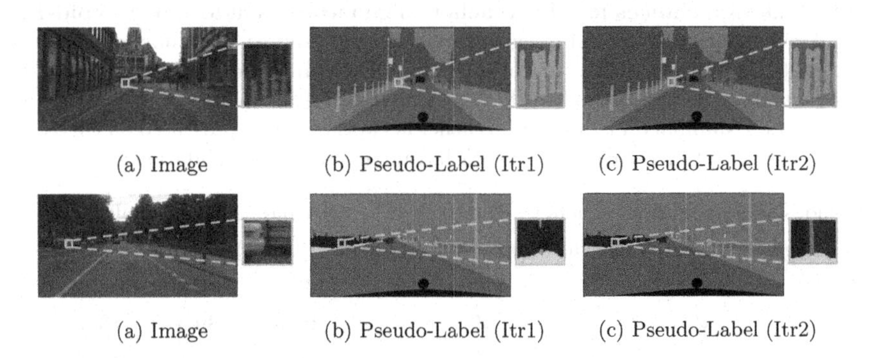

(a) Image (b) Pseudo-Label (Itr1) (c) Pseudo-Label (Itr2)

(a) Image (b) Pseudo-Label (Itr1) (c) Pseudo-Label (Itr2)

Fig. 3. Generated pseudo-labels by Teacher improves qualitatively with more iterations. We only observe subtle difference between pseudo-labeled video frames at iteration 1 and iteration 2. The results from iteration 2 capture better thin objects, as zoomed-in in the yellow regions. (Color figure online)

Fig. 4. **Segmentation results by Student on Cityscapes val set.**

Design Choices for the Teacher to Generate Pseudo-labels: When generating the pseudo-labels, there are four factors involved in our design, namely (1) assignment of void label to the ego-car region, (2) employment of test-time augmentation, (3) more Cityscapes human-labeled images, and (4) Mapillary Vistas pretraining. This "Void-Ego-Car" design, or factor (1), improves 0.7% PQ, 0.5% AP, and 0.4% mIOU. We think the wrongly generated labels in the ego car region slightly affect the model training. Without employing the test-time augmentation, (*i.e.*, multi-scale inference and left-right flipping), when generating the pseudo-labels, the performance drops by 0.8% PQ, 1.1% AP, and 0.9% mIOU. Excluding more Cityscapes human-labeled images for fine-tuning the Teacher network degrades the performance by 1.4% PQ, 1.2% AP, and 1.3% mIOU. Finally, if the Teacher network is not pretrained on the Mapillary Vistas dataset, the performance decreases by 2.2% PQ, 2.2% AP, and 1.5% mIOU (Table 3).

Table 3. **Design choices for the Teacher to generate pseudo-labels. Void-Ego-Car:** Assignment of void label to ego-car regions. **Test-Aug:** Test-time augmentation (*i.e.*, multi-scale inputs and left-right flips). **Val-Fine:** Inclusion of Cityscapes val-fine set for training the Teacher network. **MV-Pretrained:** Employment of a pretrained checkpoint on Mapillary Vista for the Teacher network. Results presented on Cityscapes validation set. Note the Student network has no access to the validation set.

Pseudo-label generation scheme				Validation set results		
Void-Ego-Car	Test-Aug	Val-Fine	MV-Pretrained	PQ (%)	AP (%)	mIOU (%)
✓	✓	✓	✓	67.5	39.8	83.7
	✓	✓	✓	66.8	39.3	83.3
✓		✓	✓	66.7	38.7	82.8
✓	✓		✓	66.1	38.6	82.4
✓	✓	✓		65.3	37.6	82.2

Design Choices for Training the Student: In Table 4, we report the results when training the Student network with different training set splits. The baseline Student network, trained with Cityscapes train-fine, attains the performance of 63.1% PQ, 35.2% AP, and 80.1% mIOU. Using the pseudo-labeled train-sequence, the performance is improved by 2.2% PQ, 2.4% AP, and 2.1% mIOU. Mixing human-labeled train-fine and pseudo-labeled train-sequence slightly degrades the performance. We think it is because of the inconsistent annotations between human-labeled and pseudo-labeled images, since train-fine is a subset of train-sequence. Finally, adding more pseudo-labeled images (train-sequence and train-extra) improves the result to 66.9% PQ, 40.2% AP, and 84.2% mIOU.

Table 4. Design choices for training the Student. We experiment with different training splits for training the Student. Results presented on Cityscapes validation set.

Training set for student			Validation set results		
Train-fine	Train-sequence	Train-extra	PQ (%)	AP (%)	mIOU (%)
✓			63.1	35.2	80.1
	✓		65.3	37.6	82.2
✓	✓		65.2	37.3	82.0
	✓	✓	66.9	40.2	84.2

Vary Human-Labeled Images and Fix Pseudo-labeled Images: In Fig. 5, we explore the semi-supervised setting with different amounts of human-labeled images but fixed amount of pseudo-labeled images. In particular, the Teacher network has *only* been trained with different numbers of Cityscapes train-fine images (*i.e.*, no other human-labeled images, such as Mapillary Vistas). The generated pseudo-labels (on Cityscapes train-sequence and train-extra) are used to train another Student network. Both Teacher and Student networks employ the Xception-71 as backbone. For comparison, we also show the performance of the supervised setting with the same amount of human-labeled images. As shown in the figure, we observe (1) the semi-supervised learning setting consistently improves over the fully supervised setting in all three metrics (PQ, AP, and mIOU) as more human-labeled images are exploited, (2) when using only 40% of the human-labeled images, our semi-supervised learning method could reap 98.9%, 97.2%, and 98.6% performance from its fully supervised counterparts in PQ, AP, and mIOU, respectively, and (3) when using 100% of the human-labeled images, our semi-supervised learning method attains 65.2% PQ, 38.6% AP, and 82% mIOU, comparable to the fully supervised counterpart with a Mapillary Vistas pretrained checkpoint (65.3% PQ, 38.8% AP, and 82.5% mIOU in [15]).

Fix Human-Labeled Images and Vary Pseudo-labeled Images: In Fig. 6, we explore the semi-supervised setting with different amounts of pseudo-labeled

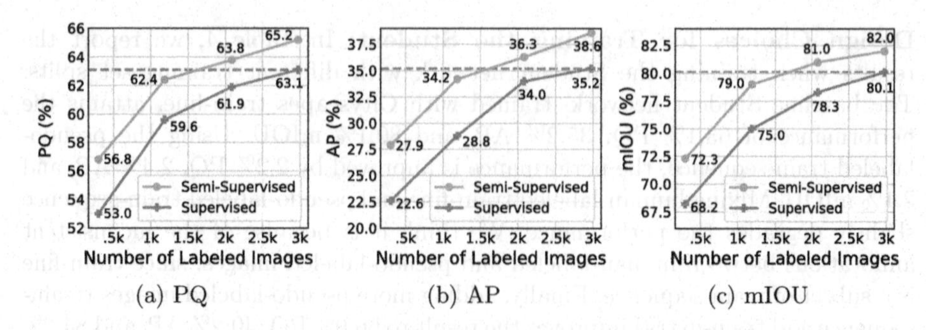

(a) PQ (b) AP (c) mIOU

Fig. 5. Semi-supervised learning with a fraction of the original labels may match supervised segmentation performance. Results presented on Cityscapes validation data. We vary the numbers of human-annotated images from train-fine set. With only 40% of labeled data, our semi-supervised learning method attains 98.9% PQ, 97.2% AP, and 98.6% mIOU of the performance of their fully supervised counterparts.

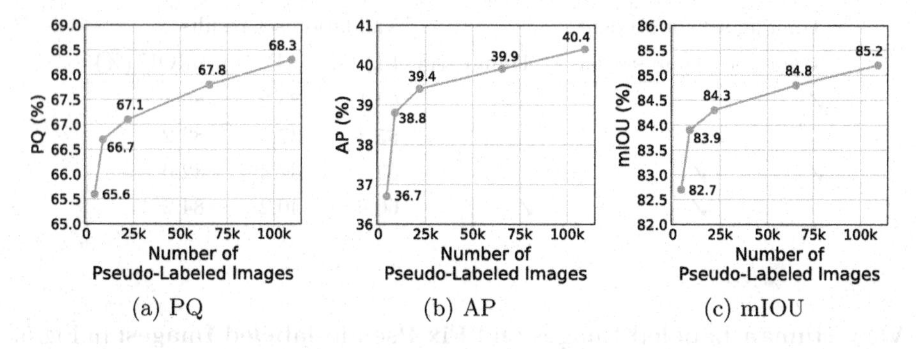

(a) PQ (b) AP (c) mIOU

Fig. 6. Increasing the amount of unlabeled data improves segmentation performance for PQ, AP, and mIOU. Results presented on Cityscapes validation data.

images. In particular, the Teacher network will generate different numbers of pseudo-labeled images for training the Student network. As shown in the figure, we observe consistent improvement in all three metrics when more and more pseudo-labeled images are included in the training.

Training Method: In Table 5, we experiment with the effect of different training methods: supervised, semi-supervised, and iterative semi-supervised learning. We employ our most powerful backbone, WR-41, attempting to push the envelope of performance. We observe a significant improvement of semi-supervised learning over supervised learning by 5.1% PQ, 4.4% AP, and 5.2% mIOU, mostly because of the small Cityscapes dataset. Adopting the iterative semi-supervised learning further improves the performance by 1.3% PQ, 1% AP, and 0.2% mIOU. We think there is more room for improving PQ and AP, since the mIOU result is starting to be saturated, as demonstrated in the public leader-board (*i.e.*, differences between top-performing models are about 0.1%).

Table 5. Comparisons with different training methods. Results presented on Cityscapes validation set. The proposed WR-41 is used as the network backbone. Semi-supervised learning (*i.e.*, only iterate once in our framework) significantly improves the performance, while iterative semi-supervised learning further improves the results.

Training method	PQ (%)	AP (%)	mIOU (%)
Supervised	64.0	38.0	80.7
Semi-supervised	69.1	42.4	85.9
iterative semi-supervised	70.4	43.4	86.1

Transfer Learning from Cityscapes to Mapillary Vistas: The transfer learning from the large-scale Mapillary Vistas to Cityscapes has been shown to be effective in the literature [15,61,95] and in our work as well, since both datasets contain street-view images. In Table 6, we experiment with the other direction of transfer learning from Cityscapes to Mapillary Vistas. The baseline model with WR-41 backbone, pretrained only on ImageNet [69], attains the performance of 37.1% PQ, 16.7% AP, and 56.2% mIOU. If we pretrain the model on the original Cityscapes trainval-fine set, we observe a slight degradation. Interestingly, when we further pretrain the model on the generated pseudo-labels, we observe a small amount of 0.7% improvement in PQ. We think the improvement gained from Cityscapes pretraining is marginal, mainly because the Cityscapes images are mostly taken in Germany, while Mapillary Vistas contains more diverse images.

Table 6. Transfer learning from Cityscapes to Mapillary Vistas. Results presented on Mapillary Vistas validation data. We experiment with the network pretrained on ImageNet [69], Cityscapes [17] labeled data (trainval-fine set), and Cityscapes pseudo-labeled data (trainval-sequence, and train-extra).

Training Set		Val Set		
Labeled	Pseudo-Labeled	PQ (%)	AP (%)	mIOU (%)
		37.1	16.7	56.2
✓		36.9	16.5	55.3
✓	✓	37.8	17.0	56.2

4.3 Modified Wide ResNet-38: WR-41

In this subsection, we report the experimental results with our modified Wide ResNet-38 [83,90], called WR-41, on both ImageNet [69] and Cityscapes [17].

ImageNet-1K Val Set: In Table 7, we report the results on the ImageNet-1K validation set. As shown in the table, our TensorFlow re-implementation of wide ResNet-38 (WR-38), proposed in [83], attains 20.36% Top-1 error, which is slightly worse than the one reported in the original paper. We think there are some differences between the deep learning libraries. Note however that our main focus is on the segmentation results, while ImageNet is only used for pretraining. Our proposed WR-41 achieves a slightly better performance. Employing the drop path [32] with a constant survival probability 0.8 improves the performance.

Table 7. *Single-model* error rates on ImageNet-1K validation set.

Backbone	Drop-Path	Top-1 Error	Top-5 Error
WR-38 [83] (our TF imp.)		20.36%	5.11%
WR-38 [83] (our TF imp.)	✓	19.89%	4.98%
WR-41		20.08%	4.93%
WR-41	✓	19.41%	4.68%

Table 8. Adopting Panoptic-DeepLab with our WR-41 achieves better segmentation accuracy on Cityscapes val set with fewer model parameters and fewer M-Adds than with our TensorFlow re-implemented WR-38.

Backbone	Drop-Path	Multi-Grid	Params (M)	M-Adds (B)	PQ (%)	AP (%)	mIOU (%)
WR-38 [83] (our TF imp.)			173.75	3486.32	62.4	35.3	79.1
WR-38 [83] (our TF imp.)	✓		173.75	3486.32	62.6	36.7	79.5
WR-38 [83] (our TF imp.)	✓	✓	173.75	3493.80	63.1	37.4	80.1
WR-41			147.27	3238.81	63.0	35.5	80.0
WR-41	✓		147.27	3238.81	63.6	36.8	80.4
WR-41	✓	✓	147.27	3246.40	64.0	38.0	80.7

Cityscapes Val Set: In Table 8, we report the Cityscapes validation set results when using Panoptic-DeepLab [15] with WR-38 (our TensorFlow re-implementation) and WR-41 as backbones. As shown in the table, we observe (1) using drop path [32] (constant survival probability 0.8) consistently improves the performance in both backbones, (2) the performance could be further improved by adopting the multi-grid scheme proposed in [12] (where the unit rates in the last two or three residual blocks are set to (1, 2) or (1, 2, 4) for WR-38 and WR-41, respectively), (3) using WR-41 as backbone slightly improves over WR-38, and (4) Panoptic-DeepLab with WR-41 as backbone is slightly faster (and with slightly fewer parameters) than with WR-38 because the ASPP module [11] is added on the last feature map with 2048 channels (instead of 4096 channels). Additionally, the GPU inference times (Tesla V100-SXM2) on a 1025×2049 input for WR-38 and WR-41 are 437.9 ms and 396.5 ms, respectively.

In Table 9, we report the effect of using test-time augmentation (*i.e.*, multi-scale inputs and left-right flips) and pretraining on Mapillary Vistas, when using

Table 9. Effect of test-time augmentation on Cityscapes val set. MV: Mapillary Vistas pretrained. **Flip**: Left-right flips. **MS:** Multi-scale inputs.

Method	MV	Flip	MS	PQ (%)	AP (%)	mIOU (%)
Panoptic-DeepLab (WR-41)				64.0	38.0	80.7
Panoptic-DeepLab (WR-41)		✓		64.5	39.3	81.2
Panoptic-DeepLab (WR-41)		✓	✓	65.0	40.7	81.4
Panoptic-DeepLab (X-71) [15]		✓	✓	64.1	38.5	81.5
Panoptic-DeepLab (WR-41)	✓			66.5	41.5	83.4
Panoptic-DeepLab (WR-41)	✓	✓		66.7	41.8	83.7
Panoptic-DeepLab (WR-41)	✓	✓	✓	67.3	43.4	83.8
Panoptic-DeepLab (X-71) [15]	✓	✓	✓	67.0	42.5	83.1

Table 10. Panoptic-DeepLab with proposed WR-41 backbone on Cityscapes test set. MV: Mapillary Vistas pretrained.

Method	MV	PQ (%)	AP (%)	mIOU (%)
Panoptic-DeepLab (X-71) [15]		62.3	34.6	79.4
Panoptic-DeepLab (WR-41) (ours)		63.7	36.5	81.5
Panoptic-DeepLab (X-71) [15]	✓	65.5	39.0	84.2
Panoptic-DeepLab (WR-41) (ours)	✓	66.5	40.6	84.5

Panoptic-DeepLab with WR-41 as network backbone. The performance consistently improved with test-time augmentation and pretraining on Mapillary Vistas. Additionally, adopting Panoptic-DeepLab with WR-41 slightly outperforms Panoptic-DeepLab with X-71 as reported in [15].

Cityscapes Test Set: In Table 10, we report the Cityscapes test set results when using Panoptic-DeepLab [15] with our modified WR-41. Without extra data, our Panoptic-DeepLab (WR-41) outperforms Panoptic-DeepLab (X-71) by 1.4% PQ, 1.9% AP, and 2.1% mIOU. With Mapillary Vistas pretraining, our Panoptic-DeepLab (WR-41) outperforms Panoptic-DeepLab (X-71) by 1.0% PQ and 1.6% AP, and 0.3% mIOU.

5 Conclusion

In this work, we have described an iterative semi-supervised learning method that significantly improves the performance of urban scene segmentation on Cityscapes, simultaneously tackling semantic, instance, and panoptic segmentation. This semi-supervised learning procedure effectively harnesses both unlabeled video frames and extra unlabeled images to improve the predictive performance of the model without the creation of additional architectures and learned modules. Namely, pseudo-labeled data garnered through a simple data augmentation (*i.e.*, multi-scale inputs and left-right flips) suffices to boost performance on supervised learning tasks. As a result, our model sets the new state-of-art performance at all three Cityscapes benchmarks without the need to fine-tune or any special design on each task. We hope our simple yet effective learning scheme could establish a baseline procedure to harness the abundant unlabeled video sequences and extra images for computer vision tasks.

Acknowledgments. We would like to thank the support from Google Mobile Vision and Brain.

References

1. Abadi, M., et al.: Tensorflow: a system for large-scale machine learning. In: Proceedings of the 12th USENIX Conference on Operating Systems Design and Implementation (2016)
2. Abu-El-Haija, S., et al.: YouTube-8M: a large-scale video classification benchmark. arXiv:1609.08675 (2016)
3. Arazo, E., Ortego, D., Albert, P., O'Connor, N.E., McGuinness, K.: Pseudo-labeling and confirmation bias in deep semi-supervised learning. arXiv:1908.02983 (2019)
4. Badrinarayanan, V., Galasso, F., Cipolla, R.: Label propagation in video sequences. In: CVPR (2010)
5. Bell, S., Upchurch, P., Snavely, N., Bala, K.: OpenSurfaces: a richly annotated catalog of surface appearance. ACM Trans. Graph. **32**, 1–17 (2013)
6. Budvytis, I., Sauer, P., Roddick, T., Breen, K., Cipolla, R.: Large scale labelled video data augmentation for semantic segmentation in driving scenarios. In: ICCV Workshop (2017)
7. Caba Heilbron, F., Escorcia, V., Ghanem, B., Carlos Niebles, J.: ActivityNet: a large-scale video benchmark for human activity understanding. In: CVPR (2015)
8. Castrejon, L., Kundu, K., Urtasun, R., Fidler, S.: Annotating object instances with a polygon-RNN. In: CVPR (2017)
9. Chen, L.C., et al.: Searching for efficient multi-scale architectures for dense image prediction. In: NeurIPS (2018)
10. Chen, L.C., Papandreou, G., Kokkinos, I., Murphy, K., Yuille, A.L.: Semantic image segmentation with deep convolutional nets and fully connected CRFs. In: ICLR (2015)
11. Chen, L.C., Papandreou, G., Kokkinos, I., Murphy, K., Yuille, A.L.: DeepLab: semantic image segmentation with deep convolutional nets, atrous convolution, and fully connected CRFs. In: IEEE TPAMI (2017)
12. Chen, L.C., Papandreou, G., Schroff, F., Adam, H.: Rethinking atrous convolution for semantic image segmentation. arXiv:1706.05587 (2017)
13. Chen, L.-C., Zhu, Y., Papandreou, G., Schroff, F., Adam, H.: Encoder-decoder with atrous separable convolution for semantic image segmentation. In: Ferrari, V., Hebert, M., Sminchisescu, C., Weiss, Y. (eds.) ECCV 2018. LNCS, vol. 11211, pp. 833–851. Springer, Cham (2018). https://doi.org/10.1007/978-3-030-01234-2_49
14. Cheng, B., et al.: Panoptic-DeepLab. In: ICCV COCO + Mapillary Joint Recognition Challenge Workshop (2019)
15. Cheng, B., et al.: Panoptic-DeepLab: a simple, strong, and fast baseline for bottom-up panoptic segmentation. In: CVPR (2020)

16. Chollet, F.: Xception: deep learning with depthwise separable convolutions. In: CVPR (2017)
17. Cordts, M., et al.: The cityscapes dataset for semantic urban scene understanding. In: CVPR (2016)
18. Cubuk, E.D., Zoph, B., Shlens, J., Le, Q.V.: Randaugment: practical data augmentation with no separate search. arXiv:1909.13719 (2019)
19. Dai, J., He, K., Sun, J.: Boxsup: exploiting bounding boxes to supervise convolutional networks for semantic segmentation. In: ICCV (2015)
20. Doersch, C., Gupta, A., Efros, A.A.: Unsupervised visual representation learning by context prediction. In: ICCV (2015)
21. Everingham, M., Van Gool, L., Williams, C.K., Winn, J., Zisserman, A.: The pascal visual object classes (VOC) challenge. IJCV **88**(2), 303–338 (2010)
22. Forsyth, D.A., Ponce, J.: Computer Vision: A Modern Approach. Prentice Hall Professional Technical Reference (2002)
23. Gadde, R., Jampani, V., Gehler, P.V.: Semantic video CNNs through representation warping. In: ICCV (2017)
24. Geiger, A., Lenz, P., Stiller, C., Urtasun, R.: Vision meets robotics: the KITTI dataset. Int. J. Robot. Res. **32**, 1231–1237 (2013)
25. Gidaris, S., Singh, P., Komodakis, N.: Unsupervised representation learning by predicting image rotations. In: CVPR (2018)
26. Girshick, R., Donahue, J., Darrell, T., Malik, J.: Rich feature hierarchies for accurate object detection and semantic segmentation. In: CVPR (2014)
27. Hariharan, B., Arbelaez, P., Bourdev, L., Maji, S., Malik, J.: Semantic contours from inverse detectors. In: ICCV (2011)
28. He, K., Zhang, X., Ren, S., Sun, J.: Deep residual learning for image recognition. In: CVPR (2016)
29. Hénaff, O.J., Razavi, A., Doersch, C., Eslami, S., Oord, A.v.d.: Data-efficient image recognition with contrastive predictive coding. arXiv:1905.09272 (2019)
30. Hinton, G., Vinyals, O., Dean, J.: Distilling the knowledge in a neural network. arXiv preprint arXiv:1503.02531 (2015)
31. Hong, S., Noh, H., Han, B.: Decoupled deep neural network for semi-supervised semantic segmentation. In: NeurIPS (2015)
32. Huang, G., Sun, Yu., Liu, Z., Sedra, D., Weinberger, K.Q.: Deep networks with stochastic depth. In: Leibe, B., Matas, J., Sebe, N., Welling, M. (eds.) ECCV 2016. LNCS, vol. 9908, pp. 646–661. Springer, Cham (2016). https://doi.org/10.1007/978-3-319-46493-0_39
33. Ioffe, S., Szegedy, C.: Batch normalization: accelerating deep network training by reducing internal covariate shift. In: ICML (2015)
34. Iscen, A., Tolias, G., Avrithis, Y., Chum, O.: Label propagation for deep semi-supervised learning. In: CVPR (2019)
35. Khoreva, A., Benenson, R., Hosang, J., Hein, M., Schiele, B.: Simple does it: weakly supervised instance and semantic segmentation. In: CVPR (2017)
36. Kingma, D.P., Ba, J.: Adam: a method for stochastic optimization. In: ICLR (2015)
37. Kirillov, A., Girshick, R., He, K., Dollár, P.: Panoptic feature pyramid networks. In: CVPR (2019)
38. Kirillov, A., He, K., Girshick, R., Rother, C., Dollár, P.: Panoptic segmentation. In: CVPR (2019)
39. Kornblith, S., Shlens, J., Le, Q.V.: Do better imagenet models transfer better? In: CVPR (2019)
40. Krizhevsky, A., Sutskever, I., Hinton, G.E.: Imagenet classification with deep convolutional neural networks. In: NeurIPS (2012)

41. Lake, B.M., Ullman, T.D., Tenenbaum, J.B., Gershman, S.J.: Building machines that learn and think like people. Behav. Brain Sci. (2017)
42. Lee, D.H.: Pseudo-label: the simple and efficient semi-supervised learning method for deep neural networks. In: ICML Workshop (2013)
43. Li, J., Raventos, A., Bhargava, A., Tagawa, T., Gaidon, A.: Learning to fuse things and stuff. arXiv:1812.01192 (2018)
44. Li, L.J., Fei-Fei, L.: Optimol: automatic online picture collection via incremental model learning. IJCV **88**, 147–168 (2010). https://doi.org/10.1007/s11263-009-0265-6
45. Li, Q., Arnab, A., Torr, P.H.S.: Weakly- and semi-supervised panoptic segmentation. In: Ferrari, V., Hebert, M., Sminchisescu, C., Weiss, Y. (eds.) ECCV 2018. LNCS, vol. 11219, pp. 106–124. Springer, Cham (2018). https://doi.org/10.1007/978-3-030-01267-0_7
46. Li, Q., Qi, X., Torr, P.H.: Unifying training and inference for panoptic segmentation. arXiv:2001.04982 (2020)
47. Li, Y., Qi, H., Dai, J., Ji, X., Wei, Y.: Fully convolutional instance-aware semantic segmentation. In: CVPR (2017)
48. Liang, J., Homayounfar, N., Ma, W.C., Xiong, Y., Hu, R., Urtasun, R.: Polytransform: deep polygon transformer for instance segmentation. arXiv:1912.02801 (2019)
49. Lin, T.-Y., et al.: Microsoft COCO: common objects in context. In: Fleet, D., Pajdla, T., Schiele, B., Tuytelaars, T. (eds.) ECCV 2014. LNCS, vol. 8693, pp. 740–755. Springer, Cham (2014). https://doi.org/10.1007/978-3-319-10602-1_48
50. Liu, C., et al.: Auto-DeepLab: hierarchical neural architecture search for semantic image segmentation. In: CVPR (2019)
51. Liu, S., Qi, L., Qin, H., Shi, J., Jia, J.: Path aggregation network for instance segmentation. In: CVPR (2018)
52. Liu, W., Rabinovich, A., Berg, A.C.: Parsenet: looking wider to see better. arXiv:1506.04579 (2015)
53. Luc, P., Neverova, N., Couprie, C., Verbeek, J., LeCun, Y.: Predicting deeper into the future of semantic segmentation. In: ICCV (2017)
54. Mustikovela, S.K., Yang, M.Y., Rother, C.: Can ground truth label propagation from video help semantic segmentation? In: Hua, G., Jégou, H. (eds.) ECCV 2016. LNCS, vol. 9915, pp. 804–820. Springer, Cham (2016). https://doi.org/10.1007/978-3-319-49409-8_66
55. Neuhold, G., Ollmann, T., Bulò, S.R., Kontschieder, P.: The mapillary vistas dataset for semantic understanding of street scenes. In: ICCV (2017)
56. Nilsson, D., Sminchisescu, C.: Semantic video segmentation by gated recurrent flow propagation. In: CVPR (2018)
57. Papandreou, G., Chen, L.C., Murphy, K.P., Yuille, A.L.: Weakly-and semi-supervised learning of a deep convolutional network for semantic image segmentation. In: ICCV (2015)
58. Papandreou, G., Zhu, T., Chen, L.-C., Gidaris, S., Tompson, J., Murphy, K.: PersonLab: person pose estimation and instance segmentation with a bottom-up, part-based, geometric embedding model. In: Ferrari, V., Hebert, M., Sminchisescu, C., Weiss, Y. (eds.) Computer Vision – ECCV 2018. LNCS, vol. 11218, pp. 282–299. Springer, Cham (2018). https://doi.org/10.1007/978-3-030-01264-9_17
59. Pathak, D., Krahenbuhl, P., Darrell, T.: Constrained convolutional neural networks for weakly supervised segmentation. In: ICCV (2015)
60. Pinheiro, P.O., Collobert, R., Dollár, P.: Learning to segment object candidates. In: NeurIPS (2015)

61. Porzi, L., Bulò, S.R., Colovic, A., Kontschieder, P.: Seamless scene segmentation. In: CVPR (2019)
62. Porzi, L., Hofinger, M., Ruiz, I., Serrat, J., Bulo, S.R., Kontschieder, P.: Learning multi-object tracking and segmentation from automatic annotations. In: CVPR (2020)
63. Qi, H., et al.: Deformable convolutional networks - COCO detection and segmentation challenge 2017 entry. In: ICCV COCO Challenge Workshop (2017)
64. Radosavovic, I., Dollár, P., Girshick, R., Gkioxari, G., He, K.: Data distillation: towards omni-supervised learning. In: CVPR (2018)
65. Real, E., Shlens, J., Mazzocchi, S., Pan, X., Vanhoucke, V.: YouTube-BoundingBoxes: a large high-precision human-annotated data set for object detection in video. In: CVPR (2017)
66. Ren, S., He, K., Girshick, R., Sun, J.: Faster R-CNN: towards real-time object detection with region proposal networks. In: NeurIPS (2015)
67. Riloff, E., Wiebe, J.: Learning extraction patterns for subjective expressions. In: EMNLP (2003)
68. Rosenberg, C., Hebert, M., Schneiderman, H.: Semi-supervised self-training of object detection models. WACV/MOTION (2005)
69. Russakovsky, O., et al.: ImageNet large scale visual recognition challenge. IJCV **115**, 211–252 (2015). https://doi.org/10.1007/s11263-015-0816-y
70. Russell, B.C., Torralba, A., Murphy, K.P., Freeman, W.T.: LabelMe: a database and web-based tool for image annotation. IJCV **77**, 157–173 (2008). https://doi.org/10.1007/s11263-007-0090-8
71. Scudder, H.: Probability of error of some adaptive pattern-recognition machines. IEEE Trans. Inf. Theor. **11**, 363–371 (1965)
72. Shi, W., Gong, Y., Ding, C., Ma, Z., Tao, X., Zheng, N.: Transductive semi-supervised deep learning using min-max features. In: Ferrari, V., Hebert, M., Sminchisescu, C., Weiss, Y. (eds.) ECCV 2018. LNCS, vol. 11209, pp. 311–327. Springer, Cham (2018). https://doi.org/10.1007/978-3-030-01228-1_19
73. Souly, N., Spampinato, C., Shah, M.: Semi supervised semantic segmentation using generative adversarial network. In: ICCV (2017)
74. Sun, C., Shrivastava, A., Singh, S., Gupta, A.: Revisiting unreasonable effectiveness of data in deep learning era. In: ICCV (2017)
75. Sun, P., et al.: Scalability in perception for autonomous driving: Waymo open dataset. arXiv:1912.04838 (2019)
76. Tang, Y., Wang, J., Gao, B., Dellandréa, E., Gaizauskas, R., Chen, L.: Large scale semi-supervised object detection using visual and semantic knowledge transfer. In: CVPR (2016)
77. Voigtlaender, P., et al.: Mots: multi-object tracking and segmentation. In: CVPR (2019)
78. Wang, H., Zhu, Y., Green, B., Adam, H., Yuille, A., Chen, L.C.: Axial-DeepLab: stand-alone axial-attention for panoptic segmentation. arXiv:2003.07853 (2020)
79. Wang, P., et al.: Understanding convolution for semantic segmentation. arXiv:1702.08502 (2017)
80. Wei, Y., et al.: STC: a simple to complex framework for weakly-supervised semantic segmentation. In: IEEE TPAMI (2016)
81. Wei, Y., Xiao, H., Shi, H., Jie, Z., Feng, J., Huang, T.S.: Revisiting dilated convolution: a simple approach for weakly-and semi-supervised semantic segmentation. In: CVPR (2018)

82. Wu, J., Yildirim, I., Lim, J.J., Freeman, B., Tenenbaum, J.: Galileo: perceiving physical object properties by integrating a physics engine with deep learning. In: NeurIPS (2015)
83. Wu, Z., Shen, C., Van Den Hengel, A.: Wider or deeper: revisiting the ResNet model for visual recognition. Pattern Recogn. **90**, 119–133 (2019)
84. Xie, Q., Hovy, E., Luong, M.T., Le, Q.V.: Self-training with noisy student improves imagenet classification. arXiv:1911.04252 (2019)
85. Xiong, Y., Liao, R., Zhao, H., Hu, R., Bai, M., Yumer, E., Urtasun, R.: UPSNet: a unified panoptic segmentation network. In: CVPR (2019)
86. Yalniz, I.Z., J'egou, H., Chen, K., Paluri, M., Mahajan, D.: Billion-scale semi-supervised learning for image classification. arXiv:1905.00546 (2019)
87. Yang, T.J., et al.: DeeperLab: single-shot image parser. arXiv:1902.05093 (2019)
88. Yarowsky, D.: Unsupervised word sense disambiguation rivaling supervised methods. In: ACL (1995)
89. Yuan, Y., Chen, X., Wang, J.: Object-contextual representations for semantic segmentation. arXiv:1909.11065 (2019)
90. Zagoruyko, S., Komodakis, N.: Wide residual networks. In: BMVC (2016)
91. Zhai, X., Oliver, A., Kolesnikov, A., Beyer, L.: S4l: self-supervised semi-supervised learning. In: ICCV (2019)
92. Zhao, H., Shi, J., Qi, X., Wang, X., Jia, J.: Pyramid scene parsing network. In: CVPR (2017)
93. Zheng, Z., Zheng, L., Yang, Y.: Unlabeled samples generated by GAN improve the person re-identification baseline in vitro. In: ICCV (2017)
94. Zhu, X., Xiong, Y., Dai, J., Yuan, L., Wei, Y.: Deep feature flow for video recognition. In: CVPR (2017)
95. Zhu, Y., et al.: Improving semantic segmentation via video propagation and label relaxation. In: CVPR (2019)
96. Zhu, Y., et al.: Improving semantic segmentation via self-training. arXiv:2004.14960 (2020)

Spatially Aware Multimodal Transformers for TextVQA

Yash Kant[1], Dhruv Batra[1,3], Peter Anderson[1,2P], Alexander Schwing[4], Devi Parikh[1,3], Jiasen Lu[1], and Harsh Agrawal[1(✉)]

[1] Georgia Institute of Technology, Atlanta, Georgia
hagrawal9@gatech.edu
[2] Google, Columbus, USA
[3] Facebook AI Research (FAIR), New York City, USA
[4] University of Illinois, Urbana, Champaign, USA

Abstract. Textual cues are essential for everyday tasks like buying groceries and using public transport. To develop this assistive technology, we study the TextVQA task, i.e., reasoning about text in images to answer a question. Existing approaches are limited in their use of spatial relations and rely on fully-connected transformer-based architectures to implicitly learn the spatial structure of a scene. In contrast, we propose a novel spatially aware self-attention layer such that each visual entity only looks at neighboring entities defined by a spatial graph. Further, each head in our multi-head self-attention layer focuses on a different subset of relations. Our approach has two advantages: (1) each head considers local context instead of dispersing the attention amongst all visual entities; (2) we avoid learning redundant features. We show that our model improves the absolute accuracy of current state-of-the-art methods on TextVQA by 2.2% overall over an improved baseline, and 4.62% on questions that involve spatial reasoning and can be answered correctly using OCR tokens. Similarly on ST-VQA, we improve the absolute accuracy by 4.2%. We further show that spatially aware self-attention improves visual grounding.

Keywords: VQA · TextVQA · Self-attention

1 Introduction

The promise of assisting visually-impaired users gives us a compelling reason to study Visual Question Answering (VQA) [3] tasks. A dominant class of questions (~20%) asked by visually-impaired users on images of their surroundings involves

J. Lu—Work was partially done as a member of PRIOR @ Allen Institute for AI.

Electronic supplementary material The online version of this chapter (https://doi.org/10.1007/978-3-030-58545-7_41) contains supplementary material, which is available to authorized users.

© Springer Nature Switzerland AG 2020
A. Vedaldi et al. (Eds.): ECCV 2020, LNCS 12354, pp. 715–732, 2020.
https://doi.org/10.1007/978-3-030-58545-7_41

reading text in the image [5]. Naturally, the ability to reason about text in the image to answer questions such as "Is this medicine going to expire?", "Where is this bus going?" is of paramount importance for these systems. To benchmark a model's capability to reason about text in the image, new datasets [7,31,36] have been introduced for the task of Text Visual Question Answering (TextVQA).

Answering questions involving text in an image often requires reasoning about the relative spatial positions of objects and text. For instance, many questions such as "What is written on the player's jersey?" or "What is the next destination for the bus?" ask about text associated with a particular visual object. Similarly, the question asked in Fig. 1, "What sponsor is to the right of the players?", explicitly asks the answerer to look to the `right` of the players. Unsurprisingly, ∼13% of the questions in the TextVQA dataset use one or more spatial prepositions[1].

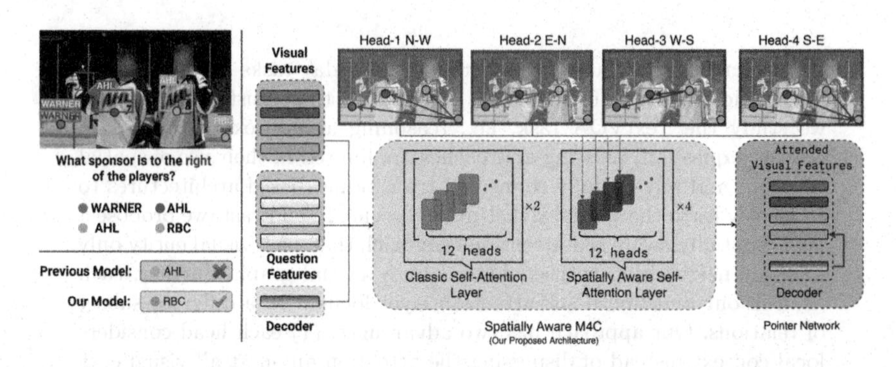

Fig. 1. (a) On questions that require spatial reasoning (∼13% of the TextVQA dataset), compared to previous approaches [14,36], our model can reason about spatial relations between visual entities to answer questions correctly. (b) We construct a spatial-graph that encodes different spatial relationships between a pair of visual entities and use it to guide the self-attention layers present in multi-modal transformer architectures.

Existing methods for TextVQA reason jointly over 3 modalities – the input question, the visual content and the text in the image. LoRRA [36] uses an off-the-shelf Optical Character Recognition (OCR) system [9] to detect OCR tokens and extends previous VQA models [2] to select single OCR tokens from the images as answers. The more recently proposed Multimodal Multi-Copy Mesh (M4C) model [14] captures intra- and inter-modality interactions over all the inputs – question words, visual objects and OCR tokens – by using a multi-modal transformer architecture that iteratively decodes the answer by choosing words from either the OCR tokens or some fixed vocabulary. The superior performance of M4C is attributed to the use of multi-head self-attention layers [42] which has become the defacto standard for modeling vision and language tasks [10,26,27,37,41].

[1] We use several prepositions such as 'right', 'top', 'contains', *etc.* to filter questions that involve spatial reasoning.

While these approaches take advantage of detected text, they are limited in how they use spatial relations. For instance, LoRRA [36] does not use any location information while M4C [14] merely encodes the absolute location of objects and text as input to the model. By default, self-attention layers are fully-connected, dispersing attention across the entire global context and disregarding the importance of the local context around a certain object or text. As a result, in existing models the onus is on them to *implicitly* learn to reason about the relative spatial relations between objects and text. In contrast, in the Natural Language Processing community, it has proven beneficial to *explicitly* encode semantic structure between input tokens [39,45,47]. Moreover, while multiple independent heads in self-attention layers model different context, each head independently looks at the same global context and learns redundant features [27] that can be pruned away without substantially harming a model's performance [29,44].

We address the above limitations by proposing a novel spatially aware self-attention layer for multimodal transformers. First, we follow [22,48] to build a spatial graph to represent relative spatial relations between all visual entities, i.e., all objects and OCR tokens. We then use this spatial graph to guide the self-attention layers in the multimodal transformer. We modify the attention computation in each head such that each entity attends to just the neighboring entities as defined by the spatial graph, and we restrict each head to only look at a subset of relations which prevents learning of redundant features.

Empirically, we evaluate the efficacy of our proposed approach on the challenging TextVQA [36] and Scene-Text VQA (ST-VQA) [7] datasets. We first improve the absolute accuracy of the baseline M4C model on TextVQA by 3.4% with improved features and hyperparameter optimization. We then show that replacing the fully-connected self-attention layers in the M4C model with our spatially aware self-attention layers improves absolute accuracy by a further 2.2% (or 4.62% for the ~14% of TextVQA questions that include spatial prepositions and has a majority answer in OCR tokens). On ST-VQA our final model achieves an absolute 4.2% improvement in Average Normalized Levenshtein Similarity (ANLS). Finally, we show that our model is more visually grounded as it picks the correct answer from the list of OCR tokens 8.8% more often than M4C.

2 Related Work

Models for TextVQA: Several datasets and methods [7,14,31,36] have been proposed for the TextVQA task – i.e., answering questions which require models to explicitly reason about text present in the image. LoRRA [36] extends Pythia [15] with an OCR attention branch to reason over a combined list of answers from a static vocabulary and detected OCR tokens. Several other models have taken similar approaches to augmenting existing VQA models with OCR inputs [6,7,31]. Building on the success of transformers [42] and BERT [11], the Multimodal Multi-Copy Mesh (M4C) model [14] (which serves as our baseline) uses a multimodal transformer to jointly encode the question, image and text and employs an auto-regressive decoding mechanism to perform multi-step answer decoding. However, these methods are limited in how they leverage the relative spatial relations between visual entities such as objects and OCR tokens. Specifically, early models [6,7,31] proposed for the TextVQA task did not encode any explicit spatial information while M4C [14] simply adds a location embedding of the absolute location to the input feature. We improve the performance of these models by proposing a general framework to effectively utilize the relative spatial structure between visual entities within the transformer architecture.

Multimodal Representation Learning for Vision and Language: Recently, several general architectures for vision and language [10,14,21,23,26, 27,32,33,37,40,41,50] were proposed that reduce architectural differences across tasks. These models (including M4C) typically fuse vision and language modalities by applying either self-attention [4] or co-attention [28] mechanisms to capture intra- and inter-modality interactions. They achieve superior performance on many vision and language tasks due to their strong representation power and their ability to pre-train visual grounding in a self-supervised manner. Similar to M4C, these methods add a location embedding to their inputs, but do not explicitly encode relative spatial information (which is crucial for visual reasoning). Our work takes the first step towards modeling relative spatial locations within the multimodal transformer architecture.

Leveraging Explicit Relationships for Visual Reasoning: Prior work has used Graph Convolutional Nets (GCN) [18] and Graph Attention Networks (GAT) [43] to leverage explicit relations for image captioning [48] and VQA [22]. Both these methods construct a spatial and semantic graph to relate different objects. Although our relative spatial relations are inspired from [22,48], our encoding differs greatly. First, [22,48] looks at all the spatial relations in every attention head, whereas each self attention head in our model looks at different subset of the relations, i.e., each head is only responsible for a certain number of relations. This important distinction prevents spreading of attention over the entire global context and reduces redundancy amongst multiple heads.

Context Aware Transformers for Language Modeling: Related to the use of spatial structure for visual reasoning tasks, there has been a body of work on modeling the underlying structure in input sequences for language modeling tasks. Previous approaches have considered encoding the relative position difference between sentence tokens [35] as well as encoding the depth of each word in a parse tree and the distance between word pairs [45]. Other approaches learn to adapt the attention span for each attention head [39,47], rather than explicitly modeling context for attention. While these methods work well for sequential input like natural language sentences, they cannot be directly applied to our task since our visual representations are non-sequential.

3 Background: Multimodal Transformers

Following the success of transformer [42] and BERT [11] based architectures on language modeling and sequence-to-sequence tasks, multi-modal transformer-style models [10,14,21,23,26,27,32,33,37,40,41,50] have shown impressive results on several vision-and-language tasks. Instead of using a single input modality (i.e., text), multiple modalities are encoded as a sequence of input tokens and appended together to form a single input sequence. Additionally, a type embedding unique to each modality is added to distinguish amongst input token of different modalities.

The core building block of the transformer architecture is a self-attention layer followed by a feed-forward network. The self-attention layer aims at capturing the direct relationships between the different input tokens. In this section, we first briefly recap the attention computation in the multi-head self-attention layer of the transformer and highlight some issues with classical self-attention layers.

3.1 Self-attention Layer

A self-attention (SA) layer operates on an input sequence represented by N d_x-dimensional vectors $X = (\mathbf{x}_1, \ldots, \mathbf{x}_N) \in \mathbb{R}^{d_x \times N}$ and computes the attended sequence $\tilde{X} = (\tilde{\mathbf{x}}_1, \ldots, \tilde{\mathbf{x}}_N) \in \mathbb{R}^{d_x \times N}$. For this, self-attention employs h independent attention heads and applies the attention mechanism of Bahdanau *et al.* [4] to its own input. Each head in a self-attention layer transforms the input sequence X into query $Q^h = [\mathbf{q}_1^h, \ldots, \mathbf{q}_N^h] \in \mathbb{R}^{d_h \times N}$, key $K^h = [\mathbf{k}_1^h, \ldots, \mathbf{k}_N^h] \in \mathbb{R}^{d_h \times N}$, and value $V = [\mathbf{v}_1^h, \ldots, \mathbf{v}_N^h] \in \mathbb{R}^{d_h \times N}$ vectors via learnable linear projections parameterized by $W_Q^h, W_K^h, W_V^h \in \mathbb{R}^{d_x \times d_h}$:

$$(\mathbf{q}_i^h, \mathbf{k}_i^h, \mathbf{v}_i^h) = (\mathbf{x}_i W_Q^h, \mathbf{x}_i W_K^h, \mathbf{x}_i W_V^h) \quad \forall i \in [1, \ldots, N].$$

Generally, d_h is set to d_x/H. Each attended sequence element $\tilde{\mathbf{x}}_i^h$ is then computed via a weighted sum of value vectors, i.e.,

$$\tilde{\mathbf{x}}_i^h = \sum_{j=1}^{n} \alpha_{ij}^h \mathbf{v}_j^h. \tag{1}$$

The weight coefficient α_{ij}^h is computed via a Softmax over a compatibility function that compares the query vector $\mathbf{q}_i^h$ with key vectors of all the input tokens $\mathbf{k}_j^h$, $j \in [1, \ldots, N]$:

$$\alpha_{ij} = \text{Softmax}\left(\frac{\mathbf{q}_i^h (\mathbf{k}_j^h)^T}{\sqrt{d_h}}\right). \tag{2}$$

The computation in Eq. (1) and Eq. (2) can be more compactly written as:

$$\texttt{head}_h = \mathcal{A}^h(Q^h, K^h, V^h) = \text{Softmax}\left(\frac{Q^h (K^h)^T}{\sqrt{d_h}}\right) V^h \quad \forall h = [1, \ldots, H]. \tag{3}$$

The output of all heads are then concatenated followed by a linear transformation with weights $W^O \in \mathbb{R}^{(d_h \cdot H) \times d_x}$. Therefore, in the case of multi-head attention, we obtain the attended sequence $\tilde{X} = (\widetilde{x}_i, \dots, \widetilde{x}_N)$ from

$$\tilde{X} = \mathcal{A}(Q, K, V) = [\mathtt{head}_1, \dots, \mathtt{head}_H] \, W^O. \tag{4}$$

Application to Multi-modal Tasks: For multi-modal tasks, the self-attention is often modified to model cross-attention from one modality U_i to another modality U_j as $\mathcal{A}(Q_{U_i}, K_{U_j}, V_{U_j})$ or intra-modality attention $\mathcal{A}(Q_{U_i}, K_{U_i}, V_{U_i})$. Note, U_i, U_j are simply sets of indices which are used to construct sub-matrices. Some architectures like M4C [14] use the classical self-attention layer to model attention between tokens of all the modalities as $\mathcal{A}(Q_U, K_U, V_U)$ where $U = U_1 \cup U_2 \cup \dots \cup U_M$ is the union of all M input modalities.

3.2 Limitations

The aforementioned self-attention layer exposes two limitations: (1) self-attention layers model the global context by encoding relations between every single pair of input tokens. This disperses the attention across every input token and overlooks the importance of semantic structure in the sequence. For instance, in the case of language modeling, it has proven beneficial to capture local-context [47] or the hierarchical structure of the input sentence by encoding the depth of each word in a parsing tree [45], (2) multiple heads allow self-attention layers to jointly attend to different context in different heads. However, each head independently looks at the entire global information and there is no explicit mechanism to ensure that different attention heads capture different context. Indeed, it has been shown that the heads can be pruned away without substantially hurting a model's performance [29,44] and that different heads learn redundant features [27].

4 Approach

To address both limitations, we extend the self-attention layer to utilize a graph over the input tokens. Instead of looking at the entire global context, an entity attends to just the neighboring entities as defined by a relationship graph. Moreover, heads consider different types of relations which encodes different context and avoids learning redundant features. In what follows, we introduce the notation for input token representations. Next, we formally define the heterogeneous graph over tokens from multiple modalities which are connected by different edge types. Finally, we describe our approach to adapt the attention span of each head in the self-attention layer by utilizing this graph. While our framework is general and easily extensible to other tasks, we present our approach for the TextVQA task.

4.1 Graph over Input Tokens

Let us define a directed cyclic heterogeneous graph $\mathcal{G} = (X, \mathcal{E})$ where each node corresponds to an input token $\mathbf{x}_i \in X$. $\mathcal{E}$ is a set of all edges $e_{i \to j}, \forall \mathbf{x}_i, \mathbf{x}_j \in X$. Additionally, we define a mapping function $\Phi_x : X \to \mathcal{T}^x$ that maps a node $\mathbf{x}_i \in X$ to one of the modalities. Consequently the number of node types is equal to the number of input modalities, i.e., $|\mathcal{T}^x| = M$. We also define a mapping function $\Phi_e : \mathcal{E} \to \mathcal{T}^e$ that maps an edge $e_{i \to j} \in \mathcal{E}$ to a relationship type $t_l \in \mathcal{T}^e$.

We represent the question as a set of tokens, i.e., $X^{\text{ques}} = \{\mathbf{x} \in X : \Phi_x(\mathbf{x}) = \text{ques}\}$. The visual content in the image is represented via a list of object region features $X^{\text{obj}} = \{\mathbf{x} \in X : \Phi_x(\mathbf{x}) = \text{obj}\}$. Similarly, the list of OCR tokens present in the image is referred to as $X^{\text{ocr}} = \{\mathbf{x} \in X : \Phi_x(\mathbf{x}) = \text{ocr}\}$. Following M4C, the model decodes multi-word answer $Y^{\text{ans}} = (\mathbf{y}_1^{\text{ans}}, \ldots, \mathbf{y}_T^{\text{ans}})$ for T time-steps.

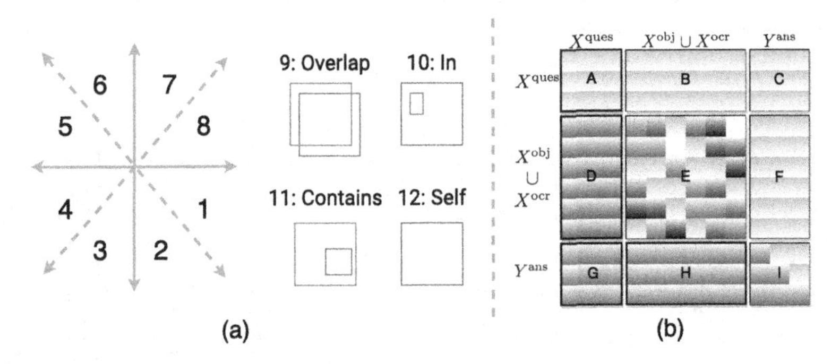

Fig. 2. (a) The spatial-relations graph encodes twelve type of relations between two object or OCR tokens $r_i, r_j \in \mathcal{R}$. (b) Denotes the attention mask between different data modalities. In our spatially aware self-attention layer, object and OCR tokens attend to each other based on a subset of spatial relations $\mathcal{T}^h \subseteq \mathcal{T}^{\text{spa}}$. They also attend to question tokens via t_{imp} relation. Any input token $x \in X$ do not attend to answer token $y^{\text{ans}} \in Y$ while y^{ans} can attend to tokens in X as well as previous answer tokens $y_{<t}^{\text{ans}}$.

Spatial Relationship Graph: Answering questions about text in the image involves reasoning about the spatial relations between various OCR tokens and objects present in the image. For instance, the question "What is the letter on the player's hat?" requires to first detect a hat in the image and then reason about the '`contains`' relationship between the letter and the player's hat.

To encode these spatial relationships between all the objects X^{obj} and OCR tokens X^{ocr} present in the image, i.e., all the regions $r \in \mathcal{R} = X^{\text{obj}} \cup X^{\text{ocr}}$, we construct a spatial graph $G_{\text{spa}} = (\mathcal{R}, \mathcal{E}_{\text{spa}})$ with nodes corresponding to the union of all objects and OCR tokens. The mapping function $\Phi_{spa} : \mathcal{E}_{\text{spa}} \to \mathcal{T}^{\text{spa}}$ assigns a spatial relationship $t_l \in \mathcal{T}^{\text{spa}}$ to an edge $e = (r_i, r_j) \in \mathcal{E}_{\text{spa}}$. The mapping function utilizes the rules introduced by Yao *et al.* [48] which we illustrate in Fig. 2(a). We use a total of twelve types of spatial relations (e.g.,

$\langle r_i - \texttt{contains} - r_j \rangle$, $\langle r_i - \texttt{is-inside} - r_j \rangle$ as well as a '`self-relation`'). Note that G_{spa} is symmetric, i.e., for every edge $e_{i \to j}$ there is a reverse edge $e_{j \to i}$.

Implicit Relationship between Objects, OCR and Question Tokens: For the TextVQA task, different types of spatial relations might be useful for different question types. For instance, a question asking about 'what is written on the player's jersey' might focus on the `contains` relationship, whereas a question asking about 'what sponsor is to the right of the player' might utilize the `right` relationship. Thus, to inject semantic information from the question into the object and OCR representation, we allow object and OCR tokens to attend to question tokens. In our general framework, we accomplish this via a bipartite graph $G_{\text{imp}}(\mathcal{R}, X^{\text{ques}}, \mathcal{E}_{\text{imp}})$ connecting all the object and OCR tokens $r_i \in \mathcal{R}$ to all question tokens $\mathbf{x}_j \in X^{\text{ques}}$ via an implicit edge $e_{i \to j}$ of type t_{imp}. Thus, by attending to question tokens, each object and OCR token learns to implicitly incorporate useful semantic information from the question into its representation.

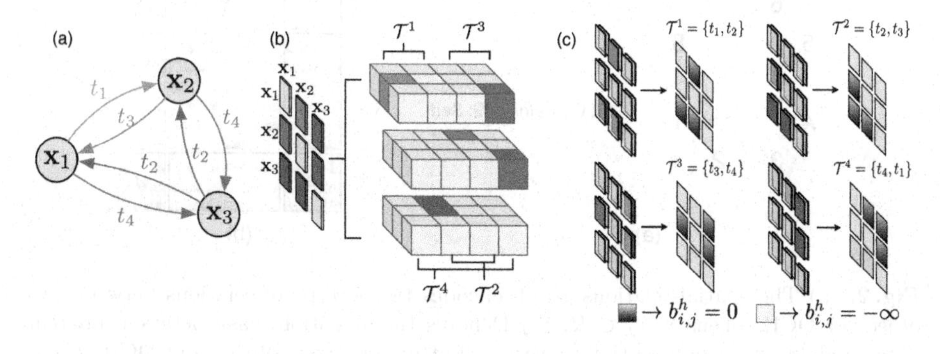

Fig. 3. (a) Spatially aware attention layer uses a spatial graph to guide the attention in each head of the self-attention layer. (b) The spatial graph is represented as a stack of adjacency matrices, each for a given relationship t_e (c) Each head indexed by h looks at a subset of relationships $\mathcal{T}^h$ defined by the size of the context ($c = 2$ here), e.g. $\texttt{head}_1$ looks at a two types of relation ($\mathcal{T}^1 = \{t_1, t_2\}$). When edge $e_{i \to j} \in \mathcal{T}^h$ (black box), bias term is set to $\beta^h_{i,j} = 0$, otherwise when $e_{i \to j} \notin \mathcal{T}^h$ (white blocks), $e_{i \to j} \notin \mathcal{T}^h = -\infty$.

4.2 Spatially aware Self-Attention Layer

As mentioned in Sect. 3, attention in a single-head h of a self-attention layer can be computed by the compatibility function defined in Eq. (2). The compatibility function computes a similarity between a query $\mathbf{q}^h_i$ corresponding to input $\mathbf{x}_i$, and the key vector $\mathbf{k}^h_j$ of input token $\mathbf{x}_j$. Within a single head we want attention to only look at relevant tokens. We model it by allowing each head to focus on only a subset of edge types $\mathcal{T}^h \subseteq \mathcal{T}^e$. In other words, we want each token $\mathbf{x}_i$ to only focus on tokens $\mathbf{x}_j$ when they are connected via an edge $e_{i \to j}$ of type $\Phi_e(e_{i \to j}) \in \mathcal{T}^h$.

In the context of TextVQA, we use the combination of two graphs, $G_{\text{spa}} \cup G_{\text{imp}}$, defined over tokens from all the input data modalities $\mathbf{x} \in X$. The subset of relations $\mathcal{T}^h$ each head h attends to is subset of c spatial relationships between $(\mathbf{x}_i, \mathbf{x}_j)$ and one implicit relationship between question and image tokens, i.e.,

$$\mathcal{T}^h = \{t_{\text{imp}}, t_h, t_{h+1}, \cdots, t_{(h+c) \bmod |\mathcal{T}^{\text{spa}}|}\}, \quad t \in \mathcal{T}^e = \mathcal{T}^{\text{spa}} \cup t_{\text{imp}}.$$

When $c > 1$, multiple heads are aware of a given spatial relationship and we are encouraging the models to jointly attend to information from different representation subspaces [42]. When $c = 1$, each head only focuses on a one type of spatial relationship. Similarly when $c = |\mathcal{T}^{spa}| + 1$, each head attends to all the spatial relationships as well as the implicit relationship t_{imp}. We empiricaly observed that $c = 2$ works best for our setting.

As illustrated in Fig. 3, to weigh the attention in each head based on the subset of spatial relationships $\mathcal{T}^h$, we introduce a bias term defined as

$$b_{i,j}^h = \begin{cases} \beta_{t_l}^h & t_l \in \mathcal{T}^h, \quad \mathbf{x}_i, \mathbf{x}_j \in X \\ -\infty & \text{otherwise} \end{cases}, \tag{5}$$

to modify the computation of the attention weights α_{ij}^h over different tokens. Specifically, we compute attention weights as follows[2]:

$$\alpha_{ij}^h = \text{Softmax}\left(\frac{\mathbf{q}_i^h(\mathbf{k}_j^h)^T + b_{i,j}^h}{\sqrt{d_h}}\right). \tag{6}$$

Intuitively, as illustrated in Fig. 3 if there is no edge $e_{i\to j}$ of type $t_l \in \mathcal{T}^h$ between nodes $\mathbf{x}_i$ and $\mathbf{x}_j$, then the compatibility score $\mathbf{q}_i^h(\mathbf{k}_j^h)^T + b_{i,j}^h$ is negative infinity and the attention weights α_{ij}^h become zero. Otherwise, the attention weights can be modulated based on the specific edge type $t_l = \Phi_e(e_{i\to j})$ by learning a bias term for each edge type $\beta_{t_l}^h \in \{\beta_{t_1}^h, \ldots, \beta_{|\mathcal{T}^e|}^h\}$. Alternatively, we can set $\beta_{t_l}^h$ to zero if we do not want to modulate attention based on the edge type between a pair of tokens. Classical self-attention layers described in Sect. 3.1 are hence a special case which is obtained when $|\mathcal{T}^e| = 1$ and G is a fully connected graph.

Specifically, for TextVQA, as illustrated in Fig. 2(b), all object and OCR tokens attend to each other based on that subset of relations $\mathcal{T}^h$ that a head is responsible for. Since we want the representations of object and OCR tokens to contain information about the question, all object and OCR tokens attend to all question tokens via the edge of type t_{imp}. For simplicity, we don't learn this relation-dependent bias and run all our experiments with $\beta_{t_l}^h$ set to zero.

Importantly, our graph-aware self attention layer overcomes the aforementioned two limitations of classical self-attention layers. First, each head is able to focus on a subset of relations $\mathcal{T}^h \subseteq \mathcal{T}^e$. Consequently, the attention is not distributed over the entire sequence of tokens and each token gathers information from only a specific subset of tokens. Second, we are forcing each head to look at a different context which prevents the heads from learning redundant features.

[2] If for a given $\mathbf{x}_i$, $\Phi_e(e_{i\to j}) \notin \mathcal{T}^h, \forall j \in [1, N]$, then we explicitly set $\alpha_{i,j}^h = 0$,.

Causal Attention for Answer Tokens: During decoding, the M4C model generates answer tokens $\mathbf{y}_t^{\text{ans}}$ $\forall t$ one step at a time. Inspired by the success of several text-to-text models [12,25], the M4C architecture uses a causal attention mask where $\mathbf{y}_t^{\text{ans}}$ attends to all question, image, and OCR tokens $\mathbf{x} \in X$ along with entries in the answer $\mathbf{y}_{<t}^{\text{ans}}$ prior to time t. We follow [14,36] to generate the answer tokens. During decoding, at each step the model transforms the predicted token from the previous step to a d-dimensional vector z_t. We use z_t to compute similarity with all OCR-tokens and vocabulary words and pick the most similar one. We iteratively decode the answer over 12 time steps.

4.3 Implementation Details

Following M4C [14], the input to our multimodal transformer consists of three different modalities – 1) 20 Question tokens, 2) 100 Object tokens, and 3) 50 OCR tokens. Below, we briefly describe the construction of each of the modal features. For a detailed discussion we refer the reader to M4C [14].

Question Features: We encode the question text using three layers of a BERT [11] model pre-trained on English Wikipedia and Book-Corpus [51].

Object Features: We encode the object regions by extracting features from a ResNeXT-152 [46] based Faster R-CNN model [34] trained on Visual Genome [19] with attribute loss. We then add an absolute-location embedding to these features by using the bounding box coordinates.

OCR Features: Similarly, for OCR, we extract region features using the same object detector and we append an embedding obtained from FastText [8] and PHOC features [1] of the ocr-text. We also add an absolute-location embedding by using the bounding box coordinates of the OCR token similar to object features.

5 Experiments

We evaluate our model on the TextVQA dataset [36] and the ST-VQA dataset [7]. Our model outperforms previous work by a significant margin and sets the new state-of-the-art on both datasets.

5.1 Evaluation on TextVQA Dataset

The TextVQA dataset [36] contains 28,408 images from the Open Images dataset [20], with human-written questions asking about text in the image. Following VQAv2 [3], each question in the TextVQA dataset has 10 free-response answers, and the final accuracy is measured via soft voting of the 10 answers (VQA Accuracy). Following the M4C model [14], we collect the top 5000 frequent words from the answers in the training set as our answer vocabulary. We compare our method with the recent proposed LoRRA [36], M4C [14] and 2019 TextVQA Challenge leaderboard entires [24,38].

Table 1. Results on TextVQA [36] dataset. We compare our model (rows 11–13) against the prior works (row 1–5) and the improved baselines (rows 6–10). † Indicates our ablations for improved baseline. †† Indicates the best model from improved baseline.

	Method	Structure	OCR system	DET backbone	w/ST-VQA	Beam size	Accu. on val	Accu. on test
1	LoRRA [36]	–	R-ml	ResNet	✗	–	26.5	27.6
2	DCD [24]	–	–	–	–	–	31.4	31.4
3	MSFT [38]	–	–	–	–	–	32.9	32.4
4	M4C [14]	4N	R-en	ResNet	✗	1	39.4	39.0
5	M4C [14]	4N	R-en	ResNet	✓	1	40.5	40.4
6	M4C [14]†	4N	G	ResNet	✗	1	41.8	–
7	M4C [14]†	4N	G	ResNeXt	✗	1	42.0	–
8	M4C [14]†	6N	G	ResNeXt	✗	1	42.7	–
9	M4C [14]†	6N	G	ResNeXt	✓	1	43.3	–
10	M4C [14]††	6N	G	ResNeXt	✓	5	43.8	42.4
11	SA-M4C (ours)	2N→4S	G	ResNeXt	✗	1	43.9	–
12	SA-M4C (ours)	2N→4S	G	ResNeXt	✓	1	45.1	–
13	SA-M4C (ours)	2N→4S	G	ResNeXt	✓	5	**45.4**	**44.6**

Improved M4C Baseline (M4C†). To establish a strong baseline we further improve M4C by replacing the Rosetta-en OCR system with the Google OCR system[3] which we qualitatively find to be more accurate, detecting text with higher recall and having fewer spelling errors. This improves the performance from 39.4% to 41.8% (Rows 4 and 6 in Table 1). Next, we replace the ResNet-101 [13] backbone of the Faster R-CNN [34] feature-extractor with a ResNeXt-151 backbone [46] as recommended by [27]. This further improves the performance from 41.8% to 42.0 % (Rows 6 and 7). Finally, we add two additional transformer layers (Row 8), jointly train M4C on ST-VQA [7] (Row 9) and use beam search decoding (Row 10) to establish the final improved baselines 43.8% and 42.4% on validation and test set respectively.

Our Results (SA-M4C). Our model consist of 2 normal self-attention layers and 4 spatially aware self-attention layers (2N→4S). As shown in Table 1 Row 13, our model is 2.2% (absolute) higher than it's counterpart in Row 10 and 4.4% better than the baseline M4C model (Row 5). Note that the improved M4C model in Row 10 and our method use the same input features, equal number of transformer layers and have the same number of parameters. Next, we perform model ablations to analyze the source of the gains in our method.

Model Structure Ablations. We answer the question, "How many spatially aware self-attention layers are helpful?", by incrementally replacing the self-attention layers in M4C with the proposed spatially aware self-attention layers.

[3] https://cloud.google.com/products/ai/.

Table 2. Ablations on TextVQA.

Method	Struc.	Context	Accu. (val)
1 M4C [14]†	6N	–	42.70
2 SA-M4C (ours)	4N→2S	1	43.19
3 SA-M4C (ours)	2N→4S	1	43.80
4 M4C-Random	2N→4S	1	42.09
5 M4C-Top-9 [49]	2N→4T	–	43.26
6 M4C-ReGAT [22]	2N→4Re	–	43.20
7 SA-M4C (ours)	2N→4S	2	**43.90**

Table 3. Results on ST-VQA dataset.

Method	Struc.	Beam size	VQA Accu.	ANLS on val	ANLS on test
1 SAN+STR [7]	–	–	–	–	0.135
2 VTA [6]	–	–	–	–	0.282
3 M4C [14]	4N	1	38.05	0.472	0.462
4 M4C [14]†	6N	1	40.71	0.499	-
5 SA-M4C (ours)	2N→4S	1	42.12	0.510	-
6 SA-M4C (ours)	2N→4S	5	**42.23**	**0.512**	**0.504**

Table 2 Row 1, 2 and 3 show that the performance improves as we replace normal self-attention layers with spatially aware self-attention. We achieve the best performance after replacing 4 out of 6 self-attention layers (43.19% *vs.* 43.80%). It's important to note that keeping the bottom self-attention layer is critical to model attention across modalities since attention for question tokens are masked in spatially aware self-attention.

Span of Spatially Aware Self-attention Head. Recall that the context-size parameters (c) is the number of relationships $|\mathcal{T}^h|$ each attention head looks at, and controls the sparsity of each head in spatially aware self-attention. When $c > 1$, multiple heads are aware of a given spatial relationships which jointly attend information from different representation subspaces. Sweeping over the context-size (c) we find that $c = 2$ works the best (Row 7 in Table 2).

Comparing with Other Methods that Induce Sparsity into Transformer. We further compare our approach with other formulations [22,49] that induce sparsity in the Transformer architectures as well as randomly mask attention heads. We describe each setting as follows.

- **Random masking (M4C-Random).** We randomly initialize a spatial graph by assigning an edge of a given type between two nodes including a no-edge with equal probability. We use this graph as input to our spatially aware self-attention layer. Through this comparison, we want to establish the importance of spatial graph induced sparsity vs random sparsity in self-attention layers. We report this baseline by averaging across 5 different seeds.
- **Top-k Attention (M4C-Top-k).** Instead of masking the attention weights based on a graph, we explicitly select the *top-k* attention weights and mask the rest [49]. We use $k = 9$ which corresponds to inducing the same level of sparsity as our baseline model. This helps to establish the need of guiding the attention based on spatial relationships.
- **Graph Attention (M4C-ReGAT).** We implement ReGAT-based attention layer Li *et al.* [22] which endows Graph Attention Network [43] encoding with spatial information by adding a bias term specific to each relation.[4]. The

[4] We use the code released by the authors https://github.com/linjieli222/VQA_ReGAT/.

goal is to establish the improvements of our spatially aware self-attention layer compared to prior work.

Table 2 Row 4 - Row 7 show these comparisons. We observe that random masking decreases the performance on TextVQA dataset by 1.2% which verifies the importance of a correct spatial relationship graph. By selecting the *top-k* connections (M4C-Top-9), we observe an improvement of 0.66% compared to M4C-Random. However, M4C-Top-9 still underperforms compared to our proposed SA-M4C model by 0.64%. Similarly, our proposed SA-M4C model outperforms the graph attention version (M4C-ReGAT) by 0.7%.

5.2 Evaluation on ST-VQA

We also report results on the ST-VQA [7] dataset which is another recently proposed dataset for the TextVQA task. ST-VQA contains 18,921 training and 2,971 test images sourced from several datasets [5,16,17,19,30]. Following M4C, we report results on the Open Dictionary (Task-3) as it matches the TextVQA setting where no answer candidates are provided at test time.

The ST-VQA dataset adopts Average Normalized Levenshtein Similarity (ANLS) defined as $1 - d_L(a_{\text{pred}}, a_{\text{gt}})/\max(|a_{\text{pred}}|, |a_{\text{gt}}|)$ averaged over all questions. a_{pred} and a_{gt} refer to prediction and ground-truth answers respectively while d_L is edit distance. The metric truncates scores lower than 0.5 to 0 before averaging. We use both VQA accuracy and ANLS as the evaluation metric to facilitate comparison with prior work.

For training and validation on ST-VQA we use the same splits used by M4C [14] generated by randomly selecting 17,028 images for training and the remaining 1,893 for validation. We train the improved baseline model and our best model (spatially aware self-attention) on ST-VQA and report results in Table 3. Following prior works [7,14] we show VQA Accuracy and ANLS both on validation set and only the latter on the test set. On the validation set our improved baseline achieves an accuracy of 40.71% and an ANLS of 0.499 improving by 2.66% and 0.027 absolute. Further, the final model with spatially aware self-attention layers achieves an accuracy of 42.23% and an ANLS of 0.512 improving by 1.52% and 0.013 in absolute gains on the validation set. On the test set, our best model achieves state-of-the-art performance of 0.504 ANLS.

6 Analysis

Spatial Reasoning: We look at the source of improvements in our model both quantitatively and qualitatively. First, we look at the performance of our model on subset of questions from TextVQA validation dataset that involve spatial reasoning. For this, we carefully curate a list of spatial-prepositions (see Supplementary for detail), and filter questions based on occurrence of one or more of these spatial-prepositions. After applying this filter, We observe that $\sim$14% of the questions (709/5000) are retained. On this subset D_{spa}, our model perform

2.83% better than M4C. Since, OCR tokens can answer only ∼65% of the questions in the validation set, we also look at the subset of questions that require spatial reasoning and has a majority answer in OCR tokens. On this subset $D_{\text{spa+ocr}}$ (409/5000 questions), our model performs 4.62% better than M4C.

Visual Grounding: As a proxy to analyze visual grounding of our model, we look at instances in which models predict the answer using the list of OCR tokens without relying on the vocabulary. Our model picks an answer from the list of OCR tokens on 368/701 questions from the D_{spa} subset, and achieves 52.85% accuracy. This greatly improves the performance over M4C which only achieves 44.05% accuracy on a similar number (398/709) of questions that were answered using OCR tokens. The increase in performance is similar on $D_{\text{spa+ocr}}$ where we achieve a score of 67.95% on 260/401 questions compared to 59.27% achieved by M4C over 273/401 questions.

Qualitative Analysis: In Fig. 4, we can qualitatively see how our models can reason about relative positions of object and text in the image. Our model picks the correct answer by reasoning about relations like '`right`', '`top-left`'. Our model can also reason about spatial relations between object ('green square') and text ('lime'). In the last row, we show instances where based on the type of spatial relationship mentioned in the question, our model changes the answer.

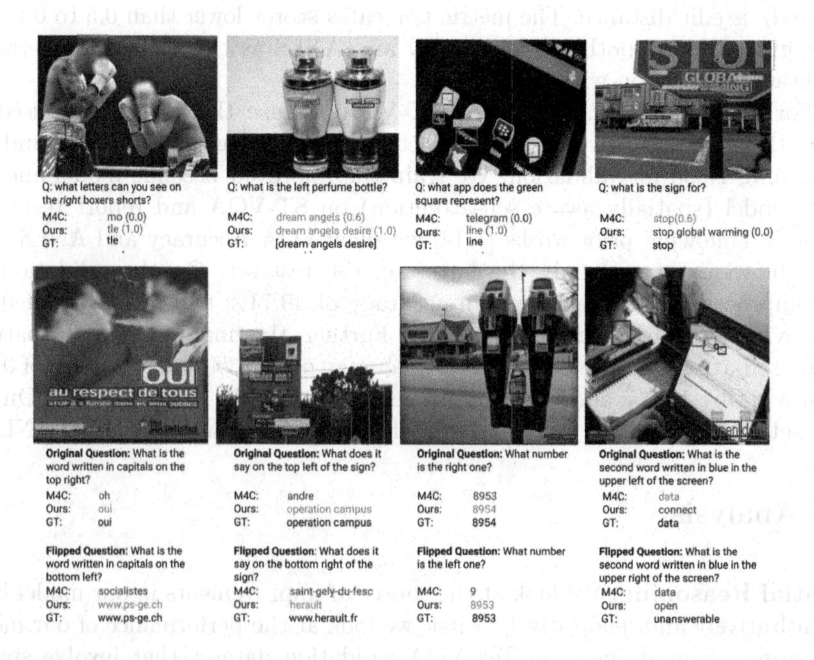

Fig. 4. Qualitative Examples: Top row shows the output of M4C and our method on several image-question pairs. The bottom row show examples where we flipped the spatial relation in the original question to see whether the models change their answers.

Potential Sources of Error: While our model improves performance by encoding spatial relationships, we are still far away from human baseline. The model is not robust to spelling mistakes in the OCR tokens. Secondly, these models have trouble generating the stop condition during decoding. Finally, while our model can encode relative spatial relationships, for reasoning about absolute positions in the image, our model can benefit from stronger cues about absolute locations.

7 Conclusion

We developed a spatially aware self-attention layer in which each input entity only looks at neighboring entities as defined by a spatial graph. This allows each input to focus on a local context instead of dispersing attention amongst all other entities. Each head also focuses on a different subset of the spatial relations which avoids learning redundant features. We apply our general framework on the task of TextVQA by constructing a spatial graph between object and OCR tokens and utilizing it in the spatially aware self-attention layers. We found this graph-based attention to significantly improve results achieving state-of-the-art performance on the TextVQA and ST-VQA dataset. Finally, we present our analysis showing how our method improves visual grounding.

Acknowledgements. The Georgia Tech effort was supported in part by NSF, AFRL, DARPA, ONR YIPs, ARO PECASE, Amazon. The views and conclusions contained herein are those of the authors and should not be interpreted as necessarily representing the official policies or endorsements, either expressed or implied, of the U.S. Government, or any sponsor.

References

1. Almazán, J., Gordo, A., Fornés, A., Valveny, E.: Word spotting and recognition with embedded attributes. IEEE Trans. Pattern Anal. Mach. Intell. **36**, 2552–2566 (2014)
2. Anderson, P., et al.: Bottom-up and top-down attention for image captioning and visual question answering. In Proceedings of the IEEE Conference on Computer Vision and Pattern Recognition, pp. 6077–6086 (2018)
3. Antol, S., et al.: VQA: visual question answering. In: Proceedings of the IEEE International Conference on Computer Vision, pp. 2425–2433 (2015)
4. Bahdanau, D., Cho, K., Bengio, Y.: Neural machine translation by jointly learning to align and translate. In: ICLR (2015)
5. Bigham, J.P., et al.: VizWiz: nearly real-time answers to visual questions. In: W4A (2010)
6. Biten, A.F., et al.: ICDAR 2019 competition on scene text visual question answering. arXiv preprint arXiv:1907.00490 (2019)
7. Biten, A.F., et al.: Scene text visual question answering. In: Proceedings of the IEEE International Conference on Computer Vision, pp. 4291–4301 (2019)
8. Bojanowski, P., Grave, E., Joulin, A., Mikolov, T.: Enriching word vectors with subword information. Trans. Assoc. Comput. Linguist. **5**, 135–146 (2017)

9. Borisyuk, F., Gordo, A., Sivakumar, V.: Rosetta. In: Proceedings of the 24th ACM SIGKDD International Conference on Knowledge Discovery & Data Mining (2018)
10. Chen, Y.-C., et al.: UNITER: learning universal image-text representations. arXiv preprint arXiv:1909.11740 (2019)
11. Devlin, J., Chang, M.-W., Lee, K., Toutanova, K.: Pre-training of deep bidirectional transformers for language understanding. In: NAACL-HLT, Bert (2019)
12. Dong, L., et al.: Unified language model pre-training for natural language understanding and generation. In: NeurIPS (2019)
13. He, K., Zhang, X., Ren, S., Sun, J.: Deep residual learning for image recognition. In: 2016 IEEE Conference on Computer Vision and Pattern Recognition (CVPR) (2016)
14. Hu, R., Singh, A., Darrell, T., Rohrbach, M.: Iterative answer prediction with pointer-augmented multimodal transformers for TextVQA. arXiv preprint arXiv:1911.06258 (2019)
15. Jiang, Y., Natarajan, V., Chen, X., Rohrbach, M., Batra, D., Parikh, D.: Pythia v0. 1: the winning entry to the VQA challenge 2018. arXiv preprint arXiv:1807.09956 (2018)
16. Karatzas, D., et al.: ICDAR 2015 competition on robust reading. In: 2015 13th International Conference on Document Analysis and Recognition (ICDAR), pp. 1156–1160 (2015)
17. D. Karatzas, F., et al.: ICDAR 2013 robust reading competition. In: 2013 12th International Conference on Document Analysis and Recognition, pp. 1484–1493 (2013)
18. Kipf, T., Welling, M.: Semi-supervised classification with graph convolutional networks. ArXiv, abs/1609.02907 (2016)
19. Krishna, R., et al.: Visual genome: connecting language and vision using crowdsourced dense image annotations. Int. J. Comput. Vis. **123**, 32–73 (2016)
20. Kuznetsova, A., et al.: The open images dataset v4: Unified image classification, object detection, and visual relationship detection at scale. arXiv:1811.00982 (2018)
21. Li, G., Duan, N., Fang, Y., Jiang, D., Zhou, M.: Unicoder-VL: a universal encoder for vision and language by cross-modal pre-training. arXiv preprint arXiv:1908.06066 (2019)
22. Li, L., Gan, Z., Cheng, Y., Liu, J.: Relation-aware graph attention network for visual question answering. ArXiv, abs/1903.12314 (2019)
23. Li, M.H., Yatskar, L., Yin, D., Hsieh, C.-J., Chang, K.-W.: VisualBert: a simple and performant baseline for vision and language. arXiv preprint arXiv:1908.03557 (2019)
24. Lin, Y., Zhao, H., Li, Y., Wang, D.: Dcd_zju, TextVQA challenge 2019 winner (2014). https://visualqa.org/workshop.html
25. Peter, J., et al.: Generating wikipedia by summarizing long sequences. In: ICLR (2018)
26. Lu, J., Batra, D., Parikh, D., Lee, S.: ViLBERT: pretraining task-agnostic visiolinguistic representations for vision-and-language tasks. ArXiv, abs/1908.02265 (2019)
27. Lu, J., Goswami, V., Rohrbach, M., Parikh, D., Lee, S.: 12-in-1: multi-task vision and language representation learning. arXiv preprint arXiv:1912.02315 (2019)
28. Lu, J., Yang, J., Batra, D., Parikh, D.: Hierarchical question-image co-attention for visual question answering. In: Advances in Neural Information Processing Systems, pp. 289–297 (2016)

29. Michel, P., Levy, O., Neubig, G.: Are sixteen heads really better than one? ArXiv, abs/1905.10650 (2019)
30. Mishra, A., Alahari, K., Jawahar, C.V.: Image retrieval using textual cues. In: ICCV (2013)
31. Mishra, A., Shekhar, S., Singh, A.K., Chakraborty, A.: OCR-VQA: visual question answering by reading text in images. In: Proceedings of the International Conference on Document Analysis and Recognition (2019)
32. Murahari, V., Batra, D., Parikh, D., Das, A.: Large-scale pretraining for visual dialog: a simple state-of-the-art baseline. arXiv preprint arXiv:1912.02379 (2019)
33. Qi, D., Su, L., Song, J., Cui, E.D.B., Bharti, T., Sacheti, A.: ImageBert: cross-modal pre-training with large-scale weak-supervised image-text data. ArXiv, abs/2001.07966 (2020)
34. Ren, S., He, K., Girshick, R., Sun, J.: Faster R-CNN: towards real-time object detection with region proposal networks. In:NuerIPS, pp. 91–99 (2015)
35. Shaw, P., Uszkoreit, J., Vaswani, A.: Self-attention with relative position representations. ArXiv, abs/1803.02155 (2018)
36. Singh, A., et al.: Towards VQA models that can read. In: 2019 IEEE/CVF Conference on Computer Vision and Pattern Recognition (CVPR), pp. 8309–8318 (2019)
37. Su, W., ZhuX., Cao, Y., Li, B., Lu, L., Wei, F., Dai, J.: VL-BERT: pre-training of generic visual-linguistic representations. arXiv preprint arXiv:1908.08530 (2019)
38. Anonymous submission. Msft_vti, textvqa challenge 2019 top entry (post-challenge) (2014). https://evalai.cloudcv.org/web/challenges/challenge-page/244/
39. Sukhbaatar, S., Grave, E., Bojanowski, P., Joulin, A.: Adaptive attention span in transformers. In: ACL (2019)
40. Sun, C., Myers, A., Vondrick, C., Murphy, K., Schmid, C.: Videobert: a joint model for video and language representation learning. In: Proceedings of the IEEE International Conference on Computer Vision, pp. 7464–7473 (2019)
41. Tan, H., Bansal, M.: LXMERT: learning cross-modality encoder representations from transformers. arXiv preprint arXiv:1908.07490 (2019)
42. Vaswani, A., et al.: Attention is all you need. In: NIPS (2017)
43. Veličković, P., Cucurull, G., Casanova, A., Romero, A., Liò, P., Bengio, Y.: Graph attention networks. In: International Conference on Learning Representations (2018)
44. Voita, E., Talbot, D., Moiseev, F., Sennrich, R., Titov, I.: Analyzing multi-head self-attention: specialized heads do the heavy lifting, the rest can be pruned. In: ACL (2019)
45. Wang, X., Tu, Z., Wang, L., Shi, S.: Self-attention with structural position representations. In: EMNLP/IJCNLP (2019)
46. Xie, S., Girshick, R., Dollar, P., Tu, Z., He, K.: Aggregated residual transformations for deep neural networks. In: 2017 IEEE Conference on Computer Vision and Pattern Recognition (CVPR) (2017)
47. Yang, B., Tu, Z., Wong, D.F., Meng, F., Chao, L.S., Zhang, T.: Modeling localness for self-attention networks. In: EMNLP (2018)
48. Yao, T., Pan, Y., Li, Y., Mei, T.: Exploring visual relationship for image captioning. In: Ferrari, V., Hebert, M., Sminchisescu, C., Weiss, Y. (eds.) Computer Vision – ECCV 2018. LNCS, vol. 11218, pp. 711–727. Springer, Cham (2018). https://doi.org/10.1007/978-3-030-01264-9_42
49. Zhao, G., Lin, J., Zhang, Z., Ren, X., Su, Q., Sun, X.: Explicit sparse transformer: concentrated attention through explicit selection. ArXiv, abs/1912.11637 (2019)

50. Zhou, L., Palangi, H., Zhang, L., Hu, H., Corso, J.J., Gao, J.: Unified vision-language pre-training for image captioning and VQA. ArXiv, abs/1909.11059 (2019)
51. Zhu, Y., et al.: Aligning books and movies: towards story-like visual explanations by watching movies and reading books. In: IEEE International Conference on Computer Vision (ICCV), pp. 19–27 (2015)

Every Pixel Matters: Center-Aware Feature Alignment for Domain Adaptive Object Detector

Cheng-Chun Hsu[1(✉)], Yi-Hsuan Tsai[2], Yen-Yu Lin[1,3], and Ming-Hsuan Yang[4,5]

[1] Academia Sinica, Taipei, Taiwan
hsu06118@citi.sinica.edu.tw
[2] NEC Labs America, Texas, USA
[3] National Chiao Tung University, Hsinchu, Taiwan
[4] UC Merced, Merced, USA
[5] Google Research, Cambridge, USA

Abstract. A domain adaptive object detector aims to adapt itself to unseen domains that may contain variations of object appearance, viewpoints or backgrounds. Most existing methods adopt feature alignment either on the image level or instance level. However, image-level alignment on global features may tangle foreground/background pixels at the same time, while instance-level alignment using proposals may suffer from the background noise. Different from existing solutions, we propose a domain adaptation framework that accounts for each pixel via predicting pixel-wise objectness and centerness. Specifically, the proposed method carries out center-aware alignment by paying more attention to foreground pixels, hence achieving better adaptation across domains. We demonstrate our method on numerous adaptation settings with extensive experimental results and show favorable performance against existing state-of-the-art algorithms. Source codes and models are available at https://github.com/chengchunhsu/EveryPixelMatters.

1 Introduction

As a key component to image analysis and scene understanding, object detection is essential to many high-level vision applications such as instance segmentation [4,10–12], image captioning [20,37,38], and object tracking [18]. Although significant progress on object detection [9,26,29] had been made, an object detector that can adapt itself to variations of object appearance, viewpoints, and backgrounds [2] is always in demand. For example, a detector used for autonomous driving is required to work well under diverse weather conditions, even if training data may be acquired under some particular weather conditions.

Electronic supplementary material The online version of this chapter (https://doi.org/10.1007/978-3-030-58545-7_42) contains supplementary material, which is available to authorized users.

© Springer Nature Switzerland AG 2020
A. Vedaldi et al. (Eds.): ECCV 2020, LNCS 12354, pp. 733–748, 2020.
https://doi.org/10.1007/978-3-030-58545-7_42

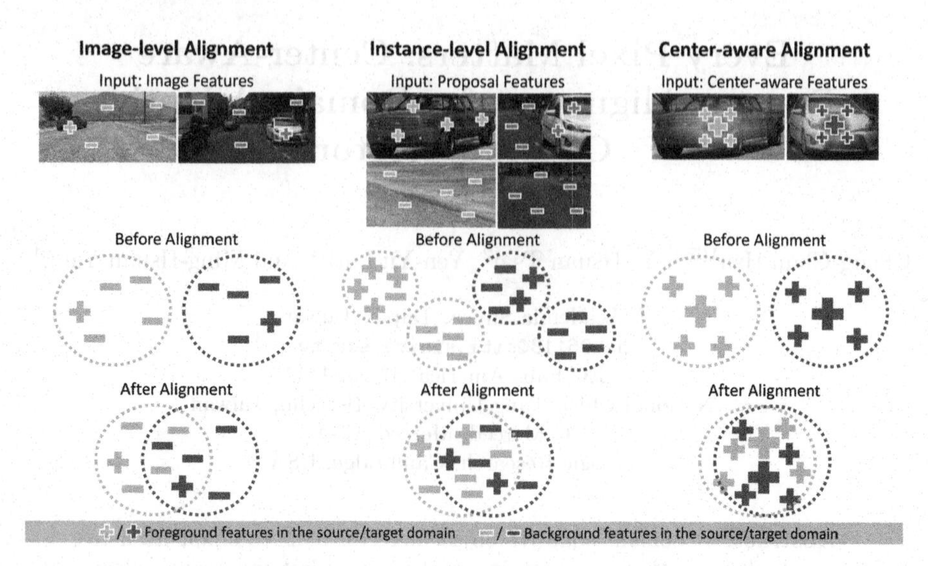

Fig. 1. Comparisons between different alignment methods. 1) For image-level alignment, it considers both foreground/background pixels, which may lead to noisy alignment and focus more on background pixels. 2) Instance-level alignment is performed on proposals, in which the pooled feature on all the pixels within the proposal could mix foreground/background signals. In addition, proposals in the target domain may contain much more background pixels due to the domain gap. 3) The proposed center-aware alignment focuses on foreground pixels with higher confidence scores of objectness and centerness, i.e., those marked by larger "+" showing higher centerness response, which play a crucial role to reduce the confusion during alignment.

To address this challenge, *unsupervised domain adaptation* (UDA) methods [7,28,31,35,36] have been developed to adapt models trained on an annotated source domain to another unlabeled target domain. Adopting a similar strategy to the classification task [36] using adversarial feature alignment, numerous UDA methods for objection detection [1,2,14–16,21,22,32] are proposed to reduce the domain gap across source and target domains. However, such alignment is usually performed on the image level that adapts global features, which is less effective when the domain gap is large [5,32]. To improve upon global alignment[1], existing methods [2,14,41] adapt instance-level distributions that pool features of all the pixels within a proposal. However, since pixel distributions are unknown in the target domain, the proposal extracted from the target domain could contain many background pixels. As a result, this may significantly confuse the alignment procedure when adapting instance-level features of target proposals to the source distribution that contains mostly foreground pixels (Fig. 1).

In this paper, we propose to take every pixel into consideration when aligning feature distributions across two domains. To this end, we design a module to estimate pixel-wise objectness and centerness of the entire image, which allows

[1] In this paper, we use image-level alignment and global alignment interchangeably.

our alignment process to focus on foreground pixels, instead of the proposal that may contain tangled foreground/background pixels as considered in the prior work. In order to predict the pixel-wise information, we revisit the object detection framework and adopt fully-convolutional layers. As a result, our method aims to align the centered discriminative part of the objects across domains, namely the regions with high objectness scores and close to the object centers (see Fig. 1). Thereby, these regions are less sensitive to irrelevant background pixels in the target domain and facilitate distribution alignment. To the best of our knowledge, we make the first attempt to leverage pixel-wise objectness and centerness for domain adaptive object detection.

To validate the proposed method, we conduct extensive experiments on three benchmark settings for domain adaptation: Cityscapes [3] → Cityscapes Foggy [33], Sim10k [17] → Cityscapes, and KITTI [8] → Cityscapes. The experimental results show that our center-aware feature alignment performs favorably against existing state-of-the-art algorithms. Furthermore, we provide ablation study to demonstrate the usefulness of each component in our method. The major contributions of this paper are summarized as follows. First, we propose to discover discriminative object parts on the pixel level and better handle the domain adaptation task for object detection. Second, center-aware distribution alignment with its multi-scale extension is presented to account for object scales and alleviate the unfavorable effects caused by cluttered backgrounds during adaptation. Third, comprehensive ablation studies validate the effectiveness of the proposed framework with center-aware feature alignment.

2 Related Work

In this section, we review a few research topics relevant to this work, including object detection and domain adaptive object detection.

2.1 Object Detection

Object detection studies can be categorized into anchor-based and anchor-free detectors. Anchor-based detectors compile a set of anchors to generate object proposals, and formulate object detection as a series of classification tasks over the proposals. Faster-RCNN [30] is the pioneering anchor-based detector, where the region proposal network (RPN) is employed for proposal generation. Owing to its effectiveness, RPN is widely adopted in many anchor-based detectors [25,26].

Anchor-free detectors skip proposal generation, and directly localize objects based on the fully convolutional network (FCN) [27]. Recently, anchor-free methods [6,23,40] leverage keypoint (i.e., the center or corners of a box) localization and achieve comparable performance with anchor-based methods. Yet, these methods require complex post-processing for grouping the detected points. To avoid such a process, FCOS [34] proposes per-pixel prediction, and directly predicts the class and offset of the corresponding object at each location on the feature map. In this work, we take advantages of the property in anchor-free methods to identify discriminate areas for the alignment procedure.

Table 1. Alignment schemes adopted by existing methods, including global alignment (*G*), instance-level alignment (*I*), low-level feature alignment (*L*), pixel-level alignment (*P*) via style transfer or CycleGAN, pseudo-label re-training (*PL*), and the proposed center-aware alignment (*CA*) that considers pixel-wise objectness and centerness. * indicates that pixel-level alignment is only applied during adapting from Sim10k to Cityscapes.

Method	*G*	*I*	*L*	*P*	*PL*	*CA*
DAF [2] CVPR'18	√	√				
SC-DA [41] CVPR'19	√	√				
SW-DA [32] CVPR'19	√		√	√*		
DAM [22] CVPR'19	√			√		
MAF [14] ICCV'19	√	√				
MTOR [1] CVPR'19	√	√				
STABR [21] ICCV'19	√				√	
PDA [15] WACV'20	√			√		
Ours	√					√

2.2 UDA for Object Detector

Che *et al.* [2] first present two alignment practices, *i.e.*, image-level and instance-level alignments, by adopting adversarial learning at image and instance scales, respectively. For image-level alignment, Saito *et al.* [32] further indicate that aligning lower-level features is more effective since global feature alignment suffers from the cross-domain variations of foreground objects and background clutter. To improve instance-level alignment, Zhu *et al.* [41] apply k-means clustering to group proposals and obtain the centroids of these clusters, which achieves a balance between global and instance-level alignment. However, their method introduces additional data-independent hyper-parameters for clustering and is not end-to-end trainable. Other variants improve feature alignment based on a hierarchical module [14], a style-transfer based method to address the source-biased issue [22], a teacher-student scheme to explore object relations [1], and a progressive alignment scheme [15].

While the above methods are based on two-stage detectors, Kim *et al.* [21] propose a one-stage adaptive detector for faster inference, via a hard negative mining technique for seeking more reliable pseudo-labels. However, their method only partially alleviates the issues brought by background and does not consider every pixel during feature alignment to reduce the domain gap. We also note that all aforementioned methods are based on anchors, in which performing instance-level alignment would be sensitive to inaccurate proposals in the target domain and the mixture of foreground/background pixels in a proposal. In contrast, we address these drawbacks by predicting pixel-wise objectness and proposing center-aware feature alignment, which only focuses on the discriminative parts of objects at the pixel scale. In Table 1, we summarize the alignment methods used in the aforementioned techniques for domain adaptive object detection.

3 Proposed Method

In this section, we first describe global feature alignment, and then introduce the proposed center-aware alignment that utilizes pixel-wise objectness and ceterness. To improve the performance, we further incorporate multi-scale alignment that takes object scale into account during adaptation.

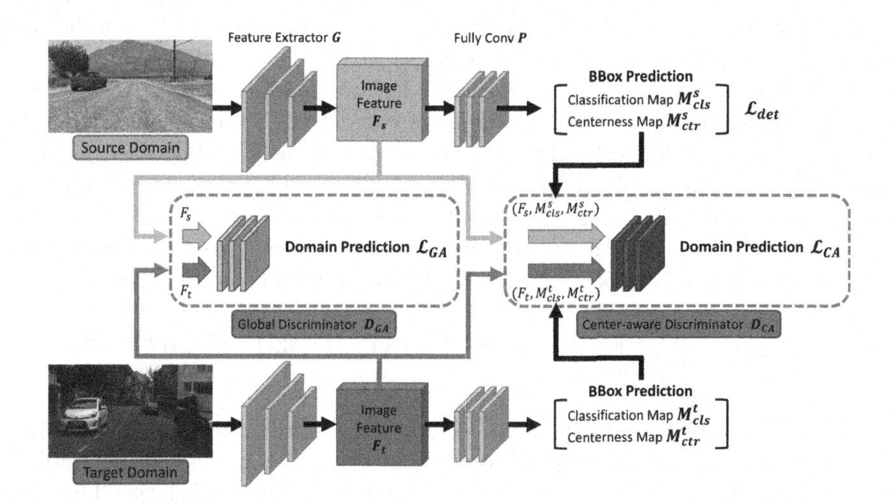

Fig. 2. Proposed framework for domain adaptive object detection. Given the source and target images, we feed them to a shared feature extractor G to obtain their features F. Then, the global alignment on these features is performed via a global discriminator D_{GA} and a domain prediction loss $\mathcal{L}_{GA}$. Next, we pass the feature through the fully-convolutional module P to produce the classification and centerness maps. These maps and the feature F are utilized to generate the center-aware features. Finally, we use a center-aware discriminator D_{CA} and another domain prediction loss $\mathcal{L}_{CA}$ to perform the proposed center-aware feature alignment. Note that the bounding box prediction loss $\mathcal{L}_{det}$ is only operated on source images using their corresponding ground-truth bounding boxes.

3.1 Algorithm Overview

Given a set of source images I_s, their ground-truth bounding boxes B_s, and unlabeled target images I_t, our goal is to predict bounding boxes B_t on the target image. To this end, we propose to utilize two alignment schemes that complement each other: global alignment that accounts for image-level distributions and the proposed center-aware alignment that focuses more on foreground pixels. The overall procedure is illustrated in Fig. 2. Given a shared feature extractor G across domains, we first extract features $F = G(I)$ and perform global alignment via using a global discriminator and a domain prediction loss. Second, followed by G, a fully-convolutional module P is adopted to predict pixel-wise objectness and centerness maps. Through combining these maps with the feature F, we employ another center-aware discriminator and its domain prediction loss to perform center-aware alignment.

3.2 Global Feature Alignment

The goal of global alignment is to align the feature maps on the image level to reduce the domain gap. To this end, we apply the adversarial alignment technique [2] via utilizing a global discriminator D_{GA}, which aims to identify whether the pixels on each feature map come from the source or the target domain.

Particularly, given the K-dimensional feature map $F \in \mathbb{R}^{H \times W \times K}$ of the spatial resolution $H \times W$ from the feature extractor G, the output of D_{GA} is a domain classification map that has the same size as F, while each location represents the domain label corresponding to the same location on F. Note that we set the domain label z of source and target domain as 1 and 0, respectively. Therefore, the discriminator can be optimized by minimizing the binary cross-entropy loss. For a location (u, v) on F, the loss function can be written as

$$\mathcal{L}_{GA}(I_s, I_t) = -\sum_{u,v} z \log(D_{GA}(F_s)^{(u,v)}) + (1-z)\log(1 - D_{GA}(F_t)^{(u,v)}). \quad (1)$$

To perform adversarial alignment, we apply the gradient reversal layer (GRL) [7] to feature maps of both source/target images, in which the sign of the gradient is reversed when optimizing the feature extractor via the GRL layer. Then the mechanism works as follows. The loss for the discriminator is minimized via (1), while the feature extractor is optimized by maximizing this loss, in order to deceive the discriminator. We also note that most existing methods (those in Table 1) utilize such global alignment that focuses on image-level distributions (i.e., more background pixels in reality). We also use global alignment in our framework to complement the proposed center-aware alignment that focuses on foreground pixels.

3.3 Center-Aware Alignment

As mentioned in Sect. 1 and Table 1, existing methods [2,14,41] for instance-level alignment are based on proposals, and thus these approaches may suffer from the background effect. In order to address this issue, we propose a center-aware alignment method that allows us to focus on discriminative object regions. To this end, we adopt a center-aware discriminator D_{CA} for aligning features in the high-confidence area on the pixel level.

Definition. With a designed fully-convolutional network P (as detailed in Sect. 3.5) and feature map $F \in \mathbb{R}^{H \times W \times K}$ from the feature extractor G, we pass F through P, and obtain a classification output $M_{cls} \in \mathbb{R}^{H \times W \times C}$ and a class-agnostic centerness output $M_{ctr} \in \mathbb{R}^{H \times W}$, where C is the number of categories. Each location on the classification and centerness maps indicates corresponding objectness and centerness scores, respectively.

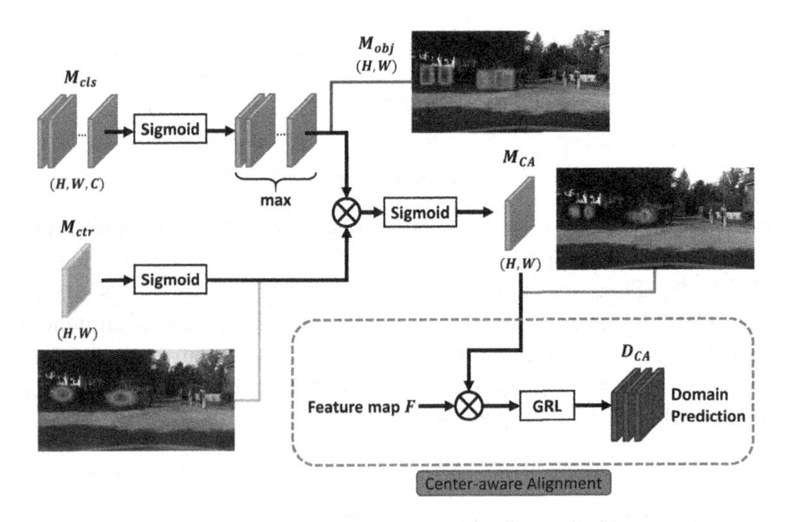

Fig. 3. Proposed center-aware alignment. Given the classification output M_{cls}, we first convert it to a class-agnostic map M_{obj}, which is then merged with the centerness output M_{ctr} into a center-aware map M_{CA} via (2) to identify potential object locations. Next, we use this map M_{CA} as the guidance to weight the global feature map F. Finally, this weighted feature map serves as the input to the center-aware discriminator D_{CA} to enable the proposed center-aware alignment in the feature space via (3).

Discover Object Region. In order to find the confident area containing foreground objects, we utilize two cues derived from our object detector as mentioned above: 1) a class-agnostic map of the objectness scores and 2) a centerness map that highlight object centers, so that the alignment can focus more on object parts. First, the objectness map can be obtained from the classification output M_{cls}. To obtain the class-agnostic map, we apply the *sigmoid* activation on each channel and take the *max* operation over categories. Similarly, the final class-agnostic centerness map is obtained via applying the *sigmoid* activation on the centerness output M_{ctr}. Overall, the final map M_{CA} to guide our center-aware alignment is calculated as follows:

$$\begin{aligned} M_{obj} &= \max_{c}(\sigma(M_{cls})), \\ M_{CA} &= \sigma(\delta\, M_{obj} \odot \sigma(M_{ctr})), \end{aligned} \tag{2}$$

where σ represents the *sigmoid* activation and $\odot$ denotes the element-wise product, i.e., Hadamard product, on the spatial maps. Since the values in M_{obj} and $\sigma(M_{ctr})$ are ranged from 0 to 1, a scaling factor δ is introduced for preventing the value from being too small after the multiplication. The factor δ is set to 20 in all experiments.

Perform Alignment. With the center-aware map M_{CA}, we are able to highlight the area where alignment on the pixel level should pay attention. To use this map as the guidance to our center-aware alignment, we multiply it by the feature map F and then feed it into the center-aware discriminator D_{CA}:

$$\mathcal{L}_{CA}(I_s, I_t) = -\sum_{u,v} z \log(D_{CA}(M^s_{CA} \odot F_s)^{(u,v)}) + (1-z)\log(1 - D_{CA}(M^t_{CA} \odot F_t)^{(u,v)}). \tag{3}$$

We note that, since M_{CA} is a map of resolution $H \times W$, we duplicate it for K channels to compute its element-wise product with the feature map $F \in \mathbb{R}^{H \times W \times K}$. Then, we adopt a similar alignment process as described in (1) via the GRL layer. As a result, different from the global alignment method as described in Sect. 3.2, our model aligns pixel-wise features that are likely to be the object and hence mitigates the non-matching issue between foregrounds and backgrounds. The entire process of center-aware alignment is illustrated in Fig. 3.

3.4 Overall Objective for Proposed Framework

Given source images I_s, target image I_t, and the ground-truth bounding boxes B_s in the source domain, our goal is to predict bounding boxes B_t on the unlabeled target data. We have described the objective for feature alignment on both source and target images. Here, we introduce the details of the object detection objective on the source domain using I_s and B_s.

Objective for Object Detector. Motivated by the anchor-free detector [34], our fully-convolutional module P consists of the classification, centerness, and regression branches. The three branches output the objectness map M_{obj}, centerness map M_{ctr}, and regression map M_{reg}, respectively. For the classification and regression branches, their goals are to predict the classification score and the distance to the four sides of the corresponding object box for each pixel, respectively. We denote their loss functions as $\mathcal{L}_{cls}$ and $\mathcal{L}_{reg}$, which can be optimized via the focal loss [25] and IoU loss [39], respectively. For the centerness branch, it predicts the distance between each pixel and the center of the corresponding object box and can be optimized by the binary cross-entropy loss [34] denoted as $\mathcal{L}_{ctr}$. The overall objective for the detector on the source domain is:

$$\mathcal{L}_{det}(I_s, B_s) = \mathcal{L}_{cls} + \mathcal{L}_{reg} + \mathcal{L}_{ctr}. \tag{4}$$

Here, we omit the argument (I_s, B_s) of each loss function for simplicity.

Overall Objective. In order to obtain domain-invariant features across the source and target domains, we apply adversarial learning to feature maps using two discriminators, D_{GA} and D_{CA}, which perform the global alignment and center-aware alignment by minimizing the objective functions $\mathcal{L}_{GA}$ and $\mathcal{L}_{CA}$, respectively. The details can be found in Sect. 3.2 and Sect. 3.3. The overall loss function can be expressed as:

$$\mathcal{L}(I_s, I_t, B_s) = \mathcal{L}_{det}(I_s, B_s) + \alpha\mathcal{L}_{GA}(I_s, I_t) + \beta\mathcal{L}_{CA}(I_s, I_t), \tag{5}$$

where α and β are the weights used to balance the three terms.

3.5 Network Architecture and Discussions

Different from the prior work [2,14,41] that focuses on instance-level alignment, our center-aware feature alignment requires pixel-wise predictions for objectness and centerness maps, so we cannot directly adopt the network architecture in previous methods. In this section, we introduce our architecture via using a fully-convolutional module for producing pixel-wise predictions, as well as a multi-scale extension to account for the object scale during adaptation.

Network Architecture. As mentioned in Sect. 3.4, we connect feature map F with the fully-convolutional detection head P that contains three branches: the classification, centerness, and regression branches. Different from previous methods, all branches are constructed by the fully-convolutional network, so that the predictions are performed on the pixel level. Specifically, the three branches consist of four 3×3 convolutional layers, and each of them has 256 filters. For both discriminators in global and center-aware alignments, i.e., D_{GA} and D_{CA}, we use the same fully-convolutional architecture as the detection branch, in order to maintain the consistency of the output size and thus map to the original input image.

Multi-scale Alignment. We observe that such a fully-convolutional architecture is not robust to the object scale, which is crucial to the performance of feature alignment. Therefore, in the feature extractor G, we use the feature pyramid network (FPN) [24] to handle different sizes of objects. Particularly, FPN utilizes five levels of feature map, which can be denoted as F^i for $i = \{3, 4, ..., 7\}$. The feature map F^3 is responsible for the smallest objects, while the feature map F^7 focuses on the largest objects. Each of the feature maps in the pyramid, i.e., F^i, has 256 channels.

We connect each layer with one head that contains three detection branches and two discriminators, i.e., D_{GA} and D_{CA}, and thus the loss function in (5) can be extended to the feature map of each layer. As a result, we are able to align each individual feature map F^i via global and center-aware alignments via (1) and (3). It follows that each aligned layer is responsible for a certain range of object size while making the overall alignment process consistent.

How Pixel-wise Prediction Helps Feature Alignment. It is worth mentioning that we take advantage of the pixel-wise prediction for the following reasons: 1) Pixel-wise prediction does not involve any fixed anchor-related hyper-parameters to produce proposals, which could be biased to the source domain during training; 2) Pixel-wise prediction considers all the pixels during training, which helps increase the capability of the model to identify the discriminative area of target objects; 3) The alignment can be performed on the pixel level and focuses on foreground pixels, which enables the model to learn better feature alignment. Note that the proposed method only depends on pixel-wise prediction, in which our method can be also applied to other similar detection models using the fully-convolutional module.

4 Experimental Results

We first provide the implementation details, and then describe datasets and evaluation metrics. Next, we compare our method with the state-of-the-art methods on multiple benchmarks. Finally, we conduct further analysis to understand the effect of each component in our framework. All the source code and models will be made available to the public.

4.1 Implementation Details

We implement our method with the PyTorch framework. In all the experiments, we set α and β in (5) as 0.01 and 0.1, respectively. Considering that center-aware alignment involves the detection output from (2), we first pre-train the detector only with the global alignment as a warm-up stage to ensure the reliability of detection before applying center-aware alignment and training the full objective in (5). Note that we set a larger α as 0.1 during pre-training for a faster convergence. For the adversarial loss using reversed gradients via GRL, we set the weight as 0.01 and 0.02 for D_{GA} and D_{CA}, respectively. The model is trained with learning rate of 5×10^{-3}, momentum of 0.9, and weight decay of 5×10^{-4}. The input images are resized with their shorter side as 800 and longer side less or equal to 1333.

4.2 Datasets

We follow the dataset setting as described in [2] and perform experiments for weather, synthetic-to-real and cross-camera adaptations on road-scene images.

Weather Adaptation. Cityscapes [3] is a scene dataset for driving scenarios, which are collected in dry weather. It consists of 2975 and 500 images in the training and validation set, respectively. The segmentation mask is provided for each image, consisting of eight categories: *person*, *rider*, *car*, *truck*, *bus*, *train*, *motorcycle* and *bicycle*. The Foggy Cityscapes [33] dataset is synthesized from Cityscapes as foggy weather. In the experiment, we adapt the model from Cityscapes to Foggy Cityscapes for studying the domain shift caused by the weather condition.

Synthetic-to-Real. Sim10k [17] is a collection of synthesized images, which consists of 10,000 images and their corresponding bounding box annotations. We use images of Sim10k as the source domain, while Cityscapes is considered as the target domain. The adaptation from Sim10k to Cityscapes is used to evaluate the adaptation ability from synthesized to real-world images. Following the literature, only the class *car* is considered.

Cross-camera Adaptation. KITTI [8] is similar to Cityscapes as a scene dataset, except that KITTI has a different camera setup. The training set of KITTI consists of 7,481 images. We use the KITTI and Cityscapes as the source domain and target domain respectively, and evaluation the capability of cross-camera adaptation. Following the literature, only the class *car* is considered.

Table 2. Results of adapting Cityscapes to Foggy Cityscapes. The first and second groups adopt VGG-16 and ResNet-101 as the backbone, respectively. Note that results of each class are evaluated in $mAP^r_{0.5}$.

Method	Cityscapes → Foggy Cityscapes									
	Backbone	person	rider	car	truck	bus	train	mbike	bicycle	$mAP^r_{0.5}$
Baseline (F-RCNN)	VGG-16	17.8	23.6	27.1	11.9	23.8	9.1	14.4	22.8	18.8
DAF [2] CVPR'18		25.0	31.0	40.5	22.1	35.3	20.2	20.0	27.1	27.6
SC-DA [41] CVPR'19		33.5	38.0	48.5	26.5	39.0	23.3	28.0	33.6	33.8
MAF [14] ICCV'19		28.2	39.5	43.9	23.8	39.9	33.3	**29.2**	33.9	34.0
SW-DA [32] CVPR'19		29.9	**42.3**	43.5	24.5	36.2	32.6	30.0	35.3	34.3
DAM [22] CVPR'19		30.8	40.5	44.3	**27.2**	38.4	**34.5**	28.4	32.2	34.6
Ours (w/o adapt.)		30.5	23.9	34.2	5.8	11.1	5.1	10.6	26.1	18.4
Ours (GA)		38.7	36.1	53.1	21.9	35.4	25.7	20.6	33.9	33.2
Ours (CA)		41.3	38.2	56.5	21.1	33.4	26.9	23.8	32.6	34.2
Ours (GA+CA)		**41.9**	38.7	**56.7**	22.6	**41.5**	26.8	24.6	**35.5**	**36.0**
Oracle		47.4	40.8	66.8	27.2	48.2	32.4	31.2	38.3	41.5
Ours (w/o adapt.)	ResNet-101	33.8	34.8	39.6	18.6	27.9	6.3	18.2	25.5	25.6
Ours (GA)		39.4	41.1	54.6	23.8	42.5	31.2	25.1	35.1	36.6
Ours (CA)		40.4	**44.9**	**57.9**	24.6	**49.6**	32.1	25.2	34.3	38.6
Ours (GA+CA)		**41.5**	43.6	57.1	**29.4**	44.9	**39.7**	**29.0**	**36.1**	**40.2**
Oracle		44.7	43.9	64.7	31.5	48.8	44.0	31.0	36.7	43.2

4.3 Overall Performance

We compare our method with existing state-of-the-art approaches in Table 2 and Table 3, while the results evaluated by other metrics are provided in Table 4. We present two baselines: proposal-based Faster R-CNN [30] and our fully-convolutional detector denoted as "Ours (w/o adapt.)", both without adaptation. In all the tables, we denote global alignment and center-aware alignment as "GA" and "CA", respectively. To understand how much domain gap our model reduces, we also present the "Oracle" results, in which the model is trained and tested on the target domain using our model. Moreover, we consider two backbone architectures as our feature extractor: VGG-16 [19] or ResNet-101 [13].

Weather Adaptation. In Table 2, we notice that our baseline without adaptation performs similarly (i.e., around 18%) to the F-RCNN baseline, using the VGG-16 backbone. After adaptation, our method (GA + CA) improves our baseline by 17.6% and performs the best compared to other methods in $mAP^r_{0.5}$, especially against the ones [2,14,41] that adopt both global and instance-level alignments. Overall, for both architectures, we consistently show that using the proposed center-aware alignment performs better than global alignment, and combining both is complementary and achieves the best performance.

Table 3. Results of adapting Sim10k/KITTI to Cityscapes. The first and second groups adopt VGG-16 and ResNet-101 as the backbone, respectively. The symbol * indicates that additional training images generated via pixel-level adaptation are used.

		Sim10k	KITTI
Method	Backbone	$mAP^r_{0.5}$	$mAP^r_{0.5}$
Baseline (F-RCNN)	VGG-16	30.1	30.2
DAF [2] $_{CVPR'18}$		39.0	38.5
MAF [14] $_{ICCV'19}$		41.1	41.0
SW-DA [32] $_{CVPR'19}$		42.3	-
SW-DA* [32] $_{CVPR'19}$		47.7	-
SC-DA [41] $_{CVPR'19}$		43.0	42.5
Ours (w/o adapt.)		39.8	34.4
Ours (GA)		45.9	39.1
Ours (CA)		46.6	41.9
Ours (GA+CA)		**49.0**	**43.2**
Oracle		69.7	69.7
Ours (w/o adapt.)	ResNet-101	41.8	35.3
Ours (GA)		50.6	42.3
Ours (CA)		51.1	43.6
Ours (GA+CA)		**51.2**	**45.0**
Oracle		70.4	70.4

Table 4. More mAP metrics of adapting Sim10k/KITTI to Cityscapes using ResNet-101 as the backbone.

Method	Sim10k → Cityscapes						KITTI → Cityscapes					
	mAP	$mAP^r_{0.5}$	$mAP^r_{0.75}$	mAP^r_S	mAP^r_M	mAP^r_L	mAP	$mAP^r_{0.5}$	$mAP^r_{0.75}$	mAP^r_S	mAP^r_M	mAP^r_L
Ours (w/o adapt.)	23.1	41.8	22.4	5.1	26.8	46.6	15.9	35.3	12.8	1.5	17.8	36.5
Ours (GA)	26.4	50.6	25.2	5.7	26.3	57.3	18.8	42.3	14.7	5.0	24.5	35.9
Ours (CA)	26.8	51.1	26.3	**7.5**	27.9	54.6	20.3	43.6	17.3	4.1	25.4	40.8
Ours (GA+CA)	**28.6**	**51.2**	**27.4**	7.1	**30.2**	**58.3**	**22.2**	**45.0**	**20.0**	**5.3**	**28.1**	**43.1**
Oracle (ResNet-101)	44.6	70.4	46.2	15.7	49.2	79.2	44.6	70.4	46.2	15.7	49.2	79.2

Synthetic-to-Real. In the left part of Table 3, we show that our final model (GA+CA) using the VGG-16 backbone performs favorably against existing methods. We note that, compared to a recent method, SW-DA* [32], that adds the augmented data into training via the pixel-level adaptation technique, our result is still better than theirs. We also notice that the improvement from GA-only to GA+CA using the ResNet-101 backbone is not significant. However, we will show that more performance gain can be achieved when using other mAP metrics with a higher standard later.

Cross-camera Adaptation. In the right part of Table 3, we show that our method achieves favorable performance against others, and adding CA consistently improves the results, e.g.., 8.8% and 9.7% gain compared to the baseline without adaptation, using VGG-16 or ResNet-101, respectively.

More Discussions. Although the CA-only model performs competitively against the GA-only model, they essentially focus on different tasks. For global alignment, it tries to align image-level distributions, which is necessary to help reduce the domain gap but may focus too much on background pixels. For our center-aware alignment, we focus more on pixels that are likely to be the foreground, in which the alignment process considers foreground distributions more. As such, they act as a different role, in which combing both is complementary to further improve the performance (i.e., GA+CA).

In addition, in Table 2, we notice that the performance of some categories that are underrepresented such as *truck* and *mbike* is lower than that of other categories. One reason is that these categories contain less foreground pixels in the source domain, in which our center-aware alignment may pay less attention to them. One could adopt a stronger backbone (e.g.., ResNet-101 in Table 2) to improve the performance or use the category prior that allows the model to focus more on those underrepresented categories, which is not in the scope of this work and could be one future work.

4.4 More Results and Analysis

In this section, we provide detailed analysis in the proposed method with more mAP measurements. In addition, we visualize our center-aware maps and more results are provided in the supplementary material.

More mAP Metrics. In Table 4, we show more mAP metrics than $\text{mAP}^r_{0.5}$, to analyze where our method helps the detector adapting to different scenarios. On the Sim10k case, as discussed in Sect. 4.3, we observe that our full model using ResNet-101 does not improve $\text{mAP}^r_{0.5}$ a lot compared with the GA-only model. However, we show that under a more challenging case, e.g., $\text{mAP}^r_{0.75}$, mAP^r_S and mAP^r_M, adding CA improves results over GA-only by 2.2%, 1.4%, and 3.9%, respectively. It validates the usefulness of our center-aware alignment for challenging adaptation cases. Similar observations could be found in the KITTI case. Such measurements also suggest an interesting aspect for domain adaptive object detection to better understand its challenges.

Multi-scale Alignment. To verify the effectiveness of our multi-scale alignment scheme, we conduct an ablation study on Sim10k $\rightarrow$ Cityscapes using the ResNet-101. In Table 5, we compare results using all the scales ($F^3 \sim F^7$), three scales ($F^5 \sim F^7$) via removing the bottom two scales, three scales ($F^3 \sim F^5$) via removing the top two scales, and a single scale F^5. Note that, we choose the single scale as F^5 since it is the middle scale, which has the most influential impact. We show that adding more scales gradually improves the performance on all the metrics, which validates the usefulness of our proposed multi-scale

Fig. 4. Comparisons of response maps on Sim10k-to-Cityscapes. The maps on the first row are extracted from the feature layer F^3 which focuses on smaller objects, while the second row is for the feature layer F^6. After adding the proposed center-aware alignment, the model could focus more on the objects and reduce background noises.

Table 5. Ablation study of our multi-scale alignment using ResNet-101.

Sim10k → Cityscapes						
Aligned Scale	mAP	$\text{mAP}^r_{0.5}$	$\text{mAP}^r_{0.75}$	mAP^r_S	mAP^r_M	MAP^r_L
w/o adapt.	23.1	41.8	22.4	5.1	26.8	46.6
F^5	24.2	48.9	22.4	5.7	24.0	52.4
$F^3 \sim F^5$	26.2	48.7	25.0	6.9	28.7	53.0
$F^5 \sim F^7$	26.1	49.2	25.8	6.2	26.8	54.8
$F^3 \sim F^7$	**28.6**	**51.2**	**27.4**	**7.1**	**30.2**	**58.3**

alignment method. Moreover, $F^3 \sim F^5$ is responsible for smaller objects, in which the $\text{mAP}^r_S/\text{mAP}^r_M$ results are better than the $F^5 \sim F^7$ ones. In contrast, mAP^r_L is better for $F^5 \sim F^7$ as it handles larger objects. This indicate that our multi-scale alignment is effective for handling various size of objects.

Qualitative Analysis. We first show some example results of the response map that our method tries to localize the object. In Fig. 4, the baseline without adaptation has difficulty to find any object centers, while our global alignment method is able to localize some objects. Adding the proposed center-aware alignment enables our method to discover more object centers at different object scales. We also note that, each scale in our model may focus on a different size of object, e.g., the upper example in Fig. 4 may miss larger objects. However, those objects missing at a smaller scale could be identified at another scale.

5 Conclusions

In this paper, we propose a center-aware feature alignment method to tackle the task of domain adaptive object detection. Specifically, we propose to generates pixel-wise maps for localizing object regions, and then use them as the guidance for feature alignment. To this end, we develop a method to discover center-aware regions and perform the alignment procedure via adversarial learning that allows the discriminator to focus on features coming from the object region. In addition, we design the multi-scale feature alignment scheme to handle different object sizes. Finally, we show that incorporating global and center-aware alignments improves domain adaptation for object detection and achieves state-of-the-art performance on numerous benchmark datasets and settings.

Acknowledgment. This work was supported in part by the Ministry of Science and Technology (MOST) under grants MOST 107-2628-E-009-007-MY3, MOST 109-2634-F-007-013, and MOST 109-2221-E-009-113-MY3, and by Qualcomm through a Taiwan University Research Collaboration Project. M.-H. Yang is supported in part by NSF CAREER Grant 1149783.

References

1. Cai, Q., Pan, Y., Ngo, C.W., Tian, X., Duan, L., Yao, T.: Exploring object relation in mean teacher for cross-domain detection. In: CVPR (2019)
2. Chen, Y., Li, W., Sakaridis, C., Dai, D., Gool, L.V.: Domain adaptive faster r-cnn for object detection in the wild. In: CVPR (2018)
3. Cordts, M., et al.: The cityscapes dataset for semantic urban scene understanding. In: CVPR (2016)
4. Dai, J., He, K., Sun, J.: Instance-aware semantic segmentation via multi-task network cascades. In: CVPR (2016)
5. Dai, S., Sohn, K., Tsai, Y.H., Carin, L., Chandraker, M.: Adaptation across extreme variations using unlabeled domain bridges. arXiv preprint arXiv:1906.02238 (2019)
6. Duan, K., Bai, S., Xie, L., Qi, H., Huang, Q., Tian, Q.: Centernet: Keypoint triplets for object detection. In: ICCV (2019)
7. Ganin, Y., Lempitsky, V.: Unsupervised domain adaptation by backpropagation. In: ICML (2015)
8. Geiger, A., Lenz, P., Urtasun, R.: Are we ready for autonomous driving? the kitti vision benchmark suite. In: CVPR (2012)
9. Girshick, R.: Fast r-cnn. In: ICCV (2015)
10. Hariharan, B., Arbeláez, P., Girshick, R., Malik, J.: Simultaneous detection and segmentation. In: ECCV (2014)
11. Hariharan, B., Arbeláez, P., Girshick, R., Malik, J.: Hypercolumns for object segmentation and fine-grained localization. In: CVPR (2015)
12. He, K., Gkioxari, G., Dollar, P., Girshick, R.: Mask R-CNN. In: ICCV (2017)
13. He, K., Zhang, X., Ren, S., Sun, J.: Deep residual learning for image recognition. In: CVPR (2016)
14. He, Z., Zhang, L.: Multi-adversarial faster-rcnn for unrestricted object detection. In: ICCV (2019)
15. Hsu, H.K., et al.: Progressive domain adaptation for object detection. In: WACV (2020)
16. Inoue, N., Furuta, R., Yamasaki, T., Aizawa, K.: Cross-domain weakly-supervised object detection through progressive domain adaptation. In: CVPR (2018)
17. Johnson-Roberson, M., Barto, C., Mehta, R., Sridhar, S.N., Rosaen, K., Vasudevan, R.: Driving in the matrix: Can virtual worlds replace human-generated annotations for real world tasks? In: ICRA (2017)

18. Kang, K., et al.: T-cnn: Tubelets with convolutional neural networks for object detection from videos. In: TCSVT (2018)
19. Karen, S., Andrew, Z.: Very deep convolutional networks for large-scale image recognition. In: ICLR (2015)
20. Karpathy, A., Li, F.F.: Deep visual-semantic alignments for generating image descriptions. In: CVPR (2015)
21. Kim, S., Choi, J., Kim, T., Kim, C.: Self-training and adversarial background regularization for unsupervised domain adaptive one-stage object detection. In: ICCV (2019)
22. Kim, T., Jeong, M., Kim, S., Choi, S., Kim, C.: Diversify and match: A domain adaptive representation learning paradigm for object detection. In: CVPR (2019)
23. Law, H., Deng, J.: Cornernet: Detecting objects as paired keypoints. In: ECCV (2018)
24. Lin, T.Y., Dollar, P., Girshick, R., He, K., Hariharan, B., Belongie, S.: Feature pyramid networks for object detection. In: CVPR (2017)
25. Lin, T.Y., Goyal, P., Girshick, R., He, K., Dollar, P.: Focal loss for dense object detection. In: ICCV (2017)
26. Liu, W., et al.: Ssd: Single shot multibox detector. In: ECCV (2016)
27. Long, J., Shelhamer, E., Darrell, T.: Fully convolutional networks for semantic segmentation. In: CVPR (2015)
28. Long, M., Cao, Y., Wang, J., Jordan, M.I.: Learning transferable features with deep adaptation networks. In: ICML (2015)
29. Redmon, J., Divvala, S., Girshick, R., Farhadi, A.: You only look once: Unified and real-time object detection. In: CVPR (2016)
30. Ren, S., He, K., Girshick, R., Sun, J.: Faster r-cnn: Towards real-time object detection with region proposal networks. In: NeurIPS (2015)
31. Saito, K., Watanabe, K., Ushiku, Y., Harada, T.: Maximum classifier discrepancy for unsupervised domain adaptation. In: CVPR (2018)
32. Saito1, K., Ushiku, Y., Harada, T., Saenko, K.: Strong-weak distribution alignment for adaptive object detection. In: CVPR (2019)
33. Sakaridis, C., Dai, D., Gool, L.V.: Semantic foggy scene understanding with synthetic data. In: IJCV (2018)
34. Tian, Z., Shen, C., Chen, H., He, T.: Fcos: Fully convolutional one-stage object detection. In: ICCV (2019)
35. Tsai, Y.H., Sohn, K., Schulter, S., Chandraker, M.: Domain adaptation for structured output via discriminative patch representations. In: ICCV (2019)
36. Tzeng, E., Hoffman, J., Saenko, K., Darrell, T.: Adversarial discriminative domain adaptation. In: CVPR (2017)
37. Wu, Q., Shen, C., Wang, P., Dick, A., van den Hengel, A.: Image captioning and visual question answering based on attributes and external knowledge. In: TPAMI (2018)
38. Xu, K., et al.: Show and attend and tell: Neural image caption generation with visual attention. In: ICML (2015)
39. Yu, J., Jiang, Y., Wang, Z., Cao, Z., Huang, T.: Unitbox: An advanced object detection network. In: ACMMM (2016)
40. Zhou, X., Zhuo, J., Krähenbühl, P.: Bottom-up object detection by grouping extreme and center points. In: CVPR (2019)
41. Zhu, X., Pang, J., Yang, C., Shi, J., Lin, D.: Adapting object detectors via selective cross-domain alignment. In: CVPR (2019)

URIE: Universal Image Enhancement for Visual Recognition in the Wild

Taeyoung Son[1], Juwon Kang[1], Namyup Kim[1], Sunghyun Cho[2], and Suha Kwak[2](✉)

[1] Department of Computer Science and Engineering, POSTECH, Pohang, Korea
[2] Graduate School of Artificial Intelligence, POSTECH, Pohang, Korea
suha.kwak@postech.ac.kr
http://cvlab.postech.ac.kr/research/URIE/

Abstract. Despite the great advances in visual recognition, it has been witnessed that recognition models trained on clean images of common datasets are not robust against distorted images in the real world. To tackle this issue, we present a Universal and Recognition-friendly Image Enhancement network, dubbed URIE, which is attached in front of existing recognition models and enhances distorted input to improve their performance without retraining them. URIE is universal in that it aims to handle various factors of image degradation and to be incorporated with any arbitrary recognition models. Also, it is recognition-friendly since it is optimized to improve the robustness of following recognition models, instead of perceptual quality of output image. Our experiments demonstrate that URIE can handle various and latent image distortions and improve the performance of existing models for five diverse recognition tasks where input images are degraded.

Keywords: Visual recognition · Image enhancement

1 Introduction

The development of deep learning has made the great advances in visual recognition. Especially, the recent advances in this field have been concentrated on accurate and comprehensive understanding of high-quality images taken in limited environments. However, images in the real world are often corrupted by various factors including adverse weather conditions, sensor noise, under- or over-exposure, motion blur, and compression artifact. Since most of existing recognition models do not consider these factors, their performance could be easily degraded when input image is corrupted [4,11,42], which could be fatal in safety-critical applications like autonomous driving.

A straightforward way to resolve this issue is to improve the quality of input image fed to recognition models by a preprocessing method. Image restoration

Electronic supplementary material The online version of this chapter (https://doi.org/10.1007/978-3-030-58545-7_43) contains supplementary material, which is available to authorized users.

© Springer Nature Switzerland AG 2020
A. Vedaldi et al. (Eds.): ECCV 2020, LNCS 12354, pp. 749–765, 2020.
https://doi.org/10.1007/978-3-030-58545-7_43

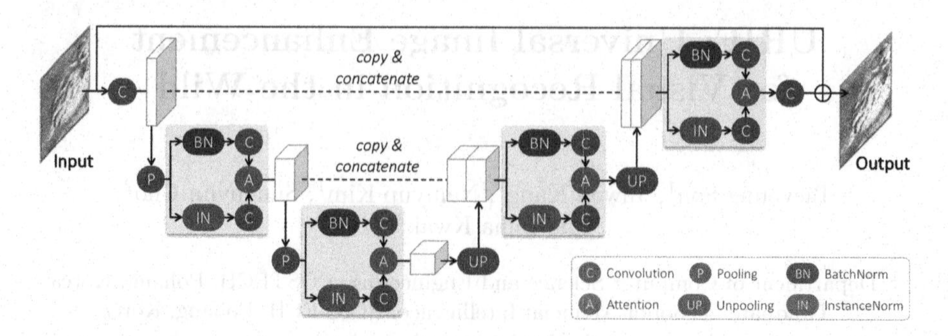

Fig. 1. Overall architecture of URIE. The Selective Enhancement Modules (SEM) are indicated by gray rectangles. Details of these modules are illustrated in Fig. 2.

seems suited to this purpose at first glance, but unfortunately, it is not in general for the following reasons. First, most of restoration methods assume that images are degraded by a single and known factor of corruption. They are thus hard to handle realistic recognition scenarios where images can be corrupted by various factors whose types and degrees are hidden. Second, they are not trained for visual recognition but for human perception. Hence input images restored by them do not always guarantee performance improvement of recognition models as demonstrated in [4,16,24,35]. Finally, recent restoration networks tend to be heavy in computation, thus could make the entire system overly expensive when integrated with recognition models.

This paper addresses the problem through a novel image enhancement model dedicated to robust visual recognition in the wild. We call our model Universal and Recognition-friendly Image Enhancement network, dubbed URIE, since we keep the following properties in our mind when developing the model. (1) *Universally applicable*: Our model aims to be attached in front of any existing recognition models to improve their robustness in any situations without retraining them. To this end, it has to deal with various factors of degradation whose types and intensities are latent, and be applicable to diverse recognition tasks and models. (2) *Recognition friendly*: Our model should be optimized for improving the performance of following recognition models, instead of making images looking plausible. (3) *Computationally efficient*: The complexity of our model has to be low to minimize the computational burden it additionally imposes.

We come up with a new network architecture and its own training strategy to implement these properties. In particular, the architecture of URIE is devised to deal with various and latent distortions efficiently and effectively. To handle diverse image distortions, URIE runs a large number of different image enhancement procedures internally by a single feed-forward path. At the same time, it has the capability to select enhancement procedures appropriate for dealing with the latent type of input distortion. These features are given by Selective Enhancement Module (SEM), a basic building block of URIE. The architecture of URIE composed of SEMs is illustrated in Fig. 1. Each SEM provides two different steps towards image enhancement. Hence, by concatenating SEMs, the network will have an exponentially large number of enhancement procedures. Moreover, each SEM has an attention mechanism that assigns a larger attention weights to its enhancement step more appropriate to deal with the input distortion than the other; this enables URIE to dynamically select useful enhancement procedures depending on the latent type of distortion.

In addition, our new training method enables URIE to be recognition-friendly as well as universal. During training, URIE is coupled to an image classification network pretrained on the original ImageNet dataset [3]. It is then trained in an end-to-end manner to improve the performance of the classifier on distorted images. As training data, we adopt ImageNet-C [11], in which images of the original ImageNet are degraded by 19 classes of distortion. This training strategy gives URIE the recognition-friendly property since the training objective is not the perceptual quality of output image but the recognition performance of the coupled classifier. In addition, we expect that training on such a large-scale dataset with a variety of corruptions allows our model to be less sensitive to the task and architecture of the following recognition model, analogously to ImageNet pretrained networks that have been used for various purposes [22, 23, 30] other than their original target.

The efficacy of URIE is evaluated on five different recognition tasks, in which recognition models are pretrained on clean images and tested with URIE on corrupted ones with no finetuning. Experimental results demonstrate that URIE improves recognition performance in the presence of image distortions including those unseen during training, and regardless of target tasks and recognition networks coupled with it. URIE is also compared with the state-of-the-art image restoration model [32], and outperforms it in terms of both recognition accuracy and computational efficiency. In summary, our contribution is four-fold:

- We introduce a new method towards robust visual recognition. Our method improves the robustness of existing recognition models without retraining them through a single and universal image enhancement module.
- We propose a novel network architecture for image enhancement that can handle various and latent types of image distortions effectively and efficiently.
- We present a new training strategy that enables our enhancement model to be universally applicable and recognition-friendly.
- In our experiments, our model improves the recognition performance of existing models substantially in the presence of various image distortions regardless of tasks and models it is tested on.

2 Related Work

2.1 Fragility of Visual Recognition Models

Fragility of deep neural networks for visual recognition has been studied to comprehend their robustness against challenging conditions that they often encounter in the real world. Diamond *et al.* [4] demonstrated that image classification networks are sensitive to noise and blur. Similarly, Pei *et al.* [24] showed that haze can degrade the performance of classification networks. On the other hand, Zendel *et al.* [42] defined various factors of image degradation appearing frequently in autonomous driving scenarios, and analyzed the robustness of existing semantic segmentation networks to them. Hendrycks and Dietterich [11] studied performance degradation of image classifiers by simulated image distortions. These results indicate that existing recognition networks trained on clean images are fragile in the presence of image distortion. Geirhos *et al.* [7] showed that one way to resolve this issue is to finetune models to distorted images but such models are not well generalized when tested on other distortion types.

Our work is motivated by these observations. We address the above problem through an image enhancement model that deals with a variety of image distortions and is optimized to improve the robustness of recognition models.

2.2 Recognition of Distorted Images

Previous methods to robust recognition of distorted images can be grouped into two classes: *direct recognition* and *recognition-friendly image enhancement.*

Direct Recognition of Corrupted Images. Classical methods in this category rely on visual features robust against image distortions like blur-invariant features [9,38]. Recent methods focus on developing new network architectures and training strategies dedicated to the robust recognition. Wang *et al.* [36] and Singh *et al.* [31] developed neural networks for classifying low resolution images, and Lee *et al.* [14] proposed a multi-task network for understanding road markings in rainy days and nighttime. Also, domain adaptation techniques have been widely adopted for corruption-aware training of recognition models. Considering different types of degradation as individual domains, they allow models for object detection and semantic segmentation to be adapted to adverse weather conditions [2,27,39] and nighttime images [26].

Recognition-Friendly Image Enhancement. Models in this category are designed and trained to enhance input images so that it becomes more informative and task-optimized for following recognizers. Diamond *et al.* [4] proposed a differentiable image processing module learned jointly with a classifier for improved classification of noisy and blurred images. Similarly, Liu *et al.* [19] developed denoising networks for classification and semantic segmentation of noisy images. Gomez *et al.* [8] studied network architectures that enhance images taken in difficult illumination conditions for robust visual odometry. Sharma *et al.* [29] investigated a way to dynamically predict image processing filters that are applied to input images and make them more classification-friendly.

The aforementioned methods of both categories are limited in terms of universality as they handle a single degradation type and/or are optimized for only one recognition task. On the other hand, our goal is to build a single enhancement model that can cope with various types of image degradation and be attached to any architectures for any recognition tasks.

2.3 Image Restoration

Most of image restoration methods focus on a single type of image distortion (*e.g.*, denoising [43], deblurring [21], dehazing [15], deraining [40], and super-resolution [5]), and consequently they are incapable of handling real world images where the type and intensity of distortion are hidden in general. A few network architectures for image restoration are designed to handle various types of degradation [18,33], but not at the same time as they have to be trained for each restoration objective. Recently, the above issue of conventional image restoration has been addressed by new architectures that can deal with various and latent factors of degradation [32,41]. However, they are not optimized for image recognition, thus may not guarantee improved performance of coupled visual recognition models. In addition, since they are computationally heavy, they will increase the complexity of overall system significantly if integrated.

3 URIE: Architecture and Training

Our final target in this paper is to improve the robustness of existing recognition models to diverse image distortions *without retraining them*. URIE is proposed to address this challenging problem by converting an input image degraded by latent distortion into another that can be recognized more reliably by any existing models. To this end, it is designed and trained to be universally applicable to various image distortions and recognition architectures, while being recognition-friendly for improved performance of the following recognition models.

Specifically, URIE is devised to derive many different enhancement procedures from a single feed-forward architecture, and at the same time, to dynamically select procedures useful to handle the input distortion. Thanks to these features, URIE can be universally applicable to a large variety of image distortions whose types and intensities are latent. This architecture is embodied by a series of basic building blocks, which we call SEMs. Furthermore, we provide a new training method that enables URIE to be recognition-friendly and universally applicable to any recongnition models.

The remainder of this section presents details of SEM, discusses overall architecture and other design points of URIE, and describes our training strategy that makes URIE universally applicable and recognition-friendly.

3.1 Selective Enhancement Module

In URIE, an image enhancement procedure is represented as a sequence of convolution layers. As a basic building block of the network, SEM provides individual steps towards image enhancement (*i.e.*, convolution layers) so that a series of such steps forms an enhancement procedure. To be specific, a single SEM presents two diverse enhancement steps, and has the capability to estimate which one among them is more appropriate to deal with the input distortion. The detailed architecture of SEM is illustrated in Fig. 2.

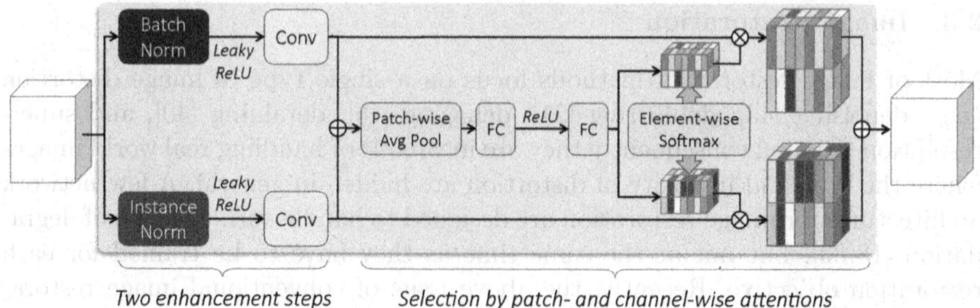

Fig. 2. Details of SEM. $\oplus$ and $\otimes$ indicate element-wise summation and multiplication between feature maps, respectively.

The two enhancement steps in SEM have separate convolution layers associated with different normalization operations, Instance Normalization (IN) [34] and Batch Normalization (BN) [12]. We adopt these two different normalization operations to learn a pair of diverse and complementary enhancement functions. IN has been used in neural style transfer to normalize the style of image while keeping its major content by normalizing statistics per image independently. Regarding a distortion type as an image style, we expect the enhancement step with IN to moderate the effect of distortion for individual images. On the other hand, the enhancement step with BN maintains the distortion information disregarded by the other, and is trained by considering a variety of distorted images together within each minibatch. We thus expect that it is complementary to its IN counterpart and useful for understanding input distortion. Empirical justification of using IN and BN can be found in the supplementary material.

The two outputs of these enhancement steps are fed to the attention module that determines which one among them is more useful to deal with the input distortion for each channel and each local patch of the input image. Let $F_{\text{in}} \in \mathbb{R}^{w\times h\times C}$ and $F_{\text{bn}} \in \mathbb{R}^{w\times h\times C}$ be the output feature maps of the two enhancement steps with resolution of $w \times h$ and C channels. They are first aggregated by element-wise summation into a single feature map, then we divide it into the 4×4 regular grid and conduct average pooling on each cell so that feature vectors for 16 local patches are obtained. The local feature vectors are then concatenated into one large vector, which is in turn fed to the following fully connected layers. The final output of the fully connected layers are two score vectors $\mathbf{s}_{\text{in}} \in \mathbb{R}^{16C}$ and $\mathbf{s}_{\text{bn}} \in \mathbb{R}^{16C}$, each of whose elements indicates how useful a specific channel of F_{in} or F_{bn} is for handling the distortion in a specific local patch.

The score vectors $\mathbf{s}_{\text{in}}$ and $\mathbf{s}_{\text{bn}}$ are used to compute attention weights. They are first reshaped into score tensors $S_{\text{in}} \in \mathbb{R}^{4\times 4\times C}$ and $S_{\text{bn}} \in \mathbb{R}^{4\times 4\times C}$, then converted to attention weight tensors A_{in} and A_{bn} of the same dimension through the element-wise softmax operation:

$$A_{\text{in}}[i,j,c] = \frac{\exp S_{\text{in}}[i,j,c]}{\exp S_{\text{in}}[i,j,c] + \exp S_{\text{bn}}[i,j,c]}, \tag{1}$$

$$A_{\text{bn}}[i,j,c] = \frac{\exp S_{\text{bn}}[i,j,c]}{\exp S_{\text{in}}[i,j,c] + \exp S_{\text{bn}}[i,j,c]}. \tag{2}$$

The spatial dimension of the two attention tensors are enlarged to $w \times h$ by bilinear interpolation to be multiplied to the outputs of the two enhancement steps, F_{in} and F_{bn}. The final output of SEM, denoted by $Y \in \mathbb{R}^{w\times h\times C}$, is then computed by weighted summation of F_{in} and F_{bn}:

$$Y = \left(A'_{\text{in}} \otimes F_{\text{in}}\right) \oplus \left(A'_{\text{bn}} \otimes F_{\text{bn}}\right), \tag{3}$$

where A'_{in} and A'_{bn} are the resized attention tensors, $\otimes$ indicates element-wise multiplication, and $\oplus$ denotes element-wise summation.

Note that the attention tensors have both spatial (*i.e.*, patch) and channel dimensions. In Eq. (3), the channel-wise attention facilitates more flexible and diverse integration of the two enhancement steps since different channels can be processed by different enhancement steps; although a SEM provides only two enhancement steps, the number of their output combinations is in principle extremely large. On the other hand, the patch-wise attention allows to process different image regions with different enhancement processes.

3.2 Overall Architecture

The overall architecture of URIE is illustrated in Fig. 1. Its structure resembles U-Net [25], but is clearly distinct from the conventional U-Net architectures in that its all layers, except for the first and the last, are not ordinary convolution layers but SEMs introduced in Sect. 3.1. The reason for following the design principle of U-Net is two-fold. First, it effectively enlarges the receptive field so as to capture high-level semantic information. This is essential for URIE that performs recognition-aware image enhancement, unlike ordinary restoration models that do not necessarily need to understand image contents. Second, it is computationally efficient since it reduces the sizes of intermediate features using pooling operations. Reducing feature resolution may degrade the perceptual quality of output as a side effect, which however is not our concern since our objective is improved recognition performance of following recognition model.

Table 1. Details of URIE. $W \times H$ denotes the input resolution.

	Output	Layer specification
(1)	$W \times H$	9×9 conv, 32, stride 1
(2)	$W/2 \times H/2$	2×2 max pool, stride 2
(3)	$W/2 \times H/2$	SEM[3×3, 64, stride 1]
(4)	$W/4 \times H/4$	2×2 max pool, stride 2
(5)	$W/4 \times H/4$	SEM[3×3, 64, stride 1]
(6)	$W/2 \times H/2$	$2\times$ bilinear upsampling
(7)	$W/2 \times H/2$	concat[(3), (6)]
(8)	$W/2 \times H/2$	SEM[3×3, 32, stride 1]
(9)	$W \times H$	$2\times$ bilinear upsampling
(10)	$W \times H$	concat[(5), (9)]
(11)	$W \times H$	SEM[3×3, 16, stride 1]
(12)	$W \times H$	3×3 conv, 3, stride 1

The key components in this architecture are SEMs, which allow URIE to handle multiple and latent factors of image degradation efficiently and effectively. Since each SEM provides two enhancement steps and a sequence of such steps can be regarded as an individual enhancement procedure, the concatenation of SEMs leads to an exponentially large number of enhancement procedures, which can handle a large variety of distortions. Moreover, through the attentions drawn by SEMs, URIE can dynamically select enhancement procedures depending on the latent type of input distortion. Since the selection is done for each local patch of the input image through the patch-wise attentions, URIE can handle different image regions with different enhancement procedures. This is more useful and realistic than applying a single enhancement function to the whole input image since individual regions of the input could be distorted by different factors in realistic scenarios, *e.g.*, images with locally different haze densities with respect to depths, and images partially over- and under-exposed at the same time.

The specification of URIE, including the size of convolution kernels and the number of channels per layer, is given in Table 1. In the table, SEM[$k \times k$, C, stride s] means that each convolution layer in the SEM consists of C number of $k \times k$ kernels with stride s. Note that we set the size of convolution kernels for the first layer especially large to capture rich information about image distortion and to enlarge the receptive field of our shallow network architecture.

3.3 Training Strategy

Our training scheme for URIE is based on two key ideas. (1) *Recognition-aware loss*: We employ a pretrained recognition network to provide URIE with supervisory signals that lead it to be recognition-friendly. Specifically, URIE is coupled with a recognition network that is pretrained on clean images and frozen during training. It is then trained to minimize a recognition loss of the pretrained model on distorted images in an end-to-end manner. (2) *Large-scale learning with distorted images*: We suggest learning URIE using a large-scale image dataset in the presence of diverse distortions so that it may learn enhancement functions insensitive to distortion types and recognition models. This is motivated by the fact that deep neural networks trained in large-scale datasets have been used for various purposes other than their original targets (*e.g.*, ImageNet pretrained classifiers used for other recognition tasks [22,23,30]).

In detail, we employ ResNet-50 [10] pretrained on the original ImageNet [3] as the coupled recognition model, and adopt the ImageNet-C dataset [11] as our training data. URIE is then trained on the ImageNet-C to minimize the cross-entropy loss of the pretrained model. The training dataset consists of images from the original ImageNet yet corrupted artificially by in total 19 different distortions with 5 intensity levels. We use only 15 distortion classes[1] among them to degrade training images, and call them *seen corruptions*. The other distortion classes[2] are kept as *unseen corruptions* for evaluating the generalization ability of our model in testing. Note that clean images from the original ImageNet are also used for training to prevent URIE from being biased to distortions.

[1] Gaussian noise, shot noise, impulse noise, defocus blur, glass blur, motion blur, zoom blur, snow, frost, fog, brightness, contrast, elastic transform, pixelation, jpeg.

[2] Speckle noise, Gaussian blur, spatter, saturation.

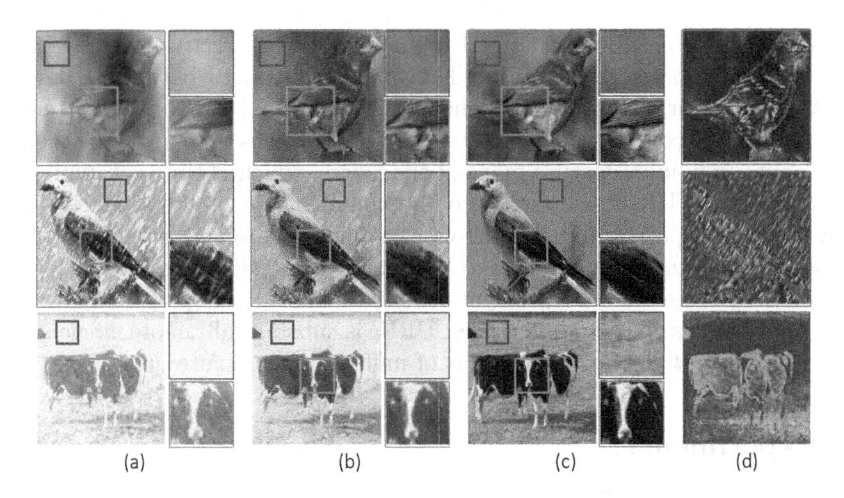

Fig. 3. Example outputs of URIE. (a) Distorted input images. (b) Outputs of URIE. (c) Ground-truth images. (d) Magnitudes of per-pixel intensity change by URIE.

3.4 Discussion

How Our Model Works. To investigate how URIE works, we apply the model trained in Sect. 3.3 to random images sampled from other datasets and corrupted by the same synthetic distortions with the ImageNet-C. Note that URIE is directly used as-is with no finetuning. Qualitative results of URIE on these images are given in Fig. 3. Interestingly, URIE does alleviate the effect of distortions to some degree although it is not trained explicitly for that purpose. More importantly, Fig. 3(b) and 3(d) demonstrate that it tends to manipulate salient local areas only (*e.g.*, object parts) and ignore remainders (*e.g.*, background). This behavior is a result of the recognition-aware learning in Sect. 3.3 that prompts the model to focus more on image regions important for recognition.

Comparison to SKNet. URIE shares a similar idea with the Selective Kernel Network (SKNet) [17] that adopts a self-attention mechanism to utilize intermediate feature maps selectively. However, the two models are clearly distinct since their overall architectures and training schemes are totally different due to their different objectives: image classification in SKNet, and image enhancement in URIE. Besides the above, their attention mechanisms are different in the following aspects. First, SKNet selects from convolution layers of different kernel sizes to implement dynamic receptive fields, but URIE selects from those with different normalization operations, IN and BN, to adapt dynamically to the latent input distortion. Second, SKNet draws only channel-wise attentions, but URIE takes both channel and spatial locations into account when drawing attentions to treat different image regions differently.

Comparison to OWAN. OWAN [32] is the current state of the art in image restoration. It is similar with URIE in that both models consider multiple and latent types of image distortion and use attentions to select appropriate image manipulation process dynamically. The major differences between URIE and OWAN lie in their techniques impelementing the ideas. OWAN draws attentions on a number of convolution layers to select useful ones among them, thus has to compute individual feature maps of the layers, which is time consuming. On the other hands, URIE selects from only two enhancement steps, but mix their outputs up in a patch- and channel-wise manner by drawing attention weights on individual channels and local patches. This scheme allows our model to enjoy an extremely large number of output combinations while keeping computational efficiency. Thanks to these differences, URIE is substantially more efficient than OWAN, *about 6 times faster* in terms of multiply-accumulate operations.

4 Experiments

In this section, we empirically verify that URIE improves performance of diverse recognition models for image classification, object detection, and semantic segmentation on real and artificially degraded images. For image restoration performance of URIE, refer to Table 5 of the supplementary material.

4.1 Training Configurations

Each training image is first resized to 256 × 256, cropped to 224 × 224, and flipped horizontally at random for data augmentation. It is then degraded on-the-fly by the synthetic distortion generator of the ImageNet-C [11], where the type and intensity of distortion are chosen at random. Note that only the 15 seen corruptions are used during training as described in Sect. 3.3. For optimization, we employ Adam [13] with learning rate 0.001, and decay the learning rate every 8 epochs by 10. Our model is trained for 30 epochs with mini-batches of size 112.

4.2 Experimental Configurations

For evaluation of our model, we employ widely used image recognition datasets, ImageNet [3], CUB [37], and PASCAL VOC [6], and generate three different versions of their test or validation sets: *Clean*–original uncorrupted images, *Seen*–images degraded by the 15 seen corruptions, and *Unseen*–images degraded by the 4 unseen corruptions. We did not generate the datasets on-the-fly but prior to evaluation so that they are fixed during testing. In addition, our model is evaluated on the collection of real haze images introduced in [24].

To verify the effectiveness of URIE, it is compared with two different image processing models. One of them is OWAN [32], the state-of-the-art image restoration model devised to deal with multiple and latent types of image distortion. OWAN is trained on the same data with ours for fair comparisons, but by using the original loss proposed in [32] since it is not straightforward to train this model with our recognition-aware loss on the large-scale dataset due to its heavy computational complexity (refer to Sect. 3.4). The other one is URIE-MSE, which

Table 2. Classification accuracy on the ImageNet dataset. The numbers in parentheses indicate the differences from the baseline. V16, R50, and R101 denote VGG-16, ResNet-50, and ResNet-101, respectively.

	OWAN			URIE-MSE			URIE		
	Clean	Seen	Unseen	Clean	Seen	Unseen	Clean	Seen	Unseen
V16	69.8 (-0.2)	31.6 (+2.0)	39.8 (+0.8)	59.5 (-10.5)	32.3 (+2.7)	38.1 (-0.9)	67.1 (-2.9)	42.4 (+12.8)	44.8 (+5.8)
R50	74.4 (-0.4)	42.6 (+2.0)	50.3 (+1.2)	66.7 (- 7.8)	44.5 (+3.9)	49.3 (+0.2)	72.9 (-1.6)	55.1 (+14.5)	56.5 (+7.4)
R101	75.9 (0.0)	47.4 (+1.5)	54.6 (+0.8)	68.8 (- 7.1)	50.0 (+4.1)	54.0 (+0.2)	74.1 (-1.8)	57.8 (+11.9)	59.4 (+5.6)

Table 3. Classification accuracy on the CUB dataset. The numbers in parentheses indicate the differences from the baseline. V16, R50, and R101 denote VGG-16, ResNet-50, and ResNet-101, respectively.

	OWAN			URIE-MSE			URIE		
	Clean	Seen	Unseen	Clean	Seen	Unseen	Clean	Seen	Unseen
V16	79.1 (0.0)	50.4 (+4.2)	42.3 (+1.7)	68.5 (-11.6)	50.5 (+4.3)	39.6 (-1.0)	77.9 (-1.2)	58.8 (+12.6)	48.9 (+8.3)
R50	84.6 (0.0)	52.8 (+3.3)	48.2 (+1.2)	76.5 (- 8.1)	58.4 (+8.9)	49.3 (+2.1)	83.7 (-0.9)	64.7 (+15.2)	54.9 (+7.7)
R101	84.9 (-0.1)	55.5 (+3.6)	48.5 (+1.8)	76.2 (- 8.8)	59.9 (+8.0)	50.5 (+3.8)	84.1 (-0.9)	67.1 (+15.2)	56.6 (+9.9)

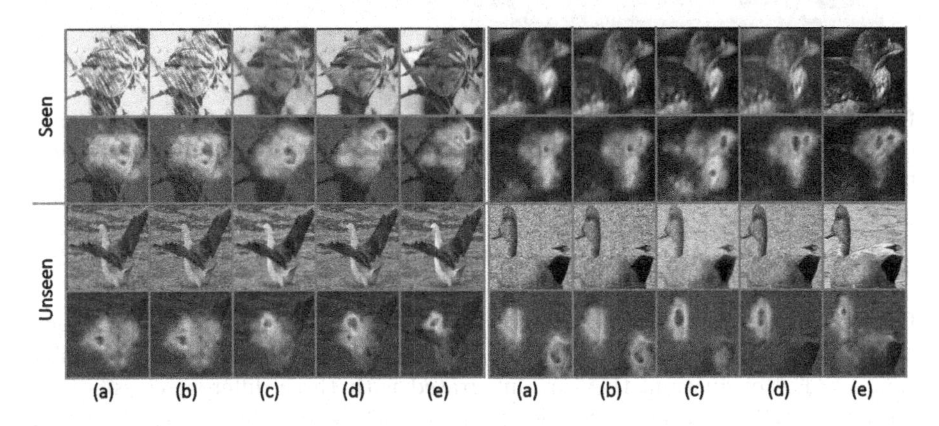

Fig. 4. Qualitative results on the CUB dataset. (a) Input distorted images. (b) OWAN. (c) URIE-MSE. (d) URIE. (e) Ground-truth images. For all images, their grad-CAMs drawn by the ResNet-50 classifier are presented alongside. Examples in the first row are degraded by seen corruptions and the others are by unseen corruptions.

is a restoration-oriented variant of URIE and trained using a Mean-Squared-Error loss instead of the recognition-aware loss. URIE and these two models are evaluated in terms of performance of coupled recognizers on the aforementioned datasets. Note that they are never finetuned to test datasets, and all the recognizers coupled with them are trained on the clean images and tested on the distorted images as-is.

Table 4. Object detection performance of SSD 300 in mAP (%) on the VOC 2007 dataset. The numbers in parentheses indicate the differences from the baseline.

OWAN			URIE-MSE			URIE		
Clean	Seen	Unseen	Clean	Seen	Unseen	Clean	Seen	Unseen
77.4 (0.0)	51.3 (+1.7)	60.2 (+0.9)	76.5 (-0.9)	52.7 (+3.1)	59.9 (+0.6)	76.5 (-0.9)	59.4 (+9.8)	62.7 (+3.4)

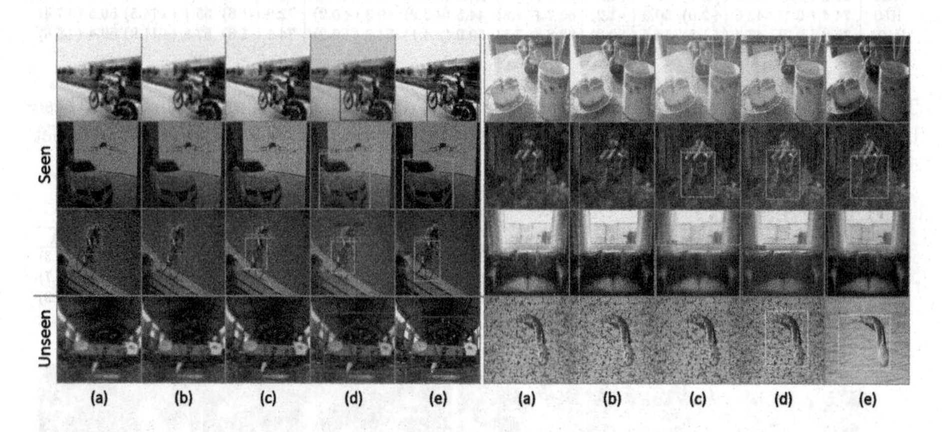

Fig. 5. Qualitative results of SSD 300 on the VOC 2007 dataset. (a) Corrupted input. (b) OWAN. (c) URIE-MSE. (d) URIE. (e) Ground-truth. Examples in the top three rows are degraded by seen corruptions and the others are by unseen corruptions.

4.3 Performance Evaluation

Classification on ImageNet. We first evaluate and compare the effectiveness of URIE and the other two models for classification on the ImageNet validation set. For evaluation, they are integrated with three different classification networks: VGG-16 [30], ResNet-50 [10] and ResNet-101 [10]. As can be seen in Table 2, the performance improvement by URIE is substantial in the case of seen corruptions and nontrivial for unseen corruptions as well, regardless of the classification network it is integrated with. These results suggest that URIE is universally applicable to diverse corruptions and to various recognition models. Also, the performance gap between URIE and the other two methods is significant, which verifies the advantage of our method. In particular, the difference between URIE and URIE-MSE in performance shows the clear advantage of our recognition-aware training. Moreover, URIE-MSE is comparable to OWAN, although it is six times more efficient as discussed in Sect. 3.4. This demonstrates the advantage of the network architecture we proposed. Unfortunately, URIE marginally degrades the performance in the case of *clean*; we suspect that the architecture of URIE could be biased to distortions while OWAN has more skip connections and better preserves input consequently.

Fine-grained Classification on CUB. Our model and the others are evaluated on the CUB validation set to examine if they work universally even in other domains where data distributions are different from that of the training data,

Table 5. Semantic segmentation performance of DeepLab v3 in mIoU (%) on the VOC 2012 dataset. The numbers in parentheses indicate the differences from the baseline.

OWAN			URIE-MSE			URIE		
Clean	Seen	Unseen	Clean	Seen	Unseen	Clean	Seen	Unseen
78.7 (-0.2)	58.2 (+3.3)	65.4(+2.1)	78.1 (-0.8)	60.1 (+5.2)	63.6 (+0.3)	78.6 (-0.3)	67.0 (+12.1)	67.2 (+3.9)

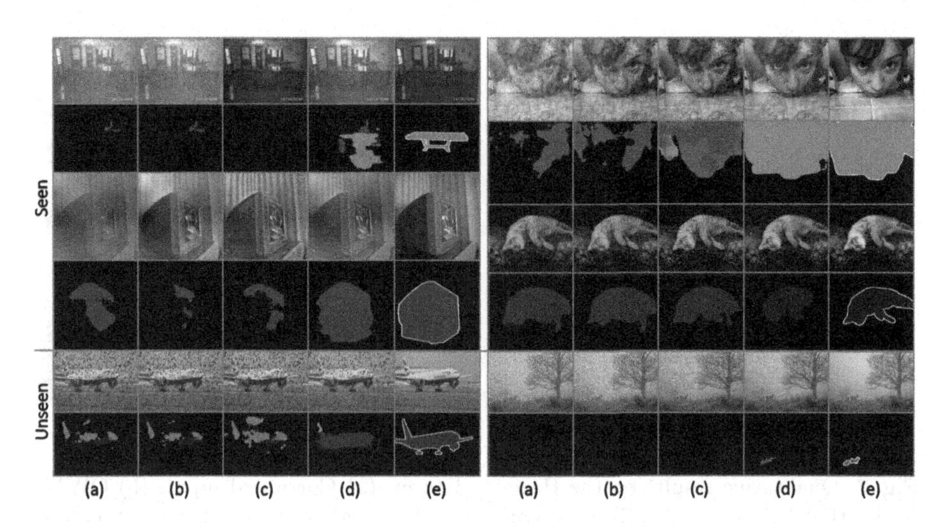

Fig. 6. Qualitative results of DeepLab v3 on the VOC 2012 dataset. (a) Corrupted input. (b) OWAN. (c) URIE-MSE. (d) URIE. (e) Ground-truth. Examples in the top two rows are degraded by seen corruptions and the others are by unseen corruptions.

i.e., ImageNet. The evaluation setting in this dataset is the same with that of the ImageNet classification. As summarized in Table 3, all models turn out to be transferable to some degree in spite of the domain gap between the ImageNet and CUB datasets; we suspect that this is an effect of the large-scale training using the ImageNet-C. However, URIE demonstrates its superiority over the other models by outperforming them substantially. Figure 3 presents qualitative results of the models and their grad-CAMs [28] computed by the ResNet-50 classifier using the ground-truth labels. In this figure, grad-CAMs computed from the results of URIE are closest to those of ground-truth images, which means that URIE best recovers image regions critical for correct classification.

Object Detection and Segmentation on PASCAL VOC. This time we evaluate the effectiveness of our model for object detection and semantic segmentation on the PASCAL VOC dataset to validate its universal applicability to totally different recognition tasks. We employ SSD 300 [20] for object detection and DeepLab v3 [1] for semantic segmentation as the recognition models. Following the convention, the object detection model is evaluated on the VOC 2007 test set while the segmentation model is tested on the VOC 2012 validation set. Table 4 and 5 summarize performance of the two recognition models, and Fig. 5 and 6 present their qualitative results when coupled with URIE and the

Table 6. Accuracy of the ResNet-50 classifier on the Haze-20 and HazeClear-20 datasets. The numbers in parentheses indicate the differences from the baseline.

OWAN		URIE-MSE		URIE	
Haze	HazeClear	Haze	HazeClear	Haze	HazeClear
58.7 (+1.7)	97.8 (-0.1)	53.4 (-3.6)	94.4 (-3.5)	60.8 (+3.8)	97.6 (-0.3)

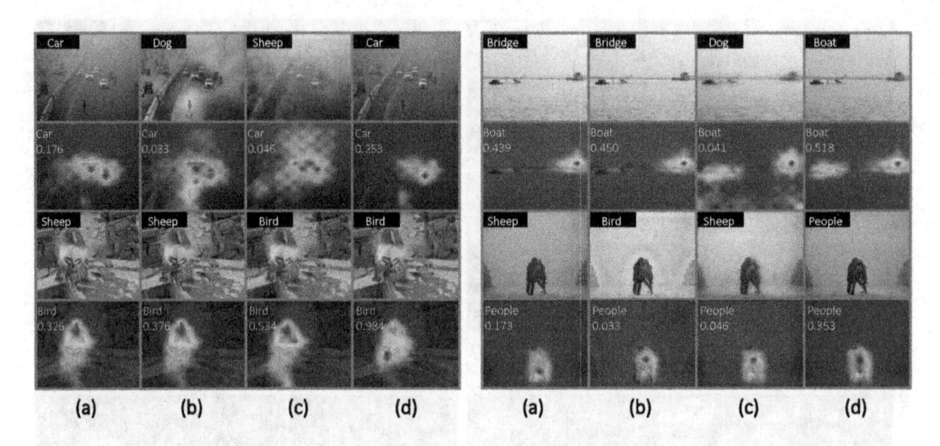

Fig. 7. Qualitative results on the Haze-20 dataset. (a) Corrupted input. (b) OWAN. (c) URIE-MSE. (d) URIE. Top-1 prediction of the ResNet-50 classifier together with its confidence score and grad-CAM are presented alongside per example.

two restoration models. In both of object detection and semantic segmentation, URIE improves recognition performance noticeably, and clearly outperforms the other two models for both seen and unseen corruptions.

Classification on Haze-20. Finally, the effectiveness of URIE is evaluated on the Haze-20 dataset [24], containing haze images of 20 object classes captured in real environments. This dataset is paired with another set, HazeClear-20, a collection of real haze-free images of the same 20 object categories. URIE and the two restoration models are applied to both of datasets and evaluated in terms of performance of a ResNet-50 classifier taking their results as input. This setting is challenging for the three models since they are trained only with synthetically distorted images. Moreover, Pei *et al.* [24] demonstrated by extensive experiments that existing dehazing algorithms do not help improve classification performance on the Haze-20 dataset. Nevertheless, as summarized in Table 6, URIE is still able to improve the classification accuracy in the presence of real-world haze, and at the same time, outperforms the two restoration methods. Figure 7 shows that URIE enhances areas around objects better than OWAN and URIE-MSE, and consequently allows the coupled classifier to predict the correct class label and draw attention most accurately.

5 Conclusion

This paper has presented a new image enhancement network dedicated to robust visual recognition in the presence of input distortion. Our network is universally applicable to various types of image distortions and many different recognition models. At the same time, it is recognition-friendly since it is optimized to improve robustness of recognition models taking its output as input. These features of our model are given by a novel network architecture and a training strategy we carefully designed. The advantages of our model have been verified throughout experiments in various settings. We however have found that our model marginally degrades performance on distortion-free images. Next on our agenda is to resolve this issue by revising the model architecture.

Acknowledgement. This work was supported by Samsung Research Funding & Incubation Center of Samsung Electronics under Project Number SRFC-IT1801-05.

References

1. Chen, L.C., Zhu, Y., Papandreou, G., Schroff, F., Adam, H.: Encoder-decoder with atrous separable convolution for semantic image segmentation. In: Proceedings of European Conference on Computer Vision (ECCV) (2018)
2. Chen, Y., Li, W., Sakaridis, C., Dai, D., Van Gool, L.: Domain adaptive faster r-cnn for object detection in the wild. In: The IEEE Conference on Computer Vision and Pattern Recognition (CVPR) (2018)
3. Deng, J., Dong, W., Socher, R., Li, L.J., Li, K., Fei-Fei, L.: ImageNet: a large-scale hierarchical image database. In: Proceedings of IEEE Conference on Computer Vision and Pattern Recognition (CVPR) (2009)
4. Diamond, S., Sitzmann, V., Boyd, S.P., Wetzstein, G., Heide, F.: Dirty pixels: Optimizing image classification architectures for raw sensor data. arXiv preprint arXiv:1701.06487 (2017)
5. Dong, C., Loy, C.C., He, K., Tang, X.: Image super-resolution using deep convolutional networks. IEEE Trans. Pattern Anal. Mach. Intell (TPAMI) **38**(2), 295–307 (2016)
6. Everingham, M., et al.: The pascal visual object classes (VOC) challenge. Int. J. Comput. Vis. **88**, 303–338 (2010). https://doi.org/10.1007/s11263-009-0275-4
7. Geirhos, R., Temme, C.R.M., Rauber, J., Schütt, H.H., Bethge, M., Wichmann, F.A.: Generalisation in humans and deep neural networks. In: Proceedings of Neural Information Processing Systems (NeurIPS) (2018)
8. Gomez, R., Zhang, Z., González-Jiménez, J., Scaramuzza, D.: Learning-based image enhancement for visual odometry in challenging hdr environments. In: Proceedings of International Conference on Robatics and Automation (ICRA) (2018)
9. Gopalan, R., Taheri, S., Turaga, P., Chellappa, R.: A blur-robust descriptor with applications to face recognition. IEEE Trans. Pattern Anal. Mach. Intell. (TPAMI) **34**(6), 1220–1226 (2012)
10. He, K., Zhang, X., Ren, S., Sun, J.: Deep residual learning for image recognition. In: Proceedings of IEEE Conference on Computer Vision and Pattern Recognition (CVPR) (2016)

11. Hendrycks, D., Dietterich, T.: Benchmarking neural network robustness to common corruptions and perturbations. In: Proceedings of International Conference on Learning Representations (ICLR) (2019)
12. Ioffe, S., Szegedy, C.: Batch normalization: Accelerating deep network training by reducing internal covariate shift. In: Proceedings of International Conference on Machine Learning (ICML) (2015)
13. Kingma, D.P., Ba, J.: Adam: a method for stochastic optimization. In: Proceedings of International Conference on Learning Representations (ICLR) (2015)
14. Lee, S., et al.: VPGNet: vanishing point guided network for lane and road marking detection and recognition. In: Proceedings of IEEE International Conference on Computer Vision (ICCV) (2017)
15. Li, B., Peng, X., Wang, Z., Xu, J., Feng, D.: Aod-net: all-in-one dehazing network. In: Proceedings of IEEE International Conference on Computer Vision (ICCV) (2017)
16. Li, S., et al.: Single image deraining: a comprehensive benchmark analysis. In: Proceedings of IEEE Conference on Computer Vision and Pattern Recognition (CVPR) (2019)
17. Li, X., Wang, W., Hu, X., Yang, J.: Selective kernel networks. In: Proceedings of IEEE Conference on Computer Vision and Pattern Recognition (CVPR) (2019) 9
18. Liu, D., Wen, B., Fan, Y., Loy, C.C., Huang, T.S.: Non-local recurrent network for image restoration. In: Proceedings of Neural Information Processing Systems (NeurIPS) (2018)
19. Liu, D., Wen, B., Liu, X., Wang, Z., Huang, T.S.: When image denoising meets high-level vision tasks: a deep learning approach. In: Proceedings of International Joint Conference on Artificial Intelligence (IJCAI) (2018)
20. Liu, W., et al.: SSD: Single shot multibox detector. In: Proceedings of European Conference on Computer Vision (ECCV) (2016)
21. Nah, S., Hyun Kim, T., Mu Lee, K.: Deep multi-scale convolutional neural network for dynamic scene deblurring. In: Proceedings of IEEE Conference on Computer Vision and Pattern Recognition (CVPR) (2017)
22. Noh, H., Hong, S., Han, B.: Learning deconvolution network for semantic segmentation. In: Proceedings of IEEE International Conference on Computer Vision (ICCV) (2015)
23. Oquab, M., Bottou, L., Laptev, I., Sivic, J.: Learning and transferring mid-level image representations using convolutional neural networks. In: Proceedings of IEEE Conference on Computer Vision and Pattern Recognition (CVPR) (2014)
24. Pei, Y., Huang, Y., Zou, Q., Lu, Y., Wang, S.: Does haze removal help cnn-based image classification? In: Proceedings of European Conference on Computer Vision (ECCV) (2018)
25. Ronneberger, O., Fischer, P., Brox, T.: U-net: Convolutional networks for biomedical image segmentation. In: Proceedings of Medical Image Computing and Computer-Assisted Intervention (MICCAI) (2015)
26. Sakaridis, C., Dai, D., Gool, L.V.: Guided curriculum model adaptation and uncertainty-aware evaluation for semantic nighttime image segmentation. In: Proceedings of IEEE International Conference on Computer Vision (ICCV) (2019)
27. Sakaridis, C., Dai, D., Hecker, S., Van Gool, L.: Model adaptation with synthetic and real data for semantic dense foggy scene understanding. In: Proceedings of European Conference on Computer Vision (ECCV) (2018)
28. Selvaraju, R.R., Cogswell, M., Das, A., Vedantam, R., Parikh, D., Batra, D.: Grad-cam: Visual explanations from deep networks via gradient-based localization. In: Proceedings of IEEE International Conference on Computer Vision (ICCV) (2017)

29. Sharma, V., Diba, A., Neven, D., Brown, M.S., Van Gool, L., Stiefelhagen, R.: Classification-driven dynamic image enhancement. In: The IEEE Conference on Computer Vision and Pattern Recognition (CVPR) (2018)
30. Simonyan, K., Zisserman, A.: Very deep convolutional networks for large-scale image recognition. In: Proceedings of International Conference on Learning Representations (ICLR) (2015)
31. Singh, M., Nagpal, S., Singh, R., Vatsa, M.: Dual directed capsule network for very low resolution image recognition. In: Proceedings of IEEE International Conference on Computer Vision (ICCV) (2019)
32. Suganuma, M., Liu, X., Okatani, T.: Attention-based adaptive selection of operations for image restoration in the presence of unknown combined distortions. In: Proceedings of IEEE Conference on Computer Vision and Pattern Recognition (CVPR) (2019)
33. Tai, Y., Yang, J., Liu, X., Xu, C.: Memnet: A persistent memory network for image restoration. In: Proceedings of International Conference on Computer Vision (ICCV). pp. 4539–4547 (2017)
34. Ulyanov, D., Vedaldi, A., Lempitsky, V.: Instance normalization: The missing ingredient for fast stylization. arXiv preprint arXiv:1607.08022 (2016) 5
35. Vidal, R.G., Banerjee, S., Grm, K., Struc, V., Scheirer, W.J.: Ug^2: a video benchmark for assessing the impact of image restoration and enhancement on automatic visual recognition. In: Proceedings of IEEE Winter Conference on Applications of Computer Vision (WACV) (2018)
36. Wang, Z., Chang, S., Yang, Y., Liu, D., Huang, T.S.: Studying very low resolution recognition using deep networks. In: Proceedings of IEEE Conference on Computer Vision and Pattern Recognition (CVPR) (2016)
37. Welinder, P., et al.: Caltech-UCSD Birds 200. Technical Report CNS-TR-2010-001, California Institute of Technology (2010)
38. Wu, Y., Ling, H., Yu, J., Li, F., Mei, X., Cheng, E.: Blurred target tracking by blur-driven tracker. In: Proceedings of IEEE International Conference on Computer Vision (ICCV) (2011)
39. Wu, Z., Suresh, K., Narayanan, P., Xu, H., Kwon, H., Wang, Z.: Delving into robust object detection from unmanned aerial vehicles: a deep nuisance disentanglement approach. In: Proceedings of IEEE International Conference on Computer Vision (ICCV) (2019)
40. Yasarla, R., Patel, V.M.: Uncertainty guided multi-scale residual learning-using a cycle spinning cnn for single image de-raining. In: Proceedings of IEEE Conference on Computer Vision and Pattern Recognition (CVPR) (2019)
41. Yu, K., Dong, C., Lin, L., Change Loy, C.: Crafting a toolchain for image restoration by deep reinforcement learning. In: Proceedings of IEEE Conference on Computer Vision and Pattern Recognition (CVPR) (2018)
42. Zendel, O., Honauer, K., Murschitz, M., Steininger, D., Fernandez Dominguez, G.: Wilddash - creating hazard-aware benchmarks. In: Proceedings of European Conference on Computer Vision (ECCV) (2018)
43. Zhang, K., Zuo, W., Chen, Y., Meng, D., Zhang, L.: Beyond a gaussian denoiser: residual learning of deep cnn for image denoising. IEEE Trans. Image Process. (TIP) **26**(7), 3142–3155 (2017)

Pyramid Multi-view Stereo Net with Self-adaptive View Aggregation

Hongwei Yi[1(✉)], Zizhuang Wei[1], Mingyu Ding[2], Runze Zhang[3], Yisong Chen[1], Guoping Wang[1], and Yu-Wing Tai[4]

[1] PKU, Beijing, China
{hongweiyi,weizizhuang,chenyisong,wgp}@pku.edu.cn
[2] HKU, Shatin, Hong Kong
myding@cs.hku.hk
[3] Tencent, Shenzhen, China
ryanrzzhang@tencent.com
[4] Kwai Inc., Beijing, China
yuwing@gmail.com

Abstract. In this paper, we propose an effective and efficient pyramid multi-view stereo (MVS) net with self-adaptive view aggregation for accurate and complete dense point cloud reconstruction. Different from using mean square variance to generate cost volume in previous deep-learning based MVS methods, our **VA-MVSNet** incorporates the cost variances in different views with small extra memory consumption by introducing two novel self-adaptive view aggregations: pixel-wise view aggregation and voxel-wise view aggregation. To further boost the robustness and completeness of 3D point cloud reconstruction, we extend VA-MVSNet with pyramid multi-scale images input as **PVA-MVSNet**, where multi-metric constraints are leveraged to aggregate the reliable depth estimation at the coarser scale to fill in the mismatched regions at the finer scale. Experimental results show that our approach establishes a new state-of-the-art on the ***DTU*** dataset with significant improvements in the completeness and overall quality, and has strong generalization by achieving a comparable performance as the state-of-the-art methods on the ***Tanks and Temples*** benchmark. Our codebase is at https://github.com/yhw-yhw/PVAMVSNet.

Keywords: Multi-view stereo · Deep learning · Self-adaptive view aggregation · Multi-metric pyramid aggregation

1 Introduction

Multi-view Stereo (MVS) aims to recover dense 3D representation of scenes using stereo correspondences as the main cue given multiple calibrated images

H. Yi and Z. Wei—Equal Contribution.

Electronic supplementary material The online version of this chapter (https://doi.org/10.1007/978-3-030-58545-7_44) contains supplementary material, which is available to authorized users.

© Springer Nature Switzerland AG 2020
A. Vedaldi et al. (Eds.): ECCV 2020, LNCS 12354, pp. 766–782, 2020.
https://doi.org/10.1007/978-3-030-58545-7_44

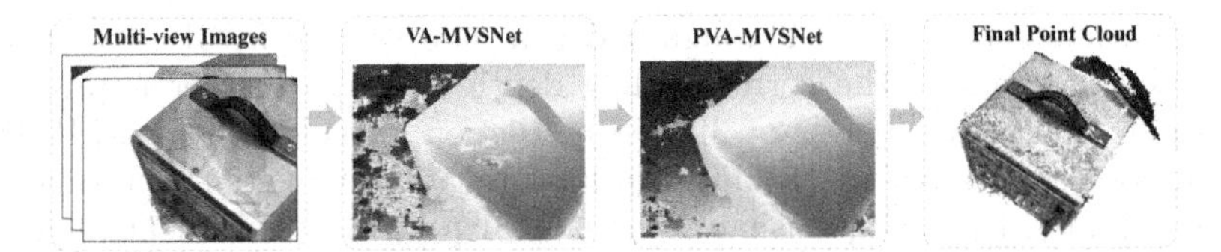

Fig. 1. VA-MVSNet performs an efficient and effective multi-view stereo with self-adaptive view aggregation to generate an accurate depth map. Cast in pyramid images additionally, PVA-MVSNet aggregates multi-scale depth maps with multi-metric constraints to boost the point cloud reconstruction with high accuracy and completeness.

[23,28,32,35]. Although they have achieved great success on MVS benchmarks [1, 22,31], many of them still have limitations in handling matching ambiguity and usually have a low completeness of 3D reconstruction. Recently, the deep neural network has made tremendous progress in multi-view stereo [17,18,42]. These methods learn and infer the information hardly obtained by stereo correspondences in order to handle matching ambiguity. However, they do not learn and utilize the following important information (Fig. 1).

First, the one-stage end-to-end deep MVS architectures [18,42,43] that directly learn from images all follow the philosophy that all view images contribute equally to the matching cost volume [13]. For instance, MVSNet [42] and R-MVSNet [43] both apply the mean square variance operation on multiple cost volumes, and DPSNet [18] selects the mean average operation. However, images from different views lead to heterogeneous image capture characteristics due to different illumination, camera geometric parameters, scene content variability, etc. Based on this observation, we propose a self-adaptive view aggregation module to learn the different significance in multiple matching volumes among images from different views. Our module benefits from the aggregated features by a self-adaptive fusion, where better element-wise matched regions are enhanced while the mismatched ones suppressed.

Second, the multi-scale information is not leveraged well to improve the robustness and completeness of 3D reconstruction. Unlike ACMM [40] where pyramid images are processed progressively to regress the depth map in a coarse-to-fine manner, we propose a novel way to aggregate multi-scale pyramid depth maps which are generated in parallel by multi-metric constraints to a refine depth map. In particular, to correct the mismatched regions at the finer depth map, we progressively aggregate the reliable depth at the coarser level to refine the finer depth map but do not introduce quantization errors benefiting from our multi-metric constraints.

To this end, we propose a novel efficient and effective pyramid multi-view stereo network with self-adaptive view aggregation, denoted as **PVA-MVSNet**. Our method constructs multi-scale pyramid images and processes them in parallel by **VA-MVSNet** to produce pyramid depth maps. To regularize 3D warping feature volumes from different views, we propose two self-adaptive element-wise view aggregation modules to learn different variance of different views in an order-independent manner. Through a depth map estimator, 3D cost volume is utilized to estimate the corresponding depth map. To further improve the robustness and completeness of 3D reconstruction generated by VA-MVSNet, our proposed multi-metric pyramid depth aggregation corrects the mismatched regions at finer depth maps using the reliable depths at coarser depth maps by checking photometric and geometric consistency.

Our main contributions are listed below:

- We propose self-adaptive view aggregation to incorporate the element-wise variances among images from different views, guiding the multiple cost volumes to aggregate a normalized one.
- We investigate to incorporate multi-scale information by our multi-metric pyramid depth maps aggregation in PVA-MVSNet, to further improve the robustness and completeness of 3D reconstruction.
- Our method establishes a new state-of-the-art on the *DTU* and a comparable performance as the state-of-the-art methods on the *Tanks and Temples*.

2 Related Work

Traditional MVS Reconstruction: Traditional MVS reconstruction algorithms can be divided into four types: voxel based [33,38], surface based [6,15], patch based [9,11] and depth map based methods [2,9,10,29,36,44]. Among those methods, the depth map based approaches are more concise and flexible. Recently, many advanced MVS algorithms estimate high quality depth maps by view selection, local propagation and multi-scale aggregation strategies. Zheng *et al.* [44] propose a depth map estimation method by solving a probabilistic graphical model. Schönberger *et al.* [29] present a new MVS system named COLMAP where geometric priors are used to better depict the probability of their graphical model. Xu *et al.* [40] propose a multi-scale MVS framework with adaptive checkerboard propagation and multi-hypothesis joint view selection to improve the performance. These works utilize predefined criteria for pixel-wise view selection, which cannot be adaptive for different scenes.

Learning Based Stereo Matching: Recently, the convolutional neural network (CNN) has made tremendous progress in many vision tasks [7,20,27,34,41], including several attempts on multi-view stereo. Early learning-based methods [8,17,19] pre-warp the images to generate plane-sweep volumes as the input. Two promising approaches [18,42] both propose the differential homography warping, which implicitly encodes multi-view camera geometries into the network and enables an end-to-end training fashion. Furthermore, R-MVSNet [43] replaces 3D-CNN in MVSNet [43] by the gated recurrent unit (GRU) to reduce memory consumption during the inference phase. Gu et al. [12] and Cheng et al. [5] both propose a cascaded MVS network through constructing coarse-to-fine cost volume which eases the memory limitation of the volume resolution in comparison with uniformly sampled cost volume [3,42,43]. P-MVSNet [24]

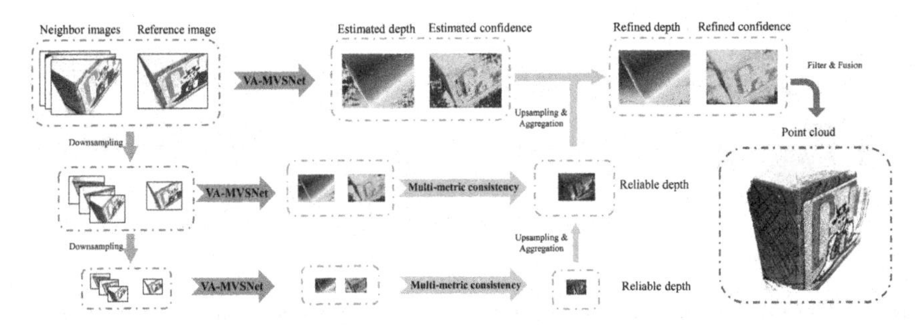

Fig. 2. Overview of PVA-MVSNet. We firstly input multi-scale pyramid images to VA-MVSNet to generate corresponding pyramid depth maps in parallel. Then we progressively replace the mismatched depths in the finer depth map with more reliable depths from a coarser level to achieve a refined depth map. Finally, we reconstruct the point cloud by filter and fusion through all estimated depth maps of the image set.

proposes a patch-wise matching module to learn the isotropic matching confidence inside the cost volume. Particularly, those methods follow the philosophy that the feature volumes from different view images contribute equally, neglecting heterogeneous image capturing characteristics due to different illumination, camera geometric parameters and scene content variability. PointMVSNet [4] is a two-stage coarse-to-fine method which needs a coarse depth map by a lower-resolution version MVSNet [42].

Based on the above analysis, we propose a self-adaptive view aggregation module to incorporate the different significance in multiple feature volumes from different views, where better element-wise matched features can be enhanced while the mismatched errors can be suppressed. To further improve the robustness and completeness of 3D reconstruction point cloud, we propose a multi-metric pyramid depth aggregation to aggregate multi-scale information in pyramid images. The mismatched depth value generated by the original image can be filled-in by the reliable depth value from the downsized image under photometric and geometric consistency.

3 Method

We first describe the overall architecture of PVA-MVSNet in Sect. 3.1. Then, we introduce the details of VA-MVSNet in Sect. 3.2. Finally, we present the multi-metric pyramid depth aggregation in Sect. 3.4.

3.1 Overall

Given a reference image $\boldsymbol{I}_{i=0}$ and $\boldsymbol{I}_{i=1,\cdots,N-1}$ neighboring images and corresponding calibrated camera parameters $\boldsymbol{Q}_{i=0}$ and $\boldsymbol{Q}_{i=1,\cdots,N-1}$, where N represents the number of multi-view images, our goal is to estimate the depth map for each reference image. Afterwards, we filter and fuse all estimated depth maps to reconstruct 3D point clouds.

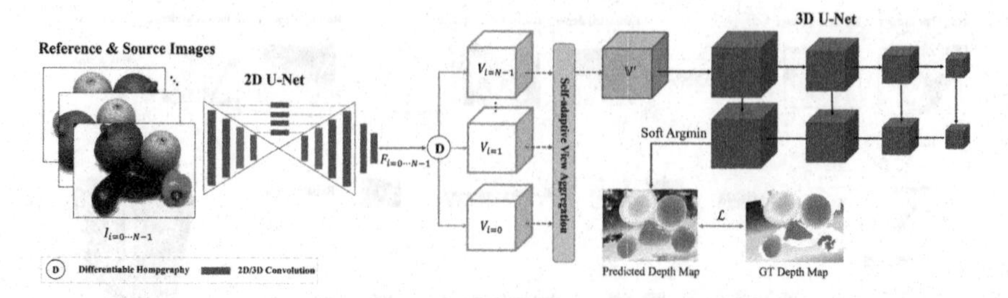

Fig. 3. The network architecture of VA-MVSNet. Multi-view images go through 2D U-Net and differentiable homography warping to generate 3D feature volumes. Cost variances in different views are encoded in self-adaptive view aggregation to aggregate the 3D cost volumes which is regularized by 3D U-Net to regress the depth map.

For the depth estimation of a reference image, our main architecture is illustrated in Fig. 2. We construct an image pyramid with K multiple scales for all images with a downsampling scale factor η. We denote k-level pyramid images and corresponding camera parameters as $I^k_{i=0,\cdots,N-1}$ and $Q^k_{i=0,\cdots,N-1}$ respectively, where $k = 0, \cdots, K-1$. The scale $k = 0$ of the pyramid represents the original image. We process each level images in the pyramid by VA-MVSNet to obtain depth maps of different scales in parallel. Then we progressively propagate the reliable depths from images with the lower resolution, which satisfy multi-metric constraints, to correct the mismatched errors of images with the higher resolution by replacements. Finally, we obtain the refined depth map of the raw image. We term our whole method PVA-MVSNet.

3.2 Self-adaptive View Aggregation

In VA-MVSNet in Fig. 3, we first design a 2D U-Net to extract $\{\boldsymbol{F}_i\}_{i=0}^{N-1}$ feature maps with larger receptive fields from the N input images. For efficient computation, the output feature map is downsampled by four to the original image size with 32 channels.

Then each feature map from different views will be warped to the reference camera frustum by the differential homography [18,42] with sampling D_i layers to build 3D plane-sweep feature volumes $\boldsymbol{V}_i$. To handle arbitrary N-view images input and the variances among images from different sources, we propose self-adaptive view aggregation to merge $\boldsymbol{V}_{i=0,\cdots,N-1}$ 3D feature volumes into one cost volume $\boldsymbol{C}$. Let W_i, H_i, D_i, C_i denote the width, height, depth sample number and channel number of the input 3D warping feature volume from image i respectively, the feature volume size can be represented as $S_i = W_i \cdot H_i \cdot D_i \cdot C_i$, the cost volume C_i aggregation can be defined as a function: $M : \underbrace{\mathbb{R}^{S_i} \times \cdots \times \mathbb{R}^{S_{N-1}}}_{N} \rightarrow \mathbb{R}^S$. In previous work [18,42,43], this is a constant function where all views contribute equally, which is the mean square error of all input feature volumes. However, it is not reasonable due to different illumination, camera position, occlusion and image content etc, where a near

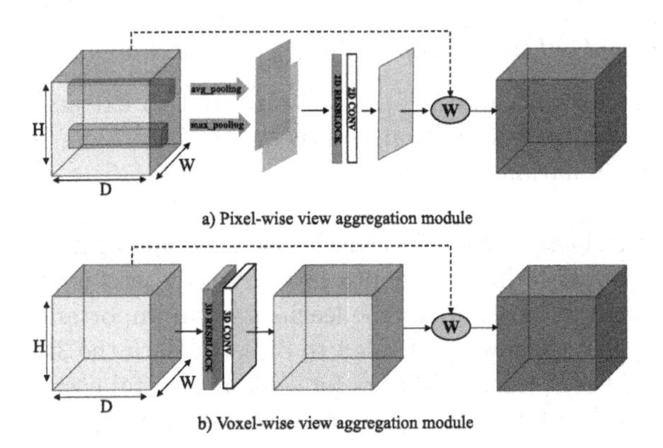

Fig. 4. Illustration of two different self-adaptive element-wise view aggregation modules, a) pixel-wise view aggregation, b) voxel-wise view aggregation.

reference image with no occlusion can provide more accurate geometric and photometric information than a far one with partial occlusion. Thus, we propose to employ self-adaptive view aggregation as this function to flexibly learn the potential different view variance from training data. To achieve this goal, we develop and investigate two different self-adaptive view aggregation modules in Fig. 4, which shows how the self-adaptive view selection incorporates the variance between different views. We introduce the attention mechanism [37,39] for guiding the network to select important matching information in different views. In the point-wise view selection, similar as ACMM [40], we consider that each pixel in the height and width dimension of 3D cost volume has different saliency but is consistent in the depth dimension. The voxel-wise view selection module is a 3D attention-guided mechanism to guide each voxel in 3D feature volumes to learn its own weight.

Pixel-Wise View Aggregation. The pixel-wise view aggregation introduces a selective weighted attention map in the height and width dimension which considers the depth number hypothesis sharing common focusing weight. Given multi-view feature volumes $V_{i=0 \cdots N-1}$, our regularized cost volumes are aggregated as $\boldsymbol{c}_{d,h,w}$:

$$\boldsymbol{v}'_{i,d,h,w} = \boldsymbol{v}_{i,d,h,w} - \boldsymbol{v}_{0,d,h,w}, \tag{1}$$

$$\boldsymbol{c}_{d,h,w} = \frac{\sum_{i=1}^{N-1}(1 + \boldsymbol{w}_{h,w}) \odot \boldsymbol{v}'_{i,d,h,w}}{N-1}, \tag{2}$$

where $\boldsymbol{w}_{h,w}$ represents a 2D weighted attention map to encode the various pixel-wise saliency among images from different sources and the reference view, and $\odot$ represents element-wise multiply operation.

To generate a 2D weighted attention map, we design a *PA-Net* in Table 1 which consists of several 2D convolutional filters and a ResNet block [14] with the squeezing 2D features from V_i' as input to learn the $\boldsymbol{w}_{h,w}$:

$$\boldsymbol{w}_{h,w} = \textit{PA-Net}(\boldsymbol{f}_{h,w}), \tag{3}$$

$$f_{h,w} = CONCAT(max\_pooling(\left\|\boldsymbol{v}'_{d,h,w}\right\|_1), avg\_pooling(\left\|\boldsymbol{v}'_{d,h,w}\right\|_1)), \quad (4)$$

where both *max_pooling* and *avg_pooling* are used to extract the highest and average cost matching information in the *depth* dimension, and $CONCAT(\cdot)$ denotes the concatenation operation.

Voxel-Wise View Aggregation. The voxel-wise view aggregation module considers that each pixel with different depth layer hypothesis d is treated differently, where each voxel in 3D feature volume learns its own importance. Based on this, we design a *VA-Net* as shown in Table 1 to directly learn the 3D weighted attention map with 3D convolutional filters for selecting useful cost information. The regularized 3D cost volumes $\boldsymbol{c}_{d,h,w}$ are aggregated by $\boldsymbol{v}'_{i,d,h,w}$:

$$\boldsymbol{c}_{d,h,w} = \frac{\sum_{i=1}^{N-1}(1 + \boldsymbol{w}_{d,h,w}) \odot \boldsymbol{v}'_{i,d,h,w}}{N-1}. \quad (5)$$

3.3 Depth Map Estimator

We design a 3D convolutional U-Net by leveraging different level information and expanding receptive fields to generate the probability volume P with a *softmax* operation along the depth dimension. The details of 3D U-Net are in the supplementary material.

To produce a continuous depth estimation, we use *soft argmin* operation [16] on the output probability volume P to estimate the depth $\boldsymbol{E}$:

$$\boldsymbol{E} = \sum_{d=d_{\min}}^{d_{\max}} d \times \boldsymbol{P}(d), \quad (6)$$

where $\boldsymbol{P}(d)$ denotes the estimated probability of all pixels for the depth hypothesis d. Following MVSNet [42], the probability map is calculated by the sum over the nearest four hypotheses in the 3D probability volume to measure the estimation quality. Comparing the estimated depth map and confidence map in Fig. 5 with [42], VA-MVSNet generates more reliable and accurate depth map with higher confidence benefiting from self-adaptive view aggregation.

Training Loss. We use the same training losses in MVSNet [42], which is the mean absolute error defined as $\mathcal{L}$:

$$\mathcal{L} = \sum_{\boldsymbol{x} \in \boldsymbol{x}_{valid}} \left\| \boldsymbol{d}(\boldsymbol{x}) - \hat{\boldsymbol{d}}(\boldsymbol{x}) \right\|_1, \quad (7)$$

where $\boldsymbol{x}_{valid}$ denotes the set of valid pixels in the ground truth, $\boldsymbol{d}(\boldsymbol{x})$ and $\hat{\boldsymbol{d}}(\boldsymbol{x})$ represent the estimated depth map and the ground truth respectively.

Table 1. The details of *PA-Net* and *VA-Net*. We denote Conv, Conv3D as 2D and 3D convolution respectively, and use GR to represent the abbreviation of group normalization and the Relu. + and & represent the element-wise addition and concatenation. K, S, F are the kernel size, stride and output channel number. N, H, W, D denote input view number, image height, image height and depth hypothesis number.

Input	Layer	Output	Output Size
PA-Net			
$f_{h,w}$	ConvGR, K = 3, S = 1, F = 16	wc_0	$W \times H \times 16$
wc_0	ResBlockGR, K = 3, S = 1, F = 16	wres_1	$W \times H \times 16$
wres_1	Conv, K = 3, S = 1, F = 1	wc_2	$W \times H \times 1$
wc_2	Sigmoid	Weight	$W \times H \times 1$
VA-Net			
v'	Conv3DGR,K=3,S=1,F=1	wc3d_0	$D \times W \times H \times 1$
wc3d_0	Conv3D, K = 3, S = 1, F = 1	wc3d_1	$D \times W \times H \times 1$
wc3d_1	Sigmoid	weight3d	$D \times W \times H \times 1$

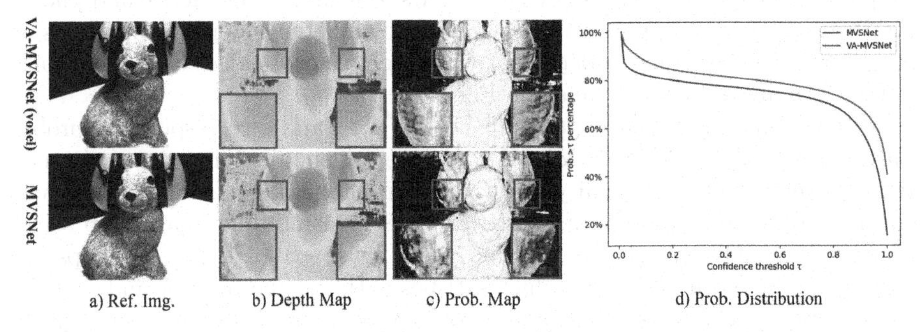

Fig. 5. Comparison on the regressed depth map, probability map and probability distribution with MVSNet [42]. (a) One reference image of Scan 12; (b) the inferred depth map; (c) the probability map; (d) the distribution of the probability map. Our self-adaptive view aggregation enhances the multi-view stereo network to generate more delicate and accurate depth estimations with higher confidence.

3.4 Multi-metric Pyramid Depth Map Aggregation

So far, our proposed network VA-MVSNet generates good-enough depth maps for the point cloud reconstruction. To further improve the robustness and completeness of 3D reconstruction, we propose a novel multi-metric pyramid depth aggregation to aggregate reliable depth estimations in a lower-resolution depth map into a higher-resolution depth map, by replacing corresponding mismatched errors.

In a higher-resolution fine estimated depth map, there are still some inaccurate depths with low confidences due to the matching ambiguity. Note that the same convolutional filter generally extracts less local-wise, but more global

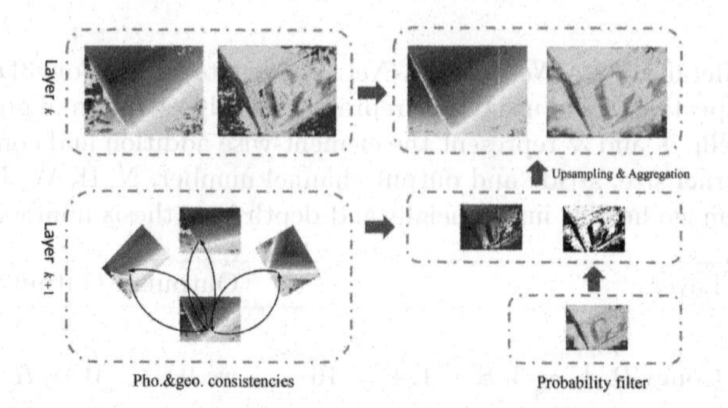

Fig. 6. Illustration of multi-metric pyramid depth map aggregation, where the reliable depth at a lower scale level $k+1$ selected by multi-metric constraints, are used to fill-in the mismatched errors at a higher scale level k by upsampling and aggregation.

information due to a larger receptive field from a downsampled image in comparison to the original image. Quite different from ACMM [40], which casts a image pyramid into VA-MVSNet to generate multi-scale depth maps in parallel, we propose to utilize multi-metric constraints, specifically, geometric and photometric consistency to progressively replace the ambiguous depth estimations at the higher scale by reliable depths at the lower scale. As a result, we optimize both depth and probability maps in Fig. 6.

Considering a pyramid depth map $D^{k=0,\cdots,K-1}$ and a corresponding probability map $P^{k=0,\cdots,K-1}$ from VA-MVSNet, we use the photometric consistency to measure the matching quality through the probability map and geometric consistency to measure the depth consistency between multiple images. To select accurate and well-matched depth value in the lower scale $k+1$ depth maps, we only select the estimated depth which satisfies both the photometric and geometric consistency. Firstly, for the photometric consistency, we expect to iteratively replace unreliable depth values with low confidence $P^k(p) < \epsilon_{low}$ at the scale k by reliable depths $P^{k+1}(p) > \epsilon_{high}$ at the downsampling scale $k+1$, where $P^k(p)$ denotes the confidence of pixel p in the probability map P^k and ϵ represent the filtering confidence threshold. After discarding mis-matched errors through the photometric consistency, we project a reference pixel p of image $\mathbf{I}_i$ to the corresponding pixel p_{proj} in the neighbor image $\mathbf{I}_j$ through $D_i(p)$ and camera parameters. In turn, we reproject p_{proj} through $D_j(p_{proj})$ back to the reference image as p_{reproj} with d_{reproj}. We remain the pixeles which satisfy the following geometric constraints in at least three neighbor views:

$$\|p - p_{reproj}\|_2 < \tau_1, \tag{8}$$

$$\|D_i(p) - d_{reproj}\|_1 < \tau_2 \cdot D_i(p). \tag{9}$$

Through our multi-metric pyramid depth map aggregation, the reliable depths at a lower scale $k+1$ can be progressively propagated to replace the mismatched depths at k scale until it leads to a final refinement at $k=0$ scale, which improves the robustness and completeness of 3D point cloud.

4 Experiments

4.1 Implementation Details

Training. We train VA-MVSNet on the DTU dataset [1], which consists of 124 different indoor scenes scanned by fixed camera trajectories in 7 different lighting conditions. Following the common practices [4,17,19,42,43], we train our network on the training split and evaluate on the evaluation part and use the same depth maps provided by MVSNet [42]. During training, the input image size is set to $W \times H = 640 \times 512$ and the number of input images $N = 5$. The depth hypotheses are sampled from 425 mm to 935 mm with depth plane number $D = 192$ in an inverse manner as illustrated in R-MVSNet [43]. We implement our network on **PyTorch** [26] and train it end-to-end for 16 epochs using *Adam* [21] with an initial learning rate 0.001 which is decayed by 0.9 every epoch. Batch size is set to 4 on 4 NVIDIA TITANX graphics cards.

Evaluation. For testing, we use $N = 7$ image views and $D = 192$ for depth plane sweeping in an inverse depth setting. We evaluate our methods on *DTU* with an original input image resolution: 1600×1184. For *Tanks and Temples*, the camera parameters are computed by OpenMVG [25] following MVSNet [42] and the input image resolution is set to 1920×1056. We use the same multi-metric constraint parameters, where $\epsilon_{low} = 0.5$, $\epsilon_{high} = 0.9$, $\tau_1 = 1$ and $\tau_2 = 0.01$.

Filtering and Fusion. We fuse all depth maps into a complete point cloud as in [10,42]. In our experiments, we only consider the reliable depth values with confidence larger than $\epsilon = 0.9$ and utilize the aforementioned geometric consistency to select those pixels occurring in more than three neighbor views. Finally, the depths are projected to 3D space and fused to produce a 3D point cloud.

Table 2. Quantitative results on the DTU evaluation dataset [1] (lower is better). Our VA-MVSNet and PVA-MVSNet (with voxel-wise view aggregation) outperform all methods in terms of completeness and overall quality with a significant improvement.

Method	Mean Distance (mm)		
	Acc	Comp	*overall*
Colmap [29]	0.400	0.664	0.532
Gipuma [10]	**0.283**	0.873	0.578
MVSNet [42]	0.396	0.527	0.462
R-MVSNet [43]	0.385	0.459	0.422
P-MVSNet [24]	0.406	0.434	0.420
PointMVSNet [4]	0.361	0.421	0.391
PointMVSNet-HiRes [4]	0.342	0.411	0.376
VA-MVSNet	0.378	0.359	0.369
PVA-MVSNet	0.379	**0.336**	**0.357**

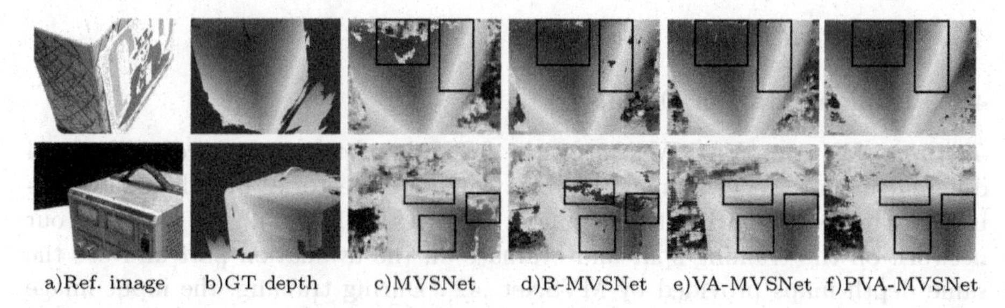

Fig. 7. Comparison of depth map estimations of *Scan* 13 *and* 11 in the *DTU* [1]. Our VA-MVSNet and PVA-MVSNet achieve more accurate, continuous and complete depth map in comparison to [42,43] methods [42,43].

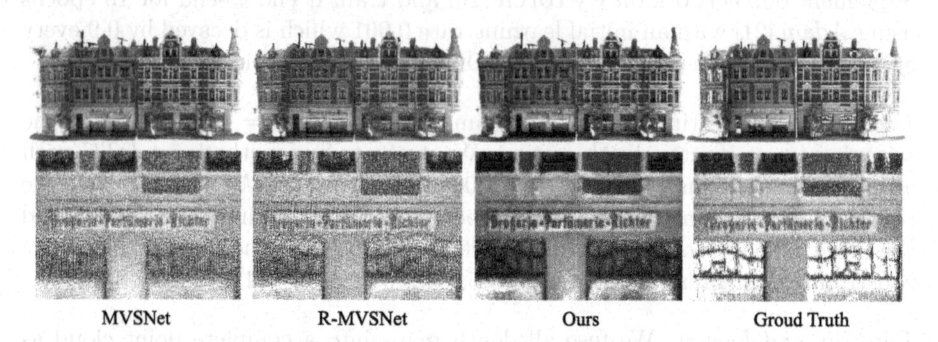

Fig. 8. Comparison of reconstruction point clouds for the model *Scan 15* in the benchmark *DTU* [1]. Our method generates denser, smoother and more complete point cloud compared with other methods [42,43].

4.2 Benchmarks Results

DTU Dataset. We evaluate our proposed method on the *DTU* [1] evaluation set. Quantitative results are shown in Table 2. The accuracy and completeness are calculated using the official matlab script provided by the *DTU* [1] dataset. The overall reconstruction quality is evaluated by calculating the average of the accuracy and completeness, as mentioned in [1,42]. While Gipuma [10] performs the best regarding to accuracy, our PVA-MVSNet and VA-MVSNet establish a new state-of-the-art both in completeness and overall quality with a significant margin compared with all previous methods [4,36,42,43]. We compare our depth maps with [42,43] in Fig. 7. VA-MVSNet predicts a more accurate, delicate and complete depth map by introducing different variances in multi-views through our proposed self-adaptive view aggregation. Moreover, PVA-MVSNet further fill-in the mismatched errors with reliable depths in the pyramid depth maps by our multi-metric pyramid depth map aggregation. Benefiting from more accurate, smooth and complete depth map estimation, our method can generate denser and more complete and delicate point clouds in Fig. 8.

Table 3. Quantitative results on the *Tanks and Temples* benchmark [22]. The evaluation metric is *f-score* which higher is better. (L.H. and P.G. are the abbreviations of *Lighthouse* and *Playground* dataset respectively).

Method	Rank	Mean	Family	Francis	Horse	L.H	M60	Panther	P.G	Train
MVSNet [42]	52.75	43.48	55.99	28.55	25.07	50.79	53.96	50.86	47.90	34.69
R-MVSNet [43]	42.62	48.40	69.96	46.65	32.59	42.95	51.88	48.80	52.00	42.38
Point-MVSNet [4]	40.25	48.27	61.79	41.15	34.20	50.79	51.97	50.85	52.38	43.06
P-MVSNet [24]	**17.00**	**55.62**	**70.04**	44.64	40.22	**65.20**	55.08	**55.17**	**60.37**	**54.29**
PVA-MVSNet	21.75	54.46	69.36	**46.80**	**46.01**	55.74	**57.23**	54.75	56.70	49.06

Fig. 9. The visualization of our partial point cloud results and the comparison with [42,43] on the Precision and Recall of *Horse* dataset on the *Tanks and Temples* [22] benchmark. The darker means the bigger error.

Tanks and Temples Benchmark. To explore the generalization of PVA-MVSNet, we compare our method **without any fine-tuning** with other baselines [3,4, 42,43] on the *Tanks and Temples*, which is a more complicated outdoor dataset. Table 3 summarizes the results. The mean *f-score* increases from 43.48 to 54.46 (larger is better, date: Mar. 5, 2020) compared with MVSNet [42], which demonstrates the efficacy and strong generalization of PVA-MVSNet. Our method outperforms Point-MVSNet [4] significantly with a higher 13% mean *f-score*, which is the best baseline on *DTU* dataset. And we achieve a comparable result with P-MVSNet [24]. The simple fusion process we adopted achieves a comparable result with P-MVSNet [24], which uses an extra *refinenet* and more depth filtering process to pursue better performance. Our partial reconstructed point clouds are shown in Fig. 9, and we compare the Precision and Recall of the *Horse* dataset with [42,43], which is provided by the *Tanks and Temples* [22] benchmark. Our method generates more accurate and complete point clouds with higher precision and recall than the others [42,43], due to the enhanced accuracy from self-adaptive view aggregation and the increased completeness and robustness from our multi-metric pyramid depth map aggregation.

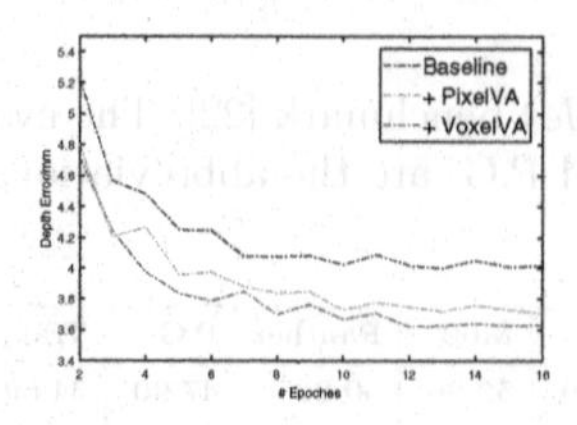

Fig. 10. Validation results of the mean average depth error with different components in VA-MVSNet during training.

Table 4. Contributions of different components in our architecture on the evaluation DTU [1].

Components	Acc	Comp	*Overall*
Baseline	0.454	0.372	0.413
+PixelVA	0.390	0.369	0.379
+VoxelVA	0.378	0.359	0.369
+PixelVA+MMP	0.392	0.341	0.366
+VoxelVA+MMP	**0.379**	**0.336**	**0.357**

4.3 Ablation Studies

In this section, we provide ablation experiments to analyze the strengths of the key components of our architecture. For following studies, to eliminate the non-learning influence, all experiments use the same consistency-check parameters in Sect. 4.1 and are tested on the *evaluation* and *validation DTU* [1] dataset.

Self-adaptive View Aggregation. As shown in Table 4, compared with our baseline method which is using the same mean square error as cost volume aggregation in MVSNet [42], both PixelVA and VoxelVA can improve the results of 3D reconstruction point cloud with a significant margin, especially on the accuracy of reconstruction quality. Specifically, the VoxelVA provides a 16.7% increase on *accuracy*, which is better than the PixelVA 14.1% due to the learning variance of the depth wise hypothesis. Besides, the VoxelVA has more parameters but less operations compared with PixelVA as denoted in Table 1. During training, as shown in Fig. 10, the depth error on *validation* dataset drops significantly by introducing our proposed novel self-adaptive view aggregation.

Number of Views. We investigate the influence of variant numbers of views N in different phases on *DTU evaluation* dataset. VA-MVSNet can process an arbitrary number of views and well leverage the variant importance in multi-views due to our proposed self-adaptive view aggregation. In the test phase, we use the model trained on 3 views to compare the reconstruction results with different numbers of views $N = 2, 3, 5, 7$. As shown in Table 5, the result with $N = 5$ achieves a great improvement compared with $N = 2, 3$, but the influence from two more extra views in $N = 7$ is quite small which can be ignored. It demonstrates that our proposed self-adaptive view aggregation can well enhance the valid information in the good neighbor views and eliminate bad information in farrer views (the neighbor views are ranked by the matching quality with the reference view in SfM [30]). In the training phase, we compare the results on the input view $N = 7$ using the models trained on $N = 3, 4, 5$. The model trained on $N = 5$ is slightly better than $N = 3$ but with more training time.

Multi-metric Pyramid Depth Aggregation. As shown in Table 4, The completeness can be averagely improved by 7.0% while introduce a negligible drop 0.39%,

Table 5. Ablation study on different number of views N in training and testing phase and different numbers of image pyramid on *DTU* [1] evaluation dataset.

Number of views		Number of pyramid	Acc. (mm)	Comp. (mm)	*Overall* (mm)
Training	Test				
$N=3$	$N=2$	\	0.415	0.467	0.441
$N=3$	$N=3$	\	0.380	0.379	0.380
$N=3$	$N=5$	\	0.381	0.361	0.371
$N=3$	$N=7$	\	0.380	0.361	0.370
$N=4$	$N=7$	\	0.380	0.359	0.370
$N=5$	$N=7$	\	0.378	0.359	0.369
$N=5$	$N=7$	$K=1$	**0.372**	0.350	0.361
$N=5$	$N=7$	$K=2$	0.378	0.341	0.360
$N=5$	$N=7$	$K=3$	0.379	**0.336**	**0.357**

Table 6. Comparisons on the time and memory cost on the *evaluation* DTU [1] dataset. MVSNet and R-MVSNet are implemented in TensorFlow while others in PyTorch.

Methods	H,W,D	Mem	Time	Overall
MVSNet	1600, 1184, 256	15.4 GB	1.18 s	0.462
R-MVSNet	1600, 1184, 512	**6.7 GB**	2.35s	0.422
PointMVSNet	1280, 960, 96	7.2 GB	1.69 s	0.391
PointMVSNet-HiRes	1600, 1152, 96	8.7 GB	5.44 s	0.376
VA-MVSNet	1600, 1184, 192	18.1 GB	**0.91** s	0.369
PVA-MVSNet	1600, 1184, 192	24.87 GB	1.01s	**0.357**

benefiting from "MMP" (multi-metric pyramid depth aggregation). As denoted in *e)* and *f)* in Fig. 7, PVA-MVSNet improves VA-MVSNet by generating more delicate and complete depth maps. To better analyse the improvement from different pyramid level image, we explore the influence by different numbers of image pyramid in Table 5. The $K=1$ level pyramid image improves both accuracy and completeness with a big margin. A trade-off between accuracy and completeness is achieved by using more pyramid images $k=2$ and $k=3$, it leads to reconstructed 3D point cloud with better overall quality.

4.4 Runtime and Memory Performance

Given time and memory performance in Table 6, all methods are tested on GeForce RTX 2080 Ti. VA-MVSNet runs fast at a speed of 0.91 s/view, even if it runs with the biggest memory consumption. Unlike PointMVSNet [4], multi-scale pyramid images can be processed independently in parallel. Therefore, with little extra time about 0.1s for multi-metric pyramid depth aggregation, the performance of 3D point cloud reconstruction increases significantly from 0.369 to 0.357 in PVA-MVSNet on *DTU* [1] dataset.

5 Conclusion

We present a novel pyramid multi-view stereo network with the self-adaptive view aggregation. The proposed VA-MVSNet dynamically selects the element-wise feature importance while suppresses the mismatching cost, which is quite efficient and effective. Casting in multi-scale pyramid images, benefiting from utilizing multi-metric constraint, PVA-MVSNet estimates a refined depth map for further improving the robustness and completeness of 3D reconstruction. Experimental results demonstrate that our proposed method PVA-MVSNet establishes a new state-of-the-art on the *DTU* dataset and shows great generalization by achieving a comparable performance as other state-of-the-art methods on *Tanks and Temples* benchmark without any fine-tuning.

Acknowledgements. This project was supported by the National Key R&D Program of China (No. 2017YFB1002705, No. 2017YFB1002601) and NSFC of China (No. 61632003, No. 61661146002, No. 61872398).

References

1. Aanæs, H., Jensen, R.R., Vogiatzis, G., Tola, E., Dahl, A.B.: Large-scale data for multiple-view stereopsis. IJCV **120**(2), 153–168 (2016)
2. Campbell, N.D.F., Vogiatzis, G., Hernández, C., Cipolla, R.: Using multiple hypotheses to improve depth-maps for multi-view stereo. In: Forsyth, D., Torr, P., Zisserman, A. (eds.) ECCV 2008. LNCS, vol. 5302, pp. 766–779. Springer, Heidelberg (2008). https://doi.org/10.1007/978-3-540-88682-2_58
3. Chen, R., Han, S., Xu, J., Su, H.: Point-based multi-view stereo network. In: ICCV (2019)
4. Chen, R., Han, S., Xu, J., Su, H.: Point-based multi-view stereo network. arXiv preprint arXiv:1908.04422 (2019)
5. Cheng, S., Xu, Z., Zhu, S., Li, Z., Li, L.E., Ramamoorthi, R., Su, H.: Deep stereo using adaptive thin volume representation with uncertainty awareness. In: CVPR (2020)
6. Cremers, D., Kolev, K.: Multiview stereo and silhouette consistency via convex functionals over convex domains. PAMI **33**(6), 1161–1174 (2010)
7. Dosovitskiy, A., et al.: FlowNet: learning optical flow with convolutional networks. In: ICCV (2015)
8. Flynn, J., Neulander, I., Philbin, J., Snavely, N.: DeepStereo: learning to predict new views from the world's imagery. In: CVPR (2016)
9. Furukawa, Y., Ponce, J.: Accurate, dense, and robust multiview stereopsis. PAMI **32**(8), 1362–1376 (2009)
10. Galliani, S., Lasinger, K., Schindler, K.: Massively parallel multiview stereopsis by surface normal diffusion. In: ICCV (2015)

11. Goesele, M., Snavely, N., Curless, B., Hoppe, H., Seitz, S.M.: Multi-view stereo for community photo collections. In: ICCV (2007)
12. Gu, X., Fan, Z., Zhu, S., Dai, Z., Tan, F., Tan, P.: Cascade cost volume for high-resolution multi-view stereo and stereo matching. In: CVPR (2020)
13. Hartmann, W., Galliani, S., Havlena, M., Van Gool, L., Schindler, K.: Learned multi-patch similarity. In: ICCV (2017)
14. He, K., Zhang, X., Ren, S., Sun, J.: Deep residual learning for image recognition. In: CVPR (2016)
15. Hiep, V.H., Keriven, R., Labatut, P., Pons, J.P.: Towards high-resolution large-scale multi-view stereo. In: CVPR (2009)
16. Honari, S., Molchanov, P., Tyree, S., Vincent, P., Pal, C., Kautz, J.: Improving landmark localization with semi-supervised learning. In: CVPR (2018)
17. Huang, P.H., Matzen, K., Kopf, J., Ahuja, N., Huang, J.B.: DeepMVS: learning multi-view stereopsis. In: CVPR (2018)
18. Im, S., Jeon, H.G., Lin, S., Kweon, I.S.: DPSNet: end-to-end deep plane sweep stereo. In: ICLR (2019)
19. Ji, M., Gall, J., Zheng, H., Liu, Y., Fang, L.: SurfaceNet: an end-to-end 3d neural network for multiview stereopsis. In: ICCV (2017)
20. Kendall, A., et al.: End-to-end learning of geometry and context for deep stereo regression. In: ICCV (2017)
21. Kingma, D.P., Ba, J.: Adam: a method for stochastic optimization. In: ICLR (2014)
22. Knapitsch, A., Park, J., Zhou, Q.Y., Koltun, V.: Tanks and temples: Benchmarking large-scale scene reconstruction. TOG **36**(4), 78 (2017)
23. Lhuillier, M., Quan, L.: A quasi-dense approach to surface reconstruction from uncalibrated images. PAMI **27**(3), 418–433 (2005)
24. Luo, K., Guan, T., Ju, L., Huang, H., Luo, Y.: P-MVSNet: learning patch-wise matching confidence aggregation for multi-view stereo. In: ICCV (2019)
25. Moulon, P., Monasse, P., Marlet, R., et al.: OpenMVG. An open multiple view geometry library (2014)
26. Paszke, A., et al.: Automatic differentiation in PyTorch. In: NeurIPS Autodiff Workshop (2017)
27. Ren, S., He, K., Girshick, R., Sun, J.: Faster R-CNN: towards real-time object detection with region proposal networks. In: NeurIPS (2015)
28. Schonberger, J.L., Frahm, J.M.: Structure-from-motion revisited. In: CVPR (2016)
29. Schönberger, J.L., Zheng, E., Frahm, J.-M., Pollefeys, M.: Pixelwise view selection for unstructured multi-view stereo. In: Leibe, B., Matas, J., Sebe, N., Welling, M. (eds.) ECCV 2016. LNCS, vol. 9907, pp. 501–518. Springer, Cham (2016). https://doi.org/10.1007/978-3-319-46487-9_31
30. Schönberger, J.L., Frahm, J.M.: Structure-from-motion revisited. In: CVPR (2016)
31. Schops, T., et al.: A multi-view stereo benchmark with high-resolution images and multi-camera videos. In: CVPR (2017)
32. Seitz, S.M., Curless, B., Diebel, J., Scharstein, D., Szeliski, R.: A comparison and evaluation of multi-view stereo reconstruction algorithms. In: CVPR (2006)
33. Sinha, S.N., Mordohai, P., Pollefeys, M.: Multi-view stereo via graph cuts on the dual of an adaptive tetrahedral mesh. In: ICCV (2007)
34. Song, X., Zhao, X., Hu, H., Fang, L.: EdgeStereo: a context integrated residual pyramid network for stereo matching. In: Jawahar, C.V., Li, H., Mori, G., Schindler, K. (eds.) ACCV 2018. LNCS, vol. 11365, pp. 20–35. Springer, Cham (2019). https://doi.org/10.1007/978-3-030-20873-8_2

35. Strecha, C., Von Hansen, W., Van Gool, L., Fua, P., Thoennessen, U.: On benchmarking camera calibration and multi-view stereo for high resolution imagery. In: CVPR (2008)
36. Tola, E., Strecha, C., Fua, P.: Efficient large-scale multi-view stereo for ultra high-resolution image sets. Mach. Vis. Appl. **23**(5), 903–920 (2012)
37. Vaswani, A., et al.: Attention is all you need. In: NeurIPS (2017)
38. Vogiatzis, G., Esteban, C.H., Torr, P.H., Cipolla, R.: Multiview stereo via volumetric graph-cuts and occlusion robust photo-consistency. PAMI **29**(12), 2241–2246 (2007)
39. Xu, K., et al.: Show, attend and tell: Neural image caption generation with visual attention. In: ICML (2015)
40. Xu, Q., Tao, W.: Multi-scale geometric consistency guided multi-view stereo. In: CVPR (2019)
41. Yang, G., Zhao, H., Shi, J., Deng, Z., Jia, J.: SegStereo: exploiting semantic information for disparity estimation. In: ECCV (2018)
42. Yao, Y., Luo, Z., Li, S., Fang, T., Quan, L.: MVSNet: depth inference for unstructured multi-view stereo. In: ECCV (2018)
43. Yao, Y., Luo, Z., Li, S., Shen, T., Fang, T., Quan, L.: Recurrent MVSNet for high-resolution multi-view stereo depth inference. In: CVPR (2019)
44. Zheng, E., Dunn, E., Jojic, V., Frahm, J.M.: PatchMatch based joint view selection and depthmap estimation. In: CVPR (2014)

SPL-MLL: Selecting Predictable Landmarks for Multi-label Learning

Junbing Li[1], Changqing Zhang[1(✉)], Pengfei Zhu[1], Baoyuan Wu[2], Lei Chen[3], and Qinghua Hu[1]

[1] Tianjin Key Lab of Machine Learning, College of Intelligence and Computing, Tianjin University, Tianjin, China
{lijunbing,zhangchangqing,zhupengfei,huqinghua}@tju.edu.cn

[2] Tencent AI lab, The Chinese University of Hong Kong, Shenzhen, China
wubaoyuan1987@gmail.com

[3] Nanjing University of Posts and Telecommunications, Nanjing, China
chenlei@njupt.edu.cn

Abstract. Although significant progress achieved, multi-label classification is still challenging due to the complexity of correlations among different labels. Furthermore, modeling the relationships between input and some (dull) classes further increases the difficulty of accurately predicting all possible labels. In this work, we propose to select a small subset of labels as landmarks which are easy to predict according to input (predictable) and can well recover the other possible labels (representative). Different from existing methods which separate the landmark selection and landmark prediction in the 2-step manner, the proposed algorithm, termed Selecting Predictable Landmarks for Multi-Label Learning (SPL-MLL), jointly conducts landmark selection, landmark prediction, and label recovery in a unified framework, to ensure both the representativeness and predictableness for selected landmarks. We employ the Alternating Direction Method (ADM) to solve our problem. Empirical studies on real-world datasets show that our method achieves superior classification performance over other state-of-the-art methods.

Keywords: Multi-label learning · Predictable landmarks · A unified framework

1 Introduction

Multi-label classification jointly assigns one sample with multiple tags reflecting its semantic content, which has been widely used in many real-world applications. In document classification, there are multiple topics for one document; in computer vision, one image may contain multiple types of object; in emotion analysis, there may be combined types of emotions, *e.g.*, relaxing and quiet. Though plenty of multi-label classification methods [12,19,21,22,27,31,32,37] have been proposed, multi-label classification is still a recognized challenging task due to the complexity of label correlations, and the difficulty of predicting all labels.

© Springer Nature Switzerland AG 2020
A. Vedaldi et al. (Eds.): ECCV 2020, LNCS 12354, pp. 783–799, 2020.
https://doi.org/10.1007/978-3-030-58545-7_45

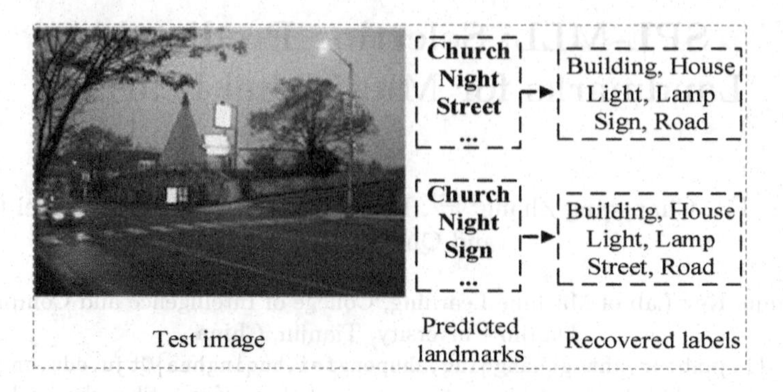

Fig. 1. Why landmarks should be predictable? With a test image, although the other related labels (*e.g.*, "street", "road") can be inferred from "sign", the landmark label "sign" itself is difficult to accurately predict (bottom row). While based on the image, the label "street" is more predictable, and accordingly, more related labels are correctly inferred (top row).

In real-world applications, labels are usually correlated, and simultaneously predicting all possible labels is usually rather difficult. Accordingly, there are techniques aiming to reduce the label space. The representative strategy is landmark based multi-label classification. Landmark based methods first select a small subset of representative labels as landmarks, where the landmark labels are able to establish interdependence with other labels. The algorithm [1] employ group-sparse learning to select a few labels as landmarks which can reconstruct the other labels. However, this method divides the landmark selection and landmark prediction as two separate processes. The method [3] performs label selection based on randomized sampling, where the sampling probability of each class label reflects its importance among all the labels. Another representative strategy is label embedding [7,13,24,39,40], which transforms label vectors into low-dimensional embeddings, where the correlations among labels can be implicitly encoded.

Although several methods have been proposed to reduce the dimensionality of label space, there are several limitations left behind these methods. First, existing landmark-based methods usually separate the landmark label selection and landmark prediction into two independent steps. Therefore, even the other labels could be easily recovered from the selected landmarks, but the landmarks themselves may be difficult to be accurately predicted with input (as shown in Fig. 1). Second, the label-embedding-based methods usually project the original label space into a low-dimensional embedding space, where embedded vectors can be used to recover the full-label set. Although the embedded vectors may be easy to predict, the embedding way (*i.e.*, dimensionality reduction) may cause information loss, and the label correlations are implicitly encoded thus lack interpretability. Considering the above issues, we jointly conduct landmarks selection, landmarks prediction and full-label recovery in a unified framework, and accordingly, propose a novel multi-label learning method, termed *Selecting Predictable*

Landmarks for Multi-Label Learning (***SPL-MLL***). The overview of SPL-MLL is shown in Fig. 2. The main advantages of the proposed algorithm include: (1) compared with existing landmark-based multi-label learning methods, SPL-MLL can select the landmarks which are both representative and predictable due to the unified objective; (2) compared with the embedding methods, SPL-MLL is more interpretable due to explicitly exploring the correlations with landmarks.

The contributions of this work are summarized as:

- We propose a novel landmark-based multi-label learning algorithm for complex correlations among labels. The landmarks bridge the intrinsic correlations among different labels, while also reduce the complexity of correlations and possible label noise.
- To the best of our knowledge, SPL-MLL is the first algorithm which simultaneously conducts landmark selection, landmark prediction, and full-label recovery in a unified objective, thus taking both representativeness and predictability for landmarks into account. This is quite different from the 2-step manner separating landmark selection and prediction.
- Extensive experiments on benchmark datasets are conducted, validating the effectiveness of the proposed method over state-of-the-arts.

2 Related Work

Generally, existing multi-label methods can be roughly categorized into three lines based on the order of label correlations [37]. The first-order strategy [4,36] tackles multi-label learning problem in the label-by-label manner, which ignores the co-existence of other labels. The second-order strategy [8,10,11] conducts multi-label learning problem by introducing the pairwise relations between different labels. For high-order strategy [14,22,28,30], multi-label learning problem is solved by establishing more complicated label relationships, which makes these approaches tend to be quite computationally expensive.

In order to reduce label space, there are approaches based on label embedding, which searches a low-dimensional subspace so that correlations among labels can be implicitly expressed [7,13,15,16,18,23,24,34,39,40]. Based on the low-dimensional latent label space, one can effectively reduce computation cost while performing multi-label prediction. The representative embedding based methods include: label embedding via random projections [13], principal label space transformation (PLST) [24] and its conditional version (CPLST) [7]. Beyond considering linear embedding functions, there are several approaches employing standard kernel functions (*e.g.*, low-degree polynomial kernels) for nonlinear label embedding. The work in [33] proposes a novel DNN architecture of Canonical-Correlated Autoencoder (C2AE), which is a DNN-based label embedding framework for multi-label classification, which is able to perform feature-aware label embedding and label-correlation aware prediction.

To explore label correlations, there are several landmark based multi-label classification models aiming to reduce the label space [1,3,5,40]. They usually first select a small subset of labels as landmarks, which are supposed to be representative and able to establish interdependency with other labels. The work in [1] models landmark selection with group-sparsity technique. Following the assumption in [1], the method in [3] alleviates this problem of computation cost by proposing an valid label selection method based on randomized sampling, and utilizes the leverage score in the best rank-k subspace of the label matrix to obtain the sampling probability of each label. It is noteworthy that these methods separate the landmark selection and landmark prediction in a 2-step manner, which can not simultaneously guarantee the representativeness and predictability of landmarks.

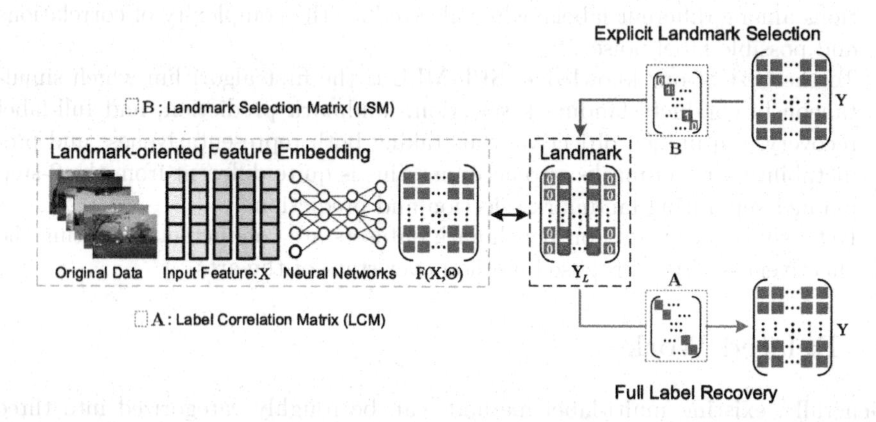

Fig. 2. Overview of SPL-MLL. The key component of our model is the landmark selection strategy, which induces the explicit landmark label matrix $\mathbf{Y}_L$. The matrix $\mathbf{B}$, termed as landmark selection matrix, is used to construct the landmark label matrix explicitly, while the matrix $\mathbf{A}$ is used to reconstruct all possible labels from landmarks. Benefitting from the explicit landmark label matrix $\mathbf{Y}_L$, the input is also able to be taken into account to ensure the predictable property for landmarks.

3 Our Algorithm: Selecting Predictable Landmarks for Multi-label Learning

For clarification, we first provide the definitions for symbols and variables used through out this paper. Let $\mathcal{X} = \mathbb{R}^D$ and $\mathcal{Y} = \{0,1\}^C$ denote the feature space and label space, where D and C are the dimensionality of feature space and label space, respectively. Given training data with the form of instance-label pairs $\{\mathbf{x}_i, \mathbf{y}_i\}_{i=1}^N$, accordingly, the feature matrix can be represented as $\mathbf{X} \in \mathbb{R}^{N \times D}$, and the label matrix is represented as $\mathbf{Y} \in \mathbb{R}^{N \times C}$. The goal of multi-label learning is to learn a model $f : \mathcal{X} \rightarrow \mathcal{Y}$, to predict possible labels accurately for new coming instances. Motivated by the landmark strategy, we propose a novel algorithm for multi-label learning, termed SPL-MLL, *i.e.*, *Selecting Predictable Landmarks for Multi-Label Learning.* SPL-MLL consists of two key components, *i.e.*, *Explicit Landmark Selection* and *Predictable Landmark Classification.*

3.1 Explicit Landmark Selection

Different from the 2-step manner [1] which only focuses on selecting landmarks that are most representative, our goal is to select landmarks which are both representative and predictable. There are two designed matrixes which are the keys to realize this goal. The first matrix is the ***label correlation matrix (LCM)*** $\mathbf{A}$ used for recovering other labels with landmarks. In self-representation manner, the matrix $\mathbf{A} \in \mathbb{R}^{C \times C}$ is obtained which captures the correlation among labels and explores the interdependency between landmark labels and the others. In the work [1], the landmarks are selected implicitly. Specifically, the underlying assumption is $\mathbf{Y} = \mathbf{YA}$, where $\mathbf{A}$ is constrained by minimizing $||\mathbf{A}||_{2,1}$ to enforce the reconstruction of $\mathbf{Y}$ mainly base on a few labels, *i.e.*, landmarks. In our model, although the linear self-representation manner is also introduced in our model, we try to obtain the landmark label matrix explicitly in the objective function.

The second critical matrix is the ***landmark selection matrix (LSM)*** $\mathbf{B}$. Note that, in the work [1], there is no explicit landmark label matrix constructed and the selection result is implicitly encoded in $\mathbf{A}$ due to its sparsity in row. Both $\mathbf{YA}$ and $\mathbf{Y}$ ($\mathbf{YA} \approx \mathbf{Y}$) in [1] are full-label matrix. Different from [1], since we aim to jointly conduct landmark selection and learn a model to predict these selected landmarks instead of all labels, we need to explicitly derive a label matrix encoding the landmarks. To this end, we introduce the matrix $\mathbf{B} \in \mathbb{R}^{C \times C}$ which is a diagonal matrix, and each diagonal element is either 0 or 1, *i.e.*, $B_{ii} \in \{0, 1\}$. Then, we can obtain the explicit landmark label matrix $\mathbf{Y}_L$ with $\mathbf{Y}_L = \mathbf{YB}$. In this way, the columns corresponding to the landmarks in $\mathbf{Y}_L$ unchanged while the elements of other columns (corresponding to non-landmark labels) will be 0. It is noteworthy that $\mathbf{B}$ is learned in our model instead of being fixed in advance. Accordingly, the explicit landmark selection objective to minimize is induced as:

$$
\begin{aligned}
\mathbf{\Gamma}(\mathbf{B}, \mathbf{A}) &= \|\mathbf{Y} - \mathbf{Y}_L\mathbf{A}\|_F^2 + \mathbf{\Omega}(\mathbf{B}) \\
&= \|\mathbf{Y} - \mathbf{YBA}\|_F^2 + \mathbf{\Omega}(\mathbf{B}), \\
& s.t. \ B_{ij} = 0, \ i \neq j; \ B_{ij} \in \{0, 1\}, \ i = j.
\end{aligned} \tag{1}
$$

Since it is difficult to strictly ensure the diagonal property for $\mathbf{B}$, a soft constraint $\mathbf{\Omega}(\mathbf{B})$ is introduced as follows:

$$
\mathbf{\Omega}(\mathbf{B}) = \lambda_1 \|\mathbf{B} - \mathbf{I}\|_F^2 + \lambda_2 \|\mathbf{B}\|_{2,1}, \tag{2}
$$

where the structure sparsity $||\mathbf{B}||_{2,1} = \sum_{i=1}^{C} \sqrt{\sum_{j=1}^{C} B_{ij}^2}$ is used to select a few landmarks, and the approximation to the identity matrix $\mathbf{I}$ ensures the labels corresponding to landmarks unchanged. The regularization parameter λ_1 and λ_2 control the degree of diagonal and sparsity property for $\mathbf{B}$, respectively. Notice that the label correlation matrix $\mathbf{A}$ is learned automatically without constraint, the underlying assumption for the correlation is sparse (similar to the existing work [1]) which is jointly ensured by the sparse landmark selection matrix $\mathbf{B}$. Then, we can obtain the explicit landmark label matrix $\mathbf{Y}_L = \mathbf{YB}$, and train a prediction model exactly for the landmarks.

3.2 Predictable Landmark Classification

Now, we firstly consider learning the classification model for accurately predicting landmarks instead of all possible labels. Beyond label correlation, modeling $\mathcal{X} \rightarrow \mathcal{Y}$ is also critical in multi-label classification. However, the traditional landmark-based multi-label classification algorithms usually separate landmark selection and landmark prediction, which may result in unpromising classification accuracy because the selected landmarks may be representative but difficult to be predicted (see Fig. 1). Recall that the goal of our model is to recover full labels with landmark labels, so our classification model only focuses on predicting landmarks $\mathbf{Y}_L$ based on $\mathbf{X}$ instead of full labels $\mathbf{Y}$. Accordingly, our predictable landmark classification objective to minimize is as follows:

$$\begin{aligned} \boldsymbol{\Phi}(\mathbf{B}, \boldsymbol{\Theta}) &= \|\boldsymbol{f}(\mathbf{X}; \boldsymbol{\Theta})\mathbf{B} - \mathbf{Y}_L\|_F^2 \\ &= \|(\boldsymbol{f}(\mathbf{X}; \boldsymbol{\Theta}) - \mathbf{Y})\mathbf{B}\|_F^2, \end{aligned} \tag{3}$$

where $\boldsymbol{f}(\cdot; \boldsymbol{\Theta})$ is the neural networks (parameterized by $\boldsymbol{\Theta}$) used for feature embedding and conducting classification for landmarks, which is implemented by fully connected neural networks.

3.3 Objective Function

Based on above considerations, a novel landmark-based multi-label classification algorithm, *i.e.*, Selecting Predictable Landmarks for Multi-Label Learning (SPL-MLL), is induced, which jointly learns landmark selection matrix, label correlation matrix, and landmark-oriented feature embedding in a unified framework. Specifically, the objective function of SPL-MLL for us to minimize is as follows:

$$\begin{aligned} \mathcal{L}(\mathbf{B}, \boldsymbol{\Theta}, \mathbf{A}) &= \boldsymbol{\Phi}(\mathbf{B}, \boldsymbol{\Theta}) + \boldsymbol{\Gamma}(\mathbf{B}, \mathbf{A}) \\ &= \|(\boldsymbol{f}(\mathbf{X}; \boldsymbol{\Theta}) - \mathbf{Y})\mathbf{B}\|_F^2 \\ &+ \|\mathbf{Y} - \mathbf{YBA}\|_F^2 + \lambda_1\|\mathbf{B} - \mathbf{I}\|_F^2 + \lambda_2\|\mathbf{B}\|_{2,1}. \end{aligned} \tag{4}$$

It is noteworthy that the critical role of matrix $\mathbf{B}$, which bridges the landmark selection and landmark classification model. With this strategy, the proposed model jointly selects predictable landmark labels, captures the correlations among labels, and discovers the nonlinear correlations between features and landmarks, accordingly, promotes the performance of multi-label prediction.

Algorithm 1: Algorithm of SPL-MLL

Input: Feature matrix $\mathbf{X} \in \mathbb{R}^{N \times D}$, label matrix $\mathbf{Y} \in \mathbb{R}^{N \times C}$, parameters λ_1, λ_2.
Initialize: $\mathbf{B} = \mathbf{I}$, initialize randomly $\mathbf{A}$.
while *not converged* **do**
    Update the parameters $\boldsymbol{\Theta}$ of $\boldsymbol{f}(\cdot; \boldsymbol{\Theta})$;
    Update $\mathbf{B}$ by Eq. (5);
    Update $\mathbf{A}$ by Eq. (6);
end
Output: $\boldsymbol{f}(\cdot; \boldsymbol{\Theta}), \mathbf{B}, \mathbf{A}$.

3.4 Optimization

Since the objective function of our SPL-MLL is not jointly convex for all the variables, we optimize our objective function by employing Alternating Direction Minimization(ADM) [17] strategy. To optimize the objective function in Eq. (4), we should solve three subproblems with respect to $\mathbf{\Theta}$, $\mathbf{B}$ and $\mathbf{A}$, respectively. The optimization is cycled over updating different blocks of variables. We apply the technique of stochastic gradient descent for updating $\mathbf{\Theta}$, $\mathbf{B}$ and $\mathbf{A}$. The details of optimization are demonstrated as follows:

- Update networks. The back-propagation algorithm is employed to update the network parameters.
- Update $\mathbf{B}$. The gradient of $\mathcal{L}$ with respect to $\mathbf{B}$ can be derived as:

$$\begin{aligned}\frac{\partial \mathcal{L}}{\partial \mathbf{B}} &= 2(\boldsymbol{f}(\mathbf{X};\mathbf{\Theta}) - \mathbf{Y})^T(\boldsymbol{f}(\mathbf{X};\mathbf{\Theta}) - \mathbf{Y})\mathbf{B} \\ &- 2\mathbf{Y}^T(\mathbf{Y} - \mathbf{YBA})\mathbf{A}^T + 2\lambda_1(\mathbf{B} - \mathbf{I}) + 2\lambda_2\mathbf{DB},\end{aligned} \tag{5}$$

  where $\mathbf{D}$ is a diagonal matrix with $D_{ii} = \frac{1}{2||\mathbf{B}_i||}$. Accordingly, gradient descent is employed based on Eq. (5).
- Update $\mathbf{A}$. The gradient of $\mathcal{L}$ with respect to $\mathbf{A}$ can be derived as:

$$\frac{\partial \mathcal{L}}{\partial \mathbf{A}} = -2\mathbf{B}^T\mathbf{Y}^T(\mathbf{Y} - \mathbf{YBA}). \tag{6}$$

  then $\mathbf{A}$ is updated by applying gradient descent based on Eq. (6). The optimization procedure of SPL-MLL is summarized as Algorithm 1.

Once the model of SPL-MLL is obtained, it can be easily applied for predicting the labels of test samples. Specifically, given a test input $\mathbf{x}$, it will be first transformed into $\boldsymbol{f}(\mathbf{x};\mathbf{\Theta})$, followed by utilizing the learned mappings $\mathbf{B}$ and $\mathbf{A}$ to predict its all possible labels with $\mathbf{y} = \boldsymbol{f}(\mathbf{x};\mathbf{\Theta})\mathbf{BA}$.

4 Experiments

4.1 Experiment Settings

We conduct experiments on the following benchmark multi-label datasets: emotions [26], yeast [8], tmc2007 [6], scene [4], espgame [29] and pascal VOC 2007 [9]. Specifically, emotions and yeast are used for music and gene functional classification, respectively; tmc2007 is a large-scale text dataset, while scene, espgame and pascal voc 2007 belong to the domain of image. The description of features for emotions, yeast, tmc2007 and scene could be referred in [4,6,8,26]. For espgame and pascal voc 2007, the local descriptor DenseSift [20] is used. These datasets can be found in Mulan[1] and LEAR websites[2]. The detailed statistics information of each dataset is listed in Table 1. We employ the standard partitions for training and testing sets (see footnote 1, 2).

[1] http://mulan.sourceforge.net/datasets-mlc.html.

[2] http://lear.inrialpes.fr/people/guillaumin/data.php.

For the proposed SPL-MLL, we utilize neural networks for feature embedding and classification. The networks consists of 2 layers: for the first and second fully connected layer, 512 and 64 neurons are deployed, respectively. A leaky ReLU activation function is employed with the batch size being 64. In addition, we initialize the matrix $\mathbf{B}$ with $\mathbf{B} = \mathbf{I}$ which captures the most sparse correlation among labels and is beneficial to landmark selection. The regularization parameters, i.e., λ_1 and λ_2 are both fixed as 0.1 for all datasets and promising performance is obtained. In our experiments, we set the constraint $\mathbf{B}_{ij} = 0, i \neq j$ in each iteration of optimization. This strictly guarantees the diagonal property and can provide clear interpretability for landmarks. The experimental results show that both convergence of our model and promising performance are achieved with this constraint. Five diverse metrics are employed for performance evaluation. For *Hamming loss* and *Ranking loss*, smaller value indicates better classification quality, while larger value of *Average precision*, *Macro-F1* and *Micro-F1* means better performance. These evaluation metrics evaluate the performance of multi-label predictor from various aspects, and details of these evaluation metrics can be found in [37]. 10-fold cross-validation is performed for each method, which randomly holds 1/10 of training data for validation during each fold. We repeat each experiment 10 times and report the averaged results with standard derivations.

Table 1. Statistics of datasets.

Dataset	#instances	#features	#labels	Cardinality	Domain
Emotions	593	72	6	1.9	Music
Scene	2407	294	6	1.1	Image
Yeast	2417	103	14	4.2	Biology
tmc2007	28596	500	22	2.2	Text
Espgame	20770	1000	268	4.7	Image
Pascal VOC 2007	9963	1000	20	1.5	Image

4.2 Experimental Results

Comparison with State-of-the-Art Multi-label Classification Methods. We compare our algorithm with both baseline and state-of-the-art multi-label classification methods. The binary relevance (BR) [27] and label powerset (LP) [4] act as baselines. We also compare ours with two ensemble methods, *i.e.*, ensemble of pruned sets (EPS) [21] and ensemble of classifier chains (ECC) [22], second-order approach - calibrated label ranking (CLR) [10] and high-order approach - random k-labelsets (RAkEL) [28], the lazy multi-label methods based on k-nearest neighbors (ML-kNN)[36] and feature-aware approach - multi-label manifold learning (MLML) [12], labeling information enrichment approach - Multi-label Learning with Feature-induced labeling information Enrichment(MLFE) [38] and robust approach for data with hybrid noise - hybrid

Table 2. Comparing results (mean ± std.) of multi-label learning algorithms. ↓ (↑) indicates the smaller (larger), the better. The values in red and blue indicate the best and the second best performances, respectively. • indicates that ours is better than the compared algorithms.

Datasets	Methods	Ranking loss ↓	Hamming loss ↓	Average precision ↑	Micro-F1 ↑	Macro-F1 ↑
Emotions	BR [27]	0.309 ± 0.021•	0.265 ± 0.015•	0.687 ± 0.017•	0.592 ± 0.025•	0.590 ± 0.016•
	LP [4]	0.345 ± 0.022•	0.277 ± 0.010•	0.661 ± 0.018•	0.533 ± 0.016•	0.504 ± 0.019•
	ML-kNN [36]	0.173 ± 0.015•	0.209 ± 0.021•	0.794 ± 0.016•	0.650 ± 0.031•	0.607 ± 0.033•
	EPS [21]	0.183 ± 0.014•	0.208 ± 0.010•	0.780 ± 0.017•	0.664 ± 0.012•	0.655 ± 0.018•
	ECC [22]	0.198 ± 0.021•	0.228 ± 0.022•	0.766 ± 0.014•	0.617 ± 0.013•	0.597 ± 0.019•
	RAkEL [28]	0.217 ± 0.026•	0.219 ± 0.013•	0.766 ± 0.031•	0.634 ± 0.023•	0.618 ± 0.036•
	CLR [10]	0.199 ± 0.024•	0.255 ± 0.012•	0.762 ± 0.024•	0.614 ± 0.037•	0.601 ± 0.038•
	MLML [12]	0.184 ± 0.015•	0.197 ± 0.013•	0.719 ± 0.018•	0.661 ± 0.039•	0.650 ± 0.047•
	MLFE [38]	0.181 ± 0.012•	0.217 ± 0.020•	0.782 ± 0.013•	0.674 ± 0.026•	0.663 ± 0.021•
	HNOML [35]	0.173 ± 0.012•	0.192 ± 0.005•	0.784 ± 0.011•	0.672 ± 0.014•	0.660 ± 0.029•
	Ours (linear)	0.172 ± 0.006	0.184 ± 0.015	0.798 ± 0.011	0.686 ± 0.013	0.675 ± 0.031
	Ours	0.170 ± 0.004	0.175 ± 0.021	0.815 ± 0.014	0.698 ± 0.021	0.687 ± 0.024
Yeast	BR [27]	0.322 ± 0.011•	0.253 ± 0.004•	0.614 ± 0.008•	0.569 ± 0.014•	0.386 ± 0.011•
	LP [4]	0.408 ± 0.008•	0.282 ± 0.005•	0.566 ± 0.008•	0.519 ± 0.023•	0.361 ± 0.025•
	ML-kNN [36]	0.171 ± 0.006	0.218 ± 0.004•	0.757 ± 0.011•	0.636 ± 0.012•	0.357 ± 0.021•
	EPS [21]	0.205 ± 0.003•	0.214 ± 0.005•	0.731 ± 0.017•	0.625 ± 0.015•	0.372 ± 0.014•
	ECC [22]	0.187 ± 0.007•	0.209 ± 0.009•	0.745 ± 0.012•	0.618 ± 0.013•	0.369 ± 0.017•
	RAkEL [28]	0.250 ± 0.005•	0.232 ± 0.005•	0.710 ± 0.009•	0.632 ± 0.009•	0.430 ± 0.012•
	CLR [10]	0.187 ± 0.005•	0.222 ± 0.005•	0.745 ± 0.008•	0.628 ± 0.012•	0.400 ± 0.018•
	MLML [12]	0.178 ± 0.002•	0.224 ± 0.005•	0.757 ± 0.009•	0.641 ± 0.014•	0.443 ± 0.025•
	MLFE [38]	0.169 ± 0.021	0.227 ± 0.010•	0.754 ± 0.012•	0.646 ± 0.013•	0.415 ± 0.011•
	HNOML [35]	0.179 ± 0.007•	0.222 ± 0.004•	0.757 ± 0.011•	0.648 ± 0.006•	0.421 ± 0.016•
	Ours (linear)	0.172 ± 0.003	0.210 ± 0.008	0.769 ± 0.006	0.659 ± 0.012	0.443 ± 0.016
	Ours	0.171 ± 0.004	0.201 ± 0.006	0.786 ± 0.005	0.667 ± 0.011	0.451 ± 0.023
Scene	BR [27]	0.236 ± 0.017•	0.136 ± 0.004•	0.715 ± 0.011•	0.609 ± 0.014•	0.616 ± 0.025•
	LP [4]	0.219 ± 0.010•	0.149 ± 0.006•	0.722 ± 0.010•	0.585 ± 0.016•	0.592 ± 0.011•
	ML-kNN [36]	0.093 ± 0.009•	0.095 ± 0.008•	0.851 ± 0.016•	0.718 ± 0.015•	0.719 ± 0.024•
	EPS [21]	0.113 ± 0.007•	0.103 ± 0.017•	0.825 ± 0.013•	0.686 ± 0.018•	0.688 ± 0.018•
	ECC [22]	0.103 ± 0.010•	0.104 ± 0.012•	0.832 ± 0.015•	0.668 ± 0.017•	0.671 ± 0.016•
	RAkEL [28]	0.106 ± 0.005•	0.106 ± 0.005•	0.829 ± 0.007•	0.636 ± 0.023•	0.644 ± 0.019•
	CLR [10]	0.106 ± 0.003•	0.138 ± 0.003•	0.817 ± 0.006•	0.612 ± 0.026•	0.620 ± 0.025•
	MLML [12]	0.079 ± 0.004•	0.098 ± 0.013•	0.862 ± 0.010•	0.728 ± 0.029•	0.729 ± 0.029•
	MLFE [38]	0.079 ± 0.002•	0.094 ± 0.003•	0.858 ± 0.013•	0.732 ± 0.021•	0.734 ± 0.019•
	HNOML [35]	0.103 ± 0.005•	0.110 ± 0.003•	0.832 ± 0.108•	0.733 ± 0.011•	0.736 ± 0.013•
	Ours (linear)	0.073 ± 0.003	0.083 ± 0.006	0.861 ± 0.005	0.738 ± 0.012	0.742 ± 0.021
	Ours	0.067 ± 0.003	0.074 ± 0.004	0.884 ± 0.005	0.746 ± 0.016	0.753 ± 0.024
Espgame	BR [27]	0.266 ± 0.003•	0.019 ± 0.002•	0.221 ± 0.001•	0.205 ± 0.004•	0.116 ± 0.001•
	LP [4]	0.496 ± 0.003•	0.031 ± 0.001•	0.055 ± 0.004•	0.109 ± 0.003•	0.060 ± 0.002•
	ML-kNN [36]	0.238 ± 0.001•	0.017 ± 0.002	0.255 ± 0.003•	0.039 ± 0.002•	0.020 ± 0.001•
	EPS [21]	0.380 ± 0.001•	0.017 ± 0.001	0.200 ± 0.003•	0.083 ± 0.002•	0.065 ± 0.001•
	ECC [22]	0.230 ± 0.001•	0.020 ± 0.002•	0.282 ± 0.001•	0.245 ± 0.004•	0.123 ± 0.001•
	RAkEL [28]	0.343 ± 0.001•	0.019 ± 0.001•	0.211 ± 0.003•	0.150 ± 0.003•	0.059 ± 0.001•
	CLR [10]	0.196 ± 0.001	0.019 ± 0.001•	0.305 ± 0.003	0.266 ± 0.004•	0.143 ± 0.001•
	MLML [12]	0.317 ± 0.000•	0.019 ± 0.003•	0.086 ± 0.002•	0.103 ± 0.003•	0.060 ± 0.002•
	MLFE [38]	0.312 ± 0.012•	0.020 ± 0.001•	0.268 ± 0.011•	0.260 ± 0.003•	0.134 ± 0.004•
	HNOML [35]	0.221 ± 0.001•	0.019 ± 0.003•	0.271 ± 0.003•	0.263 ± 0.006•	0.132 ± 0.004•
	Ours (linear)	0.223 ± 0.002	0.017 ± 0.001	0.289 ± 0.002	0.269 ± 0.002	0.143 ± 0.002
	Ours	0.220 ± 0.003	0.016 ± 0.001	0.291 ± 0.002	0.276 ± 0.004	0.149 ± 0.001

(*continued*)

Table 2. (*continued*)

Datasets	Methods	Ranking loss ↓	Hamming loss ↓	Average precision ↑	Micro-F1 ↑	Macro-F1 ↑
tmc2007	BR [27]	0.037 ± 0.007•	0.031 ± 0.004•	0.899 ± 0.025•	0.834 ± 0.014•	0.719 ± 0.011•
	LP [4]	0.324 ± 0.018•	0.041 ± 0.006•	0.594 ± 0.012•	0.791 ± 0.008•	0.721 ± 0.004•
	ML-kNN [36]	0.031 ± 0.006•	0.058 ± 0.004•	0.844 ± 0.017•	0.682 ± 0.003•	0.493 ± 0.002•
	EPS [21]	0.021 ± 0.004•	0.033 ± 0.005•	0.927 ± 0.007•	0.829 ± 0.009•	0.722 ± 0.010•
	ECC [22]	0.017 ± 0.006•	0.026 ± 0.003•	0.925 ± 0.006•	0.862 ± 0.014•	0.763 ± 0.007•
	RAkEL [28]	0.038 ± 0.008•	0.024 ± 0.002•	0.923 ± 0.005•	0.870 ± 0.011•	0.756 ± 0.006•
	CLR [10]	0.018 ± 0.005•	0.034 ± 0.004•	0.923 ± 0.011•	0.825 ± 0.013•	0.711 ± 0.011•
	MLML [12]	0.018 ± 0.001•	0.021 ± 0.001•	0.921 ± 0.002•	0.865 ± 0.011•	0.769 ± 0.008•
	MLFE [38]	0.021 ± 0.002•	0.022 ± 0.001•	0.924 ± 0.013•	0.873 ± 0.015•	0.771 ± 0.011•
	HNOML [35]	0.023 ± 0.002•	0.017 ± 0.001•	0.919 ± 0.003•	0.858 ± 0.014•	0.762 ± 0.016•
	Ours (linear)	0.015 ± 0.003	0.013 ± 0.002	0.937 ± 0.007	0.912 ± 0.008	0.781 ± 0.005
	Ours	0.012 ± 0.004	0.011 ± 0.001	0.945 ± 0.007	0.944 ± 0.007	0.792 ± 0.010

noise-oriented multilabel learning (HNOML) [35]. We try our best to tune the parameters of all the above compared methods to the best performance according to the suggested ways in their literatures.

Table 3. Performance comparisons with approaches based on label space reduction.

Datasets	tmc2007		Espgame	
Methods/Metrics	Micro-F1↑	Macro-F1↑	Micro-F1↑	Macro-F1↑
MOPLMS [1]	0.556 ± 0.012	0.421 ± 0.013	0.032 ± 0.006	0.025 ± 0.005
ML-CSSP [3]	0.604 ± 0.014	0.432 ± 0.015	0.035 ± 0.004	0.023 ± 0.006
PBR [7]	0.602 ± 0.034	0.422 ± 0.025	0.021 ± 0.008	0.014 ± 0.003
CPLST [7]	0.643 ± 0.027	0.437 ± 0.031	0.042 ± 0.005	0.023 ± 0.004
FAIE [18]	0.605 ± 0.011	0.458 ± 0.015	0.072 ± 0.008	0.026 ± 0.003
Deep CPLST	0.786 ± 0.021	0.601 ± 0.031	0.074 ± 0.004	0.016 ± 0.002
Deep FAIE	0.604 ± 0.016	0.435 ± 0.029	0.121 ± 0.011	0.024 ± 0.003
LEML [34]	0.704 ± 0.013	0.616 ± 0.022	0.148 ± 0.004	0.082 ± 0.001
SLEEC [2]	0.607 ± 0.031	0.586 ± 0.011	0.226 ± 0.016	0.108 ± 0.009
DC2AE [33]	0.808 ± 0.017	0.757 ± 0.027	0.256 ± 0.013	0.121 ± 0.009
Ours	**0.944 ± 0.007**	**0.792 ± 0.010**	**0.276 ± 0.004**	**0.149 ± 0.001**

As shown in Table 2, we report the quantitative experimental results of different methods on the benchmark datasets. Because above comparison methods are not based on neural networks, for fair comparisons, we also report the results of our model using the linear projections instead of neural networks for feature embedding. For each algorithm, the averaged performance with standard deviation are reported in terms of different metrics. As for each metric, "↑" indicates the larger the better while "↓" indicates the smaller the better. The red number and blue number indicate the best and the second best performances, respectively. According to Table 2, several observations are obtained as follows: 1) Compared with other multi-label classification methods, our algorithm achieves competitive performance on all the five benchmark datasets. For example, on emotions, scene and tmc2007, our SPL-MLL ranks as the first in terms of all metrics. 2) Compared with BR and LP, our SPL-MLL obtains much better performance on all datasets. The reason may be that these methods lack of sufficient ability to explore complex correlations among labels. 3) Compared with the three ensemble methods EPS, ECC and RAkEL, our algorithm always performs better, which further verifies the effectiveness of our SPL-MLL. 4) We also note that the performances of ML-kNN, CLR and MLML are also competitive, and the performances of CLR are slightly better than ours on espgame in terms of some metrics. However, the performances of ours are more stable and robust for different datasets. For example, CLR performs unpromising on emotions, yeast, and scene. 5) Furthermore, compared with the latest and most advanced approaches MLFE and HNOML, our model outperforms them on all datasets in terms of most metrics. In short, our proposed SPL-MLL achieves promising and stable performance compared with state-of-the-art multi-label classification methods.

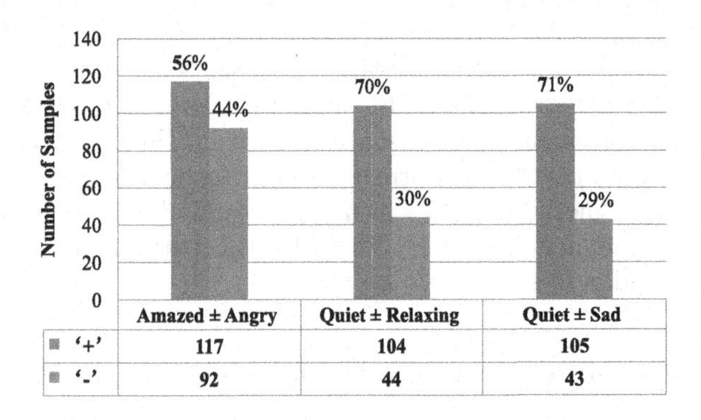

Fig. 3. Visualization of the number of co-occurrence labels on emotions. '+' and '−' denote co-occurrence and no co-occurrence of two labels, respectively.

Comparison with Label Space Reduction Methods. We compare our method with two typical landmark selection methods which conduct landmark selection with group-sparsity technique (MOPLMS) [1] and an efficient randomized sampling procedure (ML-CSSP) [3]. Moreover, we compare our method with the label embedding multi-label classification methods, which jointly reduce the label space and explore the correlations among labels. Specifically, we conduct comparison with the following label embedding based methods: Conditional Principal Label Space Transformation (CPLST) [7], Feature-aware Implicit Label space Encoding (FaIE) [18], Low rank Empirical risk minimization for Multi-Label Learning (LEML) [34], Sparse Local Embeddings for Extreme Multi-label Classification (SLEEC) [2], and the baseline method of partial binary relevance (PBR) [7]. Furthermore, we replace the linear regressors in CPLST and FAIE with DNN regressors, and name them as Deep CPLST and Deep FAIE,

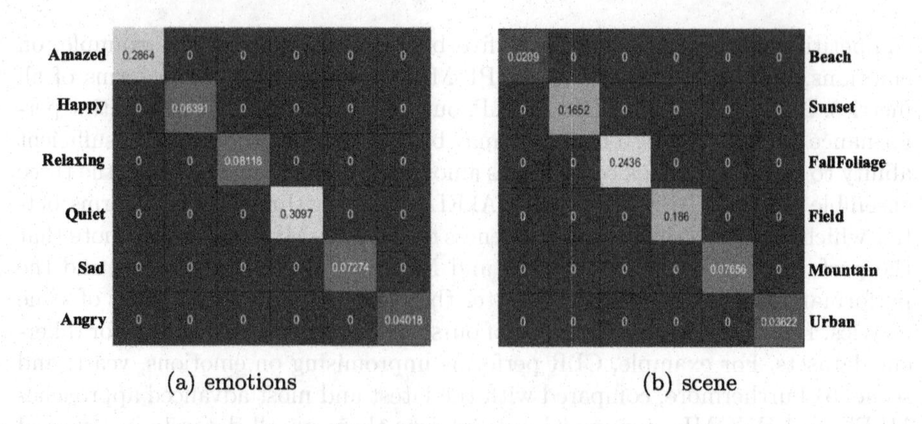

Fig. 4. Visualization of the landmark selection matrix **B**.

respectively. The work in [33] proposes a novel DNN architecture of Canonical-Correlated Autoencoder (C2AE), which can exploit label correlation effectively. Since some methods (e.g., C2AE) reported the results in terms of Micro-F1 and Macro-F1 [25], we also provides results of different approaches in terms of these two metrics for convenient comparison as shown in Table 3. According to the results, it is observed that the performance of our model is much better than the landmark selection methods [1,3] which separate landmark selection and prediction in 2-step manner. Moreover, our SPL-MLL performs superiorly against these label embedding methods.

Table 4. Ablation studies for our model on different setting on pascal VOC 2007.

Methods	Ranking loss ↓	Hamming loss ↓	Average precision ↑	Micro-F1 ↑	Macro-F1 ↑
MLFE [38]	0.232 ± 0.013	0.162 ± 0.012	0.565 ± 0.022	0.436 ± 0.026	0.357 ± 0.011
HNOML [35]	0.227 ± 0.012	0.123 ± 0.008	0.593 ± 0.023	0.443 ± 0.024	0.368 ± 0.019
NN-embeddings	0.324 ± 0.016	0.266 ± 0.011	0.431 ± 0.013	0.308 ± 0.011	0.287 ± 0.016
Ours(NN + separated)	0.243 ± 0.014	0.194 ± 0.015	0.521 ± 0.024	0.384 ± 0.009	0.311 ± 0.017
Ours(joint + linear)	0.192 ± 0.011	0.095 ± 0.012	0.608 ± 0.021	0.516 ± 0.025	0.422 ± 0.024
Ours(joint + NN)	$\mathbf{0.184 \pm 0.012}$	$\mathbf{0.083 \pm 0.013}$	$\mathbf{0.616 \pm 0.018}$	$\mathbf{0.586 \pm 0.018}$	$\mathbf{0.495 \pm 0.017}$

Ablation Studies. To investigate the advantage of our model on jointly conducting landmark selection, landmark prediction and label recovery in a unified framework, we further conduct comparison and ablation experiments on pascal VOC 2007. Specifically, we conduct ablation studies for our model under the following settings: (1) NN-embeddings: the features are directly encoded by neural network for full label recovery without landmark selection and landmark prediction; (2) Ours (NN + separated): our model is still based on the landmark selection strategy, but separates the landmark selection and landmark prediction in the 2-step manner like the work [1]; (3) Ours (joint + linear): our model employs the linear projections instead of neural networks for feature embedding. To further validate the performance improvement from our model, we also report the results of the latest and most advanced approaches MLFE [38] and HNOML [35]. The comparision results are shown in Table 4, which validates the superiority of conducting landmark selection, landmark prediction and label recovery in a unified framework.

Insight for Selected Landmarks. To investigate the improvement of SPL-MLL, we visualize the landmark selection matrix **B** on emotions and scene. As illustrated in Fig. 4, the values in yellow on the diagonal are much larger than the values in other colors, where the corresponding labels are selected landmarks. For emotions, "Amazed" and "Quiet" are most likely to be landmark labels, and "Amazed" is often accompanied by "Angry" in music, "Quiet" tends to occur simultaneously with "Relaxing" or "Sad". Thus, we can utilize the selected landmark labels to recover other related labels effectively. Similar, for scene, the label "FallFoliage" and "Field" are most likely to be landmark labels.

As shown in Fig. 3, we count the number of those samples with or without "Angry" when having "Amazed", which is represented as "Amazed ± Angry", and similarly we obtain "Quiet ± Relaxing" and "Quiet ± Sad". According to Fig. 3, it is observed that when the "Amazed" ("Quiet") emotion occurs, the probability that "Angry" ("Relaxing" and "Sad") occur simultaneously is 56% (70% and 71%). This statistics further support the reasonability of the selected landmark labels.

Fig. 5. Example predictions on espgame and scene.

Result Visualization and Convergence Experiment. For intuitive analysis, Fig. 5 shows some representative examples from espgame and scene. The correctly predicted landmark labels from our model are in red, while the labels in green, gray and black indicate the successfully predicted, missed predicted and wrongly predicted labels. Generally, although multi-label classification is rather challenging especially for the large label set, our model achieves competitive results. We find that a few labels of some samples are not correctly predicted, and the possible reasons are as follows. First, a few labels on some samples do not obviously correlate with other labels, which makes it difficult to accurately recover given selected landmark labels. Second, a few labels for some samples are associated with very small parts in images, making it difficult to predict accurately even taking the feature of images into account in our model. For example, the image labeled with "Kitchen" and "Dining" as landmark labels has the following labels predicted correctly: "Table", "Chair", "Room", "Light", "Restaurant" and "Door". However, there are labels: "Sun" and "Flower" failed to be predicted. The main reasons is that the label "Sun" and "Flower" may be not strongly correlated with the selected landmark labels in the dataset.

There are a few landmarks failed to be predicted for some samples, even though our model aims to select predictable landmark labels. For example, for the rightmost picture in the bottom of Fig. 5, we predict successfully "Sky" and "Man" as landmarks while not able to obtain the more critical landmark label "Parachuting", which leads to failure prediction for "Falling", "Fall", "Skydiving", "Parachuting". It can be seen that the parachute is rather difficult to predict due to the strong illumination.

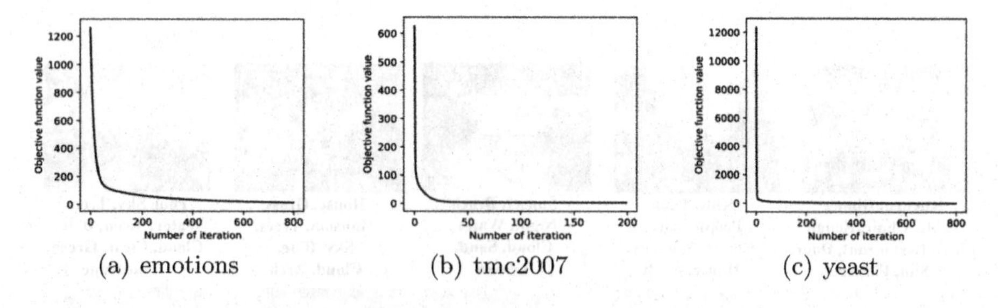

(a) emotions (b) tmc2007 (c) yeast

Fig. 6. Convergence experiment.

Figure 6 gives the convergence experiments on emotion, yeast and tmc2007. Obviously, the results demonstrate that our method can converge within a small number of iterations.

5 Conclusions and Future Work

In this paper, we proposed a novel landmark-based multi-label classification algorithm, termed *SPL-MLL: Selecting Predictable Landmarks for Multi-Label Learning*. SPL-MLL jointly takes the representative and predictable properties for landmarks in a unified framework, avoiding separating landmark selection/prediction in the 2-step manner. Our key idea lies in selecting explicitly the landmarks which are both representative and predictable. The empirical experiments clearly demonstrate that our algorithm outperforms existing state-of-the-art methods. In the future, we will consider the end-to-end manner to extend our model for image annotation with large label set.

Acknowledgements. This work is supported by the National Natural Science Foundation of China (Nos. 61976151, 61732011 and 61872190).

References

1. Balasubramanian, K., Lebanon, G.: The landmark selection method for multiple output prediction. In: International Conference on Machine Learning (2012)
2. Bhatia, K., Jain, H., Kar, P., Varma, M., Jain, P.: Sparse local embeddings for extreme multi-label classification. In: Advances in Neural Information Processing Systems, pp. 730–738 (2015)
3. Bi, W., Kwok, J.: Efficient multi-label classification with many labels. In: International Conference on Machine Learning, pp. 405–413 (2013)
4. Boutell, M.R., Luo, J., Shen, X., Brown, C.M.: Learning multi-label scene classification. Pattern Recogn. **37**(9), 1757–1771 (2004)
5. Boutsidis, C., Mahoney, M.W., Drineas, P.: An improved approximation algorithm for the column subset selection problem. In: Proceedings of the Twentieth Annual ACM-SIAM Symposium on Discrete Algorithms, pp. 968–977. SIAM (2009)
6. Charte, F., Rivera, A., del Jesus, M., Herrera, F.: Multilabel classification. Problem analysis, metrics and techniques book repository
7. Chen, Y.N., Lin, H.T.: Feature-aware label space dimension reduction for multi-label classification. In: Advances in Neural Information Processing Systems, pp. 1529–1537 (2012)
8. Elisseeff, A., Weston, J.: A kernel method for multi-labelled classification. In: Advances in neural Information Processing Systems, pp. 681–687 (2002)
9. Everingham, M., Van Gool, L., Williams, C.K., Winn, J., Zisserman, A.: The pascal visual object classes (VOC) challenge. Int. J. Comput. Vis. **88**(2), 303–338 (2010)
10. Fürnkranz, J., Hüllermeier, E., Mencía, E.L., Brinker, K.: Multilabel classification via calibrated label ranking. Mach. Learn. **73**(2), 133–153 (2008)
11. Ghamrawi, N., McCallum, A.: Collective multi-label classification. In: Proceedings of the 14th ACM International Conference on Information and Knowledge Management, pp. 195–200. ACM (2005)
12. Hou, P., Geng, X., Zhang, M.L.: Multi-label manifold learning. In: Thirtieth AAAI Conference on Artificial Intelligence (2016)
13. Hsu, D.J., Kakade, S.M., Langford, J., Zhang, T.: Multi-label prediction via compressed sensing. In: Advances in Neural Information Processing Systems, pp. 772–780 (2009)

14. Ji, S., Tang, L., Yu, S., Ye, J.: A shared-subspace learning framework for multi-label classification. ACM Trans. Knowl. Discov. Data (TKDD) **4**(2), 8 (2010)
15. Jia, X., Zheng, X., Li, W., Zhang, C., Li, Z.: Facial emotion distribution learning by exploiting low-rank label correlations locally. In: Proceedings of the IEEE Conference on Computer Vision and Pattern Recognition, pp. 9841–9850 (2019)
16. Li, X., Guo, Y.: Multi-label classification with feature-aware non-linear label space transformation. In: Twenty-Fourth International Joint Conference on Artificial Intelligence (2015)
17. Lin, Z., Liu, R., Su, Z.: Linearized alternating direction method with adaptive penalty for low-rank representation. In: Advances in Neural Information Processing Systems, pp. 612–620 (2011)
18. Lin, Z., Ding, G., Hu, M., Wang, J.: Multi-label classification via feature-aware implicit label space encoding. In: International Conference on Machine Learning, pp. 325–333 (2014)
19. Liu, J., Chang, W.C., Wu, Y., Yang, Y.: Deep learning for extreme multi-label text classification. In: Proceedings of the 40th International ACM SIGIR Conference on Research and Development in Information Retrieval, pp. 115–124. ACM (2017)
20. Lowe, D.G.: Distinctive image features from scale-invariant keypoints. Int. J. Comput. Vis. **60**(2), 91–110 (2004)
21. Read, J., Pfahringer, B., Holmes, G.: Multi-label classification using ensembles of pruned sets. In: 2008 Eighth IEEE International Conference on Data Mining, pp. 995–1000. IEEE (2008)
22. Read, J., Pfahringer, B., Holmes, G., Frank, E.: Classifier chains for multi-label classification. Mach. Learn. **85**(3), 333 (2011)
23. Ren, T., Jia, X., Li, W., Zhao, S.: Label distribution learning with label correlations via low-rank approximation. In: Proceedings of the 28th International Joint Conference on Artificial Intelligence, pp. 3325–3331. AAAI Press (2019)
24. Tai, F., Lin, H.T.: Multilabel classification with principal label space transformation. Neural Comput. **24**(9), 2508–2542 (2012)
25. Tang, L., Rajan, S., Narayanan, V.K.: Large scale multi-label classification via metalabeler. In: Proceedings of the 18th International Conference on World Wide Web, pp. 211–220. ACM (2009)
26. Trohidis, K., Tsoumakas, G., Kalliris, G., Vlahavas, I.P.: Multi-label classification of music into emotions. In: ISMIR, vol. 8, pp. 325–330 (2008)
27. Tsoumakas, G., Katakis, I.: Multi-label classification: an overview. Int. J. Data Warehous. Min. (IJDWM) **3**(3), 1–13 (2007)
28. Tsoumakas, G., Katakis, I., Vlahavas, I.: Random k-labelsets for multilabel classification. IEEE Trans. Knowl. Data Eng. **23**(7), 1079–1089 (2011)
29. Von Ahn, L., Dabbish, L.: Labeling images with a computer game. In: Proceedings of the SIGCHI Conference on Human Factors in Computing Systems, pp. 319–326. ACM (2004)
30. Wu, B., Chen, W., Sun, P., Liu, W., Ghanem, B., Lyu, S.: Tagging like humans: diverse and distinct image annotation. In: Proceedings of the IEEE Conference on Computer Vision and Pattern Recognition, pp. 7967–7975 (2018)
31. Wu, B., Jia, F., Liu, W., Ghanem, B.: Diverse image annotation. In: Proceedings of the IEEE Conference on Computer Vision and Pattern Recognition, pp. 2559–2567 (2017)
32. Wu, B., Jia, F., Liu, W., Ghanem, B., Lyu, S.: Multi-label learning with missing labels using mixed dependency graphs. Int. J. Comput. Vis. **126**(8), 875–896 (2018)

33. Yeh, C.K., Wu, W.C., Ko, W.J., Wang, Y.C.F.: Learning deep latent space for multi-label classification. In: Thirty-First AAAI Conference on Artificial Intelligence (2017)
34. Yu, H.F., Jain, P., Kar, P., Dhillon, I.: Large-scale multi-label learning with missing labels. In: International Conference on Machine Learning, pp. 593–601 (2014)
35. Zhang, C., Yu, Z., Fu, H., Zhu, P., Chen, L., Hu, Q.: Hybrid noise-oriented multi-label learning. IEEE Trans. Cybern. **50**, 2837–2850 (2019)
36. Zhang, M.L., Zhou, Z.H.: ML-KNN: a lazy learning approach to multi-label learning. Pattern Recogn. **40**(7), 2038–2048 (2007)
37. Zhang, M.L., Zhou, Z.H.: A review on multi-label learning algorithms. IEEE Trans. Knowl. Data Eng. **26**(8), 1819–1837 (2014)
38. Zhang, Q.W., Zhong, Y., Zhang, M.L.: Feature-induced labeling information enrichment for multi-label learning. In: Thirty-Second AAAI Conference on Artificial Intelligence (2018)
39. Zhang, Y., Schneider, J.: Maximum margin output coding. arXiv preprint arXiv:1206.6478 (2012)
40. Zhou, T., Tao, D., Wu, X.: Compressed labeling on distilled labelsets for multi-label learning. Mach. Learn. **88**(1–2), 69–126 (2012)

Unpaired Image-to-Image Translation Using Adversarial Consistency Loss

Yihao Zhao, Ruihai Wu, and Hao Dong(✉)

Hyperplane Lab, CFCS, Computer Science Department, Peking University, Beijing, China
{zhaoyh98,wuruihai,hao.dong}@pku.edu.cn

Abstract. Unpaired image-to-image translation is a class of vision problems whose goal is to find the mapping between different image domains using unpaired training data. Cycle-consistency loss is a widely used constraint for such problems. However, due to the strict pixel-level constraint, it cannot perform shape changes, remove large objects, or ignore irrelevant texture. In this paper, we propose a novel adversarial-consistency loss for image-to-image translation. This loss does not require the translated image to be translated back to be a specific source image but can encourage the translated images to retain important features of the source images and overcome the drawbacks of cycle-consistency loss noted above. Our method achieves state-of-the-art results on three challenging tasks: glasses removal, male-to-female translation, and selfie-to-anime translation.

Keywords: Generative adversarial networks · Dual learning · Image synthesis

1 Introduction

Learning to translate data from one domain to another is an important problem in computer vision for its wide range of applications, such as colourisation [34], super-resolution [7,13], and image inpainting [11,25]. In unsupervised settings where only unpaired data are available, current methods [2,6,10,15,20,27,31,33,35] mainly rely on shared latent space and the assumption of cycle-consistency. In particular, by forcing the translated images to fool the discriminator using a classical adversarial loss and translating those images back to the original images, cycle-consistency ensures the translated images contain enough information from the original input images. This helps to build a reasonable mapping and generate high-quality results.

The original version of this chapter was revised: The acknowledgment section has been modified. The correction to this chapter is available at https://doi.org/10.1007/978-3-030-58545-7_47

Electronic supplementary material The online version of this chapter (https://doi.org/10.1007/978-3-030-58545-7_46) contains supplementary material, which is available to authorized users.

© Springer Nature Switzerland AG 2020, corrected publication 2020
A. Vedaldi et al. (Eds.): ECCV 2020, LNCS 12354, pp. 800–815, 2020.
https://doi.org/10.1007/978-3-030-58545-7_46

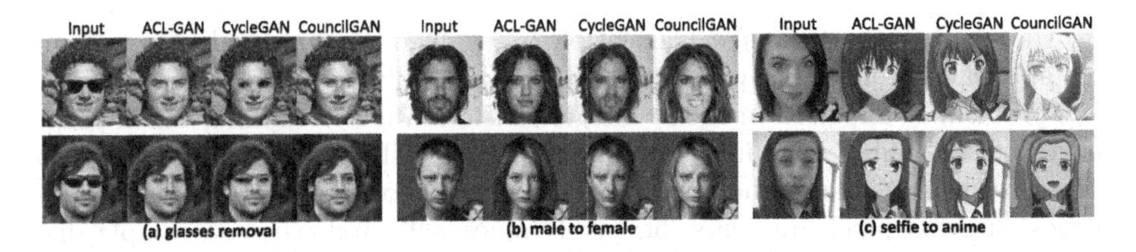

Fig. 1. Example results of our ACL-GAN and baselines. Our method does not require cycle consistency, so it can bypass unnecessary features. Moreover, with the proposed adversarial-consistency loss, our method can explicitly encourage the generator to maintain the commonalities between the source and target domains.

The main problem of cycle-consistency is that it assumes the translated images contain all the information of the input images in order to reconstruct the input images. This assumption leads to the preservation of source domain features, such as traces of large objects (*e.g.*, the trace of glasses in Fig. 1(a)) and texture (*e.g.*, the beard remains on the face in Fig. 1(b)), resulting in unrealistic results. Besides, the cycle-consistency constrains shape changes of source images (*e.g.*, the hair and the shape of the face are changed little in Fig. 1(b) and (c)).

To avoid the drawbacks of cycle-consistency, we propose a novel method, termed ACL-GAN, where ACL stands for *adversarial-consistency loss.* Our goal is to maximise the information shared by the target domain and the source domain. The adversarial-consistency loss encourages the generated image to include more features from the source image, from the perspective of distribution, rather than maintaining the pixel-level consistency. That's to say, the image translated back is only required to be similar to the input source image but need not be identical to a specific source image. Therefore, our generator is not required to preserve all the information from the source image, which avoids leaving artefacts in the translated image. Combining the adversarial-consistency loss with classical adversarial loss and some additional losses makes our full objective for unpaired image-to-image translation.

The main contribution of our work is a novel method to support unpaired image-to-image translation which avoids the drawbacks of cycle-consistency. Our method achieves state-of-the-art results on three challenging tasks: glasses removal, male-to-female translation, and selfie-to-anime translation.

2 Related Work

Generative adversarial networks (GANs) [8] have been successfully applied to numerous image applications, such as super-resolution [17] and image colourisation [34]. These tasks can be seen as "translating" an image into another image. Those two images usually belong to two different domains, termed source domain and target domain. Traditionally, each of the image translation tasks is solved by a task-specific method. To achieve task-agnostic image translation, Pix2Pix [12] was the first work to support different image-to-image translation tasks using a single method. However, it requires paired images to supervise the training.

Unpaired image-to-image translation has gained a great deal of attention for applications in which paired data are unavailable or difficult to collect. A key problem of unpaired image-to-image translation is determining which properties of the source domain to preserve in the translated domain, and how to preserve them. Some methods have been proposed to preserve pixel-level properties, such as pixel gradients [5], pixel values [29], and pairwise sample distances [3]. Recently, several concurrent works, CycleGAN [35], DualGAN [33], DiscoGAN [15], and UNIT [20], achieve unpaired image-to-image translation using cycle-consistency loss as a constraint. Several other studies improve image-to-image translation in different aspects. For example, BicycleGAN [36] supports multi-modal translation using paired images as the supervision, while Augmented CycleGAN [1] achieves that with unpaired data. DRIT [18,19] and MUNIT [10] disentangle domain-specific features and support multi-modal translations. To improve image quality, attention CycleGAN [23] proposes an attention mechanism to preserve the background from the source domain. In this paper, we adopt the network architecture of MUNIT but we do not use cycle-consistency loss.

To alleviate the problem of cycle-consistency, a recent work, CouncilGAN [24], has proposed using duplicate generators and discriminators together with the council loss as a replacement for cycle-consistency loss. By letting different generators compromise each other, the council loss encourages the translated images to contain more information from the source images. By contrast, our method does not require duplicate generators and discriminators, so it uses fewer network parameters and needs less computation. Moreover, our method can generate images of higher quality than CouncilGAN [24], because our adversarial-consistency loss explicitly makes the generated image similar to the source image, while CouncilGAN [24] only compromises among generated images. Details are described in Sect. 3.2.

3 Method

Our goal is to translate images from one domain to the other domain and support diverse and multi-modal outputs. Let X_S and X_T be the source and target domains, X be the union set of X_S and X_T (*i.e.*, $X = X_S \cup X_T$), $x \in X$ be a single image, $x_S \in X_S$ and $x_T \in X_T$ be the images of different domains. We define p_X, p_S and p_T to be the distributions of X, X_S and X_T. $p_{(a,b)}$ is used for joint distribution of pair (a, b), where a and b can be images or noise vectors. Let Z be the noise vector space, $z \in Z$ be a noise vector and $z \sim \mathcal{N}(0, 1)$.

Our method has two generators: $G_S : (x, z) \rightarrow x_S$ and $G_T : (x, z) \rightarrow x_T$ which translate images to domain X_S and X_T, respectively. Similar to [10], each generator contains a noise encoder, an image encoder, and a decoder, in which the noise encoder is used only for calculating the identity loss. The generators receive input pairs (image, noise vector) where the image is from X. In detail, the image encoder receives images sampled from X. The noise vector z obtained from the noise encoder is only for identity loss, while for other losses, the noise vector z is randomly sampled from the standard normal distribution, $\mathcal{N}(0,1)$. The noise vector z and the output of the image encoder are forwarded to the decoder and the output of the decoder is the translated image.

Moreover, there are two kinds of discriminators D_S/D_T and $\hat{D}$. $\hat{D}$ is a consistency discriminator. Its goal is to ensure the consistency between source images and translated images, and this is the core of our method. The goal of D_S and D_T is to distinguish between real and fake images in a certain domain. Specifically, the task of D_S is to distinguish between X_S and $G_S(X)$, and the task of D_T is to distinguish between X_T and $G_T(X)$.

The objective of ACL-GAN has three parts. The first, *adversarial-translation loss*, matches the distributions of generated images to the data distributions in the target domain. The second, *adversarial-consistency loss*, preserves significant features of the source images in the translated images, *i.e.*, it results in reasonable mappings between domains. The third, identity loss and bounded focus mask, can further help to improve the image quality and maintain the image background. The data are forwarded as shown in Fig. 2, and the details of our method are described below.

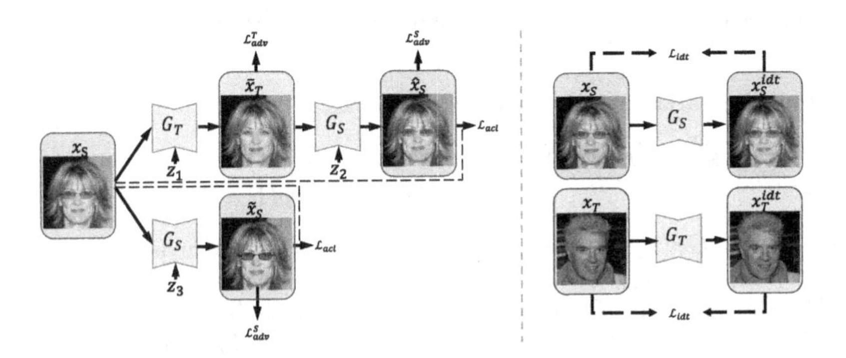

Fig. 2. The training schema of our model. (Left) our model contains two generators: $G_S : (X, Z) \rightarrow X_S$ and $G_T : (X, Z) \rightarrow X_T$ and three discriminators: D_S, D_T for $\mathcal{L}_{adv}^S, \mathcal{L}_{adv}^T$ and $\hat{D}$ for $\mathcal{L}_{acl}$. D_S and D_T ensure that the translated images belong to the correct image domain, while $\hat{D}$ encourages the translated images to preserve important features of the source images. The noise vectors z_1, z_2, z_3 are randomly sampled from $\mathcal{N}(0,1)$. (Right) $\mathcal{L}_{idt}$ encourages to maintain features, improves the image quality, stabilises the training process and prevents mode collapse, where the noise vector is from the noise encoder. The blocks with the same colour indicate shared parameters.

3.1 Adversarial-Translation Loss

For image translation between domains, we utilise the classical adversarial loss, which we call *adversarial-translation loss* in our method, to both generators, G_S and G_T, and discriminators, D_S and D_T. For generator G_T and its discriminator

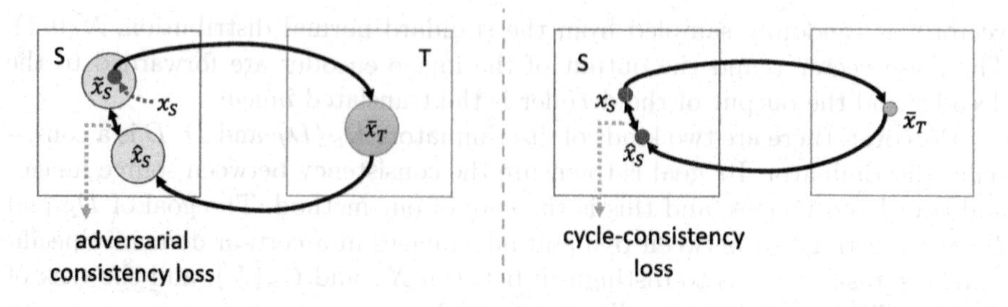

Fig. 3. The comparison of adversarial-consistency loss and cycle-consistency loss [35]. The blue and green rectangles represent image domains S and T, respectively. Any point inside a rectangle represents a specific image in that domain. (Right): given a source image x_S, cycle-consistency loss requires the image translated back, $\hat{x}_S$, should be identical to the source image, x_S. (Left): given a source image x_S, we synthesise multi-modal images in its neighbourhood distribution $\tilde{x}_S$, the distribution $\bar{x}_T$ in the target domain X_T, and the distribution $\hat{x}_S$ translated from $\bar{x}_T$. The distributions are indicated by the blue and green circles. Instead of requiring the image translated back, $\hat{x}_S$, to be a specific image, we minimise the distance between the distributions of $\hat{x}_S$ and $\tilde{x}_S$, so that a specific $\hat{x}_S$ can be any point around x_S. By doing so, we encourage $\bar{x}_T$ to preserve the features of the original image x_S. (Color figure online)

D_T, the adversarial-translation loss is as follows:

$$\begin{aligned}\mathcal{L}^T_{adv}(G_T, D_T, X_S, X_T) = \mathbb{E}_{x_T \sim p_T}[log D_T(x_T)] \\ + \mathbb{E}_{\bar{x}_T \sim p_{\{\bar{x}_T\}}}[log(1 - D_T(\bar{x}_T))]\end{aligned} \tag{1}$$

where $\bar{x}_T = G_T(x_S, z_1)$ and $z_1 \sim \mathcal{N}(0,1)$. The objective is $min_{G_T} max_{D_T}$ $\mathcal{L}^T_{adv}(G_T, D_T, X_S, X_T)$.

The discriminator D_S is expected to distinguish between real images of domain X_S and translated images generated by G_S. The generator G_S tries to generate images, $\hat{x}_S$ and $\tilde{x}_S$, that look similar to images from domain X_S. Therefore, the loss function is defined as:

$$\begin{aligned}\mathcal{L}^S_{adv}(G_S, D_S, \{\bar{x}_T\}, X_S) = \mathbb{E}_{x_S \sim p_S}[log D_S(x_S)] \\ + (\mathbb{E}_{\hat{x}_S \sim p_{\{\hat{x}_S\}}}[log(1 - D_S(\hat{x}_S))] \\ + \mathbb{E}_{\tilde{x}_S \sim p_{\{\tilde{x}_S\}}}[log(1 - D_S(\tilde{x}_S))])/2\end{aligned} \tag{2}$$

where $\hat{x}_S = G_S(\bar{x}_T, z_2)$, $\tilde{x}_S = G_S(x_S, z_3), z_2, z_3 \sim \mathcal{N}(0,1)$ and the objective is $min_{G_S} max_{D_S} \mathcal{L}^S_{adv}(G_S, D_S, \{\bar{x}_T\}, X_S)$. In summary, we can define the adversarial-translation loss as follows:

$$\mathcal{L}_{adv} = \mathcal{L}^T_{adv}(G_T, D_T, X_S, X_T) + \mathcal{L}^S_{adv}(G_S, D_S, \{\bar{x}_T\}, X_S) \tag{3}$$

3.2 Adversarial-Consistency Loss

The $\mathcal{L}_{adv}$ loss described above can encourage the translated image, $\bar{x}_T$, to be in the correct domain X_T. However, this loss cannot encourage the translated image

$\bar{x}_T$ to be similar to the source image x_S. For example, when translating a male to a female, the facial features of the female might not be related to those of the male. To preserve important features of the source image in the translated image, we propose the adversarial-consistency loss, which is realised by a consistency discriminator $\hat{D}$. The consistency discriminator impels the generator to minimise the distance between images $\tilde{x}_S$ and $\hat{x}_S$ as shown in Fig. 3. The "real" and "fake" images for $\hat{D}$ can be swapped without affecting performance. However, letting $\hat{D}$ only distinguish $\hat{x}_S$ and $\tilde{x}_S$ does not satisfy our needs, because the translated images $\hat{x}_S$ and $\tilde{x}_S$ need only to belong to domain X_S; they are not required to be close to a specific source image. Therefore, the consistency discriminator $\hat{D}$ uses x_S as a reference and adopts paired images as inputs to let the generator minimise the distances between the joint distributions of $(x_S, \hat{x}_S)$ and $(x_S, \tilde{x}_S)$. In this way, the consistency discriminator $\hat{D}$ encourages the image translated back, $\hat{x}_S$, to contain the features of the source image x_S. As $\hat{x}_S$ is generated from $\bar{x}_T$, this can encourage the translated image $\bar{x}_T$ to preserve the features of the source image x_S.

The input noise vector z enables multi-modal outputs, which is essential to make our method work. Without multi-modal outputs, given a specific input image x_S, the $\tilde{x}_S$ can have only one case. Therefore, mapping $(x_S, \hat{x}_S)$ and $(x_S, \tilde{x}_S)$ together is almost equivalent to requiring $\hat{x}_S$ and x_S to be identical. This strong constraint is similar to cycle-consistent loss whose drawbacks have been discussed before. With the multi-modal outputs, given a specific image x_S, the $\tilde{x}_S$ can have many possible cases. Therefore, the consistency discriminator $\hat{D}$ can focus on the feature level, rather than the pixel level. That is to say, $\hat{x}_S$ does not have to be identical to a specific image x_S. For example, when translating faces with glasses to faces without glasses, the $\tilde{x}_S$ and $\hat{x}_S$ can be faces with different glasses, *e.g.* the glasses have different colours and frames in x_S, $\tilde{x}_S$ and $\hat{x}_S$ in Fig. 2. Thus, $\bar{x}_T$ need not retain any trace of glasses at the risk of increasing $\mathcal{L}_{adv}^{T}$, and $\mathcal{L}_{acl}$ can still be small.

The adversarial-consistency loss (ACL) is as follows:

$$\begin{aligned}\mathcal{L}_{acl} &= \mathbb{E}_{(x_S,\hat{x}_S)\sim p(X_S,\{\hat{x}_S\})}[log\hat{D}(x_S,\hat{x}_S)] \\ &+ \mathbb{E}_{(x_S,\tilde{x}_S)\sim p(X_S,\{\tilde{x}_S\})}[log(1-\hat{D}(x_S,\tilde{x}_S))]\end{aligned} \tag{4}$$

where $x_S \in X_S$, $\bar{x}_T = G_T(x_S, z_1)$, $\hat{x}_S = G_S(\bar{x}_T, z_2)$, $\tilde{x}_S = G_S(x_S, z_3)$.

3.3 Other Losses

Identity Loss. We further apply identity loss to encourage the generators to be approximate identity mappings when images of the target domain are given to the generators. Identity loss can further encourage feature preservation, improve the quality of translated images, stabilise the training process, and avoid mode collapse because the generator is required to be able to synthesise all images in the dataset [26,35]. Moreover, the identity loss between the source image x_S and the reconstruction image x_S^{idt} can guarantee that x_S is inside the distribution of $\tilde{x}_S$ as shown in Fig. 3.

We formalise two noise encoder networks, $E_S^z : X_S \rightarrow Z$ and $E_T^z : X_T \rightarrow Z$ for G_S and G_T, respectively, which map the images to the noise vectors. Identity loss can be formalised as:

$$\mathcal{L}_{idt} = \mathbb{E}_{x_S \sim p_S}[||x_S - x_S^{idt}||_1] + \mathbb{E}_{x_T \sim p_T}[||x_T - x_T^{idt}||_1] \tag{5}$$

where $x_S^{idt} = G_S(x_S, E_S^z(x_S))$ and $x_T^{idt} = G_T(x_T, E_T^z(x_T))$.

Bounded Focus Mask. Some applications require the generator to only modify certain areas of the source image and keep the rest unchanged. We let the generator produce four channels, where the first three are the channels of RGB images and the fourth is called bounded focus mask whose values are between 0 and 1. The translated image x_T can be obtained by the formula: $x_T = x\prime_T \odot x_m + x_S \odot (1 - x_m)$, where $\odot$ is element-wise product, x_S is the source image, $x\prime_T$ is the first three output channels of the generator and x_m is the bounded focus mask. We add the following constraints to the generator which is one of our contributions:

$$\begin{aligned} \mathcal{L}_{mask} = \delta[(\max\{\sum_k x_m[k] - \delta_{max} \times W, 0\})^2 \\ + (\max\{\delta_{min} \times W - \sum_k x_m[k], 0\})^2] \\ + \sum_k \frac{1}{|x_m[k] - 0.5| + \epsilon} \end{aligned} \tag{6}$$

where δ, δ_{max} and δ_{min} are hyper-parameters for controlling the size of masks, $x_m[k]$ is the k-th pixel of the mask and W is the number of pixels of an image. The ϵ is a marginal value to avoid dividing by zero. The first term of this loss limits the size of the mask to a suitable range. It encourages the generator to make enough changes and maintain the background, where δ_{max} and δ_{min} are the maximum and minimum proportions of the foreground in the mask. The minimum proportion is essential for our method because it avoids $\tilde{x}_S$ being identical to x_S under different noise vectors. The last term of this loss encourages the mask values to be either 0 or 1 to segment the image into a foreground and a background [24]. In the end, this loss is normalised by the size of the image.

3.4 Implementation Details

Full Objective. Our total loss is as follows:

$$\mathcal{L}_{total} = \mathcal{L}_{adv} + \lambda_{acl}\mathcal{L}_{acl} + \lambda_{idt}\mathcal{L}_{idt} + \lambda_{mask}\mathcal{L}_{mask} \tag{7}$$

where $\lambda_{acl}, \lambda_{idt}, \lambda_{mask}$ are all scale values that control the weights of different losses. We compare our method against ablations of the full objective in Sect. 4.2 to show the importance of each component.

Network Architecture. For generator and discriminator, we follow the design of those in [10]. Specifically, a generator consists of two encoders and one decoder. Besides the image encoder and decoder that together form an auto-encoder architecture, our model employs the noise encoder, whose architecture is similar to the style encoder in [10]. Meanwhile, our discriminators use a multi-scale technique [32] to improve the visual quality of synthesised images.

Training Details. We adopt a least-square loss [22] for $\mathcal{L}_{adv}$ (Eq. 3) and $\mathcal{L}_{acl}$ (Eq. 4). This loss brings more stable training process and better results compared with [8]. For all the experiments, we used Adam optimiser [16] with $\beta_1 = 0.5$ and $\beta_2 = 0.999$. The batch size was set to 3. All models were trained with a learning rate of 0.0001 and the learning rate dropped by a factor of 0.5 after every $100K$ iterations. We trained all models for $350K$ iterations. The discriminators update twice while the generators update once. For training, we set $\delta = 0.001$, $\epsilon = 0.01$ and $\lambda_{idt} = 1$. The values of λ_{acl}, λ_{mask}, δ_{min} and δ_{max} were set according to different applications: in glasses removal $\lambda_{acl} = 0.2, \lambda_{mask} = 0.025, \delta_{min} = 0.05, \delta_{max} = 0.1$; in male-to-female translation $\lambda_{acl} = 0.2, \lambda_{mask} = 0.025, \delta_{min} = 0.3, \delta_{max} = 0.5$; in selfie-to-anime translation $\lambda_{acl} = 0.5, \lambda_{mask} = \delta_{min} = \delta_{max} = 0$. For fair comparison, we follow the same data augmentation method described in CouncilGAN [24].

4 Experiments

4.1 Experimental Settings

Datasets. We evaluated ACL-GAN on two different datasets, CelebA [21] and selfie2anime [14]. CelebA [21] contains 202,599 face images with 40 binary attributes. We used the attributes of gender and with/without glasses for evaluation. For both attributes, we used 162,770 images for training and 39,829 images for testing. For gender, the training set contained 68,261 images of males and 94,509 images of females and the test set contained 16,173 images of males and 23,656 images of females. For with/without glasses, the training set contained 10,521 images with glasses and 152,249 images without glasses and the test set contained 2,672 images with glasses and 37,157 images without glasses. Selfie2anime [14] contains 7,000 images. The size of the training set is 3,400 for both anime images and selfie images. The test set has 100 anime images and 100 selfie images.

Metrics. Fréchet Inception Distance (FID) [9] is an improvement of Inception Score (IS) [28] for evaluating the image quality of generative models. FID calculates the Fréchet distance with mean and covariance between the real and the fake image distributions. Kernel Inception Distance (KID) [4] is an improved measure of GAN convergence and quality. It is the squared Maximum Mean Discrepancy between inception representations.

Baselines. We compared the results of our method with those of some state-of-the-art models, including CycleGAN [35], MUNIT [10], DRIT++ [18,19], StarGAN [6], U-GAT-IT [14], Fixed-Point GAN [30], and CouncilGAN [24]. All those methods use unpaired training data. MUNIT [10], DRIT++ [18,19], and CouncilGAN [24] can generate multiple results for a single input image and the others produce only one image for each input. Among those methods, only CouncilGAN [24] does not use cycle-consistency loss. Instead, it adopts N duplicate generators and $2N$ duplicate discriminators (N is set to be 2, 4 or 6 in their paper). This requires much more computation and memory for training. We set N to be 4 in our comparison.

The quantitative and qualitative results of the baselines were obtained by running the official public codes, except for the CouncilGAN, where the results were from our reproduction.

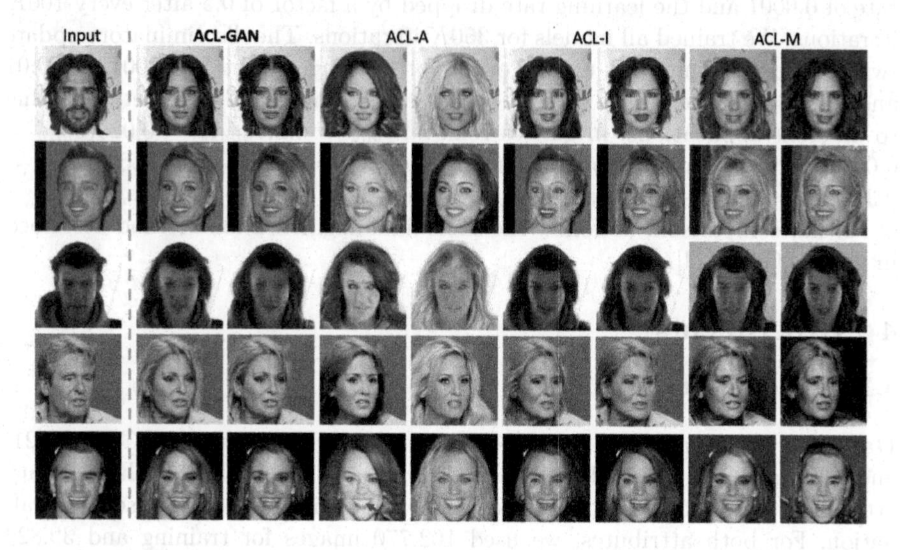

Fig. 4. Ablation studies. The male-to-female translation results illustrate the importance of different losses. From left to right: input images; ACL-GAN (with total loss); ACL-A (without $\mathcal{L}_{acl}$); ACL-I (without $\mathcal{L}_{idt}$); ACL-M (without $\mathcal{L}_{mask}$).

Table 1. Quantitative results of different ablation settings on male-to-female translation. For both FID and KID, lower is better. For KID, mean and standard deviation are listed. The ACL-GAN with total loss outperforms other settings.

Model	$\mathcal{L}_{acl}$	$\mathcal{L}_{idt}$	$\mathcal{L}_{mask}$	FID	KID
ACL-A	–	✓	✓	23.96	0.023 ± 0.0003
ACL-I	✓	–	✓	18.39	0.018 ± 0.0004
ACL-M	✓	✓	–	17.72	0.017 ± 0.0004
ACL-GAN	✓	✓	✓	**16.63**	**0.015 ± 0.0003**

4.2 Ablation Studies

We analysed ACL-GAN by comparing four different settings: 1) with total loss (ACL-GAN), 2) without adversarial-consistency loss $\mathcal{L}_{acl}$ (ACL-A), 3) without identity loss $\mathcal{L}_{idt}$ (ACL-I), and 4) without bounded focus mask and its corresponding loss $\mathcal{L}_{mask}$ (ACL-M). The results of different settings are shown in Fig. 4 and the quantitative results are listed in Table 1.

Qualitatively, we observed that our adversarial-consistency loss, $\mathcal{L}_{acl}$, successfully helps to preserve important features of the source image in the translated image, compared with the setting of ACL-A which already has identity loss and bounded focus mask. For example, as the red arrows on results of ACL-A show, without $\mathcal{L}_{acl}$, the facial features, *e.g.* skin colour, skin wrinkles, and teeth, are difficult to maintain, and the quantitative results are the worst. We found that ACL-GAN with bounded focus mask has better perceptual quality, *e.g.*, the background can be better maintained, compared with ACL-M. The quantitative results indicate that our bounded focus mask improves the quality of images because the mask directs the generator to concentrate on essential parts for translation. Even though it is difficult to qualitatively compare ACL-I with the others, ACL-GAN achieves better quantitative results than ACL-I.

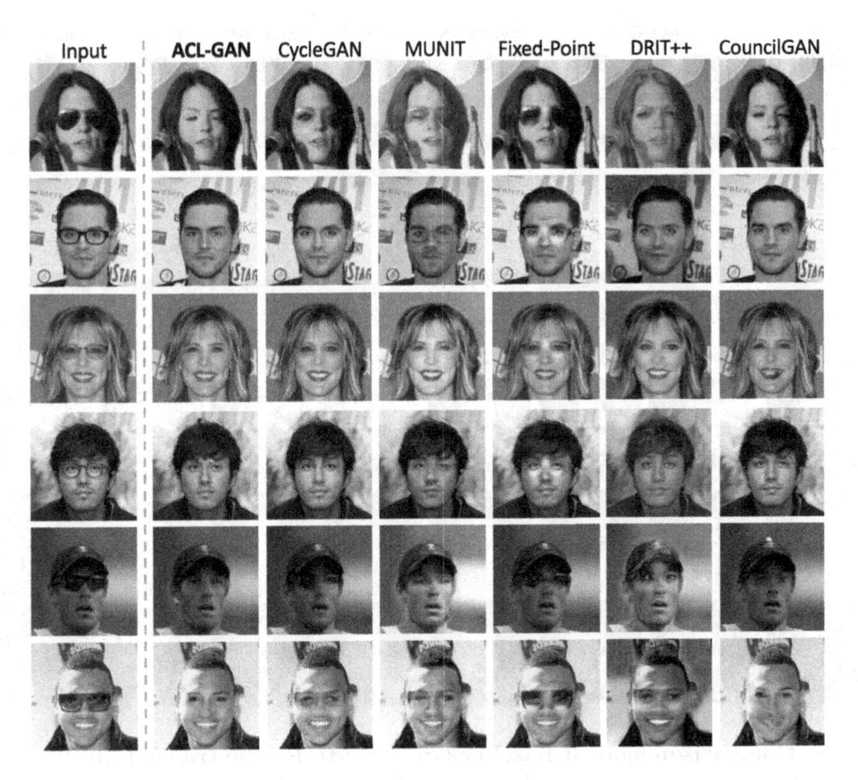

Fig. 5. Comparison against baselines on glasses removal. From left to right: input, our ACL-GAN, CycleGAN [35], MUNIT [10], Fixed-Point GAN [30], DRIT++ [18,19], and CouncilGAN [24].

4.3 Comparison with Baselines

We compared our ACL-GAN with baselines on three challenging applications: glasses removal, male-to-female translation, and selfie-to-anime translation. These three applications have different commonalities between the source and target domains, and they require changing areas with different sizes. Glasses removal requires changing the smallest area, while the selfie-to-anime translation requires changing the largest area. Table 2 shows the number of parameters of our method and baselines. Table 3 shows the quantitative results for different applications and our method outperforms the baselines. Figure 5, 6 and 7 show the results of our method and baselines.

Glasses Removal. The goal of glasses removal is to remove the glasses of a person in a given image. There are two difficulties in this application. First, the area outside the glasses should not be changed, which requires the generator to identify the glasses area. Second, the glasses hide some information of the face, such as the eyes and eyebrows. The sunglasses in some images make it even more difficult because the eyes are totally occluded and the generator is expected to generate realistic and suitable eyes.

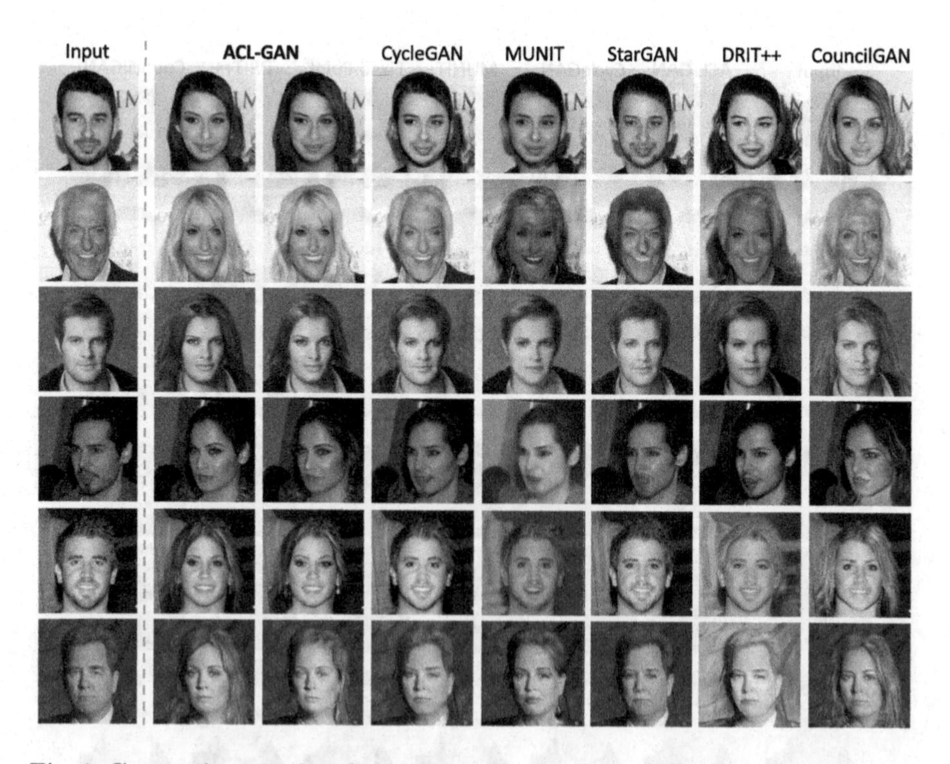

Fig. 6. Comparison against baselines on male-to-female translation. We show two translated image of our model under different noise vectors. From left to right: input, our ACL-GAN, CycleGAN [35], MUNIT [10], StarGAN [6], DRIT++ [18,19], and CouncilGAN [24].

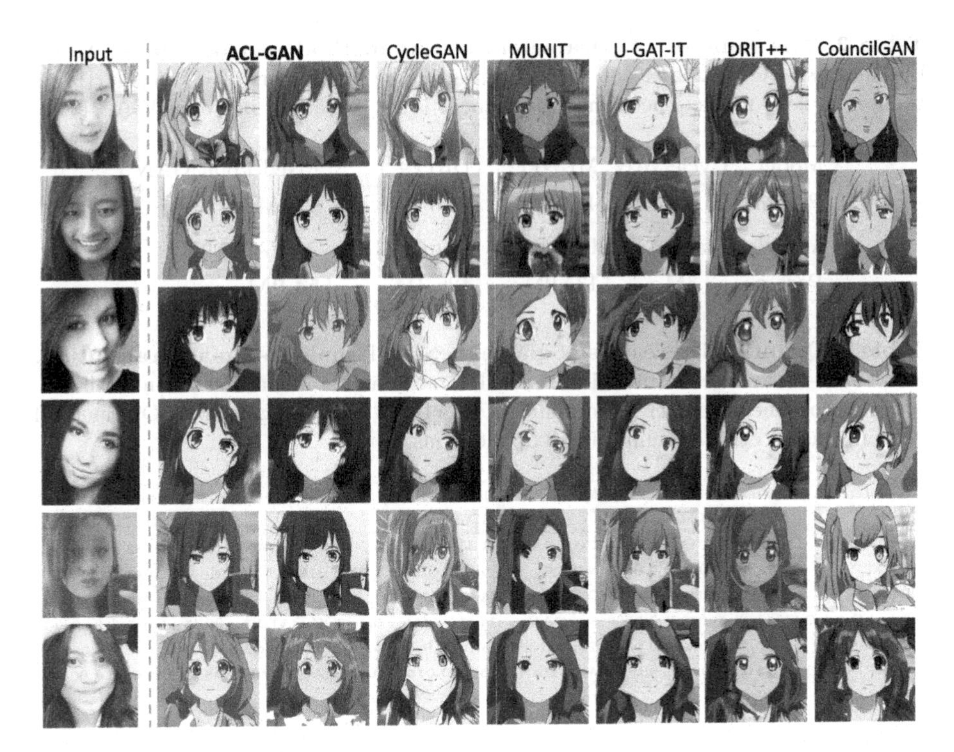

Fig. 7. Comparison against baselines on selfie-to-anime translation. From left to right: input, our ACL-GAN, CycleGAN [35], MUNIT [10], U-GAT-IT [14], DRIT++ [18,19], and CouncilGAN [24].

Figure 5 shows the results of ACL-GAN and baselines for glasses removal. Our results leave fewer traces of glasses than [10,18,19,30,35] because cycle-consistency is not required. The left columns of Table 3 show the quantitative results of glasses removal. Our method outperforms all baselines on both FID and KID. It is interesting that our method can outperform CouncilGAN and MUNIT, which use the same network architecture as ACL-GAN. Compared with MUNIT, our method does not require the image translated back to be a specific image so that no artefacts remain. Compared with CouncilGAN, it might be that, rather than requiring multiple duplicate generators to compromise each other by minimising the distances between their synthesised images, our method explicitly encourages the generated image to be similar to the source image.

Male-to-Female Translation. The goal of male-to-female translation is to generate a female face when given a male face, and these two faces should be similar except for gender. Comparing with glasses removal, this task does not require to remove objects, but there are three difficulties of this translation task. First, this task is a typical multi-modal problem because there are many possible ways to translate a male face to a female face *e.g.,* different lengths of hair. Second, translating male to female realistically requires not only changing

Table 2. The number of parameters of our method and baselines. We set the image size to be 256 × 256 for all methods. The parameters of U-GAT-IT [14] is counted in light mode. The number of councils is set to be four in CouncilGAN [24].

Model	CycleGAN	MUNIT	DRIT++	Fixed-Point GAN
Params	28.3M	46.6M	65.0M	53.2M
Model	StarGAN	U-GAT-IT	CouncilGAN	**ACL-GAN**
Params	53.3M	134.0M	126.3M	**54.9M**

Table 3. Quantitative results of glasses removal, male-to-female translation, and selfie-to-anime translation. For KID, mean and standard deviation are listed. A lower score means better performance. U-GAT-IT [14] is in light mode. Our method outperforms all other baselines in all applications.

Model	Glasses removal		Model	Male to female		Model	Selfie to anime	
	FID	KID		FID	KID		FID	KID
CycleGAN	48.71	0.043 ± 0.0011	CycleGAN	21.30	0.021 ± 0.0003	CycleGAN	102.92	0.042 ± 0.0019
MUNIT	28.58	0.026 ± 0.0009	MUNIT	19.02	0.019 ± 0.0004	MUNIT	101.30	0.043 ± 0.0041
DRIT++	33.06	0.026 ± 0.0006	DRIT++	24.61	0.023 ± 0.0002	DRIT++	104.40	0.050 ± 0.0028
Fixed-Point GAN	44.22	0.038 ± 0.0009	StarGAN	36.17	0.034 ± 0.0005	U-GAT-IT	99.15	0.039 ± 0.0030
CouncilGAN	27.77	0.025 ± 0.0011	CouncilGAN	18.10	0.017 ± 0.0004	CouncilGAN	98.87	0.042 ± 0.0047
ACL-GAN	**23.72**	**0.020 ± 0.0010**	ACL-GAN	**16.63**	**0.015 ± 0.0003**	ACL-GAN	**93.58**	**0.037 ± 0.0036**

the colour and texture but also shape, *e.g.*, the hair and beard. Third, the paired data is impossible to acquire, *i.e.*, this task can only be solved by using unpaired training data.

Figure 6 compares our ACL-GAN with baselines for male-to-female translation. For each row, two images of ACL-GAN are generated with two random noise vectors. This shows the diversity of the results of our method, which is the same in Fig. 7. Our results are more feminine than baselines with cycle-consistency loss [6,18,19,35] as our faces have no beard, longer hair and more feminine lips and eyes. We found that MUNIT [10] can generate images with long hair and no beard, which may contribute to the style latent code. However, the important features, *e.g.* the hue of the image, cannot be well-preserved. Quantitatively, the middle columns of Table 3 show the effectiveness of our method and it can be also explained by the lack of cycle-consistency loss [35] and compromise between duplicated generators [24].

Selfie-to-Anime Translation. Different from the previous two tasks, generating an animated image conditioned on a selfie requires large modifications of shape. The structure and style of the selfie are changed greatly in the target domain, *e.g.*, the eyes become larger and the mouth becomes smaller. This may lead to the contortion and dislocation of the facial features.

Figure 7 shows the results of ACL-GAN and baselines for selfie-to-anime translation. The results of methods with cycle-consistency are less in accordance with the style of anime. In contrast, without cycle-consistency loss, our method can generate images more like an anime, *e.g.* the size and the layout of facial features are better-organized than baselines. The adversarial-consistency loss helps to preserve features, *e.g.*, the haircut and face rotation of the source images. U-GAT-IT [14] utilises an attention module and AdaIN to produce more visually pleasing results. The full mode of U-GAT-IT has 670.8M parameters, which is more than twelve times the number of parameters our model has. For fairness, we evaluated U-GAT-IT in light mode for comparison. Our method can still outperform U-GAT-IT when it is in light mode, which still uses more than twice as many parameters as ours as shown in Table 2.

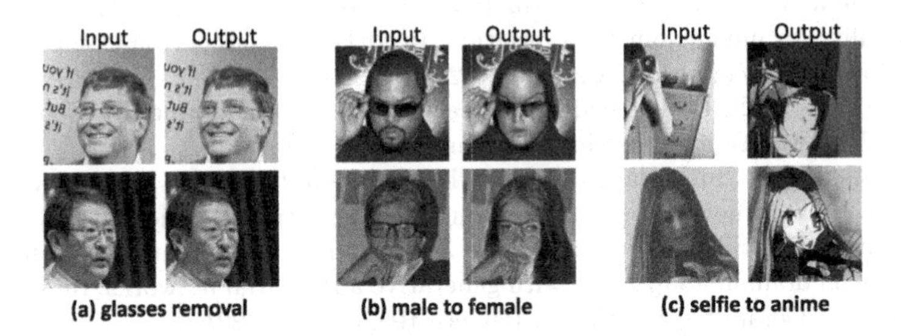

Fig. 8. Typical failure cases of our method. (a) Some glasses are too inconspicuous to be identified and they can not be completely removed by the generator. (b) We also found that the glasses might be partly erased when translating a male to a female and this is because of the imbalance inherent to the datasets. (c) The generated anime image may be distorted and blurred when the face in the input image is obscured or too small, because it is far from the main distribution of the selfie domain.

5 Limitations and Discussion

In this paper, we present a novel framework, ACL-GAN, for unpaired image-to-image translation. The core of ACL-GAN, adversarial-consistency loss, helps to maintain commonalities between the source and target domains. Furthermore, ACL-GAN can perform shape modifications and remove large objects without unrealistic traces.

Although our method outperforms state-of-the-art methods both quantitatively and qualitatively on three challenging tasks, typical failure cases are shown in Fig. 8. We also tried learning to translate real images from domain X_S to X_T and from domain X_T to X_S synchronously, but the performance was decreased, which may be caused by the increase of network workload. In addition, our method is not yet able to deal with datasets that have complex backgrounds (*e.g.* horse-to-zebra) well. Thus, supporting images with a complex background is an interesting direction for future studies. Nevertheless, the proposed method is simple, effective, and we believe our method can be applied to different data modalities.

Acknowledgements. This work was supported by the funding from Key-Area Research and Development Program of Guangdong Province (No.2019B121204008), start-up research funds from Peking University (7100602564) and the Center on Frontiers of Computing Studies (7100602567). We would also like to thank Imperial Institute of Advanced Technology for GPU supports.

References

1. Almahairi, A., Rajeswar, S., Sordoni, A., Bachman, P., Courville, A.: Augmented cycleGAN: learning many-to-many mappings from unpaired data. In: International Conference on Machine Learning (2018)
2. Anoosheh, A., Agustsson, E., Timofte, R., Gool, L.V.: ComboGAN: unrestrained scalability for image domain translation. In: IEEE Conference on Computer Vision and Pattern Recognition Workshops (2018)
3. Benaim, S., Wolf, L.: One-sided unsupervised domain mapping. In: Conference on Neural Information Processing Systems (2017)
4. Bińkowski, M., Sutherland, D.J., Arbel, M., Gretton, A.: Demystifying mmd GANs. In: International Conference on Learning Representations (2018)
5. Bousmalis, K., Silberman, N., Dohan, D., Erhan, D., Krishnan, D.: Unsupervised pixel-level domain adaptation with generative adversarial networks. In: IEEE Conference on Computer Vision and Pattern Recognition (2017)
6. Choi, Y., Choi, M., Kim, M., Ha, J.W., Kim, S., Choo, J.: StarGAN: unified generative adversarial networks for multi-domain image-to-image translation. In: IEEE Conference on Computer Vision and Pattern Recognition (2018)
7. Dong, C., Loy, C.C., He, K., Tang, X.: Image super-resolution using deep convolutional networks. IEEE Trans. Pattern Anal. Mach. Intell. **38**, 295–307 (2016)
8. Goodfellow, I., et al.: Generative adversarial nets. In: Conference on Neural Information Processing Systems (2014)
9. Heusel, M., Ramsauer, H., Unterthiner, T., Nessler, B., Hochreiter, S.: GANs trained by a two time-scale update rule converge to a local nash equilibrium. In: Conference on Neural Information Processing Systems (2017)
10. Huang, X., Liu, M.Y., Belongie, S., Kautz, J.: Multimodal unsupervised image-to-image translation. In: European Conference on Computer Vision (2018)
11. Iizuka, S., Simo-Serra, E., Ishikawa, H.: Globally and locally consistent image completion. ACM Trans. Graph. **36**, 1–14 (2017)
12. Isola, P., Zhu, J., Zhou, T., Efros, A.A.: Image-to-image translation with conditional adversarial networks. In: IEEE Conference on Computer Vision and Pattern Recognition (2017)
13. Kim, J., Lee, J.K., Lee, K.M.: Accurate image super-resolution using very deep convolutional networks. In: IEEE Conference on Computer Vision and Pattern Recognition (2016)
14. Kim, J., Kim, M., Kang, H., Lee, K.H.: U-GAT-it: unsupervised generative attentional networks with adaptive layer-instance normalization for image-to-image translation. In: International Conference on Learning Representations (2020)
15. Kim, T., Cha, M., Kim, H., Lee, J.K., Kim, J.: Learning to discover cross-domain relations with generative adversarial networks. In: International Conference on Machine Learning (2017)
16. Kingma, D.P., Ba, J.: Adam: a method for stochastic optimization. International Conference on Learning Representations (2014)
17. Ledig, C., et al.: Photo-realistic single image super-resolution using a generative adversarial network. In: IEEE Conference on Computer Vision and Pattern Recognition (2017)

18. Lee, H.Y., et al.: DRIT++: diverse image-to-image translation via disentangled representations. Int. J. Comput. Vis., 1–16 (2020)
19. Lee, H., Tseng, H., Huang, J., Singh, M., Yang, M.: Diverse image-to-image translation via disentangled representations. In: European Conference on Computer Vision (2018)
20. Liu, M., Breuel, T.M., Kautz, J.: Unsupervised image-to-image translation networks. In: Conference on Neural Information Processing Systems (2017)
21. Liu, Z., Luo, P., Wang, X., Tang, X.: Deep learning face attributes in the wild. In: IEEE International Conference on Computer Vision (2015)
22. Mao, X., Li, Q., Xie, H., Lau, R.Y., Wang, Z., Smolley, S.P.: Least squares generative adversarial networks. In: IEEE Conference on Computer Vision and Pattern Recognition (2017)
23. Mejjati, Y.A., Richardt, C., Tompkin, J., Cosker, D., Kim, K.I.: Unsupervised attention-guided image-to-image translation. In: Conference on Neural Information Processing Systems (2018)
24. Nizan, O., Tal, A.: Breaking the cycle - colleagues are all you need. In: arXiv preprint arXiv:1911.10538 (2019)
25. Pathak, D., Krahenbuhl, P., Donahue, J., Darrell, T., Efros, A.A.: Context encoders: feature learning by inpainting. In: IEEE Conference on Computer Vision and Pattern Recognition (2016)
26. Rosca, M., Lakshminarayanan, B., Warde-Farley, D., Mohamed, S.: Variational approaches for auto-encoding generative adversarial networks. In: arXiv preprint arXiv:1706.04987 (2017)
27. Royer, A., et al.: XGAN: unsupervised image-to-image translation for many-to-many mappings. In: International Conference on Machine Learning (2018)
28. Salimans, T., Goodfellow, I., Zaremba, W., Cheung, V., Radford, A., Chen, X.: Improved techniques for training GANs. In: Conference on Neural Information Processing Systems (2016)
29. Shrivastava, A., Pfister, T., Tuzel, O., Susskind, J., Wang, W., Webb, R.: Learning from simulated and unsupervised images through adversarial training. In: IEEE Conference on Computer Vision and Pattern Recognition (2017)
30. Siddiquee, M.M.R., et al.: Learning fixed points in generative adversarial networks: From image-to-image translation to disease detection and localization. In: IEEE International Conference on Computer Vision (2019)
31. Taigman, Y., Polyak, A., Wolf, L.: Unsupervised cross-domain image generation. In: International Conference on Learning Representations (2017)
32. Wang, T.C., Liu, M.Y., Zhu, J.Y., Tao, A., Kautz, J., Catanzaro, B.: High-resolution image synthesis and semantic manipulation with conditional GANs. In: IEEE Conference on Computer Vision and Pattern Recognition (2018)
33. Yi, Z., Zhang, H., Tan, P., Gong, M.: DualGAN: unsupervised dual learning for image-to-image translation. In: IEEE International Conference on Computer Vision (2017)
34. Zhang, R., Isola, P., Efros, A.A.: Colorful image colorization. In: Leibe, B., Matas, J., Sebe, N., Welling, M. (eds.) ECCV 2016. LNCS, vol. 9907, pp. 649–666. Springer, Cham (2016). https://doi.org/10.1007/978-3-319-46487-9_40
35. Zhu, J., Park, T., Isola, P., Efros, A.A.: Unpaired image-to-image translation using cycle-consistent adversarial networks. In: IEEE International Conference on Computer Vision (2017)
36. Zhu, J., Zhang, R., Pathak, D., Darrell, T., Efros, A.A., Wang, O., Shechtman, E.: Toward multimodal image-to-image translation. In: Conference on Neural Information Processing Systems (2017)

Correction to: Unpaired Image-to-Image Translation Using Adversarial Consistency Loss

Yihao Zhao, Ruihai Wu, and Hao Dong

Correction to:
Chapter "Unpaired Image-to-Image Translation Using Adversarial Consistency Loss" in: A. Vedaldi et al. (Eds.): *Computer Vision – ECCV 2020*, LNCS 12354, https://doi.org/10.1007/978-3-030-58545-7_46

In the originally published version of this chapter, the acknowledgment section has been modified.

The updated version of this chapter can be found at
https://doi.org/10.1007/978-3-030-58545-7_46

© Springer Nature Switzerland AG 2020
A. Vedaldi et al. (Eds.): ECCV 2020, LNCS 12354, p. C1, 2020.
https://doi.org/10.1007/978-3-030-58545-7_47

Author Index